Electrical Machines

Electrical machines convert energy existing in one form to another, usable, form. These machines can broadly be divided into three categories: generators, motors and transformers. Generators convert mechanical energy into electrical energy, motors convert electrical energy to mechanical energy, and transformers change the voltage level in ac system and are considered to be the backbone of a power system.

Electrical machines play an important role in domestic appliances, commercial devices and industrial applications. It is important for students of electrical and electronics engineering to learn the basic concepts of transformers, motors, generators and magnetic circuits. This book explains the design of transformers, decoding of generators and performance of electrical motors through descriptive illustrations, solved examples and mathematical derivations. Construction, working principles and applications of various electrical machines are discussed in detail. In addition, it offers an engrossing discussion on special purpose machines, which is useful from an industrial prospective in building customised machines. The text contains hundreds of worked examples and illustrations and more than a thousand self-assessment exercises. It is an ideal textbook for undergraduate students of electrical and electronics engineering.

S. K. Sahdev is Associate Dean at the Faculty of Technology and Science at Lovely Professional University, Jalandhar. He has more than thirty-five years of teaching experience. In addition, he has helped industrial units to set-up electrical laboratories for testing and developing their products. He has authored six books. His areas of interest include electrical machines, electric drives, power electronics and power systems.

Electrical Machines

S. K. Sahdev

CAMBRIDGE
UNIVERSITY PRESS

Shaftesbury Road, Cambridge CB2 8EA, United Kingdom

One Liberty Plaza, 20th Floor, New York, NY 10006, USA

477 Williamstown Road, Port Melbourne, VIC 3207, Australia

314–321, 3rd Floor, Plot 3, Splendor Forum, Jasola District Centre, New Delhi – 110025, India

103 Penang Road, #05–06/07, Visioncrest Commercial, Singapore 238467

Cambridge University Press is part of Cambridge University Press & Assessment, a department of the University of Cambridge.

We share the University's mission to contribute to society through the pursuit of education, learning and research at the highest international levels of excellence.

www.cambridge.org
Information on this title: www.cambridge.org/9781108431064

First published 2018

A catalogue record for this publication is available from the British Library

ISBN 978-1-108-43106-4 Paperback

Additional resources for this publication at www.cambridge.org/9781108431064

This book is dedicated to
my Parents,
wife – Ritu Sahdev,
son – Rohit Sahdev,
daughter-in-law – Robina Sahdev
and
grandsons – Arnav and Adhiraj

Contents

Preface

This book on 'Electrical Machines' has been written for under-graduate students of Electrical Engineering (EE) and Electrical & Electronics Engineering (EEE) belonging to various Indian and Foreign Universities. It will also be useful to candidates appearing for AMIE, IETE, GATE, UPSC Engineering Services and Civil Services Entrance Examinations.

We know that electrical energy has a wide range of applications where electrical machines play a vital role in industrial production and many other areas of science and technology. Accordingly, this book has been designed so that it be useful not only to students pursuing courses in electrical engineering but also for practising engineers and technicians.

'Electrical Machines' is taught at various universities under different titles such as Electrical Machines-I, Electrical Machines-II, DC Machines and Transformers, Electromagnetic Energy Conversion Devices, Special Purpose Machines, etc. All the topics in such courses have been covered in this single unit. As such, the book covers the revised syllabi of all Indian and Foreign Universities.

Generally, students find Electrical Machines to be one of the most difficult subjects to understand, despite the availability of a large number of text books in this field. Keeping this fact in mind, this text has been developed in a systematic manner giving more emphasis on basic concepts.

Each chapter of the book contains much needed text, supported by neat and self-explanatory diagrams to make the subject self-speaking to a great extent. A large number of solved and unsolved examples have been added in various chapters to enable students to attempt different types of questions in examination without any difficulty. Section Practice Problems have been added in all the chapters to maintain regular study and understanding. At the end of each chapter sufficient objective type questions, short-answer questions, test questions and unsolved examples have been added to make the book a complete and comprehensive unit in all respects.

The author lays no claim to original research in preparing the text. Materials available in the research work of eminent authors have been used liberally. But the author claims that he has organised the subject matter in very systematic manner. He also claims that the language of the text is lucid, direct and easy to understand.

Although every care has been taken to eliminate errors, however it is very difficult to claim perfection. I hope this book will be useful to its users (students, teachers and professionals). I shall be very grateful to the readers (students and teachers) and users of this book if they point out any mistake that might have crept in. Suggestions for the improvement of the book will be highly appreciated.

Acknowledgements

There are several people to whom I would like to express my sincere thanks. First of all, I would like to thank Mr Ashok Mittal (Hon'ble Chancellor), Mrs Rashmi Mittal (Hon'ble Pro-chancellor), Mr H. R. Singla (Director General) of Lovely Professional University, Jalandhar, who have inspired me to develop the text in the shape of a book. I would also like to thank Dr Lovi Raj Gupta, Executive Dean, (LFTS) of Lovely Professional University, who has encouraged and helped me in preparing the text.

Secondly, I would like to thank the entire executive staff, faculty and students of Lovely Professional University and Punjab Technical University for their support, collaboration and friendship.

I would like to thank all my friends, particularly Dr Manjo Kumar, Principal, DAV Institute of Engineering and Technology, Jalandhar; Dr Sudhir Sharma, HOD, Electrical, DAV Institute of Engineering and Technology, Jalandhar; Mr D. S. Rana, HOD, Electrical who have been involved, either directly or indirectly, in the successful completion of this book.

I owe my family members, relatives, friends and colleagues (Professor Bhupinder Verma, Mr R. K. Sharma, Mr Satnam Singh, Mr Amit Dhir and Ms Meenakshi Gupta) a special word of thanks for their moral support and encouragement.

I express my gratitude to the Publisher 'Cambridge University Press' and its Associate Commissioning Editor Ms Rachna Sehgal for guidance and support in bringing out the text in the shape of a book.

CHAPTER

Electro Magnetic Circuits 1

Chapter Objectives

After the completion of this unit, students/readers will be able to understand:

- ✓ What is magnetic field and its significance?
- ✓ What is a magnetic circuit?
- ✓ What are the important terms related to magnetism and magnetic circuits?
- ✓ What are the similarities and dissimilarities between magnetic and electric circuits?
- ✓ How series and parallel magnetic circuits are treated?
- ✓ What is leakage flux and how it affects magnetic circuits?
- ✓ What is magnetic hysteresis and hysteresis loss?
- ✓ What is electromagnetic induction phenomenon?
- ✓ What are Faraday's laws of electromagnetic induction?
- ✓ What are self- and mutual inductances and what is their significance?
- ✓ What is the effective value of inductances when these are connected in series–parallel combination?
- ✓ What are electromechanical energy conversion devices?
- ✓ How does torque develop by the alignment of two fields?
- ✓ What are the factors on which torque depends?
- ✓ How to determine the direction of torque or induced emf in rotating machines?

Introduction

It is always advantageous to utilise electrical energy since it is cheaper, can be easily transmitted, easy to control and more efficient. The electrical energy is generally generated from natural resources such as water, coal, diesel, wind, atomic energy, etc. From these sources, first mechanical energy is produced by one way or the other and then that mechanical energy is converted into electrical energy by suitable machines. For the utilisation of electrical energy, it is again converted into other forms of energy such as mechanical, heat, light etc. It is a well-known fact that the electric drives have been universally adopted by the industry due to their inherent advantages. The energy conversion devices are always required at both ends of a typical electrical system. The devices or machines which convert mechanical energy into electrical energy and vice-versa are called **electro–mechanical energy conversion devices.**

The operation of all the electrical machines such as *DC* machines, transformers, synchronous machines, induction motors, etc., rely upon their magnetic circuits. The closed path followed by the magnetic lines of force is called a *magnetic circuit.* The operation of all the electrical devices (e.g., transformers, generators, motors, etc.) depends upon the magnetism produced by their magnetic circuits. Therefore, to obtain the required characteristics of these devices, their magnetic circuits have to be designed carefully.

In this chapter, we shall focus our attention on the basic fundamentals of magnetic circuits and their applications as electromechanical energy conversion devices.

1.1 Magnetic Field and its Significance

The region around a magnet where its poles exhibit a force of attraction or repulsion is called **magnetic field.**

The existence of the magnetic field at a point around the magnet can also be determined by placing a magnetic needle at that point as shown in Fig. 1.1. Although magnetic lines of force have no real existence and are purely imaginary, yet their concept is very useful to understand various magnetic effects. It is assumed (because of their effects) that the magnetic lines of force possess the following important properties:

(*i*) The direction of magnetic lines of force is from N-pole to the S-pole outside the magnet. But inside the magnet their direction is from S-pole to N-pole.
(*ii*) They form a closed loop.
(*iii*) Their tendency is to follow the least reluctance path.
(*iv*) They act like stretched cords, always trying to shorten themselves.
(*v*) They never intersect each other.
(*vi*) They repel each other when they are parallel and are in the same direction.
(*vii*) They remain unaffected by non-magnetic materials.

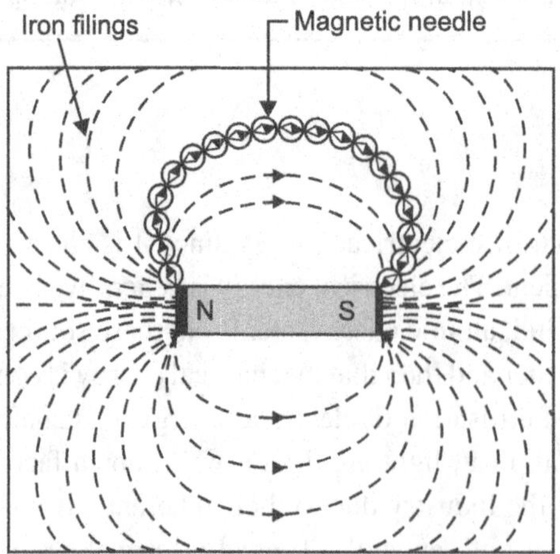

Fig. 1.1 Magnetic field around a bar magnet

1.2 Magnetic Circuit and its Analysis

The closed path followed by magnetic flux is called a **magnetic circuit.**

A magnetic circuit usually consists of magnetic materials having high permeability (e.g., iron, soft steel, etc.). In this circuit, magnetic flux starts from a point and finishes at the same point after completing its path.

Figure 1.2 shows a solenoid having N turns wound on an iron core (ring). When current I ampere is passed through the solenoid, magnetic flux ϕ Wb is set-up in the core.

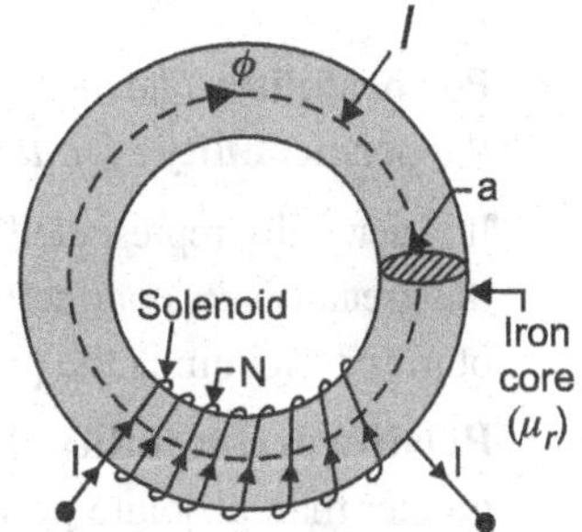

Fig. 1.2 Magnetic circuit

Let l = mean length of magnetic circuit in m;

a = area of cross-section of core in m^2;

μ_r = relative permeability of core material.

Flux density in the core material, $B = \dfrac{\phi}{a}$ Wb/m^2

Magnetising force in the core material.

$$H = \frac{B}{\mu_0 \mu_r} = \frac{\phi}{a \mu_0 \mu_r} \text{ AT/m}$$

According to work law, the work done in moving a unit pole once round the magnetic circuit (or path) is equal to the ampere-turns enclosed by the magnetic circuit.

i.e., $Hl = NI$ or $\dfrac{\phi}{a \mu_0 \mu_r} \times l = NI$ or $\phi = \dfrac{NI}{(l / a \mu_0 \mu_r)}$ Wb

The above expression reveals that the amount of flux set-up in the core is

(*i*) directly proportional to N and I i.e., NI, called *magnetomotive force* (*mmf*). It shows that the flux increases if either of the two increases and *vice-versa.*

(*ii*) inversely proportional to $l/a \mu_0 \mu_r$ called *reluctance* of the magnetic path. In fact, reluctance is the opposition offered to the magnetic flux by the magnetic path. The lower is the reluctance, the higher will be the flux and *vice-versa.*

Thus, $$\text{Flux} = \frac{\text{m.m.f}}{\text{reluctance}}$$

It may be noted that the above expression has a strong resemblance to Ohm's law for electric current (I = emf/resistance). The mmf is analogous to emf in electric circuit, reluctance is analogous to resistance and flux is analogous to current. Because of this similarity, the above expression is sometimes referred to as *Ohm's law of magnetic circuits.*

1.3 Important Terms

While studying magnetic circuits, generally, we come across the following terms:

1. **Magnetic field:** The region around a magnet where its poles exhibit a force of attraction or repulsion is called *magnetic field.*

2. **Magnetic flux (ϕ):** The amount of magnetic lines of force set-up in a magnetic circuit is called *magnetic flux*. Its unit is weber (Wb). It is analogous to *electric current I* in electric circuit.
3. *The* **magnetic flux density** *at a point is the flux per unit area at right angles to the flux at that point.*

   It is, generally, represented by letter '*B*'. Its unit is Wb/m^2 or Tesla, i.e.,

$$B = \frac{\phi}{A} \text{Wb / m}^2 \quad \text{or} \quad \text{T (1 Wb/m}^2 = 1 \times 10^4 \text{ Wb/cm}^2)$$

4. **Permeability:** *The ability of a material to conduct magnetic lines of force through it is called the* **permeability** *of that material.*

   It is generally represented by μ (*mu*, a Greek letter). The greater the permeability of a material, the greater is its conductivity for the magnetic lines of force and *vice-versa*. The permeability of air or vacuum is the poorest and is represented as μ_0 (where $\mu_0 = 4\pi \times 10^{-7}$ H/m).

   Relative permeability: The absolute (or actual) permeability μ of a magnetic material is much greater than absolute permeability of air μ_0. The relative permeability of a magnetic material is given in comparison with air or vacuum.

   Hence, the ratio of the permeability of material μ to the permeability of air or vacuum μ_0 is called the relative permeability μ_r of the material.

   i.e.,
$$\mu_r = \frac{\mu}{\mu_0} \quad \text{or} \quad \mu = \mu_0 \mu_r$$

   Obviously, the relative permeability of air would be $\mu_0/\mu_0 = 1$. The value of relative permeability of all the non-magnetic materials is also 1. However, its value is as high as 8000 for soft iron, whereas, its value for mumetal (iron 22% and nickel 78%) is as high as 1,20,000.
5. **Magnetic field intensity:** The force acting on a unit north pole (1 Wb) when placed at a point in the magnetic field is called the magnetic intensity of the field at that point. It is denoted by *H*. In magnetic circuits, it is defined as mmf per unit length of the magnetic path. It is denoted by *H*, mathematically,

$$H = \frac{\text{m.m.f}}{\text{length of magnetic path}} = \frac{NI}{l} \text{AT / m}$$

6. **Magnetomotive force (mmf):** The magnetic pressure which sets-up or tends to set-up magnetic flux in a magnetic circuit is called *magnetomotive force*. As per work law it may be defined as under:

   The work done in moving a unit magnetic pole (1 Wb) once round the magnetic circuit is called *magnetomotive force*. In general

$$\text{mmf} = NI \text{ ampere-turns (or AT)}$$

   It is analogous to *emf* in an electric circuit.
7. **Reluctance (S):** The opposition offered to the magnetic flux by a magnetic circuit is called its *reluctance*.

   It depends upon length (*l*), area of cross-section (*a*) and permeability ($\mu = \mu_0 \mu_r$) of the material that makes up the magnetic circuit. It is measured in AT/Wb.

$$\text{Reluctance, } S = \frac{l}{a\,\mu_0\,\mu_r}$$

It is analogous to *resistance* in an electric circuit.

8. **Permeance:** It is a measure of the ease with which flux can be set-up in the material. It is just reciprocal of reluctance of the material and is measured in *Wb/AT* or *henry.*

$$\text{Permeance} = \frac{1}{\text{reluctance}} = \frac{a\,\mu_0\,\mu_r}{l}\ \text{Wb/AT} \quad \text{or} \quad \text{H}$$

It is analogous to *conductance* in an electric circuit.

9. **Reluctivity:** It is specific reluctance and analogous to *resistivity* in electric circuit.

1.4 Comparison between Magnetic and Electric Circuits

Although magnetic and electric circuits have many points of similarity but still they are not analogous in all respects. A comparison of the two circuits is given below:

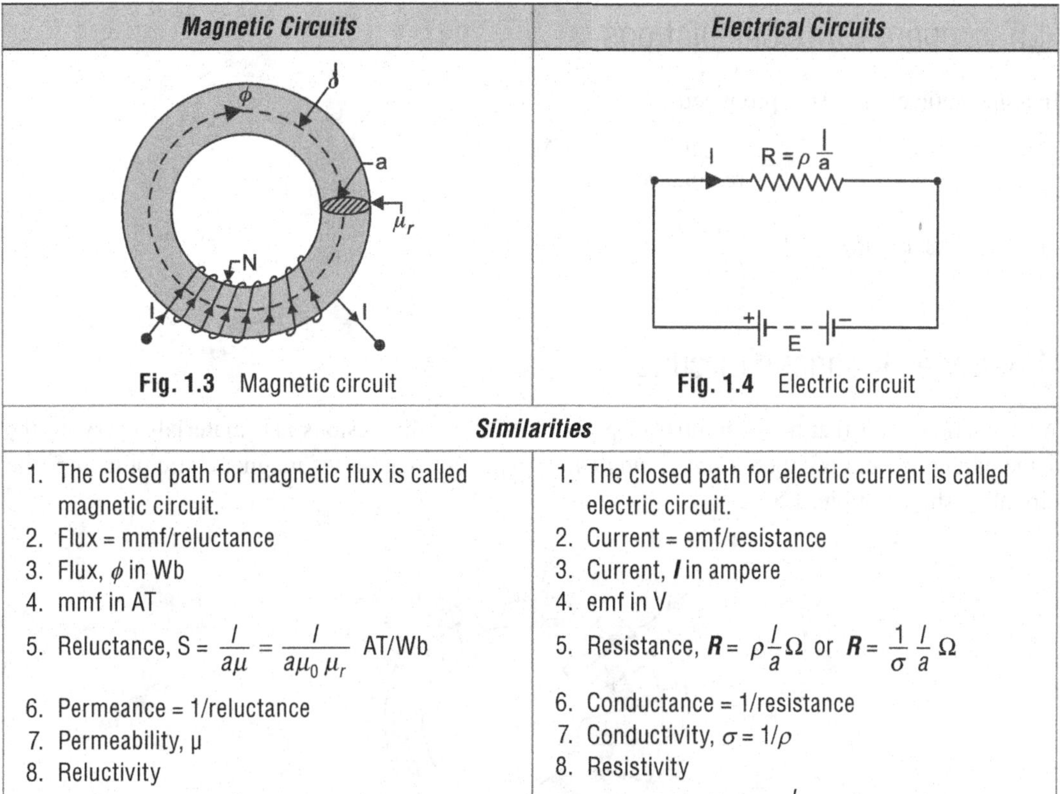

Magnetic Circuits	*Electrical Circuits*
Fig. 1.3 Magnetic circuit	**Fig. 1.4** Electric circuit
Similarities	
1. The closed path for magnetic flux is called magnetic circuit.	1. The closed path for electric current is called electric circuit.
2. Flux = mmf/reluctance	2. Current = emf/resistance
3. Flux, ϕ in Wb	3. Current, ***I*** in ampere
4. mmf in AT	4. emf in V
5. Reluctance, $S = \frac{l}{a\mu} = \frac{l}{a\mu_0\,\mu_r}$ AT/Wb	5. Resistance, $\boldsymbol{R} = \rho\frac{l}{a}\,\Omega$ or $\boldsymbol{R} = \frac{1}{\sigma}\frac{l}{a}\,\Omega$
6. Permeance = 1/reluctance	6. Conductance = 1/resistance
7. Permeability, μ	7. Conductivity, $\sigma = 1/\rho$
8. Reluctivity	8. Resistivity
9. Flux density, $\boldsymbol{B} = \frac{\phi}{a}$ Wb/m^2	9. Current density, $\boldsymbol{J} = \frac{I}{a}$ A/m^2
10. Magnetic intensity, $\boldsymbol{H} = \boldsymbol{NI/l}$	10. Electric intensity, $\boldsymbol{E} = \boldsymbol{V/d}$

Dissimilarities	
1. In fact, the magnetic flux does not flow but it sets-up in the magnetic circuit (basically molecular poles are aligned).	1. The electric current (electrons) actually flows in an electric circuit.
2. For magnetic flux, there is no perfect insulator. It can be set-up even in the non-magnetic materials like air, rubber, glass etc. with reasonable mmf	2. For electric current, there are large number of perfect insulators like glass, air, rubber, etc., which do not allow it to follow through them under normal conditions.
3. The reluctance (***S***) of a magnetic circuit is not constant rather it varies with the value of ***B***. It is because the value of μ_r changes considerably with the change in ***B***.	3. The resistance (***R***) of an electric circuit is almost constant as its value depends upon the value of ρ which is almost constant. However, the value of ρ and ***R*** may vary slightly if temperature changes.
4. Once the magnetic flux is set-up in a magnetic circuit, no energy is expanded. However, a small amount of energy is required at the start to create flux in the circuit.	4. Energy is expanded continuously, so long as the current flows through an electric circuit. This energy is dissipated in the form of heat.

1.5 Ampere-turns Calculations

In a magnetic circuit, flux produced,

$$\phi = \frac{\text{m.m.f.}}{\text{reluctance}} = \frac{NI}{l/a\,\mu_0\,\mu_r}$$

or AT required, $$NI = \frac{\phi l}{a\mu_0\mu_r} = \frac{B}{\mu_0\mu_r}l = Hl$$

1.6 Series Magnetic Circuits

A magnetic circuit that has a number of parts of different dimensions and materials carrying the same magnetic field is called a *series magnetic circuit.* Such as series magnetic circuit (composite circuit) is shown in Fig. 1.5.

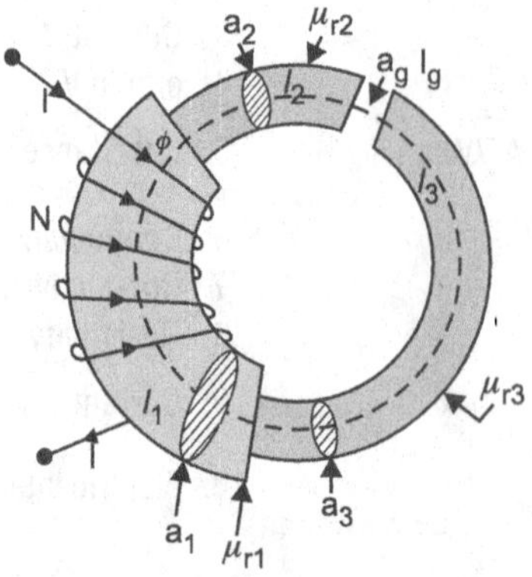

Fig. 1.5 Series magnetic circuit

Total reluctance of the magnetic circuit,

$$S = S_1 + S_2 + S_3 + S_g$$

$$= \frac{l_1}{a_1 \mu_0 \mu_{r1}} + \frac{l_2}{a_2 \mu_0 \mu_{r2}} + \frac{l_3}{a_3 \mu_0 \mu_{r3}} + \frac{l_g}{a_g \mu_0}$$

$$\text{Total mmf} = \phi S$$

$$= \phi\left(\frac{l_1}{a_1 \mu_0 \mu_{r1}} + \frac{l_2}{a_2 \mu_0 \mu_{r2}} + \frac{l_3}{a_3 \mu_0 \mu_{r3}} + \frac{l_g}{a_g \mu_0}\right)$$

$$= \frac{B_1 l_1}{\mu_0 \mu_{r1}} + \frac{B_2 l_2}{\mu_0 \mu_{r2}} + \frac{B_3 l_3}{\mu_0 \mu_{r3}} + \frac{B_g l_g}{\mu_0}$$

$$= H_1 l_1 + H_2 l_2 + H_3 l_3 + H_g l_g$$

1.7 Parallel Magnetic Circuits

A magnetic circuit which has two or more than two paths for the magnetic flux is called a *parallel magnetic circuit.* Its behaviour can be just compared to a parallel electric circuit.

Figure 1.6 shows a parallel magnetic circuit. A current carrying coil is wound on the central limb *AB*. This coil sets-up a magnetic flux ϕ_1 in the central limb which is further divided into two paths i.e., (*i*) path *ADCB* which carries flux ϕ_2 and (*ii*) path *AFEB* which carries flux ϕ_3.

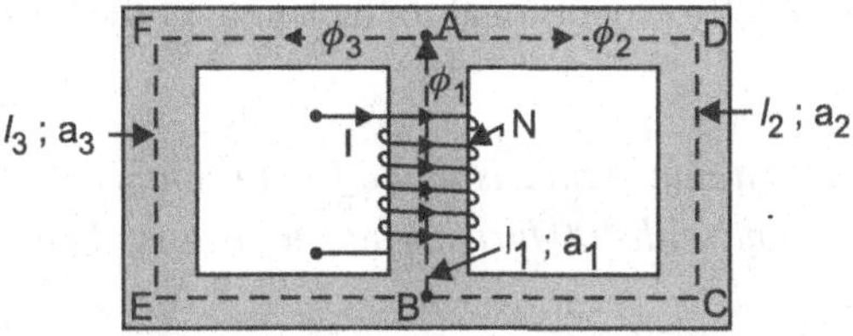

Fig. 1.6 Parallel magnetic circuit

It is clear that $\phi_1 = \phi_2 + \phi_3$

The two magnetic paths *ADCB* and *AFEB* are in parallel. The ATs required for this parallel circuit is equal to the ATs required for any one of the paths.

If S_1 = reluctance of path *BA* i.e., $l_1/a_1 \mu_0 \mu_{r1}$

S_2 = reluctance of path *ADCB* i.e., $l_2/a_2 \mu_0 \mu_{r2}$

S_3 = reluctance of path *AFEB* i.e., $l_3/a_3 \mu_0 \mu_{r3}$

∴ Total mmf required = mmf required for path *BA* + mmf required path *ADCB* or path *AFEB*.

i.e., Total mmf or $AT = \phi_1 S_1 + \phi_2 S_2 = \phi_1 S_1 + \phi_3 S_3$

1.8 Leakage Flux

The magnetic flux which does not follow the intended path in a magnetic circuit is called *leakage flux.*

When some current is passed through a solenoid, as shown in Fig. 1.7, magnetic flux is produced by it. Most of this flux is set-up in the magnetic core and passes through the air gap (an intended path). This flux is known as *useful flux* ϕ_u. However, some of the flux is just set-up around the coil and is not utilised for any work. This flux is called *leakage flux* ϕ_l.

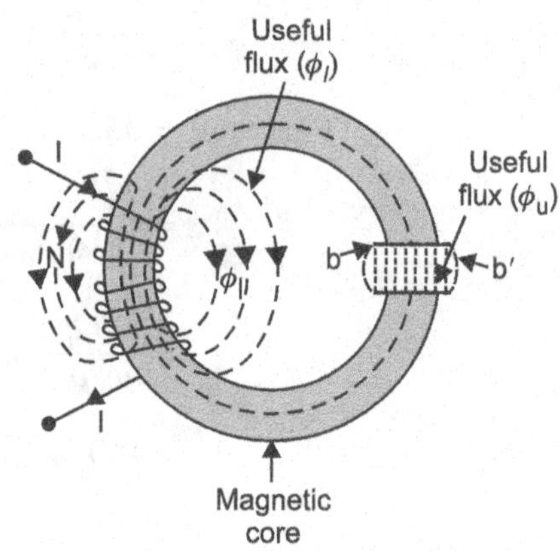

Fig. 1.7 Leakage flux

Total flux produced by the solenoid.

$$\phi = \phi_u + \phi_l$$

Leakage co-efficient or leakage factor: The ratio of total flux (ϕ) produced by the solenoid to the useful flux (ϕ_u) set-up in the air gap is known as *leakage co-efficient.* It is generally represented by letter 'λ'.

$$\therefore \quad \text{Leakage co-efficient, } \lambda = \frac{\phi}{\phi_u}$$

Fringing: It may be seen in Fig. 1.7 that the useful flux when sets-up in the air gap, it tends to bulge outwards at b and b' since the magnetic lines set-up in the same direction repel each other. This increases the effective area in the air gap and decreases the flux density. This effect is known as *fringing.* The fringing is directly proportional to the length of the air gap.

Example 1.1

An iron ring of 400 cm mean circumference is made from round iron of cross-section 20 cm². Its permeability is 500. If it is wound with 400 turns, what current would be required to produce a flux of 0·001 Wb?

Solution:

The magnetic circuit is shown in Fig. 1.8.

Mean length of magnetic path, l_m = 400 cm = 4 m

Area of X-section of iron ring, $a = 20 \times 10^{-4}$ m²

Absolute permeability, $\mu_0 = 4\pi \times 10^{-7}$

Now mmf = flux × reluctance

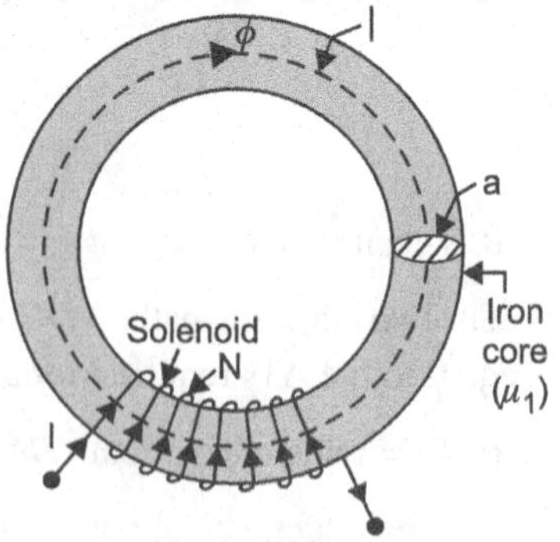

Fig. 1.8 Magnetic circuit

$$NI = \phi \times \frac{l_m}{a\mu_0\mu_r}$$

$$400 \times I = 0{\cdot}001 \times \frac{4}{20 \times 10^{-4} \times 4\pi \times 10^{-7} \times 500}$$

$$\therefore \quad \text{Current, } I = \frac{0.001 \times 4}{20 \times 10^{-4} \times 4\pi \times 10^{-7} \times 500 \times 400} = \mathbf{7{\cdot}958} \; (Ans.)$$

<u>**Example 1.2**</u>

An electromagnet has an air gap of 4 mm and flux density in the gap is 1·3 Wb/m². Determine the ampere-turns for the gap.

Solution:

Here, l_g = 4mm = 0·4 cm = 4 × 10^{-3} m; B_g = 1·3 Wb/m²

Ampere-turns required for the gap

$$= H_g \times l_g = \frac{Bg}{\mu_0} \times l_g = \frac{1 \cdot 3}{4\pi \times 10^{-7}} \times 4 \times 10^{-3} = \textbf{4136·83 AT} \ (Ans.)$$

<u>**Example 1.3**</u>

A coil of insulated wire of 500 turns and of resistance 4 Ω is closely wound on iron ring. The ring has a mean diameter of 0·25 m and a uniform cross-sectional area of 700 mm². Calculate the total flux in the ring when a DC supply of 6V is applied to the ends of the winding. Assume a relative permeability of 550.

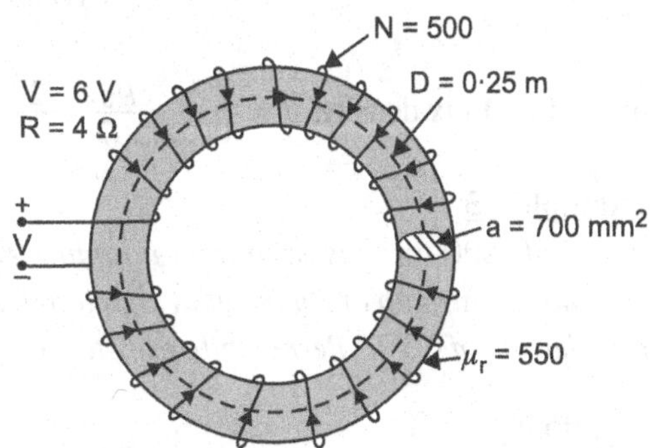

Fig. 1.9 Magnetic circuit

Solution:

Mean length of iron ring, $l = \pi D = \pi \times 0{\cdot}25 = 0{\cdot}25\,\pi$ m

Area of cross-section, α = 700 mm² = 700 × 10^{-6} m²

Current flowing through the coil,

$$I = \frac{\text{Voltage applied across coil}}{\text{Resistance of coil}}$$

$$= \frac{6}{5} = 1 \cdot 5\text{A}$$

Total flux in the ring,
$$\phi = \frac{NI}{l / a\mu_0\mu_r} = \frac{NI \times a\mu_0\mu_r}{l}$$

$$= \frac{500 \times 1 \cdot 5 \times 700 \times 10^{-6} \times 4\pi \times 10^{-7} \times 550}{0 \cdot 25\,\pi} = \textbf{0·462 mWb} \ (Ans.)$$

<u>**Example 1.4**</u>

What are the similarities between electrical circuits and magnetic circuits? An iron ring of mean length 50 cm and relative permeability 300 has an air gap of 1 mm. If the ring is provided with winding of 200 turns and a current of 1 A is allowed to flow through, find the flux density across the airgap.

Solution:

Here, l_i = 50 cm = 0·5 m; μ_r = 300; l_g = 1mm = 0·001 m; N = 200 turns; I = 1 A

$$\text{Ampere-turns required for air gap} = \frac{B}{\mu_0} l_g$$

$$\text{Ampere-turns required for iron ring} = \frac{B}{\mu_0\ \mu_r}\, l_i$$

or $$\text{Total ampere-turns required} = \frac{B}{\mu_0} l_g + \frac{B}{\mu_0\ \mu_r} l_i \quad ...(i)$$

$$\text{Ampere-turns provided by the coil} = N\,I = 200 \times 1 = 200 \quad ...(ii)$$

Equating eqn. (*i*) and (*ii*), we get,

or $$200 = \frac{B}{\mu_0}\left(l_g + \frac{l_i}{\mu_r}\right) = \frac{B}{\mu_0}\left(0.01 + \frac{0.5}{300}\right)$$

$$= \frac{B}{\mu_0}(0 \cdot 001 + 0 \cdot 00167) = \frac{B}{\mu_0} \times 0 \cdot 00267$$

or $$\text{Flux density, } B = \frac{200 \times \mu_0}{0 \cdot 00267} = \frac{200 \times 4\pi \times 10^{-7}}{0 \cdot 00267} = \mathbf{0{\cdot}09425\ T}\ (\textit{Ans.})$$

Example 1.5

A coil of 1000 *turns is wound on a laminated core of steel having a cross-section of* 5 *cm*^{2}. *The core has an air gap of* 2 *mm cut at right angle. What value of current is required to have an air gap flux density of* 0·5 *T? Permeability of steel may be taken as infinity. Determine the coil inductance.*

Solution:

Here, N = 1000 turns; a= 5 cm^2 = 5 × 10^{-4} m^2;

$$l_g = 2 \text{ mm} = 2 \times 10^{-3} \text{ m}; B = 0{\cdot}5 \text{ T}; \mu_r = \infty$$

Total ampere-turns required,

$$AT = \frac{B}{\mu_0} l_g + \frac{B}{\mu_0\ \mu_r} l_i = \frac{0 \cdot 5}{4\pi \times 10^{-7}} \times 2 \times 10^{-3} + 0 = 796$$

$$\left(\text{As } \mu_r = \infty; \frac{B}{\mu_0 \mu_r} \times l_i = 0\right)$$

Current required, $$I = \frac{AT}{I} = \frac{796}{1000} = \mathbf{0{\cdot}796\ A}\ (\textit{Ans.})$$

Inductance of coil, $$L = \frac{N\phi}{I} = \frac{N \times B \times a}{I} = \frac{1000 \times 0 \cdot 5 \times 5 \times 10^{-4}}{0.796}$$

$$= \mathbf{0{\cdot}314\ H}\ (\textit{Ans.})$$

Example 1.6

A flux density of 1·2 Wb/m^2 *is required in 2 mm air gap of an electro-magnet having an iron path 1 metre long. Calculate the magnetising force and current required if the electro magnet has 1273 turns. Assume relative permeability of iron to be 1500.*

Solution:

$$\text{Flux density, } B = 1{\cdot}2 \text{ Wb/m}^2$$

Relative permeability of iron, $\mu_r = 1500$

No. of turns, $N = 1273$

Length of iron path, $l_i = 1\text{m}$

Length of air gap, $l_g = 2\text{mm} = 0{\cdot}002$ m

Magnetising force for iron $H_i = \dfrac{B}{\mu_0\mu_r} = \dfrac{1\cdot 2}{4\pi\times 10^{-7}\times 1500} = 636{\cdot}6$ AT/m

Magnetising force for air gap, $H_g = \dfrac{B}{\mu_0} = \dfrac{1\cdot 2}{4\pi\times 10^{-7}} = 954900$ AT/m

ATs required for iron path $= H_i l_i = 636{\cdot}6 \times 1 = 636{\cdot}6$

ATs required for air gap $= H_g l_g = 954900 \times 0{\cdot}002 = 1909{\cdot}8$

Total ampere-turns $= 636{\cdot}6 + 1909{\cdot}8 = 2546{\cdot}4$

Current required, $I = \dfrac{\text{Total } ATs}{N} = \dfrac{2546\cdot 4}{1273} = \mathbf{2A}$ (*Ans.*)

Example 1.7

Estimate the number of ampere-turns necessary to produce a flux of 100000 lines round an iron ring of 6 cm² cross section and 20 cm mean diameter having an air gap 2 mm wide across it Permeability of the iron may be taken 1200. Neglect the leakage flux outside the 2 mm air gap.

Solution:

The magnetic circuit is shown in Fig. 1.10

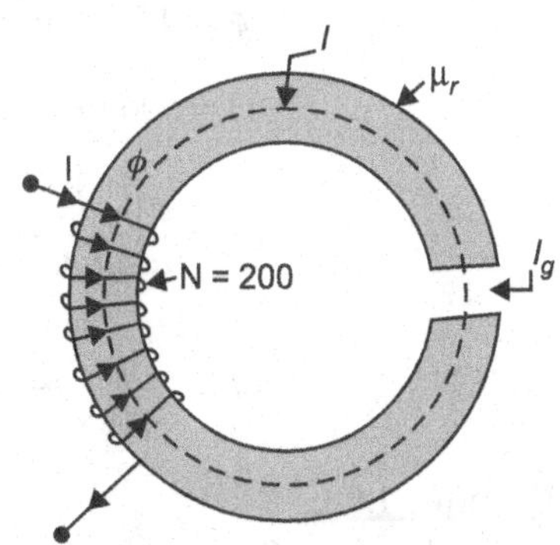

Fig. 1.10 Magnetic circuit

Area of cross section of the ring, $a = 6\text{ cm}^2 = 6 \times 10^{-4}\text{ m}^2$

Mean diameter of the ring, $D_m = 20\text{ cm} = 0{\cdot}2$ m

Length of air gap, $l_g = 2\text{ mm} = 2 \times 10^{-3}$ m

Flux set up in the ring, $\phi = 100{,}000$ lines

$= 100{,}000 \times 10^{-8}$

$= 0{\cdot}001$ Wb

Relative permeability of iron, $\mu_r = 1200$

Mean length of ring, $l_m = \pi D = \pi \times 0{\cdot}2$

$= 0{\cdot}6283$ m

Length of air gap, $l_g = 0{\cdot}002$ m

Length of iron path, $l_i = 0{\cdot}6283 - 0{\cdot}002$

$= 0{\cdot}6263$ m

Now, mmf = flux × reluctance

Ampere-turns required for iron path,

$$AT_i = \phi \times \frac{l_i}{a\mu_0\mu_r} = 0\cdot 001 \times \frac{0\cdot 6263}{6\times 10^{-4}\times 4\pi\times 10^{-7}\times 1200}$$

$= 692{\cdot}21$ AT

Ampere-turns required for air gap,

$$AT_g = \phi \times \frac{l_g}{a\mu_0} = 0 \cdot 001 \times \frac{0 \cdot 002}{6 \times 10^{-4} \times 4\pi \times 10^{-7}} = 2652{\cdot}58 \text{ AT}$$

Total ampere-turns required to produce the given flux

$$= AT_i + AT_g = 692{\cdot}21 + 2652{\cdot}58 = \mathbf{3344{\cdot}79\ AT}\ (Ans.)$$

Example 1.8

A wrought iron bar 30 *cm long and* 2 *cm in diameter is bent into a circular shape as given in Fig.* 0.11. *It is then wound with* 500 *turns of wire. Calculate the current required to produce a flux of* 0·5 *mWb in magnetic circuit with an air gap of* 1 *mm;* u_r (iron) = 4000 (*assume constant*).

Solution:

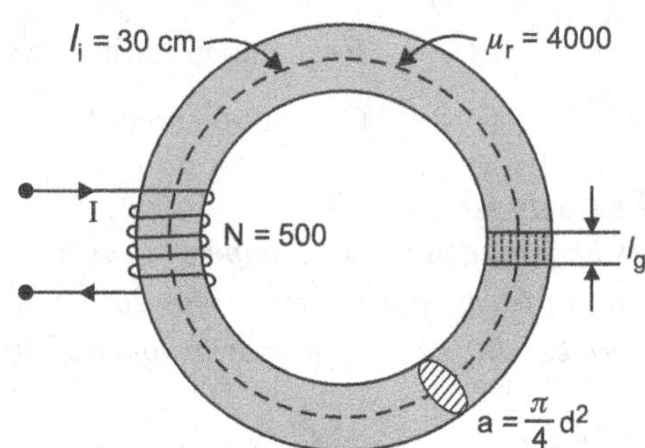

Fig. 1.11 Magnetic circuit

Here, $l_i = 30$ cm = 0·3 m;

Diameter, $d = 2$ cm

$$\therefore \quad \text{Area, } a = \frac{\pi}{4}d^2 = \frac{\pi(2)^2}{4} \times 10^{-4} \text{ m}^2$$

$$= \pi \times 10^{-4} \text{ m}^2$$

$$\phi = 0{\cdot}5 \text{ mWb} = 0{\cdot}5 \times 10^{-3} \text{ Wb}$$

$$N = 500 \text{ turns}$$

$$NI = \frac{\phi}{a}\left[\frac{l_i}{\mu_0\mu_r} + \frac{l_g}{\mu_0}\right]$$

$$I = \frac{0 \cdot 5 \times 10^{-3}}{500 \times \pi \times 10^{-4}}\left[\frac{0 \cdot 3}{4\pi \times 10^{-7} \times 400} + \frac{0 \cdot 001}{4\pi \times 10^{-7}}\right] = \mathbf{4{\cdot}433\ A}\ (Ans.)$$

Example 1.9

A circular ring 20 *cm in diameter has an air gap* 1 *mm wide cut in it. The area of a cross-section of the ring is* 3·6 *cm²*. *Calculate the value of direct current needed in a coil of* 1000 *turns uniformly wound round the ring to create a flux of* 0·5 *mWb in the air gap. Neglect fringing and assume relative permeability for the iron as* 650.

Solution:

Here, Area of cross-section of the ring, $a = 3{\cdot}6 \text{ cm}^2 = 3{\cdot}6 \times 10^{-4} \text{ m}^2$

No. of turns of the coil, $N = 1000$

Flux set-up, $\phi = 0{\cdot}5$ m Wb $= 0{\cdot}5 \times 10^{-3}$ Wb

Relative permeability of iron, $\mu_r = 650$

Length of air gap, $l_g = 1 \text{ mm} = 1 \times 10^{-3}$ m

Mean diameter of ring = 20 cm = 20×10^{-2} m

$\therefore$ Length of iron path $l_i = \pi D = \pi \times 20 \times 10^{-2}$ m $= 62{\cdot}83 \times 10^{-4}$ m

$$\text{Reluctance of iron path} = \frac{l_i}{\mu_0 \mu_r a}$$

$$= \frac{62 \cdot 83 \times 10^{-2}}{4\pi \times 10^{-7} \times 650 \times 3 \cdot 6 \times 10^{-4}} = 213669 \text{ AT/Wb.}$$

$\therefore$ AT required for iron path = $0{\cdot}5 \times 10^{-3} \times 213669 = 1068{\cdot}3$ AT

$$\text{Reluctance of air gap} = \frac{l_g}{\mu_0 a} = \frac{1 \times 10^{-3}}{4\pi \times 10^{-7} \times 3 \cdot 6 \times 10^{-4}} = 2210485 \text{ AT/Wb}$$

$\therefore$ AT required for air gap = $0{\cdot}5 \times 10^{-3} \times 2210485 = 1105{\cdot}2$ AT

$\therefore$ Total AT = $(AT)_i + (AT)_{gap} = 1068{\cdot}3 + 1105{\cdot}2 = 2173{\cdot}5$ AT

$\therefore$ Current $I = \dfrac{\text{Total AT}}{N} = \dfrac{2173 \cdot 5}{1000}$ = **2·1735 A** (*Ans.*)

Example 1.10

A coil is wound uniformly with 300 turns over a steel ring of relative permeability 900 having a mean diameter of 20 cm. The steel ring is made of bar having circular cross-section of diameter 2 cm. If the coil has a resistance of 50 ohm and is connected to 250 V DC supply, calculate (i) the mmf of the coil, (ii) the field intensity in the ring, (iii) reluctance of the magnetic path, (iv) total flux and (v) permeance of the ring.

Solution:

The magnetic circuit is shown in Fig. 1.12.

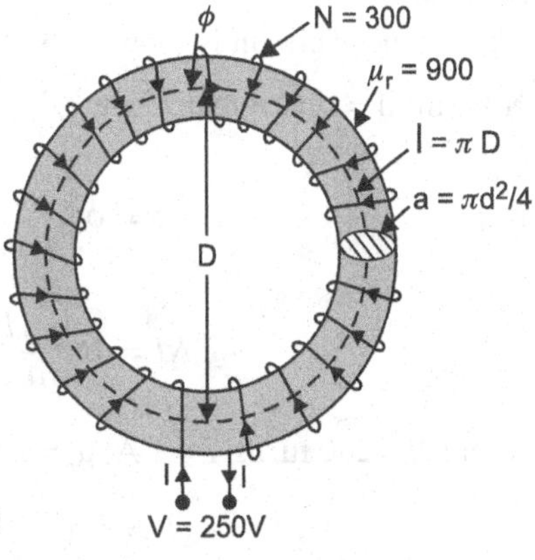

$$\text{Current through the coil, } I = \frac{V}{R} = \frac{250}{50} = 5 \text{ A}$$

(*i*) mmf of the coil = $NI = 300 \times 5$ = **1500 AT** (*Ans.*)

(*ii*) Field intensity $H = \dfrac{NI}{l}$

where, $l = \pi D = 0{\cdot}2\,\pi$ metre

$\therefore$ $H = \dfrac{1500}{0 \cdot 2\pi}$ = **2387·3 AT/m** (*Ans.*)

Fig. 1.12 Magnetic circuit

(*iii*) Reluctance of the magnetic path, $S = \dfrac{1}{a\mu_0\mu_r}$

where, $a = \dfrac{\pi}{4}d^2 = \dfrac{\pi}{4} \times (0 \cdot 02)^2 = \pi \times 10^{-4}$ m^2; $\mu_r = 900$

$\therefore$ $S = 0{\cdot}2\pi/\pi \times 10^{-4} \times 4\pi \times 10^{-7} \times 900$

= **$17{\cdot}684 \times 10^5$ AT/Wb** (*Ans.*)

(*iv*) Total flux,

$$\phi = \frac{\text{m.m.f.}}{S} = \frac{1500}{17 \cdot 684 \times 10^5} = \mathbf{0{\cdot}848\ m\ Wb}\ (Ans.)$$

(*v*) Permeance = $1/S = 1/17{\cdot}684 \times 10^5 = \mathbf{5{\cdot}655 \times 10^{-7}\ Wb/AT}$ (*Ans.*)

Example 1.11

Calculate the relative permeability of an iron ring when the exciting current taken by the 600 turn coil is 1.2 A and the total flux produced is 1 m Wb. The mean circumference of the ring is 0·5 m and the area of cross-section is 10 cm².

Solution:

$$NI = \frac{\phi \times l}{a\mu_0\mu_r}$$

$$\therefore \quad \mu_r = \frac{\phi \times l}{a\,\mu_0\,NI}$$

where $N = 600$ turns; $I = 1{\cdot}2$ A; $\phi = 1$ m Wb $= 1 \times 10^{-3}$ Wb; $l = 0{\cdot}5$ m; $a = 10$ cm² $= 10 \times 10^{-4}$ m²

$$\therefore \quad \mu_r = \frac{1 \times 10^{-3} \times 0 \cdot 5}{10 \times 10^{-4} \times 4\pi \times 10^{-7} \times 600 \times 1 \cdot 2} = \mathbf{552{\cdot}6}\ (Ans.)$$

Example 1.12

An iron ring of mean length 1 m has an air gap of 1 mm and a winding of 200 turns. If the relative permeability of iron is 500 when a current of 1 A flows through the coil, find the flux density.

Solution:

The magnetic circuit is shown in Fig. 1.13.

Now, mmf = flux × reluctance

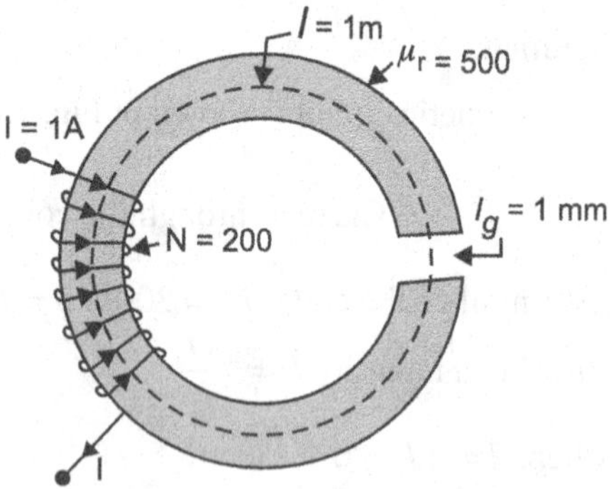

Fig. 1.13 Magnetic circuit

i.e.,

$$NI = \phi\left(\frac{l_i}{a\,\mu_0\,\mu_r} + \frac{l_g}{a\,\mu_0}\right)$$

or

$$NI = B\left(\frac{l_i}{\mu_0\,\mu_r} + \frac{l_g}{\mu_0}\right)$$

where $N = 200$ turns; $I = 1$ A; $\mu_r = 500$

$$l_g = 1 \text{ mm} = 0{\cdot}001 \text{ m};$$

$$l_i = (1 - 0{\cdot}001) = 0{\cdot}999 \text{ m}$$

$$\therefore \quad 200 \times 1 = B\left(\frac{0 \cdot 999}{4\pi \times 10^{-7} \times 500} + \frac{0 \cdot 001}{4\pi \times 10^{-7}}\right)$$

or

$$B = \frac{200}{2385 \cdot 73} = \mathbf{0{\cdot}0838\ Wb/m^2}\ (Ans.)$$

Example 1.13

A rectangular magnetic core shown in Fig. 1.14 (a). has square cross-section of area 16 cm^2. An air-gap of 2 mm is cut across one of its limbs. Find the exciting current needed in the coil having 1000 turns wound on the core to create an air-gap flux of 4 m Wb. The relative permeability of the core is 2000.

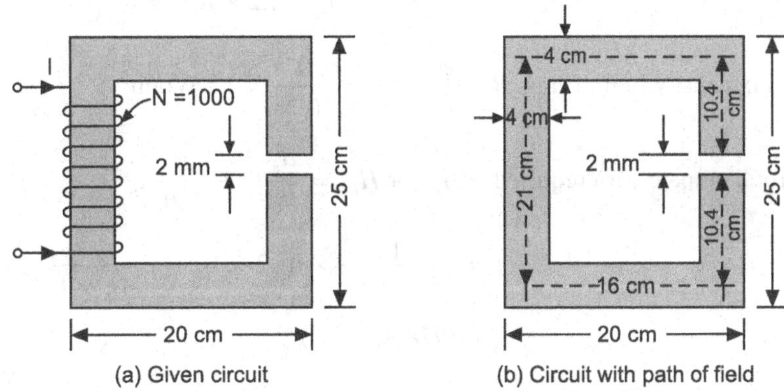

(a) Given circuit (b) Circuit with path of field

Fig. 1.14 Magnetic circuit

Solution:

Here, Area of x-section, $a = 16\ cm^2 = 16 \times 10^{-4}\ m^2$; $l_g = 2\ mm = 2 \times 10^{-3}$ m

No. of turns, $N = 1000$; Flux, $\phi = 4$ m Wb $= 4 \times 10^{-3}$ Wb; $\mu_r = 2000$

$$\text{Flux density required, } B = \frac{\phi}{a} = \frac{4 \times 10^{-3}}{16 \times 10^{-4}} = 2{\cdot}5 \text{ T}$$

$$\text{Each side of the cross-section} = \sqrt{16} = 4 \text{ cm}$$

$$\text{Length of iron-path, } l_i = \left(25 - 2 \times \frac{4}{2} + 20 - 2 \times \frac{4}{2}\right) \times 2 - 0 \cdot 2$$

$$= 73{\cdot}8 \text{ cm} = 0{\cdot}738 \text{ m}$$

$$\text{Total ampere-turns required} = \frac{B}{\mu_0} l_g + \frac{B}{\mu_0\ \mu_r} l_i$$

$$= \frac{2 \cdot 5 \times 2 \times 10^{-3}}{4\pi \times 10^{-7}} + \frac{2 \cdot 5 \times 0 \cdot 738}{4\pi \times 10^{-7} \times 2000}$$

$$= 3979 + 734 = 4713$$

$$\text{Exciting current required, } I = \frac{\text{Total ampere-turns}}{N} = \frac{4713}{1000} = \mathbf{4{\cdot}713\ A}\ (\textit{Ans.})$$

Example 1.14

An iron ring of 10 cm^2 area has a mean circumference of 100 cm. If has a saw cut of 0·2 cm wide. A flux of one mWb is required in the air gap. The leakage factor is 1.2. The flux density of iron for relative permeability 400 is 1.2 Wb/m^2. Calculate the number of ampere-turns required.

Solution:

$$\text{Flux density in air gap, } B_g = \frac{\phi_g}{a} = \frac{1 \times 10^{-3}}{10 \times 10^{-4}} = 1 \text{ Wb/m}^2$$

Flux in iron ring, $\phi_i = \lambda \times \phi_g$ (where λ is leakage factor)

$$= 1.2 \times 1 \times 10^{-3} = 1.2 \times 10^{-3} \text{ Wb}$$

$$\text{Flux density in iron ring, } B = \frac{\phi}{a} = \frac{1 \cdot 2 \times 10^{-3}}{10 \times 10^{-4}} = 1{\cdot}2 \text{ Wb/m}^2$$

$$\text{Total ampere-turn required} = H_g l_g + H_i l_i = \frac{B_g}{\mu_0} l_g + \frac{B_i}{\mu_0 \mu_r} l_i$$

$$= \frac{1}{4\pi \times 10^{-7}} \times 0 \cdot 2 \times 10^{-2} + \frac{1 \cdot 2}{4\pi \times 10^{-7} \times 400} \times 1$$

= **3978·87** (*Ans.*)

Example 1.15

A steel ring 10 *cm mean radius and of circular cross-section* 1 *cm in radius has an air gap of* 1 *mm length. It is wound uniformly with 500 turns of wire carrying current of* 3 *A. Neglect magnetic leakage. The air gap takes* 60% *of the total mmf Find the total reluctance.*

Solution:

Total mmf = NI = 500 × 3 = 1500 ATs

Mmf for air gap = 60% of total mmf = 0·6 × 1500 = 900 ATs

$$\text{Reluctance of air gap, } S_g = \frac{l_g}{a\mu_0}$$

where, l_g = 1 mm = 1 × 10^{-3} m; $a = \pi \times (0{\cdot}01)^2 = \pi \times 10^{-4}$ m^2;

$$\therefore \quad S_g = \frac{1 \times 10^{-3}}{\pi \times 10^{-4} \times 4\pi \times 10^{-7}} = \frac{10^8}{4\pi^2} \text{ ATs/Wb}$$

$$\text{Flux in the air gap, } \phi_g = \frac{\text{M.M.F. of iron}}{\text{reluctance}} = \frac{900}{10^8} \times 4\pi^2 = 36\pi^2 \times 10^{-6} \text{ Wb}$$

Mmf for iron = Total mmf – air mmf = 1500 – 900 = 600 ATs

Flux in the iron ring, $\phi_i = \phi_g = 36\,\pi^2 \times 10^{-6}$ Wb

(*since there is no magnetic leakage*)

$$\text{Reluctance of iron ring, } S_i = \frac{\text{M.M.F. of iron}}{\phi_i} = \frac{600}{36\pi^2 \times 10^{-6}} = \frac{10^8}{6\pi^2} \text{ AT/Wb}$$

$$\text{Total reluctance, } S = S_g + S_i = \frac{10^8}{4\pi^2} + \frac{10^8}{6\pi^2} = \frac{10^8}{\pi^2}\left(\frac{1}{4} + \frac{1}{6}\right)$$

$$= \mathbf{4{\cdot}22 \times 10^6\ ATs/Wb}\ (\textit{Ans.})$$

Example 1.16

Determine magnetomotive force, magnetic flux, reluctance and flux density in case of a steel ring 30 cm mean diameter and a circular cross-section 2 cm in diameter has an air gap 1 mm long. It is wound uniformly with 600 turns of wire carrying a current of 2.5 A. Neglect magnetic leakage. The iron path takes 40% of the total magnetomotive force.

Solution:

Mmf of the magnetic circuit = $NI = 600 \times 2{\cdot}5 =$ **1500 ATs** (*Ans.*)

As iron path takes 40% of the total mmf, the reluctance of iron in 40% and the rest of the reluctance (60%) is of air path.

$$\therefore \quad \frac{S_a}{S_i} = \frac{60}{40} = \frac{3}{2} = 1{\cdot}5$$

$$\text{Reluctance of air path, } S_a = \frac{l_a}{a\,\mu_0}$$

where $l_a = 1 \times 10^{-3}$ m; $a = \frac{\pi}{4}(2)^2 \times 10^{-4} = \pi \times 10^{-4}$ m^2;

$$\therefore \quad S_a = \frac{1 \times 10^{-4}}{\pi \times 10^{-4} \times 4\pi \times 10^{-7}} = 2{\cdot}533 \times 10^6 \text{ ATs/Wb}$$

$$\text{Reluctance of iron path, } S_i = \frac{S_a}{1 \cdot 5} = \frac{2 \cdot 533 \times 10^6}{1 \cdot 5} = 1{\cdot}688 \times 10^6 \text{ ATs/Wb}$$

$$\text{Total reluctance} = S_a + S_i = (2{\cdot}533 + 1{\cdot}688) \times 10^6 \text{ ATs/Wb}$$

$$= \mathbf{4{\cdot}221 \times 10^6\ ATs/Wb}\ (\textit{Ans.})$$

$$\text{Magnetic flux, } \phi = \frac{\text{m.m.f.}}{\text{reluctance}} = \frac{1500}{4 \cdot 221 \times 10^6} = \mathbf{0{\cdot}3554\ m\ Wb}\ (\textit{Ans.})$$

$$\text{Flux density, } B = \frac{\phi}{a} = \frac{0 \cdot 3554 \times 10^{-3}}{\pi \times 10^{-4}} = \mathbf{1{\cdot}131\ Wb/m^2}\ (\textit{Ans.})$$

Example 1.17

An iron ring is made up of three parts; $l_1 = 10$ cm, $a_1 = 5$ *cm*^{2}; $l_2 = 8$ cm, $a_2 = 3$ cm^2; $l_3 = 6$ cm, $a_3 = 2{\cdot}5$ cm^2. *It is wound with a* 250 *turns coil. Calculate current required to produce flux of* 0·4 *mWb.* $\mu_1 = 2670$, $\mu_2 = 1050$, $\mu_3 = 600$.

Solution:

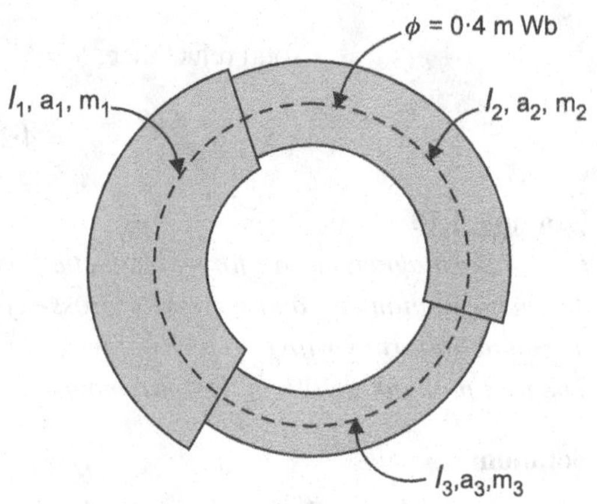

Fig. 1.15 Series magnetic circuit as per data

Flux density $B_1 = \frac{\phi}{a_1} = \frac{0 \cdot 4 \times 10^{-3}}{5 \times 10^{-4}}$

$= 0{\cdot}8 \text{ Wb/m}^2$

$B_2 = \frac{\phi}{a_2} = \frac{0 \cdot 4 \times 10^{-3}}{3 \times 10^{-4}}$

$= 1{\cdot}33 \text{ Wb/m}^2$

$B_3 = \frac{\phi}{a_3} = \frac{0 \cdot 4 \times 10^{-3}}{2 \cdot 5 \times 10^{-4}}$

$= 1{\cdot}6 \text{ Wb/m}^2$

Total Ampere-turns required

$$AT = \frac{B_1}{\mu_0 \mu_1} l_1 + \frac{B_2}{\mu_0 \mu_2} l_2 + \frac{B_3}{\mu_0 \mu_3} l_3$$

$$= \frac{1}{4\pi \times 10^{-7}} \left[\frac{0 \cdot 8}{2670} \times 0 \cdot 10 + \frac{1 \cdot 33}{1050} \times 0 \cdot 08 + \frac{1 \cdot 6}{600} \times 0 \cdot 06 \right]$$

$$= 231{\cdot}92$$

Current required to produce given flux,

$$I = \frac{AT}{N} = \frac{231 \cdot 92}{250} = \mathbf{0{\cdot}928\ A}\ (Ans.)$$

Example 1.18

The ring shaped core shown in Fig. 1.16 is made of a material having a relative permeability of 1000. The flux density in the smallest area of cross-section is 2 T. If the current through the coil is not to exceed 1·5 A, compute the number of turns of the coil.

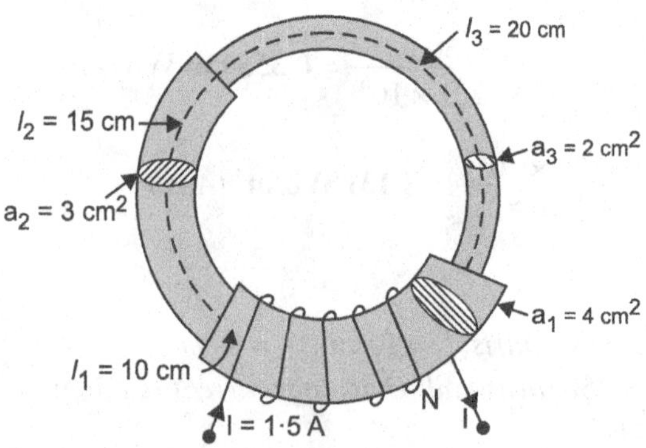

Fig. 1.16 Series magnetic circuit as per data

Solution:

Flux in the core, $\phi = B \times A = 2 \times 2 \times 10^{-4} = 4 \times 10^{-4}$ Wb

Total reluctance of the magnetic path,

$$S = S_1 + S_2 + S_3$$

$$= \frac{l_1}{a_1 \mu_0 \mu_r} + \frac{l_2}{a_2 \mu_0 \mu_r} + \frac{l_3}{a_3 \mu_0 \mu_r}$$

$$= \frac{1}{\mu_0 \mu_r}\left(\frac{l_1}{a_1} + \frac{l_2}{a_2} + \frac{l_3}{a_3}\right)$$

$$= \frac{1}{4\pi \times 10^{-7} \times 1000}\left(\frac{0 \cdot 1}{4 \times 10^{-4}} + \frac{0 \cdot 15}{3 \times 10^{-4}} + \frac{0 \cdot 2}{2 \times 10^{-4}}\right)$$

$$= 13{\cdot}926 \times 10^5 \text{ AT/Wb}$$

Total mmf required, $NI = \phi S$

or $$N \times 1{\cdot}5 = 4 \times 10^{-4} \times 13{\cdot}926 \times 10^5$$

$\therefore$ No. of turns, N = **371·36** (*Ans.*)

Example 1.19

The magnetic frame shown in Fig. 1.17 is built-up of iron of square cross-section, 3 cm side. Each air gap is 2 mm wide. Each of the coil is wound with 1000 turns and the exciting current is 1·0 A. The relative permeability of part A and part B may be taken as 1000 and 1200 respectively.

Calculate, (i) reluctance of part A; (ii) reluctance of part B; (iii) reluctance of two air gaps; (iv) total reluctance of the complete magnetic circuit; (v) mmf produced and (vi) flux set-up in the circuit.

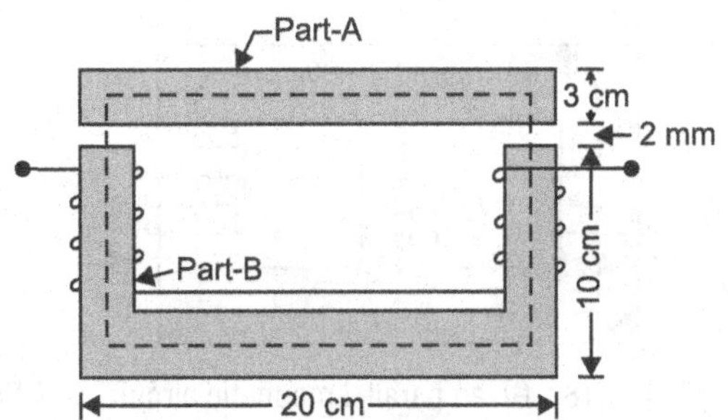

Fig. 1.17 Series magnetic circuit

Solution:

(*i*) Reluctance of path A, $S_A = \frac{l_A}{a\mu_0\mu_{rA}}$

where, $l_A = 20 - (1{\cdot}5 + 1{\cdot}5) + (1{\cdot}5 + 1{\cdot}5) = 20$ cm $= 0{\cdot}2$ m

$a = 3 \times 3 = 9$ cm$^2 = 9 \times 10^{-4}$ m^2; $\mu_{rA} = 1000$;

$\therefore$ $$S_A = \frac{0 \cdot 2}{9 \times 10^{-4} \times 4\pi \times 10^{-7} \times 1000}$$ = **176839 AT/Wb** (*Ans.*)

(*ii*) Reluctance of path *B*, $S_B = \frac{l_B}{a\mu_0\mu_{rB}}$

where, $l_B = (20 - 1{\cdot}5 - 1{\cdot}5) + 2\,(10 - 1{\cdot}5) = 34$ cm $= 0{\cdot}34$ m; $\mu_{rB} = 1200$

$$S_B = \frac{0\cdot 34}{9\times 10^{-4}\times 4\pi\times 10^{-7}\times 1200} = \mathbf{250521\ AT/Wb}\ (Ans.)$$

(*iii*) Reluctance of two air gaps, $S_g = \dfrac{l_g}{a\mu_0}$

where, $l_g = 2 + 2 = 4$ mm $= 4\times 10^{-3}$ m

$\therefore$ $$S_g = 4\times 10^{-3}/9\times 10^{-4}\times 4\pi\times 10^{-7} = \mathbf{3536776\ AT/Wb}\ (Ans.)$$

(*iv*) Total reluctance of the composite magnetic circuit,

$$S = S_A + S_B + S_g = 176839 + 250521 + 3536776$$

$$= \mathbf{3964136\ AT/Wb}\ (Ans.)$$

(*v*) Total mmf $= NI = (2\times 1000)\times 1 =$ **2000 AT** (*Ans.*)

(*vi*) Flux set-up in the circuit, $\phi = \dfrac{\text{m.m.f.}}{\text{reluctance}} = \dfrac{2000}{3964136} = \mathbf{0{\cdot}5045\ m\ Wb}\ (Ans.)$

Example 1.20

A magnetic core made of annealed sheet steel has the dimensions as shown in Fig. 1.18. The cross-section everywhere is 25 cm². The flux in branches A and B is 3500 m Wb, but that in the branch C is zero. Find the required ampere-turns for coil A and for coil C. Relative permeability of sheet steel is 1000.

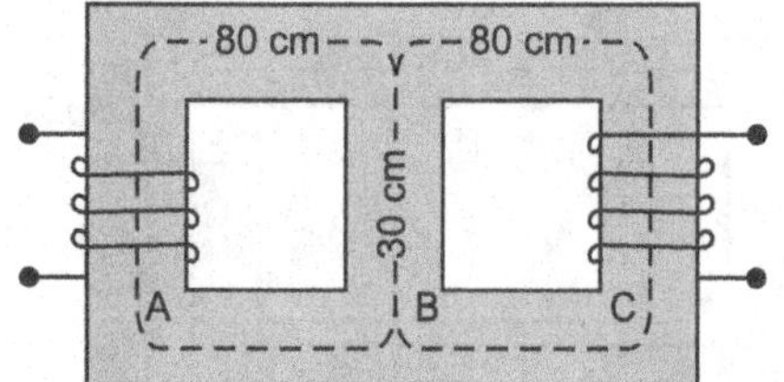

Fig. 1.18 Given parallel magnetic circuit

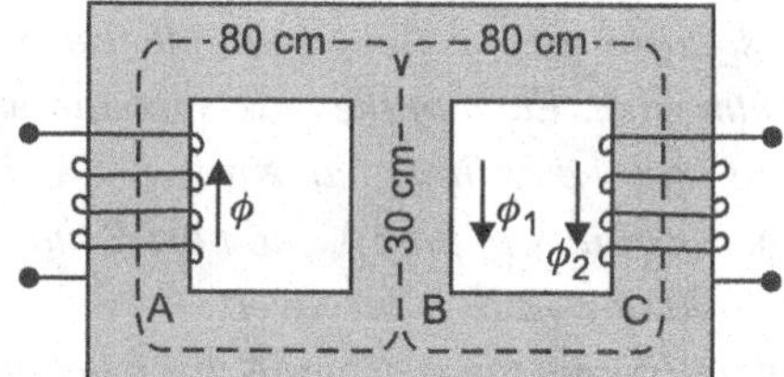

Fig. 1.19 Flux distribution in the parallel magnetic circuit

Solution:

The given magnetic circuit is a parallel circuit. To determine the ATs for coil '*A*', the flux distribution is shown in Fig. 1.19.

Since path '*B*' and '*C*' are in parallel with each other w.r.t. path '*A*',

$\therefore$ mmf for path '*B*' = mmf for path *C*,

i.e., $$\phi_1 S_1 = \phi_2 S_2$$

i.e., $$\frac{3500\times 10^{-6}\times 30\times 10^{-2}}{a\mu_0\mu_r} = \phi_2\times\frac{80\times 10^{-2}}{a\mu_0\mu_r}$$

$\therefore$ $$\phi_2 = 1312{\cdot}5\times 10^{-6}\ \text{Wb}$$

Total flux in the path '*A*', $\phi = \phi_1 + \phi_2$

$$= (3500 + 1312{\cdot}5) \times 10^{-6} = 4812{\cdot}5 \times 10^{-6}\ \text{Wb}$$

Actual (resultant) flux in path '*A*' $= \phi - \phi_2 = 3500 \times 10^{-6}$ Wb

∴ ATs required for coil '*A*' = ATs for path '*A*' + ATs for path '*B*' or '*C*'

$$= \frac{3500 \times 10^{-6}}{4\pi \times 10^{-7} \times 1000 \times 25 \times 10^{-4}}(0 \cdot 8 + 0 \cdot 3)$$

= **1225·5** (*Ans.*)

To neutralise the flux in section '*C*', the coil produces flux of 1312·5 μ Wb in opposite direction.

∴ ATs required for coil '*C*' = ATs for path '*C*' only

$$= \frac{1312 \cdot 5 \times 10^{-6} \times 0 \cdot 8}{4\pi \times 10^{-7} \times 1000 \times 25 \times 10^{-4}} = \mathbf{334{\cdot}22}\ (\textit{Ans.})$$

1.9 Magnetisation or B-H Curve

The graph plotted between flux density *B* and magnetising force H of a material is called the *magnetisation or B–H Curve* of that material.

The general shape of the *B–H* Curve of a magnetic material is shown in Fig. 1.20. The shape of the curve is non-linear This indicates that the relative permeability ($\mu_r = B/\mu_0 H$) of a magnetic material is not constant but it varies. The value of μ_r largely depends upon the value of flux density. Its shape is shown in Fig. 1.21 (for cast steel).

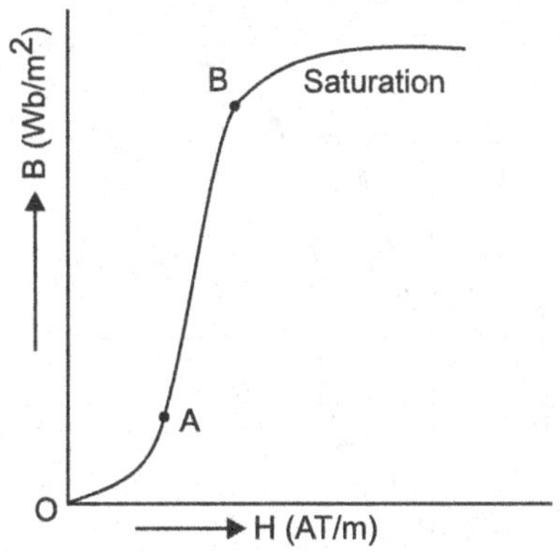

Fig. 1.20 B-H curve of a magnetic material

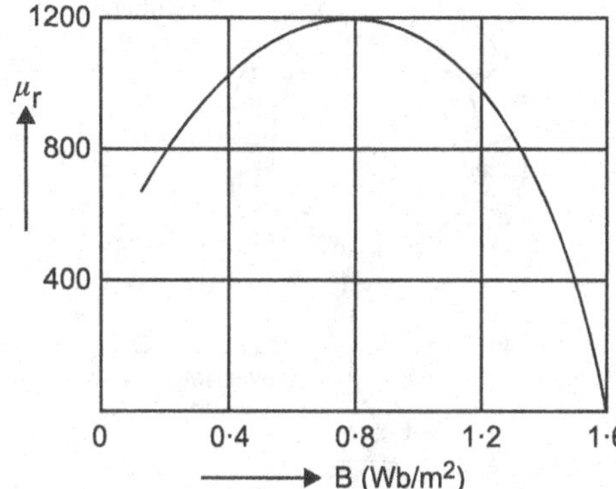

Fig. 1.21 μ_r-B curve

The *B–H* curves of some of the common magnetic materials are shown in Fig. 1.22. The *B–H* curve for a non-magnetic material is shown in Fig. 1.23. It is a straight line curve since $B = \mu_0 H$ or $B \propto H$ as the value of μ_0 is constant.

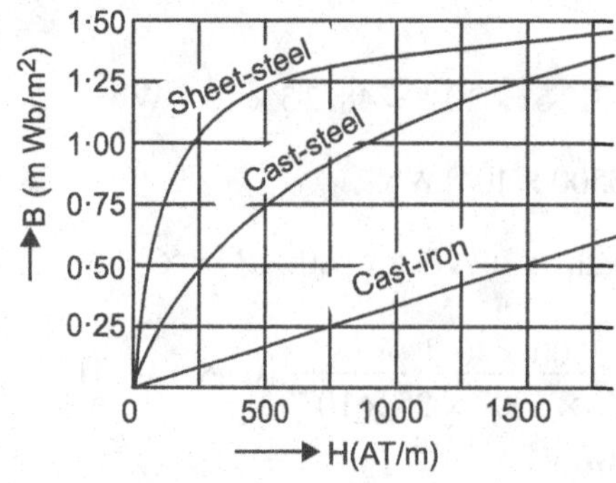

Fig. 1.22 B-H curve for different magnetic materials

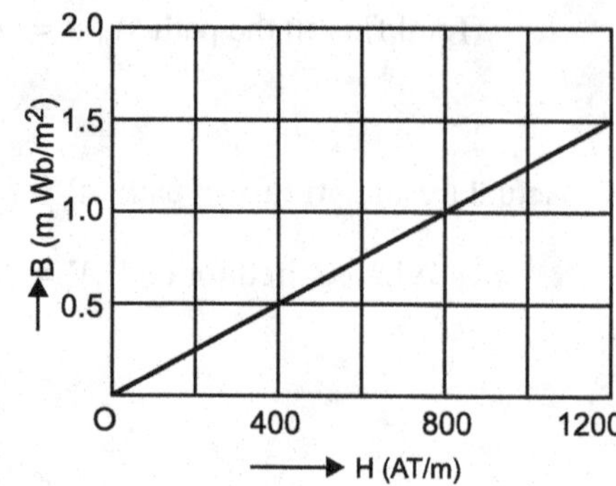

Fig. 1.23 B-H curve for a non-magnetic materials

1.10 Magnetic Hysteresis

When a magnetic material is magnetised first in one direction and then in the other (i.e., one cycle of magnetisation), it is found that flux density *B* in the material lags behind the applied magnetising force *H*. This phenomenon is known as *magnetic hysteresis.*

Hence, *the phenomenon of flux density B lagging behind the magnetising force H in a magnetic material is called* **magnetic hysteresis.**

'Hysteresis' is the term derived from the Greek word *hysterein* meaning to lag behind. To understand the complete phenomenon of magnetic hysteresis, consider a ring of magnetic material on which a solenoid is wound uniformly as shown in Fig. 1.24. The solenoid in connected to a DC source through a double pole double throw reversible switch (position '1').

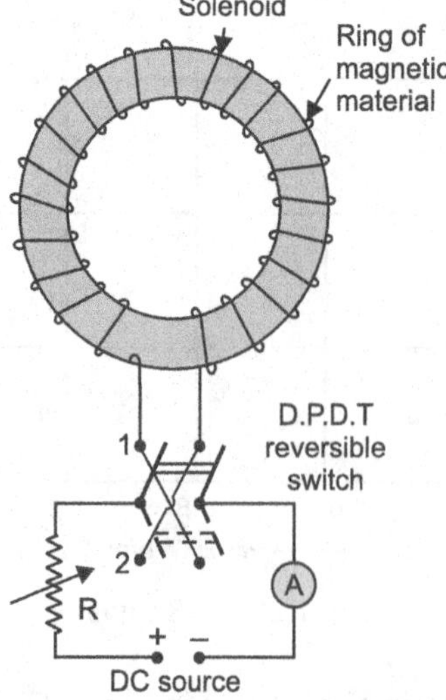

Fig. 1.24 Circuit to trace hysteresis loop

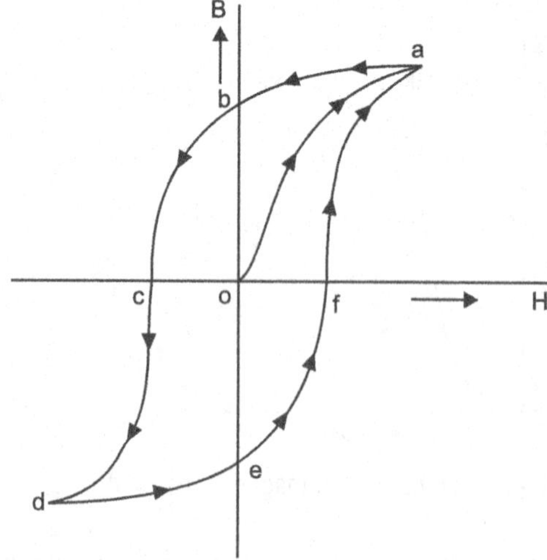

Fig. 1.25 Hysteresis loop

When the field intensity H is increased gradually by increasing current in the solenoid (by decreasing the value of R), the flux density B also increases until saturation point a is reached and curve so obtained is *oa*. If now the magnetising force is gradually reduced to zero by decreasing current in the solenoid to zero. The flux density does not become zero and the curve so obtained is *ab* as shown in Fig. 1.25. When magnetising force H is zero, the flux density still has value *ob*.

Residual Magnetism and Retentivity

This value of flux density 'ob' retained by the magnetic material is called **residual magnetism** *and the power of retaining this residual magnetism is called* **retentivity** *of the material.*

To demagnetise the magnetic ring, the magnetising force H is reversed by reversing the direction of flow of current in the solenoid. This is achieved by changing the position of double pole, double throw switch (i.e., position '2'). When H is increased in reverse direction, the flux density starts decreasing and becomes zero and curve follows the path *bc*. Thus residual magnetism of the magnetic material is wiped off by applying magnetising force *oc* in opposite direction.

Coercive Force

This value of magnetising force oc required to wipe off the residual magnetism is called **coercive force.**

To complete the loop, the magnetising force H is increased further in reverse direction till saturation reaches (point '*d*') and the curve follows the path *cd*. Again H is reduced to zero and the curve follows the path *de*. Where *oe* represents the residual magnetism. Then H is increased in the positive direction by changing the position of reversible switch to position '1' and increasing the flow of current in the solenoid. The curve follows the path of *efa* and the loop is completed. Again *of* is the magnetising force utilised to wipe off the residual magnetism *oe*.

Hence, *cf* is the total coercive force required in one cycle of magnetisation to wipe off the residual magnetism.

Since the meaning of hysteresis is lagging behind, and in this case flux density B always lags behind the magnetising force, H, therefore, loop (*abcdefa*) so obtained is called *hysteresis loop*.

1.11 Hysteresis Loss

When a magnetising force is applied, the magnetic material is magnetised and the molecular magnets are lined up in a particular direction.

However, when the magnetising force in a magnetic material is reversed, the internal friction of the molecular magnets opposes the reversal of magnetism, resulting in hysteresis. To overcome this internal friction of the molecular magnets (or to wipe off the residual magnetism), a part of the magnetising force is used. The work done by the magnetising force against this internal friction of molecular magnets produces heat. This energy, which is wasted in the form of heat due to hysteresis, is called *hysteresis loss*.

Hysteresis loss occurs in all the magnetic parts of electrical machines where there is reversal of magnetisation. This loss results in wastage of energy in the form of heat. Consequently, it increases the temperature of the machine which is undesirable. Therefore, a suitable magnetic material is selected for the construction of such parts, e.g., silicon steel is most suitable in which hysteresis loss is minimum.

1.12 Importance of Hysteresis Loop

The shape and size of hysteresis loop of a magnetic material largely depends upon its nature. For a particular location, the choice of the magnetic material depends upon the shape and size (i.e., area) of its hysteresis loop. The hysteresis loops of some of the common magnetic materials are shown in Fig. 1.26.

(i) **Hard steel:** The hysteresis loop for hard steel is shown in Fig. 1.26 (*a*). This loop has larger area which indicates that this material will have more hysteresis loss. Therefore, it is never used for the construction of machine parts. However, its loop shows that the material has high retentivity and coercivity. Therefore, it is more suitable for making permanent magnets.

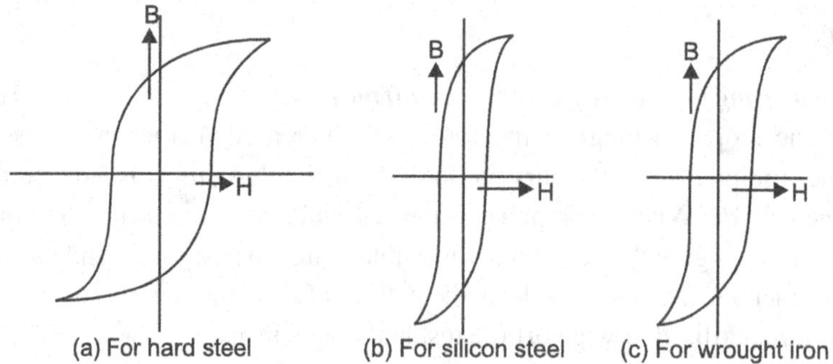

Fig. 1.26 Hysteresis loop for different magnetic materials

(ii) **Silicon steel:** The hysteresis loop for silicon steel is shown in Fig. 1.26 (*b*). This loop has smallest area which indicates that this material will have small hysteresis loss. Therefore, it is most suitable for the construction of those parts of electrical machines in which reversal of magnetisation is very quick e.g., armature of DC machines, transformer core, starter of induction motors etc.

(iii) **Wrought iron:** Figure 1.26 (*c*) shows the hysteresis loop for wrought iron. This loop shows that this material has fairly good residual magnetism and coercivity. Therefore, it is best suited for making cores of electromagnets.

Section Practice Problems

Numerical Problems

1. An iron ring has a cross-sectional area of 400 mm^2 and a mean diameter of 25 cm. It is wound with 500 turns. If the value of relative permeability is 500, find the total flux set-up in the ring. The coil resistance is 400 Ω and the supply voltage is 200 V. **(Ans.** *0·08 m Wb*)

2. An iron ring of mean diameter 22 cm and cross-section 10 cm^2 has an air gap 1 mm wide. The ring is wound uniformly with 200 turns of wire. The permeability of ring material is 1000. A flux of 0·16 mWb is required in the gap. What current should be passed through the wire? (**Ans.** *1·076 A*)

3. An iron ring has cross-section 3 cm^2 and a mean diameter of 25 cm. An air gap of 0·4 mm has been made by saw cut across the section. The ring is wound with 200 turns through which a current of 2 A is passed. If the total flux is 21×10^{-5} weber, find μ for iron assuming no leakage. (**Ans.** *2470*)

4. An iron ring has a mean circumferential length of 60 cm with an air gap of 1 mm and a uniform winding of 300 turns. When a current of 1 A flows through the coil, find the flux density. The relative permeability of iron is 300. Assume $\mu_0 = 4\pi \times 10^{-7}$ H/m. (**Ans.** *0·1256 T*)

5. In the magnetic circuit shown in Fig. 1.27, a coil of 500 turns is wound on the central limb. The magnetic path from A to B by way of outer limbs have a mean length of 100 cm each and an effective cross-sectional area of 2·5 cm^2. The central limb is 25 cm long and 5 cm^2 cross-sectional area. The air gap is 0·8 cm long. Calculate the current flowing through the coil to produce a flux of 0·3 m Wb in the air gap. The relative permeability of the core material is 800 (neglect leakage and fringing).

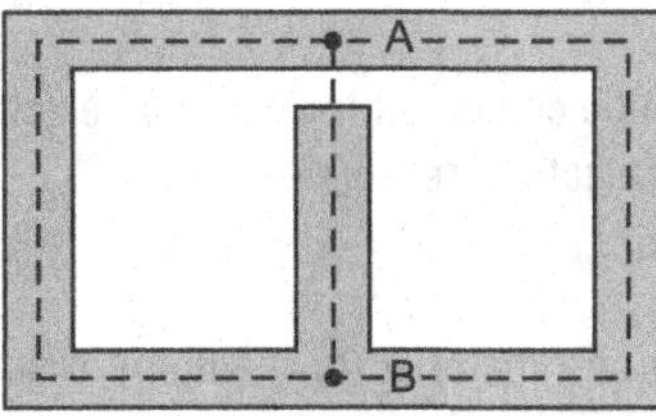

Fig. 1.27:

Short Answer Type Questions

Q.1. What is a magnet?

Ans. A substance that attracts pieces of iron is called a magnet.

Q.2. What are permanent and temporary magnets?

Ans. A magnet that retains magnetism permanently is called a permanent magnet. Whereas a magnet in which magnetism remains temporarily is called a temporary magnet. A wire wound soft iron piece becomes a temporary magnet when a *DC* is passed through the wire.

Q.3. What are magnetic poles?

Ans. The ends of a magnet from where the magnetic lines of force appear to emit or enter are called magnetic poles, these are identified as north and south poles respectively.

Q.4. What do you understand by magnetic field?

Ans. The space occupied by the magnetic lines of force around a magnet is called the magnetic field.

Q.5. What do you mean by magnetic axis?

Ans. Magnetic axis is the imaginary line joining the two poles of a magnet. It is also called as the magnetic equator.

Q.6. Define and explain a magnetic circuit.

Ans. A complete closed path followed by a group of magnetic lines force is called a magnetic circuit. In a magnetic circuit, the magnetic flux leaves from north pole, passes through the circuit and returns to the north pole.

Q.7. Mention at least four properties of magnetic lines of force.

Ans. **1.** Magnetic lines always emanate from north pole and terminate at south pole outside the magnet whereas they are set-up from south to north pole inside the magnet.

2. The magnetic lines of force do not intersect each other.

3. The magnetic lines of force set-up in one direction have a repulsive force between them and therefore do not intersect.

4. The magnetic lines of force always follow the least reluctance path.

Q.8. Explain the term MMF.

Ans. The force which drives the magnetic flux through a magnetic circuit is called the magnetomotive force. It is produced by passing electric current through wire of number of turns. It is measured in ampere-turns (*AT*).

MMF in a magnetic circuit can be compared to *EMF* in an electric circuit. Both are pressure. *MMF* is magnetic pressure and *EMF* is electric pressure.

Q.9. Define relative permeability.

Ans. The relative permeability (μ_r) of a material (or medium) is defined as the ratio of the flux density produced in that material or medium to the flux density produced in vacuum by the same magnetising force.

$$\mu_r = \frac{B}{B_0}$$

where, B – Flux density in the material or medium in tesla

B_0 – Flux density in vacuum

Q.10. Define permeance of magnetic circuit.

Ans. Permeance is the reciprocal of reluctance. It is the ease of the magnetic material with which magnetic flux is set-up in it. It is equivalent to conductance in an electric circuit. Its unit is weber per ampere-turn.

$$\text{Permeance} = \frac{1}{\text{Reluctance}}$$

$$= \frac{1}{l / \mu_0 \mu_r} = \frac{a \mu_0 \mu_r}{l}$$

where, l = length of the magnetic circuit

a – area of cross section of the magnetic circuit

μ_0 – absolute permeability of the magnetic circuit.

μ_r – relative permeability

Q.11. Give similarities of electric and magnetic circuits.

Ans.

Se. No.	*Magnetic Circuit*	*Electric Circuit*
1.	The closed path for magnetic flux is called magnetic circuit.	The closed path for electric current is called electric circuit.
2.	Flux (ϕ) = $\frac{\text{MMF}}{\text{Reluctance}}$	Current (I) = $\frac{\text{emf}}{\text{Resistance}}$
3.	*MMF* (Ampere-turns)	emf (volt)
4.	Reluctance (S) = $\frac{l}{a\mu_0\mu_r}$	Resistance (R) = $\rho\frac{l}{a}$
5.	Reluctivity	Resistivity
6.	Permeance	Conductance
7.	Flux density $B = \frac{\phi}{a}$ Wb/m^2	Current density (J) = $\frac{I}{a}$ ampere/m^2
8.	Magnetic field intensity $H = \frac{NI}{l}$ AT/m	Electric field intensity $E = \frac{V}{d}$ volt/m

Q.12. What is a composite magnetic circuit?

Ans. A composite magnetic circuit consists of different magnetic materials. The magnetic materials have different permeabilities and length it may also have an air gap. Each path will have its own reluctance. Since the materials are in series, the total reluctance is given by the sum of individual reluctances.

Q.13. State 'Ohms law' of a magnetic circuit.

Ans. The 'ohms law' of a magnetic circuit is given by

$$\text{Flux} = \frac{\text{MMF}}{\text{Reluctance}} \quad \text{i.e.,}\ \phi = \frac{NI}{S}$$

The above equation is similar to that of the ohm's law in electric circuit. Flux is analogous to current, *MMF* to *EMF* and reluctance to resistance in electric circuit.

Q.14. Define leakage factor.

Ans. Leakage factor is defined as the ratio of total flux to the useful flux.

Q.15. Why is it necessary to keep air gaps in magnetic circuits as small as possible?

Ans. Usually, the ampere-turns (*AT*) required for the airgap is much greater than that required for the magnetic circuit. It is because the reluctance of air is very high as compared to that offered by iron. Therefore, it is always preferred to keep air gaps in magnetic circuits as small as possible.

Q.16. Why does leakage occur in a magnetic circuit?

Ans. In a parallel magnetic circuit, a large amount of flux follows the intended path. At the same time a small amount of flux leaks through the surrounding air since air is not a magnetic insulator. Therefore, leakage of flux takes place easily. The leakage flux is useless, it is harmful in an electrical machine.

Q.17. What is magnetic fringing?

Ans. While crossing an air gap the magnetic lines of force tend to bulge out. The reason is that the magnetic lines of force repel each other when passing through non-magnetic material. This phenomenon is known as fringing.

Q.18. What is hysteresis in a magnetic material?

Ans. The phenomenon due to which magnetic flux density (B) lags behind the magnetic field intensity (H) in a magnetic material is called *magnetic hysteresis.*

Q.19. Give the units of *MMF*, reluctance, flux and give the relation between them.

Ans.

Quantity	*Unit*
MMF	Ampere-Turn
Flux	Weber
Reluctance	AT/weber

$$\text{Reluctance} = \frac{MMF}{\text{Flux}}$$

1.13 Electro Magnetic Induction

The phenomenon by which an emf is induced in a circuit (and hence current flows when the circuit is closed) when magnetic flux linking with it changes is called *electro-magnetic induction.*

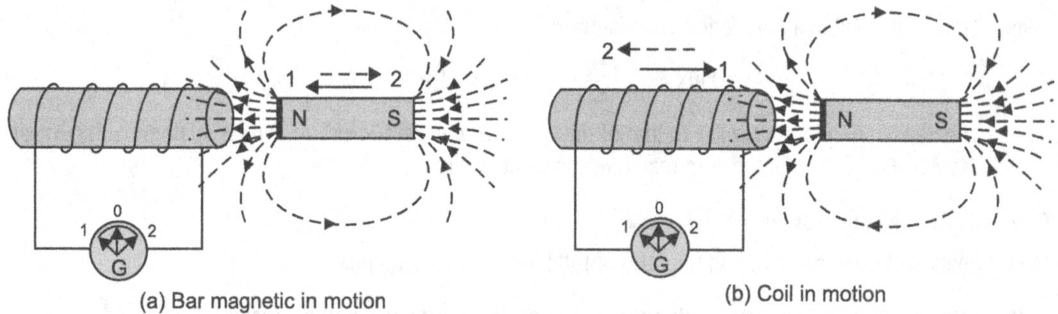

(a) Bar magnetic in motion

(b) Coil in motion

Fig. 1.28 Electromagnetic induction

For illustration, consider a coil having a large number of turns to which galvanometer is connected. When a permanent bar magnet is taken nearer to the coil or away from the coil, as shown in Fig. 1.28 (*a*), a deflection occurs in the needle of the galvanometer. Although, the deflection in the needle is opposite is two cases.

On the other hand, if the bar magnet is kept stationary and the coil is brought nearer to the magnet or away from the magnet, as shown in Fig. 1.28 (*b*), again a deflection occurs in the needle of the galvanometer. The deflection in the needle is opposite in the two cases.

However, if the magnet and the coil both are kept stationary, no matter how much flux is linking with the coil, there is no deflection in the galvanometer needle.

The following points are worth noting:

(*i*) The deflection in the galvanometer needle shows that emf is induced in the coil. This condition occurs only when flux linking with the circuit changes i.e., either magnet or coil is in motion.

(*ii*) The direction of induced emf in the coil depends upon the direction of magnetic field and the direction of motion of coil.

1.14 Faraday's Laws of Electromagnetic Induction

Michael Faraday summed up conclusions of his experiments regarding electro-magnetic induction into two laws, known as *Faraday's laws of electro-magnetic induction.*

First Law: This law states that "*Whenever a conductor cuts across the magnetic field, an emf is induced in the conductor.*"

or

"*Whenever the magnetic flux linking with any circuit (or coil) changes, an emf is induced in the circuit.*"

Figure 1.29 shows a conductor placed in the magnetic field of a permanent magnet to which a galvanometer is connected. Whenever, the conductor is moved upward or downward i.e., across the field, there is deflection in the galvanometer needle which indicates that an emf is induced in the conductor. If the conductor is moved along (parallel) the field, there is no deflection in the needle which indicates that no emf is induced in the conductor.

For the second statement, consider a coil placed near a bar magnet and a galvanometer connected across the coil, as shown in Fig. 1.30. When the bar magnet (N-pole) is taken nearer to the coil [see Fig. 1.30 (*a*)], there is deflection in the needle of the galvanometer. If now the bar magnet (*N*-pole) is taken away from the coil [see Fig. 1.30 (*b*)], again there is deflection in the needle of galvanometer but in opposite direction. The deflection in the needle of galvanometer indicates that emf is induced in the coil.

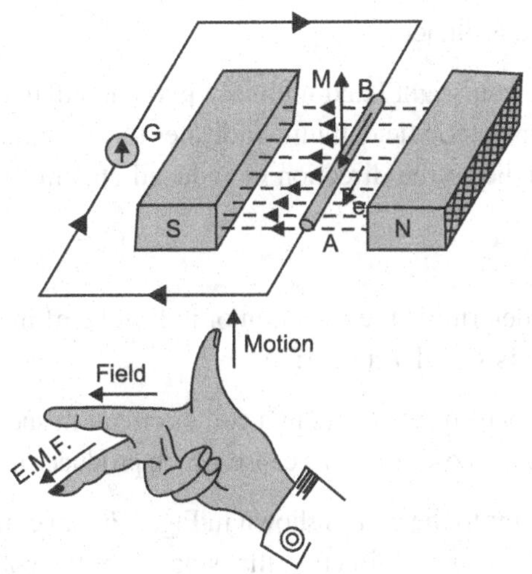

Fig. 1.29 Conductor moving in the field

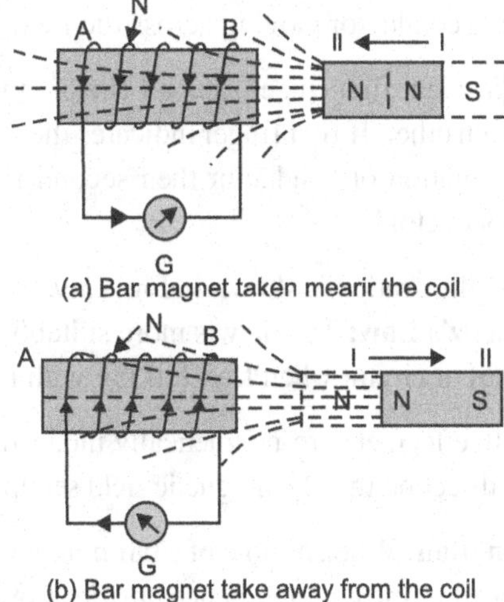

Fig. 1.30 Coil is stationary but bar magnet (field) is moving

Second Law: This law states that "*The magnitude of induced emf in a coil is directly proportional to the rate of change of flux linkages.*

Rate of change of flux linkages $= \dfrac{N(\phi_2 - \phi_1)}{t}$ Wb-turns/s

where, N = No. of turns of the coil; $(\phi_2 - \phi_1)$ = change of flux in Wb

t = time in second for the change

∴ According to Faraday's second law of electro-magnetic induction;

Induced emf, $$e \propto \frac{N(\phi_2 - \phi_1)}{t}$$

$$e = \frac{N(\phi_2 - \phi_1)}{t} \quad \text{(taking proportionality constant, as unity)}$$

In differential form, $e = N\dfrac{d\phi}{dt}$ volt

Usually, a minus sign is given to the right-hand side expression which indicates that emf is induced in such a direction which opposes the cause (i.e., change in flux) that produces it (according to Lenz's law).

$$e = -N\frac{d\phi}{dt} \text{ volt}$$

1.15 Direction of Induced emf

The direction of induced emf and hence current in a conductor or coil can be determined by either of the following two methods:

1. **Fleming's Right Hand Rule:** This rule is applied to determine the direction of induced emf in a conductor moving across the field and is stated as under;

   "Stretch, first finger, second finger, and thumb of your right hand mutually perpendicular to each other. If first finger indicates the direction of magnetic field, thumb indicates the direction of motion of conductor then second finger will indicate the direction of induced emf in the conductor."

   Its illustration is shown in Fig. 1.29.
2. **Lenz's Law:** This law is more suitably applied to determine the direction of induced emf in a coil or circuit when flux linking with it changes. It is stated as under:

   "In effect, electro-magnetically induced emf and hence current flows in a coil or circuit in such a direction that the magnetic field set up by it, always opposes the very cause which produces it."

Explanation: When *N*-pole of a bar magnet is taken nearer to the coil as shown in Fig. 1.30, an emf is induced in the coil and hence current flows through it in such a direction that side '*B*' of the coil attains North polarity which opposes the movement of the bar magnet. Whereas, when *N*-pole of the bar magnet is taken away from the coil as shown in Fig. 1.30, the direction of emf induced in the coil is reversed and side 'B' of the coil attains South polarity which again opposes the movement of the bar magnet.

1.16 Induced emf

When flux linking with a conductor (or coil) changes, an emf is induced in it. This change in flux linkages can be obtained in the following two ways:

(*i*) By either moving the conductor and keeping the magnetic field system stationary or moving the magnetic field system and keeping the conductor stationary, in such a way that conductor cuts across the magnetic field (as in case of DC and AC generators). The emf induced in this way is called *dynamically induced emf*

(*ii*) By changing the flux linking with the coil (or conductor) without moving either coil or field system. However, the change of flux produced by the field system linking with the coil is obtained by changing the current in the field system (solenoid), as in transformers. The emf induced in this way is called *statically induced emf*

1.17 Dynamically Induced emf

By either moving the conductor keeping the magnetic field system stationary or moving the field system keeping the conductor stationary so that flux is cut by the conductor, the emf thus induced in the conductor is called *dynamically induced emf*

Mathematical Expression

Considering a conductor of length l metre placed in the magnetic field of flux density B Wb/m^2 is moving at right angle to the field at a velocity υ metre/second as shown in Fig. 1.31 (*a*). Let the conductor be moved through a small distance dx metre in time dt second as shown in Fig. 1.31 (*b*).

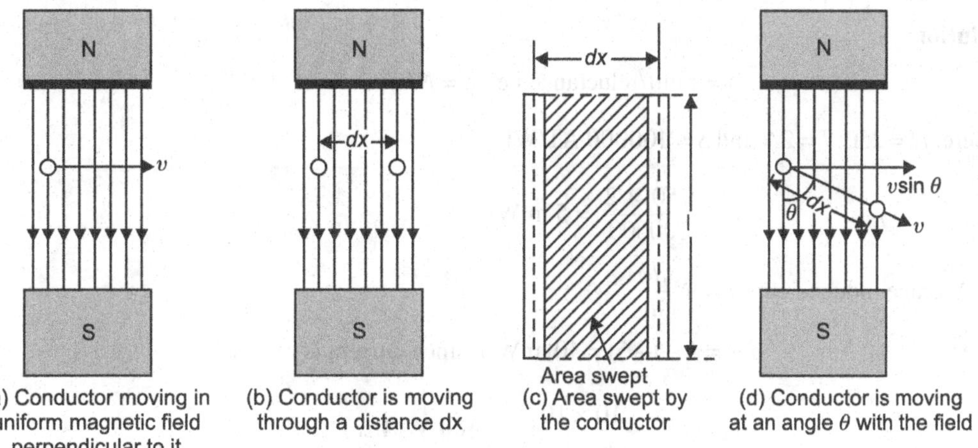

(a) Conductor moving in uniform magnetic field perpendicular to it
(b) Conductor is moving through a distance dx
(c) Area swept by the conductor
(d) Conductor is moving at an angle θ with the field

Fig. 1.31 Dynamically induced emf

Area swept by the conductor, $A = l \times dx$

Flux cut by the conductor, $\phi = B \times A = B\, l\, dx$

According to Faraday's Law of electro-magnetic induction;

$$\text{Induced emf, } e = \frac{\text{flux cut}}{\text{time}} = \frac{\phi}{dt} = \frac{B\, l\, dx}{dt} = B\, l\, \upsilon \qquad \text{(since } dx/dt = \upsilon \text{ (velocity))}$$

Now, if the conductor is moved at an angle θ with the direction of magnetic field at a velocity υ metre/second as shown in Fig. 1.31 (*d*). A small distance covered by the conductor in that direction is *dx* in time *dt* second. Then the component of distance perpendicular to the magnetic field, which produces emf, is $dx \sin \theta$.

$\therefore$ Area swept by the conductor, $A = l \times dx \sin \theta$

Flux cut by the conductor, $\phi = B \times A = B\, l\, dx \sin \theta$

$$\text{Induced emf } e = \frac{Bl\, dx \sin \theta}{dt} = B\, l\, \upsilon \sin \theta$$

Example 1.21

A coil of 500 turns in linked with a flux of 2 mWb. If this flux is reversed in 4 ms, calculate the average emf induced in the coil.

Solution:

$$\text{Average induced emf, } e = N \frac{d\phi}{dt}$$

where, $N = 500$ turns; $d\,\phi = 2 - (-2) = 4\ m\ Wb$; $dt = 4 \times 10^3\ s$;

$$\therefore \qquad e = 500 \times \frac{4 \times 10^{-3}}{4 \times 10^{-3}} = \mathbf{500\ V}\ (Ans.)$$

Example 1.22

A coil of 250 turns is wound on a magnetic circuit of reluctance 100000 AT/Wb. If a current of 2 A flowing in the coil is reversed in 5 ms, find the average emf induced in the coil.

Solution:

$$\phi = \text{mmf/reluctance i.e., } \phi = NI/S$$

where, $N = 250$; $I = 2$A and $S = 100000$ AT/Wb.

$$\therefore \qquad \phi = \frac{250 \times 2}{100000} = 5 \text{ m Wb}$$

$$\text{Average induced emf } e = N \frac{d\phi}{dt}$$

where, $d\,\phi = 5 - (-5) = 10$ m Wb (since current is reversed)

$$e = 250 \times \frac{10 \times 10^{-3}}{5 \times 10^{-3}} = \mathbf{500\ V}\ (Ans.)$$

1.18 Statically Induced emf

When the coil and magnetic field system both are stationary but the magnetic field linking with the coil changes (by changing the current producing the field), the emf thus induced in the coil is called *statically induced emf*

The statically induced emf may be:

(*i*) Self induced emf (*ii*) Mutually induced emf

(*i*) **Self induced emf:** The emf induced in a coil due to the change of flux produced by it linking with its own turns is called *self induced emf* as shown in Fig. 1.32.

The direction of this induced emf is such that it opposes the cause which produces it (Lenz's law) i.e., change of current in the coil.

Since the rate of change of flux linking with the coil depends upon the rate of change of current in the coil. Therefore, the magnitude of self induced emf is directly proportional to the rate of change of current in the coil. Therefore, the magnitude of self induced emf is directly proportional to the rate of change of current in the coil, i.e.,

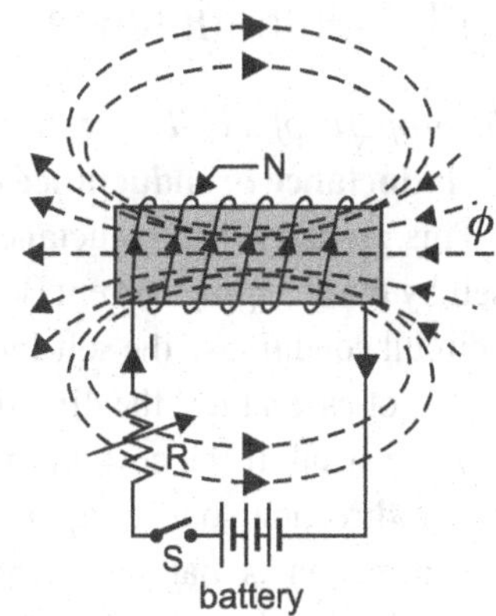

Fig. 1.32 Flux produced by coil linking with its own turns

$$e \propto \frac{dI}{dt} \quad \text{or} \quad e = L\frac{dI}{dt}$$

where L is a constant of proportionality and is called *self inductance or co-efficient of self inductance or inductance of the coil.*

(*ii*) **Mutually induced emf**

The emf induced in a coil due to the change of flux produced by another (neighbouring) coil, linking with it is called *mutually induced emf* as shown in Fig. 1.33.

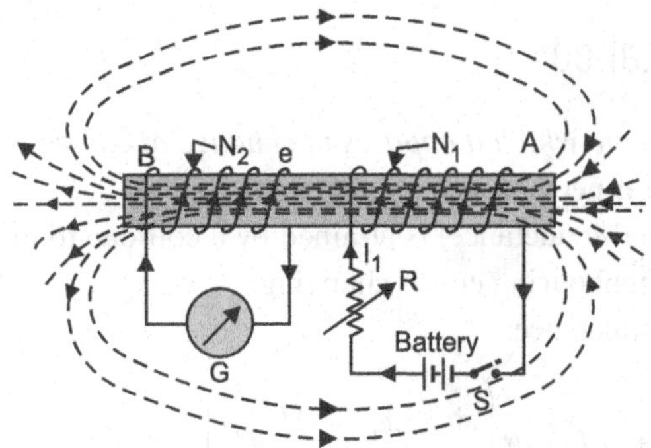

Fig. 1.33 Flux produced by coil-A linking with coil-B

Since the rate of change of flux linking with coil '*B*' depends upon the rate of change of current in coil '*A*'. Therefore, the magnitude of mutually induced emf is directly proportional to the rate of change of current in coil '*A*', i.e.,

$$e_m \propto \frac{dI_1}{dt} \text{ or } e_m = M\frac{dI_1}{dt}$$

where M is a constant of proportionality and is called *mutual inductance* or *co-efficient of mutual inductance.*

1.19 Self Inductance

The property of a coil due to which it opposes the change of current flowing through itself is called **self inductance** *or* **inductance** *of the coil.*

This property (i.e., inductance) is attained by a coil due to self-induced emf produced in the coil itself by the changing current flowing through it. If the current in the coil is increasing (by the change in circuit conditions), the self-induced emf is produced in the coil in such a direction so as to oppose the rise of current i.e., the direction of self-induced emf is opposite to that of the applied voltage. On the other hand, if the current in the coil is decreasing, the self-induced emf is produced in the coil in such direction so as to oppose the fall of current i.e., the direction of self-induced emf is in the same direction as that of the applied voltage. In fact, self-inductance does not prevent the change of current but it delays the change of current flowing through a coil.

It may be noted that this property of the coil only opposes the changing current (i.e., alternating current). However, it does not affect the steady (i.e., direct) current when flows through it. In other words, the self-inductance of the coil (by virtue of its geometrical and magnetic properties) will exhibit its presence to the alternating current but it will not exhibit its presence to the direct current.

Expressions for self inductance:

$$L = \frac{e}{dI / dt}\left(\text{since } e = L\frac{dI}{dt}\right)$$

$$= \frac{N\phi}{I}\left(\text{since } e = N\frac{d\phi}{dt} = L\frac{dI}{dt}\right) = \frac{N^2}{l / a\mu_0\mu_r}\left(\text{since } \phi = \frac{NI}{l / a\mu_0\mu_r}\right)$$

1.20 Mutual Inductance

The property of one coil due to which it opposes the change of current in the other (neighbouring) coil is called **mutual-inductance** *between the two coils.*

This property (i.e., mutual-inductance) is attained by a coil due to mutually induced emf in the coil while current in the neighbouring coil is changing.

Expression for mutual inductance:

$$M = \frac{e_m}{dI_1 / dt}\left(\text{since } e_m = M\frac{dI_1}{dt}\right)$$

$$= \frac{N_2\phi_{12}}{I_1}\left(\text{since } e_m = N_2\frac{d\phi_{12}}{dt} = M\frac{dI_1}{dt}\right)$$

$$= \frac{N_1N_2}{l / a\mu_0\mu_r}\left(\text{since } \phi_{12} = \frac{N_1I_1}{l / a\mu_0\mu_r}\right)$$

1.21 Co-efficient of Coupling

When current flows through one coil, it produces flux (ϕ_1). The whole of this flux may not be linking with the other coil coupled to it as shown in Fig. 32. It may be reduced, because of leakage flux ϕ_l, by a fraction k known as co-efficient of coupling.

Thus, the **fraction** *of magnetic flux produced by the current in one coil that links with the other is known as* **co-efficient of coupling (k)** *between the two coils.*

If the flux produced by one coil completely links with the other, then the value of k is *one* and the coils are said to be magnetically tightly coupled. Whereas, if the flux produced by one coil does not link at all with the other, then the value of k is *zero* and the coils are said to be magnetically isolated.

Mathematical expression: Considering the magnetic circuit shown in Fig. 1.34. When current I_1 flows through coil-1;

$$L_1 = \frac{N_1\phi_1}{I_1} \text{ and } M = \frac{N_2\phi_{12}}{I_1} = \frac{N_2k\phi_1}{I_1} \qquad ...(i) \left(\because \phi_{12} = k\phi_1\right)$$

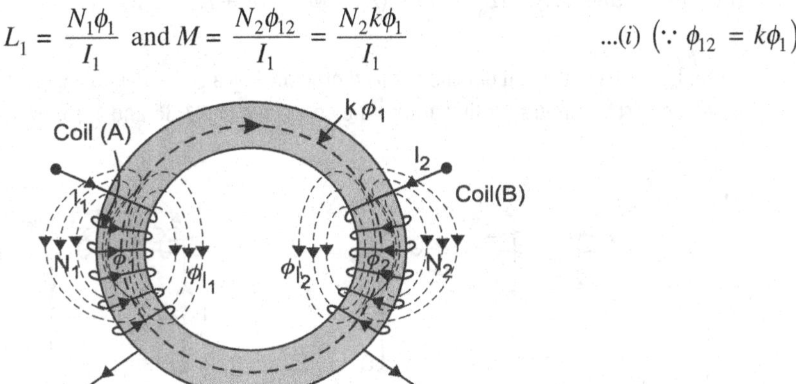

Fig. 1.34 Flux produced by one coil linking with the other

Now considering coil-2 carrying current I_2;

$$L_2 = \frac{N_2\phi_2}{I_2} \text{ and } M = \frac{N_1\phi_{21}}{I_2} = \frac{N_1k\phi_2}{I_2} \qquad ...(ii) \left(\because \phi_{21} = k\phi_2\right)$$

Multiplying equation (*i*) and (*ii*), we get,

$$M \times M = \frac{N_2k\,\phi_1}{I_1} \times \frac{N_1k\,\phi_2}{I_2}$$

or

$$M^2 = k^2\,\frac{N_1\,\phi_1}{I_1} \times \frac{N_2\,\phi_2}{I_2} = k^2\,L_1\,L_2$$

or

$$M = k\sqrt{L_1L_2} \qquad ...(iii)$$

The above expression gives a relation between mutual-inductance between the two coils and their respective self inductances.

Expression (*iii*) can also be written as,

$$k = \frac{M}{\sqrt{L_1L_2}}$$

1.22 Inductances in Series and Parallel

Consider two coils magnetically coupled having self-inductance of L_1 and L_2 respectively, and a mutual-inductance of M henry. The two coils, in an electrical circuit, may be connected in different ways giving different values of resultant inductance as given below:

Inductances in series: The two coils may be connected in series in the following two ways:

(*i*) When their fields (or mmfs.) are *additive* i.e., their fluxes are set-up in the same direction as shown in Fig. 1.35 In this case, the inductance of each coil is *increased* by M i.e.,

Total inductance, $L_T = (L_1 + M) + (L_2 + M) = L_1 + L_2 + 2M$

(*ii*) When their fields (or mmfs.) are *subtractive* i.e., their fluxes are set-up in opposite direction as shown in Fig. 1.36. In this case, the inductance of each coil is *decreased* by M, i.e.,

Total inductance, $L_T = (L_1 - M) + (L_2 - M) = L_1 + L_2 - 2M$

Note: It may be noted that direction of field produced by a coil is denoted by placing a *Dot* at the side at which current enters (or flux enters the core), see Fig. 1.35 and 1.36.

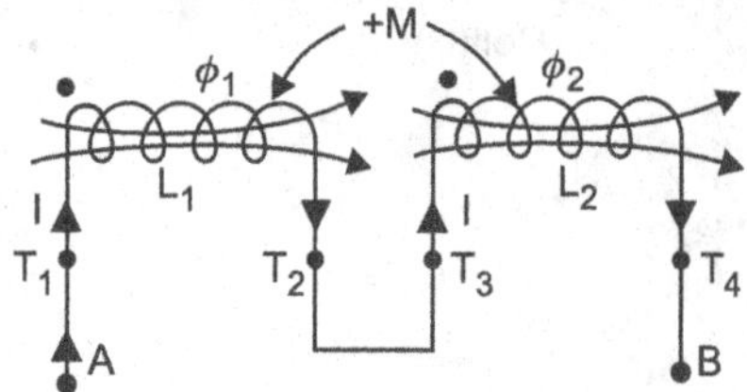

Fig. 1.35 Inductances in series having field in same direction

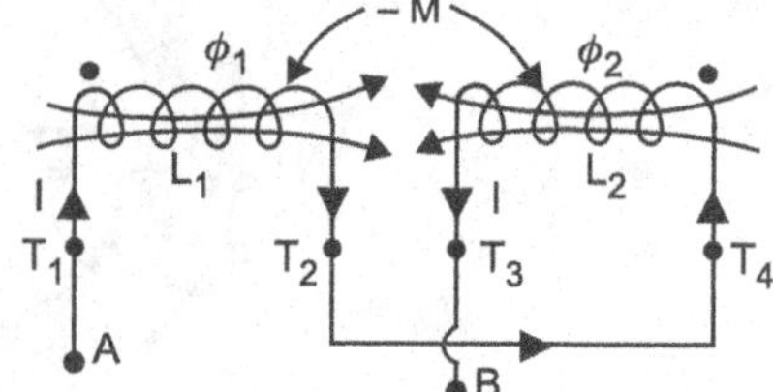

Fig. 1.36 Inductances in series with field in opposite direction

Inductances in parallel: The two coils may be connected in parallel in the following two ways:

(*i*) When the fields (or mmfs.) produced by them are in the same direction as shown in Fig. 1.37.

$$\text{Total inductance, } L_T = \frac{L_1 L_2 - M^2}{L_1 + L_2 - 2M}$$

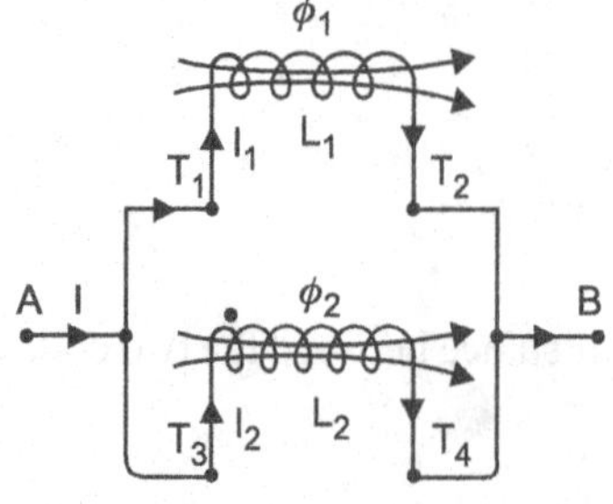

Fig. 1.37 Inductances connected in parallel fields are in same direction

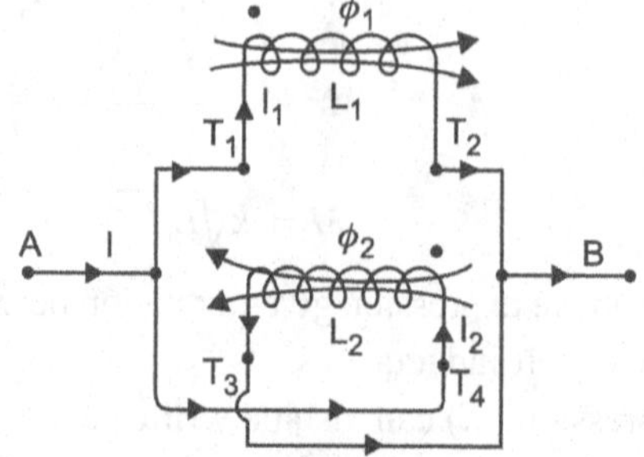

Fig. 1.38 Inductances in parallel fields are in opposite direction

(*ii*) When the fields (or mmfs.) produced by them are in the opposite direction as shown in Fig. 1.38.

$$\text{Total inductance, } L_T = \frac{L_1 L_2 - M^2}{L_1 + L_2 + 2M}$$

Example 1.23

A coil has 1500 turns. A current of 4 A causes a flux of 8 mWb to link the coil. Find the self inductance of the coil.

Solution:

$$\text{Inductance of the coil, } L = \frac{N\phi}{I}$$

where $N = 1500$; $\phi = 8 \times 10^{-3}$ Wb and $I = 4$ A.

$$\therefore \quad L = \frac{1500 \times 8 \times 10^{-3}}{4} = \mathbf{3\ H}\ (\textit{Ans.})$$

Example 1.24

Calculate the value of emf induced in circuit having an inductance of 700 micro-henry if the current flowing through it varies at a rate of 5000 A per second.

Solution:

Inductance of the coil, $L = 700 \times 10^{-6}$ H

Rate of change of current, $\frac{dI}{dt} = 5000$ A/s

$\therefore$ Magnitude of emf induced in the coil,

$$e = L\frac{dI}{dt} = 700 \times 10^{-6} \times 5000 = \mathbf{3{\cdot}5\ V}\ (\textit{Ans.})$$

Example 1.25

An air cored solenoid has 300 turns its length is 25 cm and its cross section is 3 cm^2. Calculate the self-inductance is henry.

Solution:

No. of turns of the solenoid, $N = 300$

Length of solenoid, $l = 25$ cm $= 0{\cdot}25$ m

Area of cross section, $a = 3$ cm^2 $= 3 \times 10^{-4}$ m^2

For air core, $\mu_r = 1$

$$\text{Inductance of the solenoid,} \quad L = \frac{N^2}{l/a\,\mu_0\mu_r} = \frac{N^2}{l}a\,\mu_0$$

$$= \frac{300 \times 300}{0 \cdot 25} \times 3 \times 10^{-4} \times 4\pi \times 10^{-7}$$

$$= \mathbf{0{\cdot}1375\ mH}\ (\textit{Ans.})$$

Example 1.26

Calculate the inductance of toroid, 25 cm mean diameter and 6.25 cm² circular cross-section wound uniformly with 1000 turns of wire. Hence calculate the emf induced when current in it increases at the rate of 100 A/second.

Solution:

Inductance of the toroid, $L = \frac{N^2}{l} \times a\mu_0\mu_r$

where, No. of turns, $N = 1000$ turns

Mean length $l = \pi D = 0{\cdot}25\ \pi$ m;

Area of cross-section, $a = 6{\cdot}25 \times 10^{-4}$ m² and

Relative permeability, $\mu_r = 1$

$\therefore$ $L = (1000)^2 \times 6{\cdot}25 \times 10^{-4} \times 4\pi \times 10^{-7} \times 1/0{\cdot}25\pi$

$= \mathbf{1\ mH}$ (*Ans.*)

Induced emf, $e = L\frac{dI}{dt} = 1 \times 10^{-3} \times 100 = \mathbf{0{\cdot}1\ V}$ (*Ans.*)

Example 1.27

The iron core of a choke has mean length 25 cm with an air gap of 1 mm. The choke is designed for an inductance of 15 H when operating at a flux density of 1 Wb/m². The iron core has a relative permeability of 3000 and 8 cm² area of cross-section. Determine the required number of turns of the coil.

Solution:

Inductance of the coil, $L = N^2/S_T$

where S_T is the total reluctance of the magnetic circuit

$$S_T = \frac{l_i}{a\,\mu_0\,\mu_r} + \frac{l_g}{a\,\mu_0}$$

$$= \frac{0\cdot 25}{8\times10^{-4}\times4\pi\times10^{-7}\times3000} + \frac{1\times10^{-3}}{8\times10^{-4}\times4\pi\times10^{-7}} = 1077612 \text{ AT/Wb}$$

Now $N = \sqrt{LS_T} = \sqrt{15\times1077612} = \mathbf{4020{\cdot}5\ turns}$ (*Ans.*)

Example 1.28

Two coils have a mutual inductance of 0·6 H. If current in one coil is varied from 4 A to 1 A in 0·2 second, calculate (i) the average emf induced in the other coil and (ii) the change of flux linking the later assuming that it is wound with 150 turns.

Solution:

Mutually induced emf, $e_m = M\frac{dI_1}{dt}$

where, $M = 0{\cdot}6$ H; $dI_1 = 4 - 1 = 3$ A and $dt = 0{\cdot}2$ s

$\therefore$ $e_m = 0{\cdot}6 \times 3/0{\cdot}2 = \mathbf{9\ V}$ (*Ans.*)

Now, $$e_m = N_2 \frac{d\phi_{12}}{dt}$$

$\therefore$ Change of flux with second coil, $$d\phi_{12} = \frac{e_m \times dt}{N_2} = \frac{9 \times 0 \cdot 2}{150} = \mathbf{12\ mWb}\ (\textit{Ans.})$$

Example 1.29

Two coils having 100 *and* 50 *turns respectively are wound on a core with* μ = 4000 μ_0. *Effective core length* = 60 *cm and core area* = 9 *cm*^{2}. *Find the mutual inductance between the coils.*

Solution:

We know that, Mutual Inductance $$M = \frac{N_1 N_2 \mu a}{l}$$

where, $N_1 = 100$; $N_2 = 50$; $\mu = 4000\ \mu_0$; $l = 60$ cm $= 60 \times 10^{-2}$ m; $a = 9$ cm^2 $= 9 \times 10^{-4}$ m^2

$$\therefore \quad M = \frac{100 \times 50 \times 4000\ \mu_0 \times 9 \times 10^{-4}}{60 \times 10^{-2}}$$

$$= \frac{100 \times 50 \times 4000 \times 4\mu 0 \times 9 \times 10^{-4}}{60 \times 10^{-2}} = \mathbf{37{\cdot}7\ mH}\ (\textit{Ans.})$$

Example 1.30

A wooden ring has mean diameter of 150 mm and a cross-sectional area of 250 mm^2. *It is wound with 1500 turns of insulated wire. A second coil of 900 turns in wound on the top of the first. Assuming that all flux produced by the first coil links with the second, calculate the mutual inductance.*

Solution:

Mutual-inductance, $$M = \frac{N_1 N_2}{l} a \mu_0 \mu_r$$

where, $N_1 = 1500$; $N_2 = 900$; $l = \pi D = 0{\cdot}15\ \pi$ m; $a = 250 \times 10^{-6}$ m^2; $\mu_r = 1$

$$\therefore \quad M = \frac{1500 \times 900}{0.15\pi} \times 250 \times 10^{-6} \times 4\pi \times 10^{-7} \times 1 = \mathbf{0{\cdot}9\ m\ H}\ (\textit{Ans.})$$

Example 1.31

Two coils A and B of 600 and 1000 turns respectively are connected in series on the same magnetic circuit of reluctance 2 × *10*^{6} *AT/Wb. Assuming that there is no flux leakage, calculate (i) self-inductance of each coil; (ii) mutual inductance between the two coils.*

What would be the mutual inductance if the co-efficient of coupling is 75%.

Solution:

Self inductance of coil A, $$L_1 = N_1^2 / S$$

where $N_1 = 600$ and $S = 2 \times 10^6$ AT/Wb.

$\therefore$ $L_1 = (600)^2/2 \times 10^6 = \mathbf{0{\cdot}18\ H}$ (*Ans.*)

Similarly, $L_2 = (1000)^2/2 \times 10^6 = \mathbf{0{\cdot}5\ H}$ (*Ans.*)

Mutual inductance, $M = k\sqrt{L_1 L_2}$

when $k = 1;\ M = 1\ \sqrt{0.18 \times 0.5} = \mathbf{0{\cdot}3\ H}$ (*Ans.*)

when $k = 0{\cdot}75;\ M = 0{\cdot}75\ \sqrt{0.18 \times 0.5} = \mathbf{0{\cdot}225\ H}$ (*Ans.*)

Example 1.32

Two air-cored coils are placed close to each other so that 80% of the flux of one coil links with the other. Each coil has mean diameter of 2 cm and a mean length of 50 cm. If there are 1800 turns of wire on one coil, calculate the number of turns on the other coil to give a mutual inductance of 15 mH.

Solution:

Reluctance, $S = \dfrac{l}{a\,\mu_0\,\mu_r} = \dfrac{0\cdot 5}{\frac{\pi}{4}\times(0\cdot 02)^2\times 4\pi\times 10^{-7}\times 1} = 1{\cdot}2665 \times 10^9$ AT/Wb

Now $L_1 = N_1^2\ /\ S$ and $L_2 = N_2^2\ /\ S$

$\therefore$ $\sqrt{L_1 L_2} = N_1\ N_2/S$

Also $M = k\sqrt{L_1 L_2} = k\,N_1\ N_2/S$

where $M = 15 \times 10^{-3}$ H; $N_1 = 1800$; $k = 0{\cdot}8$;

$\therefore$ $15 \times 10^{-3} = 0{\cdot}8 \times 1800 \times N_2/1{\cdot}2665 \times 10^9$

or $N_2 = \dfrac{15\times 10^{-3}\times 1\cdot 2665\times 10^9}{0\cdot 8\times 1800}$ = **13193 turns** (*Ans.*)

Example 1.33

Two coils with negligible resistance and of self inductance of 0·2 H and 0·1 H respectively are connected in series. If their mutual inductance is 0·1 H, determine the effective inductance of the combination.

Solution:

Total inductance of the two coils when connected in series;

$$L = L_1 + L_2 \pm 2\,M = 0{\cdot}2 + 0{\cdot}1 \pm 2 \times 0{\cdot}1 = \mathbf{0{\cdot}5\ H} \text{ or } \mathbf{0{\cdot}1\ H}\ (\textit{Ans.})$$

Example 1.34

The combined inductance of two coils connected in series is 0·6 H and 0·1 H depending upon the relative direction of currents in the coils. If one of the coils when isolated has a self inductances of 0·2 H, calculate the mutual inductance of the coils and the self inductance of the other coil.

Solution:

The combined inductance of the two coils when connected in series;

(*a*) having their field additive = $L_1 + L_2 + 2M = 0{\cdot}6$ ...(*i*)

(*b*) having their fields subtractive = $L_1 + L_2 - 2M = 0{\cdot}1$ ...(*ii*)

Subtracting equation (*ii*) from (*i*), we get,

$$4M = 0{\cdot}5 \text{ or } M = \mathbf{0{\cdot}125\ H}\ (Ans.)$$

From equation (*ii*); $L_1 + L_2 - 2 \times 0{\cdot}125 = 0{\cdot}1$ or $L_1 + L_2 = 0{\cdot}35$ H

Self inductance of one coil, $L_1 = 0{\cdot}2$ H

$\therefore$ Self inductance of second coil, $L_2 = 0{\cdot}25 - 0{\cdot}2 =$ **0·15 H** (*Ans.*)

Example 1.35

Two coils of self inductance 120 mH and 250 mH and mutual inductance of 100 mH are connected in parallel. Determine the equivalent inductance of combination if (i) mutual flux helps the individual fluxes and (ii) mutual flux opposes the individual fluxes.

Solution:

(*i*) When mutual flux helps the individual fluxes;

$$L_T = \frac{L_1 L_2 - M^2}{L_1 + L_2 - 2M} = \frac{120 \times 250 - (100)^2}{120 + 250 - 2 \times 100} = \mathbf{117{\cdot}65\ m\ H}\ (Ans.)$$

(*ii*) When mutual flux opposes the individual fluxes.

$$L_T = \frac{L_1 L_2 - M^2}{L_1 + L_2 + 2M} = \frac{120 \times 250 - (100)^2}{120 + 250 + 2 \times 100} = \mathbf{35{\cdot}088\ m\ H}\ (Ans.)$$

1.23 Energy Stored in a Magnetic Field

When some electrical energy is supplied to a coil, it is spent in two ways:

(*i*) A part of it is spent to meet $I^2 R$ loss which is dissipated in the form of heat and cannot be recovered.

(*ii*) The remaining part is used to create magnetic field around the coil and is stored in the magnetic field. When this field collapses, the stored energy is released by the coil and is returned to the circuit.

The energy stored in the magnetic field is given by the expression:

Energy stored in magnetic field = $\frac{1}{2} L I^2$

Example 1.36

A solenoid of 1 m in length and 10 cm in diameter has 5000 turns. Calculate the energy in the magnetic field when a current of 2 A flows in the solenoid.

Solution:

Inductance of the solenoid, $L = \frac{N^2}{l} \times a\mu_0\mu_r$

where, $N = 5000$; $a = \pi d^2/4 = 25\pi \times 10^{-4}$ m²; $l = 1$ m; $\mu_r = 1$.

$\therefore$ $L = (5000)^2 \times 25\pi \times 10^{-4} \times 4\pi \times 10^{-7} \times 1/1 = 0{\cdot}2467$ H

Energy stored $= \frac{1}{2}LI^2 = \frac{1}{2} \times 0\cdot2467 \times (2)^2 =$ **0·4934 J** *(Ans.)*

1.24 AC Excitation in Magnetic Circuits

To magnetise the magnetic circuits of electrical devices such as transformers, *AC* machines, electromagnetic relays, etc., *AC* supply is used. The magnetisation of magnetic circuits is called their excitation.

The magnetic circuits are never excited by *DC* supply, because in case of *DC* excitation the steady-state current is determined by the impressed voltage and resistance of the circuit. The inductance of the coil comes into picture only during transient period i.e., when the current is building-up or decaying during switching (*ON* or *OFF*) instants. The magnetic flux in the magnetic circuit adjusts itself in accordance with this steady value of current so that the relationship imposed by magnetisation (*B-H*) curve in satisfied. However, with *AC* excitation, inductance comes into picture even at steady-state condition. As a result for most of the magnetic circuits (not for all) the flux is determined by the impressed voltage and frequency. Then the magnetisation current adjusts itself in accordance with this flux so that the relationship imposed by the magnetisation (*B-H*) curve is satisfied.

Usually, for economic reasons, the normal working flux density in a magnetic circuit is kept beyond the linear portion of the magnetisation curve thus accurate analysis cannot be predicted for determining self inductance. However, for all practical purposes the parameters of the magnetic circuit are considered to be constant.

The reactive effect of the alternatively flux set-up by the exciting current can readily be shown as per Faraday's laws i.e.,

$$e = N\frac{d\phi}{dt}$$

Consider a magnetic core which is excited by a coil (winding) having *N* turns and carrying a current of *i* ampere as shown in Fig. 1.39. A magnetic flux ϕ is set-up by the exciting current *i*. Let the magnetic flux ϕ varies sinusoidally with respect to time *t*, then its instantaneous value is given by the relation;

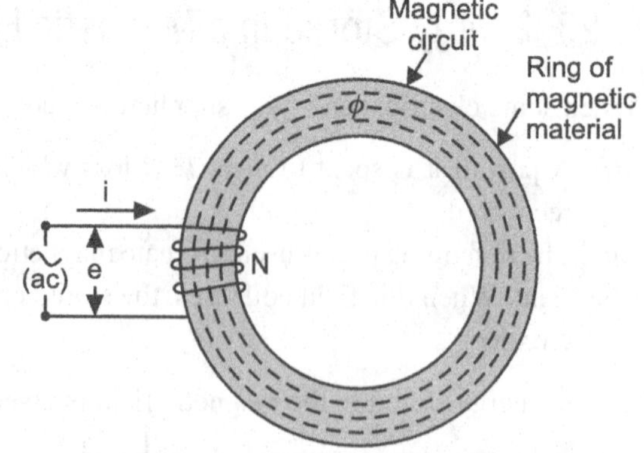

Fig. 1.39 Magnetic core excited by AC

$$\phi = \phi_m \sin \omega t = \phi_m \sin 2\pi ft \qquad ...(i)$$

where, ϕ_m = maximum value of alternating flux

f = frequency of supply impressed across the coil.

The induced *emf* in the coil

$$e = N\frac{d\phi}{dt} = N\frac{d}{dt}(\phi_m \sin 2\pi\, ft)$$

$$= 2\pi f N\,\phi_m \cos 2\pi ft = 2\pi f N\,\phi_m \sin\left(2\pi ft + \frac{\pi}{2}\right) \quad ...(ii)$$

The value of induced *emf* will be maximum when cos $2\pi ft = 1$, therefore

$$E_m = 2\pi f N\,\phi_m$$

Its effective or *r.m.s.* value,

$$E_{rms} = \frac{E_m}{\sqrt{2}} = \frac{2\pi\, fN\phi_m}{\sqrt{2}} = 4{\cdot}44\, f N\,\phi_m$$

Equation (*i*) and (*ii*) reveal that the induced *emf* leads the flux and hence the exciting current by $\frac{\pi}{2}$ radian or 90°. This induced *emf* and the coil resistance drop oppose the applied voltage. In case of electrical machines, usually the drop in resistance in only a few percent of applied voltage and therefore, neglected for close approximation. Thus, the induced *emf* E and applied voltage V may be considered equal in magnitude.

Example 1.37

For the AC excited magnetic circuit shown in Fig. 1.40. *Calculate the excitation current and induced emf of the coil to produce a core flux of* 0·6 *sin* 314 *t mWb.*

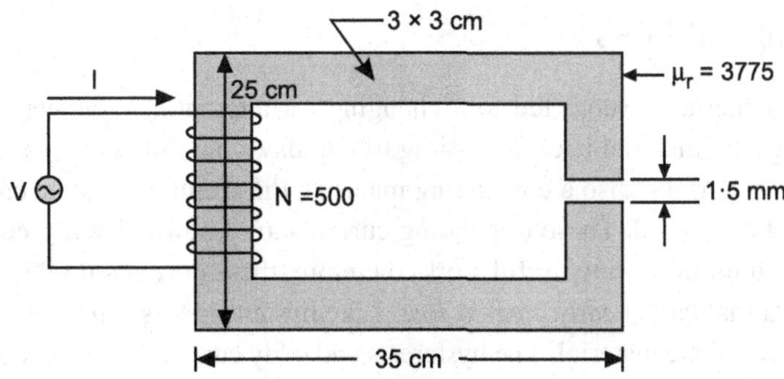

Fig. 1.40 Magnetic core excited by AC

Solution:

Here, $\phi = \phi_m \sin \omega t = 0{\cdot}6 \sin 314\, t$ mWb

Maximum value of flux, $\phi_m = 0{\cdot}6$ mWb $= 6 \times 10^{-4}$ Wb

Area of x-section, $a = 3 \times 3 = 9$ cm$^2 = 9 \times 10^{-4}$ m^2

$$\text{Flux density, } B_m = \frac{\phi_m}{\alpha} = \frac{6 \times 10^{-4}}{9 \times 10^{-4}} = 0.667T$$

$$\text{Length of air gap, } l_g = 1{\cdot}5 \text{ mm} = 1{\cdot}5 \times 10^{-3} \text{ m}$$

$$\text{Length of iron path, } l_i = \left(25 - 2\times\frac{3}{2} + 35 - 2\times\frac{3}{2}\right)\times 2 - 0.15 = 107\cdot 85\text{cm}$$

$$= 1{\cdot}0785 \text{ m}$$

$$\text{Total ampere-turns required, } AT_{\text{m}} = \frac{B_m}{\mu_0} l_g + \frac{B_m}{\mu_0\mu_r} l_i$$

$$= \frac{0\cdot 667\times 1\cdot 5\times 10^{-3}}{4\pi\times 10^{-7}} + \frac{0\cdot 667\times 1\cdot 0785}{4\pi\times 10^{-7}\times 3775}$$

$$= 796{\cdot}3 + 151{\cdot}7 = 948$$

Maximum value of excitation current required,

$$I_{\text{m}} = \frac{AT_m}{N} = \frac{948}{500} = 1{\cdot}896 \text{ A}$$

RMS value of excitation current,

$$I_{\text{rms}} = \frac{I_m}{\sqrt{2}} = \frac{1\cdot 896}{1\cdot 414} = \mathbf{1{\cdot}34\ A}\ (Ans.)$$

RMS value of induced emf in the coil, $E_{\text{rms}} = 4{\cdot}44 f N \phi_{\text{m}}$

$$= 4\cdot 44\times\frac{314}{2\pi}\times 500\times 0\cdot 6\times 10^{-3} = \mathbf{66{\cdot}57\ V}\ (Ans.)$$

1.25 Eddy Current Loss

When a magnetic material is subjected to a changing (or alternating) magnetic field, an emf is induced in the magnetic material itself according to Faraday's laws of electro-magnetic induction. Since the magnetic material is also a conducting material, these emfs. circulate currents within the body of the magnetic material. These circulating currents are known as **eddy currents**. As these currents are not used for doing any useful work, therefore, these currents produce a loss ($i^2 R$ loss) in the magnetic material called *eddy current loss*. Like hysteresis loss, this loss also increases the temperature of the magnetic material. The hysteresis and eddy current losses in a magnetic material are called *iron losses* or *core losses* or *magnetic losses*.

A magnetic core subjected to a changing flux is shown in Fig. 1.41. For simplicity, a sectional view of the core is shown. When changing flux links with the core itself, an emf is induced in the core which sets-up circulating (eddy) currents (i) in the core as shown in Fig. 1.41 (*a*). These currents produce eddy current loss ($i^2 R$), where i is the value of eddy currents and R is resistance of eddy current path. As the core is a continuous iron block of large cross-section, the magnitude of i will be very large and hence greater eddy current loss will result.

To reduce the eddy current loss, the obvious method is to reduce magnitude of eddy currents. This can be achieved by splitting the solid core into thin sheets (called laminations) in the planes parallel to the magnetic field as shown in Fig. 1.41 (*b*). Each lamination is insulated from the other by a fine layer of insulation (varnish or oxide film). This arrangement reduces the area of each section and hence the induced emf It also increases the resistance of eddy currents path since the area through which the currents can pass is smaller. This loss can further be reduced by using a magnetic material having higher value of resistivity (like silicon steel).

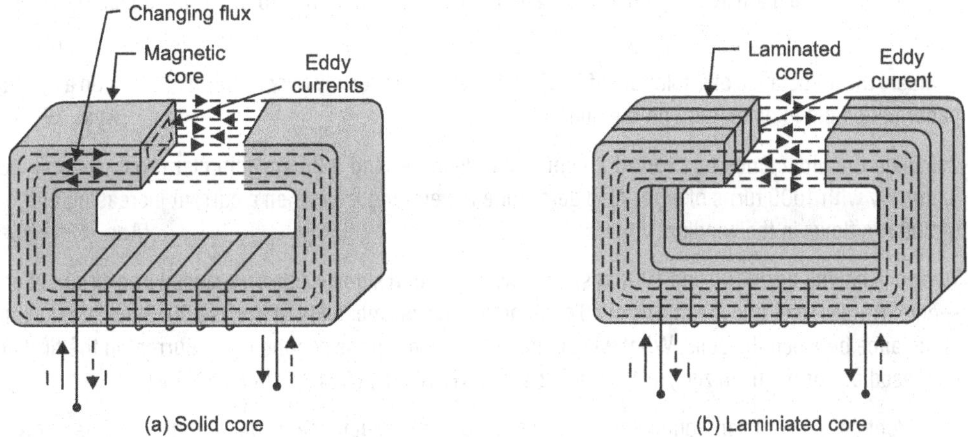

Fig. 1.41 Production of eddy currents

Useful Applications of Eddy Currents

It has been seen that when the affects of eddy currents (production on heat) are not utilised, the power or energy consumed by these currents is known as eddy current loss. However, there are the places where eddy currents are used to do some useful work e.g., in case of *Induction heating*. In this case, an iron shaft is placed as a core of an inductive coil. When high frequency current is passed through the coil, a large amount of heat is produced at the outer most periphery of the shaft by eddy currents. The amount of heat reduces considerably when we move towards the centre of the shaft. This is because outer periphery of the shaft offers low resistance path to eddy currents. This process is used for surface hardening of heavy shafts like axils of automobiles.

Eddy current effects are also used in electrical instrument e.g., providing damping torque in permanent magnet moving coil instruments and braking torque in case of induction type energy meters.

Mathematical Expression for Eddy Current Loss

Although it is difficult to determine the eddy current power loss from the current and resistance values. However, experiments reveal that the eddy current power loss in a magnetic material can be expressed as:

$$P_e = K_e B_m^2 t^2 f^2 \ V \text{ watt}$$

where, K_e = co-efficient of eddy current, its value depends upon the nature of magnetic material;

B_m = maximum value of flux density in Wb/m^2;

t = thickness of lamination in m;

f = frequency of reversal of magnetic field in Hz;

V = volume of magnetic material in m^3.

Section Practice Problems

Numerical Problems

1. A current of 8 A through a coil of 300 turns produces a flux of 4 m Wb. If this current is reduced to 2 A in 0·1 second, calculate average emf induced in the coil, assuming flux to be proportional to current.
   (**Ans.** *9 V*)
2. Estimate the inductance of a solenoid of 2500 turns wound uniformly over a length of 0·5 m on a cylindrical paper tube 4 cm in diameter. The medium is air. (**Ans.** *19·74 m H*)
3. Calculate the inductance of a toroid 25 cm mean diameter and 6·25 cm^2 circular cross-section wound uniformly with 1000 turns of wire. Also determine the emf induced when a current increasing at the rate of 200 A/s flows in the winding. (**Ans.** *1 m H; 0·2 V*)
4. Two coils having turns 100 and 1000 respectively are wound side by side on a closed iron circuit of cross-sectional area 8 cm^2 and mean length 80 cm. The relative permeability of iron is 900. Calculate the mutual inductance between the coils. What will be the induced emf in the second coil if current in the first coil is increased uniformly from zero to 10 A in 0·2 second? **(PTU.)** (**Ans.** *0·113 H; 56·5 V*)
5. Two identical coils, when connected in series have total inductance of 24 H and 8 H depending upon their method or connection. Find (*i*) self-inductance of each coil and (*ii*) mutual inductance between the coils. (**Ans.** *8H; 4 H*)
6. Two coils of self inductance 100 mH and 150 mH and mutual inductance 80 mH are connected in parallel. Determine the equivalent inductance of combination if (*i*) mutual flux helps the individual fluxes (*ii*) mutual flux opposes the individual fluxes. (**Ans.** *95·56 mH; 20·97 mH*)
7. A current of 20 A is passed through a coil of self-inductance 800 m H. Find the magnetic energy stored. If the current is reduced to half, find the new value of energy stored and the energy released back to the electrical circuit. (**Ans.** *160 J, 40 J, 120 J*)

Short Answer Type Questions

Q.1. What do you understand by electromagnetic induction?

Ans. The phenomenon by which an emf is induced in an electric circuit (or conductor) when it cuts across the magnetic field or when magnetic field linking with it changes is called *electro-magnetic inductions*. This phenomenon was discovered by Faraday.

Q.2. Define Faraday's laws of electromagnetic induction.

Ans. I Law: Whenever the flux linking with a coil or circuit changes, an emf is induced in it

II Law: The magnitude of induced emf in a coil is directly proportional to rate of change of flux linkages.

i.e.,

$$e = -N\frac{d\phi}{dt}$$

where e – Induced emf in volt

N – Number of turns

$\frac{d\phi}{dt}$ – rate of change of flux.

Minus sign due to Lenz's law.

Q.3. State Fleming's Right hand rule as well as Fleming's Left hand rule.

Ans. **Fleming's Right hand rule:** This law states that if one stretches the thumb, fore finger and middle finger of the right hand at right angles to each other in such a way that the thumb points in the direction of the motion of the conductor, the fore finger in the direction of the flux (from north to south) then the middle finger will indicate the direction of the induced emf in the conductor.

Fleming's Left Hand Rule: Stretch thumb, fore-finger and middle finger of your left hand at right angles to each other such that the fore finger points the direction of magnetic field (from north to south) and the middle finger gives the direction of current in the conductor, then the thumb will indicate the direction in which the force will act on the conductor.

Q.4. State right hand cork screw rule.

Ans. Hold the cork and place the right handed screw over it, now rotate it in such a way that it advances in the direction of flow of current. The direction in which the screw is rotated gives the direction of magnetic lines of force around the current carrying conductor.

Q.5. What do you mean by dynamically induced emf?

Ans. Either the magnetic field is stationary and the conductor is moving or the conductor is stationary and the flux is moving so that flux is cut by the conductor, an emf is induced in it. The emf induced in this way is known as dynamically induced emf. (e.g., emf generated in a *DC* generator).

Q.6. Define self inductance and give its unit.

Ans. The property of a coil due to which it opposes the change of current flowing through it is called *self inductance.* It refers to the ability of the coil to induce emf when current through it changes. Its unit is Henry.

Q.7. Distinguish between self induced and mutually induced emf.

Ans. The emf induced in a coil due to change of flux produced by it linking with its own turns is called *self induced emf.*

The emf induced in a coil due to change of flux produced by the neighbouring coil linking with it is called *mutually induced emf.*

Q.8. Does inductance play any role in *DC* circuit?

Ans. The role of inductance is only during the opening and closing of a *DC* circuit. It delays the rise of current during closing and delays the fall of current during opening. If in a *DC* circuit, the current reaches at its steady value, the circuit does not exhibit any inductance.

Q.9. What is a closed circuit?

Ans. The complete path for the flow of current through the load is known as a closed circuit.

Q.10. What is a short circuit?

Ans. If the supply mains are connected directly by a piece of wire without any load, it is known as short circuit. In these circuits, the value of the current is much greater than in the closed circuit. So the fuse melts.

Q.11. What is the type of energy being stored in a capacitor?

Ans. When a capacitor is charged, an electro-static field is set-up in between the plates. Thus electrical energy is stored in the electrostatic field set-up between the plates of the capacitor

1.26 Electro-mechanical Energy Conversion Devices

(Motors and Generators)

A *device* (machine) which *makes possible the conversion of energy from electrical to mechanical form or from mechanical to electrical form is called an* **Electromechanical energy conversion device** or **Electromechanical transducer.**

Depending upon the conversion of energy from one from to the other, the electro-mechanical device can be named as motor or generator.

1. **Motor:** *An electro-mechanical device (electrical machine) which converts electrical energy or power (EI) into mechanical energy or power (ω T) is called a* **motor**.

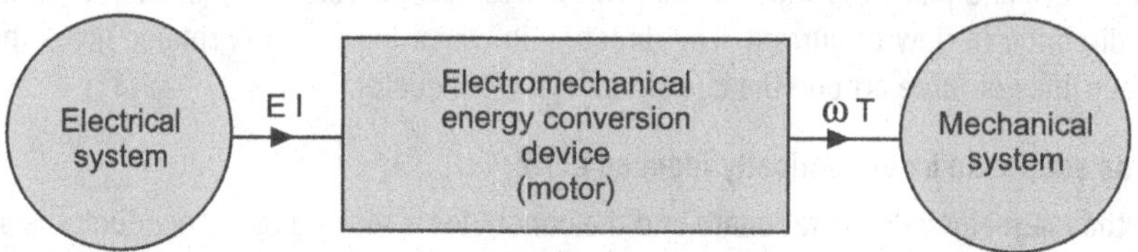

Fig. 1.42 Motor

Electric motors are used for driving industrial machines e.g., hammer presses, drilling machines, lathes, shapers, blowers for furnaces etc., and domestic appliances e.g., refrigerators, fans, water pumps, toys, mixers etc. The block diagram of energy conversion, when the electro-mechanical device works as a motor, is shown in Fig. 1.42.

2. **Generator:** An *electro-mechanical device (electrical machine) which converts mechanical energy or power* (*ω T*) into electrical energy or power (*EI*) is called **generator.**

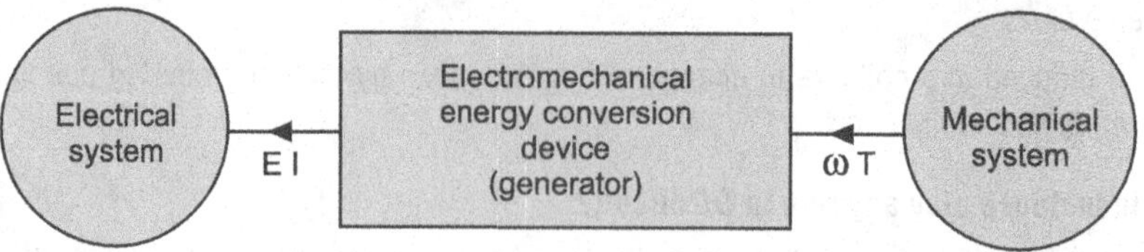

Fig. 1.43 Generator

Generators are used in hydro-electric power plants, steam power plants, diesel power plants, nuclear power plants and in automobiles. In the above said power plants various natural sources of energy are first converted into mechanical energy and then it is converted into electrical energy with the help of generators.

The block diagram of energy conversion, when the electro-mechanical device works as a generator, is shown in Fig. 1.43.

The same electro-mechanical device is capable of operating either as a motor or generator depending upon whether the input power is electrical or mechanical (see Fig. 1.44). Thus, the motoring and generating action is reversible

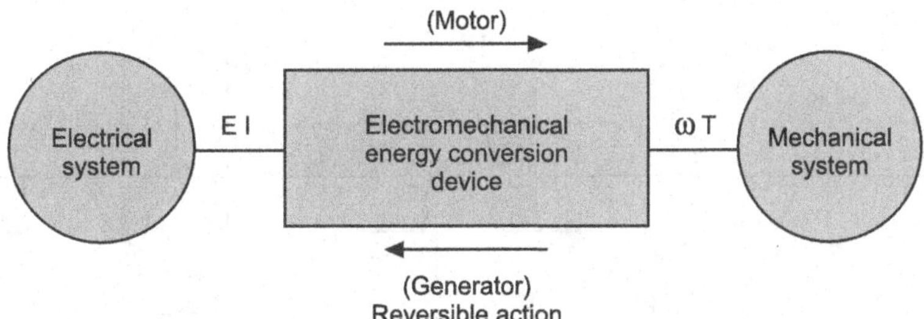

Fig. 1.44 Same machine can work as a generator or motor

The conversion of energy either from electrical to mechanical or from mechanical to electrical takes place through magnetic field. During conversion, whole of the energy in one form is not converted in the other useful form. In fact, the input power is divided into the following three parts;

(*i*) Most of the input power is converted into useful output power.
(*ii*) Some of the input power is converted into heat losses (I^2R) which are due to the flow of current in the conductors, magnetic losses (hysteresis and eddy current losses) and friction losses.
(*iii*) A small portion of input power is stored in the magnetic field of electro-mechanical device.

1.27 Torque Development by the Alignment of Two Fields

To understand the process of torque development by the alignment of two fields, the following cases may be considered:

(*i*) Soft iron piece placed in the magnetic field.
(*ii*) Permanent magnet placed in the magnetic field.
(*iii*) Electromagnet placed in the magnetic field.

1.27.1 Soft Iron Piece Placed in the Magnetic Field

Consider a soft iron piece, capable of free rotation, placed in the magnetic field of two permanent magnets (see Fig. 1.45). The magnetic lines of force are set up in the soft iron piece as shown in Fig. 1.45. The molecular poles get aligned parallel to the magnetic field due to magnetic induction. The soft iron piece obtains the polarities as marked in Fig. 1.45. This is the stable position of the soft iron piece. Torque produce in this case is zero.

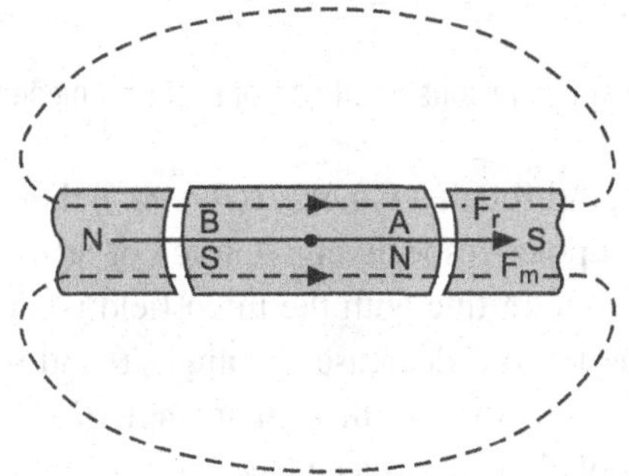

Fig. 1.45 Stable position of soft iron piece

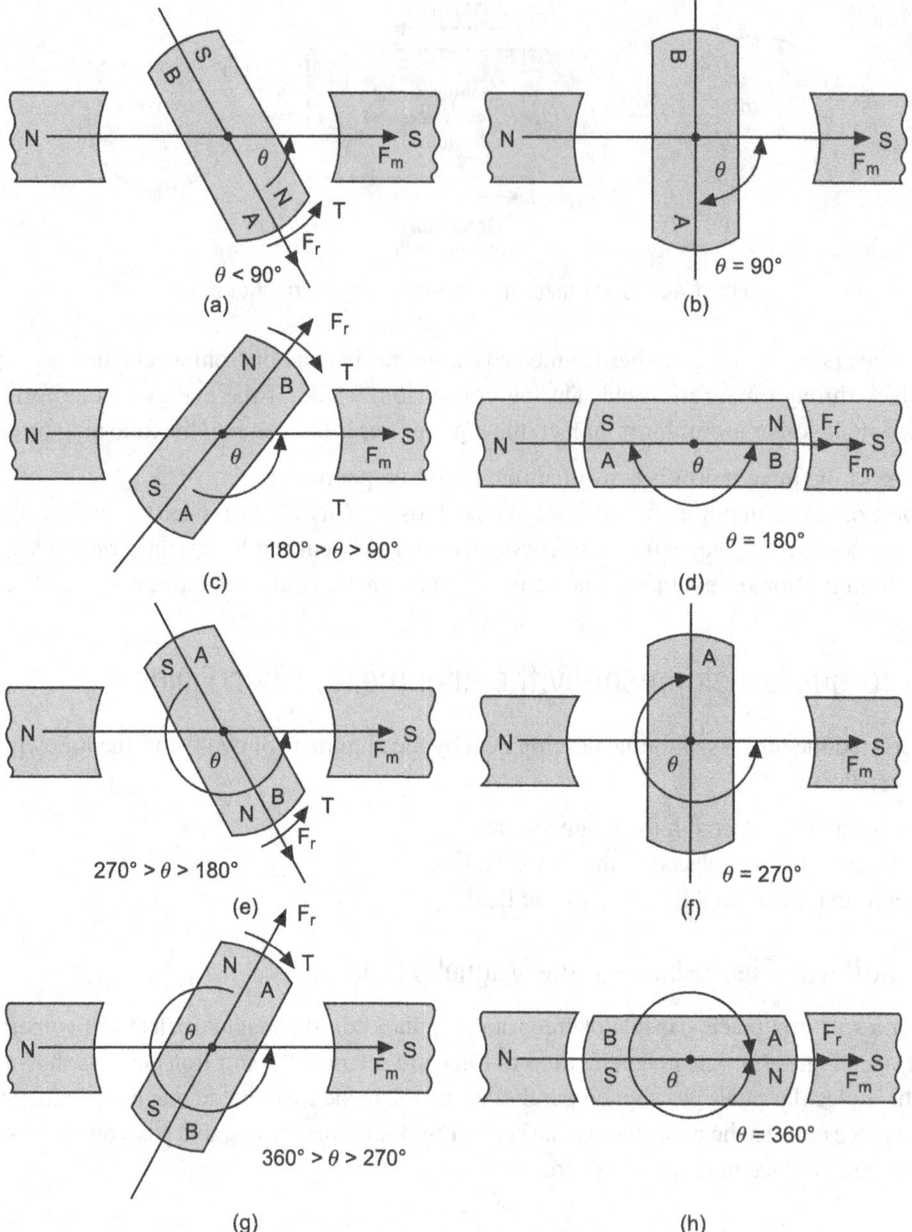

Fig. 1.46 Torque development at various positions of soft iron piece placed in the magnetic field

If the soft iron piece is rotated through an angle θ ($\theta < 90°$), then by magnetic induction ends *A* and *B* become North and South poles respectively. A force of attraction acts on the two ends and the soft iron piece will try to come in line with the main field i.e., the position of least reluctance path. This anticlockwise torque tries to decrease the angle θ and is considered as negative. [See Fig. 1.46 (*a*)]. When soft iron piece is rotated through an angle $\theta = 90°$, an equal force of attraction and repulsion acts on each end of short iron piece [see Fig. 1.46 (*b*)], therefore, torque produced is zero. This is the unstable position of the soft iron piece, because a slight change in angle θ in either direction will create a torque in that direction. When soft iron piece is rotated through an angle θ ($\theta > 90°$), then by magnetic induction ends A and B become South and North poles respectively. A force of attraction acts on the ends and soft iron piece will try to come in line with the main field. This clockwise torque tries to increase the angle θ and is considered as positive [See Fig. 1.46 (*c*)].

When soft iron piece is rotated through an angle $\theta = 180°$, then by magnetic induction ends A and B will obtain South and North polarity respectively [see Fig. 1.46 (*d*)]. There is a force of attraction at ends A as well as at B of soft iron piece which being equal and opposite cancel each other. In this position, torque produced is zero. This is the stable position, because any change in angle θ will create a torque which will tend to restore it position.

Following the similar explanation, the torque produced in the soft iron piece for various positions between 180° to 360° i.e., $180° < \theta < 360°$ can be determined.

When $270° > \theta > 180°$, torque produce is negative (anticlockwise) as shown in Fig. 1.46 (*e*).

When $\theta = 270°$, torque produced is zero, and it is an unstable position as shown in Fig. 1.46 (*f*).

When$360° > \theta > 270°$, torque produced is positive (clockwise) as shown in Fig. 1.46 (*g*).

When $\theta = 360°$, torque produced is zero and it is a stable position.

In fact, when a soft iron piece is placed in the magnetic field, by magnetic induction iron piece is magnetised. The magnetised iron piece produces its own field, the axis of that field is shown by arrow head F_r. The axis of the main magnetic field is shown by the arrowhead F_m. The rotor field F_r tries to come in line with F_m due to which torque develops. Hence, it can be said that torque is developed by the alignment of two fields.

The angle between two magnetic fields on which torque depends is called **torque angle.** This torque angle is measured with respect to the direction to the direction of rotation of soft iron piece.

Thus, it is concluded that the torque is a function of torque angle θ. The variation of torque with respect to angle θ is shown in Fig. 1.47.

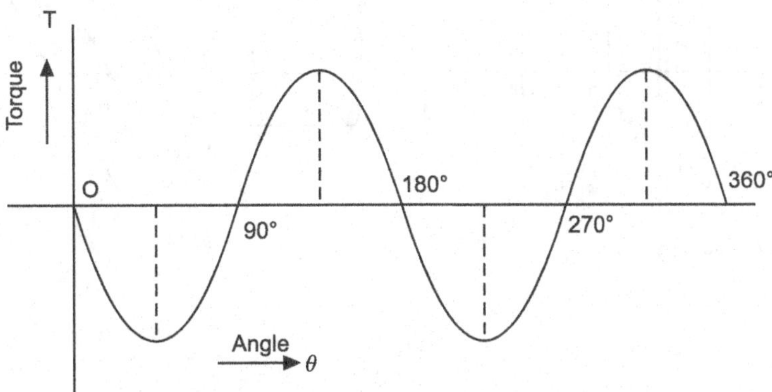

Fig. 1.47 Wave diagram of torque produced in soft iron piece

1.27.2 Permanent Magnet Placed in the Magnetic Field

Consider a permanent magnet capable of free rotation, placed in the magnetic field of two permanent magnets as shown in Fig. 1.48 (*a*). There is a force of attraction on north and south pole of the rotating magnet, which being equal and opposite cancel each other. In this position, torque produced is zero because the field of rotating magnet *Fr* and the field of stationary permanent magnet *Fm* are in line with each other.

When the rotating magnet in rotated through in angle θ (θ being less than 90°, equal to 90° more than 90° but less than 180°), its north pole will be attracted towards the south pole and south pole will be attracted towards the north pole of permanent stationary magnets [see Fig. 1.48 (*b*), (*c*) and (*d*) respectively]. In other words, we can say that the rotor field F_r tries to come in line with main field F_m and torque is developed. This anticlockwise torque is considered as negative, because it is decreasing the torque angle θ.

When the permanent rotating magnet is rotated through an angle $\theta = 180°$, the two field F_r and F_m are in line with each other but acting in opposite direction [see Fig. 1.48 (*e*)], therefore, torque developed is zero but this is the unstable position because slight change in angle θ in either direction will create a torque in that direction and the rotor will not regain its original position.

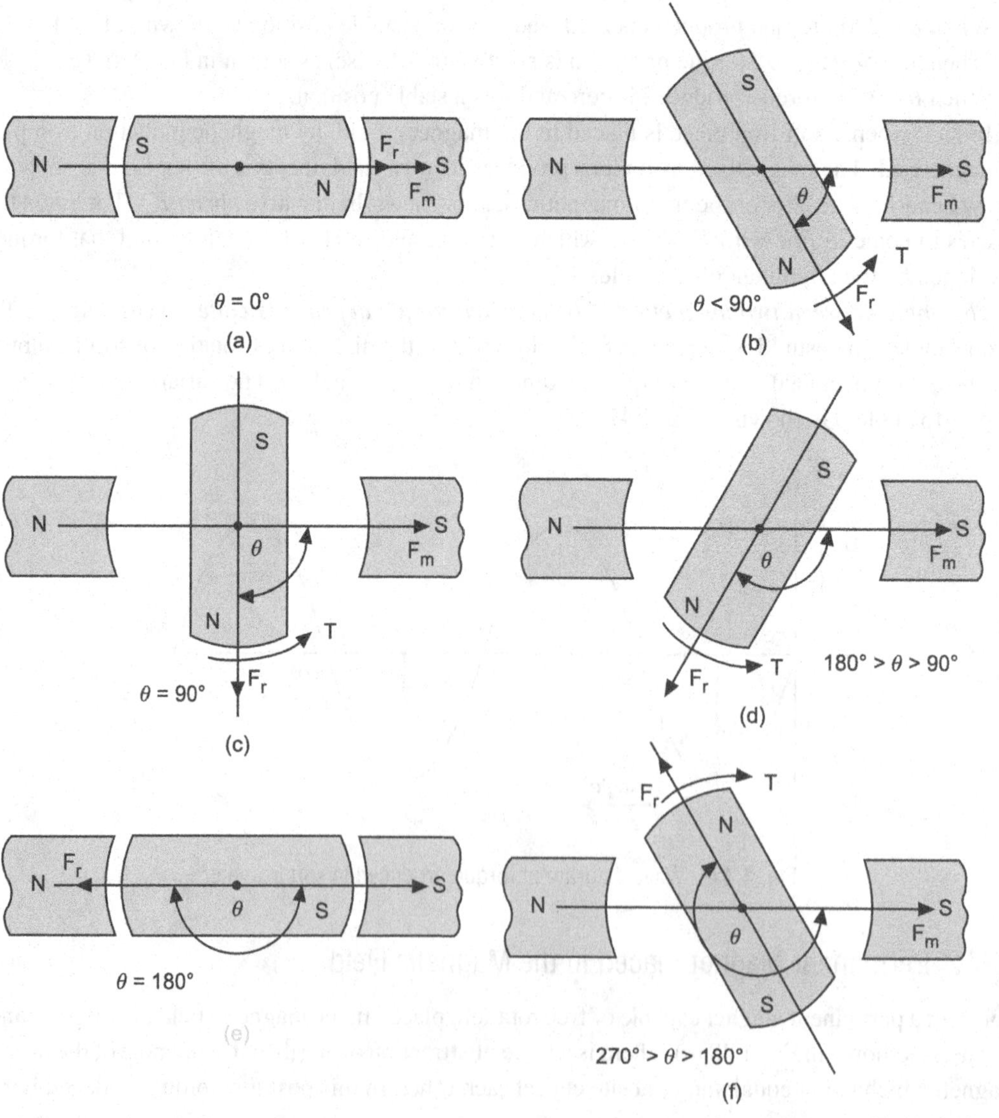

Fig. 1.48. *Contd.*

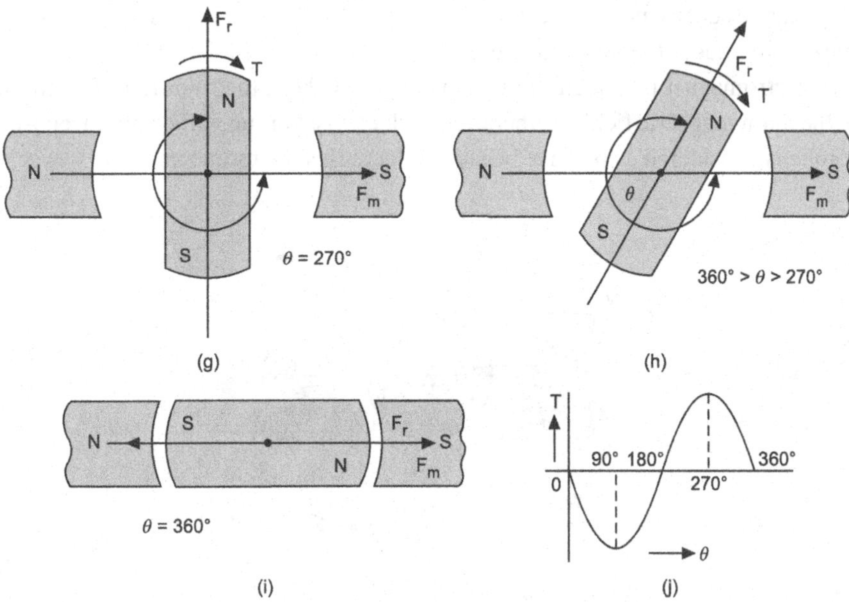

Fig. 1.48 Permanent magnet in magnetic field

When the permanent rotating magnet is rotated through an angle θ more than 180°, but less than 360° (i.e., θ is less than 270°, equal to 270°, more than 270°), its north and south poles will be attracted towards the south and north poles of the permanent stationary magnets respectively [see Figs. 1.48 (*f*), (*g*) and (*h*) respectively]. In other words, F_r will try to come in line with F_m and thus torque is developed. This clockwise torque is considered as positive, because it is increasing the torque angle θ.

The maximum negative or positive torque is produced on the rotating permanent magnet when angle $\theta = 90°$ or $\theta = 270°$ (–90°) respectively, because at these positions, there is maximum force of attraction or repulsion acting on the rotating magnet.

When rotating, permanent magnet is rotated through an angle $\theta = 360°$, two field F_r and F_m are in line with each other [see Fig. 1.48 (*i*)]. Therefore, the torque developed is zero. This is the stable position because any change in angle θ will develop a torque which tends to restore its original position.

Thus, it is concluded that the torque is produced due to the **alignment of two fields.**

The angle between the two magnetic fields on which torque depends is called **Torque angle.** This torque angle is measured with respect to the direction of rotation of rotating magnet.

The torque produced in the rotating magnet is a function of torque angle θ. The variation of torque with respect angle θ is shown graphically in Fig. 1.48 (*j*).

1.27.3 Electromagnet Placed in the Magnetic Field

Consider an electromagnet, free to rotate about its axis, is placed in the magnetic field of permanent magnets. When the field produced by the electromagnet *Fr* is in the same direction as that of the main field produced by the stationary permanent magnets *Fm*, the torque produced is zero [see Fig. 1.49 (*a*)].

When the electromagnet is rotated through an angle $\theta = 90°$, the axis of the rotor field F_r will make an angle of $\theta = 90°$ with the main field F_m. The rotor field F_r will try to come in line with the main field F_m, and produces an anticlockwise torque [see Fig. 1.49 (*b*)]. This torque is considered as negative because it reduces the torque angle θ.

When the electromagnet is rotated through an angle $\theta = 270°$, the rotor field F_r will try to come in line with the main magnetic field F_m, therefore, a clockwise torque will be produced [see Fig. 1.49 (*c*)], This torque is considered as positive because it increases the torque angle θ.

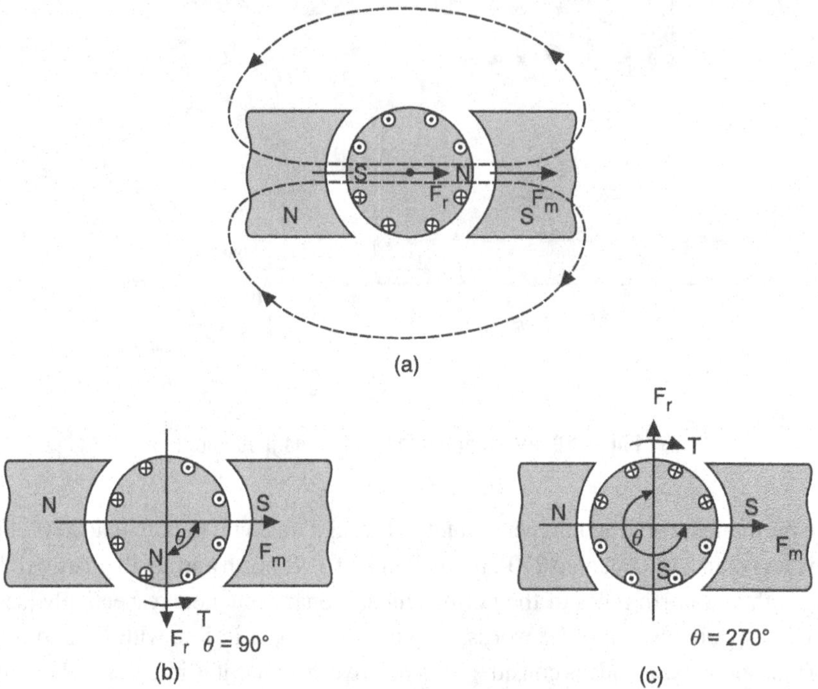

Fig. 1.49 Electromagnet in the field of permanent magnets

When torque angle θ is between zero and 180°, the torque produced is negative, whereas when θ is between 180° to 360°, the torque produced is positive. When torque angle $\theta = 90°$ or $\theta = 270°$ (–90°), torque produced is maximum negative or maximum positive respectively.

Thus, it is conclude that the variation of torque with respect to torque angle θ is similar to that in the case of permanent rotating magnet placed in the magnetic field of permanent stationary magnet[see Fig. 1.48 (*j*)].

The torque produced in this case is called an electromagnetic torque (T_e).

1.28 Production of Torque

The torque produced by the alignment of two fields (i.e., rotor field and stationary main field) varies in magnitude and direction depending upon the torque angle θ. Let us see, the effect of torque angle θ on the torque produced in the following cases:

(*i*) In case of permanent magnet
(*ii*) In case of electromagnet.

1.28.1 In Case of Permanent Magnet

Consider a permanent magnet A, which is free to rotate about its axis. Let it be placed in the magnetic field of another permanent magnet B as shown in Fig. 1.50. Let,

θ = angle between the axis of two fields F_m and F_r.

l = length of magnet A.

r = radius of circle in which rotation takes place.

F = force acting on north and south pole of magnet A.

Torque = Force × Perpendicular distance.

In the right angled triangle '*oab*', $ab = oa \sin \theta$

Distance perpendicular to force, $ab = r \sin \theta$

$\therefore$ $$\text{Torque} = 2F \times r \sin \theta$$

or $$T = 2F \times \frac{l}{2} \sin \theta [\_ r = l/2)$$

or $$T = Fl \sin \theta \text{ Where } F \text{ is force of attraction, } F = \frac{m_1 m_2}{4\pi\mu_0\mu_r d^2}$$

or $$T = K \sin \theta \{\text{where } K = F \times l \text{ is constant}\}$$

or $$T \alpha \sin \theta$$

The maximum torque will be produced when $\theta = 90°$ as said earlier.

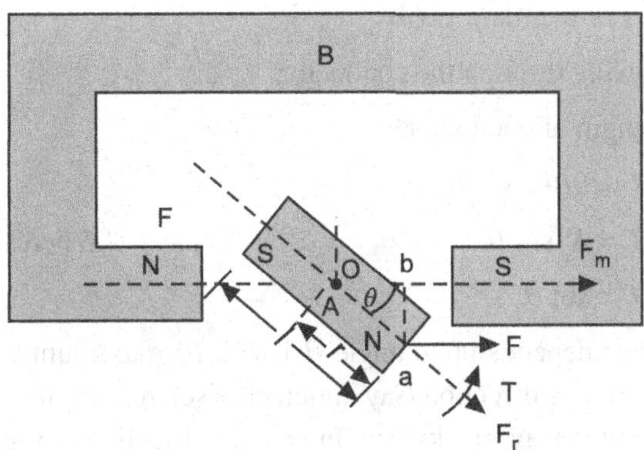

Fig. 1.50 Torque developed at an instant when a permanent magnet is placed in the uniform magnetic field

1.28.2 In Case of Electromagnet

Consider an electromagnet, having only one coil carrying current. The axis of the field produced by electromagnet Fr and axis of the main field produced by the permanent stationary magnet Fm are shown in Fig. 1.51. The angle between the two fields is q. Due to alignment of two fields torque is developed.

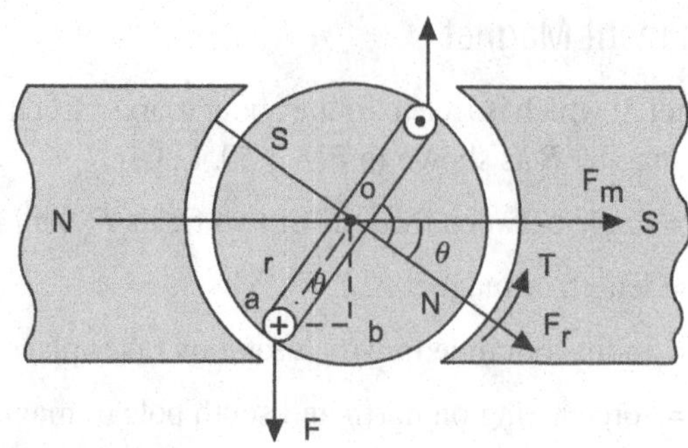

Fig. 1.51 Torque produced at an instant when an electromagnet is placed in a uniform magnetic field.

The production of torque can also be explained by the concept of electromagnetic force acting on the current carrying conductor placed in the magnetic field.

Let, F = Force acting on the two conductors.

r = radius of circle in which conductor rotates.

θ = angle between the field F_m and F_r.

Torque = Force × Perpendicular distance.

In a right angle triangle, angle $aob = \theta$

Distance perpendicular to force, $ab = oa \sin\theta$

$= r \sin\theta$

Total torque acting on the two conductors, $T = 2F_r \sin\theta$

Where B = Flux density of the main field.

I = Current flowing through the conductor.

l = Effective length of conductor.

$\therefore \quad T = 2BIlr \sin\theta$

or $\quad T = K_l \sin\theta$ [Where, $K_l = 2BIlr$ is a constant]

or $\quad T \propto \sin\theta$.

The magnitude of torque depends upon angle θ, it will be maximum when $\theta = 90°$. When θ is positive torque is produce in one direction (say anticlockwise), but when it is negative, the torque is produced in the other direction (say, clockwise). In fact, the direction of torque depends upon B and i. When either of the two is reversed the direction of torque is reversed but if both are reversed the direction of torque remains the same.

The direction of force action on a current carrying conductor when placed in the magnetic field can be determined by applying *Fleming's left hand rule.*

Fleming's Left Hand Rule: *Stretch first finger, second finger and thumb of left hand mutually perpendicular to each other. If first finger indicates the direction of main magnetic field, second finger indicates the direction of flow of current through the conductor, then thumb will indicate the direction of fore acting on the conductor.*

1.29 Production of Unidirectional Torque

As discussed earlier that by the alignment of two fields torque develops but the torque produced is not unidirectional. The unidirectional or continuous torque can be obtained by applying any one of the following methods.

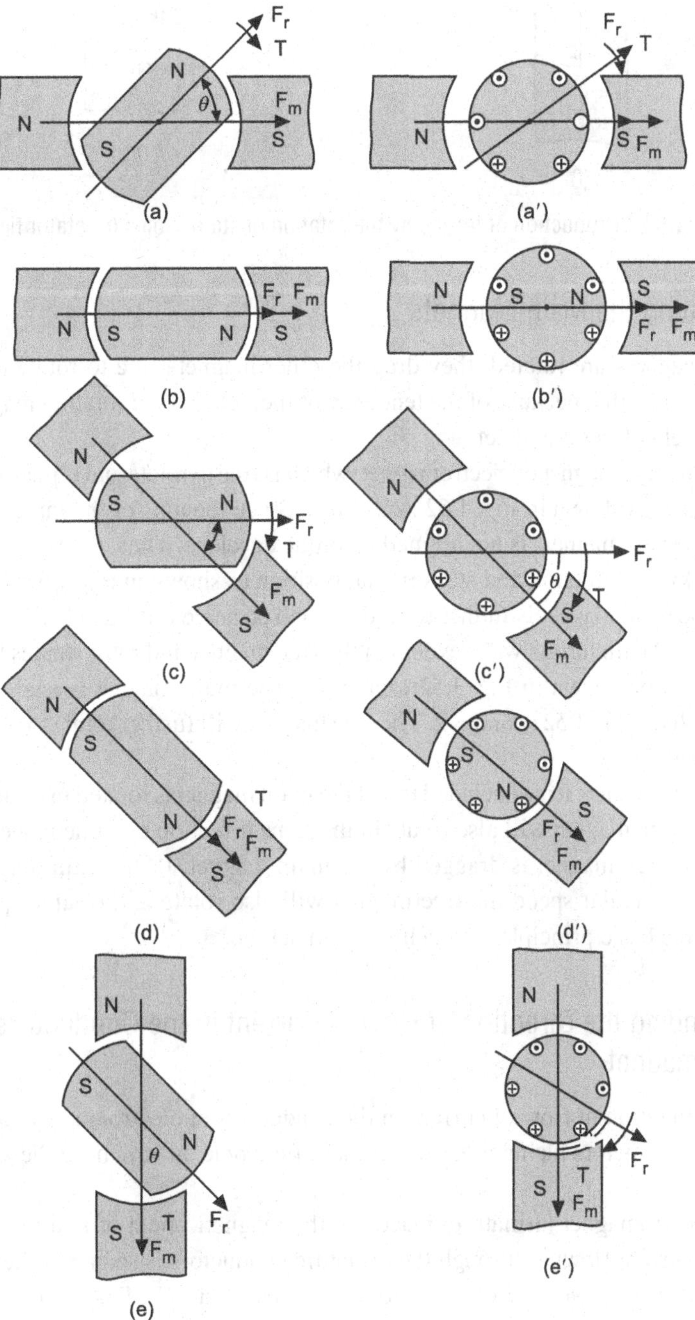

Fig. 1.52. *Contd.*

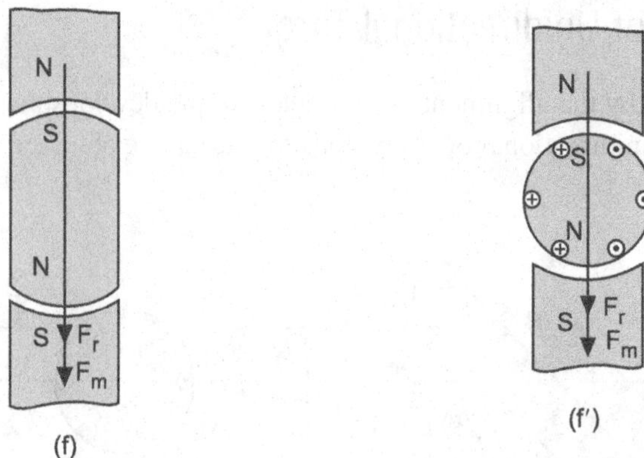

Fig. 1.52 Production of torque by the rotation of stator poles (or stator field)

1.29.1 By Rotating the Main Magnets

When the main magnets are rotated, they drag the other magnets free to rotate (or the armature, electromagnet) along with it because of the tendency of the field of free rotating magnet or armature to align with the field of main magnet.

Consider a permanent magnet or electromagnet, which is free to rotate and is placed in the magnetic field of main magnet as shown in Fig. 1.52 (*a* and *a*')]. Since the axis of the magnetic fields F_r and F_m produced by the two magnets is not aligned, a torque develops. Thus, the free magnet rotates by an angle θ in clockwise direction and stays in that position as shown in Fig. 1.52 (*b* and *b*′).

Suppose that the main magnet is further rotated through an angle θ in the clockwise direction [see Fig. 1.52 (*c* and *c*')]. Again torque will develop in the free magnet and rotor rotates through an angle θ to obtain the position shown in Fig. 1.52 (*d* and *d*'). The main magnet is again rotated through some more angle θ (see Fig. 1.52 (*e* and *e*')]. The free magnet will further rotate to obtain the position shown in Fig. (*f* and *f*').

Thus, from the above facts it is concluded that if the main magnet is rotated in a particular direction by some angle, the free magnet will also rotate in the same direction by same angle. In other words, we can say that the free magnet is dragged by the main magnet. If the main magnet is rotated in any direction at a particular speed the free magnet will also rotate at the same speed in the same direction. This is the basic principle of synchronous machine.

1.29.2 By Changing the Direction of Flow of Current in the Conductors of Electromagnet

By changing the direction of flow of current in the conductors of electromagnet (armature) in such a manner that the conductors facing a particular main field pole, always have the same direction of flow of current.

Consider an electromagnet (armature) placed in the magnetic field of main poles, as shown in Fig. 1.53 (*a*). The current flowing through the armature conductors is such that the axis of the field produced by armature F_r is 90° out of phase to that of main field F_m. Due to the alignment of two fields, torque is developed in the armature and it obtains the position as shown in Fig. 1.53 (*b*). At this position rotor (armature) will stop rotating.

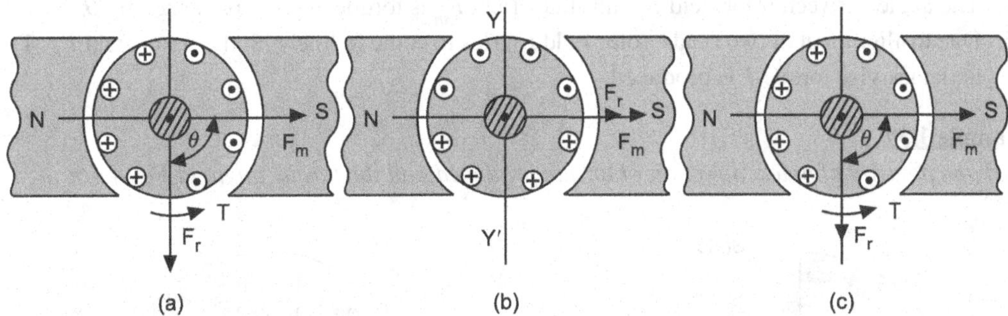

Fig. 1.53 Produced of unidirectional torque by reversing the current in armature conductors

If the direction of flow of current in the armature conductors is reversed just when they cross the *Y-Y'* axis (magnetic neutral axis) displaced by an angle $\theta = 90°$ to that of main field F_m. The torque will always be exerted on the armature. Thus a continuous torque can be obtained. This is the basic principle of DC machines. The direction of flow of current in the armature conductors is reversed by providing commutator on DC machines.

Example 1.38

In Fig. 1.54 draw the magnetic lines of force produced by the stator. Mark the polarity of poles, direction of stator and rotor field, torque angle and direction of the torque produced.

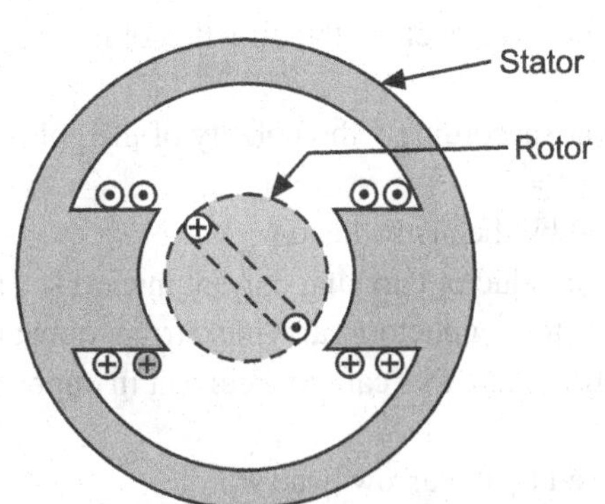

Fig. 1.54 Given figure

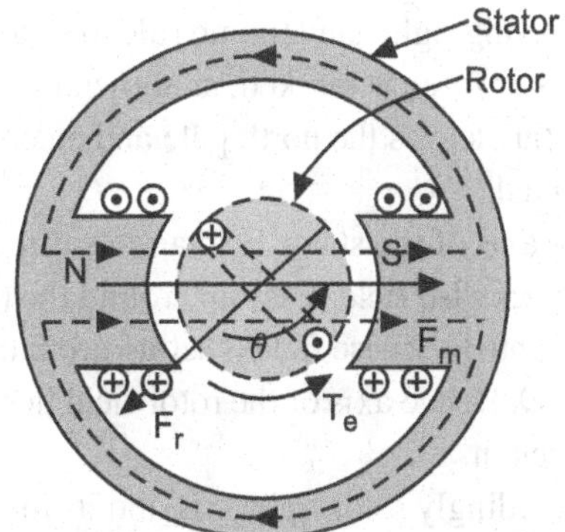

Fig. 1.55 Position of the axis of stator and rotor field

Solution:

For solution, the following steps may be considered;

1. Stator magnetic field is drawn as shown in Fig. 1.55 by applying right hand thumb rule to stator current carrying conductors.
2. We know that flux leaves the N-poles and enters the S-pole, accordingly polarity of the poles is marked in Fig. 1.55.

3. The axis and direction of stator field and rotor field are marked by arrows F_m and F_r, respectively (the axis of the rotor field lies perpendicular to the plane of coil or it lies in between the conductors carrying current in opposite direction).
4. The angle between rotor field F_r and stator field F_m is torque angle represented by θ.
5. Due to alignment of two fields, rotor field F_r tries to come in line with the stator field F_m. Thus anticlockwise torque T is produced.

Example 1.39

Mark the polarity of poles, direction of torque produced and the torque angle in Fig. 1.56.

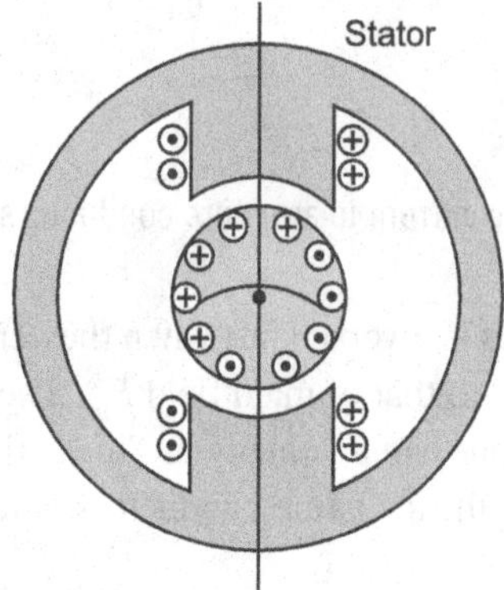

Fig. 1.56 Given figure

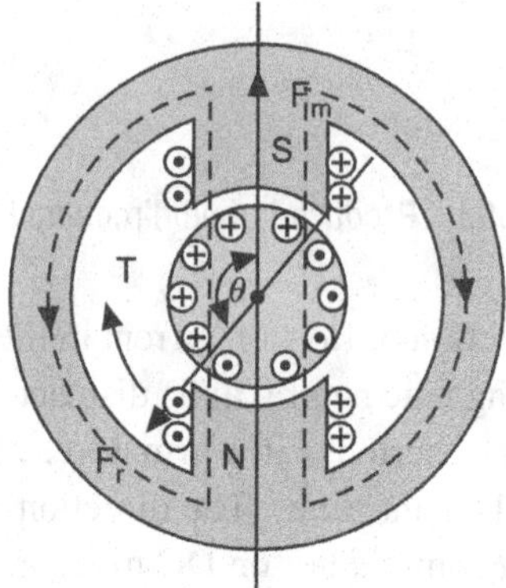

Fig. 1.57 Position of the axis of stator and rotor field

Solution:

Look at Fig. 1.56, the various quantities are marked adopting the following steps:

1. Applying right hand thumb rule to stator current carrying conductors, the direction of magnetic lines of force is marked, as shown in Fig. 1.57.
2. As flux leaves the north pole and enters and south pole, accordingly the polarity of the poles is marked.
3. The axis of the stator field and its direction is marked by the arrow head F_m.
4. A clockwise field is set-up around the bunch of rotor conductor carrying current inward i.e., ⊕ and anticlockwise field is set-up around the bunch of rotor conductors carrying current outward i.e., ⊙. Hence axis of the rotor field lies in between the conductors carrying current in opposite direction.
5. Accordingly rotor field axis and its direction is marked by the arrow head F_r.
6. The angle between rotor field and stator field is called torque angle and is marked as θ.
7. The rotor field F_r tries to come in line with the stator field F_m due to which torque T develops in clockwise direction as marked by the arrowhead.

Example 1.40

To obtain clockwise torque, mark the direction of flow of current in both the pole windings and on the rotor conductors in Fig. 1.58.

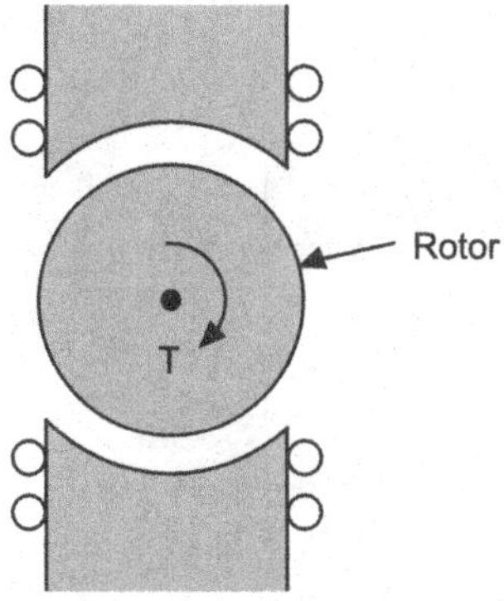

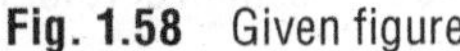

Fig. 1.58 Given figure

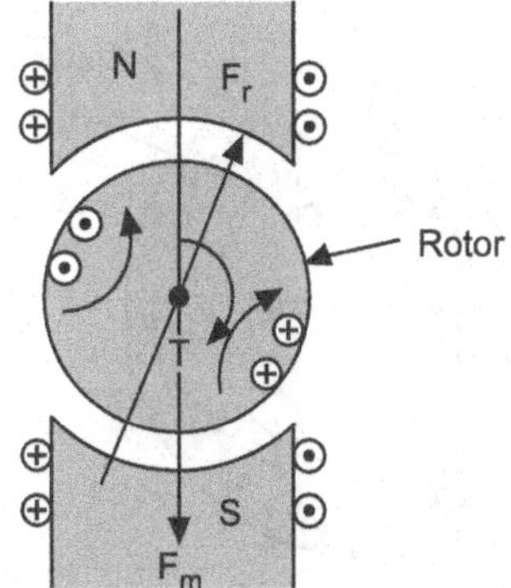

Fig. 1.59 Position of the axis of stator and rotor fields

Solution:

1. Let the direction of flow of current in pole windings be as marked in Fig. 1.59
2. Applying right hand thumb rule, mark the polarity of poles, the axis of stator field and its direction as F_m.
3. Mark the axis of rotor field perpendicular to the plane of rotor coils (see Fig. 1.59).
4. To obtain clockwise torque mark the direction to rotor field F_r as shown in Fig. 1.59.
5. To obtain the said direction of rotor field, mark the direction of flow of current in rotor conductors applying right hand thumb rule, as shown in Fig. 1.59.

Example 1.41

A hollow cylindrical stator contains a solid rotor. Both are made of soft iron. One coil carrying current is placed on the stator as shown in Fig. 1.60. Ignore the coil shown on the rotor. Mark the poles developed on the stator. Draw the lines of force. If another coil carrying current is placed on the rotor as shown in Fig. 1.60. Will the arrangement produce any torque? Give reasons. If the stator is made to rotate at N rpm what will the rotor do?

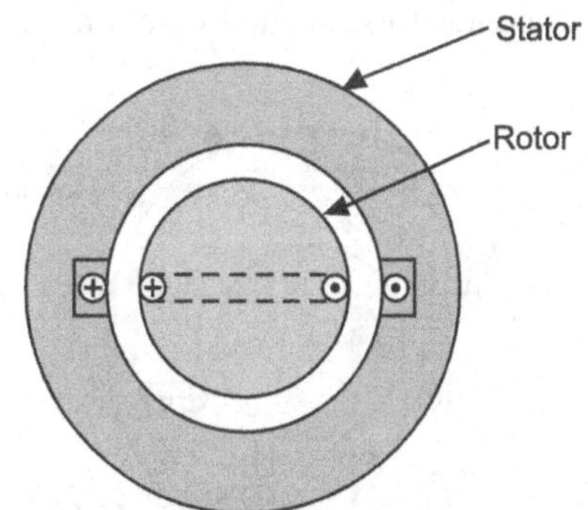

Fig. 1.60 Given figure

Solution:

Ignoring the rotor coil and applying right hand thumb rule to the current carrying conductors of stator the field produced by them is shown in Fig. 1.61. We know that magnetic lines of force leave the north pole and enter the south pole, accordingly, the polarity of stator poles is marked in Fig. 1.61. Accordingly the direction of stator field (main field) F_m is marked in Fig. 1.61 and 1.62.

If a current carrying coil is placed on the rotor as shown in Fig. 1.62, the rotor will also become an electromagnet. Applying right hand thumb rule to the rotor current carrying conductors the direction of rotor field F_r is also marked in Fig. 1.62.

The two fields F_r and F_m are in line with each other or torque angle $\theta = 0$, therefore, the torque produced in the rotor by this arrangement will be zero.

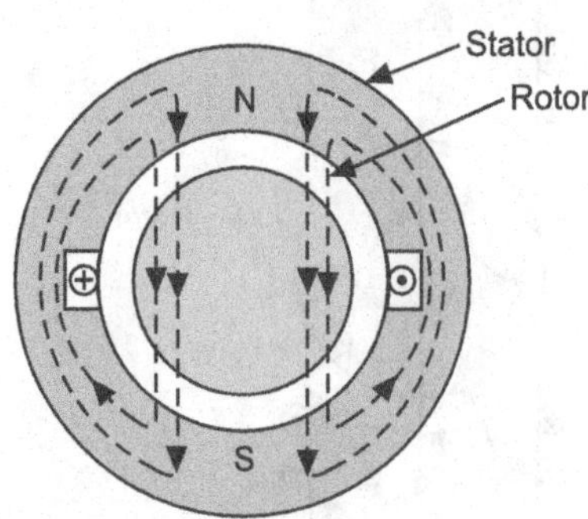

Fig. 1.61 Position of stator fields

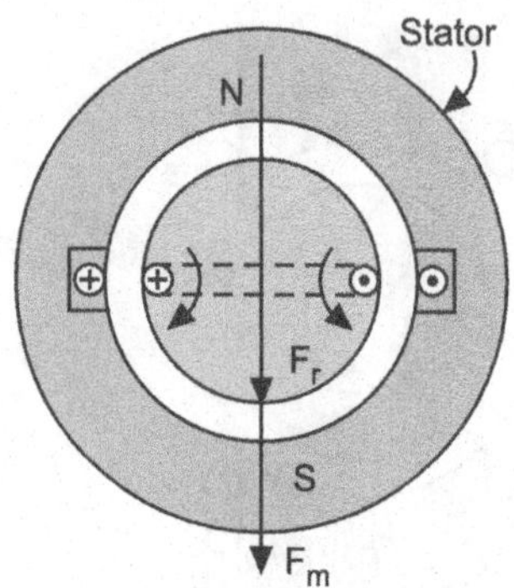

Fig. 1.62 Position of rotor field

If the stator or main field is made to rotate at N rpm in any direction the rotor field will be dragged by the main field. Thus, the rotor will also rotate in the same direction and at the same speed i.e., N rpm

Example 1.42

Fig. 1.63 shows a rotor carrying conductors placed inside a stator having four electro-magnetic poles. Referring to this figure, answer the following questions by means of neat diagram.

(a) For the direction of current in the field coils as shown, mark the polarity of each pole.

(b) For the direction of current in the field coils as shown, mark the direction of current in the rotor conductors for clockwise rotation of rotor.

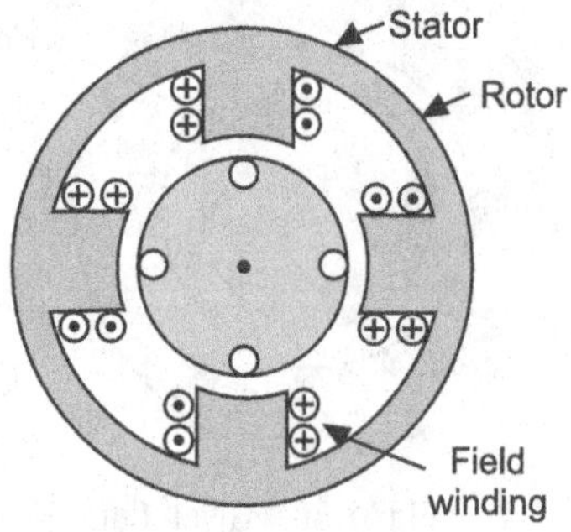

Fig. 1.63 Given figure

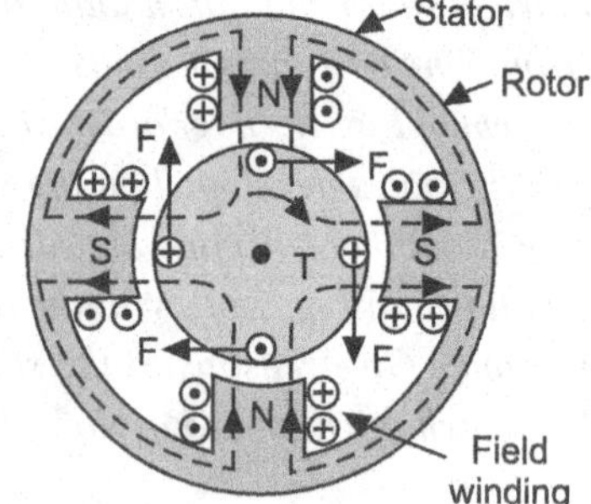

Fig. 1.64 Stator field and current in rotor conductors

Solution:

Applying right hand thumb rule to the stator current carrying conductors the direction of magnetic lines of force is marked. Flux leaves the north pole and enters the south pole, accordingly, the polarity of the poles is marked in Fig. 1.64

To obtain clockwise rotation the direction of force, which should act on the rotor conductors is marked. By applying Fleming's left hand rule, the direction of flow current in the rotor conductors is marked. The direction of flow of current in rotor conductors which are under the influence of N-pole is outward and which are under the influence of S-pole is inward (see Fig. 1.64).

1.30 emf Induced in a Rotating Coil Placed in a Magnetic Field

Consider a coil *AB* placed on the outer periphery of a soft iron solid cylindrical rotor. The stator poles carry the exciting coils. When current flows through exciting coils flux is set up as shown in Fig. 1.65. The rotor is rotating in clockwise direction at constant angular velocity of ω radians/sec. The direction of linear velocity (perpendicular to the plane of coil) acting on conductor *A* is shown in Fig. 1.66, which makes an angle θ with the direction of field. The component of velocity perpendicular to the field is $v_p = v \sin \theta$.

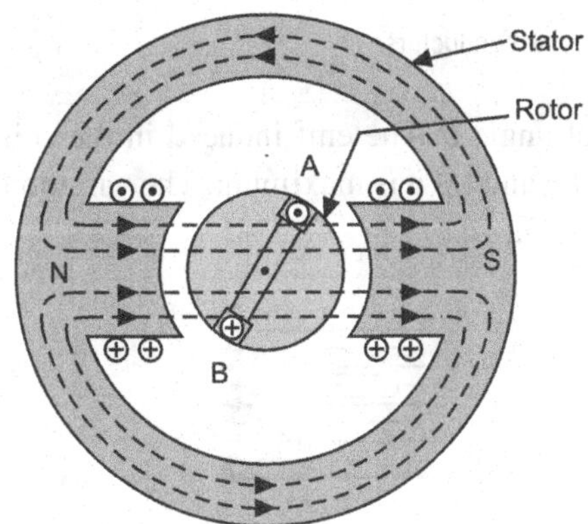

Fig. 1.65 Position of stator field

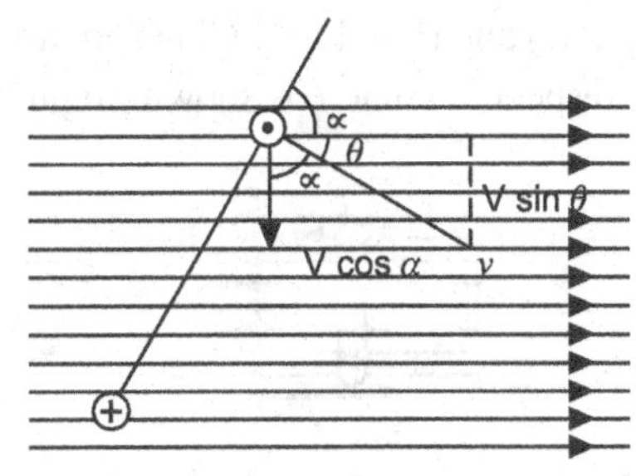

Fig. 1.66 Motion of conductor with respect to field

The emf induced in conductor *A*, $e = Blv \sin \theta$

Where, B = flux density in the rotor in Tesla

l = effective length of the conductor in metre

As the coil has two conductors only,

∴ emf induced in the coil at this instant,

$$e = 2\,Blv \sin \theta$$

If coil has *N* number of turns then,

emf induced in coil = $2NBlv \sin \theta$

If the angle between the plane of the coil and direction of magnetic field is α, then component of velocity perpendicular to the field $v_p = v \cos \alpha$ (see Fig. 1.66).

∴ The emf induced in the coil, $e = 2\,N\,Blv \cos \alpha$

The direction of induced emf in conductor *A* and *B* can be determined by applying Fleming's Right Hand Rule or Lenz's Law.

By applying Fleming's Right Hand Rule to conductor *A*, which is moving downward, the induced emf is out of the plane of paper [⊙], whereas conductor *B* is moving upward and the induced emf is into the plane of paper [⊕].

According to Lenz's Law, we can say that when coil is rotating in clockwise direction the flux linkages are decreasing [see Fig. 1.67 (*a*)]. Thus the emf induced in the coil would have the direction such that the magnetic flux produced by its current tends to oppose the decreasing flux linkages i.e., the coil will set up the field which will increases the flux linking with the coil [see Fig. 1.67 (*b*)].

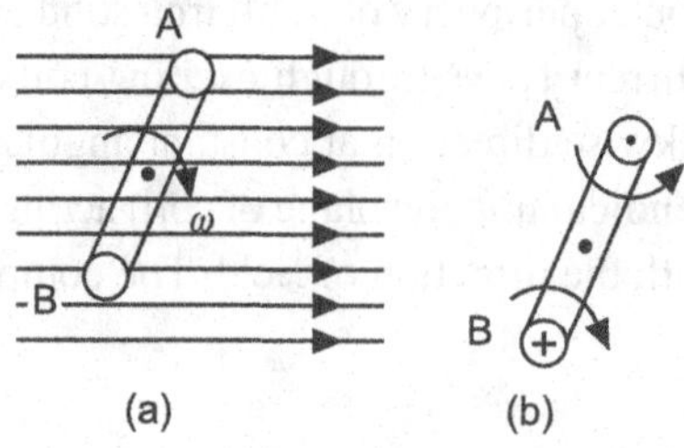

Fig. 1.67 Field produced by the rotor conductors

The magnitude of induced emf depends upon sine of angle θ. The emf induced in the coil at position (*a*), (*e*) and (*i*) in Fig. 1.68 is zero and at position (*c*) and (*g*) it is maximum. The variation of emf with respect to angle θ is shown graphically in Fig. 1.69.

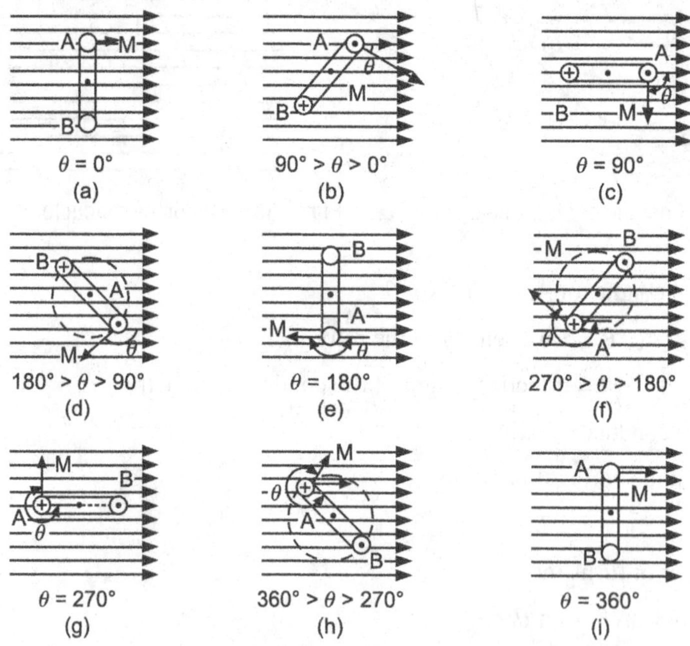

Fig. 1.68 emf induced in a coil of various instants

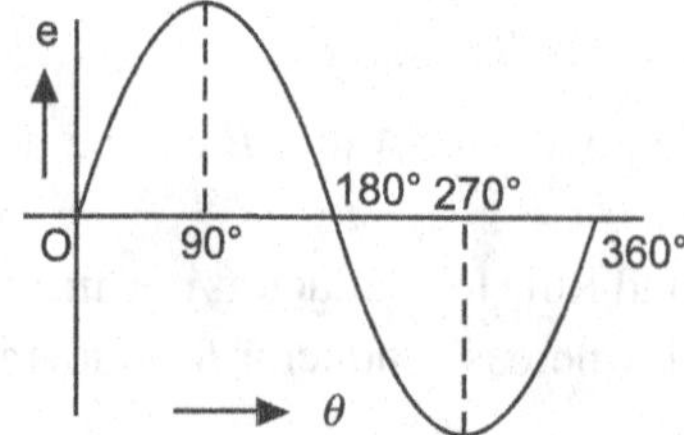

Fig. 1.69 Wave diagram of induced emf

The above arguments regarding the magnitude and direction of the emf induced are equally applicable if the main magnetic field is rotating and the coil is kept stationary.

Example 1.43

Mark the direction of emf in the conductor of the rotor of fig 1.70. when the rotor moves:

(*i*) *in the clockwise direction;*

(*ii*) *in the anticlockwise direction*

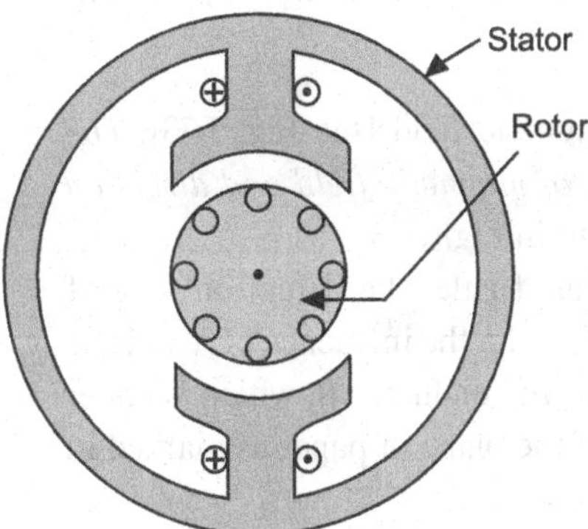

Fig. 1.70 Given figure

Solution:

Applying right hand thumb rule to the current carrying conductors of stator, the direction of magnetic lines of force is shown in Fig. 1.71 and 1.72. Accordingly the polarity of the poles is marked.

(*i*) When rotor is rotating in clockwise direction see Fig. 1.71, the direction emf induced in the rotor conductors is marked by applying Fleming's right hand rule to the conductors directly under the influence of north and south pole. The emf induced, in all the conductors under the influence of north-pole is in one direction i.e., ⊕ whereas, the emf induced in all the conductors under the influence of south pole is in other direction i.e., ⊙.

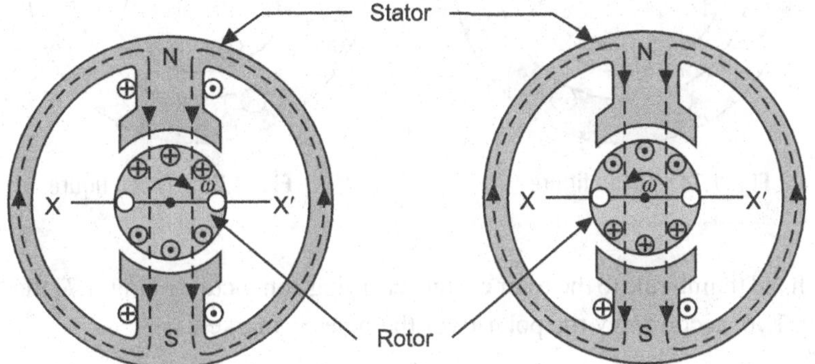

Fig. 1.71 Induced emf in rotor conductors for clockwise rotor

Fig. 1.72 Induced emf in rotor conductor for anticlockwise rotor

(*ii*) Similarly, by applying Fleming's right hand rule to the rotor conductors rotating in anticlockwise direction, the direction of induced emf is marked in them as shown in Fig. 1.72.

The emf induced in the rotor conductors placed on the magnetic neutral axis (i.e., *X*-axis) is zero, because at this position the conductor move parallel to the magnetic lines of force.

Example 1.44

A coil is rotating in clockwise direction in a two pole magnetic field. With the help of a neat diagram show what will be the direction of the induced emf in the coil? Show on the diagram the polarity of the poles and the direction of rotating of the coil.

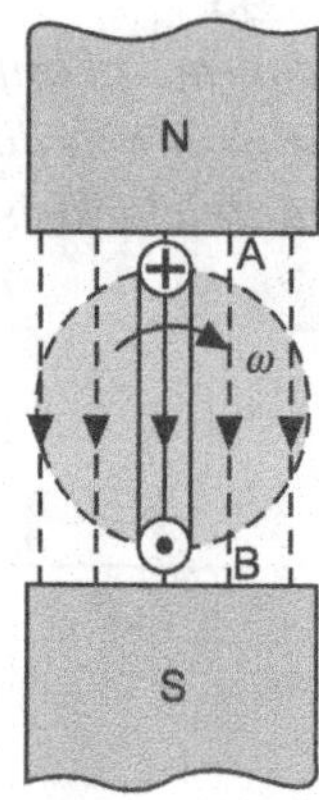

Fig. 1.73 Direction of main field and direction of induced emf

Solution:

A coil AB is placed in the magnetic field [see Fig. 1.73]. *The polarity of the poles, direction of magnetic field and direction of rotation (clockwise)* is marked in Fig. 1.73.

Applying Fleming's right hand rule, the direction of emf induced in conductor *A*, which is under the influence of *N*-pole, is into the plane of paper. Whereas, in conductor *B*, which is under the influence of *S*-pole, is out of the plane of paper as marked in Fig. 1.73.

Example 1.45

The rotors of Fig. 1.74 and 1.75 are each rotating in anti-clockwise direction, Mark in each case:

(i) *The direction of emf induced in the stator coils.*

(ii) *The coils sides which belong to the same stator coils.*

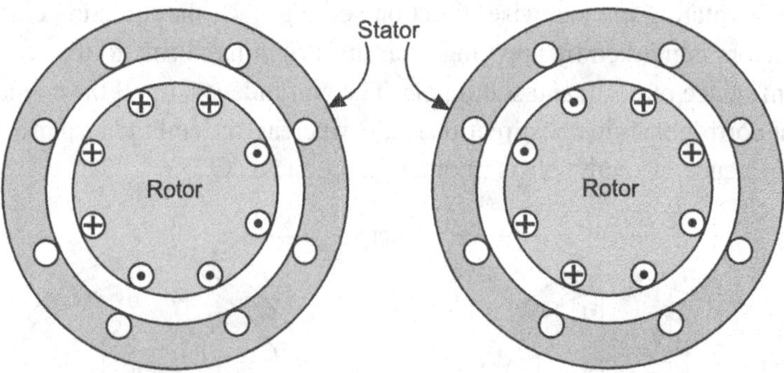

Fig. 1.74 Given figure

Fig. 1.75 Given figure

Solution:

Applying right hand thumb rule to the rotor current carrying conductors of Fig. 1.74 the field produced is shown in Fig. 1.76, accordingly the polarity of the poles is marked.

(*i*) The rotor is rotating in anticlockwise direction, therefore the relative motion of stator conductors with respect to the field is clockwise. Applying Fleming's right hand rule to the stator conductors the direction of induced emf is marked (see Fig. 1.76 and 1.77). In all the conductors under the influence of north pole, the induced emf is in one direction i.e., out of the plane of paper. Whereas in the conductors which are under the influence of south pole the induced emf is in the other direction i.e., into the plane of paper.

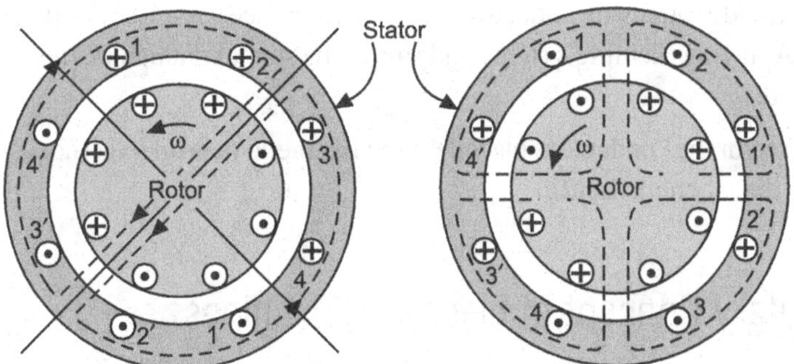

Fig. 1.76 Stator emf

Fig. 1.77 Stator emf

(*ii*) Each coil has two coil sides. If one coil side is placed under north pole, the other side of the same coil is placed under south pole, The position of the two sides must be the same under the two poles (i.e., if one side is placed at the centre of *N*-pole then the other side must also be placed at the centre of consecutive *S*-pole). The four coils in Fig. 1.76 and 1.77 are marked as 1-1′, 2-2′, 3-3′, and 4-4′.

Example 1.46

Fig 1.78 shows 6 coils placed in the slots of the stator. The rotor is carrying the field winding. Mark the direction of the emf induced in the conductors of stator.

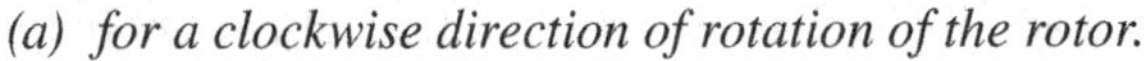

(*a*) *for a clockwise direction of rotation of the rotor.*

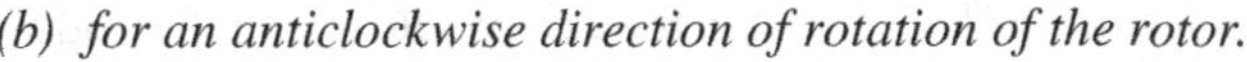

(*b*) *for an anticlockwise direction of rotation of the rotor.*

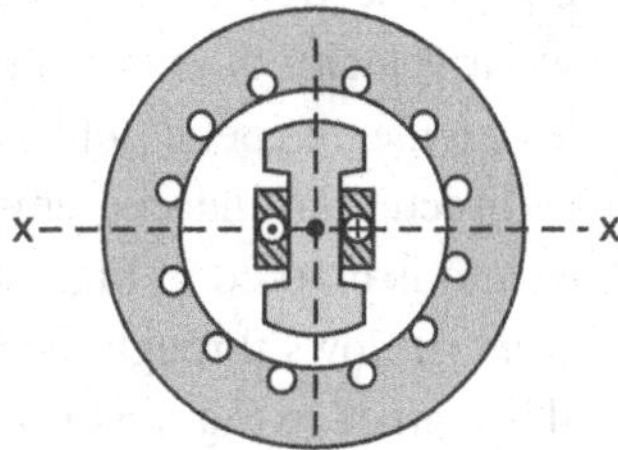

Fig. 1.78 Given figure

Solution:

Applying right hand thumb rule to the current carrying conductors of rotor, the field produced by it is shown in Fig. 1.79 and 1.80. Accordingly, the polarity of rotor poles is marked.

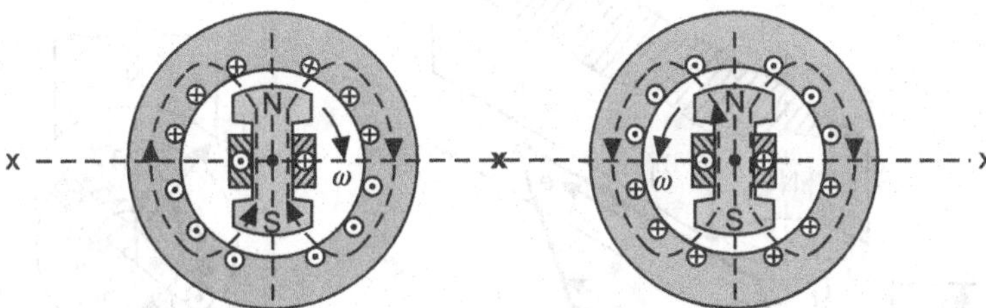

Fig. 1.79 Direction of induced emf in stator conductions for clockwise rotation

Fig. 1.80 Direction of induced emf in stator conductions for anti-clockwise rotation

(*i*) When rotor (field) rotates in clockwise direction see Fig. 1.79) flux is cut by the stator conductors. Applying Fleming's right hand rule to the stator conductors the direction of induced emf is marked.

Note: While applying Fleming's right hand rule the direction of rotation of conductors with respect to the field is considered. In the above case rotation of conductors is anticlockwise with respect to the field.

(*ii*) When rotor (field) rotates in anticlockwise direction (see Fig. 1.80) flux is cut by the stator conductors. Applying Fleming's right hand rule to the stator conductors, the direction of induced emf is marked.

(The emf induced in the conductors placed on the magnetic neutral axis (i.e., *X*-axis) will be zero as they do not cut the magnetic field).

1.31 Elementary Concept of Electrical Machines

(Energy conversion in electrical machines)

By electrical machine, here we mean a generator or motor. We shall discuss the operation of machine as a generator and motor.

1.31.1 Operation of Machine as a Generator (Conversion of Mechanical Energy into Electric Energy)

A coil is placed in a constant stationary magnetic field. Let it be rotated in clockwise direction at an angular velocity of ω radians per second by some outside driving mechanical torque T_m. The coil sides cut the magnetic field (or the flux linking with coil changes) and emf (*e*) is induced in the coil. The direction of induced emf can be determined by applying Fleming's right hand rule or Lenz's Law and is marked in Fig. 1.81 (*a*). The coil is connected to an external load resistor *R*, therefore current (*i*) flows through the coil and external load resistor. The direction of flow of current in the coil is marked in Fig. 1.81 (*a*) and (*b*). When current flows through the coil conductors, they produce their own magnetic field. The direction of this rotor field is marked by arrowhead F_r. The rotor field F_r tries to come in line with the main field F_m and an electromagnetic torque T_e is produced in the opposite direction to that of the rotation.

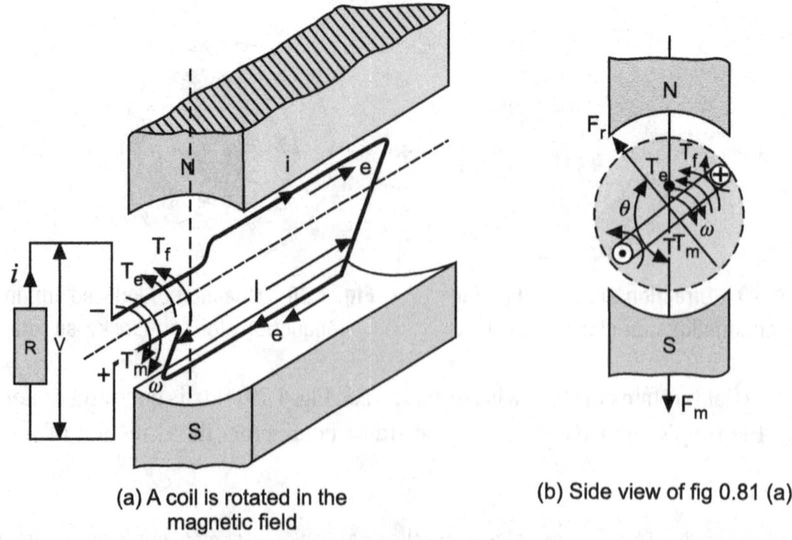

(a) A coil is rotated in the magnetic field

(b) Side view of fig 0.81 (a)

Fig. 1.81 Concept of generator

If the coil circuit is not closed, no current would flow through the coil and hence no electromagnetic torque will be developed (i.e., $T_e = 0$), under such a condition the opposition is only due to frictional torque (neglecting iron losses). Therefore, the mechanical torque T_m applied must be sufficient to overcome the frictional torque. It may be noted that frictional torque always acts in opposite direction to the direction of rotation.

i.e., $$T_m = T_f \text{ (at no-load)}$$

When the load resistance is connected, current flows and electromagnetic torque is produced in opposite direction to that of mechanical torque. Under this condition the mechanical torque T_m must be sufficient to overcome the electromagnetic torque and frictional torque (iron losses neglected).

Thus, $$T_m = T_e + T_f \quad ...(i)$$

or $$\omega T_m = \omega T_e + \omega T_f \quad ...(ii)$$

or $$\omega T_e = \omega T_m - \omega T_f$$

where, ωT_m = Mechanical input power.

ωT_f = Power losses due to friction.

ωT_e = Mechanical power developed in the rotor which is converted into electrical power.

When conductors move perpendicular to the magnetic field,

Induced emf in the coil, $e = 2\,B\,l\,v$

or $$e\,i = 2\,B\,i\,l\,v = 2\,F\,v \qquad [\because F = B\,i\,l]$$

$$= 2\text{F} \times \text{radius of coil} \times \frac{v}{\text{radius of coil}}$$

$$= T_e \times \omega \qquad [\because T = 2F \times \text{radius and } \omega = \frac{v}{\text{radius}}].$$

$$ei = \omega T_e \quad ...(iii)$$

If r is the internal resistance of the coil,

$$i = \frac{e}{r+R} \text{ or } e = ir + iR$$

multiplying both sides by i, we get,

$$ei = i^2r + i^2R \quad ...(iv)$$

$$ei - i^2r = vi$$

where, ei = electrical power generated,

i^2r = power lost in the resistance of coil, called copper losses,

vi or i^2R = electrical power output to the load.

Thus, we conclude that out of the input power ($\omega\, T_m$) only $\omega\, T_e$ is the mechanical power which is converted into electrical power ($e\,i$). After subtracting the copper losses (i^2r), the electrical power available at the load is only i^2R. This is how the conversion of power takes place in electrical machine working as a generator. The power flow diagram (neglecting iron losses) for the generator is shown in Fig. 1.82.

Mechanical power converted into electrical power.

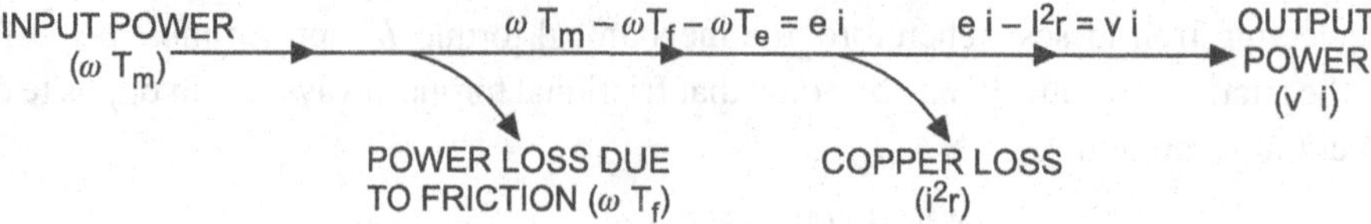

Fig. 1.82 Power flow in generator action

1.31.2 Operation of Machine as a Motor

Now consider a coil placed in the constant stationary field. The coil is connected to a battery [see Fig. 1.83 (a)]. The current i flows through the coil conductors in the direction shown in Fig. 1.83 (a) and (b). Current carrying conductors produce axis of magnetic field Fr in the direction as marked. The rotor field Fr tries to come in line with the main field Fm. Thus, electromagnetic torque is developed on the coil in anticlockwise direction. The coil is free to rotate, therefore, it starts rotating in anticlockwise direction (say at w rad/sec).

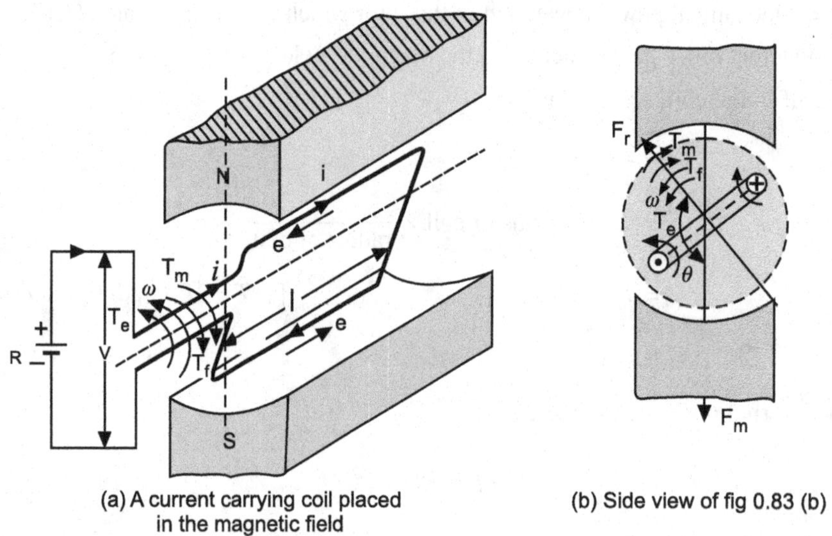

(a) A current carrying coil placed in the magnetic field

(b) Side view of fig 0.83 (b)

Fig. 1.83 Concept of motor

When the coil rotates in anticlockwise direction, flux is cut by the conductors (coil sides), therefore, an emf is induced in them in the direction marked in Fig. 1.83 (*a*), [direction is found by applying Fleming's right hand rule], which is opposite to the direction of flow of current or applied voltage *V*.20

If *r* is resistance of coil,

Applied voltage, $V = e + ir$

Multiplying both sides by *i*, we get

$$V i = ei + i^2 r$$

or

$$V i - i^2 r = ei$$

where, Vi = electrical power input,

i^2r = power lost in the resistance of the coil called copper losses,

ei = electrical power developed in the rotor which is converted into mechanical power.

that is, $ei = \omega T_e$ [from equation (*iii*)]

Thus, out of the input power (Vi), the copper losses (i^2r) are subtracted and only power ($e\ i$) is developed in the rotor which is converted into mechanical power.

Friction acts in opposite direction to that of rotation, therefore, the direction of frictional torque is in clockwise direction as marked in Fig. 1.83 (*b*). When mechanical load is applied, the mechanical torque T_m tries to reduce the speed of rotor, therefore, it is also acting in clockwise direction as marked in the figure.

Thus, the electromagnetic torque must be sufficient to meet with the sum of frictional torque and mechanical torque i.e.,

$$T_e = T_m + T_f$$

or $$\omega T_e = \omega T_m + \omega T_f$$

or $$\omega T_e - \omega T_f = \omega T_m.$$

The actual output mechanical power is only ωT_m, which is available at the shaft. This power is obtained after subtracting mechanical losses (ωT_f) from the power developed (ωT_e) in the rotor. The power flow diagram (neglecting iron losses) for the motor is shown in Fig. 1.84.

Electrical power converted into mechanical power.

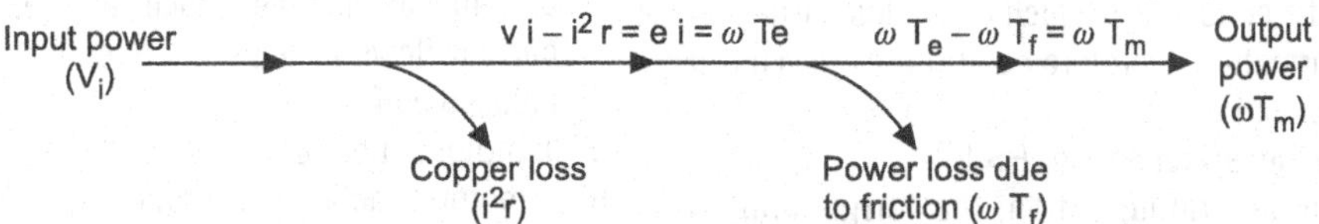

Fig. 1.84 Power flow in motor action

Conclusion

From the above discussion, the following conclusions are drawn;

(*i*) In generating action, the induced emf produces the armature current, therefore, *e* and *i* both are in same direction, whereas, in motoring action, the induced emf opposes the conduction and current flows in opposite direction to induced emf

(*ii*) In motoring action, the electromagnetic torque produces the rotation, (T_e and ω both are in same direction), whereas, in generating operation, the electromagnetic torque opposes the rotation (T_e acts in opposite direction to ω).

(*iii*) In generating action the torque angle θ is leading (F_r leads F_m by an angle θ with respect to direction of rotation), whereas in motoring action, the torque angle θ is lagging (F_r lags behind F_m by an angle θ with respect to direction of rotation).

(*iv*) In generating action, the rotation is due to mechanical torque, therefore, T_m and ω are in same direction, whereas in motoring action rotation is opposite to mechanical torque.

(*v*) In both generating and motoring action, electromagnetic torque always acts in opposite directions to mechanical torque (i.e., T_e always acts in opposite direction to T_m).

(*vi*) In both the cases, frictional torque acts in opposite direction to rotation (i.e., T_f always acts in opposite direction to ω).

Generator action	*Motor action*
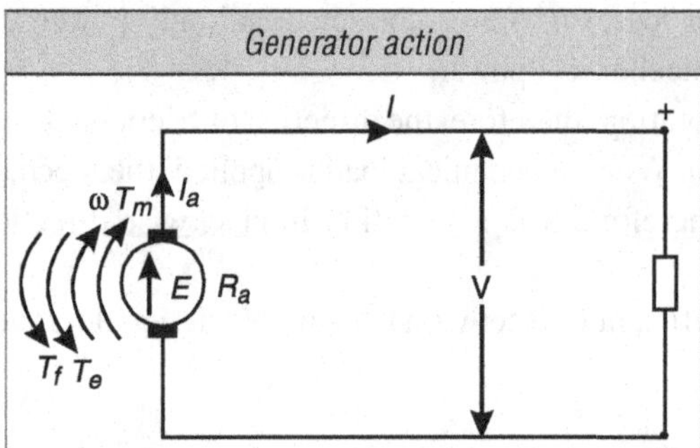	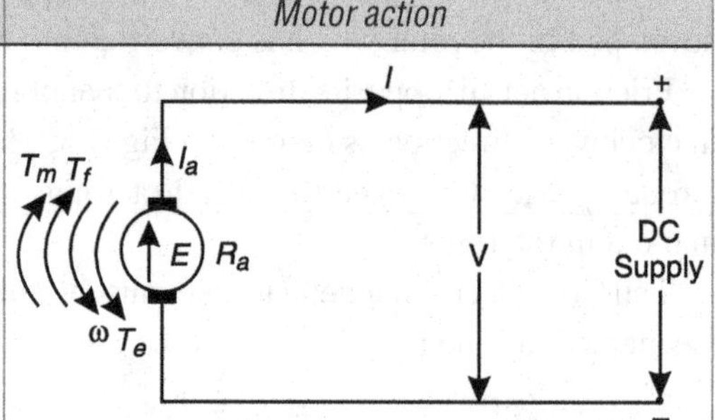
1. In generator action, the rotation is due to mechanical torque, therefore, T_m and ω are in the same direction.	1. In motor action, the rotation is due to electromagnetic torque, therefore, T_e and ω are in the same direction.
2. The frictional torque T_f acts in opposite direction to rotation ω.	2. The frictional torque T_f acts in opposite direction to rotation ω.
3. Electromagnetic torque T_e acts in opposite direction to mechanical torque T_m so that $\omega T_m = \omega T_e + \omega T_f$.	3. Mechanical torque T_m acts in opposite direction to electromagnetic torque T_e so that $\omega T_e = \omega T_m + \omega T_f$.
4. In generating action, and emf is induced in the conductors which circulates current in the armature, therefore *e* and *i* both are in the same direction.	4. In motoring action, current is impressed to the armature against the induced emf (*e*), therefore current flows in opposite direction to that of induced emf.
5. In generator action, $E > V$	5. In motor action, $E > V$
6. In generating action, the torque angle θ is leading.	6. In motoring action, the torque angle θ is legging.
7. In generating action, mechanical energy is converted into electrical energy.	7. In motoring action, electrical energy is converted into mechanical energy.

Example 1.47

The cross-sectional view of a rectangular coil rotating in a magnetic field is shown in Fig. 1.85. Determine;

(*i*) *The direction of induced emf in the coil-sides for the direction of rotation as shown in the figure.*

(*ii*) *The direction of the electromagnetic torque developed under the above condition, and also show the torque angle.*

(*iii*) *The direction of the current in the coil-sides if the circuit is closed through an external resistance.*

(*iv*) *The direction in which external mechanical torque is applied.*

(*v*) *Whether the device behaves as a motor or generator.*

(*vi*) *In which direction the frictional torque will act?*

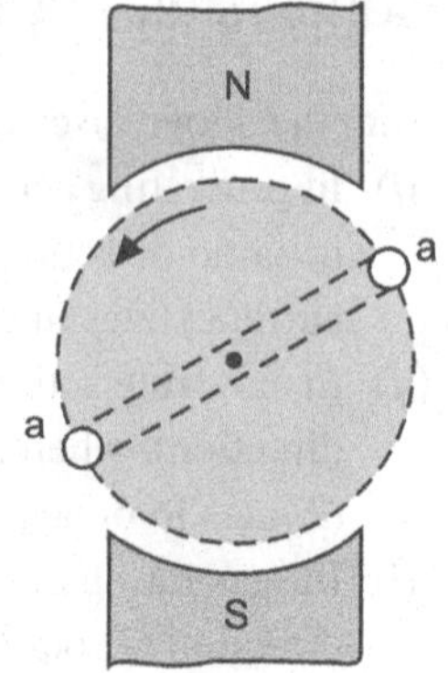

Fig. 1.85 Given figure

Solution:

(*i*) The coil is rotating in anticlockwise direction in the magnetic field, applying Fleming's right hand rule, the direction of induced emf in the coil sides in marked in Fig. 1.86 and 1.87.

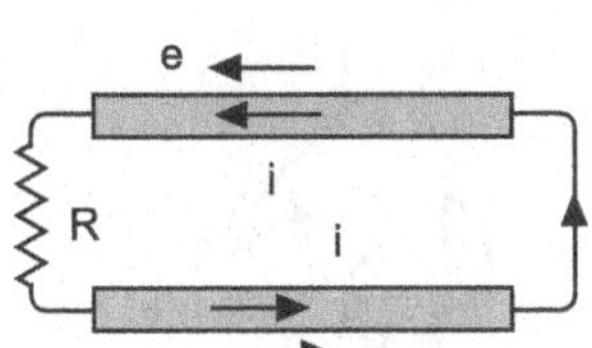

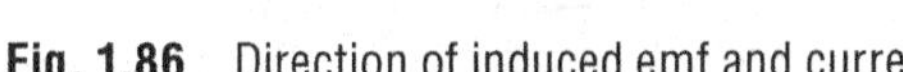

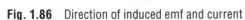

Fig. 1.86 Direction of induced emf and current

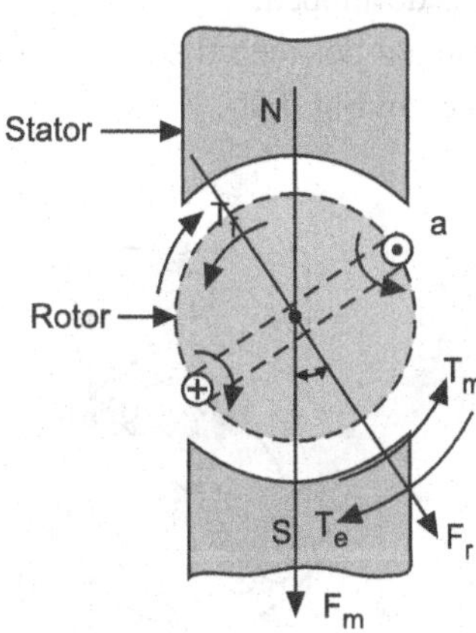

Fig. 1.87 Different quantities marked

(*ii*) When current flows through the coil sides (rotor), field is set up around the conductors, the direction of the rotor field is marked by arrow head F_r. Rotor field (F_r) tries to come in line with main field F_m, thus electromagnetic torque T_e is produced in clockwise direction as marked by the arrow head in Fig. 1.87.

(*iii*) When the circuit is closed through an external resistance, the direction of flow of current in the coil sides is shown in Fig. 1.87.
The angle between rotor field F_r and main field F_m is the torque angle θ as marked in the figure.

(*iv*) For energy conversion the mechanical torque T_m always acts in opposite direction to that of electro-magnetic torque T_e. Therefore, the direction of external mechanical torque applied is anticlockwise as marked in the figure.

(*v*) In the coil sides the direction of flow of current is same as that of the induced emf therefore, the machine acts as a generator.

(*vi*) The friction always acts in opposite direction to the direction of rotation, therefore, the direction of frictional torque is clockwise as marked by the arrow head in Fig. 1.87.

Example 1.48

The cross sectional view of a rectangular coil rotating in a magnetic field in shown in Fig. 1.88. The direction of rotation and the direction of the current in the coil sides is shown.

(*i*) *Determine the direction of the induced emf in the coil.*

(*ii*) *Mark the direction of the electromagnetic torque developed.*

(*iii*) *Show the torque angle.*

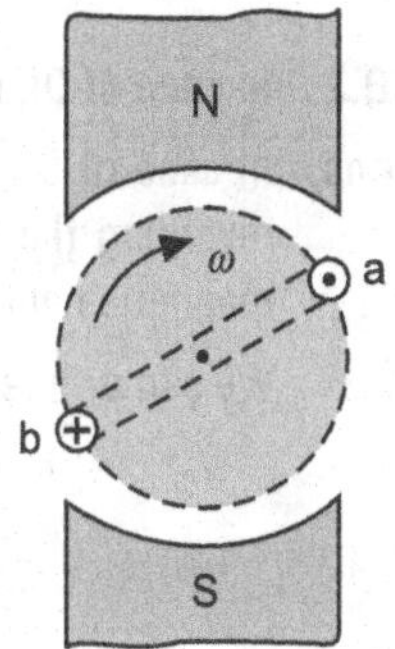

Fig. 1.88 Given figure

Solution:

(*i*) When coil rotates in the magnetic field, flux is cut by the conductors and an emf is induced. The direction of induced emf is marked in Fig. 1.89 which is determined by Fleming's right hand rule.

(*ii*) The direction of axis of main magnetic field (F_m) is marked. Depending upon the direction of flow of current in coil sides, the direction of axis of rotor field (F_r) is marked. Rotor field tries to come in line with main field, thus a clockwise electromagnetic torque (T_e), as marked in Fig. 1.90 is developed.

(*iii*) The angle between the axis of rotor field and main field (θ) is called torque angle which is marked in Fig. 1.90.

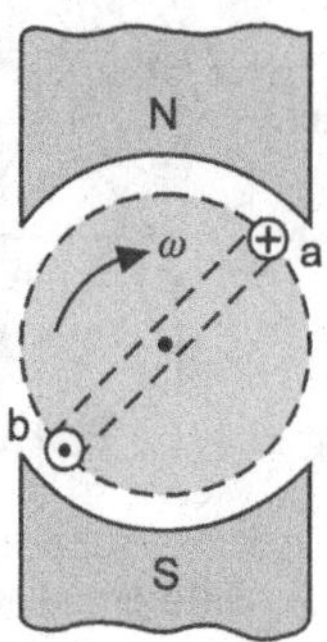

Fig. 1.89 Induced emf in the coil

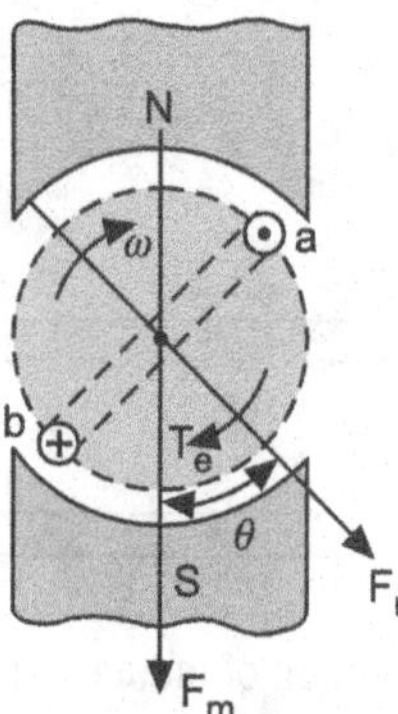

Fig. 1.90 Position of other quantities marked

Section Practice Problems

Numerical Problems

1. A conductor of 0·4 metre length is moved at a uniform speed of 5 m s^{-1} at right angle to its length and to a magnetic field. Calculate the density of the magnetic field if the emf generated in the conductor is 2V. For the same density, what will be the induced emf, if the conductor is moved at an angle of 60° with the magnetic field. (**Ans:** 1 *Wb/m*^{2}, 1·732*V*)

Short Answer Type Questions

Q.1. What do you mean by electromechanical energy conversion devices?

Ans. A device or machine that converts electrical energy into mechanical energy or vice-versa is called an electro-mechanical energy conversion device

Q.2. In case of DC motors, how alternating torque is converted into unidirectional continuous torque?

Ans. In case of DC motors, the alternating torque is converted into unidirectional continuous torque by reversing the direction of flow of current in the armature conductors when they pass through the magnetic neutral axis (using a commentator)

We can differentiate generator and motor on the basis of following:

Review Questions

1. Define the terms mmf, magnetic flux and magnetic reluctance and establish the relation which holds between these quantities for a magnetic circuit.

2. Explain the terms: permeability, reluctance and permeance.
3. Make comparison between magnetic and electric circuits.
4. What are the similarities between electrical circuits and magnetic circuits?
5. How is B-H curve of ferromagnetic material different from that of non-magnetic material? Name all the salient regions of B-H curve of magnetic material.
6. Explain hysteresis loss.
7. State and explain Faraday's laws of electro-magnetic induction.
8. Write down the expression for dynamically induced emf developed in a conductor of length *l metre* moving with a velocity of *v metre per second* in a uniform magnetic flux density of *B Wb/m*^{2}.
9. Explain the term self and mutual inductance.
10. Define co-efficient of coupling and show that $k = M/\sqrt{L_1 L_2}$.
11. Derive the expression for the equivalent inductance when two coupled coils are connected in (*a*) series (*b*) parallel.
12. Explain what you mean by eddy currents.
13. Mark the current in the conductors of the rotor in Fig. 1.91 and 1.92 to obtain an anticlockwise torque.

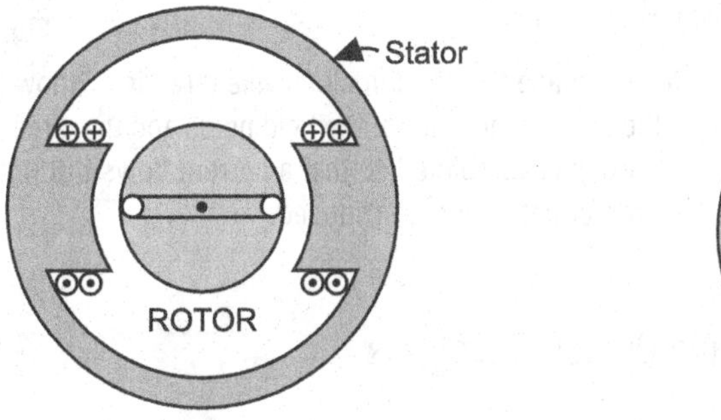

Fig. 1.91:

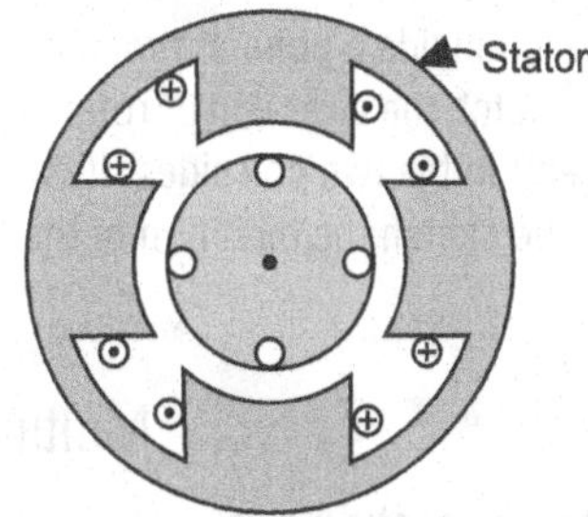

Fig. 1.92:

14. Will the rotor in Fig. 1.93 tend to rotate? If yes show the torque angle and direction of rotation.

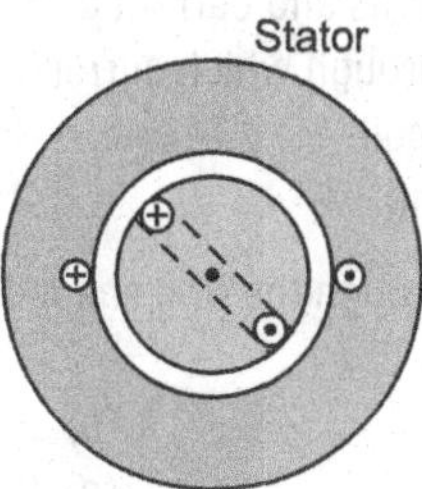

Fig. 1.93:

15. What do you mean by a motor and generator?

16. Fig. 1.94 shows a stator with 4 coils whereas, Fig. 1.95 shows a rotor with 6 coils eventually distributed over the rotor periphery. Mark the direction of emf in these conductors for a clockwise rotation. Mark also the coil sides which belong to the same coil.

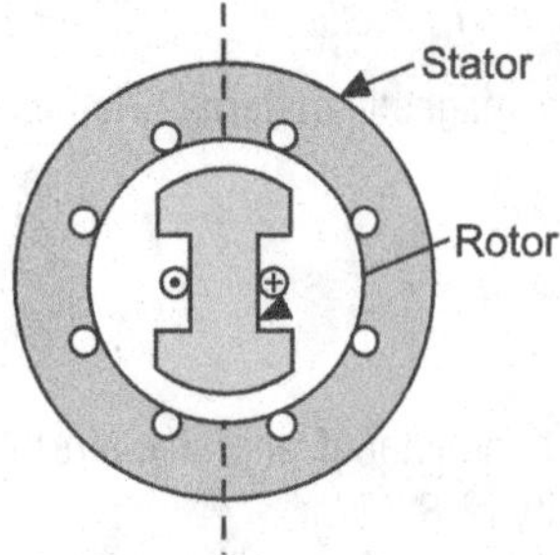

Fig. 1.94:

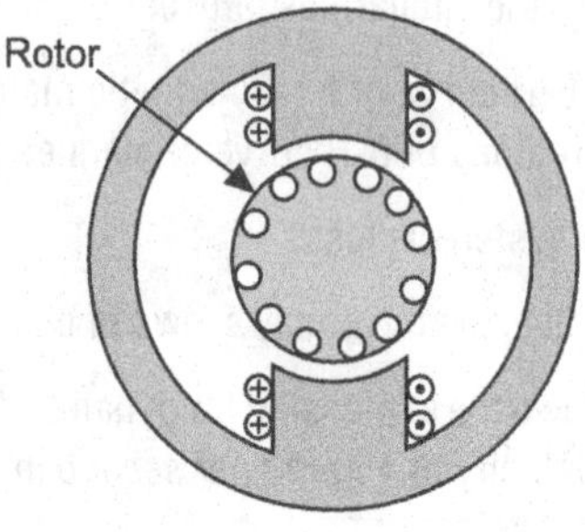

Fig. 1.95:

17. A cross-sectional view of a rectangular coil *a b* rotating in a magnetic field is shown in Fig. 1.96. Determine:

(*i*) The direction of current in the coil sides, if the circuit is closed through an external resistance, for the direction of rotation as shown in the figure, mark

(*ii*) The direction the electromagnetic torque.

(*iii*) Torque angle.

(*iv*) The direction in which external mechanical torque is applied.

(*v*) The direction in which the frictional torque will act.

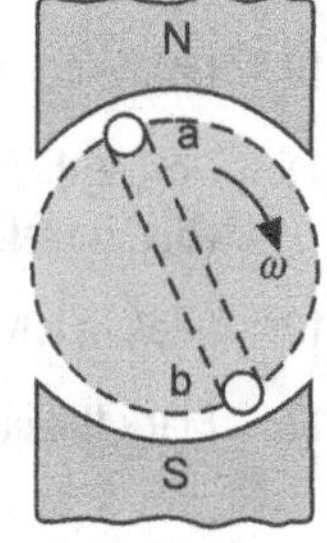

Fig. 1.96:

18. A coil placed inside a constant magnetic field is made to rotate in clockwise direction. Show by means of a neat sketch the direction of rotation of the coil, the polarity of the field poles and the direction of the induced emf in the two coil sides. If the coil circuit is completed so that a current flows through the coil, show on the sketch the direction of the electromagnetic torque produced.

Multiple Choice Questions

1. The perfect magnetic insulator is

(*a*) copper (*b*) iron

(*c*) rubber (*d*) none of above.

2. The ampere-turns are

(*a*) the product of the number of turns and current of the coil

(*b*) the number of turns of a coil through which current is flowing

(*c*) the currents of all turns of the coil

(*d*) the turns of transformer winding.

3. What will be the current passing through the ring shaped air cored coil when number of turns is 800 and ampere-turns are 3200.

(*a*) 0·25 (*b*) 2·5

(*c*) 4·0 (*d*) 0·4

4. The magnetic reluctance of a material

(*a*) increases with increasing cross-sectional area of material.

(*b*) does not vary with increasing the cross-sectional area of material.

(*c*) decreases with increasing cross-sectional area of material.
(*d*) decreases with increasing the length of material.

5. Mmf is analogous to
(*a*) electric current in electric circuit
(*b*) current density in conductor
(*c*) electromotive force

6. The magnetic strength of an electromagnet can be increased by
(*a*) increasing current in the solenoid
(*b*) increasing the number of turns of the solenoid
(*c*) both a and b
(*b*) none of above.

7. Hysteresis loss in a magnetic material depends upon
(*a*) area of hysteresis loop
(*b*) frequency of reversed of field
(*c*) volume of magnetic material
(*d*) all a, b and c.

8. The Steinmetz expression to determine the hysteresis loss in the magnetic material is
(*a*) $P_h = \eta B_{max}^{1.6}$
(*b*) $P_h = \eta B_{max}^{1.6} f$
(*c*) $P_h = \eta B_{max}^{1.6} f V$
(*d*) all, a, b and c

9. The direction of electro-magnetically induced emf is determined by
(*a*) Flemings' right hand rule
(*b*) Lenz's law
(*c*) Right hand thumb rule
(*d*) both a and b.

10. The energy store in the magnetic field is
(*a*) given by the relation $LI^2/2$
(*b*) directly proportional to the square of current flowing through the coil
(*c*) directly proportional to the inductance of the coil
(*d*) all a, b and c.

11. The self inductance of a solenoid of N-turns is proportional to
(*a*) N
(*b*) N^2
(*c*) 1/N
(*d*) $1/N^2$.

12. To reduce eddy current loss in the core of magnetic material
(*a*) the core is laminated
(*b*) the magnetic material used should have high resistivity
(*c*) both a and b
(*d*) none of above.

13. If two coils having self-inductances L_1 and L_2 and a mutual inductance M are connected in series with opposite polarity, then the total inductance of the combination will be
(*a*) $L_1 + L_2 + 2M$
(*b*) $L_1 - L_2 - 2M$
(*c*) $L_1 - L_2 + 2M$
(*d*) $L_1 + L_2 - 2M$

14. An inductive coil of 10 H develops a counter voltage of 50 V. What should be the rate of change of current in the coil
(*a*) 5 A/s
(*b*) 0·2 A/s
(*c*) 1 A/s
(*d*) 500 A/s

15. A machine that converts electrical energy into mechanical energy is called
(*a*) electric generator
(*b*) electric motor
(*c*) both a and b
(*d*) none of these

16. In electromechanical energy conversion devices, the angle between rotor field and main magnetic field is called.

(*a*) mechanical angle (*b*) electrical angle
(*c*) either a or b (*d*) torque angle

17. In an electro-mechanical device, when both the direction of rotation and direction of electromagnetic torque are in same direction the machine works as a

(*a*) generator (*b*) motor
(*c*) both a and b (*d*) none of these

18. In an electro-mechanical device, when both the direction of rotation and direction of mechanical torque are in same direction, the machine works as a

(*a*) generator (*b*) motor
(*c*) both a and b (*d*) none of these

19. In an electro-mechanical device, when induced emf (e) acts in opposite direction to the direction of flow of current (i) in the armature conductors, the machine works as a

(*a*) generator (*b*) motor
(*c*) both a and b (*d*) none of these

Keys to Multiple Choice Questions

1. d	**2.** a	**3.** c	**4.** c	**5.** c	**6.** c	**7.** d	**8.** c	**9.** d	**10.** d
11. b	**12.** c	**13.** d	**14.** a	**15.** b	**16.** d	**17.** b	**18.** a	**19.** b	

CHAPTER

Single-Phase Transformers 2

Chapter Objectives

After the completion of this unit, students/readers will be able to understand:

- ✓ What is a transformer and its necessity in power system?
- ✓ How a transformer transfers electric power from one circuit to the other i.e., what is basic principle of a transformer.
- ✓ What is core of a transformer, what is its material, why cruciform cores are considered better than square or rectangular cross-section?
- ✓ What are transformer windings, how their material and their size selected. How these wound and placed over the core?
- ✓ Why bushings are necessary for transformers?
- ✓ What is the construction of small rating transformers?
- ✓ How does a transformer behave? When is it considered to be an ideal one?
- ✓ What are the various factors on which emf induced in a transformer winding depends?
- ✓ How does a transformer behave when it is at no-load?
- ✓ What is exciting current and how is it affected by magnetisation?
- ✓ Why is there sudden inrush of magnetising current when a transformer is connected to the lines although it may be at no-load?
- ✓ How does a transformer behave when it is loaded?
- ✓ How to draw phasor diagrams of a transformer to represent various alternating quantities (neglecting resistance and reactance ampere-turns balance)?
- ✓ How does resistance of transformer windings affect its performance?
- ✓ What are mutual and leakage fluxes in a transformer?
- ✓ How does inductive reactance appear in a transformer due to leakage fluxes?
- ✓ How to determine equivalent resistance and reactance of transformer windings on its either side?
- ✓ How to draw an equivalent circuit of a transformer and how to simplify it?
- ✓ What is voltage regulation of a trans-former?
- ✓ How power factor affects the regulation of a transformer?
- ✓ What are the major losses in a transformer?
- ✓ What are the effects of voltage and frequency variation on iron losses in a transformer?
- ✓ How is efficiency of a transformer calculated at various loads?
- ✓ How to determine the conditions at which a transformer works at maximum efficiency?

- ✓ How efficiency of a transformer is affected by load and p.f.?
- ✓ How to conduct various tests such as polarity test, voltage ratio test, open circuit test, short circuit test, back-to-back test etc., on a transformer?
- ✓ How transformers are classified on the basis of their application?
- ✓ What is the necessity of putting the transformers in parallel and what conditions are to be maintained while putting them in parallel?
- ✓ When the transformers are put in parallel, how do they share the load?
- ✓ What is an auto-transformer?
- ✓ How can auto-transformer is different to a potential divider?
- ✓ What are the advantage and disadvantages of an auto-transformer in comparison to an ordinary two-winding transformer?
- ✓ Why cannot ordinary two-winding transformers be replaced by auto-transformers?
- ✓ How to draw equivalent circuit and phasor diagram of an auto-transformer?
- ✓ Can we convert a two winding transformer into an auto-transformer?
- ✓ What are the major applications of auto-transformers?

Introduction

Transformer ***is considered to be a*** **backbone** ***of a power system.***

For generation, transmission and distribution of electric power, AC system is adopted instead of DC system because voltage level can be changed comfortably by using a transformer. For economic reasons, high voltages are required for transmission whereas, for safety reasons, low voltages are required for utilisation. Transformer is an essential part of power system. Hence, *it is rightly said that* **transformer** *is a* ***backbone*** *of a power system.* In this chapter, we shall discuss the general features and principle of operation of single-phase transformers.

2.1 Transformer

A **transformer** *is a static device that transfers AC electrical power from one circuit to the other at the same frequency but the voltage level is usually changed.*

The block diagram of a transformer is shown in Fig. 2.1(*a*). When the voltage is raised on the output side ($V_2 > V_1$), the transformer is called a step up transformer, whereas, the transformer in which the voltages is lowered on the output side ($V_2 < V_1$) is called a step down transformer.

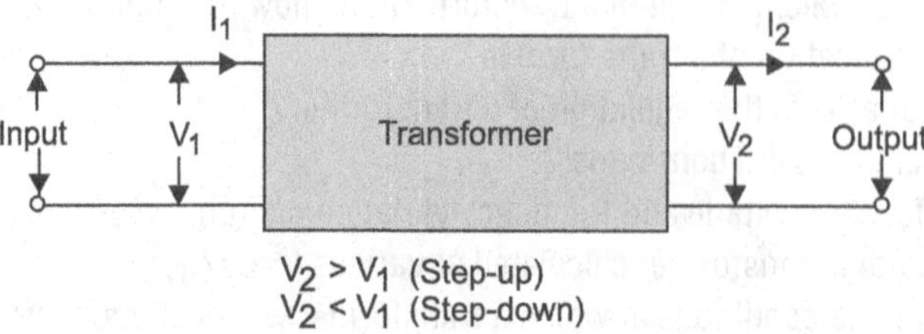

Fig. 2.1 (a) Block diagram of a single-phase transformer

Necessity

In our country, usually electric power is generated at 11 kV. The voltage level is raised to 220 kV, 400 kV or 750 kV by employing step-up transformers for transmitting the power to long distances. Then to feed different areas, as per their need, the voltage level is lowered down to 66 kV, 33 kV or 11 kV by employing step-down transformers. Ultimately for utilisation of electrical power, the voltage is stepped down to 400/230 V for safety reasons.

Thus, transformer plays an important role in the power system. The pictorial view of a power transformer is shown in Fig. 2.2 (*b*). The important accessories are labelled on it.

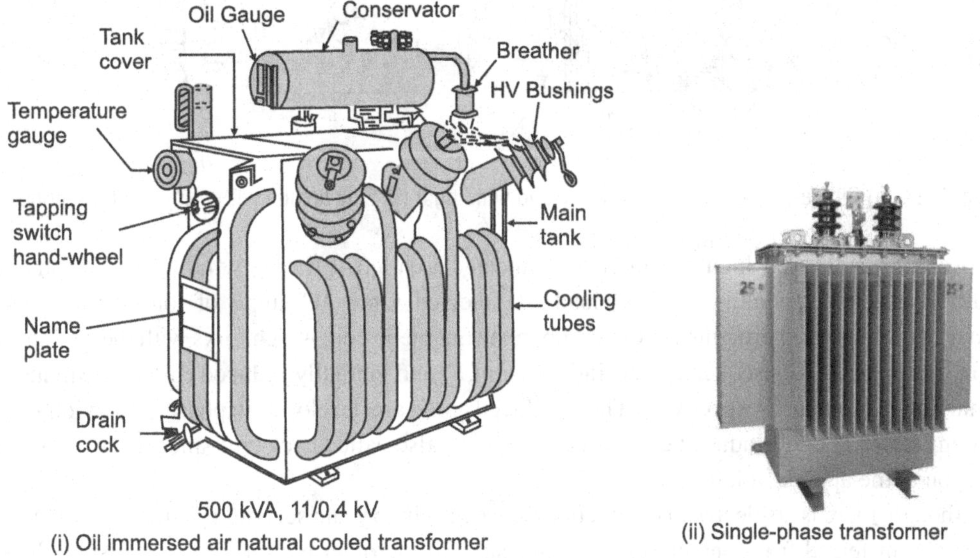

Fig. 2.1 (b) Transformer

Applications

Main applications of the transformers are given below:

(*a*) To change the level of voltage and current in electric power systems.
(*b*) As impedance-matching device for maximum power transfer in low-power electronic and control circuits.
(*c*) As a coupling device is electronic circuits
(*d*) To isolate one circuit from another, since primary and secondary are not electrically connected.
(*e*) To measure voltage and currents; these are known as instrument transformers.

Transformers are extensively used in AC power systems because of the following reasons:

1. Electric energy can be generated at the most economic level (11–33 kV)
2. Stepping up the generated voltage to high voltage, extra high voltage EHV (voltage above 230 kV), or to even ultra high voltage UHV (750 kV and above) to suit the power transmission requirement to minimise losses and increase transmission capacity of lines.
3. The transmission voltage is stepped down in many stages for distribution and utilisation for domestic, commercial and industrial consumers.

2.2 Working Principle of a Transformer

The basic principle of a transformer is **electromagnetic induction.**

A single-phase transformer consists of two windings placed over a laminated silicon steel core. The winding having less number of turns is called low-voltage winding and the winding having more number of turns is called high voltage winding.

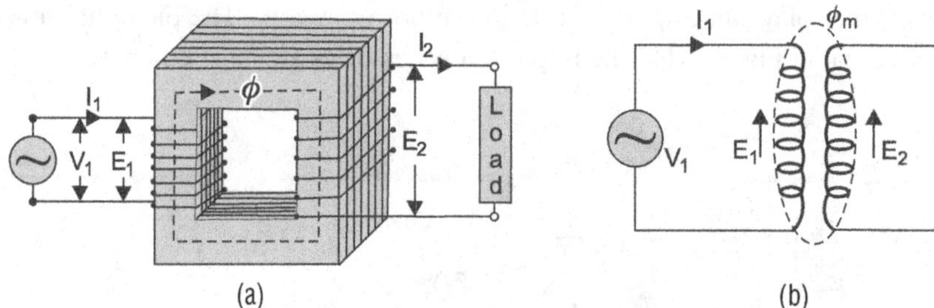

Fig. 2.2 (a) Single-phase transformer (core and windings) (b) Flux linking with primary and secondary

Also, the winding to which AC supply is connected is called a primary winding and the other one is called a secondary winding to which load is connected. Once AC supply of voltage V_1 is given to primary winding, an alternating flux is set-up in the magnetic core which links with the primary and secondary winding. Consequently, self-induced emf E_1 and mutually-induced emf E_2 are induced in primary and secondary, respectively. These induced emf's are developed in phase opposition to V_1 as per Lenz's law. The self-induced emf in the primary is also called back emf since it acts in opposite direction to the applied voltage.

Although, there is no electrical connection between primary and secondary winding, still electric power is transferred from one circuit (primary side) to the other circuit (secondary side). It is all because of magnetic coupling, i.e., the alternating flux which is set-up in the core linking with both the windings. The magnitude of induced emf in a coil depends upon rate of change of flux linkages i.e., $e \propto N$. since, the rate of change of flux for both the winding is the same, the magnitude of induced emf in primary and secondary will depend upon their number of turns, i.e., primary induced emf $E_1 \propto N_1$ and secondary induced emf $E_2 \propto N_2$. When $N_2 > N_1$, the transformer is called a step-up transformer, on the other hand, when $N_2 < N_1$ the transformer is called step-down transformer.

Turn ratio: The ratio of primary to secondary turns is called turn ratio, i.e., turn ratio $= N_1/N_2$.

Transformation ratio: The ratio of secondary voltage to primary voltage is called voltage transformation ratio of the transformer. It is represented by K.

$$K = \frac{E_2}{E_1} = \frac{N_2}{N_1} \quad \text{(since } E_2 \propto N_2 \text{ and } E_1 \propto N_1\text{)} \quad \ldots(2.1)$$

2.3 Construction of Transformer

The following are the major elements of a transformer:

(*i*) Magnetic circuit mainly comprises of transformer core having limbs and yokes.
(*ii*) Electric circuits mainly comprises of windings, insulation and bushings.
(*iii*) Tank mainly comprises of cooling devices, conservator and ancillary apparatus.

The construction of a transformer depends largely on its size and the duty which it is to perform. Designers make necessary changes as per the requirement. In practice, continuous improvements are being made in the construction of transformers.

2.3.1 Core Material

For core construction sheets of alloy steels are used. The main constituents of alloy steel are silicon and carbon in small quantities which increases the permeability at low flux densities and reduces the hysteresis loss to large extent. Addition of silicon also reduces the eddy currents to some extent because it increases the resistivity. In addition to this, these constituents also increase the mechanical strength of the core.

Now-a-days cold rolled grain oriented steel (*CRGOS*) sheets are used for core construction because of their excellent magnetic properties in the direction of rolling. This material allows flux density as high as 2.8 Wb/m^2.

2.3.2 Core Construction

In order to reduce eddy current loss, the core of the transformer is laminated. Since eddy current losses are proportional to the square of the thickness of laminations, every effort is made to reduce their thickness as small as possible. But there is a practical limit beyond which the thickness of laminations cannot be decreased further on account of mechanical considerations. This practical limit of thickness of each lamination is 0.3 mm. The laminations are usually made of 0.33 to 0.5 mm thick. These laminations are insulated from each other by a thin layer of oxide coating or varnish.

Core Cross-Section

Small transformers have rectangular section limbs with rectangular coils or square section limbs with circular coils as shown in Figs. 2.3(*a*) and (*b*).

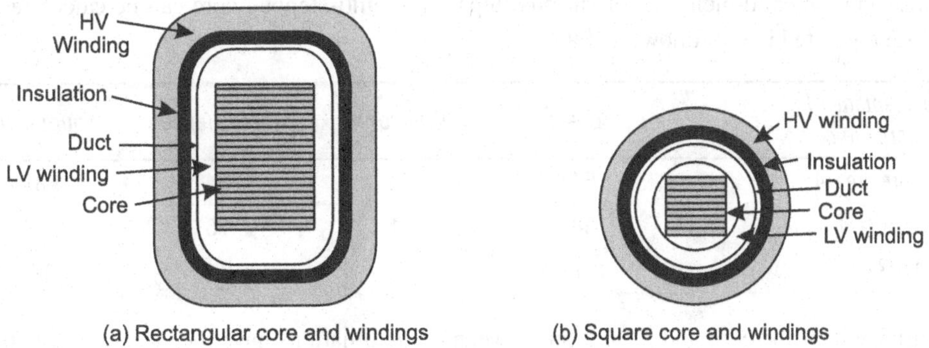

(a) Rectangular core and windings (b) Square core and windings

Fig. 2.3 Core and windings

As the size of the transformer increases, it becomes wasteful to use rectangular cores. For this purpose the cores are square shaped as shown in Fig. 2.4(a). The circle represents the inner surface of the tubular former carrying the windings. This circle is known as the **circumscribing circle.** Clearly a lot of useful space is wasted, the length of circumference of circumscribing circle being large in a comparison to its cross-section. This means that the length of mean turn of winding is increased giving rise to higher I^2R losses and conductor cost.

Circular coils are preferred for winding a transformer as they can be easily wound on machines on a former, conductors can easily be bent and winding does not bulge out due to radial forces developed during operation.

With larger transformers, cruciform cores, which utilise the space better, are used as shown in Fig. 2.4(b). As the space utilisation is better with cruciform cores, the diameter of circumscribing circle is smaller than with square cores of the same area. Thus the length of mean turn of copper is reduced with consequent reduction in cost of copper. It should be kept in mind that two different sizes of laminations are used in cruciform cores.

With large transformers, further steps are introduced as shown in Figs. 2.4(c) and (d) to utilise the core space which reduces the length of mean turn with consequent reduction in both cost of copper and copper loss.

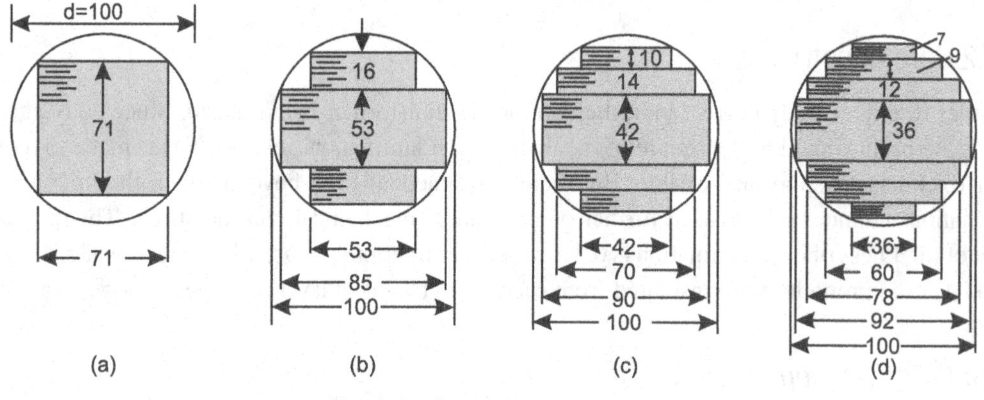

Fig. 2.4 Cruciform core section

By increasing the number of steps, the area of circumscribing circle is more effectively utilised. The most economical dimensions of various steps for a multi-stepped core can be calculated. The results are tabulated in the following Table:

Area percentage of circumscribing circle	*Square*	*Cruciform*	*Three stepped*	*Four stepped*
Gross core area A_{gc}	64	79	84	87
Net core area $A_i = kiA_{gc}$	58	71	75	78
Ratio A_i/d^2	0.45	0.56	0.6	0.62

Laminations of the core and insulation between the laminations have the effect of reducing the effective or net area of the core. The net core cross-sectional area is about 10% less than the gross-sectional area due to the laminations of the core and insulation such as paper or varnish. The ratio of the net cross-sectional area to gross cross sectional area is called the iron space factor or simply iron factor K_i.

$$K_i = \frac{A_i}{A_{gc}} \quad \text{...(2.2)} \quad (K_i < \text{unity nearly 0.9 to 0.99})$$

A_i = Iron Area or iron's net cross-sectional area of the core

A_{gc} = Overall Area or gross cross-sectional area of the core.

2.3.3 Transformer Winding

Transformer windings may be classified into two groups *viz.* concentric winding and sandwiched winding. Concentric windings are used in core type transformers as shown in Fig. 2.5(*a*). Sandwiched windings are almost exclusively used in shell type transformers as shown in Fig. 2.5(*b*).

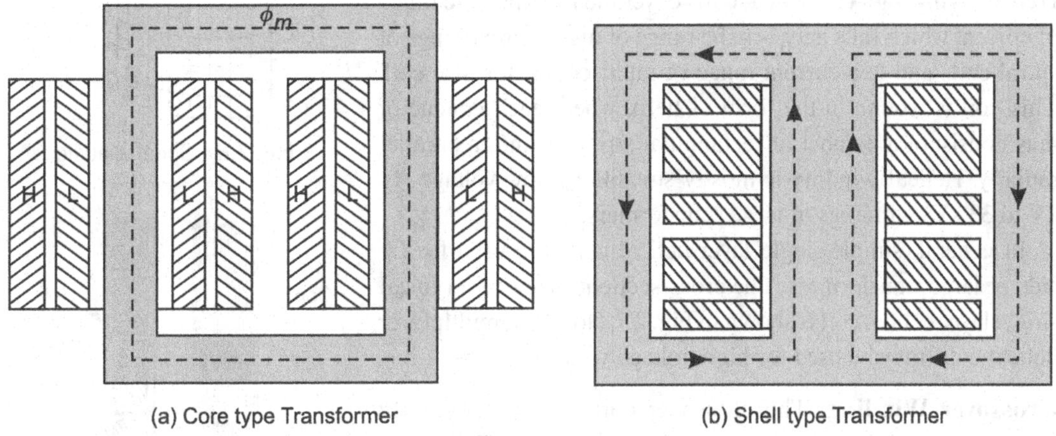

(a) Core type Transformer (b) Shell type Transformer

Fig. 2.5 Single-phase transformer (sectional view)

The positioning of the *H.V.* and *L.V.* windings with respect to the core is also very important from the point of view of insulation requirement. The low-voltage winding is placed nearer to the core in the case of concentric windings and on the outside positions in the case of sandwiched windings as shown in Figs. 2.5(*a*) and (*b*) on account of less and easier insulation facilities.

Concentric Windings

The concentric windings may be classified into four major groups, *viz.*

1. Spiral windings.
2. Helical winding.
3. Crossover winding.
4. Continuous disc winding.

Spiral Winding: The spiral coils are suitable only for windings which carry current more than 100 A. Usually these are employed for *LV* windings. However, these are also used for *HV* windings when the winding is to carry current more than 100 A.

Spiral coils normally consists of layers wound continuously from top to bottom of the coil and a composite conductor of coil may consist of a number of square or rectangular strips in parallel. These coils are mechanical very strong and are usually former wound or wound directly on to solid insulating cylinder.

The spiral coils may be wound in single layer or multilayer. In small transformers, sometimes, it is difficult to arrange them satisfactorily in single layer, then these are wound in multilayer. A typical two layer spiral coil is shown in Fig. 2.6.

When more than is required to be arranged in radial direction, it becomes necessary to introduce transpositioning throughout the length of winding to keep the resistance and leakage reactance of each conductors to be the same.

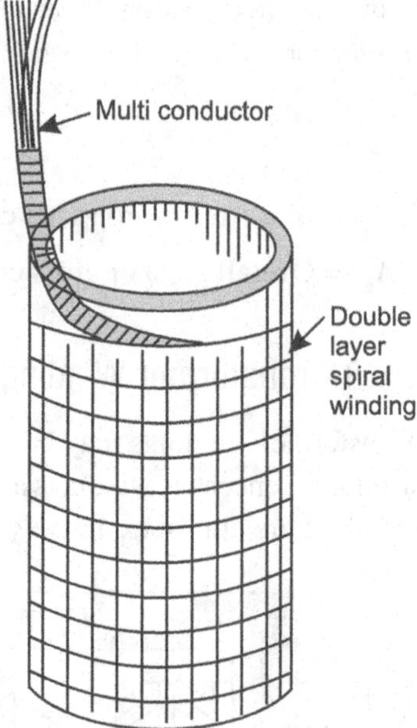

Fig. 2.6 Spiral winding

Helical Winding: The helical coils cover the intermediate range of current which falls between the range of high current range of spiral coils and low current range of multi-conductor disc coil. This coil is wound in the form of helix where each conductor may consist of a number of rectangular strips wound in parallel radially. Helical winding is mostly suitable for low-voltage (11 kV to 33 kV) windings of large transformers.

In general, simple single layer helical coils are used for *LV* side of large transformers. The cross-sectional view of a typical single layer helical coil is shown in Fig. 2.7. However, multiplayer helical coils may be used for high-voltage windings.

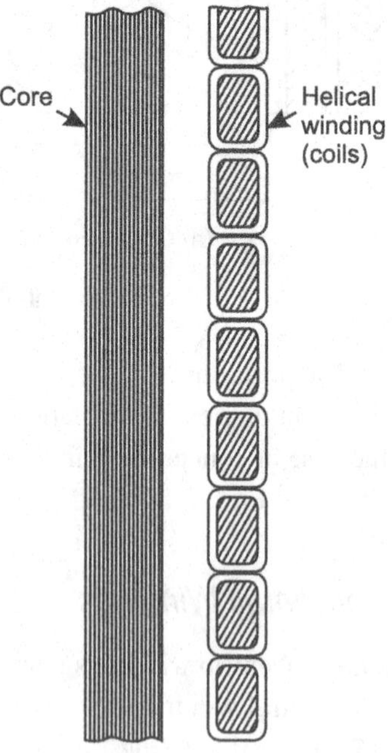

Fig. 2.7 Helical winding

Crossover Winding: The crossover coils are suitable only for windings which carry current less than 20 A. These are largely used for high voltage windings of comparatively small transformers. These coils are usually wound on formers. Generally, each coil consists of several layers and each layer carries several turns. The conductor may be of round (wire) or rectangular (strip) shape, insulated with cotton or paper covering.

The cross-sectional and isometric view of a typical crossover coil is shown in Fig. 2.8(*a*) and (*b*) respectively. The complete winding assembly employing crossover coils consists of a number of such coils connected in series. This assembly in placed over the limb of the core. The coils are usually spaced apart by means of insulating key sector to facilitate free oil circulation for cooling.

Continuous Disc Winding: The continuous disc coils are most suitable for high voltage (low current) large power transformers. As the name suggests, these coils consist of a number of discs. Each disc consist of a number of turns wound radially over one another from inside outwards and outside inwards alternately and conductors pass uninterruptedly from disc to disc. Generally, rectangular strip conductors are used and they are wound on the flat side so that each disc becomes mechanically very strong.

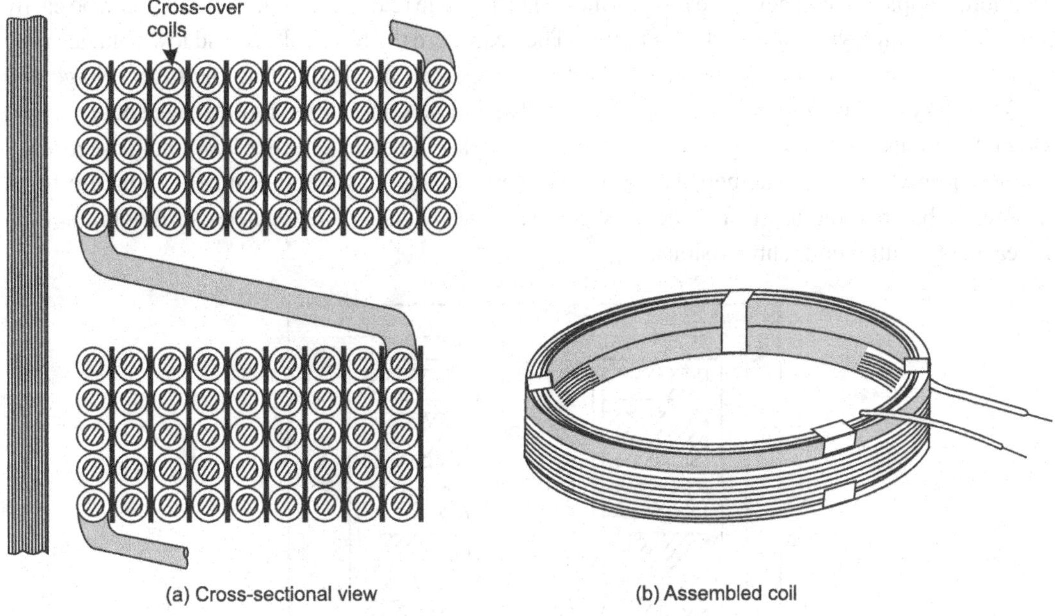

(a) Cross-sectional view (b) Assembled coil

Fig. 2.8 Cross-over winding

The cross-sectional and isometric view of a typical disc coil is shown in Figs. 2.9(*a*) and (*b*), respectively.

While using multiple strips, transpositioning of conductors has to be taken core so that uniform resistance and inductance can be ensured. The discs are wound on to an insulating cylinder and a spacer is provided between the discs when arranged vertically to ensure proper circulation of oil.

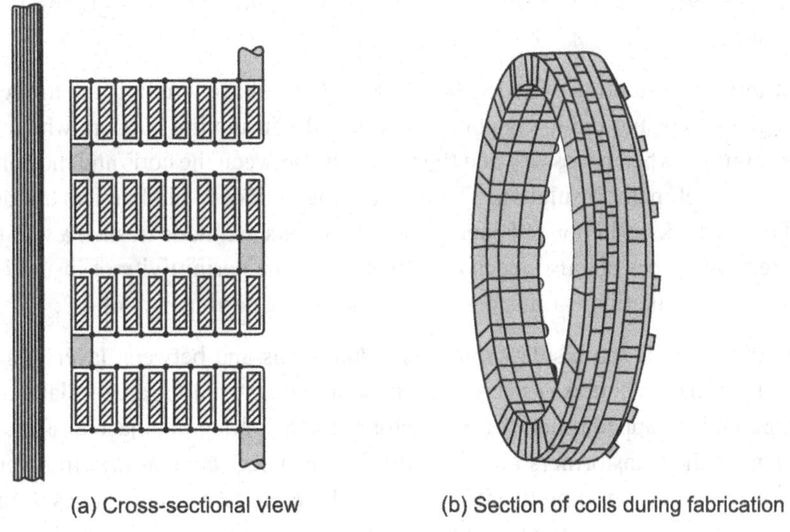

(a) Cross-sectional view (b) Section of coils during fabrication

Fig. 2.9 Continuous disc winding

Sandwiched Winding

The schematic diagram of a sandwiched winding is shown in Fig. 2.10. Sandwiched windings are commonly employed for shell-type transformers. The leakage reactance of the windings can be easily controlled by employing sandwiched winding. The nearer are the high voltage and low voltage coils, the less is the leakage flux. Leakage can be further reduced by subdividing the high voltage and low voltage coils. The high voltage sections lie between two consecutive low voltage sections. The two end sections are low-voltage sections and contain half the turns of other low-voltage sections.

Each normal section, whether high voltage or low voltage, has equal number of ampere-turns in order to balance the mmf of adjacent sections. Lower values of reactance can be obtained by increasing the number of subdivisions.

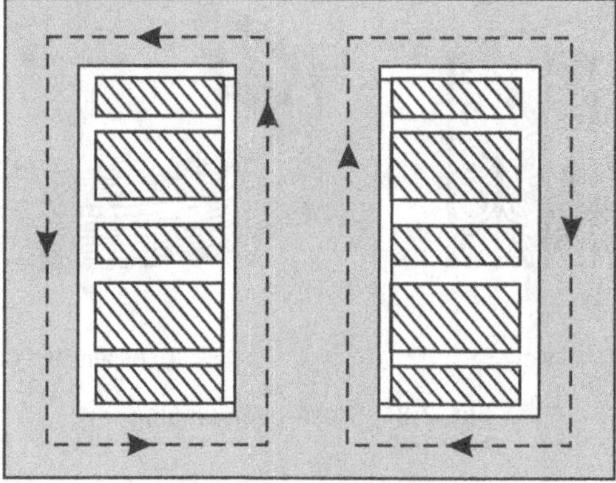

Fig. 2.10 Sandwiched winding

2.3.4 Insulation

The insulation employed in a transformer may be classified into two major groups e.g., major and mirror insulation.

Major insulation: The insulating cylinders between the low voltage winding and the core and those between the high voltage and the low voltage windings, the insulating barriers which are inserted between adjacent limbs when necessary and the insulation between the coils and the core yokes etc., fall into the category of major insulation. These insulating cylinders and barriers are usually made of pressboard or synthetic resin bonded paper. These cylinders are made in such a way that they are not only excellent insulators but also mechanically strong. The permittivity of insulating cylinders is about 4. The major insulation is transformer oil which has a permittivity of 2.2.

Minor Insulation: This is the insulation on individual turns and between layers. The conductor insulation may be of paper, cotton or glass tape, the latter being used for air insulated transformers only. The insulation is wrapped round the conductor without overlap of adjacent turns.

The windings of the transformers are generally impregnated. Now-a-days transformer oil has greatly replaced varnish as an impregnant. In general there are three main stages in preparing the windings for oil impregnation. In the first stage the pre-drying and shrinking of the coils are done.

In the second stage, further drying is carried out until the required insulation resistance is obtained. These two stages are usually accomplished by circulation of hot air at atmospheric pressure in an oven while making sure that the oxidation of the winding does not take place. In the final stage, the vapour and gas are removed from the assembled core and the windings prior to oil impregnation. This process must be carried out at a temperature of approximately 100°C and a vacuum of at least 0.5 mm. The windings of the power transformers are generally impregnated after assembly on the core.

Note: Never provide excessive insulation, it not only increases the cost but it also makes heat transfer poor.

2.3.5 Bushings

The core and winding assembly is placed in the transformer tank. The winding terminals are to be connected to the external conductors (transmission lines) or bus-bars where conductor has to pass through the top cover of the tank which is earthed on account of safety.

These outgoing terminals of a transformer are provided with **bushings.**

For system voltages upto 66 kV, non-condenser bushings such as porcelain or oil-filled bushings are used whereas for system voltages more than 66 kV, condenser bushings are employed for economic reasons.

The oil filled bushings carry a hollow tube conductor placed in the centre of a hollow porcelain cylinder. The space between the conductor and the porcelain is filled with insulating oil. Normally, there is a glass chamber at the top of the bushing to indicate the oil level which also acts as an expansion chamber for the oil when the temperature of the bushing rises. Under the influence of the electric field, dust or metallic particles present in the oil have a tendency to align themselves in the direction of the electric field, giving rise to paths of low dielectric strength which may lead to breakdown of the bushing. To break up these chains, sometimes concentric bakelite tubes are used as insulating barriers in between the porcelain and the conductor. Fig. 2.11(*a*) shows a typical oil filled bushing.

The body of a condenser type bushing shown in Fig. 2.11(*b*) is formed by inserting aluminium foil layers into a paper winding at predetermined radii. The conducting layers take the form of co-axial cylinders in such a way that the electrical stress in the radial direction does not exceed 7.5 kV/mm for synthetic resin bonded paper (s.r.b.p.) or 20 kV/mm for oil impregnated paper (o.i.p.) and the axial stress does not exceed 0.4 kV/mm for air and 0.65 kV/mm for oil. The capacitances between adjacent pair of conducting surfaces are known as partial capacitances between adjacent pair of conducting surfaces are known as partial capacitances. The thickness of the partial capacitances and the values of the partial capacitances are made equal by properly proportioning the axial length of the conducting layers. Consequently, the insulating layers are equally stressed and the dielectric stresses are kept within the limits for the insulating material used. A typical condenser bushing is shown in Fig. 2.12. For use in outdoor substations, the bushing is covered by a porcelain rain-shed which is outwardly very similar to a porcelain bushing. The space between the rain shed and the bushing assembly is filled with insulating oil. The rain sheds are designed to increase the creepage length taking into consideration the electric field distribution and thus reduce the tendency to flashover under dry as well as wet conditions.

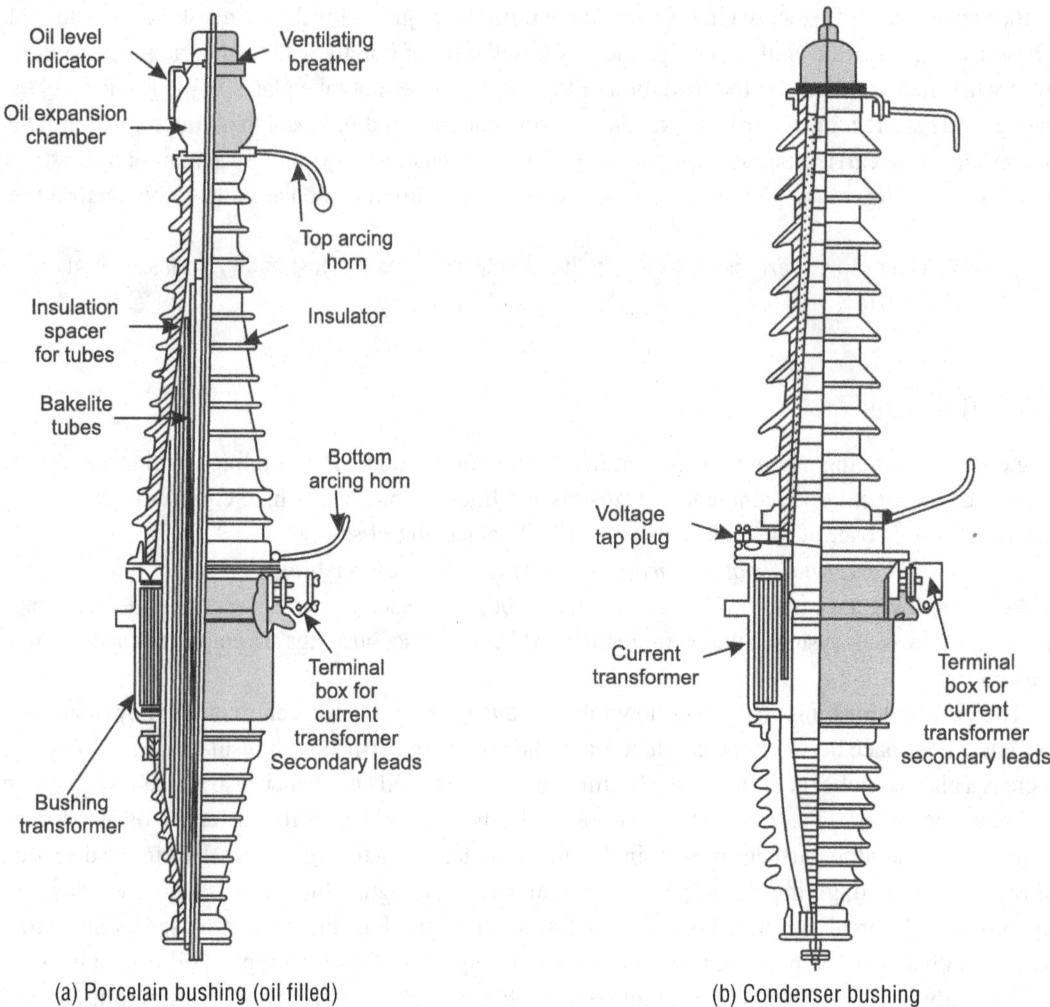

(a) Porcelain bushing (oil filled) (b) Condenser bushing

Fig. 2.11 Transformer bushings

The bushings are mounted in localised projections from the transformer tank known as turrets and current transformers are often housed in turrets around the flange barrels of the bushings. The size of the transformer tank is also greatly influenced by the portion of the lower end of the bushing inserted into the tank.

2.3.6 Transformer Tank

In modern transformers, depending upon the size, the following types of tanks are mainly used:

(*i*) Plain Sheet Steel Tank
(*ii*) Tubed Tanks
(*iii*) Corrugated Tanks
(*iv*) Radiator Tanks
(*v*) Tanks with Separate Coolers.

2.4 Simple Construction of Single-phase Small Rating (SAY 2 kVA) TRANSFORMERS

On the basis of core construction and the arrangement to windings, the transformers are named as (*i*) core type transformers (*ii*) shell type and Berry-type transformers.

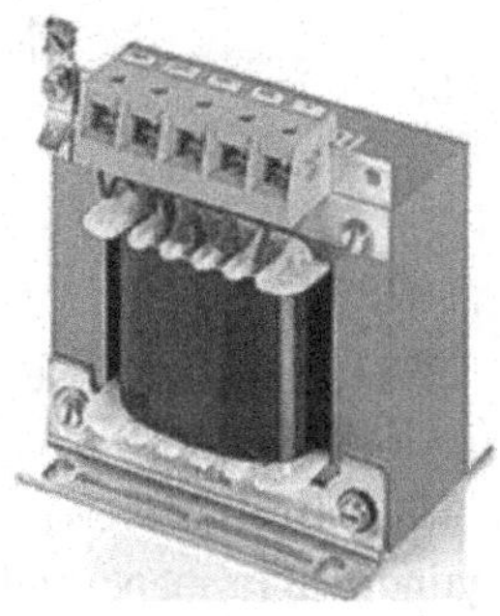

Fig. 2.12 Single-phase small transformers used with electrical gadgets

Core-type Transformers

In Such transformers the magnetic core is built up of laminations to form a rectangular frame. The laminations are cut in the form of L-shape strips as shown in Fig. 2.13(a). To eliminate high reluctance continuous joint, the laminations are placed alternately as shown in Fig. 2.13(b).

The upper horizontal portion of the core is known a ***yoke*** *and the vertical portion, as shown in Fig. 2.14(b), which carries windings is called* ***limb.*** *Usually, the cross-sectional area of yoke is kept 15 to 20% more than the limbs because it reduces the flux density and consequently reduces the iron losses.*

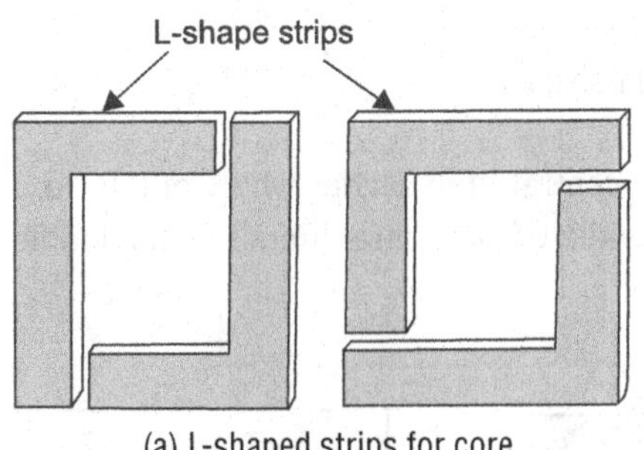

(a) L-shaped strips for core

(b) Joints kept alternately

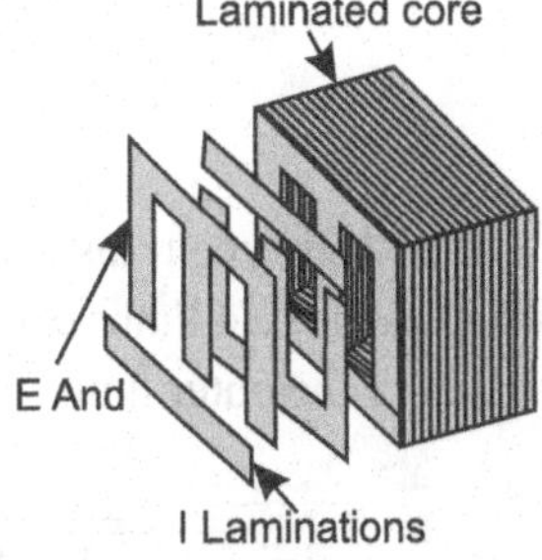

(c) Staggering of stampings

Fig. 2.13

In actual transformer construction, the primary and secondary windings are interleaved to reduce the leakage flux. Half of each winding is placed side by side or concentrically on either limb or leg of the core as shown in Fig. 2.14. However, for simplicity, the two windings are shown in Fig. 2.2(*a*) located on separate limbs of the core.

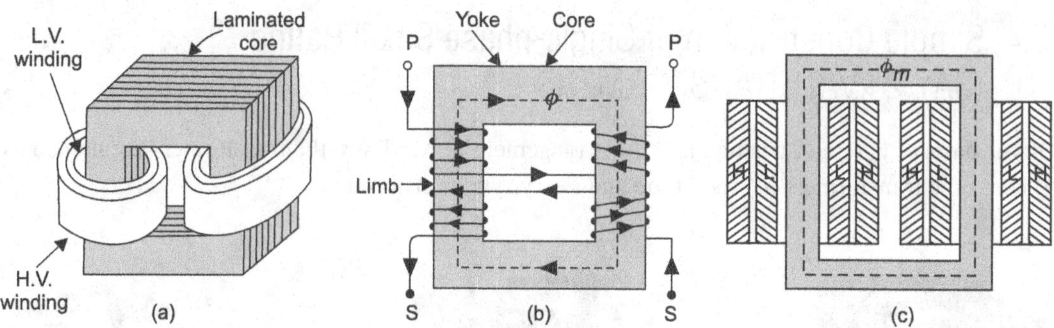

Fig. 2.14 Placing of coils over core-type transformers

While placing these windings, an insulation layer (Bakelite former) is provided between core and lower winding and between the two windings. To reduce the insulation, low voltage winding is always placed nearer the core as shown in Fig. 2.14(a). The windings used are form wound (usually cylindrical in shape) and the laminations are inserted later on.

Shell-type Transformers

In such transformers, each lamination is cut in the shape of E's and I's as shown in Fig. 2.15. To eliminate high reluctance continuous joint, the laminations are placed alternately as shown in Fig. 2.16(a).

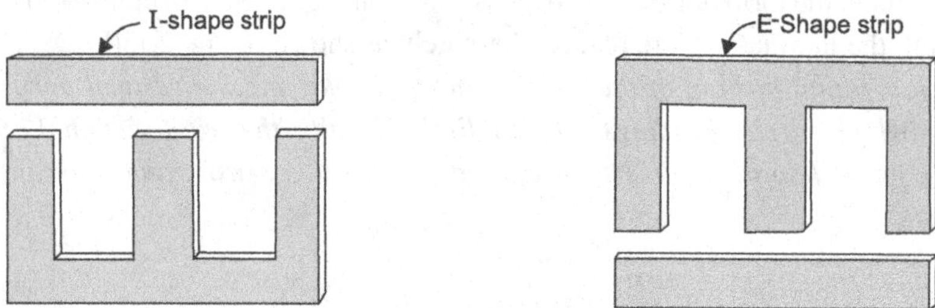

Fig. 2.15 Laminations of E and I-shapes.

In a shell-type transformer, the core has three limbs. The central limb carries whole of the flux, whereas the side limbs carry half of the flux. Therefore, the width of the central limb is about double to that of the outer limbs.

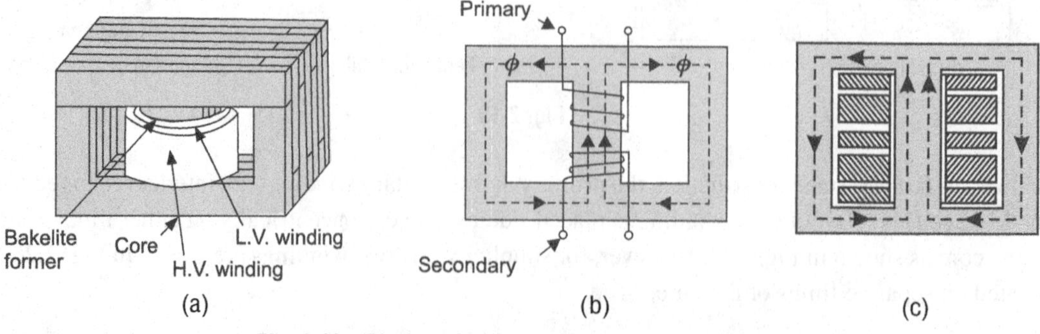

Fig. 2.16 Placing of coils over shell-type transformer

Both the primary and secondary windings are placed on the central limb side by side or concentrically (see Fig. 2.16). The low voltage winding is placed nearer the core and high voltage winding is placed outside the low voltage winding to reduce the cost of insulation placed between core and low voltage winding. In this case also the windings are form wound is cylindrical shape and the core laminations are inserted later on.

The whole assembly i.e., core and winding is then usually placed in tank filled with transformer oil. The transformer oil provides better cooling to the transformer and acts as a dielectric medium between winding and outer tank which further reduces the size of outer tank of the transformer.

Comparison between Core-type and Shell type Transformers:

Sr. No.	*Core-type transformer*	*Shell-type transformer*
1.	The windings surround a considerable portion of the core.	The *core* surrounds considerable portion of the windings.
2.	Windings are of *form*-wound, and are of cylindrical-type.	Winding are of *sandwich-type.* The coils are first wound in the form of pancakes, and complete winding consists of stacked discs.
3.	More suitable for *high voltage* transformers.	More suitable and economical for *low voltage* transformers.
4.	Mean length of coil turn is *shorter.*	Mean length of coil turn is *longer.*
5.	Core has *two limbs* to carry the windings.	Core has *three or more limbs* but the central limb carries the windings.

Berry-type Transformer

A berry type transformer is a specially designed shell type transformer and is named after its designer. The transformer core consists of laminations arranged in groups which radiate from the centre as shown (as top view) in Fig. 2.17.

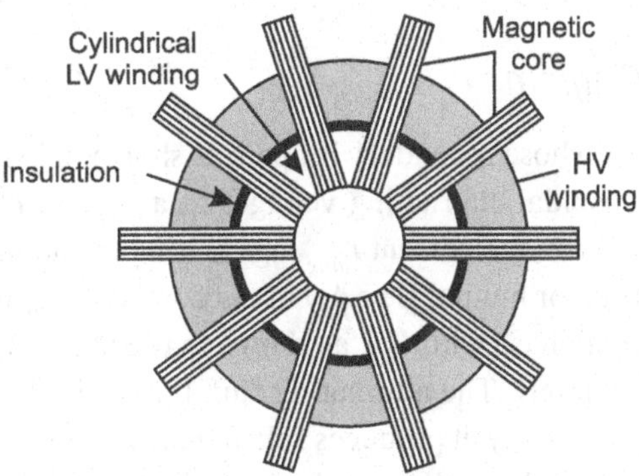

Fig. 2.17 Berry type transformer

In order to avoid possible insulating damage due to movement of strips and winding, the core and coil of the transformer are provided with rigid mechanical bracing. Good bracing reduces vibration and the objectionable noise-a humming sound - during operation.

2.5 An Ideal Transformer

To understand the theory, operation and applications of a transformer, it is better to view a transformer first as an ideal device. For this, the following assumptions are made:

(*i*) Its coefficient of *coupling (k)* is *unity.*
(*ii*) Its primary and secondary windings are *pure inductors* having *infinitely large* value.
(*iii*) Its leakage flux and *leakage inductances* are *zero.*
(*iv*) Its self and mutual inductances are zero having *no reactance* or *resistance.*
(*v*) Its efficiency is 100 percent having *no loss* due to resistance, hysteresis or eddy current.
(*vi*) Its transformation ratio (or turn ratio) is equal to the ratio of its secondary to primary terminal voltage and also as the ratio of its primary to secondary current.
(*vii*) Its core has permeability (μ) of *infinite* value.

Thus, a transformer is said to be an ideal one, if it has no ohmic resistances, moreover whole of the flux set-up in the core is considered to be linking with its primary and secondary turns, i.e., it carries no leakage flux. In other words, a transformer is said to be ideal one, when it has no losses, i.e., copper or iron losses. In actual practice, such a transformer cannot exist, but to begin with it may be considered so.

For an ideal transformer, the output must be equal to input (since it has no losses), therefore.

i.e., $$E_2 I_2 \cos\phi = E_1 I_1 \cos\phi \quad \text{or} \quad E_2 I_2 = E_1 I_2 \quad \text{or} \quad \frac{E_2}{E_1} = \frac{I_1}{I_2}$$

since $$E_2 \propto N_2;\ E_1 \propto N_1 \text{ and } E_1 \cong V_1;\ E_2 \cong V_2$$

$$\frac{V_2}{V_1} = \frac{E_2}{E_1} = \frac{N_2}{N_1} = \frac{I_1}{I_2} = K \text{ (transformation ratio)} \qquad \text{...(2.3)}$$

Hence, primary and secondary currents are inversely proportional to their respective turns.

Transformation ratio: The ratio of secondary to primary turns is called transformation ratio. It is usually devoted by letter 'K'.

Behaviour and Phasor Diagram

Consider an ideal transformer whose secondary is open, as shown in Fig. 2.18(a). When its primary winding is connected to sinusoidal alternating voltage V_1, a current I_{mag} flows through it. Since the primary coil is pure inductive, the current I_{mag} lags behind the applied voltage V_1 by 90°. This current sets up alternating flux (or mutual flux ϕ_m) in the core and magnetises it. Hence it is called magnetising current. Flux is in phase with I_{mag} as shown in the phasor diagram and wave diagram in Figs. 2.18(*b*) and (*c*), respectively. The alternating flux links with both primary and secondary windings. When it links with primary, it produces self-induced *emf* E_1 in opposite direction to that of applied voltage V_2. When it links with secondary winding, it produces mutually-induced *emf* E_2 in opposite direction to that of applied voltage. Both the *emf*s. E_1 and E_2 are shown in phasor diagram in Fig. 2.18(*b*).

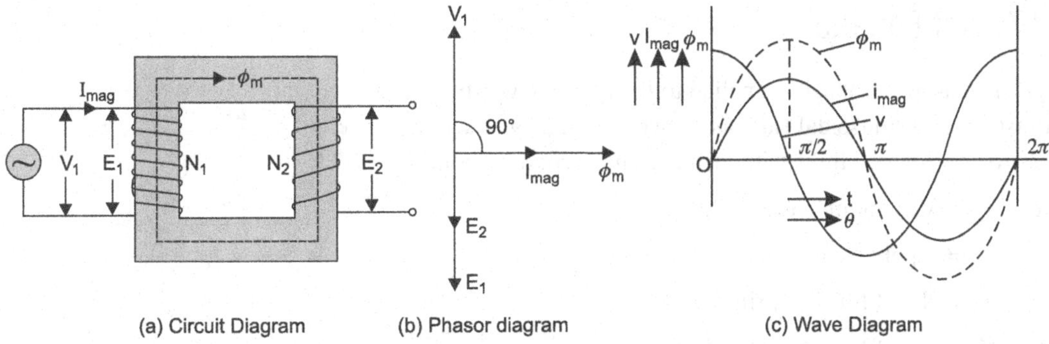

(a) Circuit Diagram (b) Phasor diagram (c) Wave Diagram

Fig. 2.18 Ideal transformer

A transformer is analogous to mechanical gear drive because of the following facts:

Sr. No.	*Mech. Gear Drive*	*Transformer*
1.	It transfers mechanical power from one shaft to the other shaft.	It transfers electrical power from one circuit to the other.
2.	There is perfect ratio between the number of teeth and the speeds of the two gears, i.e., $\frac{T_1}{T_2} = \frac{N_2}{N_1}$ where T_1 = No. of teeth of gear 1, T_2 = No. of teeth of gear 2, N_2 = Speed of gear 2, N_1 = Speed of gear 1	There is perfect ratio between the number of turns and the induced emf or current of the two windings. i.e., $\frac{N_2}{N_1} = \frac{E_2}{E_1} = \frac{I_1}{I_2}$ where N_2 = No. of secondary turns, N_1 = No. of primary turns, E_2 = EMF in secondary, E_1 = EMF is primary, I_1 = Current in primary, I_2 = Current in secondary
3.	Power is transferred through mechanical mesh.	Power is transferred through magnetic flux.

2.6 Transformer on DC

A transformer cannot work on DC. The basic working principle of a transformer is electro-magnetic induction, *i.e,* when flux linking with a coil changes an emf is induced in it. If DC is applied to one of the winding (primary) of a transformer, it will set a constant magnetic field in magnetic core. Hence no emf will be induced either in primary or secondary. Then electric power cannot be transformed from one circuit (primary) to the other (secondary).

Moreover, if rated DC voltage is applied to its primary, high current will be drawn by it since there is any counter (self) induced emf which limits the current. Consequently, heaving heat will be produced and winding insulation will burn.

Hence, a transformer cannot work on DC and it is never put-on rated DC supply.

2.7 emf Equation

When sinusoidal voltage is applied to the primary winding of a transformer, a sinusoidal flux, as shown in Fig. 2.19. is set up in the iron core which links with primary and secondary winding.

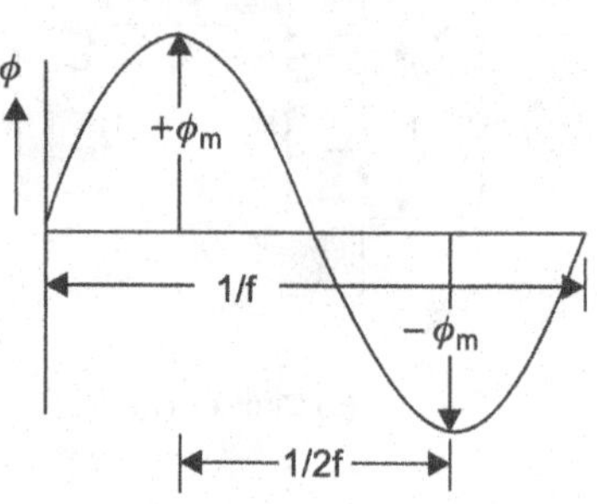

Fig. 2.19 Wave diagram of flux

Let, ϕ_m = Maximum value of flux in Wb;

f = supply frequency in Hz (or c/s);

N_1 = No. of turns in primary;

N_2 = No. of turns in secondary.

As shown in Fig. 2.19, flux changes from + ϕ_m to – ϕ_m in half a cycle i.e., $\frac{1}{2f}$ second,

$$\text{Average rate of change of flux} = \frac{\phi_m - (-\phi_m)}{1/2f} = 4f\,\phi_m \text{ Wb/s}$$

Now, the rate of change of flux per turn is the average induced emf per turn in volt.

$\therefore$ Average emf induced per turn = $4\,f\,\phi_m$ volt

For a sinusoidal wave, $\frac{\text{R.M.S. value}}{\text{Average value}}$ = Form factor = 1·11

$\therefore$ *R.M.S.* value of emf induced/turn, $E = 1.11 \times 4f\,\phi_m = 4.44f\,\phi_m$ volt

Since primary and secondary have N_1 and N_2 turns, respectively.

$\therefore$ *R.M.S.* value of emf induced in primary,

$$E_1 = (\text{emf induced/turn}) \times \text{No. of primary turns}$$

$$= 4{\cdot}44\,N_1 f\,\phi_m \text{ volt} \qquad \text{...(2.4)}$$

Similarly, r.m.s. value of emf induced in secondary,

$$E_2 = 4{\cdot}44\,N_2 f\,\phi_m \text{ volt} \qquad \text{...(2.5)}$$

From eq. (*i*), we get, $\frac{E_1}{N_1} = 4.44\,f\phi_m$ volt/turn ...(2.6)

From eq. (*ii*), we get, $\frac{E_2}{N_2} = 4.44\,f\phi_m$ volt/turn ...(2.7)

Equation (2.6) and (2.7) clearly show that emf induced per turn on both the sides i.e., primary and secondary is the same.

Again, we can find the voltage ratio,

$$\frac{E_2}{E_1} = \frac{4{\cdot}44 N_2\, f\phi_m}{4{\cdot}44 N_1\, f\phi_m} \text{ or } \frac{E_2}{E_1} = \frac{N_2}{N_1} = K \text{ (transformation ratio)}$$

Equation (i) and (ii) can be written in the form of maximum flux density B_m using relation,

$$\phi_m = B_m \times A_i \text{ (where } A_i \text{ is iron area)}$$

$\therefore$ $$E_1 = 4.44\, N_1 f B_m A_i \text{ volt} \quad \ldots(2.8)$$

and $$E_2 = 4.44\, N_2 f B_m A_i \text{ volt} \quad \ldots(2.9)$$

Example 2.1

What will be the number of primary and secondary turn of a single-phase 2310/220V, 50 Hz transformer which has in emf of 13V per turn approximately.

Solution:

Here, $$\frac{E_1}{N_1} = \frac{E_2}{N_2} = 13 \text{ V (given)}; E_1 = 2310 \text{ V}; E_2 = 220 \text{ V}$$

$\therefore$ $$\text{Primary turns, } N_1 = \frac{E_1}{13} = \frac{2310}{13} = 177.69 \cong \mathbf{178} \; (Ans.)$$

$$\text{Secondary turns, } N_2 = \frac{E_2}{13} = \frac{220}{13} = 16.92 \cong \mathbf{17} \; (Ans.)$$

Example 2.2

A power transformer has 1000 primary turns and 100 secondary turns. The cross-sectional area of the core is 6 sq. cm and the maximum flux density while in operation is 10000 Gauss. Calculate turns per volt for the primary and secondary windings.

Solution:

Here, $N_1 = 1000; N_2 = 100; A_i = 6 \text{ cm}^2 = 6 \times 10^{-4} \text{ m}^2$;

$$B_m = 10000 \text{ gauss} = 10000 \times 10^{-8} \times 10^4 = 1 \text{ tesla}$$

We know, $$E_1 = 4.44\, N_1 f B_m A_i = 4.44 \times 1000 \times 50 \times 1 \times 6 \times 10^{-4} = 133.2 \text{ V}$$

$$E_2 = 4.44\, N_2 f B_m A_i = 4.44 \times 100 \times 50 \times 1 \times 6 \times 10^{-4} = 13.32 \text{ V}$$

$$\text{On primary side, number of turns/volt} = \frac{N_1}{E_1} = \frac{1000}{133.2} = \mathbf{7.5} \; (Ans.)$$

$$\text{On secondary side, number of turns/volt} = \frac{N_2}{E_2} = \frac{100}{13.32} = \mathbf{7.5} \; (Ans.)$$

The number of turns per volt or voltage per turn on primary and secondary remains the same.

Example 2.3

The primary and secondary of a 25 kVA transformer has 500 and 40 turns, respectively. If the primary is connected to 3000 V, 50 Hz mains, calculate (i) primary and secondary currents at full load; (ii) The secondary emf and (iii) The maximum flux in the core. Neglect magnetic leakage, resistance of the winding and the primary no-load current in relation to the full load current.

Solution:

(*i*) At full load, $I_1 = \dfrac{25 \times 10^3}{3000} = \mathbf{8{\cdot}33}$ (*Ans.*)

Now $\dfrac{I_1}{I_2} = \dfrac{E_2}{E_1} = \dfrac{N_2}{N_1}$

secondary current, $I_2 = \dfrac{N_1}{N_2} \times I_1 = \dfrac{500}{40} \times 8{\cdot}33 = \mathbf{104{\cdot}15\ A}$ (*Ans.*)

(*ii*) Secondary emf, $E_2 = \dfrac{N_2}{N_1} \times E_1 = \dfrac{40}{500} \times 3000 = \mathbf{240\ V}$ (*Ans.*)

(*iii*) Using relation, $E_1 = 4{\cdot}44\, N_1 f\, \phi_m$

$$3300 = 4{\cdot}44 \times 500 \times 50\, x\, \phi_m$$

or $\phi_m = \dfrac{3000}{4.44 \times 500 \times 50} = \mathbf{27\ mWb}$ (***Ans***)

Example 2.4

The emf per turn of an 11 kV/415 V, 50 Hz single-phase core type transformer is 15 V. The maximum flux density in the core is 2.5 T. Find number of primary and secondary turns and net cross sectional area of core.

Solution:

Here, $E_1 = 11000$ V; $E_2 = 415$ V; $f = 50$ Hz; $B_m = 2.5$ T

$$\text{EMF/turn} = \frac{E_1}{N_2} = \frac{E_2}{N_2} = 15$$

$$\text{No. of primary turns, } N_1 = \frac{E_1}{15} = \frac{11000}{15} = \mathbf{733.33}\ (\textit{Ans.})$$

$$\text{No. of secondary turns, } N_2 = \frac{E_2}{15} = \frac{415}{15} = \mathbf{27.67}\ (\textit{Ans.})$$

Now, $E_1 = 4.44\, N_1 f A_i B_m$

$$\frac{E_1}{N_1} = 4.44 f A_i B_m \text{ or } 15 = 4.44 \times 50 \times A_i \times 2.5$$

$\therefore$ Net area, $A_i = \dfrac{15}{4.44 \times 50 \times 1.5} = 0.045045\ \text{m}^2 = \mathbf{450.45\ cm^2}$ (*Ans.*)

Example 2.5

A single phase 50 Hz core type transformer has rectangular cores 30 × 20 cm. and the maximum allowable density is 1.05 tesla. Find the number of turns per limb on the high and low voltage sides for a voltage ratio of 3300/200 volt. Take iron factor as 0.93.

Solution:

$$\text{Gross-Cross-Sectional Area} = 30 \times 20 = 600\ \text{cm}^2$$

$$A_{gc} = 600 \times 10^{-4}\ \text{m}^2$$

The iron factor is to be taken into consideration as the laminations are insulated from each other and

$$K_i = \frac{\text{Net Area of Cross-Section}}{\text{Gross Area of Cross-Section}} = \frac{A_i}{A_{gc}}$$

$$A_i = K_i \times A_{gc}$$

$$= 0.93 \times 600 \times 10^{-4} = 558 \times 10^{-4}\ \text{m}^2$$

$$\text{Emf induced per turn} = 4.44 f B_{max} \cdot A_i$$

$$= 4.44 \times 50 \times 1.05 \times 558 \times 10^{-4} = \mathbf{13\ volt}\ (Ans.)$$

$$\text{Primary turns} = \frac{3300}{13} = \mathbf{254\ turns}\ (Ans.)$$

$$\text{Secondary turns} = \frac{200}{13} = \mathbf{16\ turns}\ (Ans.)$$

Example 2.6

The Secondary of a 500 kVA, 4400/500 V, 50 Hz, single-phase transformer has 500 turns. Determine (i) emf per turn, (ii) primary turns, (iii) secondary full load current (iv) maximum flux (v) gross cross-sectional area of the core for flux density of 1.2 tesla and iron factor is 0.92 (vi) if the core is of square cross-section find the width of the limb.

Solution:

Here, Rating= 500 kVA; $E_1 = V_1 = 4400$ V; $E_2 = V_2 = 500$ V; $f = 50$ Hz; $N_2 = 500$; $B_m = 1.2$ T;

$k_i = 0.92$

(*i*) Emf per turn $= \dfrac{E_2}{N_2} = \dfrac{500}{500} = \mathbf{1.0\ V/turn}$ (*Ans.*)

(*ii*) Emf per turn $= \dfrac{E_2}{N_2} = \dfrac{E_1}{N_1} = \dfrac{4400}{N_1} = 1.0$

$\therefore$ Primary turns, $N_1 = \dfrac{4400}{1.0} = \mathbf{4440}$ (*Ans.*)

(*iii*) Secondary full load current, $I_2 = \dfrac{kVA \times 1000}{V_2} = \dfrac{500 \times 1000}{500} = \mathbf{1000\ A}$ (*Ans.*)

(*iv*) Maximum flux, $\phi_m = \dfrac{E_2}{4.44 \times N_2 \times f} = \dfrac{500}{4.44 \times 500 \times 50} = \mathbf{4.5\ mWb}$ (*Ans.*)

(*v*) Iron area of the core, $A_i = \dfrac{\phi_m}{B_m} = \dfrac{4.5 \times 10^{-3}}{1.2} = 37.54 \times 10^{-4}\ \text{m}^2 = 37.54\ \text{cm}^2$

Gross area of the core, $A_g = \dfrac{A_i}{k_i} = \dfrac{37.54}{0.92} = \mathbf{40.8\ cm^2}$ (*Ans.*)

(*vi*) Width of squared limb $= \sqrt{A_g} = \sqrt{40.8} = \mathbf{6.39\ cm}$ (*Ans.*)

Example 2.7

A 100 kVA, 3300/200 volt, 50 Hz single phase transformer has 40 turns on the secondary, calculate:

(i) the values of primary and secondary currents.

(ii) the number of primary turns.

(iii) the maximum value of the flux.

If the transformer is to be used on a 25 Hz system, calculate.

(iv) the primary voltage, assuming that the flux is increased by 10%

(v) the kVA rating of the transformer assuming the current density in the windings to be unaltered.

Solution:

(*i*) Full load primary current, $I_1 = \frac{100 \times 1000}{3300} = \mathbf{30.3\ A}$ (*Ans.*)

Full load secondary current, $I_2 = \frac{100 \times 1000}{200} = \mathbf{500\ A}$ (*Ans.*)

(*ii*) No. of Primary turns, $N_1 = N_2 \times \frac{E_1}{E_2} = 40 \times \frac{3300}{200} = \mathbf{660}$ (*Ans.*)

(*iii*) We know $E_2 = 4.44 f \phi_{max} N_2$ volt

$$200 = 4.44 \times 50 \times \phi_{max} \times 40$$

$\therefore$ $\phi_{max} = \frac{200}{4.44 \times 50 \times 40} = \mathbf{0.0225\ Wb}$ (*Ans.*)

(*iv*) As the flux is increased by 10% at 25 Hz

$\therefore$ Flux at 25 Hz, $\phi_m = 0.0225 \times 1.1 = 0.02475$ Wb

$\therefore$ Primary voltage $= 4.44 \times N_1 \times f' \times \phi_m$ volt

$= 4.44 \times 660 \times 25 \times 0.02475 = \mathbf{1815\ volt}$ (*Ans.*)

(*v*) For the same current density, the full load primary and secondary currents remain unaltered.

$\therefore$ kVA rating of the transformer $= \frac{30.3 \times 1815}{1000} = \mathbf{55\ kVA}$ (*Ans.*)

Section Practice Problems

Numerical Problems

1. A sinusoidal flux 0.2 Wb (max.) links with 55 turns of a transformer secondary coil. Calculate the r.m.s. value to the induced emf in the secondary coil. The supply frequency is 50 Hz. (**Ans.** *244.2 V*)
2. The primary and secondary turns of a single-phase transformer are 400 and 1100, respectively. The net cross sectional area of the core is 60 cm^2. When its primary is connected 500V, 50 Hz supply, calculate the value of maximum flux density in the core and the emf induced in secondary winding. Draw the vector diagram representing the condition. (**Ans.** *0·938 Tesla, 1250 V*)

3. A single-phase 200 kVA, 3300/240 volt, 50 Hz. Transformer carries 80 turns on its secondary. What will be its. (i) primary and secondary current on full load; (ii) the maximum value of flux; (iii) the number of primary turns. (**Ans.** *60.6 A; 833.3 A; 1100 turns*)

4. A 3300/250 V, 50 Hz, single-phase transformer has an effective cross sectional area of 125 cm^2. It has 70 turns on its low-voltage side. Calculate (*a*) the value of the maximum flux density (*b*) the number of turns on the high-voltage winding. (**Ans.** *1.287 Wb/m²; 924*)

5. The secondary of a 100 kVA, 3300/400 V, 50 Hz, one-phase transformer carries 110 turns. Determine the approximate values of the primary and secondary full-load currents, the maximum value of flux in the core and the number of primary turns. How does the core flux vary with load? (**Ans.** *30.3 A; 250 A; 16.4 mWb; 907*)

6. A 125 kVA transformer having a primary voltage of 2000 V at 50 Hz has 182 primary turns and 40 secondary turns. Neglecting losses, calculate (*a*) the full load primary and secondary current (*b*) the no-load secondary induced emf. [**Ans.** (*a*) 62.5 A, 284.4 A (*b*) 439.5 V]

Short Answer Type Questions

Q.1. What essentially is a transformer? What are the broad areas of applications of transformer?

Ans. **Transformer** is a static device that transfers AC electrical power from one circuit to the other at the same frequency but the voltage level is usually changed.

Its broad area of application is in *electrical power system.* At generating stations it is used to step-up the voltage, for economic reasons, to transmit electric power and at various sub-stations it is used to step-down the voltage for economical and safety reasons.

Q.2. Why does a transformer have iron core?

Ans. Iron core provides an easy (low reluctance) path for the magnetic lines of force.

Q.3. Why is the transformer core laminated?

Ans. The core of a transformer is laminated in order to reduce eddy current losses.

Q.4. What do you mean by iron space factor with respect to transformer core?

Ans. The ratio of the net cross-sectional area to gross cross sectional area of the transformer core is called the iron space factor or simply iron factor *Ki*.

$$Ki = Ai / Agc$$

Q.5. When a transformer is connected to the supply, how its windings are named?

Ans. When a transformer is connected to the supply, the windings are named as primary and secondary winding.

The winding to which supply is connected, is known as primary winding and the winding to which load is connected, is known as secondary winding.

Q.6. Why circular coils are always preferred over rectangular coils for winding a transformer?

Ans. Circular coils are preferred for winding a transformer as they can be easily wound on machines on a former, conductors can easily be bent and winding does not bulge out due to radial forces developed during operation.

Q.7. Why Sandwiched winding arrangement is preferred in large transformers?

Ans. Sandwiched winding provides better magnetic coupling and cooling facilities in large transformers.

Q.8. Is there a definite relation between the number of turns and voltages in transformers?

Ans. Yes, voltage is directly proportional to number of turns of the winding ($V \propto N$).

Induced emf in any winding = emf/turn × No. of turns.

Q.9. What do you mean by major and minor insulation used in transformer winding?

Ans. *Major insulation*: The insulation in the form of cylinders provided between *LV* winding and core or between *LV* and *HV* winding is termed as major insulation.

Minor insulation: The insulation provided between individual turn and between layer is termed as minor insulation.

Q.10. Why arcing horns are provided across the transformer bushings?

Ans. To protect the transformer against lighting.

Q.11. What is the significance of turn ratio in a transformer?

Ans. Turn ratio determines whether a transform is to step-up or step-down the voltage.

Q.12. What is an ideal transformer?

Ans. An ideal transformer is one which has no ohmic resistance and no magnetic leakage flux.

Q.13. Can a transformer work on DC? Justify.

Ans. No, a transformer cannot work on DC. If small DC is applied to primary, constant flux is set-up in the core and no emf is induced in the secondary. Hence, no transformer action is possible.

Q.14. A transformer is said to be analogous to mechanical gear, why?

Ans. In mechanical gear drive, mechanical power is transferred from one shaft to the other through mechanical gear, whereas in transformer electrical power is transferred from one circuit to the other through magnetic flux.

In mechanical gear drive, there is perfect ratio between number of teeth and speeds, similarly in transformer there is perfect ratio between number of turns and induced emfs.

Q.15. While drawing phasor diagram of an ideal transformer, the flux vector is drawn 90° out of phase (lagging) to the supply voltage, why?

Ans. Under ideal conditions, transformer behaves as a pure inductive circuit; therefore, magnetising current or flux phasor lags behind the voltage phasor by 90°.

Q.16. Cold rolled grain oriented steel (CRGOS) is used almost exclusively in spite of its high cost, in the construction of transformer core, why?

Ans. It is because cold rolled grain oriented steel has excellent magnetic properties in the direction of rolling and hysteresis losses are very small.

Q.17. Why is the area of yoke of a transformer kept 15 to 20% more than limb?

Ans. By keeping area of yoke 15 to 20% more than the limb, the flux density in the yoke is reduced which reduces the iron losses of the transformer consequently reduces the winding material.

Q.18. What is the thickness of laminations used for transformer core?

Ans. Thickness of laminations varies from 0.35 to 0.5 mm.

2.8 Transformer on No-load

A transformer is said to be on no-load when its secondary winding is kept open and no-load is connected across it. As such, no current flows through the secondary i.e., $I_2 = 0$. Hence, the secondary winding is not causing any effect on the magnetic flux set-up in the core or on the current drawn by the primary. But the losses cannot be ignored. At no-load, a transformer draws a small current I_0 (usually 2 to 10% of the rated value). This current has to supply the iron losses (hysteresis and eddy current losses) in the core and a very small amount of copper loss in the primary (the primary copper losses are so small as compared to core losses that they are generally neglected moreover secondary copper losses are zero as I_2 is zero).

Therefore, current I_0 lags behind the voltage vector V_1 by an angle ϕ_0 (called hysteresis angle of advance) which is less than 90°, as shown in Fig. 2.20(*b*). The angle of lag depends upon the losses in the transformer. The no-load current I_0 has two components;

(*i*) One, I_w in phase with the applied voltage V_1, called active or working component. It supplies the iron losses and a small primary copper losses.

(*ii*) The other, I_{mag} in quadrature with the applied voltage V_1, called reactive of magnetising component. It produces flux in the core and does not consume any power.

From phasor (vector) diagram shown in Fig. 2.20.

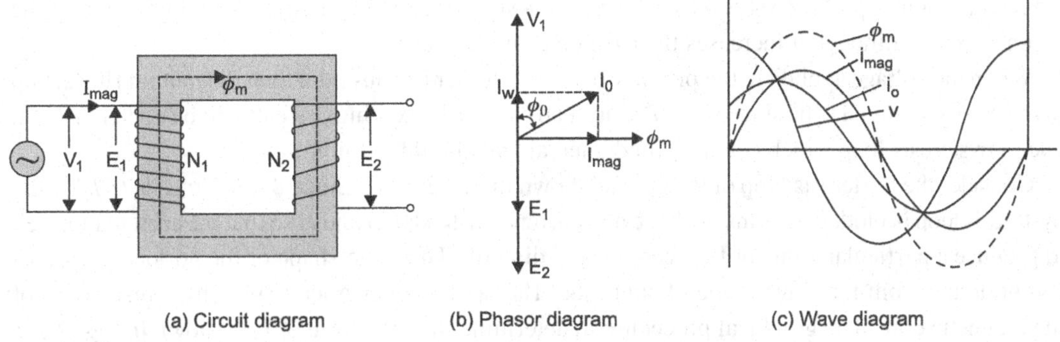

(a) Circuit diagram (b) Phasor diagram (c) Wave diagram

Fig. 2.20 Transformer on no-load

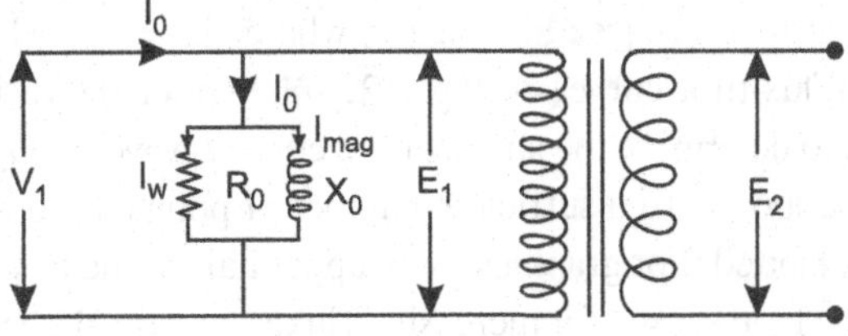

Fig. 2.21 Equivalent circuit

Working component, $I_w = I_0 \cos \phi_0$

Magnetising component, $I_{mag} = I_0 \sin \phi_0$

No-load current, $$I_0 = \sqrt{I_w^2 + I_{mag}^2} \quad \text{...(2.10)}$$

Primary p.f. at no-load, $$\cos \phi_0 = \frac{I_w}{I_0} \quad \text{...(2.11)}$$

No-load power input, $$P_0 = V_1 I_0 \cos \phi_0 \quad \text{...(2.12)}$$

$$\text{Exciting resistance, } R_0 = \frac{V_1}{I_w} \quad \text{...(2.13)}$$

$$\text{Exciting reactance, } X_0 = \frac{V_1}{I_{\text{mag}}} \quad \text{...(2.14)}$$

The equivalent circuit of a transformer at no-load is shown in Fig. 2.22. Here, R_0 represents the exciting resistance of the transformer which carries power loss component of no-load current, i.e., I_w used to meet with the no-load losses in the transformer, whereas X_0 represents the exciting reactance of the transformer which carries wattless component of no-load current, i.e., I_{mag} used to set-up magnetic field in the core.

2.9 Effect of Magnetisation on No-load (Exciting) Current

A transformer requires less magnetic material if it is operated at a higher core flux density. Therefore, from an economic point of view, a transformer is designed to operate in the saturating region of the magnetic core, although it increases the harmonics in the wave.

When the voltage applied to the primary of a transformer is sinusoidal then the mutual flux set up in the core is assumed to be sinusoidal. The no-load current I_0 (exciting current) will be non-sinusoidal due to hysteresis loop which contains fundamental and all odd harmonics.

Consider the hysteresis loop of the core as shown in Fig. 2.22(*a*). Since $\phi = BA$ and $i = (Hl/N)$, the hysteresis loop is plotted in terms of flux ϕ and current i instead of B and H so that the current required to produce a particular value of flux can be read directly. The wave-shape of the no-load current i_0 can be found from the wave shape of sinusoidal flux and $\phi - i$ characteristics (hysteresis loop) of the magnetic core. The graphical procedure to determine the waveform of i_0 is shown in Fig. 2.22.

At point O of the flux-time curve [see Fig. 2.22(*b*)], the flux is zero; this corresponds to a current OA on the hysteresis loop [see Fig. 2.22(*a*)]. At point a of the flux-time curve [see Fig. 2.22(*b*)] $\phi = aa'$; this corresponds to a current OB, [see Fig. 2.22(*a*) where OB = '2']. The corresponding value of current I_0 = '2' is plotted on flux-time curve [see Fig. 2.22(*b*)]. In brief, the various abscissas of Fig. 2.22 (*a*) are plotted as ordinates to determine the shape of the current wave on Fig. 2.22(*b*). The procedure is followed round the whole loop until a sufficient number of points are obtained. The current–time curve for the whole loop is plotted. For graphics, only upper half of the hysteresis loop is considered.

The ascending part of the loop is used for increasing fluxes, and the descending part for decreasing fluxes. The waveform of the current in Fig. 2.22(*b*) represents the magnetising component and the hysteresis component of the no-load current. It reaches its maximum at the same time as the flux wave, but the two waves do not go through zero simultaneously. Thus, the current lags behind the applied voltage by an angle slightly less than 90°.

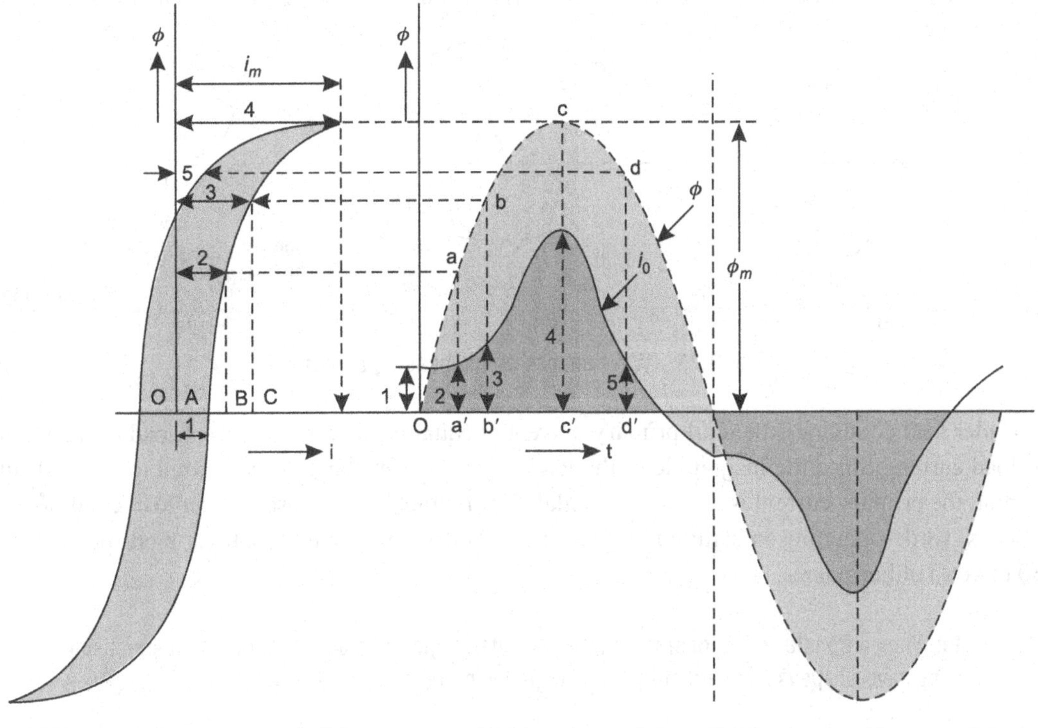

(*a*) Hysteresis loop of core material (*b*) Wave forms of sinusoidal flux f and exciting current i_0

Fig. 2.22 Wave shape of the no-load current of a transformer.

The above current is required to be modified to account for the eddy-current loss in the core and the modified current is called the *no-load current,* I_0. The corresponding current component apart from being is phase with applied voltage V_1 is sinusoidal in nature as it balances the effect of sinusoidal eddy-currents caused by sinusoidal core flux. It is, therefore, seen that eddy-currents do not introduce any harmonic in the exciting current. However, eddy-current loss reduces the angle between the applied voltage V_1 and the no-load current, I_0.

It is seen that the waveform of the no-load current contains third, fifth and higher-order odd harmonics which increase rapidly **if the maximum flux is taken further into saturation.** However, the third harmonic is the predominant one. For all practical purposes harmonics higher than third are negligible. In no-load current, the third harmonic is nearly 5 to 10% of the fundamental at rated voltage. However, it may increase to 30 to 40% when the voltage rises to 150% of its rated value.

When the load on the transformer is linear, the sinusoidal load current is so large that it swamp out the non-sinusoidal nature of the primary current. Therefore, for all practical purposes, the primary current is considered sinusoidal at all loads.

In certain 3-phase transformer-connections, third-harmonic current cannot flow, as a result the magnetising current I_{mag} is almost sinusoidal. For satisfaction of *BH* curve, the core flux ϕ must then be non-sinusoidal; it is a flat-topped wave, as shown in Fig. 2.23(*b*). This can be verified by assuming a sinusoidal magnetising current and then finding out the flux wave shape from the flux-magnetising current relationship, the normal magnetising curve, neglecting hysteresis component which is in phase with V_2. Since the flux is flat-topped, the emf which is its derivative will now be peaky with a strong third harmonic component. The various waveforms are illustrated in Fig. 2.23.

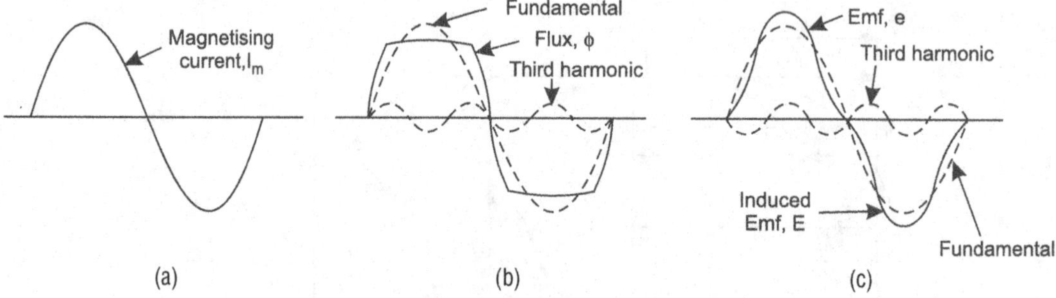

Fig. 2.23 Wave shapes of magnetising currents

Under load conditions the total primary current is equal to the phasor sum of load current and no-load current. Since the magnitude of the load current is very large as compared to the no-load current, the primary current is almost sinusoidal, for all practical purposes, under load conditions.

Thus, while operating in saturated region, the magnetising current contains more percentage (30 to 40%) of harmonics.

Note: When a transformer is operated in the saturated region of magnetisation, the magnetising current wave-shape is further distorted and contains more percentage of harmonics.

However, for economic reasons, usually transformers are designed to operate in saturated region. A transformer will require smaller quantity of magnetic material if it is designed for saturated region.

2.10 Inrush of Magnetising Current

When a transformer is initially energised, there is a sudden inrush of primary current. The maximum value attained by the flux may reach to twice the normal flux. The core is driven far into saturation with the result that the magnetising current has a **very high peak value.**

For mathematical treatment, let the applied voltage be,

$$v_1 = V_{1m} \sin (\omega t + \propto) \qquad \text{...(2.15)}$$

The secondary of the transformer is open circuited. Here $\propto$ the angle of the voltage sinusoid at $t = 0$. Suppose for the moment we neglect core losses and primary resistance, then

$$v_1 = N_1 \frac{d\phi}{dt} \qquad \text{...(2.16)}$$

where N_1 is the number of primary turns and ϕ is the flux in the core. In the steady-state

$$V_{1m} = \omega \phi_m N_1 \qquad \text{...(2.17)}$$

$$\left(\because \frac{d\phi}{dt} = \frac{d}{dt} \phi_m \sin \omega t = \omega \phi_m \cos \omega t, \text{it will be maximum when } \cos \omega t = 1 \right)$$

From eqn. (2.15) and (2.16), we get,

$$N_1 \frac{d\phi}{dt} = V_{1m} \sin (\omega t + \propto)$$

$$\frac{d\phi}{dt} = \frac{V_{1m}}{N_1} \sin (\omega t + \alpha) \quad \text{...(2.18)}$$

From eqn. (2.17) and (2.18), we get,

$$\frac{d\phi}{dt} = \omega \phi_m \sin (\omega t + \propto) \quad \text{...(2.19)}$$

Integrating the eqn. (2.19), we get,

$$\phi = -\phi_m \cos (\omega t + \propto) + \phi_c \quad \text{...(2.20)}$$

where ϕ_c is the constant of integration. Its value can be found from initial conditions at $t = 0$. Assume that when the transformer was last disconnected from the supply line, a small residual flux ϕ_r remained in the core. Thus at $t = 0$, $\phi = \phi_r$.

Substituting these values in eqn. (2.20)

$$\phi_r = -\phi_m \cos \propto + \phi_c$$

$$\therefore \quad \phi_C = \phi_r + \phi_m \cos \propto \quad \text{...(2.21)}$$

Then the eqn. (2.20) becomes

$$\phi = \underbrace{-\phi_m \cos (\omega t + \propto)}_{\substack{\text{Steady-state component} \\ \text{of flux i.e., } \phi_{ss}}} + \underbrace{\phi_r + \phi_m \cos \alpha}_{\substack{\text{transient component} \\ \text{of flux i.e., } \phi_C}} \quad \text{...(2.22)}$$

Equation (2.22) shows that the flux consists of two components, the steady-state component ϕ_{ss} and the transient component ϕ_c. The magnitude of the transient component

$$\phi_c = \phi_r + \phi_m \cos \propto$$

is a function of $\propto$, where $\propto$ is the instant at which the transformer is switched on to the supply.

If the transformer is switched on at $\propto = 0$, then $\cos \propto = 1$

$$\phi_c = \phi_r + \phi_m$$

Under this condition

$$\phi = -\phi_m \cos \omega t + \phi_r + \phi_m$$

At $\omega t = \pi$,

$$\phi = -\phi_m \cos \pi + \phi_r + \phi_m = 2\phi_m + \phi_r \quad \text{...(2.23)}$$

Thus, the core flux attains the maximum value of flux equal to $(2\,\phi_m + \phi_r)$, which is over and above twice the normal flux as shown in Fig. 2.24. This is known as **doubling effect.** Consequently, the core goes into deep saturation. The magnetising current required for producing such a large flux in the core may be as large as 10 times the normal magnetising current. Sometimes the *rms* value of magnetising current may be larger than the primary rated current of the transformer as shown in Fig. 2.25.

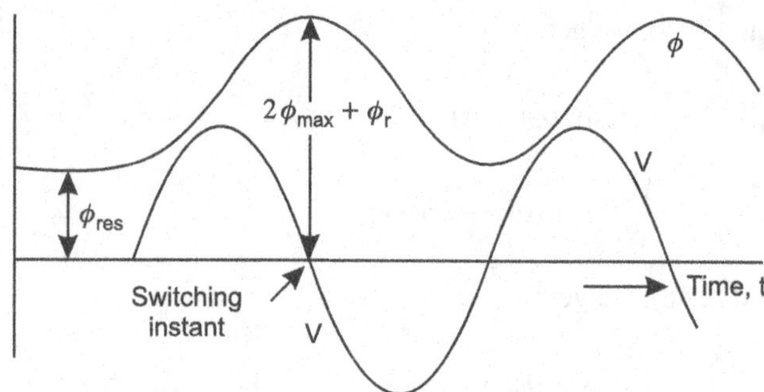

Fig. 2.24 Wave shape of inrush current

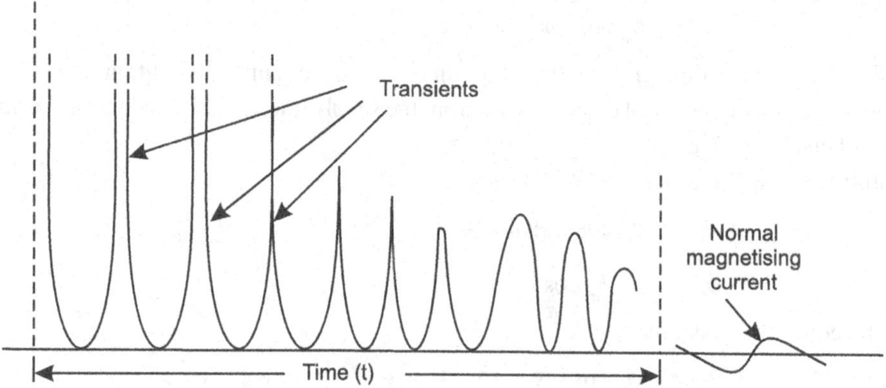

Fig. 2.25 Transients due to inrush current

This inrush current may produce electromagnetic forces about 25 times the normal value. Therefore, the windings of large transformers are strongly braced.

Inrush current may also cause (*i*) improper operation of protective devices like unwarranted tripping of relays, (*ii*) momentary large voltage drops and (*iii*) large humming due to magnetostriction of the core.

To obtain no transient inrush current, ϕ_c should be zero:

$$\phi_c = \phi_r + \phi_m \cos \alpha = 0 \text{ or } \cos \alpha = \frac{-\phi_r}{\phi_m} \quad \text{...(2.24)}$$

Since ϕ_r is usually very small $\cos \alpha = 0$ and $\alpha \cong \frac{n\pi}{2}$. ...(2.25)

In other words, if the transformer is connected to the supply line near a positive or negative voltage maximum, the current inrush will be minimised. It is usually impractical to attempt to connect a transformer at a predetermined time in the voltage cycle.

Fortunately, inrush currents do not occur as might be thought. The magnitude of the inrush current is also less than the value calculated by purely theoretical considerations. *The effects of other transformers in the system, load currents and capacitances all contribute to the* **reduction of transients.**

Example 2.8

A 230/110 V single-phase transformer has a core loss of 100 W. If the input under no-load condition is 400 VA, find core loss current, magnetising current and no-load power factor angle.

Solution:

Here, $V_1 = 230$ V; $V_2 = 110$ V; $P_i = 100$ W

Input at no-load = 400 VA

i.e., $$V_1 I_0 = 400$$

or No-load current, $$I_0 = \frac{400}{230} = 1{\cdot}739 \text{ A}$$

Core loss current, $$I_w = \frac{P_i}{V_1} = \frac{100}{230} = \mathbf{0{\cdot}4348\ A}\ (Ans)$$

Magnetising current, $$I_{mag} = \sqrt{I_0^2 - I_w^2} = \sqrt{(1{\cdot}739)^2 - (0{\cdot}4348)^2} = \mathbf{1{\cdot}684\ A}\ (Ans)$$

No-load power factor, $$\cos\phi_0 = \frac{I_w}{I_0} = \frac{0{\cdot}4348}{1{\cdot}739} = \mathbf{0{\cdot}25}\ \boldsymbol{lag}\ (Ans)$$

No-load power factor angle, $$\phi_0 = \cos^{-1} 0{\cdot}25 = \mathbf{75{\cdot}52^\circ}\ (Ans)$$

Example 2.9

A single phase, 50 kVA, 2300/230 V, 50 Hz transformer is connected to 230 V supply on the secondary side, the primary being open. The meter indicate the following readings:

Power = 230 watt

Voltage = 230 V

Current = 6·5 A

Find (i) Core loss: (ii) loss component of the current; (iii) magnetising current. Draw the phasor diagram for this condition.

Solution:

Power input at no-load, $P_0 = 230$ W

Supply voltage, $V_1 = 230$ V

Current at no-load, $I_0 = 6{\cdot}5$ A

(*i*) Since low voltage winding resistance is not given, the copper losses cannot be separated, therefore, whole of the power input will represent the iron or core losses.

∴ Core loss = **230 W** (*Ans*)

(*ii*) Using relation $$P_0 = V_1 J_0 \cos\phi_0 = V_1 I_w$$

Loss component of current, $$I_w = \frac{P_0}{V_1} = \frac{230}{230} = \mathbf{1.0\ A}\ (Ans)$$

(*iii*) Magnetising current, $I_{mag} = \sqrt{I_0^2 - I_w^2} = \sqrt{(6.5)^2 - (1.0)^2}$ = **6.423 A** (*Ans*)

Under the given condition, transformer is operated at no-load. The phasor diagram is shown in Fig. 2.11.

Where $V_1 = 230$ V; $I_0 = 6{\cdot}5$ A; $I_w = 1.0$ A; $I_{mag} = 6.423$ A; $E_1 = 230$ V; $E_2 = 2300$ V

Example 2.10

A 230V, 50 Hz transformer has 200 primary turns. It draws 5 A at 0.25 p.f lagging at no-load. Determine:

(i) Maximum value of flux in the core; (ii) Core loss; (iii) Magnetising current (iv) Exciting resistance and reactance of the transformer. Also draw its equivalent circuit.

Solution:

(*i*) Using the relation, $E_1 = 4.44\, N_1 f \phi_m$

or $230 = 4.44 \times 220 \times 50 \times \phi_m$

$\therefore$ Maximum value of flux ϕ_m = **5.18 m Wb** (*Ans*)

(*ii*) Core loss, $P_0 = V_1 I_0 \cos\phi_0 = 230 \times 5 \times 0{\cdot}25$ = **287.5 W** (*Ans*)

(*iii*) No-load p.f., $\cos\phi_0 = 0{\cdot}25$; $\sin\phi_0 = \sin\cos^{-1} 0{\cdot}25 = 0{\cdot}9682$

Magnetising current component, $I_m = I_0 \sin\phi_0 = 5 \times 0{\cdot}9682$ = **4.84A** (*Ans*)

$$\text{Exciting resistance, } R_0 = \frac{V_1}{I_w} = \frac{230}{I_0 \cos\phi_0} = \frac{230}{5 \times 0.25} = \mathbf{184\ \Omega}\ (Ans.)$$

$$\text{Exciting reactance, } X_0 = \frac{V_1}{I_{mag}} = \frac{230}{4.84} = \mathbf{47.52\ \Omega}\ (Ans.)$$

The equivalent circuit is shown in Fig. 2.26. The values of different quantities are mentioned in the solution itself.

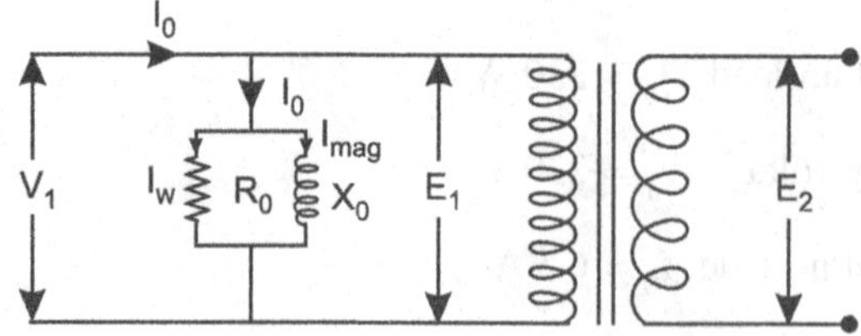

Fig. 2.26 Equivalent circuit of transformer at no-load.

Example 2.11

At open circuit, transformer of 10 kVA, 500/250 V, 50 Hz draws a power of 167 watt at 0.745 A, 500 V. Determine the magnetising current, wattful current, no-load power factor, hysteresis angle of advance, equivalent resistance and reactance of exciting circuit referred to primary side.

Solution:

Here, $V_1 = 500$ V; $I_0 = 0.745$ A; $P_0 = 167$ W

$$\text{Wattful component of current, } I_w = \frac{P_0}{V_1} = \frac{167}{500} = \mathbf{0.334\ A}\ (Ans.)$$

$$\text{Magnetising component of current, } I_{mag} = \sqrt{I_0^2 - I_w^2}$$

$$= \sqrt{(0.745)^2 - (0.334)^2} = \mathbf{0.666\ A}\ (Ans.)$$

$$\text{No-load power factor, } \cos\phi_0 = \frac{I_w}{I_0} = \frac{0.334}{0.745} = \mathbf{0.448\ lag}\ (Ans.)$$

$$\text{Hysteresis angle of advance, } \phi_0 = \cos^{-1} 0.448 = \mathbf{63.36^\circ\ lag}\ (Ans.)$$

$$\text{Exciting resistance, } R_0 = \frac{V_1}{I_w} = \frac{500}{0.334} = \mathbf{1497\ \Omega}\ (Ans.)$$

$$\text{Exciting reactance, } X_0 = \frac{V_1}{I_{mag}} = \frac{500}{0.666} = \mathbf{750\ \Omega}\ (Ans.)$$

Example 2.12

A transformer working on 2200 V, 50 Hz has 220 primary turns. The core has a mean length of magnetic path of 100 cm and cross-sectional area 1000 sq. cm. the iron having a relative permeability of 100. The iron loss is 400 watt. Calculate primary no-load current.

Solution:

Here, $V_1 = 2200$ V; $f = 50$ Hz; $N_1 = 220$ turns;

$A_i = 1000\text{ cm}^2 = 1000 \times 10^{-4} = 0.1\text{ m}^2$; $\mu_r = 1000$;

$W_i = 400$ W; $l_m = 100$ cm $= 1$ m

$$\text{Magnetising force, } H = \frac{N_1 I_{mag(\max)}}{l_m}$$

$$\text{Flux density, } B_m = \mu_0 \mu_r H = \mu_0 \mu_r \times \frac{N_1 I_{mag(\max)}}{l_m}$$

Now,

$$V_1 = 4.44\, B_m A_i N_1 f$$

$$= 4.44 \times \mu_0 \mu_r \times \frac{N_1 I_{mag(\max)}}{l_m} \times A_i \times N_1 \times f$$

or

$$I_{mag(max)} = \frac{V_1 \times l_m}{4.44 \times \mu_0 \mu_r \times N_1 \times A_i \times N_1 \times f}$$

$$I_{mag(rms)} = \frac{V_1 \times l_m}{\sqrt{2} \times 4.44 \times \mu_0 \mu_r \times N_1 A_i \times N_1 \times f}$$

$$I_{mag} = \frac{2200 \times 1}{\sqrt{2} \times 4.44 \times 4\pi \times 10^{-7} \times 1000 \times 220 \times 0.1 \times 220 \times 50}$$

$$= 2.152 \text{ A}$$

$$I_w = \frac{\text{Iron loss}}{V_1} = \frac{400}{2200} = 0.182 \text{ A}$$

$$\therefore \quad I_0 = \sqrt{(I_{mag})^2 + (I_w)^2} = \sqrt{(1.152)^2 + (0.182)^2} = \mathbf{1.166\ A}\ (\textit{Ans.})$$

Example 2.13

The iron core of a single-phase transformer has a mean length of magnetic path as 2.5m and its joints are equivalent to an air gap of 0.1 mm. The transformer has 500 and 50 turns on its primary and secondary, respectively. When a potential difference of 3000 V is applied to the primary, maximum flux density in the core is 1.2 Wb/m². Calculate (i) the cross-sectional area of the core (ii) no-load secondary voltage (iii) no-load current drawn by the primary (iv) power factor on no-load. Given that AT/m for a flux density of 1.2 Wb/m² in iron to be 500, the corresponding iron loss to be 2 watt/kg at 50 Hz and the density of iron as 7.8 g/cm³.

Solution:

Here, $N_1 = 500$; $N_2 = 50$; $E_1 = 3000$ V; $f = 50$ Hz; $B_m = 1.2$ Wb/m²

(*i*) We know, $E_1 = 4.44\, N_1 f B_m A_i$

$\therefore$ Net area of cross-section of core,

$$A_i = \frac{E_1}{4.44 N_1 f B_m} = \frac{3000}{4.44 \times 500 \times 50 \times 1.2} = 0.0225 \text{ m}^2$$

$$= \mathbf{225\ cm^2}\ (\textit{Ans.})$$

(*ii*) Transformation ratio, $K = \frac{N_2}{N_1} = \frac{50}{500} = 0.1$

No-load secondary voltage, $E_2 = KE_1 = 0.1 \times 3000 = \mathbf{300\ V}$ (*Ans.*)

(*iii*) AT for iron path, $AT_i = \text{AT/m} \times l_i = 500 \times 2.5 = 750$

AT for air gap, $AT_g = Hl_g = \frac{B_m}{\mu_0} \times l_g = \frac{1.2}{4\pi \times 10^{-7}} \times 0.1 \times 10^{-3} = 95.5$

Total $AT = AT_i + AT_g = 750 + 95.5 = 845.5$

Maximum value of magnetising current drawn by primary

$$I_{mag(max.)} = \frac{AT}{N_1} = \frac{845.5}{500} = 1.691 \text{ A}$$

Assuming the current to be sinusoidal, rms value of this current

$$I_{mag} = \frac{I_{mag(max.)}}{\sqrt{2}} = \frac{1.691}{\sqrt{2}} = 1.196 \text{ A}$$

$$\text{Volume of iron core} = \text{area of cross-section} \times \text{mean length of iron}$$
$$= 0.0225 \times 2.5 = 0.03375 \text{ m}^3$$
$$\text{Density of iron} = 7.8 \text{ g/cm}^3 = 7.8 \times 10^6 \times 10^{-3} = 7800 \text{ kg/m}^3$$
$$\text{Mass of iron core} = \text{Volume} \times \text{density} = 0.03375 \times 7800 = 263.25 \text{ kg}$$
$$\text{Total iron loss, } P_i = \text{iron loss/kg} \times \text{mass} \qquad \text{(given that iron loss/kg = 2 W)}$$
$$= 2 \times 263.25 = 526.5 \text{ W}$$

$$\text{Working component of current, } I_w = \frac{P_i}{V_1} = \frac{526.5}{3000} = 0.1755 \text{ A}$$

$$\text{No-load current, } I_0 = \sqrt{I_w^2 + I_{mag}^2} = \sqrt{(0.1755)^2 + (1.196)^2} = \mathbf{1.21\ A}\ (Ans.)$$

(*iv*) Power factor at no-load, $\cos\phi_0 = \frac{I_w}{I_0} = \frac{0.1755}{1.21} = \mathbf{0.145\ lag}$ (*Ans.*)

Section Practice Problems

Numerical Problems

1. A single phase, 50 kVA, 2300/230 V, 50 Hz transformer is connected to 230 V supply on the secondary side, the primary being open. The meter indicate the following readings: Power = 187 watt; Voltage = 230 V; Current = 6·5 A. Find (*i*) core loss; (*ii*) loss component of the current; (*iii*) magnetising current.
(**Ans.** *187 W; 0·813 A; 6·45 A*)

2. A single-phase, 230V, 50 Hz transformer draws 5A at 0.25 pf. Its primary winding carries 200 turns. Determine (*a*) the maximum value of flux in the core, (*b*) the core loss (*c*) the magnetising component of current and (*d*) exciting resistance and reactance. (**Ans.** *5.293 m Wb; 293.75 W; 4.84 A; 268 Ω, 69.2 Ω*)

3. The primary and secondary of a single-phase transformer carry 500 and 100 truns, respectively. The mean length of the flux in the iron core is 200 cm and the joints are equivalent to an air gap of 0.1 mm. If the maximum value of flux density is to be 1.1 Wb/m^2 when a potential difference of 2200 volt at 50 cycles is applied to the primary, calculate (*i*) the cross-sectional area of the core (*ii*) the secondary voltage on no-load (*iii*) the primary current and power factor on no-load.

   Assume that 400 At/m are required to produce a flux density of 1.1 Wb/m^2 and the corresponding iron loss to be 0.8 watt/kg at 50 Hz and the density of the iron to be 7.8 gm/cm^3.
(**Ans.** *183 cm^2, 440 V, 1.59 A, 0.145 lagging*)

Short Answer Type Questions

Q.1. Even at no-load, a transformer draws current from the mains. Why?

Ans. At no-load, what so ever current is drawn by a transformer that is used to meet with iron losses and to produce magnetic flux is the core.

Q.2. What do you mean by exciting resistance and exciting reactance?

Ans. At no-load, a transformer behaves as a highly inductive load connected to the system which contains some resistance and inductance in parallel. The resistance is called exciting resistance $R_0 = V_1/I_w$ and reactance is called exciting reactance, $X_0 = V_1/I_{mag}$.

Q.3. Usually, transformers are designed to operate in saturated region. Why?

Ans. If a transformer is operated in saturated region, a smaller quantity of magnetic material is required, therefore, for economic reasons, usually transformers are designed to operate is saturated region.

Q.4. What is doubling effect in transformer core?

Ans. When a transformer is initially energised, there is a sudden inrush of primary current. The maximum value attained by the flux may reach to twice the normal flux. This condition is known as *doubling effect* and the core goes into deep saturation.

Q.5. Does the magnetising current of a transformer lie in-phase with the applied voltage? Justify.

Ans. No, the phasor of magnetising current lies in quadrature i.e., lags behind the applied voltage by 90°. It is because in pure inductive circuits, current lags behind the voltage by 90°.

Q.6. What is the effect of saturation on exciting current of transformer?

Ans. When a transformer is operated in the saturated region of magnetisation, the magnetising current waveshape in further distorted and contains more percentage of harmonics.

Q.7. What are the ill-effects of inrush current of transformer?

Ans. Following are the ill-effects of inrush current in a transformer:

(*i*) Improper operation of protective devices
(*ii*) Momentary large voltage drops and
(*iii*) Large humming due to magnetostriction effect.

2.11 Transformer on Load

(Neglecting winding resistance and leakage flux)

When a certain load is connected across the secondary, a current I_2 flows through it as shown in Fig. 2.27(i). The magnitude of current I_2 depends upon terminal voltage V_2 and impedance of the load. The phase angle of secondary current I_2 with respect to V_2 depends upon the nature of load i.e., whether the load is resistive, inductive or capacitive.

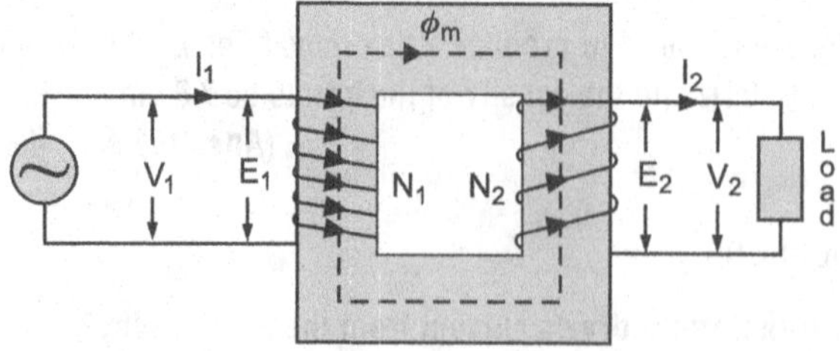

Fig. 2.27 (i) Circuit diagram for loaded transformer

The operation of the transformer on load is explained below with the help of number of diagrams;

(*i*) When the transformer is on no-load as shown in Fig. 2.27(*ii*)(*a*) it draws no-load current I_0 from the supply mains. The no-load current I_0 produces an mmf $N_1 I_0$ which sets up flux in the core.

(*ii*) When the transformer is loaded, current I_2 flows in the secondary winding. This secondary current I_2 produces an mmf $N_2 I_2$ which sets up flux ϕ_2 in the core. As per Lenz's law this flux opposes the main flux ϕ, as shown in Fig. 2.27(*ii*)(*b*).

(*iii*) As ϕ_2 is set-up in opposite direction to the main flux, the resultant flux tends to decrease and causes the reduction of self-induced emf E_1 momentarily. Thus, V_1 predominates over E_1 causing additional primary current I'_1 drawn from the supply mains. The amount of this additional current I_1' is such that the original conditions i.e., flux in the core must be restored. so that $V_1 = E_2$. The current I_1 is in phase opposition with I_2 and is called primary counter balancing current. This additional current I'_1 produces an mmf $N_1 I_1'$ which sets up flux ϕ, in the same direction as that of ϕ as shown in Fig. 2.27(*ii*)(*c*), and cancels the flux ϕ_2 set up by mmf $N_2 I_2$.

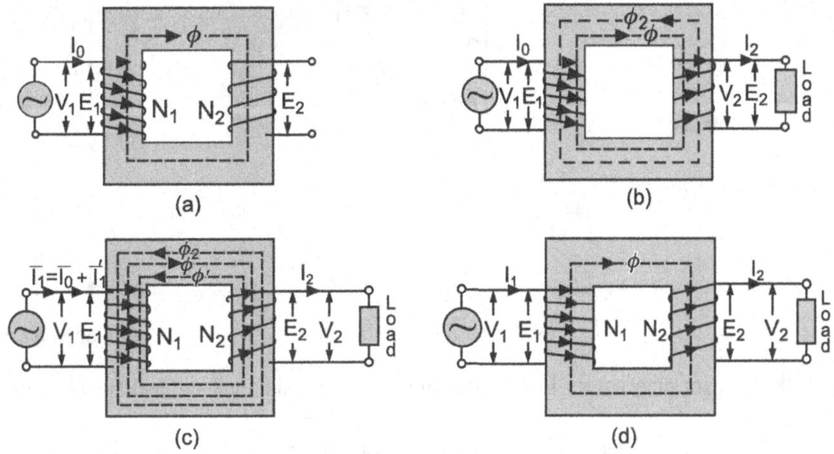

Fig. 2.27 (ii) Transformer action when load is applied to it.

Now $\qquad N_1 I_1' = N_2 I_2$ (ampere-turns balance)

$$\therefore \qquad I_1' = \frac{N_2}{N_1} I_2 = K I_2$$

(iv) Thus, the flux is restored to its original value as shown in Fig. 2.27(*ii*)(*d*). The total primary current I_1 is the vector sum of current I_0 and I_1' i.e., $I_1 = I_0 + I_1'$.

This shows that flux in the core of a transformer remains the same from no-load to full-load ; this is the reason why iron losses in a transformer remain the same from no-load to full-load

2.12 Phasor Diagram of a Loaded Transformer

(*Neglecting voltage drops in the windings; ampere-turns balance*)

Since the voltage drops in both the windings of the transformer are neglected, therefore,

$$V_1 = E_1 \text{ and } E_2 = V_2$$

While drawing the phasor diagram the following important points are to be considered.

(i) For simplicity, let the transformation ratio $K = 1$ be considered, therefore, $E_2 = E_2$.

(ii) The secondary current I_2 is in phase, lags behind and leads the secondary terminal voltage V_2 by an angle ϕ_2 for resistive, inductive and capacitive load, respectively.

(iii) The counter balancing current $I_1' = \dfrac{N_2}{N_1} I_2$ ...(2.26)

(i.e., $I_1' = K I_2$ here $K = 1 \therefore I_1' = I_2$) and is 180° out of phase with I_2.

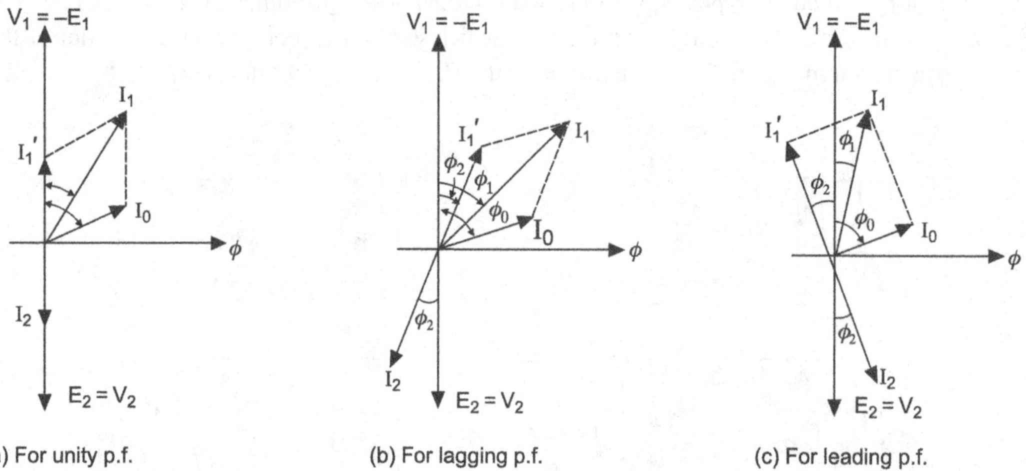

Fig. 2.28 Phasor diagram on-load (neglecting winding resistance and leakage reactance)

(iv) The total primary current I_1 is the vector sum of no-load primary current I_0 and counter balancing current I_1'.

i.e., $$\overline{I_1} = \overline{I_0} + \overline{I_1'} \text{ or } \overline{I_1} = \sqrt{(I_0)^2 + (I_1')^2 + 2I_0I_1'\cos\theta}$$ ...(2.27)

Where θ is the phase angle between I_0 and I_1'.

(v) The p.f. on the primary side is $\cos\phi_1$ which is less than the load p.f. $\cos\phi_2$ on the secondary side. Its value is determined by the relation;

$$\cos\phi_1 = \frac{I_0\cos\phi_0 + I_1'\cos\phi_2}{I_1}$$ ...(2.28)

The phasor diagrams of the transformer for resistance, inductive and capacitive loads are shown in Figs. 2.28(*a*), (*b*) and (*c*), respectively.

Alternately

The primary current I_1 can also be determined by resolving the vectors, i.e.,

$I_v = I_0 \cos\phi_0 + I'_1 \cos\phi_2$ [where $\sin\phi_0 = \sin\cos^{-1}(\cos\phi_0)$ ...(2.29)

$I_H = I_0 \sin\phi_0 + I'_1 \sin\phi_2$ and $\sin\phi_2 = \sin\cos^{-1}(\cos\phi_2)$] ...(2.30)

$$I_1 = \sqrt{(I_v)^2 + (I_H)^2}$$ ...(2.31)

Example 2.14

A 440/110 V, single phase transformer draws a no-load current of 5 A at 0.2 p.f. lagging. If a current of 120 A at 0.8 p.f lagging is supplied by the secondary, calculate the primary current and p.f.

Solution:

Transformation ratio, $K = \frac{E_2}{E_1} = \frac{110}{440} = 0.25$

Let the primary counter balancing current be I_1'.

Then $I_1' = K\,I_2 = 0.25 \times 120 = 30$ A

Now $\cos\phi_0 = 0{\cdot}2;\ \phi_0 = \cos^{-1} 0.2 = 78{\cdot}46^\circ$

$\cos\phi_2 = 0{\cdot}8;\ \phi_2 = \cos^{-1} 0.8 = 36.87^\circ$

$\theta = \phi_0 - \phi_2 = 78{\cdot}46^\circ - 36{\cdot}87^\circ = 41{\cdot}59^\circ$

$$I_1 = \sqrt{(I_0)^2 + (I_1')^2 + 2\,I_0\,I_1'\cos\theta}$$

$$= \sqrt{(5)^2 + (30)^2 + 2 \times 5 \times 30 \times \cos 41{\cdot}59''} = \mathbf{33{\cdot}9\ A}\ (Ans.)$$

Primary p.f., $\cos\phi_1 = \frac{I_1'\cos\phi_2 + I_0\cos\phi_0}{I_1}$

$$= \frac{30 \times 0{\cdot}8 + 5 \times 0{\cdot}2}{33{\cdot}9} = \mathbf{0{\cdot}7375\ lag}\ (Ans.)$$

Example 2.15

A single phase transformer with a ratio of 6600/400 V (primary to secondary voltage) takes to no-load current of 0·7 A at 0·24 power factor lagging. If a current of 120 A at a power factor of 0·8 lagging is supplied by its secondary. Estimate the current drawn by the primary winding.

Solution:

Here, $I_0 = 0{\cdot}7$A; $\cos\phi_0 = 0.24$ lag; $I_2 = 120$A; $\cos\phi_2 = 0{\cdot}8$ lag

Transformation ratio, $K = \frac{V_2}{V_1} = \frac{400}{6600} = \frac{2}{33}$

Let the primary counter balance current be I_2.

$\therefore$ $N_1 I_1' = N_2 I_2$

or $I_1' = \frac{N_2}{N_1} \times I_2 = K\,I_2 = \frac{2}{33} \times 120 = 7{\cdot}273$ A

Now, $\cos\phi_0 = 0{\cdot}24;\ \phi_0 = \cos^{-1} 0{\cdot}24 = 76{\cdot}11^\circ$

$\cos\phi_2 = 0{\cdot}8;\ \phi_2 = \cos^{-1} 0{\cdot}8 = 36{\cdot}87^\circ$

Angle between vector I_0 and I_1 [refer Fig. 2.14 (*b*)]

$$\theta = 76{\cdot}11^\circ - 36{\cdot}87^\circ = 39{\cdot}24^\circ$$

Current drawn by the primary,

$$I_1 = \sqrt{(I_0)^2 + (I_1')^2 + 2I_0I_1' \cos\theta}$$

$$= \sqrt{(0.7)^2 + (7.273)^2 + 2 \times 0.7 \times 7.273 \times \cos 39.24"} = \mathbf{7{\cdot}827\ A}\ (Ans.)$$

2.13 Transformer with Winding Resistance

In an actual transformer, the primary and secondary windings have some resistance represented by R_1 and R_2, respectively. These resistances are shown external to the windings in Fig. 2.29 To make the calculations easy the resistance of the two windings can be transferred to either side. The resistance is transferred from one side to the other in such a manner that percentage voltage drop remains the same when represented on either side.

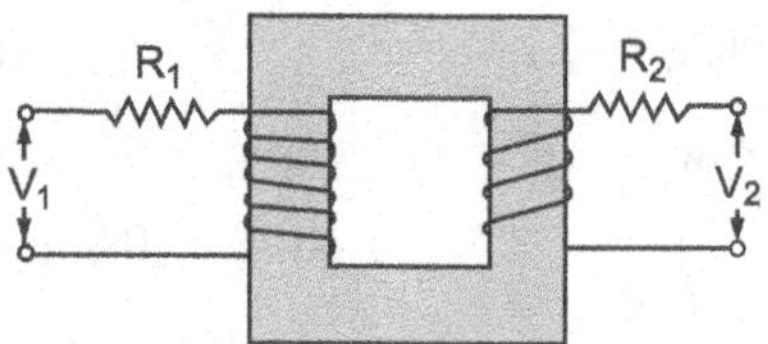

Fig. 2.29 Transformer windings with resistance

Let the primary resistance R_1 be transferred to the secondary side and the new value of this resistance be R_1' called equivalent resistance of primary referred to secondary side as shown in Fig. 2.30(*a*). I_1 and I_2 be the full load primary and secondary currents, respectively.

Then $$\frac{E_1E_2}{V_2} \times 100 = \frac{E_1E_2}{V_1} \times 100 \qquad \text{(\% voltage drops)}$$

or $$R_1' = \frac{I_1}{I_2} \times \frac{V_2}{V_1} \times R_1 = K^2R_1 \qquad ...(2.32)$$

∴ Total equivalent resistance referred to secondary.

$$R_{es} = R_2 + R_1' = R_2 + K^2 R_1 \qquad ...(2.33)$$

Now consider resistance R_2, when it is transferred to primary, let its new value be R_2' called equivalent resistance of secondary referred to primary as shown in Fig. 2.30(*c*).

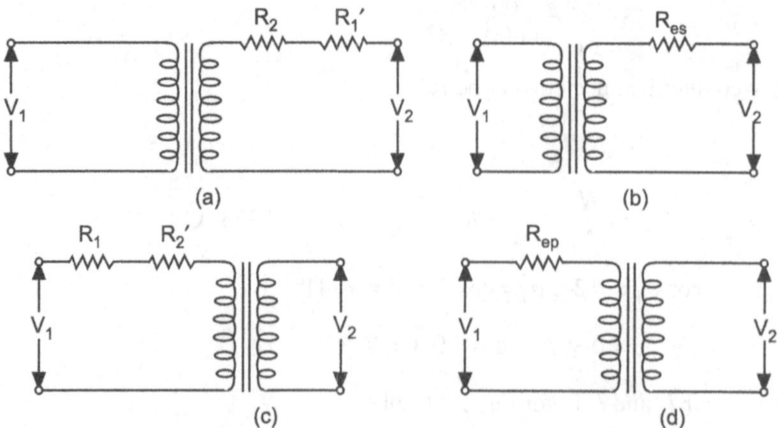

Fig. 2.30 Equivalent resistance referred to either side

Then $$\frac{I_1 R_2'}{V_1} \times 100 = \frac{I_2 R_2}{V_2} \times 100 \text{ or } R_2' = \frac{I_2}{I_1} \times \frac{V_1}{V_2} \times R_2 = \frac{R_2}{K^2} \quad \ldots(2.34)$$

∴ Total equivalent resistance referred to primary,

$$R_{ep} = R_1 + R_2' = R_1 + \frac{R_2}{K^2} \quad \ldots(2.35)$$

2.14 Mutual and Leakage Fluxes

So far, it is assumed that when AC supply is given to the primary winding of a transformer, an alternating flux is set up in the core and whole of this flux links with both the primary and secondary windings. However, in an actual transformer, both the windings produce some flux that links only with the winding that produces it.

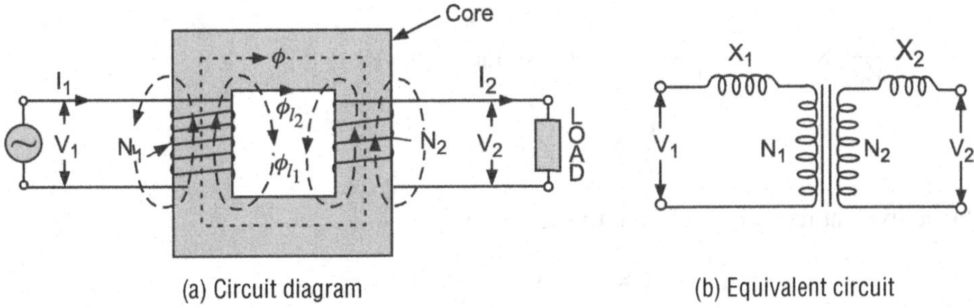

(a) Circuit diagram (b) Equivalent circuit

Fig. 2.31 Transformer primary and secondary winding reactance

The flux that links with both the windings of the transformer is called **mutual flux** *and the flux which links only with one winding of the transformer and not to the other is called* **leakage flux.**

The primary ampere turns produce some flux ϕ_{l1} which is set up in air and links only with primary winding, as shown in Fig. 2.31(a), is called *primary leakage flux.*

Similarly, secondary ampere turns produce some flux ϕ_{l2} which is set up in air and links only with secondary winding called is *secondary leakage flux.*

The primary leakage flux ϕ_{l1} is proportional to the primary current I_1 and secondary leakage flux ϕ_{l2} is proportional to secondary current I_2. The primary leakage flux ϕ_{l1} produces self-inductance L_1 $(= N_1\phi_{l1}/I_1)$ which in turn produces leakage reactance $X_1(= 2\pi f L_1)$. Similarly, secondary leakage flux ϕ_{l2} produces leakage reactance X_2 $(= 2\pi f L_2)$. The leakage reactance (inductive) have been shown external to the windings in Fig. 2.31(b).

2.15 Equivalent Reactance

To make the calculations easy the reactances of the two winding can be transferred to any one side. The reactance from one side to the other is transferred in such a manner that percentage voltage drop remains the same when represented on either side.

Let the primary reactance X_1 be transferred to the secondary and the new value of this reactance is X_1' called equivalent reactance of primary referred to secondary, as shown in Fig. 2.32(*a*).

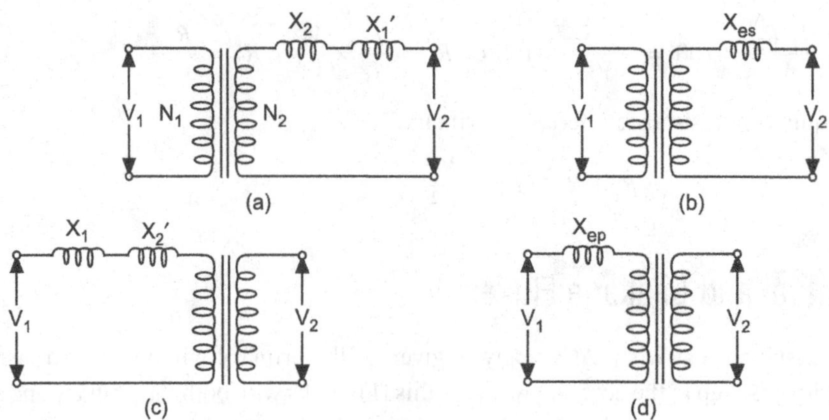

Fig. 2.32 Equivalent reactance referred to either side.

Then $$\frac{I_2 X_1'}{V_2} \times 100 = \frac{I_1 X_1}{V_1} \times 100 \text{ (\% voltage drops)}$$

or $$X_1' = \frac{I_1}{I_2} \times \frac{V_2}{V_1} \times X_1 = E^2 X_1 \qquad ...(2.36)$$

∴ Total equivalent reactance referred to secondary.

$$X_{es} = X_2 + X_1' = X_2 + K^2 X_1 \qquad ...(2.37)$$

Now, let us consider secondary reactance X_2 when it is transferred to primary side its new value is X_2' called equivalent reactance of secondary referred to primary, as shown in Fig. 2.32(*c*).

Then $$\frac{I_1 X_2'}{V_1} \times 100 = \frac{I_2 X_2}{V_2} \times 100$$

or $$X_2' = \frac{I_2}{I_1} \times \frac{V_1}{V_2} \times X_2 = \frac{X_2}{K^2} \qquad ...(2.38)$$

∴ Total equivalent reactance referred to primary.

$$X_{ep} = X_1 + X_2' = X_1 + \frac{X^2}{K^2} \qquad ...(2.39)$$

Example 2.16

A 63 kVA, 1100/220 V single-phase transformer has R_1 = 0.16 ohm, X_1 = 0.5 ohm, R_2 = 0.0064 ohm and X_2 = 0.02 ohm. Find equivalent resistance and reactance as referred to primary winding.

Solution:

Here, Transformer rating = 63 kVA; V_1 = 1100 V; V_2 = 220 V;

R_1 = 0.16 ohm; X_1 = 0.5 ohm; R_2 = 0.0064 ohm; X_2 = 0.02 ohm

Transformation ratio, $K = \frac{V_2}{V_1} = \frac{220}{1100} = 0.2$

Equivalent resistance referred to secondary side,

$$R_{es} = R_2 + R'_1 = R_2 + R_1 \times K^2 = 0.0064 + 0.16 \times (0.2)^2 = \mathbf{0.0128\ ohm}\ (Ans.)$$

Equivalent reactance referred to secondary side,

$$X_{es} = X_2 + X'_1 = X_2 + X_1 \times K^2 = 0.02 + 0.5 \times (0.2)^2 = \mathbf{0.04\ ohm}\ (Ans.)$$

Example 2.17

A 33 kVA, 2200/220V, 50Hz single phase transformer has the following parameters. Primary winding resistance $r_1 = 2.4\ \Omega$, Leakage reactance $x_1 = 6\ \Omega$ Secondary winding resistance $r_2 = 0.03\ \Omega$ Leakage reactance $x_2 = 0.07\ \Omega$. Then find Primary, Secondary and equivalent resistance and reactance.

Solution:

Here, Rating of transformer = 33 kVA; $V_1 = 2200$ V; $V_2 = 220$ V;

$$f = 50\text{ Hz};\ R_1 = 2.4\ \Omega;\ X_1 = 6\ \Omega;\ R_2 = 0.03\ \Omega;\ X_2 = 0.07\ \Omega$$

$$\text{Transformation ratio, } K = \frac{V_2}{V_1} = \frac{220}{2200} = 0.1$$

Transformer resistance referred to primary side;

$$R_{ep} = R_1 + R_2' = R_1 + \frac{R_2}{K^2} = 2.4 + \frac{0.03}{(0.1)^2} = 2.4 + 3 = \mathbf{5.4}\ \Omega\ (Ans.)$$

Transformer reactance referred to primary side;

$$X_{ep} = X_1 + X_2' = X_1 + \frac{X_2}{K^2} = 6 + \frac{0.07}{(0.1)^2} = 6 + 7 = \mathbf{13}\ \Omega\ (Ans.)$$

Transformer resistance referred to secondary side;

$$R_{es} = R_2 + R'_1 = R_2 + R_1 \times K^2 = 0.03 + 2.4(0.1)^2 = \mathbf{0.054}\ \Omega\ (Ans.)$$

Transformer reactance referred to secondary side;

$$X_{es} = X_2 + X_1' = X_2 + X_1 \times K^2 = 0.07 + 6 \times (0.1)^2 = \mathbf{0.13}\ \Omega\ (Ans.)$$

Example 2.18

A single phase transformer having voltage ratio 2500/250V (primary to secondary) has a primary resistance and reactance 1·8 ohm and 4·2 ohm, respectively. The corresponding secondary values are 0·02 and 0·045 ohm. Determine the total resistance and reactance referred to secondary side. Also calculate the impedance of transformer referred to secondary side.

Solution:

Here, $R_1 = 1{\cdot}8\ \Omega$; $X_1 = 4{\cdot}2\ \Omega$; $R_2 = 0{\cdot}02\ \Omega$; $X_2 = 0{\cdot}045\ \Omega$

$$\text{Transformation ratio, } K = \frac{V_2}{V_1} = \frac{250}{2500} = 0.1$$

Total resistance referred to secondary side,

$$R_{es} = R_2 + R_1' = R_2 + R_1 \times K^2 = 0{\cdot}02 + 1{\cdot}8 \times (0{\cdot}1)^2 = \mathbf{0{\cdot}038}\ \Omega\ (Ans.)$$

Total reactance referred to secondary side,

$$X_{es} = X_2 + X_1' = X_2 + X_1 \times K^2 = 0{\cdot}045 + 4{\cdot}2 \times (0{\cdot}1)^2 = \mathbf{0{\cdot}087}\ \Omega\ (Ans.)$$

Impedance of transformer referred to secondary side,

$$Z_{es} = \sqrt{(R_{es})^2 + (X_{es})^2} = \sqrt{(0\cdot 038)^2 + (0\cdot 087)^2}$$

$$= \mathbf{0{\cdot}095}\ \Omega\ (Ans.)$$

Section Practice Problems

Numerical Problems

1. The primary winding of a single-phase transformer having 350 turns is connected to 2.2 kV, 50 Hz supply. If its secondary winding consists of 38 turns, determine (*i*) the secondary voltage on no-load. (*ii*) The primary current when secondary current is 200 A at 0·8 p.f. lagging, if the no-load current is 5 A at 0·2 p.f. lagging. (*iii*) the p.f. of the primary current. (**Ans.** *238·8 V, 25·67 A, 0·7156 lag*)

2. The primary of a single phase, 2200/239V transformer is connected to 2.2 kV, 50 Hz supply, determine
   (*i*) the primary and secondary turns if the emf per turns 6.286 V.
   (*ii*) the primary current when the secondary current is 200 A at 0.8 p.f. lagging, if the no-load current is 5 A at 0.2 power factor lagging
   (*iii*) the power factor of the primary current. (**Ans.** *350, 38; 25.65 A; 0.715 lag*)

3. A single phase transformer takes a no-load current of 4A at a p.f. of 0.24 lagging. The ratio of turns in the primary to secondary is 4. Find the current taken by the transformer primary when the secondary supplies a load current of 240 A at a power factor 0.9 lagging. (**Ans.** *62.58A*)

4. A 5000/500 V, one-phase transformer has primary and secondary resistance of 0.2 ohm and 0.025 ohm and corresponding reactance of 4 ohm and 0.04 ohm, respectively. Determine: (*i*) Equivalent resistance and reactance of primary referred to secondary; (*ii*) Total resistance and reactance referred to secondary; (*iii*) Equivalent resistance and reactance of secondary referred to primary; (*iv*) Total resistance and reactance referred to primary.
   (**Ans.** *0·02* Ω; *0·04* Ω; *0·045* Ω; *0·08* Ω; *2·5* Ω; *4*Ω; *4·5* Ω; 8Ω)

5. A 2000/200 volt transformer has a primary resistance 2.3 ohm and reactance 4.2 ohm, the secondary resistance 0.025 ohm and reactance 0·04 ohm. Determine total resistance and reactance referred to primary side. (**Ans.** *4.8* Ω; *8.2* Ω)

6. A transformer has 500 and 100 turns on its primary and secondary, respectively. The primary and secondary resistances are 0·3 ohm and 0·1 ohm. The leakage reactances of the primary and secondary are 1·1 ohm and 0·035 ohm, respectively. Calculate the equivalent impedance referred to the primary circuit.
   (**Ans.** *3.426* Ω)

Short Answer Type Questions

Q.1. What do you know about reactance in a transformer?

Ans. The primary leakage flux linking with its own turns produce self inductance ($L_1 = N_1\phi/I_1$). The opposition offered by this inductance to the flow of primary current is called primary reactance.

Q.2. How leakage reactance of transformer can be reduced?

Ans. It can be reduced by reducing leakage flux i.e., by sandwiching the primary and secondary.

Q.3. When load current of a transformer increases, how does the input current adjust to meet the new conditions?

Ans. When load current of a transformer increases, its secondary ampere-turns (N_2I_2) set-up a flux in the core, and to counter balance this flux, primary draws extra current (I_1') from the mains so that ampere-turns are balanced i.e., $I_1'N_1 = I_2N_2$.

Q.4. How does leakage flux occur in a transformer?

Ans. When current flows through primary winding of a transformer, it produces magnetic flux. Most of the flux is set-up in the magnetic core, since it offers very low reluctance, and links with the secondary. But at the same time minute flux is set-up in air around the coil which does not link with the secondary. This magnetic flux is called primary leakage flux.

Similarly, there will be secondary leakage flux.

Q.5. Explain that "The main flux in a transformer remains practically invariable under all conditions of load."

Ans. When a transformer is connected to the supply at no-load a flux is set-up in the core called main flux (mutual flux).

When load is applied on the secondary, the secondary ampere-turns (N_2I_2) set-up a flux in the core in opposite direction to the main field which reduces the main field momentarily and hence reduces the self-induced emf E_1.

Instantly, primary draws extra current I_1' from the main to counter balance the secondary ampere-turns, such that $I_1N_1 = I_2N_2$.

Hence, the resultant flux in the core remains the same irrespective of the load.

Q.6. "The overall reactance of transformer decreases with load." Explain. **(Hint: $L = N\phi/I$)**

or

Why the transformers operate at poor *pf* when lightly loaded?

Ans. We know that total reactance of a transformer is $N\phi/I$, where I is the load current. Larger the load current, smaller will be the inductance or reactance ($2\pi fL$).

Hence, overall reactance of a transformer decreases with the increase in load.

That is why transformers operate at poor *pf* when lightly loaded.

Q.7. Does the flux in a transformer core increase with load?

Ans. No, flux in the core of a transformer remains the same from no-load to full-load.

2.16 Actual Transformer

An actual transformer has (*i*) Primary and secondary resistances R_1 and R_2, (*ii*) primary and secondary leakage reactances X_1 and X_2 (*iii*) iron and copper losses and (*iv*) exciting resistance R_0 and exciting reactance X_0. The equivalent circuit of an actual transformer is shown in Fig. 2.33.

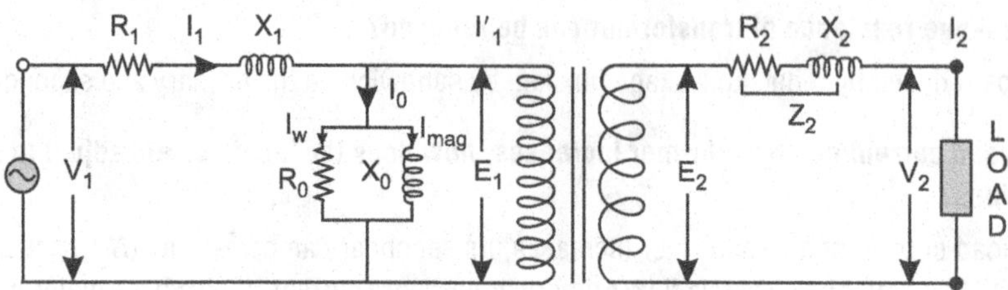

Fig. 2.33 Equivalent circuit for loaded transformer

Primary impedance,

$$\overline{Z_1} = R_1 + jX_1 \quad \text{...(2.40)}$$

Supply voltage is V_2. The resistance and leakage reactance of primary winding are responsible for some voltage drop in primary winding.

$$\therefore \quad \overline{V_1} = \overline{E_1} + \overline{I_1}(R_1 + jX_1) = \overline{E_1} + \overline{I_1}\,\overline{Z_1} \quad \text{...(2.41)}$$

where, $\overline{I_1} = \overline{I'}_1 + \overline{I_0}$

Secondary impedance, $\overline{Z_2} = R_2 + jX_2$ ...(2.42)

Similarly, the resistance and leakage reactance of secondary winding are responsible for some voltage drop in secondary winding. Hence,

$$\overline{V_2} = \overline{E_2} - \overline{I_2}(R_2 + jX_2) = \overline{E_2} - \overline{I_2}\,\overline{Z_2} \quad \text{...(2.43)}$$

The phasor (vector) diagrams of an actual transformer for resistive, inductive and capacitive loads are shown in Figs. 2.34(*a*), 2.34(*b*) and 2.34(*c*), respectively. The drops in resistances are drawn in phase with current vectors and drops in reactances are drawn perpendicular to the current vectors.

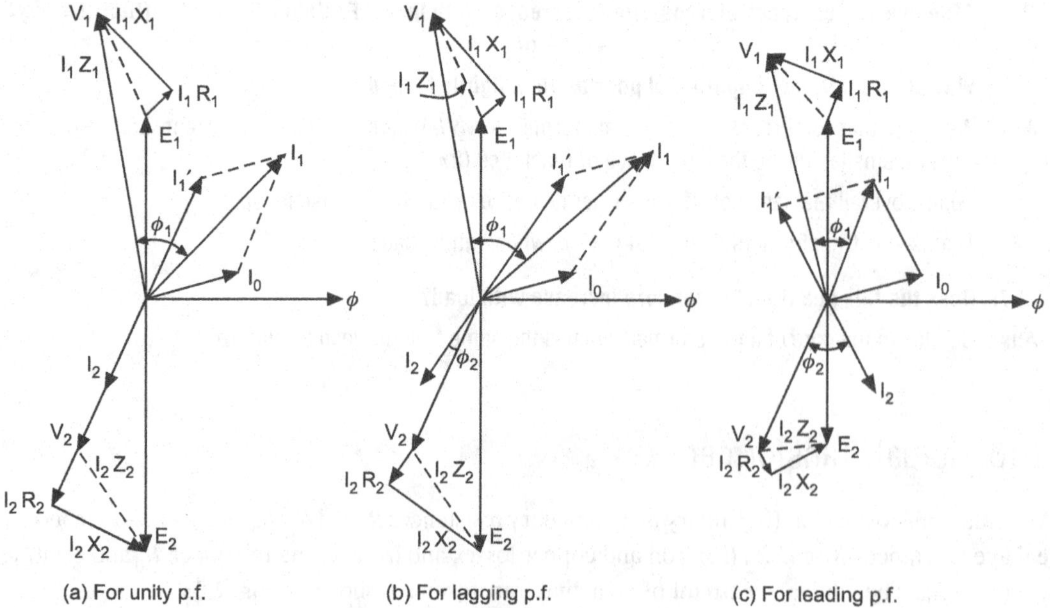

Fig. 2.34 Complete phasor diagram for loaded transformer

2.17 Simplified Equivalent Circuit

While drawing simplified circuit of a transformer, the exciting circuit (i.e., exciting resistance and exciting reactance) can be omitted.

The simplified equivalent circuit of a transformer is drawn by representing all the parameters of the transformer either on the secondary or on the primary side. The no-load current I_0 is neglected as its value is very small as compared to full load current, therefore, $I_1' = I_1$

(*i*) Equivalent circuit when all the quantities are referred to secondary.

The primary resistance when referred to secondary side, its value is $R_1' = K^2 R_1$ and the total or equivalent resistance of transformer referred to secondary, $R_{es} = R_2 + R_1'$. Similarly, the primary reactance when referred to secondary side, its value is $X_1' = K^2 X_1$ and the total or equivalent reactance of transformer referred to secondary, $X_{es} = X_2 + X_1'$. All the quantities when referred to the secondary side are shown in Fig. 2.35.

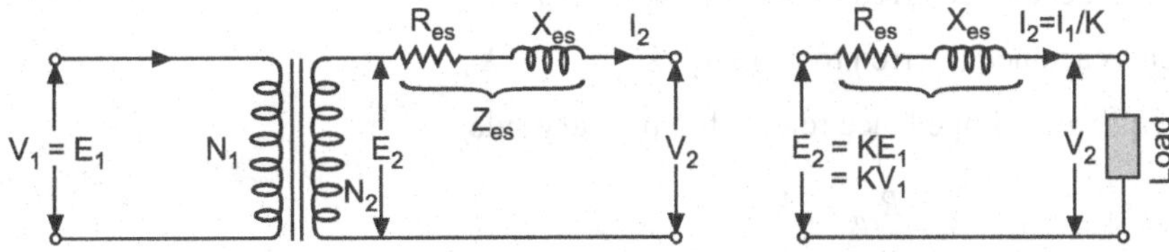

Fig. 2.35 Simplified equivalent circuit referred to secondary side

Total or equivalent impedance referred to secondary side,

$$Z_{es} = R_{es} + j\,X_{es} \qquad ...(2.44)$$

There is some voltage drop in resistance and reactance of transformer referred to secondary. Hence,

$$\overline{V_2} = \overline{E_2} - \overline{I_2}\left(R_{es} + j\,X_{es}\right) = \overline{E_2} - \overline{I_2 Z_{es}} \qquad ...(2.45)$$

Phasor Diagrams: The phasor (vector) diagrams of a loaded transformer when all the quantities are referred to secondary side for resistive, inductive and capacitive loads are shown in Figs. 2.36(*a*), 2.36(*b*) and 2.36(*c*), respectively. The voltage drops in resistances (vectors) are taken parallel to the current vector and the voltage drops in reactances (vectors) are taken quadrature to the current vector.

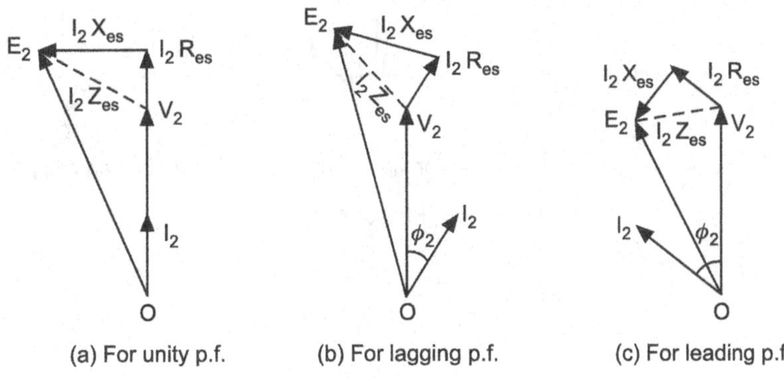

Fig. 2.36 Phasor diagram

(*ii*) Equivalent circuit when all the quantities are referred to primary.

In this case, to draw the equivalent circuit all the quantities are to be referred to primary, as shown in Fig. 2.37.

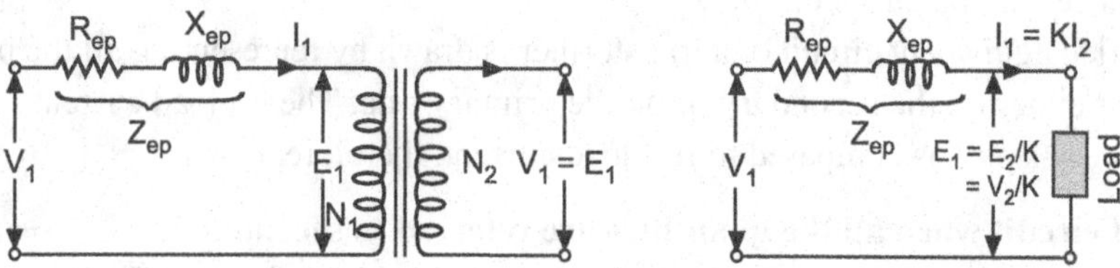

Fig. 2.37 Simplified equivalent circuit referred to primary.

Secondary resistance referred to primary, $R_2' = R_2/K^2$

Equivalent resistance referred to primary, $R_{ep} = R_1 + R_2'$

Secondary reactance referred to primary, $X_2' = X_1/K^2$

Equivalent reactance referred to primary, $X_{ep} = X_1 + X_2'$

Total or equivalent impedance referred to primary side,

$$Z_{ep} = R_{ep} + jX_{ep} \qquad ...(2.46)$$

There is some voltage drop in resistance and reactance of the transformer referred to primary side. Therefore,

$$\overline{V_1} = \overline{E_1} + \overline{I_1}\left(R_{ep} + jX_{ep}\right) = \overline{E_1} + \overline{I_1}\,\overline{Z_{ep}} \qquad ...(2.47)$$

Phasor diagrams: The phasor diagram to transformer when all the quantities are referred to primary side for different types of loads are shown in Fig. 2.38.

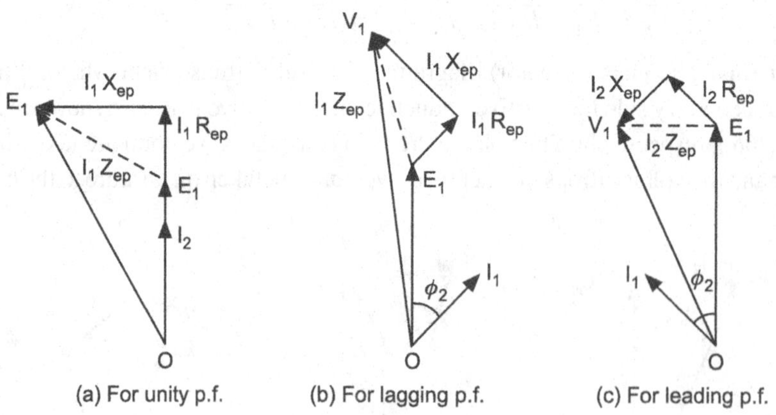

Fig. 2.38 Phasor diagram

2.18 Short Circuited Secondary of Transformer

If the secondary side of the transformer is short circuited, V_2 becomes zero which makes $E_1 = 0 \left(\text{because } E_1 = \frac{E_2}{K} = \frac{V_2}{K} \right)$. The equivalent circuit of the transformer, when all the quantities are referred to the primary side, with short circuited secondary is shown in Fig. 2.39(*a*). On short circuit the applied voltage V_1 is just utilised to circulate current $I_{1(sc)}$ in the impedance of the transformer referred to the primary side Z_{ep}. The phasor diagram of transformer under short circuit condition is shown in Fig. 2.39(*b*). The impedance of the transformer is quite small. Therefore, at normal voltage, the short circuit current $I_{1(sc)} = \frac{V_1}{Z_{ep}}$, drawn by the transformer, is very high as compared to the full load current. This current may damage the transformer. Power and distribution transformers are designed for high leakage reactance so that the short circuit current is limited between 8 times (for large transformer) to 30 times (for small transformers) of full load current.

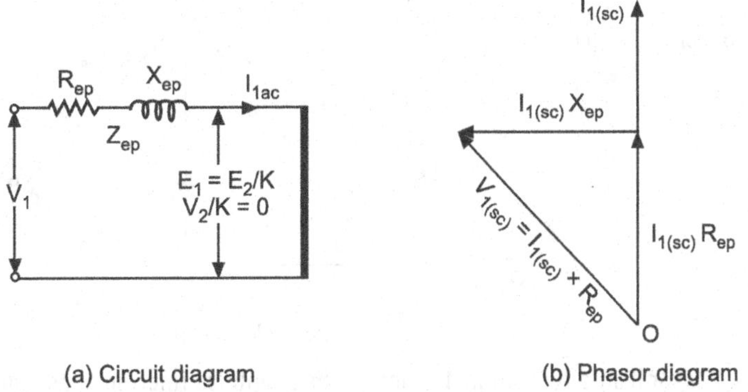

(a) Circuit diagram (b) Phasor diagram

Fig. 2.39 Transformer on short circuit.

However, if the transformer is short circuited on the secondary side intentionally to perform some test, a very low voltage $V_{1(sc)}$ (nearly 2 to 5% of its rates value) is applied at the primary terminals. Then,

$$Z_{ep} = \frac{V_{1(sc)}}{I_{1(sc)}} \qquad \ldots(2.48)$$

Power under short circuit condition

$$P_{sc} = I_{1(sc)}^2 \, R_{ep} \text{ or } R_{ep} = \frac{P_{sc}}{I_{1(sc)}^2} \qquad \ldots(2.49)$$

and

$$X_{ep} = \sqrt{\left(Z_{ep}^2\right) - \left(R_{ep}^2\right)} \qquad \ldots(2.50)$$

2.19 Expression for No-load Secondary Voltage

For a loaded transformer when all the quantities are referred to secondary side, its phasor diagram can be drawn as shown in Fig. 2.40.

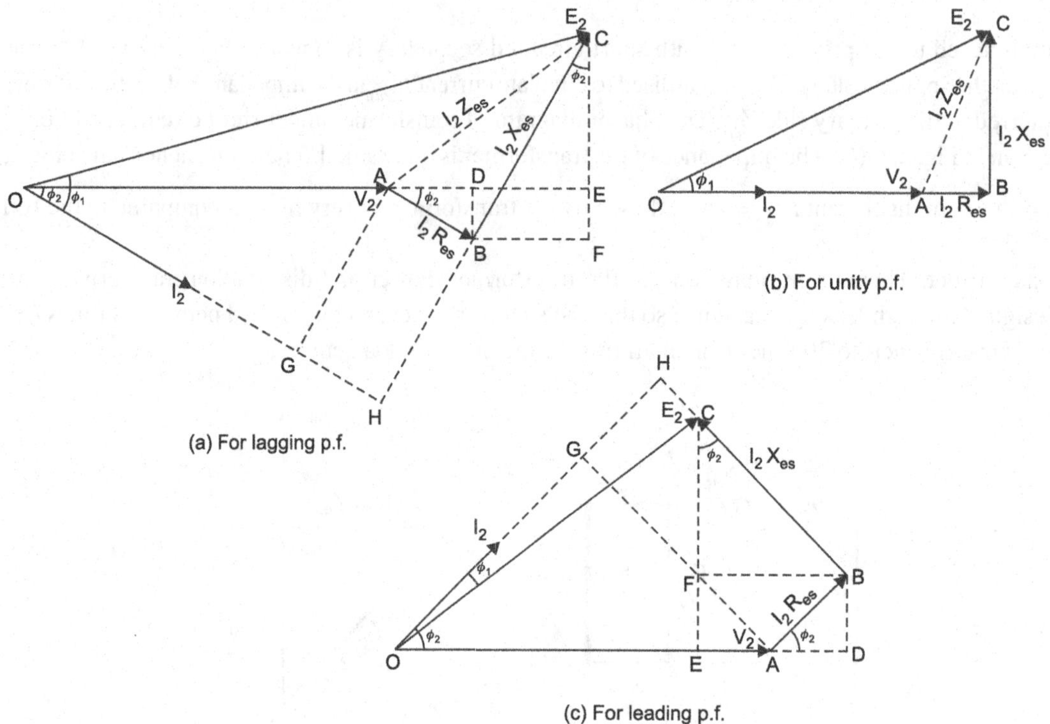

Fig. 2.40 Phasor diagram of a transformer, all the quantities referred to secondary side.

Complete the phasor diagram as shown in Fig. 2.40. From the phasor diagram we can derive the approximate as well as exact expressions for no-load secondary voltage.

(*i*) **Approximate expression;**

(*a*) *for lagging p.f.* (inductive load),

Consider right angle triangle *OEC* [see Fig. 2.40 (*a*)].

$$OC \cong OE = OA + AD + DE = OA + AD + BF$$

or

$$E_2 = V_2 + I_2 R_{es} \cos \phi_2 + I_2 X_{es} \sin \phi_2 \quad ...(2.51)$$

(*b*) *for unity p.f.* (resistive load),

Consider rt. ∠triangle OBC [see Fig. 2.40(*b*)]

$$OC \cong OB = OA + AB; \; E_2 = V_2 + I_2 R_{es} \quad ...(2.52)$$

(*c*) *for leading p.f.* (capacitive load),

Consider rt. ∠triangle *OEC* [see Fig. 2.40(c)]

$$OC \cong OE = OA + AD - DE = OA + AD - BF$$

or $$E_2 = V_2 + I_2 R_{es} \cos \phi_2 - I_2 X_{es} \sin \phi_2 \qquad \text{...(2.53)}$$

(*ii*) **Exact expression:**

(*a*) *for a lagging p.f.* (inductive load),

Consider rt. ∠triangle *OHC* [see Fig. 2.40(*a*)]

$$OC = \sqrt{(OH)^2 + (HC)^2} = \sqrt{(OG + GH)^2 + (HB + BC)^2}$$

$$= \sqrt{(OG + AB)^2 + (GA + BC)^2}$$

or $$E_2 = \sqrt{(V_2 \cos \phi_2 + I_2 R_{es})^2 + (V_2 \sin \phi_2 + I_2 X_{es})^2} \qquad \text{...(2.54)}$$

$$\text{Primary p.f., } \cos \phi_1 = \frac{OH}{OC} = \frac{OG + GH}{OC} = \frac{OG + AB}{OC} = \frac{V_2 \cos \phi_2 + I_2 R_{es}}{E_2} \qquad \text{...(2.55)}$$

(*b*) *for unity p.f.,* (resistive load)

Consider rt. ∠triangle *OBC* [see Fig. 2.40(*b*)]

$$OC = \sqrt{(OB)^2 + (BC)^2}$$

or $$OC = \sqrt{(OA + AB)^2 + (BC)^2}$$

or $$E = \sqrt{(V_2 + I_2 R_{es})^2 + (I_2 X_{es})^2} \qquad \text{...(2.56)}$$

$$\text{Primary p.f., } \cos \phi_1 = \frac{OB}{OC} = \frac{OA + AB}{OC} = \frac{V_2 + I_2 R_{es}}{E_2} \qquad \text{...(2.57)}$$

(*c*) *for leading p.f.* (capacitive load),

Consider rt. ∠triangle *OHC* [see Fig. 2.40(*c*)]

$$OC = \sqrt{(OH)^2 + (HC)^2}$$

$$= \sqrt{(OG + GH)^2 + (HB - BC)^2}$$

$$= \sqrt{(OG + AB)^2 + (GA - BC)^2}$$

or $$E_2 = \sqrt{(V_2 \cos\phi_2 + I_2 R_{es})^2 + (V_2 \sin\phi_2 - I_2 X_{es})^2} \qquad \text{...(2.58)}$$

$$\text{Primary p.f., } \cos \phi_1 = \frac{HC}{OC} = \frac{OG + GH}{OC} = \frac{OG + AB}{OC} = \frac{V_2 \cos\phi_2 + I_2 R_{es}}{E_2} \qquad \text{...(2.59)}$$

2.20 Voltage Regulation

When a transformer is loaded, with a constant supply voltage, the terminal voltage changes due to voltage drop in the internal parameters of the transformer i.e., primary and secondary resistances and inductive reactances. The voltage drop at the terminals also depends upon the load and its power factor. The change in terminal voltage from no-load to full-load at constant supply voltage with respect to no-load voltage is known as voltage regulation of the transformer.

Let, E_2 = Secondary terminal voltage at no-load.

V_2 = Secondary terminal voltage at full-load.

$$\text{Then, voltage regulation} = \frac{E_2 - V_2}{E_2} \text{ (per unit)} \quad ...(2.60)$$

$$\text{In the form of percentage, \% Reg} = \frac{E_2 - V_2}{E_2} \times 100 \quad ...(2.61)$$

When all the quantities are referred to the primary side of the transformer;

$$\text{\% Reg} = \frac{V_1 - E_1}{V_1} \times 100 \quad ...(2.62)$$

2.21 Approximate Expression for Voltage Regulation

The approximate expression for the no-load secondary voltage is

(*i*) *For inductive load*: $E_2 = V_2 + I_2 R_{es} \cos \phi_2 + I_2 X_{es} \sin \phi_2$

or $$E_2 - V_2 = I_2 R_{es} \cos \phi_2 + I_2 X_{es} \sin \phi_2$$

or $$\frac{E_2 - V_2}{E_2} \times 100 = \frac{I_2 R_{es}}{E_2} \times 100 \cos \phi_2 + \frac{I_2 - X_{es}}{E_2} \times 100 \sin \phi_2$$

where, $\frac{I_2 X_{es}}{E_2} \times 100$ = percentage resistance drop and

$\frac{I_2 X_{es}}{E_2} \times 100$ = percentage reactance drop

$$\therefore \quad \text{\% Reg} = \text{\% resistance drop} \times \cos \phi_2 + \text{\% reactance drop} \times \sin \phi_2 \quad ...(2.63)$$

Similarly

(*ii*) *For resistive load*: % Reg = % resistance drop ...(2.64)

(*iii*) *For capacitive load*:

$$\therefore \quad \text{\% Reg} = \text{\% resistance drop} \times \cos \phi_2 - \text{\% reactance drop} \times \sin \phi_2 \quad ...(2.65)$$

Condition for Zero Regulation

From the above expression, we can derive the condition at which the regulation of a transformer becomes zero, it means when the load is thrown off, the terminal voltage remains the same, i.e.,

$$0 = \text{\% resistance drop} \times \cos \phi_2 - \text{\% reactance drop} \times \sin \phi_2$$

or % reactance drop × sin ϕ_2 = % resistance drop × cos ϕ_2

$$\frac{I_2 X_{es}}{V_2} \times 100 \times \sin\phi_2 = \frac{I_2 R_{es}}{V_2} \times 100 \times \cos\phi_2$$

$$\tan\phi_2 = \frac{R_{es}}{X_{es}};\ \phi_2 = \tan^{-1}\frac{R_{es}}{X_{es}}$$

$$\text{Load } pf,\ \cos\phi_2 = \cos\tan^{-1}\frac{R_{es}}{X_{es}} \qquad \ldots(2.66)$$

The above expression reveals that the regulation of a transformer will become zero only at leading *pf* of the load that too when $\cos\phi_2 = \cos\tan^{-1}\frac{R_{es}}{X_{es}}$

Condition for Maximum Regulation

Although, this condition is never suggested for any transformer, but let us see under what condition it may occur.

Regulation will be maximum if $\frac{d}{d\phi_2}$ (regulation) = 0.

i.e.,

$$\frac{d}{d\phi_2} = \frac{I_2 R_{es}\cos\phi_2 + I_2 X_{es}\sin\phi_2}{E_2} = 0$$

or

$$-\frac{I_2 R_{es}}{E_2}\sin\phi_2 + \frac{I_2 X_{es}}{E_2}\cos\phi_2 = 0$$

or

$$\tan\phi_2 = \frac{X_{es}}{R_{es}} \text{ or } \phi_2 = \tan^{-1}\frac{X_{es}}{R_{es}}$$

This condition will occur at a lagging *pf* that too when *pf*, $\cos\phi_2 = \cos\tan^{-1}\frac{X_{es}}{R_{es}}$ ...(2.67)

2.22 Kapp Regulation Diagram

It is observed that the secondary terminal voltage of a transformer changes when the load on it changes. The change in terminal voltage (rise of fall) does not only depend upon the magnitude of load but it also depends upon pf of the load. For finding the voltage drop (or rise) which is further used to determine the regulation of the transformer, a graphical construction is employed called Kapp regulation diagram as proposed by late Dr. Kapp.

For drawing Kapp regulation diagram, it is necessary to know the equivalent resistance and reactance as referred to secondary i.e., R_{-es} and X_{es}. If I_2 is the secondary load current then, secondary terminal voltage on load, V_2, is obtained by subtracting I_2R_{es} and I_2X_{es} voltage drops vectorially from secondary no-load, E_2.

Now, E_2 is constant, hence it can be represented by a circle of constant radius *OC* as shown in Fig. 2.41. This circle is known as No-load or open-circuit emf circle. For a given load OI_2 represents the load current and is taken as the reference vector. *AB* represents I_2R_{es} and is parallel to vector OI_2, *BC* represents I_2X_{es} and is drawn at right angles to *AB*. Vector *OA* obviously represents secondary terminal voltage V_2. Since I_2 is constant, the drop triangle *ABC* remains constant in size. It is seen that end point A of V_2 lies on another circle whose centre is O'. This point O' lies at a distance of I_2X_{es} vertically below the point O and a distance of I_2R_{es} to its left as shown in Fig. 2.41.

i.e., $O'B' = AB$ and $OB' = BC$

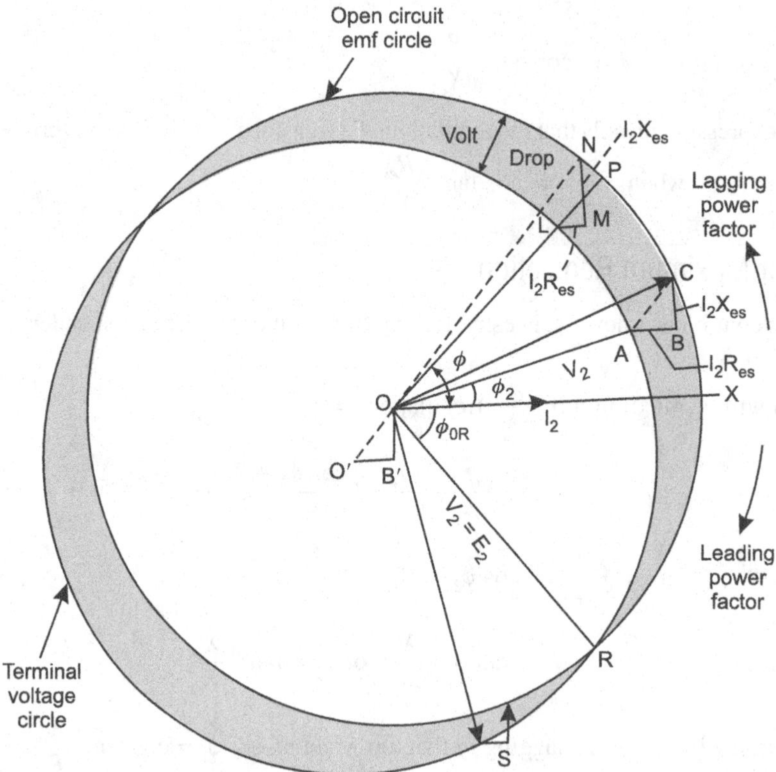

Fig. 2.41 Kapp regulation diagram of a transformer

To determine the voltage drop at full-load lagging pf cos ϕ, radius OLP inclined at an angle of ϕ with OX. $LM = I_2R_{es}$ and is drawn horizontal. $MN = I_2X_{es}$ and is drawn perpendicular to LM. Obviously ON is No-load voltage E_2. Now $ON = OP = E_2$. Similarly, OL is V_2. The voltage drop $= OP - OL = LP$.

$$\text{Hence percentage regulation} = \frac{OP - OL}{OP} \times 100 = \frac{LP}{OP} \times 100$$

It is seen that for finding voltage drop, the drop triangle LMN need not be drawn, but simply the radius OLP is sufficient to determine regulation.

The diagram shows clearly how the secondary terminal voltages falls as the angle of lag increases. Conversely, for a leading power factor, the fall in secondary terminal voltage decreases till for an angle of ϕ_{OR} leading the fall becomes zero; hence $V_2 = E_2$. For angles greater than $\phi_{OR,}$ the secondary terminal voltage V_2 (i.e., OS) becomes greater than E_2 and voltage regulation comes out to be negative.

The Kapp diagram is very useful in determining the variation of regulation with power factor. But it is not so accurate since the lengths of the sides of the impedance triangle are very small as compared to the radii of the circles. To obtain desired accuracy, the diagram has to be drawn on a sufficiently large scale.

Example 2.19

The primary winding resistance and reactance of a 10 kVA, 2000/400 V, single phase transformer is 5.5 ohm and 12 ohm, respectively, the corresponding values for secondary are 0.2 ohm and 0.45 ohm.

Determine the value of the secondary voltage at full load, 0.8 p.f. lagging, when the primary supply voltage is 2000 V.

Solution:

Transformer rating = 10 kVA = 10×10^3 VA

Primary induced voltage, $E_1 = 2000$ V;

Secondary induced voltage, $E_2 = 400$ V

Primary resistance, $R_1 = 5.5$ ohm;

Primary reactance, $X_1 = 12$ ohm

Secondary resistance, $R_2 = 0.2$ ohm; Secondary reactance,

$X_2 = 0.45$ ohm

Load p.f., $\cos \phi_2 = 0.8$ lagging

Transformation ratio, $K = \frac{E_2}{E_1} = \frac{400}{2000} = 0.2$

Primary resistance referred to secondary side,

$$R_1' = K^2 R_1 = (0.2)^2 \times 5.5 = 0.22\ \Omega$$

Equivalent resistance referred to secondary side,

$$R_{es} = R_2 + R_1' = 0.2 + 0.22 = 0.42\ \Omega$$

Primary reactance referred to secondary side,

$$X_1' = K^2 X_1 = (0.2)^2 \times 12 = 0.48\ \Omega$$

Equivalent reactance referred to secondary side,

$$X_{es} = X_2 + X_1' = 0.45 + 0.48 = 0.93\ \Omega$$

Load p.f., $\cos \phi_2 = 0.8$

∴ $\sin \phi_2 = \sin \cos^{-1} 0.8 = 0.6$

Full load secondary current, $I_2 = \frac{10 \times 10^3}{400} = 25$ A

As the primary supply voltage, $V_1 = E_1 = 2000$ V

Secondary induced voltage, $E_2 = KE_1 = 0.2 \times 2000 = 400$ V

Using the expression; $E_2 = V_2 + I_2 R_{es} \cos \phi_2 + I_2 X_{es} \sin \phi_2$

Secondary terminal voltage, $V_2 = E_2 - I_2 R_{es} \cos \phi_2 - I_2 X_{es} \sin \phi_2$

$= 400 - 25 \times 0.42 \times 0.8 - 25 \times 0.93 \times 0.6$

$= 400 - 8.4 - 13.95 =$ **377.65 V** (*Ans.*)

Example 2.20

A 15 kVA. 2300/230 V, single phase transformer has 2.5 ohm resistance and 10 ohm reactance for primary and 0.02 ohm resistance and 0.09 ohm reactance for the secondary. If the secondary terminal voltage is to be maintained at 230 V and 0.8 p.f. lagging at full load current, what should be the primary voltage?

Solution:

Transformer output = 15 kVA = 15×10^3 VA

Primary resistance, $R_1 = 2.5\ \Omega$

Primary reactance, $X_1 = 10\ \Omega$

Secondary resistance, $R_2 = 0.02\ \Omega$

Secondary reluctance, $X_2 = 0.09\ \Omega$

Secondary terminal voltage, $V_2 = 230$ V

Load power factor, $\cos \phi_2 = 0.8$ lag

Transformation ratio, $K = \dfrac{230}{2300} = 0.1$

Secondary resistance referred to primary, $R_2' = \dfrac{R^2}{K^2} = \dfrac{0.02}{(0.1)^2} = 2\Omega$

Secondary reactance referred to primary, $X_2' = \dfrac{X^2}{E^2} = \dfrac{0.09}{(0.1)^2} = 9\Omega$

Total resistance referred to primary, $R_{ep} = R_1 + R_2' = 2{\cdot}5 + 2 = 4{\cdot}5\ \Omega$

Total reactance referred to primary, $X_{ep} = X_1 + X_2' = 10 + 9 = 19\ \Omega$

Load p.f., $\cos \phi_2 = 0.8$

$\therefore$ $\sin \phi_2 = \sin \cos^{-1} 0.8 = 0{\cdot}6$

Full load primary current, $I_1 = \dfrac{15\times10^3}{2300} = 6.52\,\text{A}$

Primary induced voltage, $E_1 = \dfrac{E_2}{E} = \dfrac{V_2}{E} = \dfrac{230}{0.1} = 2300\,V$

Using the expression $V_1 = E_1 + I_1 R_{ep} \cos \phi_2 + I_1 X_{ep} \sin \phi_2$ (approx.)

$\therefore$ Primary applied voltage, $V_1 = 2300 + 6{\cdot}52 \times 4.5 \times 0.8 + 6{\cdot}52 \times 19 \times 0{\cdot}6$

$= 2300 + 23{\cdot}472 + 74{\cdot}328 =$ **2397·8 V** (*Ans.*)

Example 2.21

The turns ratio of 212 kVA single-phase transformer is 8. The resistance and reactance of the primary is 0.85 ohm and 4.8 ohm, respectively and the corresponding values for the secondary are 0.012 ohm and 0.07 ohm, respectively. Determine the voltage to be applied to the primary to obtain a current of 150 A in the secondary when the secondary terminal are short circuited. Ignore the magnetising current.

Solution:

Ratio of turns, $\frac{N_1}{N_2} = 8$

Primary resistance, $R_1 = 0{\cdot}85$;

Secondary resistance $R_2 = 0{\cdot}012\ \Omega$

Primary reactance, $X_1 = 4{\cdot}8\ \Omega$;

Secondary reactance, $X_2 = 0{\cdot}07\ \Omega$

Transformation ratio, $K = \frac{N_2}{N_1} = \frac{1}{8}$

Secondary resistance referred to primary, $R_2' = \frac{R^2}{K^2} = 0.012 \times 8 \times 8 = 0.768\ \Omega$

Equivalent resistance referred to primary $R_{ep} = R_1 + R_2' = 0{\cdot}85 + 0{\cdot}768 = 1{\cdot}618\ \Omega$

Secondary reactance referred to primary, $X_2' = \frac{X_2}{K^2} = 0.07 \times 8 \times 8 = 4{\cdot}48\ \Omega$

Equivalent reactance referred to primary, $X_{ep} = X_1 + X_2' = 4{\cdot}8 + 4{\cdot}48 = 9{\cdot}28\ \Omega$

Equivalent impedance referred to primary, $Z_{ep} = \sqrt{R_{ep}^2 + X_{ep}^2} = \sqrt{(1.618)^2 + (9.28)^2}$

$= \mathbf{9.42\ \Omega}$

Short circuit current referred to primary, $I_{1(sc)} = K\, I_{2(sc)} = \frac{1}{8} \times 150 = 18 \cdot 75\,\text{A}$

Voltage applied to the primary under short circuit condition,

$$V_{1(sc)} = I_{1(sc)} \times Z_{ep} = 18.75 \times 9.42$$

$$= \mathbf{176{\cdot}625\ V}\ (Ans.)$$

Example 2.22

A 40 kVA, 6600/250 V, 50 Hz transformer is having total reactance of 35 ohm when referred to primary side whereas its primary and secondary winding resistance is 10 ohm and 0.02 ohm, respectively.

Find full load regulation of at a p.f. 0.8 lagging.

Solution:

Rating of transformer, = 40 kVA = 40×10^3 VA

Transformation ratio, $K = \frac{250}{6600} = 0.03788$

Primary resistance, $R_1 = 10\ \Omega$

Secondary resistance, $R_2 = 0.02\ \Omega$

Total resistance, referred to primary side,

$$R_{ep} = R_1 + R_2^{'} = R_1 + \frac{R_2}{K^2} = 10 + \frac{0.02}{(0.03788)^2} = 23.94\ \Omega$$

Total reactance referred to primary side,

$$X_{ep} = 35\ \Omega$$

$$V_1 = \sqrt{(E_1 \cos\phi + I_1 R_{ep})^2 + (E_1 \sin\phi + I_1 X_{ep})^2}$$

where, $I_1 = \frac{40 \times 10^3}{6600} = 6.06$ A

$$\cos\phi = 0.8;\ \sin\phi = \sin\cos^{-1} 0.8 = 0.6$$

$\therefore$

$$V_1 = \sqrt{(6600 \times 0.8 + 6.06 \times 23.94)^2 + (6600 \times 0.6 + 6.06 \times 35)^2}$$

$$= 6843.7\ \text{V}$$

$$\%\ Reg = \frac{V_1 - E_1}{V_1} \times 100 = \frac{6843.7 - 6600}{6843.7} \times 100 = 3.56\%\ (Ans.)$$

Example 2.23

A 75 kVA single phase transformer, 6600/230 V, requires 310 V across the primary to the primary to circulate full load current on short circuit, the power absorbed being 1·6 kW. Determine the voltage regulation and the secondary terminal voltage for half full load, 0·8 p.f. lagging.

Solution:

Transformer output = 75 *kVA* = 75 × 10^3 VA

Primary induced voltage, $E_1 = 6600$ V

Secondary induced voltage, $E_2 = 230$ V

At short circuit, primary voltage, $V_{1(sc)} = 310$ V

At short circuit, power absorbed, $P_{(sc)} = 1{\cdot}6\ \text{kW} = 1{\cdot}6 \times 10^3$ W

Load p.f. $\cos\phi_2 = 0{\cdot}8$ lagging

Primary current at full load, $I_1 = \frac{75 \times 10^3}{6600} = 11.36\ A$

Primary current at short circuit $I_{1(SC)} = I_1 = 11{\cdot}36$ A

Equivalent resistance referred to primary, $R_{ep} = \frac{P_{sc}}{(I_{1sc})^2} = \frac{1.6 \times 10^3}{(11.36)^2} = 12.39\,\Omega$

Equivalent impedance referred to primary, $Z_{ep} = \frac{V_{1sc}}{I_{1sc}} = \frac{310}{11.36} = 27.29\,\Omega$

Equivalent reactance referred to primary, $X_{ep} = \sqrt{Z_{ep}^2 - R_{ep}^2} = \sqrt{(27 \cdot 29)^2 - (12 \cdot 39)^2}$

$$= 24{\cdot}32\ \Omega$$

Transformation ratio, $K = \dfrac{230}{6600}$

Equivalent resistance referred to secondary.

$$R_{es} = K^2 R_{ep} = \frac{230 \times 230 \times 12.39}{6600 \times 6600} = 0.015\Omega$$

Secondary current at full load, $I_2 = \dfrac{75 \times 10^3}{230} = 326\,A$

Secondary current at half load, $I_{2hl} = \dfrac{I_2}{2} = \dfrac{326}{2} = 163\,A$

Load p.f., $\cos\phi_2 = 0{\cdot}8\ lag$

$\therefore$ $\sin\phi_2 = \sin\cos^{-1} 0{\cdot}8 = 0{\cdot}6$

Secondary terminal voltage at half full load,

$$V_2 = E_2 - I_{2hl} R_{es} \cos\phi_2 - I_{2hl} X_{es} \sin\phi_2$$

$$= 230 - 163 \times 0{\cdot}015 \times 0{\cdot}8 - 163 \times 0{\cdot}0295 \times 0{\cdot}6$$

$$= \mathbf{225{\cdot}16\ V}\ (Ans.)$$

$$\text{Voltage regulation} = \frac{E_2 - V_2}{E_2} \times 100 = \frac{230 - 225.16}{230} \times 100$$

$$= \mathbf{2{\cdot}1\ \%}\ (Ans.)$$

Example 2.24

A 20 kVA, 2500/500 V, single phase transformer has the following parameters:

H.V. winding	*L.V. winding*
$r_i = 8\,\Omega$	$r_2 = 0.3\,\Omega$
$x_1 = 17\,\Omega$	$x_2 = 0.7\,\Omega$

Find the voltage regulation and secondary terminal voltage at full load for a power factor of

(a) 0.8 lagging *(b) 0.8 leading*

The primary voltage is held constant at 2500 V.

Solution:

Here, Rating of transformer = 20 kVA; $E_1 = 2500$ V; $E_2 = 500$ V

$$R_1 = 8\,\Omega;\ X_1 = 17\,\Omega;\ R_2 = 0.3\,\Omega;\ X_2 = 0.7\,\Omega$$

Transformer resistance referred to *LV* (secondary) side;

$$R_{es} = R_2 + R_1 \times K^2 = 0.3 + 8 \times \left(\frac{500}{2500}\right)^2 = 0.62\,\Omega$$

$$X_{es} = X_2 + X_1 \times K^2 = 0.7 + 17 \times (0.2)^2 = 2.38\,\Omega$$

Secondary full-load current, $I_2 = \dfrac{20 \times 1000}{500} = 40\text{ A}$

For p.f. 0.8 lagging, cos $\phi = 0.8$; sin ϕ = sin $\cos^{-1} 0.8 = 0.6$

$$V_2 = E_2 - I_2 R_{es} \cos \phi - I_2 X_{es} \sin \phi$$

$$= 500 - 40 \times 0.62 \times 0.8 - 40 \times 2.38 \times 0.6 = \mathbf{447.04\ V}\ (Ans.)$$

$$\% \ Reg = \frac{E_2 - V_2}{E_2} \times 100 = \frac{500 - 447.04}{500} \times 100 = \mathbf{10.59\%}\ (Ans.)$$

For p.f. 0.8 leading, cos $\phi = 0.8$; sin ϕ = sin $\cos^{-1} 0.8 = 0.6$

$$V_2 = E_2 - I_2 Res \cos \phi + I_2 X_{es} \sin \phi$$

$$= 500 - 40 \times 0.62 \times 0.8 + 40 \times 2.38 \times 0.6$$

$$= \mathbf{513.28\ V}\ (Ans.)$$

$$\% \ Reg = \frac{E_2 - V_2}{E_2} \times 100 = \frac{500 - 513.28}{500} \times 100 = \mathbf{-2.6\%}\ (Ans.)$$

Example 2.25

A 10 kVA, 500/100 V transformer has the following circuit parameters referred to primary: Equivalent resistance, R_{eq} = 0·3 ohm; Equivalent reactance, X_{eq} = 5·2 ohm. When supplying power to a lagging load, the current, power and voltage measured on primary side were 20 A, 8 kW and 500 V, respectively. Calculate the voltage on the secondary terminals under these conditions. Draw the relevant phasor diagram.

Solution:

Rating of transformer = 10 kVA = 10×10^3 VA

Power factor on primary side,

$$\cos \phi_1 = \frac{P}{V_1 I_1} = \frac{8 \times 10^3}{500 \times 20} = 0.8 \text{ kg}$$

$$\therefore \quad \sin \phi_1 = \sin \cos^{-1} 0{\cdot}8 = 0{\cdot}6$$

Primary induced emf, $E_1 = V_1 - I_1 R_{ep} \cos \phi_1 - I_1 X_{ep} \sin \phi_1$

$$= \sin 500 - 20 \times 0{\cdot}3 \times 0{\cdot}8 - 20 \times 5{\cdot}2 \times 0{\cdot}6 = 432{\cdot}8 \text{ V}$$

Transformation ratio, $K = \dfrac{100}{500} = 0 \cdot 2$

Secondary induced emf, $E_2 = E_1 \times K = 432{\cdot}8 \times 0{\cdot}2 = 86{\cdot}56$ V

Secondary terminal voltage on load,

$$V_2 = E_2 = 86{\cdot}56\ V \quad \text{(since all the parameters are referred to primary side)}$$

The equivalent circuit and phasor diagram is shown in Fig. 2.42(*b*).

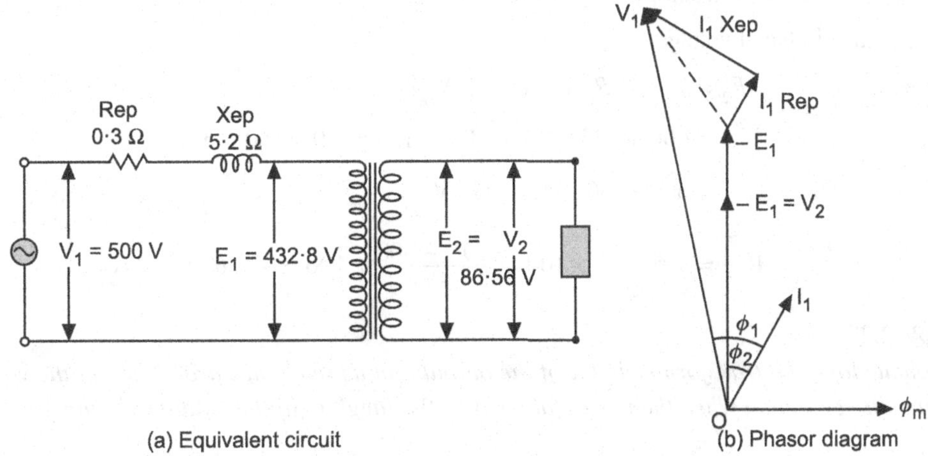

Fig. 2.42 Transformer on load.

Example 2.26

A single phase transformer with a ratio 5: 1 has primary resistance of 0·4 ohm and reactance of 1·2 ohm and the secondary resistance of 0·01 and reactance of 0·04 ohm. Determine the percentage regulation when delivering 125 A at 600 V at (i) 0·8 p.f. lagging (ii) 0·8 p.f. leading.

Solution:

Data given, $R_1 = 0{\cdot}4\ \Omega; X_1 = 1{\cdot}2\ \Omega; R_2 = 0{\cdot}01\ \Omega;$

$$X_2 = 0{\cdot}04\ \Omega;\ I_2 = 125\text{ A and } V_2 = 600\text{ V}$$

Transformation ratio, $K = 1/5 = 0{\cdot}2$

Now,

$$R_1' = K^2 R_1 = (0{\cdot}2)^2 \times 0{\cdot}4 = 0{\cdot}016\ \Omega;$$

$$R_{es} = R_2 + R_1' = 0{\cdot}01 + 0{\cdot}016 = 0{\cdot}026\ \Omega$$

$$X_1' = K^2 X_1 = (0{\cdot}2)^2 \times 1{\cdot}2 = 0{\cdot}048\ \Omega$$

$$X_{es} = X_2 + X_1' = 0{\cdot}04 + 0{\cdot}048 = 0{\cdot}088\ \Omega$$

(*i*) For p.f. $\cos\phi_2 = 0{\cdot}8$ lag; $\sin\phi_2 = \sin\cos^{-1} 0{\cdot}8 = 0{\cdot}6$

Secondary induced voltage,

$$E_2 = V_2 + I_2 R_{es}\cos\phi_2 + I_2 X_{es}\sin\phi_2$$

$$= 600 + 125 \times 0{\cdot}026 \times 0{\cdot}8 + 125 \times 0{\cdot}088 \times 0{\cdot}6$$

$$= 600 + 2{\cdot}6 + 6{\cdot}6 = 609{\cdot}2\text{ V}$$

$$\%\ Reg = \frac{E_2 - V_2}{E_2} \times 100 = \frac{609.2 - 600}{609.2} \times 100 = \mathbf{1{\cdot}51\ \%}\ (Ans.)$$

(*ii*) For p.f., $\cos\phi_2 = 0{\cdot}8$ leading; $\sin\phi_2 = 0{\cdot}6$

Secondary induced voltage,

$$E_2 = V_2 + I_2 R_{es} \cos\phi_2 - I_2 X_{es} \sin\phi_2$$

$$= 600 + 125 \times 0{\cdot}06 \times 0{\cdot}8 - 125 \times 0{\cdot}088 \times 0{\cdot}6$$

$$= 600 + 2{\cdot}6 - 6{\cdot}6 = 596 \text{ V}$$

$$\% \text{ Reg} = \frac{E_2 - V_2}{E_2} \times 100 = \frac{596 - 600}{596} \times 100 = \mathbf{-0{\cdot}67\ \%}\ (Ans.)$$

Example 2.27

If the ohmic loss of a transformer is 1% of the output and its reactance drop is 5% of the voltage, determine its regulation when the power factor is (i) 0·8 lagging (ii) 0·8 leading (iii) unity.

Solution:

Ohmic loss or resistance drop = 1%; Reactance drop = 5%

(*i*) When p.f., $\cos\phi_2 = 0{\cdot}8$ lagging; $\sin\phi_2 = \sin\cos^{-1} 0{\cdot}8 = 0{\cdot}6$

$$\% \text{ Reg} = \% \text{ resistance drop} \times \cos\phi_2 + \% \text{ reactance drop} \times \sin\phi_2$$

$$= 1 \times 0{\cdot}8 + 5 \times 0{\cdot}6 = \mathbf{3{\cdot}8\%}\ (Ans.)$$

(*ii*) When p.f., $\cos\phi_2 = 0{\cdot}8$ leading; $\sin\phi_2 = \sin\cos^{-1} 0{\cdot}8 = 0{\cdot}6$

$$\% \text{ Reg} = \% \text{ resistance drop} \times \cos\phi_2 - \% \text{ reactance drop} \times \sin\phi_2$$

$$= 1 \times 0{\cdot}8 - 5 \times 0{\cdot}6 = \mathbf{-2{\cdot}2\ \%}\ (Ans.)$$

(*iii*) When p.f. is unity % Reg = % resistance drop = **1%** (*Ans.*)

Section Practice Problems

Numerical Problems

1. A single phase transformer with a ratio 1: 2 has primary and secondary winding resistance of 0.25 ohm and 0.8 ohm, respectively and their reactances are 0.5 ohm and 2.8 ohm. Determine the no-load secondary terminal voltage of the transformer if it is delivering 10 A and 400 V at 0.8 p.f. lagging. (**Ans.** *437.64 V*)
2. A 10 kVA, 2000/400 volt single-phase transformer has the following resistances and reactances.

   Primary winding: resistance 5.0 ohm, leakage reactance 12 ohm.

   Secondary winding: resistance 0.2 ohm, leakage reactance 0.48 ohm. Determine the secondary terminal voltage at full load, 0.8 power factor lagging when the primary supply is 2000 V. (**Ans.** *377.1 V*)
3. A 230/440 V, single-phase transformer has primary and secondary winding resistance of 0.25 ohm and 0.8 ohm, respectively and corresponding reactance of 0.6 ohm and 2.8 ohm. Find the approximate secondary terminal voltage when supplying (*i*) 10 A at 0.707 p.f. lagging (*ii*) 10A at 0.707 p.f. leading.

   (**Ans.** *397.5 V; 437 V*)

4. The turn ratio of a 100 kVA transformer is 5. Its primary has a resistance and reactance of 0.3 ohm and 1.1 ohm, respectively and the corresponding values for the secondary are 0.01 ohm and 0.035 ohm. The supply voltage is 2200 V. Calculate

   (*i*) total impedance of the transformer referred to primary circuit and

   (*ii*) the voltage regulation and the secondary terminal voltage for full load having power factors (*i*) 0.8 lagging and (*ii*) 0.8 leading. (**Ans.** *2.05 Ω; 3.364%; 425.2 V; –1.54%; 446.78 V*)

5. A 17.5 kVA 66/11 kV transformer has 10% resistance and 3% leakage reactance drop. Find resistance and reactance of the transformer in ohms as referred to

   (*i*) the high voltage winding.

   (*ii*) the low voltage winding. (**Ans.** *2.52 Ω; 7.56 Ω; 0.068 Ω; 0.209 Ω*)

Short Answer Type Questions

Q.1. What is voltage regulation of a transformer?

Ans. At constant supply voltage, the change in secondary terminal voltage from no-load to full-load is called voltage regulation. It is expressed as percentage (or per unit) of no-load voltage.

$$\% \text{ regulation} = \frac{E_2 - V_2}{E_2} \times 100$$

Q.2. Why does voltage drop in a transformer?

Ans. In a transformer, voltage drop occurs due to

(*i*) Resistance of primary and secondary winding

(*ii*) Reactance of primary and secondary winding.

Q.3. Is the regulation at rated load of a transformer same at 0.8 p.f. lagging and 0.8 p.f. leading?

Ans. No, regulation at 0.8 p.f. lagging will be different to the regulation at 0.8 p.f. leading.

Q.4. What type of load should be connected to a transformer for getting zero voltage regulation?

Ans. For zero voltage regulation; $\frac{I_2 R_{es} \cos \phi_2 + I_2 X_{es} \sin \phi_2}{E_2} = 0$

or $$I_2 X_{es} \sin \phi_2 = -I_2 R_{es} \cos \phi_2$$

or $$\tan \phi = -\frac{R_{es}}{X_{es}}$$

or $$\phi_2 = -\tan^{-1} \frac{R_{es}}{X_{es}}$$

The negative sign shows that the load should be capacitive and the value *pf* should be

$$\cos \phi_2 = \cos \tan^{-1} \frac{R_{es}}{X_{es}}.$$

Q.5. Is the percentage impedance of a transformer same on primary and on secondary?

Ans. Yes, the total percentage impedance of a transformer referred to primary or secondary is the same. However, the percentage impedance of primary winding may be different to percentage impedance of secondary winding.

2.23 Losses in a Transformer

The losses which occur in an actual transformer are:

(*i*) Core or iron losses (*ii*) Copper losses

(*i*) **Core or iron losses:** When AC supply is given to the primary winding of a transformer an alternating flux is set up in the core, therefore, hysteresis and eddy current losses occur in the magnetic core.

(*a*) *Hysteresis loss:* When the magnetic material is subjected to reversal of magnetic flux, it causes a continuous reversal of molecular magnets. This effect consumes some electric power which is further dissipated in the form of heat as loss. This loss is known as hysteresis loss. ($P_h = K_h V f B_m^{2.6}$). This loss can be minimised by using silicon steel material for the construction of core.

(*b*) *Eddy current loss:* Since flux in the core of a transformer is alternating, it links with the magnetic material of the core itself also. This induces an emf in the core and circulates eddy currents. Power is required to maintain these eddy currents. This power is dissipated in the form of heat and is known as eddy current loss ($P_e = K_e V f^2 t^2 B_m^2$). This loss can be minimised by making the core of thin laminations.

It is already seen in article–8 that the flux set up in the core of the transformer remains constant from no-load to full load. Hence, iron loss is independent of the load and is known as constant losses.

(*ii*) **Copper losses:** Copper losses occur in both the primary and secondary windings due to their ohmic resistance. If I_1, I_2 are the primary and secondary currents and R_1, R_2 are the primary and secondary resistances, respectively.

Then, total copper losses = $I_1^2 R_1 + I_2^2 R_2 = I_1^2 R_{ep} = I_2^2 R_{es}$

The currents in the primary and secondary winding vary according to the load, therefore, these losses vary according to the load and are known as variable loss.

Magnetostriction and Its ill-effects

The property of ferromagnetic materials, due to which their dimensions are changed due to change of magnetic field, is called *magnetostriction effect.* In fact, the applied magnetic field changes the magnetostrictive strain until reaching its saturation value (λ). This causes change in the dimensions of ferromagnetic materials. The effect was first identified in 1842 by James Joule while observing a sample of iron.

In simple terms, as per this phenomenon, if a piece of magnetic sheet steel is magnetised it will extend itself. When the magnetisation is taken away, it goes back to its original condition. A transformer is magnetically excited by an alternating voltage and current so that it becomes extended and contracted twice during a full cycle of magnetisation.

Hence, when the core becomes magnetised, it expands then it contracts and expands again as the field is reversed. A 50 Hz transformer, therefore, expands and contracts at 100 Hz, producing a buzzing sound at that frequency.

Transformer buzzing noise (or hum) is caused by a phenomenon called **magnetostriction.**

Although, these extensions are only small dimensionally, and therefore cannot usually be seen by the naked eye. They are, however, sufficient to cause vibration resulting in a noise.

This effect also causes losses which produce heat in ferromagnetic cores.

Thus, the first cause of humming noise is magnetostriction, the other cause may be due to details of core construction, size and gauge of laminations or joints in the core

The other reasons for humming sound in a transformer may be

Loosening of laminations, mounting defects, unbalancing or insulation deterioration etc.

Methods to control transformer noise: The transformer noise can be controlled by

- using low value of flux density core,
- by proper tightening of the core by clamps, bolts etc.
- by sound insulating the transformer core from tank walls.

2.24 Effects of Voltage and Frequency Variations on Iron Losses

Power transformers are not ordinarily subjected to frequency variations and usually are subject to only modest voltage variations, but it will be interesting to note their effects.

Variation in voltage and/or frequency affects the iron losses (hysteresis and eddy current loss) in a transformer. As long as the flux variations are sinusoidal, hysteresis loss (P_h), and eddy current loss (P_e) vary according to the following relations

$$P_h \propto f(\phi_{max})^{1.6} \qquad \ldots(2.68)$$

The value 2.6 is an arbitrary value which lies between 1.5 and 2.5 depending on the grade of iron used in transformer core

and $$P_e \propto f^2\,(\phi_{max})^2 \qquad \ldots(2.69)$$

If the transformer is operated with the frequency and voltage changing in the same proportion, the flux density will remain unchanged as obvious from equation ($E = 4.44\ B_m A_i\ Nf$ or $B_m = \dfrac{E}{4.44\ A_i Nf}$ or $B_m \propto \dfrac{V}{f}$) and apparently the no-load current will also remain unaffected.

The transformer can be operated safely at frequency less than rated one with correspondingly reduced voltage. In this case iron losses will be reduced. But if the transformer is operated with increased voltage and frequency in the same proportion, the core losses may increase to a large extent which may not be tolerated. Increase in frequency with constant supply voltage will cause reduction in hysteresis loss and leave the eddy current losses unaffected. However, some increase in voltage could, therefore, be tolerated at higher frequencies, but exactly how much depends on the relative magnitude of the hysteresis and eddy current losses and the grade of iron used in the transformer core.

Example 2.28

A 1 kVA, 220/110 V, 400 Hz transformer is desired to be used at a frequency of 60 Hz. What will be the kVA rating of the transformer at reduced frequency?

Solution:

We know that $E_1 = V_1 = 4.44\,\phi_m N_1 f = 4.44\,B_m A_i N_1 f$

Assuming flux density in the core remaining unchanged, we have

$$V_1 \propto f$$

or $$\frac{V'_1}{V_1} = \frac{f'}{f}$$

or $$V'_1 = V_1 \times \frac{f'}{f} = 220 \times \frac{60}{400} = 33 \text{ volt}$$

As current rating of the transformer remains the same, the kVA rating is proportional to voltage,

∴ kVA rating of the transformer at 60 Hz,

$$\text{kVA}' = \frac{V'}{V} \times kVA = \frac{33}{220} \times 1 = \mathbf{0.15\ kVA}\ (Ans.)$$

Example 2.29

A 40 Hz transformer is to be used on a 50 Hz system. Assuming Steinmetz's coeff. as 2.6 and the losses at 40 Hz, 2.2%, 0.7% and 0.5% for copper, hysteresis and eddy currents, respectively, find (i) the losses on 50 Hz for the same supply voltage and current. (ii) the output at 50 Hz for the same total losses as on 40 Hz.

Solution:

Let W be the total power input to the transformer in both the cases in watt.

$$\text{Copper loss} = I^2 \times \text{total resistance}$$

As long as current and supply voltage remain the same, copper loss will remain the same.

∴ Copper loss at 40 Hz or 50 Hz

$$= \frac{1.2}{100} \times W = 0.012\,W \text{ watt}$$

Hysteresis loss = $\eta B_{max}{}^{1.6} f$ watt/c.c. of the magnetic material.

For the same voltage induced per turn ($E/N = 4.44 f \phi$) the product ϕf or $B_{max} f$ remains constant.

∴ $$B_{max1} f_1 = B_{\max_2} f_2$$

∴ $$\frac{B_{\max_1}}{B_{\max_2}} = \frac{f_2}{f_1} = \frac{50}{40} = 1.25$$

or $$B_{\max_2} = 0.8\,B_{\max_1}$$

∴ Hysteresis loss at 50 Hz

$$W_{h_2} = K_h = \eta B_{\max_2}^{1.6} f_2 \text{ watt/c.c.} = K_h (0.8\,B_{\max_1})^{1.6}\, 1.25\, f_1$$

$$= K_h (0.8)^{1.6} B_{\max_1}{}^{1.6} f_1 \times 1.25 = 1.25(0.8)^{1.6} \times K_h B_{\max_1}{}^{1.6} f_1 = 0.875 \times K_h B_{\max_1}{}^{1.6} f_1$$

But hysteresis loss at 40 Hz is 0.7%

$$W_{h_2} = 0.875 \times 0.7 = 0.6125\%$$

Eddy current loss, $W_e = B_{max}^2 f^2 t^2$ watt per c.c. of the magnetic material.

$$\therefore \quad \frac{W_{e_1}}{W_{e_2}} = \frac{B_{max_1}^2 f_1^2 t_1^2}{B_{max_2}^2 f_1^2 t_2^2} \quad \text{But } t_1 = t_2$$

$$\therefore \quad \frac{W_{e_1}}{W_{e_2}} = \frac{B_{max_1}^2 f_1^2}{B_{max_2}^2 f_2^2} = \left(\frac{B_{max_1}}{B_{max_2}}\right)^2 \left(\frac{f_1}{f_2}\right)^2 = \left(\frac{5}{4}\right)^2 \times \left(\frac{4}{5}\right)^2 = 1$$

$$\therefore \quad W_{e_1} = W_{e_2}$$

Hence eddy current loss will remain the same.

Thus, the three losses at 50 Hz will be

Copper loss = **1.2%**

Hysteresis loss = **0.6125%** (*Ans.*)

Eddy current loss = **0.5%**

2.25 Efficiency of a Transformer

The efficiency of a transformer is defined as the ratio of output to the input power, the two being measured in same units (either in watts or in kW).

$$\text{Transformer efficiency, } \eta = \frac{\text{output power}}{\text{input power}} = \frac{\text{output power}}{\text{output power + losses}}$$

or

$$\eta = \frac{\text{output power}}{\text{output power + iron losses + coper losses}}$$

$$= \frac{V_2 I_2 \cos\phi_2}{V_2 I_2 \cos\phi_2 + P_i + P_c} \quad \text{...(2.70)}$$

where, V_2 = Secondary terminal voltage

I_2 = Full load secondary current

$\cos\phi_2$ = *p.f.* of the load

P_i = Iron losses = Hysteresis losses + eddy current losses (constant losses)

P_c = Full load copper losses = $I^2{}_2 R_{es}$ (variable losses)

If x is the fraction of the full load, the efficiency of the transformer at this fraction is given by the relation;

$$\eta_x = \frac{x \times \textit{output at full load}}{x \times \textit{output at full load} + P_i + x^2 P_c} = \frac{x V_2 I_2 \cos\phi_2}{x V_2 I_2 \cos\phi_2 + P_i + x^2 I_2^2 R_{es}} \quad \text{...(2.71)}$$

The copper losses vary as the square of the fraction of the load.

2.26 Condition for Maximum Efficiency

The efficiency of a transformer at a given load and p.f. is expressed by the relation

$$\eta = \frac{V_2 I_2 \cos\phi_2}{V_2 I_2 \cos\phi_2 + P_i + I_2^2 R_{es}} = \frac{V_2 \cos\phi_2}{V_2 \cos\phi_2 + P_i / I_2 + I_2 R_{es}}$$

The terminal voltage V_2 is approximately constant. Thus for a given p.f., efficiency depends upon the load current I_2. In expression (*i*), the numerator is constant and the efficiency will be maximum if denominator is minimum. Thus the maximum condition is obtained by differentiating the quantity in the denominator w.r.t. the variables I_2 and equating that to zero i.e.,

$$\frac{d}{d I_2}\left(V_2 \cos\phi_2 + \frac{P_i}{I_2} + I_2 R_{es}\right) = 0$$

or

$$0 - \frac{P_i}{I_2^2} + R_{es} = 0$$

or

$$I_2^2 R_{es} = P_i \qquad \text{...(2.72)}$$

i.e., **Copper losses = Iron losses**

Thus, the efficiency of a transformer will be maximum when copper (or variable) losses are equal to iron (or constant) losses.

$$\therefore \quad \eta_{max} = \frac{V_2 I_2 \cos\phi_2}{V_2 I_2 \cos\phi_2 + 2 P_i} \qquad [\textit{since } P_c = P_i] \text{ ...(2.73)}$$

From equation (*ii*), the value of output current I_2 at which the efficiency of the transformer will be maximum is given by;

$$I_2 = \sqrt{\frac{P_i}{R_{es}}} \qquad \text{...(2.74)}$$

If x is the fraction of full load kVA at which the efficiency of the transformer is maximum.

Then, copper losses = $x^2 P_c$ (where P_c is the full load Cu losses)

Iron losses = P_i

For maximum efficiency, $x^2 P_c = P_i$; $x = \sqrt{\frac{P_i}{P_c}}$

∴ Output kVA corresponding to maximum efficiency

$$= x \times \text{full load kVA} = \text{full load kVA} \times \sqrt{\frac{P_i}{P_c}}$$

$$= \text{full load kVA} \times \sqrt{\frac{\text{iron losses}}{\text{copper losses at full load}}} \qquad \text{...(2.75)}$$

Example 2.30

A 500 kVA, 600/400V, one-phase transformer has primary and secondary winding resistance of 0.42 ohm and 0.0011 ohm, respectively. The primary and secondary voltages are 600 V and 400 V, respectively. The iron loss is 2·9 kW. Calculate the efficiency at half full load at a power factor of 0·8 lagging.

Solution:

Transformer rating, = 500 kVA

Primary resistance, $R_1 = 0{\cdot}42\ \Omega$

Secondary resistance, $R_2 = 0{\cdot}0011\ \Omega$

Primary voltage, $E_1 = 6600$ V

Secondary voltage, $E_2 = 400$ V

Iron losses, $P_i = 2{\cdot}9$ kW

Fraction of the load, $x = \frac{1}{2} = 0 \cdot 5$

Load p.f., $\cos\phi = 0{\cdot}8$ lagging

Transformation ratio, $K = \frac{E_2}{E_1} = \frac{400}{6600} = \frac{2}{33}$

Primary resistance referred to secondary,

$$R_1' = K^2 R_1 = \frac{2}{33} \times \frac{2}{33} \times 0.42 = 0.00154\,\Omega$$

Total resistance referred to secondary,

$$R_{es} = R_2 + R_1' = 0{\cdot}0011 + 0{\cdot}00154 = 0{\cdot}00264\ \Omega$$

Full load secondary current, $I_2 = \frac{kVA \times 10^3}{E_2} = \frac{500 \times 10^3}{400} = \mathbf{1250A}$

Copper losses at full load, $P_c = I_2^2\ R_{es} = (1250)^2 \times 0{\cdot}00264$

$= 4125\text{ W} = 4{\cdot}125\text{ kW}$

Efficiency of transformer at any fraction (x) of the load,

$$\eta_x = \frac{xkVA\cos\phi}{xkVA\ \cos\phi + P_i + x^2 P_c} \times 100$$

$$= \frac{0 \cdot 5 \times 500 \times 0 \cdot 8}{0 \cdot 5 \times 500 \times 0 \cdot 8 + 2 \cdot 9 + (0 \cdot 5)^2 \times 4 \cdot 125} \times 100$$

= **98·07%** (*Ans.*)

Example 2.31

A single-phase 440/110 V transformer has primary and secondary winding resistance of 0·3 ohm and 0·02 ohm, respectively. If iron loss on normal input voltage is 150 W, calculate the secondary current at which maximum efficiency will occur. What is the value of this maximum efficiency for unity power factor load?

Solution:

$$\text{Primary resistance, } R_1 = 0{\cdot}3\ \Omega$$

$$\text{Secondary resistance, } R_2 = 0{\cdot}02\ \Omega$$

$$\text{Iron losses, } P_i = 150\ \text{W}$$

$$\text{Load power factor, } \cos\phi = 1$$

$$\text{Primary induced voltage, } E_1 = 440\ \text{V}$$

$$\text{Secondary induced voltage, } E_2 = 110\ \text{V}$$

$$\text{Transformation ratio, } K = \frac{E2}{E1} = \frac{110}{440} = \frac{1}{4}$$

Primary resistance referred to secondary,

$$R_1' = K^2 R_1 = \frac{1}{4} \times \frac{1}{4} \times 0 \cdot 3 = 0 \cdot 01875\ \Omega$$

Equivalent resistance referred to secondary,

$$R_{es} = R_2 + R_1' = 0{\cdot}02 + 0{\cdot}01875 = 0{\cdot}03875\ \Omega$$

We know the condition for max, efficiency is

$$\text{Copper losses} = \text{Iron losses}$$

i.e., $$I_2^2\ R_{es} = P_i$$

Secondary current at which the efficiency is maximum,

$$I_2 = \sqrt{\frac{P_i}{R_{es}}} = \sqrt{\frac{150}{0.03875}} = \mathbf{62{\cdot}22\ A}\ (Ans.)$$

The maximum efficiency, $$\eta_{max} = \frac{I_2 V_2 \cos\phi}{I_2 V_2 \cos\phi + 2P_i} \times 100$$

$$= \frac{62.22 \times 110 \times 1}{62.22 \times 110 \times 1 + 2 \times 150} \times 100 = \mathbf{95{\cdot}8\%}\ (Ans.)$$

Example 2.32

In a 25 kVA, 2000/200 V power transformer the iron and full load copper losses are 350 W and 400 W, respectively. Calculate the efficiency at unity power factor at (i) full load and (ii) half load.

Solution:

$$\eta_x = \frac{x\ kVA \times 1000 \times \cos\phi}{x\ kVA \times 1000 \times \cos\phi + P_i + x^2\ P_c}$$

where, $\cos\phi = 1$: $P_i = 350$ W; $P_c = 400$ W

(*i*) At full-load $x = 1$

$$\therefore \quad \eta = \frac{1 \times 25 \times 1000 \times 1}{1 \times 25 \times 1000 \times 1 + 350 + 1 \times 1 \times 400} \times 100 = \mathbf{97{\cdot}087\ \%}\ (Ans)$$

(*ii*) At half-load; $x = 0{\cdot}5$

$$\therefore \quad \eta = \frac{0{\cdot}5 \times 25 \times 1000 \times 1}{0{\cdot}5 \times 25 \times 1000 \times 1 + 350 + (0{\cdot}5)^2 \times 400} \times 100 = \mathbf{96{\cdot}525\ \%}\ (Ans)$$

Example 2.33

A 220/400 V, 10 kVA, 50Hz, single-phase transformer has copper loss of 120 W at full load. If it has an efficiency of 98% at full load, unity power factor, determine the iron losses. What would be the efficiency of the transformer at half full-load at 0.8 p.f. lagging.

Solution:

$$\eta_x = \frac{x\,kVA \times 1000 \times \cos\phi}{x\,kVA \times 100 \times \cos\phi + P_i + x^2 P_c} \times 100$$

$$98 = \frac{1 \times 10 \times 1000 \times 1}{1 \times 10 \times 1000 \times 1 + P_i + 1 \times 1 \times 120} \times 100 \text{ or } P_i = \mathbf{84{\cdot}08\ W}\ (Ans)$$

when $x = 1/2$ and $\cos\phi = 0{\cdot}8$;

$$\eta_x = \frac{0{\cdot}5 \times 10 \times 1000 \times 0{\cdot}8}{0{\cdot}5 \times 10 \times 1000 \times 0{\cdot}8 + 84{\cdot}08 + (0{\cdot}5)^2 \times 120} \times 100 = \mathbf{97{\cdot}23\ \%}\ (Ans)$$

Example 2.34

A 1000 kVA, 110/220 volt, 50 Hz single phase transformer has an efficiency of 98.5% at half load and 0.8 power factor leading. Whereas its efficiency at full load unity power factor is 98.9%. Determine (i) iron loss (ii) Full load copper loss.

Solution:

Here, Rating of transformer = 1000 kVA

We know, $$\%\ \eta_x = \frac{x\,kVA \times 1000 \times p.f.}{x\,kVA \times 1000 \times p.f. + P_i + x^2 P_c} \times 100$$

(*i*) Where $\%\ \eta_{0.5} = 98.5$; $x = 0.5$; *p.f.* = 0.8 leading

$$\therefore \quad 98.5 = \frac{0.5 \times 1000 \times 1000 \times 0.8}{0.5 \times 1000 \times 1000 \times 0.8 + P_i + (0.5)^2 P_c} \times 100$$

or $$98.5 = \frac{400 \times 10^5}{4 \times 10^5 + P_i + 0.25 P_c}$$

or $P_i + 0.25\,P_c = 6100$ ...(*i*)

(*ii*) When $\%\ \eta_{fl} = 98.8$; $x = 1$; p.f. = 1

$$98.8 = \frac{1 \times 1000 \times 1000 \times 1 \times 100}{1 \times 1000 \times 1000 \times 1 + P_i + P_c}$$

$$\text{or } 10 \times 10^5 + P_i + P_c = \frac{10 \times 10^5 \times 100}{98.8} = 10.121 \times 10^5$$

$$\text{or} \qquad P_i + P_c = 12100 \qquad \ldots(ii)$$

Subtracting eq. (*i*) from (*ii*), we get

$$0.75\, P_c = 6000$$

$$\text{or} \qquad P_c = \mathbf{8000\ W}\ (Ans.)$$

From eq. (*ii*), we get, $P_i = 12100 - 8000 =$ **4100 W** (*Ans.*)

Example 2.35

A single-phase 400 kVA transformer has an efficiency of 99.13% at half load unity pf whereas it efficiency is 98.77% at full-load 0.8 pf lagging. (i) the iron loss (ii) the full load copper loss.

Solution:

Efficiency of a transformer at any fraction x of the load;

$$\eta_x = \frac{x\,kVA\cos\phi}{x\,kVA\cos\phi + P_i + x^2 P_c} \times 100$$

Case I: $x = 1$; $\cos\phi = 0{\cdot}8$; $\eta_x = 98{\cdot}77\ \%$;

$$\therefore \qquad 98{\cdot}77 = \frac{1 \times 400 \times 0{\cdot}8}{1 \times 400 \times 0{\cdot}8 + P_i + (1)^2\, P_c} \times 100$$

$$\text{or} \qquad P_i + P_c = 3{\cdot}985\text{ kW} \qquad \ldots(i)$$

Case II: $x = 0{\cdot}5$; $\cos\phi = 1$; $\eta_x = 99{\cdot}13$;

$$\therefore \qquad 99{\cdot}13 = \frac{0{\cdot}5 \times 400 \times 1}{0{\cdot}5 \times 400 \times 1 \times P_i + (0{\cdot}5)^2 P_c} \times 100$$

$$\text{or} \qquad P_i + 0{\cdot}25\, P_c = 1{\cdot}755\text{ kW} \qquad \ldots(ii)$$

Subtracting eq. (*ii*) from (*i*), we get,

$$0{\cdot}75\, P_c = 2{\cdot}23\text{ kW or } P_c = \mathbf{2{\cdot}973\ kW}\ (Ans)$$

$$\text{and} \qquad P_i = 3{\cdot}985 - 2{\cdot}973 = \mathbf{1{\cdot}012\ kW}\ (Ans)$$

Example 2.36

In a 25 kVA. 1100/400 V, single phase transformer, the iron and copper losses at full load are 350 and 400 watts, respectively. Calculate the efficiency on unity power at half load. Determine the load at which maximum efficiency occurs.

Solution:

Transformer rating = 25 kVA

Iron losses, $P_i = 350$ W

Full load copper losses, $P_c = 400$ W

Load power factor $\cos\phi = 1$

Fraction of the load $x = \frac{1}{2} = 0\cdot5$

Efficiency of transformer at any fraction of the load,

$$\eta_x = \frac{x \times kVA \times 10^3 \times \cos\phi}{x \times kVA \times 10^3 \times \cos\phi + P_i + x^2 P_c} \times 100$$

$$= \frac{0.5 \times 25 \times 10^3 \times 1}{0.5 \times 25 \times 10^3 \times 1 + 350 + (0.5)^2 \times 400} \times 100 = \mathbf{96{\cdot}52\%}\ (Ans.)$$

Output kVA corresponding to maximum efficiency

$$= \text{Rated kVA}\sqrt{\frac{P_i}{P_c}} = 25 \times \sqrt{\frac{350}{400}} = 23{\cdot}385 \text{ kVA}$$

Output power or load on maximum efficiency.

= output kVA for max efficiency × *p.f.*

= 23·385 × 1 = **23·385 kW** (*Ans.*)

Example 2.37

A 50 kVA transformer on full load has a copper loss of 600 watt and iron loss of 500 watt, calculate the maximum efficiency and the load at which it occurs.

Solution:

$$\%\,\eta = \frac{output}{output + Iron\ loss + Copper\ loss} \times 100$$

Efficiency will be maximum when: copper loss = Iron loss = 500 W

Fraction at which the efficiency is maximum, $x = \sqrt{\frac{P_i}{P_c}} = \sqrt{\frac{500}{600}} = 0{\cdot}9128$

Load at which the efficiency is maximum, i.e.,

Output = x × kVA = 0·9128 × 50 = 45·64 kVA

= 45·64 × 1 = 45·64 kW (since $\cos\phi = 1$)

$$\eta_m = \frac{45{\cdot}64 \times 1000}{45{\cdot}64 \times 1000 + 500 + 500} \times 100 = \mathbf{97{\cdot}85\%}\ (Ans)$$

Example 2.38

The iron and full-load copper losses of a 100 kVA single-phase transformer are 1 kW and 1.5 kW, respectively. Calculate the kVA loading at which the efficiency is maximum and its efficiency at this loading: (i) at unit p.f. (ii) at 8 p.f. lagging.

Solution:

Here, Rated capacity = 100 kVA; Iron loss, $P_i = 1$ kW;

Full-load copper loss, $P_c = 1.5$ kW

Output kVA corresponding to maximum efficiency

$$= x \times \text{rated kVA} = \sqrt{\frac{P_i}{P_c}} \times \text{rated kVA} = \sqrt{\frac{1}{1\cdot 5}} \times 100 = \mathbf{81.65\ kVA}\ (Ans.)$$

(*i*) At unity p.f.

$$\eta = \frac{81\cdot 65 \times 1}{81\cdot 65 \times 1 + 1 + 1} \times 100 = \frac{81\cdot 65}{83\cdot 65} \times 100 = \mathbf{97{\cdot}6\%}\ (Ans.)$$

(*ii*) At 0.8 p.f. lagging

$$\eta = \frac{81.65 \times 0.8}{81.65 \times 0.8 + 1 + 1} \times 100 = \frac{65.32}{67.32} \times = \mathbf{97.03\%}\ (Ans.)$$

2.27 Efficiency vs Load

The variation of efficiency with load is shown in Fig. 2.43 which likewise shows the constant and variable components of the total loss. It has been pointed out that with constant voltage the mutual flux of the transformer is practically constant from no-load to full load (maximum variation is from 1 to 3%). The core or iron loss is, therefore, considered constant regardless of load. Copper loss varies as the square of the load current or kVA output. The variations in copper loss with the increase in load current (or kVA) is shown in Fig. 2.43. The efficiency *vs* load curve as deduced from these is also shown in the Fig. 2.43. From the efficiency–load curve shown in Fig. 2.43 it is obvious that the efficiency is very high even at light load, as low as 10% of rated load. The efficiency is practically constant from about 20% rated load to about 20% overload. At light loads the efficiency is poor because of constant iron loss, whereas at high loads the efficiency falls off due to increase in copper loss as the square of load. From Fig. 2.43 it is also obvious that the transformer efficiency is maximum at the point of intersection of copper loss and iron loss curves i.e., when copper loss equals iron loss.

Imp. The intersection of copper loss curve and iron loss curve '*A*' gives the point of maximum efficiency. It will be seen that the efficiency changes very little over the greater part of the operating range.

2.28 Efficiency vs Power Factor

Transformer efficiency is given as

$$\eta = \frac{\text{Output}}{\text{Output} + \text{losses}} = 1 - \frac{\text{losses}}{\text{output} + \text{losses}}$$

$$= 1 - \frac{\text{losses}}{V_2 I_2 \cos\phi + \text{losses}} = 1 - \frac{\text{losses}/V_2 I_2}{\cos\phi + \text{losses} / V_2 I_2}$$

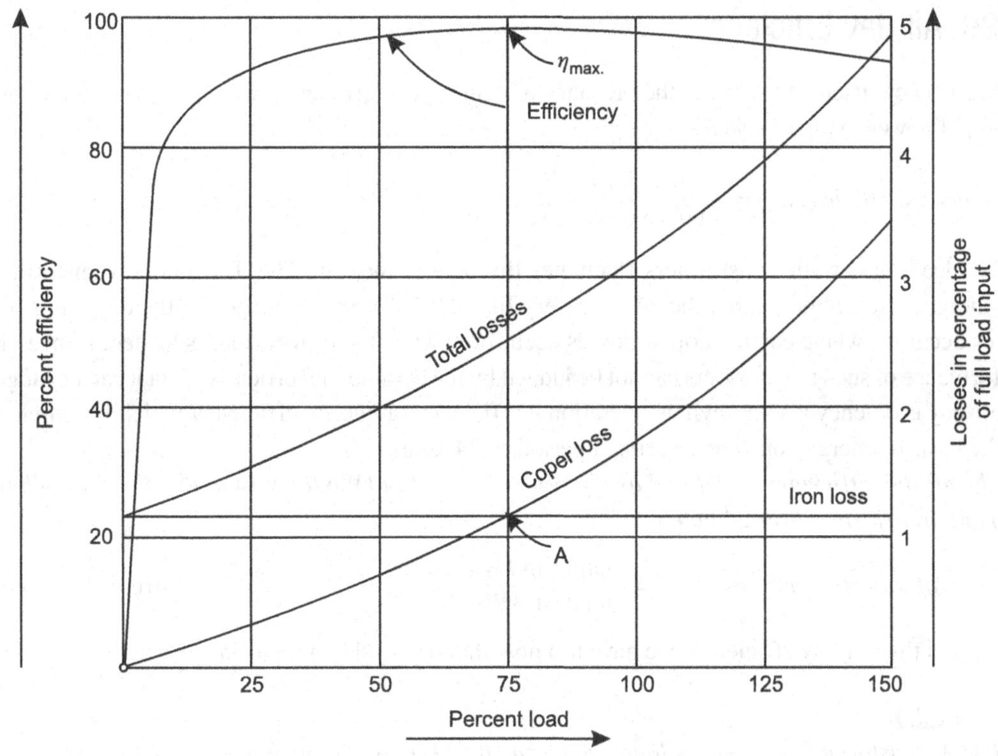

Fig. 2.43 Efficiency vs load curve of a transformer

Substituting

$$\frac{\text{losses}}{V_2 I_2} = x \text{ in above equation we get}$$

$$\eta = 1 - \frac{x}{\cos\phi + x} = 1 - \frac{x / \cos\phi}{1 + x / \cos\phi}$$

The variations of efficiency with power factor at different loads for a typical transformer are illustrated in Fig. 2.44

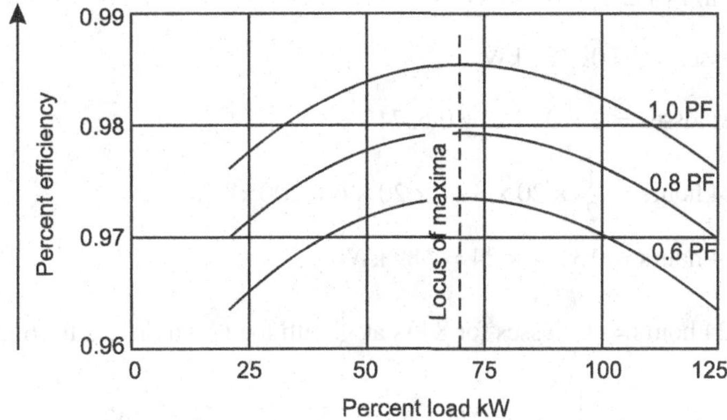

Fig. 2.44 Frequency vs load curve of a transformer

2.29 All-day Efficiency

The efficiency discussed so far is the ordinary or commercial efficiency which is given by the ratio of output power to input power i.e.,

$$\textit{Commercial efficiency, } \eta = \frac{\text{output power}}{\text{input power}}$$

The load on certain transformers fluctuates throughout the day. The distribution transformers are energised for 24 hours, but they deliver very light loads for major portion of the day. Thus iron losses occur for whole day but copper losses occur only when the transformer is loaded. Hence, the performance of such transformers cannot be judged by the commercial efficiency, but it can be judged by all-day efficiency also known as operational efficiency or energy efficiency which is computed on the basis of energy consumed during a period of 24 hours.

*The **all-day efficiency** is defined as the ratio of output in kWh (or Wh) to the input in kWh (or Wh) of a transformer over 24 hours.*

$$\therefore \quad \textit{All-day efficiency} = \eta_{all\text{-}day} = \frac{\text{output in } kWh}{\text{input in } kWh} \quad \text{...(2.76) (for 24 hours)}$$

To find this all-day efficiency, we have to know the load cycle on the transformer.

Example 2.39

A 20 kVA transformer on domestic load, which can be taken as of unity power factor, has a full load efficiency of 94.3%, the copper loss then being twice the iron loss. Calculate its all-day efficiency on the following daily cycle; no-load for 10 hours, half load for 8 hours and full load for 6 hours.

Solution:

Full load output= 20 × 1 = 20 kW

$$\text{Full load input} = \frac{\text{output}}{\eta} = \frac{20}{95{\cdot}3} \times 100 = 20{\cdot}986 \text{ kW}$$

$$\text{Total losses, } P_i + P_c = 20.986 - 20 = 0{\cdot}986 \text{ kW}$$

Now $P_c = 2\,P_i$ (given) $P_i + 2\,P_i = 0{\cdot}986$ kW

Or Iron losses, $P_i = 0{\cdot}3287$ kW

$$\text{Full load copper losses} = 2 \times 0{\cdot}3287 = 0{\cdot}6574 \text{ kW}$$

$$kWh \text{ output in 24 hours} = \frac{1}{2} \times 20 \times 8 + 1 \times 20 \times 6 = 200 \text{ kWh}$$

$$\text{Iron losses for 24 hours} = 0{\cdot}3287 \times 24 = 7{\cdot}89 \text{ kWh}$$

Copper losses for 24 hours = cu. losses for 8 hrs at $\frac{1}{2}$ full load + cu. losses for 6 hours at full load

$$= \left(\frac{1}{2}\right)^2 \times 0{\cdot}6574 \times 8 + 0{\cdot}6574 \times 6 = 5{\cdot}259 \text{ kWh}$$

input in 24 hrs = kWh output in 24 hrs + iron and cu losses in kWh for 24 hrs

$$= 200 + 7{\cdot}89 + 5{\cdot}259 = 213{\cdot}149 \text{ kWh}$$

$$\text{All-day efficiency, } \eta_{all\text{-}day} = \frac{kWh\ output\ In\ 24\ hrs}{kWh\ input\ in\ 24\ hrs} \times 100 = \frac{200}{213.149} \times 100 = \mathbf{93.83\%}\ (Ans)$$

Example 2.40

A 5 kVA single phase transformer has full-load copper loss of 100 watt and 60 watt as iron loss. The daily variation of load on the transformer is as follows:

7 AM to 1 PM — *3 kW at power factor 0.6 lagging.*
1 PM to 6 PM — *2 kW at power factor 0.8 lagging.*
6 PM to 1 AM — *5 kW at power factor 0.9 lagging.*
1 AM to 7 AM — *No-load*

Determine the all-day efficiency.

Solution:

Transformer rating = 5 kVA; P_i = 50 W; P_c = 100 W

Load variation in tabulated form is given below:

Timings	Duration in hr	Load in kW	p.f	Load in kVA = $\frac{kW}{pf}$	Fraction of Load = $\frac{\text{Actual kVA}}{\text{rated kVA}}$
7 AM to 1PM	6	3	0.6	3/0.6 = 5	5/5 = 1
1 PM to 6 PM	5	2	0.8	2/0.8 = 2.5	2.5/5 = 0.5
6 PM to 1 AM	7	5	0.9	5/0.9 = 5.55	5.55/5 = 1.11
1 AM to 7 AM	6	0	0	0	0

$$\text{kWh output in 24 hr} = 3 \times 6 + 2 \times 5 + 5 \times 7 + 0 \times 6 = 63 \text{ kWh}$$

$$\text{Iron losses in 24 hr} = P_i \text{ in kW} \times 24 = \frac{50}{1000} \times 24 = 1.2 \text{ kWh}$$

$$\text{Copper losses in 24 hr} = (1)^2 \times \frac{100}{1000} \times 6 + (0.5)^2 \times \frac{100}{1000} \times 5 + (1.11)^2 \times \frac{100}{1000} \times 7 + (0)^2 \times \frac{100}{1000} \times 6$$

$$= 0.6 + 0.125 + 0.8625 + 0 = 2.5875 \text{ kWh}$$

$$\text{All-day efficiency, } \eta_{all\text{-}day} = \frac{\text{Output in kWh in 24 hr}}{\text{Output in kWh in 24 hr} + Pi \text{ in kWh in 24 hr} + P_c \text{ in kWh is 24 hr}}$$

$$= \frac{63}{63 + 1.2 + 1.5875} \times 100 = \mathbf{95.763\%}\ (Ans.)$$

Example 2.41

A transformer has a maximum efficiency of 98% at 15 kVA at unity p.f. It is loaded as follows: 12 hrs –2 kW at p.f. 0.5; 6 hrs – 12 kW at p.f. 0.8; 6 hrs – 18 kW at p.f. 0.9, calculate all-day efficiency of the transformer.

Solution:

We know $$\eta_{max} = \frac{kVA \cos\phi}{kVA \cos\phi + 2P_i} \qquad (\because P_i = P_c)$$

or $$\frac{98}{100} = \frac{15\times 1}{15\times 1 + 2P_i}$$

or $$15 + 2P_i = \frac{15\times 100}{98} = 15.306$$

∴ Iron losses, $P_i = 0{\cdot}153$ kW

Full load copper losses, $P_c = P_i = 0{\cdot}153$ kW

During 24 hrs. the transformer is loaded as under:

Hrs.	Load in kW	P.f.	Load in kVA $\frac{kW}{p.f.}$	Fraction of load $x = \frac{\text{given load in kVA}}{\text{full load in kVA}}$
12	2	0.5	2/0.5 = 4	4/15 = 0.267
6	12	0.8	0.8 = 15	15/15 = 1
6	18	0.9	18/0.9	20/15 = 1.333

kWh output in 24 hrs $= 2\times 12 + 12\times 6 + 18\times 6 = 204$ kWh

Iron losses for 24 hrs $= 0{\cdot}153 \times 24 = 3{\cdot}672$ kWh

Copper losses for 24 hrs $= (0{\cdot}267)^2 \times 0{\cdot}153 \times 12 + (1)^2 \times 0{\cdot}153 \times 6 + (1{\cdot}333)^2 \times 0{\cdot}153 \times 6$

$= 2{\cdot}68$ kWh

Input in 24 hrs $= 204 + 3.672 + 2.68 = 210.352$ kWh

All-day efficiency, $\eta_{all\text{-}day} = \frac{204}{210{\cdot}352} \times 100$ **= 96·98 % (*Ans*)**

Section Practice Problems

Numerical Problems

1. A 500 kVA, 6600/400V transformer has primary and secondary winding resistance of 0·42 ohm and 0·0011 ohm, respectively. The iron loss is 2·9 kW. Calculate the efficiency at half full load at a power factor of 0·8 lagging. (**Ans.** *98·07%*)
2. The full load efficiency at 0.8 p.f of a 50 kVA transformer is 97.9% and at half-load 0.8 p.f. its efficiency is 97.9%. Determine iron and copper losses of the transformer. (**Ans.** *287 W; 529 W*)

3. In a 25 kVA, 1100/400 V, single phase transformer, the iron and copper loss at full load are 350 and 400 watt, respectively. Calculate the efficiency on unity power factor at half load. Determine the load on maximum efficiency. (**Ans.** *96·52%; 23·85 kW*)

4. A 100 kVA transformer supplies a lighting and power load. The iron loss is 960 W and the copper loss is 960 W at full load. The transformer is operated continuously at the rated voltage as per the following schedule in a day 100 kVA at 0.8 p.f. for 4 hrs.; 50 kVA at 0.6 p.f. for 8 hrs. and 5 kVA at 0.95 p.f. for 12 hrs. What will be the all-day efficiency of the transformer? (**Ans.** *94·5%*)

Short Answer Type Questions

Q.1. What are no-load losses occurring in the transformer?

Ans. Iron losses which are also known as magnetic losses or core losses. These losses include hysteresis loss and eddy current loss.

Q.2. Why is efficiency of a transformer high as compared to other electrical machines?

Ans. Transformer is a static device i.e., it has no rotating part, therefore, it is free of mechanical losses. Hence, it operates at higher efficiency in comparison to other electrical machines.

Q.3. Define efficiency and all-day efficiency of a transformer.

Ans. The ratio of power output (in kW) to power input (in kW) is called efficiency or commercial efficiency of a transformer, i.e.,

$$\eta = \frac{\text{Output in kW}}{\text{input in kW}}$$

The ratio of output energy (in kWh) to the input energy (in kWh) in a day of a transformer is called its all-day efficiency.

$$\eta_{all\text{-}day} = \frac{\text{Output in kWh in a day}}{\text{Input in kWh in a day}}$$

Q.4. Are transformers normally considered to be efficient devices?

Ans. Yes, normally transformers are considered as efficient devices.

Q.5. Why is the efficiency of a transformer high as much as 96%?

Ans. It is because transformers do not have rotating parts and mechanical losses do not occur.

Q.6. How can eddy current loss be reduced?

Ans. Eddy current loss can be reduced by laminating the core (0.35 mm to 0.5 mm thickness) and each lamination must be insulated from the other by an insulating layer (varnish).

Q.7. How may the iron loss be reduced to a minimum?

Ans. Iron loss can be minimised by using steel having sufficient quantity of silicon, now-a-days cold rolled grain oriented steel (*CRGOS*) is used, and the core is laminated, each lamination has a thickness 0.35 to 0.5 mm and insulated from each other.

Q.8. In a transformer, buzzing noise cannot be avoided. Justify.

Ans. Since magnetostriction phenomenon cannot be avoided, the buzzing noise produced by a transformer cannot be avoided.

2.30 Transformer Tests

All the transformers are tested before placing them in the field. By performing these tests, we can determine the parameters of a transformer to compute its performance characteristics (like voltage regulation and efficiency etc.).

Large transformers cannot be tested by direct loading because of the following reasons:

(*i*) It is almost impossible to arrange such a large load required for direct loading.
(*ii*) While performing test by direct loading, there is huge power wastage.
(*iii*) It is very inconvenient to handle the power equipment.

Therefore, to furnish the required information open circuit and short circuit tests are conducted conveniently without actually loading the transformer.

The other important tests which are conducted on a transformer are polarity test voltage ratio test and Back-to-back test.

2.31 Polarity Test

Polarity test is performed to determine the terminals with same instantaneous polarity of the two windings when terminals are not being marked. The relative polarities of the primary and secondary terminals are required to be known for

(*i*) interconnecting two or more transformers in parallel.
(*ii*) connecting three single-phase transformers while doing poly-phase transformation of power.
(*iii*) connecting windings of the same transformer in parallel or series.

For determining the relative polarity of the two windings of a transformer, the two winding are connected in series and a voltmeter is connected across them as shown in Fig. 2.45. One of the winding (preferably *HV* winding) is excited from a suitable AC voltage (less than rated value). If the polarities of the windings are as marked on the diagram, then the windings will have a subtractive polarity and the voltmeter will read the difference of E_1 and E_2 (i.e., $E_1 - E_2$). If the voltmeter reads $E_1 + E_2$ the polarity marking of one of the windings must be reversed.

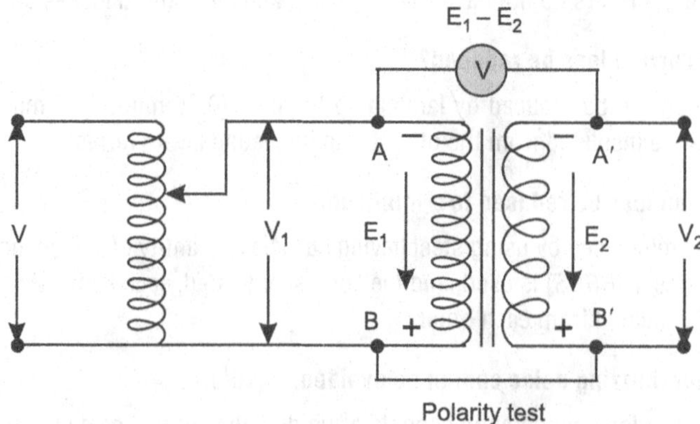

Fig. 2.45 Circuit diagram for polarity test

While performing polarity test, subtractive polarity method is preferred over additive polarity method, because in this case, the voltage between A and A' or that between B and B' is reduced. The leads connected between these terminals and two windings are not subjected to high voltage stresses. Whereas, in case of additive polarity the two windings and leads connected between AA' and BB' are subjected to high voltage stresses.

When the transformer is placed in the field, it may not be convenient to perform the above test to check the polarity. In such cases, polarity may be checked by using a battery, a switch and a DC voltmeter (*PMMC* type) which are connected in the circuit as shown in Fig. 2.46. When switch (S) is closed, the primary current increases which increases the flux linkages with both the windings inducing emf in them. The positive polarity of this induced emf in the primary is at the end to which battery is connected. The end of secondary which simultaneously acquires positive polarity which is indicated by the deflection in the (*PMMC*) voltmeter. If deflection does not occur, then open the switch. At this instant, if deflection occurs the polarity of secondary is opposite.

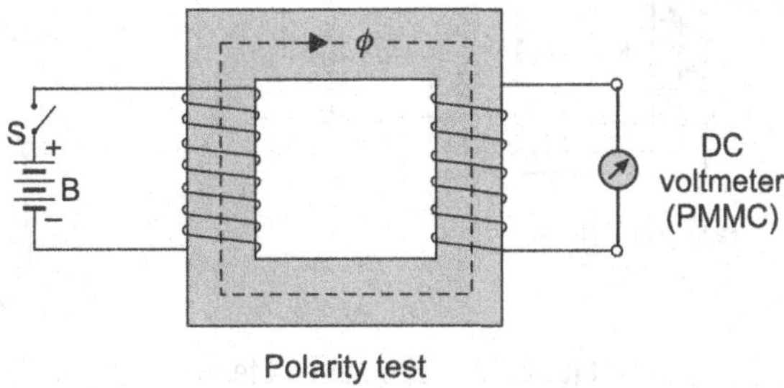

Fig. 2.46 Polarity test using PMMC

2.32 Voltage Ratio Test

The true voltage ratio is based upon turn-ratio of the two windings of a transformer. In case the two voltages are measured at no-load, their ratio is almost equal to the true value. Similarly, if the primary and secondary currents are measured on short circuit, their ratio gives true-ratio particularly if the transformer has little leakage flux and low core reluctance. voltage ratio $\frac{V_2}{V_1} = \frac{I_1}{I_2}$.

2.33 Open-circuit or No-load Test

This test is carried out at rated voltage to determine the no-load loss or core loss or iron loss. It is also used to determine no-load current I_0 which is helpful in finding the no-load parameters i.e., exciting resistance R_0 and exciting reactance X_0 of the transformer.

Usually, this test is performed on low voltage side of the transformer, i.e., all the measuring instruments such as voltage (V), wattmeter (W) and ammeter (A) are connected in low-voltage side (say primary). The primary winding is then connected to the normal rated voltage V_1 and frequency as given on the name plate of the transformer. The secondary side is kept open or connected to a voltmeter V' as shown in Fig. 2.47(*a*).

Since the secondary (high voltage winding) is open circuited, the current drawn by the primary is called no-load current I_0 measured by the ammeter *A*. The value of no-load current I_0 is very small usually 2 to 10% of the rated full-load current. Thus, the copper loss in the primary is negligibly small and no copper loss occurs in the secondary as it is open. Therefore, wattmeter reading W_0 only represents the core or iron losses for all practical purposes. These core losses are constant at all loads. The voltmeter *V'* if connected on the secondary side measures the secondary induced voltage V_2.

The ratio of voltmeter readings, $\frac{V_2}{V_1}$ gives the transformation ratio of the transformer. The phasor diagram of transformer at no-load is shown in Fig. 2.47(*b*).

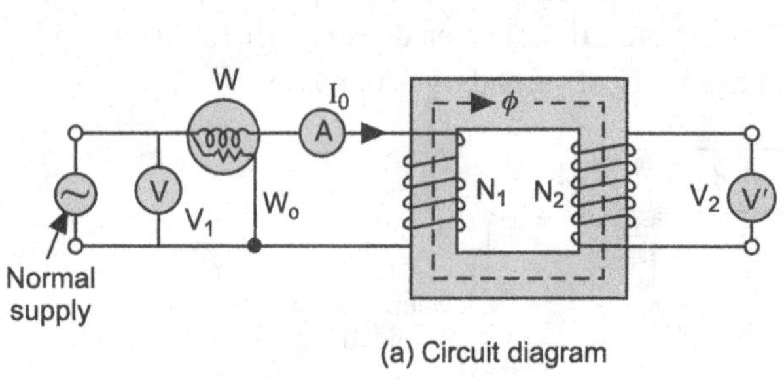

(a) Circuit diagram

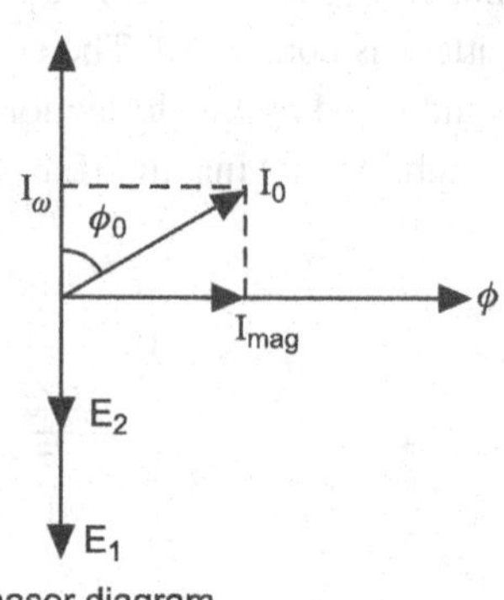

(b) Phasor diagram

Fig. 2.47 Open circuit test

Let the wattmeter reading = W_0

voltmeter reading = V_1

and ammeter reading = I_0

Then, iron losses of the transformer $P_i = W_0$

i.e., $$V_1 I_0 \cos \phi_0 = W_0$$

∴ No-load power factor, $\cos \phi_0 = \frac{W_0}{V_1 I_0}$

Working component, $I_w = \frac{W_0}{V_1}$ $(\because I_w = I_0 \cos \phi_0)$

Magnetising component $I_{mag} = \sqrt{I_0^2 - I_w^2}$

No-load, parameters, i.e.,

Equivalent exciting resistance, $R_0 = \frac{V_1}{I_w}$

Equivalent exciting reactance, $X_0 = \frac{V_1}{I_{mag}}$

The Iron losses measured by this test are used to determine transformer efficiency and parameters of exciting circuit of a transformer shown in Fig. 2.48.

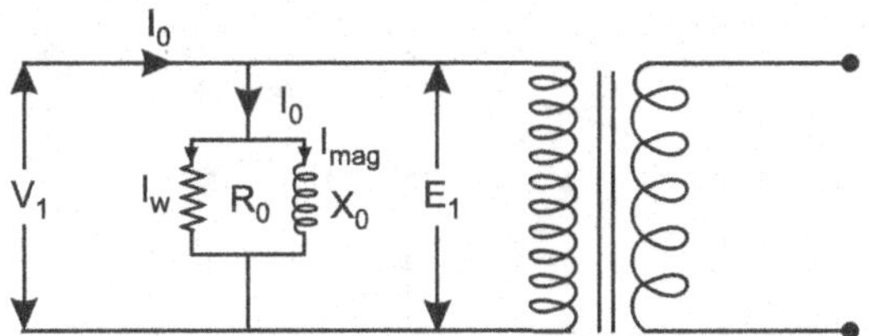

Fig. 2.48 Equivalent circuit of a transformer at no-load

2.34 Separation of Hysteresis and Eddy Current Losses

Core losses (i.e., iron losses or magnetic losses) of a transformer are constituted by (*i*) hysteresis loss and (*ii*) eddy current loss.

According to Steinmetz's empirical relations;

$$W_h = K_h Vf B_m^{1.6} \quad ...(2.77) \text{ and } W_e = K_e t^2 V f^2 B_m^2 \quad ...(2.78)$$

If thickness of laminations and volume of the core is kept constant, these losses will depend upon supply frequency and maximum flux density and hence.

$$W_h = P f B_m^{1.6} \quad ...(2.79)$$

and

$$W_e = Q f^2 B_m^2 \quad ...(2.80)$$

where P and Q are the new constants.

For a transformer, emf equation is given by the relation;

$$E = 4.44\, N f B_m A_i$$

or $B_m \propto \frac{E}{f} \propto \frac{V}{f}$ as other values are constant. ...(2.81)

For a particular value of B_m, the core losses per cycle may be represented as;

$$\frac{P_i}{f} = A + Bf \quad ...(2.82)$$

where A and B are other constants (i.e., $A = PB_m^{1.6}$ and $B = QB_m^2$)

The value of constants A and B can be determined by performing open-circuit test on the transformer at different frequencies but keeping ratio of V to f $\left(i.e.\ \frac{V}{f}\right)$ constant at every instant.

While performing this test, the applied voltage V and frequency f are varied together (by adjusting the excitation and speed of the alternator, respectively which supplies power to the transformer)

At every step, take the reading of frequency meter 'f' and wattmeter 'P_i'. Plot a curve between f and $\frac{P_i}{f}$, it will give a straight line curve as shown in Fig. 2.49.

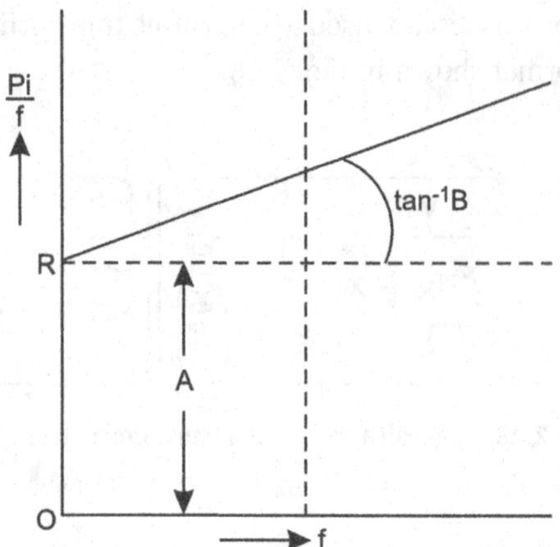

Fig. 2.49 Curve for frequency vs iron losses

Where this line intercepts the vertical axis (R) gives the value of constant A (i.e., $A = OR$), whereas, the slope of the line gives the value of constant B. Knowing the value of A and B we can separate the hysteresis and eddy current losses.

Example 2.42

The iron losses of a 400 V, 50 Hz transformer are 2500 W. These losses are reduced to 850 W when the applied voltage is reduced to 200 V, 25 Hz. Determine the eddy current loss at normal frequency and voltage.

Solution:

We know, $$E = 4.44\, N f A_i B_m$$

$$B_m \propto \frac{E}{f} \text{ (since all other quantities are constant)}$$

As $\frac{E_1}{f_1} = \frac{400}{50} = 8$ and $\frac{E_2}{f_2} = \frac{200}{25} = 8$; B_m is same in both the cases

$\therefore$ Hysteresis loss, $W_h \propto f = Pf$ (where P is a constant)

Eddy current loss, $W_e \propto f^2 = Qf^2$ (where Q is a constant)

$$\text{Total iron loss, } W_i = W_h + W_e = Pf + Qf^2 \text{ or } \frac{W_i}{f} = P + Qf \quad \text{...(2.83)}$$

When $f = 50$ Hz; $W_i = 2500$

$$\therefore \quad \frac{2500}{50} = P + Q \times 50 \text{ or } P + 50\,Q = 50 \quad \text{...(2.84)}$$

When $f = 25$ Hz; $W_i = 850$

$$\therefore \quad \frac{850}{25} = P + Q \times 25 \text{ or } P + 25Q = 34 \quad \text{...(2.85)}$$

Subtracting eq. (*iii*) from (*ii*), we get,

$$25\,Q = 16 \text{ or } Q = 0.64$$

From eq. (*ii*), we get, $P + 50 \times 0.64 = 50$; $P = 18$ ...(2.86)

Eddy current loss at normal frequency and voltage.

$$W_e = Qf^2 = 0.64 \times 50 \times 50 = \mathbf{1600\ W}\ (Ans.)$$

Example 2.43

A transformer has hysteresis and eddy current loss of 700 W and 500 W, respectively when connected to 1000 V, 50 Hz supply. If the applied voltage is raised to 2000 V and frequency to 75 Hz, find the new core losses.

Solution:

Here, $V_1 = 1000$ V; $f_1 = 50$ Hz; $W_i = 1200$ W; $W_{h1} = 700$ W

$$W_{e1} = 500 \text{ W}; V_2 = 2000 \text{ V}; f_2 = 75 \text{ Hz}$$

We know,

$$W_h \propto B_m^{1.6} \times f = PB_m^{1.6} \times f \quad \text{...(2.87)}$$

$$W_e \propto B_m^2 \times f^2 = Q\,B_m^2 \times f^2 \quad \text{...(2.88)}$$

Induced emf, $E = 4.44\,Nf\,B_m A_i$ volt

or

$$B_m \propto \frac{E}{f} \quad (\because 4.44\,NA_i \text{ are constant})$$

Substituting this value in eqs. (2.87) and (2.88), respectively, we get,

$$W_h = P\left(\frac{E}{f}\right)^{1.6} \times f = PE^{1.6} f^{-0.6} \quad \text{...(2.89)}$$

$$W_e = Q\left(\frac{E}{f}\right)^{2} \times f^2 = QE^2 \quad \text{...(2.90)}$$

Case-I: When $E = V_1 = 1000$ V and $f = f_1 = 50$ Hz

$$W_{h1} = P(V_1)^{1.6} f_1^{-0.6}$$

$$700 = P(1000)^{2.6} \times (50)^{-0.6}$$

$$700 = P \times 63096 \times 0.0956 \text{ or } P = 0.116$$

$$W_{e1} = Q \times E^2$$

$$500 = Q \times (1000)^2 \text{ or } Q = 5 \times 10^{-4}$$

Case-II: When $E = V_2 = 2000$ V and $f = f_2 = 75$ Hz

$$W_{h2} = P(V_2)^{1.6} f_2^{-0.6} = 0.116 \times (2000)^{2.6} \times (75)^{-0.6}$$

$$= 0.116 \times 192.27 \times 10^3 \times 0.075 = 1664 \text{ W}$$

$$W_{e2} = Q \times (V_2)^2 = 5 \times 10^{-4} \times (2000)^2 = 2000 \text{ W}$$

New core losses, $W_{i2} = W_{h2} + W_{e2} = 1664 + 2000 =$ **3664 W** (*Ans.*)

Example 2.44

The hysteresis and eddy current loss of a ferromagnetic sample at a frequency of 50 Hz is 25 watts and 30 watts, respectively, when the flux density of 0.75 tesla. Calculate the total iron loss at a frequency of 400 Hz, when the operating flux density is 0.3 tesla.

Solution:

At frequency, f_1= 50 Hz: 25 W; $W_{h1} = W_{e1} = 30$ W; $B_{m1} = 0.75$ tesla

frequency, $f_2 = 400$ Hz; $B_{m2} = 0.3$ tesla

$$W_{h1} = PB_{m1}^{1.6}\, f_1 \text{ or } 25 = \text{P} \times (0.75)^{2.6} \times 50$$

or
$$P = \frac{25}{(0.75)^{1.6} \times 50} = 0.7922\,; \quad W_{e1} = QB_{m1}^2 f_1^2$$

or
$$30 = Q \times (0.75)^2 \times (50)^2 \text{ or } Q = \frac{30}{(0.75)^2 \times (50)^2} = 0.02133$$

$$W_{h2} = PB_{m2}^{1.6} \times f_2 = 0.7922 \times (0.3)^{2.6} \times 400 = 46.16 \text{ W}$$

$$W_{e2} = QB_{m2}^2 \times f_2^2 = 0.02133 \times (0.3)^2 \times (400)^2 = 307.2 \text{ W}$$

Total iron losses, $P_i = W_{h2} + W_{e2} = 46.16 + 307.2 =$ **353.36 W** (*Ans.*)

Example 2.45

The following test results were obtained when a 10 kg specimen of sheet steel laminated core is put on power loss test keeping the maximum flux density and wave form factor constant.

Frequency (in Hz)	*25*	*40*	*50*	*60*	*80*
Total loss (in watt)	*18.5*	*36*	*50*	*66*	*104*

Calculate the current loss per kg at frequency of 50 Hz.

Solution:

At a given flux density and waveform factor, total iron losses are given as

$$P_i = P_h + P_e = A f + Bf^2 \text{ or } \frac{P_i}{f} = A + Bf$$

Total iron loss/cycle i.e., P_i/f for various values of frequency is given below:

f	25	40	50	60	80
P_i/f	0.74	0.9	2.0	2.1	2.3

A graph is plotted between P_i/f and f as illustrated in Fig. 2.50. From graph $A = 0.5$ and $B = 0.01$

Eddy current loss at 50 Hz = $Bf^2 = 0.01 \times (50)^2 = 25$ watt

Eddy current loss per kg at 50 Hz = $\frac{25}{10}$ = **2.5 watt** (*Ans.*)

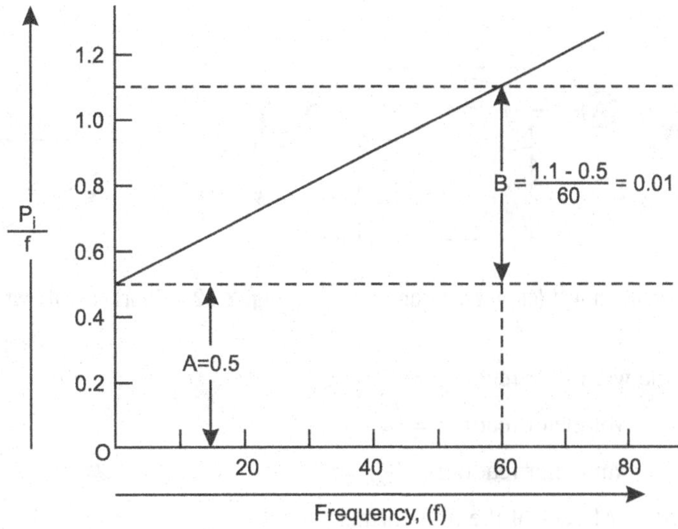

Fig. 2.50 Curve for frequency vs iron losses in a given transformer

2.35 Short Circuit Test

This test is carried out to determine the following:

(*i*) Copper losses at full load (or at any desired load). These losses are required for the calculations of efficiency of the transformer.

(*ii*) Equivalent impedance (Z_{es} or Z_{ep}), resistance (R_{es} or R_{ep}) and leakage reactance (X_{es} or X_{ep}) of the transformer referred to the winding in which the measuring instruments are connected. Knowing equivalent resistance and reactance, the voltage drop in the transformer can be calculated and hence regulation of transformer is determined.

This test is usually carried out on the high-voltage side of the transformer i.e., a wattmeter *W*, voltmeter *V* and an ammeter *A* are connected in high-voltage* winding (say secondary). The other winding (primary) is then short circuited by a thick strip or by connecting an ammeter *A'* across the terminals as shown in Fig. 2.52. A low voltage at normal frequency is applied to the high voltage winding with the help of on autotransformer so that full-load current flows in both the windings, measured by ammeters *A* and *A'*. *Low voltage is essential, failing which an excessive current will flow in both the windings which may damage them.*

Since a low voltage (usually 5 to 10% of normal rated voltage) is applied to the transformer winding, therefore, the flux set up in the core is very small about $\frac{1}{30}$th to $\frac{1}{8}$th of normal flux. The iron losses are negligibly small due to low value of flux as these losses are approximately proportional to the square of the flux. Hence, wattmeter reading W_c only represents the copper losses in the transformer windings for all practical purposes. The applied voltage V_{2sc} is measured by the voltmeter *V* which circulates the current I_{2sc} (usually full load current) in the impedance Z_{es} of the transformer to the side in which instruments are connected as shown in Fig. 2.52.

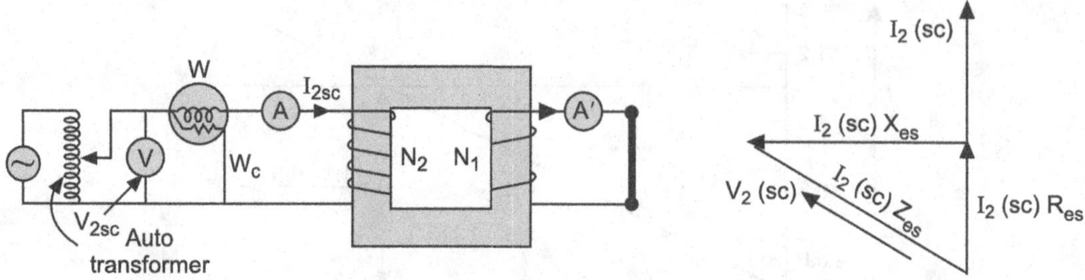

Fig. 2.51 Short circuit test (circuit diagram) **Fig. 2.52** Short circuit test (phasor diagram)

Let the wattmeter reading = W_c

voltmeter reading = V_{2sc}

and ammeter reading = I_{2sc}

Then, full load copper losses of the transformer,

$$P_c = \left(\frac{I_{2fl}}{I_{2sc}}\right)^2 Wc \quad \ldots(2.91)$$

and

$$I_{2sc}^2 R_{es} = W_C$$

Equivalent resistance referred to secondary,

$$R_{es} = \frac{W_c}{I_{2sc}^2} \quad \ldots(2.92)$$

From phasor diagram as shown in Fig. 2.52;

$$I_{2sc} Z_{es} = V_{2sc}$$

∴ Equivalent impedance referred to secondary,

$$Z_{es} = V_{2sc}/I_{2sc}$$

Equivalent reactance referred to secondary,

$$X_{es} = \sqrt{(Z_{es})^2 - (R_{es})^2}$$

* For convenience and better results, short circuit test is performed on *HV* side of the transformer; to make it clear, let us consider an example.

To perform short circuit test about 5 percent of rated voltage is required. For a 250 kVA, 2500/250V transformers, if short circuit test is performed on HV side, voltage required for the test is $2500 \times \frac{5}{100} = 125\ V$ (which is quite convenient and suitable for measurement) and the current would be $\frac{250 \times 1000}{2500} = 100\ A$.

If this test is conducted on LV side, then voltage required would be $\frac{250 \times 1000}{250} = 1000\ A$ (very high). It is observed that applied voltage is very low and current is very high, both are inconvenient to measure with ordinary instruments.

After calculating R_{es} and X_{es}, the voltage regulation of the transformer can be determined at any load and power factor.

2.36 Back-to-back Test

(*Sumpner's Test or Regenerative Test*)

Although efficiency and regulation of a transformer can be determined by performing open circuit and short circuit test but to test temperature rise, it is necessary to conduct a full-load test on a transformer. For small transformers, full-load test can be conveniently conducted, but for large transformers full-load test is very difficult. A suitable load to absorb full-load power of a large transformer may not be easily available. It will also be very expensive as a large amount of energy will be wasted in the load during the test. Therefore, large transformers can be tested for determining the maximum temperature rise by back-to-back test. This test is also called the **Regenerative test** or **Sumpner's test.**

The back-to-back test on single-phase transformers requires two identical transformers. Fig. 2.53, shows the circuit diagram for the back-to-back test on two identical single-phase transformers T_{r_1} and T_{r_2}. The primary windings of the two transformers are connected in parallel and supplied at rated voltage and rated frequency. A voltmeter, an ammeter and a wattmeter are connected to the input side as shown in Fig. 2.53.

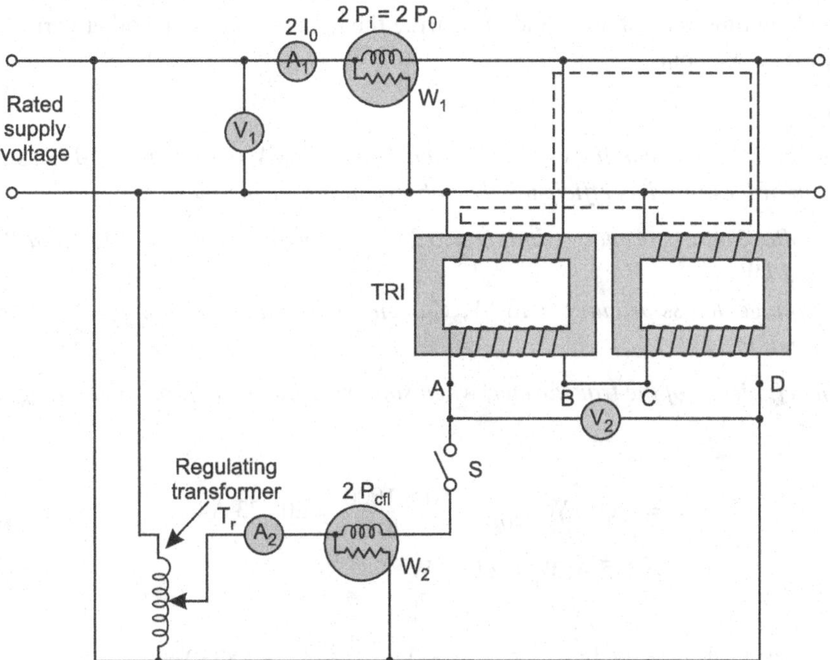

Fig. 2.53 Back-to-back test on two identical single-phase transformers

The secondary windings are connected in series with their polarities in phase opposition, which can be checked by the voltmeter V_2. The range of this voltmeter should be double the rated voltage of either transformer secondary. In order to check that the secondary windings are connected in series opposition, any two terminals (say B and C) are joined together and the voltage is measured between the remaining terminals A and D. If the voltmeter V_2 reads zero, the two secondary windings are in series opposition and terminals A and D are used for test. If the voltmeter reads a value approximately equal to twice the rated secondary voltage of either transformer, then the secondary windings are acting in the same direction. Then terminals A and C are joined and the terminals B and D are used for the test.

If the primary circuit is now closed, the total voltage across the two secondary windings in series will be zero. There will be no current in the secondary windings. The transformers will behave as if their secondary windings are open circuited. Hence, the reading of wattmeter W_1 gives the iron losses of both the transformers.

A small voltage is injected in the secondary circuit by a regulating transformer T_R excited by the main supply. The magnitude of the injected voltage is adjusted till the ammeter A_2 reads full-load secondary current. The secondary current produces full-load current to flow through the primary windings. This current will follow a circulatory path through the main bus-bars as shown by dotted line in Fig. 2.53. The reading of wattmeter W_2 will not be affected by this current. Thus, wattmeter W_2 gives the full-load copper losses of the two transformers.

The ammeter A_1 gives total no-load current of the two transformers. Thus, in this method we have loaded the two transformers to full load but the power taken from the supply is that necessary to supply the losses of both transformers.

The temperature rise of the transformers can be determined by operating these transformers back-to-back for a long time, say 48 hour, and measuring the temperature of the oil at periodic intervals of time, say every one hour.

Example 2.46

Open-circuit, and short-circuit tests were conducted on a 50 kVA, 6 360/240 V, 50 Hz, single-phase transformer in order to find its efficiency. The observations during these tests are:

O.C. test: *Voltage across primary winding = 6 360 V; Primary current = 1.0 A, and Power input = 2 kW.*

S.C. test: *Voltage across primary = 180 V; Current in secondary winding = 175 A, and Power input = 2 kW.*

Calculate the efficiency of the transformer, when supplying full-load at p.f. of 0.8 lagging.

Solution:

O.C. test: $W_i = 2\,000 \text{ W}; I_{2(fl)} = \frac{50 \times 1000}{240} = 208.33 \text{ A}$

S.C. test: $I_{2SC} = 175 \text{ A}; W_c = 2\,000 \text{ W}$

$$\therefore \text{ Cu loss at full-load, } P_c = W_c \left(\frac{I_{2(fl)}}{I_{2sc}} \right)^2 = 2000 \left(\frac{208.33}{175} \right)^2 = 2\,833 \text{ W}$$

$$\therefore \quad \text{Efficiency} = \frac{50 \times 10^3 \times 0.8 \times 100\%}{50 \times 10^3 \times 0.8 + 2\,000 + 2\,833} = \mathbf{95.33\%} \ (Ans.)$$

Example 2.47

A 15 kVA, 440/230 V, 50 Hz, single phase transformer gave the following test results:

Open Circuit (LV side)	*250 V,*	*1.8A,*	*95 W.*
Short Circuit Test (HV side)	*80 V,*	*12.0 A,*	*380 W.*

Compute the parameters of the equivalent circuit referred to LV side.

Solution:

Transformer rating = 15 kVA; $E_1 = 440$ V; $E_2 = 230$ V; $f = 50$ Hz

Open circuit test (LV side); $V_2 = 250$ V; $I_0 = 1.8$ A; $W_0 = 95$ W

Short circuit test (HV side); $V_{1(sc)} = 80$ V; $I_{1(sc)} = 12$ A; $W_c = 380$ W

From open circuit test performed on *LV* side;

$$I_w = \frac{W_0}{V_2} = \frac{95}{250} = 0.38 \text{ A}$$

$$I_{mag} = \sqrt{I_0^2 - I_w^2} = \sqrt{(1.8)^2 - (0.38)^2} = 1.75943 \text{ A}$$

$$\text{Exciting resistance, } R_0 = \frac{V_2}{I_w} = \frac{250}{0.38} = 658\ \Omega$$

$$\text{Exciting reactance, } X_0 = \frac{V_2}{I_{mag}} = \frac{250}{1.75943} = 142\ \Omega$$

From short circuit test performed on HV side;

$$Z_{ep} = \frac{V_{1(sc)}}{I_{1(sc)}} = \frac{80}{12} = 6.667\ \Omega$$

$$R_{ep} = \frac{W_c}{(I_{1(sc)})^2} = \frac{380}{(12)^2} = 2.639\ \Omega$$

$$X_{ep} = \sqrt{Z_{ep}^2 - R_{ep}^2} = \sqrt{(6.667)^2 - (2.639)^2} = 6.122\ \Omega$$

$$\text{Transformation ratio, } K = \frac{E_2}{E_1} = \frac{230}{440} = 0.5227$$

Transformer resistance and reactance referred to *LV* (secondary) side;

$$R_{es} = R_{ep} \times K^2 = 2.639 \times (0.5227)^2 = 0.7211\ \Omega$$

$$X_{es} = X_{ep} \times K^2 = 6.122 \times (0.5227)^2 = 2.673\ \Omega$$

Example 2.48

A 50 MVA, 60 Hz single-phase transformer indicates that it has a voltage rating of 8 kV: 78 kV. Open circuit test and short circuit test gave the following results:

Open Circuit Test: 8 kV, 61.9 A and 136 kW
Short Circuit Test: 650 V, 6.25 kA and 103 kW.

Determine the efficiency and voltage regulation if the transformer is operating at rated voltage and a load of 0.9 p.f. lagging.

Solution:

Here, rating of transformer = 50 MVA = 50×10^6 VA

$$V_1 = 8\text{ kV}; V_2 = 78\text{ kV}; \text{Load p.f., } \cos\phi = 0.9 \text{ lag}$$

Open circuit (*LV* side): $V_1 = 8$ kV; $I_0 = 61.9$ A; $W_0 = 136$ kW

Short circuit test (*LV* side): $V_{1(sc)} = 650$ V; $I_{1(sc)} = 6.25$ kA; $W_c = 103$ kW

From open circuit test, iron losses of the transformer,

$$P_0 = W_0 = 136\text{ kW}$$

At rated capacity, full-load current of the transformer on *LV* side,

$$I_{1(fl)} = \frac{\text{Rated capacity}}{V_1} = \frac{50 \times 10^6}{8 \times 10^3} = 6.25\text{ kA}$$

Since $I_{1(sc)} = I_{1(fl)} = 6.25$ kA, the short circuit test is performed at full load.

∴ Full load copper losses, $P_c = W_c = 103$ kW

Full-load efficiency,

$$\eta = \frac{\text{Rated kVA} \times \cos\phi}{\text{Rated kVA} \times \cos\phi + P_i \text{ in kW} + P_c \text{ in kW}}$$

$$= \frac{50 \times 10^3 \times 0.9}{50 \times 10^3 \times 0.9 + 136 + 103} = 0.9947 = \mathbf{99.47\%} \text{ (*Ans.*)}$$

Considering the data of short circuit test;

Transformer impedance referred to primary,

$$Z_{ep} = \frac{V_{1(sc)}}{I_{1(sc)}} = \frac{650}{6.25 \times 10^3} = 0.104\ \Omega$$

Transformer resistance referred to primary,

$$R_{ep} = \frac{W_c}{(I_{1(sc)})^2} = \frac{103 \times 10^3}{(6.25 \times 10^3)^2} = 0.0165\ \Omega$$

Transformer reactance referred primary,

$$X_{ep} = \sqrt{Z_{cp}^2 - R_{cp}^2} = \sqrt{(0.104)^2 - (0.0165)^2} = 0.10317\ \Omega$$

Load p.f., $\cos\phi = 0.9$; $\sin\phi = \sin\cos^{-1} 0.9 = 0.4359$

$$E_1 = V_1 - I_1 R_{ep}\cos\phi - I_1 X_{ep}\sin\phi$$

$$= 8\text{ kV} - 6.25\text{ kA} \times 0.0165 \times 0.9 - 6.25\text{ kA} \times 0.10317 \times 0.4359$$

$$= 8 - 0.0928 - 0.28107 = 7.626\text{ kV}$$

$$\text{Voltage regulation, } Reg = \frac{V_1 - E_1}{V_1} = \frac{8 - 7.626}{8} = 0.04675 \text{ (per unit)}$$

$$\% \; Reg = 0.04675 \times 100 = \mathbf{4.675\%} \; (Ans.)$$

Example 2.49

Determine the percentage efficiency and regulation at full load 0·9 p.f. lagging of a 5 kVA, 220/440 V single phase transformer. When the following test data is obtained. OC. Test – 220 V, 2 A, 100 W on L.V. side; S.C. Test – 40 V. 11·4 A, 200 W on H.V. side.

Solution:

From O.C. Test, Iron losses, $P_i = 100$ W;

From S.C. Test, Copper losses, $W_c = 200$ W (*at the load at which test is performed*)

$$\text{Full-load current on } HV \text{ side, } I_2 = \frac{kVA \times 1000}{V_2} = \frac{5 \times 1000}{440} = 11.4\text{A}$$

i.e., *SC* Test is performed at full load since $I_{2sc} = I_2$

$\therefore$ Full-load copper loss, $P_c = W_c = 200$ W

$$\text{Efficiency, } \eta = \frac{kVA \times 1000 \times \cos\phi_2}{kVA \times 1000 \times \cos\phi_2 + P_i + P_c} \times 100$$

$$= \frac{5 \times 1000 \times 0.9}{5 \times 100 \times 0.9 + 100 + 200} \times 100 = \mathbf{93{\cdot}75\%} \; (Ans.)$$

From S.C. Test:

$$R_{es} = \frac{\text{Wattmeter reading}}{(\text{Ammeter reading})^2} = \frac{200}{(11.4)^2} = 1.539\Omega$$

$$Z_{es} = \frac{\text{Volmeter reading}}{\text{Ammeter reading}} = \frac{40}{(11.4)} = 3.509\Omega$$

$$X_{es} = \sqrt{(Z_{es})^2 - (R_{es})^2} = \sqrt{(3.509)^2 - (1.539)^2} = 3.153\Omega$$

Here $\cos\phi_2 = 0{\cdot}9$; $\sin\phi_2 = \sin\cos^{-1} 0{\cdot}9 = 0{\cdot}4359$

$$E_2 = V_2 + I_2 R_{es} \cos\phi_2 + I_2 X_{es} \sin\phi_2$$

$$= 440 + 11{\cdot}4 \times 1{\cdot}539 \times 0{\cdot}9 + 11{\cdot}4 \times 3{\cdot}153 \times 0{\cdot}4359 = 471{\cdot}46 \text{ V}$$

$$\% \text{ Reg} = \frac{E_2 - V_2}{E_2} \times 100 = \frac{471.46 - 440}{471.46} \times 100 = \mathbf{6{\cdot}67\%} \; (Ans.)$$

Example 2.50

The following data was obtained when O.C. and *S.C. tests on a 5 kVA, 230/160 V, 50 Hz, transformer were performed.*

O.C. test (H.V. side) – 230V, 0·6 A, 80 watt
S.C. test (L.V. side) – 6 V, 15 A, 20 watt

Calculate the efficiency of transformer on full load at 0·8 p.f. lagging.

Solution:

From open circuit test, iron losses, $P_i = 80$ W

As short circuit test is performed on L.V. side,

$$I_{2sc} = 15 \text{ A}$$

Full load secondary current, $I_2 = \dfrac{5 \times 1000}{160} = 31.25\text{A}$

Copper losses measured at S.C. test, $W_c = 20$ W

Full load copper losses, $P_c = \left(\dfrac{I_2}{I_{2sc}}\right)^2 W_c = \left(\dfrac{31.25}{15}\right)^2 \times 20 = 86.8\text{ W}$

Efficiency of transformer at full load 0·8 p.f. lagging

$$\eta = \frac{kVA \times 1000 \times \cos\phi}{kVA \times 1000 \times \cos\phi + Pi + Pc} \times 100$$

$$= \frac{5 \times 1000 \times 0.8}{5 \times 1000 \times 0.8 + 80 + 86.8} \times 100 = \mathbf{96\%} \text{ (Ans.)}$$

Example 2.51

Following data were obtained on a 20 kVA, 50 Hz, 2000/200 volt distribution transformer

	V (Volt)	*I (Amp)*	*Power (Watt)*
O.C. Test	*200*	*4*	*120*
S.C. Test	*60*	*10*	*300*

Draw approximate equivalent circuit referred to H.V. side and L.V. side. Also calculate efficiency if the L.V. side is loaded fully at 0.8 power factor. What is maximum efficiency of the transformer at power factor and also find percentage load at which the maximum efficiency is obtained.

Solution:

Here, Rating of transformer = 20 kVA; $f = 50$ Hz; $V_1 = 2000$ V; $V_2 = 200$ V

Open circuit test is performed at rated voltage, therefore, as per data given, it is performed on *LV* side i.e., secondary side

$$\therefore \quad V_2 = 200 \text{ V}; I_0 = 4\text{A}; W_0 = 120 \text{ W}$$

Iron loss, $P_i = W_0 = 120$ W

Wattful component, $I_w = \dfrac{W_0}{V_2} = \dfrac{120}{200} = 0.6$ A

Magnetising component, $I_{mag} = \sqrt{I_0^2 - I_w^2} = \sqrt{(4)^2 - (0.6)^2} = 3.955$ A

Exciting resistance, $R_0 = \dfrac{V_2}{I_w} = \dfrac{200}{0.6} = 333.33\ \Omega$

Exciting reactance, $X_0 = \dfrac{V_2}{I_{mag}} = \dfrac{200}{3.955} = 50.57\ \Omega$

Short circuit test is performed at reduced voltage and almost at full-load current, therefore, as per data given, it is performed on HV side i.e., primary side.

$$\therefore \quad V_{1(sc)} = 60 \text{ V}; I_{1(sc)} = 10 \text{ A}; W_c = 300 \text{ W}$$

$$I_{1(sc)} = \frac{kVA \times 1000}{V_1} = \frac{20 \times 1000}{2000} = 10 \text{ A}$$

The test is performed at full-load.

Copper losses at full-load, $P_c = W_c = 300$ W

$$Z_{ep} = \frac{V_{1(sc)}}{I_{1(sc)}} = \frac{60}{10} = 6 \ \Omega$$

$$R_{ep} = \frac{W_c}{(I_{1(sc)})^2} = \frac{300}{(10)^2} = 3 \ \Omega$$

$$X_{cp} = \sqrt{Z_{ep}^2 - R_{ep}^2} = \sqrt{(6)^2 - (3)^2} = 5.2 \ \Omega$$

$$\text{Transformation ratio, } K = \frac{V_2}{V_1} = \frac{200}{2000} = 0.1$$

Exciting resistance and reactance refused to HV side

$$R'_0 = \frac{R_0}{K^2} = \frac{333.33}{(0.1)^2} = 33333 \ \Omega$$

$$X'_0 = \frac{X_0}{K^2} = \frac{50.57}{(0.1)^2} = 5057 \ \Omega$$

When resistance and reactance of transformer windings is referred to *LV* side

$$R_{es} = R_{ep} \times K^2 = 3 \times (0.1)^2 = 0.03 \ \Omega$$

$$X_{es} = X_{ep} \times K^2 = 5.2 \times (0.1)^2 = 0.052 \ \Omega$$

The approximate equivalent circuit of the transformer referred to *HV* side and *LV* side are drawn and shown in Figs. 2.54 (*a*) and (*b*), respectively. The values of various parameters are mentioned in the solution.

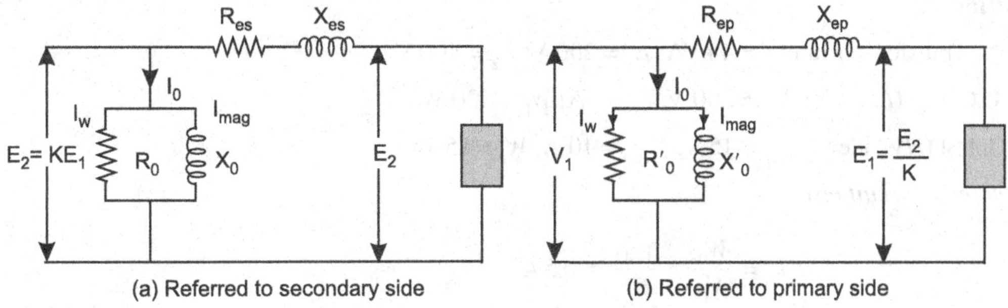

(a) Referred to secondary side (b) Referred to primary side

Fig. 2.54 Equivalent circuit

Efficiency of the transformer of full-load, 0.8 power factor,

$$\% \ \eta = \frac{kVA \times 1000 \times 0.8}{kVA \times 1000 \times 0.8 + P_i + P_c} \times 100$$

$$= \frac{20 \times 1000 \times 0.8}{20 \times 1000 \times 0.8 + 120 + 300} \times 100 = \mathbf{97.44\%} \ (Ans.)$$

Condition for maximum efficiency,

Copper loss = Iron loss

$$I_2'^2 R_{es} = 120$$

$$I_2' = \sqrt{\frac{120}{0.03}} = 63.246 \text{ A}$$

Fraction of the load at which the efficiency is maximum.

$$x = \frac{I_2'}{I_{2(fl)}} \qquad \left(I_{2(fl)}' = \frac{20 \times 1000}{200} = 100 \text{ A}\right)$$

$$= \frac{63.246}{100} = 0.63246$$

Value of max. efficiency, $\eta_{max.} = \frac{x \times kVA \times 1000 \times 0.8}{x \times kVA \times 1000 \times 0.8 + 120 + 120} \times 100$

$$= \frac{0.63246 \times 20 \times 1000 \times 0.8}{0.63246 \times 20 \times 1000 \times 0.8 + 120 + 120} \times 100$$

$$= \mathbf{97.68\%} \ (Ans.)$$

Example 2.52

Open-circuit and short-circuit tests on a 4 kVA, 200/400 V, 50 Hz, one-phase transformer gave the following test:

O.C. test:	*200 V,*	*1 A,*	*100 W*	*(on L.V. side)*
S.C. test:	*15 V,*	*10 A,*	*85 W*	*(with primary short-circuited)*

(i) Draw the equivalent circuit referred to primary: (ii) calculate the approximate regulation at the transformer at 0.8 p.f. lagging, and leading.

Solution:

Transformer rating = 4 kVA; $E_1 = 200$ V; $E_2 = 400$ V

O.C. test (*LV*, side): $V_1 = 200$ V; $I_0 = 1$ A; $W_0 = 100$ W

S.C. test (*HV* side): $V_{2(sc)} = 15$ V; $I_{2(sc)} = 10$ A; $W_c = 85$ W

From open-circuit test:

$$I_w = \frac{W_0}{V_1} = \frac{100}{200} = 0.5 \text{ A}$$

$$I_{mag} = \sqrt{I_0^2 - I_w^2} = \sqrt{(1)^2 - (0.5)^2} = 0.866 \text{ A}$$

$$R_0 = \frac{V_1}{I_w} = \frac{200}{0.5} = 400\ \Omega$$

$$X_0 = \frac{V_1}{I_{mag}} = \frac{200}{0.866} = 231\ \Omega$$

From short - circuit test:

$$R_{es} = \frac{W_c}{(I_{2(sc)})^2} = \frac{85}{(10)^2} = 0.85\ \Omega$$

$$Z_{es} = \frac{V_{2(sc)}}{I_{2(sc)}} = \frac{15}{10} = 1.5\ \Omega$$

$$X_{es} = \sqrt{Z_{es}^2 - R_{es}^2} = \sqrt{(1.5)^2 - (0.85)^2} = 1.236\ \Omega$$

$$K = \frac{400}{200} = 2$$

$$R_{ep} = \frac{R_{es}}{K^2} = \frac{0.85}{(2)^2} = 0.21\ \Omega$$

$$X_{ep} = \frac{X_{es}}{K^2} = \frac{1.236}{(2)^2} = 0.31\ \Omega$$

The equivalent circuit referred to primary side is shown in Fig. 2.55.

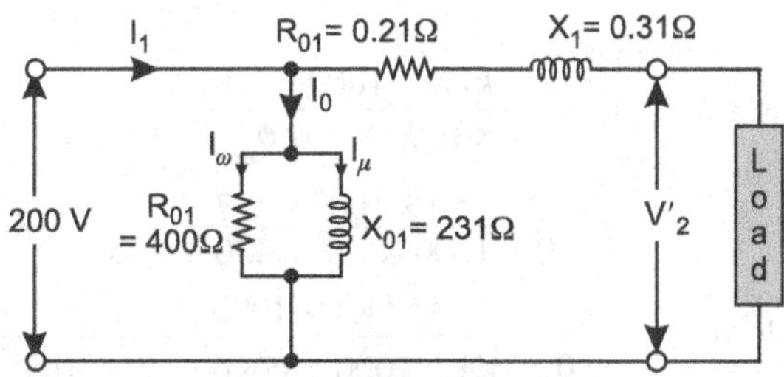

Fig. 2.55 Equivalent circuit

Power factor of the load, $\cos\phi = 0.8$; $\sin\phi = \sin\cos^{-1} 0.8 = 0.6$

Full-load secondary current, $I_2 = \dfrac{kVA \times 1000}{E_2} = \dfrac{4 \times 1000}{400} = 10$ A

(*ii*) Percentage regulation (referred to secondary) *at:*

(*a*) 0.8 p.f. lagging $= \dfrac{I_2(R_2\cos\phi + X_2\sin\phi) \times 100}{E_2}$

$$= \frac{10(0.85 \times 0.8 + 1.236 \times 0.6) \times 100}{400} = \mathbf{3.544\%}\ (Ans.)$$

(*b*) 0.8 p.f. leading $= \dfrac{10(0.85 \times 0.8 - 1.236 \times 0.6) \times 100}{400} = \mathbf{-0.154\%}$ (*Ans.*)

Example 2.53

When OC Test and SC Test were performed on a 50 kVA transformer the following results were obtained:

Open circuit tests: Primary voltage 3300 V, secondary voltage 415 V, power 430 W

Short circuit test: Primary voltage 124 V, primary current 15.3 A, primary power 525 W secondary current full load value.

Calculate:

(a) The efficiency at full-load and at half-load for 0.7 power factor.

(b) The voltage regulation for power factor 0.7: (i) lagging (ii) leading

(c) The secondary terminal voltages corresponding to (i) and (ii).

Solution:

Rating of transformer = 50 kVA; Power factor = 0.7

Open circuit test (primary): $V_1 = 3300$ V; $V_2 = 415$ V; $W_0 = 430$ W

Short circuit test (primary): $V_{1(sc)} = 124$ V; $I_{1(sc)} = 15.3$ A; $W_c = 525$ W

Short circuit test is performed at full-load secondary current,

$\therefore$ Full-load copper losses, $P_c = W_c = 525$ W

Iron losses, $P_i = W_0 = 430$ W

When p.f., $\cos\phi = 0.7$

(*a*) Full-load efficiency,

$$\eta_{fl} = \frac{kVA \times 1000 \times \cos\phi}{kVA \times 1000 \times \cos\phi + P_i + P_c}$$

$$= \frac{50 \times 1000 \times 0.7}{50 \times 1000 \times 0.7 + 430 + 525} \times 100 = \mathbf{97.34\%} \text{ (Ans.)}$$

Efficiency at half-load,

$$\eta_{0.5} = \frac{0.5\,kVA \times 1000 \times \cos\phi}{0.5\,kVA \times 1000 \times \cos\phi + P_i + (0.5)^2 P_c}$$

$$= \frac{0.5 \times 50 \times 1000 \times 0.7}{0.5 \times 50 \times 1000 \times 0.7 + 430 + (0.5)^2 \times 525} \times 100$$

$$= \mathbf{96.89\%} \text{ (Ans.)}$$

(*b*) Transformer impedance referred to primary,

$$Z_{ep} = \frac{V_{1(sc)}}{I_{1(sc)}} = \frac{124}{15.3} = 8.1\ \Omega$$

Transformer resistance referred to primary,

$$R_{ep} = \frac{W_c}{(I_{1(sc)})^2} = \frac{525}{(15.3)^2} = 2.243\ \Omega$$

Transformer reactance referred to primary,

$$X_{ep} = \sqrt{Z_{ep}^2 - R_{ep}^2}$$

$$= \sqrt{(8.1)^2 - (2.243)^2} = 7.783\ \Omega$$

At 3300 V, primary full-load current,

$$I_1 = \frac{50 \times 1000}{3300} = 15.15\ \text{A}$$

For p.f., cos ϕ = 0.7 lag; sin ϕ = sin cos^{-1} 0.7 = 0.714

$$E_1 = V_1 - I_1 R_{ep} \cos\phi - I_1 X_{cp} \sin\phi$$

$$= 3300 - 15.15 \times 2.243 \times 0.7 - 15.15 \times 7.783 \times 0.714$$

$$= 3300 - 23.787 - 84.189 = 3192\ \text{V}$$

$$\% R_{eg} = \frac{V_1 - E_1}{V_1} \times 100 = \frac{3300 - 3192}{3300} \times 100 = \mathbf{3.27\%}\ (Ans.)$$

For p.f., cos ϕ = 0.7 leading; sin ϕ = sin cos^{-1} 0.7 = 0.714

$$E'_1 = V_1 - I_1 R_{ep} \cos\phi + I_1 X_{cp} \sin\phi$$

$$= 3300 - 15.15 \times 2.243 \times 0.7 + 15.15 \times 7.783 \times 0.714$$

$$= 3300 - 23.787 + 84.189 = 3360\ \text{V}$$

$$\% R_{eg} = \frac{V_1 - E_1}{V_1} \times 100 = \frac{3300 - 3360}{3300} \times 100 = \mathbf{-1.82\%}\ (Ans.)$$

(*c*) Secondary terminal voltage at 0.7 p.f. lagging:

$$V_2 = E_2 = K \times E_1 \text{ where } K = \frac{415}{3300}$$

$$= \frac{415}{3300} \times 3192 = \mathbf{402.4\ V}\ (Ans.)$$

Secondary terminal voltage at 0.7 p.f. leading;

$$V_2' = E_2' = K \times E_1' = \frac{415}{3300} \times 3360 = \mathbf{422.5\ V}\ (Ans.)$$

Example 2.54

A 10 kVA, 2500/250 V, single phase transformer gave the following results:
Open circuit Test: 250 V, 0.8 A, 50 W
Short circuit Test: 60 V, 3A, 45 W
Determine

(a) Equivalent circuit parameters referred to LV side
(b) The efficiency at full load and 0.8 power factor lagging

(c) Load (kVA) at which the maximum efficiency occurs
(d) Voltage regulation at rated load and 0.8 pf leading
(e) Secondary terminal voltage at rated load and 0.8 pf lagging.

Solution:

Transformer rating = 10 kVA; $E_1 = 2500$ V; $E_2 = 250$ V

As per data, open circuit test is performed on *LV* side and short circuit test is performed on *HV* side.

Open circuit test (*LV* side); $V_2 = 250$ V; $I_0 = 0.8$ A; $W_0 = 50$ W

Short circuit test (*HV* side); $V_{1(sc)} = 60$ V; $I_{1(sc)} = 3$A; $W_c = 45$ W

(*a*) Open circuit test performed on *LV* side given;

$$I_0 = 0.8 \text{ A}; I_w = \frac{W_0}{V_2} = \frac{50}{250} = 0.2 \text{ A}$$

$$I_{mag} = \sqrt{I_0^2 - I_w^2} = \sqrt{(0.8)^2 - (0.2)^2} = 0.7746 \text{ A}$$

$$\text{Exciting resistance, } R_0 = \frac{V_2}{I_w} = \frac{250}{0.2} = \mathbf{1250\ \Omega}\ (\textit{Ans.})$$

$$\text{Exciting reactance, } X_0 = \frac{V_2}{I_{mag}} = \frac{250}{0.7746} = \mathbf{323\ \Omega}\ (\textit{Ans.})$$

Short circuit test performed on *HV* side gives;

$$Z_{ep} = \frac{V_{1(sc)}}{I_{1(sc)}} = \frac{60}{3} = 20\ \Omega$$

$$R_{ep} = \frac{W_c}{(I_{1(sc)})^2} = \frac{45}{(3)^2} = 5\ \Omega$$

$$X_{ep} = \sqrt{Z_{ep}^2 - R_{ep}^2} = \sqrt{(20)^2 - (5)^2} = 19.36\ \Omega$$

$$\text{Transformation ratio, } K = \frac{E_2}{E_1} = \frac{250}{2500} = 0.1$$

Parameters, when referred to *LV* side;

$$R_{es} = K^2 \times R_{ep} = (0.1)^2 \times 5 = 0.05\ \Omega$$

$$X_{es} = K^2 \times X_{ep} = (0.1)^2 \times 19.36 = 0.1936\ \Omega$$

$$(b) \text{ Full-load current on } HV \text{ side, } I_{1(fl)} = \frac{10 \times 1000}{2500} = 4\text{A}$$

$$\text{Copper loss at full-load, } P_c = \left(\frac{I_{1(fl)}}{I_{1(sc)}}\right)^2 \times W_c = \left(\frac{4}{3}\right)^2 \times 45 = 80 \text{ W}$$

$$\text{Iron loss, } P_i = W_0 = 50 \text{ W}$$

Efficiency at full-load, 0.8 p.f. lagging;

$$\eta = \frac{kVA \times 1000 \times \cos\phi}{kVA \times 1000 \times \cos\phi + P_i + P_c}$$

$$= \frac{10 \times 1000 \times 0.8}{10 \times 1000 \times 0.8 + 50 + 80} \times 100 = \mathbf{98.4\%} \ (Ans.)$$

(*c*) Fraction of load at which the efficiency is maximum;

$$x = \sqrt{\frac{P_i}{P_c}} = \sqrt{\frac{50}{80}} = 0.79$$

Load in kVA at which the efficiency is maximum

$$= x \times \text{rated kVA} = 0.79 \times 10 = \mathbf{7.9\ kVA} \ (Ans.)$$

(*d*) When p.f., $\cos\phi = 0.8$ leading; $\sin\phi = \sin\cos^{-1} 0.8 = 0.6$

$$I_{2(fl)} = \frac{10 \times 1000}{250} = 40 \text{ A}$$

$$V_2 = E_2 - I_2 R_{es} \cos\phi + I_2 X_{es} \sin\phi$$

$$= 250 - 40 \times 0.05 \times 0.8 + 40 \times 0.1936 \times 0.6$$

$$= 250 - 1.6 + 4.6464 = 253 \text{ V}$$

$$\% \ Reg = \frac{E_2 - V_2}{E_2} \times 100 = \frac{250 - 253}{250} \times 100 = \mathbf{-1.2\%} \ (Ans.)$$

(*e*) When p.f. $\cos\phi = 0.8$ lagging; $\sin\phi = \sin\cos^{-1} 0.8 = 0.6$

$$V_2 = E_2 - I_2 R_{es} \cos\phi - I_2 X_{es} \sin\phi$$

$$= 250 - 40 \times 0.05 \times 0.8 + 40 \times 0.1936 \times 0.6$$

$$= 250 - 2.6 - 4.6464 = \mathbf{243.75\ V} \ (Ans.)$$

Section Practice Problems

Numerical Problems

1. The iron losses of a transformer are 2500 W when operated on 440 V, 50 Hz; these are reduced to 850 W when operated on 220V, 25 Hz. Calculate the eddy current loss at normal frequency and voltage. (**Ans.** *1600 W*)

2. The iron losses in a transformer core at normal flux density were 30 W at 30 Hz frequency and these were changed to 54 W at 50 Hz frequency. Calculate (*a*) the hysteresis loss and (*b*) the eddy current loss at 50 Hz. (**Ans.** *44 W, 10 W*)

3. *The following test results were obtained for a 250/500 V, single-phase transformer: – Short circuit test with low voltage winding short circuited; 20 V, 12 A, 100 W; Open circuit test on low voltage side; 250 V, 1 A, 30 W. Determine the efficiency of the transformer when the output is 10 A, 500 V at 0·8 p.f. lagging.* (**Ans.** *96.4%*)

4. A 5 kVA, 230/110 V, 50 c/s transformer gave the following test results; *O.C.* test (*H.V.* side); 230 V, 0·6 A, 80 W; S.C. test (*L.V.* side); 6 V, 15 A, 20 W. Calculate the efficiency of the transformer on full-load at 0·8 p.f. lagging. Also calculate the voltage on the secondary side under full-load conditions at 0·8 p.f. leading. (**Ans.** *93·82%*, *117·4 V*)

5. A 5 kVA, 400/200 V, 50 Hz, l-phase transformer gave the following results:

No-load: 400 V, 1 A, 50 W (*LV* side)

Short-circuit: 12 V, 10 A, 40 W (*HV* side)

Calculate (*a*) the components of no-load current (*b*) the efficiency and regulation at full load and power factor of 0.8 lagging. (**Ans.** *0.125 A; 0.992 A; 97.8%; 3.13%*)

6. A transformer has copper loss of 1.5% and reactance 3.5% when tested on load. Calculate its full-load regulation at (*i*) unity power factor (*ii*) 0.8 p.f. lagging and (*iii*) 0.8 p.f. leading. (**Ans.** *1.56%; 3.32%; –0.83%*)

7. A 5 kVA, 200/400 V 50 Hz single phase transformer gave the following results.

O.C. Test	200 V,	0.7 A,	60 W	Low voltage side
S.C. Test	22 V,	120 W,	120 W	High voltage side

(*a*) Find the %age regulation when supplying full load at 0.9 p.f. lagging.

(*b*) Determine the load which gives maximum efficiency and find the value of this efficiency at unity p.f. (**Ans.** 3.08%, 4.54 kVA, 97.4%)

Short Answer Type Questions

Q.1. Why short circuit test is performed on high voltage side of transformer?

Ans. For convenience and better results, short circuit test is performed on *HV* side of a transformer.

Q.2. Why are iron losses or core losses assumed to remain constant in a power transformer from no-load to full-load?

Ans. The magnetic flux set-up in the core remains the same from no-load to full-load, hence iron losses remain constant from no-load to full-load.

Q.3. How can iron loss be measured?

Ans. Iron losses can be measured by performing no-load test on a transformer.

Q.4. The percentage leakage impedance of a one-phase 2000 V/100 V, 5 kVA transformer is 2.5. What voltage should be applied to HV side for carrying out short-circuit test at rated current?

Ans. Rated full-load on *HV* side, $I_1 = \dfrac{5 \times 1000}{2000} = 2.5\,\text{A}$

$$\% Z = \frac{I_1 Z}{V_1} \times 100$$

or $$2.5 = \frac{2.5 \times Z}{2000} \times 100$$

or $$Z = \frac{2.5 \times 2000}{2.5 \times 100} = 20\ \Omega$$

Applied voltage at short circuit, $V_{SC(1)} = I_1 Z = 2.5 \times 20 = 50\text{V}$

2.37 Classification of Transformers

The transformers are often classified according to their **applications.** Following are the important types of transformers:

(*i*) **Power Transformers:** These transformers are used to step up the voltage at the generating station for transmission purposes and then to step down the voltage at the receiving stations. These transformers are of large capacity (generally above 500 kVA). These transformers usually operate at high average load, which would cause continuous capacity copper loss, thus affecting their efficiency. To have minimum losses during 24 hours, such transformers are designed with low copper losses.

(*ii*) **Distribution Transformers:** These transformers are installed at the distribution sub-stations to step down the voltage. These transformers are continuously energised causing the iron losses for all the 24 hours, Generally the load on these transformers fluctuate from no-load to full load during this period. To obtain high efficiency, such transformers are designed with low iron losses.

(*iii*) **Instrument Transformers:** To measure high voltages and currents in power system potential transformer (*P.T.*) and current transformer (*C.T.*) are used, respectively. The potential transformers are used to decrease the voltage and current transformers are used to decrease the current up to measurable value. These are also used with protective devices.

(*iv*) **Testing transformers:** These transformers are used to step up voltage to a very high value for carrying out the tests under high voltage, e.g., for testing the dielectric strength of transformer oil.

(*v*) **Special purpose transformer:** The transformers may be designed to serve special purposes, these may be used with furnaces, rectifiers, welding sets etc.

(*vi*) **Auto-transformers:** These are single winding transformers used to step down the voltages for starting of large three-phase squirrel cage induction motors.

(*vii*) **Isolation transformer:** These transformers are used only to isolate (electrically) the electronic circuits from the main electrical lines, therefore, their transformation ratio is usually one.

(*viii*) **Impedance matching transformer:** These transformers are used at the output stage of the amplifier for impedance matching to obtain maximum output from the amplifiers.

2.38 Parallel Operation of Transformers

When the primaries and secondaries of the two or more transformers are connected separately to the same incoming and outgoing lines to share the load, the transformers are said to be connected in **parallel.**

The two single-phase transformers *A* and *B* are placed in parallel as shown in Fig. 2.56. Here the primary windings of the two transformers are joined to the supply bus-bars and the secondary windings are joined to the load through load bus-bars. Under this conditions;

$$V_1 = \text{Primary applied voltage}$$

$$V_2 = V = \text{Secondary load voltage.}$$

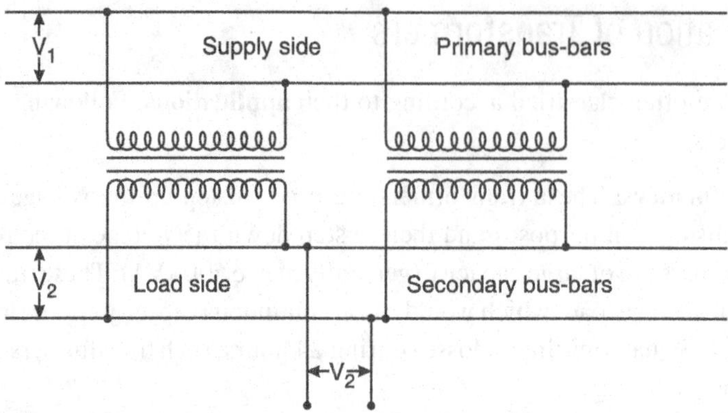

Fig. 2.56 Parallel operation of two one-phase transformers (circuit diagram)

2.39 Necessity of Parallel Operation

The following are the reasons for which transformers are put in parallel.

(*i*) When the load on the transmission lines increases beyond the capacity of the installed transformer. To overcome this problem one way is to replace the existing transformer with the new one having larger capacity (this is called augmentation of transformer) and the other way is to place one more transformer is parallel with the existing one to share the load. The *cost of replacing the transformer* is much more than placing another one in parallel with the existing one.

Hence, it is desirable to place another transformer is parallel when the electrical load on the existing transformer increases beyond its rated capacity.

(*ii*) Sometimes, the amount of power to be transformed is so high that it is not possible to build a single unit of that capacity, then we have to place two or more transformers in parallel.

Hence, parallel operation of transformers is necessary when the amount of power to be transformed is much more than that which can be handled by single unit (transformer).

(*iii*) At the grid sub stations, spare transformers are always necessary to insure the continuity of supply in case of breakdown. The size of spare transformer depends upon the size of transformers placed at the grid sub-station. Therefore, it is desirable to place transformers of smaller capacity in parallel to transform the given load which in turn *reduces the size of the spare transformer.*

Hence, it is desirable to do parallel operation of transformers if we want to keep the spare transformer of smaller size.

2.40 Conditions for Parallel Operation of One-phase Transformers

The following conditions are to be fulfilled if two or more transformers are to be operated successfully in parallel to deliver a common load.

(*i*) *Both the transformers should have same transformation ratio* i.e., *the voltage ratings of both primaries and secondaries must be identical.*

If this condition is not exactly fulfilled i.e., if the two transformers '*A*' and '*B*' have slight difference in their voltage or transformation ratios, even then parallel operation is possible. Since the transformation ratios are unequal, primary applied voltage being equal, the induced emfs in the secondary windings will not be equal. Due to this inequality of induced emfs in the secondary windings, there will be, even at no-load, some circulating current flowing from one secondary winding (having higher induced *emf*) to the other secondary windings (having lower induced *emf*). In other words, there will be circulating currents between the secondary windings and therefore between primary windings also when the secondary terminals are connected in parallel. The impedance of transformers is small, so that a small percentage voltage difference may be sufficient to circulate a considerable current and cause additional $I^2 R$ loss. When load is applied on the secondary side of such a transformer, the unequal loading conditions will occur due to circulating current. Hence, it may be impossible to take the combined full load *kVA* output from the parallel connected group without one of the transformers becoming excessively hot. For satisfactory parallel operation the circulating current should not exceed 10% of the normal load current.

(*ii*) *Both the transformers should have the same percentage impedance.*

If this condition is not exactly fulfilled, i.e., the impedance triangles at the rated *kVA*'s are not identical in shape and size, even then parallel operation will be possible, but the power factors at which the transformers operate will differ from the power factor of the load. Therefore, in this case the transformers will not share the load in proportion to their *kVA* ratings.

(*iii*) *Both the transformers must have the same polarity* i.e., *both the transformers must be properly connected with regard to their polarities.*

If this condition is not observed, the emfs in the secondary windings of the transformers which are parallel with incorrect polarity will act together in the local secondary circuits and produce the effect equivalent to a **dead short circuit.**

Polarity Check: Referring to Fig. 2.56 we have two transformers *A* and *B* with terminals of unknown polarity. Primary terminals T_1 and T_2 are connected to the primary bus bars as usual. On the secondary side, one of the terminals of the transformer *A* is joined temporarily to a terminal of transformers *B* and the other terminals of *A* and *B* transformers are connected through the **Double range voltmeter** i.e., the two secondary windings and the *double range voltage* form a closed series circuit. If the double range voltmeter reads zero, when the primary windings are energised, then the terminal which are temporarily joined are of correct polarity i.e., like polarity, but if the double range voltmeter reads twice the normal secondary voltage, these terminals are of incorrect polarity i.e., unlike polarity.

(*iv*) *In case of 3-phase transformers, the two transformers must have the same phase-sequence* i.e., *the transformers must be properly connected with regard to their phase-sequence.*

If this condition is not observed, it will have the same effect as discussed above when the polarity of two single phase transformers is not the same. Phase-sequence is also checked as discussed above.

(*v*) *In case of 3-phase transformers, the two transformers must have the connections so that there should not be any phase difference between the secondary line voltages i.e., a delta-*

star connected transformer should not be connected with a delta-delta or star-star connected transformer.

If this condition is not fulfilled, it will cause a heavy circulating current which may damage the transformers.

2.41 Load Sharing between Two Transformers Connected in Parallel

The load sharing between two transformers connected in parallel depends upon the various conditions are as discussed below:

(*i*) **When the two transformers have the same voltage ratios and their impedance voltage triangles are identical in size and shape.**

The condition for which the transformers have the same voltage ratio and impedance voltage triangles is known *ideal condition or ideal case.* Let E be the no-load secondary voltage of each transformer and V the terminal voltage. Figure 2.57 shows the equivalent circuit in which I_1 and I_2 are the currents supplied by the transformers A and B, respectively. I be the total load current lagging behind the voltage V by an angle ϕ. The impedance voltage triangles of the individual transformers are identical in shape and size and are therefore, represented by a single triangle VAB with the resistance drop (side VA) parallel to the load current vector OI as shown in Fig. 2.58. The current I_1 and I_2 in the individual transformers are both in phase with the load current I and are inversely proportional to their respective impedances.

$$\therefore \quad \bar{I} = \bar{I}_1 + \bar{I}_2 \qquad \ldots(2.93)$$

$$\text{Also} \quad \bar{I}_1 \bar{Z}_1 = \bar{I}_2 \bar{Z}_2 \qquad \ldots(2.94)$$

$$\text{or} \quad \frac{\bar{I}_1}{\bar{I}_2} = \frac{\bar{Z}_2}{\bar{Z}_1}$$

$$\text{or} \quad \bar{I}_1 = \frac{\bar{Z}_2}{\bar{Z}_1} \cdot \bar{I}_2$$

Substituting the value of I_1 in eq. (*i*), we get,

$$\bar{I} = \bar{I}_2 \frac{\bar{Z}_2}{\bar{Z}_1} + \bar{I}_2 = \bar{I}_2 \left[\frac{\bar{Z}_2 + \bar{Z}_1}{\bar{Z}_1} \right] \text{ or } \bar{I}_2 = \frac{\bar{Z}_1}{\bar{Z}_1 + \bar{Z}_2} \times \bar{I} \qquad \ldots(2.95)$$

$$\text{Similarly,} \quad \bar{I}_1 = \frac{\bar{Z}_2}{\bar{Z}_1 + \bar{Z}_2} \times \bar{I} \qquad \ldots(2.96)$$

Multiplying both sides by the common terminal voltage V, we get,

$$\bar{I}_1 \bar{V} = \frac{\bar{Z}_2}{\bar{Z}_1 + \bar{Z}_2} \bar{V} \bar{I}\text{; Similarly } \bar{I}_2 \bar{V} \frac{\bar{Z}_1}{\bar{Z}_1 + \bar{Z}_2} \bar{V} \bar{I}$$

$$\therefore \quad \overline{kVA_1} = \frac{\bar{Z}_2}{\bar{Z}_1 + \bar{Z}_2} \times \overline{kVA} \qquad \ldots(2.97)$$

and $$\overline{kVA_2} = \frac{\overline{Z_1}}{\overline{Z_1} + \overline{Z_2}} \times \overline{kVA} \quad \ldots(2.98)$$

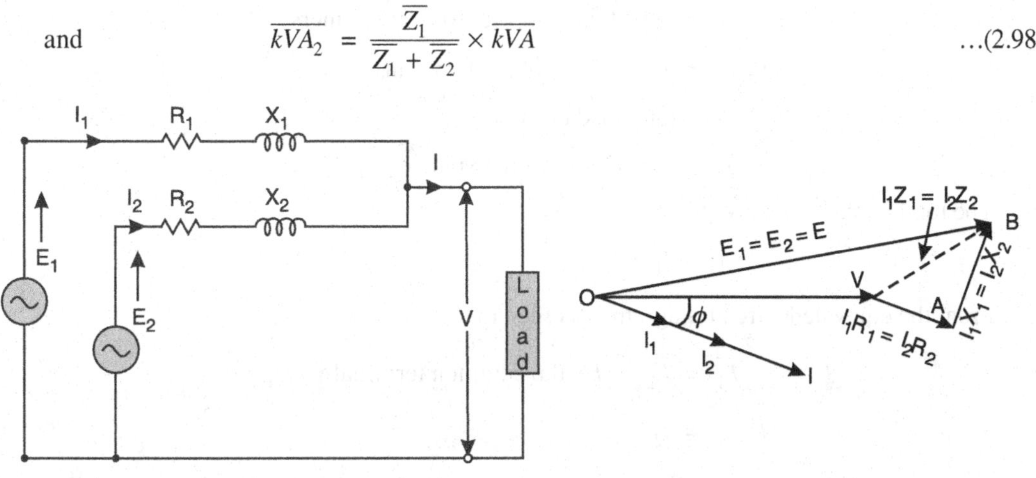

Fig. 2.57 Circuit diagram of two transformers connected in parallel

Fig. 2.58 Phasor diagram of two transformers operating in parallel (Impedance voltage triangles are identical)

(ii) When the two transformers have the same voltage ratios but different voltage triangles

In this case, no-load voltages of both secondary are equal in magnitude as well as in phase i.e., there is no phase difference between E_1 and E_2 which will only be possible if the magnetising currents of the two transformers are not very different from each other or nearly the same. Under these conditions, both sides of two transformers can be connected in parallel, and no current will circulate between them on no-load.

Figure 2.59 shows the equivalent circuit diagram when the parallel connected transformers are sharing the load current I, and it represents two impedances in parallel. The impedance voltage triangle is now represented by two triangles VAB, VA' B having common hypotenuse VB as shown in Fig. 2.60. The resistance drop sides of the triangles VA, VA' are parallel to the phasors OI_1 and OI_2 of the respective secondary currents. The sum of these vectors OI_1 and OI_2 represents the load current OI.

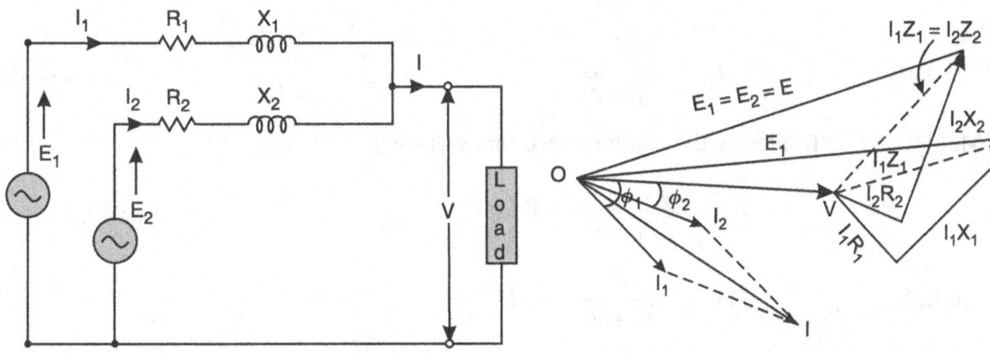

Fig. 2.59 Circuit diagram of two transformers connected in parallel

Fig. 2.60 Phasor diagram of two transformesr operating in parallel (having different voltage triangles)

$\overline{Z_1}$ and $\overline{Z_2}$ = impedances of the two transformers

$\overline{I_1}$ and $\overline{I_2}$ = the currents of the two transformers

$\bar{I}$ = total load current

$\overline{V}$ = common terminal voltage.

The total current $\bar{I}$ is given by

$$\bar{I} = \overline{I_1} + \overline{I_2}$$

From the equivalent circuit diagram, it is seen that

$$\overline{E_1} = \overline{E_2} = \overline{E}\text{, the common terminal voltage}$$

$$\therefore \quad \overline{I_1}\,\overline{Z_1} = \overline{I_2}\,\overline{Z_2} = \bar{v} \qquad \text{...(2.99)}$$

Since $\overline{Z_1}$ and $\overline{Z_2}$ are in parallel, we have the equivalent impedance $\overline{Z}$

$$\frac{1}{\overline{Z}} = \frac{1}{\overline{Z_1}} + \frac{1}{\overline{Z_2}}$$

$$\therefore \quad \overline{Z} = \frac{\overline{Z_1}\,\overline{Z_2}}{\overline{Z_1} + \overline{Z_2}}$$

∴ The total combined current

$$\bar{I} = \frac{v}{\overline{Z}} = \frac{v\left(\overline{Z_1} + \overline{Z_2}\right)}{\overline{Z_1}\,\overline{Z_2}}$$

or

$$\bar{v} = \frac{\overline{Z_1\,Z_2}}{(\overline{Z_1} + \overline{Z_2})} \times \bar{I}$$

$$\therefore \quad \overline{I_1}\,\overline{Z_1} = \frac{\overline{Z_1 Z_2}}{\overline{Z_1} + \overline{Z_2}} \times \bar{I}$$

or

$$\overline{I_1} = \frac{\overline{Z_1}}{\overline{Z_1} + \overline{Z_2}} \qquad \text{...(2.100)}$$

Similarly

$$\overline{I_2} = \frac{\overline{Z_1}}{\overline{Z_1} + \overline{Z_2}} \qquad \text{...(2.101)}$$

Multiplying both sides by the common terminal voltage $\overline{V}$, we get,

$$\overline{I_1}\,\overline{V} = \frac{\overline{Z_1}}{\overline{Z_1} + \overline{Z_2}} \times \overline{V}\,\bar{I}$$

Similarly

$$\overline{I_2}\,\overline{V} = \frac{\overline{Z_1}}{\overline{Z_1} + \overline{Z_2}} \times \overline{V}\,\bar{I}$$

Let $V \times I \times 10^{-3} = kVA$, the combined load in kVA.

$V \times I_1 \times 10^{-3} = kVA_1$ load shared by transformer 'A' in kVA.

$V \times I_2 \times 10^{-3} = kVA_2$ load shared by transformer 'B' in kVA.

$$\therefore \qquad \overline{kVA_1} = \frac{\overline{Z_2}}{\overline{Z_1} + \overline{Z_2}} \times \overline{kVA} \qquad \text{...(2.102)}$$

$$\overline{kVA_2} = \frac{\overline{Z_1}}{\overline{Z_1} + \overline{Z_2}} \times \overline{kVA} \qquad \text{...(2.103)}$$

$$\therefore \qquad \frac{kVA_1}{kVA_2} = \frac{\overline{Z_2}}{\overline{Z_1}}$$

i.e., the load shared by each transformer is inversely proportional to their impedances.

Again from equation (*ii*) or (*iii*), we get,

$$\overline{I_1} = \frac{1}{\dfrac{\overline{Z_1}}{\overline{Z_2}} + 1} \times \overline{I} \qquad \text{...(2.104)}$$

or

$$\overline{I_2} = \frac{1}{\dfrac{\overline{Z_2}}{\overline{Z_1}} + 1} \times \overline{I} \qquad \text{...(2.105)}$$

Equation (*vi*) shows that the load shared by each transformer depends upon the ratio of impedances, so the unit in which they are measured does not matter.

> **Note:** The expression, derived above are vectorial so that kVA_1 and kVA_2 are obtained in magnitude as well as in direction.

(*iii*) **When the two transformers have different voltage ratio and different voltage triangles.**

In this case, the voltage ratios or transformation ratios of the two transformers are different. It means, these no-load secondary voltages are unequal.

Let $\overline{E}_1$, $\overline{E}_2$ be the no-load secondary emfs of the two transformers and $\overline{Z}$ be the load impedance across the secondary.

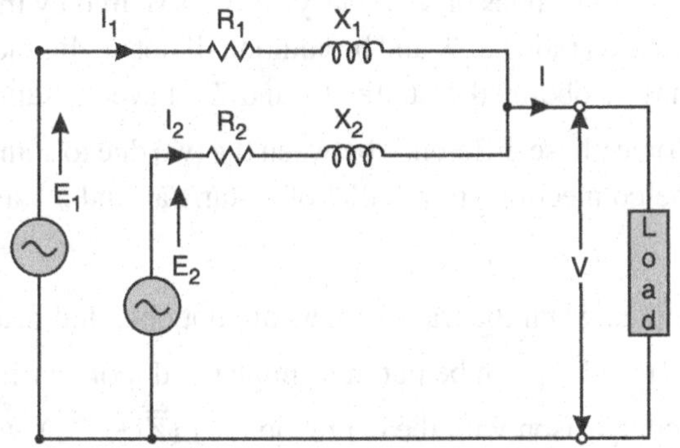

Fig. 2.61 Circuit diagram of two transformers connected in parallel

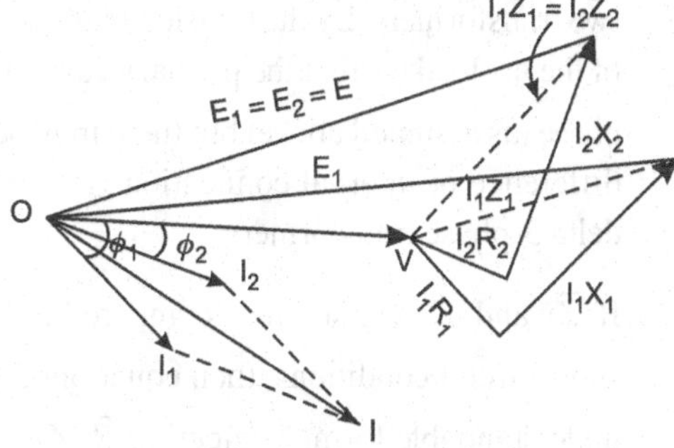

Fig. 2.62 Phasor diagram of two transformers operating in parallel having different voltage ratio and different voltage triangles.

The equivalent circuit and vector diagram are also shown in Fig. 2.61 and Fig. 2.62, respectively. It is seen that even when secondaries are on no-load, there will be some circulating-current in the secondaries because of inequality in their induced emf's. This circulating current $\overline{I}_C$ is given by

$$\overline{I}_C = (\overline{E}_1 - \overline{E}_2) / (\overline{Z}_1 + \overline{Z}_2) \quad ...(2.106)$$

As the induced emf's of the two transformers are equal to the total drops in their respective circuits.

i.e., $\overline{E}_1 = \overline{I}_1\overline{Z}_1 + \overline{V}_2 \; ; \; \overline{E}_2 = \overline{I}_2\overline{Z}_2 + \overline{V}_2$

Now $\overline{V}_2 = \overline{I}.\overline{Z}_L = (\overline{I}_1 + \overline{I}_2)\,\overline{Z}_L$ where $\overline{Z}_L$ = load impedance

$$\therefore \quad \overline{E}_1 = \overline{I}_1\overline{Z}_1 + (\overline{I}_1 + \overline{I}_2)\overline{Z}_L \quad ...(2.107)$$

$$\overline{E}_2 = \overline{I}_2\overline{Z}_2 + (\overline{I}_1 + \overline{I}_2)Z_L \quad ...(2.108)$$

$$\therefore \quad \overline{E}_1 - \overline{E}_2 = \overline{I}_1\overline{Z}_1 - \overline{I}_2\overline{Z}_2 \quad ...(2.109)$$

$$\therefore \quad \overline{I}_1 = \frac{(\overline{E}_1 - \overline{E}_2) + \overline{I}_2\overline{Z}_2}{\overline{Z}_1}$$

Substituting this value of $\overline{I}_1$ in equation (*iii*), we get,

$$\overline{E}_2 = \overline{I}_2\overline{Z}_2 + \left[\frac{(\overline{E}_1 - \overline{E}_2) + \overline{I}_2\overline{Z}_2}{Z_1} + \overline{I}_2\right]\overline{Z}_L$$

$$\therefore \quad \overline{I}_2 = \frac{\overline{E}_2\overline{Z}_1 - (\overline{E}_1 - \overline{E}_2)\overline{Z}_L}{\overline{Z}_1\overline{Z}_2 + \overline{Z}_L(\overline{Z}_1 + \overline{Z}_2)} \quad ...(2.110)$$

From the symmetry of the expression, we get

$$\overline{I}_1 = \frac{\overline{E}_1\overline{Z}_2 + (\overline{E}_1 - \overline{E}_2)\overline{Z}_L}{\overline{Z}_1\overline{Z}_2 + \overline{Z}_L(\overline{Z}_1 + \overline{Z}_2)} \quad ...(2.111)$$

The two equations (2.110) and (2.111) then give the values of secondary currents shared by the two transformers. By the division of transformation ratio i.e., *K* and by addition (if not negligible) of the no-load current the primary current may be obtained. Usually $\overline{E}_1$ and $\overline{E}_2$ have the same phase (as assumed above) but there may be some phase difference between the two due to some difference of internal connection *viz.* for the connections in parallel of a star/star and a star/delta 3-phase transformers.

If $\overline{Z}_1$ and $\overline{Z}_2$ are small as compared to $\overline{Z}_L$ i.e., when the transformers are not operated near short-circuit conditions, then equations for $\overline{I}_1$ and $\overline{I}_2$ can be put in a simpler and more easily understandable form. Neglecting $\overline{Z}_1\overline{Z}_2$ in comparison with the expression $\overline{Z}_L(\overline{Z}_1 + \overline{Z}_2)$ we have

$$\overline{I}_1 = \frac{\overline{E}_1\overline{Z}_2}{\overline{Z}_L(\overline{Z}_1 + \overline{Z}_2)} + \frac{(\overline{E}_1 - \overline{E}_2)}{\overline{Z}_1 + \overline{Z}_2} \quad ...(2.112)$$

$$\bar{I}_2 = \frac{\bar{E}_2 \bar{Z}_1}{\bar{Z}_L(\bar{Z}_1 + \bar{Z}_2)} - \frac{\bar{E}_1 - \bar{E}_2}{\bar{Z}_1 + \bar{Z}_2} \qquad \ldots(2.113)$$

The physical interpretation of second term in equations (2.112) and (2.113) is that it represents the cross-current (circulating current) between the secondaries. The first term shows how the actual load current divides between the loads. If $\bar{E}_1 = \bar{E}_2$, the ratios of the currents are inversely as the impedances (Numerical values).

Note: In the above case, it is more convenient to work with Numerical values of impedances instead of % values.

Example 2.55

A load of 500 A, at 0·8 power (lagging), at a terminal voltage of 400 V is supplied by two transformers are connected in parallel. The equivalent impedances of the two transformers referred to the secondary sides are (2 + j3) ohm and (2.5 + j5) ohm, respectively. Calculate the current and kVA supplied by each transformer and the power factor at which they operate.

Solution:

Here, $\bar{Z_1} = 2 + j3 = \sqrt{(2)^2 + (3)^2}\,\angle \tan^{-1}\frac{3}{2} = 3 \cdot 606\angle 56 \cdot 31°$

$$\bar{Z_1} = 2{\cdot}5 + j5 = \sqrt{(2 \cdot 5)^2 + (5)^2}\,\angle \tan^{-1}\frac{5}{2 \cdot 5} = 5 \cdot 59\angle 63 \cdot 44°$$

$$\bar{Z_1} + \bar{Z_2} = 2 + j3 + 2{\cdot}5 + j5 = 4{\cdot}5 + j8 = \sqrt{(4 \cdot 5)^2 + (5)^2}\,\angle \tan^{-1}\frac{8}{4 \cdot 5} = 9 \cdot 17\angle 60 \cdot 64°$$

$$\bar{I} = I\angle \cos^{-1} 0 \cdot 8 = 500\angle -36 \cdot 87°$$

$$\bar{I_1} = \frac{\bar{Z_2}}{\bar{Z_1} + \bar{Z_2}} \times \bar{I} = \frac{5 \cdot 59\angle 63 \cdot 44°}{9 \cdot 178\angle 60 \cdot 64°} \times 500\angle -36 \cdot 83° = 30 \cdot 45\angle -34 \cdot 07°$$

$I_1 =$ **304·5 A** (*Ans.*)

Power factor, $\cos \phi_1 = \cos(-34{\cdot}07°) =$ **0·8184 lag** (*Ans.*)

$$\bar{I_2} = \frac{\bar{Z_1}}{\bar{Z_1} + \bar{Z_2}} \times \bar{I} = \frac{3 \cdot 606\angle 56 \cdot 31°}{9 \cdot 178\angle 60 \cdot 64°} \times 500\angle -36 \cdot 87° = 196 \cdot 45\angle -41 \cdot 2°$$

$I_2 =$ **196·45** (*Ans.*)

Power factor, $\cos \phi_2 = \cos(-41{\cdot}2°) =$ **0·75524 lag** (*Ans.*)

Example 2.56

Two single-phase transformers connected in parallel supply a load of 1000 A at 0·8 p.f. lagging. For each transformer, the secondary emf on open circuit is 3300 V and the total leakage impedances in terms of the secondary are (0·1 + j0·2) and (0·05 + j0·4) ohm, respectively. Determine the output current for each transformer and the ratio of the kW output of the two transformers.

Solution:

$$\overline{Z_1} = 0\cdot1 + j0\cdot2 = 0\cdot2236\angle 63\cdot44^\circ$$

$$\therefore \quad \overline{Z_2} = 0\cdot05 + j0\cdot4 = 0\cdot4031\angle 82\cdot87^\circ$$

$$\overline{Z_1} + \overline{Z_2} = 0\cdot1 + j0\cdot2 + 0\cdot05 + j0\cdot4 = 0\cdot15 + j0\cdot6 = 0\cdot6185\angle 75\cdot96^\circ$$

Taking secondary terminal voltage as reference vector, the expression of the current is

$$\bar{I} = 1000\angle \cos^{-1}0.8 = 1000\angle -36.87^\circ$$

$$\overline{I_1} = \frac{\overline{Z_2}}{\overline{Z_1} + \overline{Z_2}} \times \bar{I} = \frac{0.4031\angle 82.87^\circ}{0.6185\angle 75.96^\circ} \times 100\angle -36.87^\circ = 651.7\angle -29.96^\circ$$

I_1 = **651·7 A** (*Ans.*)

Power factor, $\cos\phi_1 = \cos(-29{\cdot}96^\circ) = 0{\cdot}866$ lagging

$$\overline{I_2} = \frac{\overline{Z_1}}{\overline{Z_1} + \overline{Z_2}} \times \bar{I} = \frac{0\cdot2236\angle 63\cdot44^\circ}{0\cdot6185\angle 75\cdot96^\circ} \times 1000\angle -36\cdot87^\circ = 361\cdot5\angle -49\cdot39^\circ$$

I_2 = **361·5 A** (*Ans.*)

Power factor, $\cos\phi_2 = \cos(-49{\cdot}39^\circ) = 0{\cdot}651$ lagging

Ratio of the outputs of the two transformers.

$$\frac{kW_1}{kW_2} = \frac{VI_1\cos\phi_1}{VI_2\cos\phi_2} = \frac{3300\times651\cdot7\times0\cdot866}{3300\times361\cdot5\times0\cdot651} = \mathbf{2{\cdot}398}\ (Ans.)$$

Example 2.57

Two single-phase transformers I and II rated at 250 kVA, each are connected in parallel on both sides. Resistance and reactance drops for I and II are 1% and 6% and 1·2% and 4·8%, respectively. Calculate the load shared by each and the respective power factors, if the total load is 500 kVA at 0·8 power factor lagging. Their no-load emf's are equal.

Solution:

$$\overline{Z_1} = (1 + j6)\%; = 6\cdot083\angle 80\cdot54^\circ$$

$$\overline{Z_2} = (1{\cdot}2 + j4{\cdot}8)\% = 4\cdot948\angle 75\cdot96^\circ$$

$$\overline{Z_1} + \overline{Z_2} = (1 + j6)\% + (1\cdot2 + j4{\cdot}8)\% = (2{\cdot}2 + j10{\cdot}8)\% = 11\cdot08\angle 78\cdot48^\circ$$

$$kVA = 500\angle \cos^{-1}0\cdot8 = 500\angle -36\cdot87^\circ$$

$$\therefore \quad kVA_1 = \frac{\overline{Z_2}}{\overline{Z_1} + \overline{Z_2}}\overline{kVA} = \frac{4\cdot948\angle 75\cdot96^\circ}{11\cdot08\angle 78\cdot48^\circ} \times 500\angle -36\cdot87^\circ = 223\cdot3\angle -39\cdot39^\circ$$

kVA_1 = **223·3 kVA** (*Ans.*)

Power factor, $\cos\phi_1 = \cos(-39{\cdot}39^\circ) =$ **0·773 lag** (*Ans.*)

$$\overline{kVA_2} = \frac{\overline{Z_1}}{\overline{Z_1} + \overline{Z_2}} \times kVA = \frac{6 \cdot 083\angle 80 \cdot 54^\circ}{11 \cdot 08\angle 78 \cdot 48^\circ} \times 500\angle -36 \cdot 87^\circ = 274 \cdot 5\angle -34 \cdot 81^\circ$$

kVA_2 = **274·5 kVA** (*Ans.*)

Power factor, $\cos \phi_2 = \cos(-34{\cdot}81^\circ)$ = **0·821 lag** (*Ans.*)

Example 2.58

Two single phase transformers are connected in parallel. Both of them are having equal turns and an impedance of (0.5 + j3) ohm and (0.6 + j10) ohm with respect to the secondary. Determine how they will share a total load of 100 kW at power factor 0.8 lagging.

Solution:

Here, $\overline{Z}_1 = (0.5 + j3)$; $\overline{Z}_2 = (0.6 + j\,10)$

Load = 100 kW; Power factor, $\cos \phi = 0.8$ lag

$$\text{Load in kVA} = \frac{kW}{p.f.} = \frac{100}{0 \cdot 8} = 125$$

$$\cos \phi = 0.8;\ \phi = \cos^{-1} 0.8 = 36.87^\circ;\ \sin \phi = \sin 36.87^\circ = 0.6$$

$$\overline{kVA} = kVA\,(\cos \phi - j \sin \phi)$$

$$= 125\,(0.8 - j\,0.6) = (100 - j\,75) = 125\angle - 36 \cdot 87^\circ$$

$$\overline{Z}_1 = \sqrt{(0 \cdot 5)^2 + (3)^2}\,\angle \tan^{-1}\frac{3}{0 \cdot 5} = 3 \cdot 04\ \angle 80 \cdot 54^\circ$$

$$\overline{Z}_2 = \sqrt{(0 \cdot 6)^2 + (10)^2}\,\angle \tan^{-1}\frac{10}{0 \cdot 6} = 10 \cdot 018\angle 86 \cdot 67^\circ$$

$$\overline{Z}_1 + \overline{Z}_2 = 0.5 + j3 + 0.6 + j\,10 = 1.1 + j\,13$$

$$= \sqrt{(1 \cdot 1)^2 + (13)^2}\,\angle \tan^{-1}\frac{13}{1 \cdot 1} = 13 \cdot 05\ \angle 85 \cdot 16^\circ$$

$$kVA_1 = \frac{\overline{Z}_2}{\overline{Z}_1 + \overline{Z}_2} \times \overline{kVA} = \frac{10 \cdot 018\angle 86 \cdot 67^\circ}{13 \cdot 04\angle 85 \cdot 16} \times 125\angle - 36 \cdot 87^\circ$$

$$= 95.96\ \angle -35.36^\circ = 95.96\,(0.8155 - j\,0.5787) = 78.25 - j\,55.54$$

$$\overline{kVA_2} = \frac{\overline{Z}_1}{\overline{Z}_1 + \overline{Z}_2} \times \overline{kVA} = \frac{3 \cdot 04\ \angle 80 \cdot 54^\circ}{13 \cdot 04\angle 85 \cdot 16^\circ} \times 125\angle - 36 \cdot 87^\circ$$

$$= 29.14\ \angle -42.49 = 29.14\,(0.749 - j\,0.662) = 21.83 - j19.29$$

Load shared by transformer–I = **78.25 kW** at *p.f.* **0.8155 lag** (*Ans.*)

Load shared by transformer–II = **21.83 kW** at p.f. **0.749 lag** (*Ans.*)

Example 2.59

A transformer 'A' having an open circuit emf of 6600 V with impedance (0.3+j3) ohm referred to secondary is connected in parallel with transformer 'B' having an open circuit emf of 6400 V with impedance (0.2+j1) ohm referred to secondary side. Calculate the current delivered by each transformer to a load impedance of (8+j6) ohm.

Solution:

Here $\overline{E}_1 = 6600\angle 0^\circ = 6600 \pm j0\ ;\ \overline{E}_2 = 6400\ \angle 0^\circ = 6400 \pm j0$

$$\overline{Z}_1 = 0\cdot 3 + j3 = 3\cdot 015\angle 84\cdot 29^\circ;\overline{Z}_2 = 0\cdot 2 + j1 = 1\cdot 02\angle 78\cdot 69^\circ$$

$$\overline{Z}_1 + \overline{Z}_2 = 0\cdot 3 + j3 + 0\cdot 2 + j1 = 0\cdot 5 + j4 = 4\cdot 031\angle 82\cdot 87^\circ$$

$$\overline{Z}_L = 8 + j6 = 10\angle 36\cdot 87^\circ\ ;\ \overline{I}_1 = \frac{\overline{E}_1\overline{Z}_2 + (\overline{E}_1 - \overline{E}_2)\overline{Z}_L}{\overline{Z}_1\overline{Z}_2 + \overline{Z}_L(\overline{Z}_1 + \overline{Z}_2)}$$

$$= \frac{6600\angle 0^\circ \times 1\cdot 02\angle 78\cdot 69^\circ + (6600 - 6400)\angle 0^\circ \times 10\angle 36\cdot 87^\circ}{3\cdot 015\angle 84\cdot 29^\circ \times 1\cdot 02\angle 78\cdot 69^\circ + 10\angle 36\cdot 87 \times 4\cdot 031\angle 82\cdot 87^\circ}$$

$$= \frac{6732\angle 78\cdot 69^\circ + 2000\angle 36\cdot 87^\circ}{3\cdot 0753\angle 162\cdot 98 + 40\cdot 31\angle 119\cdot 74^\circ}$$

$$= \frac{(1320 + j6601) + (1600 + j1200)}{(-2\cdot 94 + j0\cdot 9) + (-20 + j35)} = \frac{2920 + j7801}{-22\cdot 94 + j35\cdot 9}$$

$$= \frac{8329\cdot 6\angle 69\cdot 48^\circ}{42\cdot 6\angle -53\cdot 42^\circ} = 195\cdot 53\ \angle 126\cdot 9^\circ$$

I_1 = **195.53 A** (in magnitude) (*Ans.*)

Similarly
$$\overline{I}_2 = \frac{\overline{E}_2\overline{Z}_1 - (\overline{E}_1 - \overline{E}_2)\overline{Z}_L}{\overline{Z}_1\overline{Z}_2 + \overline{Z}_L(\overline{Z}_1 + \overline{Z}_2)}$$

$$= \frac{6400\angle 0^\circ \times 3\cdot 015\angle 84\cdot 29^\circ - (6600 - 6400)\angle 0^\circ \times 10\angle 36\cdot 87^\circ}{42\cdot 6\angle -57\cdot 42^\circ}$$

$$= \frac{19296\angle 84\cdot 29^\circ - 2000\angle 36\cdot 87^\circ}{42\cdot 6\angle -57\cdot 42^\circ}$$

$$= \frac{1920 + j19200 - 1600 - j1200}{42\cdot 6\angle -57\cdot 42^\circ} = \frac{320 + j18000}{42\cdot 6\angle -57\cdot 42^\circ}$$

$$= \frac{18003\angle 88\cdot 98}{42\cdot 6\angle -57\cdot 42^\circ} = 422.6\ \angle 146.4^\circ$$

I_2 = **422.6 A** (*in magnitude*) (*Ans.*)

<u>**Example 2.60**</u>

A 500 kVA transformer is connected in parallel with a 150 kVA transformer and is supplying a load of 750 kVA at 0.8 pf lagging. Their open circuit voltages are 405 V and 415 V, respectively. Transformer A has 1% resistance and 5% reactance and transformer B has 1.5% resistance and 4% reactance. Find (a) cross-current in the secondaries on no-load and (b) the load shared by each transformer.

Solution:

It is more convenient to work with ohmic impedances and for that purpose, we will convert percentage values into Numerical values by assuming 400 volt as the terminal voltage (i.e., $V_1 = V_2 = 400$ V, this value is arbitrary but this assumption will not introduce appreciable error).

Now, $\frac{I_{1(fl)}R_1}{V_1} \times 100 = 1$ or $R_1 = \frac{V_1}{I_{(fl)} \times 100} = \frac{1}{100} \times \frac{V_1}{I_{1(fl)}}$

where, $$I_{1(fl)} = \frac{kVA_1 \times 1000}{V_1} = \frac{500 \times 1000}{400} = 1250 \text{ A}$$

$\therefore$ $$R_1 = \frac{1}{100} \times \frac{400}{1250} = 0 \cdot 0032\ \Omega$$

Similarly, $$X_1 = \frac{5}{100} \times \frac{400}{1250} = 0 \cdot 01600\ \Omega$$

Now, for transformers - B, $I_{2(fl)} = \frac{kVA_2 \times 1000}{V_2} = \frac{150 \times 1000}{400} = 375$ A

$\therefore$ $$R_2 = \frac{1 \cdot 5}{100} \times \frac{400}{375} = 0 \cdot 016\ \Omega\ ; X_2 = \frac{4}{100} \times \frac{400}{375} = 0 \cdot 0427\ \Omega$$

$\therefore$ $$\overline{Z}_1 = 0.0032 + j\,0.016 = 0.0163 \angle 78 \cdot 5^\circ$$

$$\overline{Z}_2 = 0.016 + j\,0.0427 = 0.0456 \angle 69 \cdot 46^\circ$$

$$\overline{Z}_1 + \overline{Z}_2 = 0.0032 + j0.016 + 0.016 + j0.0427 = 0.0192 + j0.0587 = 0.0618\ \angle 71 \cdot 88^\circ$$

Next step is to calculate load impedance. Let Z_L be the load impedance and V_2 be the terminal voltage which has been assumed as 400 V.

$\therefore$ $$\left(\frac{V_2}{\overline{Z}_L}\right) \times 10^{-3} = 750\angle - 36.87^\circ$$

$\therefore$ $$\overline{Z}_L = \frac{V^2 \times 10^{-3}}{750\angle - 36 \cdot 87^\circ} = \frac{400 \times 400 \times 10^{-3}}{750\angle - 36 \cdot 83}$$

$$= 0 \cdot 214\ \angle 36 \cdot 87^\circ = (0 \cdot 171 + j0 \cdot 128)\ \Omega$$

(a) $$I_C = \frac{E_1 - E_2}{Z_1 + Z_2} = \frac{(405 - 415)}{0 \cdot 0618\angle 71 \cdot 88^\circ} = -161 \cdot 8\angle - 71 \cdot 88^\circ$$

(b) $$\overline{I}_1 = \frac{\overline{E}_1\overline{Z}_2 + (\overline{E}_1 - \overline{E}_2)\overline{Z}_L}{\overline{Z}_1\overline{Z}_2 + \overline{Z}_L(\overline{Z}_1 + \overline{Z}_2)}$$

$$= \frac{405 \times 0 \cdot 0456\angle 69 \cdot 46^\circ + (405 - 415) \times 0 \cdot 214\ \angle 36 \cdot 87^\circ}{0 \cdot 0163\angle 78 \cdot 5^\circ \times 0 \cdot 0456\angle 69 \cdot 46^\circ + 0 \cdot 214\angle 36 \cdot 87^\circ \times 0 \cdot 0618\angle 71 \cdot 88^\circ}$$

$$= \frac{18 \cdot 47\angle 69 \cdot 46^\circ - 2 \cdot 14\angle 36 \cdot 87^\circ}{0 \cdot 00074\angle 147 \cdot 96^\circ + 0 \cdot 01323\angle 108 \cdot 75^\circ}$$

$$= \frac{6 \cdot 48 + j17 \cdot 29 - 1 \cdot 712 - j1 \cdot 284}{-0 \cdot 00063 + j0 \cdot 0004 - 0 \cdot 00425 + j0 \cdot 0125}$$

$$= \frac{4 \cdot 768 + j16 \cdot 01}{-0 \cdot 00488 + j0 \cdot 0129} = \frac{16 \cdot 705\angle 73 \cdot 41^\circ}{0 \cdot 0138\angle - 69 \cdot 28^\circ} = 1210\angle - 37 \cdot 3^\circ \text{A}$$

Load shared by transformer – A

$$S_1 = 400 \times 1210 \times 10^{-3} \angle{-37.3^\circ} = \mathbf{484\angle{-37.3^\circ}\ kVA}\ (Ans.)$$

$$p.f. = \cos \phi_1 = \cos 37.3° = \mathbf{0.795\ lag.}\ (Ans.)$$

$$I_2 = \frac{\overline{E}_2\overline{Z}_1 - (\overline{E}_1 - \overline{E}_2)\overline{Z}_L}{\overline{Z}_1\overline{Z}_2 + \overline{Z}_L(\overline{Z}_1\overline{Z}_2)}$$

$$= \frac{415 \times 0\cdot0163\angle78\cdot5° - (405 - 415) \times 0\cdot214\angle36\cdot87°}{0\cdot0138\angle-69\cdot28°}$$

$$= \frac{1\cdot348 + j6\cdot629 + 1\cdot712 + j1\cdot284}{0\cdot0138\angle-69\cdot28°} = \frac{3\cdot0607 + j7\cdot913}{0\cdot0138\angle-69\cdot28°}$$

$$= \frac{8\cdot4843\angle68\cdot85°}{0\cdot0138\angle-69\cdot28} = 615\angle-41\cdot87°\ \text{A}$$

Load shared by transformers - *S*,

$$S_2 = 400 \times 615 \times 10^{-3}\angle-42.87° = \mathbf{246\angle-42.87°\ kVA}\ (Ans.)$$

$$p.f. = \cos \phi_2 = \cos 42.87° = \mathbf{0.744\ lag}\ (Ans.)$$

Section Practice Problems

Numerical Problems

1. A load of 1600 kW at 0·8 p.f. is shared by two 1000 kVA transformers having equal turn ratios and connected in parallel on their primary and secondary side. The full load resistance drop is 1% and reactive drop is 6% in one of the transformers, the corresponding values in the other transformer being 1.5% and 5%. Calculate the power and power factor at which each transformer is operating.

(**Ans.** *700 kW, 0.76 lagging, 900 kW, 0.84 lag.*)

2. Two single-phase transformers supply, in parallel, a secondary load of 1000 A at 0.8 p.f. lagging. For each transformer, the secondary emf on open-circuit is 3300 V and the total leakage impedance, in terms of the secondary are (0.1 + j 0.2) and (0.05 + j 0.4) ohm, respectively. Determine the output current for each transformer and the ratio of the kW output of the two transformers. (**Ans.** *652 A; 362 A; 2.4: 1*)

3. Two transformers A and B of different ratings but equal voltage ratios share a load of 900 kVA at 0.8 power factor lagging at 400 V by operating in parallel. Transformer *A* has a rating of 600 kVA, resistive drop of 1% and reactance drop 5%. Transformer *B* has a rating of 300 kVA, resistive drop of 1.5% and reactance drop of 4%. Calculate the load shared by each transformer and the power factor at which it is working.

(**Ans.** *564 kVA at pf 0.762 lag; 336 kVA at pf 0.856 lag*)

Short Answer Type Questions

Q.1. What is the necessity of parallel operation of transformer?

Ans. (*i*) When load on the existing transformer increases beyond its rated capacity.
(*ii*) When the power to be transformed is more than the size of single unit.

Q.2. What conditions are required to be fulfilled for parallel operation of transformers?

Ans. (*i*) Both the transformers should have same transformation ratio.
(*ii*) Both the transformers should have the same percentage impedance.
(*iii*) In case of one-phase transformers, both the transformers must have same polarity.

(*iv*) In case of 3-phase transformers, both the transformers must have same phase sequence.
(*v*) In case of 3-phse transformers, the winding connections selected must be such that there should not be phase difference between the voltages of two transformers.

2.42 Auto-transformer

A transformer, in which a part of the winding is common to both the primary and secondary circuits, is called an auto-transformer. In a two winding transformer, primary and secondary windings are electrically isolated, but in an auto-transformer the two windings are not electrically isolated rather a section of the same winding acts as secondary or primary of the transformer.

Construction

The core of an auto-transformer may be rectangular [Fig. 2.63(*a*)] or circular ring-type [see Fig. 2.64(*a*)] in shape. A single winding is wound around one or two limbs of the rectangular core as shown in Fig. 2.63(*b*) or it is wound over the ring as shown in Fig. 2.64(*b*). Terminal '*B*' is taken as a common point from which one terminal for primary and one terminal of the secondary is taken out. The second terminal of the secondary is connected to point '*C*' which may be fixed or movable as shown in Figs. 2.63(*b*) and 2.64(*b*). The number of turns between *AB* are taken as N_1 and the number of turns between *BC* are taken as N_2 as shown in Figs. 2.63(*c*) and 2.64(*c*). Thus, one section of the same winding acts as a primary and the other section of the same winding acts as a secondary. When the number of secondary turns N_2 is less then the primary turns N_1 (i.e., $N_2 < N_1$) as shown in Fig. 2.63(*c*) and 64(*c*), the auto-transformer works as step-down transformer, whereas, it works as a step-up transformer if number of secondary turns N_2 is more than primary turns N_1 as shown in Figs. 2.63(*d*) and 2.64(*d*).

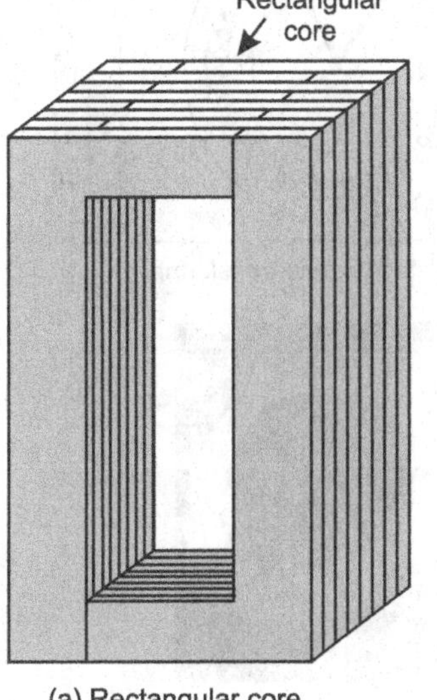

(a) Rectangular core

Rectangular laminated core

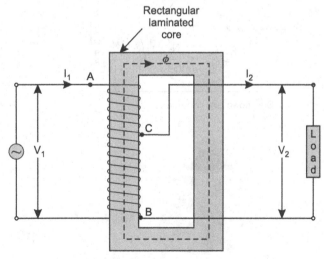

ϕ, I_1, A, I_2, C, V_1, V_2, Load, B

(b) Single-winding placed on core

Contd.

Contd.

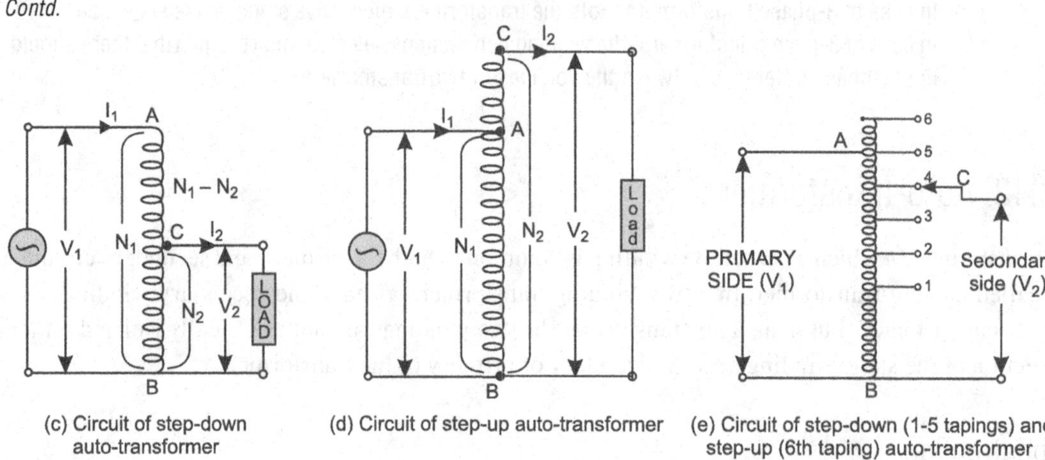

(c) Circuit of step-down auto-transformer

(d) Circuit of step-up auto-transformer

(e) Circuit of step-down (1-5 tapings) and step-up (6th taping) auto-transformer

Fig. 2.63 Single-phase rectangular core auto-transformer.

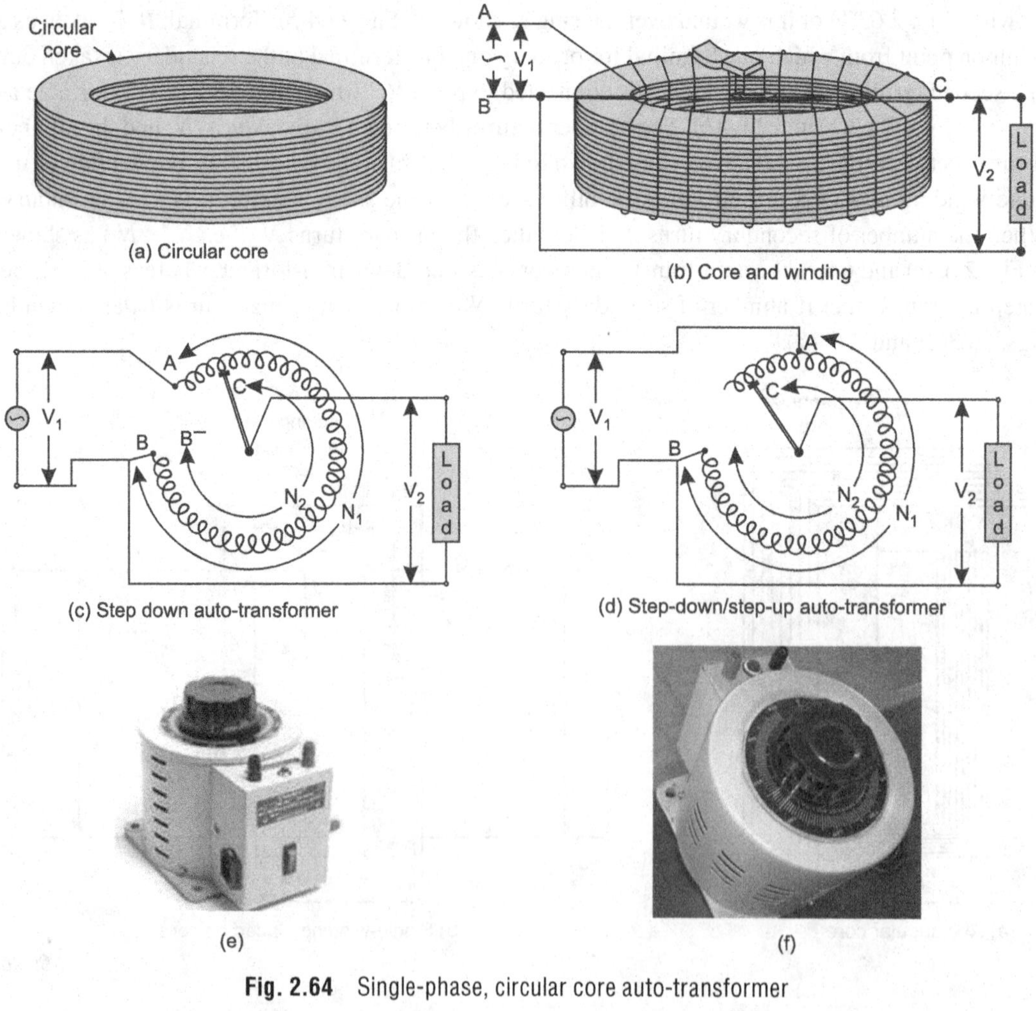

(a) Circular core

(b) Core and winding

(c) Step down auto-transformer

(d) Step-down/step-up auto-transformer

(e)

(f)

Fig. 2.64 Single-phase, circular core auto-transformer

The pictorial view of a single-phase auto-transformer used in labs is shown in Figs. 2.64(*e* and *f*). Here, point *C* is attached to a movable arm which carries a carbon brush. The brush moves over number of turns wound over a circular laminated core and its position determines the output voltage.

Working

When AC voltage V_1 is applied to winding *AB*, an exciting current starts flowing through the full winding *AB* if the internal impedance drop is neglected, then the voltage per turn in winding *AB* is V_1/N_1 and, therefore, the voltage across *BC* is $(V_1/N_1) N_2$.

When switch *S* is closed, a current I_2 starts flowing through the load and current I_1 is drawn from the source. Neglecting losses,

$$\text{Input power} = \text{Output power}$$

or

$$V_1 I_1 \cos \phi_1 = V_2 I_2 \cos \phi_2$$

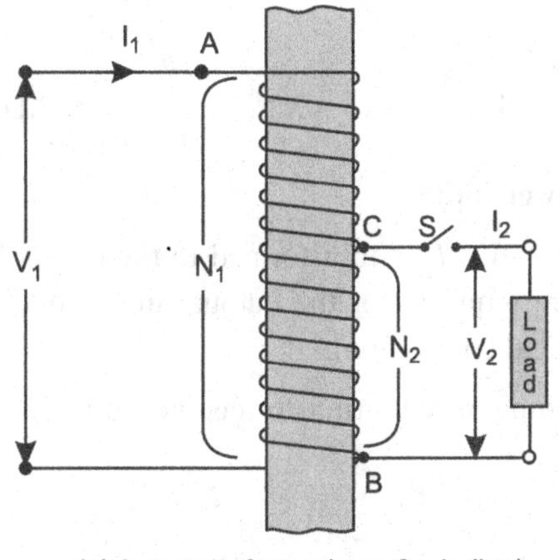

(a) Auto-transformer(core & winding)

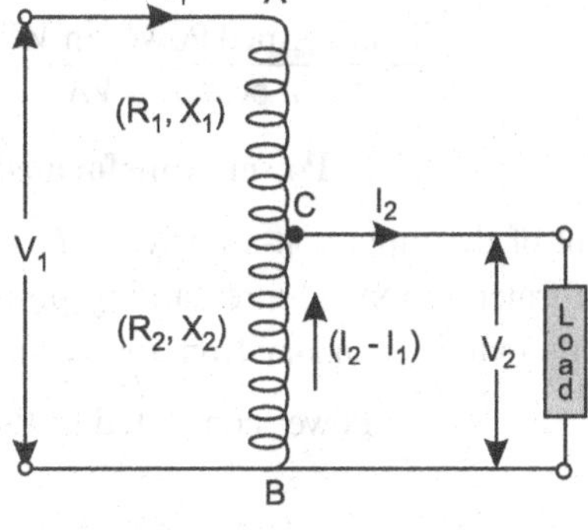

(b) Auto-transformer transformer with load

Fig. 2.65 Circuit for single-phase auto-transformer.

If internal (or leakage) impedance drops and losses are neglected, then

$$\cos \phi_1 = \cos \phi_2$$

Hence

$$V_1 I_1 = V_2 I_2$$

or

$$\frac{V_2}{V_1} = \frac{I_1}{I_2} = \frac{N_2}{N_1} = K \qquad \text{...(2.114)}$$

Here *K* is less than unity. The expression is identical to a two winding transformer.

Let at any instant the exciting current flows from *A* to *B* and it establishes a working mmf directed vertically upward in the core. When switch *S* is closed, the current in winding *BC* must flow from *B* to *C*, in order to create an mmf opposing the exciting or working mmf, as per Lenz's law. Since the working mmf in a transformer remains constant at its no-load value, the primary must draw additional current I_1 from the source, in order to neutralise the effect of current I_{BC}. In winding *AB*,

I_1 flows from A to B while in winding BC, I_2 flows from B to C. Therefore, the current in winding BC is I_1 from C to B and I_2 from B to C. Here the current I_2 is greater than I_1 (because $V_2 < V_1$) and their mmfs. are opposing each other at every instant, therefore,

$$I_{BC} = I_2 - I_1.$$

mmf of winding $AC = I_1 (N_1 - N_2) = I_1N_1 - I_1N_2$

$$= I_2N_2 - I_1N_2 = (I_2 - I_1)\, N_2 \qquad [\because I_1N_1 = I_2N_2]$$

$$= I_{BC}N_2$$

= mmf of winding CB.

It is, therefore, seen that the transformer action takes place between winding, section AC and winding section BC. In other words, the volt-amperes across winding AC are transferred by transformer action to the load connected across winding BC.

$\therefore$ Power transformed in $VA = V_{AC}I_{AC} = (V_1 - V_2)\, I_1$

Total power to be transferred or input power in $VA = V_1I_1$

$$\therefore \quad \frac{\text{Transformed Power in } VA}{\text{Input power in } VA} = \frac{(V_1 - V_2)\, I_1}{V_1I_1} = 1 - \frac{V_2}{V_1} = (1 - K) \qquad \text{...(2.115)}$$

Power transformed = (1 – *K*) × Power input

Out of the input volt-amperes V_1I_1, only $V_{AC}I_{AC} = (V_1 - V_2)\, I_1$ is transformed to the output by transformer action. The remaining power in volt-ampere required for the output, are conducted directly to the secondary from the primary (due to electrical connection).

$\therefore$ Power conducted in VA = Total power input in VA – transformed power in VA

$$= V_1I_1 - (V_1 - V_2)\, I_1 = V_2I_1$$

$$\therefore \quad \frac{\text{Power conducted in } VA}{\text{Power Input in } VA} = \frac{V_2I_1}{V_1I_1} = K \qquad \text{...(2.116)}$$

Power conducted = *K* × Power input

Hence,

$$\frac{\text{Transformed power}}{\text{Input power}} = 1 - K \qquad \text{...(2.117)}$$

and

$$\frac{\text{Conducted power}}{\text{Input power}} = K \qquad \text{...(2.118)}$$

Considering eqn. (2.4),

$$\frac{\text{Transformed power}}{\text{Input power}} = 1 - K = 1 - \frac{V_2}{V_1} = \frac{V_1 - V_2}{V_1}$$

or

$$\frac{\text{Inductively transformed power}}{\text{Total power}} = \frac{\text{High voltage} - \text{Low voltage}}{\text{High voltage}} \qquad \text{...(2.119)}$$

2.43 Auto-transformer vs Potential Divider

At first sight, an auto-transformer appears to be similar to a resistance potential divider. But this is not so, as described below.

(*i*) A resistive potential divider can't step up the voltage, whereas it is possible in an auto-transformer.

(*ii*) The potential divider has more losses and is, therefore, less efficient.

(*iii*) In a potential divider, almost entire power to load flows by conduction, whereas in auto-transformer, a part of the power is conducted and the rest is transferred to load by transformer action.

(*iv*) In a potential divider, the input current, must always be more than the output current, this is not so in an auto-transformer. If the output voltage in auto-transformer is less than the input voltage, the load current is more than the input current.

2.44 Saving of Copper in an Auto-transformer

Volume, and hence weight of copper (or aluminium), is proportional to the length and area of *X*-section of the conductor. The length of conductor is proportional to number of turns whereas area of *X*-section is proportional to the current flowing through it. Hence the weight of copper is proportional to the product of current and number of turns.

Now, with reference to Fig. 2.66(*a*), weight of copper required in an auto-transformer.

$$Wt_a = \text{weight of } Cu \text{ in section } AC + \text{weight of } Cu \text{ in section } CB$$

$$\therefore \quad Wt_a \propto I_1 (N_1 - N_2) + (I_2 - I_1) N_2 \propto I_1 N_1 + I_2 N_2 - 2 I_1 N_2 \qquad \text{...(2.120)}$$

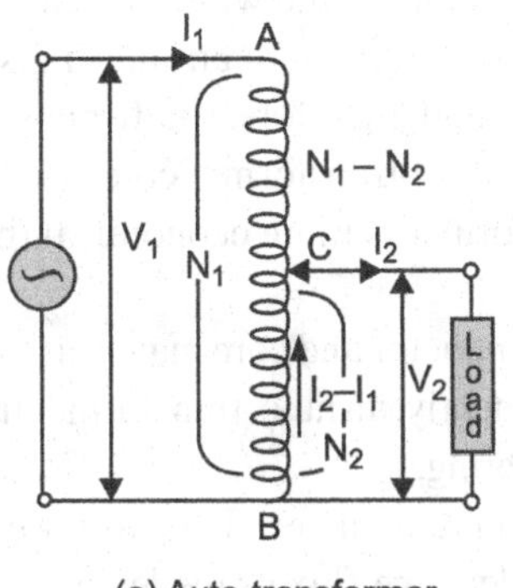

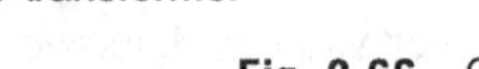
(a) Auto-transformer

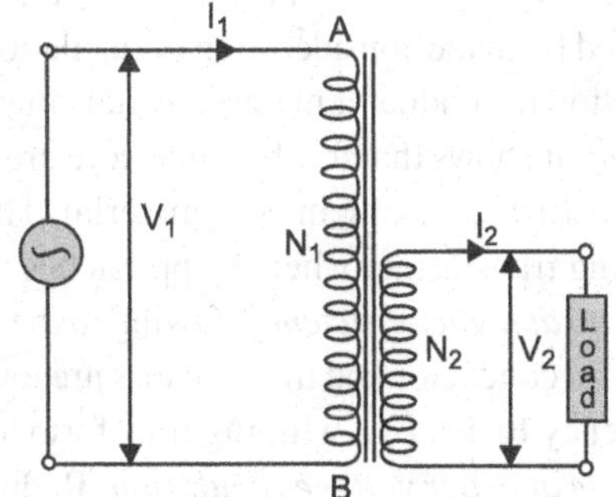

(b) Ordinary two-winding transformer

Fig. 2.66 Circuits for comparison

If an ordinary two winding transformer is to perform the same duty, then with reference to Fig. 2.66(*b*). Total weight of copper required in the ordinary transformer.

$$Wt_0 = \text{weight of } Cu \text{ on its primary} + \text{weight of } Cu \text{ on its secondary.}$$

$$\therefore \quad Wt_0 \propto I_1 N_1 + I_2 N_2 \qquad \text{...(2.121)}$$

Now, the ratio of weight of copper in auto-transformer to the weight of copper in an ordinary transformer,

$$\frac{Wt_a}{Wt_0} = \frac{I_1 N_1 + I_2 N_2 - 2 I_1 N_2}{I_1 N_1 + I_2 N_2} = \frac{I_1 N_1 + I_2 N_2}{I_1 N_1 + I_2 N_2} - \frac{2 I_1 N_2}{I_1 N_1 + I_2 N_2}$$

$$= 1 - \frac{2I_1 N_2 / I_1 N_1}{I_1 N_1 / I_1 N_1 + I_2 N_2 / I_1 N_1} = 1 - K$$

or $$Wt_a = (1 - K)\, Wt_0 \qquad ...(2.122)$$

Saving of copper affected by using an auto-transformer

= wt. of cu required in an ordinary transformer – wt. of copper required in an auto-transformer.

$= Wt_0 - Wt_a = Wt_0 - (1 - K)\, Wt_0 = K \times Wt_0$

∴ Saving = $K \times$ Wt. of copper required for two winding transformer

Hence, saving in copper increases as the transformation ratio approaches to unity, therefore, auto transformers are used when K in nearly equal to unity.

2.45 Advantages of Auto-transformer over Two-winding Transformer

1. *Quantity of conducting material required is less.* The quantity of conducting material required for an auto-transformer having same rating as that of an ordinary two winding transformer is only $(1 - K)$ times, i.e., Quantity of conducting material required for auto-transformer = $(1 - K)$ quantity of conducting material required for an ordinary two winding transformer. Thus, the cost of auto-transformer is less as compared to two winding transformer of the same rating.
2. *Quantity of magnetic material required is less.* During designing, the window dimensions are decided from the consideration of insulation and conductor material. For an auto-transformer, a reduction in conductor material means lower window area and, therefore, smaller core length is needed. It shows that for the same core area, the weight of auto-transformer core is less. Hence, there is further saving in core material. Thus auto-transformer is more economical than a two winding transformer when K approaches to unity.
3. *Operate at higher efficiency.* Owing to the reduction in conductor and core materials, the ohmic losses in conductor and the core loss are lowered. Consequently, an auto-transformer has higher efficiency than a two-winding transformer of the same rating.
4. *Operate at better voltage regulation.* Reduction in the conductor material means lower value of ohmic resistance. A part of the winding being common, leakage flux or the leakage reactance is less. In other words, an auto-transformer has lower value of leakage impedance and hence auto-transformer has lower value of leakage impedance and has better voltage regulation than a two-winding transformer of the same output.

2.46 Disadvantages of Auto-transformers

Although auto-transformers have less cost, better regulation and low losses as compared to ordinary two winding transformer of same rating. But still they are not widely used due to one major disadvantage that the secondary winding is not insulated from the primary. If an auto transformer is used to supply low voltage from a high voltage and there is a break in the secondary winding, full *primary voltage comes across the secondary terminals which may be dangerous to the operator and equipment (load).* Therefore, it is advisable not to use an auto transformer for interconnecting high voltage and low voltage system. Their use is only limited to the places where slight variation of output voltage from the input voltage is required. The other disadvantages are;

- The effective per-unit impedance of an autotransformer is smaller compared to a two-winding transformer. The reduced internal impedance results in a larger short-circuit (fault) current.
- In an autotransformer there is a loss of isolation between input and output circuits. This is particularly important in three-phase transformers where one may wish to use a different winding and earthing arrangement on each side of the transformer.

2.47 Phasor Diagram of an Auto-transformer

Considering a step-down auto-transformer shown in Fig. 2.67.

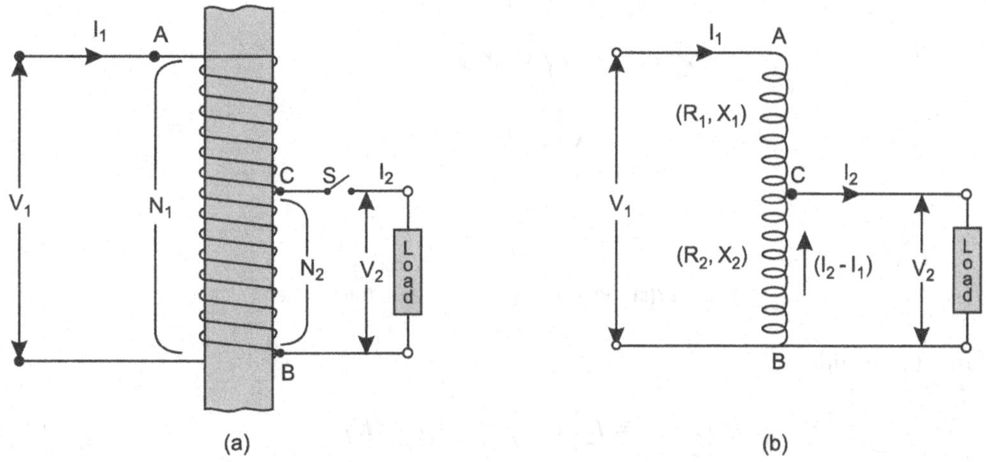

Fig. 2.67(a,b) Circuit for auto-transformer

Let V_1 = applied voltage, acting downwards (from A to B)

I_2 = supplied current flows through winding AB (A to B)

I_2 = load current which also flows through section BC (B to C)

$I_2 - I_1$ = resultant current in section BC (B to C)

ϕ = flux produced in the core linking with winding.

An emf is induced in both the sections of the winding i.e., E_{CA} and E_{BC} which act in opposite direction to the applied voltage to balance it.

The phasor diagram of an auto-transformer is shown in Fig. 2.67(*c*), where V_2 is the voltage applied across the load represented by phasor OA. Let the load is inductive, then current drawn by the load I_2 lags behind the voltage phasor V_2 by an angle ϕ_2 (where $\phi_2 = \cos^{-1}$ pf of load).

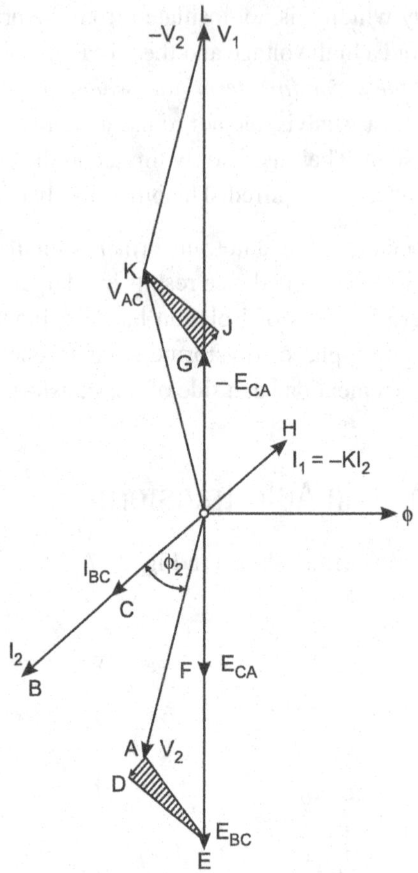

Fig. 2.67(c) Phasor diagram of an auto-transformer

Current in section *BC*.

$$I_{BC} = I_2 - I_1 = I_2\left(1 - \frac{I_1}{I_2}\right) = I_2(1 - K) \quad \text{...(2.123)}$$

where *K* is transformation ratio.

This current I_{BC} is represented by phasor *OC*. There is a voltage drop in the resistance and reactance of the winding section *BC*. To represent these drops, draw a line parallel to line *OC* from *A* and take $AD = I_2\,(1 - K)\,R_2$. Draw a perpendicular from *D* on *OC* so that $DE = I_2(1 - K)\,X_2$. Triangle *ADE* represents the voltage drop in the impedance of winding section *BC*. Here, *OE* represents the *emf* induced in the winding section *BC* i.e., E_{BC}.

The induced emf in section *CA*,

$$E_{CA} = E_{BC}\left(\frac{1-K}{K}\right) \text{ in phase with } E_{BC} \text{(vector } OF\text{)}$$

Draw this vector to the other side i.e., $OG = OF$, hence $OG = -E_{CA}$

Draw the vector $OH = I_1 = -KI_2$ by producing line *BO* to *H*. Draw impedance drop triangle *GJK* for the section *CA*, where $GJ = I_1R_1$ (parallel to I_1) and $JK = I_1X_1$ (perpendicular to I_1)

Here, phasor $OK = V_{AC}$ i.e., phasor sum of emf OG and impedance drop GK.

Finally, V_1 is the phasor sum of V_{AC} and V_{CB} i.e.,

$$V_1 = V_{AC} + V_{CB} = V_{AC} - V_2 \quad \text{...(2.124)}$$

2.48 Equivalent Circuit of an Auto-transformer

While drawing equivalent circuit of an auto-transformer, it may be considered as a two winding transformer. Where one section AC of its winding can be taken as its primary having voltage V_{AC} and current I_1 and the other section CB can be taken as its secondary having output voltage V_2 and current $I_2 - I_1$.

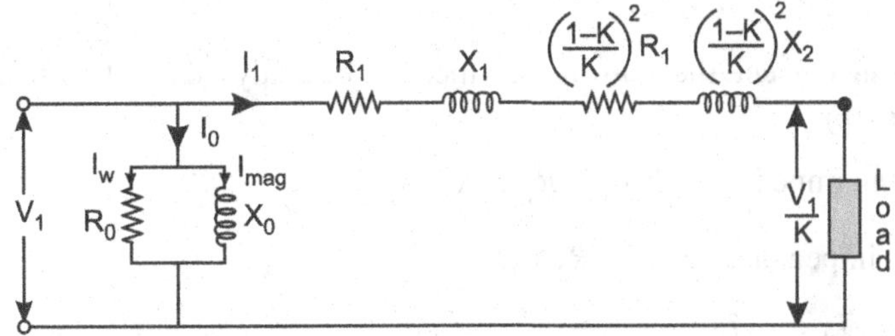

Fig. 2.68 Equivalent circuit of an auto-transformer

$$\text{Current ratio} = \frac{I_{AC}}{I_{BC}} = \frac{I_1}{I_2 - I_1} = \frac{I_1 / I_2}{1 - I_1 / I_2} = \frac{K}{1-K} \quad \text{...(2.125)}$$

Hence, the ratio of currents in sections of winding is $\left(\frac{K}{1-K}\right)$

$$\text{Ratio of induced emfs} = \frac{E_{CB}}{E_{AC}} = \frac{N_2}{N_1 - N_2} = \frac{N_2 / N_1}{1 - N_2 / N_1} = \frac{K}{1-K} \quad \text{...(2.126)}$$

The equation (2.125) and (2.126) reveal that the auto-transformer with a turn ratio of K is equivalent in its transformation to the ordinary transformer with a ratio of $\frac{K}{1-K}$.

An approximate equivalent circuit of an auto-transformer is shown in Fig. 2.68.

2.49 Simplified Equivalent Circuit of an Auto-transformer

An auto-transformer, shown in Fig. 2.69(*a*), may be considered as a two-winding transformer with winding of section AC of $(N_1 - N_2)$ turns acting as primary and the winding of section BC of N_2 turns acting as secondary. Thus, an auto-transformer behaves as a two-winding transformer primary applied voltage across AC is V_{AC} and primary current is I_1. The secondary output voltage is V_2 and the secondary winding current is $(I_2 - I_1)$. For given auto-transformer, its equivalent circuit as a two-winding transformer is shown in Fig. 2.69(*b*).

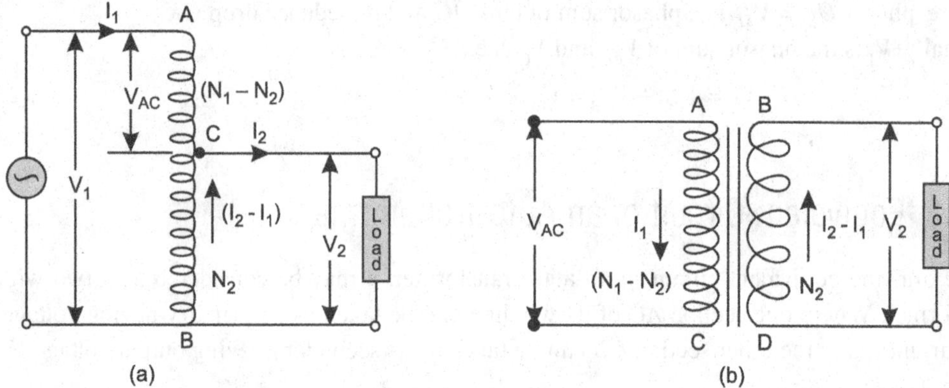

Fig. 2.69 Auto-transformer and its equivalent circuit

Neglecting exciting current, the auto-transformer can be analysed as a two-winding transformer as shown in Fig. 2.69(*b*).

Let, Primary winding impedance, $\overline{Z}_{AC} = R_1 + jX_1$

Secondary winding impedance, $\overline{Z}_{BC} = R_2 + jX_2$

Primary current, $\overline{I}_{AC} = \overline{I}_1$

Referring all the parameters on the primary side

$$R_{ep} = R_1 + R'_2 = R_1 + R_2\left(\frac{N_1 - N_2}{N_2}\right)^2 = R_1 + R_2\left(\frac{1-K}{K}\right)^2 \quad ...(2.127)$$

$$X_{ep} = X_1 + X'_2 = X_1 + X_2\left(\frac{N_1 - N_2}{N_2}\right)^2 = X_1 + X_2\left(\frac{1-K}{K}\right)^2 \quad ...(2.128)$$

Secondary voltage referred to primary side

$$V'_2 = V_2 \times \left(\frac{N_1 - N_2}{N_2}\right) = V_2\left(\frac{1-K}{K}\right)$$

$$\overline{V}_{AC} = \overline{V}'_2 + \overline{I}_1\overline{Z}_{ep}$$

$$V_{AC} = V_2\left(\frac{1-K}{K}\right) + I_1\left[\left\{R_1 + R_2\left(\frac{1-K}{K}\right)^2\right\} + j\left\{X_1 + X_2\left(\frac{1-K}{K}\right)^2\right\}\right]$$

$$V_{AC} = V_2\left(\frac{1}{K} - 1\right) + I_1(R_{ep} + jX_{ep}) \quad ...(2.129)$$

From Fig. 2.68(*a*), it reveals that

$$V_1 = V_{AC} + V_2$$

$$= V_2\left(\frac{1}{K} - 1\right) + I_1(R_{ep} + jX_{ep}) + V_2 = \frac{V_2}{K} + I_1(R_{ep} + jX_{ep}) \quad ...(2.130)$$

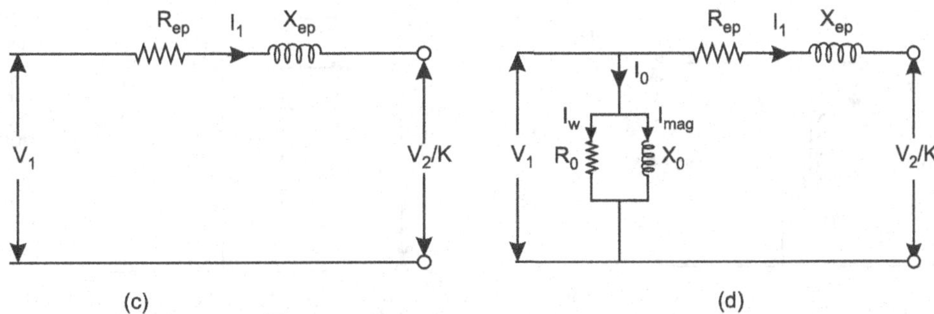

Fig. 2.69 Simplified equivalent circuit of an auto-transformer

The equivalent circuit of an auto-transformer is shown in Fig. 2.69(*c*) where exciting circuit is not considered. If exciting resistance and reactance is considered, then its equivalent circuit will be as shown in Fig. 2.69(*d*).

In Fig. 2.69(*a*), if input voltage is applied across *BC* and the load is connected across *AB*, then the transformer acts as a step-up auto-transformer.

2.50 Conversion of a Two-winding Transformer to an Auto-transformer

A conventional two-winding transformer with its polarity markings is shown in Fig. 2.70(*a*). It can be converted to a step-up autotransformer by connecting the two windings electrically in series with additive or subtractive polarities. With additive polarity between the high-voltage and low-voltage sides, a step-up transformer is obtained. Whereas, with subtractive polarity a step-down transformer is obtained.

Let us consider a conventional 24 kVA, 2500/250 V transformer to be connected in autotransformer configuration.

Additive Polarity

Figure 2.70(*b*) shows the series connections of the windings with additive polarity. The circuit is redrawn in Fig. 2.70(*c*) showing common terminal of the autotransformer at the top. Figure 2.70(*d*) shows the same circuit with common terminal at the bottom. Since the polarity is additive, V_{HV} = 2500+ 250 = 2750 V and V_{LV} = 2500 V, the transformer acts as a step-up autotransformer.

Subtractive Polarity

Figure 2.70(*e*) shows the series connections of the windings with subtractive polarity. The circuit is redrawn in Fig. 2.70(*f*) with common terminal at the top. Figure 2.70(*g*) shows the same circuit with common terminal at the bottom. Since the polarity is subtractive V_{HV} = 2500 V and V_{LV} = 2500 – 250 = 2250 V, the transformer acts as a step-down autotransformer.

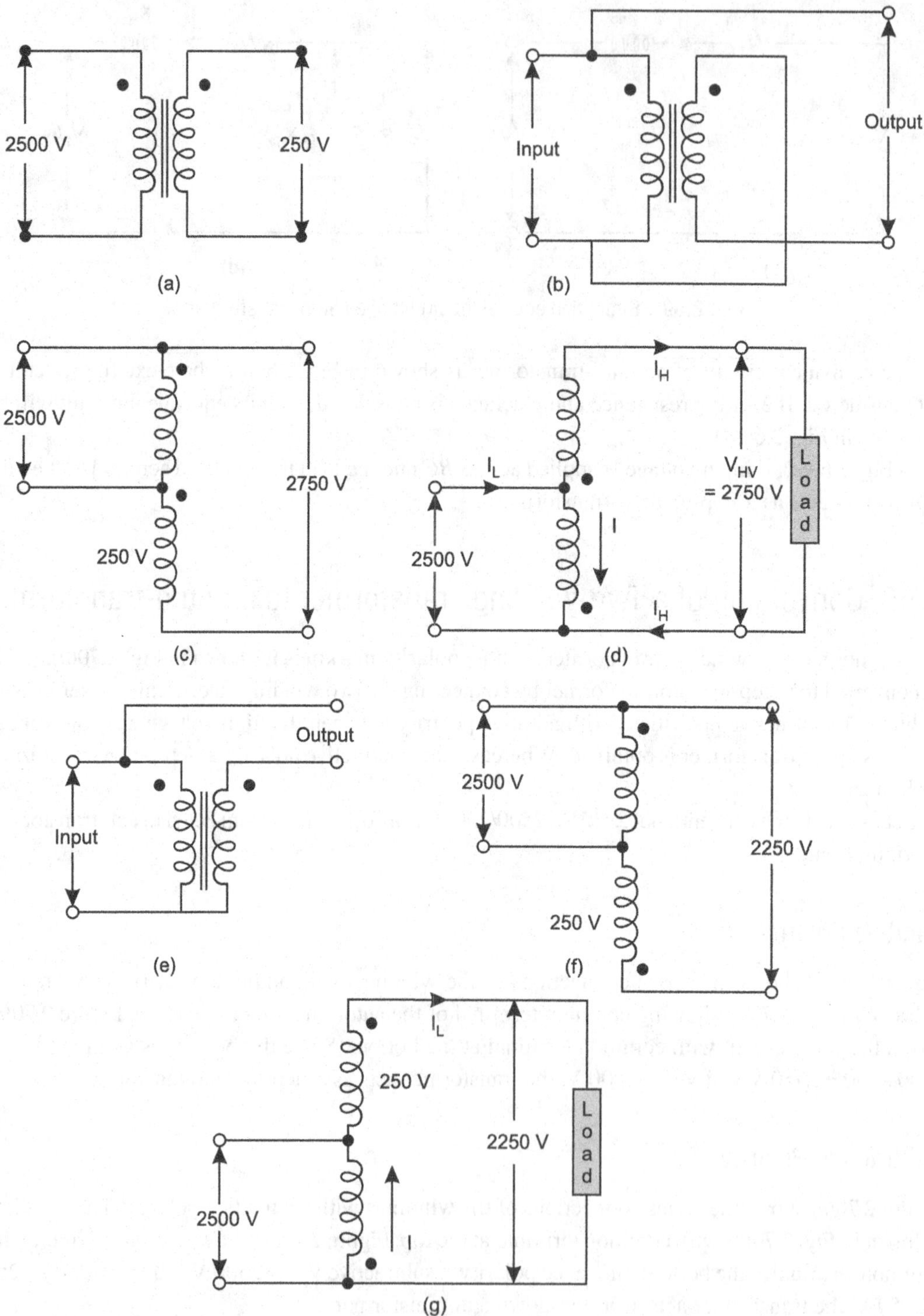

Fig. 2.70 Various electrical connections to form an auto-transformer.

2.51 Comparison of Characteristics of Auto-transformers and Two-winding Transformers

An auto-transformer can be considered as a two-winding transformer with winding *AC* as primary and winding *BC* as the secondary. Ratings and characteristics of auto-transformers and two-winding transformers so obtained, differ as discussed below:

(*i*) **Ratings.** It is seen that winding *AC* acts as the primary and winding *BC* as the secondary of a two-winding transformer, then considering input side;

$$\frac{\text{kVA rating as an auto-transformer}}{\text{kVA rating as a two-winding transformer}}$$

$$= \frac{\text{Primary input voltage } V_1 \times \text{Primary input current } I_1}{\text{Primary voltage across winding } AC \times \text{Primary current in winding } AC}$$

$$= \frac{V_1 I_1}{(V_1 - V_2)I_1} = \frac{1}{1-(V_2/V_1)} = \frac{1}{1-K}$$

Considering output side; $\dfrac{\text{kVA rating as an auto-transformer}}{\text{kVA rating as a two-winding transfer}} = \dfrac{V_2 I_2}{V_2(I_2 - I_1)}$

$$= \frac{1}{(1 - I_1/I_2)} = \frac{1}{1-K} \qquad \text{...(2.131)}$$

(*ii*) **Losses.** When a two-winding transformer is connected as an auto-transformer, the current in different sections and voltages across them remain unchanged. Therefore, losses when working as an auto-transformer are the same as the losses in a two-winding transformer. Per unit losses, however, differ.

$$\frac{\text{Per unit full-load losses as auto-transformer}}{\text{Per unit full-load losses as two-winding transformer}}$$

$$= \frac{\text{Full-load losses}}{\text{kVA rating as auto-transformer}} \times \frac{\text{kVA rating as two-winding transformer}}{\text{Full-load losses}}$$

$$= (1 - K) \qquad \text{...(2.132)}$$

(*iii*) **Impedance drop.** When a two-winding transformer is used as an auto-transformer, both *LV* and *HV* windings are utilised completely. In addition, current and voltage ratings of each winding section remain unaltered. Therefore, impedance drop at full load is the same in both the transformers. Their per unit values are, however, different. When referred to *HV* side, per unit impedance drop as an auto-transformer is with respect to voltage V_1 and for a two-winding transformer, it is with respect to $(V_1 - V_2) = V_1(1 - K)$.

$$\frac{\text{Per unit impedance drop as an auto-transformer}}{\text{Per unit impedance drop as a two-winding transformer}} = \frac{I_1 Z_1 / V_1}{I_1 Z_1 / (V_1 - V_2)} = \frac{V_1 - V_2}{V_1} = (1 - K)$$

When impedance drop is referred to *LV* side, the winding current is $(I_2 - I_1)$ as an auto-transformer and I_2 as a two-winding transformer.

$$\therefore \quad \frac{\text{Per unit impedance drop as an auto-transformer}}{\text{Per unit impedance drop as a two-winding transformer}} = \frac{(I_2 - I_1)\,Z_2\,/\,V_2}{I_2 Z_2\,/\,V_2}$$

$$= \frac{I_2 - I_1}{I_2} = (1 - K) \quad \text{...(2.133)}$$

(iv) **Voltage regulation.** Regulation in transformers is proportional to per unit impedance drop.

$$\frac{\text{Regulation as an auto-transformer}}{\text{Regulation as a two winding transformer}} = (1 - K) \quad \text{...(2.134)}$$

(v) **Short-circuit current.** Per unit short-circuit current is reciprocal of per unit impedance drop.

$$\therefore \quad \frac{\text{Per unit short-circuit current as an auto-transformer}}{\text{Per unit short-circuit current as two winding transformer}} = \frac{1}{1 - K} \quad \text{...(2.135)}$$

The value of K used in the above relations, is less than one. In general, for using these relations, the value of K for step-down or step-up auto-transformers is $K = LV/HV$.

2.52 Applications of Auto-transformers

1. Single-phase and 3-phase auto-transformers are employed for obtaining variable output voltages at the output. When used as variable ratio auto-transformers, these are known by their trade names, such as *variac, dimmerstat, autostat* etc.

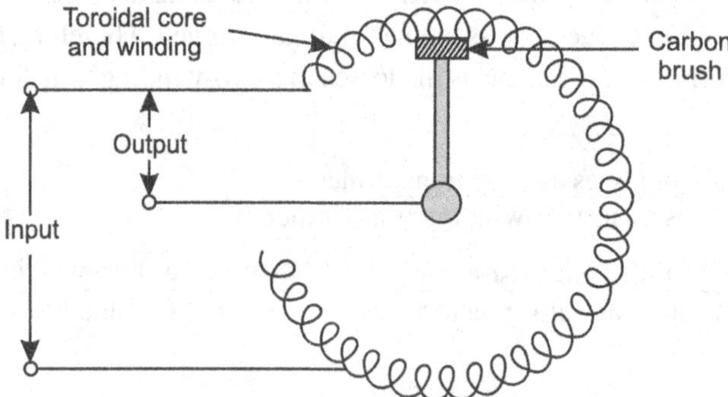

Fig. 2.71 Auto-transformer as a variac

A variable ratio auto-transformer (or variac) has a toroidal core and toroidal winding. A sliding contact with the winding is made by carbon brush, as shown in Fig. 2.71 and Fig. 2.64(*f*). The position of the sliding contact can be varied by a hand-wheel which changes output voltage. These are mostly used in laboratories.

2. Auto-transformers are also used as *boosters* for raising the voltage in an AC feeder.
3. As *furnace transformers,* for getting a convenient supply to suit the furnace winding from normal 230 V AC supply
4. Auto-transformers with a number of tappings are used for starting induction motors and synchronous motors. When auto-transformers are used for this purpose, these are known as *auto-starters.*

Example 2.61

Determine the core area, the number of turn, and the position of the tapping point for a 500 kVA, 50 Hz single phase, 6600/5000 volt auto-transformer, assuming the following approximate values: emf per turn 8 volt, maximum flux density 1.3 tesla.

Solution:

We know $E = 4.44 f B_m A_i N$

or $\frac{E}{N} = 4.44 f B_m A_i$

or $8 = 4.44 \times 50 \times 1.3 \times A_i$

or $A_i = \frac{8}{4 \cdot 44 \times 50 \times 1 \cdot 3} = 0 \cdot 02772 \text{ m}^2$

$= \mathbf{277.2\ cm^2}$ (*Ans.*)

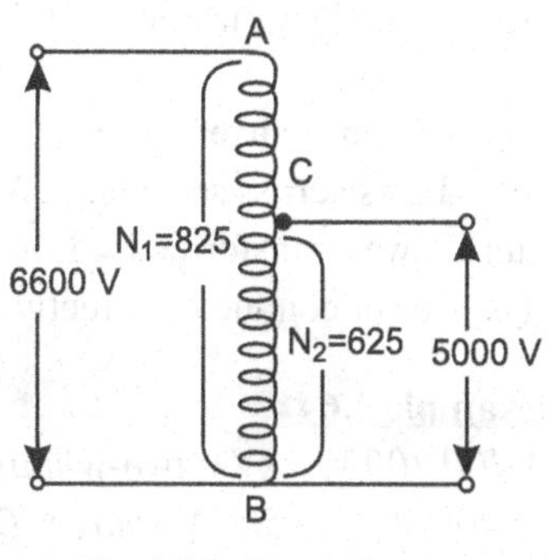

Fig. 2.72 Circuit diagram.

Turns on the primary side, $N_1 = \frac{6600}{8} = 825$

Turns on the secondary side $N_2 = \frac{5000}{8} = 625$

Hence tapping should be 200 turns from high voltage end or 625 turns from the common end as shown in Fig. 2.72.

Example 2.62

An auto-transformer having 1500 turns is connected across a 500 V AC supply. What secondary voltage will be obtained if a tap is taken at 900th turn.

Solution:

Supply voltage, $V_1 = 500$ V

Total turns, $N_1 = 1500$

Secondary turns, $N_2 = 900$

Voltage per turn $= \frac{V_1}{N_1} = \frac{500}{1500} = \frac{1}{3}$ V

Secondary voltage, V_2 = Voltage per turn $\times N_2 = \frac{1}{3} \times 900 = \mathbf{300\ V}$ (*Ans.*)

Example 2.63

An auto-transformer supplies a load of 10 kW at 250 V and at unity power factor. If the primary voltage is 500 V, determine.

(a) transformation ratio (b) secondary current (c) primary current (d) number of turns across secondary if total number of turns is 500 (e) power transformed and (f) power conducted directly from the supply mains to load.

Solution:

(*a*) Transformation ratio, $K = \frac{V_2}{V_1} = \frac{250}{500} = \mathbf{0.5}$ (*Ans.*)

(*b*) Secondary current, $I_2 = \frac{kW \times 1000}{V_2 \cos \phi} = \frac{10 \times 1000}{250 \times 1} = \mathbf{40\ A}$ (*Ans.*)

(*c*) Primary current, $I_1 = KI_2 = 0.5 \times 40 = \mathbf{20\ A}$ (*Ans.*)
(*d*) Turns across secondary, $N_2 = KN_1 = 0.5 \times 500 = \mathbf{250}$ (*Ans.*)
(*e*) Power transformed = Load × $(I - K) = 10(1 - 0.5) = \mathbf{5\ kW}$ (*Ans.*)
(*f*) Power conducted directly from supply mains = 10 – 5 = **5 kW** (*Ans.*)

Example 2.64

A 400/100 V, 5 kVA, two-winding transformer is to be used as an autotransformer to supply power at 400 V from 500 V source. Draw the connection diagram and determine the kVA output of the autotransformer.

Solution:

For a two-winding transformer $V_1I_1 = kVA \times 1000 \text{ or } I_1 = \frac{5 \times 1000}{400} = 12 \cdot 5 \text{ A}$

$$V_2I_2 = kVA \times 1000 \text{ or } I_2 = \frac{5 \times 1000}{100} = 50 \text{ A}$$

Figure 2.73 shows the use of two-winding transformer as an autotransformer to supply power at 400 V from a 500 V source.

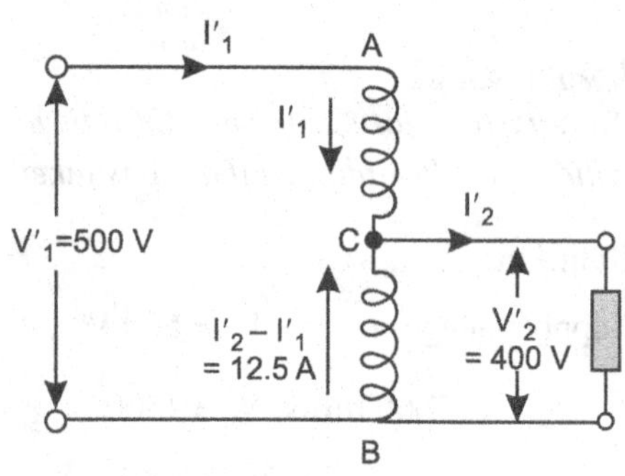

Fig. 2.73 Circuit diagram

Here $V_1' = 500$ V, $V_2' = 400$ V

Transformation ratio, $K = \frac{V_2'}{V_1'} = \frac{400}{500} = 0 \cdot 8$

$$I_1' = K \times I_2' = 0.8\, I_2'$$

Current through 400 V winding,

$$I_{BC} = I_2' - I_1' = I_2' - 0 \cdot 8\, I_2' = 0 \cdot 2 I_2'$$

Since the current rating of 400 V winding is 12.5 A

$$0 \cdot 2\, I_2' = 12.5 \text{ or } I_2' = \frac{12 \cdot 5}{0 \cdot 2} = 62 \cdot 5 \text{ A}$$

The *kVA* output of the autotransformer $= \frac{V_2' I_2'}{1000} = \frac{400 \times 62 \cdot 5}{1000} = \mathbf{25}$ (*Ans.*)

Example 2.65

The primary and secondary voltages of an auto-transformer are 250 V and 200 V, respectively. Show with the aid of a diagram the current distribution in the windings when the secondary current is 100 A and calculate the economy of copper in this particular case (in percentage).

Solution:

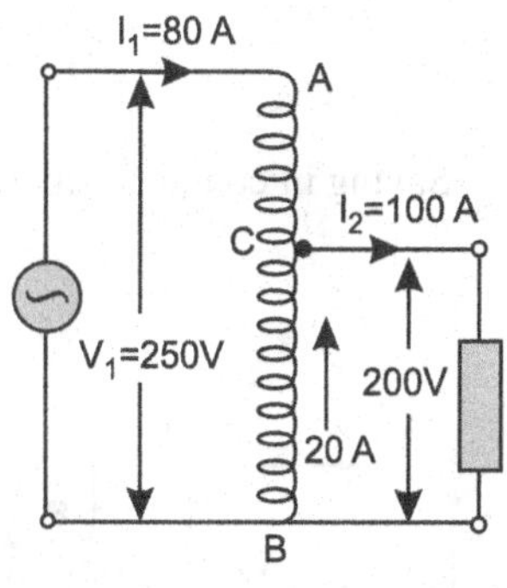

Fig. 2.74 Circuit diagram

Transformation voltage, $K = \dfrac{V_2}{V_1}$

$= \dfrac{200}{250} = 0 \cdot 8$

Secondary load current, $I_2 = 100$ A

Primary current, $I_1 = KI_2 = 0.8 \times 100 = 80$ A

The current distribution is shown in Fig. 2.65

Economy in copper = $K = 0.8$ or **80%** (*Ans.*)

Example 2.66

A 12500/2500 V transformer is rated at 100 kVA as a two-winding transformer. If the winding are connected in series to form an autotransformer, what will be the possible voltage ratios and output? Also calculate the saving in conductor material.

Solution:

$$\text{Rated current for } HV \text{ winding} = \frac{100 \times 1000}{12500} = 8 \text{ A}$$

$$\text{Rated current for } LV \text{ winding} = \frac{100 \times 1000}{2500} = 40 \text{ A}$$

It is to be noted that if the windings of the two-winding transformer are connected in series to form an autotransformer, the current in the two windings should not exceed their rated values.

To form an autotransformer, the two-windings may be connected in the following two ways:

(*a*) *First Configuration*

The winding *AC* is for 2500 V and winding *BC* for 12500 V as shown in Fig. 2.75(*a*)

Here, $V_{AC} = 2500$ V, $V_{BC} = 12500$ V

$\therefore$ $V_1 = V_{AC} + V_{BC} = 2500 + 12500 = 15000$ V

$V_2 = V_{BC} = 12500$ V

Therefore, the voltage ratio for the autotransformer of Fig. 2.75(*a*) is

$$K = \frac{V_2}{V_1} = \frac{12500}{15000} = \frac{5}{6} = \mathbf{0.833} \ (Ans)$$

By *KCL* at point *C* $I_2 = I_{AC} + I_{BC} = 40 + 8 = 48$ A

The current distribution is shown in Fig. 2.75(*a*).

kVA rating of 15000/12500 V autotransformer

$$= \frac{V_2 I_2}{1000} = \frac{12500 \times 48}{1000} = \mathbf{600\ kVA} \ (Ans.)$$

or $$= \frac{V_1 I_1}{100} = \frac{15000 \times 40}{1000} = \textbf{600 kVA } (Ans)$$

Saving in conductor material

$$K_1 = \frac{5}{6} = 0 \cdot 8333 \text{ pu} = \textbf{83.3 per cent } (Ans.)$$

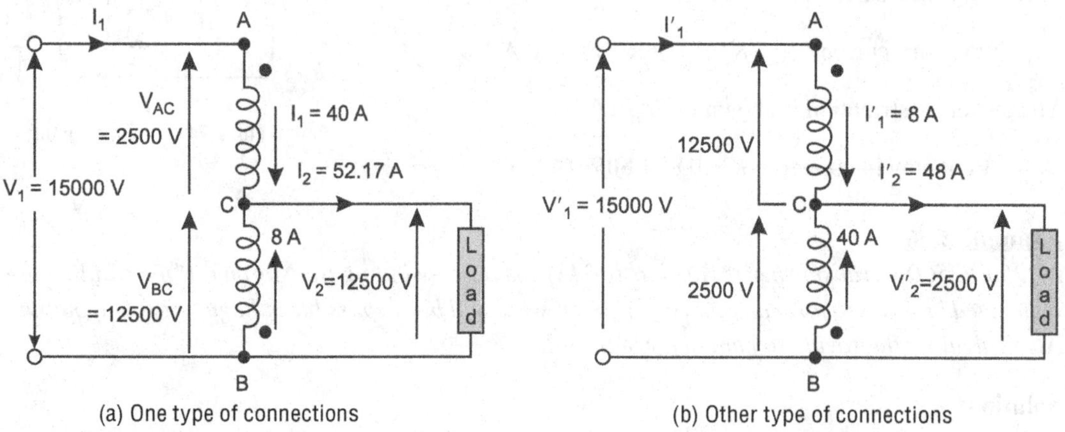

(a) One type of connections (b) Other type of connections

Fig. 2.75 Electric circuit of an auto-transformer with different configurations

(b) Second Configuration

Here the winding *AC* is for 12500 V and *BC* for 2500 V, as shown in Fig. 2.75(*b*). Therefore V_{AC} = 12500 V, V_{BC} = 2500 V

$$V_1' = V_{AC} + V_{BC} = 12500 + 2500 = 15000 \text{ V}$$

$$V_2' = V_{BC} = 2500 \text{ V}$$

$$K_2 = \frac{V_2'}{V_1'} = \frac{2500}{15000} = \frac{1}{6} = \textbf{0.167 } (Ans.)$$

By *KCL* at point *C*, $I_2' = I_{AB} + I_{BC} = 8 + 40 = 48$ A

The current distribution is shown in Fig. 2.75(*b*).

kVA rating of 15000/2500 V autotransformer

$$= \frac{V_2' I_2'}{1000} = \frac{2500 \times 48}{1000} = \textbf{120 kVA } (Ans.)$$

or $$= \frac{V_1' I_1'}{1000} = \frac{15000 \times 8}{1000} = 120 \text{ kVA}$$

Saving in conductor material = $K_2 = \frac{1}{6} = 0 \cdot 167 \text{ pu} = \textbf{16.7\%}$ (*Ans.*)

Example 2.67

A load of 100 kVA is to be supplied at 500 volt from 2500 V supply mains by an auto-transformer. Determine the current and voltage rating for each of the two windings. What would be the kVA rating of the transformer, if it were used as a two winding transformer.

Solution:

Load to be supplied = 100 kVA

Secondary voltage, V_2 = 500 V

Primary voltage, V_1 = 2500 V

$$\text{Load current, } I_2 = \frac{kVA \times 1000}{V_2} = \frac{100 \times 1000}{500} = \mathbf{200\ A}\ (Ans)$$

Assuming input equal to output (i.e., neglecting losses)

$$\text{Input current, } I_1 = \frac{kVA \times 1000}{V_1} = \frac{100 \times 1000}{2500} = \mathbf{40\ A}\ (Ans.)$$

The voltage across winding *BC* is 500 volt and current flowing through it is phasor difference of currents I_1 and I_2. Neglecting no-load current, phasor difference of currents I_1 and I_2 is equal to numerical difference of two currents I_1 and I_2 and therefore current flowing through winding *BC* is 200 – 40 = 160 A, refer Fig. 2.3

Hence voltage rating of winding *BC* = **500 V** (*Ans.*)

Current rating of winding *BC* = **200 A** (*Ans.*)

Although voltage across winding *AC* is not necessarily in phase with the voltage across *CB*. However the angle between the two voltages is usually so small that the voltages are added numerically.

Voltage across winding, *AC* = 2500 – 500 = 2000 V

Current flowing through the winding *AC* is I_1 i.e., 40 A

Hence voltage rating of winding *AC* = **200 volt** (*Ans.*)

Current rating of winding *AC* = **40 A** (*Ans.*)

The auto-transformer, when connected as two winding transformer;

The permissible voltage input = 2000 V

Permissible voltage output = 500 V

The rating of transformer as a two winding transformer

$$= \frac{\text{Permissible voltage output} \times \text{permissible current output}}{1000}$$

$$= \frac{500 \times 160}{1000} = \mathbf{80\ kVA}\ (Ans.)$$

Example 2.68

An auto-transformer is required to boost the voltage of a 500 kVA line from 3100 volt to 3300 volt. Calculate the current in the winding and the kVA rating of the two winding transformer to which it would be equivalent.

Solution:

$$\text{Primary winding voltage} = 3100 \text{ volt}$$

$$\text{Secondary winding voltage} = 3300 - 3100 = 200 \text{ volt}$$

$$\text{The input current} = \frac{500 \times 1000}{3100} = \mathbf{161.29\ A}\ (Ans.)$$

$$\text{and output current} = \frac{500 \times 1000}{3300} = \mathbf{151.51\ A}\ (Ans.)$$

$$\text{Current in the primary} = 161.29 - 151.51 = \mathbf{9.78\ A}\ (Ans.)$$

and the current in the secondary winding = 151.51 A

The kVA of the equivalent two winding transformer

$$= 500\left(1 - \frac{3100}{3300}\right) = \mathbf{30.3\ kVA}\ (Ans.)$$

Example 2.69

A 25 kVA, 2500/250 V, two-winding transformer is to be used as a step-up autotransformer with constant source voltage of 2500 V. At full load of unity power factor, calculate the power output, power transformed and power conducted. If the efficiency of the two-winding transformer at 0.8 power factor is 95 per cent, find the efficiency of the autotransformer.

Solution:

$$\text{Rated current of 2500 V winding, } I_{HV} = \frac{25 \times 1000}{2500} = 10\text{ A}$$

$$\text{Rated current of 250 V winding, } I_{LV} = \frac{25 \times 1000}{250} = 100\text{ A}$$

With the polarities shown in Fig. 2.76, the output voltage,

$$= 2500 + 250 = 2750\text{ V}$$

By *KCL* at point *C*, the input line current, $I_1 = 100 + 10 = 110$ A.

The current distribution is shown in Fig. 2.76

Output current of autotransformer, $I_2 = 100$ A.

$$kVA \text{ rating of autotransformer} = \frac{2750 \times 100}{1000} = 275\ \text{kVA}$$

Considering input side;

or $$kVA \text{ rating of autotransformer} = \frac{2500 \times 110}{1000} = 275\text{ kVA}$$

Power output at full load, unity power factor

$$= \text{kVA} \cos\phi = 275 \times 1 = \mathbf{275\ kVA}\ (Ans.)$$

Here winding *BC* acts as the primary and the winding *AC* as the secondary.

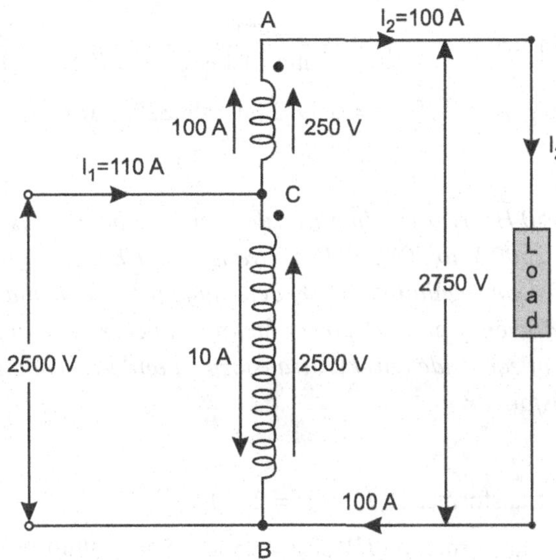

Fig. 2.76 Circuit diagram.

$$kVA \text{ transformed} = \frac{V_{AC}I_{AC}}{1000} = \frac{250 \times 100}{1000} = 25 \text{ kVA}$$

$$\text{Also, } kVA \text{ transformed} = \frac{V_{BC}I_{BC}}{1000} = \frac{2500 \times 10}{1000} = 25 \text{ kVA}$$

$$\text{Power transformed} = \text{kVA} \cos\phi = 25 \times 1 = \mathbf{25\ kW}\ (Ans.)$$

$$\text{kVA conducted} = \frac{V_{BC}I_{AB}}{1000} = \frac{2500 \times 100}{1000} = 250 \text{ kVA}$$

Alternatively

$$\text{kVA conducted} = \text{input kVA} - \text{transformed kVA}$$
$$= 275 - 25 = 250 \text{ kVA}$$

$$\text{Power conducted} = \text{kVA} \cos\phi = 250 \times 1 = \mathbf{250\ kW}\ (Ans.)$$

Calculation of efficiency

For two winding transform: Efficiency, $\eta = \dfrac{\text{output}}{\text{output + losses}}$ or output + losses = $\dfrac{\text{output}}{\eta}$

$$\therefore \quad \text{losses} = \left(\frac{1}{\eta} - 1\right) \times \text{output}$$

$$\text{Output of two-winding transformer} = 250 \times 100 \times 0.8 = 2000 \text{ W}$$

$$\text{Losses in a two-winding transformer} = \left(\frac{1}{0 \cdot 95} - 1\right) \times 2000 = 1052 \cdot 6 \text{ W}$$

Since autotransformer operates at rated voltages and rated currents, losses in autotransformer

= losses in two-winding transformer

= 1052.6 W

Efficiency of auto-transformer $\eta_a = \dfrac{\text{Output}}{\text{Output + losses}} = \dfrac{275 \times 1000 \times 0 \cdot 8}{275 \times 1000 \times 0 \cdot 8 + 1052 \cdot 6}$

$= 0.9952$ pu $=$ **99.52%** (*Ans.*)

Example 2.70

A 200 kVA, 2500/500 V, 50 Hz, two-winding transformer is to be used as an auto-transformer to step-up the voltage from 2500 V to 3000 V. If the transformer has an efficiency of 96% at 0.8 pf lagging, impedance of 4% and regulation of 3%, determine (i) voltage and current ratings of each side (ii) kVA rating (iii) efficiency at unity power factor (iv) percentage impedance (v) regulation, (vi) short-circuit current of each side (vii) kVA transformed and kVA conducted at full load, while it is used as an auto-transformer.

Solution:

Rated voltage of the auto-transformer on *LV* side = 2500 V

Rated voltage of the auto-transformer on *HV* side = 2500 + 500 = **3000** (*Ans.*)

Rated current of *HV* winding, $I_{HV} = \dfrac{200 \times 1000}{2500} = 80$ A

Rated current of *LV* winding $I_{LV} = \dfrac{200 \times 1000}{500} = 400$ A

Rated current of *HV* side of the auto-transformer

= Rated current 500 V side of the two-winding transformer

= **400 A** (*Ans.*)

Applying *KCL* at point *C*, input line current, $I_1 = 400 + 80 =$ **480 A** (*Ans.*)

The current distribution is shown in Fig. 2.77

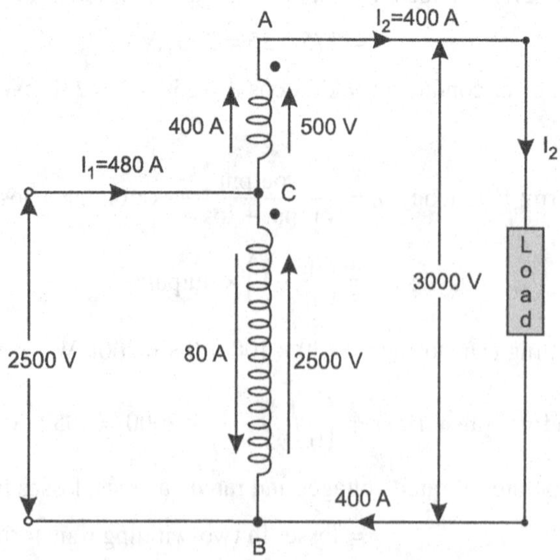

Fig. 2.77 Circuit diagram

(*ii*) kVA rating of the auto-transformer = $\frac{3000 \times 400}{1000}$ = **1200 kVA** (*Ans.*)

$$\text{Transformation ratio, } K = \frac{LV \text{ side voltage}}{HV \text{ side voltage}} = \frac{2500}{3000} = 0 \cdot 833$$

(*iii*) Power output of two-winding transformer at 0.8 pf lagging

$$= 200 \times 0.8 = 160 \text{ kW}$$

$$\text{Full-load losses} = \frac{\text{Output}}{\text{Efficiency}} - \text{output} = \frac{160}{0 \cdot 96} - 160 = 6 \cdot 667 \text{ kW}$$

Since the auto-transformer operates at rated voltages and rated currents, the losses remain constant i.e., losses of autotransformer = 6.667 kW

(*iv*) Efficiency of auto-transformer for an output of 1200 × 1 kW is

$$\eta = \frac{\text{Output}}{\text{Output} + \text{Losses}} \times 100 = \frac{1200}{1200 + 6 \cdot 667} \times 100 = \mathbf{99.45\%} \text{ (Ans.)}$$

(*v*) Percent impedance of two-winding transformer = 4%

The ohmic drop at full load is the same in both cases. If the impedance drop is referred to *HV* side of the auto-transformer (see Fig. 2.77), the pu (per unit) or percent drop in the auto-transformer is with respect to V_1 while that in the two-winding transformer it is with respect to

$$V_1 - V_2 = V_2 (1 - K)$$

If the impedance drop is referred to *LV* side of the auto-transformer, then the winding current is

$$I_2 - I_1 = I_2 (1 - K),$$

So the impedance drop with respect to I_2 is to be reduced by $(1 - K)$ times in the case of an auto-transformer.

Therefore pu or percent impedance drop in an auto-transformer

= $(1 - K) \times pu$ or percent impedance drop as a two-winding transformer

= (1 – 0.833) × 4 = **0.667%** (*Ans.*)

(*v*) As regulation is proportional to pu or percent impedance drop,

So regulation as an auto-transformer = $(1 - K)$ × regulation as a two-winding transformer

= (1 – 0.833) × 3 = **0.5%** (*Ans.*)

(*vi*) Short-circuit current as an auto-transformer

$$= \frac{1}{\text{pu impedance of the auto-transformer}}$$

$$= \frac{1}{0.00667} = 150 \text{ pu}$$

So short-circuit current on *HV* side = $\frac{150 \times 400}{1000}$ = **60 kA** (*Ans.*)

Short-circuit current on *LV* side = $\frac{150 \times 480}{1000}$ = **72 kA** (*Ans.*)

(*vi*) Here winding *BC* acts as the primary and the winding *AC* as the secondary

$$\text{So kVA transformed} = \frac{80 \times 2500}{1000} \text{ or } \frac{400 \times 500}{1000} = \mathbf{200\ kVA}\ (Ans.)$$

$$\text{kVA conducted} = 1200 - 200 = 1000 \text{ kVA } (Ans.)$$

Example 2.71

Determine the values of the currents flowing in the various branches of a 3 phase, star connected auto-transformer loaded with 400 kW at 0.8 power factor lagging and having a ratio of 440/550 volt. Neglect voltage drops, magnetising current and all losses in the transformer.

Solution:

$$\text{Primary current} = \frac{kW \times 1000}{\sqrt{3} \times E_{L1} \times \cos\phi}$$

$$I_1 = \frac{400 \times 1000}{\sqrt{3} \times 440 \times 0 \cdot 8} = \mathbf{656\ A}$$

$$\text{Secondary line current} = \frac{kW \times 100}{\sqrt{3} \times E_{L_2} \times \cos\phi} = \frac{400 \times 1000}{\sqrt{3} \times 550 \times 0 \cdot 8} = \mathbf{525\ A}$$

Note: Current in sections $R'R$, $Y'Y$, $B'B = I_2 = 525$ A
and current in section OR', Oy', $OB' = I_1 - I_2 = 656 - 525 = 131$ A

All the currents are shown with their directions in the circuit diagram shown in Fig. 2.78.

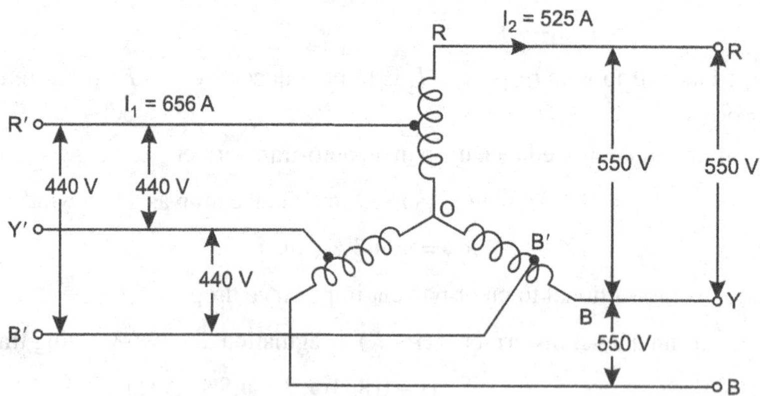

Fig. 2.78 Circuit diagram of a 3-phase auto-transformer.

Section Practice Problems

Numerical Problems

1. An auto-transformer is used to step down from 240 volts to 200 volt. The complete winding consist of 438 turns and secondary delivers a current of 15 ampere. Determine (*i*) Secondary turns (*ii*) the primary current (*iii*) the current in the secondary winding. Neglect the effect of the magnetising current.

(**Ans.** *365; 12.5 A; 2.8 A*)

2. An auto-transformer supplies a load of 3 kW at 115 volt at a power factor of unity. If the primary voltage applied is 230 volt. Calculate (*i*) the power transformed (*ii*) the power conducted directly from the supply lines to the load. (**Ans.** *1.5 kW, 1.5 kW*)

3. The primary and secondary voltages of an auto-transformer are 440 and 352 volt, respectively. Calculate the value of the currents in distribution with the help of the diagram. Calculate the economy of copper in this case. (**Ans.** I_2 = *100 A*; I_1 = *80 A; 80%*)

4. An 11000/22000 volt transformer is rated at 100 kVA as a two winding transformer. If the winding are connected in series to form an auto-transformer what, will be the voltage ratio and output? (**Ans.** *6; 118 kVA*)

5. An autotransformer supplies a load of 5 kW at 110 V at unity power factor. If the applied primary voltage is 220 V, calculate the power transferred to the load (*a*) inductively, (*b*) conductively. (**Ans.** *2.5 kW; 2.5 kW*)

6. The primary and secondary voltages of an autotransformer are 500 V and 400 V, respectively. Show with the aid of a diagram the current distribution in the windings when the secondary current is 100 A. Calculate the economy in the conductor material. (**Ans.** *80%*)

7. A 2200/220 V transformer is rated at 15 kVA as a two-winding transformer. It is connected as an autotransformer with low-voltage winding connected additively in series with high-voltage winding so that the output voltage is 2200 V. The autotransformer is excited from a 2420 V source. The autotransformer is loaded so that the rated currents of the windings are not exceeded. Find

(*a*) the current distribution in the windings,
(*b*) *kVA* output,
(*c*) *kVA* transferred conductively and inductively from input to output,
(*d*) saving in conductor material as compared to a two-winding transformer of the same *kVA* rating.

(**Ans.** *68.2A, 6.82 A, 75.02 A; 165 kVA; 150 kVA, 15 kVA, 15 kVA; 90.9%*)

8. An 11500/2300 V transformer is rated at 100 kVA as two-winding transformer. If the two windings are connected in series to form an auto-transformer what will be the voltage ratio and output? (**Ans.** *0.833; 600 kVA; 0.167; 120 kVA*)

9. An auto-transformer supplies a load of 5 kW at 125 V and at unity power factor. If the primary voltage is 250 V, determine.
(*a*) transformation ratio (*b*) secondary current (*c*) primary current (*d*) number of turns across secondary if total number of turns is 250 (*e*) power transformed and (*f*) power conducted directly from the supply mains to load. (**Ans.** *0.5; 40 A; 20 A; 125; 2.5 kW; 2.5 kW*)

10. A 25 kVA, 2000/200 V, 2-winding transformer is to be used as a step-up autotransformer with constant source voltage of 2000 V. At full load of unity power factor, calculate the power output, power transformed and power conducted. If the efficiency of the two-winding transformer at 0.8 power factor is 95 per cent, find the efficiency of the autotransformer. (**Ans.** *275 kW; 25 kW; 250 kW; 99.52%*)

11. A 200 kVA, 2300/460 V, 50 Hz, two-winding transformer is to be used as an auto-transformer to step-up the voltage from V to 2760 V. If the transformer has an efficiency of 96% at 0.8 pf lagging, impedance of 4% and regulation of 3%, determine (*i*) voltage and current ratings of each side (*ii*) kVA rating (*iii*) efficiency at unity power factor (*iv*) percentage impedance (*v*) regulation, (*vi*) short-circuit current of each side (*vii*) kVA transformed and kVA conducted at full load, while it is used as an auto-transformer.
(**Ans.** *86.96 A; 434.8 A; 521.76 A; 1200 kVA; 99.45%, 0.667%; 0.5%; 65.22 kA; 78.264 kA; 200 kVA; 1000 kVA*)

Review Questions

1. Explain what is a transformerand its necessity in power system?
2. Explain the working principle of a transformer.
3. State why the core of a transformer is laminated?
4. State why silicon steel in selected for the core of a transformer?
5. Give the constructional details of a core-type transformer.
6. In a transformer explain how power is transferred from one winding to the other.
7. Show that $(E_1/E_2) = (I_2/I_1) = (T_1/T_2)$ in a transformer.
8. What will you expect if a transformer is connected to a DC supply?
9. Derive an expression for the emf induced in a transformer winding. Show that the emf induced per turn in primary is equal to the emf per turn in secondary.
10. Explain the behaviour of a transformer on no-load.
11. Explain that "The main flux in a transformer remains practically invariable under all conditions of load".
12. Draw and explain the phasor diagram of a loaded transformer (neglecting voltage drop due to resistance and leakage reactance).
13. Draw the phasor diagram and equivalent circuit of a single-phase transformer.
14. Draw a neat phasor diagram showing the performance of a transformer on-load.
15. Draw and explain the phasor diagram of single-phase transformer connected to a lagging p.f. load.
16. "The overall reactance of transformer decreases with load." Explain.
    (**Hint:** $L = N\phi/I$; when I increases, L decreases)
17. What do you mean by voltage regulation of a transformer?
18. What are the various losses in a transformer? Where do they occur and how do they vary with load?
19. Define efficiency of a transformer and find the condition for obtaining maximum efficiency.
20. Write short note on 'All-day efficiency of a transformer".
21. Distinguish between 'power efficiency' and 'all-day efficiency' of a transformer.
22. What information can be obtained from the open circuit test of transformer? How can you get these informations?
23. How open circuit and short circuit tests are performed on a single-phase transformer. Draw circuit diagram for each test. Also mention uses of these tests.
24. Define an autotransformer. How does the current flow in different parts of its windings?
25. Give the constructional features of an auto-transformer. State the applications of autotransformers.
26. What are the applications of autotransformers?
27. Derive an expression for the saving of copper in an autotransformer as compared to an equivalent two-winding transformer.
28. Explain the working principle and construction of an auto-transformer.

29. Define an auto-transformer. Distinguish clearly the difference between a resistive potential divider and an auto-transformer.

30. What is an auto-transformer? State its merits and demerits over the two winding transformer.

31. Give the constructional features of an auto-transformer.

Draw and explain the phasor diagram of an auto-transformer on load.

32. What is an autotransformer? State its merits and demerits over the two-winding transformer. Give the constructional features and explain the working principle of a single-phase autotransformer.

33. Derive an expression for saving in conductor material in an autotransformer over a two-winding transformers of equal rating. State the advantages and disadvantages of autotransformer over two-winding transformers.

34. In an auto transformer how the current flows in different parts of its windings? Derive an expression for the saving of copper in an auto-transformer as compared to an equivalent two winding transformer.

35. Derive an expression for the approximate relative weights of conductors material in an autotransformer and a two-winding transformer, the primary voltage being V_1 and the secondary voltage V_2. Compare the weights of conductor material when the transformation ratio is 3. Ignore the magnetising current.

[Ratio = $1 - (V_2/V_1)$; 2/3]

36. What are the reasons of higher efficiency of autotransformers as compared to conventional transformers?

37. What is meant by the terms transformed voltamperes and conducted voltamperes in an autotransformer? Show that two windings connected as an autotransformer will have greater *VA* rating than when connected as a two-winding transformer.

38. Show that in case of an auto-transformer

$$\frac{\text{Inductively transferred power}}{\text{Total power}} = \frac{\text{High voltage-low voltage}}{\text{High voltage}}$$

39. If an autotransformer is made from a two-winding transformer having a turns ratio $\frac{T_1}{T_2} = n$, show that

$$\frac{\text{magnetising current as an autotransformer}}{\text{magnetising current as a two-winding transformer}} = \frac{n-1}{n}$$

$$\frac{\text{shortcircuit current as an autotransformer}}{\text{short-circuit current as a two-winding transformer}} = \frac{n}{n-1}$$

Multiple Choice Questions

1. The phase relationship between the primary and secondary voltages of a transformer is
(*a*) 90 degree out of phase
(*b*) in the same phase
(*c*) 180 degree out of phase
(*d*) 270 degree out of phase

2. Transformer core is laminated
(*a*) because it is difficult to fabricate solid core
(*b*) because laminated core provides high flux density
(*c*) to reduce eddy current losses.
(*d*) to avoid hysteresis losses.

3. The induced emf in the transformer secondary will depend on
(*a*) frequency of the supply only.
(*b*) Number of turns in secondary only.
(*c*) Frequency and flux in core.
(*d*) Frequency, number of secondary turns and flux in the core.

4. A transformer with output of 250 kVA at 3000 V, has 600 turns on its primary and 60 turns on secondary winding. What will be the transformation ratio of the transformer?
(*a*) 10 (*b*) 0·1
(*c*) 100 (*d*) 0·01

5. The primary applied voltage in an ideal transformer on no-load is balanced by
(*a*) primary induced emf (*b*) secondary induced emf
(*c*) secondary voltage (*d*) iron and copper losses

6. If R_1 is the resistance of primary winding of the transformer and K is transformer ratio then the equivalent primary resistance referred to secondary will be
(*a*) KR_1^2 (*b*) KR_1
(*c*) K^2R_1 (*d*) R_1/K^2

7. The eddy current loss in the transformer occurs in the
(*a*) primary winding (*b*) core
(*c*) secondary winding (*d*) none of the above.

8. Which of the following electrical machines has the highest efficiency?
(*a*) DC generator (*b*) AC generator
(*c*) transformer (*d*) induction motor.

9. If the iron losses and full load copper losses are given then the load at which the efficiency of a transformer is maximum, is given by
(*a*) full load × $\dfrac{\textit{iron loss}}{\textit{f.l.cu loss}}$ (*b*) full load × $\sqrt{\dfrac{\textit{iron loss}}{\textit{f.l.cu loss}}}$
(*c*) full load × $\left(\dfrac{\textit{iron loss}}{\textit{f.i. cu. loss}}\right)$ (*d*) full load × $\sqrt{\dfrac{\textit{f.l.cu. loss}}{\textit{iron loss}}}$

10. Kapp-regulation diagram is used to determine
(*a*) Iron losses in a transformer (*b*) copper losses in a transformer
(*c*) Voltage regulation of a transformer (*d*) Efficiency of a transformer

11. In an auto transformer, there are
(*a*) always two windings (*b*) one winding only without taps
(*c*) one winding with taps taken out (*d*) two windings put one upon the other.

12. The marked increase in kVA capacity produced by connecting a 2 winding transformer as an autotransformer is due to
(*a*) increase in turn ratio
(*b*) increase in secondary voltage
(*c*) increase in transformer efficiency
(*d*) establishment of conductive link between primary and secondary.

13. The kVA rating of an ordinary two-winding transformer is increased when connected as an auto-transformer because

(*a*) transformation ratio is increased
(*b*) secondary voltage is increased
(*c*) energy is transferred both inductively and conductivity
(*d*) secondary current is increased

14. In an auottransformer a break occurs at the point *P*. The value of V_2 will be

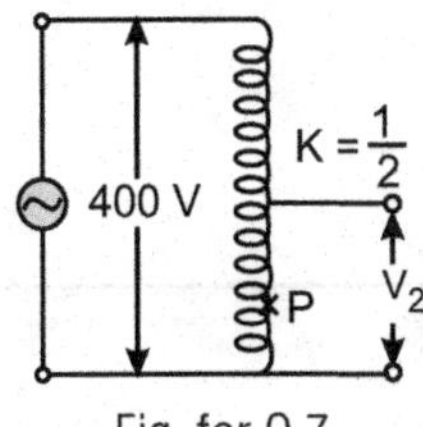

Fig. for Q.7

(*a*) 200 V (*b*) 100 V
(*c*) 50 V (*d*) 400 V

15. The saving in Cu achieved by converting a two-winding transformer into an auto-transformer is determined by

(*a*) voltage transformation ratio (*b*) load on the secondary
(*c*) magnetic quality of core material (*d*) size of the transformer core.

Keys to Multiple Choice Questions

1. c	**2.** c	**3.** d	**4.** b	**5.** b	**6.** c	**7.** c	**8.** c	**9.** b	**10.** c
11. c	**12.** d	**13.** c	**14.** d	**15.** a					

CHAPTER

Three-Phase Transformers 3

Chapter Objectives

After the completion of this unit, students/readers will be able to understand:

- ✓ Constructional features of a three-phase transformer.
- ✓ Advantages of using three-phase trans-formers.
- ✓ How to determine relative primary and secondary of a 3-phase transformer?
- ✓ How to check the polarities of windings?
- ✓ How are the windings of three-phase transformers are connected and what are their relative merits and demerits?
- ✓ What are the advantage of using zig-zag connection?
- ✓ What is the necessity of connecting transformers in parallel?
- ✓ Which necessary conditions are to be fulfilled before connecting three-phase transformers in parallel?
- ✓ How to calculate the load sharing when the transformers are operated in parallel?
- ✓ What is a three-winding transformer? What is the use of third winding?
- ✓ How tertiary winding suppresses harmonic voltages and prevents neutral potential to oscillate to undesirable extent?
- ✓ What are off-load and on-load tap changers?
- ✓ What are open delta or Vee-Vee connections?
- ✓ What are Scott connections?
- ✓ How is load shared in Scott-connected transformers?
- ✓ Why is cooling of transformers necessary?

Introduction

Three phase system is invariably adopted for generation, transmission and distribution of electrical power due to economical reasons. Usually, power is generated at the generating stations at 11 kV (or 33 kV), whereas, it is transmitted at 750 kV, 400 kV, 220 kV, 132 kV or 66 kV due to economical reasons. At the receiving stations, the voltage level is decreased and power is transmitted through shorter distances. While delivering power to the consumers, the voltage level is decreased to as low as 400V (line value) for safety reasons.

Thus to increase the voltage level at the generating stations step-up transformers and to decrease the voltage level at the receiving stations, step down transformers are employed.

3.1 Merits of Three-phase Transformer over Bank of Three Single-phase Transformers

The voltage level in three-phase system at the generating stations and at the receiving stations can be changed either by employing a bank of three single-phase transformers (inter connecting them in star or delta) or by employing one three phase transformer. Generally, one three phase transformer is preferred over a bank of three single phase transformers because of the following reasons.

(*i*) It requires smaller quantity of iron and copper. Hence, its cost is nearly 15% lesser than a bank of three single phase transformers of equal rating.
(*ii*) It has smaller size and can be accommodated in smaller tank and hence needs smaller quantity of oil for cooling.
(*iii*) Because of smaller size, it occupies less space; moreover it has less weight.
(*iv*) It needs less number of bushings.
(*v*) It operates at slightly better efficiency and regulation.

These transformers suffer from the following disadvantage.

(*i*) It is more difficult and costly to repair three-phase transformers.
(*ii*) It is difficult to transport single large unit of three-phase transformer than to transport three single phase transformers individually.

The advantages of three-phase transformer (such as lower cost, lower weight, lower space requirement etc.) over weighs its disadvantages and hence are invariably employed in the power system to step-up or step-down the voltage level.

3.2 Construction of Three-phase Transformers

Form construction point of view, three-phase transformers are also classified as
(*i*) Core type transformers (*ii*) Shell type transformers

Core Type Transformers

In core type three-phase transformers, the core has three limbs of equal area of cross-section. Three limbs are joined by two horizontal (top and bottom) members called yokes. The area of cross- section of all the limbs and yokes is the same since at every instant magnitude of flux set-up in each part is the same. The core consists of laminations of silicon steel material having oxide film coating on both the sides for insulation. The laminations are usually of *E* and *I* shape and are staggered alternately to decrease reluctance of magnetic path and increase the mechanical strength.

The complete section of a three-phase core type transformer with its plan is shown in Fig. 3.1. This type of transformers is usually wound with circular cylindrical coils. The low voltage* (*LV*) winding

* When *LV* winding is placed nearer the core, less insulation is required between the core and the *LV* winding in comparison to that if *HV* winding is placed nearer the core and hence reduces the cost of construction.

is wound nearer the core and high voltage (*HV*) winding is wound over low voltage winding as shown in Fig. 3.1. Insulation is always provided between the core and low voltage winding and between low voltage winding and high voltage winding.

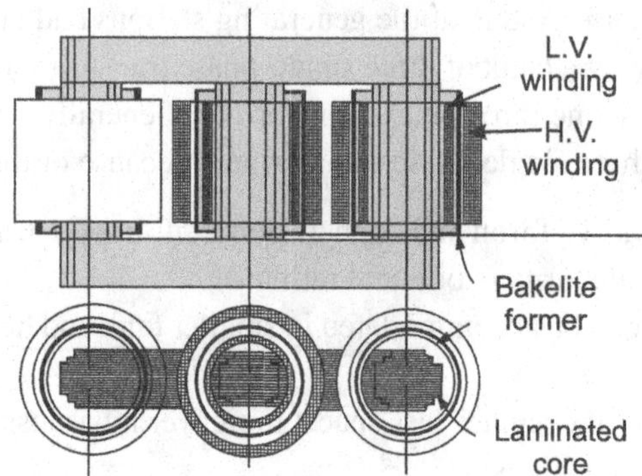

Fig. 3.1 Construction of a three-phase core type transformer

The core construction for very large capacity three-phase transformers is slightly changed. In this case, the core consists of three main limbs on which windings are arranged and two additional limbs at the sides without winding are formed as shown in Fig. 3.2. This arrangement allows decreasing the height of the yoke and consequently decreases the overall height of the core. However, the length increases. This facilitates the transportation of transformers by rail. In this arrangement, the magnetic circuits for each phase are virtually independent:

In either case, the magnetic circuits of the three-phases are somewhat unbalanced, the middle phase having less reluctance than the outer two. This causes the magnetising current of middle phase slightly less than the outers. But during operation, the magnetising current is so small that it does not produce any noticeable effect.

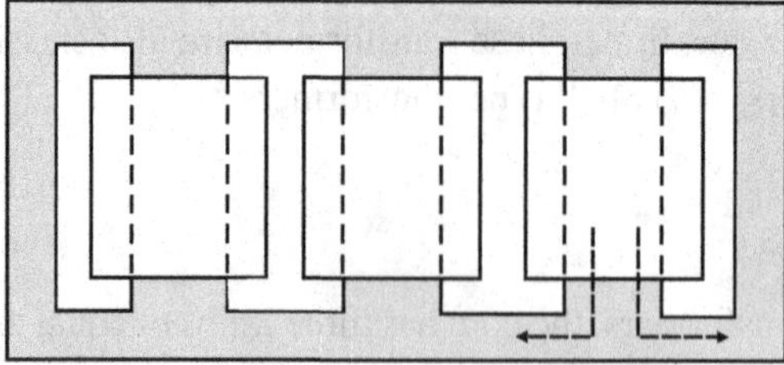

Fig. 3.2 Core and winding of core-type three-phase transformer

Shell Type Transformers

In shell type transformers, the core construction is such that the windings are embedded in the core instead of surrounding the iron as shown in Fig. 3.3. The area of cross section of the central limbs is double to that of the side limbs and horizontal members.

The low voltage and high voltage windings of the three-phases are wound on the central limbs. These windings are placed vertically in the three portions as shown in Fig. 3.3. The low voltage (*LV*) winding is always placed neater to the core and high voltage (*HV*) winding is placed over the low voltage winding for economic reasons. To obtain uniform distribution of flux in the core, usually second winding placed on the central limb is wound in the reverse direction as shown in Fig. 3.4.

Note: The detailed construction of core, winding, bushing, etc., has already been dealt with in chapter-4.

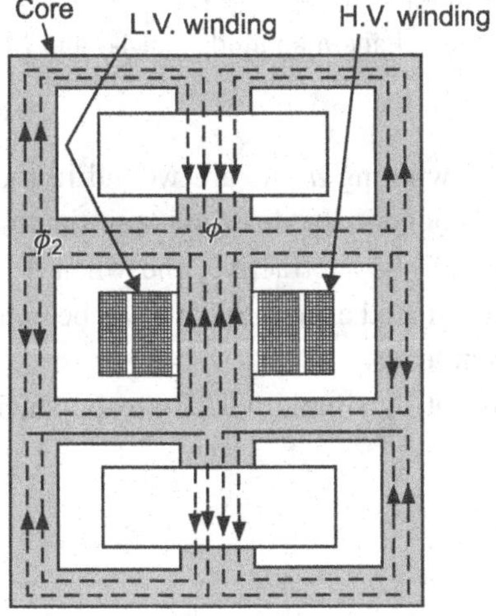

Fig. 3.3 Construction of a three-phase shell-type transformer

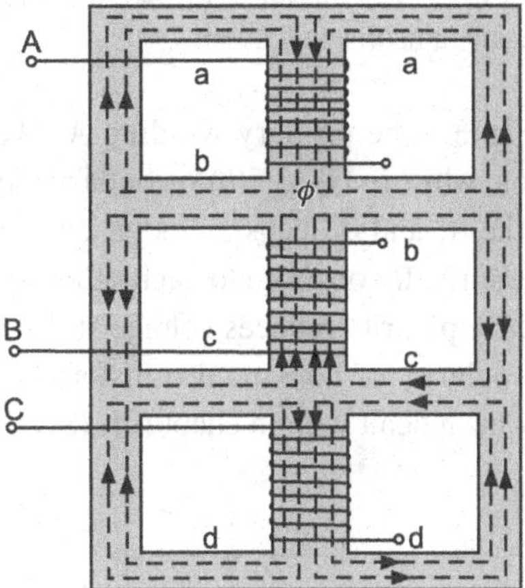

Fig. 3.4 Construction of a three-phase shell-type transformer

3.3 Determination of Relative Primary and Secondary Windings in Case of Three-phase Transformer

At the place of manufacturing, all the primary and secondary winding terminals are traceable. Otherwise, all the primary and secondary winding terminals belonging to each other are to be determined.

To determine which secondary belongs to which primary, proceed as follows:

Short circuit all the phases except one primary and a probable secondary. Connect a voltmeter across the secondary and circulate a small direct current in the primary. A momentary deflection of voltmeter on making and breaking of primary current confirms that the secondary corresponds to the primary chosen otherwise check the other secondary. The test is repeated for other windings too. For this test, the terminals of all the windings must be open and available for testing.

3.4 Polarity of Transformer Windings

In order to connect windings of the same transformer in parallel or series or in star/delta, or to interconnect two or more transformers in parallel, or to connect single phase transformers for poly-phase transformation of voltages, it is necessary to know the polarity of the different winding terminals. The polarity of the transformer terminals is usually indicated by means of standard marking scheme followed by manufactures – the scheme varies from country to country.

Transformer polarity marking designates the relative instantaneous directions of current and voltage in the transformer leads. In one system high voltage leads are indicated by letter *A* (or *H*) and low voltage leads by *a* (or *X*), and tertiary, if any by (*A*) [or *Y*], each with a subscript 1, 2, 3 etc., depending upon the number of leads. The *A-a* scheme is shown in Fig. 3.5

> **Note:** As per Indian standards based on *IEC* (International Electro-technical Commission) document, in single-phase power transformers 1.1 is used for A_1 (or H_1), 1.2 for A_2 (or H_2), 2.1 for a_1 (or X_1), 2.2 for a_2 (or X_2), 3.1 for (A_1) or Y_1, etc., the voltage magnitudes in *A-a* and (*A*) being in decreasing order.

In Fig. 3.5 the primary winding A_1–A_2 and secondary winding a_1–a_2 are wound in the same direction, while in Fig. 3.5(*b*) secondary winding $a_2 - a_1$ is wound in the direction opposite to A_1–A_2. When the A_1 and a_1 leads are adjacent, the polarity is said to be subtractive, and when A_1 and a_1 are diametrically opposite to each other, the polarity is designated as additive. It may be noted that subtractive polarity reduces voltage stress between adjacent leads.

The polarity of unmarked transformers can be found out by simple polarity tests which have already been dealt with in chapter-1. (Section 1.31).

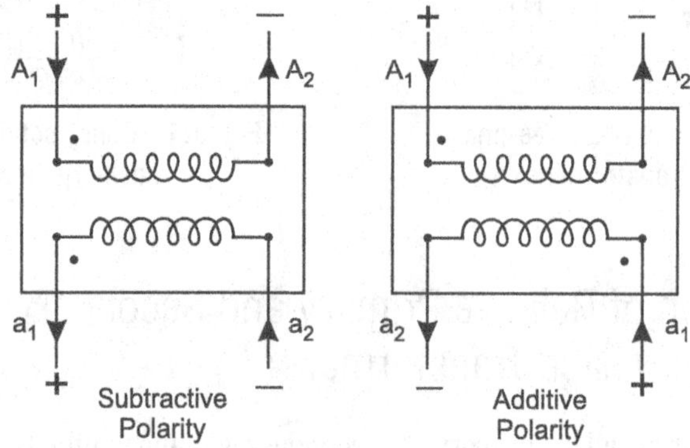

Fig. 3.5 Transformer polarity designation

3.5 Phasor Representation of Alternating Quantities in Three-phase Transformer Connections

Before to study the connections of three-phase transformers, it is necessary to learn the characteristics of balanced three-phase systems as well as the conventions followed to designate currents and voltages of a three-phase system. If the supply voltage is balanced, the voltages can be represented by a voltage triangle shown in Fig. 3.6(*a*), where *A*, *B*, *C* are the nomenclatures of the three lines of the system and *N* stands for the neutral or star point of the system. *A*, *B*, *C* are the three vertices of the equilateral triangle *ABC* and *N* is the circum centre of the triangle. The voltages and currents with double subscripts notation are represented by phasors. For example, V_{ab}, I_{ab} represent voltage of point *a* with respect to point *b* and current flowing from point *a* to point *b*, respectively. With the arrow pointing towards *A* the line represents voltage phasor $V_{AN.}$ Thus the line-to-line voltage (also called as line voltage) phasor and line-to-neutral voltage (also known as phase voltage) phasors can be drawn from the voltage triangle shown in Fig. 3.6(*a*). The phasor diagram is shown in Fig. 3.6(*b*) where phase voltage phasors which are equal in magnitude are displaced from each other by 120°. The line voltage phasors are also equal in magnitude and displaced from each other by 120°, but the magnitude of a line voltage phasor is $\sqrt{3}$ times the magnitude of a phase voltage phasor. Further it may be noted that the set of line voltage phasors is displaced from the set of phase voltage phasors by 30°.

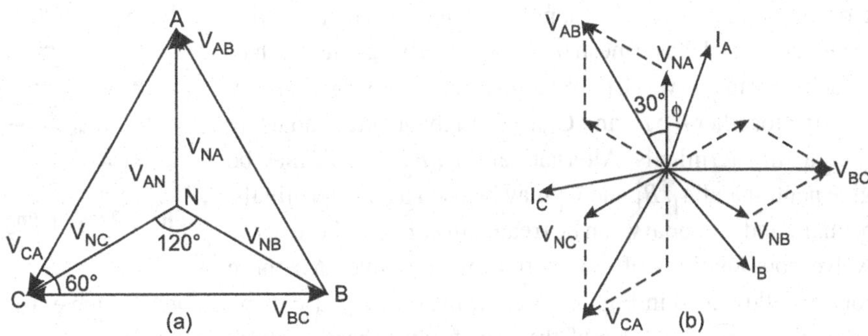

Fig. 3.6 Phasors of a balanced three-phase system

If the lines carry balanced load then the magnitudes of the line currents I_A, I_B and I_C are equal and displaced with respect to each other by 120°. They are equally displaced from the corresponding phase voltages by an angle ϕ as shown in Fig. 3.6(*b*), the power factor being cos ϕ. In this case, the power factor is lagging as currents lag behind the phase voltages.

Note: As per Indian Standards Specifications, the terminals of a three-phase transformer are marked as U, V, W with numerics 1, 2, 3, etc.

Important relations in three-phase connected system:

Star-connections | *Delta-connections*

$E_L = \sqrt{3}\ E_{ph}$ ...(3.1); $I_L = I_{ph}$ ...(3.2) | $E_L = E_{ph}$ ...(3.3); $I_L = \sqrt{3}\ I_{ph}$ ...(3.4)

Power, $P = \sqrt{3}\ V_L I_L \cos\phi = 3\ V_{ph} I_{ph} \cos\phi$ ...(3.5) (*both for star and delta*)

In three-phase transformers

Primary | *Secondary*

$E_{ph1} = 4.44 f \phi_m N_1$ ...(3.6) | $E_{ph2} = 4.44 f \phi_m N_2$ ...(3.7)

$= 4.44 f B_m A_i N_1$ ...(3.8) | $= 4.44 f B_m A_i N_2$ ...(3.9)

$$\text{Transformation ratio, } K = \frac{E_{ph2}}{E_{ph1}} = \frac{N_2}{N_1} \quad ...(3.10)$$

3.6 Three-phase Transformer Connections

As mentioned earlier, the standard nomenclature for phases is *A*, *B*, *C*. Capital letters *A*, *B*, *C* are used for *HV* windings, lower case letters *a*, *b*, *c* for *LV* winding and (*A*), (*B*), (*C*) for a tertiary windings if provided. Neutral terminals precede line terminals. Each winding has two ends designated by the subscript numbers 1, 2. However, if there are intermediary tapping (or there are sections of the same winding) these are numbered in order of their separation from end 1, as illustrated in Fig. 3.7.

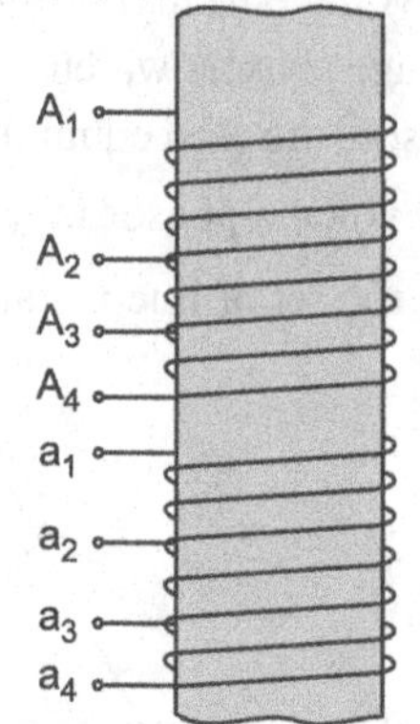

Fig. 3.7 Winding terminals

The interconnection of the phase windings to give a three-phase, three-wire or three-phase, four wire supply provides three alternative modes of connection (*i*) delta or mesh (*ii*) star and (*iii*) zig-zag. Each of these can be achieved in two ways. For Example 3.a star-connection may be had by joining together A_1, B_1 and C_1, to form the neutral and using A_2, B_2 and C_2 as line terminals. Alternatively A_2, B_2, and C_2 may be joined to give neutral and A_1, B_1 and C_1 may be used as line terminals.

Since primary and secondary can be treated in either of these ways, at least twelve combinations of connection are possible. All these combinations are illustrated in Fig. 3.8. According to the phase displacement which exists between the line voltages on the two sides of the transformer these combinations can be arranged in the following four main groups:

Group 1: With 0° phase displacement : Connections with symbolic notation Yy0, Dd0, Dz0

Group 2: With 180° phase displacement : Connections with symbolic notation Yy6, Dd6, Dz6

Group 3: With 30° lag phase displacement : Connections with symbolic notation Dy1, Yd1, Yz1

Group 4: With 30° lead phase displacement : Connections with symbolic notation Dy11, Yd11, Yz11

Among the above, the most popular method of connecting transformers is the delta–star connections (*Dy*11). The connection diagram illustrates the relative position of the terminals in the terminal box and the arrangement of internal connections. The *HV* winding is connected to a three-wire supply having phase sequence *A*, *B*, *C* and at the instant chosen the potential of *A* is at its maximum positive value. Thus the line voltage on the *HV* side will form an equilateral triangle as illustrated. The line terminals are A_1, B_1 and C_1 and the corners of the triangles are so marked. It is to be noted that the phase *A* of the transformer lies between A_1 and B_1, phase *B* between B_1 and C_1 and phase *C* between C_1 and A_1 owing to internal connections used.

In case of *LV* winding, since $V_{a_1a_2}$ is in phase with $V_{A_1A_2}$, the line na_1 is, therefore, drawn parallel to B_1A_1 in order to establish the direction of phase–voltages phasor. Similarly nb_1 and nc_1 depend on C_1B_1 and A_1C_1 and the phasor diagram for *LV* shows the voltages of the terminals a_1, b_1 and c_1 with respect to *n*.

From the phasor diagrams of induced voltages for *HV* and *LV* sides it is observed that the voltage of the *HV* line A_1 is at maximum value, the *LV* line a_1 has been advanced by 30°. In fact, the transformer connections have produced this displacement of 30° for all lines.

All the other methods of connections in the same main group no. 4 give a similar 30° advance which is the basis of classification.

In case of group no. 1 there is no phase displacement, whereas group no. 2 gives phase displacement of 180°. In group no. 3, the line voltage of *LV* winding lags by 30°.

For parallel operation of transformers the essential requirement is that the transformer connections belong to the same main group so that there may not be any phase displacement between the line voltages.

Group No.	Phase Displacement	Symbolic Notation	Winding connections HV	Winding connections LV	Phasor Diagrams of Induced Voltage HV	Phasor Diagrams of Induced Voltage LV
1	0°	Yy0	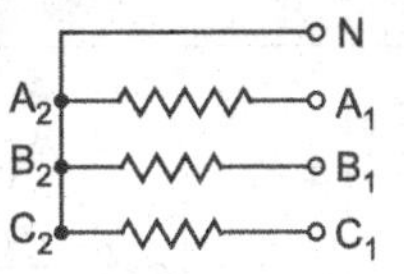	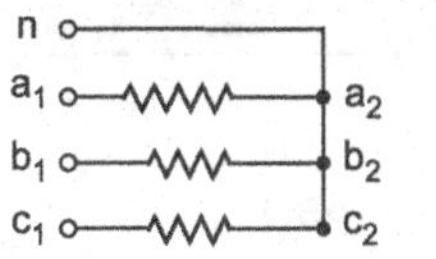	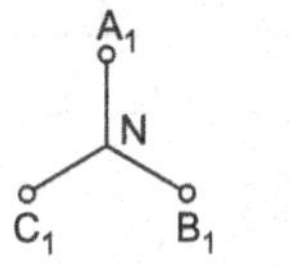	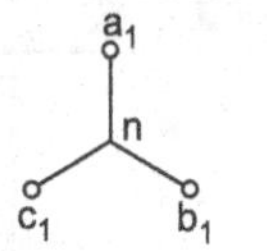
		Dd0	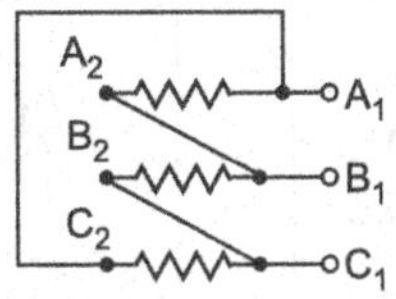	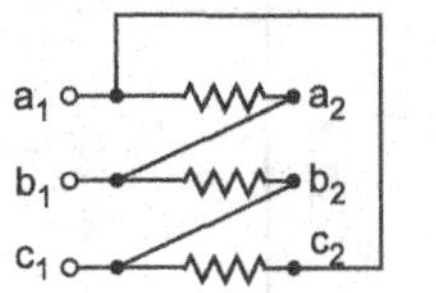	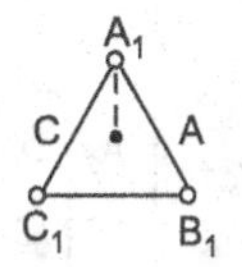	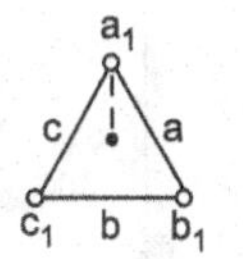
		Dz0	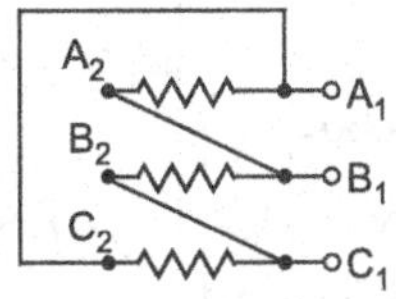	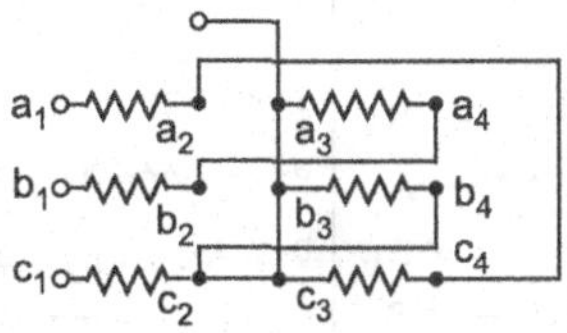	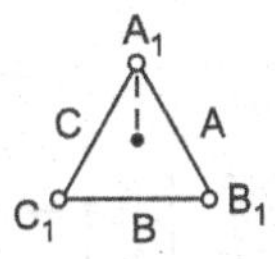	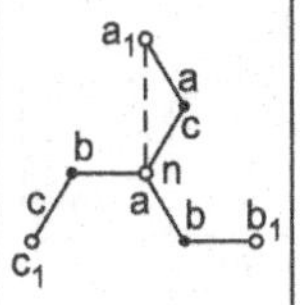
2	180°	Yy6	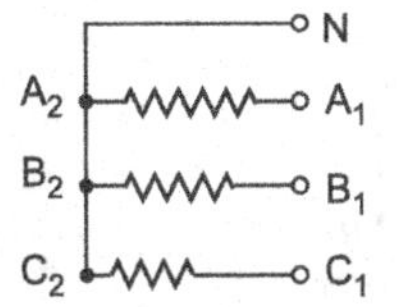	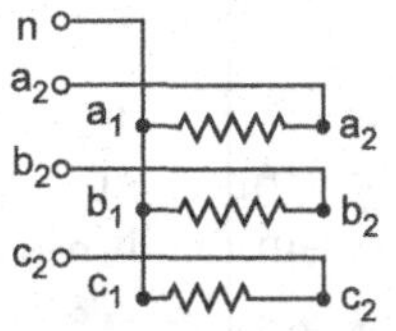	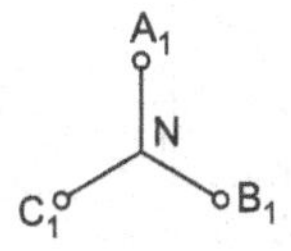	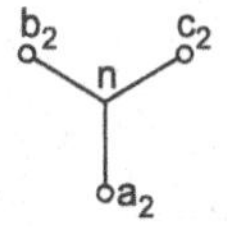
		Dd6	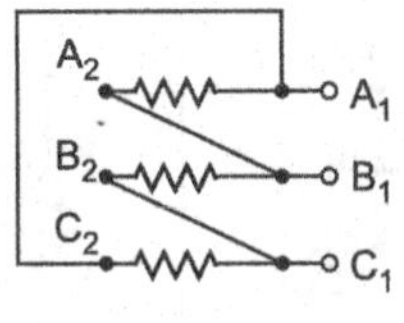	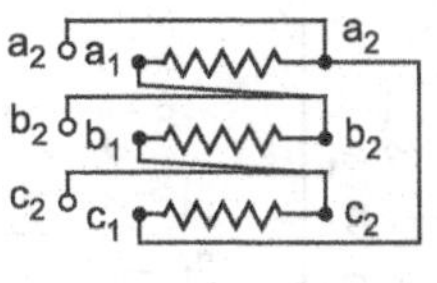	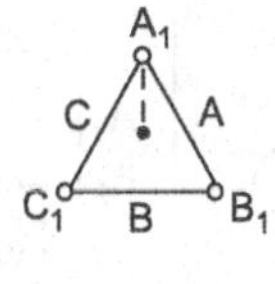	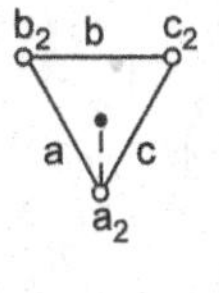
		Dz6	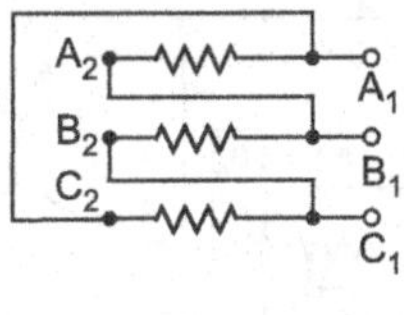	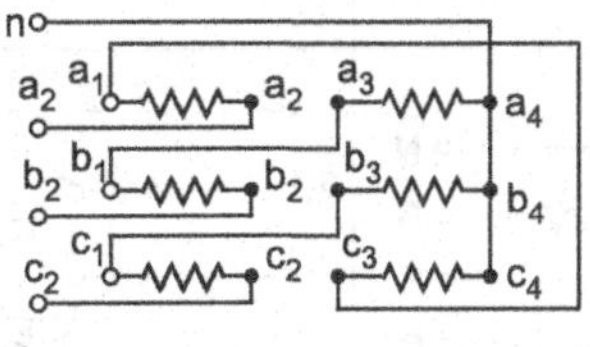	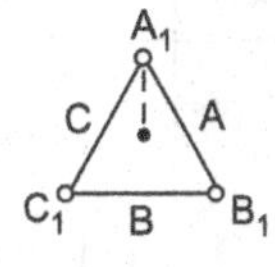	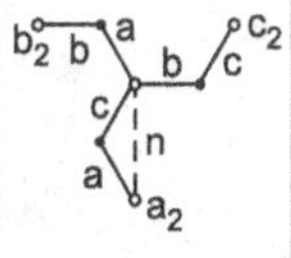

Fig. 3.8. Contd.

Fig. 3.8. Contd.

Group No.	Phase Displacement	Symbolic Notation	Winding connections		Phasor Diagrams of Induced Voltage	
			HV	LV	HV	LV
3	30°	Dy1	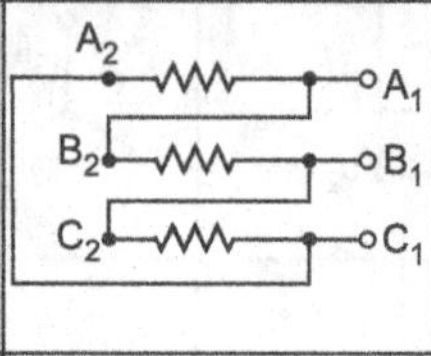	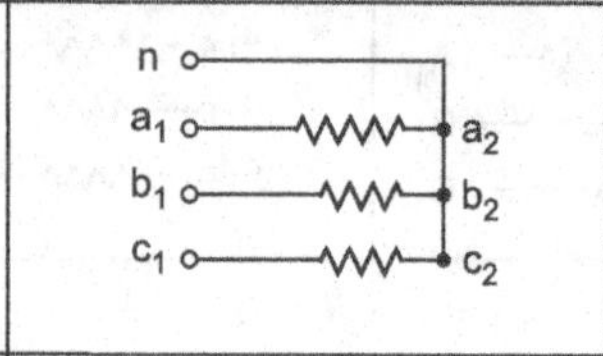	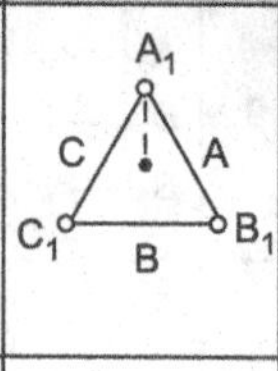	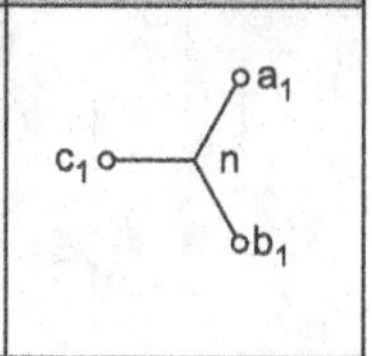
		Yd1	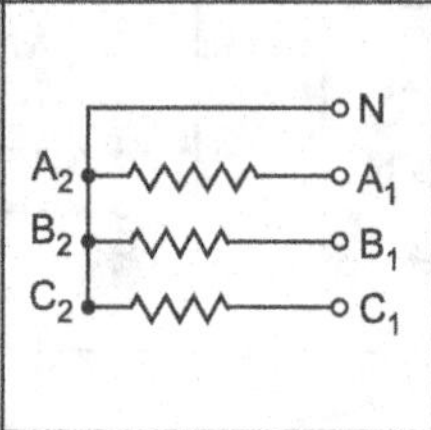	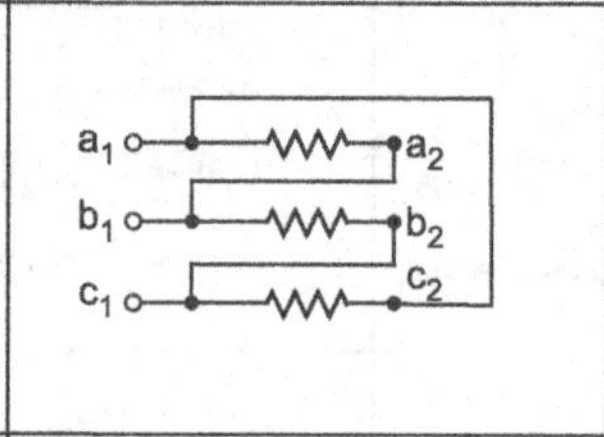	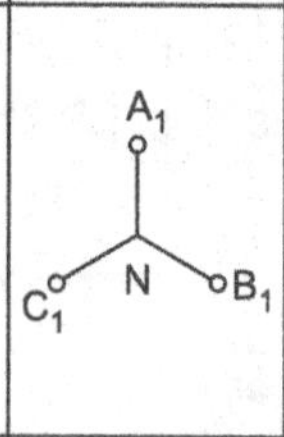	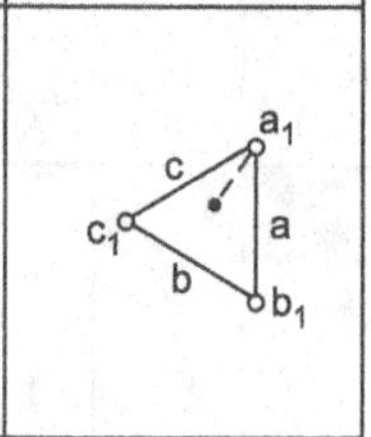
		Yz1	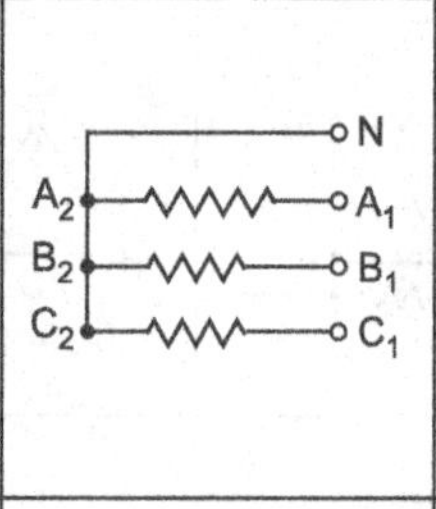	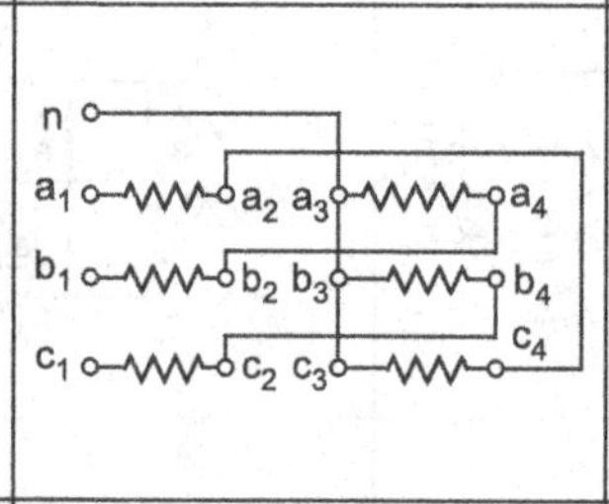	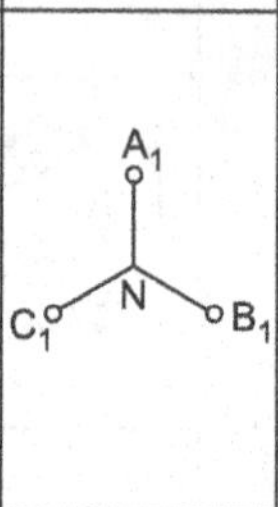	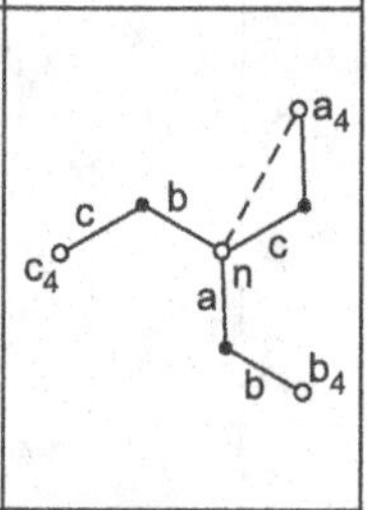
4	+30°	Dy11	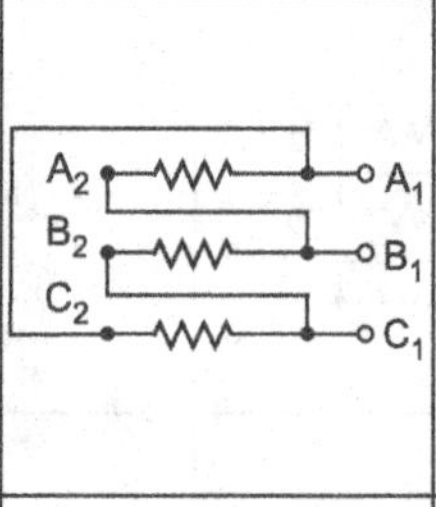	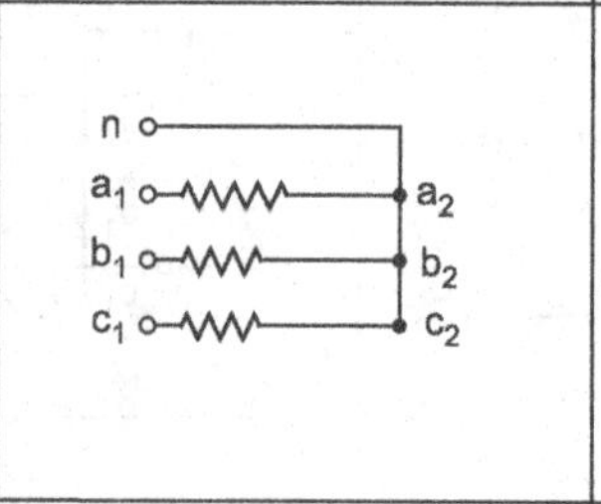	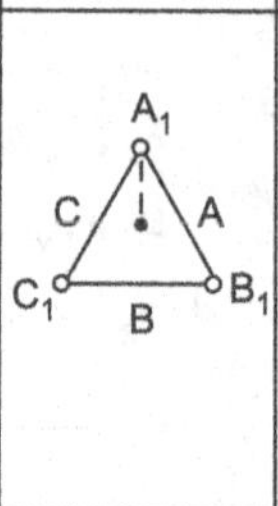	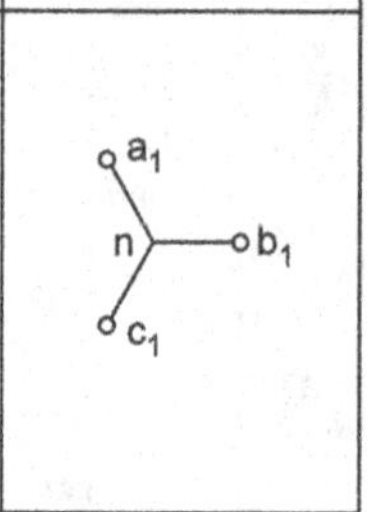
		Yd11	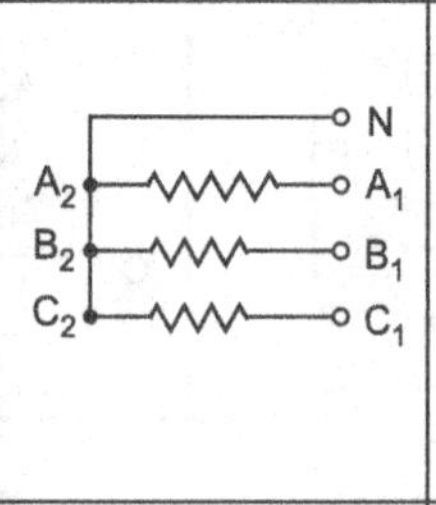	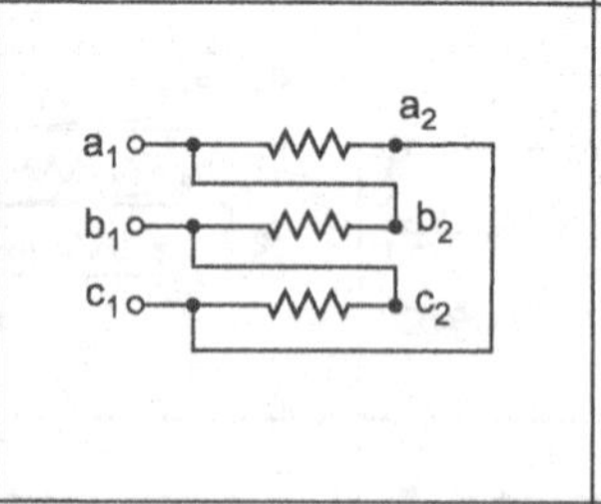	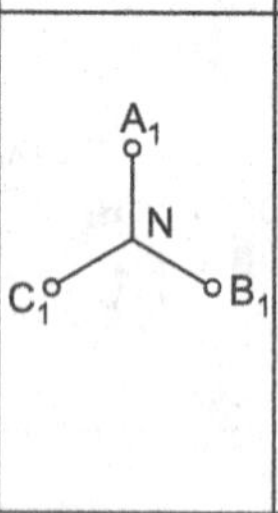	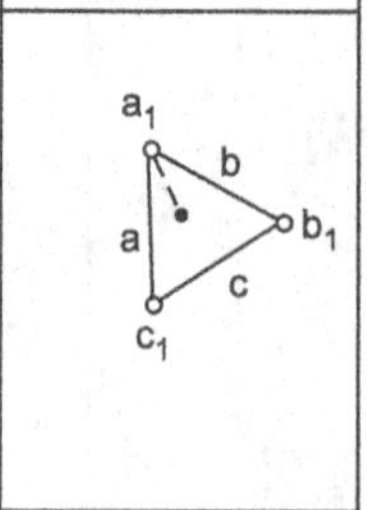
		Yz11	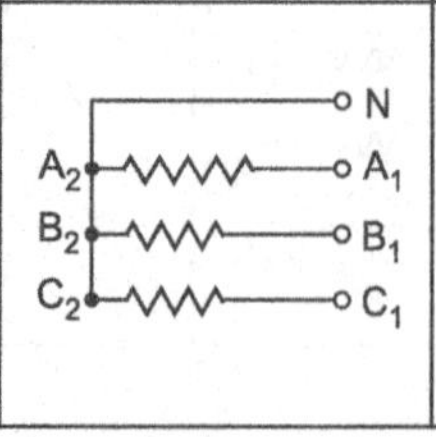	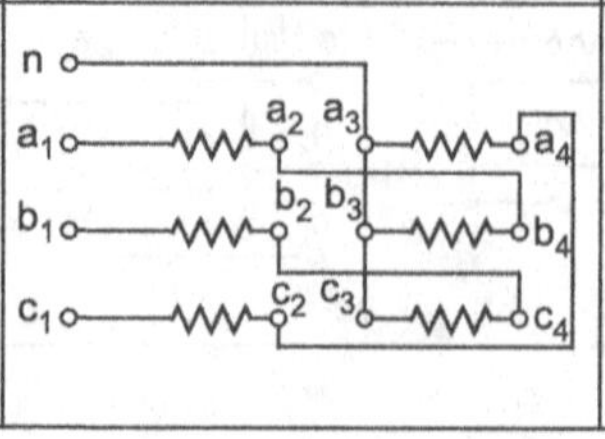	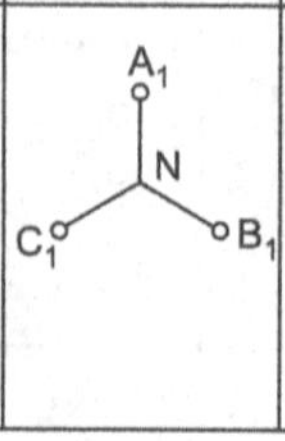	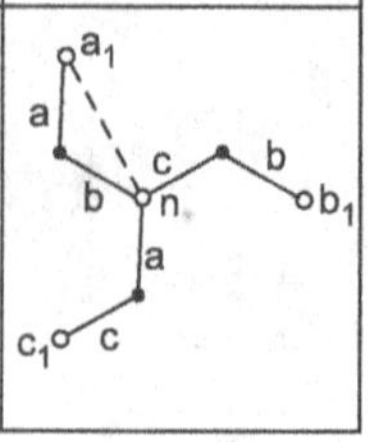

Fig. 3.8 Various winding connections of three-phase transformers

With three-phase transformers, it is possible to obtain any desired phase shift if the primary and/or secondary windings are divided into a sufficient number of sections. The zig-zag connection is one of the example of sectionalised winding. By using these connections, the effect of third harmonics in line to neutral voltages and line to line voltages is reduced. In these connections, half sections are connected in opposition which gives the larger total fundamental voltage and causes the co-phasial third harmonics to cancel. The fundamental line-to-neutral voltage is 0.866 times the arithmetic sum of the two section voltages (phase voltages in zig-zag, connection being composed of two half-voltages with a phase difference of 60°). This causes the reduction in voltage and to compensate the same 15 percent more turns are required for a given total voltage per phase. Consequently, it necessitates an increase in the frame size over that normally used for the given rating. However, the advantages of zig-zag arrangement may offset the cost. The other important advantages of zig-zag connections are unbalanced loads on the secondary side are distributed better on the primary side. The zig-zag/star connection has been employed where delta connections were mechanically weak (on account of large number of turns and small copper sections) in *HV* transformers. These are also preferred for rectifiers.

3.7 Selection of Transformer Connections

When a transformer is to be placed in a power system to step-up or step down the voltage, it is selected as per its connections which are having some peculiar characteristics as explained below:

3.7.1 Star-Star (Yy0 or Yy6) Connections

The star-star connections of a three-phase transformer are shown in Figs. 3.9.1(*a* and *b*) and 9.2(*a* and *b*). Their phasor diagrams are also shown in Figs. 3.9.1(*c* and *d*) and 3.9.2 (*c* and *d*). It may be seen that line voltage is $\sqrt{3}$ times the phase voltage and there is a phase difference of 30° between them. In Fig. 3.9.1 the secondary voltage is in phase with the primary, whereas in Fig. 3.9.2 the secondary voltage system is 180° out of phase from the primary, the former is designated as Yy0 and the latter is designated as Yy6 for three-phase transformers.

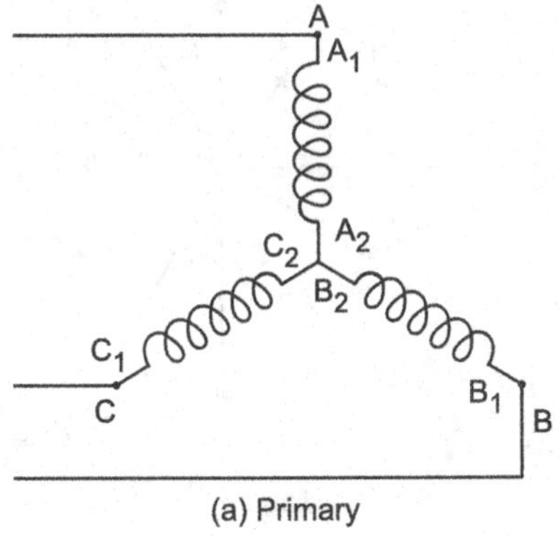

(a) Primary

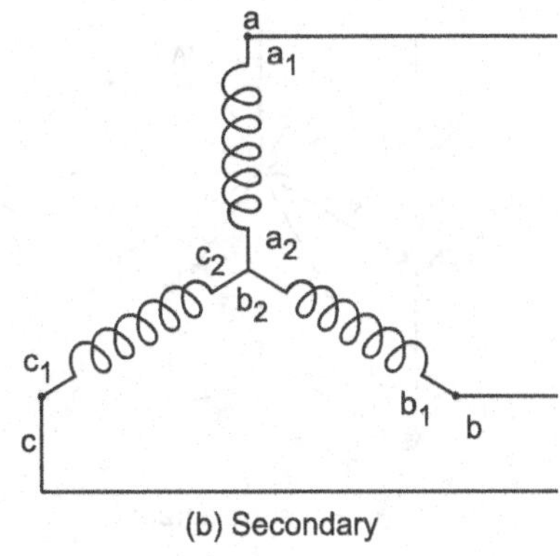

(b) Secondary

Fig. 3.9.1. Contd.

Fig. 3.9.1. Contd.

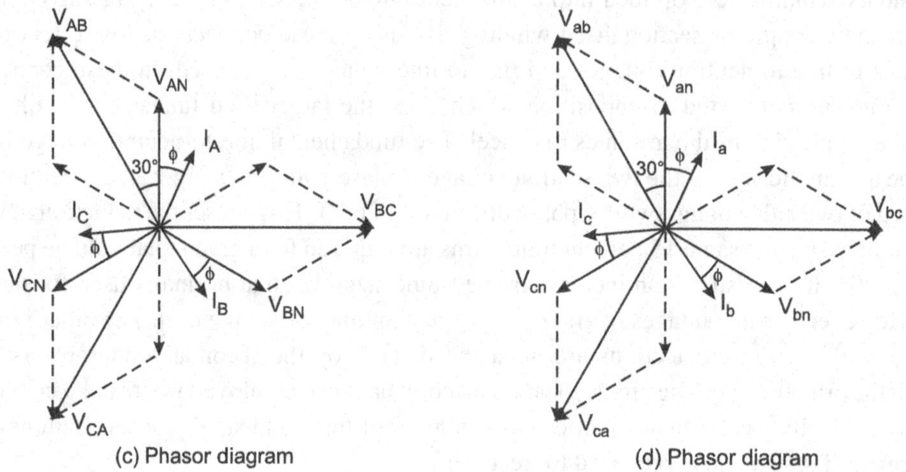

(c) Phasor diagram (d) Phasor diagram

Fig. 3.9.1 Star-star (Yy0) connection and their phasor diagram (0° phase shift)

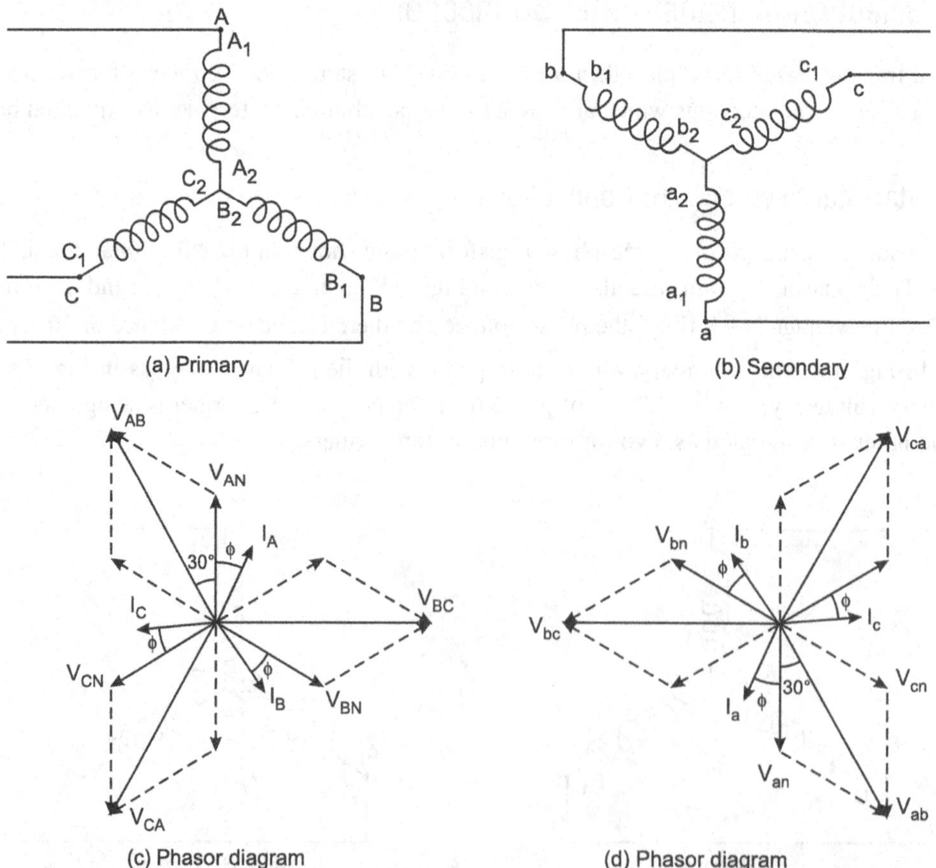

(a) Primary (b) Secondary

(c) Phasor diagram (d) Phasor diagram

Fig. 3.9.2 Star-star (Yy6) connection and their phasor diagram (180° phase shift)

Star-star connected three-phase transformers are operated with grounded neutrals, i.e., the neutral of the primary is connected to the neutral of the power source. If the neutral is kept in isolation, the unbalanced load on the secondary will shift the position of neutral which changes the magnitude of phase voltages. A grounded neutral in the primary prevents this unsatisfactory operation.

With an isolated neutral the third-harmonic components in the magnetising currents of the three primary windings are in phase and as such they have no path. As the path for third harmonic current is absent, the phase voltages become non-sinusoidal though the line voltages are sinusoidal.

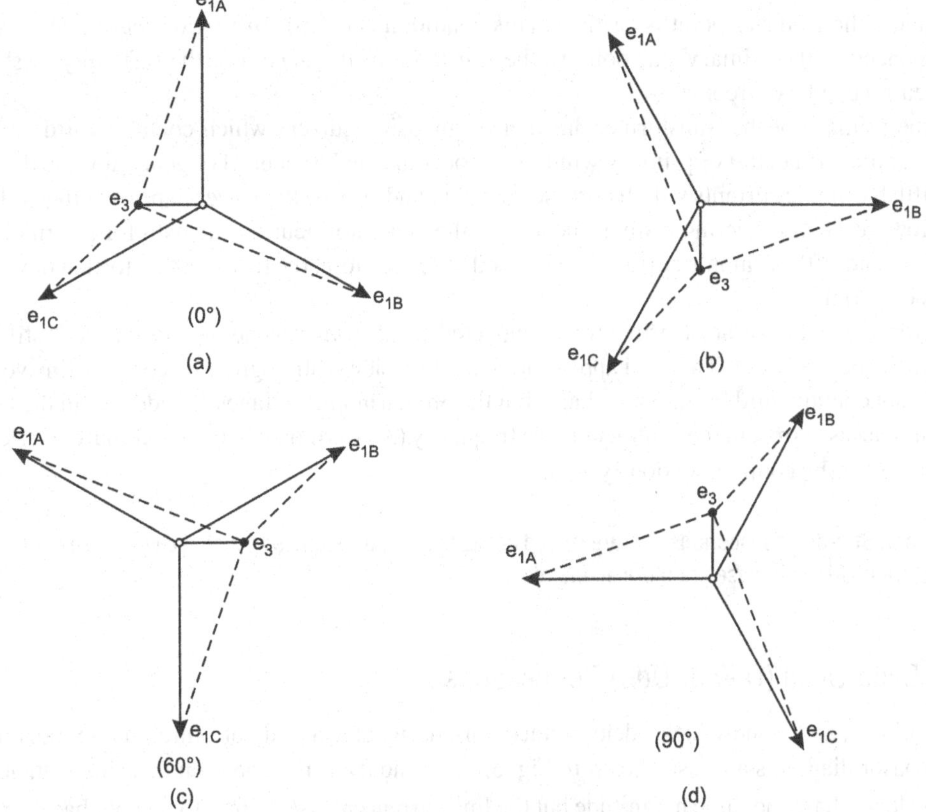

Fig. 3.10 Oscillating neutral at different instants

For illustration, consider that the fundamental phasors for the balanced phase voltages are e_{1A}, e_{1B} and e_{1C} each of magnitude e_1, while the third harmonic voltage for each phase is e_3. At a certain time their relative positions are as shown in Fig. 3.10. As the third harmonic phase voltage has the frequency different from that of the fundamental (the frequency of third harmonic is thrice the fundamental frequency) phase voltage, their relative positions vary with time, that is, position of N' changes. This phenomenon is known as oscillating neutral. The maximum voltage at any phase is $\overline{e}_1 + \overline{e}_3$. Thus harmonic voltage is undesirable as it develops high stress in the insulation of the windings. The use of a grounded neutral or a tertiary delta winding will allow a path for the third-harmonic current and thus produces a sinusoidal flux and a sinusoidal phase voltage.

The star-star transformer connections are most economical. These are employed in small current and high voltage transformers. The advantages and disadvantages of such an arrangement are given below:

Advantages:

(*i*) The number of turns per phase and the amount of insulation is minimum because phase voltage is only $\frac{1}{\sqrt{3}}$ times to that of line voltage.

(*ii*) There is no phase displacement between the primary and secondary voltages.

(*iii*) It is possible to provide a neutral connection since star point is available on both the sides.

Disadvantages:

(*i*) Under unbalanced load conditions on the secondary side, the phase voltages of load side change unless the load star point is earthed. This condition is called shifting of neutral. However by connecting the primary star point to the star point of the generator the difficulty of shifting neutral can be overcome.

(*ii*) The primary of the transformer draws a magnetising current which contains third and fifth harmonic. If neutral of primary winding is not connected to neutral of generator, the third and fifth harmonic currents will distort the core flux and change the wave shape of output voltages. However, by connecting primary neutral to the generator neutral, the path for return of these third and fifth harmonic currents is provided and, therefore, the trouble of distortion of voltages is overcome.

(*iii*) Even if neutral point of primary is connected to neutral of generator or earthed, still third harmonic may exist. This will appear on secondary side. Although the secondary line voltages do not contain third harmonic voltage; but the 3rd harmonic voltages are additive in the neutral and causes current in the neutral of triple frequency (3rd harmonic) which will cause interference to the nearby communication system.

> **Note:** Star-star connections are rarely used because of the difficulties associated with the exciting current although these are more economical.

3.7.2 Delta-Delta (Dd0 or Dd6) Connections

Figure 3.11.1(*a* and *b*) shows delta-delta connections, designated as Dd0 of a three-phase transformer. Their phasor diagrams are also shown in Fig. 3.11.1(*c* and *d*). It may be seen that line voltages and phase voltages have the same magnitude but the line currents are $\sqrt{3}$ times to that of phase currents, i.e., $I_A = \sqrt{3} I_{AB}$.

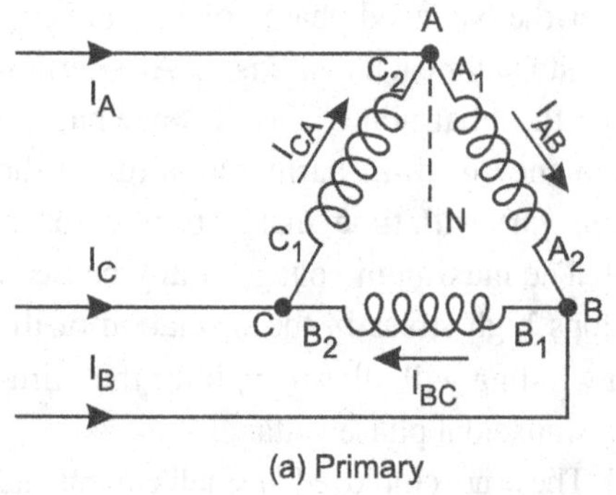

(a) Primary

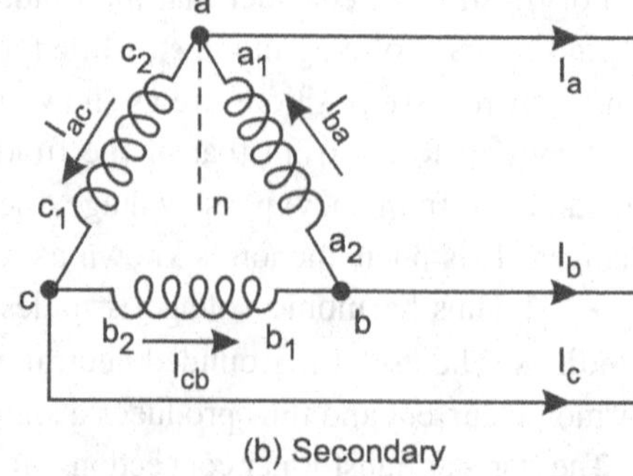

(b) Secondary

Fig. 3.11.1. Contd.

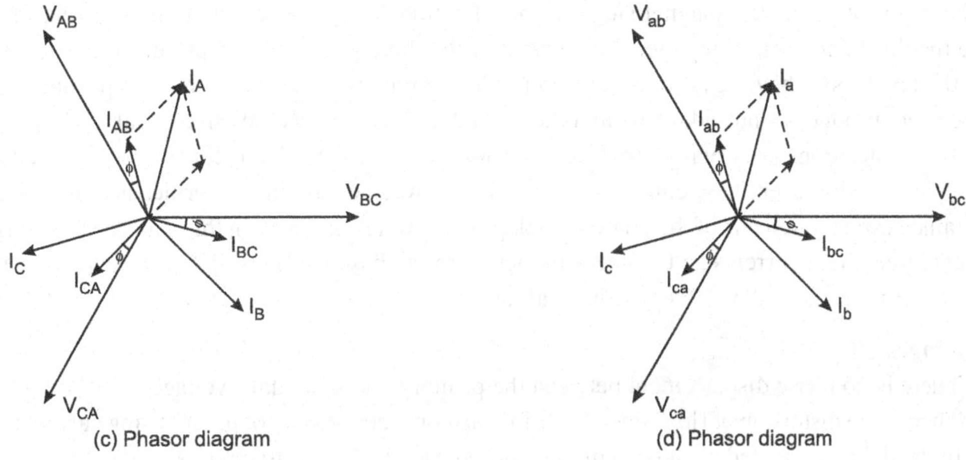

(c) Phasor diagram (d) Phasor diagram

Fig. 3.11.1 Delta-delta (Dd0) connection of transformer (0° phase shift)

If the connections of the phase windings are reversed on either side as shown in Fig. 3.11.2(*a* and *b*) we may have the phase difference of 180° between the primary and secondary systems as shown in Fig. 3.11.2(*c* and *d*). These connections are designated as Dd6.

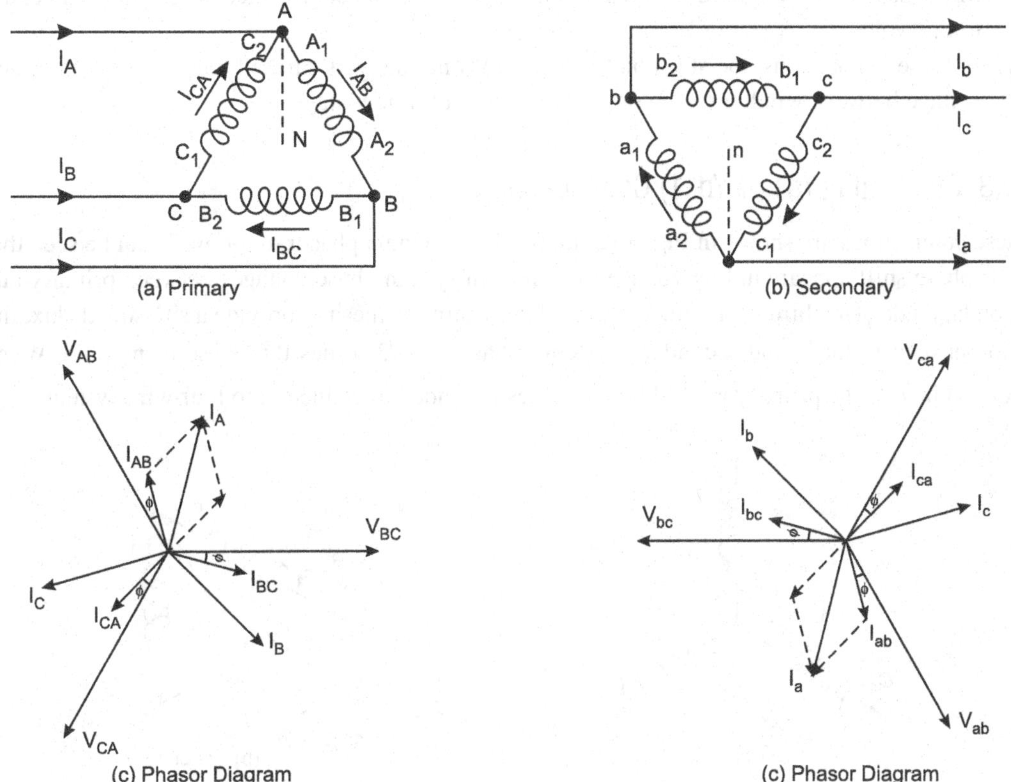

(c) Phasor Diagram (c) Phasor Diagram

Fig. 3.11.2 Delta-delta (Dd6) connection of transformer and their phasor diagram (180° phase shift)

The primary draws the magnetising current of a transformer which contains third harmonic. Since the third harmonic components of current of the three phases are displaced from each other by 120° × 3 = 360°, they are all in phase and produce circulating current within the primary delta. This current produces sinusoidal flux and the secondary voltage becomes sinusoidal.

These connections are usually employed in power system where large currents on low voltages are employed. Moreover, these connections are suitable when continuity of service is required to be maintained even though one of the phases develops fault. When operated in this way, the transformer delivers three phase currents and voltages in their correct phase relationship but the capacity of the transformer is reduced to 57.7% of the normal rating.

Advantages:

(*i*) There is no phase displacement between the primary and secondary voltages.

(*ii*) There is no distortion of flux, since the third harmonic component of magnetising current flows in the delta-connected primary winding without flowing in the line wires.

(*iii*) For winding, conductors with smaller diameter are required as cross-section of conductor is reduced because the phase current is $\frac{1}{\sqrt{3}}$ times of line current.

(*iv*) No difficulty is experienced even though the load is unbalanced on the secondary side.

Disadvantages:

(*i*) More insulation is required in comparison to star-star connections since phase voltage in equal to line voltage.

(*ii*) In these connections, the star point is absent, if one line gets earthed due to fault, maximum voltage between windings and core will become full line voltage.

3.7.3 Star-Delta (Yd1 or Yd11) Connections

These connections are shown in Fig. 3.12.1(*a*, *b*). By drawing a phasor diagram, it can be seen that a 30° phase shift appears in line voltage as well as in system phase voltages between primary and secondary side. The third-harmonic currents flow within the mesh to provide a sinusoidal flux, the ratio between primary and secondary system voltages is $\sqrt{3}$ times the phase turns ratio. When operated in *Y*-Δ, the primary neutral is sometimes grounded to connect it to four-wire system.

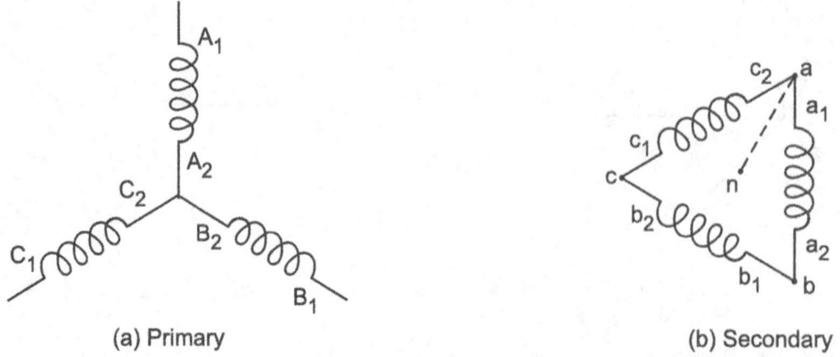

Fig. 3.12.1 Phase shift 30° lead

Figure 3.12.1 shows phase shift of 30° lag whereas Fig. 3.12.2(*a* and *b*) shows phase shift of 30° lead. If the transformer is poly-phase and the high voltage side is star connected, the former connection is termed as Yd1 and the latter Yd11.

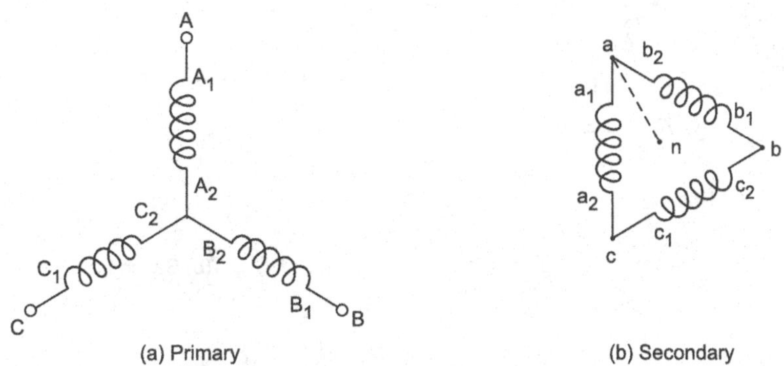

Fig. 3.12.2 Phase shift 30° lag

Usually, the transformers with these connections are used where the voltage is to be stepped down. For example, at the receiving end of a transmission line. In this type of transformer connections, the neutral of the primary winding is earthed. In this system line voltage ratio is $\frac{1}{\sqrt{3}}$ times of transformer turn-ratio and secondary line voltages have a phase shift of ± 30° with respect to primary line voltages. On the *HV* side of the transformer insulation is stressed only to the extent of 57.7% of the line voltage and, therefore, there is some saving in the cost of insulation.

3.7.4 Delta-Star (Dy1 or Dy11) Connections

The Δ-*Y* connections and phasor diagrams of a three-phase transformer supplying a balance load are shown in Figs. 3.13.1 and 3.13.2. It may be noted that the secondary system phase voltages i.e., V_{an} etc., lag the primary system phase voltages V_{AN} etc., by 30°. The ratio of primary to secondary line voltages is $1 / \sqrt{3}$ times the transformation ratio for the individual phase windings. No difficulty arises due to third harmonic currents as a delta connection allows a path for these currents.

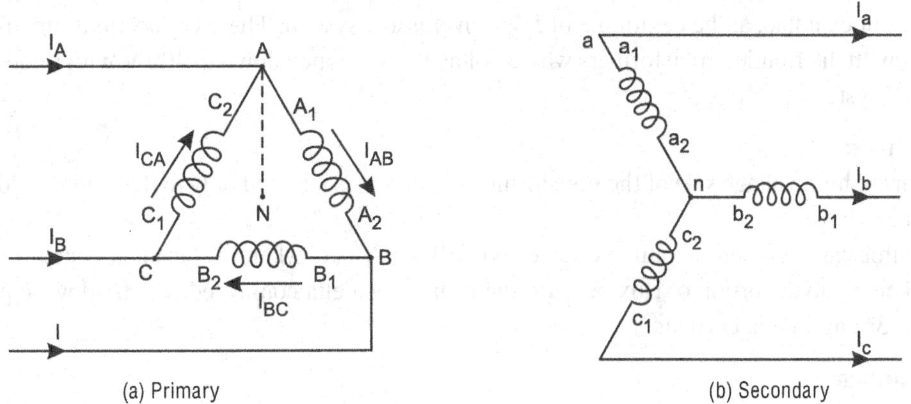

Fig. 3.13.1 Delta–Star connection of transformer (Phase shift 30° lag)

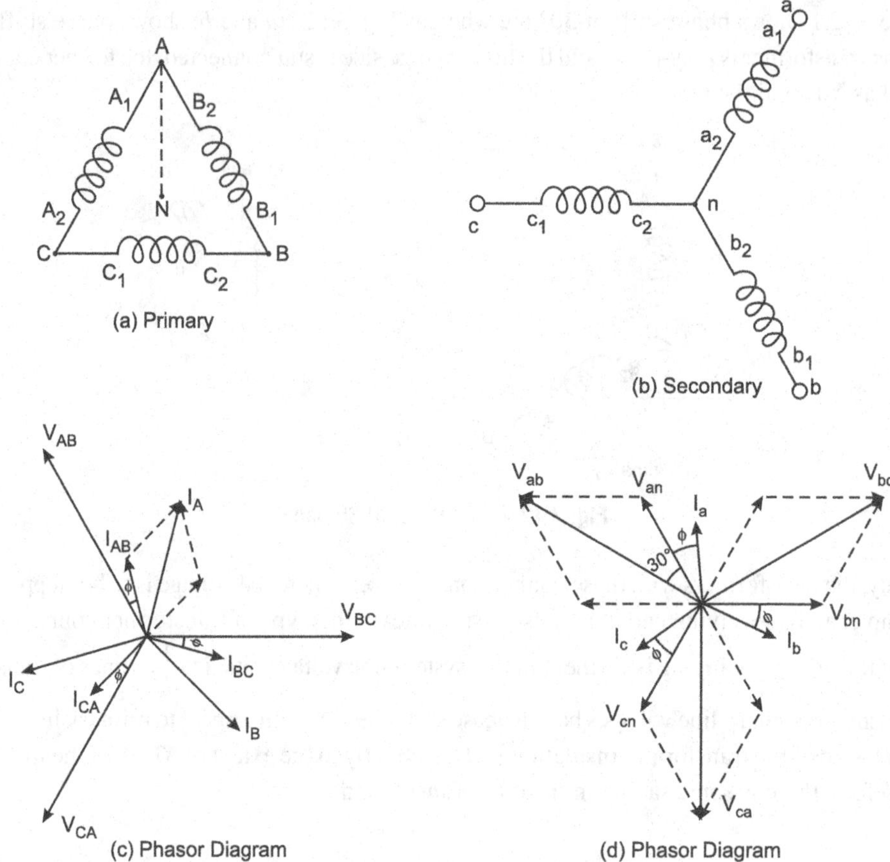

Fig. 3.13.2 Delta–star connection of transformer (Phase shift 30° lead)

The use of such connections permits a grounded neutral on the secondary side to provide a three-phase four-wire supply system. By reversing the connections on either side, the secondary system voltage can be made to lead the primary system by 30° as shown in Fig. 3.13.2. If the transformer is three-phase and the high voltage side is delta connected, the transformer is designated as Dy1 and the latter Dy11.

The transformers with these connections are employed where it is necessary to step up the voltage. For example, at the beginning of *HV* transmission system. These connections are also very popular with distribution transformers where voltages are stepped down to 400 V with three-phase, four-wire system.

Advantages:

(*i*) On the high voltage side of the transformer, insulation is stressed only to the extent of 57.7% of line voltage.

(*ii*) In this case, the neutral point is stable and will not "float" when load is unbalanced.

(*iii*) There is no distortion of flux because the primary is delta connected which allows a path for the 3rd harmonic currents.

Disadvantages:

(*i*) In this scheme of connection the line voltage ratio is $\sqrt{3}$ times of transformer turn-ratio.

(*ii*) The secondary line voltages have a phase shift of ± 30° with respect to primary line voltages.

The neutral of the secondary is grounded to provide three-phase, four-wire system and this scheme of connections is widely used in distribution systems because it can be used to serve both the three phase power equipment and single phase lighting circuits. In such case the 11 kV/400 V transformer connections provide a four-wire secondary, with neutral converted to the 4th wire. The three-phase equipment is connected to the line wires to operate at 400 V while the lighting equipment is connected between one of the line wires and neutral to operate at 230 volt.

Note: Star–delta or delta–star connected transformers cannot be operated is parallel with star-star or delta-delta connected transformers even though the voltage ratios are correctly adjusted as there will be a 30° phase difference between corresponding voltages on the secondary side.

3.7.5 Delta-Zigzag Connections

In this case, the primary is connected in delta and each secondary winding is divided in two halves such as $a_1 - a_2, a_3 - a_4; b_1 - b_2, b_3 - b_4$ and $c_1 - c_2, c_3 - c_4$. The six halves are then connected to form a zigzag connection as shown in Figs. 3.14.1 and 3.14.2. A_1, a_1, a_3 are of same polarity; similarly, B_1, b_1, b_3 and C_1, c_1, c_3 are of same polarity respectively. If the voltage of each half of the transformer secondary is V_2, the phase voltage of the secondary system is $\sqrt{3}V_2$ and the voltage is $\sqrt{3} \times \sqrt{3}\, V_2 = 3V_2$. If I_2 be the rated current of each secondary winding, the volt-ampere available from the three-phase transformer will be $3\sqrt{3}\, V_2 V_1$ while the sum of the individual winding capacities is $6V_2 I_2$. Thus only 0.866 of the combined individual capacity is available at the output.

There is no phase difference between phase voltages of the primary and secondary systems in the connection of Fig. 3.14.1(*a*) whereas a phase shift of 180° occurs if the secondary is connected as shown in Fig. 3.14.2. In poly-phase transformers if the delta is on the high voltage side, which is usually the case, the connections of Fig. 3.14.1 are termed as Dz0 and that of Fig. 3.14.2 are termed as Dz6.

Delta-zigzag connections are useful for supply of power to rectifiers.

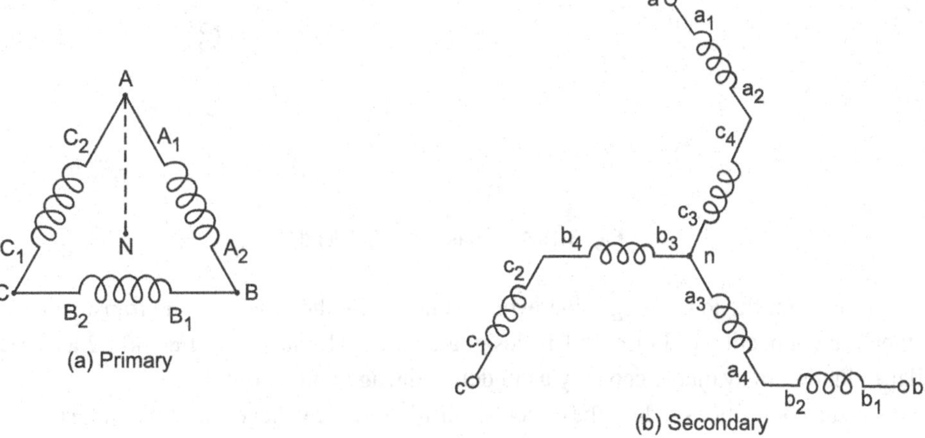

Fig. 3.14.1 0° Phase shift

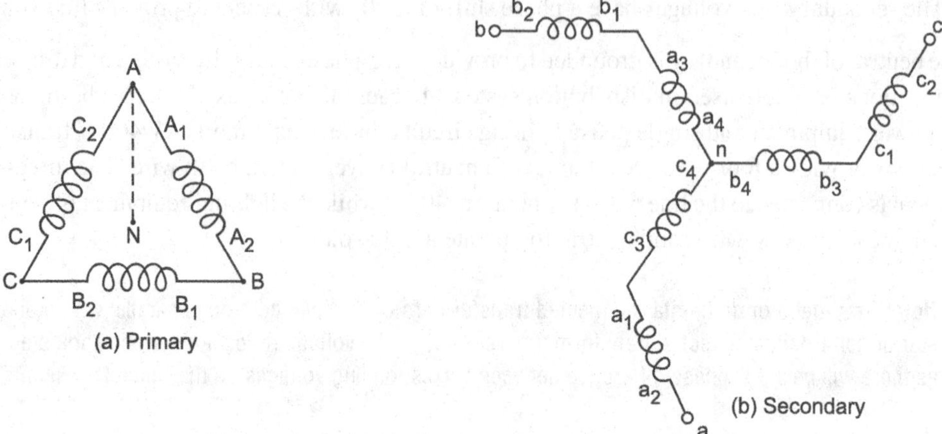

Fig. 3.14.2 Delta–zigzag connection of transformer

3.7.6 Star-Zigzag Connection

In this case, the primary is connected in star and each secondary which is made of two helps of $a_1 - a_2$, $a_3 - a_4$; $b_1 - b_2$ etc. is connected in *zig-zag.* Zigzag connections have already been explained while discussing delta–zigzag connections. In Fig. 3.15.1 the phase voltage of the secondary system leads the primary voltage by 30° while in Fig. 3.15.2 they lag by 30°.

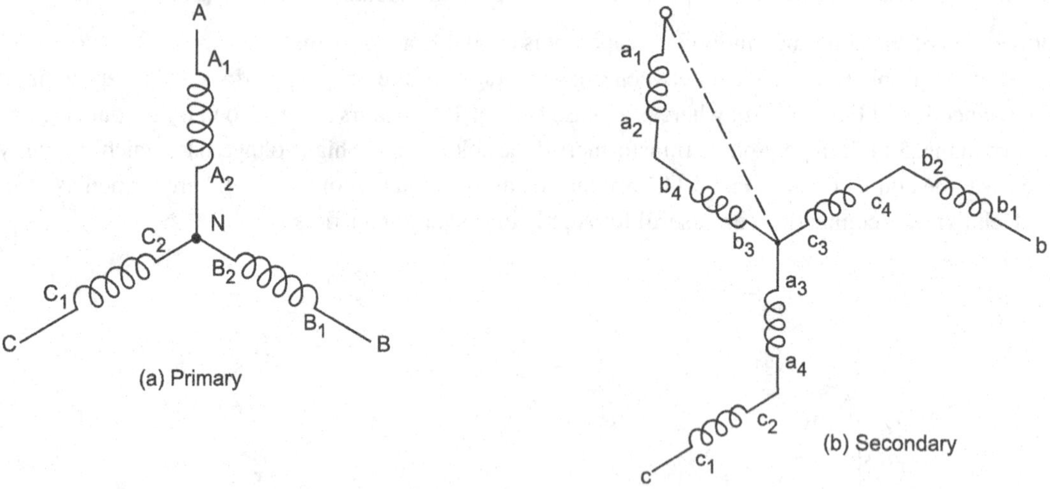

Fig. 3.15.1 Phase shift 30° lead

If the zigzag is on the low voltage side, which is normally the case, and the transformer is poly-phase, then the connection of Fig. 3.15.1 is designated as Yz11 and that of Fig. 3.15.2 as Yz1. There is similar reduction in available capacity as in delta–zigzag connections.

These zig-zag connections reduce the effect of third-harmonic voltage and at the same time permits unbalanced loading even though the primary is star-connected with isolated neutral.

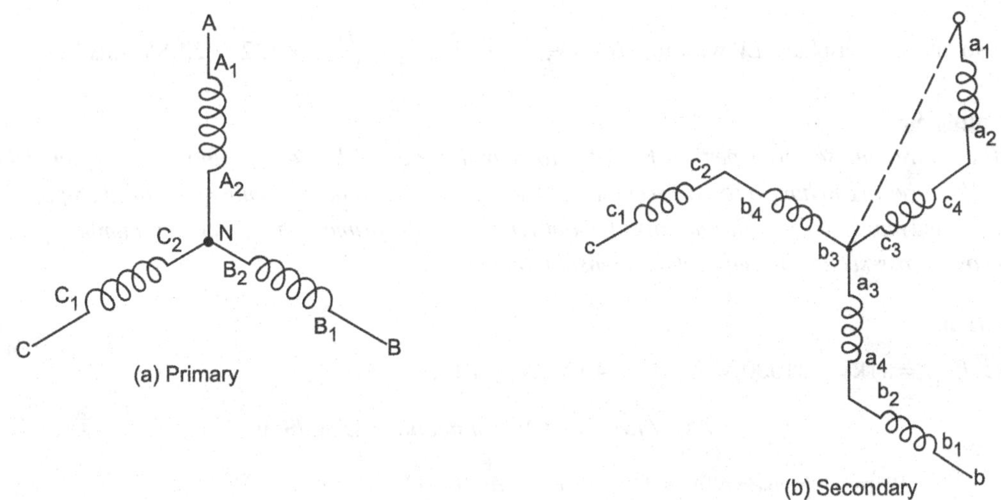

Fig. 3.15.2 Phase shift 30° lag

Example 3.1

Find the turn-ratio (primary to secondary) of a 11000/400 volt, delta/star connected, three phase transformer.

Solution:

Primary phase voltage, $E_{Ph1} = E_{L1} = 11000$ V (delta-connected primary)

Secondary phase voltage, $E_{Ph2} = \frac{E_{L2}}{\sqrt{3}} = \frac{400}{\sqrt{3}} = 231$ V (star-connected secondary)

Turn-ratio (primary to secondary), $\frac{N_1}{N_2} = \frac{E_{ph1}}{E_{ph2}} = \frac{11000}{231} = \mathbf{47.62}$ (*Ans.*)

Example 3.2

A three-phase 50 Hz transformer core has a cross-section of 400 cm² (gross). If the flux density be limited to 1.2 Wb/m², find the number of turns per phase on high and LV side winding. The voltage ratio is 2200/220 V, the HV side being connected in star and LV side in delta. Consider stacking factor as 0.9.

Solution:

Net iron cross-section, $A_i = 400 \times 0.9 = 360 \text{ cm}^2 \text{ or } 0.036 \text{ m}^2$

HV side phase voltag7e, $E_{ph1} = \frac{2200}{\sqrt{3}} = 1270 \cdot 17$ V

No. of turn per phase on *HV* winding, $N_1 = \frac{E_{ph1}}{4 \cdot 44\, B_m \times A_i \times f} = \frac{1270.17}{4 \cdot 44 \times 1 \cdot 2 \times 0 \cdot 036 \times 50}$

$= \mathbf{132}$ (*Ans.*)

LV side phase voltage, $E_{ph2} = 220$ V

$$\text{No. of turn per phase } LV \text{ winding, } N_2 = \frac{E_{ph2}}{E_{ph1}} \times N_1 = \frac{220}{1270 \cdot 17} \times 132 = \mathbf{22.86} \text{ (Ans.)}$$

Example 3.3

A 50 Hz, three-phase core type transformer is to be built for an 11 kV /440 V ratio, connected in delta-star. The cores are to have a square section and the coils are of circular. Taking an induced emf of 15 V per turn and maximum core flux density of about 1.1 T. Find the primary and secondary number of turns and cores cross-sectional area neglecting insulation thickness.

Solution:

Here, $E_{1\ (L)} = 11\text{kV} = 11000$ V; $E_{2\ (L)} = 440$ V; $\mathrm{B}_m = 1.1$ T

EMF/ turn = 15 V; *connections*-Delta/Star

$$\text{Primary phase voltage, } E_{1(\text{ph})} = E_{1\ (\text{L})} = 11000 \text{ V}$$

$$\text{Primary turns/phase, } N_1 = \frac{11000}{15} = \mathbf{733 \cdot 3} \text{ (Ans)}$$

$$\text{Secondary phase voltage, } E_{2(\text{ph})} = \frac{E_{2\ (L)}}{\sqrt{3}} = \frac{440}{\sqrt{3}} = 254 \text{ V}$$

$$\text{Secondary turns/phase, } N_2 = \frac{254}{15} = \mathbf{17} \text{ (Ans)}$$

Now,
$$E_{1\ (ph)} = 4{\cdot}44 f B_m A N_1$$

$$11000 = 4.44 \times 50 \times 1.1 \times A \times 733.3$$

$\therefore$
$$\text{Core–sectional area, } A = 0.06143 \text{ m}^2 = \mathbf{614.3 \text{ cm}^2} \text{ (Ans.)}$$

Example 3.4

A three-phase, 50 Hz transformer of shell type has cross-sectional area of core as 400 cm². If the flux density is limited to 1.2 Tesla, find the number of turns per phase on high voltage and low voltage side. The voltage ratio is 11000/400 V, the higher voltage side being connected in star and low voltage side in delta. Also determine the transformation ratio.

Solution:

Here, $f = 50$ Hz; $A = 400 \text{ cm}^2 = 400 \times 10^{-4} \text{ m}^2$; $B_{\text{m}} = 1.2$ T

$E_{1(L)} = 11000$ V; $E_{2(L)} = 400$ V; Connections - Star/Delta

Primary phase voltage,
$$E_{1ph} = \frac{E_{1(L)}}{\sqrt{3}} = \frac{11000}{\sqrt{3}} = 6351 \text{ V}$$

Now,
$$E_{1ph} = 4{\cdot}44 f B_m A N_1$$

$\therefore$
$$\text{Primary turns/phase, } N_1 = \frac{E_{1(ph)}}{4 \cdot 44 f B_m A}$$

$$= \frac{6351}{4 \cdot 44 \times 50 \times 1 \cdot 2 \times 400 \times 10^{-4}} = \mathbf{596}\ (Ans.)$$

Transformation, ratio, $K = \frac{E_{2(ph)}}{E_{1(ph)}} = \frac{400}{11000/\sqrt{3}} = \frac{\sqrt{3} \times 400}{11000} = \mathbf{0{\cdot}06298}\ (Ans.)$

Secondary turns/phase, $N_2 = KN_1 = 0{\cdot}06298 \times 596 = \mathbf{37{\cdot}54}\ (Ans.)$

Example 3.5

A three-phase step down transformer is connected to 6600 volt mains and takes a current of 24 amperes. Calculate the secondary line voltage, line current and output for the following connections (i) Delta-delta (ii) Star-star (iii) Delta-star (iv) Star-delta. The ratio of turns of per phase is 12. Neglect losses.

Solution:

Ratio of turns per phase = 12

Transformation ratio, $K = \frac{1}{12}$

(*i*) *Delta-Delta connections:*

In Delta connections, line voltage = phase voltage.

Primary line voltage = 6600

Primary phase voltage = 6600 V

Secondary phase voltage $= \frac{6600}{12} = 550$ V

Secondary line voltage = **550 V** (*Ans*)

Line current $= \sqrt{3} \times$ phase current

$\therefore$ Primary phase current $= \frac{24}{\sqrt{3}}$

Secondary phase current $= \frac{24}{\sqrt{3}} \times 12$

Secondary line current $= \sqrt{3} \times \frac{24}{\sqrt{3}} \times 12 = \mathbf{288\ A}$ (*Ans*)

Output $= \sqrt{3}\, V_L\, I_L = \frac{\sqrt{3} \times 550 \times 288}{1000}$ kVA= **274.36 kVA** (*Ans*)

(*ii*) *Star-Star Connections*:

In star connections, line voltage $= \sqrt{3} \times$ phase voltage

Also, line current = phase current.

Primary phase voltage $= \frac{6600}{\sqrt{3}}$

$$\text{Secondary phase voltage} = \frac{6600}{\sqrt{3} \times 12}$$

$$\text{Secondary line voltage} = \sqrt{3} \times \frac{6600}{\sqrt{3} \times 12} = \mathbf{550\ V}\ (\textit{Ans})$$

$$\text{Primary line current} = 24\ \text{A}$$

$$\text{Secondary line current} = 24 \times 12 = \mathbf{288\ A}\ (\textit{Ans})$$

$$\text{Output in kVA} = \frac{\sqrt{3} \times 550 \times 288}{1000} = \mathbf{274.36\ kVA}\ (\textit{Ans})$$

(*iii*) *Delta-Star Connections*:

$$\text{Primary phase voltage} = 6600\ \text{volts}$$

$$\text{Secondary phase voltage} = \frac{6600}{12} = 550\ \text{V}$$

$$\text{Secondary line voltage} = \sqrt{3} \times 550 = \mathbf{952{\cdot}63\ V}\ (\textit{Ans})$$

$$\text{Primary phase current} = \frac{24}{\sqrt{3}}$$

$$\text{Secondary phase current} = \frac{24}{\sqrt{3}} \times 12 = 116{\cdot}27\ \text{A}$$

$$\text{Secondary line current} = \mathbf{116.27\ A}\ (\textit{Ans})$$

$$\text{Output in kVA} = \frac{\sqrt{3} \times 952 \cdot 63 \times 166 \cdot 27}{1000} = \mathbf{274{\cdot}36\ kVA}\ (\textit{Ans})$$

(*iv*) *Star-Delta connections*:

$$\text{Primary phase voltage} = \frac{6600}{\sqrt{3}}$$

$$\text{Secondary phase voltage} = \frac{6600}{\sqrt{3}} \times \frac{1}{12} = \mathbf{317{\cdot}54\ V}\ (\textit{Ans})$$

$$\text{Secondary line voltage} = \mathbf{317{\cdot}54\ V}\ (\textit{Ans})$$

$$\text{Primary line current} = \text{primary phase current} = \mathbf{24\ A}\ (\textit{Ans})$$

$$\text{Secondary phase current} = 24 \times 12 = 288\ \text{A}$$

$$\text{Secondary line current} = \sqrt{3} \times 288 = \mathbf{498{\cdot}83\ A}\ (\textit{Ans})$$

$$\text{Output in kVA} = \frac{\sqrt{3} \times 317 \cdot 54 \times 498 \cdot 83}{1000} = \mathbf{274.36\ kVA}\ (\textit{Ans})$$

Example 3.6

A three-phase, 50 Hz, core type transformer is required to be built for a 10000/500 V ratio, connected in star/mesh. The cores are to have a square section and the coils are to be circular. Taking an induced emf of about 15 V per turn and maximum core density of about 1.1 tesla find the cross-sectional dimensions of the core and the number of turns per phase.

Solution:

$$\text{Emf per turn} = 15$$

$$\text{The primary phase voltage, } E_{ph1} = \frac{10000}{\sqrt{3}} \text{ volt}$$

$$\therefore \quad \text{Primary turns, } N_1 = \frac{10000}{\sqrt{3}} \times \frac{1}{15} = \mathbf{384} \ (Ans.)$$

$$\therefore \quad \text{Secondary turns, } N_2 = N_1 \times K$$

$$\text{Where } K = \text{Transformation ratio} = \frac{E_{ph2}}{E_{ph1}} = \frac{500}{\frac{10000}{\sqrt{3}}} = \frac{\sqrt{3}}{20}$$

$$\therefore \quad N_2 = 384 \times \frac{\sqrt{3}}{20} = 33.1; \mathbf{34} \ (Ans.)$$

Also from the fundamental equation $E_{ph} = 4.44 f.\, \phi_{m.}\, N_{ph} = 4.44 f B_{m.} A_i N_{ph}$ volt

$$\frac{E_{ph}}{N_{ph}} = 15 = 4.44 \times 50 \times 1.1 \times A_i$$

$$\therefore \quad A_i = \frac{15}{4 \cdot 44 \times 50 \times 1 \cdot 1} = 612 \times 10^{-4} \text{ m}^2 = \mathbf{612\ cm^2} \ (Ans.)$$

Gross-cross-sectional area of the core,

$$A_{gc} = \frac{A_i}{K_i}$$

$$\therefore \quad A_{gc} = \frac{612}{0 \cdot 9} = 680 \text{ cm}^2 \text{ (assuming } K_i \text{ to be 0.9)}$$

$$\text{Side of square} = \sqrt{A_{gc}} = \sqrt{680} = \mathbf{26\ cm} \ (Ans.)$$

Example 3.7

A three-phase transformer, rated at 1000 kVA, 11/3.3 kV has its primary star-connected and secondary delta connected. The actual resistances per phase of these windings are, primary 0.375 ohm, secondary 0.095 and the leakage reactances per phase are primary 9.5 ohm, secondary 2 ohm. Calculate the voltage at normal frequency which must be applied to the primary when the secondary terminals are short circuited. Calculate also the power under these conditions.

Solution:

$$\text{Primary phase voltage, } E_{ph1} = \frac{11000}{\sqrt{3}} = 6352 \text{ V}$$

$$\text{Transformation ratio, } K = \frac{E_{ph2}}{E_{ph1}} = \frac{3300}{6352} = 0 \cdot 5195$$

$$\text{Primary full load current, } I_{ph1} = I_{L1} = \frac{1000 \times 1000}{\sqrt{3} \times 11000} = 52 \cdot 49 \text{ A}$$

Total resistance referred to primary per phase,

$$R_{ep} = R_1 + \frac{R_2}{K^2} = 0 \cdot 375 + \frac{0 \cdot 095}{(0 \cdot 5195)^2} = 0 \cdot 727 \text{ ohm}$$

Total reactance referred to primary per phase,

$$X_{ep} = X_1 + \frac{X_2}{K^2} = 9 \cdot 5 + \frac{2}{(0 \cdot 5195)^2} = 16 \cdot 9 \text{ ohm}$$

$$\text{Total impedance referred to primary, } Z_{ep} = \sqrt{R_{ep}^2 + X_{ep}^2}$$

$$Z_{ep} = \sqrt{(0 \cdot 727)^2 + (16 \cdot 9)^2} = 16.95 \text{ ohm per phase}$$

Phase voltage applied to primary when the secondary in short circuited,

$$V_{1sc(ph)} = I_{ph1} Z_{ep}$$

$$= 52.49 \times 16.95 = 889 \text{ V}$$

Voltage applied at the terminals, $E_{L1} = \sqrt{3}\, E_{ph1} = \sqrt{3} \times 889 =$ **1540 V** (*Ans.*)

Power input when secondary is short circuited,

$$P_c = 3 I_{ph1}^2 R_{ep} = 3 \times (52.49)^2 \times 0.727$$

= 5964 W = **5.964 kW** (*Ans.*)

Example 3.8

A 33/66 kV, 5 MVA, three-phase star-connected transformer with short circuited secondary passes full-load current with 7% primary potential difference and losses are 30 kW. With full potential difference on the primary and the secondary open circuited, the losses are 15 kW. What will be the efficiency of the transformer at full-load and 0.8 power factor?

Solution:

At short-circuit full-load copper losses, $P_c = 30$ kW

At open-circuit, iron losses, $P_i = 15$ kW

Total losses = $P_i + P_c = 15 + 30 = 45$ kW = 0.045MW

At 0.8 p.f., rated output = $5 \times 0.8 = 4$ MW

$$\text{Full-load efficiency, } \eta = \frac{\text{rated output}}{\text{rated output + losses}} \times 100$$

$= \dfrac{4}{4 + 0 \cdot 045} \times 100 =$ **98.88%** (*Ans.*)

Example 3.9

A 2 MVA three-phase, 33/6.6 kV, delta/star transformer has a primary resistance of 8 ohm per phase and a secondary resistance of 0.08 ohm per phase. The percentage impedance is 7%. Calculate the secondary terminal voltage, regulation and efficiency at full load 0.75 power factor lagging when the iron losses are 15 kW.

Solution:

Here, transformer rating = 2 MVA = 2×10^6 VA

$$\text{Transformation ratio, } K = \frac{E_{ph2}}{E_{ph1}} = \frac{6 \cdot 6 / \sqrt{3}}{33} = 0 \cdot 1155$$

Referring to secondary side,

$$\text{Secondary phase current, } I_{ph2} = I_{L2} = \frac{2 \times 10^6}{\sqrt{3} \times 6 \cdot 6 \times 10^3} = 175 \text{ A}$$

Percent impedance drop = 7% of E_{ph2}

$$= \frac{7}{100} \times \frac{6 \cdot 6 \times 1000}{\sqrt{3}} = 266 \cdot 7 \text{ V}$$

or,
$$I_{ph2} Z_{es} = 266.7$$

$\therefore$
$$\text{Impedance/phase, } Z_{es} = \frac{266 \cdot 7}{175} = 1.524\ \Omega$$

Resistance per phase referred to secondary side,

$$R_{es} = R_2 + R_1 + K^2 = 0.08 + 8 \times (0.1155)^2$$

$$= 0.1867\ \Omega$$

$$\text{Reactance/phase, } X_{es} = \sqrt{Z_{es}^2 - R_{es}^2} = \sqrt{(1 \cdot 524)^2 - (0 \cdot 1867)^2}$$

$$= 1.51\ \Omega$$

Power factor, $\cos \phi_2 = 0.75$; $\sin \phi_2 = \sin \cos^{-1} 0.75 = 0.6613$

$$E_{ph2} = \frac{6 \cdot 6 \times 1000}{\sqrt{3}} = 3810 \text{ V}$$

Secondary phase voltage after drop in resistance and reactance.

$$V_{ph2} = E_{ph2} - I_{ph2} R_{es} \cos \phi_2 - I_{ph2} X_{es} \sin \phi_2$$

$$= 3810 - 175 \times 0.1867 \times 0.75 - 175 \times 1.51 \times 0.6613$$

$$= 3810 - 24.5 - 174.5 = 3611 \text{ V}$$

Secondary terminal voltage, $V_{L2} = \sqrt{3} \times V_{ph2} = \sqrt{3} \times 3611 =$ **6254 V** *(Ans.)*

$$\text{Voltage regulation, } \% \, Reg = \frac{E_{ph2} - V_{ph2}}{E_{ph2}} \times 100$$

$$= \frac{3810 - 3611}{3810} \times 100 = \mathbf{5.22\%} \; (Ans.)$$

$$\text{Full-load coper losses, } P_c = 3I_{ph2}^2 \, R_{es} = 3 \times (175)^2 \times 0.1867$$

$$= 17153 \text{ W} = 17.153 \text{ kW}$$

$$\text{Efficiency at full-load, } \eta = \frac{MVA \times 1000 \times \cos\phi}{MVA \times 1000 \times \cos\phi + P_i + P_c} \times 100$$

$$= \frac{2 \times 1000 \times 0.75}{2 \times 1000 \times 0 \cdot 75 + 15 + 17 \cdot 153} \times 100$$

$$= \mathbf{97.9\%} \; (Ans.)$$

Example 3.10

The percentage impedance of a three-phase, 11000/400 V, 500 kVA, 50 Hz transformer is 4.5%. Its efficiency at 80% of full-load, unity power factor is 98.8%. Load power factor is now varied while the load current and the supply voltage are held constant at their rated values. Determine the load power factor at which the secondary terminal voltage is minimum.

Solution:

$$\text{Output at 80\% of full load and unity of} = 500 \times 0.8 \times 1.0 = 400 \text{ kW}$$

$$\text{Input at 80\% of full load and unity pf} = \frac{400}{\eta} = \frac{400}{0 \cdot 988} = 404 \cdot 8583 \text{ kW}$$

$$\text{Total losses at 80\% of full load} = \text{Input} - \text{output}$$

$$= 404.8583 - 400$$

$$= 4.8583 \text{ kW}$$

$$\therefore \quad \text{Copper losses at 80\% full load} = \frac{4 \cdot 8583}{2} = 2.429 \text{ kW}$$

$$\text{Full load copper losses, } P_c = \frac{2 \cdot 429}{(0 \cdot 8)^2} = 3 \cdot 796 \text{ kW}$$

$$\text{Percentage resistance} = \text{Percentage copper loss of full load output}$$

$$= \frac{3 \cdot 7955}{500} \times 100 = 0.7591\%$$

$$\text{Percentage impedance} = 4.5\% \text{ (given)}$$

$$\therefore \quad \text{Percentage reactance} = \sqrt{(4 \cdot 5)^2 - (0 \cdot 7591)^2} = 4.4355\%$$

For terminal voltage to be minimum, the drop or percentage regulation will be maximum i.e., $v_r \cos\phi + v_x \sin\phi$ will be maximum or $$\frac{d}{d\phi}(v_r \cos\phi + v_x \sin\phi) = 0$$

or $$-v_r \sin\phi + v_x \cos\phi = 0$$

or $$\phi = \tan^{-1}\frac{v_x}{v_r} = \tan^{-1}\frac{4 \cdot 4355}{0 \cdot 7591} = 80.3°$$

Thus load power factor for minimum voltage, $\cos\phi = \cos 80.3° =$ **0.1687** (lag) (*Ans.*)

Section Practice Problems

Numerical Problems

1. A three-phase 50 Hz transformer of shell type has an iron cross-section of 400 sq. cm (gross). If the flux density be limited by 1.2 tesla, find the number of turns per phase on high and low voltage windings. The voltage ratio is 11000/550, the high voltage winding being connected in star and low voltage winding in mesh. (**Ans.** *670, 58*)

2. A 440 V, three-phase supply is connected through a transformer of 1: 1 ratio which has its primary connected in Delta and secondary in star to a load consisting of three 11 ohm resistors connected in delta. Calculate the currents (*i*) in the transformer windings (*ii*) in the resistors (*iii*) in the line to the supply (*iv*) the load. Find also (*v*) the power supplied and (*vi*) power dissipated by each resistor. (**Ans.** *120 A, 120 A, 69.3A, 207.3A, 144 kW, 48 kW*)

3. A 120 kVA, 6000/400 V, Y/Y, 3-ϕ, 50 Hz transformer has an iron loss of 1800 W. The maximum efficiency occur at 3/4 full load. Find the efficiency of the transformer at (*i*) full load and 0.8 power factor and (*ii*) the maximum efficiency at unity power factor. (**Ans.** *95%, 96.15%*)

4. A 100 kVA, three-phase 50 Hz 3300/400 V transformer is delta-connected on the *HV* side and star-connected on *LV* side. The resistance of the *HV* winding is 3.5 Ω per phase and that of the *LV* winding 0.02 Ω per phase. Calculate the iron losses of the transformer at normal voltage and frequency if its full-load efficiency be 95.8% at 0.8 pf (lag). (**Ans.** *1.2 kW*)

Short Answer Type Questions

Q.1. What are the advantages of using three-phase transformers over a bank of three one-phase transformers?

Ans. (*i*) It requires smaller quantity of iron and copper. Hence, its cost is nearly 15% lesser than a bank of three single phase transformers of equal rating.
(*ii*) It has smaller size and can be accommodated in smaller tank and hence needs smaller quantity of oil for cooling.
(*iii*) It has less weight and occupies less space.
(*iv*) It needs less number of bushings.
(*v*) It operates at slightly better efficiency and regulation.

Q.2. Draw the connection diagram for delta-Y connection.

Ans. The Δ-*Y* connections of a three-phase transformer supplying a balance load are shown in Fig. Q2. In these connections, the secondary system phase voltages lag the primary system phase voltages by

30°. The ratio of primary to secondary line voltages is $1/\sqrt{3}$ times the transformation ratio for the individual phase windings. No difficulty arises due to third harmonic currents as a delta connection allows a path for these currents.

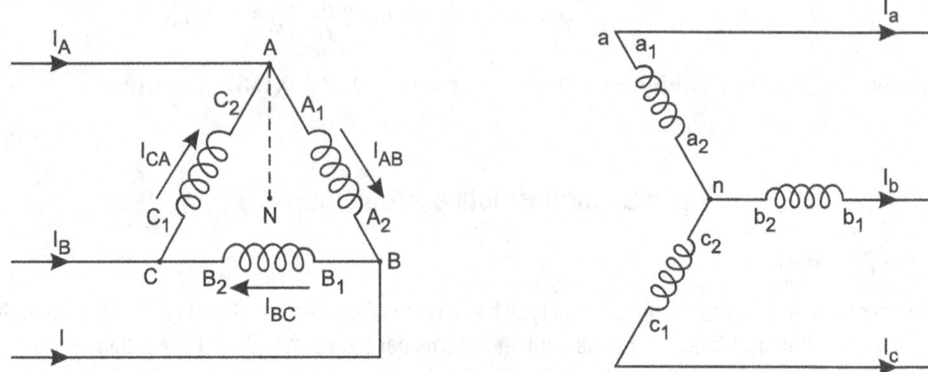

Fig. Q.2 Delta–star connection of transformer (Phase shift 30° lag)

Q.3. A star/delta transformer has a phase current transformation ratio (star phase: 1 delta phase). Its star/delta line current transformation ratio will be given by....................

Ans. In star connections, $I_{L_1} = I_{ph_1}$ or $I_{ph_1} = I_{L_1}$

In delta connections, $I_{L_2} = \sqrt{3} I_{ph_2}$ or $I_{ph_2} = \dfrac{I_{L_2}}{\sqrt{3}}$

$$\text{Ratio} = \frac{I_{ph_1}}{I_{ph_2}} = 1 \text{ or } \frac{I_{L1}}{I_{L2}/\sqrt{3}} = 1 \text{ or } \frac{I_{L1}}{I_{L2}} \times \sqrt{3} = 1 \text{ or } \frac{I_{L1}}{I_{L2}} = \frac{1}{\sqrt{3}}$$

Q.4. Star-star connected transformers are rarely used, why?

Ans. (*i*) Under unbalanced load conditions on the secondary side, the phase voltages of load side change unless the load star point is earthed. This condition is called shifting of neutral.

(*ii*) The primary of the transformer draws a magnetising current which contains third and fifth harmonic. If neutral of primary winding is not connected to neutral of generator, the third and fifth harmonic currents which distort the core flux and change the wave shape of output voltages.

(*iii*) Even if neutral point of primary is connected to neutral of generator or earthed, still third harmonic may exist.

Star-star connections are rarely used because of the difficulties associated with the exciting current although these are more economical.

3.8 Parallel Operation of Three-phase Transformers

When the primaries and secondaries of the two or more transformer are connected separately to the same incoming and outgoing lines to share the load, the transformers said to be connected in **parallel.**

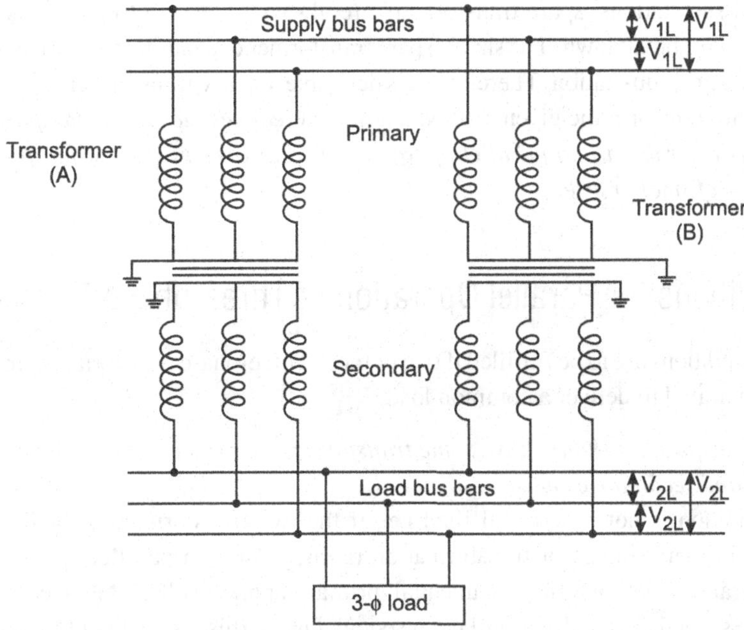

Fig. 3.16 Connections for parallel operation of two three-phase transformers

The two three-phase transformers *A* and *B* are placed in parallel as shown in Fig. 3.16. Here the primary windings of the two transformers are connected to the supply bus-bars and the secondary windings are connected to the load through load bus-bars. Under this conditions

$$V_{1L} = \text{Primary applied voltage}$$

$$V_{2L} = V_L = \text{Secondary load voltage.}$$

3.9 Necessity of Parallel Operation of Three-phase Transformers

As explained earlier, following are the reasons for which transformers are put in parallel.

(*i*) When the load on the transmission lines increases beyond the capacity of the installed transformer. To overcome this problem one way is to replace the existing transformer with the new one having larger capacity (this is called augmentation of transformer) and the other way is to place one more transformer is parallel with the existing one to share the load. The *cost of replacing the transformer* is much more than placing another one in parallel with the existing one.

Hence, it is desirable to place another transformer is parallel when the electrical load on the existing transformer increases beyond its rated capacity.

(*ii*) Sometimes, the amount of power to be transformed is so high that it is not possible to build a single unit of that capacity, then we have to place two or more transformers in parallel.

Hence, parallel operation of transformers is necessary when the amount of power to be transformed is much more than that which can be handled by single unit (transformer).

(*iii*) At the grid sub stations, spare transformers are always necessary to insure the continuity of supply in case of breakdown. The size of spare transformer depends upon the size of transformers placed at the grid sub-station. Therefore, it is desirable to place transformers of smaller capacity in parallel to transform the given load which in turn *reduces the size of the spare transformer. Hence, it is desirable to do parallel operation of transformers if we want to keep the spare transformer of smaller size.*

3.10 Conditions for Parallel Operation of Three-phase Transformers

The following conditions are to be fulfilled if two or more Three-phase transformers are to be operated successfully in parallel to deliver a common load.

(*i*) *Both the transformers should have same transformation ratio i.e., the voltage ratings of both primaries and secondaries must be identical.*
If this condition is not exactly fulfilled i.e., if the two transformers '*A*' and '*B*' have slight difference in their voltage or transformation ratios, even then parallel operation is possible. Since the transformation ratios are unequal, primary applied voltage being equal, the induced emfs in the secondary windings will not be equal. Due to this inequality of induced emfs in the secondary windings, there will be, even at no-load, some circulating current flowing from one secondary winding (having higher induced *emf*) to the other secondary windings (having lower induced *emf*). In other words, there will be circulating currents between the secondary windings and therefore between primary windings also when the secondary terminals are connected in parallel. The impedance of transformers is small, so that a small percentage voltage difference may be sufficient to circulate a considerable current and cause additional $I^2 R$ loss. When load is applied on the secondary side of such a transformer, the circulating current will tend to produce unequal loading conditions. Hence, it may be impossible to take the combined full load *kVA* output from the parallel connected group without one of the transformers becoming excessively hot. For satisfactory parallel operation the circulating current should not exceed 10% of the normal load current.

(*ii*) *Both the transformers should have the same percentage (or per unit) impedance.*
If this condition is not exactly fulfilled, i.e., the impedance triangles at the rated *kVA*'s are not identical in shape and size, even then parallel operation will be possible, but the power factors at which the transformers operate will differ from the power factor of the load. Therefore, in this case the transformers will not share the load in proportion to their *kVA* ratings.

(*iii*) *Both three-phase transformers must have the same phase-sequence* i.e., *the transformers must be properly connected with regard to their phase-sequence.*
If this condition is not observed, the secondary of one transformer will act as a load on the other, i.e., the secondaries will be under short circuit condition, they will be heated up and damage the insulation quickly This condition must be fulfilled by all means.

(*iv*) *In case of three-phase transformers, the two transformers must have such connections that there should not be any phase displacement between the secondary line voltages.*
The primaries and secondaries of three-phase transformers may be connected in different form of connections. These connections produce various magnitudes and phase displacements.

The magnitudes can be adjusted by changing the tapings but phase displacement cannot be compensated. Therefore the following types of connections are permissible for connecting three-phase transformers in parallel.

Transformer-I:	Yy	Dd	Yy	Yd	Yd
Transformer-II:	Yy	Dd	Dd	Dy	Yz

However, transformers with (α + 30°) and (–30°) angle can also be connected in parallel but only after reversing the connections of either primary or secondary.

3.11 Load Sharing between Three-phase Transformers Connected in Parallel

The load sharing between two transformers connected in parallel depends upon the various conditions as discussed in chapter-2, Art No. 2-41. The only difference is that, in case of three-phase transformers per phase impedance is to be considered while determining the load sharing. If percentage impedances of the two transformers having different rating are given, their values have to be converted as per base kVA for calculating load sharing, i.e., % Resistance or % Reactance at base kVA = $\frac{\text{Base kVA}}{\text{Rated kVA}}$ × % R or % X at rated kVA.

Example 3.11

Two three-phase transformers each of 100 kVA are connected in parallel. One transformer has (per-phase) resistance and reactance of 1% and 4% respectively and the other has (per-phase) resistance and reactance of 1.5% and 6% respectively. Calculate the load shared by each transformer and their pf when the total load to be shared is 120 kVA, 0.8 p.f. lagging.

Solution:

Here, Load = 120 kVA at p.f. 0.8 lagging

$$\overline{S} = 120 \angle \cos^{-1} 0.8 = 120 \angle -36 \cdot 87°$$

Percentage impedance of transformer-I,

$$\overline{Z}_1 = (1 + j4) = 4 \cdot 12 \angle 76°$$

Percentage impedance of transformer-II,

$$\overline{Z}_2 = (1.5 + j6) = 6 \cdot 18 \angle 76°$$

$$\overline{Z}_1 + \overline{Z}_2 = (1 + j4) + (1.5 + j6) = (2.5 + j10) = 10 \cdot 3 \angle 76°$$

Load shared by transformer-I,

$$\overline{S}_1 = \frac{\overline{Z}_2}{\overline{Z}_1 + \overline{Z}_2} \times \overline{S}$$

$$= \frac{6 \cdot 18 \angle 76°}{10 \cdot 3 \angle 76°} \times 120 \angle -36 \cdot 87°$$

$= 72\angle -36\cdot 87° \text{ kVA}$

$=$ **57.6 kW at 0.8 p.f. lagging** (*Ans.*)

$$\text{Load shared by transformer-II, } \bar{S}_2 = \frac{\bar{Z}_1}{\bar{Z}_1 + \bar{Z}_2} \times \bar{S} = \frac{4\cdot 12\angle 76°}{10\cdot 3\angle 76°} \times 120\angle -36\cdot 87°$$

$= 48\angle -36\cdot 87° kVA$

$=$ **38.4 kW at 0.8 p.f. lagging** (*Ans.*)

Example 3.12

Two 1000 kVA and 500 kVA, three-phase transformers are operating in parallel. The transformation ratio is same for both i.e., *6600/400, delta-star. The equivalent secondary impedances of the transformers are (0.001 + j0.003) ohm and (0.0028 + j0.005) ohm per phase respectively. Determine the load shared and pf of each transformer if the total load supplied by them is 1200 kVA at 0.866 pf lagging.*

Solution:

Impedance of 1000 kVA transformer, $\bar{Z}_1 = (0.001 + j0.003) = 0.003162 \angle 71.57° \text{ ohm}$

Impedance of 500 kVA transformer, $\bar{Z}_2 = (0.0028 + j0.005) = 0.00573 \angle 60.75° \text{ ohm}$

$$\bar{Z}_1 + \bar{Z}_2 = (0.001 + j0.003) + (0.0028 + j0.005) = (0.0038 + j0.008)$$

$$= 0.00886 \angle 64.59° \text{ ohm}$$

$$\text{Load supplied, } \bar{S} = 1200 \angle -\cos^{-} 0.866 = 1200 \angle -30° \text{ kVA}$$

Load shared by 1000 kVA transformer,

$$\bar{S}_1 = \bar{S} \times \frac{\bar{Z}_2}{\bar{Z}_1 + \bar{Z}_2} = 1200\angle -30° \times \frac{0\cdot 00573\angle 60\cdot 75}{0\cdot 00886\angle 64\cdot 59}$$

$= 776 \angle -33.84° \text{ kVA}$

$=$ **645 kW at *pf* 0.83 lagging** (*Ans.*)

$$\text{Load shared by 500 kVA transformer, } \bar{S}_2 = \bar{S} \times \frac{\bar{Z}_1}{\bar{Z}_1 + \bar{Z}_2} = 1200\angle -30° \times \frac{0\cdot 003162\angle 71\cdot 57°}{0\cdot 00886\angle 64\cdot 59°}$$

$= 428 \angle -23° \text{ kVA}$

$=$ **394 kW at *pf* 0.92 lagging** (*Ans.*)

Example 3.13

Two 400 kVA and 800 kVA transformers are connected in parallel. One of them (400 kVA transformer) has 1.5% resistive and 5% reactive drops whereas the other (800 kVA transformer) has 1% resistive and 4% reactive drops. The secondary voltage of each transformer is 400 V on load. Determine how they will share a load of 600 kVA at a pf of 0.8 lagging.

Solution:

Let the base kVA be 800 kVA

Percentage impedance of 400 kVA transformer at base kVA

$$\bar{Z}_1 = \frac{800}{400} \times (1 \cdot 5 + j5) = 3 + j10 = 10.44 \angle 73.3^\circ$$

Percentage impedance of 800 kVA transformer at base kVA

$$\bar{Z}_2 = \frac{800}{800} \times (1 + j4) = 1 + j4 = 4.123 \angle 76^\circ$$

$$\bar{Z}_1 + \bar{Z}_2 = (3 + j10) + (1 + j4) = 4 + j14 = 14.56 \angle 74^\circ$$

Load to be shared, $\bar{S} = 600 \angle -\cos^{-1} 0.8 = 600 \angle -36.87^\circ$ kVA

Load shared by 400 kVA transformer, $\bar{S}_1 = \bar{S} \times \frac{\bar{Z}_2}{\bar{Z}_1 + \bar{Z}_3}$

$$= 600 \angle -36 \cdot 87^\circ \times \frac{4 \cdot 123 \angle 76^\circ}{14 \cdot 56 \angle 74^\circ} = 169.9 \angle -34.87^\circ$$

= **139.4 kW at pf 0.82 lagging** (*Ans.*)

Load shared by 800 kVA transformer, $\bar{S}_2 = \bar{S} \times \frac{\bar{Z}_1}{\bar{Z}_1 + \bar{Z}_2}$

$$= 600 \angle -36 \cdot 87^\circ \times \frac{10 \cdot 44 \angle 73 \cdot 3^\circ}{14 \cdot 56 \angle 74^\circ} = 430.2 \angle -37.57^\circ$$

= **340.6 kW at pf 0.792 lagging** (*Ans.*)

3.12 Three Winding Transformers (Tertiary Winding)

Transformers usually have third winding in addition to the normal primary and secondary winding called **tertiary winding**, and the transformer in called triple wound or 3-winding transformer. The third winding is placed on the core because of any of the following reasons:

(*i*) In star-star connected transformers to suppress harmonic voltages, to allow sufficient earth fault current to flow for operation of protective devices and to limit voltage unbalancing when transformer is supplying an unsymmetrical load.

(*ii*) To prevent neutral potential, in star/star connected transformer, from oscillating to undesirable extent.

(*iii*) To supply an additional load at a voltage different from that of the primary and secondary.

(*iv*) To supply power to phase compensating devices, such as condensers,* operated at a voltage that is different from both primary and secondary voltage.

(*v*) To measure voltage of an HV testing transformer.

* Usually, in all extensive high voltage systems, regulation of voltage at certain points is provided by means of synchronous condensers. These condensers are connected to the high voltage system through transformers. From economic considerations, the condenser connections are made through a third winding placed in the transformers.

Usually, tertiary windings are delta-connected to provide path for zero sequence currents in the case of single line or double line to ground faults. This low reactance path reduces voltage unbalancing caused by these unbalanced ground faults.

The **tertiary winding** *is called as an* **auxiliary winding** *when it is employed for supplying an additional small load at a different voltage. On the other hand, it is also called as* **stabilising winding** *when it is employed to limit the short-circuit current.*

Operating Principle of 3-winding Transformer

The *operating principle* of a 3-winding (or triple wound) transformer is essentially the same as that of a two-winding transformer. The primary winding of a three winding transformer acts as the magnetising winding, and its current produces the main magnetic flux. The flux links the secondary and tertiary windings and induces emfs in them in proportion to their number of turns respectively. When loads are connected across the secondary and tertiary windings, currents I_2 and I_3 flow in them. The ampere-turns are balanced as given below:

$$I_1N_1 = (-I_2 N_2) + (-I_3N_3) + I_0 N_1 \quad \ldots(3.11)$$

or

$$I_1 = I_2' + I_3' + I_0 \qquad \left(\text{since } I_2' = -I_2\frac{N_2}{N_1} \text{ and } I_3' = -I_3\frac{N_2}{N_1}\right) \quad \ldots(3.12)$$

Thus in a 3-winding transformer, power is transferred to secondary and tertiary winding simultaneously.

Usually, both secondary and tertiary windings are not fully loaded at the same time, moreover the currents I_2' and I_3' will not be in phase at the same instant. Therefore, the primary winding is usually designed for a lower load than the sum of the rated powers of the secondary and tertiary windings.

Construction

In a *three-winding transformer*, there are three sets of windings (primary, secondary and tertiary) placed on the core. Three winding transformers may be either single three-phase units or three one-phase units connected in a three-phase units placed in the same tank or three one-phase separate units connected in a three-phase bank. In this transformer, three windings are operated at three different voltages termed as high voltage, medium voltage and low voltage windings.

The two possible winding arrangements starting from core outwards are : (*i*) *LV*, *MV* and *HV* and (*ii*) *MV*, *LV* and *HV* as shown in Fig. 3.17.

In two-winding transformers, the kVA ratings of both the primary and the secondary windings are the same and same is the rating of transformer but it is not so in case of 3-winding transformers. The kVA rating of the 3-winding transformer is considered to be equal to the largest kVA rating of any of its windings (i.e., primary winding).

The rating of tertiary winding depends upon its intended application. If it is provided for supply of an additional load, the winding is designed and calculated on the same basis as the primary and secondary. When it is employed only for balancing of loads and controlling short-circuit currents, it carries current only for short-duration, and its rating depends mainly on its heat-capacity. In practice, the x-section of the winding wire is generally determined by the fault conditions irrespective of the fact for what application it is going to be used because it carries maximum current during this condition.

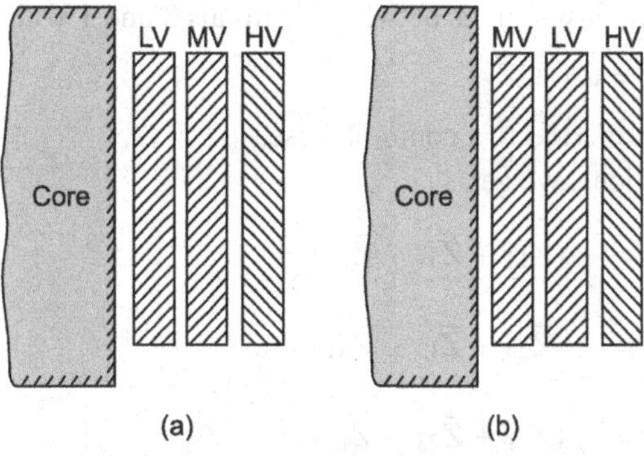

Fig. 3.17 Winding arrangement

Equivalent Circuit

The equivalent circuit can be drawn where each winding is represented by its own resistance and leakage reactance referred to a common voltage (base voltage). The equivalent circuit is shown in Fig. 3.18, neglecting circuit. The resistances and reactances can be converted to per unit values on the basis of an assumed common or base kVA.

Referring to circuit shown in Fig. 3.18, neglecting exciting current I_0, we have $\bar{I}_1 + \bar{I}_2 + \bar{I}_3 = 0$.

The impedances for such an equivalent circuit can be determined from the data of three short-circuit tests performed on the transformer.

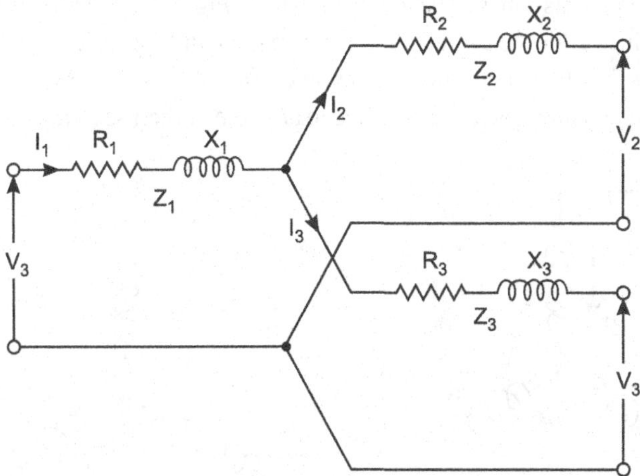

Fig. 3.18 Equivalent circuit of three-winding transformer

Short circuit impedance of windings between terminals 1 and 2 with winding 3 open (i.e. impedance of equivalent circuit when the terminals of circuit 1 and 2 are short-circuited and the terminals of circuit 3 are open).

$$\bar{Z}_{12} = \bar{Z}_1 + \bar{Z}_2$$

Short circuit impedance of windings between terminals 2 and 3 with winding 1 open

$$\bar{Z}_{23} = \bar{Z}_2 + \bar{Z}_3$$

and short circuit impedance of windings between terminals 3 and 1 with winding 2 open,

$$\bar{Z}_{31} = \bar{Z}_3 + \bar{Z}_1$$

All the impedances are referred to a common base.

Solving the above equations, we get,

$$\bar{Z}_1 = \frac{1}{2}(\bar{Z}_{12} + \bar{Z}_{31} - \bar{Z}_{23}) \quad \text{...(3.13)}$$

$$\bar{Z}_2 = \frac{1}{2}(\bar{Z}_{23} + \bar{Z}_{12} - \bar{Z}_{31}) \quad \text{...(3.14)}$$

and

$$\bar{Z}_3 = \frac{1}{2}(\bar{Z}_{31} + \bar{Z}_{23} - \bar{Z}_{12}) \quad \text{...(3.15)}$$

The core-loss, exciting or magnetising impedance and turn-ratio can be determined by performing open-circuit test on any of the three windings.

3.12.1 Stabilisation Provided by Tertiary Winding in Star-Star Transformer

Star-star transformers may comprise of single-phase units, or three-phase shell or five-limb core-type units. All these transformer units suffer from the following disadvantages:

(*i*) They cannot supply unbalanced loads between line and neutral, i.e., zero sequence loads. (*ii*) Their phase voltages may get distorted by third-harmonic emfs.

The reason is that mmf due to third-harmonic and zero sequence currents have an iron path.

The zero sequence impedance is so high between primary and secondary windings that it causes large voltage unbalancing. This can well be understood from Fig. 3.19(*a*). The single-phase load or fault current I_2 tends to induce I_1 in the corresponding primary winding (phase-*A*), but in the absence of a neutral the current I_1 has to return through the other two phases *B* and *C*. The reactance of these windings, which is magnetising reactance here due to absence of their secondary currents, is very high.

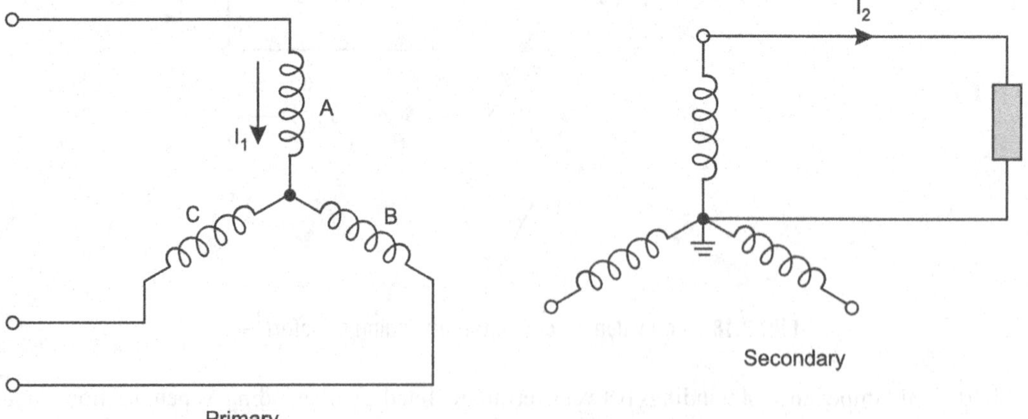

Fig. 3.19(a) Distribution of currents in windings due to single-phase load or ground fault

The voltages of these phases *B* and *C* rise and become nearly equal to the primary line voltage. On the other hand, the voltage of the faulty or loaded phase *A* reduces to a very low value. It develops a serious voltage unbalancing.

By the use of a delta connected tertiary winding, zero sequence currents are provided with low reactance paths, this prevents the voltage unbalancing in case of unbalanced loads or ground faults. Figure 3.19(*b*) shows how currents are distributed in the windings of star-star-delta connected transformer in case of single-phase loads or faults. It is assumed that each winding has equal turns per phase. It is seen that no current has to encounter magnetising impedance magnetically linked balancing current paths are available. Thus voltage unbalancing is prevented.

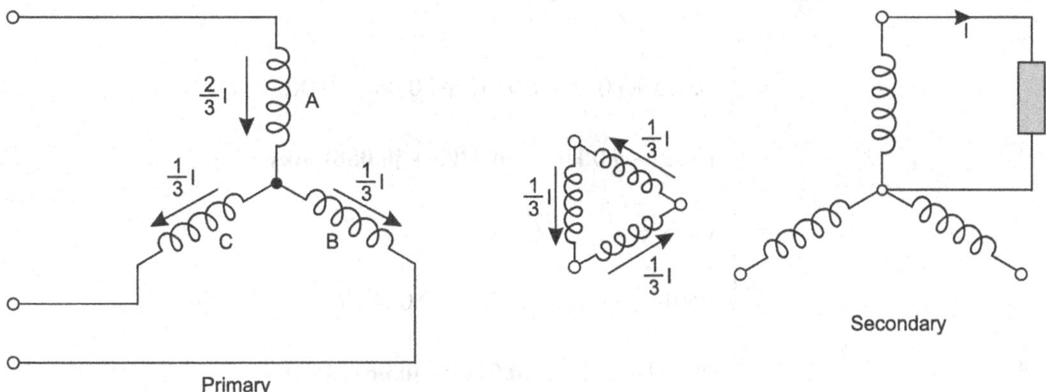

Fig. 3.19(b) Distribution of unbalanced current in winding when a connected tertiary is placed

The two disadvantages mentioned above are reduced to some extent in the three-limbs of core type transformers, because zero sequence flux is forced out of the core limb to high-reluctance air and tank leading to reduction in flux and zero sequence impedance thereby reducing voltage unbalance considerably. The third-harmonic flux also has similar high reluctance path and so magnitude of third harmonic flux is small. The distortion in voltage wave shape is small and transformer has more or less sinusoidal flux and emf when the input voltage is sinusoidal. It is nevertheless usual to provide a delta-connected tertiary even to the three-limb core type transformer if the primary and secondary are connected star-star. Further the third harmonic fluxes have been found to cause losses in tanks.

Example 3.14

The short circuit tests gave the following pu values of a 3-winding transformer.

$\overline{Z}_{12} = (0.012 + j0.064)$; $\overline{Z}_{23} = (0.025 + j0.064)$; $\overline{Z}_{31} = (0.016 + j0.12)$

The open circuit test on the primary gives the following values pu

$$\overline{Y}_0 = (0.02 - j\,0.05)$$

The secondary is supplying full rated power at 0.8 pf lagging and tertiary is kept open. Find the impedance pu of each winding.

Solution:

Here,
$$\overline{Z}_{12} = (0.012 + j\,0.064)$$
$$\overline{Z}_{23} = (0.025 + j\,0.064)$$
$$\overline{Z}_{31} = (0.016 + j\,0.12)$$

$$\overline{Z}_1 = \frac{1}{2}(\overline{Z}_{12} + \overline{Z}_{31} - \overline{Z}_{23})$$

$$= \frac{1}{2}[0.012 + j\,0.064 + 0.016 + j\,0.12 - 0.025 - j\,0.064\,]$$

$$= \frac{1}{2}\ (0.003 + j\,0.12) = \mathbf{(0.0015 + j0.06)}\ (Ans.)$$

$$\overline{Z}_2 = \frac{1}{2}(\overline{Z}_{23} + \overline{Z}_{12} - \overline{Z}_{31})$$

$$= \frac{1}{2}[0.025 + j\,0.064 + 0.012 + j\,0.064 - 0.016 - j\,0.12]$$

$$= \frac{1}{2}\ (0.021 + j\,0.116) = \mathbf{(0.0105 + j0.058)}\ (Ans.)$$

$$\overline{Z}_3 = \frac{1}{2}\ (\overline{Z}_{31} + \overline{Z}_{23} - \overline{Z}_{12}\ )$$

$$= \frac{1}{2}\ [0.016 + j\,0.12 + 0.025 + j\,0.064 - 0.012 - j\,0.064]$$

$$= \frac{1}{2}\ (0.029 + j0.12) = \mathbf{(0.0145 + j0.06)}\ (Ans.)$$

Example 3.15

An 11 kV/400 V, three-phase, delta/star transformer supplies a load of 100 kW at unity pf between line R and neutral. It is also supplying a balanced load of 500 kW at 0.8 pf lagging. Determine the current magnitude in each primary winding and each input line. State assumptions if any and draw the relevant diagram.

Solution:

Secondary phase voltage, $V_{ph2} = \dfrac{400}{\sqrt{3}} = 231\text{ V}$

Primary phase voltage, $V_{ph1} = V_{L1} = 11000\text{ V}$

Transformation ratio, $K = \dfrac{V_{ph2}}{V_{ph1}} = \dfrac{231}{11000}$

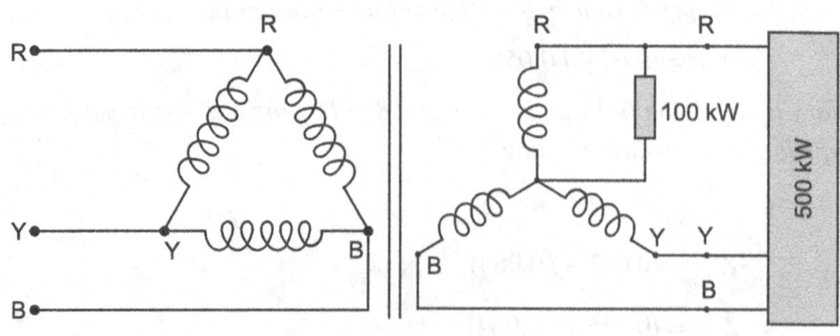

Fig. 3.20 Circuit diagram

Current in red phase of secondary due to single phase load of 100 kW at unity *pf.*

$$= \frac{100 \times 1000}{231} = 433 \text{ A} \qquad \text{(in phase with respective phase voltage)}$$

Current in all phases due to balanced load of 500 kW at 0.8 pf lagging

$$= \frac{500 \times 1000}{3 \times 231 \times 0 \cdot 8} = 902 \text{ A}$$

(lagging behind their respective phase voltages by $\phi = \cos^{-1} 0.8 = 36.87°$)

Assuming phase voltage V_{RY} on primary as reference phasor,

$$\bar{V}_{RY} = 11000\angle 0°\text{V} \; ; \; \bar{V}_{YB} = 11000\angle -120°\text{ V} \; ; \; \bar{V}_{BR} = 11000\angle 120°\text{ V}$$

Current in primary phase *RY* due to single phase load of 100 kW unity *pf*

$$I_{RY(1)} = 433 \times \frac{231}{11000} = 9 \cdot 1\text{ A in phase with voltage } V_{RY}$$

$$\bar{I}_{RY(1)} = 9.1\angle 0° \text{ A}$$

Current in primary phase *RY* due to three-phase balance load of 500 kW at 0.8 pf lagging.

$$I_{RY(2)} = 902 \times \frac{231}{11000} = 18 \cdot 95\text{ A lagging } V_{RY} \text{ by } 36.87°$$

$$\therefore \quad \bar{I}_{RY(2)} = 18.95 \angle -36.87° \text{A}$$

$$\text{Current in phase } RY,\ \bar{I}_{RY} = 9.1 \angle 0° + 18.95 \angle -36.87°$$

$$= 9.1 + (15.16 - j\,11.37) = (24.26 - j11.37)$$

$$= \mathbf{26.8 \angle -25.1° \text{ A}} \; (Ans.)$$

$$\text{Current in phase } YB,\ \bar{I}_{YB} = 18.95\angle -120° - 36.87° = 18.95 \angle -156.87° = (-17.43 - j7.44) \text{ A}$$

$$= \mathbf{18.95 \angle 23.11° \text{ A}} \; (Ans.)$$

$$\text{Current in phase } BR,\ \bar{I}_{BR} = 18.95\angle 120° - 36.87° = 18.95 \angle 83.13° = (2.67 + j18.81) \text{ A}$$

$$= \mathbf{19\angle 81.92°} \; (Ans.)$$

$$\text{Line current, } \bar{I}_R = \bar{I}_{RY} - \bar{I}_{BR} = (24 \cdot 26 - j11 \cdot 37) - (2 \cdot 67 + j18 \cdot 81)$$

$$= (21.59 - j30.18) = \mathbf{37.1 \angle -54.42° \text{A}} \; (Ans.)$$

$$\text{Line current, } \bar{I}_Y = \bar{I}_{YB} - \bar{I}_{RY} = (-17.43 - j7.44) - (24.26 - j11.37)$$

$$= (-41.69 - j18.81) = \mathbf{45.74 \angle 24.28° \text{ A}} \; (Ans.)$$

$$\text{Line current, } \bar{I}_B = \bar{I}_{BR} - \bar{I}_{YB} = (2.67 + j18.81) - (-17.43 - j7.44)$$

$$= (20.1 + j26.25) = \mathbf{33.1\angle 52.56°} \; (Ans.)$$

Example 3.16

A 3300/400/110 V star-star-delta transformer takes a magnetising current of 6A and a balanced load of 500 kVA at 0.8 pf lagging and 200 kVA at 0.6 pf leading on the tertiary. Determine the primary current and its p.f.

Solution:

Here, $I_{mag} = 6$ A; $\bar{I}_{mag} = 0 - j6 = 6\angle{-90°}$

Secondary current, $I_{ph2} = I_{L2} = \dfrac{kVA \times 1000}{\sqrt{3}V_{L2}} = \dfrac{500 \times 1000}{\sqrt{3} \times 400} = 720$ A

Transformation ratio, $K = \dfrac{V_{ph2}}{V_{ph1}} = \dfrac{400/\sqrt{3}}{3300/\sqrt{3}} = \dfrac{4}{33}$

Primary phase current, $I'_{ph1} = KI_{ph2} = \dfrac{4}{33} \times 720 = 87 \cdot 5\,\text{A}$ at 0.8 pf, lagging

$$\bar{I}'_{ph1} = \bar{I}'_{ph1}\angle - \cos^{-1} 0 \cdot 8 = 87 \cdot 5\angle - 36 \cdot 87°$$

$$= (70 - j52.5)\ \text{A}$$

Tertiary winding current (phase value) $= I_{Tph} = \dfrac{200 \times 1000}{3 \times 110} = 606$ A at 0.6 pf leading

Transformation ratio for tertiary, $K_T = \dfrac{110}{3300/\sqrt{3}} = \dfrac{1}{10\sqrt{3}}$

Primary phase current due to tertiary, $I'_{T1} = K_T \times I_{Tph} = \dfrac{1}{10\sqrt{3}} \times 606 = 35$A at 0.6 pf leading

$$\bar{I}'_{T1} = I'_{T1}\angle\cos^{-1} 0.6 = 35\angle 53.13° = (21 + j28)\ \text{A}$$

Total primary current, $I_1 = \bar{I}_{mag} + \bar{I}'_{ph1} + \bar{I}'_{T1}$

$$= 0 - j6 + 70 - j52.5 + 21 + j28 = 91 - j30.5$$

$$= \sqrt{(91)^2 + (30 \cdot 5)^2}\angle - \tan^{-1}\frac{30 \cdot 5}{91} = 96\angle - 16 \cdot 53°$$

= **96 A at pf 0.948 lagging** (*Ans.*)

Example 3.17

A star/star/delta connected transformer has secondary load of 40A at 0.8 pf lagging and tertiary has load of 30A at 0.71 pf lagging. The ratio of turns is 10 : 2 : 1 for primary, secondary and tertiary respectively. Calculate the primary current and pf.

Solution:

Transformation ratio, $K = \dfrac{N_2}{N_1} = \dfrac{2}{10} = \dfrac{1}{5}$

Transformation ratio for tertiary, $K_T = \dfrac{N_T}{N_1} = \dfrac{1}{10}$

Primary current due to secondary, $I'_1 = KI_2 = \frac{1}{5} \times 40 = 8A$

$$\bar{I}'_1 = I'_1 \angle \cos^{-1} 0.8 = 8\angle - 36 \cdot 87^\circ$$

$$= 8 (\cos 36.87^\circ - j \sin 36.87^\circ)$$

$$= (6.4 - j4.8) \text{ A}$$

Primary current due to tertiary, $\bar{I}'_{1T} = K_T I_{T_{ph}} = \frac{1}{10} \times \frac{30}{\sqrt{3}} = \sqrt{3}\text{A}$

$$\bar{I}'_{1T} = I'_{1T} \angle -\cos^{-1} 0.71$$

$$= 1.732 \angle -45^\circ$$

$$= 1.732 (\cos 45^\circ - j \sin 45^\circ)$$

$$= (1.23 - j1.23) \text{ A}$$

(Magnetising current is not given, hence neglected)

Total primary current, $\bar{I}_1 = y\bar{I}'_1 + \bar{I}'_{1T} = (6 \cdot 4 - j4 \cdot 8) + (1 \cdot 23 - j1 \cdot 23)$

$$= (7.63 - j6.03) \text{ A} = \sqrt{(7 \cdot 63 - j6 \cdot 03)} \text{ A}$$

$$= \sqrt{(7 \cdot 63)^2 + (6 \cdot 03)^2} \angle - \tan \frac{6 \cdot 03}{7 \cdot 63}$$

$$= 9.73 \angle 38.32^\circ$$

= **9.73 A at pf 0.7846 lagging** (*Ans.*)

3.13 Tap-changers on Transformers

All the electrical equipments connected at the consumers end are designed to operate satisfactorily at a particular voltage level. Therefore, it is essential to supply the electrical energy to the consumers at a level which must fall within the prescribed limits. However, in the power system, due to change in load (may be seasonal or otherwise), the transformer output voltage on the consumer's terminal may change beyond the permissible limits. This can be controlled by providing tap-changing transformers. Taps can be provided either on the primary or on the secondary.

Secondary output voltage can be regulated by either changing the number of turns in the primary or secondary on the basis of following principle.

Let V_1, N_1 and V_2, N_2 be primary and secondary quantities respectively. If N_1 is decreased, the emf per turn on primary $\left(= \frac{V_1}{N_1}\right)$ increases which increases secondary terminal voltage $\left(V_2 = \frac{V_1}{N_1} \times N_2\right)$. One the other hand, if N_2 is increased keeping N_1 constant, the secondary terminal voltage ($V_2 \propto N_2$) still increases.

Thus, secondary terminal voltage can be increased either by decreasing the primary turns or by increasing the secondary turns and vice-versa.

Choice between Primary and Secondary to provide Taps to Regulate Output Terminal Voltage of the Transformer

While selecting the side to provide taps to regulate secondary output voltage, we always try to maintain voltage per turn, as far as possible, constant. If the primary voltage per turn decreases, the core flux decreases which results in poor utilisation of core, although it reduces the core losses. On the other hand, if the primary voltage per turn increases, the core flux increases which results in magnetic saturation of the core. It also increases the core losses.

In the transformers located at the generating stations, the primary voltage has to be kept constant, consequently the taps are provided on the secondary side. However, if transformers are energised from a variable source, as at the receiving end of a transmission line (receiving sub-stations), the taps are usually provided on the primary side.

The other factors which may also be taken into consideration are given below:

(*i*) Transformers with large turns ratio, are tapped on the *HV* side since this enables a smoother control of the output voltage. If in such transformers, the taps are provided on the *LV* side, it varies the output voltage to a large step which is usually undesirable.

(*ii*) Tap-changing gear provided on the *HV* side are to handle low currents, although more insulation has to be provided.

(*iii*) It is difficult to tap the *LV* winding, since it is placed next to the core. Whereas, the *HV* winding, placed outside the *LV* winding, is easily accessible and can, therefore, be taped more easily.

3.14 Types of Tap-changers

There are two types of tap-changers, called

(*i*) No-load (or off-load) tap-changers and

(*ii*) On load tap-changers

The working of these tap-changers is explained below with the help of schematic diagrams.

3.14.1 No-load (or Off-load) Tap-changer

These tap-changers are used for seasonal voltage variations. The schematic diagram of a no-load tap-changer is shown in Fig. 3.21(*a*). A winding is tapped and its leads are connected to six studs marked 1 to 6. The studs are stationary and are arranged in a circle. The face plate carrying the studs can be mounted any where on the transformer, say on the yoke or on any other convenient place (say a separate box). The rotatable arm '*A*' is attached to a hand wheel which is kept outside the tank and can be rotated.

Sometimes, in case of large transformers, the rotatable arm is rotated with the help of a motor (with gear drive) and the controls of the motor are placed on the panel board although hand wheel is also provided for manual operation.

The active number of turns of the winding which remain in the circuit depends upon the position of rotating arm. If the winding is tapped at 2% interval, then 'at various positions of the rotatable arm '*A*', the winding in the circuit will be as under:

(*i*) at studs 1–2: Full winding is in the circuit.
(*ii*) at studs 2–3: 98% of the winding is in the circuit.
(*iii*) at studs 3–4: 96% of the winding is in the circuit.
(*iv*) at studs 4–5: 94% of the winding is in the circuit and
(*v*) at studs 5–6: 92% of the winding is in the circuit.

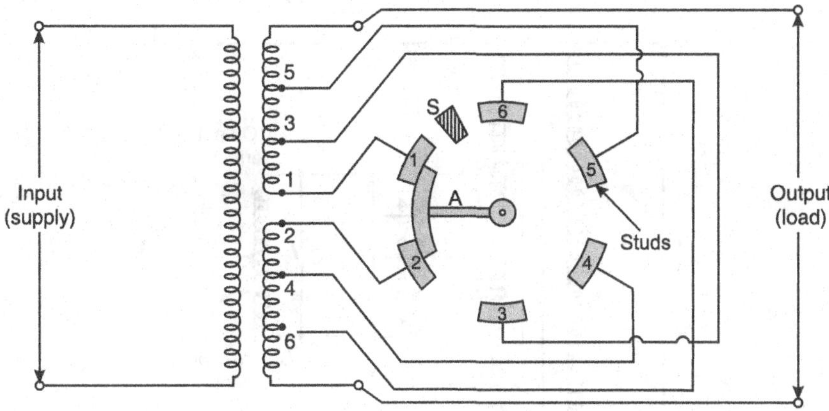

Fig. 3.21 (a) Off-load Tap-changer

A stopper *S* is placed in between the stud 1 and 6. It fixes the final position of the arm '*A*' at stud 5–6. Moreover, it prevents the clockwise rotation of the arm '*A*' from stud-1. It prevents the connections of stud 1 and 6 through arm '*A*'. If stud 1 and 6 are connected then only the lower part of the winding is cut out of circuit which is undesirable from mechanical-stress considerations.

In this case, the tap-changing is carried out only after the transformer is disconnected from the supply. For instance, let the arm '*A*' is at stub 1 and 2 and the whole winding is in the circuit. Now, if we want to reduce the winding to 96%, we have to rotate the arm in anticlockwise direction to bridge stud 3 and 4 through arm '*A*'. While doing so, the transformer is disconnected from the supply, arm is rotated to the desired position and then the transformer is energised.

This tap-changer is never operated on load. If it would be operated on load, there would be heavy sparking at the studs when arm '*A*' is separated from them. It may damage the tap-changer and the transformer winding.

3.14.2 On-load Tap-changer

This tap-changer is used for daily or short period voltage regulations. The output voltage can be regulated with the help of this tap-changer without any supply interruptions. During the operation of an on-load tap-changer the following points must be kept in mind;

(*i*) Never open the main circuit during the operation of tap-changer otherwise dangerous sparking will occur, and
(*ii*) No part of the tapped winding should get short circuited.

One form of an on-load tap-changer provided with a centre-tapped reactor is shown in Fig. 3.21(*b*). The function of the reactor is to prevent the short-circuiting of the tapped winding. The switches 1, 2, 3, 4 and 5 are connected to the correspondingly marked taps.

During normal operation switch *S* is closed [see Fig. 3.21(*b*)], switches 2, 3, 4 and 5 are opened and switch 1 is closed. The entire winding is in the circuit. The two halves of the reactor carry half of the total current in opposite directions. Since the whole reactor is wound in the same direction, the mmf produced by the two halves is opposite to each other. Since these mmfs are equal, therefore, the net mmf is practically zero. Hence, the reactor is almost non-inductive and the impedance offered by it is very small, consequently, the voltage drop in the centre-tapped reactor is negligible.

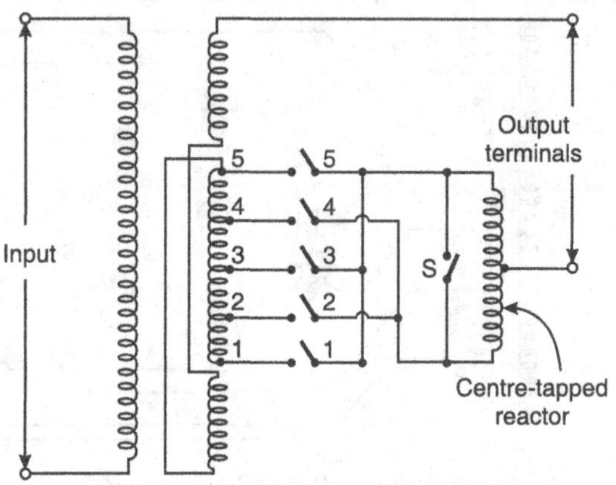

Fig. 3.21 (*b*) On-load tap changer

When a change in voltage is required, the following sequence of operations is adopted.

(*i*) *Open the switch S*: By opening this switch, total current flows through the upper half of the reactor and there is more voltage drop. Since reactor is to carry full load current momentarily, the reactor must be designed accordingly.

(*ii*) *Close the switch 2*: When switch 2 is closed, the winding between taps 1 and 2 is connected across the reactor. Since the impedance offered by the reactor is high for a current flowing in only one direction, the local circulating current flowing through the reactor and tapped winding is quite small.

Thus, reactor prevents the tapped winding from getting short circuited. The terminal voltage at this instant is mid-way between the potentials of tappings 1 and 2.

(*iii*) *Open the switch I*: The entire current now flows through the lower half of the reactor which causes more voltage drop.

(*iv*) *Close the switch S*: By closing the switch, current is divided equally in the upper and lower part of the reactor which causes almost negligible voltage drop.

The same sequence of operations is repeated if the tapping is to be changed from stud 2 to 3.

Section Practice Problems

Numerical Problems

1. Two transformers are connected in open delta and deliver a balanced three phase load of 250 kW at 440 volt and a power factor 0.8. Calculate.

(*a*) Secondary line current.
(*b*) kVA load on each transformer
(*c*) The power delivered by each transformer
(*d*) If the third transformer having the same rating as each of the other two is added to form a Δ-bank, what total load can be handled. (**Ans.** *410 A; 180.4 kVA; 179 kW, 71 kW; 541.2 kVA*)

2. A 500 kW load is to be supplied at 2300 V by two transformers connected in open delta. The p.f. of the load is 0.86. Calculate the current and kVA rating required by each of the transformers. What kVA load could be carried at some time in future if third transformer would be added to give delta connection. (**Ans.** *146 A; 865.5 kW; 1006.8 kVA*)

3. The short-circuit tests gave the following pu values of a 3-winding transformer $Z_{12} = (0.01 + j0.06)$; $Z_{23} = (0.02 + j0.06)$ and $Z_{31} = (0.016 + j0.1)$

The open-circuit test on the primary side gave the following values per unit $Y_0 = (0.02 - j0.05)$

The secondary is supplying full rating at 0.8 pf lagging and tertiary open.

Find the impedance of each winding. (**Ans.** $0.003 + j0.05$; $0.007 + j0.01$; $0.013 + j0.05$)

4. A 3300/400/110 V star-star-delta transformer take a magnetising current of 6A and a balanced three-phase load of 750 kVA at pf 0.8 lagging and 200 kVA at 0.6 *pf* leading on the tertiary. Determine the primary current and its *pf*. (**Ans.** 138.8 A at 0.913 pf lagging)

Short Answer Type Questions

Q.1. What is the necessity of parallel operation of three-phase transformers?

Ans. (*i*) It is desirable to place another transformer is parallel when the electrical load on the existing transformer increases beyond its rated capacity.

(*ii*) Parallel operation of transformers is necessary when the amount of power to be transformed is much more than that which can be handled by single unit (transformer).

(*iii*) It is desirable to do parallel operation of transformers if we want to keep the spare transformer of smaller size.

Q.2. What do you expect if star-delta transformer is connected in parallel with a star-star transformer?

Ans. The phasor displacement cannot be compensated and the transformers will come under direct short circuit condition.

Q.3. Why a tertiary winding is also called as auxiliary winding and stabilising winding?

Ans. The **tertiary winding** is called as an **auxiliary winding** when it is employed for supplying an additional small load at a different voltage. On the other hand, it is also called as **stabilising winding** when it is employed to limit the short-circuit current.

Q.4. What are on-load tap changing transformers?

Ans. **No-load (or off-load) top-changer.** These tap-changers are used for seasonal voltage variations. The schematic diagram of a no-load tap-changer is shown in Fig. Q.7). A winding is tapped and its leads are connected to six studs marked 1 to 6. The studs are stationary and are arranged in a circle. The face plate carrying the studs can be mounted anywhere on the transformer, say on the yoke or on any other convenient place (say a separate box). The rotatable arm '*A*' is attached to a hand wheel which is kept outside the tank and can be rotated.

By changing the position of arm '*A*', tapings can be changed or the voltage can be changed.

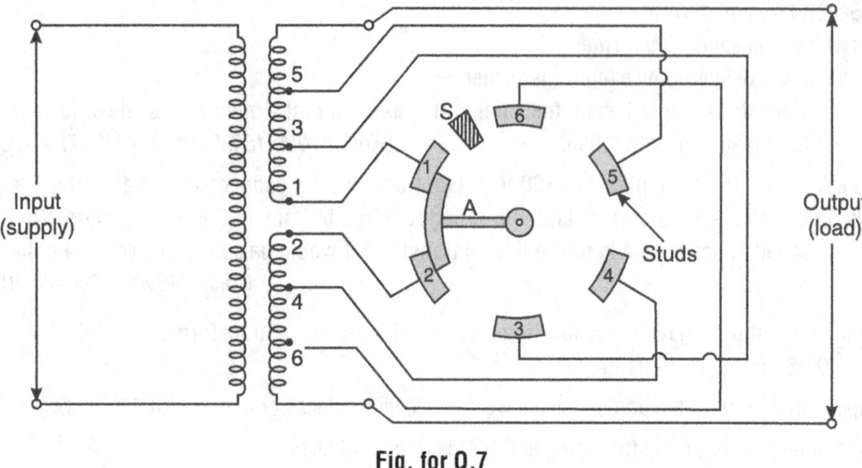

Fig. for Q.7

Q.5. Give limitations of on load tap changing transformer.

Ans. No-load tap changes is never operated on load. If it would be operated on load, there would be heavy sparking at the studs when arm '*A*' is separated from them. It may damage the tap-changer and the transformer winding.

3.15 Transformation of Three-phase Power with Two Single-phase Transformers

Three-phase power can be transformed by means of only two single phase transformers or by means of only two windings placed on primary and secondary of a three-phase transformer. This can be done by two methods, namely (*i*) Open-delta connection method and (*ii*) T-connection method. Both of these methods result in slightly unbalanced output voltage under load because of the unsymmetrical relations. However, this problem is not considered to be a serious problem in commercial transformers.

3.16 Open-Delta or V-V Connections

The delta-delta connections of three transformers for a three-phase system are shown in Fig. 3.22(*a*). Let us suppose that in one of the transformers, a fault develops due to which it is removed. If the primaries are connected to three-phase supply as shown in Fig. 3.22(*b*) then three equal three-phase voltages will be available at the secondary terminals at no-load. This method of transforming three-phase power by means of only two one-phase transformers is called **open delta** or **Vee-Vee** connections.

The basis of operation of open-delta connections is because of the fact that the vector sum of any two of the line voltages in a balanced three-phase system is equal to the third line voltage. Thus even though one of the transformer has been removed, the voltage between the terminals of the secondary to which load has been connected remains unchanged. The following are the points which favour the use of open-delta or vee-vee connections.

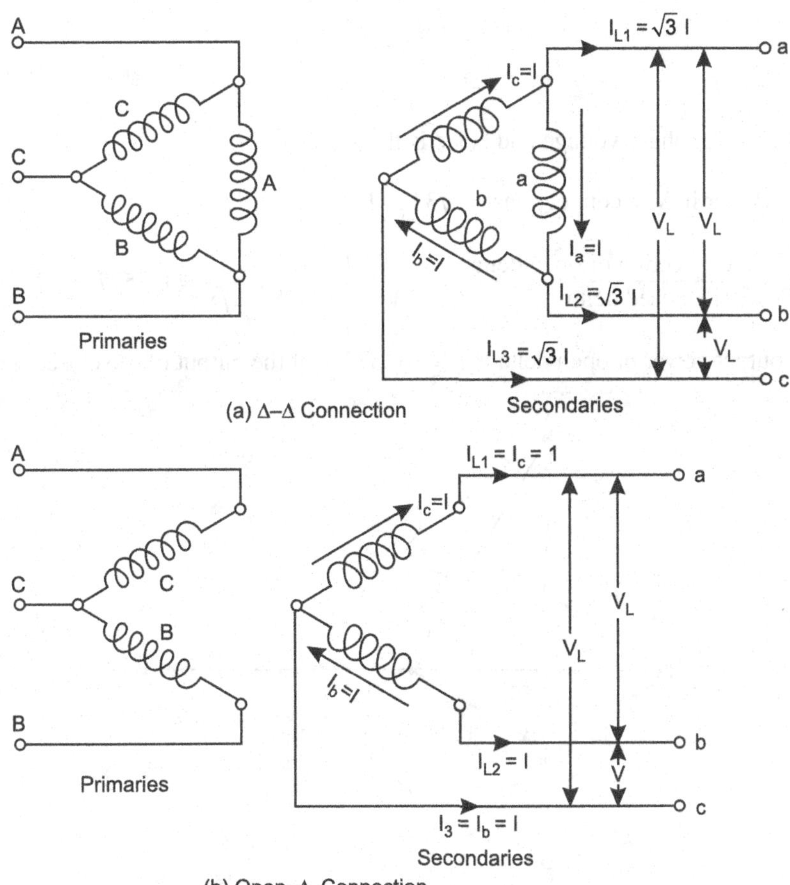

Fig. 3.22 Δ–Δ and open Δ connections

(*i*) When the three phase load is comparatively small, so that the installation does not warrant a three phase transformer bank.

(*ii*) When one of the transformers in a Δ-Δ bank fails, so that the service may be continued until the faulty transformer is repaired or good one is substituted.

(*iii*) In a new installation advantage is taken of the open-delta or V_{ee} connections by installing initially only two transformers of the capacity to meet the present maximum demand. When the load on the system increases to the expected full load, a third transformer is added to close the delta.

Hence open-delta or Vee-connections are used when it is anticipated that in future load will increase necessitating the closing of the open-delta at some later stage.

The phasor diagram of an open delta or $V_{cc} - V_{cc}$ connections is shown in Fig. 3.23.

It is important to note that the total load that can be handled by *V-V* connections is not two-third of the capacity of a delta-delta bank but is only 57.7% of it. Mathematically, it is proved as below:

$$\text{Power in delta arrangement} = \sqrt{3}V_L\ I_L\ \cos\phi$$

$$= \sqrt{3}\,V_L \sqrt{3}\,I\ \cos\phi$$

$$\therefore \quad I_L = \sqrt{3}I$$

$$= 3V_L\, I \cos\phi$$

where V_L is the line or phase voltage and $\cos\phi$ is the p.f.

$$\text{Power in V-V connections} = \sqrt{3}\, V_L \; I \; \cos\phi$$

$$\therefore \quad \frac{\text{Power in V-V connection or open delta}}{\text{Power in closed delta}} = \frac{\sqrt{3}\, V_L \; I \; \cos\phi}{3V_L \; I \; \cos\phi} = \frac{1}{\sqrt{3}} = 0 \cdot 577 = 57 \cdot 7\% \quad \ldots(3.16)$$

Therefore, output power of open delta is $\frac{1}{\sqrt{3}}$ or 57.7% of the output of the closed delta.

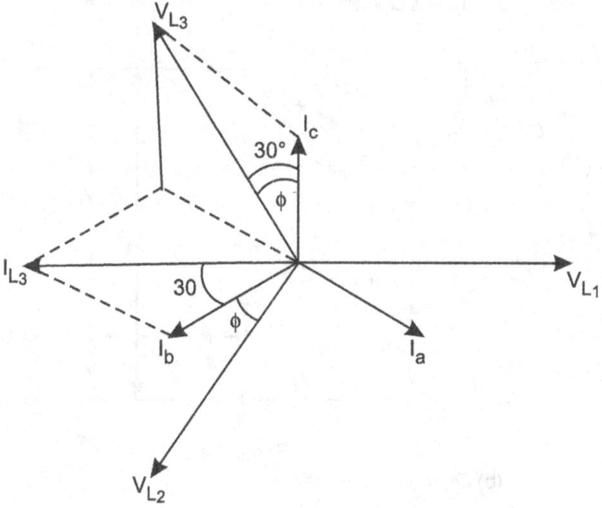

Fig. 3.23 Phasor diagram for open delta connection

Note: It is assumed that the power factor in both the systems is same but in actual practice the power factor in the *V-V* connections is lesser than the actual power factor of the load.

The two transformers constitute 66.6 per cent of the installed capacity of the three, but they are able to deliver only 57.7% of the three in open delta connections. For example, if the three transformers, each having rating of 33.33 kVA are connected to form closed delta bank, the rating of the bank would be 100 kVA. But if one of these transformers is removed, the rating of the resulting open Δ bank is not 66.67 kVA, but it will be only 57.7 kVA. The total load that can be carried by an open delta transformer bank is not two-thirds of the capacity of a Δ–Δ bank, but it is only 57.7 percent of it. This is a reduction of about 14% $\left(\frac{66.67 - 57 \cdot 7}{66 \cdot 67} \times 100\right)$ from its normal rating or in other words to obtain two-thirds of the output of the original closed Δ, each transformer in the open delta must have a rating of $\frac{66 \cdot 67}{57 \cdot 7}$ or 115% of the original rating. The reduction in rating is caused because of the average power factor at which the transformers operate which is less than that of actual power factor of the load. This average power factor is the ratio of $\frac{57 \cdot 8}{66 \cdot 67}$ (or 86.6%) and is always equal to 86.6% of the balanced load power factor. In other words, we can say that this is the power factor at which the two transformers operate when the load is of unity p.f.

If the transformer of *B*-phase is removed from the bank on each primary and secondary as shown in Fig. 3.24(*a*) the same line voltages V_{AB}, V_{BC}, V_{CA}, V_{ab}, V_{bc} and V_{ca} will appear in the systems as $V_{bc} = -(V_{ca} + V_{ab})$. Obviously the same secondary phase voltages V_{an} etc., will appear and there is no difficulty in supplying three-phase load. The balanced three-phase currents I_a, I_b, I_c are supplied to the load at a power factor of cos ϕ lag [Fig. 3.24(*b*)]. The currents in the secondary windings are $I_{ba} = -I_b$, $I_{ac} = I_c$. On the primary side the line currents and winding currents are related as $I_B = -I_{AB}$, $I_C = I_{CA}$ and $I_A = -(I_B + I_C)$. It may be seen that one of the windings operates at a power factor of cos (30° + θ) and the other at cos (30° – θ).

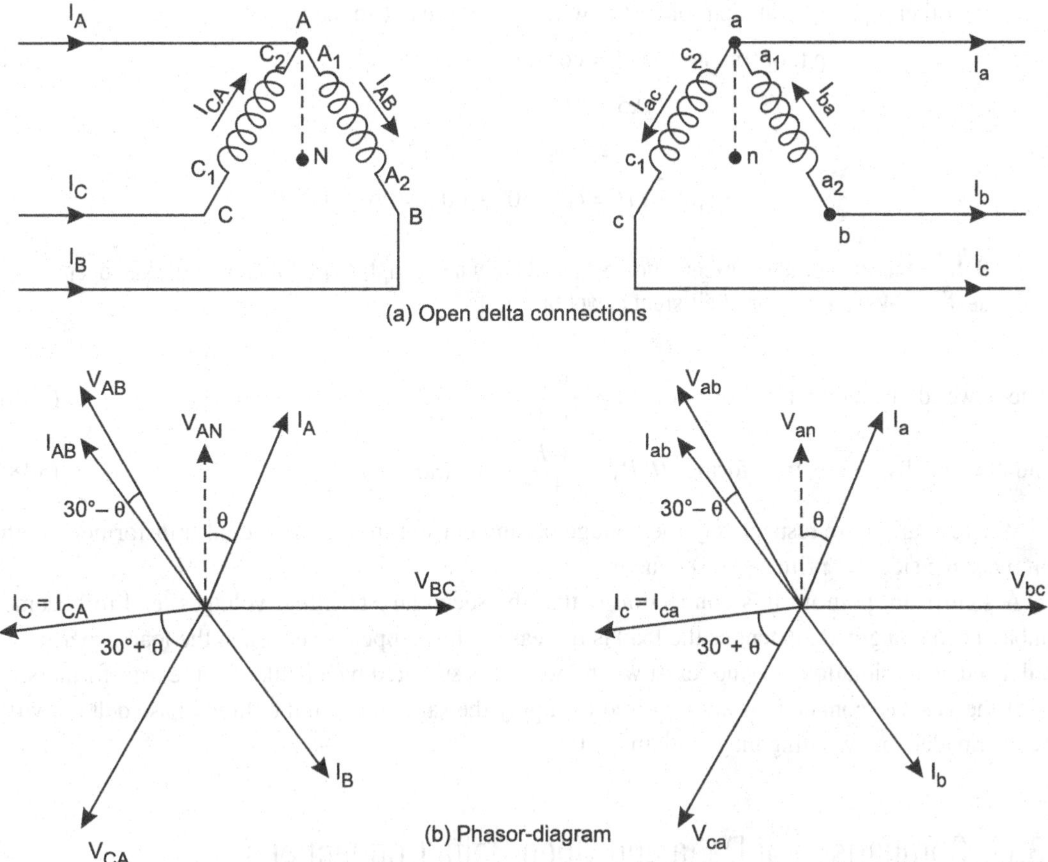

Fig. 3.24 Open-delta connections and phasor diagram

The performance of two transformer connected in open delta will change as per the load p.f. as mentioned below:

(*i*) When the load p.f. is unity, the transformers operate at the same p.f.

$$\text{P.f. of Transformer } I = \cos(30° - 0) = \cos 30° = 0.866$$

$$\text{P.f. of Transformer } II = \cos(30° + 0) = \cos 30° = 0.866$$

$$\because \quad \cos\phi = 1$$

$$\therefore \quad \phi = 0$$

(*ii*) When the load power factor is 0.866, one transformer will operate at unity p.f. while the p.f. of the other will be 0.5.

$$\text{p.f. of transformer } I = \cos(30° - 30°) \text{ where } \phi = \cos^{-1} 0.866 = 30°$$

$$= \cos 0° = \text{unity}$$

$$\text{p.f. of transformer } II = \cos(30° + 30°) = \cos 60° = 0.5$$

(*iii*) When the load p.f. is 0.5, one transformer will operate at zero p.f. and delivers, **no power,** while the other, operating at a p.f. of 0.866, will take up the entire load.

$$\text{p.f. of transformer } I = \cos(30° - 60)$$

$$\cos^{-1} 0.5 = 60°$$

$$= \cos 30° = 0.866$$

$$\text{p.f. of } II = \cos(30° + 60°) = \cos 90° = 0$$

Note: Hence it shows that except for a balanced unity p.f. load, the two transformers in an open delta or *V-V*-bank operate at different power factors.

The power delivered by transformer *I*, $P_1 = \dfrac{VI}{1000} \times \cos(30 - \phi)$ ...(3.17)

and power delivered by transformer *II*, $P_2 = \dfrac{VI}{1000} \times \cos(30 + \phi)$ ...(3.18)

Where *V* and *I* are respectively the voltage of, and current in, the individual transformers, both on the same side i.e., primary or secondary.

A significant point in this connection is that the secondary terminal voltages tend to become unbalanced to a greater extent as the load is increased, this happens even when the load is perfectly balanced. This situation does not exist when the load is supplied by a bank of three transformers.

If the *Vee-Vee* connections are required to supply the same load on the three phase delta, it will be overloaded and winding may be damaged.

3.17 Comparison of Delta and Open Delta Connections

(*i*) When similar transformers are used, the voltages given by the delta and Vee connections are the same and their outputs are proportional to their line currents.

(*ii*) In balanced delta connection the line current is $\sqrt{3}$ of the phase current whereas in Vee-connection the line current is the same as the phase current.

(*iii*) With non-inductive balanced load, each transformer of the delta connection carries one-third of the total load at unity power factor. Under the same conditions each transformer of Vee-connection carries one half of the load at a p.f. 0.866.

(*iv*) The ratio of the power in Vee-Vee to Power in delta connection is 0.577.

3.18 T-T Connections or Scott Connections

The other method of transformation of three-phase power from one circuit to another by using two transformers is *T-T* connections These connections were proposed by C.F. Scott therefore, these are frequently called as *Scott connections.* This scheme of connections is also used for three-phase to two-phase, and vice-versa transformation.

In this arrangement, two transformers are used. One of them must have at least two primary and two secondary coils so that a centre tap may be brought out from each side (or the transformer should have centre-tap primary and centre-tap secondary). This transformer is called *main transformer.* The other transformer must have primary and secondary windings having number of turns 0.866 times of the respective turns on the main transformer. This transformer is called *teaser transformer.* The current ratings of the two transformers should be the same. The connections are shown in Fig. 3.25(*a*). It may be noted that one end of both the primary and secondary of the teaser transformer are connected to the centre taps on both primary and secondary of the main transformers respectively whereas, the two ends of main transformer (*A* and *B*) and 86.6% tapping point (*C*) on teaser transformer are connected to the three-phase supply on the primary side.

The two ends (*a* and *b*) of the secondary of main transformer and 86.6% tapping point (*c*) of secondary of teaser transformer are taken to connect the load, as shown in Fig. 3.25(*a*).

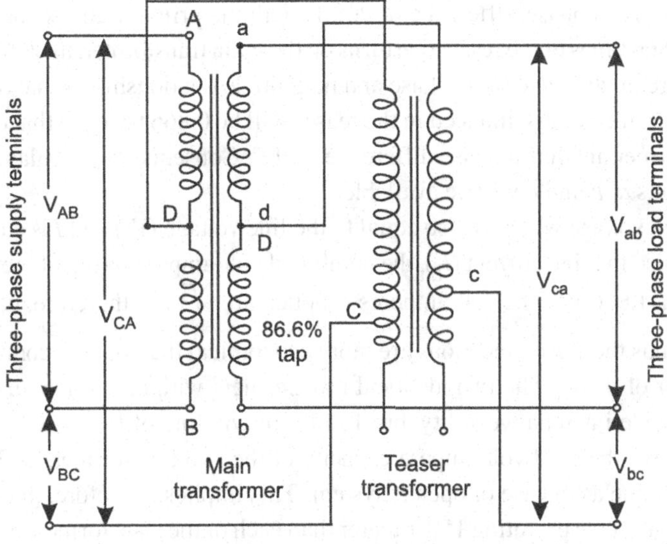

Fig. 3.25(a) Scott connections

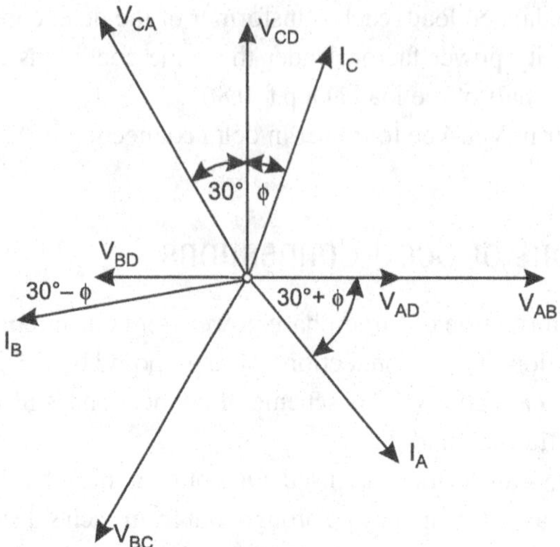

Fig. 3.25(b) Phasor diagram of T-T connected transformer

Since primary and secondary of teaser transformer are connected to the centre tap of the primary and secondary of the main transformer respectively giving a shape of English letter '*T*', as shown in Fig. 3.26 (*a* and *b*) the connections are known as *T-T* connections.

Let V_{AB}, V_{BC} and V_{CA} be the applied voltages across the line terminals, then

$$V_{CD} = V_{CA} + V_{AD} = V \angle 120° + \frac{1}{2} \angle 0°$$

$$= V(-0.5 + j\,0.866) + 0.5\text{ V} = j\,0.866\text{ V} = 0.8666\text{ V} \angle 90° \qquad \ldots(3.19)$$

The above expression shows that voltage applied across the teaser is 0.866 times of that applied across the main and has a phase difference of 90°. If both the primary and secondary of the teaser transformer have 0.866 times of the respective turns on the main transformer, then the induced voltages in the secondary circuit will have same phase and magnitude relationship as that of applied voltages on the primary. Thus, the voltage induced in the teaser will be 0.866 times to that of induced voltage in the main transformer and has a phase difference of 90°. Consequently, a balanced 3-ϕ system of voltages across points *a*, *b* and *c* will be available.

If *V* is the applied voltage which is also equal to the line voltage (V_L) and *I* is current rating of the winding, which is also the line current (I_L), the combined volt-ampere rating of the two transformers is *VI* + 0.866 *VI* i.e., 1.866 *VI*. The volt-amperes supplied are $\sqrt{3}\,VI$, therefore, utility factor is $\sqrt{3}$ /1.866 i.e., 0.928. Thus the *T-T* connections are more economical than open Δ connections which are having utility factor of 0.866. The two identical transformers with centre-tap and 86.6 tap may be used. This will provide interchangeability, but 13.14% of winding of teaser will remain idle. With identical transformers, the total volt-ampere capacity of the two transformers is 2*VI* and the utility factor is 0.866, the same as in case of open Δ system. This implies, therefore, that each transformer in a T-T connection must have a rating 15% greater than each of the transformers employed in a three transformer bank.

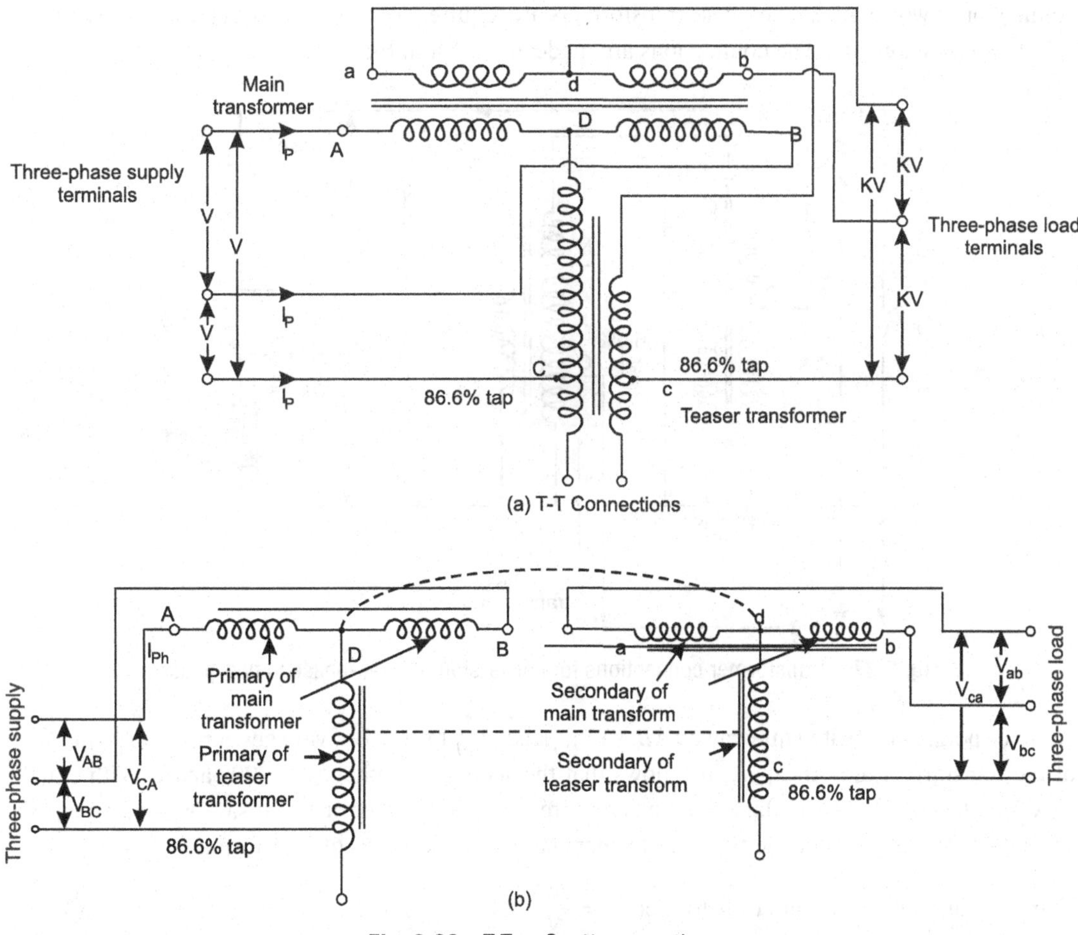

Fig. 3.26 T-T or Scott connections

For any balanced load of power factor cos ϕ (lag), one of the two halves of the main transformer operates at a pf of cos $(30^\circ + \phi)$ and the other at cos $(30^\circ - \phi)$. This is similar to the condition as in open Δ. Mainly because of this (i.e., different loading effects) the voltages on the secondary side tend to become unbalanced to a greater extent with the increase in load.

It may be noted that this arrangement provides the three-phase, four-wire system. Three phase power loads may be connected between lines *a*, *b* and *c* while lighting load may be connected between *ad* and *bd*. There is a further advantage of availability of neutral on the teaser transformer. This permits a true three-phase four-wire system with the use of two transformers, which we could not get in open Δ system.

3.19 Conversion of Three-phase to Two-phase and vice-versa

In some cases such as for electric furnaces, it is desirable to work with two-phase currents. From the power supply system, three-phase AC supply is available, therefore, it is necessary to convert three-phase supply to two-phase supply. This can be achieved by using Scott-connections. In these connections, two identical one-phase transformers are required, one of them must have centre tapped primary and secondary. The connections are made as shown in Fig. 3.27.

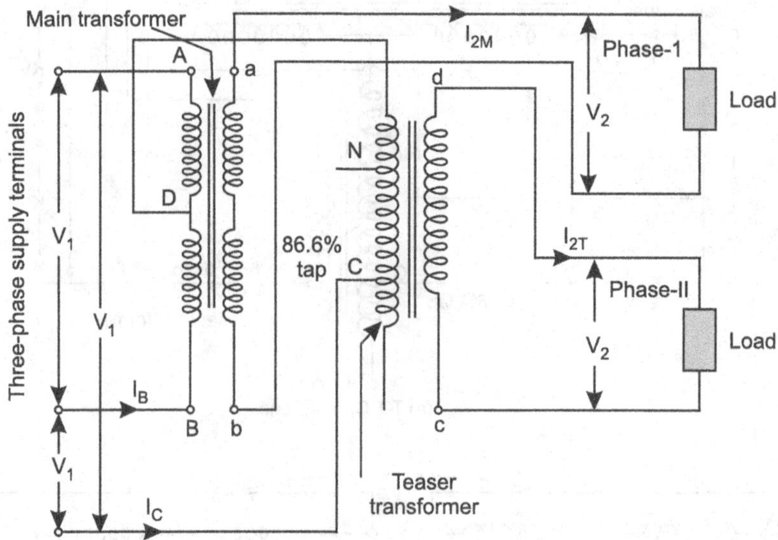

Fig. 3.27 Transformer connections for conversion of three-phase to two-phase

Since point D is located midway on AB, so V_{CD} leads V_{AB} by 90° i.e., voltages across the primaries of the transformer are 90° apart. It follows that the secondary voltages are 90° apart. With equal fluxes in the core, the secondary windings require an equal number of turns (say, N_2) to give equal secondary voltages. Hence the two transformers have unequal ratios of transformation.

$$\text{Transformation ratio of main transformer, } K = \frac{N_2}{N_1} \quad \ldots(3.20)$$

$$\text{Transformation ratio of teaser transformer} = \frac{N_2}{0.866\ N_1} = 1 \cdot 15\text{K} \quad \ldots(3.21)$$

It is to be noted that point D is not the neutral point of the primary system, as its voltage with respect to line terminals A, B and C are not equal to $\frac{V_1}{\sqrt{3}}$. Let N be such a point that the voltages from point N to points A, B and C are equal. Point N, therefore, is the neutral of the primary system.

The neutral point on the three phase side can be located on the teaser transformer. The neutral must have a voltage of $\frac{V}{\sqrt{3}}$ i.e., 0.577 to C and since the voltage C to D is 0.866 V, the neutral point N will be 0.866 V – 5.77 V from D or 0.289 V i.e., a number of turns below D equivalent to 28.9 per cent of the primary turns in main transformer. Since 0.289 is one third of 0.866, the neutral point is one third the way down the teaser transformer winding from D to C or point N divides the teaser winding in the ratio of 1: 2.

For determination of primary currents let us neglect the magnetising current and consider load component currents I_{2M} and I_{2T} only.

Considering ampere-turn balance of teaser transformer, $I_C \times 0.866\ N_1 = I_{2T}\ N_2$

or $$I_C = I_{2T} \times \frac{N_2}{0.866\,N_1} = 1 \cdot 15\ KI_{2T} \quad \text{...(3.22)}$$

For the main transformer $$\frac{I_A N_1}{2} - \frac{I_B N_1}{2} = I_{2M} N_2$$

$$\text{or } I_A - I_B = 2I_{2M}\frac{N_2}{N_1} = 2KI_{2M} \quad \text{...(3.23)}$$

With a balanced two phase load, I_{2T} and I_{2M} are equal in magnitude but 90° apart, hence

$$I_{2T} = j\,I_{2M}$$

$$\text{or } I_{2M} = \frac{I_{2T}}{j} = -j\,I_{2T} = \frac{-j\,I_C}{1.15\,K} \quad \text{...(3.24)}$$

Substituting the value of I_{2M} in expression (*iv*), we get,

$$I_A - I_B = 2K \times \left(\frac{-jI_C}{1.15K}\right) = \frac{-2jI_C}{1.15}$$

or $$I_A - I_B + j\sqrt{3}I_C = 0 \quad \text{...(3.25)}$$

As phasor sum of three line currents is zero

∴ $$\bar{I}_A + \bar{I}_B + \bar{I}_C = 0 \quad \text{...(3.26)}$$

Solving equations (*vi*) and (*vii*), we get

$$I_A = I_B(-0.5 + j\,0.866) = I_B \angle 120° \quad \text{...(3.27)}$$

$$\text{and } I_C = I_B(-0.5 - j\,0.866) = I_B \angle -120° \quad \text{...(3.28)}$$

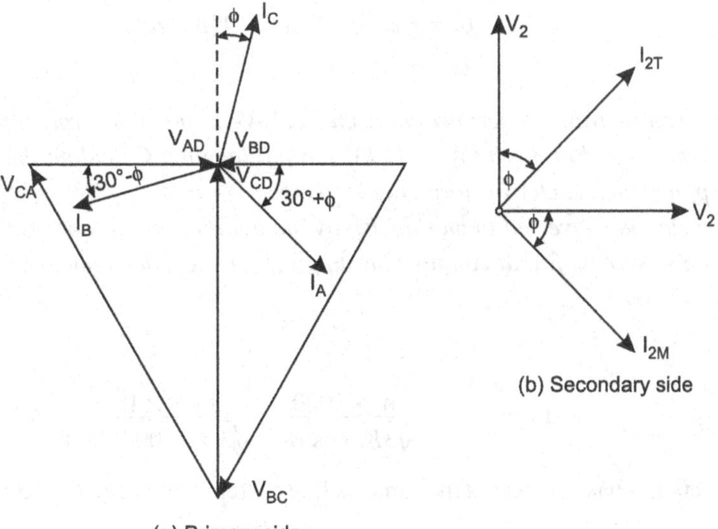

Fig. 3.28 Phasor diagram

Thus I_A, I_B and I_C are equal in magnitude but are 120° apart from each other and so they give a balanced system.

Phasor diagram for voltages and currents, neglecting the transformer impedance and magnetising current, is shown in Fig. 3.28. The load currents are lagging behind their respective secondary voltages by an angle ϕ. One half of the main transformer primary winding AB, has an angle of lag of $(30° - \phi)$ while the other has angle of lag of $(30° + \phi)$ where teaser primary has an angle of lag of ϕ. This is the reason why the volt-ampere capacity of the primary winding of main transformer must be greater than that of its secondary.

Example 3.18

A balanced three-phase load of 1000 kW and 0.8 p.f. lagging is supplied by Vee-connected transformers. Calculate the line and phase currents and the power factor at which each transformer is working. The working voltage is 3.3 kV.

Solution:

$$\text{Line current, } I_L = \frac{kW \times 1000}{\sqrt{3} \times 3.3 \times 1000 \times 0.8} = 218.8 \text{ A}$$

Phase current of the transformer is equal to the line current for Vee connections

$$I_1 = I_2 = I_{ph} = 218.8 \text{ A}$$

$$\text{Load pf, } \cos\phi = 0.8;\ \phi = \cos^{-1} 0.8 = 36.87°$$

The current I_1 will make an angle of $\phi_1 = 30 + \phi = 30 + 36°.87° = 66°.87°$ with phase voltage of respective transformer

$\therefore$ $\cos\phi_1 = \cos 66.87° = \mathbf{0.393\ lagging}$ (*Ans.*)

Current I_2 will make an angle of $\phi_2 = 30° - \phi = 30° - 36.87°$

$= -6.87°$ with phase voltage of respective transformer.

$\therefore$ $\cos\phi_2 = \cos(-6.87°) = \mathbf{0.993}$ (*Ans.*)

Example 3.19

Two identical single phase transformers are connected in V-V-connection across three-phase mains and deliver a balanced load of 3048 kW at 11 kV, 0.8 p.f. lagging. Calculate the line and phase currents and the pf at which each transformer is working.

The V-V-connection is converted to mesh/mesh by the addition of an identical unit. Calculate the additional load of same p.f. that can now be supplied for the same temp. rise. Also calculate the phase and line currents.

Solution:

$$\text{Line current, } I_L = \frac{kW \times 1000}{\sqrt{3} E_L \cos\phi} = \frac{3048 \times 1000}{\sqrt{3} \times 11000 \times 0.8} = 200 \text{ A}$$

In Vee-Vee connection, phase current is the same as line current and is equal to 200 A in this case.

Load p.f. $\cos\phi = 0.8$

$\therefore$ $\phi = \cos^{-1} 0.8 = 36.87°$

$$\text{p.f. of transformer-I} = \cos(30° - \phi) = \cos(30° - 36.87°)$$

$$= \cos 6.87° = \mathbf{0.993\ lagging}\ (Ans.)$$

$$\text{p.f. of transformer-II} = \cos(30° + 36.87°)$$

$$= \cos 66.87° = \mathbf{0.393\ lagging}\ (Ans.)$$

Power supplied by the delta for same temperature rise i.e., for same current is

$$= 3\, E_{ph}\, I_{ph} \cos\phi$$

$$= \frac{3 \times 11 \times 1000 \times 200 \times 0.8}{1000} = 5280 \text{ kW}$$

Note: In delta connection line voltage is the same as phase voltage.

$\therefore$ Extra power supplied by delta = 5280 – 3048 = **2232 kW** (*Ans.*)

$$I_{ph} = \mathbf{200\ A}\ (Ans.)$$

$\therefore$ line current = $200 \times \sqrt{3}$ = **346 A** (*Ans.*)

Example 3.20

Two-transformers are connected in open delta and supply a balanced three-phase load of 300 kW at 400 volt and a p.f. of 0.866, determine:

(*a*) *the secondary line current.*

(*b*) *the kVA load on each transformer*

(*c*) *the power delivered by the individual transformers.*

If a third transformer having the same rating i.e., 200 kVA is added to form a delta bank, what will be the total load that can be supplied.

Solution:

Load supplied = 300 kW; Secondary line voltage, V_{L2} = 400 V

Load pf, $\cos\phi$ = 0.866

(*a*) Secondary line current $I_{L2} = \dfrac{kW \times 1000}{\sqrt{3} \times V_{L2} \times \cos\phi} = \dfrac{300 \times 1000}{\sqrt{3} \times 400 \times 0.866}$ = **500 A** (*Ans.*)

(*b*) kVA load on each transformer = $\dfrac{V_{L2} \times I_{L2}}{1000} = \dfrac{400 \times 500}{1000}$ = **200** (*Ans.*)

(*c*) For a power factor $\cos\phi = 0.866$; $\phi = \cos^{-1} 0.866 = 30°$

Power delivered by one transformer, P_1 = kVA cos (30° – ϕ) = 200 cos 0° = **200 kW** (*Ans.*)

Power delivered by the other transformer,

$$P_2 = \text{kVA} \cos(30° + \phi)$$

$$= 200 \cos 60° = \mathbf{100\ kW}\ (Ans.)$$

The output of three transformers, each of 200 kVA rating, when connected in Δ bank

$$= 3 \times 200 = \mathbf{600\ kVA}\ (Ans.)$$

Example 3.21

A three-phase 150 kW balanced load at 1000V, 0.866 p.f. lagging is supplied from 2000 volt, three-phase main through (i) three single-phase transformers connected in (i) delta-delta and then (ii) in Vee-Vee. Find the current in the windings of each transformer and the p.f. at which they operate in each case.

Solution:

Delta-Delta Connection

$$\text{Secondary line current, } I_{L2} = \frac{kW \times 1000}{\sqrt{3}\, V_L \cos\phi} = \frac{150 \times 1000}{\sqrt{3} \times 1000 \times 0.866} = 100 \text{ A}$$

$$\text{Secondary phase current, } I_{ph2} = \frac{100}{\sqrt{3}} = \mathbf{57.7\ A}\ (\textit{Ans.})$$

Turns ratio,
$$K = \frac{N_2}{N_1} = \frac{E_{ph2}}{E_{ph1}} = \frac{1000}{2000} = \frac{1}{2}$$

$$\therefore \quad \text{Primary phase current, } I_{ph1} = 57.5 \times \frac{1}{2} = \mathbf{28.85\ A}\ (\textit{Ans.})$$

p.f at which each transformer is operating is **0.866 lagging** (*Ans.*)

Vee-Vee Connection

Let I_L' be the secondary line current which is also the phase current in open delta. Then

$$I_L' = \frac{kW \times 1000}{\sqrt{3}\, V_L \cos\phi} = \frac{150 \times 1000}{\sqrt{3} \times 1000 \times 0.866} = 100 \text{ A}$$

$$\text{Secondary phase current, } I_{ph}' = 100 \text{ A}$$

$$\text{Primary phase current} = 100 \times \frac{1}{2} = 50\text{A}$$

$$\text{Load p.f. } \cos\phi = 0.866;\ \phi = \cos^{-1} 0.866 = 30°$$

$$\text{p.f. of Transformer } I = \cos(30 - \phi) = \cos(30 - 30°) = \cos 0 = 1 \text{ (unity)}$$

$$\text{p.f. of Transformer } II = \cos(30 + \phi) = \cos(30° + 30°) = \cos 60° = 0.5 \text{ lagging}$$

Hence when the load p.f. is 0.866 the pf of transformer-I is Unity and II is 0.5 as already explained.

Example 3.22

Three 1100/110 V transformers are connected ΔΔ and supply a lighting load of 120 kW. If one of these transformers is damaged and hence removed for repairs, what currents will be flowing in each transformer when (i) the three transformers were in service (ii) the two transformers are in service.

Solution:

(*i*) When three transformers were in service

$$\text{Load } P = 120 \text{ kW}$$

$$\text{Power factor, } \cos\phi = 1.0 \text{ (for lighting load)}$$

$$\text{Primary line voltage, } V_{L1} = 1100 \text{ volt}$$

$$\text{Secondary line voltage, } V_{L2} = 110 \text{ volt}$$

$$\text{Secondary line current, } I_{L2} = \frac{kW \times 1000}{\sqrt{3} \times V_{L2} \times \cos\phi} = \frac{120 \times 1000}{\sqrt{3} \times 110 \times 1} = 630 \text{ A}$$

$$\text{Primary line current, } I_{L1} = \frac{120 \times 1000}{\sqrt{3} \times 1100 \times 1} = 63 \text{ A}$$

For delta connections,

$$\text{Secondary phase current, } I_{ph2} = \frac{630}{\sqrt{3}} = \mathbf{363.7\ A}\ (Ans.)$$

$$\text{Primary phase current, } I_{ph1} = \frac{63}{\sqrt{3}} = \mathbf{36.37\ A}\ (Ans.)$$

(*ii*) When only two transformers are in service

$$\text{Secondary line current, } I'_{L2} = \frac{kW \times 1000}{\sqrt{3} \times V_{L2} \times \cos\phi} = \frac{120 \times 1000}{\sqrt{3} \times 110 \times 1} = 630 \text{ A}$$

$$\text{Secondary phase current, } I'_{ph2} = \text{Secondary line current} = \mathbf{630\ A}\ (Ans.)$$

$$\text{Primary phase current, } I'_{ph1} = \text{Secondary phase current} \times \text{transformation ratio}$$

$$= 630 \times \frac{110}{1100} = \mathbf{63\ A}\ (Ans.)$$

Example 3.23

Three 1100/110 volt transformers are, connected in delta/delta. This bank of transformers is supplying a lighting load of 100 kW. One of these transformers is damaged and removed for repairs.

(*i*) *What currents were flowing in each transformer when the three transformers were in service.*

(*ii*) *What current flows in each of the two transformers after third has been removed.*

(*iii*) *What is the kVA output of each transformer if the two transformers, connected in open-delta, carry full load with normal heating.*

Solution:

(*i*) All the three transformers in service with **Delta/Delta Connection**

Power factor for lighting load is always unity

$$\therefore \quad \text{Secondary line current, } I_{L2} = \frac{kW \times 1000}{\sqrt{3} V_{L2} \cos\phi} = \frac{100 \times 1000}{\sqrt{3} \times 110 \times 1} = 525 \text{ A}$$

$$\text{Secondary phase current, } I_{ph2} = \frac{I_L}{\sqrt{3}} = \frac{525}{\sqrt{3}} = \mathbf{303\ A}\ (Ans.)$$

$$\text{Turns ratio} = \frac{V_{ph2}}{V_{ph1}} = \frac{110}{1100} = \frac{1}{10}\ (\Delta/\Delta \text{ connected})$$

$$\therefore \quad \text{Primary phase current, } I_{ph1} = I_{ph2} \times \frac{1}{10} = 303 \times \frac{1}{10} = \mathbf{30.3\ A}\ (Ans.)$$

(*ii*) When one transformer is removed the connections are changed to Vee-Vee-Connections.

$\therefore$ Secondary line current, I_{L2} = **525 A** = I'_{ph2} (*Ans.*)

Primary phase current, $I'_{ph1} = 525 \times \frac{1}{10}$ = **52.5 A** (*Ans.*)

(*iii*) Power output of Delta/Delta connection is 100 kW

$\therefore$ Power output of open-delta = 0.577 × 100 = **57.7 kW** (*Ans.*)

When the load p.f. is unity, the both the transformers in Vee-Vee connection operate at the same p.f. 0.866.

p.f. of transformer-I = cos (30 – 0) = 0.866 $(\because \phi = \cos^{-1} = 0°)$

p.f. of transformer-II = cos (30 + 0) = 0.866

kW load taken by each transformer = $\frac{57.7}{2}$

$\therefore$ kVA rating of each transformer = $\frac{57.7}{2} \times \frac{1}{\text{p.f.}} = \frac{57.7}{2} \times \frac{1}{0.866}$ = **33.33 kVA** (*Ans.*)

Example 3.24

The primary and secondary windings of two transformers each rated 220 kVA, 11/2 kV and 50 Hz are connected in open delta. Find (i) the kVA load that can be supplied from this connection; (ii) currents on HV side if a delta connected three phase load of 250 kVA, 0.8 pf (lag) 2 kV is connected to the LV side of the connections.

Solution:

(*i*) The kVA load that can be supplied by two transformers, each having rating of 220 kVA

= 2 × kVA rating of each transformer × 0.866

= 2 × 220 × 0.866 = **381 kVA** (*Ans.*)

Secondary line current, $I_{L2} = \frac{kVA \times 1000}{\sqrt{3} \times V_{L2}} = \frac{220 \times 1000}{\sqrt{3} \times 2000} = 63.5$ A

Secondary phase current, I_{ph2} = Secondary line current = 63.5 A

Primary phase current (current on *HV* side)

= Secondary phase current × transformation ratio

$= 63.5 \times \frac{2}{11}$ = **11.54 A** (*Ans.*)

Example 3.25

Two single phase furnaces A and B are supplied at 100 V by means of Scott-connected transformers from a three-phase, 6 kV system. Furnace A is supplied from the teaser transformer. Calculate the line currents on the three phase side when (i) each furnace takes 600 kW at p.f. 0.8 (lag). (ii) furnace A takes 500 kW at unity power factor and furnace B 600 kW at 0.8 pf (lag).

Solution:

$$\text{Secondary voltage, } V_2 = 100 \text{ V}$$

$$\text{Primary main voltage, } V_1 = 6 \text{ kV} = 6000 \text{ V}$$

$$\text{Transformation ratio, } K = \frac{V_2}{V_1} = \frac{100}{6000} = \frac{1}{60}$$

(*i*) Secondary currents, $I_{2T} = I_{2M} = \dfrac{600 \times 1000}{100 \times 0.8} = 7500 \text{ A}$

Teaser primary current, $I_{IT} = 1.15\, KI_{2\text{M}} = 1.15 \times \dfrac{1}{60} \times 7500 = 144 \text{ A}$

Considering Fig. 3.29.

$$\text{Current in line } C, I_C = 144 \text{ A}$$

Current in primary of main transformer,

$$I_1 = KI_{2M} = \frac{1}{60} \times 7500 = 125 \text{ A}$$

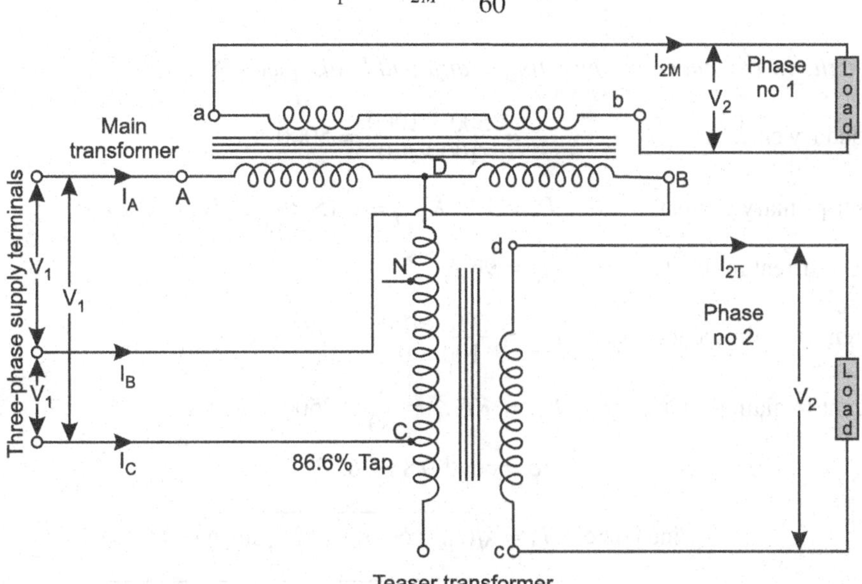

Fig. 3.29 Circuit diagram

The current in lines A and B of the primary of main transformer will be the phasor sum of current $KI_{2\text{M}}$ and one half of the teaser primary current I_C. As the power factor is the same on both secondaries, the line currents are in quadrature. [See Fig. 3.30(*a*)]

$$\therefore \quad \text{Line currents, } I_A = I_B = \sqrt{I_1^2 + \left(\frac{1}{2} I_C\right)^2} = \sqrt{(125)^2 + \left(\frac{144}{2}\right)^2}$$

$$= \mathbf{144.25 \text{ A}} \; (\textit{Ans.})$$

$$\text{Phase angle between phasor } I_C \text{ and } I_A = 90° + \tan^{-1}\frac{72}{125} = 120°$$

Similarly, phase difference exists between phasors I_C and I_B and phasors I_A and I_B.

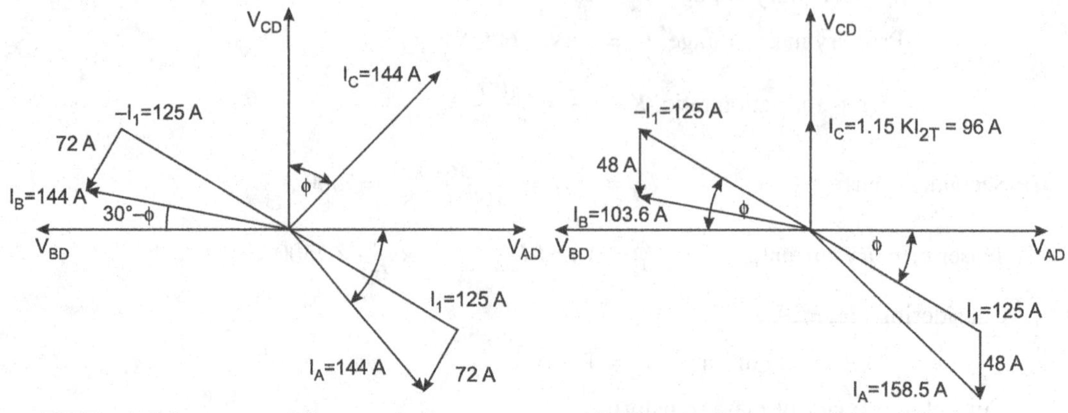

Fig. 3.30 Phasor diagram

(ii) *When the two furnaces are operating at different loads. [see Fig. 3.30(b)]*

Secondary current, $$I_{2T} = \frac{500 \times 1000}{100 \times 1.0} = 5000 \text{ A}$$

Teaser primary current, $$I_{1T} = 1.15\,K\,I_{2T} = 1.15 \times \frac{1}{60} \times 5000 = 96 \text{ A}$$

Hence current in line C, $$I_C = 96 \text{ A}$$

Current in main secondary, $$I_{2M} = \frac{600 \times 1000}{100 \times 0.8} = 7500 \text{ A}$$

Current in main primary, $$I_{1M} = KI_{2M} = \frac{1}{60} \times 7500 = 125 \text{ A}$$

and $$\phi = \cos^{-1} 0.8 = 36.87°$$

$$\text{Line current, } I_A = \sqrt{(I_{1M} \cos \phi)^2 + (I_{1M} \sin \phi + 0.5 I_{1T})^2}$$

$$= \sqrt{(125 \times 0.8)^2 + (125 \times 0.6 + 0.5 \times 96)^2}$$

= **158.5 A** (*Ans.*)

$$\text{Line current, } I_B = \sqrt{(I_{1M} \cos \phi)^2 + (I_{1M} \sin \phi - 0.5\, I_{1T})^2}$$

$$= \sqrt{(125 \times 0.8)^2 + (125 \times 0.6 - 0.5 \times 96)^2}$$

= **103.6 A** (*Ans.*)

Example 3.26

Two one-phase furnaces A and B are supplied at 80 V by means of a Scott-connected transformer combination from a three-phase 6600 V system. Calculate the line currents on the three-phase side when the furnace A takes 500 kW at unity p.f. and B takes 800 kW at 0.7 p.f. lagging. Draw the corresponding vector diagrams.

Solution:

$$\text{Secondary voltage, } V_2 = 80 \text{ V}$$

$$\text{Primary main voltage, } V_1 = 6600 \text{ V}$$

$$\text{Transformation ratio, } K = \frac{V_2}{V_1} = \frac{80}{6600}$$

Let furnace A be supplied by the teaser

$$\text{Teaser secondary current, } I_{2T} = \frac{kW \times 1000}{V_2 \times \cos \phi} = \frac{500 \times 1000}{80 \times 1} = 6250 \text{ A}$$

$$\text{Teaser primary current, } I_{1T} = 1.5 \times K \times I_{2T} = 1.5 \times \frac{80}{6600} \times 6250 = 87.6 \text{ A}$$

Current in the secondary of main transformer,

$$I_{2M} = \frac{800 \times 1000}{80 \times 0.7} = 14286 \text{ A}$$

Current in the primary main transformer,

$$I_{2M} = I_{2M} \times K = 14286 \times \frac{80}{6600} = 173 \text{ A}$$

From the phasor diagram shown in Fig. 3.31, I_A is given as,

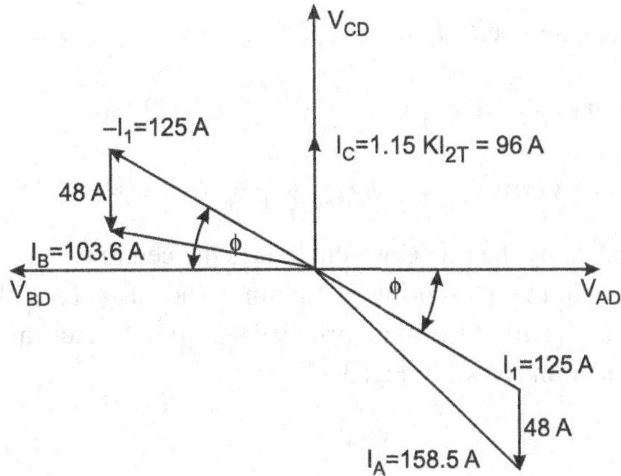

Fig. 3.31 Phasor diagram

$$I_A = \sqrt{(I_{1M} \cos \phi)^2 + (I_{1M} \sin \phi + 0.5\ I_{1T})^2}$$

$$= \sqrt{(173 \times 0.7)^2 + (173 \times 0.714 + 43.8)^2}$$

$$= \mathbf{207\ A}\ (Ans.)$$

$$\text{Line current, } I_B = \sqrt{(I_{1M}\cos\phi)^2 + (I_{1M}\sin\phi - 0.5I_{1T})^2}$$

$$= \sqrt{(173\times 0.7)^2 + (173\times 0.714 - 43.8)^2}$$

$$= \mathbf{145\ A}\ (Ans.)$$

Example 3.27

Two single-phase electric furnaces are supplied power at 80 volt from a three-phase, 11 kV system by means of two single-phase Scott-connected transformers, with similar secondary windings. When the load on one transformer is 500 kW and on the other is 800 kW, what current will flow in each of the three-phase lines at unity p.f.?

Solution:

Let load of 500 kW be connected across the teaser secondary winding and 800 kW across main secondary winding.

Secondary voltage, $V_2 = 80$ V; Primary main voltage, $V_1 = 11000$ V

$$\text{Transformation ratio } K = \frac{V_2}{V_1} = \frac{80}{11000} = \frac{8}{1100}$$

$$\text{Teaser secondary current, } I_{2T} = \frac{\text{Load in kW} \times 1000}{\text{Voltage}} = \frac{500\times 1000}{80} = 6250\text{ A}$$

$$\text{Teaser primary current, } I_{1T} = \frac{1}{0.866}\frac{N_2}{N_1}\times I_{2T} = 1.15\,KI_{2T}$$

$$= 1.15\times\frac{8}{1100}\times 6250 = \mathbf{52.5\ A.}$$

$$\text{Current in the line } C,\ I_C = I_{1T} = 52.5\text{ A}$$

$$\text{Main secondary current, } I_{2M} = \frac{800\times 1000}{80} = 10000\text{ A}$$

$$\text{Main primary current, } I_{1M} = KI_{2M} = \frac{8}{1100}\times 10000 = 72.8\text{ A}$$

The current I_{1T} is supplied by the line C which divides the centre tap and its components 0.5 I_{1T} flow in opposite directions in the main primary winding. The current I_{1M} flows in the primary of the main transformer. In each half of the main primary winding the current I_{1M} is 90° out of phase with 0.5 I_{1T} (see phasor diagram shown in Fig. 3.32).

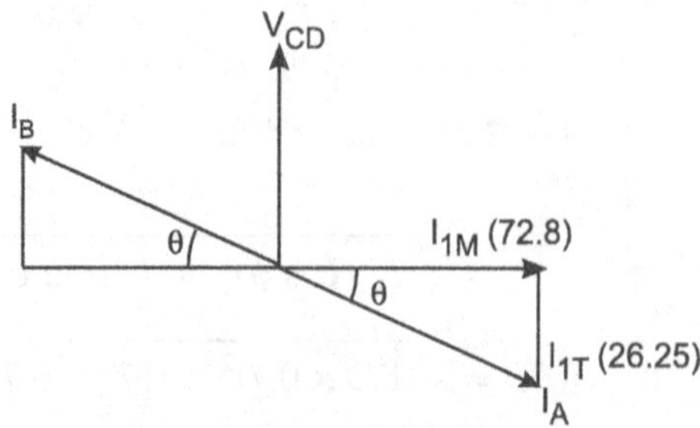

Fig. 3.32 Phasor diagram

Now current in the line A and B

$$\therefore \quad I_A = I_B = \sqrt{(I_{1M})^2 + (0.5\, I_{1T})^2}$$

$$= \sqrt{(72.8)^2 + (0.5 \times 52.5)^2} = \mathbf{77.3\ A.}\ (Ans.)$$

$$\tan\theta = \frac{26.25}{72.8} = 0.360 \text{ or } \theta = \tan^{-1} 0.360 = \mathbf{19.8°}\ (Ans.)$$

Example 3.28

It is desired to transform 2000 V, 400 kVA three phase power to two-phase power at 500 V by Scott-connected transformers. Determine the voltage and current ratings of both primary and secondary of each transformer. Neglect the transformer no-load currents.

Solution:

$$\text{Primary voltage, } V_1 = 2000 \text{ V}$$

$$\text{Secondary voltage, } V_2 = 500 \text{ V}$$

$$\text{Transformation ratio, } K = \frac{V_2}{V_1} = \frac{500}{2000} = \frac{1}{4}$$

For Teaser Transformer

$$\text{Voltage rating of secondary, } V_{2T} = V_2 = 500 \text{ V}$$

$$\text{Voltage rating of primary, } V_{1T} = V_1 = 0.866\, V_1 = 0.866 \times 2000 = \mathbf{1732\ V}\ (Ans.)$$

$$\text{Current rating of secondary} = I_{2T} = \frac{\text{Load on teaser transformer in kVA} \times 1000}{V_2}$$

$$= \frac{400 \times 1000}{2 \times 500} = \mathbf{400\ A}\ (Ans.)$$

$$(\because \text{ Load on teaser} = \frac{1}{2} \times \text{total load})$$

$$\text{Current rating of primary} = I_{1T} = 1.15\, K\, I_{2T} = 1.15 \times \frac{1}{4} \times 400 = \mathbf{115\ A}\ (Ans.)$$

For Main Transformer

$$\text{Voltage rating of secondary} = V_2 = 500 \text{ V } (Ans.)$$

$$\text{Voltage rating of primary} = V_1 = 2000 \text{ V } (Ans.)$$

$$\text{Current rating of secondary} = I_{2M} = I_{2T} = 400 \text{ A}$$

$\because$ load is same on both transformers

$$\text{Main primary current, } I_{1M} = K\, I_{2M} = \frac{1}{4} \times 400 = 100 \text{ A}$$

$$\text{Current rating of primary, } I_A \text{ or } I_B = \sqrt{I_{1M}^2 + (0.5\, I_{1T})^2}$$

$$\because \quad \text{pf is same on both secondaries} = \sqrt{(100)^2 + (0.5 \times 115)^2} = \mathbf{115.4\ A}\ (Ans.)$$

Example 3.29

A balanced three-phase, 100 kW load at 400 V and 0.8 pf lagging is to be obtained from a balanced two phase 1100 V lines. Determine the kVA rating of each unit of the Scott-connected transformer.

Solution:

Secondary line voltage, $V_2 = 400$ V

Primary voltage, $V_1 = 1100$ V

Voltage rating of the secondary of main transformer = Supply line voltage = 400 V

Voltage rating of the secondary of teaser transformer = 0.866 V_2 = 0.866 × 400 = 346.4 V

Voltage rating of primary of each transformer = V_1 = 1100 V

Current rating of primary of each transformers,

$$I_1 = \frac{\text{Load in kW} \times 1000}{2 \times V_1 \times pf}$$

$$= \frac{100 \times 1000}{2 \times 1100 \times 0.8} = 56.82 \text{ A}$$

Current rating of secondary of each transformer,

$$I_2 = \frac{\text{Load in kW} \times 1000}{\sqrt{3}\, V_2 \times \cos\phi}$$

$$= \frac{100 \times 1000}{\sqrt{3} \times 400 \times 0.8} = 180.42 \text{ A}$$

Volt-ampere rating of primary and secondary of teaser transformer is the same

i.e., $1100 \times 56.82 \times 10^{-3}$ = **62.5 kVA** (*Ans.*)

Volt-ampere rating of secondary main transformer = $\frac{400 \times 180.42}{1000}$ = **72.17 kVA** (*Ans.*)

Volt-ampere rating of primary of main transformer = $\frac{1100 \times 56.82}{1000}$ = **62.5 kVA** (*Ans.*)

Since, two identical transformers are used to provide inter-changeability, both must have the same rating i.e., **72.17 kVA** (*Ans.*)

Example 3.30

Two single phase Scott-connected transformers supply a three-phase four-wire 50 Hz distribution system with 400 V between lines. The hv windings are connected to a two-phase 6600 V (per phase) system. The core area is 200 cm², while the maximum allowable flux density is 1.2 T. Determine the number of turns on each winding and the point to be tapped for the neutral wire on the three-phase side.

Solution:

Maximum flux density, $B_m = 1.2$ T

Maximum allowable flux, $\phi_m = B_m \times$ core area

$= 1.2 \times 0.02 = 0.024$ Wb

Voltage on primary (*HV*) side, $V_1 = 6600$ V

Number of turns on *HV* side of both the transformers,

$$N_1 = \frac{V_1}{4.44\ f\phi_m} = \frac{6600}{4.44 \times 50 \times 0.024} = \mathbf{1239}\ (Ans.)$$

Number of turns of *lv* side of main transformer,

$$N_2 = \frac{V_2}{4.44 f\phi_m} = \frac{400}{4.44 \times 50 \times 0.024} = \mathbf{75}\ (Ans.)$$

Number of turns on *LV* side of teaser transformer = $0.866\ N_2 = 0.866 \times 75 =$ **65** (*Ans.*)

$$\text{Number of turns between } CN = \frac{2}{3} \times 65 = \mathbf{43}\ (Ans.)$$

3.20 Difference between Power and Distribution Transformers

In power system, the voltage level is raised at the generating station and it is decreased at the various sub-stations for economic reasons. The transformers used in the power system are called either power transformers or distribution transformers depending upon the duty performed by them. They basically differ from design point of view, as explained below:

(*i*) **Power Transformers:** These transformers are used to step up the voltage at the generating station for transmission purposes and then to step down the voltage at the receiving stations. These transformers are of large sizes (generally above 500 kVA). These transformers usually operate at high average load, which would cause continuous heavy copper loss, thus affecting their efficiency. To have minimum losses during 24 hours, such transformers are designed with *low copper losses.*

(*ii*) **Distribution Transformers:** These transformers are installed at the distribution sub-stations to step down the voltage. These transformers are continuously energised causing the iron losses for all the 24 hours, when generally the load fluctuates from no-load to full-load during this period. To obtain high efficiency, such transformers are designed with *low iron losses.*

3.21 Cooling of Transformers

When a transformer is connected to the mains, heat is produced in the transformer due to various losses. This increases its temperature. In fact, output of the transformer is limited by the rise in temperature. Therefore, some means are provided to cool down the transformer so that the rise in temperature may not go beyond the permissible limits.

According to specifications, the rise in temperature of a transformer when working at the rated output shall not exceed 45°C to 60°C.

Most of the electrical machines have rotating part, therefore, it is easy to cool down such machines by providing fan on their shaft and ducts in their construction. But in case of transformers there is no rotating part, therefore, it is difficult to cool down the inner parts of a transformer. However, depending upon the size of a transformer, various methods have been evolved to dissipate heat produced in the transformer. These are mentioned below:

1. **Natural Cooling**
   (*a*) Air Natural cooling (*AN*)
   (*b*) Oil immersed Natural cooling (*ON*)
   (*c*) Oil immersed Forced oil circulation Natural cooling (*OFN*)
2. **Artificial Cooling (Air)**
   (*a*) Oil immersed Forced oil circulation with air Blast cooling (*OFB*)
   (*b*) Oil immersed air Blast cooling (*OB*)
   (*c*) Air Blast cooling (*AB*)
3. **Artificial Cooling (Water)**
   (*a*) Oil immersed Water cooling (*OW*)
   (*b*) Oil immersed Forced oil circulation with Water cooling (*OFW*)
4. **Mixed Cooling**
   (*a*) Oil immersed Natural cooling with alternative additional air Blast cooling (*ON/OB*)
   (*b*) Oil-immersed Natural cooling with alternative additional Forced oil circulation (*ON/OFN*)
   (*c*) Oil-immersed Natural cooling with alternative additional Forced oil circulation air Blast cooling (*ON/OFB*)
   (*d*) Oil-immersed Natural cooling with alternative additional forced oil circulation and Water cooling (*ON/OFW*)

3.22 Methods of Transformer Cooling

Some of the common methods of transformer cooling are explained below:

1. **Air Natural Cooling (*AN*)**

   Small transformers of a few kVA rating are cooled by the Natural air surrounding the core and winding of the transformer. The heat produced in the transformer due to losses dissipated by conduction, convection and radiation.

   The transformers cooled by this method are used in laboratories and small appliances like radios, rectifiers etc.

2. **Oil immersed Natural cooling (*ON*)**

   In this case, the assembly of core and winding (assembled unit) of a transformer is placed in a steel tank filled with pure insulating oil as shown in Fig. 3.33. The cooling is affected by natural circulation (convection currents) of the oil through the cooling ducts provided between *coils* and between *coils and core.*

   For small transformers, the tanks are usually smooth surfaced but for large sizes the tanks are made of corrugated sheets as shown in Fig. 3.34. Sometimes fins are attached to the other surface as shown in Fig. 3.35 to improve cooling.

   In most of the pole mounted transformers upto the rating of 200 kVA number of round or elliptical tubes are provided with the tank externally as shown in Fig. 3.36. The transformer oil has large coefficient of expansion, therefore, convection currents develop which cause the oil to circulate through the external tubes as shown in Fig. 3.36. The cooling can further be improved by using a radiator in place of external cooling tubes.

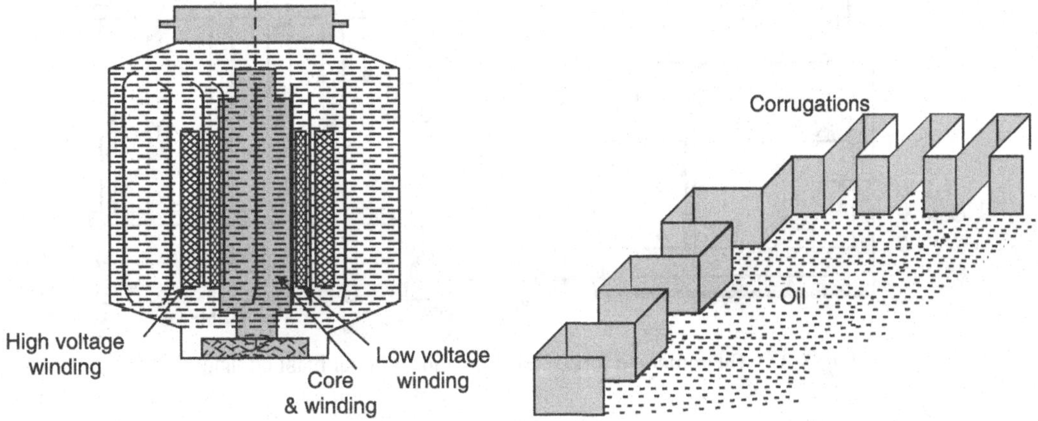

Fig. 3.33 Oil natural cooling

Fig. 3.34 Corrugated tank

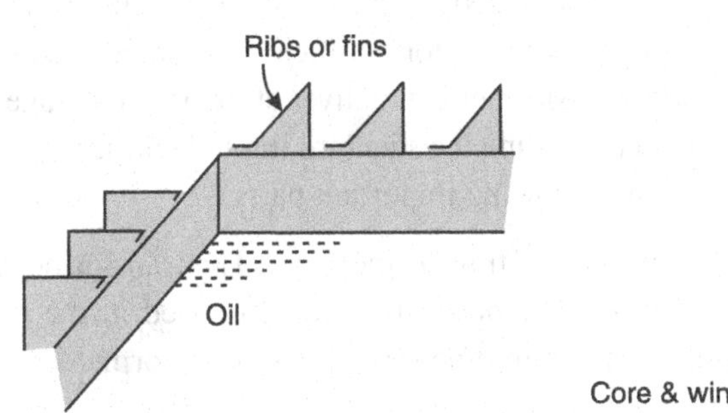

Fig. 3.35 Tank with fins

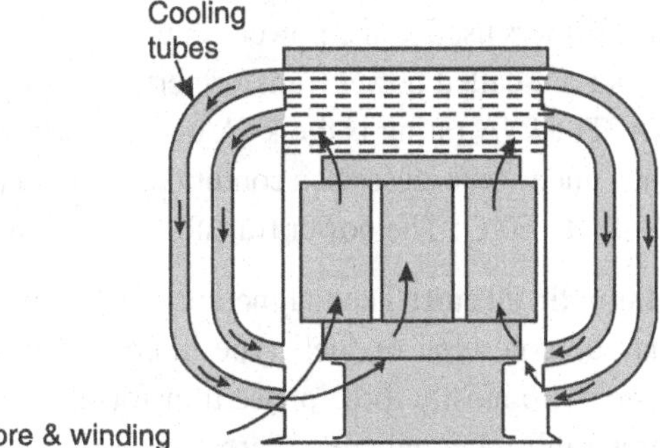

Fig. 3.36 Tank with cooling tubes

3. **Oil immersed Forced oil-circulation Natural cooling (*OFN*)**

   In this type, core and coils are immersed in oil and the cooling is effected basically by forced oil circulation through motor driven oil pump as shown in Fig. 3.37. This method of cooling is one of the latest to be adopted and is employed with the radiators separately attached with the main tank of the transformer. A pump forces the oil through the ducts provided between winding and back to the radiators. The transformer tank in this case is also made plain, the radiators being relied upon to give necessary cooling.

4. **Oil immersed Forced oil circulation with air Blast cooling (*OFB*)**

   This is achieved by means of fans blowing under the radiator as shown in Fig. 3.37. The radiators may be placed at some little distance from the transformer tank. Fans are placed under the radiators in such a way as to blow air through them and thus increases the cooling.

   The method in which forced air cooling is adopted in addition to the oil-forced cooling is called *OFB* method i.e., oil immersed forced oil circulation with air blast cooling.

   OFN method raises the rating by one-third and *OFB* by another third.

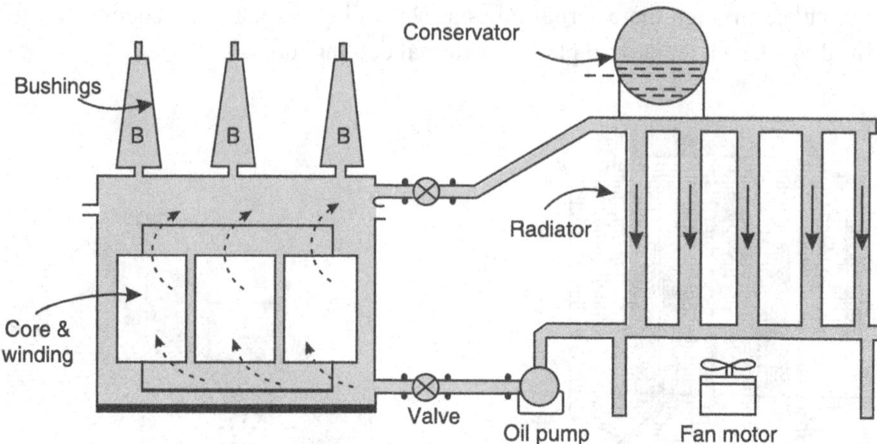

Fig. 3.37 Oil immersed forced oil circulation with air blast cooling

3.23 Power Transformer and its Auxiliaries

The transformers used in the power system for transfer of electric power or energy from one circuit to the other are called power transformers. The rating of a transformer includes voltage, frequency and kVA. The kVA rating is the kVA output that a transformer can deliver at the rated voltage and frequency under general service conditions without exceeding the standard limit of temperature rise (usually 45° to 60°C) The power transformer has the following important parts :

1. **Magnetic circuit:** The magnetic circuit comprises of transformer core. The transformer core may be core type or shell type in construction. The power transformers used in the power system are mostly three phase transformers. In a core type three-phase transformer core has three limbs of equal area of cross-section.
2. **Electrical circuit:** In three phase transformers there are three primary (*H.V.*) windings and three secondary (*L.V.*) windings. Whole of the *L.V.* winding is wound over one limb next to the core, then whole of the *H.V.* winding is wound over the *L.V.* winding. In between the *L.V.* winding and *H.V.* winding and between core and *L.V.* winding insulation is provided.
3. **Transformer oil:** Transformer oil is a mineral oil obtained by fractional distillation of crude petroleum. The oil is used only in the oil cooled transformers. The oil not only carries the heat produced due to losses in the transformer, by convection from the windings and core to the transformer tank, but also has even more important function of insulation.

   When transformer delivers power, heat is produced due to the iron and copper losses in the transformer. This heat must be dissipated effectively otherwise the temperature of the winding will increase. The raise in temperature further increases the losses. Thus, the efficiency of the transformer will decrease. As there is no rotating part in the transformer, it is difficult to cool down the transformer as compared to rotating machines. Various methods are adopted to cool down the transformers of different rating. The common methods are air natural cooling, oil immersed natural cooling, oil immersed forced oil circulation natural cooling, oil immersed forced oil circulation with air blast cooling, oil immersed forced oil circulation with water cooling etc.

Generally, for cooling of distribution transformers, oil immersed natural cooling method is adopted. Cooling tubes or small cooling radiators are used with the main tanks, as shown in Fig. 3.38, to increase the surface area for the dissipation of heat.

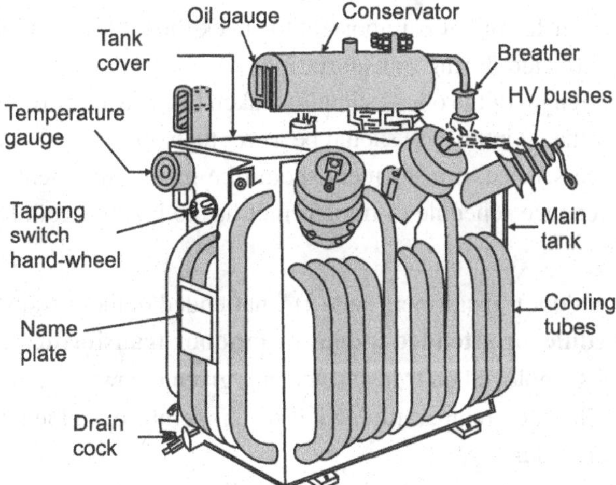

Fig. 3.38 Pictorial view of a 200 kVA, 11 kV/400 V oil immersed natural cooled distribution transformer

4. **Tank Cover:** A number of parts are arranged on the tank cover of which most important are :

   (i) **Bushing:** The internal winding of the transformer are connected to the lines through copper rods or bars which are insulated from the tank cover, these are known as bushings. Upto 33 kV ordinary porcelain bushing can be used. Above this voltage oil filled bushings or condenser bushing are employed.

   (ii) **Oil conservator tank:** Oil conservator is also known as an oil expansion chamber. It is a small cylindrical air tight and oil tight vessel. The oil conservator is connected with a tube to the main transformer tank at the tank cover. This tank is partially filled with oil. The expansion and contraction of oil, changes the oil level in the conservator.

   (iii) **Breather:** The transformer oil should not be allowed to come in contact with atmospheric air, since a small amount of moisture causes a great decrease in the dielectric strength of transformer oil. All the tank fittings are made air tight. When oil level in the oil conservator tank changes due to expansion and contraction of oil because of change of oil temperature, air moves in and out of the conservator. This action is known as breathing.

   The breathed air is made to pass through an apparatus called breather to abstract moisture. Breather, contains Silica gel or some other drying agent such as *calcium chloride.* This ensures that only dry air enters the transformer tank.

   (iv) **Buchholz Relay:** This is installed in between the main tank and the oil conservator. It is a gas relay which gives warning of any fault developing inside the transformer, and if the fault is dangerous, the relay disconnects the transformer circuit. This relay is installed in the transformer having capacity more than 750 kVA.

All the important parts of a 200 kVA, 11 kV/400 V oil immersed natural cooled distribution transformer are shown in Fig. 3.38.

3.24 Maintenance Schedule of a Transformer

A power or distribution transformer when despatched from a factory, its tank is filled with oil upto a certain level in the conservation tank. Above the oil level, dry air or inert gas under pressure is filled. On receipt of the transformer, it is important to check the oil level and the dielectric strength of the oil which may be affected during transportation.

To check the electric strength of the oil, its sample is taken from near the top and bottom of the tank. The dielectric strength of the oil in no case should be lower than 30 kV for a 4 mm gap.

For a long trouble-free service, transformers should be given due attention regularly. For this purpose, a proper maintenance schedule of transformer has to be prepared. Different maintenance schedule is prepared for

(*i*) Attended indoor or outdoor transformers and (*ii*) Unattended outdoor transformers.

(*i*) **Maintenance schedule for attended indoor or outdoor transformers.**

The different checks employed on transformers are given below:

(*a*) *Hourly check*: Check oil temperature, winding temperature, ambient temperature, load and voltage after very hour.

The temperature is kept within the permissible limits by adjusting the load.

(*b*) *Daily Check*:

(*i*) Check oil level daily. If low, fill moisture free oil.

(*ii*) Check the colour of the silica gel in the breather daily. Its colour should be blue. If its colour is pink replace it.

(*c*) *Quarterly check*: Check the working of cooling fans and pumps quarterly. If not working properly, get them repaired.

(*d*) *Half-yearly*:

(*i*) Check dielectric strength, if below 30 kV for 4 mm then filter it or replace it.

(*ii*) Check bushes, insulators and cable boxes half-yearly. If found defective replace them.

(*e*) *Yearly*: (*i*) Check oil alarm circuits, relays, contacts, earth resistance and lightning arrestors.

(*f*) *Five yearly*:

(*i*) Carry out overall inspection of the transformer including lifting of coils and core.

(*ii*) Cleaning of transformer with moisture free transformer oil.

(*ii*) **Maintenance schedule of outdoor transformers.**

These transformers are checked after a certain period as given below:

(*a*) *Quarterly*:

(*i*) Check oil level and condition of silica gel.

(*ii*) Clean bushings, check conditions of switching and alarm circuits etc.

(*b*) *Yearly*:

(*i*) Check dielectric strength of oil.

(*ii*) Check earth resistance and lightning arrestors etc.

(*c*) *Five-yearly*:

(*i*) Carryout overall inspection of the transformer.

(*ii*) Cleaning of transformer with moisture free transformer oil.

3.25 Trouble Shooting of a Transformer

For any type of internal fault of the transformer, the winding insulation and bushings are tested. For this insulation test between transformer winding and core or tank is carried out. Similarly, to test insulation of bushings, insulation test between each terminal of transformer and tank is carried out. When fault is located, it is removed.

Section Practice Problems

Numerical Problems

1. Two transformers each of 80 kVA are connected in parallel. One has resistance and reactance of 1% and 4% respectively and the other has resistance and reactance of 1.5% and 6% respectively. Calculate the load shared by each transformer and the corresponding power factor when the total load shared is 100 kVA 0.8 pf lagging. (**Ans.** *48 kW, 0.8 pf lagging; 32 kW, 0.8 pf lagging*)

2. A 500 kVA transformer with 1.5% resistive and 5% reactive drops is connected in parallel with a 1000 kVA transformer with 1% resistive and 4% reactive drops. The secondary voltage of each transformer is 400 V on load. Determine how they share a load of 500 kVA at a pf of 0.8 lagging. (**Ans.** *116 kW, 0.82 (lag); 284 kW, 0.7923 (lag)*)

3. The primary and secondary windings of two transformers each rated 250 kVA, 11/2 kV and 50 Hz are connected in open delta. Find (*i*) the kVA load that can be supplied from this connection; (*ii*) currents on HV side if a delta connected three phase load of 250 kVA, 0.8 pf (lag) 2 kV is connected to the LV side of the connection. (**Ans.** *43.3 kVA; 13A*)

4. Two single phase furnaces working at 100 volt are connected to 6600 volt, three-phase mains through Scott-connected transformers. Calculate the current in each line of the three-phase mains, when the power drawn by furnace I is 600 kW and the other 900 kW respectively at a power factor of 0.71 lagging. Neglect losses in transformers. (**Ans.** *84500 A; 148 A; 12680 A; 192.1 A; 205.8 A*)

Short Answer Type Questions

Q.1. Compare delta and open delta connections.

Ans. Comparison of Delta and Open Delta Connections:

(*i*) When similar transformers are used, the voltages given by the delta and Vee connections are the same and their outputs are proportional to their line currents.

(*ii*) In balanced delta connection the line current is $\sqrt{3}$ of the phase current whereas in Vee-connection the line current is the same as the phase current.

(*iii*) With non-inductive balanced load, each transformer of the delta connection carries one-third of the total load at unity power factor. Under the same conditions each transformer of Vee-connection carries one half of the load at a p.f. 0.866.

(*iv*) The ratio of the power in Vee-Vee to Power in delta connection is 0.577.

Q.2. Why Scott connections are also known as T-connections?

Ans. Since primary and secondary of teaser transformer are connected to the centre tap of the primary and secondary of the main transformer respectively giving a shape of English letter '*T*', as shown in Fig. Q.3 (*a* and *b*) the connections are known as *T-T* connections.

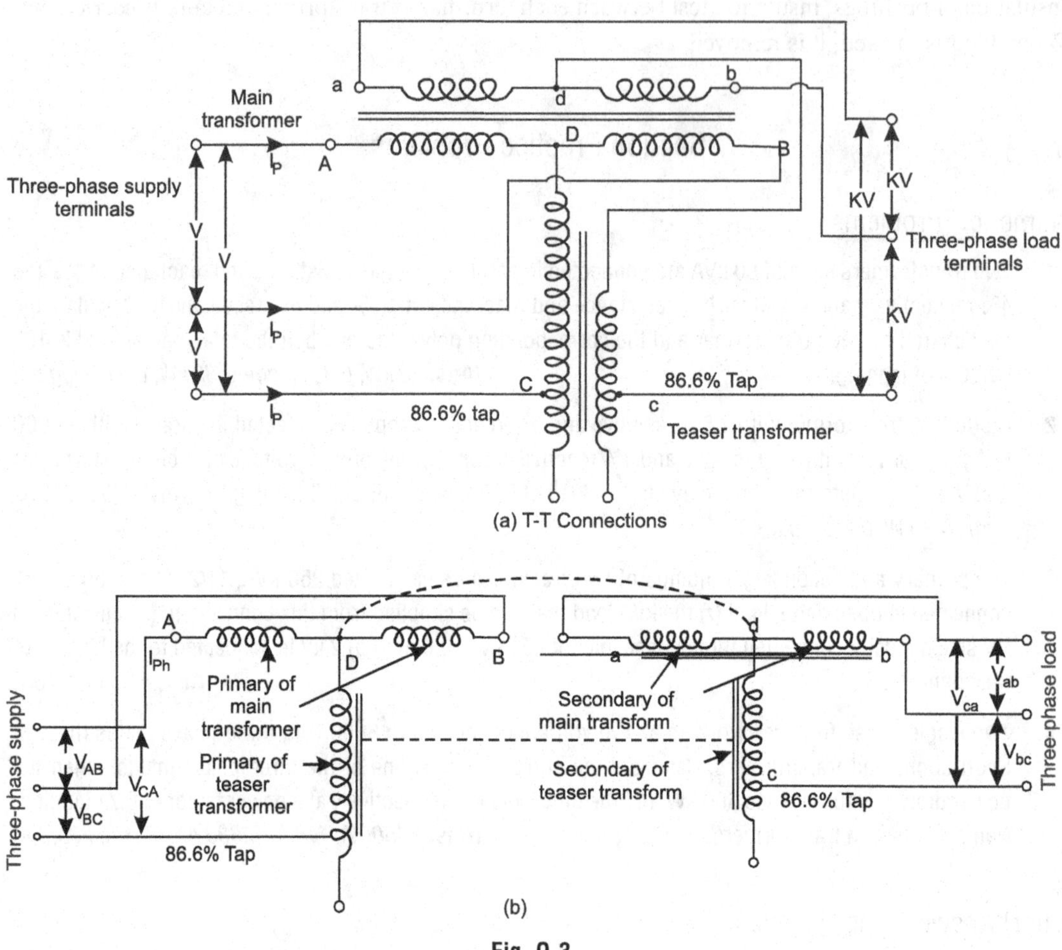

Fig. Q.3

Q.3. What is the function of a breather is a transformer?

Ans. **Breather:** The transformer oil should not be allowed to come in contact with atmospheric air, since a small amount of moisture causes a great decrease in the dielectric strength of transformer oil. All the tank fittings are made air tight. When oil level in the oil conservator changes, air moves in and out of the conservator. This action is known as breathing.

The breathed air is made to pass through an apparatus called breather to abstract moisture. Breather, contains *Silica-gel* or some other drying agent such as *calcium chloride*. This ensures that only dry air enters the transformer tank.

Review Questions

1. A single unit of three-phase transformer is preferred over three units of one-phase transformers to transfer power in three-phase circuits, why?
2. *LV* winding is always placed nearer the core, why?
3. Parallel operation of the transformers in the power system is necessary, why?
4. Distribution transformers are designed for smaller iron losses, why?
5. Tap-changers are usually employed on the *HV* side, why?
6. No-load tap changers are only operated at off load, why?
7. A reactor is employed with on-load tap changer, why?
8. It is difficult to cool down transformer than to cool down an induction motor or alternator, why?
9. In power transformer, breather is necessary, why?
10. In distribution transformers, cooling tubes are provided, why?
11. How can you differentiate between power and distribute transformers.
12. What is the importance of cooling of transformer?
13. Write a short note on a star-Delta transformer.
14. Write a short note on earthing of transformer.
15. What is a three-phase transformer? Give its merits over a bank of three single-phase transformers.
16. Explain the construction of a three-phase (*i*) core type and (*ii*) shell type transformers with the help of neat sketches.
17. Explain the following connections of three-phase transformers
    (*i*) Star-star
    (*ii*) Delta-delta
    (*iii*) Delta-star
    (*iv*) Star-delta giving the merits of each one of them.
18. What do you understood by parallel operation of transformers? What is its necessity?
19. State and explain the conditions necessary for parallel operation of two single-phase transformers.
20. State and explain the conditions necessary for parallel operation of two three-phase transformers.
21. Explain with the help of a phasor diagram the load shared by each transformer, when two are connected in parallel having same voltage ratios and same percentage impedances.
22. Explain with the help of a phasor diagram the load shared by each transformer when two are connected in parallel having same voltage ratios but different impedance triangles.
23. What is a tap-changer? Where and how they are placed with the transformers?
24. Explain the working of an off-load tap-changer with the help of a neat-schematic diagram.
25. Explain the working of an on-load tap-changer with the help of neat schematic diagram.
26. What is the necessity of cooling of transformer? Explain with neat sketch oil immersed forced oil circulation natural cooling.

27. Write short note on

(*i*) Oil immersed natural cooling

(*ii*) Oil immersed forced oil circulation with air blast cooling.

Multiple Choice Questions

1. The transformation ratio of a three-phase transformer is given by the relation.

(*a*) $\frac{E_{2(L)}}{E_{1(L)}}$ (*b*) $\frac{E_{2(ph)}}{E_{1(ph)}}$

(*c*) $\frac{E_{2(L)}}{E_{1(ph)}}$ (*d*) $\frac{E_{2(ph)}}{E_{1(L)}}$

2. Usually *HV* winding is placed

(*a*) next to the core (*b*) over the low voltage winding

(*c*) both *a* and *b* (*d*) none of these

3. In parallel operation of two transformers if the transformers have different percentage impedances.

(*a*) they will share the load as per their rating.

(*b*) the transformers will be on dead short circuiting and will be damaged.

(*c*) They will not share the load as per their rating but there will be no damage to the transformers.

(*d*) none of these

4. The seasonal voltage adjustments on the transformer are made through

(*a*) off-load tap-changer (*b*) on load tap-changer

(*c*) both *a* and *b* (*d*) none of these

5. Cooling of transformer

(*a*) reduces losses (*b*) increases efficiency

(*c*) increase life (*d*) all of these

Keys to Multiple Choice Questions

1. b	**2.** b	**3.** c	**4.** a	**5.** d

CHAPTER

4

DC Generator

Chapter Objectives

After the completion of this unit, students/readers will be able to understand:

- ✓ What is a DC generator and how does it convert mechanical energy into electrical energy?
- ✓ What are the main constructional features of a DC machine, material used for different parts and their function?
- ✓ How armature winding is designed and placed in the armature slots? Give details of lap and wave winding design.
- ✓ What are the factors on which emf induced in the armature depends?
- ✓ What are the factors on which torque developed in the armature depends?
- ✓ What is armature reaction and how does it affect the performance of machine?
- ✓ What is commutation? What is meaning of bad and good commutation?
- ✓ What are causes of sparking at the brushes and how can it be reduced or removed?
- ✓ What are inter-poles and compensating winding and how are these helpful in improving commutation?
- ✓ What are different types of DC generators on the basis of field excitation?
- ✓ How is voltage built-up in DC shunt generator and what is critical resistance?
- ✓ What are the performance characteristics of DC generators and their field of applications?
- ✓ What are the losses in a DC generator?
- ✓ What is efficiency of a DC generator and under what conditions does it become maximum?

Introduction

A DC machine is an electro-mechanical energy conversion device. When it converts mechanical power (ωT) into DC electrical power (EI), it is known as a DC generator. On the other hand, when it converts DC electrical power into mechanical power it is known as a DC motor.

Although battery is an important source of DC electric power, but it can supply limited power. There are some applications where large quantity of DC power is required (such as in chemical and metal extraction plants, for electroplating and electrolysis processes etc.), at such places DC generators are used to deliver power.

In short, we can say that DC machines have their own role in the field of engineering. In this chapter, we shall focus our attention on the common topics of DC generators.

4.1 DC Generator

An electro-mechanical energy conversion device (or electrical machine) that converts mechanical energy or power (ωT) into DC electrical energy or power (EI) is called **DC generator.**

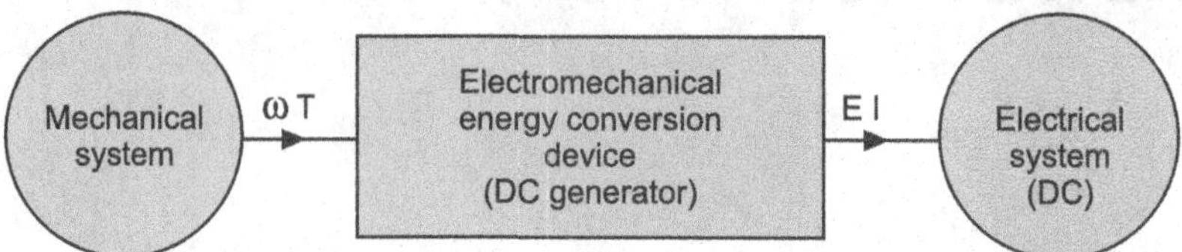

Fig. 4.1 Block diagram of electro-magnetic energy conversion (Generator action)

Working Principle

The basic principle of a DC generator is *electro-magnetic induction* i.e.,

"*When a conductor cuts across the magnetic field, an* **emf** *is induced in it.*"

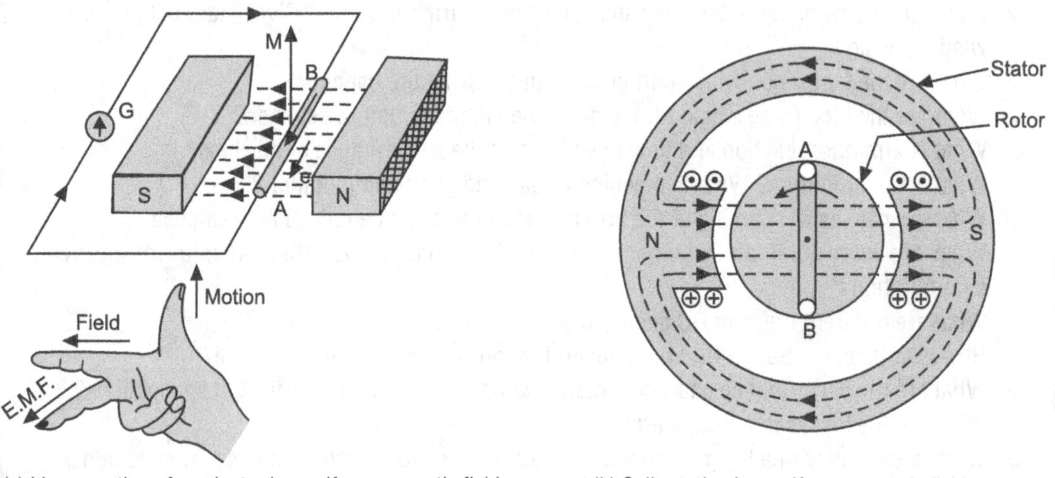

(a) Linear motion of conductor in a uniform magnetic field (b) Coil rotating in a uniform magnetic field

Fig. 4.2 Generation of emf

Consider Fig. 4.2(*a*), here, when a conductor is moved vertically upward or downward, the deflection in the galvanometer clearly shows that an *emf* is induced in the conductor since flux is cut by the conductor. But, when it is moved horizontally (left or right), there is no deflection in the galvanometer which shows that no emf is induced in the conductor since flux cut is zero and conductor moves just parallel to the magnetic lines of force.

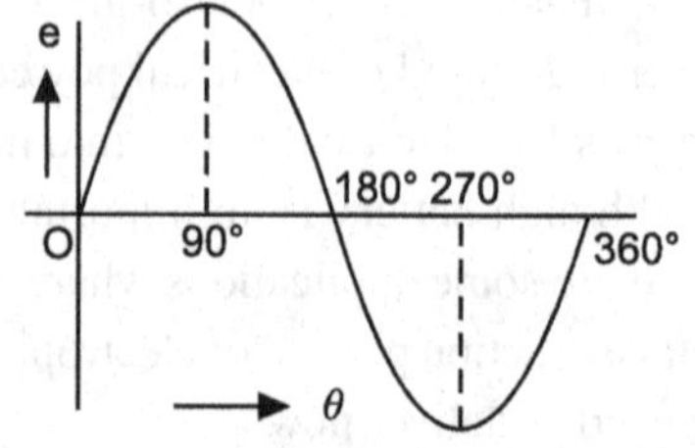

Fig. 4.2(c) Wave shape of induced emf

In fact, in a generator, a coil is rotated at a constant speed of ω radians per second in a strong magnetic field of constant magnitude as shown in Fig. 4.2(*b*). An *emf* is induced in the coil by the phenomenon of *dynamically induced emf* (e =

Blv sin *θ*; *e* ∝ sin *θ*). The magnitude and direction of induced emf changes periodically depending upon sine of angle *θ*. The wave shape of the induced *emf* is shown in Fig. 4.2(*c*), which is AC for internal as well as external load.

This AC is converted into DC with the help of commutator, as explained in the Articles to follow.

Thus, the working principle of a DC generator is **electro-magnetic induction.**

4.2 Main Constructional Features

The complete assembly of various parts in a scattered form of a DC machine is shown in Fig. 4.3. The essential parts of a DC machine are described below:

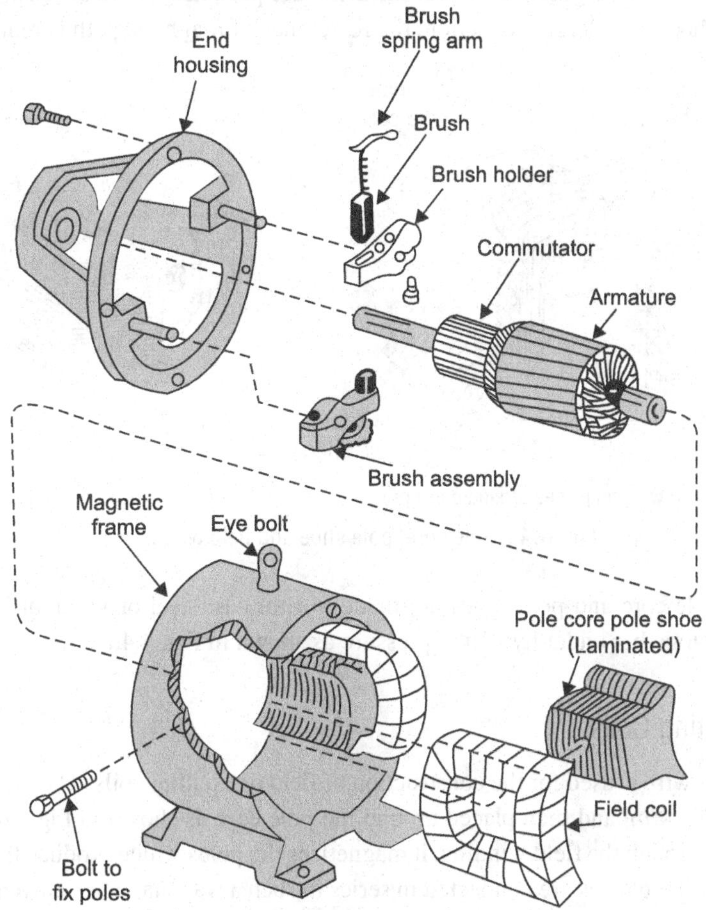

Fig. 4.3 Disassembled parts of a DC machine

1. Magnetic Frame or Yoke

The outer cylindrical frame to which main poles and inter poles are fixed is called yoke. It also helps to fix the machine on the foundation. It serves two purposes:

(*i*) It provides mechanical protection to the inner parts of the machine.
(*ii*) It provides a low reluctance path for the magnetic flux.

The yoke is made of cast iron for smaller machines and for larger machines, it is made of cast steel or fabricated rolled steel since these materials have better magnetic properties as compared to cast iron.

2. Pole Core and Pole Shoes

The pole core and pole shoes are fixed to the magnetic frame or yoke by bolts. They serve the following purposes:

(*i*) They support the field or exciting coils.
(*ii*) They spread out the magnetic flux over the armature periphery more uniformly.
(*iii*) Since pole shoes have larger X-section, the reluctance of magnetic path is reduced.

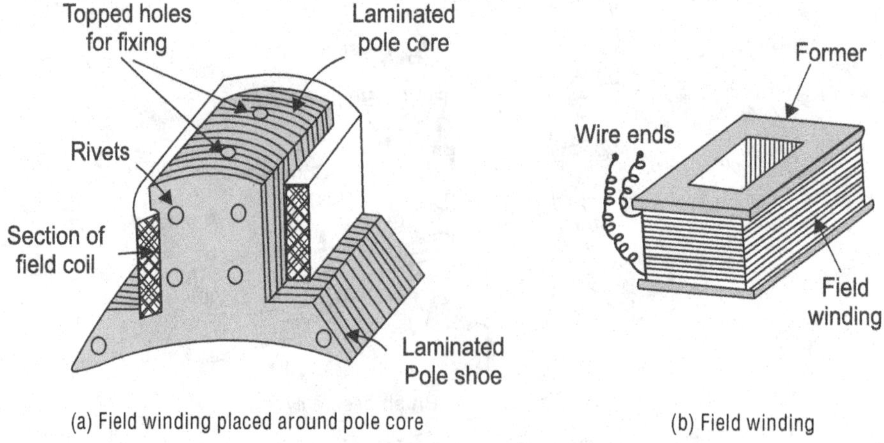

(a) Field winding placed around pole core (b) Field winding

Fig. 4.4 Pole core, pole shoe and field winding

Usually, the pole core and pole shoes are made of thin cast steel or wrought iron laminations which are riveted together under hydraulic pressure as shown in Fig. 4.4(*a*).

3. Field or Exciting Coils

Enamelled copper wire is used for the construction of field or exciting coils. The coils are wound on the former [see Fig. 4(*b*)] and then placed around the pole core as shown in Fig. 4(*a*). When direct current is passed through the field winding, it magnetises the poles which produce the required flux. The field coils of all the poles are connected in series in such a way that when current flows through them, the adjacent poles attain opposite polarity as shown in Fig. 4.5.

4. Armature Core

It is cylindrical is shape and keyed to the rotating shaft. At the outer periphery slots are cut, as shown in Fig. 4.6, which accommodate the armature winding. The armature core shown in Fig. 4.6, serves the following purposes:

(*i*) It houses the conductors in the slots.
(*ii*) It provides an easy path for magnetic flux.

Since armature is a rotating part of the machine, reversal of flux takes place in the core, hence hysteresis losses are produced. To minimise these losses silicon steel material is used for its construction. When it rotates, it cuts the magnetic field and an emf is induced in it. This emf circulates eddy currents which results in eddy current loss in it. To reduce these losses, armature core is laminated, in other words we can say that about 0.3 to 0.5 mm thick stampings are used for its construction. Each lamination or stamping is insulated from the other by varnish layer (see Fig. 4.6).

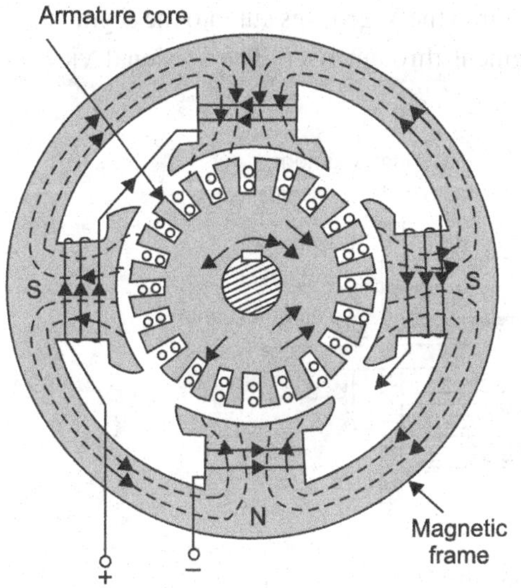

Fig. 4.5 Magnetic circuit of DC machine

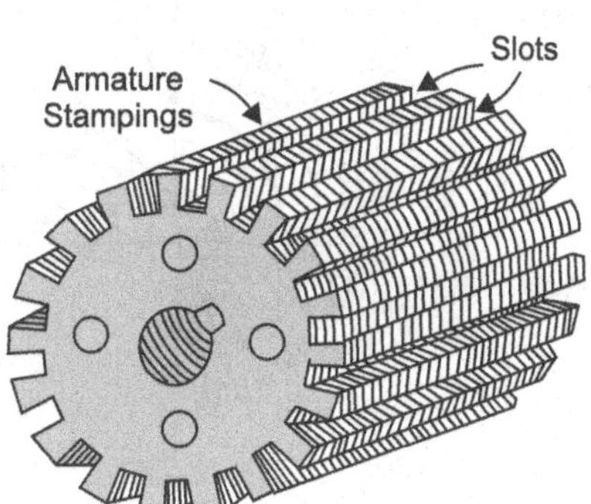

Fig. 4.6 Armature core

5. Armature Winding

The insulated conductors housed in the armature slots are suitably connected. This is known as armature winding. The armature winding acts as the heart of a DC machine. It is a place where one form of power is converted to the other form i.e., in case of generator, mechanical power is converted into electrical power and in case of motor, electrical power is converted into mechanical power. On the basis of connections, there are two types of armature windings named (*i*) Lap winding and (*ii*) Wave winding (detailed discussions in the coming articles).

(*i*) *Lap winding*: In this winding, the connections are such that the number of parallel paths is equal to number of poles. Thus, if machine has P poles and Z armature conductors, then there will be P parallel, paths, each path will have Z/P conductors in series. In this case, the number of brushes is equal to the number parallel paths. Out of which half the brushes are positive and the remaining (half) are negative.

(*ii*) *Wave winding*: In this winding, the connections are such that the numbers of parallel paths are only two irrespective of the number of poles. Thus, if machine has Z armature conductors, there will be only two parallel paths each having $Z/2$ conductors in series. In this case, the number of brushes is equal to two i.e., number of parallel paths.

6. Commutator

It is an important part of a DC machine and serves the following purposes:

(*i*) It connects the rotating armature conductors to the stationary external circuit through brushes.

(*ii*) It converts the alternating current induced in the armature conductors into unidirectional current in the external load circuit in generator action, whereas, it converts the alternating torque into unidirectional (continuous) torque produced in the armature in motor action.

The commutator is of cylindrical shape and is made up of wedge-shaped hard drawn copper segments. The segments are insulated from each other by a thin sheet of mica. The segments are held together by means of two V-shaped rings that fit into the V-grooves cut into the segments. Each armature coil is connected to the commutator segment through riser. The sectional view of the commutator assembly is shown in Fig. 4.7.

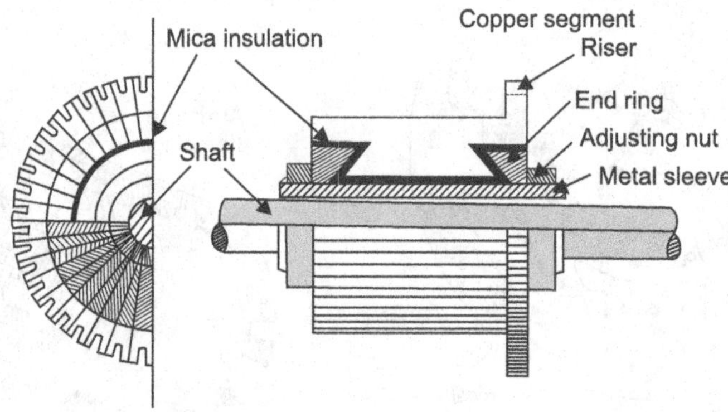

Fig. 4.7 Commutator

7. Brushes

The brushes are pressed upon the commutator and form the connecting link between the armature winding and the external circuit. They are usually made of high grade carbon because carbon is conducting material and at the same time in powdered form provides lubricating effect on the commutator surface. The brushes are held in particular position around the commutator by brush holders and rocker.

8. Brush Rocker

It holds the spindles of the brush holders. It is fitted on to the stationary frame of the machine with nut and bolts. By adjusting its position, the position of the brushes over the commutator can be adjusted to minimise the sparking at the brushes.

9. End Housings

End housings are attached to the ends of the main frame and support bearings. The front housing supports the bearing and the brush assemblies whereas the rear housing usually supports the bearing only.

10. Bearings

The bearings may be ball or roller bearings these are fitted in the end housings. Their function is to reduce friction between the rotating and stationary parts of the machine. Mostly high carbon steel is used for the construction of bearings as it is very hard material.

11. Shaft

The shaft is made of mild steel with a maximum breaking strength. The shaft is used to transfer mechanical power from or to the machine. The rotating parts like armature core, commutator, cooling fan etc. are keyed to the shaft.

4.3 Simple Loop Generator and Function of Commutator

For simplicity, consider only one coil *AB* placed in the strong magnetic field. The two ends of the coil are joined to slip rings *A*′ and *B*′ respectively. Two brushes rest on these slip rings as shown in Fig. 4.8.

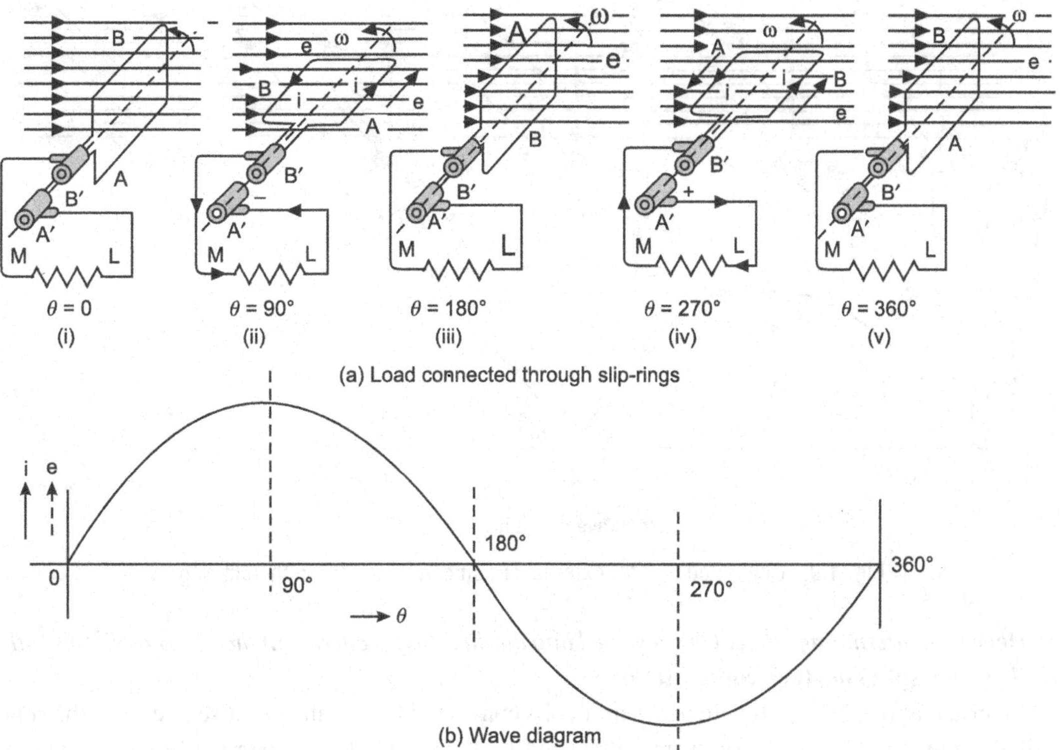

Fig. 4.8 Generated emf for external circuit connected through slip rings

When this coil is rotated in counter clockwise direction at an angular velocity of ω radius per second, the magnetic flux is cut by the coil and an emf is induced in it. The position of the coil at various instants is shown in Fig. 4.8(*a*) and the corresponding value of the induced emf and its direction is shown in Fig. 4.8(*b*). The induced emf is alternating and the current flowing through the external resistance is also alternating i.e., at second instant current flows in external resistance from *M* to *L*, whereas, at fourth instant it flows from *L* to *M* as shown in Fig. 4.8(*b*).

Commutator Action

Now, consider that the two ends of the coil are connected to only one slip ring split into two parts (segment) i.e., *A*″ and *B*″. Each part is insulated from the other by a mica layer. Two brushes rest on these parts of the ring as shown in Fig. 4.9(*a*).

In this case when the coil is rotated is counter clockwise direction at an angular velocity of ω radians per second, the magnetic flux is cut by the coil and an emf is induced in it. The magnitude of emf induced in the coil at various instants will remain the same as shown in Fig. 4.8(*b*).

However, the flow of current in the external resistor or circuit will become unidirectional i.e., at second instant the flow of current in the external resistor is from *M* to *L* as well as the flow of current in the external resistor is from *M* to *L* in the fourth instant, as shown in Fig. 4.9(*a*). Its wave shape is shown in Fig. 4.9(*b*).

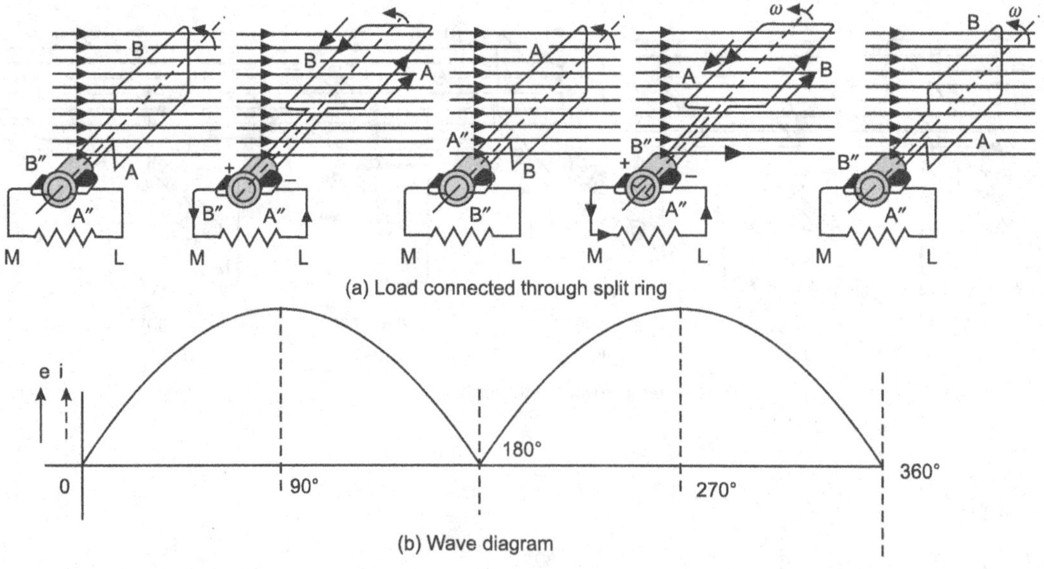

Fig. 4.9 Generated emf for external circuit connected through split ring

Hence, *an alternating current is converted into unidirectional current in the external circuit with the help of a* **split ring** (i.e., **commutator**).

In an actual machine, there are number of coils connected to the number of segments of the ring called commutator. The emf or current delivered by these coils to the external load is shown in Fig. 4.10(*a*). The actual flow of current flowing in the external load is shown by the firm line which fluctuates slightly. The number of coils placed on the armature is even much more than this and a pure direct current is obtained at the output as shown in Fig. 4.10(*b*).

Thus, *in actual machine working as a generator, the function of commutator is to convert the* **alternating current** *produced in the armature into* **direct current** *in the external circuit.*

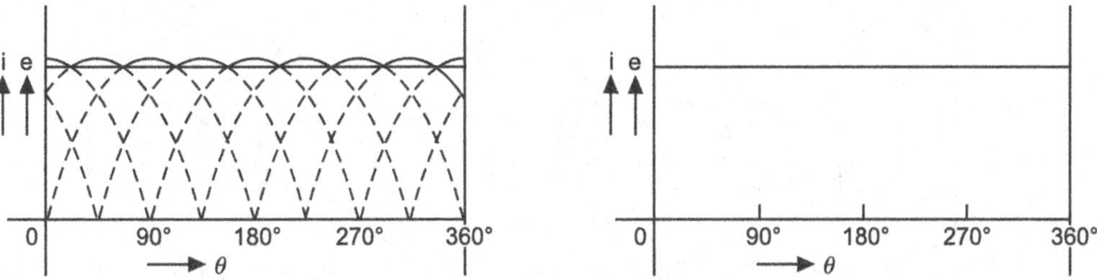

(a) Wave shape of output delivered by number of coils (b) True wave shape of output delivered by a dc generator

Fig. 4.10 Wave shape of output delivered by a DC generator

4.4 Connections of Armature Coils with Commutator Segments and Location of Brushes

Consider an armature which has four coils 1, 2, 3 and 4 equally spaced in the armature slots as shown in Fig. 4.11. The number of commutator segments is equal to number of coils. When the armature is rotated clockwise, the direction of induced emf and hence the current in coil sides 3′, 1, 4′, 2 is downward and in coil sides 1′, 3, 2′, 4 is upward. The coils should be connected in such a way that the emf induced in the two sides of the same coil be added up as shown in Fig. 4.12, The coil sides 1–4′, 1′–2, 2′–3 and 3′–4 are connected to the commutator segments namely 1, 2, 3 and 4 as shown in Fig. 4.13.

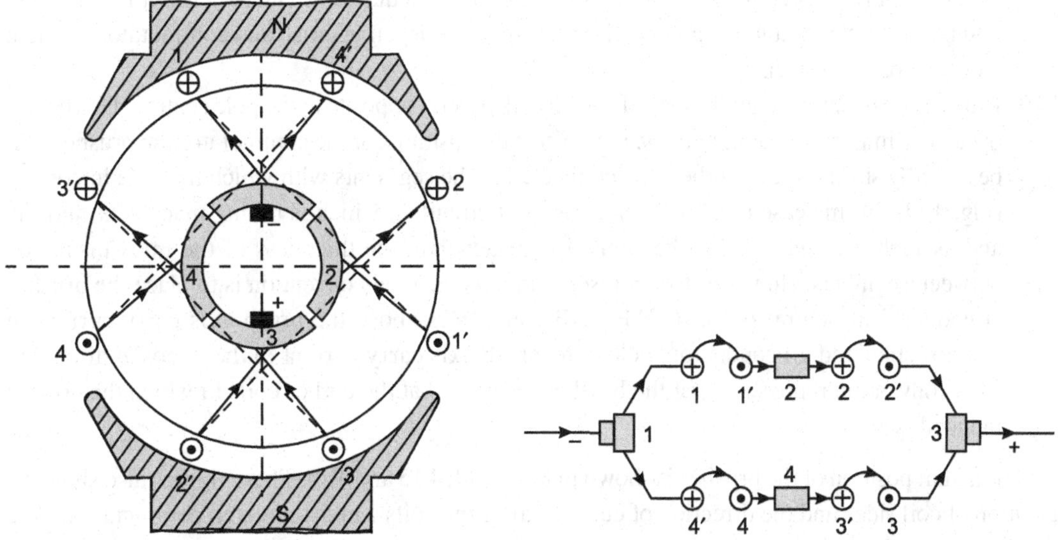

Fig. 4.11 Coils and commutator segments **Fig. 4.12** Electrical connections

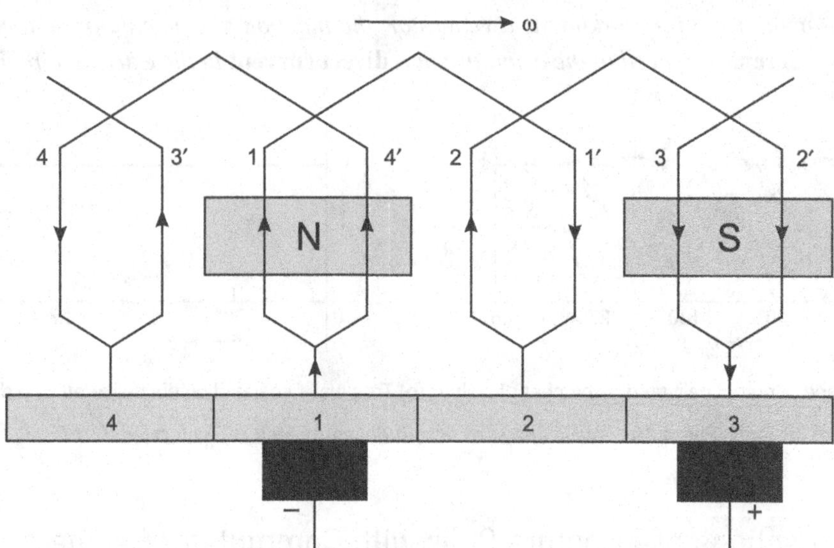

Fig. 4.13 Developed windings diagram for 2-pole, 4-coil, 4-segment DC machine.

The next point of discussion is to determine how many brush sets are required and where these are to be placed with respect to the poles. The brushes are to collect or deliver the current and arc placed at such a position that sparking is minimum at the brushes. The rules to be followed for determining the position of brushes are given below:

(*i*) Place the brush at each meeting point or each separating point of two emf's. The brushes at the meeting point are of positive polarity and those at the separating point are of negative polarity. At segment 1, current is separated towards conductor 1 and 4′, hence it is a separating point. Whereas, current is coming towards segment 3 from conductor 2′ and 3, hence it is a meeting point. So, the commutator segment 1 is the position of negative brush and commutator segment 3 of the positive brush.

(*ii*) Brushes are generally equally spaced and placed directly opposite to the pole centres. This brings out a very important point, namely, that in certain positions of the commutator the brushes will be actually short circuiting the coils connected to the segments with which they are in contact (Fig. 4.14). In this case only two coils are short circuited, which is taking place continuously and as such the width of brushes must be greater than the thickness of the mica insulation between segments. However, for representation, the general convention is to place the brushes at geometrical neutral plane (G.N.P.). All conductors above the brush axis carry current in one direction and all conductors below the brush axis carry current in the opposite direction. This convention represents that the brushes are placed at the coil or coils in which the voltage induced is zero.

The actual position of the brushes is shown in Figs. 4.14, 4.15 and 4.16. These figures also show the position of coil sides and the direction of currents after the coils 2 and 4 undergo commutation. The negative brush short circuits the segments 1 and 4 whereas positive brush short circuit the segments 2 and 3. The coils 4 and 2 are short circuited by the brushes and hence no emf should induce in these coils. The armature winding forms a closed circuit and consists of two parallel paths. When a coil under goes commutation no emf is induced in it since it passes through magnetic neutral axis (*MNA*), and the coil is short-circuited by the brushes, hence no sparking will take place.

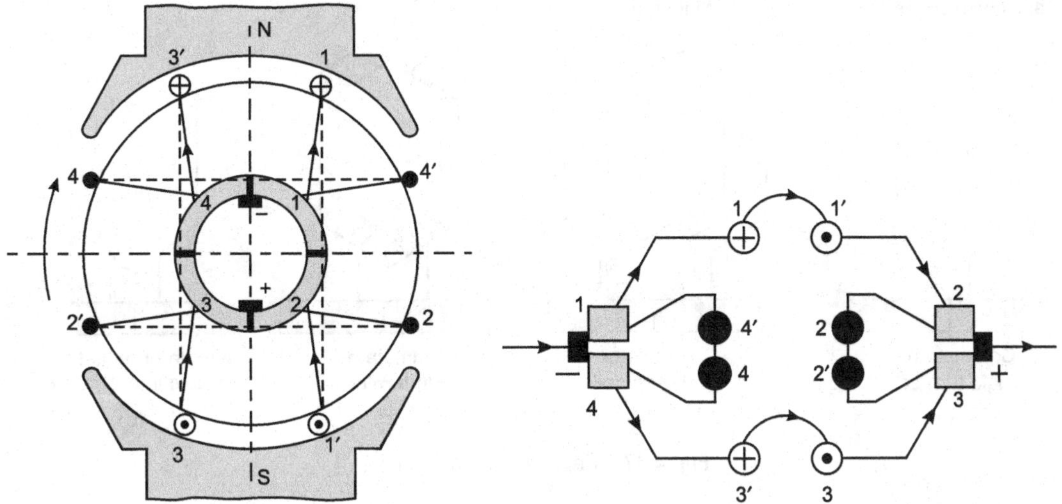

Fig. 4.14 Brush position

Fig. 4.15 Brush position and electrical connections

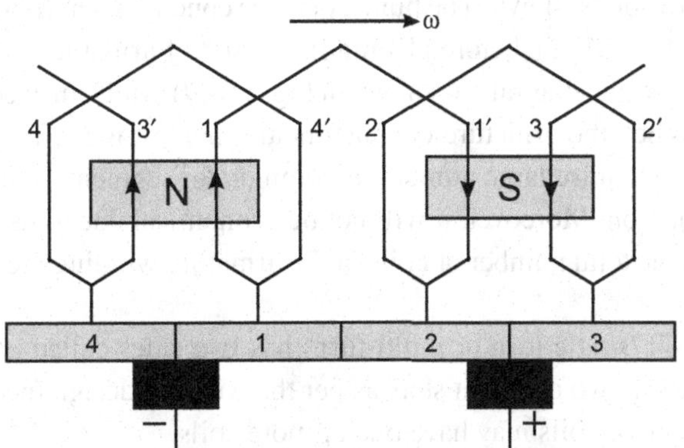

Fig. 4.16 Brush position at an instant

4.5 Armature Winding

At the outer periphery of an armature core, slots are cut. In these slots number of conductors are placed which are connected with each other in proper arrangement forming series–parallel paths depending upon the requirement. This arrangement of connections is known as *armature winding*. To understand the armature winding schemes, it is desirable to have an idea about the following terms.

1. **Conductor:** The length of wire embedded in armature core and lying within the magnetic field is called the conductor (see Fig. 4.17(*a*), where *AB* is a conductor). It may be having one or more parallel strands. Total number of conductors in the armature winding are represented by the symbol *Z*.

2. **Turn:** Two conductors lying in a magnetic field connected in series at the back, as shown in Fig. 4.17(*a*), so that emf induced in them is additive is known as a turn.

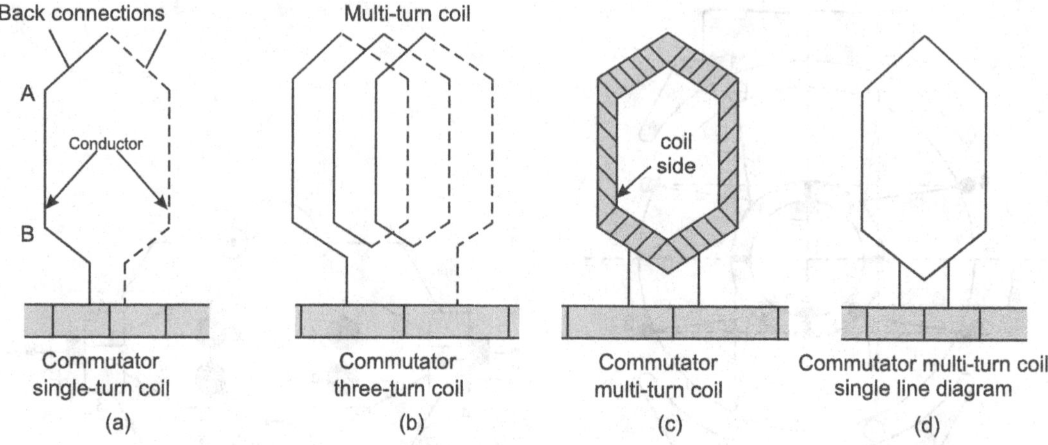

Fig. 4.17 Representation of a coil.

3. **Coil.** A coil may be a single turn coil having only two conductors, as shown in Fig. 4.17(*a*), or it may be a multi-turn coil having more than two conductors as shown in Fig. 4.17(*b*). In Fig. 4.17(*b*), a three-turn coil is shown. The bunch of three conductors may be wrapped by the cotton tape, as shown in Fig. 4.17(*c*), before placing in the slot of armature. A multi-turn coil can be represented by single line diagram as shown in Fig. 4.17(*d*). Multi-turn coils are used to develop higher voltages. When the armature conductors are more, it is not feasible to use single turn coils because it will require large number of commutator segments and if used it will not give spark-less commutation. Moreover, it will not be economical due to use of more copper in the end connections. The total number of coils in the armature winding are represented by symbol '*C*'.
4. **Coil side.** Each coil (single turn or multi-turn) has two sides called *coil sides.* Both the coil sides are embedded in two different slots as per the winding design (nearly a pole pitch apart).
5. **Coil group.** A group of coils may have one or more coils.
6. **Winding.** When number of coil groups are arranged on the armature in a particular fashion as per the design, it is called an armature winding.
7. **Inductance effect.** All the coils have some inductance effect as current is changing in them. Due to inductance effect the flow of current is opposed causing reduction in resultant output voltage. Over-hanging end connections have some adverse effect due to inductance.

Terms Related to Winding Design

8. **Front-end and Back-end connectors.** A wire that is used to connect end of a coil at the front to the commutator segment is called a front-end connector.

   Whereas, a wire that is used to connect one coil side to the other coil side at the back is called back-end connector.

9. **Pole pitch.** In general, it is defined as number of armature slots per pole. It may also be defined as the number of armature conductors per pole. If there are 36 conductors for 4 poles then pole pitch will be equal to 36/4 i.e., 9 conductors per pole.
10. **Back pitch.** The distance in terms of number of armature conductors between first and last conductor of the same coil i.e., the distance between two coil sides of the same coil is called *back pitch.* It is also called the *coil span* or *coil spread* and is denoted by Y_b, as shown in Figs. 4.18(*a*), (*b*) and (*c*)

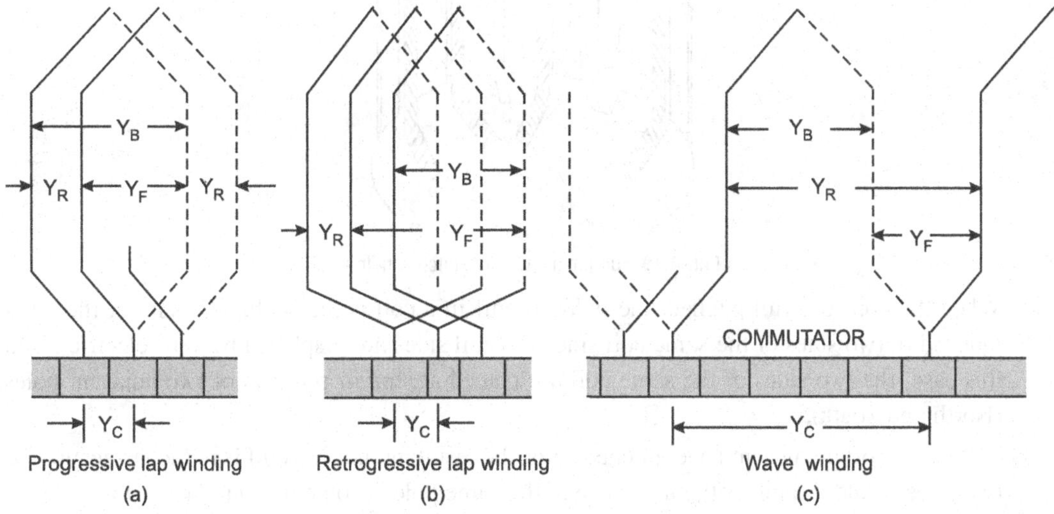

Fig. 4.18 Terms used in coils

11. **Front pitch.** The distance in terms of number of armature conductors or number of slots between second conductor of one coil and the first conductor of the next coil which are connected to the same commutator segment on the front is called *front pitch.* It is denoted by Y_F, as shown in Figs. 4.18 (*a*), (*b*) and (*c*).
12. **Resultant pitch.** The distance in terms of number of armature conductors or number of slots between the start of one coil and the start of the next coil to which it is connected is called resultant *pitch.* It is denoted by Y_R as shown in Figs. 4.18(*a*), (*b*) and (*c*).
13. **Commutator pitch.** The distance measured in terms of commutator segments between the segments to which the two ends of a coil are connected is called *commutator pitch.* It is denoted by Y_C as shown in Figs. 4.18(*a*), (*b*) and (*c*).
14. **Coil span or coil pitch.** The distance in terms of number of armature conductors or number of slots between two sides of the same coil is called *coil span or coil pitch.*

    The coil span may be equal to, less than or more than pole pitch. Accordingly, the coils are knows as full pitched, short pitched or over pitched coils respectively.

    For instance, consider Fig. 4.19. where pole pitch is say 4. One side of coil-*A* is placed is slot No. 1 and the other side is placed in slot No. 5, then the coil span is 5-1 = 4 which is equal to pole pitch hence coil-*A* is called a full-pitched coil. Whereas, in case of coil-*B*, the coil span is 6–3 = 3 which is less than pole pitch, this coil is called a short pitched coil.

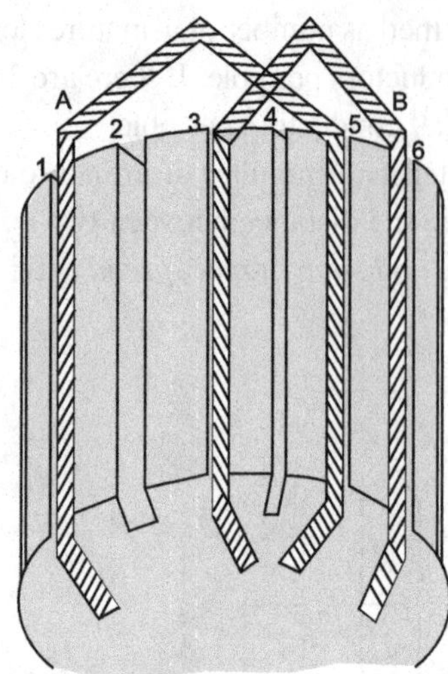

Fig. 4.19: Full-pitch and short-pitch winding

When the coils are full pitched, the induced emf in a coil is the arithmetic sum of the emfs induced in two sides of the same coil since the coil sides are displaced by 180° electric al. In this case, the two sides of the same coil are placed at similar position of two adjacent poles (North and South).

In case of short pitched and over pitched coils, the resultant induced emf is reduced because the two sides would fall under the influence of the same pole at some instant. At this instance the induced emf in the two sides will oppose each other causing reduction in resultant emf (phasor difference).

The advantage of short pitched winding is that in this case the copper used for end connections is reduced substantially which reduces the cost of machine. If also improves the commutation (reduction of sparking at brushes) because inductance of overhang connections is reduced. Moreover, it reduces the copper losses and improves the efficiency to some extent. Hence, many a times short pitched winding is used.

15. **Degree of Re-entrant of an armature winding.** There can be single re-entrant or multi re-entrant armature winding. While tracing through a winding once, if all armature conductors are included on returning to the starting point, the winding is called single re-entrant winding. It will be double re-entrant if only half of the conductors are included in tracing through the winding once.

4.6 Types of Armature Winding

The continuous or closed armature windings are of two types:

1. Gramme-ring winding   2. Drum Winding

The gramme-ring type of armature winding is an early form of armature winding. This winding is replaced by more efficient drum type winding. The gramme-ring type of armature winding was having following disadvantages.

Disadvantages

(*i*) Only half the coil was available to link with the pole flux. The other half of the portion of winding was lying inside the core which was used only as connectors and so there was wastage of copper.

(*ii*) As each turn was to pass through the centre of the core, therefore, it was difficult to wind and require more labour, hence was costlier.

(*iii*) The maintenance and repairs were more costly.

(*iv*) Insulation of winding was also difficult.

(*v*) For the same pole flux and armature velocity the emf induced in ring winding was half of that induced in the drum winding having same number of coils.

(*vi*) Construction was having a large air gap, so stronger field excitation was required to produce the required flux.

4.7 Drum Winding

In this type of winding, the conductors are housed in slots cut over the armature surface and connected to one another at the front and back through connectors. It has the following two main advantages.

1. Whole of the copper used is active except end connectors, i.e., it cuts flux and, therefore, is active in generating emf.
2. The coils can be pre-formed and insulated before placing on the armature, which reduces the construction cost.

The drum winding may be either single layer or double layer winding. In single layer winding, only one conductor or one coil side is placed in each armature slot. It is rarely used because of its more cost.

In two-layer winding, as shown in Fig. 4.20, there are two conductors or coil sides per slot arranged in two layers. Mostly it is employed because of economy reasons. These coils are usually are wound on machine driven formers which give them proper shape. The turns may be bound together with cotton tape. The ends are left bare to solder them to the commutator segments. The coils are then dipped into some insulating compound such as asphalt and are then dried. If the machine is to be operated at high temperatures, other materials, such as mica and paper tape, fibre glass tape, and silicon impregnated insulation may be used.

Before placing the coils in the slots, liners of leatheroid are placed to ensure mechanical protection is provided to the coils. After dropping the coils into the slots, wedges of wood or hard fibre are driven to hold them in place, as shown in Figs. 4.20(*a*) and (*b*).

To obtain maximum voltage a full pitched winding is used. However, the span may be reduced to as much as eight-tenth of the pole pitch without any serious reduction in the induced emf. When this is done, the winding is called short pitch (or fractional pitch) winding.

Usually, one side of every coil lies in the top of one slot and the other side lies in the bottom of some other slot at a distance of approximately one pole pitch along the armature. Thus, at least two coil sides occupy each slot. Such windings, in which two coil sides occupy each slot are most commonly used for all medium sized machines. In case of large machines, more coil-sides can be placed in a single slot as shown in Figs. 4.20(*b*) and 4.20(*d*). Placing of several elements in a single slot gives fewer slots than segments which have got following advantages.

(*i*) By providing more conductors in one slot, the number of slots decreases. As the number of slots is reduced the armature core teeth become mechanically stronger to bear the stresses.

(*ii*) In this arrangement, the number of commutator segments is increased, the voltage between segments adjacent to each other decreases and the number of turns of wire in the coil or coils connected to adjacent segments also decreases. The result is that there is less sparking at the commutator because of the improved commutation.

(*iii*) The overall cost of construction is reduced.

In general there are two types of drum type armature windings: (*i*) lap winding (*ii*) wave windings.

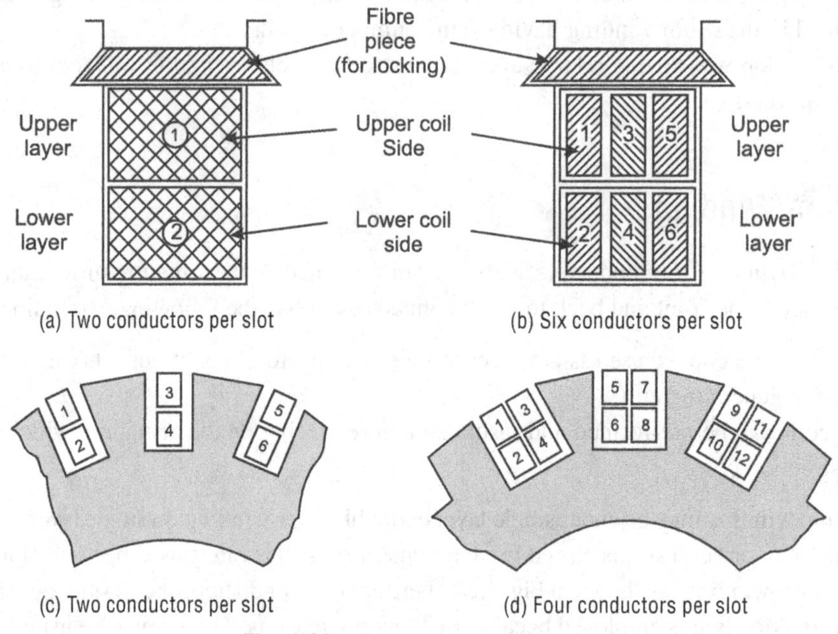

Fig. 4.20 Double-layer winding

4.8 Lap Winding

Single turn lap winding is shown in Figs. 4.21(*a*) and (*b*). In lap winding a coil side under one pole is connected directly to the coil side of another coil which occupies nearly the corresponding position under the next pole and the finish end of other coil is connected to a commutator segment and to the start end of the adjacent coil situated under the same pole and all coils are connected similarly forming a closed loop. Since the sides of successive coils overlap each other, it is called a lap winding. Lap winding may be further classified as a simplex (single) or multiplex (double or triple) windings.

Simplex lap winding: In this winding, there are as many parallel paths as there are field poles on the machine.

Double or duplex lap winding: In this case, two similar simplex windings are placed in alternate slots on the armature and connected to alternate commutator segments. Thus, each winding carries half of the armature current.

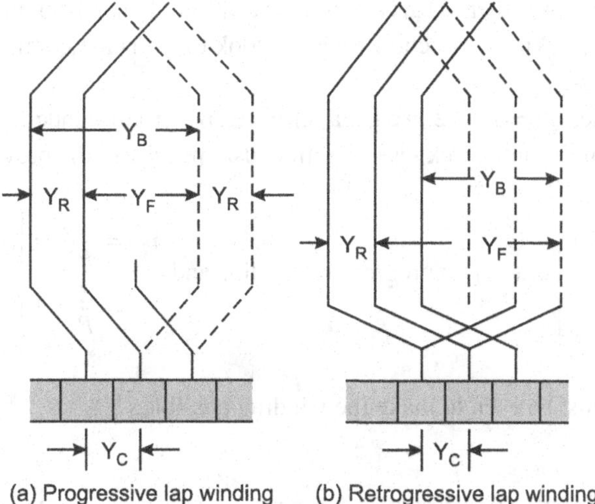

(a) Progressive lap winding (b) Retrogressive lap winding

Fig. 4.21 Lap winding connections

Triple or triplex lap winding: In this case, three similar simplex windings are placed to occupy every third slot and connected to every third commutator segment. Thus, each winding carries one third of the armature current. Similarly, there can be multiplex lap winding having even more than three simplex winding as per the requirement.

The above explanation clearly shows that the sole purpose of employing multiplex lap winding is to increase the number of parallel** paths enabling the armature to carry a large total current, at the same time reducing the conductor current to improve commutation conditions.

Following points regarding lap winding should be carefully noted:

1. The back and front pitches must be odd. But they cannot be equal. They differ by 2 or some multiple thereof.
2. The coil or back pitch Y_B must be nearly equal to a pole pitch $\left(\frac{Z}{P}\right)$.
3. The average pitch $Y_A = \frac{Y_B + Y_F}{2}$. It is equal to pole pitch $= \frac{Z}{P}$.
4. Commutator pitch $Y_C = \pm 1$ for simplex winding (in general, $Y_C = \pm m$ where m is the multiplicity of winding)
5. Resultant pitch Y_R is always even, being the arithmetical difference of two odd numbers, i.e., $Y_R = Y_B - Y_F$. It is equal to 2 m where m is multiplicity of winding.
6. For a two-layer winding, the number of slots and number of commutator segments are equal to the number of coils (i.e., half the number of coil sides).
7. The number of parallel paths in the armature = mP where m is the multiplicity of winding and P the number of poles.

Taking the first condition, we have $Y_B = Y_F \pm 2$.

* In duplex lap winding the number of parallel circuits is twice the number of poles and in triplex lap winding, the number of parallel circuits is three times the number of poles.

(*a*) If $Y_B > Y_F$ i.e., $Y_B = Y_F + 2$, then we get a progressive or right-handed winding i.e., a winding which progresses in the clockwise direction when we look towards the commutator end. Obviously, $Y_C = +1$, in this case.

(*b*) In case $Y_B < Y_F$ i.e., $Y_B = Y_F - 2$, we get a retrogressive or left-handed winding i.e., a winding which advances in the anti-clockwise direction just opposite to the previous case. In this case, $Y_C = -1$.

(*c*) Hence, $$\left.\begin{aligned} Y_F &= \frac{Z}{P} - 1 \\ Y_B &= \frac{Z}{P} + 1 \end{aligned}\right] \quad \text{...(4.1) for progressive winding and} \quad \left.\begin{aligned} Y_F &= \frac{Z}{P} + 1 \\ Y_B &= \frac{Z}{P} - 1 \end{aligned}\right] \quad \text{...(4.2) for retrogressive}$$

winding

Obviously, Z/P must be even to make the winding possible.

4.9 Numbering of Coils and Commutator Segments in Developed Winding Diagram

The developed winding diagram can be obtained by removing the armature periphery, cutting it along the slot and laying it out flat so that the slots and conductors can be viewed to trace out the armature winding. While drawing developed winding diagrams, we will designate only the coil sides with number (not individual turns). The upper side of the coil will be represented by a firm continuous line, whereas the lower side will be represented by a dotted line. The consecutive number i.e., 1, 2, 3,….. etc. will be assigned to the coil sides such that odd numbers are assigned to the top conductors and even numbers to the lower sides for a two-layer winding. The commutator segments will also be numbered consecutively, the number of the segments will be equal to the number of coils and the upper side will be connected to it.

The readers will be able to understand how to develop armature winding diagram by going through the following examples:

Example 4.1

Draw the developed winding diagram of a progressive lap winding for 4 pole, 16 slot single layer showing therein position of poles, direction of motion, direction of induced emf and position of brushes.

Solution:

Here, Number of poles, $P = 4$

Number of coil sides, $Z = 16$

$$\text{Pole pitch, } Y_P = \frac{Z}{P} = \frac{16}{4} = 4$$

For progressive winding, $Y_B = Y_P + 1 = 4 + 1 = 5$

$$Y_F = Y_B - 2\text{ m} = 5 - 2 \times 1 = 3 \text{ (here } m = 1)$$

The connections are to be made as per the following table:

Back End Connections	Front End Connections
1 to (1 + 5) = 6	6 to (6 – 3) = 3
3 to (3 + 5) = 8	8 to (8 – 3) = 5
5 to (5 + 5) = 10	10 to (10 – 3) = 7
7 to (7 + 5) = 12	12 to (12 – 3) = 9
9 to (9 + 5) = 14	14 to (14 – 3) = 11
11 to (11 + 5) = 16	16 to (16 – 3) = 13
13 to (13 + 5) = 18(18–16=2)	18 to (18 – 3) = 15
15 to (15 + 5) = 20(20–16=4)	4 to (4 – 3) = 1

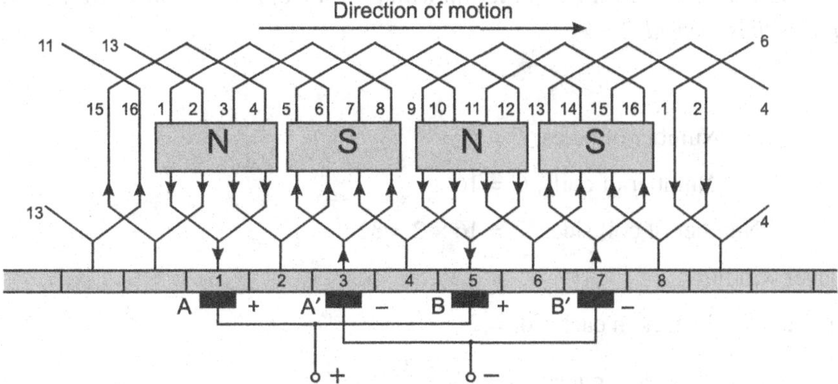

Fig. 4.22 Developed winding diagram

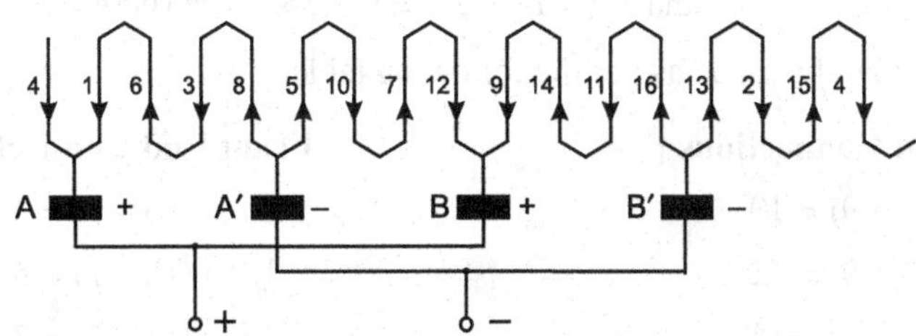

Fig. 4.23 Spiral winding diagram

Steps to draw developed winding diagram

1. Draw 16 coil-sides and number them.
2. Since it is a single layer winding draw all the 16 coil-sides by firm continuous lines.
3. Draw four poles *N, S, N* and *S* alternately, each pole covers the space of 16/4 = 4 coil sides. Consider that the poles are below the coil sides.
4. Mark the direction of motion (or rotation) of the conductors, say it is from left to right as marked.
5. Connect conductor 1 to 6 at the back and 6 to 3 at the front and so on as per the table.

6. Mark the direction of flow of current in the coil-sides i.e., downward in the coil sides which are in the influence of north-pole and upward for south-pole.
7. Connect each front end of the coil to each commutator segment as marked in Fig. 4.22.
8. Place four brushes at four commutator segments where either current is meeting (segment No. 1 and 5) or separating (segment No. 3 and 7).
9. The brushes where current is meeting are joined together and a positive terminal is obtained. The other two brushes where current is separating are joined together and a negative terminal is obtained.
10. The series, spiral winding connections are shown in Fig. 4.23, the brushes are placed as explained above.

Example 4.2

Draw a developed winding diagram of a simple two-layer lap-winding for a four-pole generator with 16 coils. Also draw the equivalent ring diagram with position of brushes and diagram representing parallel circuits thus formed.

Solution:

Here, Number of poles, $P = 4$

Number of coils, $C = 16$

Number of coil sides, $Z = 16 \times 2 = 32$

It is a two-layer winding,

$\therefore$ Number of coil sides in each slot = 2

$$\text{Pole pitch, } Y_P = \frac{Z}{P} = \frac{32}{4} = 8$$

For progressive winding, $Y_B = Y_P + 1 = 8 + 1 = 9$

and $Y_F = Y_B - 2\text{m} = 9 - 2 \times 1 = 7$ (here $m = 1$)

The connections are to be made as per the following table:

Back End Connections	Front End Connections
1 to (1 + 9) = 10	10 to (10 – 7) = 3
3 to (3 + 9) = 12	12 to (12 – 7) = 5
5 to (5 + 9) = 14	14 to (14 – 7) = 7
7 to (7 + 9) = 16	16 to (16 – 7) = 9
9 to (9 + 9) = 18	18 to (18 – 7) = 11
11 to (11 + 9) = 20	20 to (20 – 7) = 13
13 to (13 + 9) = 22	22 to (22 – 7) = 15
15 to (15 + 9) = 24	24 to (24 – 7) = 17
17 to (17 + 9) = 26	26 to (26 – 7) = 19
19 to (19 + 9) = 28	28 to (28 – 7) = 21

21 to (21 + 9) = 30   30 to (30 – 7) = 23

23 to (23 + 9) = 32   32 to (32 – 7) = 25

25 to (25 + 9) = 34(34–32=2)   34 to (34 – 7) = 27

27 to (27 + 9) = 36(36–32=4)   36 to (36 – 7) = 29

29 to (29 + 9) = 38(38–32=6)   38 to (38 – 7) = 31

31 to (31 + 9) = 40(40–32=8)   8 to (8– 7) = 1

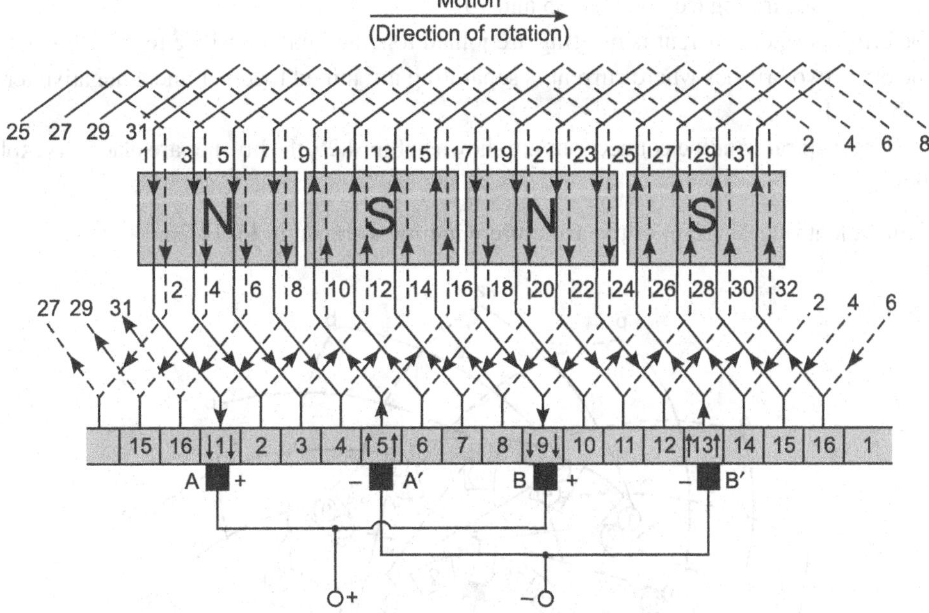

Fig. 4.24 Developed winding diagram

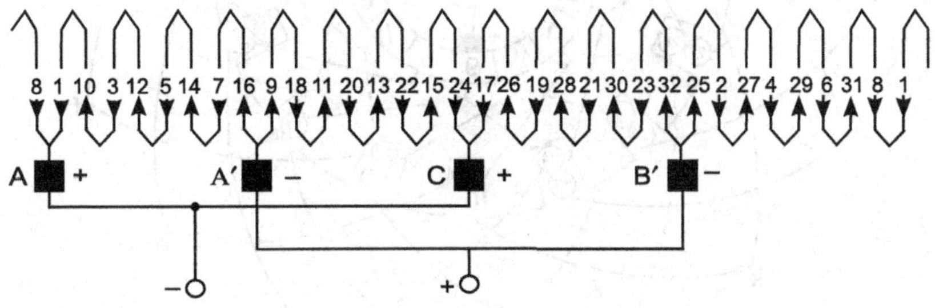

Fig. 4.25 Spiral winding diagram

Steps to draw developed winding diagram

1. Draw 16 × 2 = 32 coil-sides and number them.
2. Since it is a two-layer winding draw all the upper (odd numbered) coil-sides by firm continuous lines and all the lower (even numbered) coil sides by dotted lines.

3. Draw four poles *N, S, N* and *S* alternately, each pole covers the space of 32/4 = 8 coil sides. Consider that the poles are below the coil sides.
4. Mark the direction of motion (or rotation) of the conductors, say it is from left to right as marked.
5. Connect conductor 1 to 10 at the back and 10 to 3 at the front and so on as per the table.
6. Mark the direction of flow of current in the coil-sides i.e., downward in the coil sides which are in the influence of N-pole and upward for the S-pole.
7. Connect each front end of the coil to each commutator segment as marked in Fig. 4.24.
8. Place four brushes at four commutator segments where either current is meeting (segment No. 1 and 9) or separating (segment No. 5 and 13).
9. The brushes where current is meeting are joined together and a positive terminal is obtained. The other two brushes where current is separating are joined together and a negative terminal is obtained.
10. The series, spiral winding connections are shown in Fig. 4.25, the brushes are placed as explained above.

The equivalent ring diagram of the armature winding is shown in Fig. 4.26.

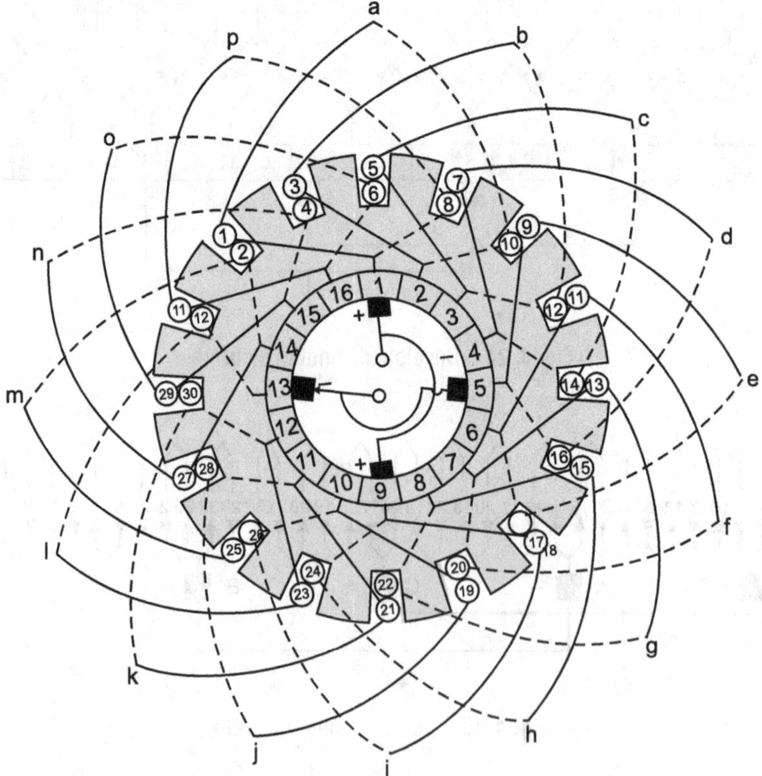

Fig. 4.26 Ring winding diagram

When brushes of same polarity are connected together, then all the armature conductors are divided into four parallel paths as shown in Fig. 4.27.

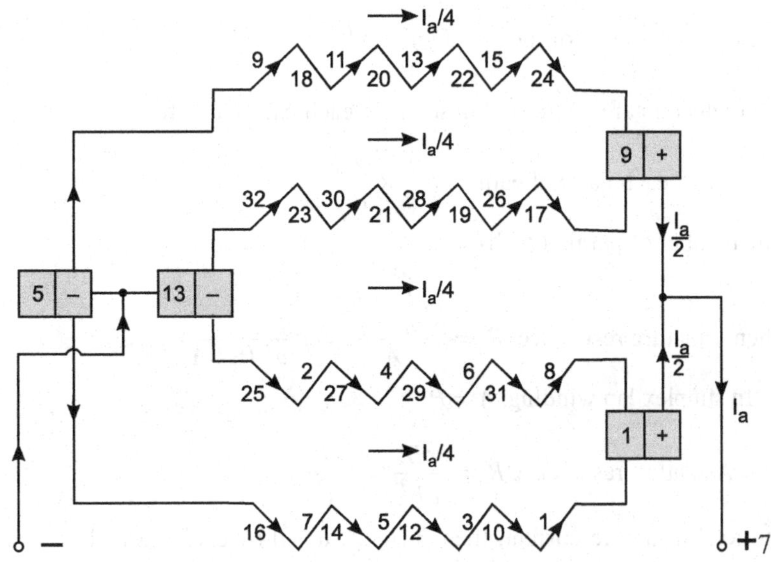

Fig. 4.27 Brush position and parallel paths

4.10 Characteristics of a Simplex Lap Winding

From the above examples it is concluded that

1. A simplex lap winding is a singly re-entrant closed winding because winding may start from any coil side (here it starts from coil side No. 1) returns back to the same coil side after connecting all the coil sides of the armature once and only once.
2. Both back pitch (Y_B) and front pitch (Y_F) of all the coils remain the same and odd number.
3. Total number of brushes is equal to the number of poles. These brushes are connected to the coil sides which lie in between the poles and have no emf induced in them at that particular instant.
4. There are as many parallel paths in the armature as the number of poles (see Fig. 4.27). That is why such a winding is also known as '*Multiple Circuit*' or '*parallel*' winding.

   In general, number of parallel paths in armature, $A = mP$ where m is the multiplicity of the lap winding. For simplex lap winding $m = 1$ ($\therefore A = P$), for duplex lap winding $m = 2$ and so on.
5. The emf developed across the positive (+ve) and negative (–ve) brushes is equal to the emf generated in any one of the parallel paths.

   Generated emf, E_g = emf generated per parallel path

   $$= \text{Average emf/coil side} \times \frac{Z}{P}$$

   where $\frac{Z}{P}$ is the number of coil sides connected in series per parallel path.
6. If l is the length of each armature conductor (or coil side), including overhang, a is the area of cross-section and ρ is the resistivity of conductor material, then

Resistance of each conductor (or coil side) = $\rho \frac{l}{a}$

Since $\frac{Z}{P}$ conductors are connected in series in each parallel path.

Resistance of each parallel path = $\rho \frac{l}{a} \times \frac{Z}{P}$

Say the number of parallel paths = A

$$\text{Then armature resistance, } R_a = \frac{\rho \frac{l}{a} \times \frac{Z}{P}}{A} = \rho \frac{l}{a} \cdot \frac{Z}{P \times A}$$

In simplex lap winding, $A = P$

$$\therefore \quad \text{Armature resistance, } R_a = \frac{\rho l Z}{a P^2} = \frac{\rho l Z}{a A^2} \qquad \dots (4.3)$$

7. If I_a is the total armature current, then current carried by each parallel path or conductor is I_a/P (in general I_a/A).

Example 4.3

Design and draw a 2 layer progressive duplex lap winding for a 6 pole DC generator with 24 slots, each slot having 2 coil sides. Indicate the direction of motion, position of poles, position of brushes with their polarities and parallel paths formed there-of.

Solution:

Here, Number of poles, $P = 6$

Number of slots = 24

Number of coil sides, $Z = 24 \times 2 = 48$ (2 layer winding)

For duplex winding, $m = 2$

$$\text{Pole pitch, } Y_P = \frac{Z}{P} = \frac{48}{6} = 8$$

For progressive winding, $Y_B = Y_P + 1 = 8 + 1 = 9$

and $Y_F = Y_B \pm 2m = 9 - 2 \times 2 = 5$ (since $m = 2$)

The connections are to be made as per the following table.

Steps to draw developed winding diagram

1. Draw $24 \times 2 = 48$ coil-sides and number them.
2. Since it is a two-layer winding draw all the upper (odd numbered) coil-sides by firm continuous lines and all the lower (even numbered) coil sides by dotted lines.
3. Draw six poles *N, S, N, S, N* and *S* alternately, each pole covers the space of 48/6 = 8 coil sides. Consider that the poles are below the coil sides.
4. Mark the direction of motion (or rotation) of the conductors, say it is from left to right as marked.

Winding-I		*Winding-II*	
Back-end Connections	*Front-end Connections*	*Back-end Connections*	*Front-end Connections*
1 to (1 + 9) = 10	10 to (10 – 5) = 5	3 to (3 + 9) = 12	12 to (12 – 5) = 7
5 to (5 + 9) = 14	14 to (14 – 5) = 9	7 to (7 + 9) = 16	16 to (16 – 5) = 11
9 to (9 + 9) = 18	18 to (18 – 5) = 13	11 to (11 + 9) = 20	20 to (20 – 5) = 15
13 to (13 + 9) = 22	22 to (22 – 5) = 17	15 to (15 + 9) = 24	24 to (24 – 5) = 19
17 to (17 + 9) = 26	26 to (26 – 5) = 21	19 to (19 + 9) = 28	28 to (28 – 5) = 23
21 to (21 + 9) = 30	30 to (30 – 5) = 25	23 to (23 + 9) = 32	32 to (32 – 5) = 27
25 to (25 + 9) = 34	34 to (34 – 5) = 29	27 to (27 + 9) = 36	36 to (36 – 5) = 31
29 to (29 + 9) = 38	38 to (38 – 5) = 33	31 to (31 + 9) = 40	40 to (40 – 5) = 35
33 to (33 + 9) = 42	42 to (42 – 5) = 37	35 to (35 + 9) = 44	44 to (44 – 5) = 39
37 to (37 + 9) = 46	46 to (46 – 5) = 41	39 to (39 + 9) = 48	48 to (48 – 5) = 43
41 to (41 + 9) = 50=2	50 to (50 – 5) = 45	43 to (43 + 9) = 52 = 4	52 to (52 – 5) = 47
45 to (45 + 9) = 54=6	6 to (6 – 5) = 1	47 to (47 + 9) = 56 = 8	8 to (8 – 5) = 3

5. It is a duplex winding having two entrants, start the connections of one winding from conductor (or coil side) number 1. Connect 1 to 10 at the back and 10 to 5 at the front and so on as per the table.

   For second winding start from coil side No. 3. Connect 3 to 12 at the back and 12 to 7 at the front and so on as per the table.
6. Mark the direction of flow of current in the coil-sides i.e., downwards in the coil sides which are under the influence of North-pole and upward for South pole.
7. Connect each front end of the coil to each commutator segment as marked in Fig. 4.28.

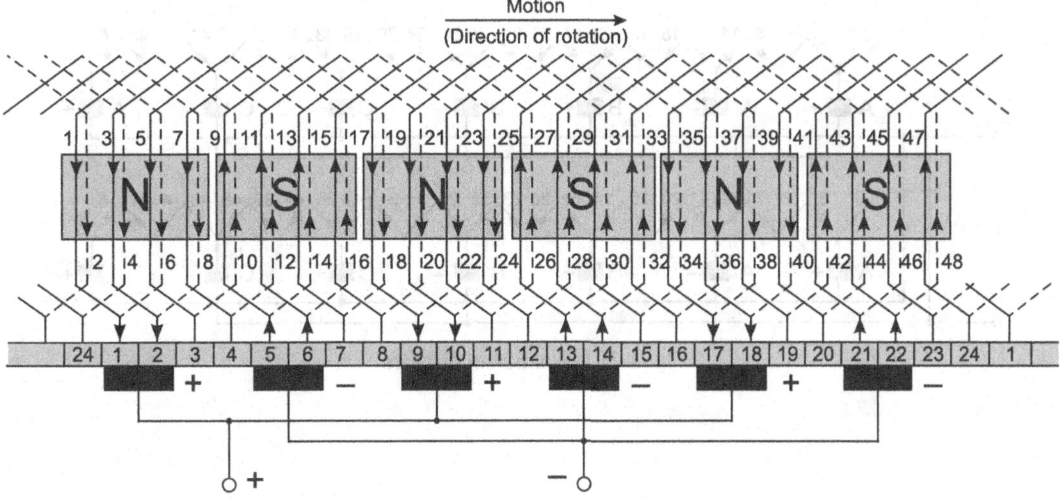

Fig. 4.28 Developed winding diagram

8. Place six brushes at six places, each brush is to cover two commutator segments, as shown in Fig. 4.28. Brushes are placed where either current is meeting (segment No. 1–2; 9–10 and 17–18) or separating (segment No. 5–6; 13–14; 21–22).
9. The three brushes where current is meeting are joined together and a positive terminal is obtained. The other three brushes where current is separating are joined together and a negative terminal is obtained.
10. The series, spiral winding connections of the two windings with separate entrant are shown in Fig. 4.29, the brushes are placed as explained above.

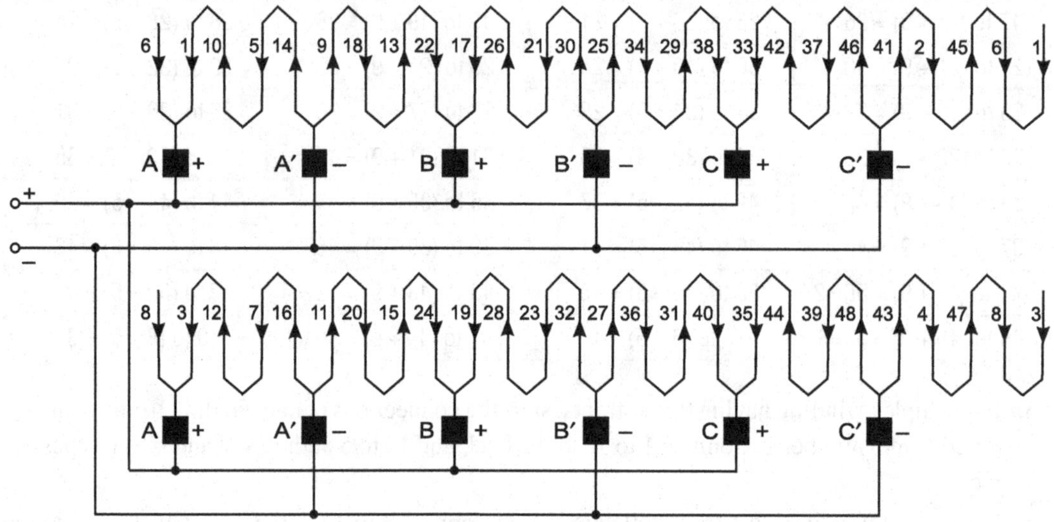

Fig. 4.29 Spiral winding diagram

11. The parallel paths formed by each winding are drawn in Figs. 4.30 and 4.31. Here, each winding is having six parallel paths. In all there are 12 (i.e., $A = mP$ *when* $m = 2$ and $P = 6$ in this case) parallel paths.

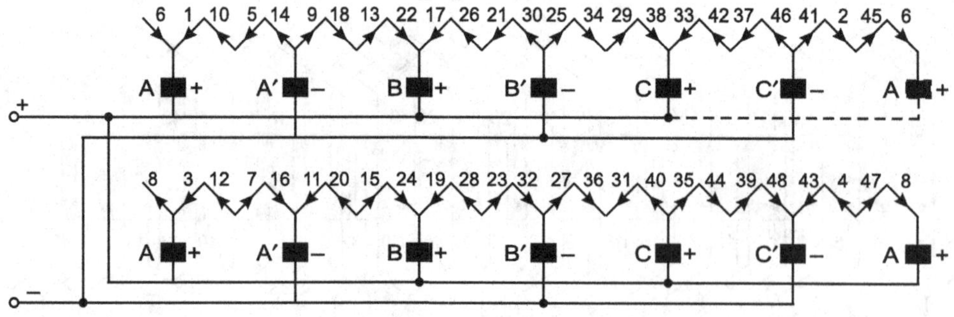

Fig. 4.30 Position of brushes

4.11 Characteristics of a Multiplex Lap Winding

From the winding diagram for duplex lap winding, the following conclusions have been drawn:

1. There are two independent winding, each closed on itself. Accordingly, the winding is said to be a *doubly re-entrant duplex winding.*

   It is important to note that that in duplex winding, if the coils are not multiple of 2, there will not be two independent windings and in such cases, the winding will close only once, then it is known as singly re-entrant duplex winding.

   Similarly, for triplex winding, the number of coils must be a multiple of 3 and so on.
2. Total number of brushes is equal to the number of poles but each brush must have a span of two commutator segments to collect current provided by two windings, as shown in Fig. 4.28. Similarly, for triplex winding each brush must have a span of three segments and so on.
3. The number of parallel paths for each winding is equal to the number of poles (i.e., 6 in this case). Therefore, total number of parallel paths for the armature is mP where m is the multiplicity, in this case $m = 2$, therefore the number of parallel paths = $2 \times 6 = 12$, as shown in Fig. 4.31.

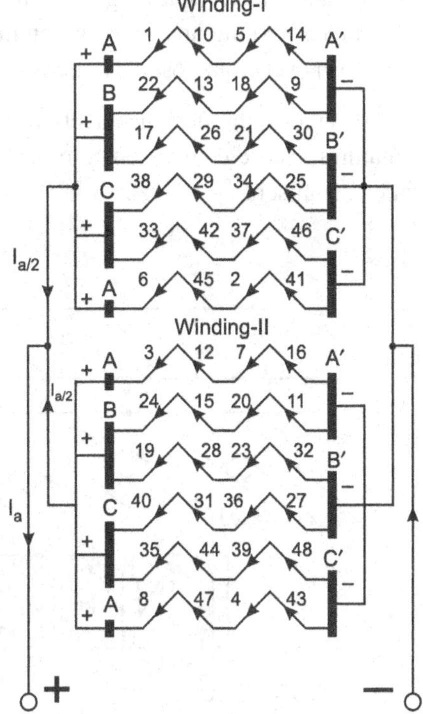

Fig. 4.31 Parallel paths

4. The number of conductors in each parallel path are Z/mP (here $m = 2$)
   In this case, No. of conductors connected in series in each parallel path, say,

$$Z_S = \frac{Z}{mP} = \frac{48}{2 \times 6} = 4$$

5. It is obvious that multiplex lap winding is used in the machines where we require high current to be delivered at small voltages.

4.12 Equalising Connections and their Necessity

In simplex lap winding, the number of parallel paths are equal to the number of poles and all the conductors of any one parallel path lie under one pair of poles. If the flux produced by all the poles is exactly the same, the emf induced in each parallel path and hence the current carried by each one of them will be equal. But in spite of best efforts, this condition cannot be obtained due to different properties of steel used for construction of poles and variation of air-gap between the poles and armature. Hence, there is always a slight difference in the induced emf in various parallel paths which results in large circulating currents in the armature winding. This circulating current flows from one parallel path to the other from one commutator segment to the other through brushes. This circulating current has the following ill-effects:*

1. Armature circulating current heats-up the armature winding.
2. It causes heavy sparking at the brushes which not only heats-up the machine but also causes undue burning and wear of commutator and brushes. If it is carried too far, it may result in flash over from positive to negative brushes.

To overcome this detrimental effect, usually, equalising connections are provided in all lap-wound armatures. These connections are made with the help of very low resistance (almost zero) wires called equalisers.

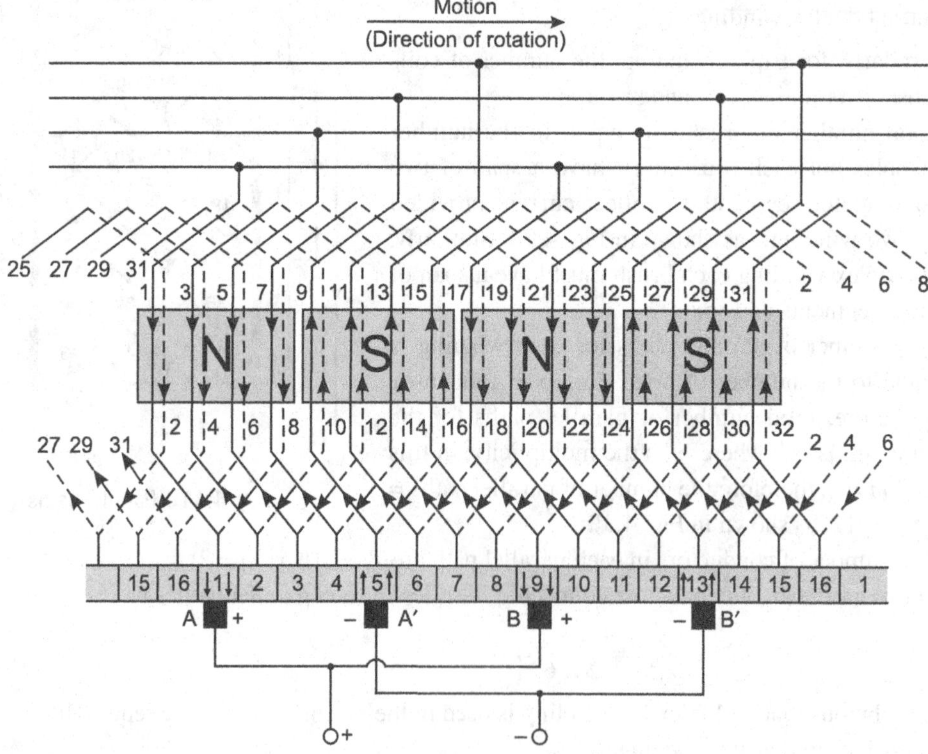

Fig. 4.32 Developed winding diagram

An equaliser is a copper wire having very low resistance and is used to connect the points of different parallel paths of armature winding which should have same potential under ideal conditions. These equaliser rings carry the circulating current and relieve the brushes to carry these circulating current. Under this condition, the brushes are to carry only the load current.

For making equaliser connections, the number of coils should be multiples of the number of poles and at the same time, the number of coils per pole should be divisible by a small number 2, 3 or 4.

* The current flowing in various paths may also differ due to difference in their resistances, but this would not pose serious effects.

For instance, consider a four-pole, 16-slot double layer, lap wound armature winding having 32 coil sides. Its developed winding diagram is shown in Fig. 4.32. Here, the number of coil sides per pole are 8. The equaliser connections are shown in Fig. 4.32. The following points are worth noting:

1. That every 3rd coil is connected to an equaliser. Each coil-side (or coil) connected to the equaliser occupies the same position relative to the poles such that same emf is induced in all the coils at every instant.
2. The number of connections to each equalising ring is equal to the number of pair of poles.
3. The number of equalising rings is equal to the number of coils under one pair of poles (i.e., 8/2 = 4).
4. Here, alternate coils have been connected to the equalising ring and as such it is said that the winding is equalised by 50%. If all the coils would have been connected to the equalisers it would have been said that the winding is 100% equalised.

Example 4.4

A six-pole DC armature with lap connected winding has 72 slots and 2 coil-sides per slot. Determine the winding pitches and connections to the 6 equalising rings.

Solution:

$$\text{Number of poles, } P = 6$$

$$\text{Number of slots} = 72$$

$$\text{Number of coil sides, } Z = 72 \times 2 = 144$$

$$\text{Number of parallel path, } A = P = 6 \text{ (for simplex lap winding)}$$

$$\text{Pole pitch, } Y_P = \frac{Z}{P} = \frac{144}{6} = 24$$

$$\text{Back pitch, } Y_B = Y_P + 1 = 24 + 1 = 25$$

$$\text{Front pitch, } Y_F = Y_B - 2 = 25 - 2 = 23$$

Distance between the coils having same potential i.e.,

$$Y_{eq} = \frac{\text{Total number of coils}}{\text{Pair of poles}} = \frac{Z/2}{P/2} = \frac{144/2}{6/2} = 24$$

$$\text{Total number of tapings} = \text{No. of equaliser rings} \times \text{pair of poles}$$

$$= n_r \times \frac{P}{2} = 6 \times \frac{6}{2} = 18$$

$$\text{Distance between adjacent tapings} = \frac{\text{Total No. of coils}}{\text{Total No. of taps}} = \frac{144/2}{18} = 4$$

Hence, the equaliser rings are to be arranged as per the following table:

Ring No.	I	II	III	IV	V	VI
Coil No.	1	5	9	13	17	21
	25	29	33	37	41	45
	49	53	57	61	65	69

Example 4.5

Calculate the different pitches for 4 pole, lap wound DC machine having 24 slots each has 2 coils-sides.

Solution:

Here, $P = 4$; No. of slots = 24; No. of coil sides/slot = 2

$$\therefore \quad Z = 24 \times 2 = 48; \text{Pole pitch, } Y = \frac{Z}{P} = \frac{48}{4} = 12$$

Back pitch, $Y_B = Y_{B+1} = 12 + 1 = 13$; Front pitch, $Y_P = Y_P - 1 = 12 - 1 = 11$

Resultant pitch, $Y_R = Y_B - Y_F = 13 - 11 = 2$; commutator pitch, $Y_C = 1$

4.13 Simplex Wave Winding

A simplex* wave winding is shown in Fig. 4.33. In lap winding a coil side under one pole is connected directly to a coil side at the back which occupies almost corresponding positioned coil side under the next pole and the finish end of that coil is connected to a commutator segment and to the start end of the adjacent coil situated under the same original pole. But in the wave winding the coil side is not connected back rather progresses forward to another coil side placed under the next pole, as shown in Fig. 4.33. In this way, the winding progresses, passing successively every North and South pole till it returns to the coil side from where it was started. It clearly shows that the connections of wave winding always progresses in the same direction around the armature instead of moving forward and backward alternately like that of lap winding. As the shape of winding is wavy, it is named as *wave winding.*

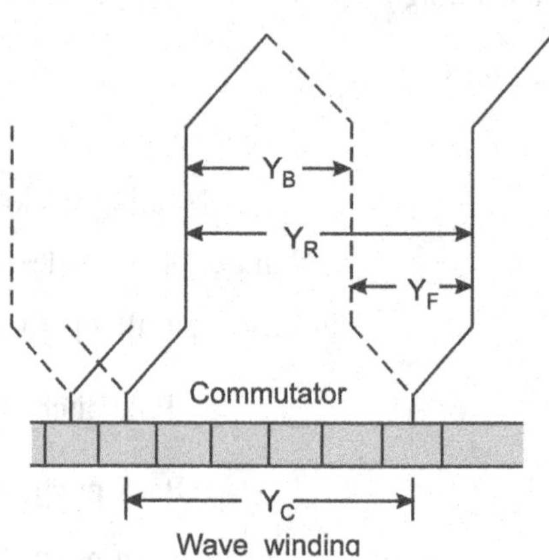

Fig. 4.33 Wave winding

The wave winding may be classified as progressive wave winding or retrogressive wave winding.

Progressive wave winding. If after completing one round of the armature, the winding falls in a slot to the right of its starting point the winding is known as *progressive wave winding.*

Retrogressive wave winding. If after completing one round of the armature. The winding falls in a slot to the left of its starting point, the winding is known as *retrogressive wave winding.*

The wave winding may also be a multiplex winding (double, triple or more).

A simplex wave winding has only two parallel paths irrespective of the number of pair of poles of the machine. However, a multiplex wave winding has 2 *m* parallel paths, where *m* is the multiplicity.

* Like lap winding, a wave winding may also be multiplex winding i.e., duplex, triplex and so on...

Regarding wave winding, the following points are worth noting:

(*a*) *Simplex wave winding*

1. Both back and front pitches must be odd.
2. Both back and front pitches must be nearly equal to pole pitch. They may be equal to each other or may differ by 2 i.e.,

$$Y_F = Y_B \text{ or } Y_F = Y_B \pm 2$$

3. Commutator pitch, $Y_C = Y_A$ (average pitch)
4. An average pitch Y_A must be an integer given by the relation

$$Y_A = \frac{Y_B + Y_F}{2} = \frac{Z \pm 2}{P} = \frac{Z/2 \pm 1}{P/2} = \frac{\text{No. of commutator segments} - 1}{\text{No. of pair of poles}}$$

The wave winding will be closed itself only if the value of average pitch (Y_A) comes out to be a whole number (i.e., integer) and agrees to the above formulae.

It is important to note here that in formulae $Y_A = \frac{Z-2}{P}$, the value ±2 has been taken due to the reason that after one round of armature the winding falls either short or go ahead by two conductors from the starting conductor. If it is not done and the average pitch is taken as *Z/P* then after one round of armature the winding would have been closed without including all coil-sides.

The positive sign provides progressive winding whereas negative sign provides retrogressive winding.

It is mentioned that average pitch (Y_A) must be a whole number, it shows that this winding is not possible with any number of coil-sides. For example, with 32 conductors in a four-pole machine, $Y_A = \frac{32 \pm 2}{4} = 8\frac{1}{2}$ or $7\frac{1}{2}$ is a fractional number, the wave winding is not possible.

5. Resultant pitch, $Y_R = Y_B + Y_F$
6. (*a*) For even number of pair of pole machines i.e., the machines which are having 2, 4, 6,...... pair of poles;

   (*i*) the number of coils must be odd.

   (*ii*) the number of commutator segments must be odd.

   (*iii*) the average pitch, $Y_A = \frac{Y_B + Y_F}{2}$ may be odd or even.

   (*b*) For odd number of pair of pole machines i.e., the machines which are having 3, 5, 7..... pair of poles;

   (*i*) the number of coils may be odd or even

   (*ii*) the number of commutator segments may be odd or even.

   (*iii*) the average pitch must be odd for even number of coils but it must be even for odd number of coils.

   (*c*) *Multiplex wave winding*

   (*i*) $Y_B = Y_F \pm 2m$

Positive sign for progressive winding and negative for retrogressive winding.

(*ii*) $Y_A = \frac{Z \pm 2m}{P}$ which must be an integer.

(*iii*) In case of duplex winding

(*a*) If No. of pair of poles is add and average pitch is odd, it must have odd number of commutator segments and coils.

If average pitch is even machine must have an even number of commutator segments and coils.

(*b*) If No. of pair of poles is even, the number of commutator segments and coils must be even irrespective whether the average pitch is odd or even.

If average pitch is odd, the winding will be singly re-entrant and if even doubly re-entrant.

4.14 Dummy Coils

While designing wave winding, a particular number of coil-sides are required. To accommodate these coil-sides, the stemming with particular number of slots are needed which may not be available with standard armature punching. In such cases, some coil-sides (or conductors) are placed in the vacant slots. These coil sides are not connected to other coil-sides of the winding and do not participate in the conversion of power. Such coil-sides or coils are called *dummy coil*. These are employed only to make the mechanical balancing of the armature.

Thus, *the coils placed on the armature which do not participate in the conversion of power but are employed only to make mechanical balancing of the armature are known as* **dummy coils.**

Example 4.6

Draw a developed winding diagram showing position of poles, direction of rotation, polarity and position of brushes of a simplex two-layer wave winding for a four-pole DC generator with 30 armature conductors. Hence point out the characteristics of a simple wave winding.

Solution:

Here, $P = 4; Z = 30$

$$\text{Average pitch, } Y_A = \frac{Z \pm 2}{P} = \frac{30 \pm 2}{4} = 8^{**} \text{ or } 7$$

If Y_A is taken odd i.e., 7, then $Y_B = Y_F$

Starting from 1st coil-side, connect coil-side 1 to 1 + 7 = 8 at the back and 8 to 8 + 7 = 15 at the front through commutator segment 8. The first coil side is connected to commutator segments 1 at the front. Thus $Y_C = 8 - 1 = 7$. The coil side 15 is further connected to coil side 15 + 7 = 22 at the back and 22 to 22 + 7 = 29 at the front and so on, as per the table given below, till winding is closed.

* If Y_A is taken as 8, then $Y_B = 9$ and $Y_F = 7$ or $Y_F = 9$ and $Y_B = 7$. However, it is important to note that if $Y_C = 7$ the armature will rotate in one direction and if $Y_C = 8$, it will rotate in opposite direction.

Back-end Connections	Front-end Connections	Through Segment No.
1 to (1 + 7) = 8	8 to (8 + 7) = 15	8
15 to (15 + 7) = 22	22 to (22 + 7) = 29	15
29 to (29 + 7) = 36 = 6	6 to (6 + 7) = 13	7
13 to (13 + 7) = 20	20 to (20 + 7) = 27	14
27 to (27 + 7) = 34 = 4	4 to (4 + 7) = 11	6
11 to (11 + 7) = 18	18 to (18 + 7) = 25	13
25 to (25 + 7) = 32 = 2	2 to (2 + 7) = 9	5
9 to (9 + 7) = 16	16 to (16 + 7) = 23	12
23 to (23 + 7) = 30	30 to (30 + 7) = 37 = 7	4
7 to (7 + 7) = 14	14 to (14 + 7) = 21	11
21 to (21 + 7) = 28	28 to (28 + 7) = 35 = 5	3
5 to (5 + 7) = 12	12 to (12 + 7) = 19	10
19 to (19 + 7) = 26	26 to (26 + 7) = 33 = 3	2
3 to (3 + 7) = 10	10 to (10 + 7) = 17	9
17 to (17 + 7) = 24	24 to (24 + 7) = 31 = 1	1

The developed winding diagram, as per table, is shown in Fig. 4.34.

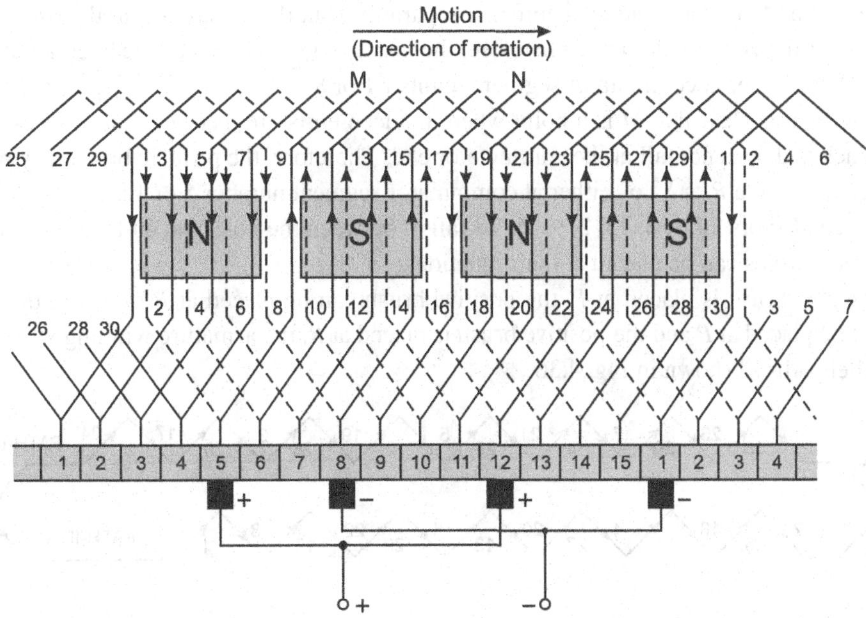

Fig. 4.34 Developed winding diagram

The table shows that winding connections started from coil side number 1 finishes at the same coil side, thus, form a closed circuit.

The ring or spiral winding diagram of a wave wound armature is shown in Fig. 4.35.

Brush Position

1. The direction of induced *emf* in various coil-sides is marked as per the following grouping (see Fig. 4.34)

Downward	Upward	Downward	Upward
(N-Pole)	(S-Pole)	(N-Pole)	(S-Pole)
1–8	9–16	17–23	24–30
(8)	(8)	(7)	(7)

2. The direction of induced emf in various coil-sides is also marked in the spiral winding diagram shown in Fig. 4.35.

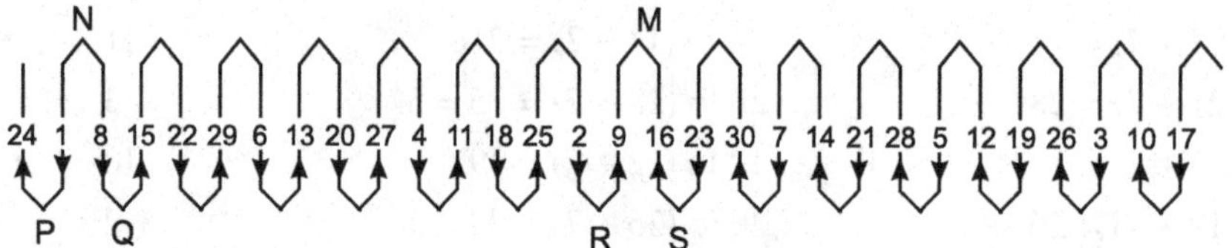

Fig. 4.35 Spiral winding

3. Figure 4.35 shows that whole winding is divided into two portions i.e., between *N* to *M* and to *M* to *N*.
4. Here *N* is a point where induced emfs are separating, but this position is at the back and does not fall at the commutator end. Therefore, the negative brush may be placed either at position *P* or *Q* i.e., either at commutator segment number 1 or 8.
5. Similarly, it is clear that *M* is a point where induced emfs are meeting, but this position is at the back and does not fall at the commutator end. Therefore, the positive brush may be placed either at position *R* or *S* i.e., either at commutator segment number 5 or 12.
6. Here, conductors 24, 1, 8, 15, 2, 9, 16, 23 fall in between the poles, no emf is induced in them and the brushes can be placed at their junctions.
7. If negative brush is placed at *P*, the positive brush can be placed at *R* or *S*. Let the negative brush be placed at *P* and the positive brush is placed at *R*, the armature winding will form two parallel paths as shown in Fig. 4.36.

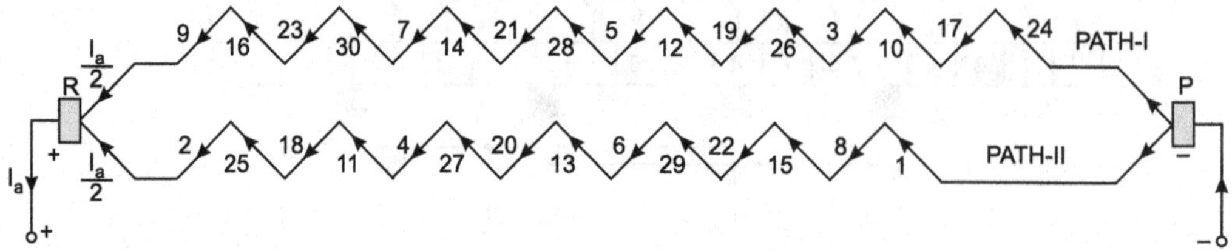

Fig. 4.36 Parallel paths

8. If in all, 4 brushes (two positive and two negative) are used and are placed in both alternative positions, then the loops lying between the brushes of same polarity are being short circuited as shown in Fig. 4.37, but the number of parallel paths remain only two.

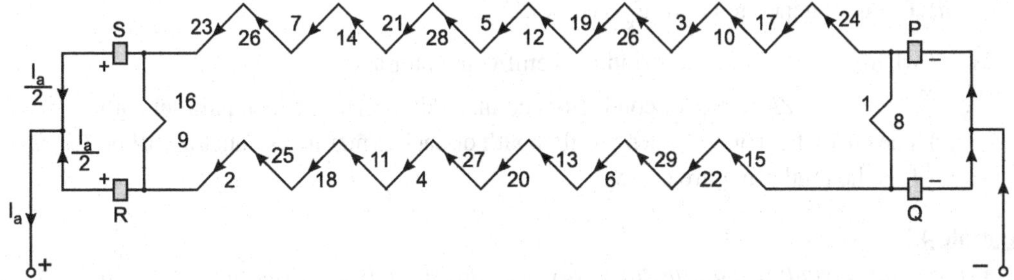

Fig. 4.37 Path formation with four brushes

9. The ring diagram with brush position is shown in Fig. 4.38.

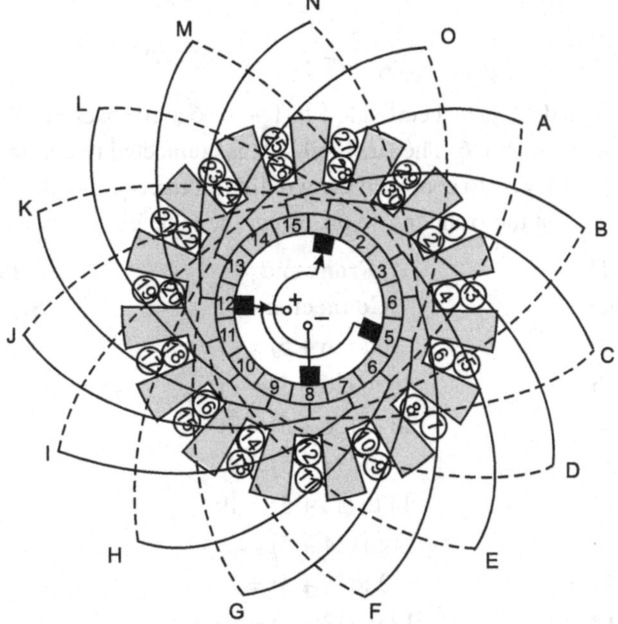

Fig. 4.38 Ring diagram

From the above explanation, it is summarised that:

(*i*) Only two brushes are necessary (at position *P* and *R*, see Fig. 4.36), although their number may be kept equal to the number of poles (at position *P*–*Q*, negative and *R*–*S*, positive; see Fig. 4.38).

Note: *Only two brushes are necessary, however, parallel brushes may be added to make their number equal to the number of poles. The parallel brush arrangement reduces the current per brush set which reduces the axial length of commutator and consequently reduces the cost of commutator or machine. This arrangement is used in large machines, for small machines only two brushes (one set) is sufficient.*

(*ii*) Wave wound armature winding has only two parallel paths irrespective of the number of poles of the machine. That is why, this winding is sometimes called '*two circuit winding*' or '*series winding*'.

(*iii*) The total generated emf, $E_g = e_{av} \times Z/2$

where e_{av} = average induced emf/conductor and

$Z/2$ = No. of conductors connected in series in each parallel path.

(*iv*) The current carried by each parallel path or each armature conductor will be $I_a/2$ where I_a is the total armature current.

Example 4.7

Draw a developed winding diagram for a wave wound, six-pole DC armature having 16 slots with 2 coil-sides per slot, each coil has single turn.

Solution:

Here, $P = 6; Z = 16 \times 2 = 32$

$$\text{Average pitch, } Y_A = \frac{Z \pm 2}{P} = \frac{32 - 2}{6} = 5 \text{ (add number)}$$

Starting from 1st coil-side, connect coil-side 1 to 1 + 5 = 6 at the back and 6 to 6 + 5 = 11 at front through commutator segment No. 6. The first coil-side is connected to commutator segment No. 1 at the front. Thus $Y_C = 6 - 1 = 5$. To complete the winding, connect coil-side 11 to 11 + 5 = 16 at the back and 16 to 16 + 5 = 21 at the front and so on, as per the table given below:

Back-end Connections	Front-end Connections	Through Segment No.
1 to (1 + 5) = 6	6 to (6 + 5) = 11	6
11 to (11 + 5) = 16	16 to (16 + 5) = 21	11
21 to (21 + 5) = 26	26 to (26 + 5) = 31	16
31 to (31 + 5) = 36 = 4	4 to (4 + 5) = 9	5
9 to (9 + 5) = 14	14 to (14 + 5) = 19	10
19 to (19 + 5) = 24	24 to (24 + 5) = 29	15
29 to (29 + 5) = 34 = 2	2 to (2 + 5) = 7	4
7 to (7 + 5) = 12	12 to (12 + 5) = 17	9
17 to (17 + 5) = 22	22 to (22 + 5) = 27	14
27 to (27 + 5) = 32	32 to (32 + 5) = 37 = 5	3
5 to (5 + 5) = 10	10 to (10 + 5) = 15	8
15 to (15 + 5) = 20	20 to (20 + 5) = 25	13
25 to (25 + 5) = 30	30 to (30 + 5) = 35 = 3	2
3 to (3 + 5) = 8	8 to (8 + 5) = 13	7
13 to (13 + 5) = 18	18 to (18 + 5) = 23	12
23 to (23 + 5) = 28	28 to (28 + 5) = 33 = 1	1

The table shows that winding connections stated from coil-side No. 1 finishes at the same coil side, thus, form a closed circuit.

The ring or spiral winding diagram of a wave wound armature is shown in Fig. 4.40.

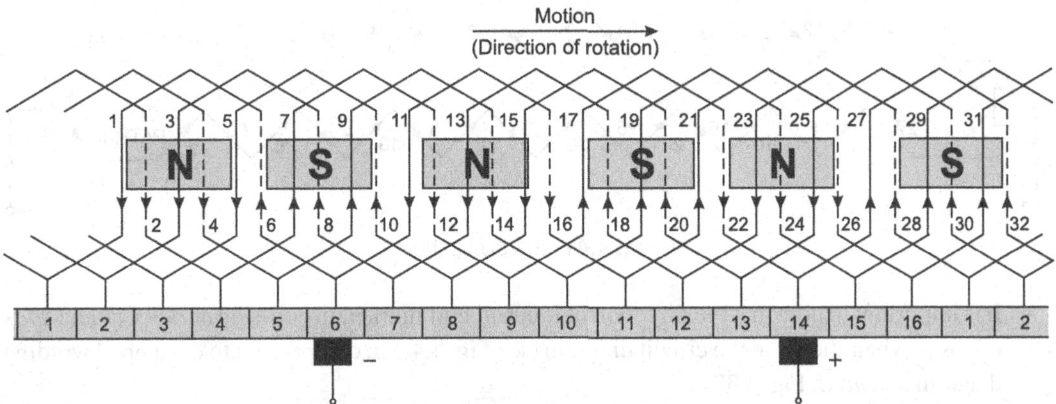

Fig. 4.39 Developed winding diagram

Brush Position

1. The direction of induced *emf* in various coil-sides is marked as per the following grouping (see Fig. 4.39):

Downward (N-Pole)	Upward (S-Pole)	Downward (N-Pole)	Upward (S-Pole)	Downward (N-Pole)	Upward (S-Pole)
1–5	6–10	11–16	17–21	22–26	27–32
(5)	(5)	(6)	(5)	(5)	(6)

2. The direction of induced emfs in various coil-sides is also marked in the spiral winding diagram, shown in Fig. 4.40.

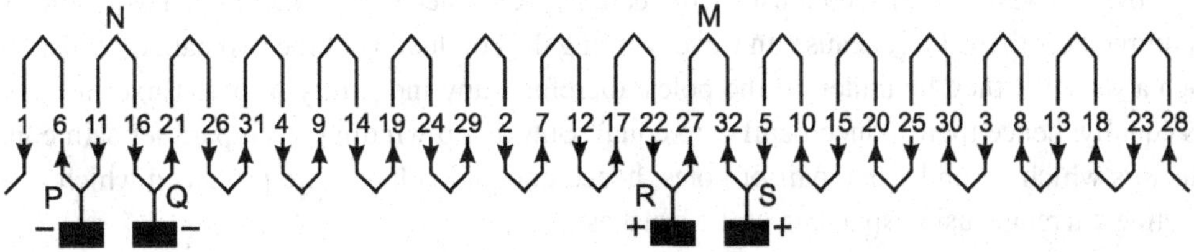

Fig. 4.40 Spiral winding

3. Figure 4.40 shows that whole winding is divided into two portions between *N* to *M* and *M* to *N*.
4. From Fig. 4.40, it is clear that point *N* is a position where induced emfs are separating, but this position is at the back and does not fall at the commutator end. Therefore, negative brush may be placed either at position *P* or *Q* i.e., either at commutator segment number 6 or 11.
5. Similarly, from Fig. 4.40, it is clear that point *M* is a position where induced emfs are meeting, but this position is at the back and does not fall at the commutator end. Therefore, positive brush may be placed either at position *R* or *S* i.e., either at commutator segment number 14 or 3.
6. If negative brush is placed at *P*, the corresponding position of positive brush will be *R*. However, if the negative brush is placed at *Q*, the corresponding position of positive brush will be *S*.
7. The two parallel paths thus formed are shown in Fig. 4.41.

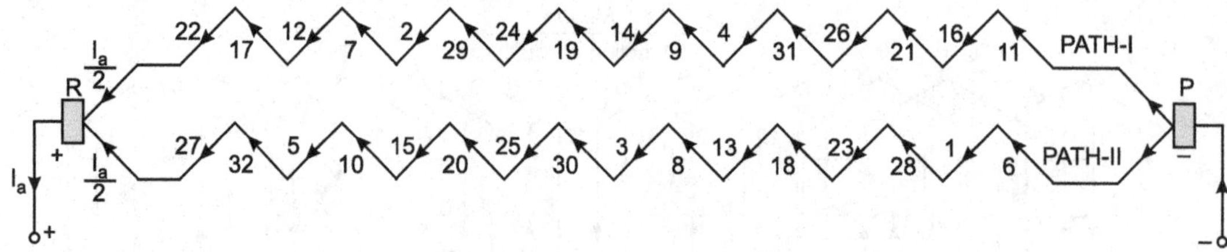

Fig. 4.41 Parallel paths

8. It is important to note that the direction of induced emf marked in coil-side number 11 and 27 is opposite when shown in the circuit diagram (see Fig. 4.41) in comparison to developed winding diagram shown in Fig. 4.39.

   In fact, these two coil-sides, at this particular instant, are passing through magnetic neutral axis and the induced emf in these conductors is zero. Hence, they do not affect the system.

4.15 Applications of Lap and Wave Windings

The lap winding contains more number of parallel paths and provides large current. Hence, this winding is applied to the generators which are to deliver more current. Wave winding is more suitable for small generators specially these are meant for 500–600 V circuits.

The main advantage of wave winding is that it gives more emf than lap winding for given number of poles and armature conductors, whereas, the lap winding would require large number of conductors for the same emf. This will result in higher cost of winding and less utilisation of space in the armature slots.

Moreover, in wave winding, equalising connections are not necessary, whereas in lap winding these are required definitely. It is because, in wave winding the conductors of the two paths are distributed in such a way that they lie under all the poles, therefore, any inequality of pole fluxes affects two paths equally, hence their induced emfs are equal. But, in lap winding, each parallel path contains conductors which lie under one pair of poles, hence unequal voltages are produced which set-up a circulating current causing sparking at the brushes.

Thus, in general practice, wave winding is preferred, the lap winding is only used for low-voltage, high-current machines.

Section Practice Problems

Numerical Problems

1. Draw a developed winding diagram of progressive lap winding for 4 pole, 24 slot with one coil side per slot, single layer winding showing there in position of poles, direction of motion, direction of induced emf and position of brushes.
2. Design and draw a 2 layer progressive duplex lap winding for a 4 pole DC generator with 16 slots. Show the position of brushes and their polarity. Also draw a diagram representing number of parallel paths thus formed.
3. Prepare a winding diagram for a four-pole wave connected armature of a DC generator having 22 coil sides.

Short Answer Type Questions

Q.1. Define statically and dynamically induced emf.

Ans. When flux linking with the coil changes on emf is induced in the coil is called statically induced emf, in this case both coil and magnetic field system are stationary.

When a constructor cuts across the magnetic field an emf is induced in the conductor is called dynamically induced emf, in this case either conductor on the magnetic field system is moving (or rotating).

Q.2. Why is commutator employed in DC machines?

Ans. Commutator in employed in DC machines:

1. To convert AC into DC, in generator action, to deliver DC electric power to the load.
2. To convert DC into AC, in motor action, to develop unidirectional torque in armature.

Q.3. Define coil pitch and coil span.

Ans. *Coil pitch or Coil span*: The distance between two sides of the same coil is called coil pitch or coil span.

The coil-pitch is usually expressed in terms of number of armature conductor or number of slots between two sides of the same coil.

The coil span is usually expressed in terms of angular displacement (electrical angle) between the two sides of the same coil.

Q.4. Define resultant pitch.

Ans. The distance in terms of number of armature conductor or number of slots between the start of one coil and the start of next coil to which it is connected is called resultant pitch.

Q.5. Explain the difference between integral slot and fractional slot winding.

Ans. The winding in which the number of armature slots per pole give an integer (whole number), the winding is called integral slot winding.

The winding in which the number of armature slots per pole gives fractional value, the winding is called fractional slot winding.

Q.6. Define the factors and state benefits of short pitched coils.

Ans. The advantage of short pitched winding is that in this case the copper used for end connections is reduced substantially which reduces the cost of machine. If also improves the commutation (reduction of sparking at brushes) because inductance of overhang connections is reduced. Moreover, it reduces the copper losses and improves the efficiency to some extent. Hence, many a times short pitched winding is used.

Q.7. What are equaliser rings and why are they used?

Ans. *Equaliser rings and their use:* An equaliser is a copper wire having very low resistance and is used to connect the points of different parallel paths of armature winding which should have same potential under ideal conditions. These equaliser rings carry the circulating current and relieve the brushes to carry these circulating current. Under this condition, the brushes are to carry only the load current.

Q.8. What is the basic principle on which a generator operates?

Ans. A generator operates on the basic principle of production of dynamically induced emf (electro-magnetic induction).

Q.9. Why is the pole shoe section of a DC machine made larger than pole core?

Ans. The pole shoe is made larger because it is to support the field coils and spreads out the flux over the armature periphery more uniformly. The larger *x*-section area reduces the reluctance of the magnetic path.

Q.10. Is it necessary to use a large number of coils and commutator segments with the coils evenly distributed around the surface of the armature of a DC machine?

Ans. Yes, it is necessary it provides practically constant unidirectional voltage at the brushes.

Q.11. Why the armature of a DC machine is made of laminated silicon steel?

Ans. To reduce hysteresis and eddy current loss.

Q.12. In large DC machines electro-magnets are preferred over permanent magnets, why?

Ans. On account of their greater magnetic effect and field strength regulation.

Q.13. In small DC machine, cast iron yokes are preferred, why?

Ans. Because of their low cost.

Q.14. Why fractional pitch winding is preferred over full pitch winding?

Ans. To effect saving in the copper of end connections and to have improved commutation by reducing inductance effect of overhang.

Q.15. What for brushes are employed in DC generator.

Ans. The brushes are employed to collect current from the commutator and supply it to the external load circuit.

Q.16. For what type of DC machine, wave winding is employed?

Ans. For low current and high voltage rating machine.

Q.17. Which are the different types of armature windings commonly used in DC machines?

Ans. Lap and wave windings.

Q.18. For what type of DC machine lap winding is employed?

Ans. For high current and low voltage rating machine.

Q.19. Why equaliser rings are used?

Ans. Equaliser rings are used in lap wound machines to avoid unequal distribution of current at the brushes, thus reducing sparking at the brushes.

Q.20. In which type of armature winding equaliser connections are used?

Ans. In lap winding.

4.16 emf Equation

Let, P = Number of poles of the machine.

ϕ = Flux per pole in Wb

Z = Total number of armature conductors.

N = Speed of armature in *rpm*

A = Number of parallel paths in the armature winding.

In one revolution of the armature; Flux cut by one conductor $= P\phi$ Wb

Time taken to complete one revolution, $t = 60/N$ second

∴ Average induced emf in one conductor,

$$e = \frac{P\phi}{t} = \frac{P\phi}{60/N} = \frac{P\phi N}{60} \text{ volt}$$

The number of conductors connected in series in each parallel path = Z/A.

∴ Average induced emf across each parallel path or across the armature terminals,

$$E = \frac{p\phi N}{60} \times \frac{Z}{A} = \frac{PZ\phi N}{60A} \text{ volt} \quad \ldots(4.4)$$

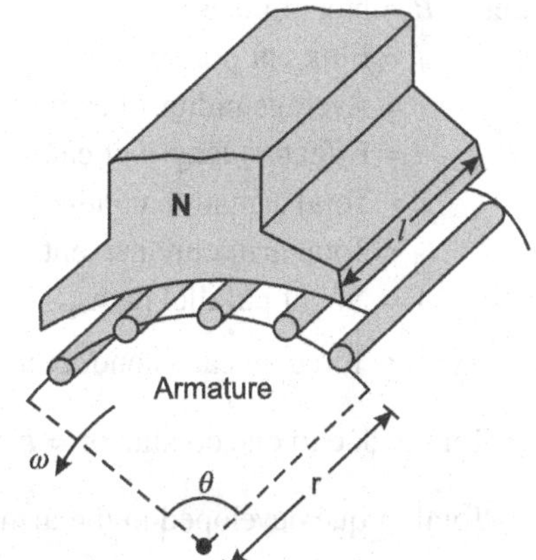

Fig. 4.42 A portion of a DC machine

or $$E = \frac{PZ\phi n}{A} \text{ where } n \text{ in speed in } r.p.s.$$

i.e., $$n = \frac{N}{60}$$

For a given machine, the number of poles and number of conductors per parallel path (Z/A) are constant.

∴ $$E = K\phi n \text{ where } K = \frac{PZ}{A} \text{ is a constant or } E \propto \phi n$$

or $$E = K_1\phi N \text{ where } K_1 = \frac{PZ}{60A} \text{ is another constant or } E \propto \phi N \quad \ldots(4.5)$$

or $$E \propto \phi\,\omega \text{ where } \omega = \frac{2\pi N}{60} \text{ is the angular velocity in radian/second}$$

Thus, we conclude that the induced emf is directly proportional to flux per pole and speed. Moreover, the polarity of the induced emf depends upon the direction of magnetic field and the direction of rotation. If either of the two is reversed, the polarity of induced emf i.e., brushes is reversed, but when both are reversed the polarity does not change.

This induced emf is fundamental phenomenon to all DC machines whether they are working as generator or motor. However, when the machine is working as a generator, this induced emf is called generated emf and is represented as E_g, i.e., $E_g = \frac{PZ\phi N}{60A}$ volt.

Whereas, in case the machine is working as a *motor*, this induced emf is called back emf as it acts opposite to the supply voltage V. Then $E_b = \frac{PZ\phi N}{60A}$ volt.

4.17 Torque Equation

We know that when a current carrying conductor is placed in the magnetic field a force is exerted on it which exerts turning moment or torque ($F \times r$) (see Fig. 4.43). This torque is produced due to electro-magnetic effect, hence is called *electromagnetic torque.*

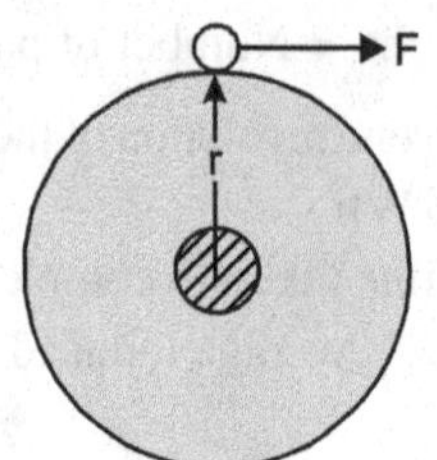

Fig. 4.43 Conductor placed on armature of radius 'r'

Let P = No. of poles.
ϕ = Flux per pole in Wb.
r = Average radius of armature in metre.
l = Effective length of each conductor in metre.
Z = Total armature conductors.
I_a = Total armature current.
A = No. of parallel paths.

Average force on each conductor, $F = Bil$ newton

Torque due to one conductor = $F \times r$ newton metre

Total torque developed in the armature,

$$T = ZFr \text{ newton metre}$$

or
$$T = ZB\,i\,l\,r$$

Now, current in each conductor, $i = I_a/A$

*Average flux density, $B = \phi/a$

where 'a' is the X- sectional area of flux path at radius r.

Obviously,
$$a = \frac{2\pi rl}{P}\text{m}^2 \;\therefore\; B = \frac{\phi P}{2\pi r l} \text{ tesla}$$

Substituting these values in equation (i), we get

$$T = Z \times \frac{\phi p}{2\pi r l} \times \frac{I_a}{A} \times l \times r \text{ or } T = \frac{PZ\phi I_a}{2\pi \text{A}} \text{Nm} \qquad \ldots(4.6)$$

Alternately; The power developed in the armature is given as

$$EI_a = \omega T \quad \text{or} \quad EI_a = \frac{2\pi N}{60} \times T$$

or
$$\frac{\phi ZNP}{60A} \times I_a = \frac{2\pi N}{60} \times T \quad \text{or} \quad T = \frac{PZ\,\phi N}{2\pi A}\text{Nm } (\textit{As above})$$

For a particular machine, the number of poles (P), number of conductors per parallel path (Z/A) are constant.

$$\therefore \qquad T = K\frac{\varphi \text{I}_\alpha}{2\pi} \text{ where } K = \frac{PZ}{A} \text{ is a constant}$$

The constant K for a given machine is the same for the emf equation as well as the torque equation.

Also, $T = K_2\,\phi\,I_a$ where $K_2 = \dfrac{PZ}{2\pi A}$ is another constant. Thus, $T \propto \phi\,\text{I}_a$ ...(4.7)

* If B = flux density in tesla; θ = arc angle subtended by the pole shoe; l = length of pole shoe; r = radius of pole shoe. Then, $\phi = B \times a = B \times 2\pi r \dfrac{\theta}{360} \times l$ weber

Thus, we conclude that torque produced in the armature is directly proportional to flux per pole and armature current. Moreover, the direction of electromagnetic torque developed in the armature depends upon the direction of flux or magnetic field and the direction of flow of current in armature conductors. The direction of torque produced and hence the direction of rotation is reversed if either of the two is reversed. But when both are reversed the direction of torque does not change.

Example 4.8

An eight pole lap wound DC generator has 960 conductors, a flux of 40 m Wb per pole and is driven at 400 rpm. Find OC emf.

Solution:

Open circuit emf, $E_g = \frac{\phi ZNP}{60A}$

where, $\phi = 40$ m Wb $= 40 \times 10^{-3}$ Wb; $Z = 960$; $N = 400$ rpm; $P = 8$

$$A = P = 8 \text{ (lap winding)}$$

$$\therefore \quad E_g = \frac{40 \times 10^{-3} \times 960 \times 400 \times 8}{60 \times 8} = \mathbf{256\ V}\ (Ans.)$$

Example 4.9

A 4-pole, DC machine is having 500 wave wound conductors and running at 1000 rpm. The flux per pole is 30 m Wb. What will be the voltage induced in the armature winding.

Solution:

Here, $P = 4$; $A = 2$ (Wave wound); $Z = 728$; $N = 1800$ rpm;

$\phi = 35$ m Wb $= 35 \times 10^{-3}$ Wb

Generated voltage, $E_g = \frac{\phi ZNP}{60A} = \frac{35 \times 10^{-3} \times 728 \times 1800 \times 4}{60 \times 2} = \mathbf{1528.8\ V}$ (*Ans.*)

Example 4.10

A 4-pole, DC machine has 144 slots in the armature with two coil-sides per slot, each coil has two turns. The flux per pole is 20 m Wb, the armature is lap wound and if rotates at 720 rpm, what is the induced emf (i) across the armature (ii) across each parallel path?

Solution:

Here, $P = 4$; $A = P = 4$ (Lap wound); $\phi = 20$ m Wb $= 20 \times 10^{-13}$ Wb; $N = 720$ rpm

No. of slots = 144 with 2 coil sides per slot and each coil has two turns

$$\therefore \quad Z = 144 \times 2 \times 2 = 576$$

Induced emf across armature, $E_g = \frac{\phi ZNP}{60A} = \frac{20 \times 10^{-3} \times 576 \times 720 \times 4}{60 \times 4} = \mathbf{138.24\ V}$ (*Ans.*)

Voltage across each parallel path = E_g = **138.24 V** (*Ans.*)

Example 4.11

A six-pole machine has an armature with 90 slots and 8 conductors per slot, the flux per pole is 0.05 Wb and rms at 1000 rpm. Determine induced emf if winding is (i) lap connected (ii) wave connected.

Solution:

Here, $P = 6$; $\phi = 0.05$ Wb; $N = 1000$ rpm

No. of slots = 90 each slot with 8 conductors

$$\therefore \quad Z = 90 \times 8 = 720$$

(*i*) *When lap connected*: $A = P = 6$

$$\text{Induced emf, } E_g = \frac{\phi ZNP}{60A} = \frac{0.05 \times 720 \times 1000 \times 6}{60 \times 6} = \mathbf{600\ V}\ (Ans.)$$

(*ii*) *When wave connected*: $A = 2$

$$\text{Induced emf, } E_g = \frac{\phi ZNP}{60A} = \frac{0.05 \times 720 \times 1000 \times 6}{60 \times 2} = \mathbf{1800\ V}\ (Ans.)$$

Example 4.12

A DC generator carries 600 conductors on its armature with lap connections. The generator has 8 poles with 0.06 Wb useful flux. What will be the induced emf at its terminals if it is rotated at 1000 rpm? Also determine the speed at which it should be driven to induce the same voltage with wave connections?

Solution:

Here, $P = 8$; $Z = 600$; $\phi = 0.06$ Wb; $N = 1000$ rpm.

$$A = P = 8 \text{ (when lap wound)}$$

$$\text{Induced emf, } E_g = \frac{\phi ZNP}{60A} = \frac{0.06 \times 600 \times 1000 \times 8}{60 \times 8} = \mathbf{600\ V}\ (Ans.)$$

when wave wound, let the speed be N′ rpm but $E_g = 600$ V

$$\text{Now, } N' = \frac{E_g \times 60A}{\phi ZP} = \frac{600 \times 60 \times 2}{0.06 \times 600 \times 8} = \mathbf{250\ rpm}\ (Ans.)$$

Example 4.13

A wave wound armature of an eight-pole generator has 51 slots. Each slot contains 16 conductors. The voltage required to be generated is 300 V. What would be the speed of coupled prime mover if flux per pole is 0.05 Wb.

If the armature is rewound as lap wound machine and run by same prime mover, what will be the generated voltage.

Solution:

Here, $P = 8$; $\phi = 0.05$ Wb; No. of slots = 51; conductors per slot = 16

$$\therefore \quad Z = 51 \times 16 = 816$$

When the machine is wave wound, $A = 2$ and $E_g = 300$ V

Now, $$E_g = \frac{\phi ZNP}{60A} \text{ or } 300 = \frac{0.05 \times 816 \times N \times 8}{60 \times 2}$$

$\therefore$ $$\text{Speed, } N = \frac{300 \times 60 \times 2}{0.05 \times 816 \times 8} = \mathbf{110.3\ rpm}\ (Ans.)$$

When the machine is rewound as lap winding, $A = P = 8$ and $N = 110.3$ rpm.

$$E_g = \frac{0.05 \times 816 \times 110.3 \times 8}{60 \times 8} = \mathbf{75\ V}\ (Ans.)$$

Example 4.14

A six-pole lap wound armature rotating at 350 rpm is required to generate 260 V. The effective flux per pole is about 0.05 Wb. If the armature has 120 slots, determine the suitable number of conductors per slot and hence determine the actual value of flux required to generate the same voltage.

Solution:

Here, $P = 6$; $A = P = 6$; $N = 350$ rpm; $E_g = 260$ V; $\phi = 0.05$ Wb

Now, $$E_g = \frac{\phi ZNP}{60A} \text{ or } 260 = \frac{0.05 \times Z \times 350 \times 6}{60 \times 6}$$

or $$Z = \frac{260 \times 60 \times 6}{0.05 \times 350 \times 6} = \frac{260 \times 24}{7}$$

$$\text{No. of conductors/slot} = \frac{Z}{\text{No. of slots}} = \frac{260 \times 24}{7 \times 120} = 7.43 \cong 8 \text{ (an integer)}$$

For 8 conductors/slot, $Z = 120 \times 8 = 960$

$$\text{Actual value of flux required, } \phi = \frac{E_g \times 60A}{ZNP} = \frac{260 \times 60 \times 6}{960 \times 350 \times 6} = \mathbf{0.0464\ Wb}\ (Ans.)$$

Example 4.15

The emf generated by a 4 pole DC generation is 400 V, when the armature is driven at 1200 rpm. Calculate the flux per pole if the wave wound generator has 39 slots having 16 conductors per slot.

Solution:

Induced emf, $$E_g = \frac{\phi ZNP}{60A}$$

where, $P = 4$; $E_g = 400$ V; $N = 1200$ rpm; $Z = 39 \times 16 = 624$; $A = 2$ (wave winding)

$\therefore$ $$\textit{Flux per pole, } \phi = \frac{E_g \times 60A}{ZNP} = \frac{400 \times 60 \times 2}{624 \times 1200 \times 4} = 0{\cdot}016 \text{ Wb} = \mathbf{16\ mWb}\ (Ans.)$$

Example 4.16

The armature of a four-pole 250 V, lap wound generator has 500 conductors and rms of 400 rpm. Determine the useful flux per pole. If the number of turns in each field coil is 1000, what is the average induced emf in each field coil on breaking its connection if the magnetic flux set-up by it dies away completely in 0.1 second?

Solution: Here, $P = 4$; $E_g = 250$ V; $Z = 500$; $A = P = 4$; $N = 400$ rpm

$$\phi = \frac{E_g \times 60A}{ZNP} = \frac{260 \times 60 \times 4}{500 \times 400 \times 4} = \mathbf{0.075\ Wb}\ (Ans.)$$

No. of turns of exciting winding, $N_t = 1000$

$$e = N_t \frac{d\phi}{dt} = 1000 \times \frac{0.075 - 0}{0.1} = \mathbf{750\ V}\ (Ans.)$$

Example 4.17

A four-pole generator has an induced emf of 250 V when driven at 500 rpm. The armature is lap wound and has 600 conductors. The radius of the pole shoe is 20 cm and it subtends an angle of 60°. Calculate the flux density in the air-gap if the length of pole shoe is 18 cm.

Solution:

Here, $P = 4$; $E_g = 250$ V; $N = 500$ rpm; $A = P = 4$; $Z = 600$

$$\text{Flux per pole, } \phi = \frac{E_g \times 60A}{ZNP} = \frac{250 \times 60 \times 4}{600 \times 500 \times 4} = 0.05\ \text{Wb}$$

$$\text{Pole shoe arc} = 2\pi r \times \frac{\theta}{360} = 2\pi \times 0.2 \times \frac{60}{360} = 0.21\ \text{m}$$

$$\text{Pole shoe area} = \text{Pole shoe arc} \times l = 0.21 \times 0.18 = 0.0378\ \text{m}^2$$

$$\text{Flux density in air gap} = \frac{\phi}{\text{area}} = \frac{0.05}{0.0378} = \mathbf{1.323\ tesla}\ (Ans.)$$

Example 4.18

A four-pole wave wound DC generator has 51 slots on its armature and each slot has 24 conductors. The flux per pole is 0·01 weber. At what speed must the armature rotate to give an induced emf of 220 V. What will be the emf developed if the winding is lap connected and the armature rotates at the same speed.

Solution:

$$\text{Induced emf, } E_g = \frac{\phi ZNP}{60A}$$

where, $\phi = 0{\cdot}01$ Wb; $Z = 51 \times 24 = 1224$; $E = 220$ V; $P = 4$; $A = 2$ (wave winding).

$$\therefore \qquad 220 = \frac{0 \cdot 01 \times 1224 \times N \times 4}{60 \times 2}$$

$$\text{or} \qquad N = \frac{220 \times 60 \times 2}{0 \cdot 01 \times 1224 \times 4} = \mathbf{539{\cdot}21\ rpm}\ (Ans.)$$

For lap winding, $A = P = 4$;

$$E_g = \frac{0 \cdot 01 \times 1224 \times 539 \cdot 21 \times 4}{60 \times 4} = \mathbf{110\ V}\ (Ans.)$$

Example 4.19

A four-pole, wave wound DC armature has a bore diameter of 84 cm. It has 600 conductors and the ratio of pole arc to pole pitch is 0.8. If the armature is running at 360 rpm and the flux density in the air gap is 1.2 tesla, determine the induced emf in the armature if the effective length of armature is 20 cm.

Solution:

Here $P = 4$; $A = 2$ (wave wound); $Z = 600$; $N = 360$ rpm

Armature bore diameter, $D = 84$ cm $= 0.84$ m

$$\text{Pole pitch} = \frac{\pi \times \text{bore diameter}}{\text{No. of poles}} = \frac{\pi \times 0.84}{4} = 0.66 \text{ m}$$

$$\text{Pole shoe arc} = \text{Pole pitch} \times \frac{\text{Pole arc}}{\text{Pole pitch}} = 0.66 \times 0.8 = 0.528 \text{ m}$$

$$\text{Pole shoe area, } a = \text{Pole shoe arc} \times \text{Pole length}$$

$$= 0.528 \times 0.2 = 0.1056 \text{ m}^2$$

$$\text{Flux per pole, } \phi = B \times a = 1.2 \times 0.1056 = 0.12672 \text{ Wb}$$

$$\text{Induced emf, } E_g = \frac{\phi ZNP}{60A} = \frac{0.12672 \times 600 \times 360 \times 4}{60 \times 2} = \mathbf{912.4\ V}\ (Ans.)$$

Example 4.20

A four-pole DC generator has rated armature current of 200 A. If the armature is lap-wound, determine the current flowing through each parallel path of the armature. What will be its value if the armature is wave wound?

Solution:

Here, $P = 4$; $A = P = 4$ (lap winding); $I_a = 200$ A

For lap-wound maczhine, current in each parallel path

$$= \frac{I_a}{A} = \frac{200}{4} = \mathbf{50\ A}\ (Ans.)$$

For wave-wound machine, current in each parallel path

$$= \frac{I_a}{A} = \frac{200}{2} = \mathbf{100\ A}\ (Ans.)$$

Example 4.21

A four-pole lap-connected DC machine has an armature resistance of 0.15 ohm. The total armature conductors are 48. Find the armature resistance of the machine when rewound for wave connections.

Solution:

Here, $P = 4$; $R_a = 0.15$ ohm; $Z = 48$

For lap-winding, $A = P = 4$

$\therefore$ Resistance of each parallel path, $R_1 = R_a \times A = 0.15 \times 4 = 0.6\ \Omega$

$$\text{No. of conductors in each parallel path} = \frac{Z}{A} = \frac{48}{4} = 12$$

$$\text{Resistance of each conductor} = \frac{0.6}{12} = 0.05\ \Omega$$

For wave-winding, $A = 2$

$$\text{No. of conductors in each parallel path} = \frac{48}{2} = 24$$

$$\text{Resistance of each parallel path} = 24 \times 0.05 = 1.2\ \Omega$$

$$\text{Armature resistance, } R_a = \frac{1.2}{2} = \mathbf{0.6\ \Omega}\ (Ans.)$$

Example 4.22

A DC machine running at 750 rpm has an induced emf of 200 V. Calculate the speed at which the induced emf will be 250 V (ii) the percentage increase in main field flux for an induced emf of 250 V at a speed of 700 rpm.

Solution:

Here, $E_1 = 200$ V; $N_1 = 750$ rpm

(*i*) $E_2 = 250$ V; $N_2 =?$

(*ii*) $E_3 = 250$ V, $N_3 = 700$ rpm, $\phi_3 =?$

We know $E \propto N\phi$

(*i*) $\therefore$ $$\frac{N_2}{N_1} = \frac{E_2}{\phi_2} \times \frac{\phi_1}{E_1} \text{ as } \phi_2 = \phi_1$$

$$N_2 = \frac{250}{200} \times 750 = \mathbf{937.5\ rpm.}$$

(*ii*) $$\frac{N_3}{N_1} = \frac{E_3}{\phi_3} \times \frac{\phi_1}{E_1} \text{ or } \frac{\phi_3}{\phi_1} = \frac{E_3}{E_1} \times \frac{N_1}{N_3} = \frac{250}{200} \times \frac{750}{700} = 1.34$$

$\therefore$ %age increase in flux = **1.34%** (*Ans.*)

Example 4.23

A 50 HP, 400 V, 4 pole, 1000 rpm, DC motor has flux per pole equal to 0.027 Wb. The armature having 1600 conductors is wave connected. Calculate the gross torque when the motor takes 70 ampere.

Solution:

$$\text{Torque developed, } T = \frac{P\phi Z I_a}{2\pi A}$$

Where, $P = 4$; $\phi = 0.027$ Wb; $Z = 1600$; $I_a = 70$A; $A = 2$ (wave connected)

$$T = \frac{4 \times 0.027 \times 1600 \times 70}{2 \times \pi \times 2} = \mathbf{963\ Nm}\ (Ans.)$$

Example 4.24

The induced emf in a DC machine is 220 volts at a speed of 1500 rpm. Calculate the electromagnetic torque developed at an armature current of 20 A.

Solution:

Here $E = 220$ V; $N = 1500$ rpm; $I_a = 20$ A

Power developed in the armature

$$\omega T_e = E_b I_a$$

or $$T_e = \frac{E_b I_a}{\omega} = \frac{E_b I_a}{2\pi N / 60} = \frac{220 \times 20 \times 60}{2\pi \times 1500} = \mathbf{23.87\ Nm}\ (Ans.)$$

Example 4.25

The electromagnetic torque developed in a DC machines is 80 Nm for an armature current of 20 A. Find the torque for a current of 30 A. What is the induced emf for a speed of 900 rpm?

Solution:

$T_{e1} \propto \phi_1 I_{a1}$ and $T_{e2} \propto \phi_2 I_{a2}$

or $$\frac{T_{e2}}{T_{e1}} = \frac{I_{a2}}{I_{a1}}$$ (Assuming flux remains constant)

$$T_{e2} = 80 \times \frac{30}{20} = \mathbf{120\ Nm}\ (Ans.)$$

Now $$T_{e2} = \frac{k}{2\pi}\phi_2 I_{a2}$$

$\therefore$ $$k\phi_2 I_{a2} = 2\pi T_{e2} \text{ or } k\phi_2 = \frac{2\pi \times 120}{30} = 8\pi$$

$$\text{Now } E = kn\phi = \frac{8\pi \times 900}{60} = \mathbf{376.8\ volt}\ (Ans.)$$ $(\because \phi_2 = \phi_1 = \phi)$

4.18 Armature Reaction

When a DC generator is loaded, a current flows through the armature conductor in the same direction as that of the induced (or generated) emf The armature conductors carrying current, produce their own magnetic field called armature field.

The effect of armature field produced by the armature current carrying conductors on the main magnetic field is known as **armature reaction.**

Let us see the effect of armature field on the main magnetic field when the generator is loaded.

Consider a bipolar generator. At no-load, no current flows through the armature conductors and the flux distribution in the armature is shown in Fig. 4.44. The vector OF_m represents the mmf produced by the main field. It is observed that the Magnetic Neutral Axis (*MNA*), which are perpendicular to the main field passing through the armature, and the Geometrical Neutral Axis (*GNA*) coincide with each other. The brushes (B_1 and B_2) are always placed at *MNA*. Here, they are shown as touching the armature conductors directly, but in reality they touch the commutator segments connected to these conductors.

Geometrical Neutral Axis: *The line passing through the geometrically central point between the two adjacent opposite magnetic poles is called* **geometrical neutral axis** (*GNA*).

Magnetic Neutral Axis: *The line passing through the magnetically neutral position between the two adjacent opposite magnetic poles is called* **magnetic neutral axis (MNA).** When a conductor (or coil) passes through these axis, no emf is induced in the conductor (or coil).

When load is applied to the generator, current* flows through the armature conductors which sets up armature field as shown in Fig. 4.45. The vector OF_A represents the mmf produced by the armature field.

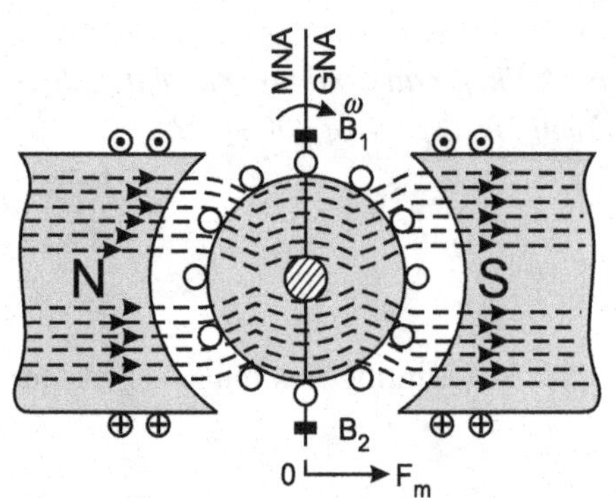

Fig. 4.44 Main field produced by main poles

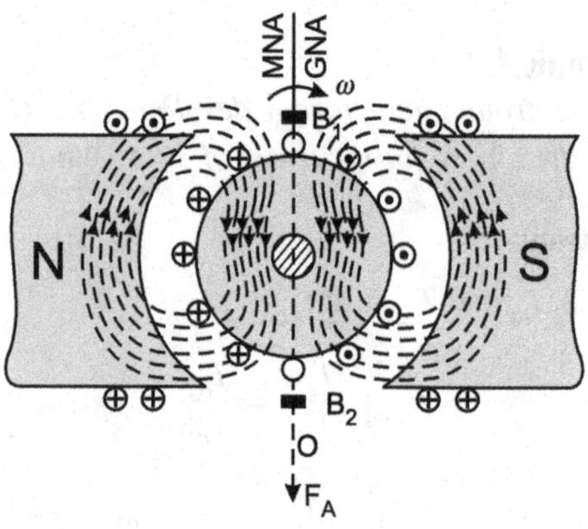

Fig. 4.45 Field produced by armature conductors

This armature flux interacts with the main flux and a resultant flux is set up in the armature as shown in Fig. 4.46. It can be observed that the resultant flux is no longer uniform. It is concentrated (becomes stronger) at the trailing pole tips and is rare (becomes weaker) at the leading pole tips.

The resultant mmf is shown by the vector OF which is the vector sum of OF_m and OF_A. Thus, the *MNA* are shifted to new position displaced from its original position by an angle θ.

The new position of magnetic neutral axis i.e., the shifting of axis by an angle θ depends upon the magnitude of load applied on the generator. Larger the load, larger will be the shift or larger will be the value of angle θ. It means the shifting of *MNA* is not constant, it varies and depends upon the magnitude of load applied on the machine. Moreover, the shift is in the direction of rotation. (in generating action).

As per the new position of *MNA*, the distribution of armature flux is shown in Fig. 4.47. The vector OF_{AR} represents the new position of mmf producing resultant armature field. This armature field has two component (*i*) OF_C which is perpendicular to the main mmf OF_m and produces the cross magnetising effect. (*ii*) OF_D which opposes the main mmf OF_m and produces the demagnetising effect.

* The direction of flow of current in armature conductors is determined by Flemming's right hand rule. The direction of induced emf and direction of flow of current is the same.

Thus the armature magnetic field produces.

(*i*) Cross magnetising effect which creates a magnetic field in between the two adjacent opposite poles where brushes are placed for commutation.

(*ii*) Demagnetising effect which weaken the main magnetic field and changes the flux distribution such that at trailing pole tips the flux is strengthened and at leading pole tips the flux is weakened.

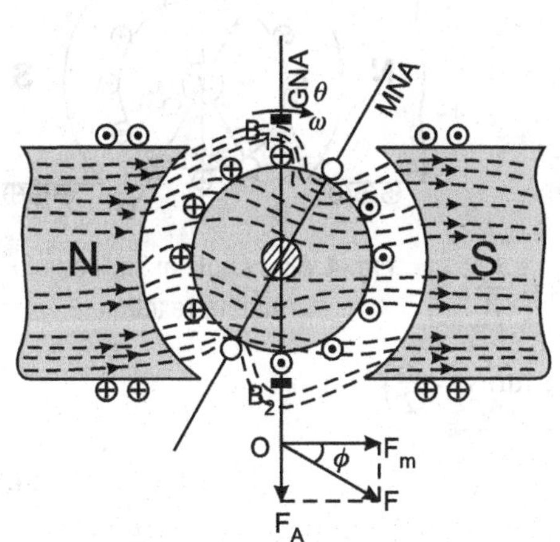

Fig. 4.46 Resultant field

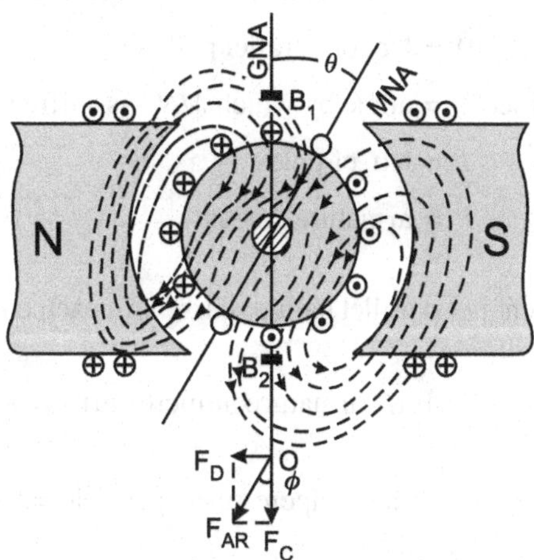

Fig. 4.47 New position of MNA

Ill-effects of armature reaction

The above two effects caused by the armature reaction lead to poor commutation (increases sparking at the brushes or at the commutator surface) and increases iron losses. Let us see how it happens;

(*i*) *Sparking at brushes:* During commutation i.e., when a coil is short circuited by the brushes through commutator segments should have zero emf induced in it. That is why the brushes are usually placed along the interpolar axis where the flux cut by the coil is zero and no emf is induced in it. But due to armature reaction the magnetic neutral axis (*MNA*) are shifted and the coil which undergoes commutation induces some emf causing sparking at the brushes. At heavy loads, the induced emf in the commutating coil may be so high that it may produce a spark that may spread around the commutator surface forming a ring of fire. By all means, it has to be avoided otherwise it would damage the commutator surface and brushes.

(*ii*) *Iron losses:* The flux density in the leading and trailing pole tips is changing due to change in load on the generator or due to armature reaction. This change in flux density causes more iron losses in the pole shoes.

4.19 Calculations for Armature Ampere-turns

Armature reaction causes demagnetising and cross-magnetising effect. Due to armature reaction, let the *MNA* be shifted ahead by an angle *θ*. The new position of *MNA* is represented by line *AB*. Draw another line *CD* making the same angle *θ* with *GNA* but in opposite direction as shown in Fig.

4.48. It may be observed that all the conductors lying between *AOC* and *BOD* carry current in such a direction *mmf* produced by them in opposite to the main *mmf.* These conductors cause demagnetising effect and these turns are known as *demagnetising turns.*

All other turn lying between angle *AOD* and *COB* produce cross magnetising *mmf* and are called *cross-magnetising turns.*

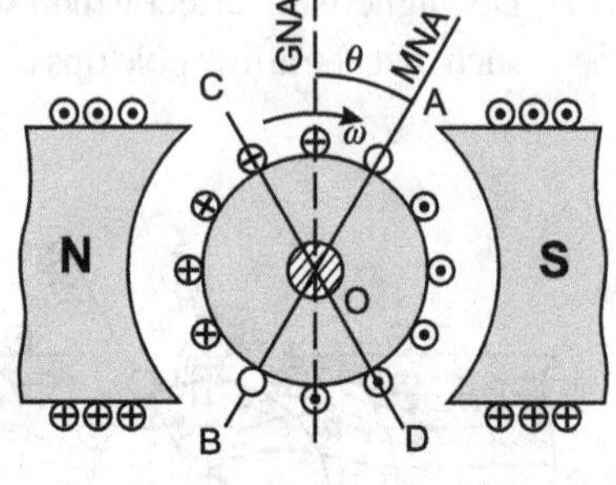

Fig. 4.48 Angle of shift due to armature reaction

Let Z = total number of armature conductors

A = No. of parallel paths

θ^* (mech.) = angle by which *MNA* is shifted

P = No. of poles

I_a = armature current

Current per parallel path or current in each conductor, $I_c = \frac{I_a}{A}$

$$\text{Total armature ampere-turns} = \frac{I_c Z}{2} \quad \left(\text{since turns} = \frac{Z}{2}\right)$$

$$\text{Total ampere-turns per pole} = \frac{I_c Z}{2P} \qquad \ldots(4.8)$$

Demagnetising ampere-turns = ampere-turns lying with in the angle (*AOC* + *BOD*)

$$= \frac{I_c Z}{2} \cdot \left(\frac{2\theta + 2\theta}{360}\right) = \frac{I_c Z}{2} \cdot \frac{4\theta}{360} = I_c Z \frac{2\theta}{360}$$

$$\text{Demagnetising ampere-turns/pole, } AT_d = \frac{I_c Z}{2} \cdot \frac{2\theta}{360} = I_c Z \cdot \frac{\theta}{360} \qquad \ldots(4.9)$$

$$\text{Cross magnetising ampere-turns/pole, } AT_c = \frac{I_c Z}{2P} - \frac{I_c Z\theta}{360} = I_c Z\left(\frac{1}{2P} - \frac{\theta}{360}\right) \qquad \ldots(4.10)$$

Example 4.26

A six-pole generator has 600 wave wound conductors on its armature. It delivers 100 ampere at full-load. If the brush lead is 6° mechanical, calculate the armature demagnetising and cross-magnetising ampere-turns per pole.

Solution:

Here, $P = 4$; $A = 2$ (wave-wound); $Z = 600$; $I_a = 100$ A; $\theta = 6°$

$$\text{Current in each conductor, } I_c = \frac{I_a}{A} = \frac{100}{2} = 50 \text{ A}$$

$$\text{Demagnetising ampere-turns/pole, } AT_d = I_c Z \cdot \frac{\theta}{360} = 50 \times 600 \times \frac{6}{360} = \mathbf{500} \; (Ans.)$$

* In a bipolar machine mechanical angle is 360° and the electrical angle is also 360°, whereas, in four-pole machine mech. angle is 360° but the electrical angle is 720° i.e., electrical angle = 2 × mech. angle or electrical angle = $\frac{P}{2}$ mech. angle.

Cross-magnetising ampere-turns/pole, $AT_c = I_c Z.\left(\frac{1}{2P} - \frac{\theta}{360}\right)$

$$= 50 \times 600 \times \left(\frac{1}{2 \times 6} - \frac{6}{360}\right) = \mathbf{2000}\ (Ans.)$$

Example 4.27

A 250 kW, 500 V, 4 pole lap wound armature has 720 conductors. It is given a brush lead of 3° mechanical from its geometrical neutral axis (GNA). Calculate demagnetising and cross magnetising ampere-turns per pole. Neglect shunt field current.

Solution:

Here, Load = 250 kW; V = 500 V; P = 4; Z = 720; θ = 3° (mech.)

$$\text{Load current, } I_L = \frac{250 \times 1000}{500} = 500 \text{ A}$$

Armature current $I_a = I_L$ = 500 A (shunt field current neglected)

No. of parallel paths, $A = P$ = 4 (lap wound)

$$I_c = \frac{I_a}{A} = \frac{500}{4} = 125 \text{ A}$$

Demagnetising ampere-turns/pole,

$$AT_d = I_c Z.\frac{\theta}{360} = 125 \times 720 \times \frac{3}{360} = \mathbf{750}\ (Ans.)$$

Cross-magnetising ampere-turns/pole,

$$AT_c = I_c Z\left(\frac{1}{2P} - \frac{\theta}{360^\circ}\right) = 125 \times 720 \times \left(\frac{1}{2 \times 4} - \frac{3}{360}\right) = \mathbf{10500}\ (Ans.)$$

Example 4.28

A four-pole motor has a wave wound armature with 720 conductors. The brushes are displaced backward through 4 degree mechanical from GNA. If the total armature current is 80A, calculate demagnetising and cross-magnetising ampere-turns per pole.

Solution:

Here, P = 4; A = 2; Z = 720; θ = 4; I_a = 80 A

$$\text{Current in each conductor, } I_c = \frac{I_a}{A} = \frac{80}{2} = 40A$$

$$\text{Demagnetising ampere-turns/pole, } AT_d = I_c Z.\frac{\theta}{360} = 40 \times 720 \times \frac{4}{360} = \mathbf{320}\ (Ans.)$$

$$\text{Cross-magnetising ampere-turns/pole, } AT_c = I_c Z\left(\frac{1}{2P} - \frac{\theta}{360}\right)$$

$$= 40 \times 720 \times \left(\frac{1}{2 \times 4} - \frac{4}{360}\right) = \mathbf{3280}\ (Ans.)$$

Example 4.29

A 250 V, 10 kW, 8 pole DC generator has single turn coils. The armature is wave-wound with 90 commutator segments. If the brushes are shifted by 2 commutator segments at full-load, calculate (i) total armature reaction ampere-turns (ii) demagnetising ampere-turns and (iii) cross-magnetising ampere-turns.

Solution:

Here, Load = 10 kW; $V = 250$ V; $P = 8$; $A = 2$ (wave wound)

No. of commutator segments = 90

Brush shift = 2 commutator segments

$$\text{Load current; } I_L = \frac{10 \times 1000}{250} = 40 \text{ A}$$

$$\text{Armature current, } I_a = I_L = 40 \text{ A}$$

$$\text{Current per conductor, } I_c = \frac{I_a}{2} = \frac{40}{2} = 20\text{A}$$

Total armature conductors, Z = No. of commutator segment × No. of parallel path = 90 × 2 = 180

$$\text{Brush shift, } \theta = \frac{2}{90} \times 360 = 8° \text{ (mech.)}$$

(*i*) Total armature-reaction ampere-turns/pole $= \dfrac{I_c Z}{2P} = \dfrac{40 \times 180}{2 \times 8} = \mathbf{450}$ (*Ans.*)

(*ii*) Demagnetising ampere-turns/pole $= I_c Z . \dfrac{\theta}{360} = 40 \times 180 \times \dfrac{8}{360} = \mathbf{160}$ (*Ans.*)

(*iii*) Cross-magnetising ampere-turns/pole $= \dfrac{I_c Z}{2P} - \dfrac{I_c Z.\theta}{360} = 450 - 160 = \mathbf{290}$ (*Ans.*)

Example 4.30

A 150 kW, 250 V, 6 pole lap wound generator has 600 conductors on its armature. Due to armature reaction if the MNA are shifted by 18° electrical, determine (i) demagnetising ampere-turns (ii) cross-magnetising ampere-turns (iii) series turns required to balance the demagnetising component neglecting magnetic leakage.

Solution:

Here, Load = 150 kW; $V = 250$ V; $P = 6$; $A = P = 6$; $Z = 600$; $\theta = 18°$ electrical

$$\text{Load current, } I_L = \frac{150 \times 1000}{250} = 600 \text{ A}$$

$$\text{Armature current, } I_a = I_L = 600 \text{ A}$$

$$\text{Current per conductor, } I_c = \frac{I_a}{A} = \frac{600}{6} = 100 \text{ A}$$

$$\text{Brush shift, } \theta = \frac{\text{electrical angle}}{P/2} = \frac{18}{6/2} = 6° \text{ (mechanical)}$$

(*i*) Demagnetising ampere-turns, $AT_d = I_c Z.\dfrac{\theta}{360°} = 100 \times 600 \times \dfrac{6}{360}$ = **1000** (*Ans.*)

(*ii*) Cross-magnetising ampere-turns, $AT_c = I_c Z.\left(\dfrac{1}{2P} - \dfrac{\theta}{360}\right)$

$$= 100 \times 600 \times \left(\frac{1}{2 \times 6} - \frac{6}{360}\right) = \mathbf{4000}\ (Ans.)$$

(*iii*) Series turns required to balance the demagnetising component = $\dfrac{AT_d}{I_a} = \dfrac{1000}{600}$

$$= \frac{5}{3} = \mathbf{1.67}\ (Ans.)$$

4.20 Commutation

In a DC machine, one of the major function is the delivery of current from the armature (rotating part) to the external circuit (stationary part) or vicenegativersa. This operation is conducted with the help of brushes and commutator.

During this operation one of the armature coil moves from the influence of one pole to the other and consequently the current in this coil is reversed. While moving from one pole to the other the coil is short circuited by the brushes through commutator segments for fraction of a second (say about $\dfrac{1}{500}$ second). This operation is called *commutation.*

Thus, *the process in which a coil is short circuited by the brushes through commutator segments while it passes from the influence of one pole to the other is called* **commutation.** In this process the current in the coil is reversed.

The duration for which a coil remains short circuited is called **commutation period.**

Explanation

For better understanding, consider a machine having ring winding, a part of which is shown in Fig. 4.49. Assume that the width of brush is equal to the width of commutator segment and the insulation between the segments is very thin (negligibly small). The current per conductor is I_c and the armature is rotating in such a direction, that coils are moving from left to right. Let the coil '*B*' undergoes commutation. Step-wise explanation in given below:

1. As shown in Fig. 4.49 (*a*), the brush is in contact with commutator segment '*a*' and collects current $2I_c$ coming equally from both the sides.
2. As the armature is moving, in the first step as shown in Fig. 4.49(*b*), the brush contact with segment '*b*' starts increasing and contact with segment '*a*' starts decreasing. Consequently, the current flowing towards the brush via segment '*b*' starts increasing* and through segment '*a*' starts decreasing.

* It may be noted that at first instant the current in segment 'b' is zero but at second instant its value is $I_c - x$ and at the next instant it will become I_c which will be move than $(I_c - x)$. Similarly, at first instant current in segment 'a' was 2 I_c (i.e., $I_c + I_c = 2I_c$) which is reduced to $I_c + x$ and at the next instant it will be reduced to I_c only.

It may be noted that current in coil 'B' decreases from I_c to x.

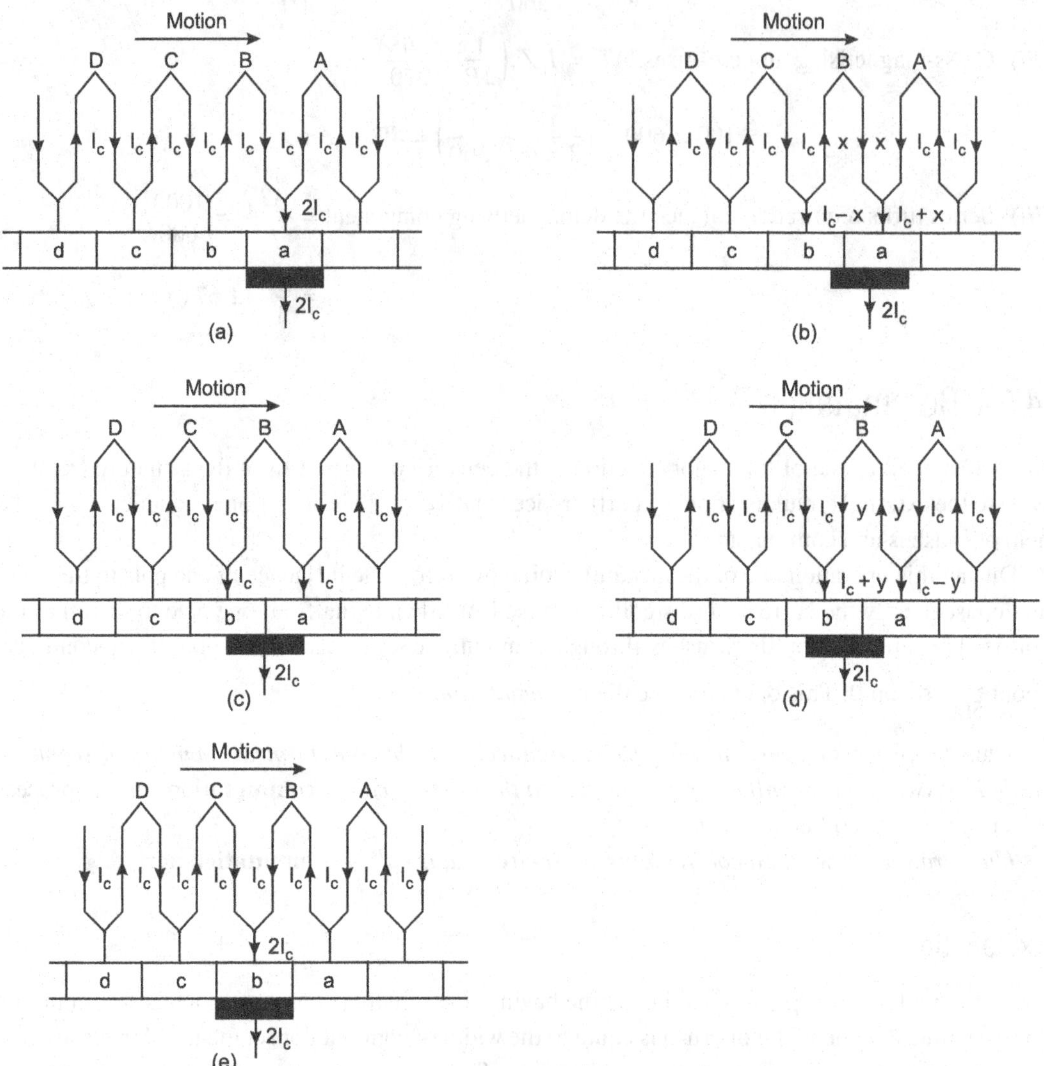

Fig. 4.49 Commutation during brush shifts from segment 'a' to 'b'

3. At the next instant, as shown in Fig. 4.49(*c*), the brush is at the centre of both the segments '*b*' and '*a*' and covers half of the area of both the segments. At this instant brush is drawing equal current (I_c) from both the segments '*b*' and '*a*' and its total value is $2I_c$.

   It may be noted that current in the coil '*B*' at this instant reduces to zero.

4. Further at the next instant, as shown in Fig. 4.49(*d*), larger area of segment '*b*' has come in contact with the brush than segment '*a*'. Accordingly, brush draws more current ($I_c + y$) from segment '*b*' and draws smaller current ($I_c - y$)* from segment '*a*'.

---

* It may be noted that at each instant current drawn by the brush is $2I_c$. First instant, $2I_c$; second instant ($I_c - x + I_c + x = 2I_c$); third instant ($I_c + I_c = 2I_c$); fourth instant ($I_c + y + I_c - y = 2I_c$).

It may be noted that current in the coil '*B*' has reversed and starts increasing.

5. At the next (final) instant, as shown in Fig. 4.49(*e*), the brush completely comes in contact with segment '*b*' and draws equal current I_c from both the sides.

   It may be noted that current in the coil '*B*' has totally reversed and obtains its rated value I_c.

Thus, the commutation process is completed for coil '*B*'. The same process continues for the next coils to come (i.e., coil *C*, *D* and so on..........).

The reversal of current through a coil undergoing commutation may occur in any of the following manners (see Fig. 4.50):

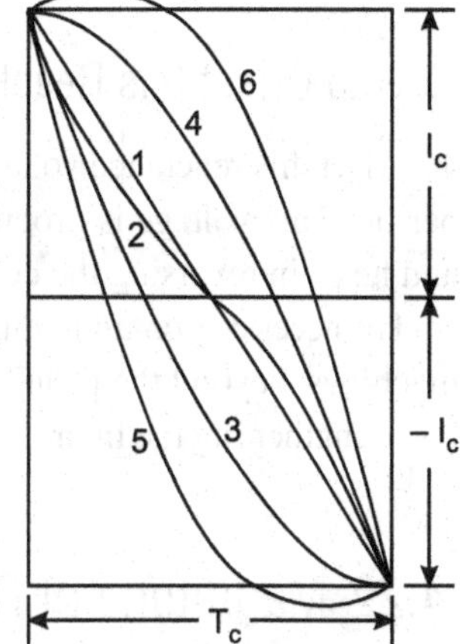

Fig. 4.50 Curves for reversal of current

Curve-I. Shows *linear commutation,* where current changes from $+I_c$ to $-I_c$ uniformly. It is an ideal commutation and provides uniform current density at the brush contact surface. In this case, the power loss at the brush contact is minimum.

Curve-2. Shows *sinusoidal commutation.* It results in satisfactory commutation.

Curve-3. Shows an *accelerated* or *over-commutation* where the current attains its final value with a zero (quick) rate of change at the end of commutation period. Usually, it provides a satisfactory commutation.

Curve 4. Shows a *retarded* or *under-commutation* where the final rate of change of current is very high. In such conditions sparking at the trailing edge of the brush is inevitable.

Curve 5. Shows sharply *accelerated commutation* where the current may reach to its proper final value without a spark but it involves very high localised current densities at the brush contact leading to sparking and heating which further leads to deterioration of brushes.

Curve 6. Shows sharply *retarded commutation* where current reaches to its final value which causes excessive current density under brushes resulting in sparking at the brushes.

Conclusion

From the above discussion, it is concluded that for *satisfactory commutation,* the current in the coil which undergoes commutation must be reversed completely during its commutation period T_c.

4.21 Cause of Delay in the Reversal of Current in the Coil going through Commutation and its Effect

The main cause of delay in the reversal of current in the coil undergoing commutation is its inductive property. It is apparent that the coils placed in the armature may have large number of turns embedded in the magnetic core having high permeability, due to which it possesses appreciable amount of self-inductance.

The inductive property of the coil undergoing commutation gives rise to a voltage called *reactance voltage.* This voltage opposes the reversal of current in it. Although, this voltage is very small, it produces a large current in the coil whose resistance is very small under short circuit. Due to this, heavy **sparking** occur at the brushes.

Thus, it is observed that even though the brushes are set at such a position that coils undergoing short-circuit are in the magnetic neutral axis where they are not cutting any flux and hence no emf is induced in them due to rotation of armature but there will be an emf induced due to self-inductance which causes severe sparking at the brushes.

How to Offset this Reactance Voltage?

To offset this reactance voltage, a voltage is produced in each coil as it passes through commutating period. This voltage is produced in the coil in such a direction that it offsets the reactance voltage and helps in reversing the current in the coil. Such a voltage is called the *commutating emf.*

The necessary commutating emf can be produced either (*i*) by shifting the brushes in non-interpolar machines, beyond the position of *MNA* which is selected to neutralise the armature reaction. or (*ii*) by strengthening the inter-poles more than that is necessary to neutralise the armature reaction flux.

4.22 Magnitude of Reactance Voltage

An approximate value for inductance (*L*) of a coil placed in the armature can be determined by Hobert's formula. Hobart gave an experimental rule for determining inductance of a coil, i.e.,

$$L = n^2 (8\, l_e + 0.8\, l_f) \times 10^{-6} \text{ henry} \quad \ldots(4.11)$$

where, l_e = embedded length of coil in metre

l_f = free length (over hang) of coil in metre

n = number of turns of a coil

Induced emf in a coil due to self-inductance, i.e.,

$$\text{Reactance voltage} = L\frac{di}{dt} \quad \ldots(4.12)$$

where, di = Change of current during commutation

$= I_c - (-I_c) = 2I_c$

$dt = T_c$ (time for which the coil is short circuited)

$$T_c = \frac{W_b - W_m}{v_c} \text{ second} \quad \ldots(4.13)$$

Here, W_b = brush width in metre

W_m = width (or thickness) of mica insulation

v_c = linear velocity of commutator $\frac{\pi DN}{60}$

Hence, reactance voltage $= n^2 (8l_e + 0.8l_f).\dfrac{2I_c}{T_c} \times 10^{-6}$ volt ...(4.14)

The above voltage is developed when only one coil is short circuited by the brush width. If brush width is such that it is short circuiting *m* coils, then

$$\text{reactance voltage} = mn^2(8l_e + 0.8l_f).\frac{2I_c}{T_e} \times 10^{-6}\text{ V} \quad ...(4.15)$$

Example 4. 31

A four-pole lap-wound DC machine has an armature of 20 cm diameter and runs at 1500 rpm. If the armature current is 120 A, thickness of the brush is 10 mm and the self-inductance of each coil is 0.15 mH, determine the average emf induced in each coil during commutation.

Solution:

Here, $P = 4$; $A = P = 4$; $D = 20$ cm $= 0.2$ m; $I_a = 120$ A;

$$L = 0.15\text{ mH} = 0.15 \times 10^{-3}\text{ H};\ W_b = 10\text{ mm} = 0.01\text{ m}$$

$$\text{Linear velocity, } v_c = \frac{\pi DN}{60} = \frac{\pi \times 0.2 \times 1500}{60} = 15.71\text{ ms}^{-1}$$

$$\text{Commutation time, } T_c = \frac{W_b - W_m}{V_c} = \frac{0.01 - 0}{15.71} = (W_m = 0\text{ not given})$$

$$= 636 \times 10^{-3} = 0.636\text{ ms}$$

$$\text{Current per conductor, } I_c = \frac{I_a}{A} = \frac{120}{4} = 30\text{ A}$$

$$\text{Average reactance voltage} = L.\frac{2I_c}{T_c} = 0.15 \times 10^{-3} \times \frac{2 \times 30}{0.636 \times 10^{-3}}$$

$= \mathbf{14.15\ V}$ (*Ans.*)

Example 4. 32

Calculate the reactance voltage for a six-pole wave connected generator if its speed is 300 rpm, diameter of commutator 1.4 metre, number of commutator segments 440, brush width 3 cm, length of each conductor 1 m, length of core 40 cm, turns per commutator segment 2 and armature current is 500 ampere.

Solution:

Here, $P = 6$; $A = 2$; $N = 300$ rpm; $D = 1.4$ m; $W_b = 3$ cm; $I_a = 500$ A; $n = 2$

Linear or peripheral velocity,

$$v_c = \frac{\pi DN}{60} = \frac{\pi \times 1.4 \times 300}{60} = 22\text{ ms}^{-1}$$

$$\text{Period of commutation, } T_c = \frac{W_b - W_m}{v_c} = \frac{0.03 - 0}{22} = \frac{0.03}{22}\text{ second}$$

$$\text{Current per conductor, } I_c = \frac{I_a}{A} = \frac{500}{2} = 250 \text{ A}$$

Length of conductor = 1 m

Embedded length of conductor = 40 cm = 0.4 m

Free length of conductor = 1 – 0.4 = 0.6 m

Embedded length per turn, $l_e = 2 \times 0.4 = 0.8$ m

Free length per turn, $l_f = 2 \times 0.6 = 1.2$ m

$$\text{Width of each segment} = \frac{\pi D}{\text{No. of segments}} = \frac{\pi \times 1.4}{440} = 0.01 \text{ m}$$

Brush width, W_b = 3 cm = 0.03 m

$$\text{No. of segments short circuited by the brush, } m = \frac{0.03}{0.01} = 3$$

No. of turns per coil, $n = 2$

$\therefore$ Self-inductance, $L = mn^2 (8\, l_e + 0.8\, l_f) \times 10^{-6}$H

$= 3 \times 2^2 \times (8 \times 0.8 + 0.8 \times 1.2) \times 10^{-6}$ H

$= 88.32 \times 10^{-6}$ H

$$\text{Reactance voltage} = L \times \frac{2I_c}{T_c} = 88.32 \times 10^{-6} \times \frac{2 \times 250}{0.032 / 22} = \mathbf{32.38\ V}\ (Ans.)$$

4.23 Good Commutation and Poor Commutation

Good Commutation

Good commutation means no sparking at the brushes and the commutator surface remains unaffected during continuous operation of DC machines. Efforts are made to obtain good commutation.

Poor Commutation (Causes of sparking at brushes)

A machine is said to have poor commutation if there is sparking at the brushes and the commutator surface gets damaged during its operation. Sparking at the brushes results in overheating at the commutator brush contact and pitting of commutator. These effects are cumulative, since any sparking impairs the contact which increases heating and further aggravates the situation.

Poor commutation may be caused by mechanical or electrical conditions. The mechanical conditions for poor commutation may be due to uneven surface of the commutator, non-uniform brush pressure or vibrations of brushes in the brush holder etc. By making proper mechanical arrangements, the sparking due to mechanical conditions can be avoided (or eliminated).

The electrical conditions for poor commutation is development of emf in the coils undergoing commutation which may be due to armature reaction or inductance effect of the coil.

The production of emf due to armature reaction in the coil undergoing commutation can be eliminated by providing interpoles or compensating winding on the machine. The production of emf due to inductance can be counter balanced either (*i*) by strengthening the interpoles more than that is necessary to neutralise the armature reaction flux or (*ii*) by shifting the brushes beyond the position of *MNA* which is selected to neutralise the armature reaction.

4.24 Interpoles and their Necessity

The narrow poles placed in between the main poles of a DC machine are called interpoles or commutating poles.

Necessity

Interpoles are provided in between the main poles of DC machine and are energised to such an extent that they must neutralise the armature field produced by the armature winding when machine is loaded. At the same time they must neutralise the emf induced due to inductance in the coil undergoing commutation.

For illustration,

Consider a coil *ab* placed on the armature, rotating in a uniform magnetic field. At this instant, when the induced emf in the coil *ab* which undergoes short circuit must be zero as shown in Figs. 4.51 (*a*) and (*b*). Then no current flows and there is no sparking at the brushes which are slipping from one commutator segment to the next.

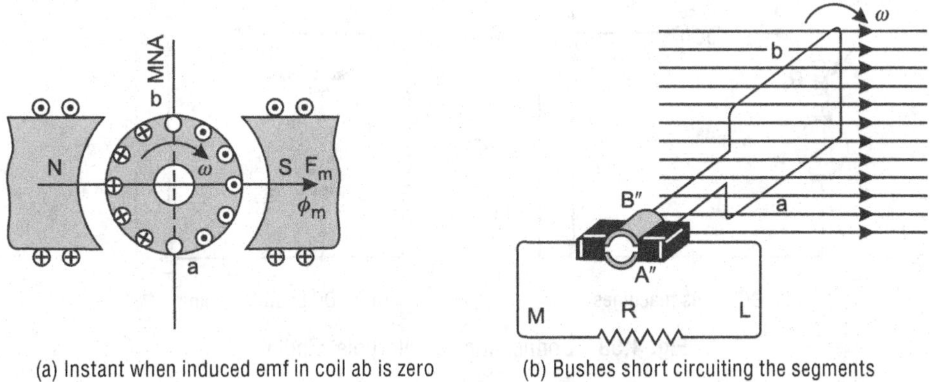

(a) Instant when induced emf in coil ab is zero

(b) Bushes short circuiting the segments

Fig. 4.51 Armature rotating in uniform magnetic field

When load is applied, the armature conductors also carry current and produce their own magnetic field ϕ_A in the direction shown in Fig. 4.52(*a*). Thus coil *ab* undergoing commutation, this coil is short circuited for a small period. Although it is moving parallel to main field ϕ_m but at the same time it is cutting the armature field ϕ_A at right angles. Hence, an emf is induced in the short circuited coil *ab* as shown in Fig. 4.52(*a*). This produces lot of current in the coil and hence causes sparking at the brushes.

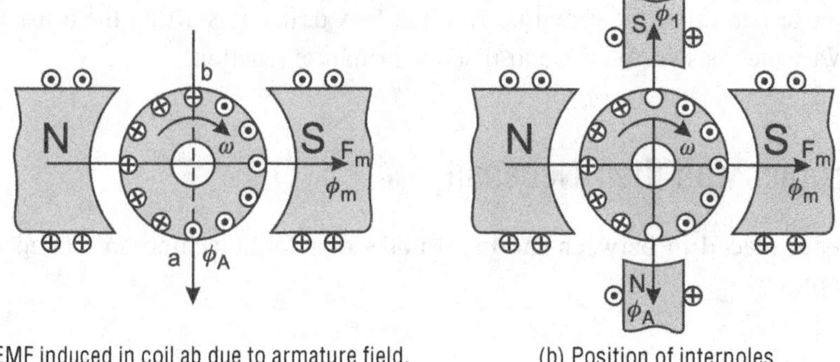

(a) EMF induced in coil ab due to armature field.

(b) Position of interpoles

Fig. 4.52 Interpoles neutralising the armature field

The sparking at the brushes can be avoided by neutralising the armature field. For this, interpoles are provided in the DC machines which are placed in between the main poles as shown in Fig. 4.52(*b*). The flux produced by the interpoles, i.e., ϕ_i must be equal to that of armature field ϕ_A. As ϕ_A is proportional to I_a, therefore, the winding of the interpoles is connected in series with the armature as shown in Fig. 4.53(*a*) and 4.53(*b*). Usually, the interpoles are tapered in order to ensure that there may not be saturation at the root of the pole at heavy overloads. At the same time the air-gap under the interpoles is kept larger (1.5 to 1.7 times) than that under the main poles in order to avoid saturation in the interpoles.

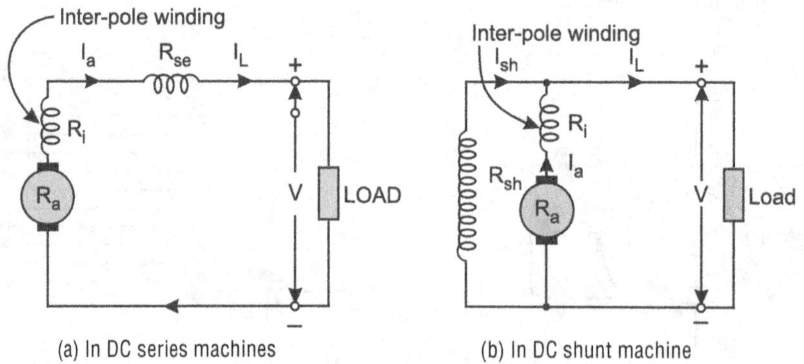

(a) In DC series machines

(b) In DC shunt machine

Fig. 4.53 Connections of interpole winding

4.25 Compensating Winding and its Necessity

An armature reaction produces a demagnetising and cross-magnetising effect. The demagnetising effect is compensated by incorporating a *few extra turns* to the main-field winding, whereas, to neutralise the cross-magnetising effect a compensating winding is used.

In this case, a number of conductors or coils are embedded in the slots of the pole shoes and are connected in series with the armature winding in such a way that current flowing through these conductors or coils sets up a magnetic field which neutralises the cross-magnetising effect of armature field. This winding is known as *compensating winding,* as shown in Fig. 4.54.

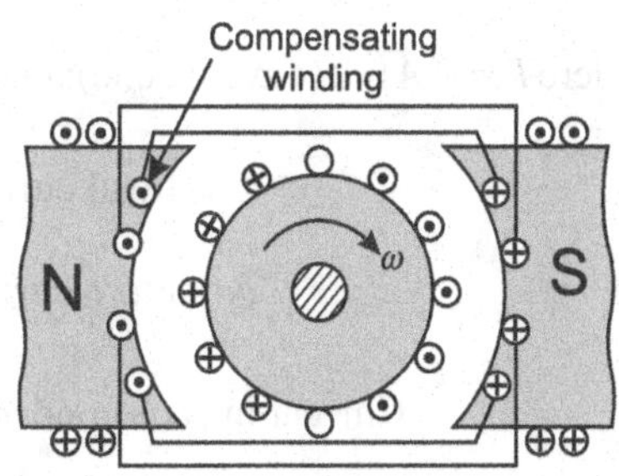

Fig. 4.54 Compensating winding

When current flows through this winding, it sets up magnetic field which is equal and opposing to the cross-magnetising effect of armature field and neutralises it. Thus, no emf is induced in the coil passing through *MNA* and hence sparking at the brushes is eliminated. Thus, a sparkless or good commutation is obtained.

Example 4.33

Estimate the number of turns needed on each commutating pole of six-pole generator delivering 200kW at 200V, given that the number of armature conductors is 540 and the winding is lap connected interpole air-gap is 1.0 cm and the flux density in the interpole air-gap is 0.3 Wb/m². Neglect the effect of iron parts of the circuit and of leakage.

Solution:

Here, Load = 200 kW; $V = 200$ V; $Z = 540$; $A = P = 6$

$$l_{ig} = 1.0 \text{ cm} = 0.01 \text{ m}; \; B_{ig} = 0.3 \text{ Wb/m}^2$$

$$\text{Load current, } I_L = \frac{kW \times 1000}{V} = \frac{200 \times 1000}{200} = 1000 \text{ A}$$

$$\text{Armature current, } I_a = I_L = 1000\text{A}$$

$$\text{Current per conductor, } I_c = \frac{I_a}{A} = \frac{1000}{6} = 166.7 \text{ A}$$

$$\text{Armature reaction ampere turns per pole} = \frac{I_c Z}{2P} = \frac{166.7 \times 540}{2 \times 6} = 7500$$

$$\text{Ampere-turns for air gap flux per pole, } AT_{ig} = \frac{l_{ig}\, B_{ig}}{\mu_0} = \frac{0.01 \times 0.3}{4\pi \times 10^{-7}} = 2386$$

Total ampere-turns required for each interpole, $AT_{ip} = 7500 + 2386 = 9886$

$$\text{No. of turns for each interpole} = \frac{AT_{ip}}{I_a} = \frac{9886}{1000} = \mathbf{9.886} \; (Ans.)$$

Example 4.34

A four-pole, 20 kW, 400 V DC generator has a wave wound armature with 600 conductors. The mean flux density in the air gap of interpole is 0.5 tesla on full-load. If the air gap under interpole is 4 mm, calculate the number of turn required on each interpole.

Solution:

Here $P = 4$; $A = 2$ (wave-wound); load = 20 kW; $V = 400$ V; $Z = 600$; $B_{ig} = 0.5$ T; $l_{ig} = 4$ mm $= 4 \times 10^{-3}$ m

$$\text{Load current, } I_L = \frac{kW \times 1000}{V} = \frac{20 \times 1000}{400} = 50 \text{ A}$$

$$\text{Armature current, } I_a = I_L = 50 \text{ A}$$

$$\text{Current in each conductor, } I_c = \frac{I_a}{A} = \frac{50}{2} = 25 \text{ A}$$

$$\text{Armature reaction ampere-turns per pole} = \frac{I_c Z}{2P} = \frac{25 \times 600}{2 \times 4} = 1875$$

$$\text{Ampere-turns for air-gap, } AT_{ig} = \frac{B_{ig} l_{ig}}{\mu_0} = \frac{0.5 \times 4 \times 10^{-3}}{4\pi \times 10^{-7}} = 1591$$

$$\text{Total interpole ampere-turns, } AT_{ip} = 1875 + 1591 = 3466$$

$$\text{Number of turns on each interpole} = \frac{AT_{ip}}{I_a} = \frac{3466}{50} = \mathbf{69.32} \text{ (Ans.)}$$

Example 4.35

A four-pole, lap wound DC generator carries 360 armature conductors and delivers a full-load current of 400 A. Determine the number of turns of each Interpole if the air-gap flux of the interpole is 0.5 tesla and the effective air-gap length is 3 mm. Assuming that the number of ampere-turns required for the remaining magnetic circuit is only one-tenth of the air-gap.

Solution:

Here, $P = 4$; $A = P = 4$ (lap winding); $Z = 360$; $I_a = 400$ A

$$\text{Current per conductor, } I_c = \frac{I_a}{A} = \frac{400}{4} = 100\text{A}$$

$$\text{Armature reaction ampere-turns per pole, } AT = \frac{I_c Z}{2P} = \frac{100 \times 360}{2 \times 4} = 4500$$

Inter-pole air gap, $l_{ig} = 3$ mm $= 3 \times 10^{-3}$ m

Flux density in the inter-pole air gap, $B_{ig} = 0.5$ T

$$\text{Ampere-turns required for inter-pole air gap} = \frac{B_{ig} l_{ig}}{\mu_0} = \frac{0.5 \times 3 \times 10^{-3}}{4\pi \times 10^{-7}} = 1193.2$$

$$\text{Ampere-turns required for remaining magnetic circuit} = 10\% \text{ of } 1193.2 = \frac{10}{100} \times 1193.2 = 119.3$$

Total ampere-turns required for the interpole = 4500 + 1312.5 = 5812.5

$$\text{Number of turn on each interpole} = \frac{5812.5}{400} = \mathbf{14.53} \text{ (Ans.)}$$

Example 4.36

A four-pole, 25 kW, 250 V wave-wound generator has 42 slots with 10 conductors per slot. The brushes are given a lead of 8° mechanical when the generator delivers full-load current. Calculate the number of turns on the compensating winding if the ratio of pole arc to pole pitch is 0.8.

Solution:

Here, $P = 4$; $A = 2$ (wave-winding); Load = 25 kW; $V = 250$ V; $Z = 42 \times 10 = 420$; $\theta = 8°$ (mech.)

$$\text{Load current, } I_L = \frac{kW \times 1000}{V} = \frac{25 \times 1000}{250} = 100\text{ A}$$

$$\text{Armature current, } I_a = I_L = 100\text{ A}$$

$$\text{Current per conductor, } I_c = \frac{I_a}{A} = \frac{100}{2} = 50\text{A}$$

$$\text{Total armature reaction ampere-turns} = \frac{I_c Z}{2P} = \frac{50 \times 420}{2 \times 4} = 2625$$

Ampere-turns per pole for compensating winding

$$= \frac{I_c Z}{2P} \times \frac{\text{Pole arc}}{\text{Pole pitch}} = 2625 \times 0.8 = 2100$$

$$\text{Turns per pole for compensating winding} = \frac{\text{ATs for each pole face}}{I_a} = \frac{2100}{100} = \mathbf{21} \text{ (Ans.)}$$

4.26 Methods of Improving Commutation

Commutation may be improved by employing following methods.

1. **By use of high resistance brushes.** High resistance carbon brushes help the current to be reversed in the coil undergoing commutation and reduces sparking at the brushes.
2. **By shifting of brushes.** In this method, brushes are shifted to the new position of *MNA* so that no *emf* be induced in the coil undergoing commutation. Thus, the sparking at the brushes is eliminated. But in this case, the position of *MNA* changes with the change in load on the machine and simultaneously the position of brushes cannot be changed.

   Hence, this method is employed in the machine which we do not have interpoles and the load on the machine remain almost constant.
3. **By use of interpoles or commutating poles.** In this method, narrow poles are placed in between the main poles of a DC machine which re-energised to such an extent that they neutralise the field produced by the armature under load. Hence, no *emf* is induced in the coil which undergoes commutation.
4. **By use of compensating winding.** In this method, a number of conductors or coils are embedded in the slots provided at the pole shoes faces and carry current of such a magnitude and direction that field produced by them neutralises the armature field and improve commutation.

Section Practice Problems

Numerical Problems

1. A four-pole, lap wound DC machine is having 500 conductors on its armature and running at 1000 rpm. The flux per pole is 30 m Wb Calculate the voltage induced in the armature winding. What will be induced emf if the armature is rewound for wave winding? (**Ans.** *250 V; 500 V*)

2. A 4-pole machine has an armature with 90 slots and 8 conductors per slot, the flux per pole is 0.05 Wb and runs at 1200 rpm. Determine induced emf if winding is (i) lap connected (ii) wave connected. (**Ans.** *720 V; 1440 V*)

3. A wave wound armature of a six-pole generator has 51 slots. Each slot contains 20 conductors. The voltage required to be generated is 250 V. What would be the speed of coupled prime mover if flux per pole is 0.07 Wb.
If the armature is rewound as lap wound machine and run by same prime mover, what will be the generated voltage. (**Ans.** *70 rpm; 83.3 V*)

4. The armature of a four-pole 200 V, lap wound generator has 400 conductors and rms of 300 rpm. Determine the useful flux per pole. If the number of turns in each field coil is 900, what is the average induced emf in each field coil on breaking its connection if the magnetic flux set-up by it dies away completely in 0.1 second? (**Ans.** *0.1 Wb; 900 V*)

5. A four-pole, lap wound DC armature has a bore diameter of 70 cm. It has 540 conductors and the ratio of pole arc to pole pitch is 0.72. If the armature is running at 500 rpm and the flux density in the air gap is 1.2 tesla, determine the induced emf in the armature if the effective length of armature is 20 cm. (**Ans.** *427.68 V*)

6. The armature resistance of a four-pole, lap wound DC machine is 0.18 ohm. What will be the value of armature resistance if the machine is rewound for wave connections. (**Ans.** *0.72 Ω*)

7. A current of 72A is supplied by a four-pole lap wound DC generator with 480 conductors on its armature. The brushes are given an actual lead of 12° (mechanical). Calculate the cross-magnetising ampere-turns per pole. (**Ans.** *792*)

8. A four-pole DC generator supplies a current of 100A at full load. It has wave wound armature with 722 conductors. If the brush lead is 8° calculate the armature demagnetising and cross-magnetising ampere-turns per pole. (**Ans.** *802; 3710.5*)

9. A four-pole wave wound DC machine has an armature of 25 cm diameter and runs at 1200 rpm. If the armature current is 140 A, thickness of brush is 12 mm and the self inductance of each armature coil is 0.14 mH, calculate the average emf induced in each coil during commutation. (**Ans.** *25.64 V*)

10. Determine the number of turns on each commutating pole of a 6-pole machine if the flux density in the air gap of the commutating pole is 0.5 Wb/m^2 at full-load and the effective length of the airgap is 4 mm. The full load current is 600 A and the armature is lap wound with 540 conductors. Assume the number of ampere-turns required for the remainder of the magnetic circuit to be one-tenth of that of the air-gap. (**Ans.** *10.4*)

11. A 420 V, four-pole, 25 kW DC generator has a wave-connected armature winding with 846 conductors. The mean flux density in the air-gap under the interpoles is 0.5 Wb/m^2 on full load and the radial gap length is 0.4 cm. Calculate the number of turns required on each interpole. (**Ans.** *79.62*)

Short Answer Type Questions

Q.1. What is armature reaction?

Ans. The effect of magnetic field set up by the armature current carrying conductors on the distribution of flux set-up by the main poles of a DC machine is called armature reaction.

Q.2. Mention two unwanted effects of armature reaction?

Or

State the effects of armature reaction in DC machines.

Ans. Demagnetisation or weakening of main field and cross-magnetisation or distortion of main field.

Q.3. How will you define magnetic neutral plane (mnp)?

Ans. The plane through the axis, along which no emf is induced in the armature conductors, is known as magnetic neutral plane.

Q.4. Give the relation between electrical and mechanical angle in case of rotating machine.

Ans. In a multipolar machine, the electrical angle completed in one revolution is $\frac{P}{2} \times 360°$, whereas mechanical angle is only 360°.

$$\therefore \qquad \theta_{elect.} = \frac{P}{2}\,\theta_{mech}$$

Q.5. In a DC machine without interpole to get improved commutation, whether the brush shift should be varied with change in load or brush shift should be fixed?

Ans. Brush shift has to be varied with the change in load in a DC machine so as to lie along *MNA* to provide sparkless commutation.

Q.6. Explain bad commutation.

Ans. A machine is said to have poor or bad commutation if there is sparking at the brushes and commutator surface. The bad commutation may be caused by mechanical or electrical conditions. The mechanical conditions may be due to uneven commutator surface non uniform brush pressure, vibration of brushes in the brush holder etc. The electrical conditions include development of *emf* in the coil undergoing commutation due to armature reaction or due to inductance.

Q.7. What is linear commutation?

Ans. When current in a coil undergoing commutation changes from + I_C to – I_C uniformly, it is called linear commutation. It is an ideal commutation.

Q.8. Name the factor that opposes reversal of current in a coil undergoing commutation.

Ans. Self-inductance of the coil undergoing commutation opposes the reversal of current.

Q.9. Why are the interpoles of a DC machine tapered?

Ans. Interpoles are tapered in order to ensure that there may not be saturation at the root of the pole at heavy overloads.

Q.10. What is the function of compensating winding in a DC machine?

Ans. The compensating winding is used to neutralise the effect of armature reaction outside the influence of the interpoles. It also maintains uniform flux distribution under the faces of the main poles.

Q.11. In what way compensating winding is connected to the armature?

Ans. The compensating winding is connected in series with the armature in a manner so that the direction of current through the compensating winding conductors in any pole face will be opposite the field produced by the armature current carrying conductors.

Q.12. What is the role of interpoles and compensating winding in DC machine?

Ans. Both are used to neutralise the armature field produced by the current carrying conductors of the armature when machines are loaded.

Q.13. What are disadvantages of armature reaction?

Ans. Armature reaction produces demagnetising and cross-magnetising effect on the main magnetic field.

Because of armature reaction, the *MNA* are shifted due to which an emf is induced in the coil undergoing commutation which causes heavy sparking at brushes.

4.27 Types of DC Generators

D.C. generators are generally classified according to the methods of their field excitation. On the basis of this criteria, they can be classified as:

1. Separately excited DC generators
2. Self excited DC generators – these are further classified as:
   (*i*) Shunt wound DC generators
   (*ii*) Series wound DC generators
   (*iii*) Compound wound DC generators.
      (*a*) Long shunt compound wound generators
      (*b*) Short shunt compound wound generators.

Except the above, there are also permanent magnet type DC generators. In these generators, no field winding is placed around the poles. These machines have fairly constant magnetic field. Although these machines are very compact but are used only in small sizes like dynamos in automobiles etc. The main disadvantages of these machine is that the flux produced by the magnets deteriorates with the passage of time which changes the characteristics of the machine.

4.28 Separately-excited DC Generators

A DC generator in which current is supplied to the field winding from an external DC source is called a separately excited DC generator. The flux produced by the poles depends upon the field current with in the unsaturated region of magnetic material of the poles (i.e., $\phi \propto I_f$), but in the saturated region, the flux remains constant. Its conventional diagram is shown in Fig. 4.55.

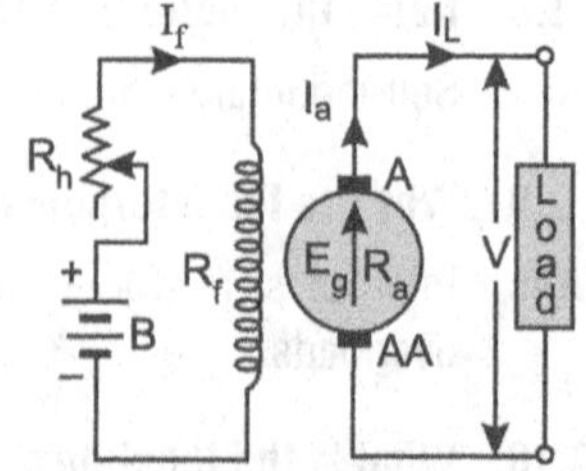

Fig. 4.55 Circuit diagram for separately excited DC generator

Important relations: Here, $I_a = I_L$...(4.16)

where I_a is armature current and I_L is the line current.

Terminal voltage, $V = E_g - I_a R_a$

If contact brush drop per brush (υ_b) is known,

$$V = E_g - I_a R_a - 2\,\upsilon_b. \quad \text{...(4.17)}$$

Power developed $= E_g I_a$;

Power output $= VI_L = VI_a$

4.29 Self-excited DC Generators

A DC generator whose field winding is excited by the current supplied by the generator itself is called a *self-excited DC generator.*

In a self-excited *DC* generator, the field coils may be connected in parallel with the armature, in series with the armature or partly in series and partly in parallel with the armature winding. Accordingly, the self-excited generators may be classified as

(*i*) Shunt wound generators
(*ii*) Series wound generators
(*iii*) Compound wound generators.

(i) Shunt Wound Generators

In a shunt wound generator, the field winding is connected across the armature winding forming a parallel or shunt circuit. Therefore, full terminal voltage is applied across the field winding. A very small current I_{sh} flows through it because this winding has many turns of fine wire having very high resistance R_{sh} (of the order of 100 ohm). Its conventional diagram is shown in Fig. 4.56.

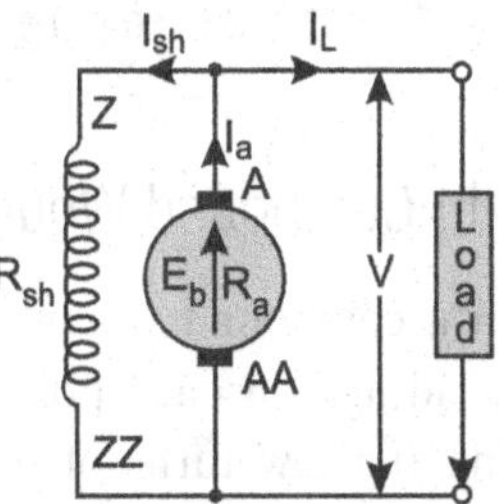

Fig. 4.56 Circuit diagram for DC shunt generator

Important relations:

Shunt field current, $I_{sh} = V/R_{sh}$ ...(4.18)

Where R_{sh} is the shunt field winding resistance. The field current I_{sh} is practically constant at all loads, therefore, the DC shunt machine is considered to be constant flux machine.

Armature current, $I_a = I_L + I_{sh}$ ...(4.19)

Terminal voltage, $V = E_g - I_a R_a$

Including brush contact drop, $V = E_g - I_a R_a - 2\upsilon_b$ ...(4.20)

Power developed $= E_g I_a$; Power output $= VI_L$

(ii) Series Wound Generators

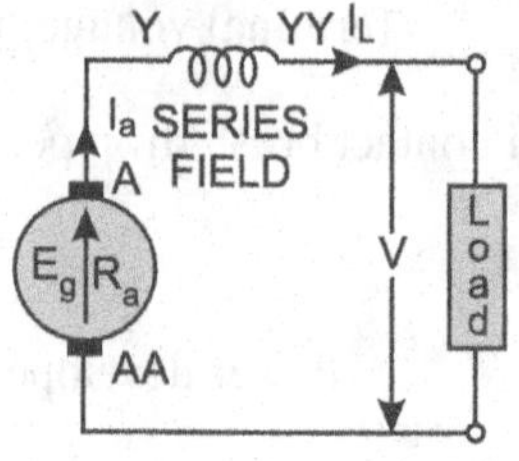

Fig. 4.57 Circuit diagram for DC series generator

In a series wound generator, the field winding is connected in series with the armature winding forming a series circuit. Therefore, full line current I_L or armature current I_a flows through it. Since the series field winding carries full load current, it has a few turns of thick wire having low resistance (usually of the order of less than one ohm). Its conventional diagram is shown in Fig. 4.57.

Important relations:

Series field current, $I_{se} = I_L = I_a$...(4.21)

Series field winding resistance = R_{se}

Terminal voltage, $V = E_g - I_a R_a - I_{se} R_{se} = E_g - I_a (R_a + R_{se})$

Including brush contact drop, $V = E_g - I_a (R_a + R_{se}) - 2\upsilon_b$ ...(4.22)

Power developed = $E_g I_a$; Power output = $VI_L = VI_a$

Note: The flux developed by the series field winding is directly proportional to the current flowing through it (i.e., $\phi \propto I_{se}$). But it is only true before magnetic saturation, after saturation flux becomes constant even if the current flowing through it is increased.

(iii) Compound Wound Generators

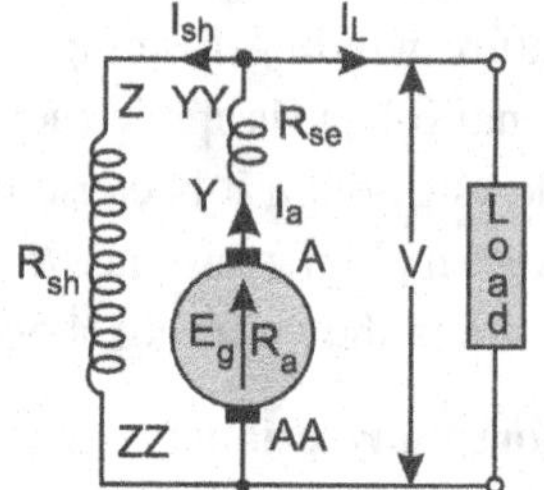

Fig. 4.58 Circuit diagram for long shunt DC compound generator

In a compound wound generator, there are two sets of field windings on each pole. One of them is connected in series (having few turns of thick wire) and the other is connected in parallel (having many turns of fine wire) with armature. A compound wound generator may be;

(*a*) *Long shunt* in which the shunt field winding is connected in parallel with the combination of both armature and series field winding. The conventional diagram of lone shunt compound generator is shown in Fig. 4.58.

Important relations

Shunt field current, $I_{sh} = \dfrac{V}{R_{sh}}$

Series field current, $I_{se} = I_a = I_L + I_{sh}$ ...(4.23)

Terminal voltage, $V = E_g - I_a R_a - I_{se} R_{se}$

$= E_g - I_a (R_a + R_{se})$

Including brush contact drop, $V = E_g - I_a (R_a + R_{se}) - 2\, v_b$ ...(4.24)

Power developed = $E_g I_a$; Power output = VI_L

(*b*) *Short shunt* in which the shunt field winding is connected in parallel with only armature winding. The conventional diagram of short shunt compound generator is shown in Fig. 4.59.

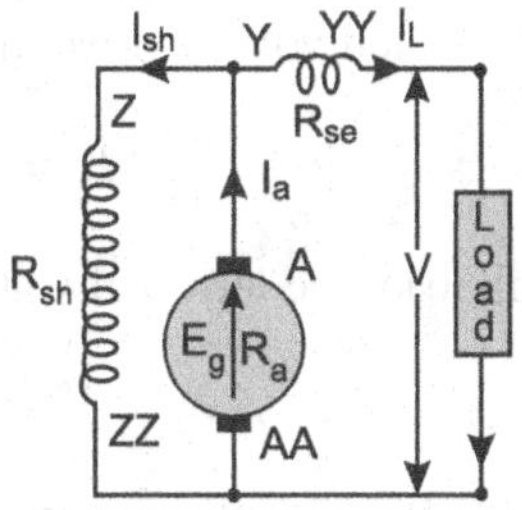

Fig. 4.59 Circuit diagram for short shunt DC compound generator

Important relations

Series field current, $I_{se} = I_L$

Shunt field current, $I_{sh} = \dfrac{V + I_L R_{se}}{R_{sh}}$

$$= \frac{E_g - I_a R_a}{R_{sh}} \qquad \text{...(4.25)}$$

$$I_a = I_L + I_{sh}$$

Terminal voltage, $V = E_g - I_a R_a - I_L R_{se}$

Including brush contact drop, $V = E_g - I_a R_a - I_L R_{se} - 2v_b$ ...(4.26)

Power developed $= E_g I_a$; Power output $= VI_L$

Cumulatively and Differentially Compound-wound Generators

In compound wound DC generators, the field is produced by the shunt as well as series winding. Generally the shunt field is stronger than the series field. When the series field *assist* the shunt field, the generator is called as **cumulatively compound wound generator** [see Fig. 4.60(*a*)]. However, when the series field *opposes* the shunt field, the generator is known as **differentially compound wound generator** [see Fig. 4.60(*b*)].

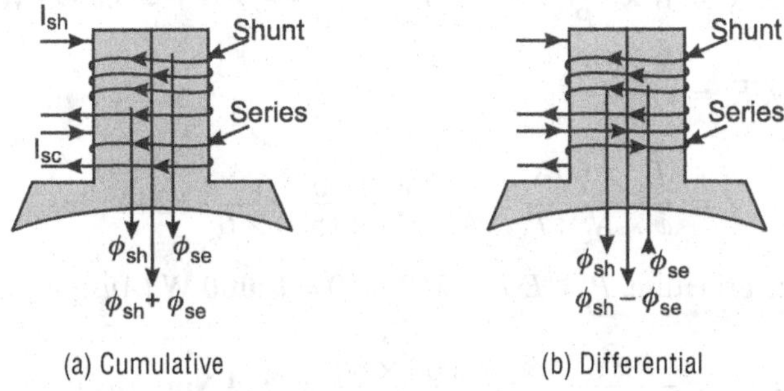

(a) Cumulative (b) Differential

Fig. 4.60 Winding position and direction of flow of current in shunt and series winding

4.30 Voltage Regulation of a DC Shunt Generator

At no-load, the voltage at the terminals of a shunt generator is maximum and is called no-load generated emf When load is applied on the generator, the terminal voltage decreases due to drop in the armature circuit.

Generated voltage or voltage at the terminals at no-load

$$E_g = E_0 \frac{P\phi ZN}{60A}$$

At full load, the terminal voltage is

$$V = E_g - I_{a(fl)} R_a - V_b$$

Where, $I_{a(fl)}$ = Full-load armature current

R_a = Armature resistance

V_b = Total voltage drop at the brushes

The rise in terminal voltage from full-load to no-load at constant speed of a DC generator is called its ***voltage regulation.*** It is expressed as a percentage of full-load terminal voltage i.e.,

$$\% \text{ Voltage regulation} = \frac{E_0 - V}{V} \times 100 \qquad \text{...(4.27)}$$

Example 4.37

A 12 kW, six-pole DC generator develops an emf of 240 at 1500 rpm. The armature has a lap connected winding. The average flux density over the pole pitch is 1.0 T. The length and diameter of the armature is 30 cm and 25 cm respectively. Calculate flux per pole, total number of active armature conductors. Power generated in the armature and torque developed when the machine is delivering 50 A current to the load.

Solution:

Here, $P = 6$; $V = 240$ V; $N = 1500$ rpm; $A = P = 6$;

$B = 1.0$ T; $D = 0.25$ m; $l = 0.3$ m; $I_a = I_L = 50$ A

Flux per pole, $\phi = B \times \frac{\pi D}{P} \times l = 1.0 \times \frac{\pi \times 0.25}{6} \times 0.3 =$ **0.0393 Wb** (*Ans.*)

Now, $E_g = \frac{\phi ZNP}{60\text{ A}}$

$\therefore$ $Z = \frac{E_g \times 60\text{A}}{\phi \times N \times P} = \frac{240 \times 60 \times 6}{0.0393 \times 1500 \times 6} =$ **244** (*Ans.*)

Power developed in the armature, $P_g = E_g I_a = 240 \times 50 =$ **12000 W** (*Ans.*)

Torque developed, $T = \frac{P_g}{\omega} = \frac{P_g}{2\pi N / 60} = \frac{12000 \times 60}{2\pi \times 1500} =$ **76.4 Nm** (*Ans.*)

Example 4.38

A 12-pole DC shunt generator has 50 slots on its armature with 12 conductors per slot with wave winding. The armature and field winding resistance is 0.5 ohm and 60 ohm respectively. The generator is supplying a resistive load of 15 ohm at terminal voltage of 300 V when running at a speed of 625 rpm. Find the armature current, the generated emf and the flux per pole.

Solution:

The conventional diagram is shown in Fig. 4.61.

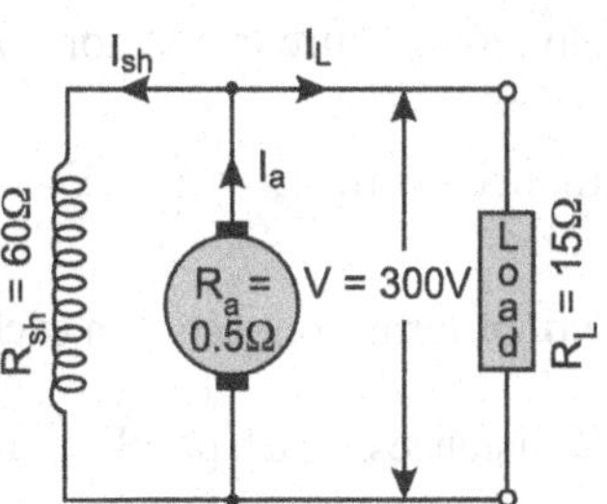

Fig. 4.61 Circuit diagram for shunt generation

$$I_L = \frac{V}{R_L} = \frac{300}{15} = 20A;$$

$$I_{sh} = \frac{V}{R_{sh}} = \frac{300}{60} = 5 \text{ A}$$

Armature current, $I_a = I_L + I_{sh} = 20 + 5 = \mathbf{25\ A}$ (*Ans.*)

Generated emf, $E_g = V + I_a R_a$

$= 300 + 25 \times 0{\cdot}5 = \mathbf{312{\cdot}5\ V}$ (*Ans.*)

Now, $$E_g = \frac{\phi ZNP}{60A}$$

where, $Z = 50 \times 12 = 600$; $N = 625$ rpm; P = 12; A = 2;

$$\therefore \quad 312{\cdot}5 = \frac{f \times 600 \times 625 \times 12}{60 \times 2}$$

or $\phi = \mathbf{8{\cdot}33\ m\ Wb}$ (*Ans.*)

Example 4.39

A four-pole shunt generator with lap connected armature has field and armature resistance of 50 Ω and 0.1 Ω respectively. The generator is supplying a load of 2.4 kW at 100 V. Calculate the armature current, current in each conductor and generated emf.

Solution:

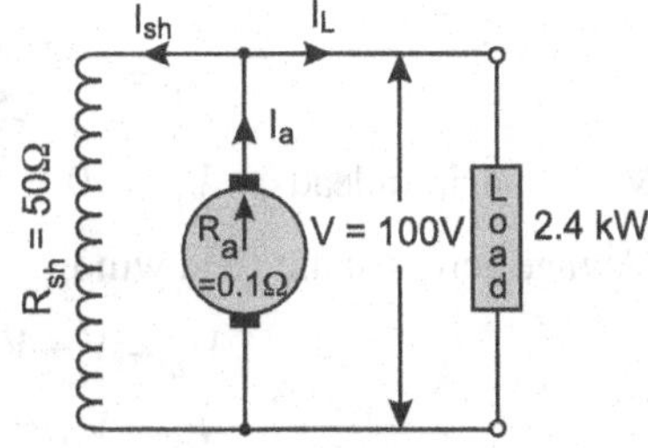

Fig. 4.62 Circuit diagram

Load current, $I_L = \frac{2.4 \times 1000}{100} = 24$ A

Shunt field current; $I_{sh} = \frac{V}{R_{sh}} = \frac{100}{50} = 2$A

Armature current, $I_a = I_L + I_{sh} = 24 + 2 = \mathbf{26\ A}$ (*Ans.*)

Current in each conductor, $I_c = \frac{I_a}{A} = \frac{26}{6} = \mathbf{4.33\ A}$ (*Ans.*)

Generated emf, $E_g = V + I_a R_a$

$= 100 + 26 \times 0.1 = \mathbf{102.6\ V}$ (*Ans.*)

Example 4.40

The useful flux of a four-pole, lap wound DC generator is 0.07 Wb. Its armature carries 220 turns and each turn has 0.004 ohm resistance, if its armature current is 50 A, running at a speed of 900 rpm, determine its terminal voltage.

Solution:

Here, $P = 4$; $A = 4$ (*lap – wound*); $\phi = 0{\cdot}07$ Wb; $N = 900$ rpm; $I_a = 50$ A

No. of armature conductors, $Z = 2 \times \text{No. of turns} = 2 \times 220 = 440$

Induced emf, $$E = \frac{PZ\phi N}{60A} = \frac{4 \times 400 \times 0\cdot07 \times 900}{60 \times 4} = 462 \text{ V}$$

No. of turns connected in each parallel path $= \frac{220}{4} = 55$

Resistance of each parallel path $= 55 \times 0{\cdot}004 = 0{\cdot}22$ ohm

Four resistances, each of 0.22 ohm, are connected in parallel as shown in Fig. 4.63(*a*).

$\therefore$ Armature resistance, $R_a = \frac{0\cdot22}{4} = 0\cdot055\,\Omega$

Terminal voltage, $V = E - I_a R_a$ (generator action)

$= 462 - 50 \times 0{\cdot}055$

$=$ **459·25 V** (*Ans.*)

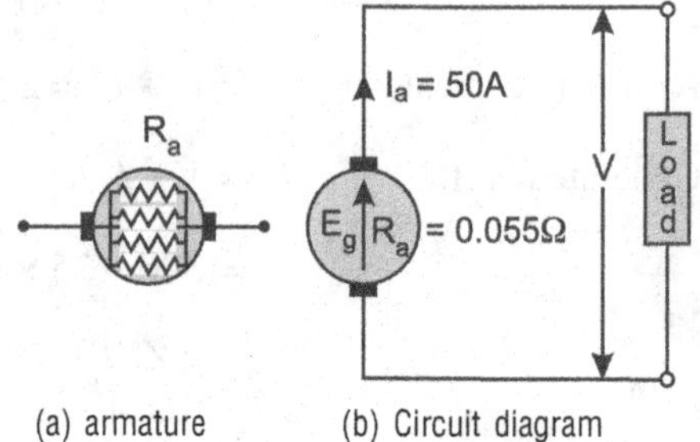

Fig. 4.63 Circuit diagram

Example 4.41

A four-pole DC shunt generator with a wave wound armature having 390 conductors has to supply a load of 500 lamps each of 100 W at 250 V. Allowing 10 V for the voltage drop in the connecting leads between the generator and the load and brush drop of 2 V. Calculate the speed at which the generator should be driven. The flux per pole is 30 m Wb and the value of $R_a = 0{\cdot}05\,\Omega$ *and* $R_{sh} = 65\,\Omega$.

Solution:

The conventional circuit diagram of the DC shunt generator is shown in Fig. 4.64.

Total load $= 500 \times 100$ W

$$I_L = \frac{500 \times 100}{250} = 200 \text{ A}$$

Voltage drip in leads, $V_L = 10$ V

Voltage across shunt field winding,

$$V_{sh} = V + V_L = 250 + 10 = 260 \text{ V}$$

$$I_{sh} = V_{sh}/R_{sh} = 260/65 = 4 \text{ A}$$

$$I_a = I_L + I_{sh} = 200 + 4 = 204 \text{ A}$$

Armature drop $= I_a R_a = 204 \times 0{\cdot}05 = 10{\cdot}2$ V

Total brush drop, $2\upsilon_b = 2$ V

Generated emf, $E_g = V + I_a R_a + V_L + 2\upsilon_b$

$= 250 + 10{\cdot}2 + 10 + 2 = 272{\cdot}2$ V

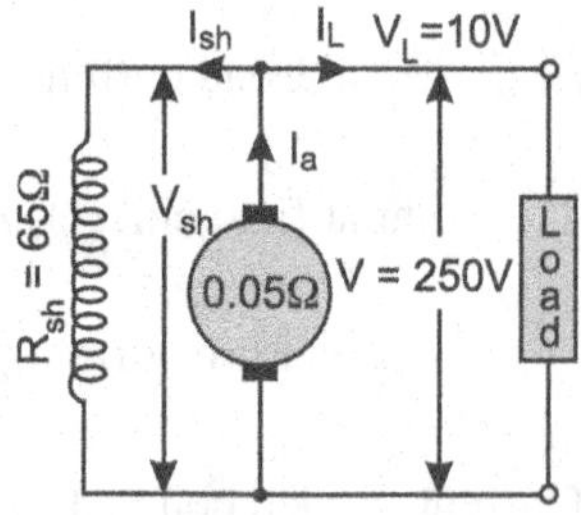

Fig. 4.64 Circuit diagram

Now, $$E_g = \frac{P\phi NZ}{60\,A} \text{ or } 272{\cdot}2 = \frac{4 \times 30 \times 10^{-3} \times N \times 390}{60 \times 2}$$

or $$N = \frac{272\cdot2 \times 60 \times 2}{4 \times 30 \times 10^{-3} \times 390} = \textbf{698 rpm} \text{ (Ans.)}$$

<u>**Example 4.42**</u>

The armature of a four-pole DC shunt generator has 378 wave connected conductors. The armature and shunt winding resistance of the generator is 1 ohm and 100 ohm respectively. The flux per pole is 0.02 Wb. If a load resistance of 10 ohm is connected across the armature terminals and the generator is driven at 1000 rpm, calculate the power absorbed by the load.

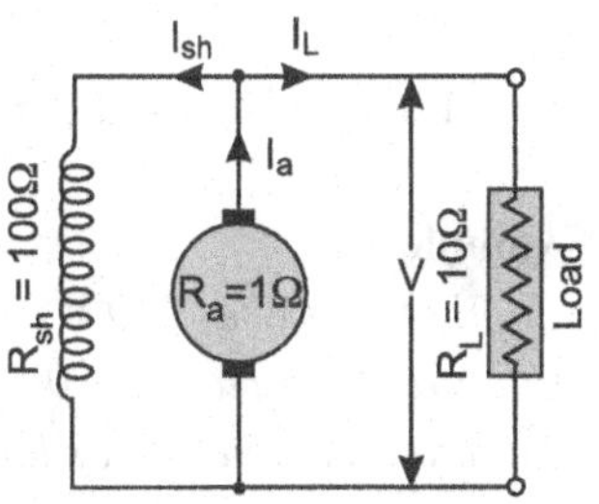

Fig. 4.65 Circuit diagram

Solution:

The conventional circuit is shown in Fig. 4.65.

$$\text{Generated emf, } E_g = \frac{PZ\phi N}{60\,A} = \frac{4 \times 378 \times 0 \cdot 02 \times 1000}{60 \times 2} = 252\text{ V}$$

$$\text{Line current, } I_L = \frac{V}{R_L} = \frac{V}{10} \text{ (where } V \text{ is terminal voltage)}$$

$$\text{Shunt field current, } I_{sh} = \frac{V}{R_{sh}} = \frac{V}{100}$$

$$\text{Armature current, } I_a = I_L + I_{sh} = \frac{V}{10} + \frac{V}{100} = (0 \cdot 11\,V)$$

$$\text{Using the relation, } E_g = V + I_a R_a;$$

$$252 = V + 0{\cdot}11\,V \times 1{\cdot}0$$

$\therefore$ Terminal voltage, $V = 227$ volt

$$\therefore \quad \text{Load current, } I_L = \frac{V}{R_L} = \frac{227}{10} = \mathbf{22{\cdot}7\ A}$$

Power absorbed by the load, $P = VI_L = 227 \times 22{\cdot}7 =$ **5·153 kW** (*Ans.*)

<u>**Example 4.43**</u>

A load of 7.5 kW at 230V is supplied by a short-shunt cumulatively compound DC generator. If the armature, series and shunt field resistances are 0·4, 0·3 and 100 ohm respectively. Calculate the induced emf and the load resistance.

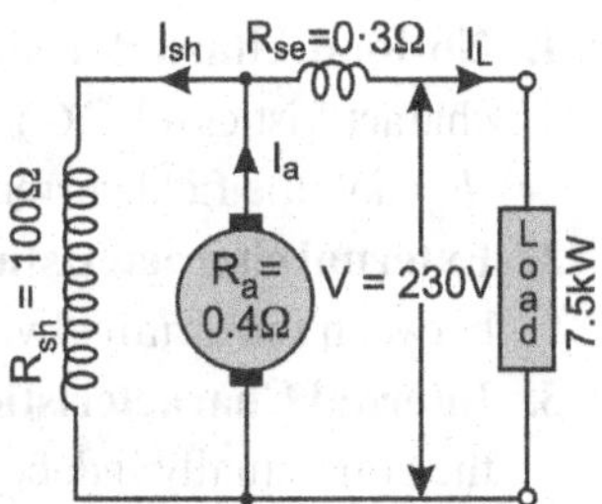

Fig. 4.66 Circuit diagram

Solution:

Refer to Fig. 4.66.

$$I_L = \frac{7 \cdot 5 \times 1000}{230} = 32 \cdot 61\text{ A}$$

$$I_{sh} = \frac{V + I_L R_{se}}{R_{sh}} = \frac{230 + 32 \cdot 61 \times 0 \cdot 3}{100} = 2 \cdot 39\text{ A}$$

$$I_a = I_L + I_{sh} = 32{\cdot}61 + 2{\cdot}39 = 35\text{ A}$$

$$\text{Induced emf, } E_g = V + I_L R_{se} + I_a R_a$$

$= 230 + 32{\cdot}61 \times 0{\cdot}3 + 35 \times 0{\cdot}4 =$ **253·78 V** (*Ans.*)

$$\text{Load resistance, } R_L = \frac{V^2}{P} = \frac{(230)^2}{7\cdot 5\times 1000} = \mathbf{7{\cdot}053\ ohm}\ (Ans.)$$

Example 4.44

A load of 20 kW at 230 V is supplied by a compound DC generator. If the series, shunt field and armature resistances are 0·05, 115 and 0.1 ohm respectively. Calculate the generated emf when the generator is connected as long shunt.

Solution:

$$\text{Load} = 20\text{ kW} = 20\times 10^3\text{ W}$$

$$V = 230\text{ V};\ R_a = 0{\cdot}1\ \Omega;\ R_{se} = 0{\cdot}05\ \Omega;\ R_{sh} = 115\ \Omega$$

Line current, $I_L = \frac{20\times 10^3}{230} = 86\cdot 96\text{ A}$

Shunt field current, $I_{sh} = \frac{V}{R_{sh}} = \frac{230}{115} = 2\text{A}$

Armature current, $I_a = I_L + I_{sh} = 86{\cdot}96 + 2 = 88{\cdot}96\text{ A}$

Generated emf, $E_g = V + I_a R_a + I_a R_{se}$

$$= 230 + 88{\cdot}96\times 0{\cdot}1 + 88{\cdot}96\times 0{\cdot}05$$

$$= \mathbf{243{\cdot}3\ V}\ (Ans.)$$

4.31 Characteristics of DC Generators

To determine the relation between different quantities of a DC generator, the following are the important characteristics of DC generators:

1. **No-load characteristics.** It is also known as magnetic characteristics or open-circuit characteristics (*O.C.C.*). It shows the relation between the no-load generated emf in the armature (E_0) and the field current (i.e., exciting current) I_f, at a specified speed.
2. **External characteristics.** It is also called the performance characteristics. It shows the relation between the terminal voltage (V) and the load current I_L.
3. **Internal Characteristics.** It is also known as total characteristics. It gives the relation between the emf actually induced in the armature (E_g) and the armature current I_a.

4.32 No-load Characteristics of DC Generators or Magnetisation Curve of DC Generator

It shows the relation between the no-load generated emf in the armature (E_0) and the field current I_f, at a specified speed. To obtain this characteristics, proceed as follows:

Open the field winding of the generator and connect it to a separate DC source through a rheostat as shown in Fig. 4.67. Connect an ammeter in the field circuit and a voltmeter across the armature. Reduce the field current to zero and run the armature at a specified speed. Get the reading of voltmeter and mark the point '*a*' on the graph. To plot the characteristics take field current I_f along X-axis and no-load generated emf (E_0) along Y-axis. Increase the field current in steps and get the corresponding voltmeter readings. Plot these values on the graph. The curve thus obtained (shown in Fig. 4.68) shows the no-load characteristics or open circuit characteristics (*O.C.C.*) of the generator.

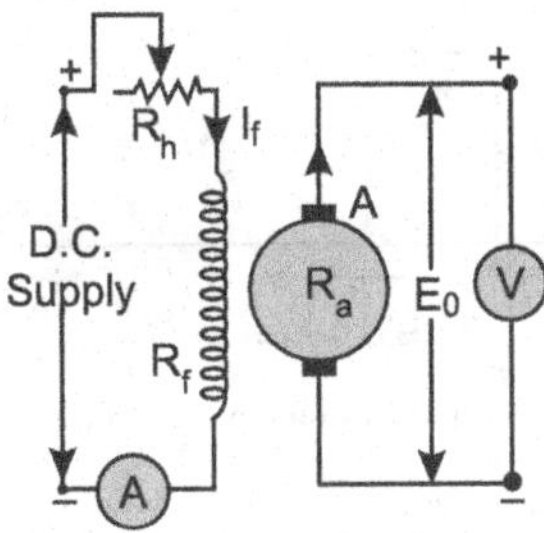

Fig. 4.67 Circuit diagram

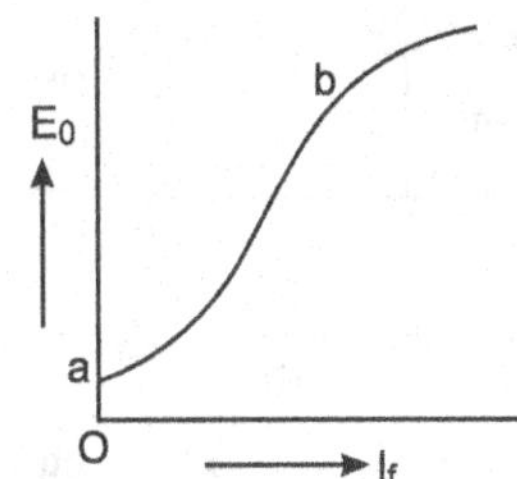

Fig. 4.68 No-load characteristics

Analysis of the curve

While analysing the curve, the following points are worth noting:

1. The curve starts from point '*a*' instead of '*O*' when the field current is zero. It is because of the residual magnetism of the poles.
2. The initial part of the curve (*ab*) is almost a straight line because at this stage the magnetic material is unsaturated and it has high permeability.
3. After point '*b*' the curve bends and the generated emf (E_0) becomes almost constant. It is because after point '*b*', the poles (magnetic material) starts getting saturated.

4.33 Voltage Build-up in Shunt Generators

The shunt generator is a self-excited DC generator whose field winding is supplied current from the output of the generator itself. But question arises how it can supply current to the field winding before the voltage being generated? And if the field current is not supplied, how can the voltage be generated? Let us find out its answer from the following explanation.

The open circuit characteristics of a DC shunt generator is shown in Fig. 4.69(*b*). The shunt field resistance is represented by a straight line *OX*. When armature is rotated at a constant speed of ω rad/sec, the small *residual flux* of the poles is cut by the armature conductors, and very small *emf* (*oa*) is induced in the armature. If now key (*K*) connected in the shunt field winding, as shown in Fig. 4.69(*a*), is closed, current *ob* flows in the field winding. This current increases the flux produced by the poles and voltage generated in the armature is increased to *oc* which further increases the field current to *od* which further builds up the voltage. This building up action comes to an end at point *f* where the *o.c.c.* intersects the shunt field resistance line *OX*. It is, because beyond this point, the induced voltage is less than that required to maintain the corresponding field current. Thus, the final current in the field winding is *ef* and the final voltage build up by the generator for a given *O.C.C.* is *oe* as shown in Fig. 4.69(*b*).

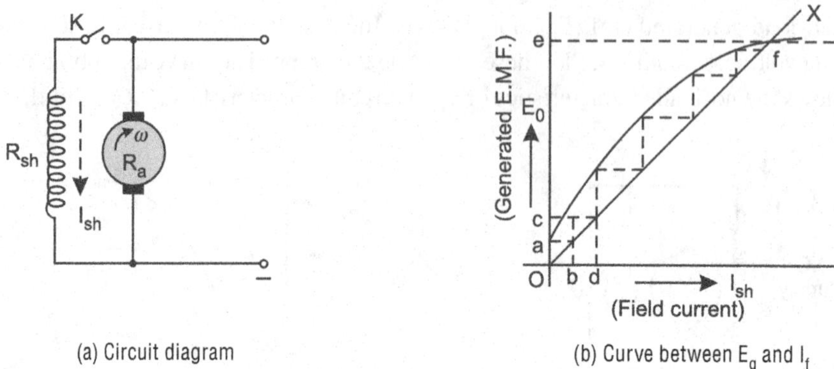

(a) Circuit diagram

(b) Curve between E_g and I_f

Fig. 4.69 Open circuit voltage built-up

4.34 Critical Field Resistance of a DC Shunt Generator

The open circuit characteristic of a DC shunt generator are shown in Fig. 4.70. The line *OX* is drawn in such a way that its slope gives the field winding resistance, i.e.,

$$R_{sh} = \frac{OB(in\ volt)}{OC(in\ ampere)}$$

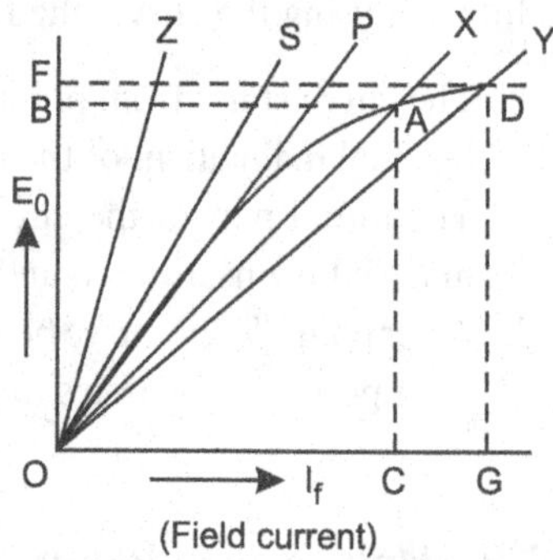

Fig. 4.70 Critical resistance

In this case, the generator can build up a maximum voltage *OB* with a shunt field resistance R_{sh}. A line *OY* represents a smaller resistance. With this resistance, the generator can build up a maximum voltage *OF* which is slightly more than *OB*. If the field resistance is increased, the slope of the resistance line increases. Consequently the maximum voltage which the generator can build up, at a specified speed, decreases. If the value of R_{sh} is increased to such an extent that the resistance line does not cut the no-load characteristics at all (*OZ*), then it is apparent that the voltage will not be built-up (i.e., the generator fails to excite).

If the resistance line (*OP*) just coincide with the slope of the curve, at this value of field resistance, the generator will *just excite*. This resistance, given by the tangent to the *O.C.C.* is called the **critical resistance** at a specified speed.

Thus, *the slope of the tangent drawn on the O.C.C. is called* **critical resistance.**

Critical resistance of a field winding. It is that maximum value resistance of a field winding which is required to build-up voltage in a generator. If the value of field resistance is more than this value, the generator would not build-up the voltage.

Critical load resistance. The minimum value of load resistance on a DC shunt generator with which it can be in position to build-up is called its critical load resistance.

Critical speed of a DC shunt generator. It is the speed of a DC shunt generator at which shunt field resistance will represent the critical field resistance.

4.35 Load Characteristics of Shunt Generator

It is also called external or performance characteristics of shunt generator. It shows relation between the terminal voltage V on load and the load current I_L. To obtain this characteristics, proceed as follows:

Connect an ammeter A_1 and rheostat in the field circuit and an ammeter A_2 and voltmeter V on the load side as shown in Fig. 4.71. Apply a variable load across the terminals. At start switch off the load and run the generator at rated speed. No-load emf (generated voltage E_g) will appear across the voltmeter. Then switch on the load through switch S and increase the load gradually keeping field current (ammeter reading A_1) constant with the help of rheostat R_h. Take the readings of voltmeter V and ammeter A_2 at various instants and plot the curve. The curve so obtained is shown in Fig. 4.71.

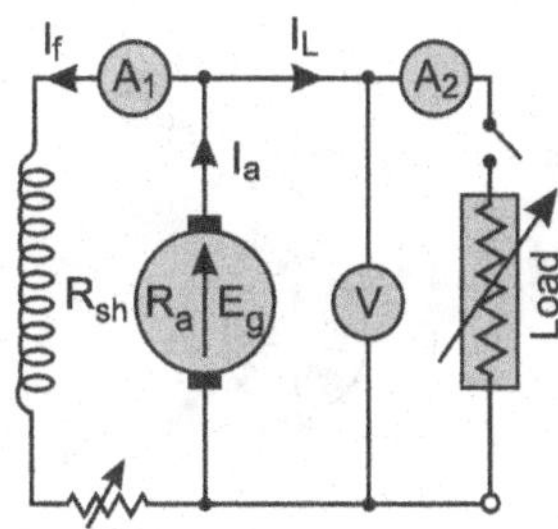

Fig. 4.71 Circuit diagram

Analysis of the Curve

While analysis the curve, the following points are highlighted:

1. At no-load, the voltage across the terminals is maximum and is considered to be equal to generated emf E_g.
2. As the load is increased gradually, the load current I_L increases but the terminal voltage decreases. The decrease in voltage is because of the following reasons:
   (*i*) Due to increase in voltage drop in the armature resistance ($I_a R_a$)
   (*ii*) Due to armature reaction, when load current or armature current I_a increases, the demagnetising effect of the armature field increases on the main field which reduced the induced emf Consequently the terminal voltage decreases.
   (*iii*) The drop in terminal voltage further causes decreases in field current. This will, in turn, causes the decrease in induced emf which reflects the drop in terminal voltage. However, the field current can be kept constant by adjusting the rheostat connected in the field circuit.
3. During initial portion of the curve AB, the tendency of the voltage drop due to armature resistance is more than armature reaction.
4. At point B these two effects neutralise each other.
5. After point B, armature reaction dominates and the curve turns back (BC portion of the curve), as shown in Fig. 4.72.
6. The point C at which the external characteristic cuts the current axis corresponds to a gradual short circuit.

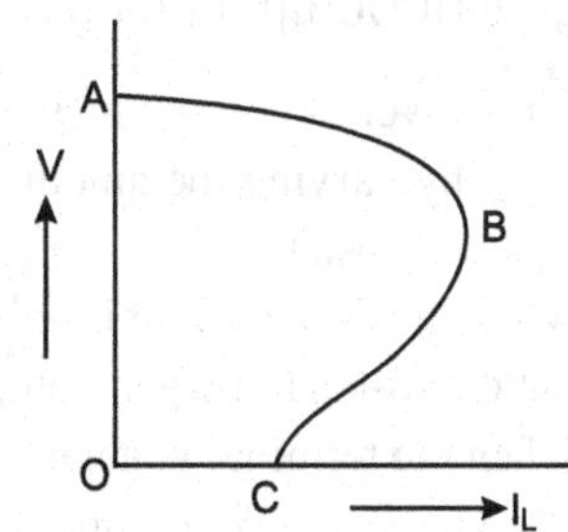

Fig. 4.72 Curve between V and I_L

4.36 Load Characteristics of Series Generators

In this generator, the field winding is connected in series with the armature and load (see Fig. 4.73). Therefore, full armature current I_a flows through it. When load increases, I_a increases which increases flux and consequently generated emf is also increased. This, correspondingly increases the terminal voltage *V*. Thus, a series generator has a rising characteristic (curve *OA*) as shown in Fig. 4.73.

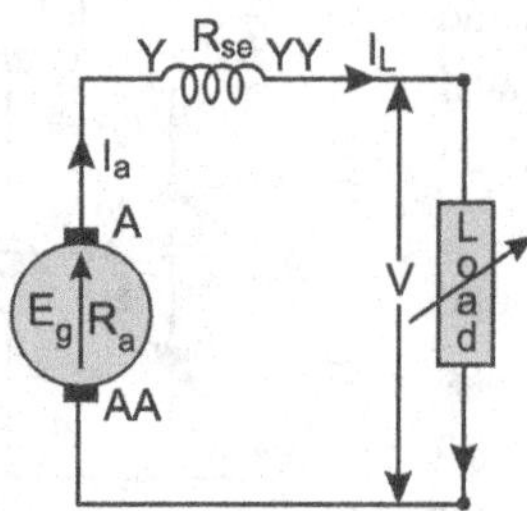

Fig. 4.73 Circuit diagram

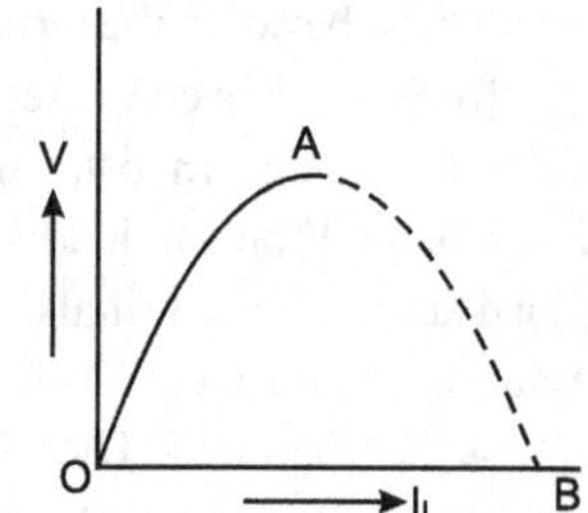

Fig. 4.74 Curve between V and I_L

However, at higher loads, the terminal voltage begins to reduce because of the excessive demagnetising effects of armature reaction. Ultimately, the terminal voltage reduces to zero at load current *OB* as shown in Fig. 4.74.

4.37 Load Characteristics of Compound Generator

There are some applications where constant terminal voltage is essential. At such places, shunt generator is not suitable, because its terminal voltage decreases with the increase in load on it.

However it can be made suitable for such applications by connecting a few turns in series with the armature as shown in Fig. 4.75. The field produced by these series turns assist the field produced by the shunt winding. Such generators are known as compound generators. In such generators when load current increases, the flux increases which increases the induced emf This extra induced emf compensates the voltage drop in the armature resistance and the demagnetising effect due to armature reaction. Hence, the terminal voltage *V* remains substantially constant.

Degree of Compounding

A cumulatively – compound wound generator is shown in Fig. 4.75. Its level of compounding can be changed by varying the amount of current passing through the series field winding by connecting a by-pass rheostat R_h.

When the field current is adjusted such that the terminal voltage *V* on full load remains the same as that on no-load, the generator is called to be *level* or *flat compounded generator* (see Fig. 4.76).

When the terminal voltage on full-load is more than its terminal voltage at no-load, the generator is called to be an over compounded generator.

On the other hand, when the terminal voltage on full-load is less than no-load voltage, the generator is called to be as under compounded generator.

However, if the field produced by the series field winding acts in opposite direction to the field produced by the shunt field winding, the generator is called to be differentially compounded (see Fig. 4.76).

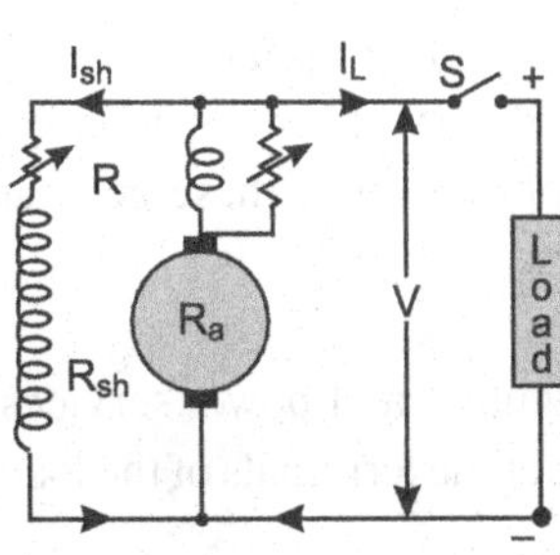

Fig. 4.75 Circuit diagram

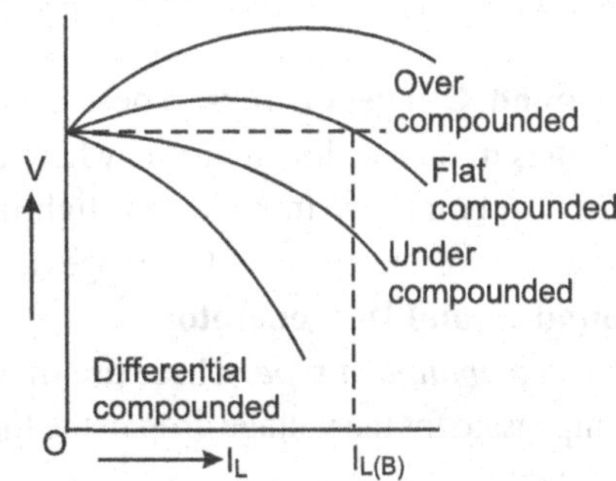

Fig. 4.76 Load characteristics of compound generator

4.38 Causes of Failure to Build-up Voltage in a Generator

There may be one or more of the following reasons due to which a generator fails to build-up voltage:

1. When the residual magnetism in the field system is destroyed.
2. When the connections of the field winding are reversed. This, in fact, destroys the residual magnetism due to which generator fails to build up voltage.
3. In case of shunt-wound generators, the other causes may be
   (*i*) the resistance of shunt field circuit may be more than the critical resistance.
   (*ii*) the resistance of load circuit may be less than critical resistance.
   (*iii*) the speed of rotation may be below the rated speed.
4. In case of series-wound generators, the other causes may be
   (*i*) the load circuit may be open: it may be due to faulty contact between brushes and commutator or commutator surface may be greasy or dirty and making no contact with the brushes.
   (*ii*) the load circuit may have high resistance.

Rectification: If the generator is not building up because of absence of residual magnetism due to any reason, the *field coils should be connected to a DC source for a small period in order to magnetise the poles.*

4.39 Applications of DC Generators

Depending upon the characteristics of various types of DC generators, their important applications are given below:

1. **Separately excited DC generators.** Although, these generators are more costly than self excited generators as they require a separate source for their field excitation. But their response to the

change in field resistance is more quick and precise. Therefore, these are employed where quick and definite response to control is important such as Ward–Leonard System of speed control.

2. **Shunt-wound DC generators.** As they provide constant terminal voltage, they are best suited for battery charging. Along with field regulators, they are also used for light and power supply purposes.
3. **Series-wound DC generators.** These generators have very few applications. Their best application is in the DC locomotives, where they supply field current for regenerative braking. They are also employed in series arc lighting. Another application of these generators is as series boosters for increasing DC voltage across the feeders.
4. **Compound-wound DC generators**
   (*i*) *Over-compounded type.* These are more suited for lighting and power services, as they compensate for the voltage drop in the lines and voltage at the terminals of the load remains constant.
   (*ii*) *Differential-compounded type.* They are usefully employed as are welding sets. In such cases, generator is practically short-circuited every time the electrode touches the metal plates to be welded.

4.40 Losses in a DC Generator

While conversion of mechanical energy into electrical energy, *a part of energy dissipated in the form of heat in surrounding air is called losses in the generator.*

These losses affect the efficiency of the generator. A reduction in these losses leads to higher efficiency. Thus, the major objective in the design of a DC machine is the reduction of these losses. The various losses occurring in a DC machine can be sub-divided as:

1. Copper losses.
2. Iron losses
3. Mechanical losses.

1. **Copper losses.** The various windings of the DC machine, made of copper, have some resistance. When current flows through them, there is power loss proportional to the square of their respective currents. *These power losses are called* **copper losses.**
   In general, the various copper losses in a DC machine are:
   (*i*) Armature copper loss = $I_a^2 R_a$
   (*ii*) Shunt field copper loss = $I_{sh}^2 R_{sh}$
   (*iii*) Series field copper loss = $I_{se}^2 R_{se}$
   (*iv*) Interpole winding copper loss = $I_i^2 R_i = I_a^2 R_i$
   (*v*) Brush contact loss = $I_a^2 R_b$ or $2I_a v_b$
   (*vi*) Compensating winding copper loss = $I_a^2 R_c$
   The brush contact drop is generally included in armature copper losses.

2. **Iron losses.** The losses which occur in the iron parts of a DC generator are known as *iron losses or core losses or magnetic loss.* These losses consist of the following:

   (*i*) **Hysteresis loss.** Whenever a magnetic material is subjected to reversal of magnetic flux, this loss occurs. It is due to retentivity property of the material. It is expressed with reasonable accuracy by the following expression: $P_h = K_h V f B_m^{1.6}$ ....(4.28)

   where, K_h = hysteresis constant in J/m^3 i.e., energy loss per unit volume of magnetic material during one magnetic reversal, its value depends upon nature of material;

   V = Volume of magnetic material in m^3;

   f = frequency of magnetic reversal in cycle/second and

   B_m = maximum flux density in the magnetic material in tesla.

   It occurs in the rotating armature. To minimise this loss, the armature core is made of silicon steel which has low hysteresis constant.

   (*ii*) **Eddy current loss.** When flux linking with the magnetic material changes (or flux is cut by the magnetic material) an emf is induced in it which circulates eddy currents through it. These eddy currents produce eddy current loss in the form of heat. It is expressed with reasonable accuracy by the expression:

   $$P_e = K_e V f^2 t^2 B_m^2 \qquad \text{...(4.29)}$$

   where, K_e = constant called co-efficient of eddy current, its value depends upon the nature of magnetic material;

   t = thickness of lamination in m;

   V, f and B_m are the same as above.

   The major part of this loss occur in the armature core. To minimise this loss, the armature core is laminated into thin sheets (0.3 to 0.5 mm) since this loss is directly proportional to the square of thickness of the laminations.

3. **Mechanical losses.** As the armature of a DC machine is a rotating part, some power is required to overcome:

   (*i*) Air friction of rotating armature (windage loss)

   (*ii*) Friction at the bearing and friction between brushes and commutator (friction loss).

   These losses are known as **mechanical losses.** To reduce these losses proper lubrication is done at the bearings.

4.41 Constant and Variable Losses

The losses in a DC generator may also be sub-divided into:

1. Constant losses; 2. Variable losses

1. **Constant losses.** The losses in a DC machine which remain the same at all loads are called *constant losses.* The constant losses in a DC machine are:

   (*i*) Iron losses; (*ii*) Mechanical losses; (*iii*) Shunt field copper losses

2. **Variable losses.** The losses in DC machine which vary with load are called *variable losses.* The variable losses in a DC machine are;

(*i*) Armature copper loss; (*ii*) Series field copper loss (*iii*) Interpole winding copper loss and (*iv*) Compensating winding copper loss.

4.42 Stray Losses

The sum of the iron losses and mechanical losses in a DC machine are known as *stray losses* i.e.,

Stray losses = Iron losses + Mechanical losses.

4.43 Power Flow Diagram

The mechanical power (ωT_m) is supplied to the generator which is converted into electrical power (VI_L). While conversion, various losses occur in the machine. The power flow diagram for a DC generator is shown in Fig. 4.77.

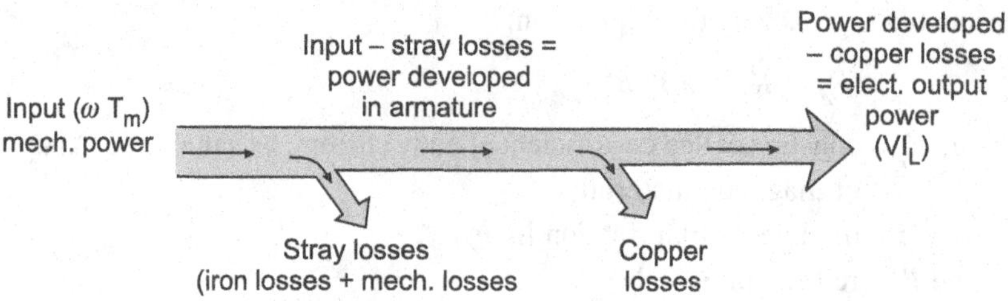

Fig. 4.77 Power flow diagram of a DC generator

4.44 Efficiency of a DC Generator

The ratio of output power to the input power of a DC generator is called its *efficiency*.

$$\text{Efficiency, } \eta = \frac{\text{output}}{\text{Input}}; \text{ where, Power output} = VI_L \text{ watt}$$

Power input = Power output + Variable losses + Constant losses

Since the shunt field current I_{sh} is very small as compared to line current,

therefore, $I_L \cong I_a$ (neglecting I_{sh})

$\therefore$ Variable losses = $I_L^2 R_a$

Constant losses = P_c (say)

Then, power output = $VI_L + I_L^2 R_a + P_c$

$$\therefore \quad \eta = \frac{VI_L}{VI_L + I_L^2 R_a + P_c} \quad \ldots(4.30)$$

4.45 Condition for Maximum Efficiency

The efficiency of a DC generator is given as:

$$\eta = \frac{\text{output}}{\text{output} + \text{variable losses} + \text{constant losses}}$$

$$= \frac{VI_L}{VI_L + I_L^2 R_a + P_c} = \frac{1}{1 + \frac{I_L R_a}{V} + \frac{P_c}{VI_L}} \qquad \text{...(4.31)}$$

Now, efficiency will be maximum when the denominator of eq. (*i*) in minimum i.e.,

$$\frac{d}{dI_L}\left(1 + \frac{I_L R_a}{V} + \frac{P_c}{VI_L}\right) = 0$$

or

$$\frac{R_a}{V} - \frac{P_c}{VI_L^2} = 0 \text{ or } \frac{R_a}{V} = \frac{P_c}{VI_L^2} \text{ or } I_L^2 R_a = P_c$$

i.e., **Variable losses = Constant losses**

Hence, *the efficiency of a DC machine will be maximum when the line current is such that constant loss is equal to the variable loss.*

The line current at which the efficiency of DC machine is maximum.

$$I_L = \sqrt{\frac{P_c}{R_a}} \qquad \text{...(4.32)}$$

Example 4.45

A DC generator is connected to a 220 V DC mains. The current delivered by the generator to the mains is 100 A. The armature resistance is 0.1 ohm. The generator is driven at a speed of 500 rpm Calculate (i) the induced emf (ii) the electromagnetic torque (iii) the mechanical power input to the armature neglecting iron, winding and friction losses, (iv) Electrical power output from the armature, (v) armature copper loss.

Solution:

(*i*) The induced emf, $E_g = V + I_a R_a = 220 + 0.1 \times 100 =$ **230 V** (*Ans.*)

(*ii*) Using the relation, $\omega T = E_g I_a$

Electromagnetic torque,

$$T = \frac{E_g I_a}{\omega} = \frac{E_g I_a}{2\pi N} \times 60 \qquad \left[\because \ \omega = \frac{2\pi N}{60}\right]$$

$$= \frac{230 \times 100 \times 60}{2\pi \times 500} = \textbf{439.27 Nm} \ (\textit{Ans.})$$

(*iii*) Neglecting iron, winding and friction losses,

$$\text{Input to armature} = \omega T \ (\text{or } E_g I_a)$$

$$= \frac{2\pi\ NT}{60} = \frac{2\pi \times 500 \times 439.27}{60} = \textbf{23000 W} \ (\textit{Ans.})$$

(*iv*) Electrical power output = $VI_a = 220 \times 100 =$ **22000 W** (*Ans.*)

(*v*) Armature copper losses = $I_a^2 R_a = (100)^2 \times 0.1 =$ **1000 W** (*Ans.*)

Example 4.46

A shunt generator supplies 195 A at 220 V. Armature resistance is 0.02 ohm, shunt field resistance is 44 ohm. If the iron and friction losses amount to 1600 watt, find (i) emf generated; (ii) copper losses; (iii) b.h.p. of the engine driving the generator.

Solution:

The conventional circuit is shown in Fig. 4.78.

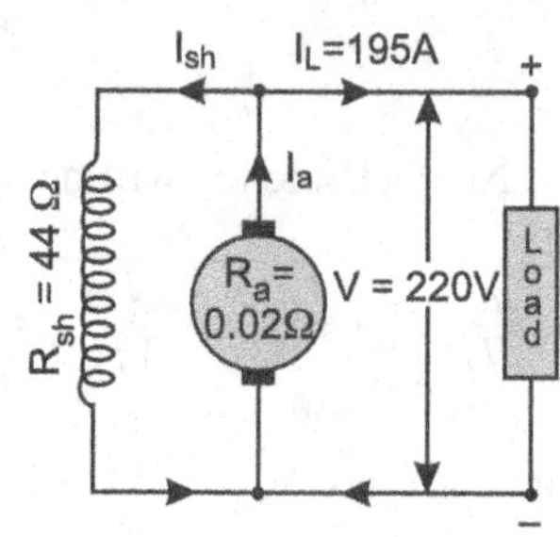

Fig. 4.78 Circuit diagram

Shunt field current,

$$I_{sh} = \frac{V}{R_{sh}} = \frac{220}{44} = 5\text{A}$$

Armature current, $I_a = I_L + I_{sh} = 195 + 5 = 200$ A

Generated or induced emf,

$$E_g = V + I_a R_a = 220 + 200 \times 0.02 = \mathbf{224\ V}\ (Ans.)$$

Armature copper loss;

$$= I_a^2 R_a = (200)^2 \times 0.02 = 800\ \text{W}$$

$$\text{Shunt field copper loss} = I_{sh}^2 R_{sh} = (5)^2 \times 44 = 1100\,\text{W}$$

$$\text{Total copper losses} = 800 + 1100 = \mathbf{1900\ W}\ (Ans.)$$

$$\text{Output power} = VI_L = 220 \times 195 = 42900\ \text{W}$$

$$\text{Input power} = 42900 + 1600 + 1900 = 46400\ \text{W}$$

$$\text{B.H.P. of the engine driving the generation} = \frac{46400}{735.5} = \mathbf{63.08\ H.P.}\ (Ans.)$$

Example 4.47

A 400 V shunt generator has full-load current of 200 A. Its armature resistance is 0.06 ohm, field resistance is 100 ohm and the stray losses are 2000 watt. Find the h.p. of prime-mover when it is delivering full load, and find the load for which the efficiency of the generator is maximum.

Solution:

The conventional circuit is shown in Fig. 4.79.

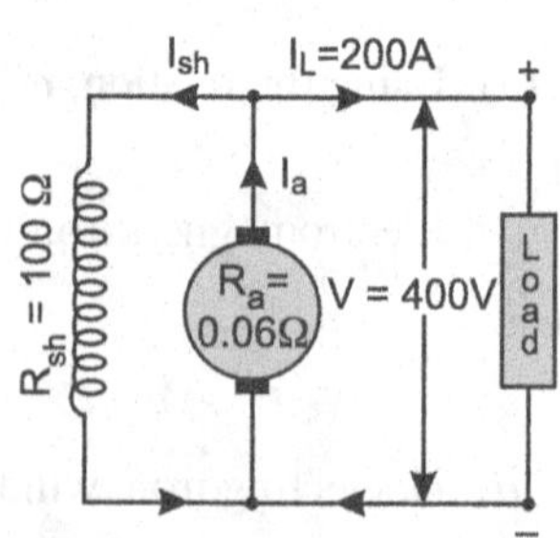

Fig. 4.79 Circuit diagram

$$\text{Shunt field current, } I_{sh} = \frac{V}{R_{sh}} = \frac{400}{100} = 4\text{A}$$

Armature current, $I_a = I_L + I_{sh} = 200 + 4 = 204$ A

Armature copper loss; $= I_a^2 R_a = (204)^2 \times 0.06 = 2497$ W

Shunt field copper loss $= I_{sh}^2 R_{sh} = (4)^2 \times 100 = 1600$ W

Total losses = 2497 + 1600 + 2000 = 6097 W

Output power $= VI_L = 400 \times 200 = 80000$ W

Input power = Output power + losses

= 80000 + 6097 = 86097 W

Horse power of prime-mover $= \dfrac{\text{Input power}}{735.5} = \dfrac{86097}{735.5}$ = **17.06 H.P.** (*Ans.*)

Constant losses = stray losses + shunt field copper loss = 2000 + 1600 = 3600 W

Condition for maximum efficiency is,

Variable losses = constant losses.

Let, I_L' be the load current at which the efficiency is maximum and armature current is I_a'

$$\therefore \quad I_a'^2 R_a = 3600 \quad \text{or} \quad I_a' = \sqrt{\frac{3600}{0.06}} = 245\text{ A}$$

Load current, $I_L' = I_a' - I_{sh} = 245 - 4 = 241$ A

Load for which the efficiency is maximum $= I_L'V = 241 \times 400 =$ **96.4 kW** (*Ans.*)

Section Practice Problems

Numerical Problems

1. The armature of a four-pole DC generator has 90 slots on its armature with 6 conductors per slot. If the armature is lap-wound, flux per pole is 6×10^{-2} Wb and running at 1000 rpm, determine its induced emf. Also determine the electrical power output of the machine if the current per conductor is 100 ampere. (**Ans.** *540 V; 216 kW*)

2. A 10 kW, six-pole DC generator develops an emf of 200 V at 1500 rpm. The armature has a lap-connected winding. The average flux density over a pole pitch is 0.9 T. The length and diameter of the armature are 0.25 m and 0.2 m respectively. Calculate (a) the flux per pole (b) the total number of active conductors in the armature and (c) the torque developed by the machine when the armature supplies a current of 60 A. (**Ans.** *23.6 mWb, 340, 76.4 Nm*)

3. The field and armature resistance of a four-pole shunt generator with lap-connected armature is 50 Ω and 0.1 Ω respectively. It is supplying a 2400 W load at a terminal voltage of 100 V. Calculate the total armature current, the current per armature path and the generated emf. (**Ans.** *26A; 6.5 A; 102.6 V*)

4. A DC shunt generator supplies a load of 100 A at 200 V. The generator has four-pole and its shunt field and armature resistance is 80 ohm and 0.1 ohm respectively. Find (i) total armature current (ii) current per conductor and (iii) generated emf. (**Ans.** *102.5 A, 25.6A; 210.25 V*)

5. An eight-pole DC shunt generator supplies a load of 12.5 ohm at 250 V. Its armature carries wave connected 778 conductors and its armature and field winding resistances are 0.24 ohm and 250 ohm respectively. Find the armature current, the induced emf and the flux per pole.

(**Ans.** *821 A; 255.04 V; 9.834 m Wb*)

6. A six-pole, 12 kW, 240 V DC machine is wave-connected. If the same machine is lap-connected, all other data remaining same, calculate its voltage, current and power ratings. (**Ans.** *80V; 150 A; 12 kW*)

7. The induced voltage of a shunt generator is 254 V. When the machine is loaded, the terminal voltage drops down to 240 V. Determine the load current, if the armature resistance is 0.04 Ω, and the field circuit resistance is 24 Ω ohm neglecting armature reaction. (**Ans.** *340 A*)

8. A four-pole, lap-wound DC shunt generator has field and armature resistances of 100 Ω and 0.1 Ω respectively. It supplies power to 50 lamps rated for 100 volt, 60 watt each. Calculate the total armature current and the generated emf by allowing a contact drop of 1 volt per brush. (**Ans.** *105.1V*)

9. Find the resistance of the load which takes a power of 5 kW from a DC shunt generator whose external characteristic is given by the equation: $V = 250 - 0.5\ I_L$. (***Ans.*** *11.48 Ω*)

10. A four-pole DC generator is having useful flux of 0.07 Wb per pole, it carries 220 turns each of 0.004 ohm resistance. The shunt and series winding resistance is 100 ohm and 0.02 ohm respectively. If the generator carries an armature current of 50 A and running at a speed of 900 rpm find its terminal voltage and power delivered by it. (**Ans.** *458.25 V; 20.814 W*)

11. A 500 V shunt generator has a full load current of 100 A and stray losses being 1·5 kW. Armature and field resistances are 0·3 and 250 ohm respectively. Calculate the input power and efficiency.

(**Ans.** *55·621 kW; 89·89%*)

Short Answer Type Questions

Q.1. What are the requirements of self-excited DC machine?

Or

What are the conditions to be fulfilled for a DC shunt generator to build-up emf.

Ans. Following are the requirements of a self-excited DC machine:

1. Its poles must have residual magnetism.
2. The field winding must be connected in such a way that field produced by the field winding must add to the residual magnetism. It should never be connected in reverse direction.
3. The field resistance should not be more than critical resistance.
4. The armature should not be rotated in reverse direction.
5. The resistance of the load circuit must not be below the critical resistance.

Q.2. Define critical resistance of DC shunt generator.

Ans. Critical resistance of a DC shunt generator. The maximum value of a shunt field winding resistance required to build-up voltage in a DC generator is called critical resistance. If the value of field winding resistance is more than this value, the generator will not be in position to built-up the voltage.

Q.3. What are the different methods by which flux is established in the magnetic circuit of a DC generator?

Ans. The magnetic flux in a DC generator, can be established by the following methods:

(*i*) by means of permanent magnets.

(*ii*) By the field coils excited by some external source

(*iii*) By the field coils excited by the generator itself.

Q.4. How can you differentiate between the excitation system of a DC series generator and that of a DC shunt generator?

Ans. In a series wound generator, the field winding is connected in series with the armature winding so that whole current flows through it and the load. In a shunt wound generator the field winding is connected across the armature winding forming a parallel or shunt circuit.

Q.5. Why is field coils of a DC shunt generator are wound with a large number of turns of fine wire.

Ans. The field coils of a DC shunt generator are wound with a large number of turns of fine wire so as to have high resistance in order to keep shunt field current as low as possible as well as to provide required mmf.

Q.6. What type of current flows in field winding and armature winding of a DC shunt generator? [A.M.I.e., Winter 2003]

Ans. In field winding, it is DC whereas in armature winding it is AC.

Q.7. What is meant by OCC of a DC generator?

Ans. *OCC* mean open circuit characteristics. The curve drawn between generated emf at no-load and the field current, when the machine is running at a specified speed is known as magnetic characteristic or open-circuit characteristic (*OCC*) of a DC generator.

Q.8. What is the critical resistance of field circuit?

Ans. The maximum value of resistance of field circuit of a DC shunt generator with which the DC shunt generator will be able to excite it is called critical resistance.

Q.9. In what respect the separately excited DC generator is superior to self excited DC generator?

Ans. The separately excited DC generator provides stable operation with any field excitation in comparison to self excited generators.

Q.10. What is critical speed in a DC shunt generator?

Ans. It is the speed of a DC generator at which shunt field resistance will represent the critical field resistance.

Q.11. A DC shunt generator is developing rated terminal voltage at some speed. For this generator,

(*i*) If only the direction of rotation is reversed, will the generator build-up?

(*ii*) If the direction of rotation and also the residual magnetism are reversed simultaneously will the generator build up?

Ans. (*i*) The generator will not build up. When the direction of rotation is reversed, the polarity of the generated emf will be reversed. the direction of field current will also be reversed and hence the residual magnetism present in the poles will be wiped off and the generator will fail to build up.

(*ii*) When the direction of rotation and the residual magnetism both are reversed simultaneously then it is obvious that a small emf will be produced initially which will further strengthen the field and hence the generator will build up voltage.

Q.12. What kind of generator can be expected to have no-load voltage and full load voltage to be same?

Ans. It can be expected in case of flat-compounded DC generator.

Q.13. If the terminal voltage of a DC generator at full load is more than its open-circuit voltage. What is the mode of excitation of the generator?

Ans. It can happen only if the machine is cumulatively compound wound.

Q.14. What type of DC generator can be used for battery charging?

Ans. DC shunt wound generator.

Q.15. What type of DC generator can be used for general power supply?

Ans. Cumulative compound wound is the most suitable.

Q.16. A DC series generator is not considered suitable for general electric supply?

Ans. Because DC series generators provide rising V–I characteristics.

Q.17. A six-pole, 12 kW, 240 V DC machine is wave-connected. If this machine is now lap-connected, all other things remaining the same, calculate the voltage and current ratings.

Ans. We know $E_g = \dfrac{\phi ZNP}{60\,A}$

If all other factors are same only the winding connections are being changed

$$E_g \propto \frac{1}{A}$$

i.e.,

$$E_{g_{wave}} \propto \frac{1}{A_{wave}} \text{ and } E_{g\,lap} = \frac{1}{A_{lap}}$$

$$\frac{E_{g_{lap}}}{E_{g_{wave}}} = \frac{A_{wave}}{A_{lap}} \text{ or } E_{g_{lap}} = \frac{2}{P} \times E_{g_{wave}} = \frac{2}{6} \times 240 = \mathbf{80\ V}\ (\textit{Ans.})$$

$$\text{Current, } I = \frac{Power}{V} = \frac{12 \times 1000}{80} = \mathbf{150\ A}\ (\textit{Ans.})$$

Review Questions

1. Name the various parts of a DC machine and give their function.
2. Explain the principle of action of a DC generator. Describe briefly its important parts.
3. What is the principle commercially adopted to transform mechanical energy to electrical energy? Illustrate in the case of a simple DC generator. How can the generator be made to supply DC?
4. Explain how commutator works in a DC machine to generate DC voltage.
5. Define: pole pitch, front pitch, back pitch, resultant pitch, average pitch and commutator pitch. Illustrate them with the help of neat sketches.

6. Explain what do you understand by (*i*) lap and wave windings (*ii*) duplex windings.
7. What is meant by duplex winding? If a duplex winding does not close after one passage around the armature, is the number of commutator segments even or odd. Explain.
8. Illustrate with suitable diagrams the patterns of (*i*) lap winding and (*ii*) wave winding of the armature of a DC machine. Explain the relative merits and the applications of the two types of windings.
9. What are the similarities and dissimilarities between lap and wave windings in a DC machine?
10. Explain why equaliser connections are used in lap windings and dummy coils are sometimes used in wave windings.
11. What are the ordinary types of armature windings for DC machines? Explain the essential difference between them and give their advantages, disadvantages and applications.
12. Highlight the factors on which the choice of type of armature winding of a DC machine will depend.
13. Find the expression for the equivalent armature resistance of a P-pole, lap wound armature, when there are total Z conductors, the length of each conductor being l cm. Assume cross-sectional area as a cm^2 and specific resistance of a conductor material as ρ ohm–cm.
14. Write short notes on the following:
    (*i*) Principle of operation of DC generator.
    (*ii*) Construction and function of commutator.
    (*iii*) Lap winding.
    (*iv*) Wave winding-merits and demerits over lap winding.
15. Derive emf equation of a DC generator (or DC machine).
16. Explain how will you distinguish between a lap and wave winding. How can you recognise the winding of a DC machine by counting its brushes?
17. What are the different types of excitation employed in DC machines?
18. Explain the principle of action of a DC shunt generator.
19. Explain what is meant by armature reaction of a DC machine. Describe different methods for minimising armature reaction.
20. Explain what you would understand by “armature reaction” in a DC generator. What are its effects?
21. How demagnetising and cross-magnetising ampere-turns per pole are calculated in a DC machine?
22. “As brushes in a DC machines are shifted, magnetising/demagnetising ampere-turns appears”. Explain with the help of a diagram.
23. Develop an expression for the demagnetising and cross magnetising armature ampere-turns in a DC generator.
24. What are the types of commutation possible in a DC machine? Explain them with graph.
25. Explain what do you mean by commutation in a DC machine. Mention the methods of improving it.
26. What do you mean by good commutation and bad commutation?
27. Briefly describe the function of interpoles in a DC machine.
28. Define *O.C.C.* (no-load characteristics) of a DC generator. Explain how it is obtained for a given generator.
29. Explain how a shunt generator builds up. What limits the voltage to which a generator can build up?

30. What is critical field resistance in a DC generator?

31. Discuss the internal and external characteristics of a shunt generator.

32. Mention the different reasons for the drop in the terminal voltage of a shunt generator when it is loaded.

33. Sketch the load characteristic of (*i*) DC shunt and (*ii*) DC series generator. Give reasons for the particular shape in each case.

34. Explain the load characteristics of a level-compounded DC generator. What do you mean by cumulative and differential compounding? Mention their applications.

35. Write a short note on external characteristics of series, shunt and compound wound DC generators.

36. Under what circumstances does a DC shunt generator fails to build up.

37. What are the various energy losses in a DC machine and how do they vary with load?

38. Derive a condition for maximum efficiency of a DC generator.

Multiple Choice Questions

1. The direction of electro-magnetic torque developed in the armature of a DC machine can be reversed by reversing the direction of
(*a*) rotation
(*b*) field flux
(c) armature current
(*d*) either field flux or armature current.

2. A thicker wire is used in the DC series field winding than DC shunt field winding in DC machines.
(*a*) to prevent mechanical vibrations.
(*b*) to produce large flux.
(*c*) because it carries the load current which is much higher than shunt field current for the same rating of DC machines.
(*d*) to provide strength.

3. The brushes for commutator are made of
(*a*) Copper
(*b*) Aluminium
(*c*) Cast iron
(*d*) Carbon.

4. In DC generators, dummy coil are provided
(*a*) to reduce eddy current losses
(*b*) to enhance flux density
(*c*) to amplify voltage
(*d*) to mechanically balance the rotor.

5. A shunt generator produces 445 A at 250V. The resistances of shunt field and armature are 50 Ω and 0.025 ohm respectively. The armature voltage drop will be
(*a*) 5 V
(*b*) 11.25 V
(*c*) 22.5 V
(*d*) 25 V

6. Eddy current loss varies
(*a*) inversely as the thickness of the laminations
(*b*) inversely as the square of the thickness of the laminations
(*c*) directly as thickness of laminations.
(*d*) directly as the square of the thickness of laminations.

7. Hysteresis loss in a DC machine depends upon
(*a*) volume and grade of iron
(*b*) maximum value of flux density
(*c*) frequency of magnetic reversals
(*d*) all of the above.

8. Iron losses in a DC machine take place in
(*a*) yoke
(*b*) commutator
(*c*) main body
(*d*) armature rotor.

9. The main requirement of a DC armature winding is that it must be
(*a*) a wave winding
(*b*) a lap winding
(*c*) a closed one
(*d*) drum winding

10. For full-pitch winding in a four-pole, 36-slot DC machine, the coil occupies ________ slot.
(*a*) 1 and 9
(*b*) 1 and 10
(*c*) 3 and 11
(*d*) 4 and 12

11. In a long-shunt DC generator, the series field winding is excited by ________ current.
(*a*) shunt
(*b*) armature
(*c*) load
(*d*) external

12. A DC generator with wave winding is used for________ current, ________ voltage.
(*a*) low, high
(*b*) low, low
(*c*) high, low
(*d*) high, high

13. The emf induced in the armature of a DC generator is inversely proportional to
(*a*) pole flux
(*b*) field current
(*c*) number of armature parallel paths
(*d*) number of dummy coils

14. The external characteristic of a shunt generator can be obtained directly from its ________ characteristics.
(*a*) internal
(*b*) open-circuit
(*c*) closed-circuit
(*d*) performance

15. The O.C.C. of a self-excited DC generator starts with some minimum voltage instead of zero due to
(*a*) residual pole flux
(*b*) high armature speed
(*c*) magnetic inertia
(*d*) high field circuit resistance

16. If residual magnetism of a shunt generator is destroyed due to any reason, it may be restored by connecting its shunt field
(*a*) to earth
(*b*) to a DC source
(*c*) to neutral
(*d*) to an AC source

17. A DC generator is considered to be an ideal one if it has ________ voltage regulation.
(*a*) low
(*b*) zero
(*c*) positive
(*d*) negative

18. In a DC generator, compensating winding is provided to
(*a*) compensate for decrease in main flux
(*b*) neutralise armature mmf
(*c*) neutralise cross-magnetising flux
(*d*) maintain uniform flux distribution.

19. Which statement is not true for the interpoles in the following.
(*a*) these are smaller in size spaced in between the main poles
(*b*) their field windings are connected in parallel with the armature so that they carry part of the armature current

(*c*) their polarity, in the case of generators is the same as that of the main pole ahead
(*d*) they automatically neutralise not only reactance voltage but cross-magnetisation as well

20. In a DC generator, the effect of armature reaction on the main field is to
(*a*) reverse it
(*b*) distort it
(*c*) reduce it
(*d*) both (*b*) and (*c*)

21. In a loaded DC generator, to reduce sparking, brushes have to be shifted.
(*a*) clockwise
(*b*) counterclockwise
(*c*) either (*a*) or (*b*)
(*d*) in the direction of rotation.

22. The commutation process in a DC gener-ator basically involves
(*a*) passage of current from moving arm-ature to a stationary load
(*b*) reversal of current in an armature coil as it crosses MNA
(*c*) conversion of reactance voltage
(*d*) all of above

Keys to Multiple Choice Questions

1. d	**2.** c	**3.** d	**4.** d	**5.** b	**6.** d	**7.** d	**8.** d	**9.** c	**10.** b
11. b	**12.** a	**13.** c	**14.** b	**15.** a	**16.** b	**17.** b	**18.** c	**19.** d	**20.** d
21. d	**22.** d								

CHAPTER

5

DC Motors

Chapter Objectives

After the completion of this unit, students/readers will be able to understand:

- ✓ What is a DC motor and how does it convert electrical energy into mechanical energy?
- ✓ What is back emf and its significance?
- ✓ What are the factors on which torque developed in a DC motor depends?
- ✓ How DC motors are classified on the basis of their field excitation?
- ✓ What are the performance characteristics of DC motors and how are they selected for different applications?
- ✓ What is the need of a starter and how do starters limit the in-rush flow of starting current and protect the motor?
- ✓ How is a four point starter for DC shunt motor different than a three point starter?
- ✓ How are the steps of a starting resistance designed?
- ✓ Which type of starter is used for starting of DC series motors?
- ✓ What are the factors on which the speed of a DC motor depends and how can we control the speed of DC motors as per the requirement?
- ✓ How is speed of electric trains controlled?
- ✓ What is electric breaking? How is it superior to mechanical braking and what are the different methods by which electric motors or drives are stopped?
- ✓ What are the various losses of a DC machine, how these are categorised?
- ✓ How to determine various losses of a DC machine with the help of different tests?
- ✓ How to calculate the efficiency of a DC machine once the losses are determined?
- ✓ How to prepare maintenance schedule of a DC machine to improve its reliability and performance?

Introduction

A machine that converts mechanical power into DC electrical power is called a DC generator. The same machine when used to convert DC electrical power into mechanical power, it known as a DC motor. From construction point of view there is no difference between a DC generator and motor.

The DC motors are very useful where wide range of speeds and perfect speed regulation is required such as electric traction.

5.1 DC Motor

An electro-mechanical energy conversion device (electrical machine) that converts DC electrical energy or power (EI) into mechanical energy or power (ωT) is called a **DC motor.**

Electric motors are used for driving industrial machines, e.g., hammers, presses, drilling machines, lathes, rollers in paper and steel industry, blowers for furnaces, etc., and domestic appliances, e.g., refrigerators, fans, water pumps, toys, mixers, etc. The block diagram of energy conversion, when the electro-mechanical device works as a motor, is shown in Fig. 5.1.

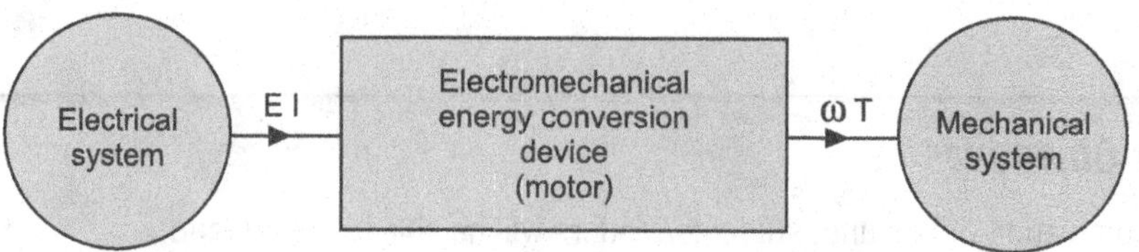

Fig. 5.1 Block diagram of electromagnetic energy conversion (motor action)

5.2 Working Principle of DC Motors

The operation of a DC motor is based on the principle that when a current carrying conductor is placed in a magnetic field, a mechanical force is experienced by it. The direction of this force is determined by *Fleming's Left Hand Rule* and its magnitude is given by the relation:

$$F = Bil \text{ newton}$$

For simplicity, consider only one coil of the armature placed in the magnetic field produced by a bipolar machine [see Fig. 5.2(*a*)]. When DC supply is connected to the coil, current flows through it which sets up its own field as shown in Fig. 5.2(*b*). By the interaction of the two fields (i.e., field produced by the main poles and the coil), a resultant field is set up as shown in Fig. 5.2(*c*). The tendency of this is to come to its original position i.e., in straight line due to which force is exerted on the two coil sides and torque develops which rotates the coil.

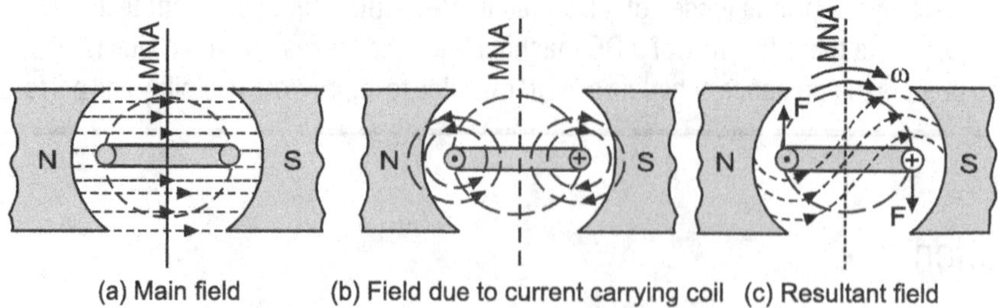

Fig. 5.2 Working principle of a motor

Alternately, it can be said that the main poles produce a field F_m. Its direction is marked in Fig. 5.3. When current is supplied to the coil (armature conductors), it produces its own field marked as F_r. This field tries to come in line with the main field and an electromagnetic torque develops in clockwise direction as marked in Fig. 5.3.

In actual machine, a large number of conductors are placed on the armature. All the conductors, placed under the influence of one pole (say, North pole) carry the current in one direction (outward). Whereas, the other conductors placed under the influence of other pole i.e., south pole, carry the current in opposite direction as shown in Fig. 5.4. A resultant rotor field is produced. Its direction is marked by the arrow-head F_r. This rotor field F_r tries to come in line with the main field F_m and torque (T_e) develops. Thus, rotor rotates.

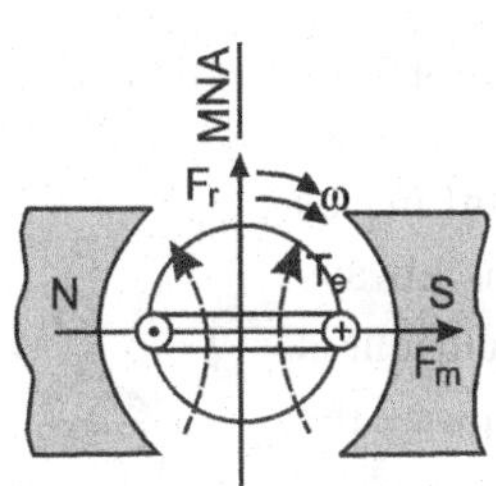

Fig. 5.3 Position of main field F_m and rotor field F_r

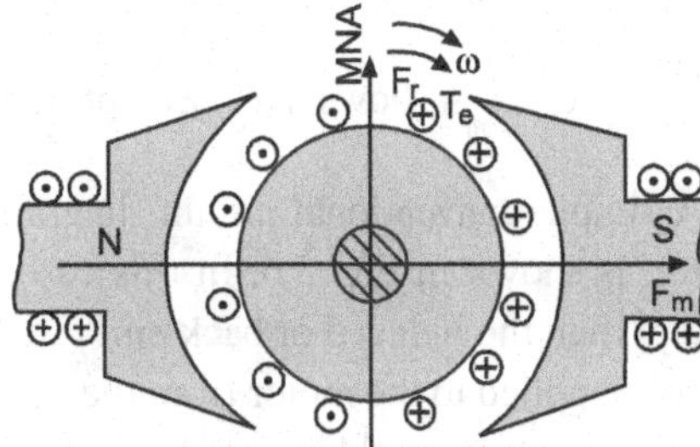

Fig. 5.4 Motor action

It can be seen that to obtain a continuous torque, the direction of flow of current in each conductor or coil side must be reversed when it passes through the magnetic neutral axis (*MNA*). This is achieved with the help of a commutator.

Function of a Commutator

The function of a commutator in DC motors is to reverse the direction of flow of current in each armature conductor when it passes through the *M.N.A.* to obtain continuous torque.

5.3 Back emf

It has been seen that when current is supplied to the armature conductors, as shown in Fig. 5.5(*a*), placed in the main magnetic field, torque develops and armature rotates. Simultaneously, the armature conductors cut across the magnetic field and an emf is induced in these conductors. The direction of this induced emf in the armature conductors is determined by Fleming's Right Hand Rule and is marked in Fig. 5.5(*b*).

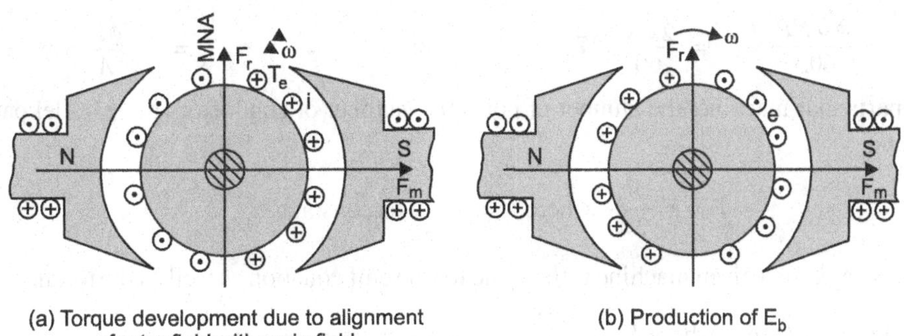

(a) Torque development due to alignment of rotor field with main field

(b) Production of E_b

Fig. 5.5 Back emf

It can be seen that the direction of this induced emf is opposite to the applied voltage. That is why this induced emf is called back emf (E_b). The magnitude of this induced emf is given by the relation;

$$E_b = \frac{PZ\phi N}{60\,A}$$

or $$E_b = \frac{ZP}{60\,A}\phi N \text{ or } E_b \propto \phi N \qquad \left(\text{since } \frac{ZP}{60\,A} \text{ are constant}\right) \quad ...(5.1)$$

Also, $N \propto \frac{E_b}{\phi}$ shows that speed of motor is inversely proportional to magnetic field or flux.

A simple conventional circuit diagram of the machine working as motor, is shown in Fig. 5.6. In this case, the supply voltage is always greater than the induced or back emf (i.e., $V > E_b$). Therefore, current is always supplied to the motor from the mains and the relation among the various quantities will be; $E_b = V - I_a R_a$.

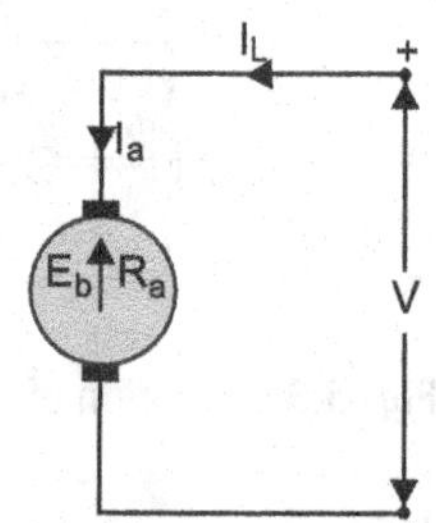

Fig. 5.6 Circuit diagram ($E_b < V$)

Significance of Back emf

The current flowing through the armature is given by the relation:

$$I_a = \frac{V - E_b}{R_a}$$

When mechanical load applied on the motor increases, its speed decreases which reduces the value of E_b. As a result the value ($V - E_b$) increases which consequently increases I_a. Hence, motor draws extra current from the mains.

Thus, **the back emf** *regulates the input power as per the extra load.*

5.4 Electro-magnetic Torque Developed in DC Motor

The electrical power which is supplied to a DC motor is converted into mechanical power. The conversion of power takes place in the armature as stated below:

The power developed in the armature is given as

$$EI_a = \omega T_e \qquad \text{or} \qquad EI_a = \frac{2\pi N}{60} \times T_e$$

or $$\frac{\phi\, ZNP}{60\text{A}} \times I_a = \frac{2\pi N}{60} \times T_e \qquad \text{or} \qquad T_e = \frac{PZ\phi N}{2\pi A}\,\text{Nm}$$

For a particular machine, the number of poles (P), number of conductors per parallel path (Z/A) are constant.

$$\therefore \qquad T = K\frac{\phi I_a}{2\pi} \text{ where } K = \frac{PZ}{A} \text{ is a constant}$$

The constant K for a given machine is the same for the emf equation as well as the torque equation.

As well as, $T = K_2\,\phi\, I_a$ where $K_2 = \frac{PZ}{2\pi A}$ is another constant or $T \propto \phi\, \text{I}_a$ ...(5.2)

Thus, it is concluded that torque produced in the armature of a DC machine is directly proportional to flux per pole and armature current. Moreover, the direction of electromagnetic torque developed in the armature depends upon the direction of flux or magnetic field and the direction of flow of current in the armature conductors. If either of the two is reversed the direction of torque produced is reversed and hence the direction of rotation. But when both are reversed the direction of torque (or rotation) does not change.

5.5 Shaft Torque

In DC motors whole of the electromagnetic torque (T_e) developed in the armature is not available at the shaft. A part of it is lost to overcome the iron and mechanical (friction and windage) losses. Therefore, shaft torque (T_{sh}) is somewhat less than the torque developed in the armature.

Thus, in case of DC motors, the actual torque available at the shaft for doing useful mechanical work is known as **shaft torque.**

Brake Horse Power (B.H.P.)

In case of motors, the mechanical power (H.P.) available at the shaft is known as brake horse power (B.H.P.). If T_{sh} is the shaft torque in Nm and N is speed in rpm then,

$$\text{Useful output power} = \omega\, T_{sh} = 2\pi N\, T_{sh}/60 \text{ watt}$$

$$\text{Output in B.H.P.} = \frac{2\pi N T_{sh}}{60\times 735.5} \qquad \ldots(5.3)$$

5.6 Comparison of Generator and Motor Action

It has been seen that the same machine can be used as a DC generator or as a DC motor. When it converts mechanical energy (or power) into electrical energy (or power), it is called a DC generator and when it is used for reversed operation, it is called a DC motor.

The comparison of generator and motor action is given below:

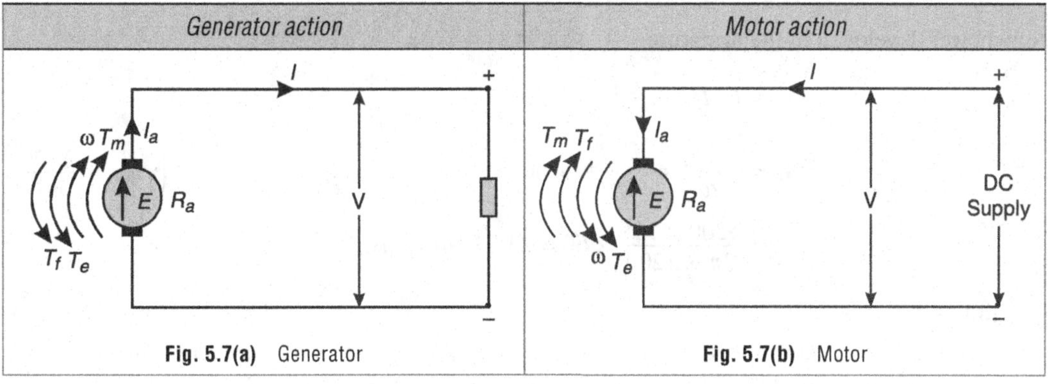

Fig. 5.7(a) Generator

Fig. 5.7(b) Motor

1. In generating action, the rotation is due to mechanical torque, therefore, T_m and ω are in the same direction.	1. In motoring action, the rotation is due to electromagnetic torque, therefore, T_e and ω are in the same direction.
2. The frictional torque T_f acts in opposite direction to rotation ω.	2. The frictional torque T_f acts in opposite direction to rotation ω.
3. Electromagnetic torque T_e acts in opposite direction to mechanical torque T_m so that $\omega T_m = \omega T_e + \omega T_f$.	3. Mechanical torque T_m acts in opposite direction to electromagnetic torque T_e so that $\omega T_e = \omega T_m + \omega T_f$.
4. In generating action, an emf is induced in the armature conductors which circulates current in the armature when load is connected to it. Hence, e and i both are in the same direction.	4. In motoring action, current is impressed to the armature against the induced emf (e), therefore current flows in opposite direction to that of induced emf.
5. In generator action, $E > V$	5. In motor action, $E < V$
6. In generating action, the torque angle θ is leading.	6. In motoring action, the torque angle θ is lagging.
7. In generating action, mechanical energy is converted into electrical energy.	7. In motoring action, electrical energy is converted into mechanical energy.

Example 5.1

A 50 HP, 400 V, 4 pole, 1000 rpm, DC motor has flux per pole equal to 0.027 Wb. The armature having 1600 conductors is wave connected. Calculate the gross torque when the motor takes 70 ampere.

Solution:

$$\text{Torque developed, } T = \frac{P\varphi Z I_a}{2\pi A}$$

Where, $P = 4$; $\phi = 0.027$ Wb; $Z = 1600$; $I_a = 70$A; $A = 2$ (wave connected)

$$T = \frac{4 \times 0 \cdot 027 \times 1600 \times 70}{2 \times \pi \times 2} = \mathbf{963\ Nm}\ (Ans.)$$

Example 5.2

The induced emf in a DC machine is 200 V at a speed of 1200 rpm. Calculate the electromagnetic torque developed at an armature current of 15 A.

Solution:

Here, $E_b = 200$ V; $N = 1200$ rpm; $I_a = 15$ A

Now power developed in the armature,

$$\omega T_e = E_b I_a$$

or

$$T_e = \frac{E_b I_a}{\omega} = \frac{E_b I_a}{2\pi N} \times 60 \qquad \left(\because \omega = \frac{2\pi N}{60}\right)$$

$$= \frac{200 \times 15}{2\pi \times 1200} \times 60 = \mathbf{23.87\ Nm}\ (Ans.)$$

Example 5.3

A four-pole DC motor has a wave-wound armature with 594 conductors. The armature current is 40 A and flux per pole is 7.5 m Wb. Calculate H.P. of the motor when running at 1440 rpm

Solution:

Torque developed, $$T = \frac{PZ\phi I_a}{2\pi A} = \frac{4 \times 594 \times 7.5 \times 10^{-3} \times 40}{2\pi \times 2} = \mathbf{56.72\ Nm}\ (Ans.)$$

Power developed = ωT watt; where $\omega = \frac{2\pi N}{60}$

$$H.P. = \frac{\text{power developed}}{735.5};\ H.P. = \frac{\omega T}{735.5} = \frac{2\pi NT}{60 \times 735.5}$$

or $$H.P. = \frac{2 \times 1440 \times 56.72}{60 \times 735.5} = \mathbf{11.63}\ (Ans)$$

Example 5.4

A DC motor has 6-poles with lap wound armature. What will be its brake horse power when it draws a current of 340 A and rotates at 400 rpm. The flux per pole is 0.05 Wb and the armature carries 864 turns, Neglect mechanical losses.

Solution:

Here, $P = 6$; $A = P = 6$ (lap wound); $I_L = 340$ A; $N = 400$ rpm,

$$\phi = 0.05 \text{ Wb; No. of turns} = 864$$

$$Z = 864 \times 2 = 1728$$

$$\text{Back emf, } E_b = \frac{\phi ZNP}{60A} = \frac{0.05 \times 1728 \times 400 \times 6}{60 \times 6} = 576 \text{ V}$$

$$\text{Armature current, } I_a = I_L = 340 \text{ A}$$

$$\text{Power developed} = E_b \times I_a = 576 \times 340 = 195840 \text{ W}$$

$$\text{Neglecting losses, brake } HP = \frac{E_b I_a}{735.5} = \frac{195840}{735.5} = \mathbf{266.27}\ (Ans.)$$

5.7 Types of DC Motors

On the basis of the connections of armature and their field winding, DC motors can be classified as;

1. **Separately excited DC motors:** The conventional diagram of a separately excited DC motor is shown Fig. 5.8. Its voltage equation will be;

$$E_b = V - I_a R_a - 2v_b \quad \text{(where } v_b \text{ is voltage drop per brush)} \quad ...(5.4)$$

2. **Self excited DC motors:** These motors can be further classified as;

(*i*) **Shunt motors:** Their conventional diagram is shown in Fig. 5.9.

Important relations:

$$I_{sh} = V/R_{sh} \quad ...(5.5)$$

$$I_a = I_L - I_{sh} \quad ...(5.6)$$

$$E_b = V - I_a R_a - 2v_b \quad \text{(where } v_b \text{ is voltage drop per brush)} \quad ...(5.7)$$

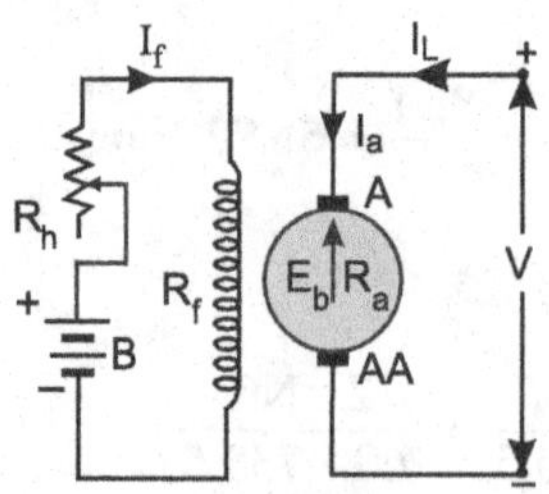

Fig. 5.8 Circuit diagram of separately excited DC motor

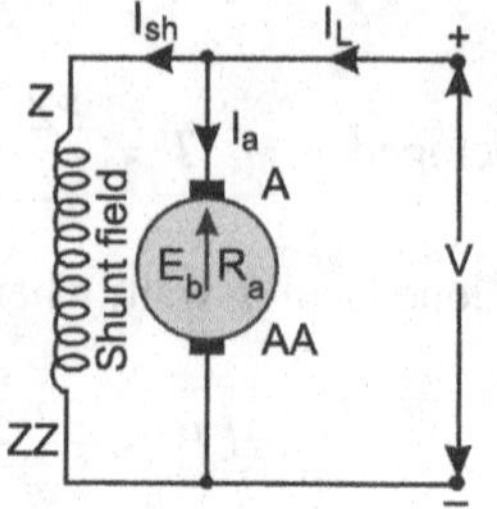

Fig. 5.9 DC shunt motor

(*ii*) **Series motor:** Its conventional diagram is shown in Fig. 5.10.
Important relations:

$$I_L = I_a = I_{se} \quad ...(5.8)$$

$$E_b = V - I_a (R_a + R_{se}) - 2v_b \quad ...(5.9)$$

(*iii*) **Compound motor:** Its conventional diagram (for long shunt) is shown in Fig. 5.11.

$$I_{sh} = \frac{V}{R_{sh}}; I_a = I_L - I_{sh;}\ E_b = V - I_a (R_a + R_{se}) - 2v_b \quad ...(5.10)$$

In all the above voltage equations, the brush voltage drop v_b is sometimes neglected since its value is very small.

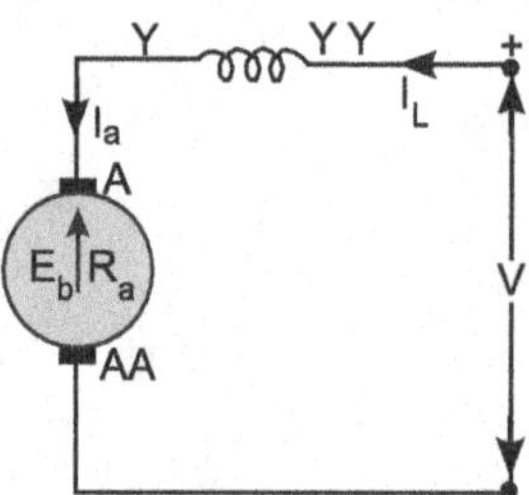

Fig. 5.10 DC series motor

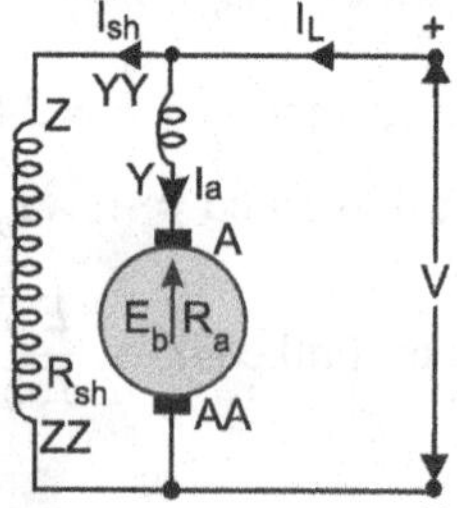

Fig. 5.11 DC compound motor

The compound motor can be further subdivided as;

(*a*) **Cumulative compound motors:** In these motors, the flux produced by both the windings is in the same direction, i.e.,

$$\phi_r = \phi_{sh} + \phi_{se}.$$

(*b*) **Differential compound motors:** In these motors, the flux produced by the series field winding is opposite to the flux produced by the shunt field winding, i.e.,

$$\phi_r = \phi_{sh} - \phi_{se}$$

Example 5.5

The armature resistance of a DC shunt motor is 0.5 ohm, it draws 20 A from 220 V mains and is running at a speed of 80 radian per second. Determine (i) Induced emf (ii) Electromagnetic torque (iii) Speed in rpm.

Solution:

Here, $V = 220$ V; $I_a = 20$ A; $R_a = 0.5\ \Omega$; $\omega = 80$ rad/s

$$\text{Induced emf, } E = V - I_a R_a = 220 - 20 \times 0.5 = \mathbf{210\ V}\ (Ans.)$$

$$\text{Electromagnetic torque, } T = \frac{EI_a}{\omega} = \frac{210 \times 20}{80} = \mathbf{52.5\ Nm}\ (Ans.)$$

$$\text{Speed in rpm, } N = \frac{60\omega}{2\pi} = \frac{60 \times 80}{2 \times \pi} = \mathbf{764\ rpm}\ (Ans.)$$

Example 5.6

A 400 V DC motor takes an armature current of 100 A when its speed is 1000 rpm If the armature resistance is 0·25 ohm, calculate the torque produced in Nm.

Solution:

Here, $V = 400$ V; $I_a = 100$ A; $R_a = 0{\cdot}25\ \Omega$; $N = 1000$ rpm

$$\text{Induced emf, } E = V - I_a R_a \text{ (motor action)}$$

$$= 400 - 100 \times 0{\cdot}25 = 375 \text{ V}$$

Using the relation, $\omega T = EI_a$ or

$$\frac{2\pi NT}{60} = EI_a \qquad \left[\text{Because } \omega = \frac{2\pi N}{60}\right]$$

$$\therefore \quad \text{Torque produced, } T = \frac{60EI_a}{2\pi N} = \frac{60 \times 375 \times 100}{2\pi \times 1000} = \mathbf{358{\cdot}1\ Nm}\ (Ans.)$$

Example 5.7

The armature and series field winding resistance of a 220 V, four-pole DC series motor is 0.75 ohm. It has 782 wave wound armature conductors. If it draws 40 A from the supply mains and has a flux of 25 mWb, determine its speed and gross torque developed.

Solution:

Here, $V = 250$ V; $P = 4$; $A = 2$ (wave winding); $Z = 782$;

$$R_a + R_{se} = 0.75\ \Omega;\ I = 40 \text{ A};\ \phi = 25 \times 10^{-3} \text{ Wb}$$

$$E = V - I(R_a + R_{se}) = 250 - 40 \times 0{\cdot}75 = 220 \text{ V}$$

$$E = \frac{\phi ZNP}{60A}$$

or

$$N = \frac{60\ AE}{\phi ZP} = \frac{60 \times 2 \times 220}{25 \times 10^{-3} \times 782 \times 4} = \mathbf{337{\cdot}6\ rpm}\ (Ans.)$$

$$T_a = \frac{\phi ZPI_a}{2\pi A} = \frac{25 \times 10^{-3} \times 782 \times 4 \times 40}{2\pi \times 2} = \mathbf{248{\cdot}9\ Nm}\ (Ans.)$$

Example 5.8

The armature and shunt field resistance of a four-pole, lap wound DC shunt motor is 0.05 ohm and 25 ohm respectively. If its armature contains 500 conductors, find the speed of the motor when it take 120 A from a DC mains of 100 V supply. Flux per pole is 2×10^{-2} Wb.

Solution:

The circuit is shown in Fig. 5.12;

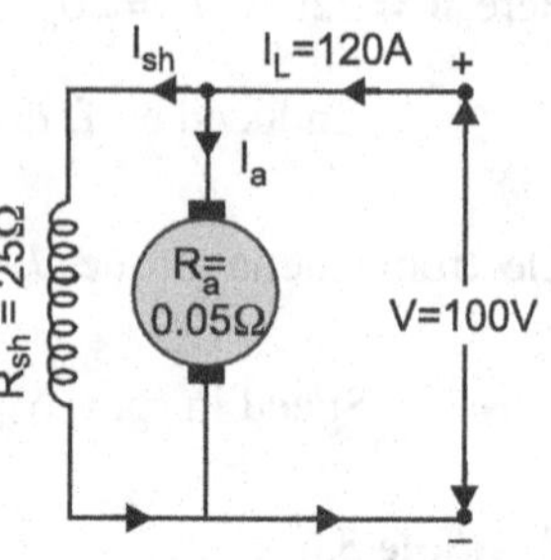

Fig. 5.12 Circuit diagram

$$I_{sh} = \frac{V}{R_{sh}} = \frac{100}{25} = 4\text{A};$$

$$I_a = I_L - I_{sh} = 120 - 4 = 116 \text{ A}$$

$$E_b = V - I_a R_a = 100 - 116 \times 0.05 = 94.2 \text{ V.}$$

Now, $$E_b = \frac{P\phi ZN}{60\,A}$$

or $$94.2 = \frac{6 \times 2 \times 10^{-2} \times 500 \times N}{60 \times 6}$$

or $$N = \mathbf{565.2\ rpm}\ (\textit{Ans.})$$

Example 5.9

A 6-pole, 440 V DC motor has 936 wave wound armature conductors. The useful flux per pole is 25 m Wb. The torque developed is 45·5 kg-m. Calculate the following, if armature resistance is 0·5 ohm; (i) Armature current (ii) Speed

Solution:

Here, $$P = 6;\ Z = 936;\ \phi = 25 \text{ m Wb} = 25 \times 10^{-3} \text{ Wb}$$

$$A = 2\ (\textit{wave wound armature});\ V = 440 \text{ V};\ R_a = 0{\cdot}5 \text{ ohm};$$

Torque developed, $$T = 45{\cdot}5 \text{ kg-m} = 45{\cdot}5 \times 9{\cdot}81 = 446{\cdot}35 \text{ Nm}$$

(*i*) Using the relation, $T = \dfrac{PZ\phi I_a}{2\pi A}$

Armature current, $$I_a = \frac{2\pi A \times T}{PZ\phi} = \frac{2\pi \times 2 \times 446 \cdot 35}{6 \times 936 \times 25 \times 10^{-3}} = \mathbf{39{\cdot}95\ A}\ (\textit{Ans.})$$

(*ii*) Induced emf, $E = V - I_a R_a$ (motor action)

$$= 440 - 39{\cdot}95 \times 0{\cdot}5 = 420 \text{ V}$$

Using the relation, $\omega T = EI_a$

or $$\frac{2\pi N}{60} \times I = EI_a$$

Speed $$N = \frac{60 \times EI2}{\pi T} = \frac{60 \times 420 \times 39 \cdot 95}{2\pi \times 446 \cdot 35} = \mathbf{359\ rpm}\ (\textit{Ans.})$$

Example 5.10

The electromagnetic torque developed in a DC machine is 80 Nm for an armature current of 30 A. What will be the torque for a current of 15 A? Assume constant flux. What is the induced emf at a speed of 900 rpm and an armature current of 15 A?

Solution:

Torque developed, $T_1 = 80$ Nm

Armature current, $I_{a1} = 30$ A

Armature current, $I_{a2} = 15$ A

Let the torque developed is T_2 Nm when the armature current is 15 A.

Now $T \propto \phi I_a$

When flux ϕ is constant, $T \propto I_a$

$$\therefore \quad \frac{T_2}{T_1} = \frac{I_{a2}}{I_{a1}}$$

$$\text{or} \quad T_2 = \frac{I_{a2}}{I} \times T1 = \frac{15}{30} \times 80 = \mathbf{40\ Nm}\ (Ans.)$$

Power developed in the armature = $E_2 I_{a_2} = \omega_2 T_2$

$$\text{where} \quad \omega_2 = \frac{2\pi N_2}{60} = \frac{2\pi \times 900}{60} = 30\pi$$

$$\therefore \quad \text{Induced emf} = E_2 = \frac{\omega_2 T_2}{I_{a_2}} = \frac{30\pi \times 40}{15} = \mathbf{251{\cdot}33\ V}\ (Ans.)$$

5.8 Characteristics of DC Motors

The performance of a DC motor can be easily judged from its characteristic curves, known as motor *characteristics*. The characteristics of a motor are those curves which show relation between the two quantities. On the basis of these quantities, the following characteristics can be obtained:

1. **Speed and Armature current i.e., N – I_a Characteristics:** It is the curve drawn between speed N and armature current I_a. It is also known as *speed characteristics*.
2. **Torque and Armature current i.e., T–I_a Characteristics:** It is the curve drawn between torque developed in the armature T and armature current I_a. It is also known as *electrical characteristic*.
3. **Speed and Torque i.e., N–T characteristics:** It is the curve drawn between speed N and torque developed in the armature T. It is also known as *mechanical characteristics*.

The following important relations must be kept in mind while discussing the motor characteristics:

$$E_b \propto N\phi \quad \text{or} \quad N \propto \frac{E_b}{\phi} \quad \text{and} \quad T \propto \phi I_a$$

5.9 Characteristics of Shunt Motors

The conventional diagram of this motor is shown in Fig. 5.13. In these motors, the shunt field current $I_{sh} = V/R_{sh}$ remains constant since the supply voltage V is constant. Hence, the flux in DC shunt motors is practically constant (although at heavy loads, somewhat flux decreases due to armature reaction).

1. $N - I_a$ characteristics

We know that, $N \propto \frac{E_b}{\phi}$

Since flux is constant; $N \propto E_b$ or $N \propto V - I_a R_a$

If the armature drop ($I_a R_a$) is negligible, the speed of the motor will remain constant for all values of load as shown by the dotted line *AB* in Fig. 5.14. But strictly speaking, as the armature current increases due to the increase of load, armature drop $I_a R_a$ increases and speed of the motor decreases slightly as shown by the straight line *A C* in Fig. 5.14 (neglecting armature reaction). Moreover, the characteristic curve does not start from a point of zero armature current because a small current, no-load armature current I_{a0}, is necessary to maintain rotation of the motor at no-load.

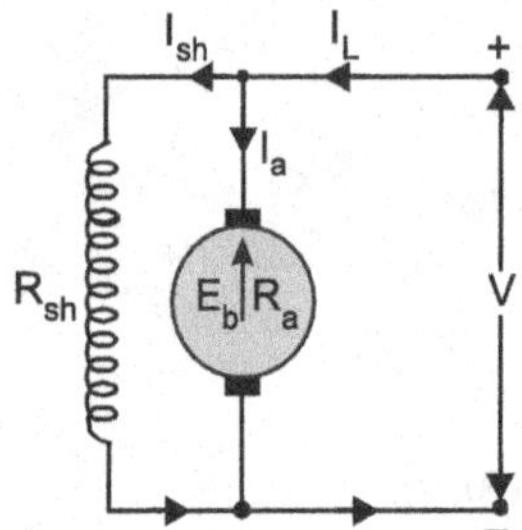

Fig. 5.13 Circuit for shunt motor

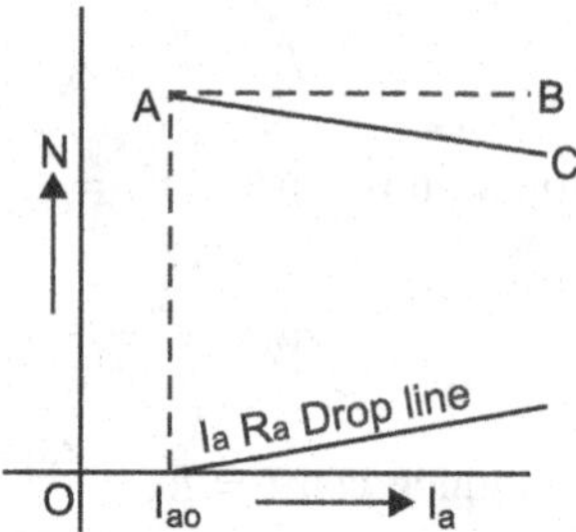

Fig. 5.14 N-I_a characteristics of shunt motor

Since there is no appreciable change in the speed of a DC shunt motor from no-load to full load that is why it is considered to be a constant speed motor. This motor is best suited where almost constant speed is required and the load may be thrown off totally and suddenly.

2. $T - I_a$ Characteristics

We know that, $T \propto \phi I_a$

Since flux is constant, $T \propto I_a$

Hence, the electrical characteristic (i.e., $T - I_a$) is a straight line passing through the origin as shown in Fig. 5.15. It is clear from the characteristic curve that a large armature current is required at the start if machine is on heavy load. Thus, shunt motor should never be started on load.

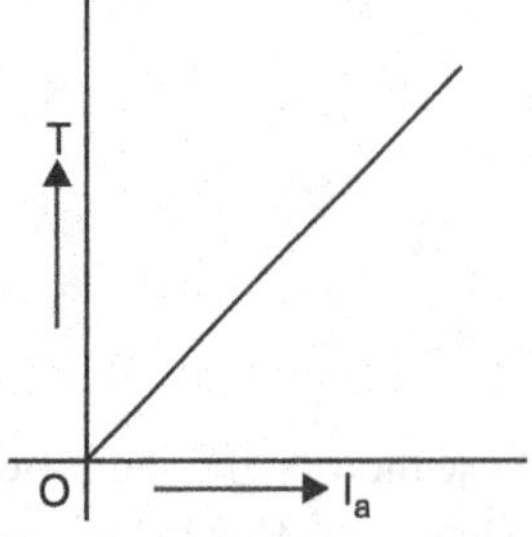

Fig. 5.15 T–I_a characteristics of shunt motor

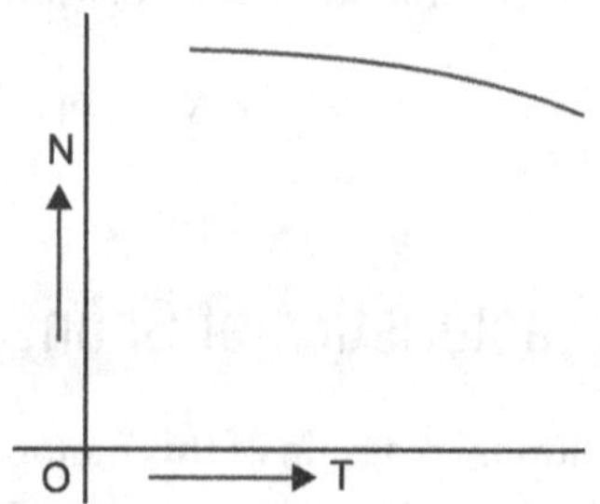

Fig. 5.16 N-T characteristics of shunt motor

3. N – T Characteristics

The $N - T$ characteristic is derived from the first two characteristics. When load torque increases, armature current I_a increases but speed decreases slightly. Thus with the increase in load or torque, the speed decreases slightly as shown in Fig. 5.16.

5.10 Characteristics of Series Motors

The conventional diagram a series motor is shown in Fig. 5.17. In these motors, the series field winding carries the armature current. Therefore, the flux produced by the series field winding is proportional to the armature current before magnetic saturation, but after magnetic saturation flux becomes constant.

1. N – I_a Characteristics

We know that
$$N \propto \frac{E_b}{\phi}$$

where, $E_b = V - I_a (R_a + R_{se})$

When armature current increases, the induced emf (back emf) E_b decreases, due to $I_a (R_a + R_{es})$ drop whereas flux ϕ increases as $\phi \propto I_a$ before magnetic saturation. However, under normal conditions $I_a (R_a + R_{se})$ drop is quite small and may be neglected.
Considering E_b to be constant,
$$N \propto \frac{1}{\phi} \propto \frac{1}{I_a}$$

Thus, before magnetic saturation, the $N - I_a$ curve follows the hyperbolic path as shown in Fig. 5.18.

In this region, the speed decreases abruptly with the increase in load or armature current.
After magnetic saturation, flux becomes constant, then
$$N \propto E_b \propto V - I_a (R_a + R_{se})$$

Thus, after magnetic saturation, the $N - I_a$ curve follows a straight line path and speed decreases slightly as shown in Fig. 5.18.

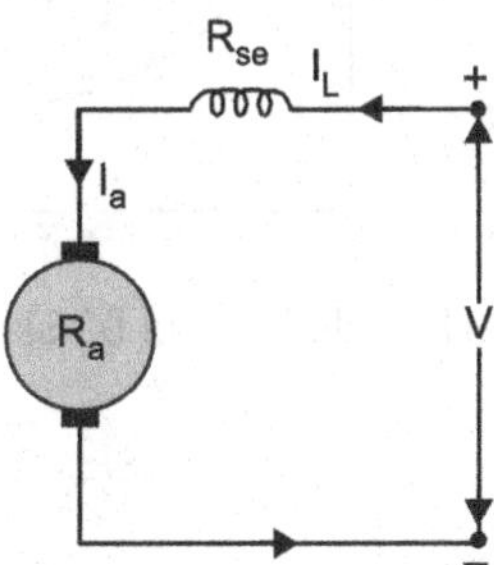

Fig. 5.17 Circuit diagram of series motor

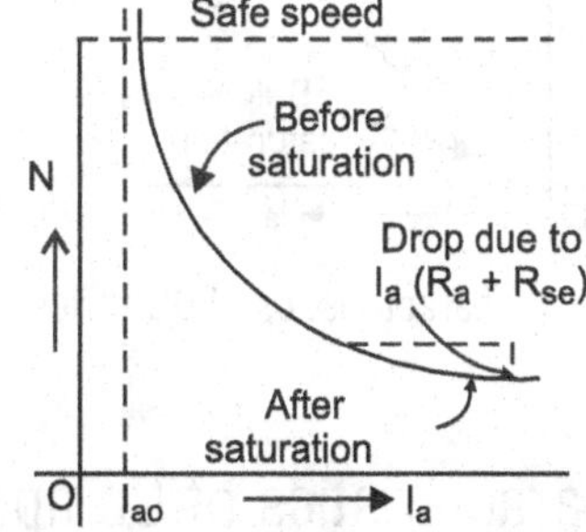

Fig. 5.18 N-Ia characteristics of DC series motor

From this characteristic, it is concluded that the series motor is a variable speed motor, i.e., its speed changes when the armature current (or load) changes. As the load on this motor decreases, speed increases. If this motor is connected to the supply without load, armature current will be very small and hence speed will be dangerously high which may damage the motor due to heavy centrifugal forces.

Therefore, a series motor is never started on no-load. However, to start a series motor, mechanical load (not belt driven load because belt slips over the pulley) is put on it first then started.

2. $T - I_a$ Characteristics

We know that, $T \propto \phi I_a$

In series motors, before magnetic saturation $\phi \propto I_a$

Hence, before magnetic saturation the electromagnetic torque produced in the armature is proportional to the square of the armature current. Therefore, this portion of the curve (*OA*) is a parabola passing through the origin as shown in Fig. 5.19.

However, after magnetic saturation, the flux ϕ becomes constant.

$$\therefore \quad T \propto I_a$$

Hence, after magnetic saturation, the curve (*AB*) becomes a straight line.

It is seen that before magnetic saturation $T \propto I_a^2$. When load is applied to this motor at start, it takes large current and heavy torque is produced which is proportional to square of this current. Thus, this motor is capable to pick up heavy loads at the start and best suited for electric traction.

3. N – T Characteristics

This characteristic is derived from the first two characteristics. At low value of load, I_a is small, torque is small but the speed is very high. As load increases, I_a increases, torque increases but the speed decreases rapidly. Thus for increasing torque, speed decreases rapidly as shown in Fig. 5.20.

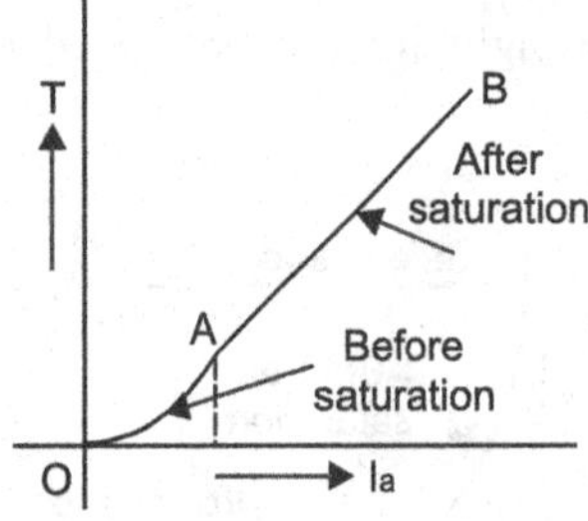

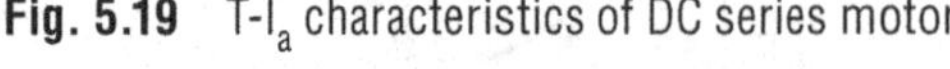
Fig. 5.19 T-I_a characteristics of DC series motor

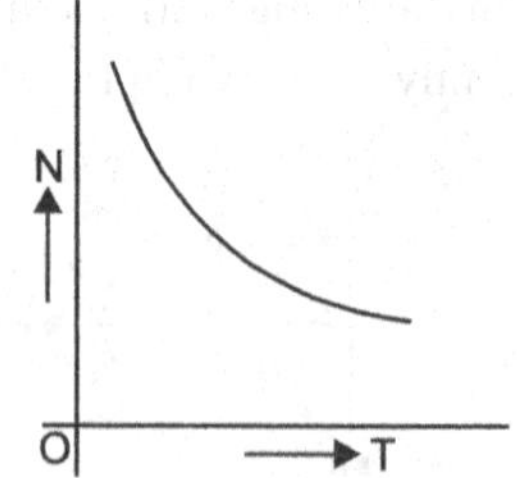

Fig. 5.20 N-T characteristics of DC series motor

5.11 Characteristics of Compound Motors

There are two types of compound wound DC motors namely; cumulative compound motors and differential compound motors. Cumulative compound motors are most common. The characteristics of these motors lies between the shunt and series motors. The $N-I_a$ characteristics, $T-I_a$ characteristics and $N-T$ characteristics are shown in Figs. 5.21(*a*), (*b*) and (*c*), respectively.

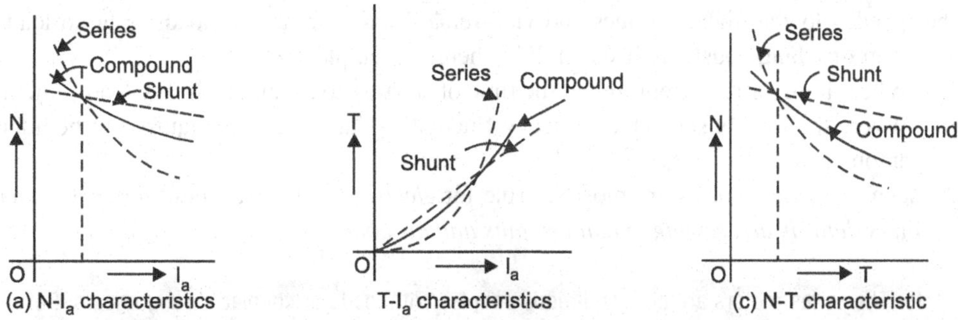

Fig. 5.21 Characteristics of cumulatively compound DC motor

However the $N-I_a$, $T-I_a$ and $N-T$ characteristics of a differentially compound motor are shown in Figs. 5.22(*a*), (*b*) and (*c*), respectively.

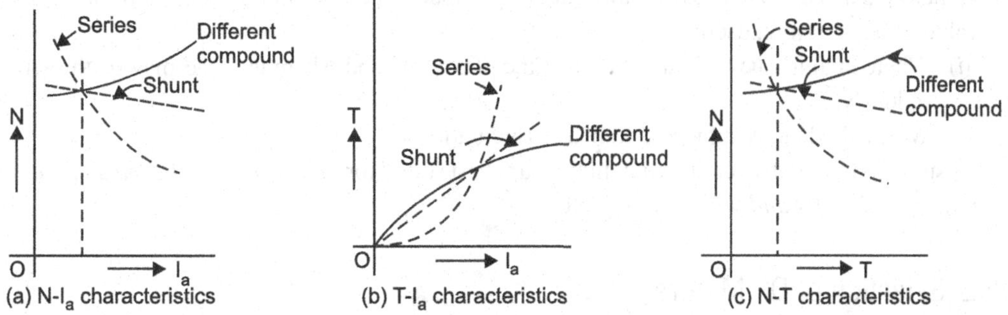

Fig. 5.22 Characteristics of differentially compound DC motor

5.12 Applications and Selection of DC Motors

5.12.1 Applications of DC Motors

As per the characteristics of DC motors, different types of DC motors are applied for different jobs as explained below:

1. **Separately excited motors:** Very accurate speeds can be obtained by these motors. Moreover, these motors are best suited where speed variation is required from very low value to high value. These motors are used in steel rolling mills, paper mills, diesel – electric propulsion of ships, etc.
2. **Shunt motors:** From the characteristics of a shunt motor we have seen that it is almost constant speed motor. It is, therefore, used;
   (*i*) Where the speed between no-load to full load has to be maintained almost constant.
   (*ii*) Where it is required to drive the load at various speeds (various speeds are obtained by speed control methods) and any one of the speed is required to be maintained almost constant for a relatively long period.

As such the shunt motors are most suitable for industrial drives such as *lathes, drills, grinders, shapers, spinning and weaving machines, line shafts in the group drive,* etc.

3. **Series motors:** The characteristics of a series motor reveal that it is variable speed motor i.e., the speed is low at higher torques and vice-versa. Moreover, at light loads or at no-load, the motor attains dangerously high speed. It is, therefore, employed:
   (*i*) Where high torque is required at the time of starting to accelerate heavy loads quickly.
   (*ii*) Where the load is subjected to heavy fluctuations and speed is required to be adjusted automatically.

   As such the series motors are most suitable for *electric traction, cranes, elevators, vacuum cleaners, hair driers, sewing machines, fans and air compressors,* etc.

> **Note:** The series motors are always directly coupled with loads or coupled through gears. Belt loads are never applied to series motor, because the belt may slip over the pulley or it may break. Then the motor will operate at light loads or at no-load and will attain dangerously high speed which may damage the motor.

4. **Compound motors:** The important characteristic of this motor is that the speed falls appreciably on heavy loads as in a series motor, but at light loads, the maximum speed is limited to safe value. It is, therefore, used;
   (*i*) Where high torque is required at the time of starting and where the load may be thrown off suddenly.
   (*ii*) Where the load is subjected to heavy fluctuations.

   As such the cumulative compound, motors are best suited for *punching and shearing machines, rolling mills, lifts and mine - hoists,* etc.

5.12.2 Selection of DC Motors

While selecting a DC motor for a particular work, one is to consider the following points:

1. *Selection of power rating*: If the size (*HP* or *kW* capacity) of a motor is more than the load to be picked-up by it then it will operate at lighter load than its rating. Hence, there will be more losses (i.e., motor will operate at poor efficiency) and unnecessarily there will be wastage of power. Thus extra expenditure will incur during operation. At the same time its initial cost will be more.

   On the other hand, if the size of the motor is less than the load to be picked-up by it, then it will be over loaded and it will be heated-up beyond its limits. Hence the motor may be damaged if operated continuously. Otherwise, also the protective devices will disconnect the motor under over-load conditions and this affect the production.

   Thus, a motor should be chosen of a size for its maximum power utilisation. During operation, it should be heated-up within the permissible temperature limits and it should never be over-heated. But at the same time it should be capable to take-up the over-loads temporarily.

2. *Characteristics of the motor:* For satisfactory operation, selection of power rating of a motor is not only the criteria, rather one should know the characteristics of a motor (i.e., behaviour of a motor under different load conditions) so that a work can be done quickly and efficiently without any breakdown. Therefore, before selection of motor, one should know the following particulars of the work

(*i*) Torque requirement during starting and running at different loads.
(*ii*) Requirement of accelerating and braking torque
(*iii*) Frequency of switching
(*iv*) Temperature at the work place
(*v*) Environmental conditions, etc.

After knowing the above particulars at the work place, a suitable motor (with proper enclosures) to meet with the requirements is selected.

5.13 Starting of DC Motors

To start a DC motor, when it is switched–*ON* to the supply with full rated voltage, it draws heavy current during starting period (more than its rated value). This excessive current overheats the armature winding and may even damage the winding insulation. Therefore, during starting period a resistance called starter in connected in series with the armature circuit to limit the starting current.

5.14 Necessity of Starter for a DC Motor

Under normal operating conditions, the voltage equation for a motor is given as

$$E_b = V - I_a R_a \quad \text{or} \quad I_a R_a = V - E_b$$

The armature current is given by the relation;

$$I_a = \frac{V - E_b}{R_a}$$

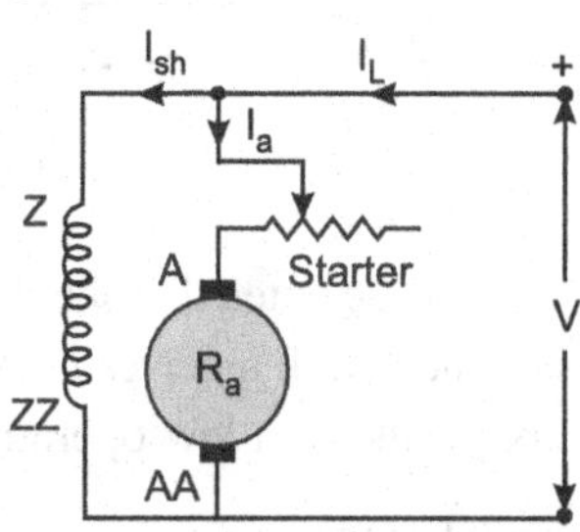

Fig. 5.23 DC shunt motor starter

When the motor is at rest, the induced emf E_b in the armature is zero ($E_b \propto N$). Consequently, if full voltage is applied across the motor terminals, the armature will draw heavy current ($I_a = V/R_a$) because armature resistance is relatively small. This *heavy starting current* has the following effects:

(*i*) It will blow out the fuses and prior to that it may damage the insulation of armature winding due to excessive heating effect if starting period is more.
(*ii*) Excessive voltage drop will occur in the lines to which the motor is connected. Thus, the operation of the appliances connected to the same line may be impaired and in some cases they may refuse to work.

To avoid this heavy current at start, a variable resistance is connected in series with the armature, as shown in Fig. 5.23, called a **starting resistance or starter,** and thus the armature current is limited to safe value $\left(I_a = \frac{V}{R_a + R}\right)$. Once the motor picks up speed, emf is built up and current is reduced $\left(I_a = \frac{V - E_{b1}}{R_a + R}\right)$. After that the starting resistance is gradually reduced. Ultimately, whole of the resistance is taken out of circuit when the motor attains normal speed.

Another important feature of a starter is that it contains protective devices such as overload protection coil (or relay) which provides necessary protection to the motor against over loading and no-volt release coil.

5.15 Starters for DC Shunt and Compound Wound Motors

The basic function of a starter is to limit the current in the armature circuit during starting or accelerating period. Starters are always rated on the basis of output power and voltage of the motor with which they are to be employed, e.g., 10 HP, 250 V shunt motor starter). A simplest type of starter is just a variable resistance (a rheostat) connected in series with the armature alone (not with the motor as a whole) as shown in Figs. 5.24(*a*) and (*b*).

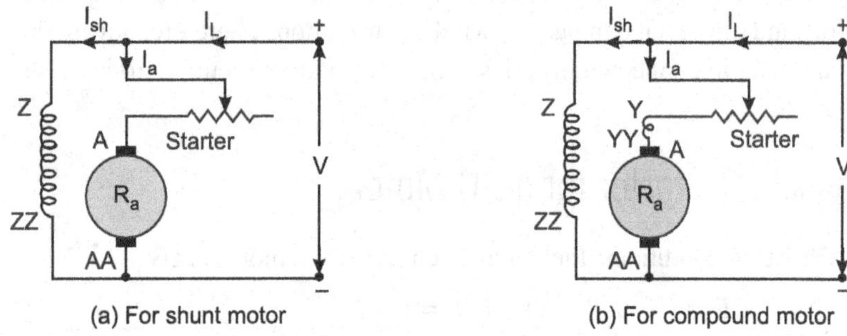

Fig. 5.24 DC Series and Compound motor starter

It may be noted that shunt field is kept independent of starting resistance. It is because when supply is connected, it receives normal rated voltage and sets-up maximum (rated) flux. A higher value of flux results in a low operating speed and a higher motor torque for a particular value of starting current since speed is inversely proportion to flux per pole $\left(N \propto \frac{1}{\phi}\right)$ whereas motor torque is proportional to product of flux per pole and armature current ($T \propto \phi I_a$). Hence, for a given load torque, the motor will accelerate quickly and reduces the starting period. Thus, the heating effect to armature winding is reduced.

For all practical application, this starter is further modified which includes protective devices such as over-load release and no-volt release. The over-load release protects the motor against over-loading i.e., when the motor is over-loaded (or short circuited) this relay brings the plunger to its *OFF* position. On the other hand, the no-volt release brings the plunger to its *OFF* position so that the motor may not start again without starter. For shunt and compound motors there are two standard types of starters named as (*a*) three point starter and (*b*) four point starter.

5.16 Three-point Shunt Motor Starter

The schematic connection diagram of a shunt motor starter is shown in Fig. 5.25. It consists of starting resistance *R* divided into several sections. The tapping points of starting resistance are connected to number of studs. The last stud of the starting resistance is connected to terminal *A* to which one terminal of the armature is connected. The + ve supply line is connected to the line terminal *L* through main switch. From line terminal, supply is connected to the starting lever *SL* through over load release coil *OLRC*. A spring *S* is placed over the lever to bring it to the off position, when supply goes off. A soft iron piece *SI* is attached with the starting lever which is pulled by the no volt release coil under normal running condition. The far end of the brass strip *BS* is connected to the terminal *Z* through a no volt release coil *NVRC*. One end of the shunt field winding in connected to *Z* terminal of the starter. An iron piece is lifted by *OLRC* under abnormal condition to short circuit the no-volt release coil. The negative supply line is connected directly to the other ends of shunt field winding and armature of the DC shunt motor.

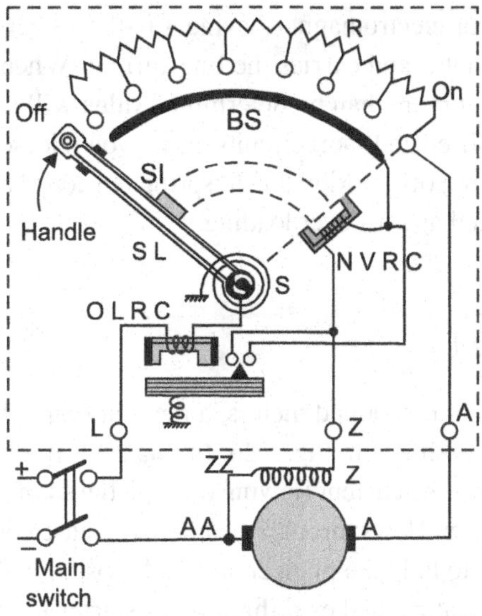

Fig. 5.25 Three-point shunt motor starter

Operation

First of all the main switch is closed with starting lever resting in off position. The handle is then turned clockwise to the first stud and brass strip. As soon as it comes in contact with first stud, whole of the starting resistance *R* is inserted in series with the armature and the field winding is directly connected across the supply through brass strip. As the handle is turned further the starting resistance is cut out of the armature circuit in steps and finally entire starting resistance is cut out of armature circuit.

No-volt Release Coil and its Function

A no-volt release coil is a small electromagnet having many turns of fine wire. It is connected in series with shunt field winding and therefore, carries a small field current. When the handle is turned to on position, the no-volt release coil is magnetised by the field current and holds the starting lever at on position. In case of failure or disconnection of the supply, this coil is demagnetised and the lever comes to the off position due to spring tension. Consequently. The motor is disconnected from the supply. If the spring with the no-volt release coil is not used the lever would remain in *ON*-position in case of supply failure. And again, when the supply comes, the motor would be connected directly to the lines without starter.

The other important advantage of connecting the no-volt release coil in series with the shunt field winding is that due to an accident if the circuit of field winding becomes open, the *NVRC* will be demagnetised and the starting lever is immediately pulled back to off position by the spring. Otherwise the motor would have attained dangerously high speed.

Over-load Release Coil and its Function

An over-load release coil is an electromagnet having small number of turns of thick wire. It is connected in series with the motor and carries the line current. When the motor is over loaded (or short circuited), a heavy current more than predetermined value will flow through it. Then, the iron piece (armature or plunger) is lifted and short circuits the no-volt release coil. Hence the starting lever is released and pulled back to the off position due to spring tension. Thus the motor is disconnected from the supply and is protected against over loading.

5.17 Four-point Starter

For speed control of DC shunt or compound motors, a rheostat (variable resistor R_h) is connected in series with the field winding, as shown in Fig. 5.26. In this case, if a three-point starter is used and the value of R_h is so adjusted that the current flowing through the shunt field winding is very small. It may be seen that the same current flows through the no-volt release coil, then the magnetic strength of the coil may be insufficient to hold the plunger at its *ON* position. This is an undesirable feature of a three-point starter. This feature makes a three-point starter unsuitable for such applications. Accordingly, a four-point starter is designed, as shown in Fig. 5.26, in which the current flowing through the no-volt release coil is made independent of the shunt field circuit. Figure 5.26(*a*) shows a four-point starter used with a shunt wound motor whereas Fig. 5.26(*b*) shows a starter used with a compound wound machine.

Operation

The working of a four-point starter is similar to a three-point starter with slight changes.

In this case, when the plunger touches the first stud, the line current is divided into the following three parts:

(*i*) First part passes through starting resistance and armature (as well as in series field for compound motors).

(*ii*) Second part passes through the field winding (and speed control resistance if applied) and

(*iii*) The third part passes through no-volt release coil and protective resistance connected in series with the coil.

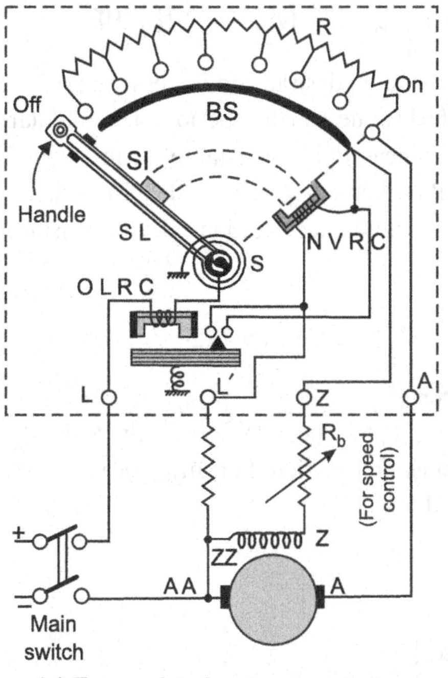

(a) Four-point shunt motor starter

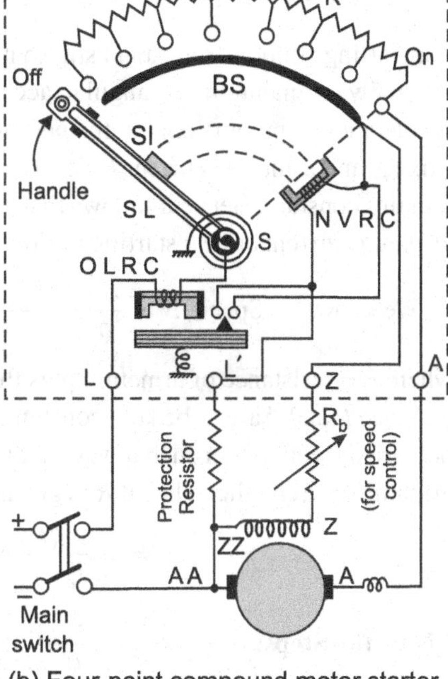

(b) Four-point compound motor starter

Fig. 5.26 Four-point DC motor starter

It is evident that is a four-point starter, the no-volt release coil circuit is independent of the field circuit. Therefore, the change of current in the field circuit do not affect the pull exerted by the holding coil which remains always sufficient to prevent the spiral spring from restoring the plunger to its *OFF* position.

While starting a motor with a four-point starter it is necessary to ensure that the field circuit is closed and the rheostat connected in series with the shunt field winding must be at zero resistance position. Moreover, whole of the starting resistance must come in series with the armature.

Whenever a shunt motor is required to be stopped, it must be stopped by opening the line switch. In fact this switch can be opened without any appreciable arc since the motor develops back emf nearly equal to applied voltage and the net voltage across the switch contacts is very small. The electromagnetic energy stored in the field does not appear at the switch but it gradually discharges through the armature. The motor should never be stopped by bringing the plunger (starting arm) back to the *OFF* position, because in such cases, when the field circuit breaks at the last stud placed near the *OFF* position, a heavy spark occurs owing to the inductive nature of the field. Usually, this sparking burns the contact.

Moreover, while stopping the motor, the value of resistance connected in the field circuit should always be reduced to zero so that speed of motor falls to its normal value. This ensures that, when the motor is started next time, it must start with a strong field and higher starting torque.

5.18 Calculation of Step Resistances Used in Shunt Motor Starter

For starting a motor from stand still to its rated speed, it is normally desirable to increase the speed gradually to maintain the angular acceleration constant during the starting period. The angular acceleration is proportional to the net torque, which is in turn nearly proportional to the product of flux (ϕ) and armature current I_a i.e., $T \propto \phi I_a$. The flux ϕ will remain constant provided line voltage remains constant. Hence it follows that substantially constant angular acceleration calls for constant armature current during starting period.

We know, Speed, $N \propto \dfrac{E_b}{\phi} \propto \dfrac{V - I_a R}{\phi}$

where R = resistance of armature plus the resistance of starter.

Since I_a and ϕ are to be kept constant, therefore, for increasing the speed (N) gradually, R should be varied (reduced) in such a way that the above relation must be satisfied. For different values of armature current, the value of R is given by the expression, d

$$R = \frac{V - KN}{I_a} \quad \text{where } K \text{ is a constant}$$

Note the Steps: $N \propto \dfrac{V - I_a R}{\phi}$ or $\phi N \propto V - I_a R$

or $KN = V - I_a R$ or $R = \dfrac{V - KN}{I_a}$

In starters, usually the value of R is changed in steps and, therefore, armature current will change in two extreme values. Accordingly, the steps of the starter are designed in such a way that armature current varies in between these limits so that torque may not change to a greater degree.

The fuse or *MCB* placed in the motor circuit is usually not larger than 150% of the motor full-load current.

Let I_1 and I_2 be the maximum and minimum value of the current drawn by the motor during starting. Let the starter has n-sections each having a resistance as $r_1, r_2, r_3 \ldots r_n$ as shown in Fig. 5.27. Let the total resistance of armature circuit when the starting arm is kept at stud No. 1, 2, 3, … n and $n + 1$ be $R_1, R_2, R_3, \ldots R_n$ and R_{n+1} (where $R_{n+1} = R_a$), respectively.

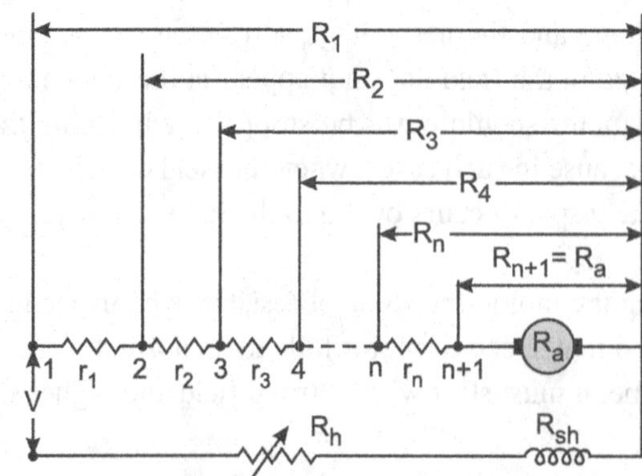

Fig. 5.27 Design of shunt motor starter

Just at the time of starting when the starting arm is brought in contact with stud-1, the motor is stationary and there is no back emf, the maximum value of current drawn by the motor,

$$I_1 = \frac{V}{R_1} \qquad ...(5.11)$$

The motor starts rotating and picks up speed N_1 which develops a back emf E_{b1} (where $E_{b1} \propto N_1$) and the current drops to I_2 (the starting arm is still at stud-1), therefore,

$$I_2 = \frac{V - E_{b1}}{R_1} \qquad ...(5.12)$$

Now the starting arm is shifted to stud-2, the speed of the motor is still N_1 and the back emf is also E_{b1} but the circuit resistance is reduced to R_2 due to this the current again attains its maximum value I_1, i.e.,

$$I_1 = \frac{V - E_{b1}}{R_2} \qquad ...(5.13)$$

The motor picks-up speed further and its value reaches to N_2 which develops a back emf E_{b2} ($E_{b2} \propto N_2$) and the current drops to I_2 (the starting arm is still at stud-2), therefore,

$$I_2 = \frac{V - E_{b2}}{R_2} \qquad ...(5.14)$$

Further, starting arm is shifted to stud-3, the speed is still N_2 and back emf is E_{b2} but the circuit resistance is reduced to R_3, thus the current again attains its maximum value I_1, i.e.,

$$I_1 = \frac{V - E_{b2}}{R_3} \qquad ...(5.15)$$

Dividing eqn. (5.13) by (5.12) and eqn. (5.15) by (5.14), we get,

$$\frac{I_1}{I_2} = \frac{V - E_{b1}}{R_2} \times \frac{R_1}{V - E_{b1}} = \frac{V - E_{b2}}{R_3} \times \frac{R_2}{V - E_{b2}}$$

or

$$\frac{I_1}{I_2} = \frac{R_1}{R_2} = \frac{R_2}{R_3} = K \text{ (say)} \qquad ...(5.16)$$

and

$$\frac{R_1}{R_2} \times \frac{R_2}{R_3} = K^2 \quad \text{or} \quad \frac{R_1}{R_3} = K^2$$

Similarly, the starting arm is shifted to next studs in sequence and ultimately reaches to stud number (n + 1), then we get,

$$\frac{R_1}{R_2} = \frac{R_2}{R_3} = \frac{R_3}{R_4} = \ldots = \frac{R_{n-1}}{R_n} = \frac{R_n}{R_a} = K$$

and

$$\frac{R_1}{R_2} \times \frac{R_2}{R_3} \times \frac{R_3}{R_4} \times \ldots \times \frac{R_{n-1}}{R_n} \times \frac{R_n}{R_a} = K^n$$

or

$$\frac{R_1}{R_a} = K^n \quad \text{or} \quad K = \left(\frac{R_1}{R_a}\right)^{\frac{1}{n}} \qquad ...(5.17)$$

If the values of armature resistance R_a, limiting currents I_1 (max. value) and I_2 (min. value) are known, the value of R_1 and K can be determined from eqn. (5.11) and (5.16) respectively. At the same time, the number of section (n) can be determined by using eqn. (5.17).

By knowing the value of R_1 and K, one can determine the values of $R_2, R_3, R_4, \ldots R_n$ by using the eqn. (*vi*) and then the value of resistance of each step can be determined. i.e., $r_1 = R_1 - R_2$; $r_2 = R_2 - R_3$; $r_3 = R_3 - R_2$; ...; $r_n = R_n - R_a$

Example 5.11

A 152 V DC shunt motor has an armature resistance of 0.3 Ω, a brush voltage drop of 2 V, the rated full load current is 70 A. Calculate (i) the current at the instant of starting as a percentage of full load current (ii) the value of starting resistance to limit the motor current at the instant of starting to 150 percent of the rated load current.

Solution:

Here, $V = 152$ V; $R_a = 0.3\ \Omega$; $I_{fl} = 70$ A

(*i*) Starting current with no-additional resistance in the armature circuit,

$$I_S = \frac{V - \text{brush contact drop}}{R_a} = \frac{152 - 2}{0.3} = 500\text{ A}$$

$$I_S \text{ as a \%age of full-load} = \frac{I_S}{I_{fl}} \times 100 = \frac{500}{70} \times 100 = 714.3$$

Let the total resistance required in the armature circuit to limit the starting current to 150 per cent of the rated load current be R_1, then

$$R_1 = \frac{V - \text{brush contact drop}}{1.5 \times I_{fl}} = \frac{152 - 2}{1.5 \times 70} = 1.43\text{ ohm}$$

(*ii*) Additional resistance required for starting

$$R = R_1 - R_a = 1.43 - 0.3 = \mathbf{1.31}\ \Omega\ (Ans.)$$

Example 5.12

A 10 b.h.p. (7.46 kW), 200-V shunt motor has full-load efficiency of 85%. The armature has a resistance of 0.238 Ω. Calculate the value of the starting resistance necessary to limit the starting current to 1.5 times the full-load current at the moment of first switching on. The shunt current may be neglected. Find also the back emf of the motor, when the current has fallen to its full-load value, assuming that the whole of the starting resistance is still in circuit.

Solution:

$$\text{Full load motor current, } I_{fl} = \frac{7.46 \times 1000}{200 \times 0.85} = 43.88\text{ A}$$

$$\text{Starting current, } I_1 = 1.5 \times 43.88 = 65.83\text{ A}$$

$$R_1 = \frac{V}{I_1} = \frac{200}{65.83} = 3.038\ \Omega;\ R_a = 0.238\ \Omega$$

$$\therefore \quad \text{Starting resistance} = R_1 - R_a = 3.038 - 0.238 = \mathbf{2.8}\ \Omega\ (Ans.)$$

Now, full-load current, $I_{fl} = 43.88 \text{ A} = I_a$

Now, $$I_a = \frac{V - E_b}{R_1}$$

$\therefore$ $$E_b = V - I_a R_1 = 200 - (43.88 \times 3.038) = \mathbf{67\ V}\ (Ans.)$$

Example 5.13

A 230 V shunt motor has an armature resistance of 0.4 Ω. The starting armature current must not exceed 45 A. If the number of sections are 5, calculate the values of resistance steps to be used in the starter.

Solution:

Number of sections of the starter, $n = 5$

Resistance of armature circuit at the starting instant

$$R_1 = \frac{V}{I_{\max}} = \frac{230}{45} = 5.11\ \Omega$$

Ratio of maximum to minimum current during starting,

$$\frac{I_{\max}}{I_{\min}} = K = \left(\frac{R_1}{R_a}\right)^{1/n} = \left(\frac{5.11}{0.4}\right)^{1/5} = 1.6645$$

Resistance, $R_2 = \frac{R_1}{K} = \frac{5.11}{1.6645} = 3.07\ \Omega$; Resistance, $R_3 = \frac{R_2}{K} = \frac{3.07}{1.6645} = 1.844\ \Omega$

Resistance, $R_4 = \frac{R_3}{4} = \frac{1.84}{1.6645} = 1.108\ \Omega$; Resistance, $R_5 = \frac{R_4}{K} = \frac{1.108}{1.6645} = 0.666\ \Omega$

Resistance of 1st to 5th section $r_1 = R_1 - R_2 = 5.11 - 3.07 = \mathbf{2.04}\ \Omega\ (Ans.)$

$r_2 = R_2 - R_3 = 3.07 - 1.844 = \mathbf{1.226}\ \Omega\ (Ans.)$

$r_3 = R_3 - R_4 = 1.844 - 1.108 = \mathbf{0.736}\ \Omega\ (Ans.)$

$r_4 = R_4 - R_5 = 1.108 - 0.666 = \mathbf{0.442}\ \Omega\ (Ans.)$

$r_5 = R_5 - R_a = 0.666 - 0.4 = \mathbf{0.266}\ \Omega\ (Ans.)$

Example 5.14

A 220-V shunt motor has an armature resistance of 0.5 Ω. The armature current of starting must not exceed 40 A. If the number of sections is 6, calculate the values of the resistor steps to be used in this starter.

Solution:

Here, $V = 220$ V; $R_a = 0.5\ \Omega$; $I_{max} = 40$ A; $n = 6$

Now, $$R_1 = \frac{220}{40} = 5.5\ \Omega,$$

$$\frac{R_1}{R_a} = K^n \text{ or } R_1 = R_a K^n$$

$\therefore \quad 5.5 = 0.4\,K^6 \text{ or } K^6 = 5.5/0.4 = 13.75$

$$6 \log_{10} K = \log_{10} 13.75 = 1.1383;\ K = 1.548$$

Now, $\quad R_2 = \dfrac{R_1}{K} = \dfrac{5.5}{1.548} = 3.553\ \Omega$

$$R_3 = \frac{R_2}{K} = \frac{3.553}{1.548} = 2.295\ \Omega$$

$$R_4 = \frac{2.295}{1.548} = 1.482\ \Omega$$

$$R_5 = \frac{1.482}{1.548} = 0.958\ \Omega$$

$$R_6 = \frac{0.958}{1.548} = 0.619\ \Omega$$

Resistance of various (1st to 6th) sections;

$r_1 = R_1 - R_2 = 5.5 - 3.553 = \mathbf{1.947}$ W *(Ans.)*

$r_2 = R_2 - R_3 = 3.553 - 2.295 = \mathbf{1.258}$ W *(Ans.)*

$r_3 = R_3 - R_4 = 2.295 - 1.482 = \mathbf{0.813}\ \Omega$ *(Ans.)*

$r_4 = R_4 - R_5 = 1.482 - 0.958 = \mathbf{0.524}\ \Omega$ *(Ans.)*

$r_5 = R_5 - R_6 = 0.958 - 0.619 = \mathbf{0.339}\ \Omega$ *(Ans.)*

$r_6 = R_6 - R_a = 0.619 - 0.4 = \mathbf{0.219}\ \Omega$ *(Ans.)*

Example 5.15

A 250 V DC shunt motor has an armature resistance of 0.125 ohm and a full load current of 150 ampere. Calculate the number and values of resistance elements of a starter to start the motor on full load, the maximum current not to exceed 200 A.

Solution:

As the motor is to start against its full load, hence the minimum current is its full load current i.e., 150 A.

$$I_{max} = 200 \text{ A};\ I_{min} = 150 \text{ A}$$

$\therefore \quad K = \dfrac{I_{max.}}{I_{min.}} = \dfrac{200}{150} = \dfrac{4}{3}$

Now $\quad R_1 = \dfrac{V}{I_{max.}} = \dfrac{250}{200} = 1.25 \text{ ohm and } R_a = 0.125 \text{ ohm}$ (given)

Now $\quad \left(\dfrac{I_{max}}{I_{min}}\right)^n = \dfrac{R_1}{R_a} \quad \text{or} \quad \left(\dfrac{4}{3}\right)^n = \dfrac{1.25}{0.125} = 10$

Taking logs on both sides, we have

$$n.\log\frac{4}{3} = \log 10 \text{ or } n \times 0.1248 = 1 \text{ or } n = \mathbf{8}\ (Ans.)$$

$R_2 = \frac{R_1}{K} = 1.25 \times \frac{3}{4} = 0.938\ \Omega$; $R_3 = \frac{R_2}{K} = 0.938 \times \frac{3}{4} = 0.703\ \Omega$

$R_4 = \frac{R_2}{K} = 0.703 \times \frac{3}{4} = 0.527\ \Omega$; $R_5 = \frac{R_4}{K} = 0.527 \times \frac{3}{4} = 0.395\ \Omega$

$R_6 = \frac{R_5}{K} = 0.395 \times \frac{3}{4} = 0.296\ \Omega$; $R_7 = \frac{R_6}{K} = 0.222 \times \frac{3}{4} = 0.222\ \Omega$

$R_8 = \frac{R_7}{K} = 0.222 \times \frac{3}{4} = 0.167\ \Omega$; $R_9 = \frac{R_8}{K} = 0.167 \times \frac{3}{4} = 0.125\ \Omega = R_a$

Resistance of various steps in the order of 1st to 8th,

$r_1 = R_1 - R_2 = 1.25 - 0.931 = \mathbf{0.312\ \Omega}$; $r_2 = R_2 - R_3 = 0.938 - 0.703 = \mathbf{0.235\ \Omega}$

$r_3 = R_3 - R_4 = 0.703 - 0.527 = \mathbf{0.176\ \Omega}$; $r_4 = R_4 - R_5 = 0.527 - 0.395 = \mathbf{0.132\ \Omega}$

$r_5 = R_5 - R_6 = 0.395 - 0.296 = \mathbf{0.099\ \Omega}$; $r_6 = R_6 - R_7 = 0.296 - 0.222 = \mathbf{0.074\ \Omega}$

$r_7 = R_7 - R_8 = 0.222 - 0.167 = \mathbf{0.055\ \Omega}$; $r_8 = R_8 - R_9 = 0.167 - 0.125 = \mathbf{9.042\ \Omega}$

Example 5.16

Find the value of the step resistance in a 6-stud starter for a 5 h.p. (3.73 kW), 200-V shunt motor. The maximum current in the line is limited to twice the full-load value. The total Cu loss is 50% of the total loss. The normal field current is 0.6 A and the full-load efficiency is found to be 88%.

Solution:

$$\text{Output} = 3730\ \text{W}$$

$$\text{Input} = \frac{\text{Output}}{\eta} = \frac{3730}{88} \times 100 = 4238\ \text{W}$$

$$\text{Total loss} = 4238 - 3730 = 508\ \text{W}$$

$$\text{Armature Cu loss} = \frac{508}{2} = 254\ \text{W}$$

$$\text{Input current} = \frac{\text{Input}}{V} = \frac{4238}{200} = 21.19\ \text{A}$$

$$\text{Armature current, } I_a = I_L - I_{sh} = 21.19 - 0.6 = 20.59\ \text{A}$$

Now $(20.59)^2 R_a = 254$

$\therefore$ $R_a = \frac{254}{(20.59)^2} = 0.5989\ \Omega$

Permissible input current = 21.19 × 2 = 42.38 A

Permissible armature current = 42.38 – 0.6 = 41.78 A

$\therefore$ $R_1 = \frac{200}{41.78} = 4.787\ \Omega;\ n = 5;\ \frac{R_1}{R_a} = K^5$

$\therefore$ $4.787 = K^5 \times 0.5989$

$\therefore \qquad K^5 = 4.787/0.5989 = 7.993$

or $\qquad 5 \log K = \log 7.993 = 0.9027$

$\therefore \qquad \log K = 0.1805; K = 1.516$

Now $\quad R_2 = \dfrac{R_1}{K} = \dfrac{4.789}{1.516} = 3.159\ \Omega$; $\qquad R_3 = \dfrac{R_2}{K} = \dfrac{3.159}{1.516} = 2.084\ \Omega$

$R_4 = \dfrac{R_3}{K} = \dfrac{2.084}{1.516} = 1.376\ \Omega$; $\qquad R_5 = \dfrac{R_4}{K} = \dfrac{1.376}{1.516} = 0.908\ \Omega$

Resistance in 1st step – $R_1 - R_2 = 4.787 - 3.159 =$ **1.628 Ω.** (*Ans.*)

Resistance in 2nd step – $R_2 - R_3 = 3.159 - 2.084 =$ **0.075 Ω.** (*Ans.*)

Resistance in 3rd step – $R_3 - R_4 = 2.084 - 1.376 =$ **0.708 Ω.** (*Ans.*)

Resistance in 4th step – $R_4 - R_5 = 1.376 - 0.908 =$ **0.468 Ω.** (*Ans.*)

Resistance in 5th step – $R_5 - R_a = 0.908 - 0.5989 =$ **0.309 Ω.** (*Ans.*)

5.19 Series Motor Starter

A series motor starter is also called a two-point starter. Its internal and external connections are shown in Fig. 5.28.

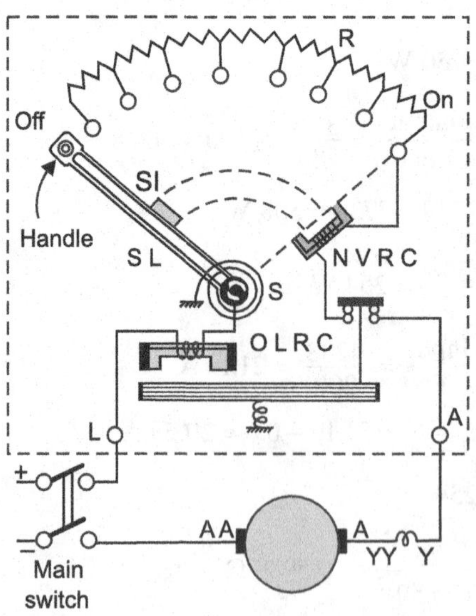

Fig. 5.28 Two-point series motor starter

Here, for starting the motor, the control arm or plunger is moved in clockwise direction from its *OFF* position to *ON* position against the spring tension. In the beginning, all the sections of the starting resistance are connected in series with the armature to limit the current to predetermined value. As the starting arm moves, various steps of the starting resistance are cut out of circuit and ultimately whole of the starting resistance is cut-out and the control arm is held in the *ON* position by an electromagnet. The hold-on (no-volt release) coil is connected in series with the armature circuit. If the motor loses its load, current drawn by the motor decreases which decreases the strength of the hold-on coil. Immediately the arm goes back to the *OFF* position due to strong spring tension; thus preventing the motor from over-speed. The control arm or plunger also goes back to the *OFF* position when supply goes *OFF* or when the supply voltage decreases appreciably. *L* and *Y* are the two points of the starter through which supply line terminal and the motor (series field) terminal is connected.

For stopping the motor, the line switch should always be opened rather than bringing the control arm back to its *OFF* position. If it is done a heavy sparking occurs at the last stud placed near the *OFF* position. This sparking occurs due to dissipation of energy stored in the magnetic field of series field winding.

Example 5.17

A 240 V series motor takes 40 A when giving its rated output at 1500 rpm. Its armature and series field resistance is 0.18 Ω and 0.12 Ω respectively. Find the external resistance which must be added to obtain rated torque (i) at starting and (ii) at 1000 rpm.

Solution:

Since torque remains the same in both the cases, the current drawn by the motor remains constant at 40 A, because $T \propto \Phi I$.

We know, current drawn by the motor, $I = \dfrac{V - E_b}{R_a + R_{se} + R}$

where R is external resistance connected in series with motor at start

$$\therefore \quad 40 = \frac{240 - 0}{0.18 + 0.12 + R} \qquad \text{(since at start } E_b = 0\text{)}$$

or
$$R = \frac{240}{40} - 0.3 = \mathbf{5.7\ \Omega}\ (Ans.)$$

$$\text{Back emf, } E_{b_1} = V - I(R_a + R_{se}) = 240 - 40 \times (0.18 + 0.12) = 228 \text{ V}$$

$$\text{Back emf, } E_{b_2} = E_{b_1} \times \frac{N_2}{N_1} = 228 \times \frac{1000}{1500} = 152 \text{ V}$$

$$\because \quad I = \frac{V - E_b}{R_a + R_{se} + R},\ 40 = \frac{240 - 152}{0.18 + 0.12 + R}$$

or
$$R = \frac{88}{40} - 0.3 = \mathbf{1.9\ \Omega}\ (Ans.)$$

Section Practice Problems

Numerical Problems

1. A DC series motor draws 50 A at 230 V. Resistance of armature and series field winding is 0.2 Ω and 0.1 Ω respectively. Calculate (*i*) brush voltage (*ii*) back emf, (*iii*) power wasted in armature, and mechanical power developed. (**Ans.** *225V, 215 V, 500 W, 10.75 kW*)

2. The field resistance and armature resistance of a 120 V DC dhunt motor is 60 ohm and 0.2 ohm respectively. The motor draws 60 A current at full-load, 1800 rpm. If the brush contact drop is 3V, find the speed of the motor at half load. (**Ans.** *1902*)

3. A 250 V DC shunt motor, on no-load draws a current of 5A and runs at 1000 rpm. The armature and field resistance of the motor is 0.025 ohm and 250 ohm respectively. If the motor draws 41 A at full-load what will be the speed of the motor. Assume that the field is wakened by 3% due to armature reaction. (**Ans.** *993.67*)

4. A DC series motor draws 80 A at 230 V and rotates at 1000 rpm. The armature and series field resistance is 0.11 ohm and 0.14 ohm respectively. Calculate the speed of the motor when it draws 20 A from the mains assuming that the field is reduced to 0.4 times the previous one. (**Ans.** *2678 rpm*)

5. A 220 V DC series motor draws full-load line current of 38 A at the rated speed of 500 rpm. The series field and armature resistance of the motor is 0.2 ohm and 0.4 ohm respectively. Considering brush drop as 3.0 V irrespective of the load and neglecting armature reaction, find:
   (*i*) The speed of the motor when the load current drops to 19 A
   (*ii*) The speed on removal of load when the motor takes only 1A from supply.
   (*iii*) The internal horse power developed in each of the above cases.
   (**Ans.** *1058 rpm; 21172 rpm; 10, 5.31, 0.294*)

6. A belt driven DC shunt generator runs at 1200 rpm delivering 10 kW at 220 V brushes. The belt breaks, following which the machine operates as a motor drawing 2 kW power. What will be its speed as a motor. The armature and field resistance are 0.25 Ω and 55 Ω respectively. Ignore armature reaction and assume the contact drop at each brush to be 1V. (**Ans.** *1100*)

7. The field and armature resistance of a DC shunt machine is 60 ohm and 0.015 ohm respectively. When working as a generator it delivers 60 kW at 240 V and 720 rpm. What will be its speed when the same machine works as a motor drawing 60 kW at 240 V assuming that contact voltage drop per brush is 1 V. (**Ans.** *686.3 rpm*)

8. A 12 hp, 230 V shunt motor has an armature circuit resistance of 0.5 Ω and a field resistance of 115 Ω. At no-load and rated voltage the speed is 1200 rpm and the armature current is 2A. If the load is applied, the speed drops to 1100 rpm. Determine the armature current, the line current and torque. (**Ans.** *40.2 A; 42.2 A; 73.2 Nm*)

9. A 250 V, four-pole shunt motor has wave winding with 500 conductors. The armature circuit resistance is 0.25 Ω, field resistance is 100 Ω and the flux per pole is 0.02 Wb. Armature reaction is neglected. If the motor draws 14.5 A from the mains, then compute speed and the torque developed in the armature. (**Ans.** *741 rpm, 38.2 Nm*)

10. The field winding and armature resistance of a 230 V DC shunt motor is 230 ohm and 0.5 ohm respectively. It draws 3 A at no-load and runs at a speed of 1000 rpm. At full load and rated voltage, the current drawn is 23 A and the armature reaction causes a drop of 2% in flux. Determine (*i*) full-load speed (*ii*) full-load torque. (**Ans.** *976 rpm, 47.15 Nm*)

11. The field and armature resistance of a 200 V DC series motor is 0.2 ohm and 0.3 ohm respectively. It draws 20 A at full-load and runs at 700 rpm. Assuming the magnetic circuit unsaturated, what will be the speed if (*i*) the load torque is increased by 44% (*ii*) the motor current is 10A? (**Ans.** *577.2 rpm, 1437 rpm*)

12. A 120 V DC shunt motor has an armature and shunt field winding resistance of 0.2 Ω and 120 Ω respectively. The brush voltage drop is 2 V. The rated full-load armature current is 75 A. Calculate the current at the instant of starting and its value in terms of percentage of full-load armature current. (**Ans.** *591 A; 788%*)

13. A 440 V, 80 A, 1200 rpm DC shunt motor has an armature resistance of 0.55 Ω. Determine the resistance for each step of the starter for accomplishing starting in 4 steps. The maximum armature current should not exceed 150% of the full-load current. Neglect shunt field current being small.
(**Ans.** *1.3847 Ω, 0.8618 Ω, 0.5364 Ω and 0.3338 Ω*)

14. A 25 hp, 220 V DC shunt motor has an armature resistance of 0.1 Ω. At full-load, it draws an armature current of 95A and runs at a speed of 600 rpm. It is to be braked by plugging. Estimate the value of the resistance which should be placed in series with it to limit the current to 130 A. What should be the initial value of the electric braking torque and value when the speed has fallen to half its full load value?
(**Ans.** *3.211 Ω; 400.5 Nm; 302.57 Nm*)

Short Answer Type Questions

Q.1. How does a DC motor differ from a DC generator in construction?

Ans. Basically, there is no difference in the construction of a DC generator and motor, however the frame of the generator can, as a rule, be open but those of motors should be either partly or totally enclosed.

Q.2. Why the emf generated in the armature of a DC motor is called the back emf?

Ans. Because this induced emf in a DC motor opposes the applied voltage, therefore, it is called back or counter emf.

Q.3. What do you understand by self excitation mode of DC machine? Name two DC machines working in this mode.

Ans. The DC machine in which field winding(s) is/are excited by the current supplied by the machine itself is said to be in its self excited made. In such machines the field coils are inter-connected with the armature winding.

The self excited machines may be (*i*) DC shunt machine (*ii*) DC series machine

Q.4. What will be the effect on direction of rotation of a DC series motor if the supply terminals are reversed?

Ans. There will be no effect, motor will rotate in the same direction, as before the reversal of line terminals.

Q.5. How can the direction of rotation of a DC shunt motor be reversed?

Ans. Direction of rotation of a DC shunt motor can be reversed by reversing the current flow through either the armature winding or the field winding.

Q.6. What do you mean by speed regulation of a DC motor?

Ans. The speed regulation of a DC motor is defined as the change in speed when the load on the motor is reduced from rated load to zero load and is expressed in percentage of rated full-load speed

Q.7. In a DC motor, the brushes are given a backward shift why?

Ans. In a DC motor, the brushes are given backward shift to neutralise the effect of armature reaction and further to reduce sparking at the commutator.

Q.8. What should be the location of interpoles in a (*i*) motor (*ii*) generator.

Ans. In case of a motors, the interpoles having the same polarity as that of main poles must be placed behind them, whereas in case of a generators, the interpoles having same polarity as that of the main pole must be placed ahead of them in the direction of rotation.

Q.9. Why large variable speed DC motors are fitted with compensating windings?

Ans. To reduce sparking at brushes due to armature reaction.

Q.10. Why is the starting torque of a DC series motor more than that of a DC shunt motor of the same rating?

Ans. In case of a DC series motor flux ϕ is directly proportional to armature current I_a (before saturation) and, therefore, torque developed $T_e \propto I_a^2$ while in case of a DC shunt motor torque developed $T_e \propto I_a$ (flux being constant). Obviously on the basis of the same kW output and speed, a DC series motor develops the torque more than that of a shunt motor. Hence, series motors are capable to pick heavy loads at start.

Q.11. What would you expect if the field winding of a DC shunt motor gets disconnected while in normal operation?

Ans. The flux will drop to almost zero value (residual value) and, therefore, the speed will increase to a tremendously high value and the motor may get damaged due to heavy centrifugal forces.

Q.12. What is the purpose and location of a series field winding?

Ans. Series field winding is connected in series with the armature and placed over the magnetic pole core to set up necessary flux in the machine for production of torque.

Q.13. A DC shunt motor runs away at heavy loads. Comment.

Ans. The field may be weakened, due to armature reaction at heavy loads, to the extent that the motor runs away.

Q.14. What is difference between cumulative compound and differential compound wound motors?

Ans. In cumulative compound wound DC motors, the field produced by series field winding assists the field produced by shunt field winding, whereas in differential compound wound motors series field opposes the shunt field.

Q.15. A long shunt DC compound motor is cumulatively compounded. Without any change in connection, it is now run as a DC generator. State whether it will now be differentially or cumulatively compounded.

Ans. When a cumulatively compounded long shunt DC motor is run as a generator, the shunt and series field currents produce flux in opposite directions and thus it will operate as a differentially compounded generator.

Q.16. What is the field of application of DC shunt motor and DC series motor.

Ans. DC shunt motors are almost constant speed motors. These motors develop medium starting torque and their speed can be adjusted by using speed control methods. Therefore, shunt motors can be used for the loads which are totally and suddenly thrown off without resulting in excessive speed. Being constant speed motors, they are best suited for driving of line shafts, machine lathes, milling machines, conveyors, fans and for all purposes where constant speed is required.

DC series motors are variable speed motors. These motors develop large starting torque but attain dangerously high speed on no-load. So these motors cannot be operated at no-load. Series motors are best suited for electric traction, elevators, cranes, vacuum cleaners, sewing machines etc.

Q.17. Identify suitable DC motors for the following applications:

(*a*) Electric traction (*b*) Vacuum cleaners (*c*) Paper making (*d*) Shearing and punching.

Ans. (*a*) DC series motor (*b*) DC series motor (*c*) DC cumulative compound motor (*d*) DC cumulative compound motor.

Q.18. What would happen if a DC motor is directly switched ON to the supply, without any starter?

Ans. If a DC motor is directly switched *ON* to the supply, it will draw heavy current (about 10 – 12 times its rated value) which may over-heat the motor or damage it, depending upon the starting period.

Q.19. Why a starter is necessary for a DC motor?

Or

Why starters are used for DC motors.

Ans. A starter is employed to limit the starting current and provide necessary safety to the motor.

Q.20. A DC shunt motor is connected to a three-point starter. What would happen if

(i) the starter handle is moved rapidly from 'off to 'on' position;

(ii) the field circuit becomes open-circuited with the motor running at no-load;

(iii) the starter handle is pulled back to stop the motor.

Ans. (*i*) When the starter handle is moved rapidly from '*OFF*' to '*ON*' position, the resistance connected in series with the armature is being cut out at a faster rate and the speed of the motor is increased at a slow rate. Because of this the armature current will increase to a larger value and the ratio of upper and lower values of currents will not be maintained.

(*ii*) With the opening of field circuit with motor running at no-load, the field current will reduce to zero. So flux collapses and reduced to zero, in the process, the speed of motor will suddenly starts increasing and continue to increase if the supply is not cut-off.

(*iii*) Sparking will occur at the last stud due to release of energy stored in the magnetic field of field winding, which may burn it.

Q.21. It is desirable that the starting lever of a motor starter should fall back to off position when the power goes off. Why?

Ans. If the starter arm remains in the "*ON*" position when supply goes off, then on restoring the power supply, the motor will start without starting resistance. Hence, motor will draw heavy current when supply is restored.

Q.22. Discuss the limitations of a three-point starter.

Ans. In a three-point starter, the no-volt release coil is connected in series with the shunt field winding. If the field current is decreased to low value for speed control then magnetic strength of the no-volt release coil decreases to such an extent that spring tension brings the starting arm back to *OFF* position. To overcome this problem a four-point starter is used.

Q.23. Why do you prefer starter to rheostat in starting DC motors?

Ans. The starter also contains protective devices, such as no-volt release, overload release, etc.

Q.24. How does four point DC shunt motor starter differs from three point starter?

Ans. In a four point starter the circuits of no-volt release coil and shunt field are made independent and so the operation of no-volt release is not effected due to variations in field current.

5.20 Speed Control of DC Motors

The speed of a DC motor is given by the relation $N \propto \frac{E_b}{\phi}$ where $E_b = V - I_a R_a$

$$\therefore \qquad N \propto \frac{V - I_a R_a}{\phi}$$

From the above equation it is clear that the speed of DC motors can be controlled;

1. By varying flux per pole ϕ. This is known as *flux* or *field control method.*
2. By varying the armature drop, i.e., by varying the resistance of armature circuit. This is known as *armature control method.*
3. By varying the applied voltage. This is known as *voltage control method.*

5.21 Speed Control of Shunt Motors

5.21.1 Field Control Method

The speed of a DC motor can be increased by weakening the field and it can be decreased by strengthening the field. In DC shunt motors, the speed adjustments can be made by field control method by employing any one of the following methods:

1. Field-rheostat control method
2. Reluctance control method
3. Field-voltage control method

1. **Field rheostat control method.** The flux produced by the shunt field winding depends upon the current flowing through it (i.e., $\phi \propto I_{sh}$ and $I_{sh} = \left(\frac{V}{R_{sh}}\right)$. When a variable resistance R is connected in series with the shunt field winding as shown in Fig. 5.29(*a*). The shunt field current $\left(I_{sh} = \frac{V}{R_{sh} + R}\right)$ is reduced and hence the flux ϕ. Consequently, the motor runs at a speed higher than the normal speed (since $N \propto 1/\phi$), as shown in Fig. 5.29(*b*) The amount of increase in speed depends upon the value of variable resistance R.

   This method is most common since very little power ($I^2_{sh} R$) is wasted in the shunt field variable resistance due to relatively small value of I_{sh}.

   In this method speed can be only increased, thus we can obtain speeds above normal. The other limitations and drawbacks of this method are:

   (*i*) Speeds below normal cannot be obtained.

(*ii*) The speeds above normal can be obtained by weakening the field. The advantage of high speed cannot be considered for increase in power output rather to compensate the power, armature has to draw extra current from the mains.

(*iii*) To obtain high speed, the field is very weak, to obtain certain load torque, armature draws extra current which may causes overheating of armature winding; poor commutation and instability.

Hence, the high speeds are restricted to certain extent. High speeds are also restricted due to mechanical considerations such as centrifugal forces developing in the armature conductors and friction occurring at the bearings.

Accordingly there is a limit to the maximum speed obtainable with this method. It is mainly because of poor communication at weak fluxes. A ratio of maximum to minimum speeds of 6 is fairly common.

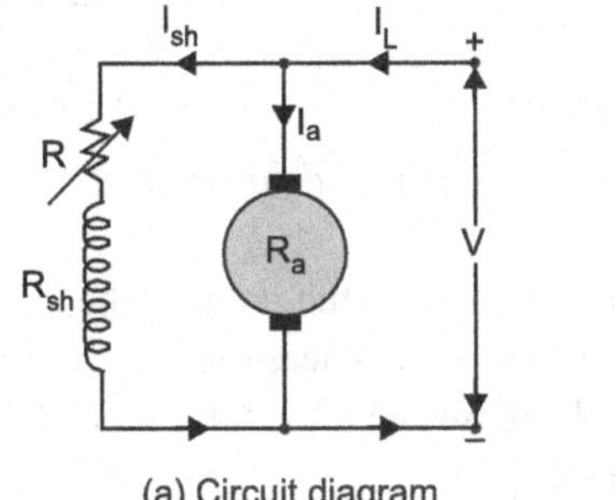

(a) Circuit diagram

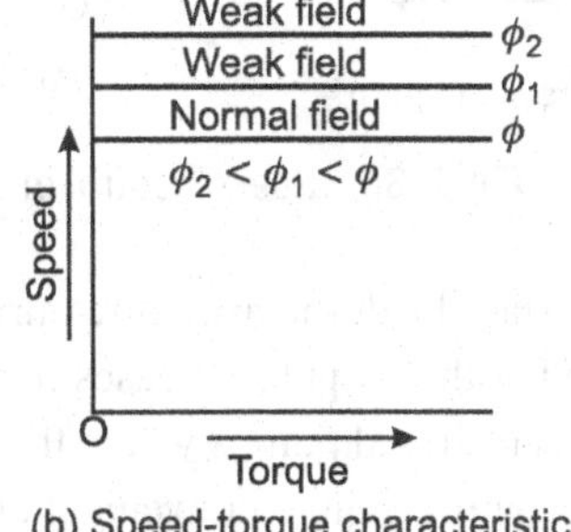

(b) Speed-torque characteristics

Fig. 5.29 Speed control of DC shunt motor by field control method

2. **Reluctance control method.** The flux or field can also be controlled by controlling the reluctance of the magnetic circuit, but this can only be obtain by employing some special mechanical features which increases the cost of machine. Therefore, practically this method is not employed.
3. **Field-Voltage control method.** The flux or field can also be controlled by supplying variable voltage to the field winding. This can only be achieved by disconnecting the field winding from the armature and supplying variable voltage from some other source. Then the motor is treated as separately excited DC motor.

5.21.2 Armature Control Method

In this method, voltage applied to the armature is changed without altering the field current. This cannot be achieved by just changing the common supply voltage because it will also affect the field simultaneously and the combined effect will have little change in the speed of shunt motor. Hence, to adjust the speed of DC shunt motors by armature control method, different arrangements are made and the methods are known as:

1. Armature-resistance control method
2. Shunted-armature control method
3. Armature terminal-voltage control method

1. **Armature resistance control method.** In a shunt motor, flux is constant when applied terminal voltage and shunt field resistance are constant. Therefore, speed of the motor is directly proportional to induced *emf* (i.e., $N \propto E_b$ and $E_b = V - I_a R_a$). The value of E_b depends upon the drop in the armature circuit. When a variable resistance is connected in series with the armature as shown in Fig. 5.30(*a*) the induced emf $[E_b = V - I_a (R_a + R)]$ is reduced and hence the speed. Thus, the motor runs at a speed lower than the normal speed as shown in Fig. 5.30(*b*).

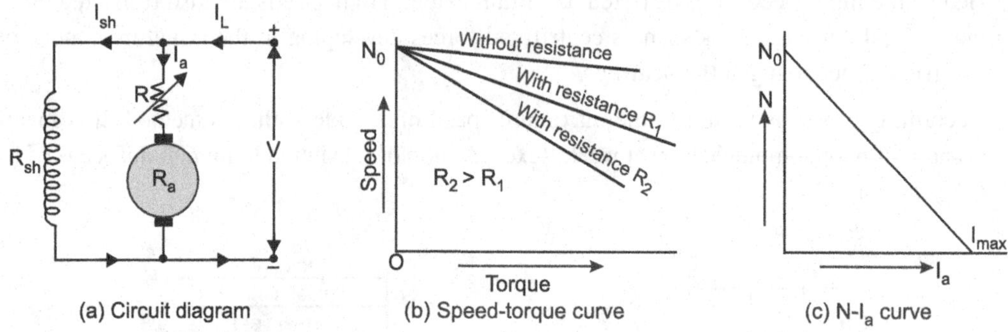

(a) Circuit diagram (b) Speed-torque curve (c) N-I_a curve

Fig. 5.30 Speed control of DC shunt motor by armature control method

For a constant load, the armature current remains the same so the input to the motor remains the same but the output decreases in proportion to the speed. Therefore, for the loads such as fans and centrifugal pumps where the load torque decreases with the fall in speed, this method of speed control is quite convenient and economical.

By this method, a wide range of speeds (below normal) can be obtained. Moreover, motor develops any desired torque over its operating range since torque depends only upon the armature current (flux remaining unchanged). The major advantage of this method is that the speed of the motor can be reduced to any low value and creeping speed (only a few rpm) can be developed so easily, as shown in Fig. 5.30(*c*).

The major disadvantage of this method of speed control is that there is heavy loss of power in the control rheostat.

2. **Shunted-armature control method.** An armature resistance control method is very simple and usually employed when speeds below normal are required. But this method suffers from the drawback that the motor speed changes with every change in load on the machine. It is because the speed variation does not depend only upon the controlling resistance but it also depends upon load current. Because of this reason constant speed cannot be obtained in case of rapidly changing loads. To improve the stability, a shunt diverted is used across the armature in addition to series resistance, as shown in Fig. 5.31(*a*). By this arrangement, the changes in armature current due to change in load torque will not be so effective to change the voltage across armature and hence the speed of the motor remains almost constant.

Figure 5.31(*b*) shows the speed-torque characteristics for shunted armature control method. In this method, the ideal no-load speed of the motor (N'_0) will be less than that obtained by conventional series resistance method (N_0). This difference occur due to the fact that even if no current flows through the motor armature, there is some current in the series resistance and the diverter. Hence, there is some voltage drop in series resistance which reduces the voltage coming across the armature. Thus, in this arrangement, the voltage coming across the armature is always less than the supply voltage. Whereas in case of conventional series resistance control method, the voltage coming across armature is *V*. The following three characteristics have been drawn in Fig. 5.31(*b*).

(*i*) Normal speed-torque characteristics – curve-*A*.

(*ii*) Speed-torque characteristics with conventional series resistance control – curve-*B*

(*iii*) Speed-torque characteristics with shunted armature control – curve-*C*.

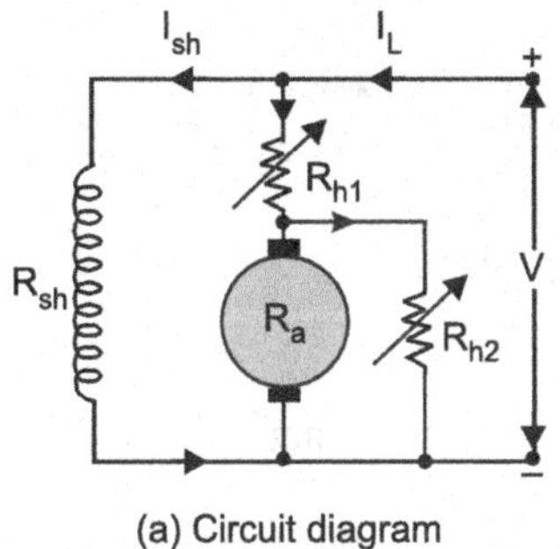

(a) Circuit diagram

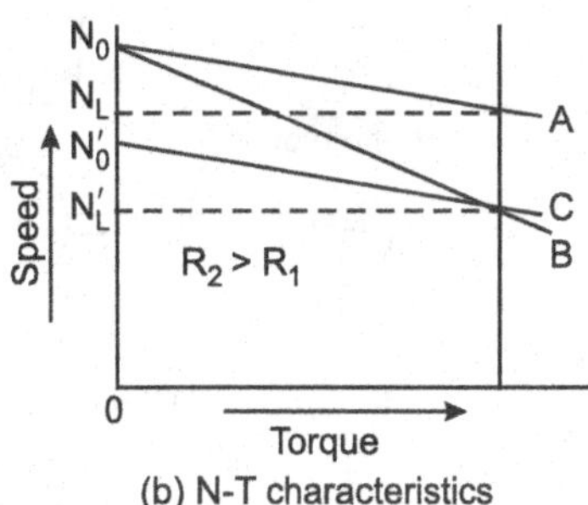

(b) N-T characteristics

Fig. 5.31 Speed control of DC shunt motors by shunted armature control method.

3. **Armature terminal-voltage control method.** In this method, speed of a DC shunt motor is changed by changing the voltage applied across the armature. This can be achieved only by providing a separate voltage source. Although this method avoids the disadvantages of poor speed regulation and low efficiency, which are the characteristics of armature-resistance control method, but it is very expensive and has very high initial cost. It is because an adjustable voltage generator or an adjustable electronic regulator is required to supply variable voltage to the armature. Both these machines are very costly.

   This method gives a large range of speeds with any desired number of speed points. It is essentially a constant-torque system because the output of the motor decreases with a decrease in applied voltage and a corresponding decrease in speed.

5.22 Speed Control of Separately Excited Motors

Sometimes, it is also called an armature voltage control method of a DC shunt motor. In this case, a separate DC source is used to supply power to the armature of DC shunt motor or separately excited motor and the field winding is excited from a different source. The most common and accurate speed control method of separately excited DC motors is Ward-Leonard system as explained below:

Ward-Leonard System

This system is used to supply variable voltage to the motor. As shown in Fig. 5.32, a DC generator *G* is mechanically coupled with a prime mover *PM* which rotates the generator at constant speed. The field winding of the DC generator is connected to a constant voltage DC supply line through a field regulator and reversing switch. The DC motor M is fed from the generator G and its field winding is connected directly to a constant DC supply line.

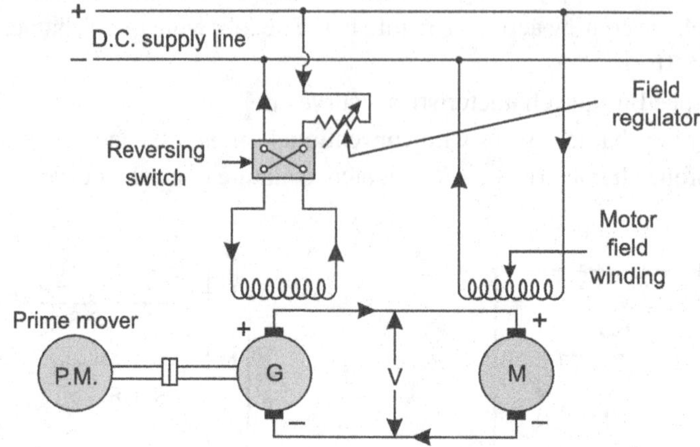

Fig. 5.32 Ward-Leonard method of speed control of DC motors

The voltage of the generator fed to the motor, can be varied from zero to its maximum value by means of its field regulator. By reversing the direction of the field current by means of the reversing switch, the polarity of the generated voltage can be reversed and hence the direction of rotation of motor M. Hence, by this method, the speed and direction of rotation both can be controlled very accurately.

The capital investment in this system is very high as two extra machines (a generator and a prime mover) are required. This system of speed control is best suited where almost unlimited speed control in either direction of rotation is required, e.g., in steel rolling mills, paper machines, elevators, cranes, diesel-electric propulsion of ships, etc.

5.23 Speed Regulation

The *speed regulation of a DC motor is defined as the change in speed from full load to no-load and is expressed as a percentage of the full load speed.*

$$\therefore \quad \%\ Speed\ regulation = \frac{\text{N.L. speed} - \text{F.L. speed}}{\text{F.L. speed}} \times 100 = \frac{N_0 - N}{N} \times 100 \qquad \ldots(5.18)$$

Example 5.18

The armature resistance of a 220 V DC generator is 0.4 ohm. It is delivering a load of 4 kW at rated terminal voltage Now the machine is operated as a motor and draws the same armature current at the same terminal voltage. In this operation, if the flux per pole is increased by 10% what will be the ratio of generator to motor

Solution:

As a generator; $\quad I_{al} = I_L = \dfrac{4 \times 1000}{220} = 18.182 \text{ A}$

$$E_g = V + I_{al} R_a = 220 + 18.182 \times 0.4 = 227.27 \text{ V}$$

As a motor; $I_a = I_{a2} = 18.182$ A

$$E_b = V - I_{a2}R_a = 220 - 18.182 \times 0.4 = 212.73 \text{ V}$$

$$\phi_2 = 1.1\ \phi_1 \text{ (given)}$$

Now $$N_1 \propto \frac{E_g}{\phi_1} \text{ and } N_2 \propto \frac{E_b}{\phi_2}$$

$$\therefore \quad \frac{N_1}{N_2} = \frac{E_g}{\phi_1} \times \frac{\phi_2}{E_b} = \frac{227.27}{\phi_1} \times \frac{1.1\phi_1}{212.73} = \mathbf{1.175} \text{ (Ans.)}$$

Example 5.19

A 250 V, 8 kW with R_a = 0.7 ohm DC shunt motor runs at 1250 rpm Its field is adjusted at no-load so that its armature carries a current of 16 A. When load is applied, the current drawn by it increases to 40 A and speed falls to 1150 rpm. How much flux per pole reduces due to armature reaction.

Solution:

At no-load; $$E_{b0} = V - I_{a0}R_a = 250 - 1{\cdot}6 \times 0{\cdot}7 = 248{\cdot}88 \text{ V}$$

At full-load; $$E_b = V - I_aR_a = 250 - 40 \times 0{\cdot}7 = 222 \text{ V}$$

Now, $$N_0 \propto \frac{E_{b0}}{\phi_0} \text{ and } N \propto \frac{E_b}{\phi}$$

Therefore $$\frac{N}{N_0} = \frac{E_b}{\phi} \times \frac{\phi_0}{E_{b0}}$$

or $$\phi = \frac{E_b}{E_{b0}} \times \frac{N_0}{N} \times \phi 0$$

or $$\phi = \frac{222}{248 \cdot 88} \times \frac{1250}{1150} \times 1 = 0 \cdot 96956$$

Reduction in flux = 1 – 0·96956 = 0·03044 = **3·044** % (*Ans.*)

Example 5.20

A fan motor whose torque is proportional to the square of the speed is driven by a separately excited DC motor. The motor draws a current of 8 A at 120 V and runs at a speed of 500 rpm. What voltage and current is to be applied to run the motor at 750 rpm. The armature resistance of the motor is 1 ohm and losses are neglected.

Solution:

When 120 V is applied to the armature

Armature current, I_{a1} = 8A

Back emf, $E_{b1} = V - I_{a1}R_a = 120 - 8 \times 1 = 112$ V

Speed, N_1 = 500 rpm

Let voltage of V volt be applied to the armature to raise the speed to 750 rpm.

Then, $T_2 = T_1 \times \left(\frac{N_2}{N_1}\right)^2 = T_1 \times \left(\frac{750}{500}\right)^2 = 2.25\ T_1$ (Since $T \propto N^2$ given)

or $I_{a2}\phi_2 = 2.25\ I_{a1}\phi_1$ (since $T \propto \phi I_a$)

Assuming flux as constant

$$I_{a2} = 2.25 \times I_{a1} = 2.25 \times 8 = \mathbf{18\ A}\ (Ans.)$$

Now, $\frac{E_{b2}}{E_{b1}} = \frac{N_2}{N_1}$

or $E_{b2} = \frac{N_2}{N_1} \times E_{b1} = \frac{750}{500} \times 112 = 168\ \text{V}$

Applied voltage, $V_2 = E_{b2} + I_{a2}R_a = 168 + 18 \times 1 = \mathbf{186\ V}\ (Ans.)$

Example 5.21

A 100 kW belt-driven shunt generator running at 300 rpm on 220 V bus bars continues to run as a motor when the belt breaks then taking 10 kW from the mains. What will be its speed? Armature resistance is 0.025 ohm, field circuit resistance is 60 ohm contact drop per brush is 1 V and ignore armature reaction.

Solution:

Shunt field current, $I_{sh} = \frac{220}{60} = 3.67\ \text{A}$

(*i*) When the machine runs as a generator (see Fig. 5.33)

$$I_{L1} = \frac{100 \times 1000}{220} = 454.54\text{A}$$

$$I_{a1} = I_{L1} + I_{sh} = 454.54 + 3.67 = 458.21\ \text{A}$$

$$E_g = V + I_{a1}R_a + 2v_b$$

$$= 220 + 458.21 \times 0.025 + 2 \times 1 = 233.455\ \text{V}$$

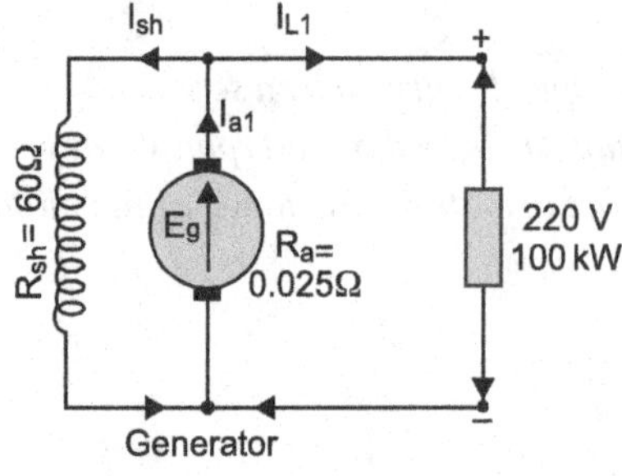

Fig. 5.33 Circuit diagram

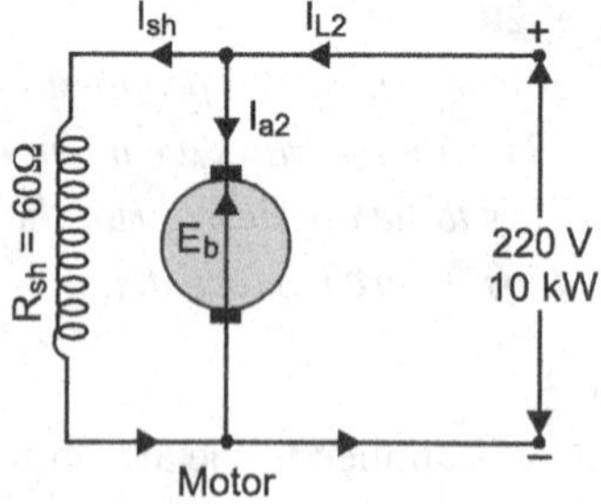

Fig. 5.34 Circuit diagram

(*ii*) When the machine runs as a motor (see Fig. 5.59.)

$$I_{L2} = \frac{10 \times 1000}{220} = 45.45\ \text{A};\ I_{a2} = I_{L2} - I_{sh} = 45.45 - 3.67 = 41.78\ \text{A}$$

$$E_b = V - I_{a2}R_a - 2\upsilon_b = 220 - 41.78 \times 0.025 - 2 \times 1 = 216.955 \text{ V}$$

In shunt machines, flux is constant;

$$\therefore \quad N_1 \propto E_g \text{ and } N_2 \propto E_b$$

or $$\frac{N_2}{N_1} = \frac{E_b}{E_g}$$

or $$N_2 = \frac{E_b}{E_g} \times N_1 = \frac{216.955}{233.455} \times 300 = \mathbf{278.8\ rpm}\ (Ans)$$

Example 5.22

A 240 V DC shunt motor has a field resistance of 400 ohm and an armature resistance of 0.1 ohm. The armature current is 50 A and the speed is 1000 rpm Calculate the additional resistance in the field to increase the speed to 1200 rpm Assume that armature current remains the same and the magnetisation curve to be a straight line.

Solution:

When no resistance is added in the shunt field winding

Shunt field current, $$I_{sh} = \frac{V}{R_{sh}} = \frac{240}{400} = 0.6 \text{ A}$$

$$E_{b1} = E_{b2} = V - I_aR_a = 240 - 50 \times 0.1 = 235 \text{ V (since } I_a \text{ is constant)}$$

Now, we know that, $$N \propto \frac{E_b}{\phi}$$

$$\therefore \quad N_1 \propto \frac{E_{b1}}{\phi_1} \text{ and } N_2 \propto \frac{E_{b2}}{\phi_2}$$

or $$\frac{N_2}{N_1} = \frac{E_{b2}}{\phi_2} \times \frac{\phi_1}{E_{b1}} = \frac{\phi_1}{\phi_2} \qquad (\because E_{b1} = E_{b2} = 235 \text{ V})$$

As the magnetisation curve is a straight line, $\phi_1 \propto I_{sh1}$ and $\phi_2 \propto I_{sh2}$

$$\therefore \quad \frac{N_2}{N_1} = \frac{I_{sh1}}{I_{sh2}}$$

or $$I_{sh2} = \frac{N_1}{N_2} \times I_{sh1} = \frac{1000}{1200} \times 0.6 = 0.5 \text{ A}$$

or $$\frac{240}{400 + R} = 0.5$$

$$\therefore \quad R = \mathbf{80\Omega}\ (Ans)$$

Example 5.23

A 250 V shunt motor runs at 1500 rpm at full load with an armature current of 15 A. The total resistance of the armature and brushes is 0.6 Ω. If the speed is to be reduced to 1200 rpm with the same armature current, calculate the amount of resistance to be connected in series with the armature and power lost in this resistor.

Solution:

During normal operating conditions;

Armature current, $I_{a1} = 15$ A

Back emf, $E_{b1} = V - I_{a1} R_a = 250 - 15 \times 0.6 = 241$ V

Speed, $N_1 = 1500$ rpm

Let a resistance R ohm be connected in series with armature, then

Speed, $N_2 = 1200$ rpm

Back emf, $E_{b2} = V - I_{a2} (R + R_a) = 250 - 15(0.6 + R) = 241 - 15$ R

Now, $\frac{E_{b2}}{E_{b1}} = \frac{N_2}{N_1}$ (Since flux is constant)

$\frac{241 - 15}{241} = \frac{1200}{1500}$ or $241 - 15R = 192.8$ or $R =$ **3.213 Ω** (*Ans.*)

Power lost $= I_a^2 R = (15)^2 \times 3.213 =$ **723 W** (*Ans.*)

Example 5.24

A DC shunt motor runs at 1000 rpm on 220 V supply. Its armature and field resistances are 0.5 ohm and 110 ohm respectively and the total current taken from the supply is 26 A. It is desired to reduce the speed to 750 rpm keeping the armature and field currents same. What resistance should be inserted in the armature circuit?

Solution:

Shunt field current, $I_{sh} = \frac{V}{R_{sh}} = \frac{220}{110} = 2\text{A}$

Armature current, $I_a = I_L - I_{sh} = 26 - 2 = 24$ A

When no resistance is connected in series with the armature,

Induced emf, $E_{b1} = V - I_a R_a = 220 - 24 \times 0.5 = 208$ V

Let the resistance connected in series with the armature be R ohm, as shown in Fig. 5.35. to reduce the speed to 750 rpm

Then, induced emf, $E_{b2} = V - I_a (R_a + R) = 220 - 24 (0{\cdot}5 + R) = (208 - 24 R)$ V ...(*i*)

Now, we know that, $N \propto \frac{E_b}{\phi}$

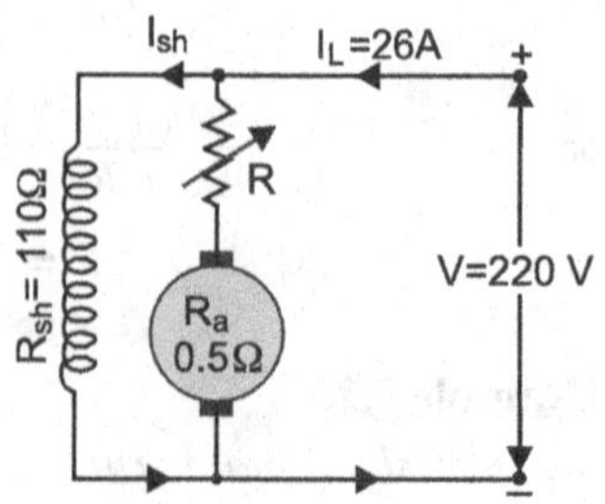

Fig. 5.35 Circuit diagram

Flux ϕ is constant because shunt field current remains the same.

∴ $N_1 \propto E_{b1}$ and $N_2 \propto E_{b2}$

Taking their ratio, $\frac{E_{b2}}{E_{b1}} = \frac{N_2}{N_1}$

or $E_{b2} = \frac{N_2}{N_1} \times E_{b1} = \frac{750}{1000} \times 208$

$= 156\text{V}$ ...(*ii*)

Equating eq. (*i*) and (*ii*), we get,

$$208 - 24R = 156 \quad \text{or} \quad R = \mathbf{2.167\Omega}\ (Ans)$$

Example 5.25

The armature resistance of a 200 V DC shunt motor is 0.5 ohm. When it draws an armature current of 30 A, it runs at a speed of 750 rpm. In order to reduce the speed to 450 rpm, what resistance must be added in the armature circuit keeping armature current constant. Further, what will be the speed of motor if its additional armature resistance is kept the same but armature current is reduced to 15 A.

Solution:

When no resistance is added in the armature circuit,

$$\text{Armature current, } I_{a1} = 30 \text{ A}$$

$$\text{Back emf, } E_{b1} = V - I_{a1} R_a = 200 - 30 \times 0.5 = 185 \text{ V}$$

$$\text{Speed, } N_1 = 750 \text{ rpm}$$

When a resistance (say *R ohm*) is connected in series with armature

$$\text{Armature current, } I_{a2} = I_{a1} = 30\text{A}$$

$$\text{Back emf, } E_{b2} = V - I_{a2}(R_a + R) = 200 - 30(0.5 + R) = (185 - 30R) \text{ volt.}$$

$$\text{Speed, } N_2 = 450 \text{ rpm}$$

$$\text{Now, } \frac{E_{b2}}{E_{b1}} = \frac{N_2}{N_1} \qquad \text{(assuming flux as constant)}$$

or
$$\frac{185 - 30R}{185} = \frac{450}{750}$$

or
$$R = \frac{185 - 0.6 \times 185}{30} = \mathbf{2.467\ \Omega}\ (Ans.)$$

When armature current (say I_{a3}) changes to 15A.

$$\text{Back emf, } E_{b3} = V - I_{a3}(R + R_a) = 200 - 15(0.5 + 2.467) = 155.5 \text{ V}$$

$$\text{Speed } N_3 = N_1 \times \frac{E_{b3}}{E_{b1}} = 750 \times \frac{155.5}{185} = \mathbf{630.4 \text{ rpm}}\ (Ans.)$$

Example 5.26

A 230 V shunt motor is taking a current of 50 A. Resistance of shunt field is 46 Ω and the resistance of the armature is 0.02 Ω. There is a resistance of 0.6 Ω in series with the armature and the speed is 900 rpm. What alternation must be made in the armature circuit to raise the speed to 1000 rpm, the torque remaining the same.

Solution:

$$\text{Line current, } I_{L1} = 50 \text{ A and } N_1 = 900 \text{ rpm}$$

$$\text{Shunt field current, } I_{sh} = \frac{V}{R_{sh}} = \frac{230}{46} = 5\text{A}$$

$$\text{Armature current, } I_{a1} = I_L - I_{sh} = 50 - 5 = 45 \text{ A}$$

$$\text{Back emf, } E_{b1} = V - I_a (R_a + R) = 230 - 45\,(0.02 + 0.6) = 202.1 \text{ V}$$

Let the series resistance be reduced from 0.6 Ω to R_1 ohm to raise the speed to 1000 rpm.

Since load torque is constant

∴ $$T_2 = T_1 \text{ or } I_{a2}\,\phi_2 = I_{a1}\,\phi_1 \qquad \text{(since } T \propto \phi I_a\text{)}$$

or $$I_{a2} = I_{a1} \times \frac{\phi_1}{\phi_2} = I_{a1} = 45 \text{ A} \qquad \text{(assuming flux unchanged)}$$

$$\text{Speed, } N_2 = 1000 \text{ rpm}$$

$$\text{Back emf, } E_{b2} = V - I_{a2}\,(R_a + R_1) = 230 - 45\,(0.02 + R_1) = 229.1 - 45\text{R}_1$$

Also $$E_{b2} = E_{b1} \times \frac{N_2}{N_1} \qquad \text{(since flux is constant)}$$

or $$229.1 - 45\,\text{R}_1 = 202.1 \times \frac{1000}{900}$$

or $$R_1 = \mathbf{0.1\ \Omega}\ (Ans.)$$

i.e., Additional resistance of 0.6 Ω will have to be reduced to 0.1 Ω in order to raise the motor speed from 900 rpm to 1000 rpm. (*Ans.*)

Example 5.2

A 250 V shunt motor has an armature current of 20 A when running at 1000 rpm against full-load torque. The armature resistance is 0.5 Ω. What resistance must be inserted in series with the armature to reduce the speed to 800 rpm at the same torque, and what will be the speed if the load torque is halved with this resistance in the circuit. Assume the flux to remain constant throughout and neglect brush contact drop.

Solution:

At normal conditions; $I_{a1} = 20$ A; $N_1 = 1000$ rpm

$$\text{Back emf, } E_{b1} = V - I_{a1} R_a = 250 - 20 \times 0.5 = 240 \text{ V}$$

Let R be the resistance connected in series with the armature circuit to reduce the speed to 800 rpm, i.e., $N_2 = 800$ rpm

$$\text{Load torque, } T_2 = T_1 \qquad \text{(given)}$$

or $$I_{a2}\,\phi_2 = I_{a1}\,\phi_1 \qquad \text{(since } T \propto \phi I_a\text{)}$$

or $$I_{a2} = I_{a1} = 20 \text{ A} \qquad \text{(flux } \phi \text{ being constant)}$$

$$\text{Back emf } E_{b2} = V - I_{a2}\,(R_a - R) = 250 - 20\,(0.5 + R) = 240 - 20\,R$$

Now, $$\frac{E_{b2}}{E_{b1}} = \frac{N_2}{N_1} \quad (\phi \text{ being constant})$$

or $$\frac{240 - 20R}{240} = \frac{800}{1000}$$

or $$R = \mathbf{2.4\ \Omega}\ (Ans.)$$

When load torque is reduced to half

i.e., $$T_3 = 0.5\ T_1 \text{ or } I_{a3}\ \phi_3 = 0.5\ I_{a1}\ \phi_1$$

or $$I_{a3} = 0.5\ I_{a1} = 0.5 \times 20 = 10 \text{ A (since } \phi \text{ is constant)}$$

$$\text{Back emf, } E_{b3} = V - I_{a3}(R + R_a) = 250 - 10\ (2.4 + 0.5) = 221 \text{ V}$$

$$\frac{N_3}{N_1} = \frac{E_{b3}}{E_{b1}} \quad (\phi \text{ being constant})$$

or $$N_3 = N_1 \times \frac{E_{b3}}{E_{b1}} = 1000 \times \frac{221}{240} = \mathbf{920.8\ rpm}\ (Ans.)$$

Example 5.28

The field winding resistance and armature resistance of a 240 V DC shunt motor is 120 ohm and 0.1 ohm respectively. It draws 24 A at rated voltage to run at 1000 rpm. Find the value of additional resistance required in the armature circuit to reduce the speed to 800 rpm when (i) the load torque is proportional to speed (ii) the load torque varies as the square of the speed.

Solution:

At normal conditions; $I_{L1} = 24$ A and $N_1 = 1000$ rpm

$$\text{Shunt field current, } I_{sh} = \frac{V}{R_{sh}} = \frac{240}{120} = 2\text{A}$$

$$\text{Armature current, } I_{a1} = I_{L1} - I_{sh} = 24 - 2 = 22 \text{ A}$$

$$\text{Back emf, } E_{b1} = V - I_{a1}R_a = 240 - 22 \times 0.1 = 237.8 \text{ V}$$

(*i*) Let R be the additional resistance required to be connected in armature circuit to reduce the speed to 800 rpm when the load torque is proportional to speed, i.e., $T_1 \propto N_1$ and $T_2 \propto N_2$

$\therefore$ $$T_2 = T_1 \times \frac{N_2}{N_1} = T_1 \times \frac{800}{1000} = 0.8\ T_1$$

or $$I_{a2}\ \phi_2 = 0.8\ I_{a1}\phi_1 \text{ (since } T \propto \phi I_a)$$

or $$I_{a2} = 0.8 \times 22 = 17.6 \text{ (since } \phi_2 = \phi_1 = \text{constant)}$$

$$E_{b2} = V - I_{a2}\ (R_a + R) = 240 - 17.6\ (0.1 + R) = 238.24 - 17.6\ R$$

Since $$\frac{E_{b2}}{E_{b1}} = \frac{N_2}{N_1},$$

or $$\frac{238.24 - 17.6R}{237.8} = \frac{800}{1000}$$

or $238.24 - 17.6\text{R} = 190.24$ or $R = \mathbf{2.73\ \Omega}$ (*Ans.*)

(*ii*) Let R_1 be the additional resistance required in the armature circuit to reduce the speed to 800 rpm when the load torque varies as the square of the speed

$$T_3 = T_1 \times \left(\frac{N_2}{N_1}\right)^2 = T_1 \times \left(\frac{800}{1000}\right)^2 = 0.64\ T_1$$

or $$I_{a3}\ \phi_3 = 0.64\ I_{a1}\ \phi_1 \qquad \text{(since } T \propto \phi I_a\text{)}$$

or $$I_{a3} = 0.64 \times 22 = 14.08 \text{ A} \qquad \text{(since } \phi_3 = \phi_1 = \text{constant)}$$

$$E_{b3} = V - I_a (R_1 + R_a) = 240 - 14.08\ (R_1 + 0.1) = 238.592 - 14.08\ R_1$$

Also, $$\frac{E_{b3}}{E_{b1}} = \frac{N_3}{N_1} \quad \text{or} \quad \frac{238.592 - 14.08\ R_1}{237.8} = \frac{800}{1000}$$

$238.592 - 14.08\ R_1 = 190.24$ or $R_1 =$ **3.434 Ω** *(Ans.)*

5.24 Speed Control of DC Series Motors

The speed of DC series motors can be controlled by any one of the following methods:

(*i*) Armature control method
(*ii*) Field control method
(*iii*) Series – parallel control method

5.24.1 Armature Control Method

By making various adjustments in the armature circuit of a DC series motor its speed can be controlled. According to the adjustments made, these methods are knows as;

1. Armature series resistance control method
2. Shunted armature control method
3. Armature terminal voltage control method.

1. **Armature series resistance control method.** In this method, a variable resistance is connected in series with the armature or motor as shown in Fig. 5.36(*a*). If the load and torque developed by the machine is constant, the speed of the motor depends upon back emf i.e., $N \propto E_b$. When no addition resistance is connected in series with the armature, then $E_{b1} = V - I_a (R_a + R_{se})$. However, when an additional resistance R is connected in series with the armature, then back emf, $E_{b2} = V - I_a (R_a + R_{se} + R)$. Obviously $E_{b2} < E_{b1}$ accordingly $N_2 < N_1$. Hence, by connecting an additional resistance is series with the armature we can obtain speeds below normal.

   The speed-torque characteristics of a DC series motor without additional resistance and with additional series resistance is shown in Fig. 5.36(*b*). The maximum range of speed control of about 3: 1 will be available depending upon the load.

   This is the most common method employed for the speed control of series motors. The power loss in the control resistance for many applications of DC series motors is not too serious, since in these applications the control is utilised for a large portion of time for decreasing the speed under light loads.

The major applications of this method of speed-control are for driving cranes, hoists, trains, etc., because such drives operate intermittently.

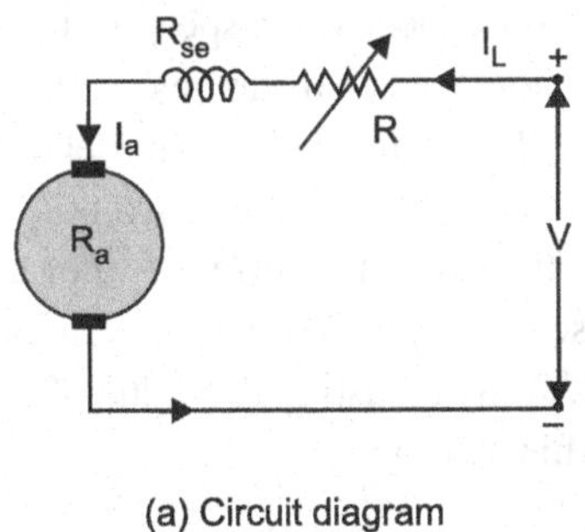

(a) Circuit diagram

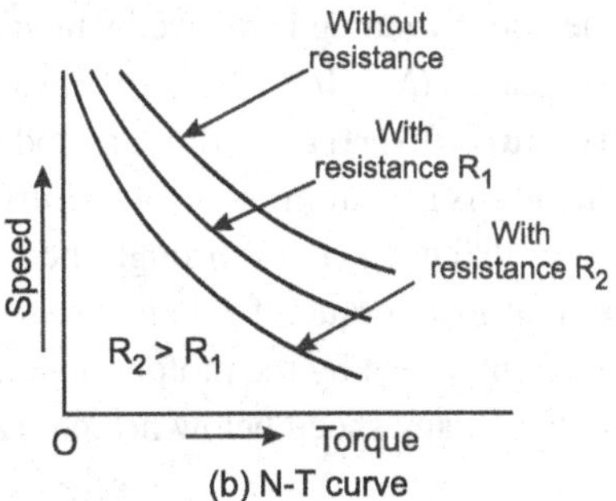

(b) N-T curve

Fig. 5.36 Speed control of DC series motors by armature control method

2. **Shunt armature control method.** In this method, combination of a rheostat R_2 shunting the armature and a rheostat R_1 in series with armature, as shown in Fig. 5.37 is used. It provides slow speeds at light loads. This is an arrangement which accomplishes the speed control both by lowering the voltage applied to the armature and by varying the flux. In fact, the voltage applied to the armature terminals is varied by varying series rheostat R_1 whereas the exciting current is varied by varying the shunting rheostat R_2 keeping the armature current constant.

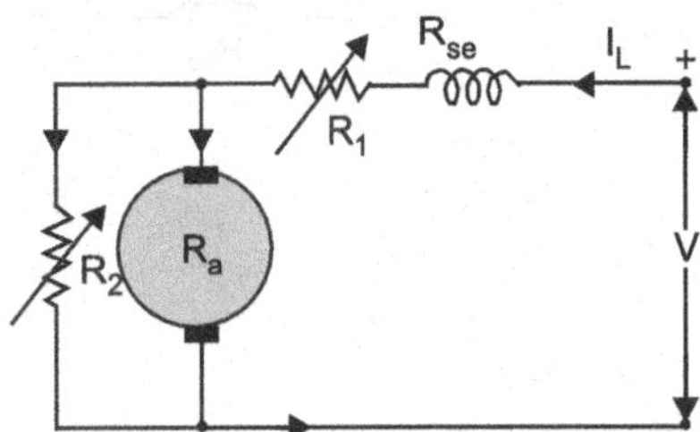

Fig. 5.37 Circuit diagram for shunted armature control method

In this method, by the insertion of armature diverter R_2, armature current reduces, then flux ϕ must increase because torque developed ($T \propto I_a$) by the armature remains the same. This causes increase in current drawn from the supply mains which increases the field and decreases the speed.

Thus, by this method, speeds below normal can be obtained. A wide range of speeds below normal can be obtained by taking different ratios of R_1 and R_2.

The application of this method is restricted to the places where speed control for short internals is required because this method is not so economical due to considerable power losses in speed controlling resistance.

3. **Armature terminal voltage control method.** In this method, a variable voltage power supply is used to change the supply voltage which changes the speed of a DC series motor. But the cost of such equipment is so high that this method is rarely applied.

5.24.2 Field Control Method

The speed of series motors can be controlled by varying the flux produced by the series field winding. The variation of flux can be brought about by anyone of the following ways;

1. **Field diverters:** In this method, a variable resistance R is connected in parallel with the series field winding as shown in Fig. 5.38. Its effect is that it diverts the path of the current I_L drawn by the motor. A part of the current I_D flows through diverter and the current flowing through the series field winding is reduced which reduces the flux ϕ. Consequently, the speed of the motor is increased ($N \propto I/\phi$). Thus by this method, only speeds above normal can be obtained.
2. **Armature diverter:** In this method, variable resistance R is connected in parallel with the armature is shown in Fig. 5.39. Its effect is that it diverts the path of the line current I_L. Some of the current I_D flows through the diverter and reduces the armature current I_a. For a given constant load torque, if I_a is reduced then ϕ must increase ($T \propto \phi I_a$). This results in increase in current drawn by the motor and a fall in speed ($N \propto 1/\phi$). By adjusting the value of diverter resistance, any speed below normal can be obtained by this method.

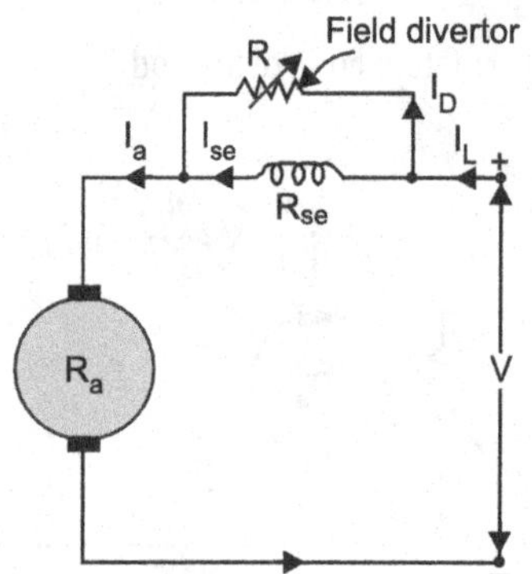

Fig. 5.38 Circuit diagram for field divertor method

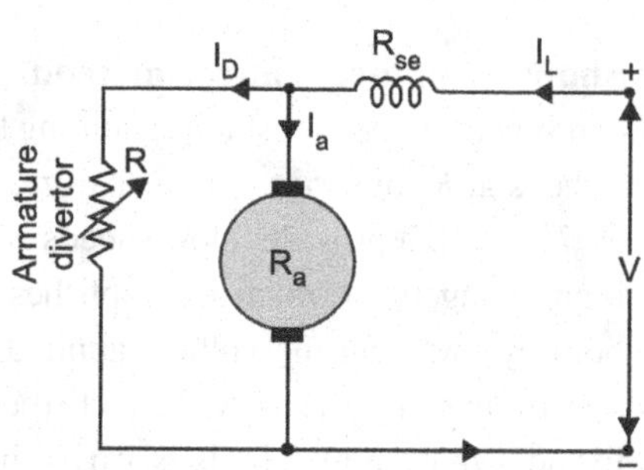

Fig. 5.39 Circuit diagram for armature diverter

3. **Tapped field control:** In this method, the number of turns of the series field winding can be changed by short circuiting a part of it as shown in Fig. 5.40. We know that flux produced by the winding depends upon the ampere-turns (i.e., $\phi \propto I_{se} \times$ *No. of turns*). As the number of turns are reduced, the speed of the motor is increased ($N \propto 1/\phi$). Thus, only speeds above normal can be obtained by this method.

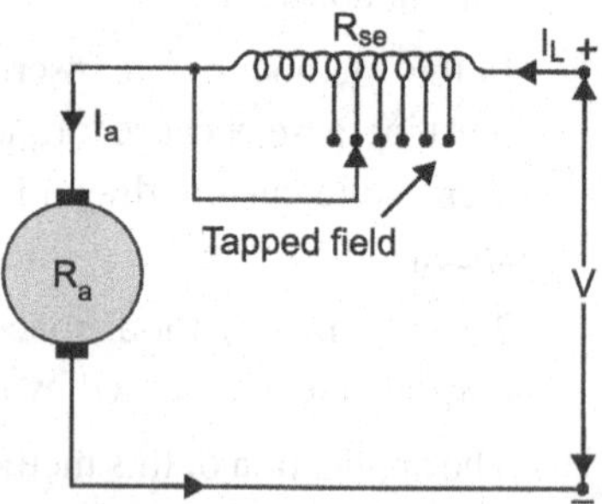

Fig. 5.40 Circuit diagram for tapped field

5.24.3 Voltage Control Method

In this method, the voltage across the series motors are changed by connecting them in series or in parallel or the combination of both. This in widely used in electric traction.

To explain this method, let us consider only two similar series motors (for simplicity) whose shafts are mechanically coupled. Firstly they are connected is series and then in parallel as shown in Figs. 5.41(*a*) and (*b*), respectively. The current flowing through each motor and the voltage across it is shown in Fig. 5.41.

When in series:

$$\text{Speed, } N_1 \propto \frac{E_{b1}}{\phi_1} \propto \frac{V/2 - I_a(R_a + R_{se})}{I_a}$$

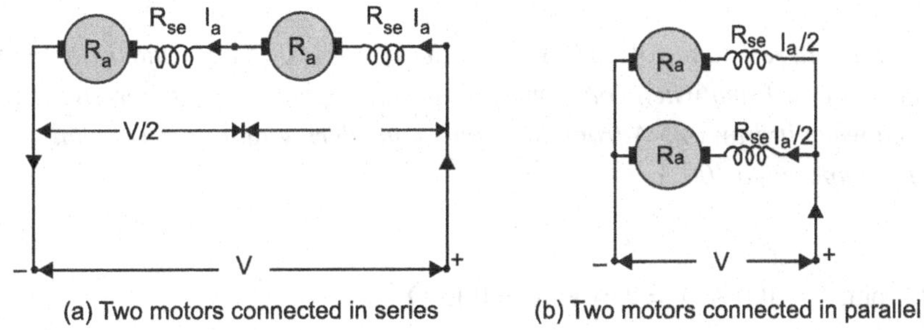

(a) Two motors connected in series (b) Two motors connected in parallel

Fig. 5.41 Voltage control method

Neglecting drops, $$N_1 \propto \frac{V/2}{I_a} \propto \frac{V}{2I_a} \quad \text{...(5.19)}$$

When in parallel: $$\text{Speed, } N_2 \propto \frac{E_{b_2}}{\varphi_2} \propto \frac{V - \frac{I_a}{2}(R_a + R_{se})}{I_a/2} \propto \frac{2V}{I_a} \quad \text{...(5.20)}$$

From eq. (*i*) and (*ii*), we get, $$\frac{N_2}{N_1} = \frac{2V}{I_a} \times \frac{2I_a}{V} = 4 \text{ or } N_2 = 4N_1 \quad \text{...(5.21)}$$

Thus, when the motors are connected in series, low speeds are obtained and when they are connected in parallel high speeds (nearly 4 times to that of first case) are obtained. Many speeds can be obtained by having more number of motors and connecting them in series, in parallel or the combination of both.

Usually, this method is employed in electric locomotives for controlling the speed of trains. In this case, combination of series-parallel and resistance control method is employed. The sequence of connections are shown in Fig. 5.42:

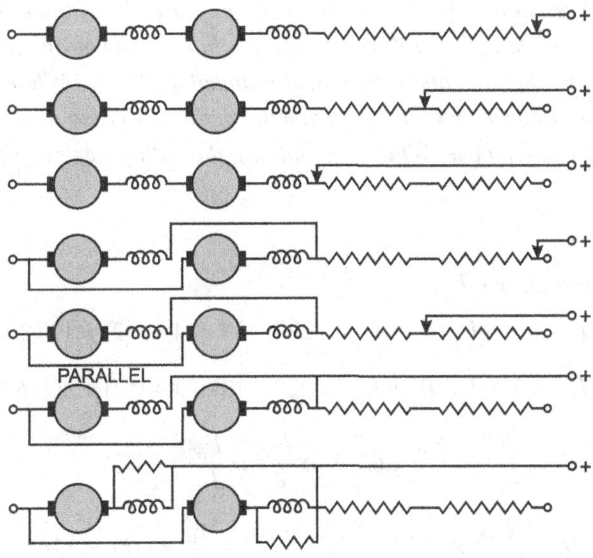

Fig. 5.42 Combination of series-parallel and resistance control of DC series motors

Example 5.29

The armature and field resistance of a 230 V DC series motor is 0.15 ohm and 0.1 ohm respectively. It runs at a speed of 800 rpm when connected to rated voltage drawing a current of 100 A. What will be its speed when it draws 25 A from the supply considering that its flux is only 45 percent at this current as compared to 100 A.

Solution:

Here $N_1 = 800$ rpm; $I_1 = 100$ A; $V = 230$ V; $R_a = 0.15\ \Omega$

$$R_{se} = 0.1\ \Omega;\ I_2 = 25\ \text{A};\ \phi_2 = 0.45\ \phi_1$$

We know, $$N \propto \frac{E_b}{\phi} \quad \text{or} \quad N_1 \propto \frac{E_{b1}}{\phi_1}$$

and $$N_2 \propto \frac{E_{b2}}{\phi_2} \quad \text{or} \quad \frac{N_2}{N_1} = \frac{E_{b2}}{\phi_2} \times \frac{\phi_1}{E_{b1}}$$

Where, $$E_{b_1} = V - I_{a_2}(R_a + R_{se})$$

$$= 230 - 100\ (0.15 + 0.1) = 230 - 25 = 205\ \text{V}$$

$$E_{b2} = V - 1_{a2}(\text{R}_a + \text{R}_{se})$$

$$= 230 - 25\ (0.15 + 0.1) = 230 - 6.25 = 223.75\ \text{V}$$

$\therefore$ $$\text{Speed, } N_2 = \frac{E_{b2}}{\phi_2} \times \frac{\phi_1}{E_{b1}} \times N_1$$

$$\frac{223.75}{0.45\phi_1} \times \frac{\phi_1}{205} \times 800 = \mathbf{1940.4\ rpm}\ (\textit{Ans.})$$

Example 5.30

The field and armature resistance of a 220 V DC series motor is 0.2 ohm and 0.4 ohm. At full-load it draws 38 A and runs at a speed of 600 rpm. Assume that brush voltage drop is 3 volt. (i) What will be the speed of motor when its intake current is reduced to 19 A (ii) What will be its speed if the load is removed and the motor draws only 1 A from the mains. (iii) Also determine the horse power developed by the motor in each case. While determining the values neglect effect of saturation and armature reaction.

Solution:

Since saturation is neglected, $\phi \propto I_a$

$$E_{b1} = V - I_{a1}(R_a + R_{se}) = 220 - 38\ (0{\cdot}4 + 0{\cdot}2) = 197{\cdot}2\ \text{V}$$

$$E_{b2} = V - I_{a2}(R_a + R_{se}) = 220 - 19\ (0{\cdot}4 + 0{\cdot}2) = 208{\cdot}6\ \text{V}$$

Now, $$N_1 \propto \frac{E_{b1}}{\phi_1} \alpha \frac{E_{b1}}{I_{a1}} \text{ and } N_2 \alpha \frac{E_{b2}}{\phi_2} \alpha \frac{E_{b2}}{I_{a2}}$$

$$N_2 = \frac{E_{b2}}{I_{a2}} \times \frac{I_{a1}}{E_{b1}} \times N_1 = \frac{208.6 \times 38}{19 \times 197.2} \times 600 = \mathbf{1269{\cdot}37\ rpm}\ (\textit{Ans.})$$

$$E_{b3} = V - I_{a3}(R_a + R_{se}) = 220 - 1\ (0{\cdot}4 + 0{\cdot}2) = 219{\cdot}4\ \text{V}$$

$$\therefore \qquad N_3 = \frac{E_{b3}}{I_{a3}} \times \frac{I_{a1}}{E_{b1}} \times N_1 = \frac{219.4 \times 38}{1 \times 197.2} \times 600 = \mathbf{25366\ rpm}\ (Ans.)$$

Internal horse power developed in each case;

$$HP_1 = \frac{E_{b1}\, I_{a1}}{735.5} = \frac{197.2 \times 38}{735.5} = \mathbf{10{\cdot}19}\ (Ans.)$$

$$HP_2 = \frac{E_{b2}\, I_{a2}}{735 \cdot 5} = \frac{208 \cdot 6 \times 19}{735 \cdot 5} = \mathbf{5{\cdot}39}\ (Ans.)$$

$$HP_3 = \frac{E_{b3}\, I_{a3}}{735 \cdot 5} = \frac{219 \cdot 4 \times 1}{735 \cdot 5} = \mathbf{0{\cdot}298}\ (Ans.)$$

Example 5.31

A series motor with unsaturated magnetic circuit and with negligible resistance when running at a certain speed, on a given load takes 50 A at 500 V. If the load torque varies as cube of the speed, find the resistance necessary to reduce the speed by 20%.

Solution:

In series motors, $T \propto \phi I_a;\ T \propto,\ I_a^2$ $\qquad (\because\ \phi \propto I_a)$

Given that, $T \propto N^3$

$$\therefore \qquad I_{a1}^2 \propto N_1^3 \text{ and } I_{a2}^2\ N_1^3$$

or
$$\frac{I_{a2}}{I_{a1}} = \left(\frac{N_2}{N_1}\right)^{1.5}$$

or
$$I_{a2} = \left(\frac{N_2}{N_1}\right)^{1.5} \times I_{a1}$$

Now, $N_2 = 0{\cdot}8\ N_1$

$$I_{a2} = (0{\cdot}8)^{1{\cdot}5} \times 50 = 35{\cdot}777 \text{ A}$$

Circuit resistance in first case, $R_1 = \dfrac{V}{I_{a1}} = \dfrac{500}{50} = 10\Omega$

Circuit resistance in second case, $R_2 = \dfrac{V}{I_{a2}} = \dfrac{500}{35.777} = 13.975\Omega$

Additional resistance required to reduce the speed;

$$R = R_2 - R_1 = 13{\cdot}975 - 10 = \mathbf{3{\cdot}975}\ \Omega\ (Ans.)$$

Example 5.32

A 240 V, DC series motor has a resistance of 0.2 ohms, when the line current is 40 A, speed is 1800 rpm. Find the resistance to be added in series with the motor (i) to make the motor run at 900 rpm when line current is 60A (ii) to limit the speed to 3600 rpm when line current is 10 A (iii) speed of motor when it is connected directly to mains and line current is 60 A. Assume flux at 60A is 1.8 times of flux at 40A.

Solution:

Here, $V = 240$ V; $R_m = 0.2\ \Omega$; $I_{a1} = 40$ A; $N_1 = 1800$ rpm

$$\text{Back emf } E_{b1} = V - I_{a1}R_m = 240 - 40 \times 0.2 = 232 \text{ V}$$

$$\text{For } I_{a1} = 40 \text{ A; flux} = \phi_1$$

$$\text{For } I_{a2} = 60 \text{ A; flux} = \phi_2 \text{ where } \phi_2 = 1.8\ \phi_1$$

(*i*) $N_2 = 900$ rpm; $I_{a2} = 60$ A; then $\phi_2 = 1.8\ \phi_1$

Let R be the resistance added in motor circuit

$$E_{b2} = V - I_{a2}(R_m + R) = 240 - 60\,(0.2 + R)$$

$$= 240 - 12 - 60\,R = 228 - 60\text{ R}$$

Now $$N \propto \frac{E_b}{\phi} \quad \text{or} \quad \frac{N_2}{N_1} = \frac{E_{b2}}{\phi_2} \times \frac{\phi_1}{E_{b1}}$$

or $$\frac{900}{1800} = \frac{(228 - 60R)}{1.8\ \phi_1} \times \frac{\phi_1}{232}$$

or $$228 - 60\text{ R} = \frac{1.8 \times 232 \times 900}{1800}$$

or $$228 - 60\text{R} = 208.8 \quad \text{or} \quad R = \mathbf{0.32\ \Omega}\ (Ans.)$$

(*ii*) $N_3 = 3600$ rpm; $I_{a3} = 10$ A; $\phi_3 \propto I_{a3}$ and $\phi_1 \propto I_{a1}$

Let R_1 be the resistance added in series with the motor

$$E_{b3} = V - I_{a3}(R_m + R_1) = 240 - 10\,(0.2 + \text{R}_1)$$

$$= 240 - 2 - 10\text{ R}_1 = 238 - 10\text{ R}_1$$

$$N \propto \frac{E_b}{\phi};\ \frac{N_3}{N_1} = \frac{E_{b3}}{\phi_3} \times \frac{\phi_1}{E_{b1}}$$

$$\frac{N_3}{N_1} = \frac{E_{b_3}}{E_{b_3}} \times \frac{I_{a1}}{I_{b_1}}$$ (Considering linear magnetisation)

$$\frac{3600}{1800} = \frac{(238 - 10R_1)}{10} \times \frac{40}{232}$$

or $$238 - 10\text{R}_1 = \frac{3600 \times 10 \times 232}{1800 \times 40}$$

or $$238 - 10\text{R}_1 = 116 \text{ or } R_1 = \mathbf{12.2\ \Omega}\ (Ans.)$$

(*iii*) $N_4 = ?$; $I_{a4} = 60$ A and $\phi_4 = 1.8\ \phi_1$

$$E_{b4} = V - I_{a4}(R_m) = 240 - 60 \times 0.2 = 228 \text{ V}$$

$$\frac{N_4}{N_1} = \frac{E_{b4}}{\phi_2} \times \frac{\phi_1}{E_{b1}}$$

or $$N_4 = \frac{228}{1.8\phi_1} \times \frac{\phi_1}{232} \times 1800$$

or $$N_4 = \mathbf{982.76\ rpm}\ (Ans.)$$

<u>**Example 5.33**</u>

A 400 V series motor has a total resistance of 0·25 ohm, when running at 1200 rpm it draws a current 25A. When a regulating resistance of 2·75 ohm is included in the armature circuit, it draws a current of 15A. Find the speed and ratio of the two mechanical outputs. Assume that the flux with 15A is 70% of that with 25A.

Solution:

$$E_{b1} = V - I_{a1}(R_a + R_{se}) = 400 - 25 \times 0{\cdot}25 = 393{\cdot}75 \text{ V}$$

$$E_{b2} = V - I_{a2}[R_a + R_{se}) + R] = 400 - 15\,[0{\cdot}25 + 2{\cdot}75] = 355 \text{ V}$$

$$\phi_2 = 0{\cdot}75\ \phi_1$$

Now, $$N_1 \propto \frac{E_{b1}}{\phi_1} \text{ and } N_2 \propto \frac{E_{b2}}{\phi_2}$$

$\therefore$ $$N_2 = \frac{Eb_2}{\phi_2} \times \frac{\phi_1}{Eb_1} \times N_1 = \frac{355}{0.75\phi_1} \times \frac{\phi_1}{393.75} \times 1200 = \mathbf{1442{\cdot}54\ rpm}\ (Ans.)$$

Mechanical power developed in two cases;

$$P_1 = E_{b1} I_{a1} = 393{\cdot}75 \times 25 = \mathbf{9843{\cdot}75\ W}\ (Ans.)$$

$$P_2 = E_{b2} I_{a2} = 355 \times 15 = \mathbf{5325\ W}\ (Ans.)$$

<u>**Example 5.34**</u>

The field and armature resistance of a 200 V DC series motor is 0.3 ohm and 0.5 ohm respectively. The motor runs at a speed of 700 rpm when drawing a current of 25 A. What value of additional series resistance is required to reduce the speed to 350 rpm keeping intake current constant.

Solution:

Motor input current, $I_1 = 25$ A

Back emf, $E_{b1} = V - I_1(R_a + R_{se}) = 200 - 25\,(0.5 + 0.3) = 180$ V

Speed, $N_1 = 700$ rpm

Let R be the resistance connected in series with the motor to reduce the speed to

$$N_2 = 350 \text{ rpm}$$

Motor input current, $I_2 = I_1 = 25$ A

Back emf, $E_{b2} = V - I_2(R + R_a + R_{se})$

$= 200 - 25\,(R + 0.5 + 0.3) = 180 - 25$ R

Since $$N \propto \frac{E_b}{\Phi},$$

$$\frac{N_2}{N_1} = \frac{E_{b2}}{E_{b1}} \times \frac{\Phi_1}{\Phi_2} = \frac{E_{b2}}{E_{b1}}$$ ($\phi_2 = \phi_1$ as field current remains the name)

or $$\frac{350}{700} = \frac{180 - 25R}{180}$$

or $$R = \mathbf{3.6\ ohm}\ (Ans.)$$

Example 5.35

The field and armature resistance of a 500 V DC series motor is 0.2 ohm and 0.3 ohm respectively. The motor runs at 500 rpm when drawing a current of 49 A. If the load torque varies as the square of the speed, determine the value of the external resistance to be added in series with the armature for the motor to run at 450 rpm Assume linear magnetisation.

Solution:

Here, $N_1 = 500$ rpm; $I_{a1} = 40$A; $V = 600$ V; $N_2 = 450$ rpm

$$T \propto N^2;\ R_a = 0.3\ \Omega;\ R_{se} = 0.2\ \Omega;\ \phi \propto I_a$$

$$E_{b1} = V - I_{a1}(R_a + R_{se}) = 600 - 40\ (0.3 + 0.2) = 580\ \text{V}$$

$$T \propto \phi I_a \text{ or } T \propto I_a^2$$

and $$T \propto N^2$$

$\therefore$ $$I_a^2 = N^2$$

and $$\left(\frac{I_{a2}}{I_{a1}}\right)^2 = \left(\frac{N_2}{N_1}\right)^2$$

or $$I_{a2} = \frac{N_2}{N_1} \times I_{a1} = \frac{450}{500} \times 40 = 36\ \text{A}$$

Let R be the resistance added in series with the armature

$$E_{b2} = V - I_{a2}(R_a + R_{sc} + R) = 600 - 36\ (0.3^0 + 0.2 + R) = 582 - 36R$$

Now $$\frac{N_2}{N_1} = \frac{E_{b2}}{\phi_2} \times \frac{\phi_1}{E_{b1}} \text{ or } \frac{N_2}{N_1} = \frac{E_{b2}}{I_{a2}} \times \frac{I_{a1}}{E_{b1}}$$

or $$\frac{450}{500} = \frac{(582 - 36R)}{36} \times \frac{40}{580}$$

or $$582 - 36R = \frac{450 \times 36 \times 580}{500 \times 40}$$

or $$582 - 36\text{R} = 469.8 \text{ or } R = \mathbf{3.117\ \Omega}\ (Ans.)$$

Example 5.36

A 240 V DC series motor takes 40 A when giving its rated output at 1200 rpm. The combined value of armature and series field resistance is 0.3 Ω. Find what resistance must be added to obtain the rated torque (a) at starting (b) at 800 rpm.

Solution:

Rated voltage, $V = 240$ V

Rated current, $I_1 = 40$ A

Rated speed, $N_1 = 1200$ rpm

Back emf, $E_{b1} = V - I_a R_m = 240 - 40 \times 0.3 = 228$ V

(*i*) At start when speed is zero, the back emf E_{b2} will be zero and rated torque will be developed when current drawn by the motor is of rated value i.e.,

Current, $I_2 = 40$ A

Back emf, $E_{b2} = V - I_2 (R_1 + R_m)$

$$\therefore \quad \text{additional resistance, } R_1 = \frac{V - E_{b2}}{I_2} - R_m = \frac{240 - 0}{40} - 0.3 = \mathbf{5.7\ ohm}\ (Ans.)$$

(*ii*) At 800 rpm, the rated torque will be developed only when current drawn by the motor is rated one i.e.,

$$I_3 = 40 \text{ A}$$

Since field current remains the same, back emf will be proportional to speed i.e.,

$$E_{b2} = E_{b_1} \times \frac{N_3}{N_1} = 228 \times \frac{800}{1200} = 152 \text{ V}$$

$$\therefore \quad \text{required additional resistance, } R_2 = \frac{V - E_{b3}}{I_3} - R_m = \frac{240 - 152}{40} - 0.3 = \mathbf{1.9\ \Omega}\ (Ans.)$$

Example 5.37

A DC series motor, connected to a 440 V supply, runs at 1500 rpm when taking a current of 50 A. Calculate the value of resistance which when inserted in series with motor will reduce the speed to 1000 rpm, the gross torque being then half the previous value. The resistance between the terminals of the motor is 0.2 Ω. Assume linear magnetisation.

Solution:

Here, $V = 440$ V; $I_1 = 50$ A; $R_m = 0.2\ \Omega$; $N_1 = 1500$ rpms; $N_2 = 1000$ rpm

Back emf, $E_{b1} = V - I_1 R_m = 440 - 50 \times 0.2 = 430$ V

Let R be the resistance connected in series to reduce the speed.

$$\text{Now } T_2 = \frac{1}{2} T_1 \text{ (given)}$$

or $$I_2 \phi_2 = \frac{1}{2} I_1 \phi_1 \qquad \text{(since } T \propto \phi I\text{)}$$

or $$I_2^2 = \frac{1}{2} I_1^2 \text{ (since } \phi \propto I\text{)}$$

or $$I_2 = \sqrt{\frac{1}{2}}\, I_1 = \frac{50}{\sqrt{2}} = 35.35 \text{ A}$$

Back emf, $E_{b2} = V - I_2(R + R_m) = 440 - 35.35\,(0.2 + R) = 432.93 - 35.35$ R

We know, $$\frac{E_{b2}}{E_{b1}} = \frac{N_2}{N_1} \times \frac{\phi_2}{\phi_1},$$

$$\frac{432.93 - 35.35R}{430} = \frac{1000}{1500} \times \frac{35.35}{50} \qquad \left(\text{since } \frac{\phi_2}{\phi_1} = \frac{I_2}{I_1} = \frac{35.35}{50}\right)$$

or $$R = \mathbf{6.51\ \Omega}\ (\textit{Ans.})$$

Example 5.38

A DC series motor has the following rating: 200 V, 20 A and 1200 rpm. Armature and series field resistances are 0.1 Ω and 0.2 Ω respectively. Magnetic circuit can be assumed to the linear.

At what speed the motor will run at rated torque if a resistance of 20 Ω is placed in parallel with the armature?

Solution:

Here, $V = 200$ V; $I_1 = 20$ A; $R_a = 0.1$ Ω; $R_{se} = 0.2$ Ω; $N_1 = 1200$ rpm; $R_D = 20$ Ω

$$\text{Back emf, } E_{b1} = V - I_1(R_a + R_{se})$$

$$= 200 - 20\ (0.1 + 0.2) = 214 \text{ V}$$

When a resistance of 20 Ω is connected parallel to armature, as shown in Fig. 5.43, let the motor input current be I_2 ampere.

$$\text{Voltage across armature} = V - I_2R_{se} = 200 - 0.2\ I_2$$

$$\text{Diverter current, } I_D = \frac{200 - 0.2I_2}{20} = 10 - 0.01\ I_2$$

Armature current, $$I_{a2} = I_2 - I_D = I_2 - (10 - 0.01\ I_2) = 1.01\ I_2 - 10$$

At rated torque i.e., $$T_2 = T_1$$

or $$I_{a2}\Phi_2 = I_{a1}\Phi_1 \text{ (since } T \propto \Phi I_a)$$

or $$I_{a2} = I_{a1} \times \frac{\Phi_1}{\Phi_2}$$

or $$1.01\ I_2 - 10 = I_1 \times \frac{I_1}{I_2} \text{ (series } \Phi \propto I)$$

or $$1.01\ I_2 - 10 = \frac{20 \times 20}{I_2}$$

or $$I_2 = 25.45 \text{ A}$$

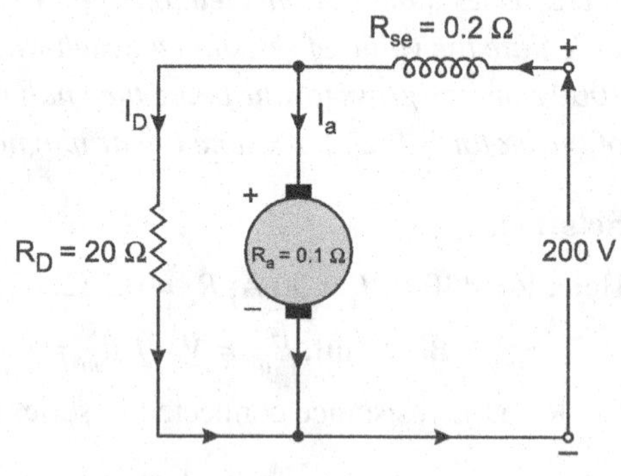

Fig. 5.43 Circuit diagram

$$\text{Armature current, } I_{a2} = 1.01\ I_2 - 10 = 1.01 \times 25.45 - 10 = 15.7 \text{ A}$$

$$\text{Back emf, } E_{b2} = V - I_2 R_{se} - I_{a2} R_a = 200 - 25.25 \times 0.2 - 15.7 \times 0.1 = 193.34 \text{ V}$$

$$\text{Then, } N_2 = N_1 \times \frac{E_{b2}}{E_{b1}} \times \frac{\Phi_1}{\Phi_2} = 1200 \times \frac{193.34}{214} \times \frac{20}{25.45} = \mathbf{852\ rpm}\ (\textit{Ans.})$$

Example 5.39

A series motor runs at 800 rpm when taking a current of 60 A at 460 V. The resistance of the armature circuit is 0.2 Ω and that of the field winding is 0.1 Ω. Calculate the speed when a 0.15 Ω diverter is connected in parallel with the field winding. Assume the torque to remain unaltered and the flux to be proportional to the field current.

Solution:

At normal conditions, when no diverter is connected

Armature current, $I_{a1} = I_{se1} = I_{L1} = 60$ A

$$E_{b1} = V - I_{a1} R_a - I_{se1} R_{se} = 460 - 60 \times 0.2 - 60 \times 0.1 = 442 \text{ V}$$

When a diverter of 0.15 Ω is connected parallel to series field winding as shown in Fig. 5.44, let the speed be N_2 and the line current be I_{L2}.

Armature current, $I_{a2} = I_{L2}$

Series field current, $I_{se2} = I_{L2} \dfrac{R_D}{R_{se} + R_D} = I_{L2} \dfrac{0.15}{0.1 + 0.15} = \dfrac{3}{5} I_{L2}$

Since torque remains the same,

$$T_2 = T_1$$

or $$I_{a2} \Phi_2 = I_{a1} \Phi_1 \quad (\text{Since } \Phi \propto I_{se})$$

or $$I_{a2} I_{se2} = I_{a1} I_{se1}$$

or $$I_{L2} \times \frac{3}{5} I_{L2} = 60 \times 60$$

or $$I_{L2} = \sqrt{\frac{3600 \times 5}{3}} = 77.46 \text{ A}$$

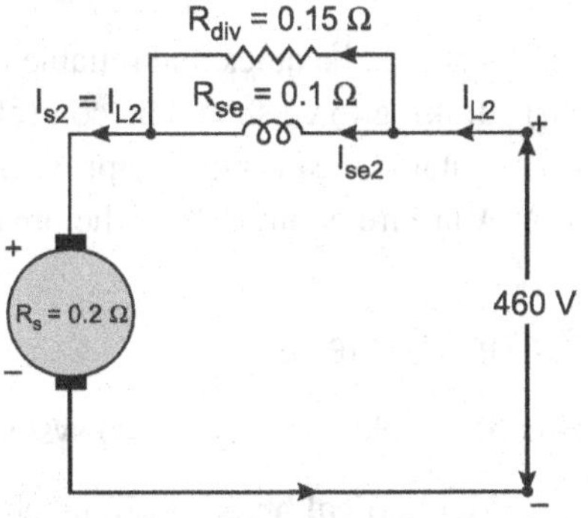

Fig. 5.44 Circuit diagram

$$I_{se2} = \frac{3}{5} \times I_{L2} = \frac{3}{5} \times 77.46 \text{ A} = 46.48 \text{ A}$$

Back emf $E_{b2} = V - I_{a2} R_a - I_{se2} R_{se} = 460 - 77.46 \times 0.2 - 46.48 \times 0.1 = 440.06$ V

$$\text{Speed, } N_2 = N_1 \times \frac{E_{b2}}{E_{b1}} \times \frac{\Phi_1}{\Phi_2}$$

$$= N_1 \times \frac{E_{b2}}{E_{b1}} \times \frac{I_{se1}}{I_{se2}} = 800 \times \frac{440.06}{442} \times \frac{46.48}{60}$$

= **617.6 rpm** (*Ans.*)

5.25 Electric Braking

When load is removed from an electric motor and supply is disconnected, it continues to run for some time due to inertia. This continuous rotation may cause damage to equipment or to the manufactured product and even loss of human life. Therefore, quick stopping of a motor is essential in such cases.

Thus, *the method by which an electric motor or drive is brought to rest quickly when load on it is removed and supply is disconnected is called* **braking.**

Braking may be employed to bring the drive to rest or it may be employed to lower down the speed when load is reduced. The device employed for braking has to absorb the kinetic and potential energy of the moving parts.

Objectives of Braking

While stopping the drive, braking may be used for any one of the following objectives:

(*i*) Reducing the time of stopping i.e., reducing the time of continuous running when drive is switched – *OFF.*

(*ii*) Stopping the drive exactly at specified point, e.g., in lifts; for safety reasons, the drive has to be stopped at a specific point.

(*iii*) For feeding back some of the energy to the supply system.

Essential Features of Braking

(*i*) It should be quick and reliable in action.

(*ii*) Braking force should be controllable.

(*iii*) Suitable means must be provided for dissipating the stored energy of the rotating parts.

(*iv*) A failure of any part of the breaking system should result in the application of brakes.

Braking Systems

Broadly, in electric drives, two systems of braking can be applied:

1. **Mechanical braking:** In mechanical braking the stored energy of rotating parts is dissipated in the form of heat where frictional torque acts against the rotation. The motor may be stopped by using a brake shoe or brake band or brake drum.
2. **Electrical braking:** In electrical braking, the kinetic energy of the motor is converted into electrical energy and is dissipated as heat in a resistance or returned to the supply system.

Advantages and Disadvantages of Electrical Braking over Mechanical Braking

Before dealing with electric braking, let us try to find out the advantages and disadvantages of electric braking over mechanical braking.

Advantages

(*i*) *Saving; no replacement cost*: In case of mechanical braking, due to friction there is excessive wear on brake lining which needs frequent and costly replacement. But in electric braking no such replacement is required hence saves money.

(*ii*) *Small maintenance cost*: In case of mechanical braking there is more wear and tear which increases the maintenance cost. There is no wear and tear in case of electric braking which reduces the maintenance cost to very low value.

(iii) *Energy saving*: In regenerative braking, a part of energy is returned to the supply this reduces the operating cost. This is only possible in electric braking.

(iv) *Reliability:* In case of mechanical braking, the heat produced at brake blocks may result in failure of brakes. But in case of electric braking, the heat produced does not affect the braking system this improves its reliability.

(v) *Smooth operation*: Mechanical braking may produce jerks but electric braking never produces jerks.

(vi) *System Capacity:* By employing electric braking the capacity of the system can be increased by way of higher speeds and handling of heavy loads.

Disadvantages or Limitations

(i) *It cannot hold the machine or load stationary:* In case of electric braking, the driving motor starts working as a generator during braking period. But at stand-still the machine (motor) cannot work as generator. Therefore, an electric brake can reduce the speed of the machine or load and bring it to almost at stand-still condition but it cannot hold it stationary. Hence, electric brake alone cannot fulfil the purpose, friction (mechanical) brake is always provided in addition to it.

(ii) *Cost:* Addition equipment employed for electric braking increases the cost.

(iii) *Choice of motor is limited:* In case of electric braking system, the motor has to function as a generator and must have suitable braking characteristics. Thus, the choice of motor is limited.

Electric braking is superior to mechanical braking since it is much quicker and eliminates the cost of maintenance of mechanical brakes. However, in order to finally bring the motor to a stand still position and hold it there, friction brakes are also essential.

5.26 Types of Electric Braking

There are three types of electric braking.

(i) Plugging

(ii) Rheostatic braking

(iii) Regenerative braking.

5.26.1 Plugging

This is also called a counter-current braking. It is simplest type of braking and involves the reversing of the connections to the armature of the motor so that it tends to run in the reverse direction. In order to stop the machine, a special device must be installed to cut off the supply as soon as it comes to rest. This method can be applied to DC motors, induction motors and synchronous motors. It is particularly suitable where the drive has to be rapidly reversed such as in planning machines, etc.

(a) **Plugging applied to DC Shunt motors:** The direction of rotation of DC shunt motor can be reversed by reversing the current flowing through the armature. The expression for the armature current is $I_a = \dfrac{V - E_b}{R_a}$.

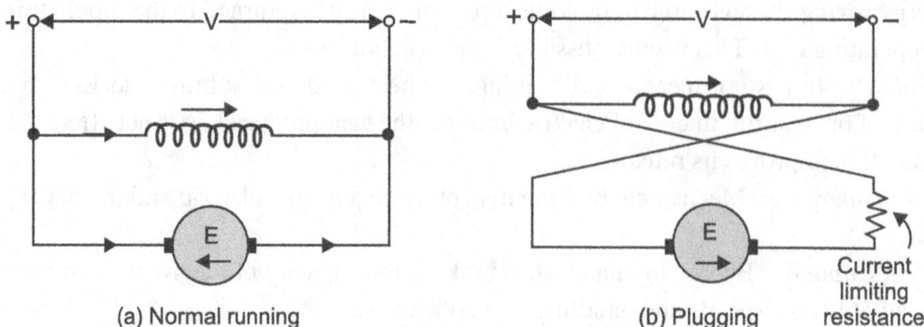

Fig. 5.45 Plugging with shunt motor

During normal running, the back *emf* E_b is opposite to the applied voltage 'V' but at the instant of reversal a voltage equal to $V + E_b$ is impressed across armature. This will cause a great rush of current (almost double the normal) in the armature circuit. To prevent this the starting resistance is reinserted in the circuit as shown in Fig. 5.45.

(*b*) **Plugging applied to DC series motors:** The method is same as in case of DC shunt motor. The changeover of connections is shown in Fig. 5.46.

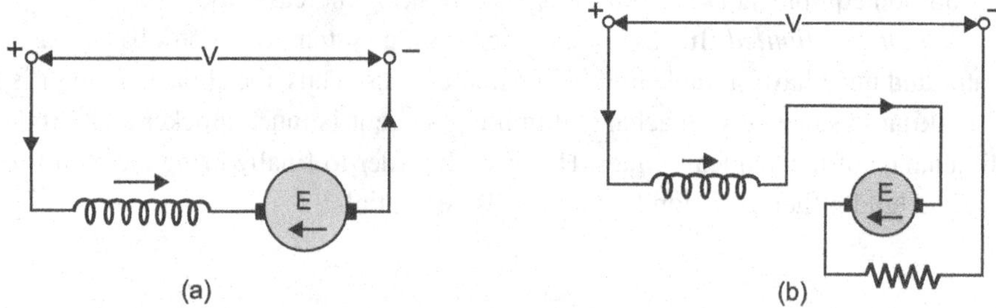

Fig. 5.46 Plugging with series motor

In both cases, the motor must be disconnected from the supply when it comes to rest otherwise it may rotate in the opposite direction.

5.26.2 Rheostatic Braking

This is also called a dynamic braking. With this method of braking the motor is disconnected from the supply and used as a generator driven by the momentum of the equipment to be braked and supplying current to resistance; DC or synchronous motors can be braked in this way. The induction motors, however, require a separate source of DC excitation during this type of braking. This type of braking is applied to different motors as explained below;

(a) **DC shunt motors:** The armature is disconnected from the supply and a resistance is connected across it, as shown in Fig. 5.47(*b*). The motor then works as a separately excited generator and the braking torque is applied by the current flowing in the resistance as shown in Fig. 5.47.

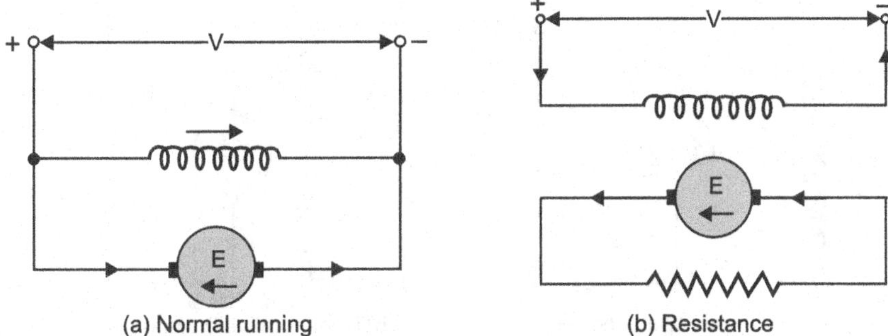

Fig. 5.47 Rheostatic braking of shunt motors

The magnitude of the braking effect can be controlled by varying either the resistance in the armature circuit or the field current. The braking is not effective if the supply fails during this period.

(b) **DC series motors:** The motor after being disconnected from the supply mains, works as a self excited series generator. Sometimes, in order to build up more field flux the connections of armature are reversed as shown in Fig. 5.48(*b*).

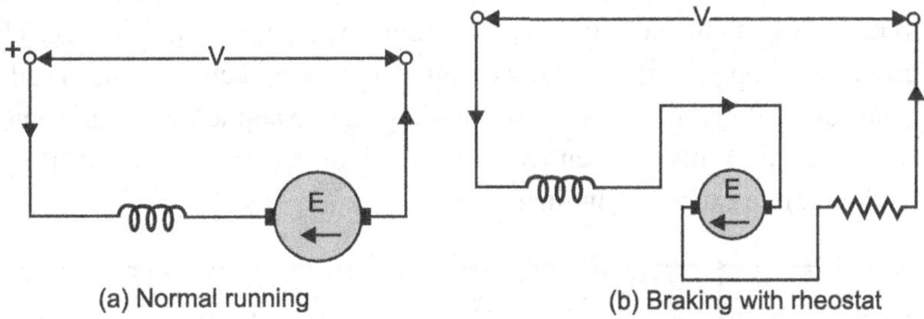

Fig. 5.48 Rheostatic braking of series motors

In such cases if the resistance is above a certain critical value equal to the initial slope of the current-voltage characteristics, the machine will not be self excited, therefore, resistance should be low enough.

5.26.3 Regenerative Braking

This type of braking is mostly employed in electric traction. The kinetic energy of the motor is converted into electrical energy, which is fed back to the supply lines instead of wasting it in a rheostat. This method can be applied to DC motors and induction motors.

(*a*) **DC shunt motors:** If the emf generated by the motor is greater than the supply voltage, power will be fed back to the supply. The emf in a shunt motor depends upon the field flux and speed. During this type of braking the field is disconnected from the armature and connected to a separate supply as shown in Fig. 5.49. The field current is so adjusted that E > V and this is returned to the mains.

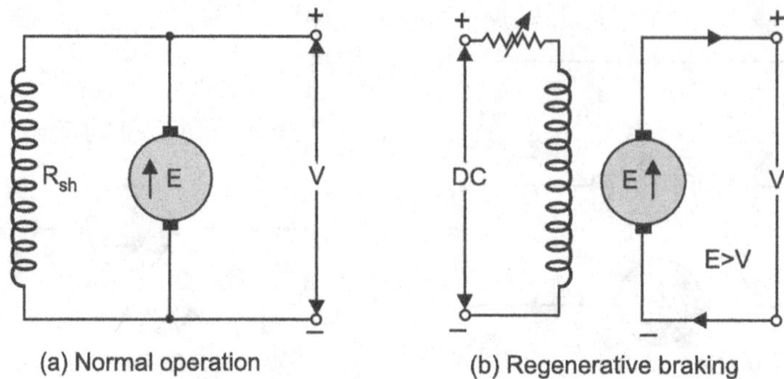

(a) Normal operation (b) Regenerative braking

Fig. 5.49 Regenerative braking with shunt motors

Alternatively, if the field excitation does not change but the load causes the speed to increase, the induced *emf* may become more than the supply voltage and power will be fed back to the supply.

(*b*) **DC series motors:** Regenerative braking cannot be applied to DC series motors in an ordinary way. For regenerative braking, the motor should work as a generator this is achieved by reversing the connections of series field. But even if the field connections are reversed at the exact moment, this method would still be not successful, because at the instant of reversal, the emf induced in the motor will be small, so the current will flow through the field winding in wrong direction, which will reverse the field and cause the motor emf to help the supply voltage. This will result in short circuit of supply. Due to these complications, this method is not used for common industrial purposes. However, regenerative braking can be applied with series motors used for traction purposes but in this case, either the field windings are modified or the machines are supplied with separate exciting windings.

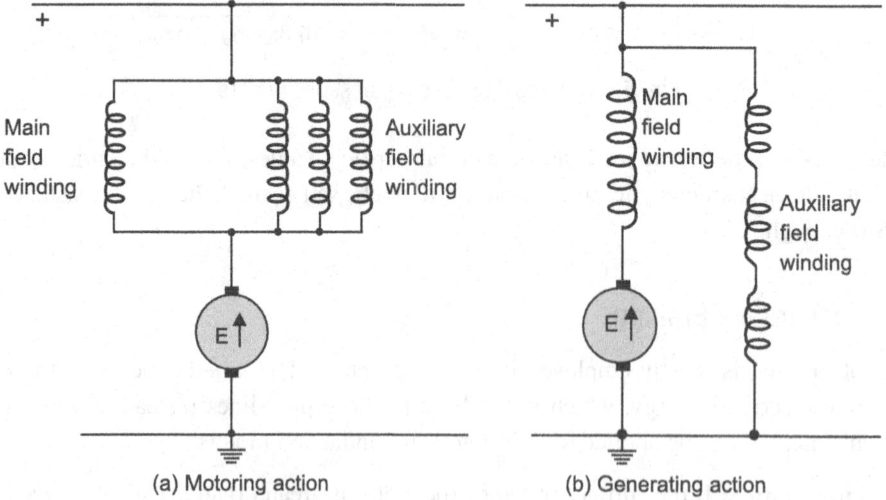

(a) Motoring action (b) Generating action

Fig. 5.50 Regenerative braking with series motors

One method of regenerative braking with series motors is French method. If there is a single series motor, it is provided with one main series field winding and 2 or 3 auxiliary field windings.

During normal running the main and auxiliary field windings are placed in parallel with each other as shown in Fig. 5.50(*a*). During regenerative braking, the auxiliary windings are placed in series with each other and combination is connected in parallel across the armature and main series field winding as shown in Fig. 5.50(*b*). In this case the machine works as a differentially compound generator and sends back power to supply lines.

In locomotives there are four to six motors and therefore, no auxiliary windings are required. The regenerative braking can be applied in this case as shown in Fig. 5.51.

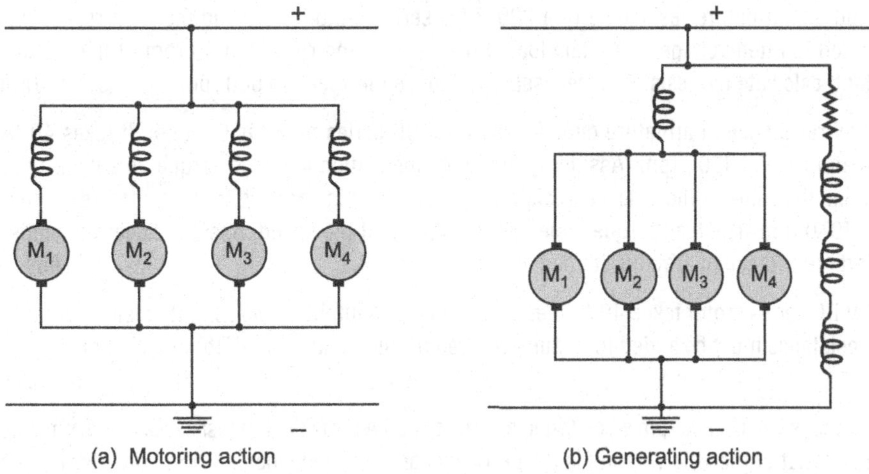

(a) Motoring action (b) Generating action

Fig. 5.51 Regenerative braking

An alternative method of regenerative braking is applied by using an exciter for controlling the excitation of the field windings during regeneration.

Section Practice Problems

Numerical Problems

1. A 220 V DC shunt motor takes armature current of 48 A and runs at 760 rpm. Find the resistance to be added in the field circuit to increase the speed to 950 rpm at an armature current of 78 A. Assume flux is proportional to field current. The armature and field winding resistance is 0.15 Ω and 240 Ω respectively. (**Ans.** *66.48* Ω)

2. A 500 V DC shunt motor is running at 700 rpm. Its speed is to be raised to 1000 rpm by weakening the field but the total torque remaining unchanged. The armature and field resistances are 0.8 Ω and 750 Ω respectively and the current at the lower speed is 12 A. Calculate the additional shunt field resistance required assuming the magnetic circuit to be unsaturated. Neglect all losses. (**Ans.** *330* Ω)

3. The armature resistance of 250 V DC shunt motor is 0.5 ohm. Its armature current is 20 A when running at 1200 rpm and develops full-load torque. What resistance must be inserted in series with the armature to reduce the speed to 600 rpm at the same torque, and what will be the speed if the load torque is halved with this resistance in the circuit. Assume the flux to remain constant throughout and neglect brush contact drop. (**Ans.** *6* Ω*, 925 6 rpm*)

4. A 230 V DC shunt motor draws a current of 50 A. Resistance of shunt field winding and the armature is 46 Ω and 0.02 Ω respectively. When there is a resistance of 0.6 Ω in series with the armature, the speed of motor is 800 rpm. What alternation must be made in the armature circuit to raise the speed to 850 rpm, the torque remaining the same.

(**Ans.** *0.6 Ω resistance connected in series with armature has to be reduced to 0.319 Ω*)

5. The armature resistance of a 220 V DC shunt motor is 0.5 ohm. Its armature draws a current of 30 A at full-load and runs at a speed of 800 rpm. If a resistance of 1.0 Ω is placed in the armature circuit, find the speed at (a) full-load torque (b) double full-load torque. (**Ans.** *683 rpm, 507 rpm*)

6. The field and armature resistance of a 220 V DC series motor is o.3 and 0.5 ohm respectively. When connected to rated voltage, at certain load it draws 25 A and runs at 1000 rpm. If the current remains constant, calculate the resistance necessary to reduce the speed to 500 rpm. (**Ans.** *3.6 Ω*)

7. The total resistance of armature circuit of a 600 V DC series motor is 0.5 ohm. It takes 40 A from the mains and runs at 1200 rpm. Assuming linear magnetisation and load torque as square of the speed, determine the value of the external resistance to be added in series with the armature for the motor to run at 1080 rpm. The load torque varies as the square of the speed. Assume linear magnetisation and take total resistance of armature circuit as 0.5 Ω. (**Ans.** *3.1167 Ω*)

8. A 240 V DC series motor takes 40 A when giving its rated output at 750 rpm. Its resistance is 0.3 Ω. Find what resistance must be added to obtain the rated torque (a) at starting (b) at 500 rpm.

(**Ans.** *5.7 Ω, 1.9 Ω*)

9. A series motor runs at 800 rpm when taking a current of 60 A at 460 V. The resistance of the armature circuit is 0.2 Ω and that of the field winding is 0.1 Ω. Calculate the speed when a 0.15 Ω diverter is connected in parallel with the field winding. Assume the torque to remain unaltered and the flux to be proportional to the field current. (**Ans.** *617.6 rpm*)

10. A DC series motor having a resistance of 0.4 Ω and runs at 2000 rpm when connected to 250 V supply and drawing a current of 20 A.

(*i*) What resistance must be added in order to limit the speed to 4000 rpm when armature current is 5A? Assume flux varies linearly.

(*ii*) What resistance must be added to limit the speed to 1000 rpm when armature current is 30 A? Assume that the flux at 30 A is 20% more than that at 20A.

(*iii*) At what speed the motor will run if it is connected directly to 250 V supply when the load is such that the current drawn is 30 A? (**Ans.** (*i*) *25.4 Ω*, (*ii*) *3.093 Ω*, (*iii*) *1640 rpm*)

Short Answer Type Questions

Q.1. List various methods of controlling speed of DC shunt motors.

Or

Name the methods used for speed control of DC motors.

Ans. (*i*) Field control (*ii*) Armature control (*iii*) Voltage control method

Q.2. What are the factors that affect the speed of a DC motor?

Ans. We know, speed, $N \propto \frac{E_b}{\phi} \propto \frac{V - I_a R_a}{\phi}$

Hence, the factors influencing the speed of a DC motor are:

(*i*) Applied or terminal voltage (*ii*) Flux per pole ϕ (*ii*) Resistance of armature circuit and (*iv*) armature current or load.

Q.3. In a field rheostat control method what factors do limit the increase of speed of DC shunt motors?

Ans. In field rheostat control method, the factors which limit the increase of speed of DC shunt motors are; (*i*) demagnetising effect of armature reaction on the field, (*ii*) commutation problems at high speeds, (*iii*) friction at bearings and (*iv*) large centrifugal forces developed at high speed.

Q.4. Which method is adopted to control the speed of a DC shunt motor above its base speed?

Ans. To obtain speeds above base speed, field control method is adopted.

Q.5. A DC shunt motor is running at rated speed with a resistance R in series with the armature, will the speed of the motor increase or decrease if additional resistance of finite value is connected (a) across R and (b) in series with R?

Ans. (*a*) When resistance *R* is connected across *R*, its effective value will decrease hence speed will increase

(*b*) On the other hand when a resistance *R* is connected in series with the existing resistance *R*, its effective value will increase, then the speed of motor will decrease.

Q.6. While controlling the speed of a DC shunt motor what should be done to achieve a constant torque drive?

Ans. We know $T \propto \phi I_a$, therefore, while controlling the speed of a DC shunt motor, the product value of field and armature current must remain the same.

Q.7. In the armature circuit of DC motor, an external resistance is introduced. How shall it affect the speed of the motor?

Ans. The speed of motor will decrease.

Q.8. List different methods of speed control employed for DC series motors.

Ans. The various methods of speed control of DC series motor are:

1. **Armature control methods.** These may be sub-divided as
   (*i*) Armature series resistance control method
   (*ii*) Shunted armature control method
   (*iii*) Armature terminal voltage control method
2. **Field control Methods.** These may be sub-divided as
   (*i*) Field diverter method.
   (*ii*) Armature diverter method
   (*iii*) Tapped field control method
3. **Voltage control Method**

Q.9. Do the series motors run faster in series combination or parallel combination?

Ans. Series motors will run faster in parallel combination.

Q.10. Why the power loss in control resistance for many applications of DC series motors is not too serious as in case of DC shunt motors?

Ans. Because in case of DC series motors, the control is utilised for long duration while reducing the speed under light load conditions and it is only employed intermittently while carrying full load.

Q.11. How plugging of a DC motor is done?

Ans. In DC motors plugging is obtained by reversing the supply terminals to the armature of the motor.

Q.12. Why rheostatic braking is economical to plugging?

Ans. In rheostatic braking, energy required from the supply to brake the motor is eliminated as compared to the plugging.

5.27 Losses in a DC Machine

A DC machine is used to convert mechanical energy into electrical energy or vice-versa. While doing so, the whole of input energy does not appear at the output but a part of it is lost in the form of heat in the surroundings. This wasted energy is called *losses in the machine.*

These losses affect the efficiency of the machine. A reduction in these losses leads to higher efficiency. Thus, the major objective in the design of a DC machine is to reduce these losses. The various losses occurring in a DC machine can be sub-divided as:

1. Copper losses.
2. Iron losses.
3. Mechanical losses

1. **Copper losses:** The various windings of the DC machine, made of copper, have some resistance. When current flows through them, there will be power loss proportional to the square of their respective currents. *These power losses are called* **copper losses.**

   In general, the various copper losses in a DC machine are:

   (*i*) Armature copper loss = $I_a^2\ R_a$

   (*ii*) Shunt field copper loss = $I_{sh}^2\ R_{sh}$

   (*iii*) Series field copper loss = $I_{se}^2\ R_{se}$

   (*iv*) Interpole winding copper loss = $I_i^2\ R_i = I_a^2\ R_i$

   (*v*) Brush contact loss = $I_a^2 Rb = 2 I_a\ v_b$

   (*vi*) Compensating winding copper loss = $I_a^2 R_c$

   The loss due to brush contact is generally included in armature copper loss.

2. **Iron losses:** *The losses which occur in the iron parts of a DC machine are called iron losses or core losses or magnetic losses.* These losses consist of the following:

   (*i*) **Hysteresis loss:** Whenever a magnetic material is subjected to reversal of magnetic flux, this loss occurs. It is due to retentivity (a property) of the magnetic material. It is expressed with reasonable accuracy by the following expression:

   $$P_h = K_h V f\ B_m^{1\cdot 6}$$

   where,

   K_h = hysteresis constant in J/m³ i.e., energy loss per unit volume of magnetic material during one magnetic reversal, its value depends upon nature of material;

V = Volume of magnetic material in m^3.

f = frequency of magnetic reversal in cycle/second and

B_m = maximum flux density in the magnetic material in Tesla.

It occurs in the rotating armature. To minimise this loss, the armature core is made of silicon steel which has low hysteresis constant.

(*ii*) **Eddy current loss:** When flux linking with the magnetic material changes (or flux is cut by the magnetic material) an emf is induced in it which circulates eddy currents through it. These eddy currents produce eddy current loss in the form of heat. It is expressed with reasonable accuracy by the expression:

$$P_e = K_e V f^2 t^2 B_m^2$$

where,

K_e = constant called co-efficient of eddy current, its value depends upon the nature of magnetic material;

t = thickness of lamination in m; V, f and B_m are the same as above.

The major part of this loss occurs in the armature core. To minimise this loss, the armature core is laminated into thin sheets (0·3 to 0·5 mm) since this loss is directly proportional to the square of thickness of the laminations.

3. **Mechanical losses:** As the armature of a DC machine is a rotating part, some power is required to overcome:

   (*i*) Air friction of rotating armature (windage loss).

   (*ii*) Friction at the bearing and friction between brushes and commutator (friction loss).

These losses are known as **mechanical losses.** To reduce these losses proper lubrication is done at the bearings.

5.28 Constant and Variable Losses

The losses in a DC machine may also be sub-divided into two categories, i.e.,

1. Constant losses; 2. Variable losses.

1. **Constant losses:** The losses in a DC machine which remain the same at all loads are called *constant losses.* The constant losses in a DC machine are:

   (*i*) Iron losses; (*ii*) Mechanical losses; (*iii*) Shunt field copper losses.

2. **Variable losses:** The losses in a DC machine which vary with load are called *variable losses.* The variable losses in a DC machine are:

   (*i*) Armature copper loss; (*ii*) Series field copper loss; (*iii*) Interpole winding copper loss.

5.29 Stray Losses

The sum of the iron loss and mechanical loss in a DC machine are known as *stray losses* i.e.,

Stray losses = Iron loss + Mechanical loss.

5.30 Power Flow Diagram

(i) When machine is working as a generation

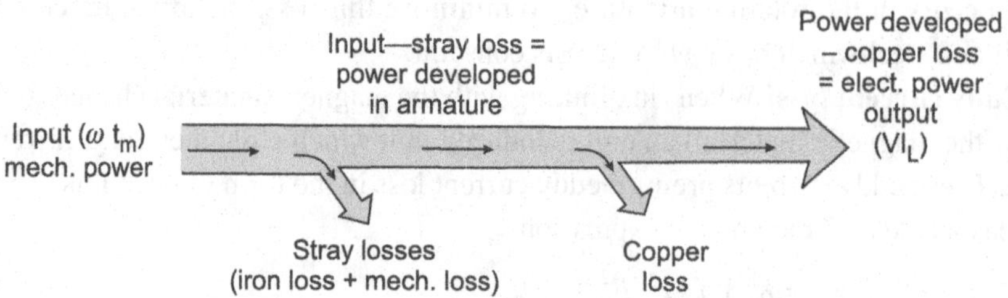

Fig. 5.52 Power flow diagram (Generator action)

The mechanical power (ωT_m) is supplied to the generator which is converted into electrical power (VI_L). While conversion, various losses occur in the machine. The power flow diagram for a DC generator is shown in Fig. 5.52.

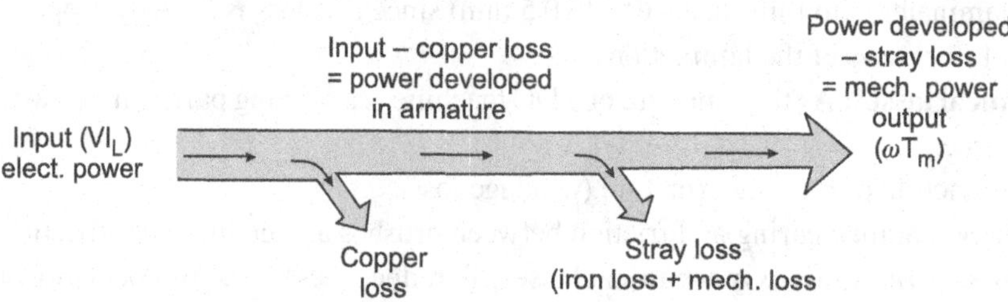

Fig. 5.53 Power flow diagram (Motor action)

Although losses in a DC machine are the same whether it works as a generator or as a motor but the flow of power is opposite. The power flow diagram for a DC motor is shown in Fig. 5.53.

5.31 Efficiency of a DC Machine

The ratio of output power to the input power of a DC machine is called its *efficiency***.

$$\text{Efficiency, } \eta = \frac{\text{Output}}{\text{Input}};$$

---

* The ratio of output to input of a DC motor is called its commercial efficiency, i.e., $\eta = \frac{\text{output}}{\text{input}}$. The ratio of mechanical power output to the power developed in the armature is called its mechanical efficiency, i.e., $h_m = \frac{\text{mech. power output}}{\text{mech. power developed in the armature}}$ The ratio of electrical power developed in the armature to the electrical power input is called electrical efficiency of the motor, i.e., $h_e = \frac{\text{power developed in the armature}}{\text{electrical power input}}$.

Also, commercial efficiency, $h = h_m \times h_e$

(*i*) *When machine is working as a generator*

Power output = VI_L watt

Power input = Power output + Variable losses + Constant losses

Since the shunt field current I_{sh} is very small as compared to line current,

We may consider, $I_L \cong I_a$ (neglecting I_{sh})

$\therefore$ Variable losses = $I_L^2 R_a$; Constant losses = P_c (say)

Then, power input = $VI_L + I_L^2 R_a + P_c$ $\quad \therefore \quad \eta = \dfrac{VI_L}{VI_L + I_L^2 R_a + P_c}$ ...(5.22)

(*ii*) *when machine is working as a motor.*

Power input = VI_L

Power output = Power input – Variable losses – constant losses = $VI_L - I_1^2 R_a - P_c$

$$\therefore \quad \eta = \frac{VI_L - I_L^2 R_a - P_c}{VI_L} \quad ...(5.23)$$

5.32 Condition for Maximum Efficiency

The efficiency of a DC generator is given as:

$$\eta = \frac{\text{output}}{\text{output + variable losses + constant loses}}$$

$$= \frac{VI_L}{VI_L + I_L^2 R_a + P_c} = \frac{1}{1 + \dfrac{I_L R_a}{V} + \dfrac{P_c}{VI_L}} \quad ...(5.24)$$

Now, efficiency will be maximum when the denominator of eq. (5.24) in minimum i.e.,

$$\frac{d}{dI_L}\left(1 + \frac{I_L R_a}{V} + \frac{P_e}{VI_L}\right) = 0$$

or $$\frac{R_a}{V} - \frac{P_e}{VI_L^2} = 0 \quad \text{or} \quad \frac{R_a}{V} = \frac{P_c}{VI_L^2} \quad \text{or} \quad I_L^2 R_a = P_c \quad ...(5.25)$$

i.e., **Variable losses = Constant losses**

Hence, the efficiency of a DC machine will be maximum when the line current is such that constant loss is equal to the variable loss.

The line current at which the efficiency of DC machine is maximum. $I_L = \sqrt{\dfrac{P_c}{R_a}}$ ...(5.26)

Example 5.40

A DC generator is connected to a 220 V DC mains. The current delivered by the generator to the mains is 100 A. The armature resistance is 0·1 ohm. The generator is driven at a speed of 500 rpm Calculate (i) the induced emf (ii) the electromagnetic torque (iii) the mechanical power input to the armature neglecting iron, windage and friction losses, (iv) Electrical power output from the armature, (v) armature copper loss.

Solution:

(*i*) The induced emf, $E_g = V + I_a R_a$

$= 220 + 0{\cdot}1 \times 100 = \mathbf{230\ V}$ (*Ans.*)

(*ii*) Using the relation, $\omega T = E_g I_a$

Electromagnetic torque, $T = \dfrac{E_g I_a}{\omega} = \dfrac{E_g I_a}{2\pi N} \times 60 \qquad \left[\because \omega = \dfrac{2\pi N}{60}\right]$

$= \dfrac{230 \times 100 \times 60}{2\pi \times 500} = \mathbf{439{\cdot}27\ Nm}$ (*Ans.*)

(*iii*) Neglecting iron, windage and friction losses,

Input to armature $= \omega T$ or $E_g I_a$

$= \dfrac{2\pi N T}{60} = \dfrac{2\pi \times 500 \times 439 \cdot 27}{60} = \mathbf{23000\ W}$ (*Ans.*)

(*iv*) Electrical power output $= V I_a = 220 \times 100 = \mathbf{22000\ W}$ (*Ans.*)

(*v*) Armature copper losses $= I_a^2 R_a = (100)^2 \times 0{\cdot}1 = \mathbf{1000\ W}$ (*Ans.*)

Example 5.41

A shunt generator supplies 195 A at 220 V. Armature resistance is 0.02 ohm, shunt field resistance is 44 ohm. If the iron and friction losses amount to 1600 watt, find (i) emf generated; (ii) copper losses; (iii) b.h.p. of the engine driving the generator. (iv) commercial, mechanical and electrical efficiency.

Solution:

The conventional circuit is shown in Fig. 5.54.

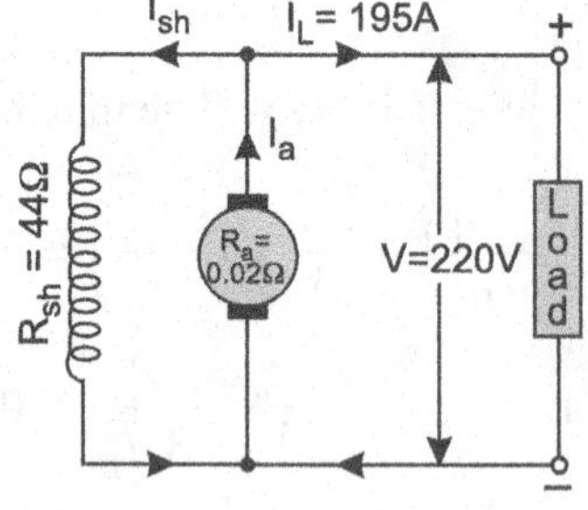

Fig. 5.54 Circuit diagram

Shunt field current, $I_{sh} = \dfrac{V}{R_{sh}} = \dfrac{220}{44} = 5A$

Armature current, $I_a = I_L + I_{sh} = 195 + 5 = 200$ A

(*i*) Generated or induced emf, $E = V + I_a R_a$

$= 220 + 200 \times 0{\cdot}02$

$= \mathbf{224\ V}$ (*Ans.*)

Armature copper loss $= i_a^2 R_a = (200)^2 \times 0 \cdot 02 = 800$ W

Shunt field copper loss $= I_{sh}^2 R_{Sh} = (5)^2 \times 44 = 1100W$

(*ii*) Total copper losses $= 800 + 1100 = \mathbf{1900\ W}$ (*Ans.*)

Output power $= V I_L = 220 \times 195 = 42900$ W

Input power $= 42900 + 1600 + 1900 = 46400$ W

(*iii*) B.H.P. of the engine driving the generator = $\frac{46400}{735 \cdot 5}$ = **63·08 H.P.** (*Ans.*)

(*iv*) Commercial efficiency, $\eta = \frac{\text{final output of machine}}{\text{input to machine}}$

$$= \frac{42900}{46400} \times 100 = \mathbf{92.45\%} \text{ (Ans.)}$$

Mechanical efficiency, $\eta_m = \frac{\text{Power developed in armature}}{\text{Power input}}$

$$= \frac{E_g I_a}{46400} \times 100 = \frac{224 \times 200}{46400} \times 100 = \mathbf{96.55\%} \text{ (Ans.)}$$

Electrical efficiency, $\eta_e = \frac{\text{Electrical Power output}}{\text{Power generated in armature}}$

$$= \frac{VI_L}{E_g I_a} \times 100 = \frac{220 \times 195}{224 \times 200} \times 100 = \mathbf{95.76\%} \text{ (Ans.)}$$

Alternatively, commercial efficiency,

$$\eta = \eta_m \times \eta_e$$

$$= \frac{96.55}{100} \times \frac{95.76}{100} \times 100 = 92.45\% \text{ (True as above)}$$

Example 5.42

A 10 kW, 250 V DC self excited shunt generator driven at 1000 rpm has armature resistance 0.15 ohm and field current 1.64 A. Find rated load, armature induced emf, torque developed and efficiency when the rotational losses are 540 watt. Assume constant speed operation.

Solution:

Here, output power = 10 kW = 10 × 1000 W

$$R_a = 0.15\ \Omega;\ I_{sh} = 1{\cdot}64\ \text{A};\ V = 250\ \text{V};\ N = 1000\ \text{rpm}$$

Load current, $I_L = \frac{10 \times 1000}{250}$ = **40 A** (*Ans.*)

Armature current, $I_a = I_L + I_{sh} = 40 + 1{\cdot}64 = 41{\cdot}64$ A

Induced or generated emf, $E_g = V + I_a R_a$

$= 250 + 41{\cdot}64 \times 0{\cdot}15 =$ **256·246 V** (*Ans.*)

Power developed in the armature, $\omega T = E_g I_a$

∴ Torque developed, $T = \frac{E_g I_a}{\omega} = \frac{256 \cdot 246 \times 41 \cdot 64 \times 60}{2\pi \times 1000}$ $\left(\text{since } \omega = \frac{2\pi N}{60}\right)$

= **101·89 Nm** (*Ans.*)

Power developed = $E_g I_a = 256{\cdot}246 \times 41{\cdot}64 = 10670$ W

$$\text{Mechanical losses} = 540 \text{ W}$$

$$\text{Power input} = 10670 + 540 = 11210 \text{ W}$$

$$\text{Power output} = 10 \text{ kW} = 10 \times 1000 \text{ W}$$

$$\text{Efficiency, } \eta = \frac{\text{output}}{\text{input}} \times 100 = \frac{10 \times 1000}{11210} \times 100 = \mathbf{89{\cdot}2\%} \text{ (Ans.)}$$

Example 5.43

A 200 V shunt motor has $R_a = 0.1$ ohm, $R_f = 240$ ohm and rotational loss = 236 W. On full load the line current is 9.8 A with the motor running at 1450 rpm. Determine

(a) the mechanical power developed
(b) the power output
(c) the load torque (d) the full load efficiency

Solution:

Here, $V = 200$ V; $R_a = 0.1\ \Omega$; $R_{sh} = 240\ \Omega$; (Mech. + Iron loss) = 236 W

$$I_L = 9.8 \text{ A}; N = 1450 \text{ rpm}$$

$$\text{Shunt field current, } I_{sh} = \frac{V}{R_{sh}} = \frac{200}{240} = 0.833 \text{ A}$$

$$\text{Armature current, } I_a = I_L - I_{sh} = 9.8 - 0.833 = 8.967 \text{ A}$$

$$E_b = V - I_a R_a = 200 - 8.967 \times 0.1 = 199.1 \text{ V}$$

(*a*) $$\text{Mechanical power developed} = E_b \times I_a = 199.1 \times 8.967 = 1785.33 \text{ W}$$

(*b*) $$\text{Power output} = \text{Mech. power developed} - \text{rotational losses}$$

$$= 1785.33 - 236 = 1549.33 \text{ W}$$

(*c*) Now $$\frac{2\pi NT}{60} = \text{Power output}$$

or $$\text{Load torque, } T = \frac{\text{Power output} \times 60}{2\pi N}$$

or $$\text{Load torque, } T = \frac{1549.33 \times 60}{2\pi \times 1450} = \mathbf{10.2 \text{ Nm}} \text{ (Ans.)}$$

(*d*) $$\text{Input to the motor} = VI_L = 200 \times 9.8 = 1960 \text{ W}$$

$$\text{Full-load efficiency, } \eta = \frac{\text{Power output}}{\text{Input}} \times 100 = \frac{1549.3}{1960} \times 100 = \mathbf{79.04\%} \text{ (Ans.)}$$

Example 5.44

A long shunt dynamo running at 1000 rpm supplies 22 kW at a terminal voltage 220 V. The resistance of armature, shunt field and series field are 0.05, 110 and 0.06 ohm respectively. The overall efficiency at above load is 88%. Find (a) copper losses (b) iron and friction losses.

Solution:

The circuit is shown in Fig. 5.55; Here, Power $P = 22$ kW; $V = 220$ V; $R_a = 0.05\ \Omega$

$R_{sh} = 110\ \Omega;\ R_{se} = 0.06\ \Omega;\ \eta = 88\%$

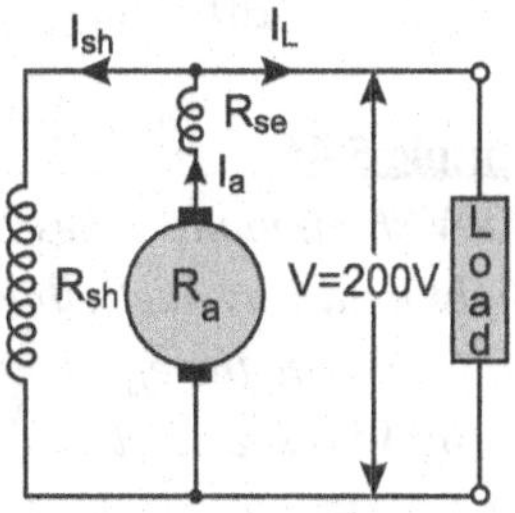

$$\text{Output current, } I_L = \frac{\text{Power output}}{V} = \frac{22 \times 1000}{220} = 100 \text{ A}$$

$$I_{sh} = \frac{V}{R_{sh}} = \frac{220}{110} = 2\text{A}$$

$$I_a = I_L + I_{sh} = 100 + 2 = 102 \text{ A}$$

$$\text{Armature copper loss} = I_a^2 R_a = (102)^2 \times 0.05 = 520.2 \text{ W}$$

$$\text{Series field copper loss} = I_{se}^2 R_{se} = (102)^2 \times 0.06 = 624.24 \text{ W}$$

$$\text{Shunt field copper loss} = I_{sh}^2 R_{sh} = (2)^2 \times 110 = 440 \text{ W}$$

$$\text{Total copper losses} = I_a^2 R_a + I_{se}^2 R_{se} + I_{sh}^2 R_{sh} = 520.2 + 624.24 + 440$$

$$= \mathbf{1584.44\ W}\ (Ans.)$$

Fig. 5.55 Circuit diagram

$$\text{Power input} = \frac{\text{Power output}}{\eta} = \frac{22 \times 1000}{0.88} = 25000 \text{ W}$$

$$\text{Total loss, } P_T = \text{Power input} - \text{power output} = 25000 - 22000 = 3000 \text{ W}$$

$$\text{Iron and friction losses, } P_S = P_T - \text{copper losses} = 3000 - 1584.44 = \mathbf{1415.56}\ (Ans.)$$

Example 5.45

A 220 V DC series motor takes 50 A. Armature resistance 0.1 ohm, series field resistance 0·08 ohm. If the iron and friction losses are equal to copper losses at this load, find the B.H.P. and efficiency.

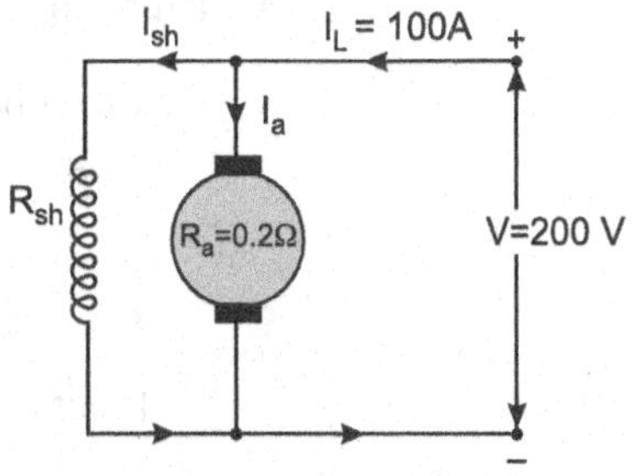

Fig. 5.56 Circuit diagram

Solution:

The circuit is shown in Fig. 5.56;

$$\text{Armature current, } I_a = I_L = 50 \text{ A}$$

$$\text{Copper losses} = I_a^2 (R_a + R_{se}) = (50)^2 \times (0{\cdot}1 + 0{\cdot}08) = 450 \text{ W}$$

$$\text{Iron and friction losses} = \text{copper losses} = 450 \text{ W}$$

$$\text{Total losses} = \text{Iron and friction losses} + \text{copper losses}$$

$$= 450 + 450 = 900 \text{ W}$$

$$\text{Input power} = V I_L = 220 \times 50 = 11000 \text{ W}$$

$$\text{Output power} = \text{Input power} - \text{Total losses}$$

$$= 11000 - 900 = 10100 \text{ W}$$

$$\therefore \quad \text{B.H.P.} = \frac{\text{output power}}{\text{Input power}} = \frac{10100}{735.5} = \mathbf{13{\cdot}732\ H.P.}\ (Ans.)$$

$$\text{Efficiency, } \eta = \frac{\text{output power}}{\text{input power}} \times 100 = \frac{10100}{11000} \times 100 = \mathbf{91{\cdot}82\ \%}\ (Ans.)$$

Example 5.46

A 200V shunt motor takes 10A when running on no-load. At higher loads the brush drop is 2V and at light load it is negligible. The stray load loss at a line current of 100 V is 50% of the no-load loss. Calculate the efficiency at a line current of 100A if armature and field resistance are 0.2 W and 100 W respectively.

Solution:

Here, $V = 200$ V; $I_{L0} = 10$ A; $V_{bf} = 2$V; $V_{b0} = 0$ V

Stray load loss = 50% of no-load loss; $R_a = 0.2\ \Omega$; $R_{sh} = 100\ \Omega$

$$\text{Input at no-load} = V \times I_{L0} = 200 \times 10 = 2000 \text{ W}$$

$$I_{sh} = \frac{V}{R_{sh}} = \frac{200}{100} = 2\text{A}$$

$$\text{Shunt field Cu loss} = I_{sh}^2 R_{sh} = (2)^2 \times 100 = 400 \text{ W}$$

$$\text{Stray loss} = \frac{50}{100} \times 2000 = 1000 \text{ W}$$

At load, armature current, $I_a = I_L - I_{sh} = 100 - 2 = 98$ A

$$\text{Armature copper loss} = I_a^2\ R_a = (98)^2 \times 0.2 = 1920.8 \text{ W}$$

$$\text{Loss at brushes} = V_{bf} \times I_a = 2 \times 98 = 196 \text{ W}$$

$$\text{Total losses} = \text{Stray loss} + \text{Armature Cu loss} + \text{Shunt field Cu loss} + \text{Brush contact loss}$$

$$= 1000 + 1920.8 + 400 + 196 = 3516.8 \text{ W}$$

$$\text{Input to motor} = V \times I_L = 200 \times 100 = 20000 \text{ W}$$

$$\text{Motor efficiency, } \eta = \frac{\text{output}}{\text{input}} = \frac{\text{Input} - \text{Losses}}{\text{Input}} \times 100$$

$$= \frac{20000 - 3516.8}{20000} \times 100 = \mathbf{82.416\%}\ (Ans.)$$

Example 5.47

A shunt generator has a full-load current of 195 A at 250 V. The stray losses are 720 W and the shunt field coil resistance is 50 Ω. It has a full-load efficiency of 88%. Find the armature resistance. Also find the current corresponding to maximum efficiency.

Solution:

$$\text{Generator rated output} = V \times I_L = 250 \times 195 = 48750 \text{ W}$$

$$\text{Input power to generator} = \frac{\text{Rated power output}}{\text{Full-load efficiency}} = \frac{48750}{0.88} = 55398 \text{ W}$$

$$\text{Total losses} = \text{Power input} - \text{power output} = 55398 - 48750 = 6648 \text{ W}$$

$$\text{Shunt field current, } I_{sh} = \frac{V}{R_{sh}} = \frac{250}{50} = 5\text{A}$$

$$\text{Shunt field copper loss} = I_{sh}^2 R_{sh} = (5)^2 \times 50 = 1250 \text{ W}$$

$$\text{Stray loss} = 720 \text{ W}$$

$$\text{Armature current, } I_a = I_L + I_{sh} = 195 + 5 = 200 \text{ A}$$

$$\text{Armature copper loss, } I_a^2 R_a = \text{Total losses} - \text{stray loss} - \text{shunt field copper loss}$$

$$= 6648 - 720 - 1250 = 4678 \text{ W}$$

or

$$\text{Armature resistance, } P_C = \frac{4678}{I_a^2} = \frac{4678}{(200)^2} = 0.11695\ \Omega \textit{ (Ans.)}$$

$$\text{Constant losses, } P_C = \text{Shunt field loss} + \text{Stray loss} = 1250 + 720 = 1970 \text{ W}$$

Armature current at which the efficiency will be maximum (say I_a'),

$$I_a' = \sqrt{\frac{P_c}{R_a}} = \sqrt{\frac{1970}{0.11695}} = 130 \text{ A}$$

Load current at which the efficiency will be maximum (say I_L'),

$$I_L' = I_a' - I_{sh} = 130 - 4 = \mathbf{126\ A} \textit{ (Ans.)}$$

Example 5.48

A 400 V shunt generator has full-load current of 200 A. Its armature resistance is 0·06 ohm, field resistance is 100 ohm and the stray losses are 2000 watt. Find the h.p. of prime-mover when it is delivering full load, and find the load for which the efficiency of the generator is maximum.

Solution:

Here, $V = 400$ V; $I_L = 200$ A; $R_a = 0.06\ \Omega$; $R_{sh} = 100\ \Omega$

$$\text{Stray losses} = 2000 \text{ W}$$

$$\text{Shunt field current, } I_{sh} = \frac{V}{R_{sh}} = \frac{400}{100} = 4\text{A}$$

$$\text{Armature current, } I_a = I_L + I_{sh} = 200 + 4 = 204 \text{ A}$$

$$\text{Armature copper loss} = I_a^2 R_a = (204)^2 \times 0.06 = 2497 \text{ W}$$

$$\text{Shunt field copper loss} = I_{sh}^2 R_{sh} = (4)^2 \times 100 = 1600 \text{ W}$$

$$\text{Total losses} = \text{Armature copper loss} + \text{shunt field copper loss} + \text{stray loss}$$

$$= 2497 + 1600 + 2000 = 6097 \text{ W}$$

$$\text{Output power} = VI_L = 400 \times 200 = 80000 \text{ W}$$

$$\text{Input power} = \text{Output power} + \text{losses} = 80000 + 6097 = 86097 \text{ W}$$

$$\text{Horse power of prime mover} = \frac{\text{Input power}}{735.5} = \frac{86097}{735.5} = \textbf{117·06 H.P.} \ (Ans.)$$

$$\text{Constant losses, } P_c = \text{stray losses} + \text{shunt field copper loss}$$

$$= 2000 + 1600 = 3600 \text{ W}$$

Condition for maximum efficiency is,

$$\text{Variable losses} = \text{Constant losses}$$

Let, I_L' be the load current at which the efficiency is maximum and armature current is I_a'.

$$I_a' = \sqrt{\frac{P_c}{R_a}} = \sqrt{\frac{3600}{0 \cdot 06}} = 245 \text{ A}$$

$$\text{Load current, } I_L' = I_a' - I_{sh} = 245 - 4 = 241 \text{ A}$$

Load for which the efficiency is maximum = $I_L'V = 241 \times 400 = 96{\cdot}4 \times 10^3$ W = **96·4 kW** (*Ans.*)

Example 5.49

A 250 V, 20 kW shunt motor running at 1500 rpm has a maximum efficiency of 85% when delivering 80% of its rated output. The resistance of the shunt field winding is 125 Ω. Determine the efficiency and speed of the motor when it draws 100 A from the mains.

Solution:

$$\text{Rated output} = 20 \text{ kW} = 20000 \text{ W}$$

$$\text{Actual output} = \text{Operating } \eta \times \text{rated output} = 0.8 \times 20000 \text{ W}$$

$$\text{Maximum efficiency, } \eta_{\max} = 85\%$$

$$\text{Efficiency of machine, } \eta, = \frac{\text{Output}}{\text{Output} + \text{losses}}$$

or

$$\text{Total losses} = \frac{1 - \eta_{\max} \text{ output}}{\eta_{\max}} = \left(\frac{1}{\eta_{\max}} - 1\right) \times \text{actual output}$$

$$= \left(\frac{1}{0.85} - 1\right) \times 0.8 \times 20000 = 2824 \text{ W}$$

At maximum efficiency, constant losses (rotational losses + shunt field loss)

$$= \text{Variable armature circuit losses, } I_a^2 R_a$$

$$= \frac{\text{Total losses}}{2} = \frac{2824}{2} = 1412 \text{ W}$$

$$\text{Input to motor, } P_{in} = \text{Motor output, } P_{out} + \text{total losses}$$

$$= 20000 \times 0.8 + 2824 = 18824 \text{ W}$$

$$\text{Input line current, } I_{L1} = \frac{P_{in}}{V} = \frac{18824}{250} = 75.3 \text{ A}$$

$$\text{Shunt field current, } I_{sh} = \frac{V}{R_{sh}} = \frac{250}{125} = 2 \text{ A}$$

$$\text{Armature, } I_{a1} = I_{L1} - I_{sh} = 75.3 - 2 = 73.3 \text{ A}$$

$$\text{Armature copper loss, } I_{a1}^2 R_a = 1412 \text{ W}$$

or $$\text{Armature circuit resistance, } R_a = \frac{1412}{I_{a1}^2} = \frac{1412}{73.3^2} = 0.263\ \Omega$$

$$\text{Now input of motor, } I_{L2} = 100 \text{ A}$$

$$\text{Armature current, } I_{a2} = I_{L2} - I_{sh} = 100 - 2 = 98 \text{ A}$$

$$\text{Armature copper loss} = I_{a2}^2 R_a = 98^2 \times 0.263 = 2{,}525 \text{ W}$$

$$\text{Motor input} = VI = 250 \times 100 = 25000 \text{ W}$$

$$\text{Motor output} = \text{Motor input} - \text{constant losses} - \text{armature copper losses}$$

$$= 25000 - 1412 - 2525 = 21063 \text{ W}$$

$$\text{Motor efficiency, } \eta = \frac{\text{Output}}{\text{Input}} \times 100 = \frac{21063}{25000} \times 100 = \mathbf{84.25\%} \ (Ans.)$$

Assuming flux remaining the same, the speed of the motor,

$$N_2 = N_1 \times \frac{E_{b2}}{E_{b1}} = N_1 \times \frac{V - I_{a2}R_a}{V - I_{a1}R_a}$$

$$= 1500 \times \frac{250 - 98 \times 0.263}{250 - 73.3 \times 0.263} = \mathbf{1458\ rpm}\ (Ans.)$$

5.33 Test Performed to Determine Efficiency of DC Machines

To determine the efficiency of a DC machine namely, three methods can be used direct method, indirect method and regenerative method.

1. **Direct Method.** In this method, full load is applied to the machine and output is directly measured. Although, this method looks to be very simple but it is very difficult and inconvenient to apply full load to large size machines. Therefore, this method is restricted only to determine the efficiency of small sized machines. *Brake test* comes under this category.
2. **Indirect Method.** In this method, the losses are determined without actually loading the machine. The power is required to supply the losses only, therefore, large machines can be tested by applying this method. Accordingly, this method is usually employed to determine the efficiency of large DC shunt and compound wound machines. The major drawback of this method is that the temperature rise of the winding on load and commutating qualities cannot be *assessed* since the machine is operated at light loads. *Swinburne's test* falls in this category.
3. **Regenerative Method.** In this method, two identical, mechanically coupled machines are used. One acts as a motor drawing power from the supply and the other acts as a generator which

feeds the generated power back to the supply. Thus motor draws power equal to the total losses of the two machines at rated speed and load. Very large machines can be tested by this method. Moreover, the machines can be operated (or tested) for long duration and their performance regarding commutation, temperature rise, etc., can be studied conveniently. *Hopkinson's* test falls in this category.

5.34 Brake Test

It is a direct method of determining the efficiency of a DC motor. This test is performed only with small motors. The brake is applied to a pulley (either air cooled or water cooled) mounted on the motor shaft as shown in Fig. 5.57. The load on the motor is increased by tightening the belt mounted on the pulley. The electrical connections are made as shown in the circuit diagram.

Let, spring balance reading on tight side = W_1 kg

Spring balance reading on loose side = W_2 kg

Motor speed = N rpm (reading taken by tachometer)

Radius of pulley = r metre

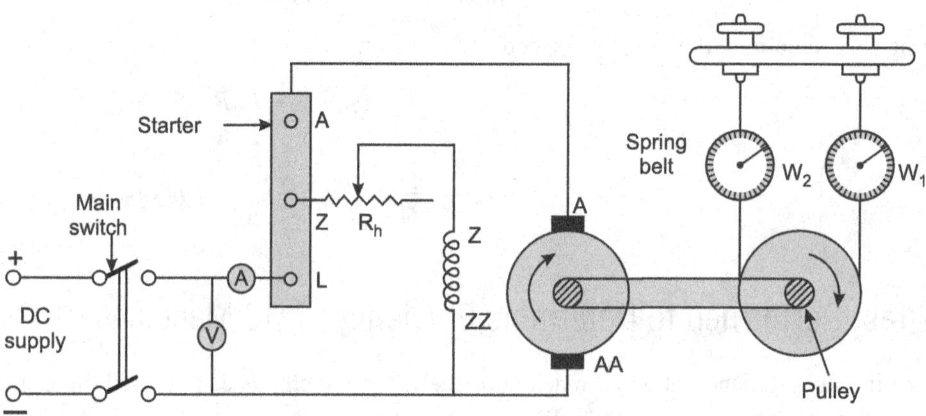

Fig. 5.57 Circuit arrangement for brake test

Thickness of belt = t metre

$$\text{Motor output} = T \times \frac{2\pi N}{60} = (W_1 - W_2)\left(r + \frac{t}{2}\right)\frac{2\pi N}{60}$$

(since T = effective pull × effective radius)

$$= (W_1 - W_2)\, r \times \frac{2\pi N}{60} \text{ kg m/s (neglect } t \text{ being small)}$$

$$= (W_1 - W)\, r \times \frac{2\pi N}{60} \times 9.81 \text{ Nm/s or watt} \quad \ldots(5.27)$$

If voltmeter reading = V volt

$$\text{Armature reading} = I \text{ ampere}$$

$$\text{Motor input} = VI \text{ watt} \quad \text{...(5.28)}$$

$$\therefore \text{ Efficiency of motor, } \eta = \frac{\text{output}}{\text{input}} = \frac{(W_1 - W_2)\, r \times \frac{2\pi N}{60} \times 9.81}{VI} \quad \text{...(5.29)}$$

This method of measuring efficiency has the following drawbacks:

(*i*) We cannot determine the efficiency of large machines because such facilities of loading are not available.

(*ii*) The output measured by this method is not accurate because belt is not offering a constant load (usually belt slips over the pulley)

A special precaution is to be observed while performing this test on a series motor. If the brakes applied fails, the motor may obtain a dangerously high speed; therefore, this test is usually applied only on shunt and compound machines.

Example 5.50

In a brake test, the DC motor took 38.5 A from a 240 V supply mains. The brake pulley of radius 35 cm had an effective load of 30 kg and the speed was 720 rpm. Find the bhp (metric) and the efficiency at the above load.

Solution:

$$\text{Torque exerted, } T = \text{Effective load} \times \text{radius of brake pulley}$$

$$= (W_1 - W_2)\, r = 30 \times 0.35 = 10.5 \text{ kg-m or } 10.5 \times 9.81 \text{ Nm}$$

$$\text{Output of motor} = T \times \frac{2\pi N}{60} = 10.5 \times 9.81 \times 2 \times \pi \times \frac{720}{60} = 7766.4 \text{ watt}$$

$$\text{BHP of motor} = \frac{\text{Output in watts}}{735.5} = \frac{7766.4}{735.5} = \mathbf{10.56} \text{ (\textit{Ans.})}$$

$$\text{Input to motor} = V \times I = 240 \times 38.5 = 9240 \text{ watt}$$

$$\text{Efficiency of motor, } \eta = \frac{\text{Output}}{\text{Input}} \times 100 = \frac{7766.4}{9240} \times 100 = \mathbf{84.05\%} \text{ (\textit{Ans.})}$$

Example 5.51

In a brake test on a DC shunt motor, the load on tight side of the brake band was 32 kg and the other side 2 kg. The motor was running at 1200 rpm; its input being 65 A at 400 V DC. The pulley diameter is 1 m. Determine the torque, output of the motor and efficiency of the motor.

Solution:

$$\text{Effective force, } F = 32 - 2 = 30 \text{ kg}$$

$$\text{Pulley radius, } r = \frac{1.0}{2} = 0.5 \text{ m}$$

$$\text{Torque exerted, } T = F \times r = 30 \times 0.5 = 15 \text{ kg-m or } 15 \times 9.81 = \mathbf{147.15 \text{ Nm}} \text{ (\textit{Ans.})}$$

$$\text{Motor output} = T \times \frac{2\pi N}{60} = \frac{147.15 \times 2\pi \times 1200}{60} = \mathbf{18499\ W}\ (Ans.)$$

$$\text{Input power to motor} = V \times I = 400 \times 65 = 26000$$

$$\text{Motor efficiency, } \eta = \frac{\text{Output}}{\text{Input}} \times 100 = 71.15\%\ (Ans.)$$

Example 5.52

In a break test on DC motor, the effective load on the brake drum was 23 kgf, the effective diameter of the drum 45 cm and the speed 960 rpm. The input of the motor was 28A at 230 V. Calculate the efficiency of the motor.

Solution:

Here, load $(W_1 - W_2) = 23$ kg f; $d_m = 45$ cm $= 0.45$ m;

$$N = 960 \text{ rpm}; I_L = 28 \text{ A}; V = 230 \text{ V}$$

$$\text{Torque developed, } T = (W_1 - W_2) \times \frac{dm}{2} = 23 \times \frac{0.45}{2} = 5.175 \text{ kg m}$$

$$= 5.175 \times 9.81 = 50.77 \text{ Nm}$$

$$\text{Motor output} = \frac{2\pi NT}{60} = \frac{2\pi \times 960 \times 50.77}{60} \text{ Nm/s or watt} = 5106 \text{ W}$$

$$\text{Motor input} = V \times I_L = 230 \times 28 = 6440 \text{ W}$$

$$\text{Motor efficiency, } \eta = \frac{\text{output}}{\text{input}} \times 100 = \frac{5106}{6440} \times 100 = \mathbf{79.28\%}\ (Ans.)$$

5.35 Swinburne's Test

Swinburne's test is performed determine the constant losses in a DC shunt machine. In this test, the machine is operated as a motor on no-load. This no-load test is also known as Swinburne's test. A voltmeter and two ammeters A_1 and A_2 are connected in the circuit as shown in Fig. 5.58. The normal rated voltage V is applied to the motor terminals. The ammeter A_1 and A_2 measure the no-load line current I_{LO} and shunt field current I_{sh} respectively. The voltmeter measures the applied voltage. As there is no output at no-load, all the power supplied to the motor, given by the product of current I_{LO} and voltage V, is being utilised to meet with losses only. The following are the losses at no-load:

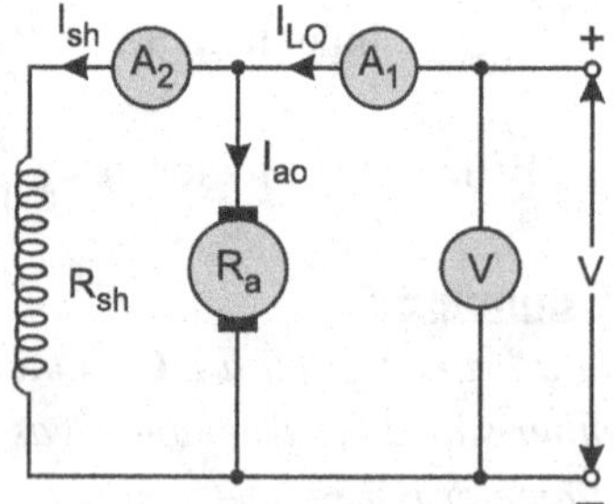

Fig. 5.58 Circuit arrangement for Swinburne's Test

(*i*) Iron losses in the core
(*ii*) Windage and friction losses at bearing and commutator.
(*iii*) Shunt field copper losses.
(*iv*) Armature copper losses at no-load (very smell)

Except the armature copper loss all other losses are the constant losses.

No-load input power to the machine $= V I_{LO}$ watt ...(5.30)

No-load armature current, $I_{a0} = I_{LO} - I_{sh}$

The resistance of armature circuit including the inter pole winding, etc., is determined (measured) by disconnecting one end of the shunt field circuit. Let its value be R_a.

Then, variable losses $= I_a^2 R_a$ watt

$\therefore$ Constant losses $P_c = (V I_{LO} - I_{a0}^2 R_a)$ watt ...(5.31)

Note: For all practical purposes the copper losses in the armature at no-load may be neglected being very small.

After determining the constant losses, the efficiency of the machine, when it is working as a motor or generator can be calculated at any load, as discussed below:

Let I_L be the line current at which efficiency is to be calculated.

(*i*) *When the machine is working as a motor*:

Armature current, $I_a = I_L - I_{sh}$ ...(5.32)

Variable or armature copper loss at load $= I_a^2 R_a$

$\therefore$ Total losses $= P_c + I_a^2 R_a$ ...(5.33)

Input power $= VI_L$

Output power = input power – total losses

$= VI_L - (P_c + I_a^2 R_a)$ ...(5.34)

$\therefore$ Efficiency, $\eta = \dfrac{output}{Input} = \dfrac{VI_L - \left(P_c + I_a^2 R_a\right)}{VI_L}$ ...(5.35)

(*ii*) *When the machine is working as a generator*;

Armature current, $I_a = I_L + I_{sh}$ ...(5.36)

Variable or armature copper loss at load $= I_a^2 R_a$

Total losses $= P_c + I_a^2 R_a$

Output power $= VI_L$

Input power = output power + total losses

$= VI_L + (P_c + I_a^2 R_a)$ ...(5.37)

$\therefore$ Efficiency, $\eta = \dfrac{VI_L}{VI_L + (P_c + I_a^2 R_a)}$ ...(5.38)

Advantages

1. Very convenient and economical method since power required is very small.
2. Stray or constant losses are determined, therefore, efficiency at any desired value of load can be determined.

Disadvantages

1. Since the test is performed at no-load, the effect of temperature rise and performance of commutator cannot be assessed properly.
2. This test cannot be performed with DC series motors because at no-load series motors obtain dangerously high speeds.
3. The change in iron losses from no-load to full-load are not accounted for, although this change is prominent due to armature reaction.

Example 5.53

A 250 V shunt motor takes 4A at no-load and resistance of armature is 0.4 Ω and that of shunt field is 125 Ω calculate: (i) output is kW (ii) Efficiency of motor when motor current is 102A.

Solution:

Here, $I_{a0} = 4$ A; $R_a = 0.4\ \Omega$; $R_{sh} = 125\ \Omega$;

$$I_{sh} = \frac{V}{R_{sh}} = \frac{250}{125} = 2A \text{ A}$$

$$I_{ao} = I_{LO} - I_{sh} = 4 - 2 = 2\text{A}$$

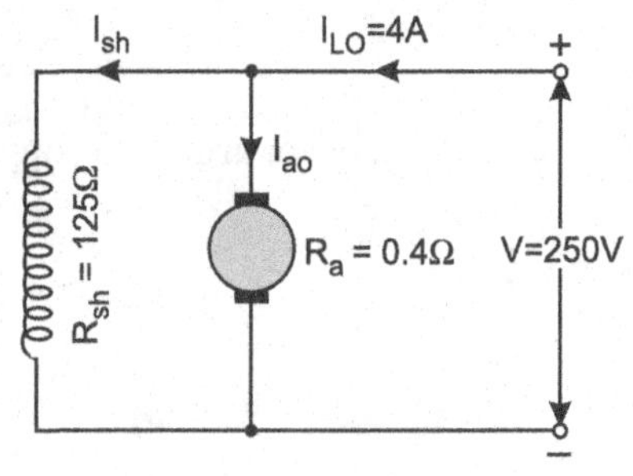

Fig. 5.59 Circuit diagram

$$\text{Copper losses at no-load} = I_{a0}^2 R_a = (2)^2 \times 0.4 = 1.6 \text{ W}$$

$$\text{Power input at no-load} = V I_{LO} = 250 \times 4 = 1000 \text{ W}$$

$$\text{Constant losses} = 1000 - 1.6 = 998.4 \text{ W}$$

$$\text{Armature full load current, } I_a = I_{L(fl)} - I_{sh} = 102 - 2 = 100 \text{ A}$$

$$\text{Armature Cu loss at full-load} = I_a^2 R_a = (100)^2 \times 0.4 = 4000 \text{ W}$$

$$\text{Total losses at full-loaded} = 4000 + 998.4 = 4998.4 \text{ W}$$

$$\text{Input at full-load} = VI_{L(fl)} = 250 \times 102 = 25500 \text{ W}$$

$$\text{Output} = \text{Input} - \text{losses} = 25500 - 4998.4 = 20501.6 \text{ W}$$

$$= \mathbf{20.5016\ kW}\ (\textit{Ans.})$$

$$\text{Efficiency of the motor, } \eta = \frac{\text{output}}{\text{input}} \times 100$$

$$= \frac{20501.6}{25500} \times 100 = \mathbf{80.4\%}\ (\textit{Ans.})$$

Example 5.54

A 50 kW, 250 V compound motor takes a current of 9A while running on no-load at rated voltage and speed. The shunt field current is 5A. The resistances of the windings when hot are:

Armature 0.1 Ω, series field 0.07 Ω and interpole 0.03 Ω. The brush drop is 2V. Determine motor output and the efficiency when the motor intake is 155 A.

Solution:

At No-load; $I_{LO} = 9A$; $I_{sh} = 5A$

$$\text{Input at no-load} = I_{LO} \times V = 9 \times 250 = 2250 \text{ W}$$

$$\text{Armature current, } I_{a0} = I_{LO} - I_{sh} = 9 - 5 = 4A$$

Considering the machine as long-shunt;

$$\text{Variable losses} = \text{Copper losses in armature circuit and at brushes}$$

$$= I_{a0}^2 (R_a + R_{se} + R_i) + I_{a0} \times \text{brush drop}$$

$$= 4^2 (0.1 + 0.07 + 0.03) + 4 \times 2 = 11.2 \text{ W}$$

$$\text{Constant losses, } P_C = \text{Input at no-load} - \text{variable losses}$$

$$= 2250 - 11.2 = 2238.8 \text{ W}$$

$$\text{At full-load; Line current, } I_{Lf} = 155 \text{ A}$$

$$\text{Power input} = I_{L1} \times V = 155 \times 250 = 38750 \text{ W}$$

$$\text{Armature current, } I_{af} = I_{Lf} - I_{sh} = 155 - 5 = 150 \text{ A}$$

$$\text{Variable losses} = I_{af}^2 (R_a + R_{se} + R_1) + I_{af} \times \text{brush drop}$$

$$= 150^2 (0.1 + 0.07 + 0.03) + 150 \times 2 = 4800 \text{ W}$$

$$\text{Total losses} = P_C + \text{variable losses} = 2238.8 + 4800 = 7038.8 \text{ W}$$

$$\text{Motor output, } P_{out} = P_{in} - \text{total losses}$$

$$= 38750 - 7038.8 = 31711.2 \text{ W or } \mathbf{31.7\ kW} \ (Ans.)$$

$$\text{Motor efficiency, } \eta = \frac{P_{out}}{P_{in}} \times 100 = \frac{31711.2}{38750} \times 100 = \mathbf{81.84\%} \ (Ans.)$$

Example 5.55

A 100 kW, 500 V shunt generator was run as a motor on no-load at its rated voltage and speed. The total current taken was 9·8 A including a shunt current of 2·7 A. The resistance of the armature circuit (including interpoles) at normal working temperature was 0·11 ohm. Calculate the efficiency at half full-load.

Solution:

$$\text{Terminal voltage, } V = 500 \text{ V}$$

$$\text{Armature resistance, } R_a = 0{\cdot}11 \text{ ohm}$$

$$\text{No-load line current, } I_{LO} = 9{\cdot}8 \text{ A}$$

$$\text{Shunt field current, } I_{sh} = 2{\cdot}7 \text{ A}$$

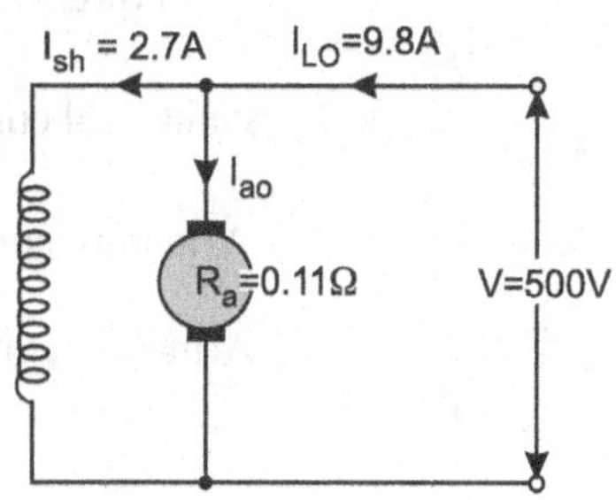

Fig. 5.60 Circuit diagram

When machine is operated at no-load as a motor as shown in Fig. 5.60.

$$\text{Input at no-load} = VI_{L0} = 500 \times 9{\cdot}8 = 4900 \text{ W}$$

No-load armature current, $I_{a0} = I_{L0} - I_{sh} = 9{\cdot}8 - 2{\cdot}7 = 7{\cdot}1 \text{ A}$

Variable or armature copper losses at no-load

$$= I_{a0}^2 \, R_a = (7{\cdot}1)^2 \times 0{\cdot}11 = 5{\cdot}55 \text{ W}$$

$$\text{Constant losses} = 4900 - 5{\cdot}55 = 4894{\cdot}45 \text{ W}$$

When the machine is operated at half full-load as a generator.

$$\text{Output power} = \frac{100}{2} = 20\text{kW} = 50 \times 10^3 \text{ W}$$

$$\text{Line current, } I_L = \frac{50 \times 10^3}{500} = 100 \text{ A}$$

Armature current at half full-load, $I_a = I_L + I_{sh} = 100 + 2{\cdot}7 = 102{\cdot}7 \text{ A}$

Variable or armature copper losses, at half full-load

$$= I_a^2 \, R_a = (102{\cdot}7)^2 \times 0{\cdot}11 = 1160{\cdot}2 \text{ W}$$

$$\text{Total losses at half full load} = \text{constant losses} + \text{variable losses}$$

$$= 4894{\cdot}45 + 1160{\cdot}2 = 6054{\cdot}65 \text{ W}$$

$$\text{Input power} = \text{output power} + \text{total losses}$$

$$= 50000 + 6054{\cdot}65 = 56054{\cdot}65 \text{ W}$$

$$\text{Efficiency, } \eta = \frac{\text{output power}}{\text{input power}} \times 100 = \frac{50000}{56054 \cdot 65} \times 100$$

$= \mathbf{89{\cdot}2\%}$ *(Ans.)*

Example 5.56

A 500 V, 25 hp, DC shunt motor takes 2.4 A while running light. The field and armature resistances are 650 Ω and 0.57 Ω respectively. Calculate full-load efficiency, assuming a brush drop of 2V.

Solution:

At No-load; $I_{L0} = 2.4$ A; $V = 500$ V

$$\text{Power input, } P_{in0} = VI_{L0} = 500 \times 2.4 = 1200 \text{ W}$$

$$\text{Shunt field current, } I_{sh} = \frac{V}{R_{sh}} = \frac{500}{650} = 0.77 \text{ A}$$

$$\text{Armature current, } I_{a0} = I_{L0} - I_{sh} = 2.4 - 0.77 = 1.63 \text{ A}$$

$$\text{Armature copper loss} = I_{a0}^2 R_a + I_{a0} \times \text{brush drop}$$

$$= (1.63)^2 \times 0.57 + 1.63 \times 2 = 4.77 \text{ W}$$

$$\text{Constant losses, } P_C = P_{in0} - \text{no-load armature copper loss}$$

$$= 1200 - 4.77 = 1195.23 \text{ W}$$

At full-load; Power output $P_{out} = 25 \times 735.5$

$= 18387.5$ W

At full-load, let the armature current be I_a ampere

$$\text{Power input, } P_{in} = V \times (I_a + I_{sh}) = 500\,(I_a + 0.77)$$

$$\text{Armature copper loss} = I_a^2 R_a + I_a \times v_b = 0.57\,I_a^2 + 2I_a$$

$$\text{Also power input, } P_{in} = P_{out} + \text{armature copper loss} + \text{constant losses}$$

or $$500\,(I_a + 0.77) = 18387.5 + 0.57\,I_a^2 + 2I_a + 1195.3 \text{ W}$$

or $$0.57\,I_a^2 - 498\,I_a + 19197.8 = 0$$

or $$I_a = \frac{498 \pm \sqrt{(498)^2 - 4 \times 0.57 \times 19197.8}}{2 \times 0.57}$$

$$= 833.26 \text{ A or } 41.05 \text{ A}$$

Since the value of current 833.26 A is impracticable

$$\text{Full-load armature current, } I_a = 41.05 \text{ A}$$

$$\text{Motor input} = VI_L = 500 \times (41.05 + 0.77) = 20910 \text{ W}$$

$$\text{Motor efficiency, } \eta = \frac{\text{Motor output}}{\text{Motor input}} \times 100$$

$$= \frac{18387.5}{20910} \times 100 = \mathbf{87.94\%} \ (\textit{Ans.})$$

5.36 Hopkinson's Test

Hopkinson's test is basically a regenerative test. It is also known as *back-to-back* test. To perform this test, two identical machines are required. These machines are mechanically coupled to each other. One of them works as a motor which acts as a prime-mover for the other machine which works as a generator.

The electrical power or energy supplied to the motor is converted into mechanical energy which is further converted into electrical energy by the second machine coupled to it, and fed back to the motor through supply system. In the process, in fact, the two machines draw electrical power or energy to meet with the losses of the two machines. Since the mechanics are identical, the losses in each machine are determined by dividing the input into two equal parts. Usually, this test is performed on large size machines at full-load for longer duration.

Circuit Arrangements and Procedure of Performing the Test

The electrical circuit arrangement is shown in Fig. 5.61.

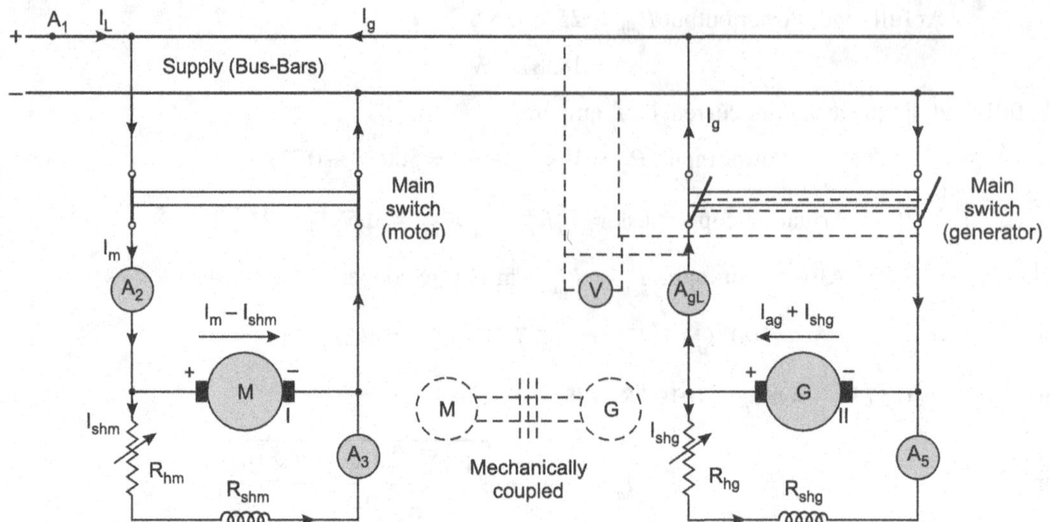

Fig. 5.61 Circuit arrangement for Hopkinson's test

Connect machine-I to the supply through its main switch (motor) and starter keeping main switch of the other machine-II open. Operate machine-I to its rated speed and adjust the field current I_{shg} of second machine such that the voltage developed by this machine (working as a generator) is 1 to 2V more than the mains voltage having same polarity as that of bus-bars. The voltage and polarity can be checked with the help of a paralleling voltmeter V. Now, connect the second machine (generator) to the mains through its main switch.

It may be noted that a machine with smaller excitation acts as a motor and a machine with larger excitation acts as a generator.

To increase the load, the excitation of machine-I (i.e., motor) is decreased gradually. With the decrease in excitation its back emf decreases due to which motor draws more current from the mains. This process goes on till the machine draws its full-load current as per the rating. During this process, the supply voltage must be maintained at its rated value.

The value of the resistances involved in the armature circuit such as armature winding resistance including brush contact, interpole winding resistance, compensating winding resistance of the two machines are measured by passing full-load current through them after the completion of whole procedure when these windings attain their final temperature at full-load.

Let various ammeters A_1, A_2, A_3, A_4 and A_5 connected in the circuit shown in Fig. 5.61 measure line current I_L, motor current I_m, motor shunt field current I_{shm}, generator current I_g and generator shunt field current I_{shg} respectively.

Where, motor current, $I_m = I_L + I_g$ (as per Kirchhoff's current law) ...(5.39)

$$\text{Motor input} = VI_m \quad \text{...(5.40)}$$

$$\text{and generator output} = VI_g \quad \text{...(5.41)}$$

where V is bus = bar voltage.

If η_m and η_g are the efficiencies of motor and generator respectively, then motor output $= VI_m \eta_m$ ...(5.42)

and generator input $= \dfrac{VI_g}{\eta_g}$ ...(5.43)

But motor output = Generator input

or $VI_m \eta_m = \dfrac{VI_g}{\eta_g}$ or $\eta_m \eta_g = \dfrac{I_g}{I_m}$ ...(5.44)

Since armature, field and stray power losses in both of the machines are considered to be equal then $\eta_m = \eta_g$

Thus, $\eta_m = \eta_g = \sqrt{\dfrac{I_g}{I_m}} = \sqrt{\dfrac{I_g}{I_g + I_L}} = \sqrt{\dfrac{I_m - I_L}{I_m}}$ ...(5.45)

This assumption is true for large machines since there is very slight difference in their armature and excitation current, but in case of small machines, the difference between the armature currents and shunt field currents is large. To obtain accurate results, armature and shunt field losses are determined separately and stray losses are assumed to be equal in both the machines.

Current supplied by the mains, $I_L = I_m - I_g$ ...(5.46)

Current drawn by motor $= I_m$

Motor shunt field current $= I_{shm}$

Motor armature current, $I_{am} = (I_m - I_{shm})$ ...(5.47)

Current supplied by generator $= I_g$

Generator shunt field current $= I_{shg}$

Generator armature current, $I_{ag} = (I_g + I_{shg})$ ...(5.48)

Terminal voltage $= V$

Total losses of both machines = Power supplied by the mains $= VI_L$

Armature copper loss in motor $= (I_m - I_{shm})^2 R_{am}$ ...(5.49)

Armature copper loss in generator $= (I_g + I_{shg})^2 R_{ag}$ ...(5.50)

Shunt field copper loss in motor $= VI_{shm}$

Shunt field copper loss in generator $= VI_{shg}$

Total copper losses $= (I_m - I_{shm})^2 R_{am} + (I_g + I_{shg})^2 R_{ag} + VI_{shm} + VI_{shg}$

Stray losses of both the machines, $P_S = VI_L - [I_m - I_{shm}]^2 R_{am} + (I_g + I_{shg})^2 R_{ag} + V(\mathrm{I_{shm}} + \mathrm{I_{shg)}}]$ ...(5.51)

Stray loss of each machine $= \dfrac{P_S}{2}$ ...(5.52)

For efficiency of motor (machine-I)

Motor input$= VI_m$ ...(5.53)

$$\text{Total losses in the motor} = \text{Armature loss} + \text{field loss} + \text{stray loss}$$

$$= (I_m - I_{shm})^2 R_{am} + VI_{shm} + \frac{P_S}{2} \quad \text{...(5.54)}$$

$$\text{Motor output} = VI_m - \left[(I_m - I_{shm})^2 R_{am} + VI_{shm} + \frac{P_S}{2}\right] \quad \text{...(5.55)}$$

$$\text{Motor efficiency} = \frac{\text{Output}}{\text{Input}} \times 100 \quad \text{...(5.56)}$$

For efficiency of generator (machine-II)

$$\text{Generator output} = VI_g \quad \text{...(5.57)}$$

$$\text{Total losses in the generator} = \text{Armature loss} + \text{field loss} + \text{stray loss}$$

$$= (I_g + I_{shg})^2 R_{ag} + VI_{shg} + \frac{P_S}{2} \quad \text{...(5.58)}$$

$$\text{Generator input} = VI_g + (I_g + I_{shg})^2 R_{ag} + VI_{shg} + \frac{P_S}{2} \quad \text{...(5.59)}$$

$$\text{Generator efficiency, } \eta_g = \frac{\text{Output}}{\text{Input}} \times 100 \quad \text{...(5.60)}$$

Thus, the efficiency of motor and generator can be determined from the data available from the test.

Advantages

(*i*) It is economical since small power is required.
(*ii*) The performance of the machines regarding commutation and temperature rise, etc., can be conveniently studied as the machines are tested at full-load, that too for long duration.
(*iii*) Accurate results are obtained since the efficiency is determined under load conditions and the stray load loss are being taken into account.

Disadvantage

The main disadvantage of this test is that there is necessity of two identical machines which are rarely available.

Example 5.57

The results of Hopkinson's test on two similar DC machines are as follows:

Line voltage 250 V, Motor armature current 23A, Generator armature current 20A, Generator field current 0.4 A. Motor armature current 0.3 A. Armature resistance of each machine 0.5 Ω. Calculate the efficiency of each machine.

Solution:

Supply voltage, $V = 250$ V

Generator armature current, $I_{og} = 20$ A

Generator shunt field current, $I_{shg} = 0.4$ A

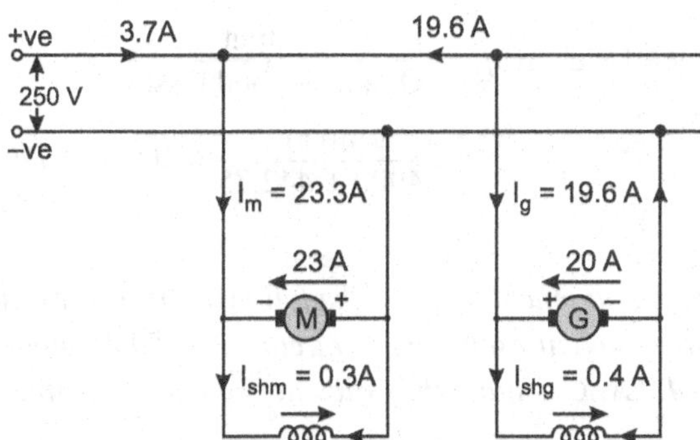

Fig. 5.62 Circuit diagram

Generator output current, $I_g = I_{ag} - I_{shg} = 20 - 0.4 = 19.6$ A

Motor armature current, $I_{am} = 23$ A

Motor shunt field current, $I_{shg} = 0.3$ A

Motor input current, $I_m = 23.3$ A

Input line current, I_L = Motor input current – generator output current

$= 23.3 - 19.6 = 3.7$ A

Input power to set $= VI_L = 250 \times 3.7 = 925$ W

Motor armature copper loss $= (I_{am})^2 R_{am} = (23)^2 \times 0.5 = 264.5$ W

Generator armature copper loss $= (I_{ag})^2 R_{ag} = (20)^2 \times 0.5 = 200$ W

Motor shunt field copper loss $= VI_{shm} = 250 \times 0.3 = 75$ W

Generator shunt field copper loss $= VI_{shg} = 250 \times 0.4 = 100$ W

Total copper loss $= 264.5 + 200 + 75 + 100 = 639.5$ W

Total stray power loss, P_S = Input to machines – total copper loss

$= 925 - 639.5 = 285.5$ W

$$\text{Stray power loss per machine} = \frac{P_S}{2} = \frac{285.5}{2} = 142.75 \text{ W}$$

Input to motor $= VI_m = 250 \times 23.3 = 5825$ W

Total losses in motor = Armature loss + field loss + stray power loss

$= 264.5 + 75 + 142.75 = 482.25$ W

$$\text{Motor efficiency, } \eta_m = \frac{\text{Input} - \text{Total losses}}{\text{Input}} \times 100 = \frac{5825 - 482.25}{5825} \times 100$$

$= 91.72\%$ (*Ans.*)

Generator output $= VI_g = 250 \times 19.6 = 4900$ W

Total losses in generator = Armature loss + field loss + stray power loss

$= 200 + 100 + 142.75 = 442.75$ W

$$\text{Generator efficiency, } \eta_g = \frac{\text{Output}}{\text{Output + Total losses}} \times 100$$

$$= \frac{4900}{4900 + 442.75} \times 100 = 91.72\% \text{ (Ans.)}$$

Example 5.58

The Hopkinson test on two shunt machines gave the following results for full load:

Line voltage, 250 V; line current excluding field currents, 50 A; motor armature current 380 A ; field currents, 5A and 4.2 A. Calculate the efficiency of each machine. Armature resistance of each machine is 0.02 ohm.

Solution.

Here, $V = 250$ V; $I_{am} = 380$ A; $I_L = 50$ A; $R_a = 0.2\ \Omega$

In Hopkinson's test, the machine working as a generator must have field current more than the machine working as a motor.

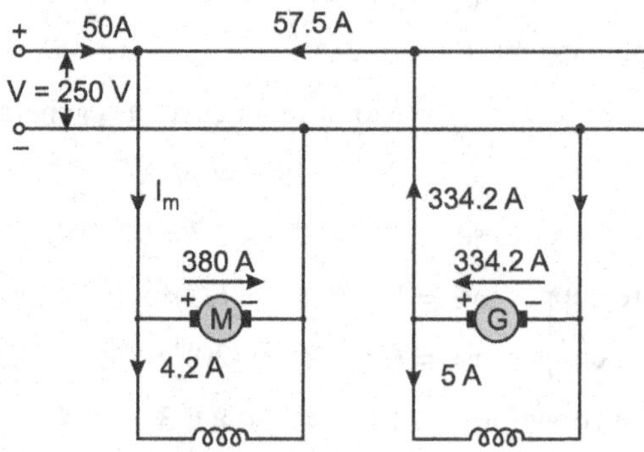

Fig. 5.63 Circuit diagram

Therefore, $I_{shg} = 5\text{A};\ I_{shm} = 4.2$ A

Current drawn by the motor, $I_m = I_{am} + I_{shm}$

$= 380 + 4.2 = 384.2$ A

Current delivered by the generator,

$$I_g = I_m - I_L$$

$$= 384.2 - 50 = 334.2 \text{ A}$$

$$I_{ag} = I_g - I_{shg}$$

$$= 334.2 - 5 = 329.2 \text{ A}$$

Armature copper loss in motor $= (I_{am})^2 R_a = (380)^2 \times 0.02 = 2888$ W

Armature copper loss in generator $= (I_{ag})^2 R_a = (329.2)^2 \times 0.02 = 2168$ W

Field copper loss in motor $= VI_{shm} = 250 \times 5 = 1250$ W

$$\text{Field copper loss in generator} = VI_{shg} = 250 \times 4.2 = 1050 \text{ W}$$

$$\text{Power supplied to the set} = VI_L = 250 \times 50 = 12500 \text{ W}$$

$$\begin{aligned}\text{Total stray losses} &= \text{Power input to set} - \text{copper losses in both the machines}\\ &= 12500 - (2888 + 2168 + 1250 + 1050)\\ &= 12500 - 7357 = 5144 \text{ W}\end{aligned}$$

Stray losses for each machine, $P_S = \dfrac{5144}{2} = 2572 \text{ W}$

$$\text{Motor input} = VI_m = 250 \times 384.2 = 96050 \text{ W}$$

$$\begin{aligned}\text{Motor output} &= \text{motor input} - \text{copper losses} - \text{stray losses}\\ &= 96050 - (2888 + 1250) - 2572 = 89340 \text{ W}\end{aligned}$$

$$\text{Efficiency, } \eta_m = \frac{\text{output}}{\text{input}} \times 100 = \frac{89340}{96050} \times 100 = 93.01\% \; (Ans.)$$

$$\text{Generator output} = VI_g = 250 \times 334.2 = 83550 \text{ W}$$

$$\begin{aligned}\text{Generator input} &= \text{generator output} + \text{copper loss} + \text{stray loss}\\ &= 83550 + (2169 + 1050) + 2572 = 89341 \text{ W}\end{aligned}$$

$$\text{Generator efficiency, } \eta_g = \frac{\text{output}}{\text{input}} \times 100 = \frac{83550}{89341} \times 100 = \mathbf{93.52\%} \; (Ans.)$$

Example 5.59

The following test results were obtained while Hopkinson's test was performed on two similar DC shunt machines:

Supply voltage = 250 V, Field current of motor = 2A, Field current of generator = 2.5 A, Armature current of generator = 60A, Current taken by the two armatures from supply = 15A, Resistance of each armature circuit = 0.2 ohm.

Calculate the efficiency of the motor and generator under these conditions of load.

Solution.

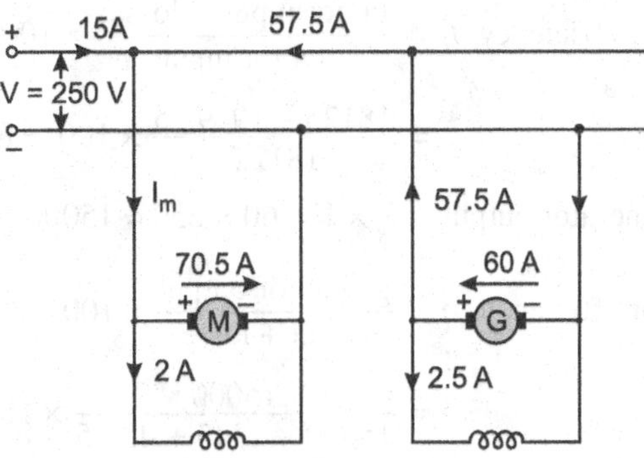

Fig. 5.64 Circuit diagram

Here, $V = 250$ V; $I_{shm} = 2$A; $I_{shg} = 2.5$ A; $I_{ag} = 60$ A

$$I_L = 15 \text{ A}; R_{am} = R_{ag} = 0.2\ \Omega$$

$$\text{Generator current, } I_g = I_{ag} - I_{shg} = 60 - 2.5 = 57.5 \text{ A}$$

$$\text{Motor current, } I_m = I_g + I_L$$

$$= 57.5 + 15 = 72.5 \text{ A}$$

$$\text{Motor armature current, } I_{am} = I_m - I_{shm}$$

$$= 72.5 - 2 = 70.5 \text{ A}$$

$$\text{Motor armature copper loss} = I_{am}^2 R_{am}$$

$$= (70.5)^2 \times 0.2 = 994.05 \text{ W}$$

$$\text{Motor shunt field Cu loss} = I_{shm} V = 2 \times 250 = 500 \text{ W}$$

$$\text{Generator armature Cu loss} = I_{ag}^2 R_{ag} = (60)^2 \times 0.2 = 720 \text{ W}$$

$$\text{Generator shunt field Cu loss} = I_{shg.} V = 2.5 \times 250 = 625 \text{ W}$$

$$\text{Copper losses of motor} = 994.05 + 500 = 1494.05 \text{ W}$$

$$\text{Copper losses of generator} = 720 + 625 = 1345 \text{ W}$$

$$\text{Total copper losses} = 1494.05 + 1345 \cong 2839.05 \text{ W (say 2839 W)}$$

$$\text{Input to the set of two machines} = VI_L = 250 \times 15$$

$$= 3750 \text{ W}$$

$$\text{Total stray losses} = 3750 - 2839$$

$$= 911 \text{ W}$$

$$\text{Stray losses of each machine} = \frac{911}{2} = 455.5 \text{ W}$$

$$\text{Total losses of motor} = \text{Cu loss} + \text{stray loss}$$

$$= 1494.05 + 455.5 = 1949.55 \text{ W}$$

$$\text{Motor input} = I_m \times V = 72.5 \times 250 = 18125 \text{ W}$$

$$\text{Motor efficiency, } \eta = \frac{\text{motor input} - \text{losses}}{\text{motor input}} \times 100$$

$$= \frac{18125 - 1949.55}{1812.5} \times 100 = \mathbf{89.24\%} \text{ (Ans.)}$$

$$\text{Generator output} = I_g \times V = 60 \times 250 = 15000 \text{ W}$$

$$\text{Generator efficiency, } \eta = \frac{\text{generator output}}{\text{output + losses}} \times 100$$

$$= \frac{15000}{15000 + 1345 + 455.5} \times 100$$

$$= \mathbf{89.28\%} \text{ (Ans.)}$$

5.37 Testing of DC Series Machines

(*Field test*)

Small DC series machines can be tested by brake test but large DC series machines cannot be tested by brake test because neither it is convenient nor possible to develop a mechanism to apply load on such large machines directly.

Moreover, DC series machines cannot be tested by Swinburne's test, because at no-load these machines obtain dangerously high speeds. In view of this*, *field test* is considered to be most suitable for determining efficiency of these machines.

Field Test

To determine efficiency of large *DC* series machines usually field test is employed. To perform this test, two** identical DC series machines are coupled mechanically, as shown in Fig. 5.65, and their fields are connected in series so that the iron losses of both the machines be made equal. One of the machine-I to which supply is given, operates as a motor and drives the other machine-II. This other machine operates as a separately excited generator. A variable load R_L is connected directly across its terminals without any switch.

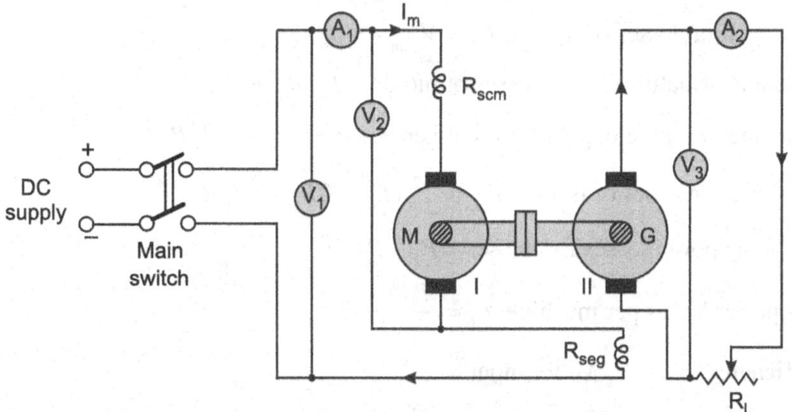

Fig. 5.65 Circuit arrangement for field test of DC series motors

Performance

The machine-I (motor) is switched-*ON* to the supply in a usual manner and the output of the other machine-II, working as a generator, is dissipated in the variable load resistor R_L. The voltage across the motor terminals V_2 is kept equal to its rated value. It is obvious that the supply voltage V_1 will be more than V_2 i.e., $V_1 = V_2 + I_m R_{seg}$. The load resistor R_L is varied till the ammeter A_1, shows full-

*Field test looks to be similar to regenerative test but it is not so because in this case the output of the generator is not feedback to the motor rather it is dissipated in a load resistor. This will be confirmed after going through the test.

**It is not difficult to have two identical DC series machines because in electric traction DC series motors are used in pairs.

load motor current. The hot resistances of various windings are measured by voltmeter-ammeter (or other suitable) method after performing the test. Note down the readings of various measuring instruments as mentioned below:

Precautions

Every effort is to be made to get accurate measurements since the accuracy of the test depends upon the accuracy of these measurements.

Let,

Supply voltage = Reading of voltmeter $V_1 = V_1$ volt

Motor input current = Reading of ammeter $A_1 = I_m$

Terminal voltage of generator = Reading of voltmeter $V_3 = V_3$

Load current or generator current = Reading of ammeter $A_2 = I_g$

Armature resistance of each machine = R_a

Series field resistance of each machine = R_{se}

$$\text{Input to the whole set} = V_1 I_m \quad \ldots(5.61)$$

$$\text{Output} = V_3 I_g \quad \ldots(5.62)$$

$$\text{Total losses of the set, } P_T = V_1 I_m - V_3 I_g \quad \ldots(5.63)$$

$$\text{Series field and armature copper losses of motor} = I_m^2 (R_a + R_{se}) \quad \ldots(5.64)$$

$$\text{Series field and armature copper losses of generator} = I_m^2 R_{se} + I_g^2 R_a \quad \ldots(5.65)$$

$$\text{Total copper losses of the set, } P_{cu} = I_m^2 (R_a + 2R_{se}) + I_g^2 R_a \quad \ldots(5.66)$$

$$\text{Stray power losses for the set} = P_T - P_{cu}$$

$$\text{Stray power losses per machine, } P_s = \frac{P_T - P_{cu}}{2} \quad \ldots(5.67)$$

Motor efficiency

$$\text{Motor input} = V_2 I_m \quad \ldots(5.68)$$

$$\text{Motor losses} = I_m^2 (R_a + R_{se}) + P_s \quad \ldots(5.69)$$

$$\text{Motor output} = V_2 I_m - I_m^2 (R_a + R_{se}) - P_s \quad \ldots(5.70)$$

$$\text{Motor efficiency, } \eta_m = \frac{V_2 I_m - I_m^2 (R_a + R_{se}) - P_s}{V_2 I_m} \quad \ldots(5.71)$$

Generator efficiency

$$\text{Generator output} = V_3 I_g \quad \ldots(5.72)$$

$$\text{Generator losses} = I_g^2 R_a + I_m^2 R_{se} + P_s \quad \ldots(5.73)$$

$$\text{Generator input} = V_3 I_g + I_g^2 R_a + I_m^2 R_{se} + P_s \quad \ldots(5.74)$$

$$\text{Generator efficiency, } \eta_g = \frac{V_3 I_g}{V_3 I_g + I_g^2 R_a + I_m^2 R_{se} + P_s} \quad \ldots(5.75)$$

Disadvantages of Field Test Method

1. Even for a small error in the measurement of the input to motor or output of generator may cause a relatively large error in computed the efficiency.
2. Whole of the power supplied to the set is wasted.

Example 5.60

The following test data is obtained after performing a field test on two identical, mechanically coupled DC series motors (with their field windings connected in series):

Motor: Armature current-50 A, Armature voltage-500 V; Field winding voltage drop 35 V

Generator: Armature current-38 A; Armature voltage-400 V; Field winding voltage drop-32 V; Resistance of each armature is 0.2 Ω.

Calculate the efficiency of each machine at this load.

Solution.

$$\text{Voltage across motor armature} = 500 \text{ V}$$

$$\text{Voltage across motor, } V_2 = 500 + 35 = 535 \text{ V}$$

$$\text{Supply voltage, } V_1 = V_2 + \text{voltage drop across generator field winding}$$

$$= 535 + 32 = 567 \text{ V}$$

$$\text{Terminal voltage of generator, } V_3 = 400 \text{ V}$$

$$\text{Generator armature current, } I_g = 38 \text{ A}$$

$$\text{Motor armature current, } I_m = 50 \text{ A}$$

$$\text{Total power input} = V_1 I_m = 567 \times 50 = 28350 \text{ W}$$

$$\text{Generator output} = V_3 I_g = 400 \times 38 = 15200 \text{ W}$$

$$\text{Total copper losses} = I_m^2 \times 0.2 + I_g^2 \times 0.2 + I_m \; (35 + 32) \text{ watt}$$

$$= (50)^2 \times 0.2 + (38)^2 \times 0.2 + 50 \times 67 = 4149.8 \text{ W}$$

$$\text{Stray losses of the set} = \text{total power input} - \text{power output} - \text{total copper losses}$$

$$= 28350 - 15200 - 4149.8 = 9000.2 \text{ W} \cong 9000 \text{ W}$$

$$\text{Stray losses per machine, } P_S = \frac{9000}{2} = 4500 \text{ W}$$

Motor Efficiency

$$\text{Input power to motor} = V_2 I_m \times 535 \times 50 = 26750 \text{ W}$$

$$\text{Total losses in motor} = 50^2 \times 0.2 + 50 \times 35 + 4500 \text{ W} = 6750 \text{ W}$$

$$\text{Motor output} = \text{Input} - \text{total losses} = 26750 - 6750 = 20000 \text{ W}$$

$$\text{Motor efficiency, } \eta_m = \frac{\text{Motor output}}{\text{Motor input}} \times 100 = \frac{20000}{26750} \times 100 = \mathbf{74.77\%} \; (Ans.)$$

Generator Efficiency

$$\text{Generator output} = V_2 I_g = 400 \times 38 = 15200 \text{ W}$$

$$\text{Total losses in generator} = (38)^2 \times 0.2 + 50 \times 32 + 4500 = 6388.8 \text{ W}$$

$$\text{Generator input} = \text{Output} + \text{total losses} = 15200 + 6388.8 = 21588.8 \text{ W}$$

$$\text{Generator efficiency, } \eta_g = \frac{\text{Generator output}}{\text{Generator input}} \times 100$$

$$= \frac{15200}{21588.8} \times 100 = \mathbf{70.4\%} \text{ (Ans.)}$$

Example 5.61

The field windings of the two similar tramway motors are connected in series and the motors are coupled to perform a test. The following test results were obtained while one machine acts as a motor and the other acts as a generator.

Motor: Armature Current: 56A, Armature Voltage: 590 V,

Voltage drop a cross field winding: 40V

Generator: Armature Current: 44A, Armature Voltage: 400V, Field Voltage drop: 40V, Resistance of each armature: 0.3 ohm.

Calculate the efficiency of motor and generator.

Solution:

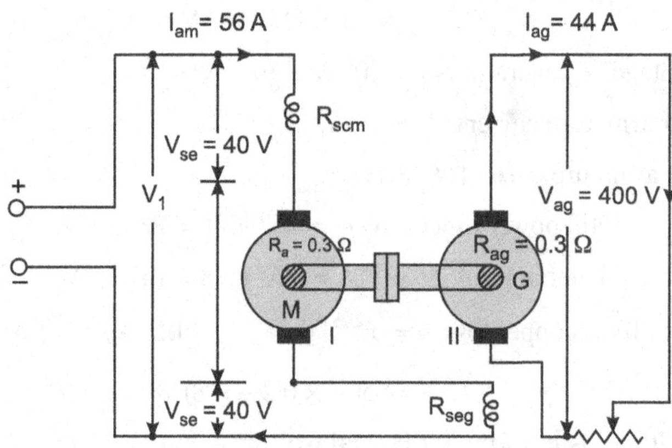

Fig. 5.66 Circuit diagram

Here, $I_{am} = 56 \text{ A}; V_{am} = 590 \text{ V}; V_{se} = 40 \text{ V}$

$$I_{ag} = 44 \text{ A}; V_{ag} = 400 \text{ V}; V_{se} = 40 \text{ V}$$

$$R_{am} = R_{ag} = 0.3\ \Omega$$

$$\text{Input to the whole set} = V_1 I_{am} = (40 + 590 + 40) \times 56 = 37520 \text{ W}$$

$$\text{Output of the set} = V_{ag} I_{ag} = 400 \times 44 = 17600 \text{ W}$$

$$\text{Total losses of the set} = \text{Input} - \text{Output} = 37520 - 17600 = 19920 \text{ W}$$

Series field and armature Cu loss of motor

$$= V_{sem} \times I_{am} + I_{am}^2 R_{am}$$

$$= 40 \times 56 + (56)^2 \times 0.3 = 3180.8 \text{ W}$$

Series field and armature Cu loss of generator

$$= V_{seg} \times I_{am} + I_{ag}^2 R_{ag}$$

$$= 40 \times 56 + (44)^2 \times 0.3 = 2820.8 \text{ W}$$

$$\text{Total Cu losses of the set} = 3180.8 + 2820.8 = 6001.6 \text{ W}$$

$$\text{Stray losses of the set} = 19920 - 6001.6 = 13918.4 \text{ W}$$

$$\text{Stray losses of each machine} = \frac{13918.4}{2} = 6959.2 \text{ W}$$

For motor efficiency

$$\text{Motor input} = (V_{am} + V_{sem}) \times I_{am} = (590 + 40) \times 56 = 35280 \text{ W}$$

$$\text{Motor losses} = 3180.8 + 6959.2 = 10140 \text{ W}$$

$$\text{Motor efficiency, } \eta_m = \frac{\text{motor input} - \text{losses}}{\text{motor input}} \times 100$$

$$= \frac{35280 - 10140}{35280} \times 100 = \mathbf{71.26\%} \text{ (Ans.)}$$

For generator efficiency:

$$\text{Generator output} = V_{ag} I_{ag} = 400 \times 44 = 17600 \text{ W}$$

$$\text{Generator losses} = 2820.8 + 6959.2 = 9780 \text{ W}$$

$$\text{Generator efficiency, } \eta_g = \frac{\text{Generator output}}{\text{Generator output} + \text{losses}} \times 100$$

$$= \frac{17600}{17600 + 9780} \times 100 = \mathbf{64.28\%} \text{ (Ans.)}$$

5.38 Inspection/maintenance of DC Machines

The basic purpose of inspection/maintenance is to ensure uninterrupted and efficient service from the machine. The frequency of routine inspection/maintenance depends upon conditions under which a machine is operating. The normal routine inspection/maintenance tasks are discussed below:

1. **Mechanical maintenance**
   - (*i*) *Bearings:* Check sound and excessive play if any. Check for proper lubrication-avoid excessive lubrication.
   - (*ii*) *General*: Check all the fixing fixtures – tight them it required.

   Check that all the cover-plates and enclosures are clean and correctly in place – if not fix them.
2. **Electrical maintenance**

   Always keep the insulation dry and in good condition for better reliability.
   - (*i*) *Commutator:* Trouble may be caused due to dirt deposits deposit, of carbon dust in the grooves between segments, rough surface or eccentricity. Check for these and clean the commutator surface with a dry soft fluff free cloth if required.
   - (*ii*) *Brushes:* Trouble may be caused due to less spring tension, wrong brush positioning, insufficient contact surface or too short a brush. Check for there and do the adjustments or changes as per the need.

(*iii*) *Winding*: Trouble may be caused due to deterioration of insulation. Check the insulation resistance of armature winding and field winding periodically – if faculty, it should be repaired.

(*iv*) *Insulation resistance*: The insulation resistance of the whole machine be checked periodically which should not be less than 1 M Ω. If its value is found to be less than this then the machine should be cleaned and dried.

(*v*) *Earthing*: Outer frame of the motor must be properly earth, the earth resistance should be checked, if found more pour water in earthing pit.

(*vi*) *Starter*: Check the functioning of the starter, its relay must be sensitive enough to protect the machine.

5.39 Faults in DC Machines

In DC machines faults may occur in field windings or in armature winding.

1. *Faults in field winding*: The faults in field winding may be

(*a*) *Open circuit fault*: Check the winding for open circuit, if found faulty, check the opening at the terminals if so tight-up the terminals otherwise replace the field winding.

(*b*) *Earth fault*: Check for the earth fault, if found remove it or replace the coil.

(*c*) *Short circuit fault:* Check for the voltage drop in each turn and locate the fault and replace the turn or repair the field winding.

2. *Faults in armature winding:* In armature winding, following faults may develop:

(*i*) *Short-circuit in coils*: The turns placed in the same slot of the armature may get short circuited due to insulation failure between the turns. Heavy local current flows and heat is produced. This may lead to open circuit. The heavy current is purely local and it does not over-load the brushes.

(*ii*) *Open-circuit in coils:* Due to open circuiting of armature coil, half of the armature which carries sound coils carries double the normal current and the other portion carries no current. This produces heavy sparking at the brushes which may damage it. Therefore machine has to be stopped and repaired immediately.

(*iii*) *Earthed coil:* When a coil in the armature comes in direct contact with armature stampings due to insulation rupturing, the fault is called earth fault. If this fault occurs at one place, the effect is not that serious but if this fault occurs at two places, the two coils are being short circuited causing a serious short circuit fault which produces heavy sparking due to unbalancing of armature currents.

5.40 Trouble Shooting in a DC Motor

For trouble shooting of a DC motor, the following steps are taken:

1. Check the fuses.
2. Check the electrical connections at the terminals
3. Check the wear and tear of the brushes.

4. Check the continuity of field winding.
5. Check the armature winding.

Section Practice Problems

Numerical Problems

1. The field and armature resistance of a 250 V DC shunt generator is 50 ohm and 0.02 ohm respectively. It delivers 195 A at rated voltage. The iron and friction losses equal 1050W. Find (*a*) emf generated (*b*) copper losses (*c*) output of the prime-mover (*d*) commercial, mechanical and electrical efficiencies.
(**Ans.** 254 V; 2050 W; 51.85 kW; 94.11%, 97.975%, 95.96%)

2. A four-pole DC shunt motor draws a current of 36 A when connected to 230 V mains and delivering 10 hp. If armature and field circuit resistances are 0.1 Ω and 300 Ω respectively, find the motor current when load is 5 hp. Neglect armature reaction effect. (**Ans.** 19627 A)

3. A 6-pole, 250 V series motor is wave-connected. There are 240 slots and each slot has four conductors. The flux per pole in 1.75×10^{-2} Wb when the motor is taking 80 A. The field resistance is 0.05 Ω, the armature resistance is 0.1 Ω and the iron and frictional loss is 0.1 kW. Calculate (*a*) speed (*b*) bhp (*c*) shaft torque (*d*) the pull in newton at the rim of the pulley of diameter 30 cm.
(**Ans.** 283.3 rpm; 25.75; 638.34 Nm; 4255.6 N)

4. In a brake test, a DC shunt motor draws 40 A from a 231 V supply mains. The brake pulley of radius 30 cm had an effective load of 35 kg and the speed was 720 rpm. Find the bhp (metric) and the efficiency at the above load. (**Ans.** 10.56; 84.05%)

5. The field resistance and armature resistance of a 400 V DC shunt motor is 200 ohm and 0.5 ohm respectively. It draws 5 A at no-load. What will be its output and efficiency when it draws 50 A from the mains. Also find the percentage change in speed from no-load to full-load. (**Ans.** 16.852 kW; 84.26%,; 5.65%)

6. A DC machine is rated at 5 kW, 250 V, 2000 rpm. Its armature resistance R_a is 1 Ω. Driven from the electrical end at 2000 rpm the no-load power input to the armature is I_a = 1.2 A at 250 V with the field winding (R_f = 250 Ω) excited by I_f = 1A. Estimate the efficiency of the machine as a 5 kW generator.
(**Ans.** 83.48%)

7. The following test results we obtained when Hopkinson test was performed on two identical DC shunt machines at full load:
Line voltage 250 V, line current excluding field currents 50 A;
Motor armature current 380 A; Field currents 5 A and 4.2 A.
Calculate the efficiency of each machine. The armature resistance of each machine = 0.02 ohm.
(**Ans.** η_m = 92.03%; η_g = 91.9%)

Short Answer Type Questions

Q.1. What do you mean by "energy loss" in a machine?

Ans. Whole of the input energy supplied to the electrical machine is not converted into useful output energy but a part of it is lost in the form of heat while doing useful work. The energy converted into heat is called the energy loss.

Q.2. What are the different losses which occur in a DC machine.

Ans. The following losses occur in a DC machine:

1. **Copper losses:** These may be
   (*i*) Armature copper loss ($I_a^2 R_a$)
   (*ii*) Series field copper loss ($I_{se}^2 R_{se}$)
   (*iii*) Interpole winding copper loss ($I_a^2 R_i$)
   (*iv*) Compensating winding copper loss ($I_a^2 R_c$)
   (*v*) Brush contact loss ($I_a v_b$)
   (*vi*) Shunt field copper loss ($I_{sh}^2 R_{sh}$)
2. **Iron-loss:** (*i*) Hysteresis loss ($K_b V f B_m^{1.6}$) (*ii*) Eddy current loss ($K_e V f^2 t^2 B_m^2$).
3. **Mechanical loss:** (*i*) Friction loss (*ii*) Windage loss

Q.3. How do the various losses occurring in a DC machine vary with the load?

Ans. The copper losses occurring in armature windings i.e., series field winding, compensating winding and interpole winding, vary as the square of the load current whereas copper losses occurring in DC shunt field winding is constant. Iron and mechanical losses are independent of load and remain constant.

Q.4. What are the components of iron loss in DC machine and in which part of the machine it occurs?

Ans. The components of iron loss are hysteresis and eddy-current losses obviously these losses occur in the iron parts of the machine such as armature core, pole shoe, etc.

Q.5. How does eddy current losses change if the thickness of laminations is increased?

Ans. Eddy current loss will increase as the square of the thickness of magnetic material laminations.

Q.6. What type of tests are performed to determine efficiency of DC motors?

Ans. Basically, direct, indirect and regenerative tests are performed to determine the efficiency of DC machines.

Q.7. Can a Swinburne's test be applied to a DC series motor? Justify.

Ans. No, Swinburne's test is performed at no-load and at no-load series motors obtain dangerously high speed and gets damaged. Therefore, Swinburne's test cannot be applied for determining efficiency of DC series motors.

Q.8. What is the main drawback of brake test?

Ans. The output of the motor cannot be measured accurately.

Q.9. Why constant losses can be considered equal to input to a machine at no-load?

Ans. Because at no-load variable losses (armature copper losses and losses in series field winding, interpole winding and compensating winding, if any) are negligible owing to very small armature current.

Q.10. State the advantages of Swinburne's test.

Ans. Advantages of Swinburne's test are:

1. Very convenient and economical method since power required is very small.
2. Stray or constant losses are determined, therefore, efficiency at any desired value of load can be determined.

Q.11. Why Hopkinson's test is also called the regenerative or back-to-back test?

Ans. The motor supplies mechanical power which is used to drive the generator while generator develops electrical power which is sent back to the system and is utilised in driving the DC motor in addition to electrical power drawn from the supply mains.

Q.12. What are the possible causes of sparking at brushes?

Ans. Sparking at brushes may occur due to

(*i*) trouble in brushes,

(*ii*) commutation difficulties

(*iii*) excessive load current.

Review Questions

1. Explain the principle of operation of a DC motor.
2. Explain with suitable diagram the working of a DC motor.
3. Explain the function of commutator in a DC motor.
4. What is back emf? Give its significance.
5. On what factors does the torque developed by a DC motor depends?
6. What is back emf? Is the back emf greater or lesser than the applied voltage? Why? By what amount do the two voltages differ. Write voltage equation of a motor.
7. A DC shunt motor operating from a constant voltage supply is running steadily on no-load. Explain how the motor will adjust itself on application of load.
8. Derive an expression for the speed of a DC motor in terms of back emf and flux per pole.
9. Explain back emf in a DC motor. Explain the effect of field current on the space of a DC motor.
10. How may the direction of rotation of a DC shunt motor be reversed? What is the effect of reversing the line terminals?
11. Derive the torque equation of a DC machine.
12. Define an expression for the torque developed by a DC motor in terms of ϕ, Z, P, A, I_a where the symbols have their usual meaning.
13. Differentiate between the motoring and generating action of a DC machine.
14. Mention the various types of DC motors and their uses.
15. Show that a shunt motor is almost constant speed motor. Draw its speed-armature current characteristics.
16. Sketch the speed-torque curve of a DC series motor and discuss its nature. What are the applications for DC series motors?
17. A DC series motor should not be started without load why?
18. Show that a series motor develops high starting torque.
19. Using characteristics, explain why a DC series motor

    (*i*) Is suitable for electric traction

    (*ii*) Should never be started without a load on it.

20. Sketch the speed-load and torque-load characteristics of DC cumulatively-compound motor and comment on the shape of the characteristics. Indicate where such a motor can be ideally used.
21. Some of the applications given below need suitable motors. Motors available are series, shunt, cumulatively compound, differential compound DC motors. Mention the motor used for the following applications; give reasons for your answer.

    Blower, shearing machines, diesel electric locomotives, cranes, hoists centrifugal pumps, elevators and rolling mills.
22. Describe the factors for the selection of DC motors for specific application.
23. List applications of DC shunt, DC series and DC compound motors.
24. How will you determine the value of resistance of each step of a DC shunt motor starter when the maximum and minimum value of current varies between I_1 and I_2?
25. Explain the function of no-volt and over load release coil in a two-point series motor starter?
26. While starting a DC motor, why does a resistor connected in series with the armature? Describe a suitable starter for starting a DC shunt motor having No-volt and overload protections.
27. Name various starters used for DC motor and explain working of any one of them with a suitable diagram.
28. Write a short note on DC motor starters.
29. Explain why a starter is required for starting a DC motor. Describe a three-point starter having no-volt and over-load protections for starting a DC shunt motor. What modification is made in a 4 point starter?
30. Explain the working principle of a 3 point starter of a DC shunt motor.
31. What may happen to a DC shunt motor connected to a three-point starter if the field excitation is kept minimum at the time of starting?
32. A DC shunt motor is connected to a three-point starter. Explain what would happen if:
    (*i*) The starter handle is moved rapidly from OFF to the ON position.
    (*ii*) The field circuit is open and an attempt is made to start the motor.
    (*iii*) The field circuit becomes open-circuited with the motor running at no-load, with the assumption that the starter is not provided with the no-volt release and the spring.
    (*iv*) The field excitation is minimum at the time of starting.
33. Why is starting current high in a DC motor? Explain the working of a four-point starter for a DC machine.
34. Write a general expression for the speed of a DC motor in terms of supply voltage and flux per pole.
35. Describe the speed control methods of DC shunt motors.
36. Describe briefly the methods of speed control used for DC series motors.
37. Explain with neat sketch how speed control of a DC shunt motor is obtained by Ward Leonard control system. How the direction of rotation of the motor is usually reversed in this method of speed control?
38. Identify the factors that predominantly affect the speed of a DC motor. Discuss two methods for varying the speed of a DC series motor for constant load torque. List one application of each of these methods.
39. Define speed regulation and percentage speed regulation of a DC motor.
40. Which of the following three motors has the poorest speed regulation: Shunt motor, series motor or cumulative compound motor? Explain.

41. Explain the series-parallel control of two identical series motors. How many speeds are possible? Are the motors run faster in series combination or parallel combination?

42. Explain the speed control of DC series motor by (*i*) field divertors (*ii*) variable resistance in series with motor.

43. With relevant circuit diagrams, describe different methods of speed control of DC series motor.

44. Explain four methods by which flux is varied to control the speed of a DC series motor, with neat sketches.

45. What is meant by braking of DC motors? Describe, briefly, the various methods of braking of DC shunt motors.

46. Why is electric braking of electric motors superior to mechanical braking? How is dynamic braking of DC shunt motor done?

47. What is regenerative braking? Why it cannot be accomplished with a series wound DC motor without modifications in the circuit diagram? Give merits and demerits?

48. What are the various losses occurring in rotating machines? Mention the method to reduce them.

49. Give an account of the losses which occur in a DC machine and explain how these losses vary with load and speed.

50. Mention the factors on which the hysteresis loss (W_b) and eddy current loss (W_e) in a DC machine depend. How these losses are reduced?

51. Draw a power flow diagram of a DC motor and define commercial, mechanical and electrical efficiency of a motor.

52. Derive condition for maximum efficiency of a DC motor.

53. What are the various tests performed on a DC Machine? Explain any two.

54. To determine no-load losses, Swinburne's test is performed. Explain it and mention its limitations.

55. Describe the Hopkinson's test for obtaining the efficiency of two similar shunt machines.

56. What is Hopkinson test? Draw a diagram and explain the procedure of Hopkinson test.

57. Classify various losses in a DC machine and indicate the factors on which these depend. Explain the regenerative method of determining the efficiency of a DC machine. List the merits and demerits of the method.

58. Explain a suitable method for determining the efficiency of a series motor.

59. With a neat circuit diagram discuss how the efficiency of DC series motors can be determined, by conducting field's test.

Multiple Choice Questions

1. Usually, the normal value of the armature resistance of a DC motor is
   (*a*) 100 Ω (*b*) 0.5 Ω
   (*c*) 10 Ω (*d*) 0.005 Ω

2. The ratio E_b/V of a DC motor represents its
   (*a*) efficiency (*b*) speed regulation
   (*c*) starting torque (*d*) Running Torque

3. The power developed in the armature of a DC motor is equal to
(*a*) armature current multiplied by back emf
(*b*) power input minus losses
(*c*) power output plus iron losses
(*d*) power output multiplied by efficiency

4. If load on a DC shunt motor is increased, its speed is decreased due primarily to
(*a*) increase in its flux
(*b*) decrease in back emf
(*c*) increase in armature current
(*d*) increase in brush drop

5. If the voltage applied across the armature of a DC shunt motor is increased by 5% keeping its load current and field constant, what will be the effect on its speed?
(*a*) decrease by about 5 per cent
(*b*) remain unchanged
(*c*) increase by about 5 per cent
(*d*) increase by 10 per cent

6. The torque available at the shaft of a DC motor is less than the torque developed in the armature because of.........losses.
(*a*) copper
(*b*) mechanical
(*c*) iron
(*d*) rotational

7. Neglecting saturation, if current taken by series motor is increased from 10 A to 13 A, the percentage increase in its torque is..... percent
(*a*) 80
(*b*) 69
(*c*) 44
(*d*) 39

8. The armature resistance of a 110 V DC motor is 0.5 ohm. What will be its armature current if the back emf is 100 V.
(*a*) 20 A
(*b*) 200 A
(*c*) 220 A
(*d*) 60 A

9. With the increase in load, the speed of a DC shunt motor
(*a*) reduces slightly
(*b*) remains constant
(*c*) increases slightly
(*d*) increases proportionately

10. For drives requiring high starting torque but only fairly constant speed such as crushers, the most suitable DC motor would be
(*a*) shunt
(*b*) series
(*c*) compound
(*d*) permanent magnet

11. DC shunt motors are most suitable to drive fans because they require
(*a*) small torque at start
(*b*) large torque at high speeds
(*c*) practically constant voltage
(*d*) both (*a*) and (*b*)

12. When a 220 V DC shunt motor draws an armature current of 10 A, it develops a torque of 50 Nm. What will be the torque developed when armature current reduces to 20 A.
(*a*) 54 N-m
(*b*) 81 N-m
(*c*) 108 N-m
(*d*) None of the above

13. One can change the speed of a DC motor by varying
(*a*) its flux per pole
(*b*) resistance of armature circuit
(*c*) applied voltage
(*d*) all of above

14. The most efficient method of increasing the speed of a 5 kW DC shunt motor would be
(*a*) armature control method
(*b*) flux control method
(*c*) Ward-Leonard method
(*d*) tapped-field control method

15. Usually, ... test is performed to determine the efficiency of a traction motor
(*a*) Field's (*b*) brake
(*c*) Hopkinson's (*d*) Swinburne's

16. The Swinburne's test is considered to be the best one because it
(*a*) is applicable both to shunt and compound motors
(*b*) needs one running test
(*c*) is very economical and convenient
(*d*) ignores any charge in iron loss

17. The major disadvantage of Hopkinson's test for finding efficiency of DC shunt motors is that it
(*a*) requires full-load power (*b*) ignores any change in iron loss
(*c*) needs one motor and one generator (*d*) requires two identical shunt machines

Keys to Multiple Choice Questions

1. b	**2.** a	**3.** a	**4.** b	**5.** c	**6.** b	**7.** b	**8.** a	**9.** a	**10.** c
11. d	**12.** c	**13.** d	**14.** b	**15.** a	**16.** c	**17.** d			

Synchronous Generators or Alternators

CHAPTER

6

Chapter Objectives

After the completion of this unit, students/readers will be able to understand:

- ✓ What are the basic principles on which working of synchronous machine depends?
- ✓ How emf is induced in a synchronous generator.
- ✓ Why stationery armature and revolving field system is preferred for large synchronous generators?
- ✓ What is the function and material of different parts of a synchronous machine?
- ✓ Where and why synchronous generators of salient or non-salient pole type construction are preferred.
- ✓ How magnetic poles of synchronous machines are excited.
- ✓ What are various types of windings used in synchronous machines, viz. single-phase and three-phase, concentrated and distributed, single-layer and double-layer, full-pitched and short-pitched, concentric, lap and wave winding?
- ✓ What is the meaning of different terms used while preparing different winding schemes viz. coil, coil pitch, pole pitch, electrical and mechanical angle, slot pitch, phase spread etc.
- ✓ What are coil span and distribution factors and their significance.
- ✓ Various factors on which induced emf in an alternator depends.
- ✓ How revolving field in set-up in the stator core of a 3-phase wound machine
- ✓ What is Ferrari's principle and its importance?
- ✓ Why the alternators are rated in kVA?
- ✓ What is armature resistance and leakage reactance?
- ✓ What is armature reaction and how it is affected by the type of load (resistive, inductive or capacitive)?
- ✓ What is the effect of armature reaction on the terminal voltage or induced emf of an alternator?
- ✓ How to draw a simplified equivalent circuit of an alternator?
- ✓ What is voltage regulation and how it is affected by the power factor of the load?
- ✓ How to perform open-circuit and short-circuit test on a synchronous generator?
- ✓ What is ampere-turn (or mmf) method for determining voltage regulation of a synchronous generator?
- ✓ What is zero power factor (or Potier method) of determining voltage regulation of a synchronous generator?
- ✓ What power is developed by a cylindrical synchronous generator?

- ✓ What is two-reactance concept for salient pole synchronous machines?
- ✓ How to determine direct axis synchronous reactance (X_d) and quadrature axis synchronous reactance (X_q)?
- ✓ What are transients in alternators?
- ✓ What is the meaning of sub-transient, transient and direct reactance?
- ✓ What are the various losses in an alternator?
- ✓ Why alternator are heated-up during working and how these are cooled down?

Introduction

In an *AC system* voltage level can be increased or decreased (as per requirement) very easily with the help of a transformer, therefore, this system is exclusively used for generation, transmission and distribution of electric power. The mechanical power or energy is converted into electrical power or energy with the help of an AC machine called *alternator* or *synchronous generator.* However, when the same machine can be used to convert electrical power or energy into mechanical power or energy, then it is known as a *synchronous motor.* Thus, the same machine can be operated as a generator or as a motor and in general, it is called as a *synchronous machine.* In fact, it is a machine which rotates only at synchronous speed ($N_S = 120\ f/P$) under all conditions. To understand the construction, working and performance of the machine its study is divided into number of main topics as chapters, to be followed.

In this chapter, the readers will go through various topics related to this machine when working as a *synchronous generator or alternator.*

6.1 General Aspects of Synchronous Machines

A machine in which the following relation is maintained for its satisfactory operation is called a *synchronous machine* (The machine may work as a generator or motor):

$$N_S = \frac{120f}{P} \quad \text{or} \quad f = \frac{PN_S}{120}$$

where N_S is the synchronous speed in *rpm;* f is the supply frequency and

P is the number of poles of the machine.

When the machine is to work as a generator, it has to run at synchronous speed (N_S) to generate power at certain frequency (f), called power frequency. In India its value is 50 Hz, whereas in the *USA* it is kept at 60 Hz.

When the machine works as a motor, it can rotate only at synchronous speed (N_S) since the magnetic poles are locked with the revolving field. If the machine fails to rotate at synchronous speed, it is palled out of step and stops.

Hence, synchronous machine (generator or motor) is a machine which only runs at synchronous speed and maintains the relation;

$$N_S = 120\ \text{f}/\text{P rpm}$$

6.2 Basic Principles

A *synchronous machine* is just an **electro-mechanical transducer** which converts mechanical energy into electrical energy or vice-versa. The fundamental phenomenon which make these conversions possible are:

(*i*) the law of electro-magnetic induction and (*ii*) law of interaction.

(*i*) *Law of electromagnetic induction:* This relates to the production of *emf*, i.e., *emf* is induced in a conductor whenever it cuts across the magnetic field (see Fig. 6.1). This is called *Faraday's first law of electromagnetic induction.*

(*ii*) *Law of interaction:* This law relates to the phenomenon of production of *force.* or *torque* i.e., whenever a current carrying conductor is placed in the magnetic field, by the interaction of the magnetic fields produced by the current carrying conductor and the main field, force is exerted on the conductor and torque is developed (see Fig. 6.2).

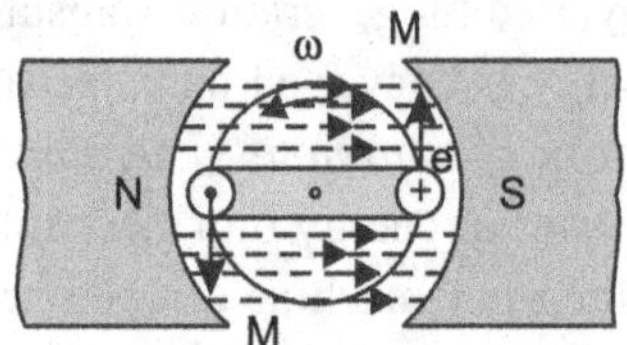

Fig. 6.1 Generator principle

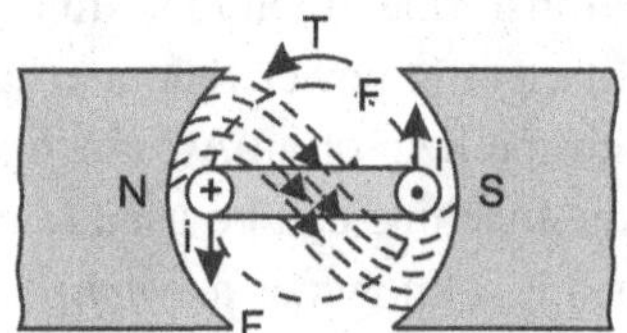

Fig. 6.2 Motor principle

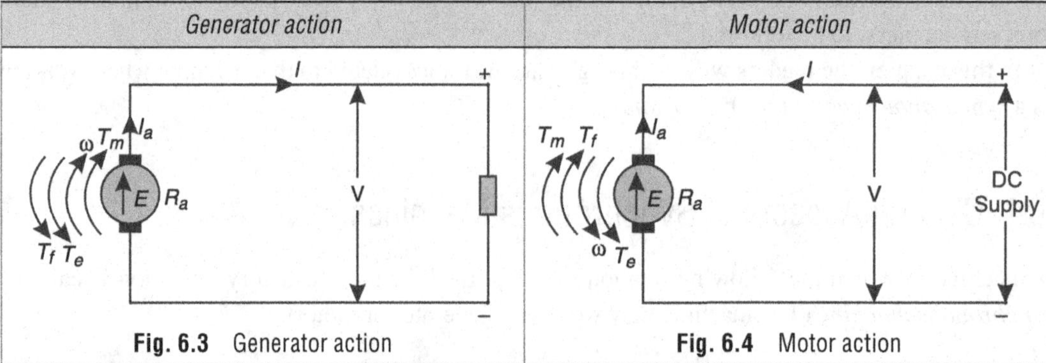

Fig. 6.3 Generator action

Fig. 6.4 Motor action

Generator action	Motor action
1. In generator action, the rotation is due to mechanical torque, therefore, T_m and ω are in the same direction.	**1.** In motoring action, the rotation is due to electromagnetic torque, therefore, T_e and ω are in the same direction.
2. The frictional torque T_f acts in opposite direction to rotation ω.	**2.** The frictional torque T_f acts in opposite direction to rotation ω.
3. Electromagnetic torque T_e acts in opposite direction to mechanical torque T_m so that $\omega T_m = \omega T_e + \omega T_f$.	**3.** Mechanical torque T_m acts in opposite direction to electromagnetic torque T_e so that $\omega T_e = \omega T_m + \omega T_f$.
4. In generator action, an emf is induced in the armature conductors which circulates current in the armature when load is connected to it. Hence, *e* and *i* both are in the same direction.	**4.** In motoring action, current is impressed to the armature against the induced emf (*e*), therefore current flows in opposite direction to that of induced emf.
5. In generator action, $E > V$	**5.** In motor action, $E < V$
6. In generator action, the torque angle θ is leading.	**6.** In motoring action, the torque angle θ is lagging.
7. In generator action, mechanical energy is converted into electrical energy.	**7.** In motoring action, electrical energy is converted into mechanical energy.

6.3 Generator and Motor Action

In generator action, an *emf* is induced in the armature conductors when they cut across the magnetic field. On closing the circuit, current flows through the armature conductors which produces another field. By the interaction of this field and main field a force is exerted on the conductor which acts is opposite direction to that of rotation. *It is this force against which the relative motion of conductors has to be maintained by the mechanical power supplied by the prime-mover, thus the* **mechanical power** is converted into **electrical power.**

In motor action, a current is supplied to the machine which flows through the armature conductors. The armature conductors produce a field which interacts with the main field. Thus, a force is exerted on the conductors and rotation takes place (i.e., torque is developed). Once rotation occurs, an *emf* is induced in the conductors due to relative motion. This *emf* acts in opposite direction to the flow of current. *The flow of current has to be maintained against this emf by applying external voltage source thus* **electrical power** is converted into **mechanical power.**

6.4 Production of Sinusoidal Alternating emf

When a conductor or coil cuts across the magnetic field an *emf* is induced in it by the phenomenon called electromagnetic induction. This can be achieved either by rotating a coil in the stationary magnetic field or by keeping the coil stationary and rotating the magnetic field. (The magnetic field can be rotated by placing the field winding on the rotating part of the machine).

For illustration see Figs. 6.5(*a*) and (*b*), two positions of a coil rotating in a stationary magnetic field are shown. Whereas, in Figs. 6.5(*c*) and (*d*), two positions of a rotating electro-magnet in a coil placed on stationary armature are shown. At first instant, the *emf* induced in the coil is zero since flux cut by the coil is zero. However, at second instant, the *emf* induced in the coil is maximum (say positive). The two instants t_1 and t_2 are marked on the wave diagram shown in Fig. 6.5(*e*). In one revolution the induced *emf* completes one cycle and its wave shape is shown in Fig. 6.5(*e*).

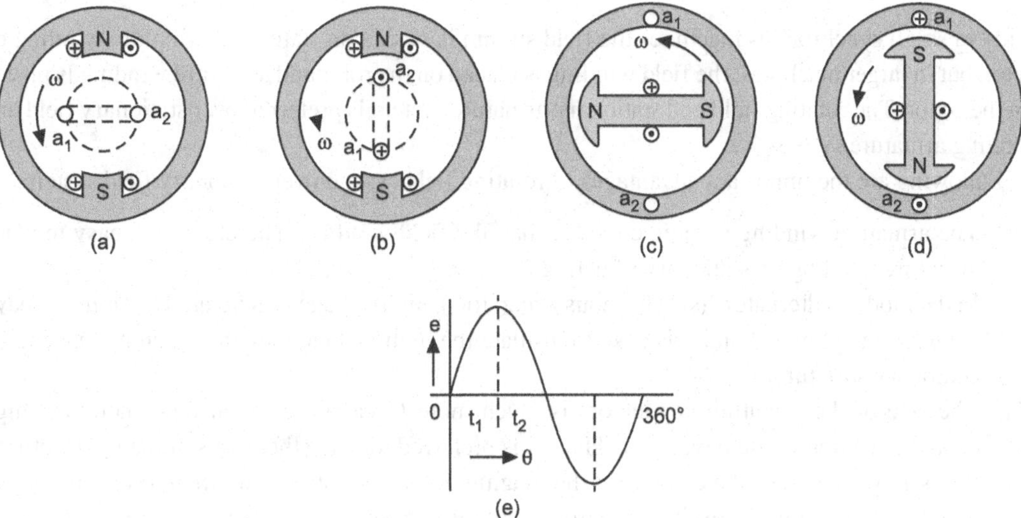

Fig. 6.5 (a to d) Generation of alternating emf, (e) Wave shape of alternating emf

6.5 Relation between Frequency, Speed and Number of Poles

In Fig. 6.6, a machine is shown having *P* number of poles on the rotor revolving at a speed at N_s *rpm* When a conductor passes through a pair of poles one cycle of *emf* is induced in it.

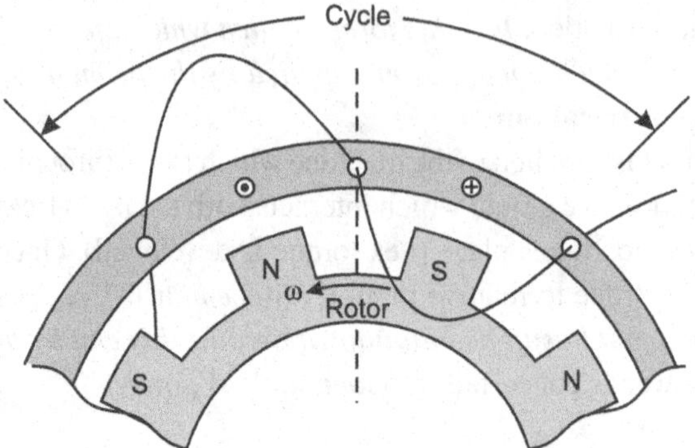

Fig. 6.6 One cycle is produced when a coil passes through a pair of poles

$$\therefore \quad \text{No. of cycle made per revolution} = \frac{P}{2}$$

$$\text{No. of revolutions made per second} = \frac{N_s}{60}$$

$$\therefore \quad \text{No. of cycles made per second} = \text{No. of cycles/revolution} \times \text{No. of revolutions/s}$$

$$f = \frac{P}{2} \times \frac{N_s}{60} = \frac{PN_s}{120} \text{ cycles/s or Hz}$$

6.6 Advantages of Rotating Field System over Stationary Field System

Only in small synchronous machines the field system is placed on stator and armature winding on rotor, but in larger machines, the field winding is placed on the rotor and armature winding is placed on the stator. The rotating field and stationary armature system is preferred over stationary field and rotating armature system.

Following are the important advantages of rotating field system over stationary field system:

(*i*) The armature winding is more complex than the field winding. Therefore, it is easy to place armature winding on stationary structure.

(*ii*) In the modern alternators (synchronous generators), high voltage is generated, therefore, heavy insulation is provided and it is easy to insulate the high voltage winding when it is placed on stationary structure.

(*iii*) The size of the armature conductors is much more to carry heavy current, therefore, high centrifugal stresses are developed. Thus, it is preferred to place them on stationary structure.

(*iv*) The size of slip rings depends upon the magnitude of flow of current, therefore, it is easy to deliver small current for excitation, through slip rings of smaller size when rotating field system is used.

(*v*) It is easier to build and properly balance high speed rotors when they carry the lighter field system.

(*vi*) The weight of rotor is small when field system is provided on rotor and as such friction losses are produced.

(*vii*) Better cooling system can be provided when the armature is kept stationary.

6.7 Constructional Features of Synchronous Machines

The important parts of a synchronous machine are given below:

1. Stator 2. Rotor 3. Miscellaneous

1. **Stator:** The outer stationary part of the machine is called stator; it has the following important parts:

   (*i*) *Stator frame:* It is the outer body of the machine made of cast iron and it protects the inner parts of the machine. It can be also made of any other strong material since it is not to carry the magnetic field. Cast iron is used only because of its high mechanical strength.

   (*ii*) *Stator Core:* The stator core is made of silicon steel material. It is made from number of stamping which are insulated from each other. Its function is to provide an easy path for the magnetic lines of force and accommodate the stator winding.

   (*iii*) *Stator Winding:* Slots are cut on the inner periphery of the stator core in which three-phase or one-phase winding is placed. Enamelled copper is used as winding material.

2. **Rotor:** The rotating part of the machine is called rotor. From construction point of view, there are two types of rotors named as
   (*i*) Salient pole type rotor; (*ii*) Non-salient pole type rotor.

   (*i*) **Salient pole type rotor:** In this case, projected poles are provided on the rotor. The cost of construction of salient pole type rotors is low, moreover sufficient space is available to accommodate field winding but these cannot bear high mechanical stresses at high speeds. Therefore, salient pole type construction is suited for medium and low speeds and are usually employed at hydro-electric and diesel power plants as synchronous generators. Since the speed of these machines (generators) is quite low, to obtain the required frequency, the machines have large number of poles as shown in Figs. 6.7 and 6.8. To accommodate such a large number of poles, these machines have larger diameter and small length.

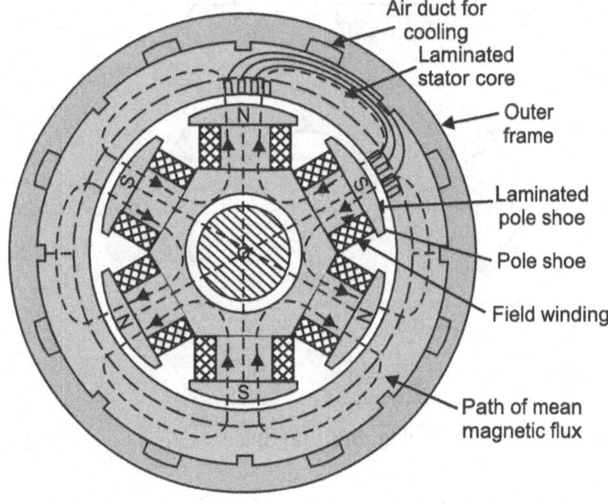

Fig. 6.7 Salient pole type alternator.

For a speed of 200 *rpm* (alternators coupled with water turbines) the diameter of the machines is as large as 14 metre and length is only 1 metre. The salient pole type rotor has the following important parts:

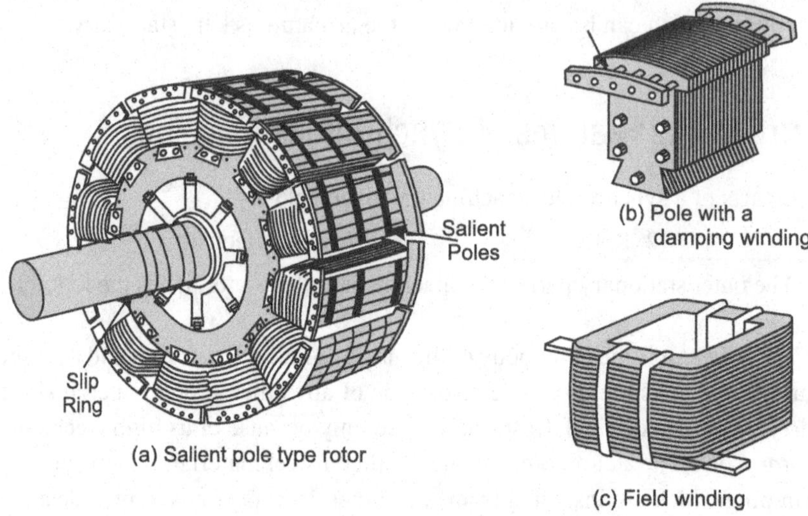

Fig. 6.8 Parts of rotor of salient pole alternator

(*a*) *Spider:* Spider is made of cast iron to provide an easy path for the magnetic flux. It is keyed to the shaft and at the outer surface, pole core and pole-shoe are keyed to it [see Fig. 6.9(a)].

(*b*) *Pole core and pole shoe:* It is made of laminated sheet material [see Figs. 6.8 (*b*) and 6.9(*b*)]. Pole core provides least reluctance path for the magnetic field and pole shoe distributes the field over the whole periphery uniformly to produce sinusoidal wave form of the generated *emf.*

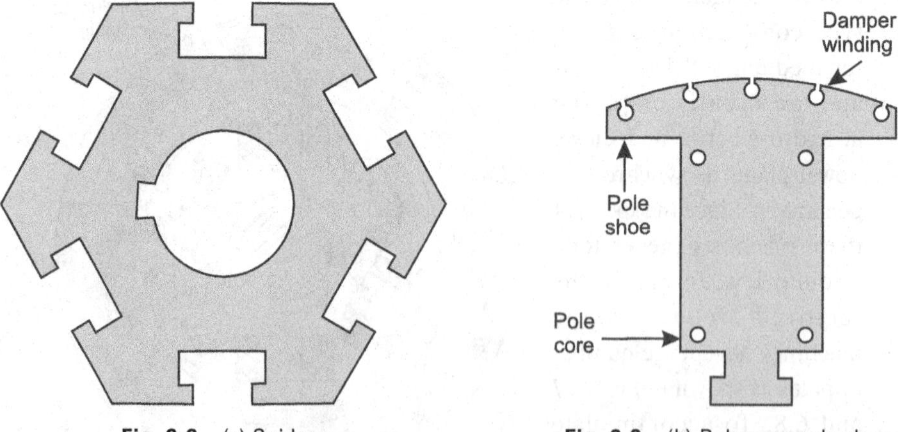

Fig. 6.9 (a) Spider

Fig. 6.9 (b) Pole core pole shoe

(*c*) *Field winding or Exciting winding:* Field winding [see Fig. 6.8 (*c*)] is wound on the former and then placed around the pole core. DC supply is given to it through slip rings. When direct current flows through the field winding, it produces the required magnetic field.

(*d*) *Damper winding:* At the outermost periphery, holes are provided [see Fig. 6.9 (*b*)] in which copper bars are inserted and short-circuited at both the sides by rings forming damper winding.

Generally, the segments on individual poles are joined together to form common rings resulting in a short-circuited squirrel cage winding similar to that used in induction machines with squirrel cage rotors. Salient pole machines are frequently provided with a damper winding on the rotor to damp rotor oscillations during transient-conditions and to facilitate smooth operation under unbalanced load conditions.

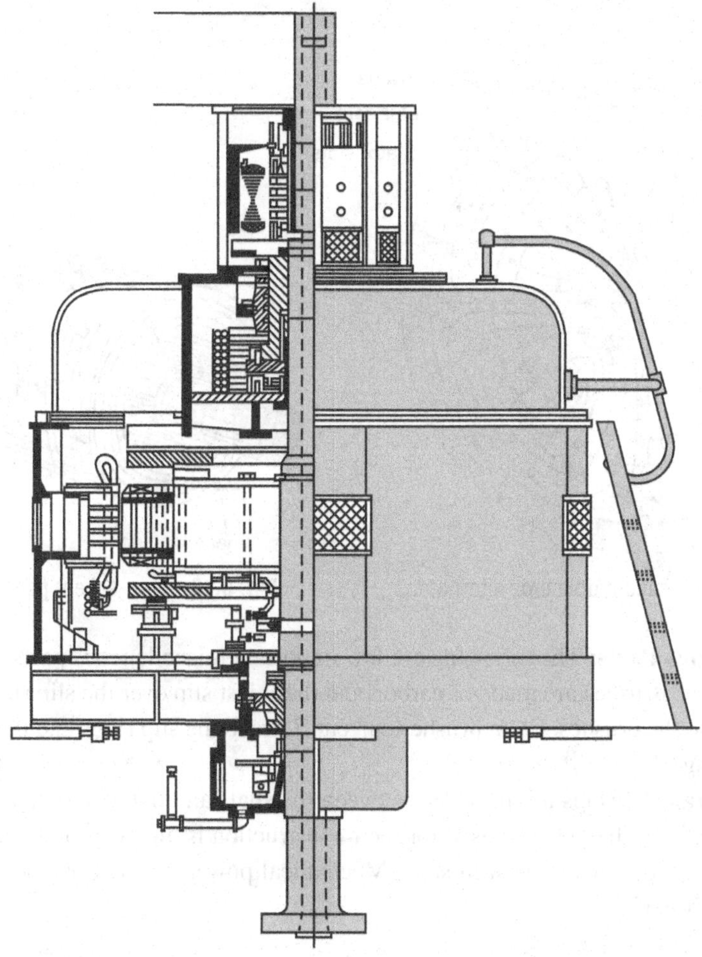

Fig. 6.10 Salient pole type alternator mounted vertically at hydro-electric power plant (half-sectional view)

At hydro-electric power plants, usually, salient pole type alternators are placed with their shafts in vertical position, as shown in Fig. 6.10.

(*ii*) Non-salient pole type rotor: A non-salient pole alternator is shown in Fig. 6.11. In this case, there are no projected poles but the poles are formed by the current flowing through the rotor (exciting) winding. Non-salient pole type construction is suited for the high speeds. The steam turbines rotate at a high speed (3000 *rpm*). When these turbines are used as prime-mover for this machine working as a generator, a small number of poles are required for given frequency. Hence, these machines have smaller diameter and larger length. Non salient pole type rotors have the following parts:

(*a*) *Rotor core:* Rotor core is made of silicon steel stampings. It is keyed to the shaft. At the outer periphery slots are cut in which exciting coils are placed. It provides an easy path to the magnetic flux.

(*b*) *Rotor winding or Exciting winding:* It is placed in rotor slots and current is passed through the winding in such a way that poles are formed according to the requirement (see Fig. 6.12).

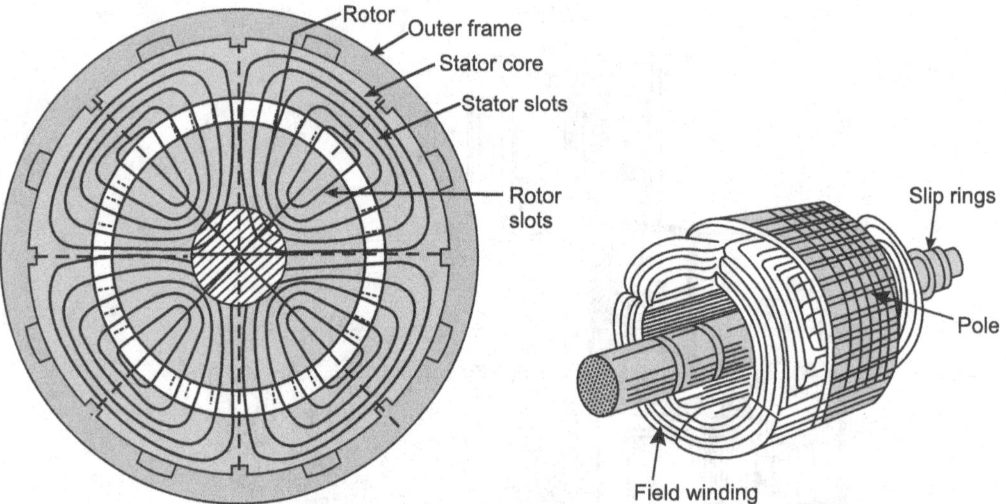

Fig. 6.11 Non-salient pole type alternator

Fig. 6.12 Non-salient pole type rotor

3. **Miscellaneous Parts:** The following are few important miscellaneous parts;

(*i*) *Brushes:* Brushes are made of carbon and these just slip over the slip rings. DC supply is given to the brushes. From brushes current flows to the slip rings and then to the exciting winding.

(*ii*) *Bearings:* Bearings are provided between the shaft and outer stationary body to reduce the friction. The material used for their construction is high carbon steel.

(*iii*) *Shaft:* Shaft is made of mild steel. Mechanical power is taken or given to the machine through shaft.

Some Special Features of Salient and Non-salient Structures

Usually the salient pole field structure has the following special features:

(*i*) These are of larger diameter and shorter length.

(*ii*) Usually, 2/3rd of the pole pitch is covered by the pole shoes (*iii*) To reduce eddy current losses, the poles are laminated.

(*iv*) The machine having such structure are employed with hydraulic turbines or with diesel engines which are usually operated at low speeds (100 to 375 rpm)

The non-salient field structure has the following special features.

(*i*) They are of smaller diameter and of very long axial length.

(*ii*) Robust construction and noiseless operation.

(*iii*) Less windage (air-resistance) loss.

(*iv*) Better in dynamic balancing.

(*v*) High operating speed (3000 rpm).

(*vi*) Nearly sinusoidal flux distribution around the periphery, and therefore, gives a better emf waveform than that obtainable with salient poles field structure.

(*vii*) There is no need of providing damper windings (except in special cases to assist in synchronising) because the solid field poles themselves act as efficient dampers.

Air-gap and its Significance

A very small air-gap increases the stray-load loss and synchronous reactance X_d. A large air-gap needs larger excitation current. Therefore, a compromise has to be made. Generally the ratio of air-gap to the pole pitch is between 0.008 and 0.02.

6.8 Excitation Systems

Since the field winding is on rotor, a special arrangement is necessary to connect DC source to the field. In small size synchronous machines, generally the field winding is excited from a separate DC source through sliprings and brushes. Sliprings are metal rings completely encircling the shaft of the machine, but insulated from it. A brush rides and slips over each slipring. The positive end of a DC voltage source is connected to one brush and negative end is connected to another brush. In large machines, various schemes are employed to supply DC excitation to the field winding. Some of the most important excitation systems are given below:

6.8.1 DC Exciters

This is a conventional method of exciting the field windings of synchronous generators. In this method, three machines namely pilot exciter, main exciter and the main 3-phase alternator are mechanically coupled and are therefore, driven by the same shaft. The pilot exciter is a DC shunt generator feeding the field winding of a main exciter. The main exciter is a separately-excited DC generator which provides the necessary current to the field winding of the main alternator through brushes and slip rings as shown in Fig. 6.13.

This conventional method of excitation suffers from cooling and maintenance problems associated with slip rings, brushes and commutator with the higher rating alternators. The modern excitation systems have been developed by eliminating the sliding contacts and brushes. This has led to develop static and brushless-excitation systems.

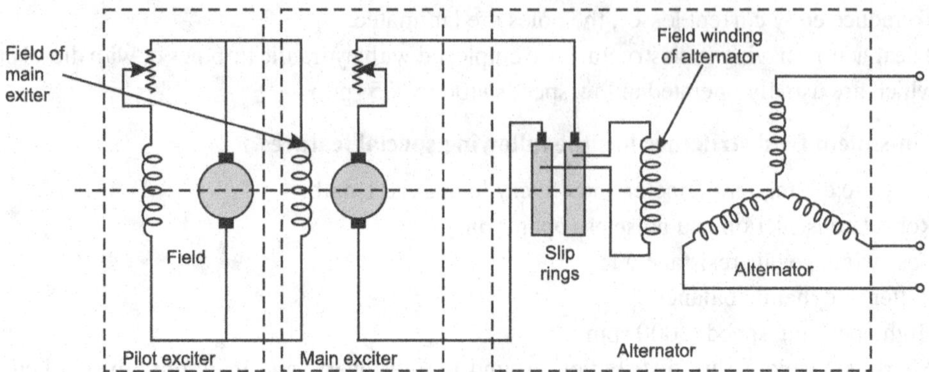

Fig. 6.13 Circuit diagram of DC exciter

6.8.2 Static Excitation System

In this method, the excitation power for the main alternator field is drawn from output terminals of the main 3-phase alternator itself. For this purpose, a three-phase transformer T_1 steps down the alternator voltage to the desired value. This three-phase voltage is fed to a three-phase full wave bridge rectifier using thyristors. The firing angle of these thyristors is controlled by means of a regulator which picks up the signal from alternator terminals through potential transformer *PT* and current transformer *CT* as shown in Fig. 6.14. The controlled DC power output from thyristor unit is delivered to the field winding of main alternator through brushes and slip rings as shown in Fig. 6.14.

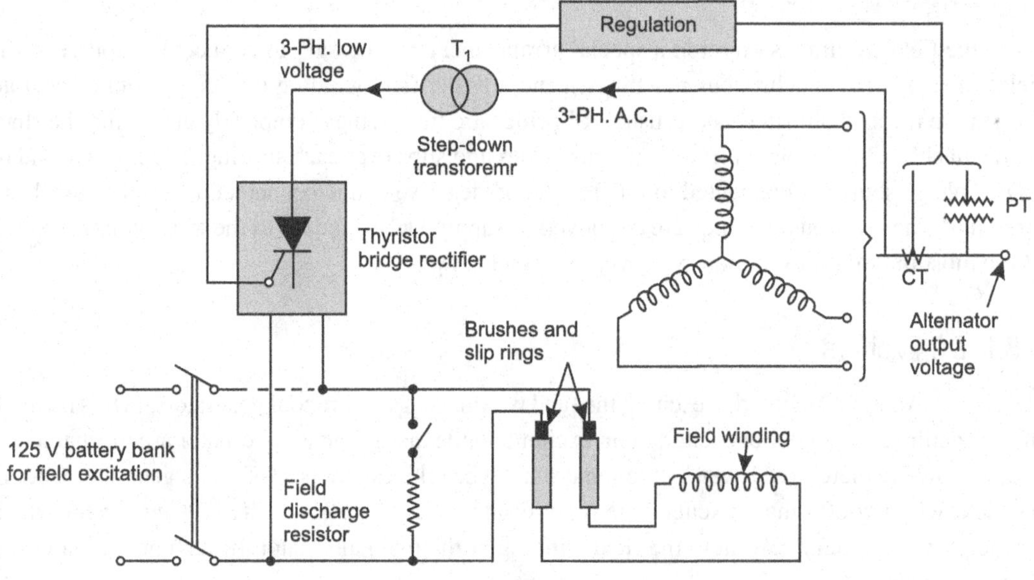

Fig. 6.14 Circuit diagram of a static exciter

While initiating the process of static excitation, first of all, field winding is switched on to the battery bank to establish field current in the alternator. The alternator speed is adjusted to rated speed. When the output voltage from alternator is built up sufficiently, the alternator field windings is disconnected from battery bank and is switched on to the thyristor bridge output. Sufficient protection devices are installed in the static excitation scheme for any possible fault in the excitation system.

The advantages of static excitation are as follows:

1. Its response time is very small about 20 m sec.
2. It eliminates the exciter windage loss and commutator, bearing and winding maintenance.
3. As the excitation energy is taken from alternator terminals itself, the excitation voltage is directly proportional to alternator's speed. This improves the overall system performance.

6.8.3 Brushless Excitation System

Figure 6.15 shows a simplified diagram of a brushless excitation system. The excitation system consists of an alternator rectifier main exciter and a pilot exciter (permanent magnet generator *PMG*). Both the main exciter and pilot exciter are driven directly from the main shaft. The main exciter has a stationary field and a rotating armature, which is directly connected, through silicon rectifiers $S_{1,}$ to the main alternator field. Thus the sliprings and brushes are eliminated. The main exciter's field is fed from a shaft driven *PMG*, having rotating permanent magnets attached to the shaft and a stationary 3-phase armature. The AC output of *PMG* is rectified by three-phase full-wave phase controlled thyristor bridges. The thyristor assembly is usually housed in removable drawers, which can be taken out easily for repair. The thyristor bridges are controlled by a set of dual firing circuits operating in parallel.

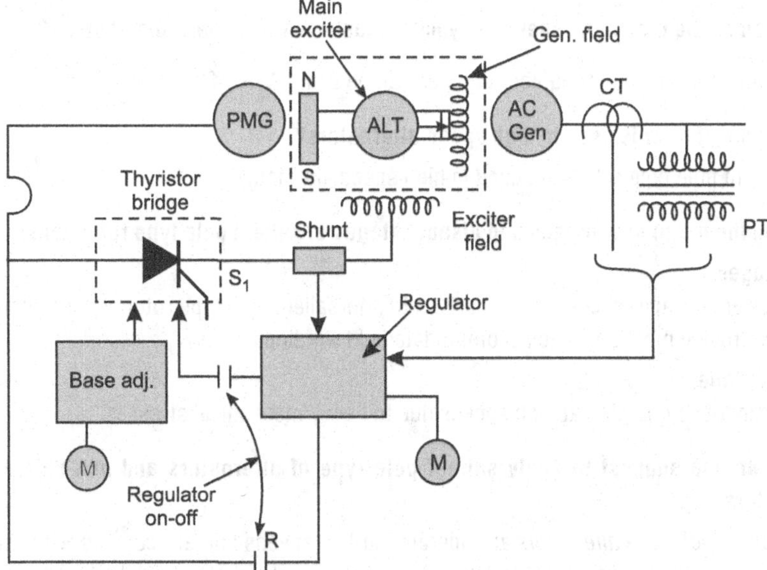

Fig. 6.15 Brushless excitation system for synchronous generator

The base excitation is controlled by an input setting to the thyristor gating circuits. This control signal is derived from the PMG via a regulated DC supply, which also serves the regulator logic circuitry. The regulator controls excitation by supplying a buck-boost control signal, which adds algebraically to the base setting. The regulator elements also comprise of solid state circuits.

This excitation system has a short time constant and a response time of less than 0.1 second.

Section Practice Problems

Short Answer Type Questions

Q.1. How will you define a synchronous machine?

Ans. An AC machine that rotates only at synchronous speed N_S is called a *synchronous machine.* Its satisfactory operation depends upon the relation.

$$N_S = \frac{120f}{P}$$

Q.2. What is a synchronous motor? State its working principle.

Ans. An *AC* machine that converts electrical power or energy into mechanical power or energy and rotates only at synchronous speed is called *synchronous motor.*

The basic principle of operation of a synchronous motor is torque development by the alignment of two fields. In this machine the two fields are magnetically locked and rotor is dragged by the stator revolving field.

Q.3. Broadly, suggest the construction of synchronous machines?

Ans. Usually large size machines have stationary armature and rotating field system because of economy and simple designing.

As for as construction of rotor is concerned, there are two types of synchronous machines namely,

(*i*) Salient pole type

(*ii*) Non-salient pole type

Q.4. What can be the maximum speed of synchronous machines operating at 50 Hz?

Ans. 3000 rpm, since minimum number of poles can be 2 and $N_s = \frac{120f}{P} = \frac{120 \times 50}{2} = 3000$ rpm.

Q.5. Which type of rotor is used in high speed alternators?

Ans. Non-salient pole type rotors are used in high speed alternators.

Q.6. Mention the major advantages and disadvantages of salient pole type rotor construction.

Ans. Advantages:

1. These are cheaper in cost in comparison to non-salient pole type rotor construction.
2. They provide more space to accommodate field winding

Disadvantages:

These cannot be operated at high speeds due to heavy mechanical stresses.

Q.7. Where do you suggest to apply salient pole type of alternators and non-salient pole type of alternators.

Ans. (*i*) *Salient pole type alternators* are operated at low speeds and are coupled with water turbines at *hydro-electric power plants* and with diesel engines at diesel power plants. These machines have large number of poles, larger diameter and smaller length.

(*ii*) *Non-salient pole type alternators* are operated at high speeds and are coupled with steam turbines at *thermal power plants.* These machines have less number of poles, smaller diameter and larger length.

Q.8. Why are the pole core and pole shoes laminated?

Ans. To reduce eddy current losses.

Q.9. Why the core of turbo-alternators in long with smaller diameter?

Ans. The turbo-alternators are operated at high speed. They carry less number of poles (only 2 or 4) which can be accommodated in smaller rotor diameter but to produce desired flux for generation of required power, core length has to be increased.

Q.10. At what voltage, usually, the field winding of an alternator is excited?

Ans. 250 V or 125 V DC.

Q.11. What are the different methods by which excitation is provided in synchronous machines?

Ans. 1. By DC exciters
2. By static excitation system.
3. By brushless excitation system.

6.9 Armature Winding

In the large synchronous machines, stationary part is the armature. On the inner periphery of the stator core, number of slots (mostly open parallel sided slots) are provided. In these slots armature winding is placed.

6.10 Types of Armature Winding

Various types of winding schemes can be adopted to wound the armature of an alternator, a few of them are given below:

1. **Single-phase and poly-phase windings:** When only one winding is placed on the armature and only one *emf* is obtained at the output, winding is called *single-phase winding.*
   When more than one windings are placed on the armature and *emfs* induced are more than one, displaced from each other by some angle, the winding is called *poly-phase winding.* Mostly three-phase winding is provided on the armature.
2. **Concentrated and distributed windings:** When one slot per pole or slots equal to the number of poles are employed, the windings thus obtained are called *concentrated windings.* Such windings give maximum induced *emfs* for given number of conductors but the wave form of induced *emf* is not exactly sinusoidal.
   When number of slots per poles are more than one, the windings thus obtained are called *distributed windings.* Such windings give slightly less than maximum induced *emf* for a given number of conductors but the wave form of induced *emf* is more sinusoidal.
3. **Single layer and double layer windings:** When only one coil side is placed in a slot, the winding is called *single layer winding.* However, when two coil sides are placed in one slot, one over the other, the winding is called *double layer winding.*
4. **Full pitched and short pitched windings:** When the two coil sides of the same coil are 180 electrical degrees apart, the winding is called *full pitch winding.* When the two sides of the same coil are less than 180 electrical degrees apart, the winding is called *short pitch winding.* The *emf* induced in each coil is maximum with full pitch winding scheme is employed whereas

emf induced in the short pitch winding is less than that. However, short pitch winding is preferred over full pitch winding because of the following reasons:

(*i*) It decreases the length at the end-connections and thus amount of copper required is saved.

(*ii*) It reduces the slot reactance and thus improves the wave shape of the generated *emf*, i.e., the generated *emf* can be made to approximately sinusoidal more easily by properly chording the winding.

(*iii*) It reduces or eliminates distorting harmonics in the wave form of generated *emf*

The only disadvantage of short pitch winding is that a few more turns are used to obtain the same voltage as it would be induced in full pitch winding.

5. **Concentric (or spiral), Lap and Wave windings:** When each group of coils under a pole is arranged into a sort of concentric shape i.e., when the current flow is traced through one such properly connected set of coils that the conductors seem to form a spiral around a portion of the core (see Fig. 6.16) the winding is called *concentric* or *chain* or *spiral winding.* This type of winding scheme is preferred for large diameter, low speed synchronous machines.

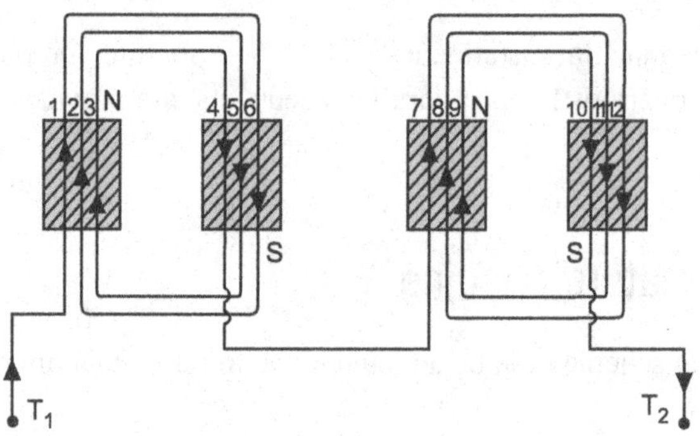

Fig. 6.16 Concentric winding

In the alternators, the lap and wave windings give the same *emf* as long as the other conditions are the same. In case of lap winding as shown in Fig. 6.17, coils or coil sides overlap the other consecutively and connections are made. Whereas in wave winding, as shown in Fig. 6.18 the coils are always forward connected. The connections of a lap winding are simpler to that of the wave winding, therefore lap winding is exclusively used.

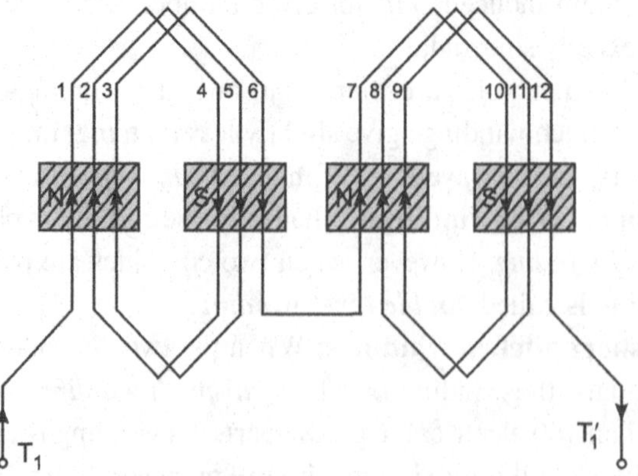

Fig. 6.17 Lap winding

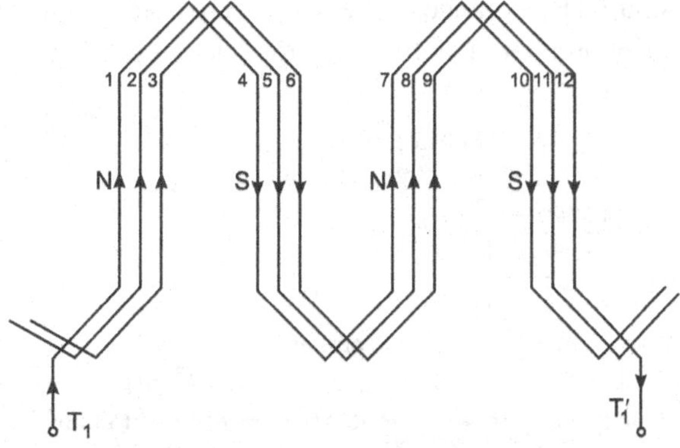

Fig. 6.18 Wave winding

6.11 Important Terms Used in Armature Winding

Some of the important terms used in the armature winding are given below:

(*i*) **Electrical angle:** When a conductor passes through a pair of poles, one cycle of *emf* is induced in it. Thus a pair of poles represents an angle of 360 electrical degrees.
There is a perfect relation between electrical and mechanical angle.

Electrical angle = Mechanical angle × Pair of poles.

(*ii*) **Pole pitch:** Distance between two neutral axis (or similar points) of adjacent poles is called *poles pitch*. The pole pitch can be expressed as number of slots per poles or electrical degrees (i.e., 180° elect.), refer to Fig. 6.20.
If S is the number of slots on the whole periphery of armature and P is the number of poles, Then,

Pole pitch = No. of slots per pole = S/P.

(*iii*) **Coil:** Two conductors placed in the two slots displaced by pole pitch (in full pitch winding) or less than pole pitch (in short pitch winding), connected at one side by the end connections form a *single turn coil* as shown in Fig. 6.19(*a*). When number of turns are connected in series and each side (coil side) is placed in the slot, it is called a multi-turn coil as shown in Figs. 6.19(*b*) and (*c*). The multi-turn coil is shown in Fig. 6.19(*d*) by a single line diagram.

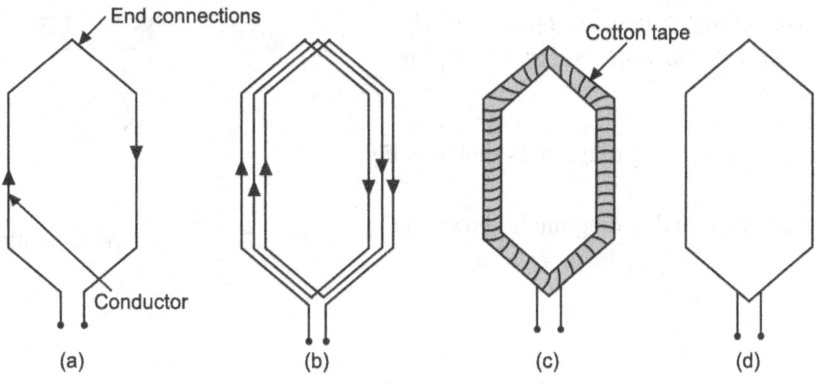

Fig. 6.19 Single and multi-turn coils

(*iv*) **Coil pitch or coil span:** The distance between two active sides of a coil is called *coil span.* It is expressed in terms of number of slots or electrical degrees. Refer to Fig. 6.20.

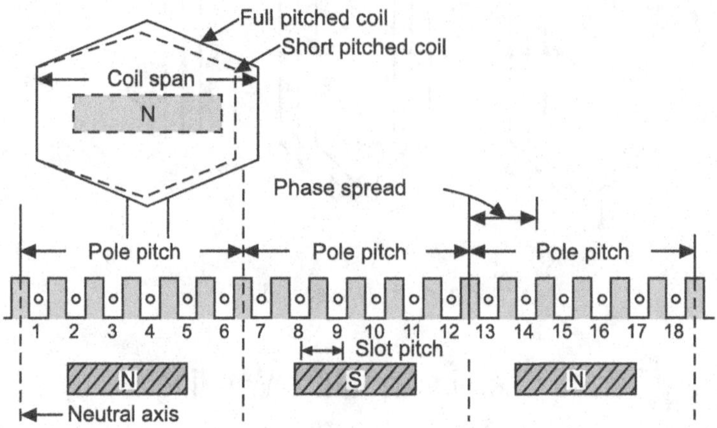

Fig. 6.20 Distributed winding

(*v*) **Slot pitch:** The distance between centre points (or similar points) of two consecutive slots or teeth is called *slot pitch.* It is expressed in electrical degrees. Refer to Fig. 6.20.

$$\text{slop pitch, } \alpha = \frac{180^\circ}{\text{No. of slots/pole}}$$

(*vi*) **Phase spread:** The angle or space of pole face over which coil sides of the same phase are spread is called *phase spread*, as shown in Fig. 6.20. In a distributed winding, the conductors of one phase under one pole are spread in number of slots so that each phase has equal distribution. In a three phase winding:

$$\text{Phase spread} = \frac{180}{3} = 60 \text{ electrical degrees}$$

or

$$\text{Phase spread} = \text{No. of slots/pole/phase}$$

Example 6.1

Draw a developed winding diagram for a 4 pole, 1-phase synchronous machine when (i) the winding is single layer concentrated in one slot (ii) the winding is single layer and distributed in three slots per pole.

Solution:

(*i*) The developed winding diagram is shown in Fig. 6.21.

(*ii*) The developed winding diagram is shown in Fig. 6.22.

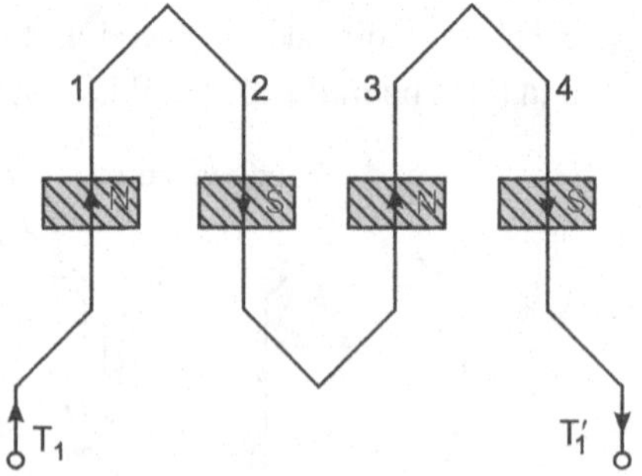

Fig. 6.21 Single layer concentrated winding

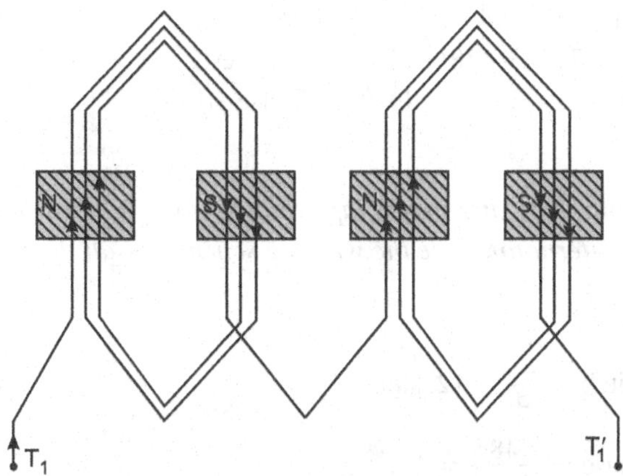

Fig. 6.22 Single layer distributed winding

Example 6.2

Draw a single-layer, full-pitch, distributed lap-winding for a 3-phase, 4-pole, 24-slot armature.

Solution:

$$\text{No. of slots per pole} = \frac{24}{4} = 6$$

$$\text{Slot pitch, } \alpha = \frac{180^\circ}{\text{No. of slots/pole}} = \frac{180^\circ}{6} = 30^\circ \text{ electrical.}$$

First phase starts with T_1 in a first slot, then second phase will start with T_2 from slot No. 5, since second phase is to start after 120 electrical degrees i.e., 120/30 = 4 slots, i.e., after a gap of 4 slots and so on.

The winding diagram is shown in Fig. 6.23.

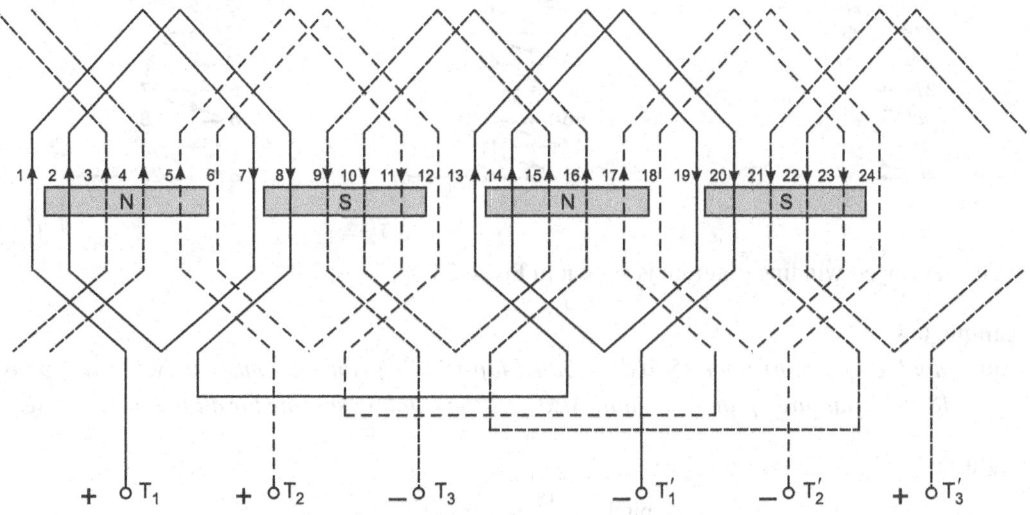

Fig. 6.23 Single-layer, full-pitch, distributed lap-winding

$T_1 - 1 - 7 - 2 - 8 - 13 - 19 - 14 - 20 - T_1'$ – Phase-I
$T_2 - 5 - 11 - 6 - 12 - 17 - 23 - 18 - 24 - T_2'$ – Phase-II
$T_3 - 9 - 15 - 10 - 16 - 21 - 3 - 22 - 4 - T_3'$ – Phase-III

Example 6.3

Draw a double-layer, full-pitch, distributed lap-winding (for one-phase only) for a 3-phase, 4-pole, 48-slot armature of an alternator. Give the winding scheme for all the three phases:

Solution:

$$\text{Pole pitch} = \frac{48}{4} = 12 \text{ slot}$$

$$\text{No. of slots/pole/phase} = \frac{48}{4 \times 3} = 4$$

For full-pitch winding, the upper conductor of first slot will be connected with lower conductor of (1 + 12 = 13) thirteenth slot and the lower conductor of 13th slot will be connected with upper conductor of (13 – 11 = 2) 2nd slot and so on, as per the following scheme. The upper conductors of each slot are represented as (1, 2, 3,…, 48) and lower conductor is represented as (1′, 2′, 3′,…, 48′)

The winding scheme for the three phases is given below:

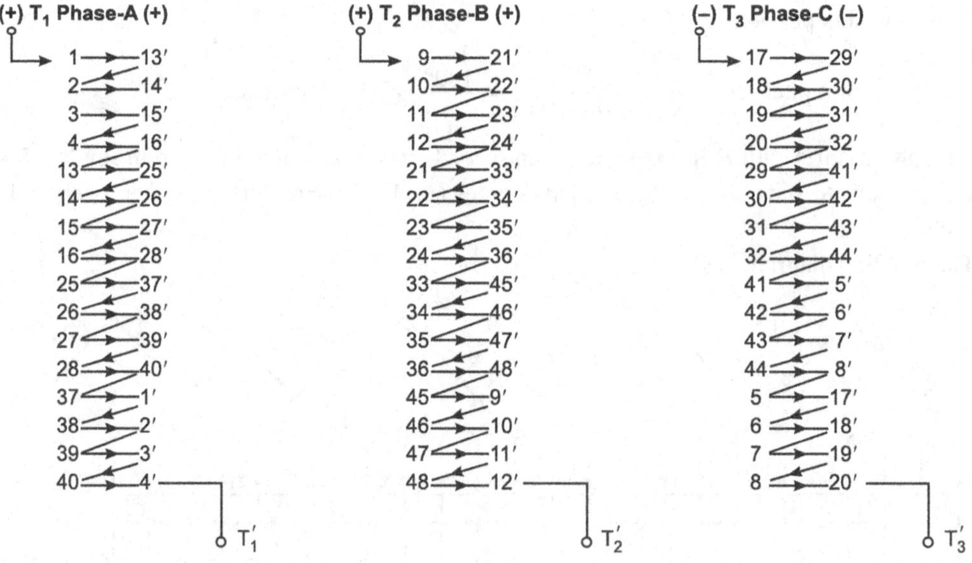

The developed winding diagram is shown in Fig. 6.24

Example 6.4

Draw a double-lay, short-pitch (5/6),distributed lap-winding (for one-phase only) for a 3-phase, 4-pole, 48 slot armature of an alternator. Also give the winding scheme for all the three phases.

Solution:

$$\text{Pole pitch} = \frac{48}{4} = 12 \text{ slot}$$

$$\text{No. of slots/pole/phase} = \frac{48}{4 \times 3} = 4 \text{ slot}$$

Winding is short pitches, coil span = $\frac{5}{6} \times 12 = 10$ slot

Upper conductor of 1st slot is to be connected with the lower conductor of (1 + 10 = 11) eleventh slot and the lower conduct or 11th slot is to be connected with upper conductor of 2nd slot (11 – 9 = 2) and so on.

The winding scheme of the three phases is given below:

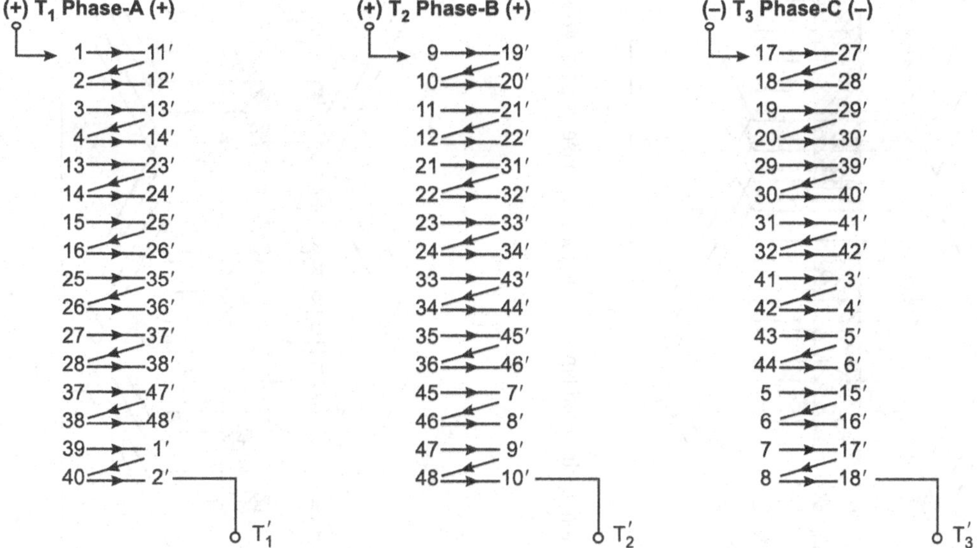

The developed winding diagram is shown in Fig. 6.25

Section Practice Problems

Numerical Problems

1. Draw a developed winding diagram for (*i*) six-pole, one-phase synchronous machine having single layer winding concentrated in one slot (*ii*) four-pole, one-phase synchronous machine having single layer winding distributed in two slots.
2. Draw a single-layer, full-pitch, distributed lap-winding for a three-phase, six-pole, 36-slot armature.
3. Draw double-layer, full-pitch, distributed lap-winding (only for one-phase) for a three-phase, four-pole, 36-slot armature of an alternation. Give the winding scheme for all the three phases.
4. Draw a double-layer, short-pitch (coil is short-pitched by one slot), distributed lap winding (only for one phase) for a three-phase, four-pole, 36-slot armature. Give the winding scheme for all the three phases.

Short Answer Type Questions

Q.1. What do you mean by one-phase and three-phase winding?

Ans. When only one winding is placed in the armature and only one emf is obtained at the output, the winding is called a single-phase winding.

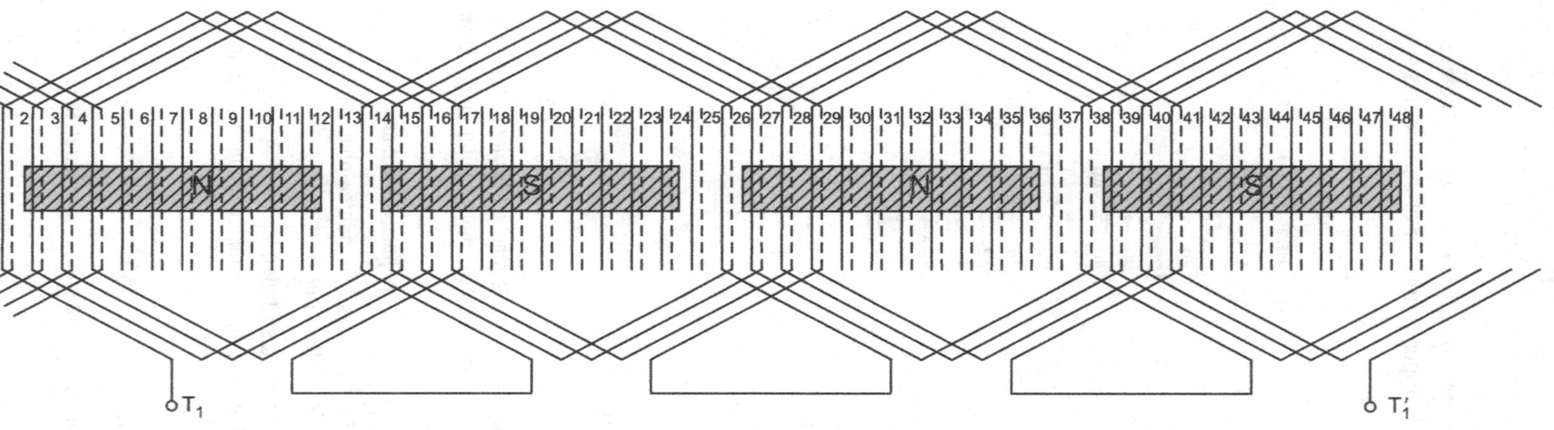

Fig. 6.24 Developed winding diagram of a double-layer, full-pitch, distributed lap-winding for a 3-phase, 4-pole, 48-slot armature

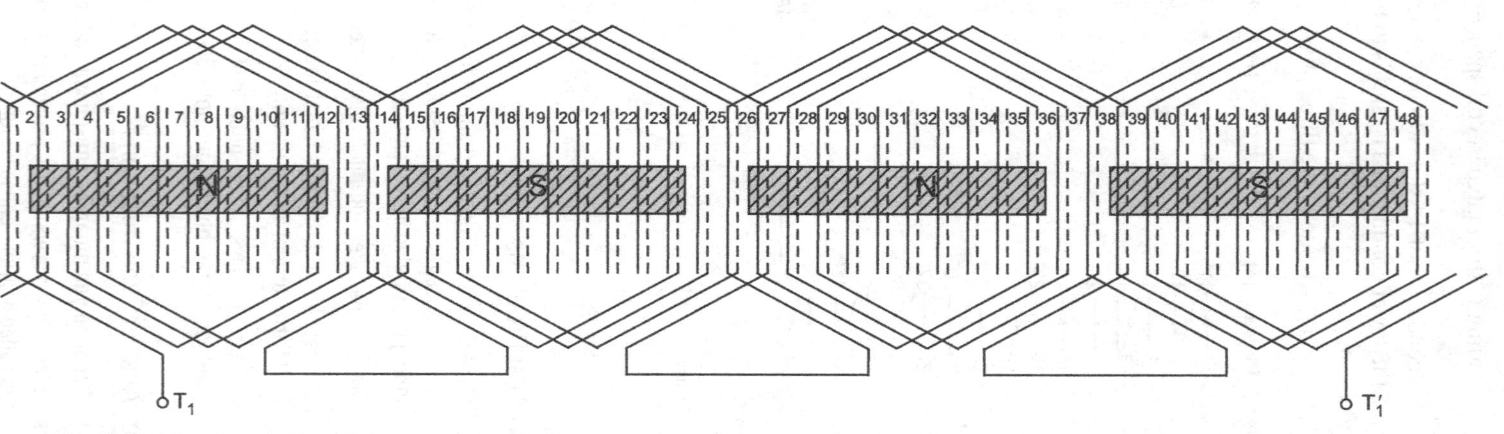

Fig. 6.25 Develop winding diagram of a double-layer, short-pitch (5/6), distributed lap-winding for a 3-phase, 4-pole, 48-slot armature.

When three windings are placed on the armature and three emfs are obtained at the output which are displaced by an angle of 120 degree electrical, the winding is called *three-phase winding.*

Q.2. What do you mean by single-layer and double-layer winding?

Ans. When only one coil side is placed in a slot, the winding is called *single layer winding.* However, when two coil sides are placed in one slot, one over the other, the winding is called *double layer winding.*

Q.3. Why do we prefer short-pitched winding?

Ans. Short pitch winding is preferred over full pitch winding because of the following reasons:

(*i*) It decreases the length at the end-connections and thus amount of copper required is saved.

(*ii*) It reduces the slot reactance and thus improves the wave shape of the generated *emf* i.e., the generated *emf* can be made to approximately sinusoidal more easily by properly chording the winding.

(*iii*) It reduces or eliminates distorting harmonics in the wave form of generated *emf*

Q.4. What is difference between electrical and mechanical angle?

Ans. A pair of poles represents an angle of 360 electrical degrees but the physical angular displacement i.e., mechanical angle depends upon number of poles on the machine. There is a perfect relation between electrical and mechanical angle.

Electrical angle = Mechanical angle × Pair of poles.

6.12 Coil Span Factor

In a full pitch winding the *coil span* or *coil pitch* is always equal to the pole pitch which is equal to 180 electrical degrees. When the coil span is less than 180 electrical degrees, the winding is called *short pitched* or *fractional pitch* or *chorded winding* as shown in Fig. 6.26.

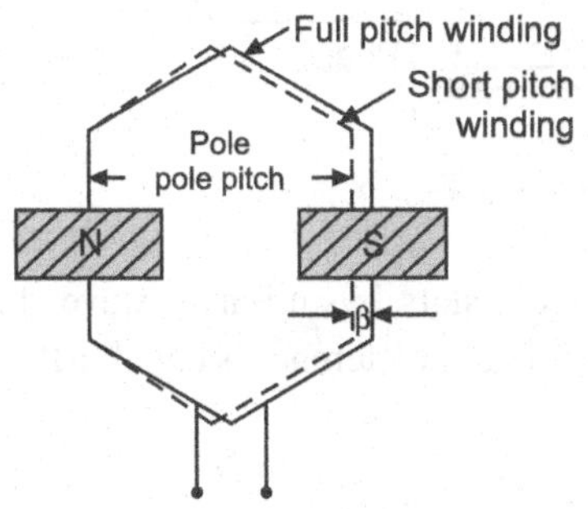

Fig. 6.26 Short pitched winding

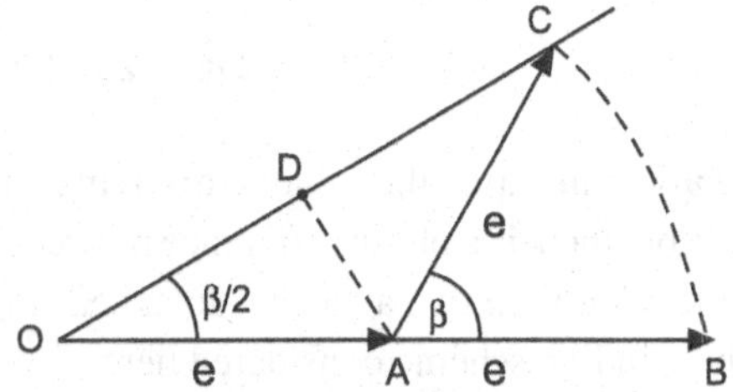

Fig. 6.27 Phasor diagram

Let '*e*' be the induced *emf* in each of the two sides of the same coil. For a full pitch winding the *emf* induced in two sides of the coil i.e., *OA* and *AB* are 180 electrical degrees apart as shown in Fig. 27. However, when the winding is short pitched by an angle β electrical degrees, the *emf* induced in the two sides of the coil are *OA* and *AC*, i.e., (180 – β) electrical degrees apart of shown in Fig. 6.27.

For a full pitch winding,

Total induced *emf* in the coil $= OA + AB = e + e = 2e$

For a short pitch winding,

$$\text{Total, induced } emf \text{ in the coil} = OC = OD + DC$$

$$= OA \cos \beta/2 + AC \cos \beta/2 \quad \text{(where } AD \text{ is } \perp \text{ on } OC\text{)}$$

$$= e \cos \beta/2 + e \cos \beta/2$$

$$= 2\, e \cos \beta/2$$

The ratio of induced emf in a coil when the winding is short pitched to the induced emf in the same coil when it is full pitched is called a **coil span factor** or **pitch factor** or **chorded factor** *and it is generally denoted be K_c. It is always less than unity.*

$$\therefore \quad \text{Coil span factor, } K_c = \frac{2e\cos\beta/2}{2e} = \cos\beta/2$$

6.13 Distribution Factor

In a concentrated winding, all the conductors of any one phase, which lie under a single pole, are placed in a single slot. When the conductors of one phase, which lie under a single pole, are placed in several slots, the winding is called distributed winding. A distributed winding is shown in Fig. 6.28 in which there are 12 slots per pole for a three-phase winding or four slots per pole per phase. In a distributed winding, the number of coils representing each phase are connected in series and is called a coil group.

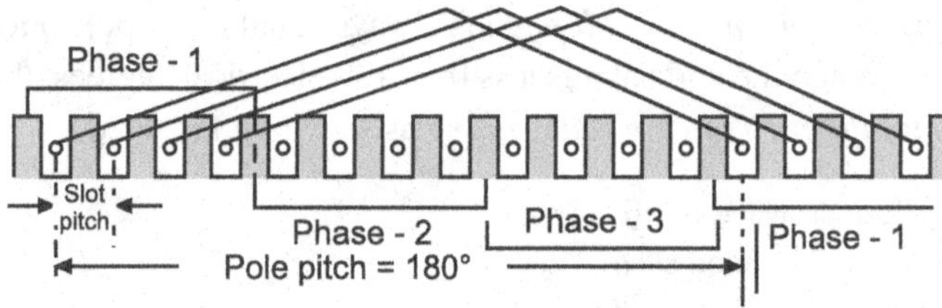

Fig. 6.28 Distributed winding

The *emfs* induced in the conductors lying in two adjacent slots is similar in wave shape and magnitude but there is a phase difference between them. This phase difference is equal to the angular displacement between two adjacent slots, i.e., slot pitch.

For the winding scheme considered here,

$$\text{Slot pitch} = \frac{180°}{\text{No. of slots/pole}} \text{electrical degrees} = \frac{180°}{12} = 15° \text{ elect.}$$

Let there be m coils connected in series in a coil group placed in adjacent slots and α be the angle between two adjacent slots in electrical degrees.

The *emfs* induced in the coil group are shown vectorially in Fig. 6.29.

Where $\quad m = \text{No. of slots/pole/phase} = 4$

and $\quad \alpha = \dfrac{180°}{\text{No. of slots/pole}} = 15° \text{ elect.}$

$$AB = BC = CD = DE$$

$$= \textit{emf induced in each coil side.}$$

$$AE = \text{Vector sum of } \textit{emfs} \text{ of } m \text{ coils.}$$

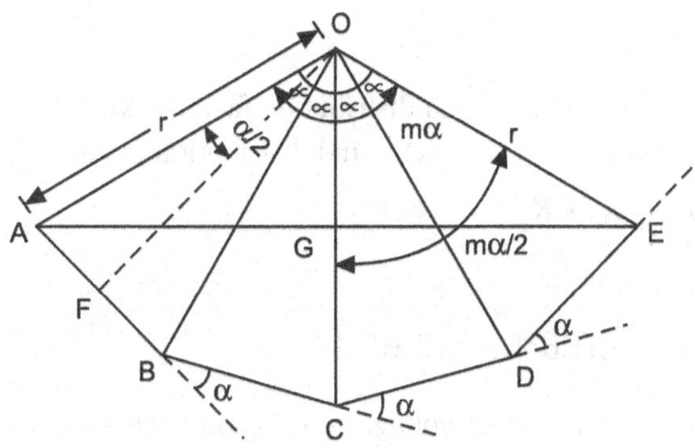

Fig. 6.29 Phasor diagram

A, *B*, *C*, *D* and *E* are the points laying on the circle of radius *r* having centre at point '*O*'. Drop a perpendicular on *AB* at *F* and on *AE* at *G* from point '*O*'.

Now $$AF = FB$$

and $$AB = AF + FB = 2AF$$

In right angle triangle *OAF*,

$$AF = r \sin \alpha/2$$

$$\therefore \quad AB = 2r \sin \alpha/2$$

Total induced emf in the coil group when all the coil sides are concentrated in one slot.

$$e_a = \text{Arithmetic sum of the induced } \textit{emf} \text{ in each coil}$$

$$= AB + BC + CD + DE = 4\,AB$$

$$= m \times AB = 2m\,r \sin \frac{\alpha}{2}$$

Now $$AG = GE$$

and $$AE = AG + GE = 2GE$$

In right angle triangle *OGE*,

$$GE = r \sin \frac{m\alpha}{2}$$

$$\therefore \quad AE = 2GE = 2r \sin \frac{m\alpha}{2}$$

Total induced emf in the coil group when the winding is distributed in number of slots to the induced emf in the coil group when the winding is concentrated in one slot is called a **distribution factor** *or* **breadth factor** *and it is generally denoted by K_d. It is always less than unity.*

$$\therefore \text{ Distribution factor, } K_d = \frac{e_v}{e_a} = \frac{2r\sin\frac{m\alpha}{2}}{2mr\sin\frac{\alpha}{2}} = \frac{\sin\frac{m\alpha}{2}}{m\sin\frac{\alpha}{2}}$$

6.14 Winding Factor

The combined effect of coil span factor and distribution factor is known as *winding factor.* In fact, winding factor is the product of coil span factor and distribution factor.

$$K_w = K_c \times K_d$$

6.15 Generation of Three-phase emf

In a three-phase system, there are equal voltages (or *emfs*) of the same frequency having a phase difference of 120°. These voltages can be produced by a three-phase *AC* generator having three identical windings (or phases) fixed on the some spindle and displaced by 120° electrical. When these windings are rotated in a stationary magnetic field as shown in Fig. 6.30(*a*) or when these windings are kept stationary and the magnetic field is rotated [see Fig. 6.30(*b*)], an *emf* is induced in each winding or phase. These *emfs* are of same magnetic and frequency but are displaced from one another by 120° electrical.

Consider three identical coils $a_1\ a_2$, $b_1\ b_2$ and c_1c_2 mounted on the rotor as shown in Fig. 6.30 (*a*) or placed on the stationary armature as shown in Fig. 6.30 (*b*). Here, a_1, b_1 and c_1 are the start terminals, whereas, a_2, b_2 and c_2 are finish terminals of three coils. It may be noted that a phase difference of 120° electrical is maintained between the corresponding start terminals a_1, b_1 and c_1. Let the three coils mounted on the same axis be rotated (or the magnetic field system be rotated keeping coils stationary) is ant i-clockwise direction at ω radians/second, as shown in Fig. 6.30(*a*) and (*b*) respectively. Three *emfs* are induced in the three coils respectively. Their magnitude and direction, at this instant are given below:

(*i*) The *emf* induced in coil $a_1\ a_2$ is zero and is increasing in the positive direction as shown by wave $a_1\ a_2$ in Fig. 6.30(*c*).

(*ii*) The coil $b_1\ b_2$ is 120° (electrical) behind the coil $a_1\ a_2$ The emf induced in this coil is negative and is becoming maximum negative as shown by the wave $b_1\ b_2$ in Fig. 6.30(c).

(*iii*) The coil c_1c_2 is 120° (electrical) behind $b_1\ b_2$ or 240° (electrical) behind $a_1\ a_2$ The *emf* induced in this coil is positive and is decreasing as shown by wave $c_1\ c_2$ in Fig. 6.30(*c*).

Phasor diagram: The emfs induced in three coils are of the same magnitude and frequency but are displaced by 120° (electrical) from each other as shown in phasor diagram [see Fig. 6.30(d)]. These can be represented by the equations:

$$e_{a1a2} = E_m \sin \omega t$$

$$e_{b1b2} = E_m \sin\left(\omega t - \frac{2\pi}{3}\right) = E_m \sin(\omega t - 120°)$$

$$e_{c1c2} = E_m \sin\left(\omega t - \frac{4\pi}{3}\right) = E_m \sin(\omega t - 240°)$$

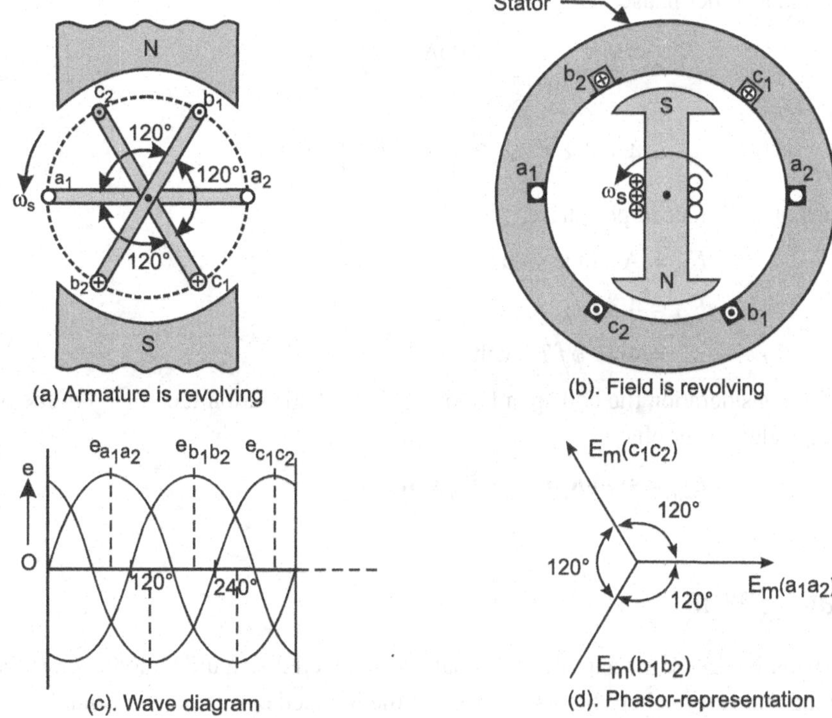

Fig. 6.30 Generation of three-phase emfs

6.16 emf Equation

Let P = No. of poles;

ϕ = Flux per pole in Wb;

N = Speed in *rpm*;

f = frequency in Hz;

Z_{ph} = No. of conductors connected in series per phase

T_{ph} = No. of turns connected in series per phase

K_c = Coil span factor;

K_d = Distribution factor

Flux cut by each conductor during one revolution = $P\,\phi$ Wb

Time take to complete one revolution = $\dfrac{60}{N}$ second

Average *emf* induced per conductor = $\dfrac{P\phi}{60/N} = \dfrac{P\phi N}{60}$

Average *emf* induced per phase,

$$= \frac{P\phi N}{60} \times Z_{ph} = \frac{P\phi N}{60} \times 2T_{ph} \qquad \left(\because T_{ph} = \frac{Z_{ph}}{2}\right)$$

$$= 4 \times \phi \times T_{ph} \times \frac{PN}{120} = 4\,\phi f T_{ph}$$

R.M.S. values of *emf* induced per phase,

$$E_{ph} = \text{Average value} \times \text{form factor}$$

$$E_{ph} = 4\,\phi f T_{ph} \times 1{\cdot}11$$

$$= 4{\cdot}44\,\phi f T_{ph} \text{ volt}$$

Taking into consideration the coil span factor (K_C) and distribution factor (K_d) of the winding. Actual *emf* induced per phase

$$E_{ph} = \mathbf{4{\cdot}44\, K_c\, K_d\, \phi f\, T_{ph}\ \textbf{volt}}$$

6.17 Wave Shape

The flux distribution in the air-gap of an alternator is not well distributed. moreover, if the winding is concentrated and full-pitched, the wave shape of the induced emf is not sinusoidal.

The wave shape can be made sinusoidal by proper designing of salient pole shoes and using short-pitch and distributed winding.

6.18 Harmonics in Voltage Wave Form

The harmonics in the output voltage wave form are developed due to non-sinusoidal wave form of the field flux. These are also developed due to variation in the reluctance of the air-gap because of slotting of stator core.

Suppression of Harmonics

Harmonics can be suppressed or eliminated by various methods as mentioned below:

(*i*) By well distributing the armature winding.
(*ii*) By using short-pitched winding.
(*iii*) By skewing the poles by one slot-pitch.
(*iv*) By using fractional slot winding.
(*v*) By using slightly larger air-gap length to increase the reluctance.

Example 6.5

What will be the number of poles of a 3-phase, 50 Hz synchronous generator running at a speed of 187.5 rpm. Which type of prime mover would you suggest for this machine?

Solution:

$$\text{Frequency}, f = 50 \text{ Hz}$$

$$\text{Speed}, N_s = 187{\cdot}5 \text{ rpm}$$

$$P = \frac{120f}{N_S} = \frac{120 \times 50}{187 \cdot 5} = \mathbf{32} \; (Ans.)$$

Since the speed of the synchronous generator is very low the prime-mover would be a *water turbine* (*hydraulic-turbine*). For such a large number of poles the machine would be a *salient-pole type*.

Example 6.6

The armature coils of a 3-phase, 4-pole, 24-slot alternator are short pitched by one slot. Determine (i) distribution factor and (ii) pitch factor.

Solution:

No. of phases = 3; No. of poles, $P = 4$

$$\text{No. of slots} = 24; \text{ No. of slots per pole} = \frac{24}{4} = 6$$

$$\text{Slot pitch}, \alpha = \frac{180°}{\text{No. of slots/pole}} = \frac{180}{6} = 30° \text{ elect.}$$

$$\text{No. of slots/pole/phase}, m = \frac{24}{4 \times 3} = 2$$

$$\therefore \text{ Distribution factor}, K_d = \frac{\sin \frac{m\alpha}{2}}{\text{m} \sin \frac{\alpha}{2}} = \frac{\sin \frac{2 \times 30}{2}}{2 \sin \frac{30}{2}} = \mathbf{0{\cdot}9659} \; (Ans.)$$

Angle by which the coil is short pitched, $\beta = \alpha = 30°$

$$\therefore \text{ Pitch factor or coil span factor}, K_c = \cos \frac{\beta}{2} = \cos \frac{30°}{2} = \mathbf{0{\cdot}9659} \; (Ans.)$$

Example 6.7

Determine the distribution and coil span factor of a 4-pole, 3-phase alternator having 36-slots which carries a short-pitched winding with a span of 8 slots.

Solution:

No. of poles, $P = 4$; No. of slots = 36

No. of phases = 3; Coil span = 8 slots (*short pitch winding*)

$$\text{Slot pitch}, a = \frac{180°}{\text{No. of slots/pole}} = \frac{180° \times 4}{35} = 20°$$

$$\text{No. of slots/pole/phase}, m = \frac{36}{4 \times 3} = 3$$

$$\text{Distribution factor}, K_d = \frac{\sin \frac{m\alpha}{2}}{m \sin \frac{\alpha}{2}} = \frac{\sin \frac{3 \times 20}{2}}{3 \sin \frac{20}{2}} = \mathbf{0{\cdot}9598} \; (Ans.)$$

$$\text{Coil span for full pitch winding} = \frac{\text{No. of slots}}{\text{No. of poles}} = \frac{36}{4} = 9$$

No. of slots by which the coil is short pitched = 9 – 8 = 1

Angle by which the coil is short pitched = $1 \times a = 20°$

$$\text{Pitch factor, } K_c = \cos\frac{\beta}{2} = \cos\frac{20}{2} = \mathbf{0{\cdot}9848}\ (Ans.)$$

Example 6.8

A 3-phase, 20-pole, 50Hz alternator has single-layer winding with full-pitch coils. The coils are connected in 60° phase group and each coil is having six turns. If the flux per pole is 0.025 Wb, determine the rms value of emf induced per phase.

Solution:

No. of poles, $P = 20$; Frequency, $f = 50$ Hz

No. of slots = 180; Flux per pole, $\phi = 0{\cdot}025$ Wb

No. of turns/coil = 6; Coil span factor, $K_c = 1$ (full-pitch winding)

Winding is single layer, full-pitched, thee-phase

$$\text{No. of slots/pole/phase, } m = \frac{180}{20 \times 3} = 3$$

$$\text{Slot pitch, } \alpha = \frac{180°}{\text{No. of slots/pole}} = \frac{180° \times 20}{180} = 20° \text{ (elect.)}$$

$$\text{Distribution factor, } K_d = \frac{\sin\frac{m\alpha}{2}}{m\sin\frac{\alpha}{2}} = \frac{\sin\frac{3\times 20}{2}}{3\sin\frac{20}{2}} = 0 \cdot 9598$$

No. of turns per phase, $T_{ph} = \frac{180 \times 6}{3} = 360$

Induced *emf*/phase, $E_{ph} = 4{\cdot}44\, K_c\, K_d\, \phi f\, T_{ph}$

$= 4{\cdot}44 \times 1 \times 0{\cdot}9598 \times 0{\cdot}025 \times 50 \times 360 = \mathbf{1917{\cdot}7\ V}$ (*Ans.*)

Example 6.9

Calculate the no-load terminal voltage of a 3-phase, 8-pole, star connected alternator running at 750 rpm having following data:

Sinusoidally distributed flux per pole = 55 m Wb

Total No. of armature slots = 72

Number of conductors/slot = 10

Distribution factor = 0·96

Assume full pitch windings.

Solution:

No. of poles, $P = 8$; Speed, $N_s = 750$ rpm

Flux, $\phi = 55 \times 10^{-3}$ Wb; No. of slots = 72

No. of conductors/slot = 10; Distribution factor, $K_d = 0{\cdot}96$

For full pitch winding,

Coil span factor, $K_c = 1$

Distribution factor is given, therefore, it is not to be calculated

$$\text{No. of turns/phase, } T_{ph} = \frac{72 \times 10}{2 \times 3} = 120$$

$$\text{Supply frequency, } f = \frac{PN_s}{120} = \frac{8 \times 750}{120} = 50 \text{ Hz}$$

Emf induced per phase,

$$E_{ph} = 4{\cdot}44\, K_c K_d f \phi T_{ph}$$

$$= 4{\cdot}44 \times 1 \times 0{\cdot}96 \times 50 \times 55 \times 10^{-3} \times 120 = 1406{\cdot}6 \text{ V}$$

Since the alternator is star connected;

$$\text{No-load terminal voltage, } E_L = \sqrt{3}\, E_{ph} = \sqrt{3} \times 1406 \cdot 6 = \mathbf{2436{\cdot}3\ V}\ (Ans.)$$

Example 6.10

Calculate the voltage per phase for a 3-phase 1500 rpm, 4-pole alternator having a double layer winding with 16 turns per coil. The slots per pole per phase are 8 and the coil span is 20 slots. The phase spread is 60° and the flux per pole is 25·8 × 10^{-3} weber.

Solution:

No. of poles, $P = 4$

Speed, $N_s = 1500$ rpm

No. of slots/pole/phase, $m = 8$

Coil span = 20 slots

Flux per pole, $\phi = 25{\cdot}8 \times 10^{-3}$ Wb

Pole pitch = No. of slots/pole = m × No. of phases = 8 × 3 = 24

No. of slots by which coil is short pitched = pole pitch – coil span = 24 – 20 = 4 slots

$$\text{Slot pitch, } \alpha = \frac{180°}{\text{No. of slots/pole}} = \frac{180°}{24} = 7 \cdot 5° \text{ elect.}$$

Angle by which the coil is short pitched,

$$\beta = 4 \times a = 4 \times 7{\cdot}5 = 30°$$

$$\therefore \quad \text{Coil span factor, } K_c = \cos\frac{\beta}{2} = \cos\frac{30}{2} = 0 \cdot 9659$$

$$\text{Distribution factor, } K_d = \frac{\sin\frac{m\alpha}{2}}{m\sin\frac{\alpha}{2}} = \frac{\sin\frac{8\times 7\cdot 5}{2}}{8\sin\frac{7\cdot 5}{2}} = \frac{0\cdot 5}{0\cdot 5232} = 0\cdot 9556$$

$$\text{Supply frequency, } f = \frac{PNs}{120} = \frac{1\times 1500}{120} = 50 \text{ Hz}$$

For double layer winding:

$$\text{No. of conductors/slot} = 16 + 16 = 32 \text{ (double layer)}$$

$$\text{No. of turns/phase, } T_{ph} = \frac{\text{No. of slots/phase} \times 32}{2}$$

$$= \frac{\text{m} \times \text{No. of poles} \times 32}{2} = \frac{8\times 4\times 32}{2} = 512$$

Emf induced per phase:

$$E_{ph} = 4{\cdot}44\, K_c\, K_d f\, \phi\, T_{ph}$$

$$= 4{\cdot}44 \times 0{\cdot}9659 \times 0{\cdot}9556 \times 50 \times 25{\cdot}8 \times 10^{-3} \times 512$$

$$= \mathbf{2706{\cdot}77 \text{ V}} \text{ (Ans.)}$$

Example 6.11

What will be the rms value of emf induced per phase in 3-phase, 6-pole, star-connected alternator having a stator with 90 slots and 8 conductors per slot? The flux per pole is 4×10^{-2} Wb and it runs at a speed of 1000 rpm. Assume full-pitched coils and sinusoidal flux distribution.

Solution:

No. of poles, $P = 6$; No. of phases $= 3$ (star connected)

No. of slots $= 90$; No. of conductors/slot $= 8$

Speed, $N_s = 1000$ rpm Flux per pole, $\phi = 4 \times 10^{-2}$ Wb

Coil span factor, $K_c = 1$ (coil is full pitched)

$$\text{No. of turns/phase, } T_{ph} = \frac{90\times 8}{2\times 3} = 120$$

$$\text{No. of slots/pole/phase, } m = \frac{90}{6\times 3} = 5$$

$$\text{Slot pitch, } \alpha = \frac{180°}{\text{No. of slots/pole}} = \frac{180° \times 6}{90} = 12° \text{ (elect.)}$$

$$\text{Distribution factor, } K_d = \frac{\sin\frac{m\alpha}{2}}{m\sin\frac{\alpha}{2}} = \frac{\sin\frac{5\times 12}{2}}{5\sin\frac{12}{2}} = 0\cdot 9567$$

$$\text{Frequency, } f = \frac{PN_s}{120} = \frac{6\times 1000}{120} = 50 \text{ Hz}$$

Generated *emf*/phase,

$$E_{ph} = 4{\cdot}44\, K_c\, K_d\, \phi f\, T_{ph}$$

$$= 4{\cdot}44 \times 1 \times 0{\cdot}9567 \times 4 \times 10^{-2} \times 50 \times 120$$

$$= \mathbf{1019{\cdot}4 \text{ V}} \text{ (Ans.)}$$

$$\text{Terminal voltage at no-load} = \sqrt{3} \times E_{ph} \text{ (star connections)}$$

$$= \sqrt{3} \times 1019 \cdot 4 = \mathbf{1765 \cdot 7\ V}\ (Ans.)$$

Example 6.12

Determine the useful flux per pole required to develop 6600 V across the terminals of a 3-phase, star connected, 50 Hz, 4-pole alternator having 60 slots with 2 conductors per slot. The coils are short pitched such that if one coil side lies in slot No.1 and the other lies in slot No.13. Assume a double-layer winding.

Solution:

Number of poles, $P = 4$; Frequency, $f = 50$ Hz

No. of slots = 60; No. of conductors/slot = 2

Terminal voltage, $E_L = 6600$ V

$$\text{Pole pitch} = \frac{\text{No. of slots}}{\text{No. of poles}} = \frac{60}{4} = 15 \text{ slots}$$

$$\text{Coil pitch} = 13 - 1 = 12 \text{ slots}$$

No. of slots by which coil is short pitched = 15 – 12 = 3

$$\text{Slot pitch, } \alpha = \frac{180°}{\text{No. of slots/pole}} = \frac{180°}{15} = 20° \text{ elect.}$$

Angle by which coil is short pitched,

$$\beta = 3 \times \alpha = 3 \times 12 = 36° \text{ elect.}$$

$$\text{Coil span factor, } K_c = \cos\frac{\beta}{2} = \cos\frac{36°}{2} = 0 \cdot 951$$

$$\text{No. of slots/pole/phase, } m = \frac{60}{4 \times 3} = 5$$

$$\text{Distribution factor, } K_d = \frac{\sin\frac{m\alpha}{2}}{m\sin\frac{\alpha}{2}} = \frac{\sin\frac{5 \times 12}{2}}{5\sin\frac{12}{2}} = \frac{0 \cdot 5}{0 \cdot 5226} = 0 \cdot 9567$$

$$\text{No. of turns/phase} = \frac{60 \times 2}{2 \times 3} = 20$$

$$\text{Induced } emf/\text{phase, } E_{ph} = \frac{E_L}{\sqrt{3}} = \frac{6600}{\sqrt{3}} = 3810 \cdot 5 \text{ V}$$

Now,

$$E_{ph} = 4{\cdot}44\, K_c\, K_d f \phi\, T_{ph}$$

$\therefore$ Useful flux/pole,

$$\phi = \frac{E_{ph}}{4 \cdot 44 K_c K_d f T_{ph}}$$

$$= \frac{3810 \cdot 5}{4 \cdot 44 \times 0 \cdot 951 \times 0 \cdot 9567 \times 50 \times 20}$$

$$= \mathbf{0 \cdot 9433\ Wb}\ (Ans.)$$

Example 6.13

A three-phase, 16 pole, 50 Hz star connected alternator has 144 slots and 10 conductors per slot. The flux per pole is 2·48 × 10⁻² weber sinusoidally distributed. The coil pitch is 2 slots short of full pitch. Find (i) speed (ii) line emf

Solution:

No. of poles, $P = 16$; Frequency, $f = 50$ Hz

No. of slots = 144; Flux/pole, $\phi = 2{\cdot}48 \times 10^{-2}$ Wb

No. of conductors/slot = 10 (*3-phase, star connected*)

Coil is short pitched by 2 slots

$$\text{Synchronous speed, } N_s = \frac{120 f}{P} = \frac{120 \times 50}{16} = \mathbf{375\ rpm}\ (\textit{Ans.})$$

$$\text{Slot pitch, } \alpha = \frac{180°}{\text{No. of slots/pole}} = \frac{180° \times 16}{144} = 20°\ (\textit{elect.})$$

Angle by which the coil is short pitched,

$$\beta = 2 \times \propto = 2 \times 20° = 40°$$

$$\text{Coil span factor, } K_c = \cos\frac{\beta}{2} = \cos\frac{40°}{2} = 0 \cdot 9397$$

$$\text{No. of slots/pole/phase, } m = \frac{144}{16 \times 3} = 3$$

$$\text{Distribution factor, } K_d = \frac{\sin\frac{m\alpha}{2}}{m\sin\frac{\alpha}{2}} = \frac{\sin\frac{3 \times 20}{2}}{3\sin\frac{20}{2}} = 0 \cdot 9598$$

$$\text{No. of turns/phase, } T_{ph} = \frac{144 \times 10}{2 \times 3} = 240$$

$$\text{Generated } \textit{emf}\text{/phase, } E_{ph} = 4{\cdot}44\ K_c\ K_d f\,\phi\,T_{ph}$$

$$= 4{\cdot}44 \times 0{\cdot}9397 \times 9598 \times 2{\cdot}58 \times 10^{-2} \times 50 \times 240$$

$$= 1191{\cdot}75 \text{ V}$$

$$\text{Line voltage, } E_L = \sqrt{3}E_{ph} = \sqrt{3} \times 1191 \cdot 75 = \mathbf{2064{\cdot}2\ V}\ (\textit{Ans.})$$

Example 6.14

A 10 MVA, 11 kV, 50 Hz, three-phase, star-connected alternator is driven at 300 rev/min. The stator winding is housed in 360 slots and has 6 conductors per slot. The coils spanning five-sixth of a pole pitch. Calculate:

(i) The sinusoidally distributed flux/pole required to give a line voltage of 11 kV on open circuit; and

(ii) the full load current per conductor.

Solution:

Rated output = 10 MVA = 10×10^6 VA

Line voltage, E_L = 11 kV = 11×10^3 V

$$\text{Speed, } N_s = 300 \text{ rpm; Frequency, } f = 50 \text{ Hz}$$

$\therefore$

$$\text{No. of slots} = 360; \text{ No. of conductors/slot} = 6$$

$$\text{Coil pitch} = \frac{5}{6} \times \text{pole pitch} = \frac{5}{6} \times 180 = 150° \text{ elect.}$$

$$\text{Angle by which coil is short pitched, } \beta = 180 - 150 = 30° \text{ elect.}$$

$$\text{Coil span factor, } K_c = \cos\frac{\beta}{2} = \cos\frac{30}{2} = 0 \cdot 9659$$

$$\text{No. of poles, } P = \frac{120 f}{N_s} = \frac{120 \times 50}{300} = 20$$

$$\text{No. of slots/pole/phase, } m = \frac{360}{20 \times 3} = 6$$

$$\text{Slot pitch, } \alpha = \frac{180°}{\text{No. of slots/pole}} = \frac{180° \times 20}{360} = 10° \text{ elect.}$$

$$\text{Distribution factor, } K_d = \frac{\sin\frac{m\alpha}{2}}{m \sin\frac{\alpha}{2}} = \frac{\sin\frac{6 \times 10}{2}}{6 \sin\frac{10}{2}} = \frac{0 \cdot 5}{0 \cdot 5229} = 0 \cdot 9561$$

$$\text{No. of turns/phase, } T_{ph} = \frac{360 \times 6}{2 \times 3} = 360$$

$$\text{Induced, } emf\text{/phase, } E_{ph} = \frac{E_L}{\sqrt{3}} = \frac{11 \times 10^3}{\sqrt{3}} = 6350 \cdot 8 \text{ V}$$

Now,

$$E_{ph} = 4{\cdot}44\, K_c\, K_d f \phi\, T_{ph}$$

$\therefore$

$$\text{Flux/pole, } \phi = \frac{E_{ph}}{4.44 \times K_c K_d f T_{ph}}$$

$$= \frac{6350 \cdot 8}{4 \cdot 44 \times 0 \cdot 9698 \times 0 \cdot 9561 \times 50 \times 360}$$

$$= \mathbf{86 \text{ m Wb}} \; (Ans.)$$

$$\text{Full load current/conductor, } I = \frac{\text{Rated output}}{\sqrt{3}\, E_L} = \frac{10 \times 10^6}{\sqrt{3} \times 11 \times 10^3} = \mathbf{524{\cdot}86 \text{ A}} \; (Ans.)$$

Example 6.15

The armature of a 6600 volt, three-phase, 20-pole, 300 rpm star-connected alternator has 180 slots. The flux per pole is 80 m Wb and the coil span is 160° electrical. Determine the number of conductors in series per phase.

Solution:

Line voltage, $E_L = 6600$ V; No. of poles, $P = 20$

Speed, $N_s = 300$ rpm; No. of slot = 180

Coil span = 160° elect.; Flux per pole, $\phi = 80 \times 10^{-3}$ Wb

$$\text{Frequency, } f = \frac{PN_s}{120} = \frac{20 \times 300}{120} = 50 \text{ Hz}$$

Angle by which coil is short pitched, β = 180 – 160 = 20° elect.

$$\text{Coil span factor, } K_c = \cos\frac{\beta}{2} = \cos\frac{20}{2} = 0 \cdot 9848$$

$$\text{No. of slots/pole/phase, } m = \frac{180}{20 \times 3} = 3$$

$$\text{Slot pitch. } a = \frac{180°}{\text{No. of slots/pole}} = \frac{180° \times 20}{180} = 20° \text{ elect.}$$

$$\text{Distribution factor, } K_d = \frac{\sin\frac{m\alpha}{2}}{m\sin\frac{\alpha}{2}} = \frac{\sin\frac{3 \times 20}{2}}{3\sin\frac{20}{2}}$$

$$= \frac{0 \cdot 5}{0 \cdot 5209} = 0 \cdot 9598$$

$$\text{Induced } emf \text{ per phase, } E_{ph} = \frac{E_L}{\sqrt{3}} = \frac{6600}{\sqrt{3}} = 3810 \cdot 5 \text{ V}$$

Now,
$$E_{ph} = 4{\cdot}44\, K_c\, K_d f\, \phi\, T_{ph}$$

∴
$$\text{No. of turns/phase, } T_{ph} = \frac{E_{ph}}{4 \cdot 44 K_c K_d f \phi}$$

$$= \frac{3810 \cdot 5}{4 \cdot 44 \times 0 \cdot 9548 \times 0 \cdot 9598 \times 50 \times 80 \times 10^{-3}} = 227$$

No. of conductors in series per phase, $Z_{ph} = 2\, T_{ph} = 2 \times 227 =$ **454** (*Ans.*)

Section Practice Problems

Numerical Problems

1. Calculate the coil span factor or the pitch factor for the following windings:
   (*a*) 54 stator slots, six-poles, when coil spans 1 to 8.
   (*b*) 42 stator slots, four-poles, coil spans 1 to 10.
   (*c*) 96 stator slots, six-poles, coil spans 1 to 13. (**Ans.** *0.94; 0.925; 0.925*)
2. Find the breadth factor for 3-ϕ winding with 2 slots per pole per phase. (**Ans.** *0.96*)
3. A star connected, three-phase 4 pole, 50 Hz alternator has a single layer winding in 24 stator slots. There are 50 turns in each coil and the flux per pole is 5 mega lines. Find the open circuit line voltage. (**Ans.** *3715 V*)
4. The field system of a 50 Hz alternator has sinusoidal flux per pole of 0.075 Wb. Find emf generated in one turn of the machine if the winding has a span of 150° (Electrical). (**Ans.** *16 V*)
5. The air gap flux of a 12 pole, 3-ϕ alternator is 0.058 weber per pole and is distributed sinusoidally over the pole. The stator has 2slots per pole per phase and 8 conductors per slot. The winding is a double layer winding with a coil span of 135° electrical apart. Find the voltage generated per phase at no-load when the machine runs at 500 rpm. (**Ans.** *1175 V*)

Short Answer Type Questions

Q.1. Why large synchronous machines are three-phase wound?

Ans. Larger *AC* machines are always 3-phase wound machines because of their high efficiency and economy.

Q.2. What is the relation between electrical and mechanical angle of an electrical machine.

Ans. *Electrical angle* = $\frac{P}{2}$ × mechanical angle

Q.3. Define coil pitch or coil span.

Ans. The distance between the two active sides of the same coil is called coil span, It may be described in number of slots or electrical angle.

Q.4. What do you mean by phase spread?

Ans. *Phase spread*: The angle or space in which coil-sides of the same phase are spread is called *phase spread.* It is represented either as number of slots or electrical angle. In 3-phase machines;

Phase spread = No. of slots/pole/phase = 60° elect.

Q.5. Define distribution factor.

Ans. *Distribution factor*: The ratio of induced *emf* in the coil group when the winding is distributed in number of slots to the induced *emf* in the same coil group when the winding is concentrated in one slot is called *distribution factor* or *breadth factor.*

$$\text{Distribution factor, } K_d = \frac{\sin \frac{m\alpha}{2}}{m \sin \frac{\alpha}{2}}$$

where, m is the phase spread = No. of slots/pole/phase and

$$\alpha \text{ is the slot pitch} = \frac{180^\circ}{\text{No. of slots/pole}}$$

6.19 Production of Revolving Field

(Physical/Graphical aspect)

A resultant magnetic field having constant magnitude and fixed polarity changes its position continuously in space is called a **revolving field.**

For simplicity, consider the stator of a two-pole synchronous machine or of an induction motor having three-phase winding represented by the concentric coils *a-a′*, *b-b′* and *c-c′* respectively as shown in Fig. 6.32 (*a*).

Let 3-phase currents having wave diagram as shown in Fig. 6.31(*a*) flows through the stator winding. Current of phase-1 flows through coil *a-a′*, current of phase-2 flows through coil *b-b′* and current of phase-3 flows through coil *c-c′* respectively. When three-phase currents flow through the three-phase winding, they produce their own magnetic fields ϕ_1, ϕ_2 and ϕ_3. The phasor diagram of the fields is shown in Fig. 6.31(*b*). The positive half cycle of the alternating current is considered as inward flow of current [cross in a circle ⊕] and negative half cycle as outward flow of current [dot in a circle ⊙]. It is marked in the start terminals (*a*, *b* and *c*) of the three coils. The direction of flow of current is opposite in the finish terminals of the same coil.

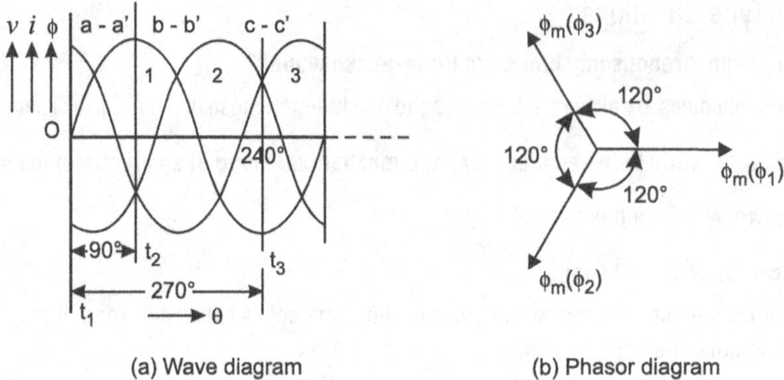

(a) Wave diagram (b) Phasor diagram

Fig. 6.31 Three-phase emfs

The three fields are given by the equation;

$$\phi_1 = \phi_m \sin \theta \text{ [where } \theta = \omega t]$$

$$\phi_2 = \phi_m \sin [\theta - 120°]$$

$$\phi_3 = \phi_m \sin [\theta - 240°]$$

At an instant t_1 when $\theta = 0$, the flow of current in the start terminals of the three coils, *a*, *b* and *c* is zero, negative and positive (i.e., zero, outward and inward) respectively whereas, in the finish terminals it is opposite to that of the start terminals as shown in Fig. 6.32(*a*). The resultant field produced by the stator winding and its direction F_m is shown in Fig. 6.32(*a*).

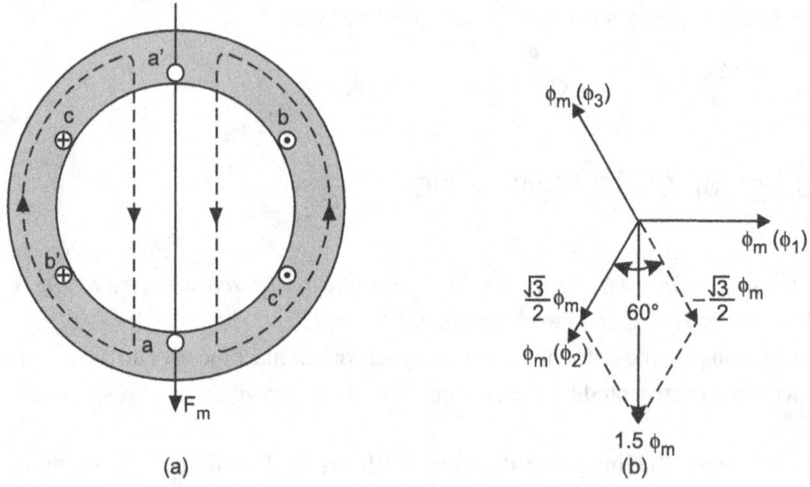

(a) (b)

Fig. 6.32 Position of resultant field at instant 't_1'

At this instant angle θ is zero, therefore,

$$\phi_1 = \phi_m \sin \theta = \phi_m \sin 0° = 0$$

$$\phi_2 = \phi_m \sin (0° - 120°) = -\frac{\sqrt{3}}{2}\phi_m$$

$$\phi_3 = \phi_m \sin(0° - 240°) = \frac{\sqrt{3}}{2}\phi_m$$

The field ϕ_1 is zero and the remaining fields ϕ_2 and ϕ_3 are shown vectorially in Fig. 6.32(*b*). Field ϕ_2 is shown in opposite direction having magnitude $\frac{\sqrt{3}}{2}\phi_m$. The resultant field ϕ_R is the vector sum of ϕ_1, ϕ_2 and ϕ_3 i.e.,

$$\phi_R = \sqrt{\phi_2^2 + \phi_3^2 + 2\phi_2\phi_3 \cos 60°}$$

or
$$\phi_R = \sqrt{\left(\frac{\sqrt{3}}{2}\phi_m\right)^2 + \left(\frac{\sqrt{3}}{2}\phi_m\right)^2 + 2\left(\frac{\sqrt{3}}{2}\phi_m\right)\left(\frac{\sqrt{3}}{2}\phi_m\right) \times \frac{1}{2}}$$

or
$$\phi_R = 1{\cdot}5\ \phi_m$$

At another instant t_2, where $\theta = 90°$, the direction of flow of current in coil-side *a* is inward and in *b* and *c* is outward. Whereas, the flow of current in the other sides of the coils is opposite, i.e., in *a*′ it is outward and in *b*′ and *c*′ it is inward as shown in Fig. 6.33(*a*). The resultant field and its direction F_m at this instant is shown in Fig. 6.33(*a*). The resultant field is rotated in anti-clockwise direction (i.e., the direction in which the supply sequence is applied to the winding, here the supply 1, 2, 3 is given to the coils *a*, *b* and *c* respectively) through an angle $\phi = 90°$ from its previous position.

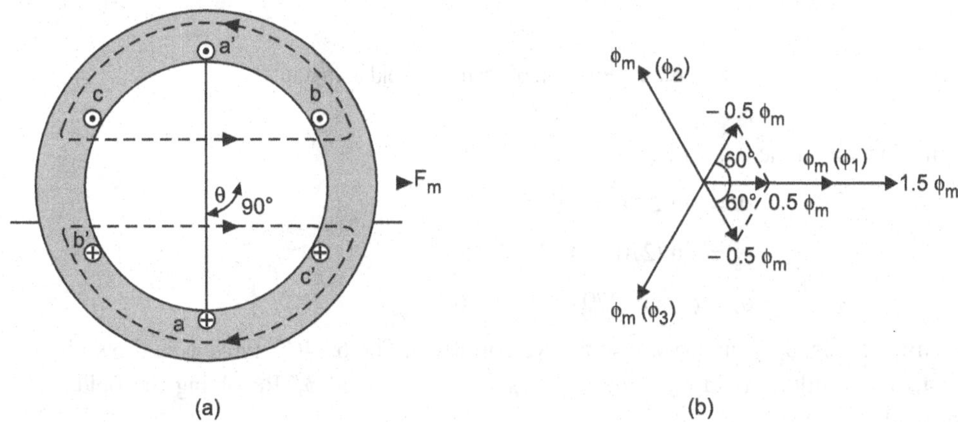

Fig. 6.33 Position of resultant field at instant 't_2'

At this instant, angle θ is 90°, therefore,

$$\phi_1 = \phi_m \sin 90° = \phi_m$$

$$\phi_2 = \phi_m \sin(90° - 120°) = -0{\cdot}5\ \phi_m$$

$$\phi_3 = \phi_m \sin(90° - 240°) = -0{\cdot}5\ \phi_m$$

The three fields, ϕ_1, ϕ_2 and ϕ_3 are shown vectorially in Fig. 6.33(*b*). Fields ϕ_2 and ϕ_3 are shown in opposite direction each having a magnitude of $0{\cdot}5\ \phi_m$. The resultant field ϕ_R is the vector sum of ϕ_1, ϕ_2 and ϕ_3. Resolving the fields along the axis of field ϕ_1.

$$\phi_R = \phi_1 + \phi_2 \cos 60° + \phi_3 \cos 60°$$

or $$\phi_R = \phi_m + 0{\cdot}5\ \phi_m \times 0{\cdot}5 + 0{\cdot}5\ \phi_m \times 0{\cdot}5$$

or $$\phi_R = 1{\cdot}5\ \phi_m$$

At some other instant t_3 where $\theta = 270°$ the direction of flow of current in start terminals of the three coils *a*, *b* and *c* is outward, inward and inward respectively, whereas in the finish terminals it is opposite to that of the start terminals as shown in Fig. 6.34(*a*). The resultant field and its direction F_m at this instant is shown in Fig. 6.34(*a*). The resultant field is rotated in anti-clockwise direction through an angle $\theta = 270°$ from its first position.

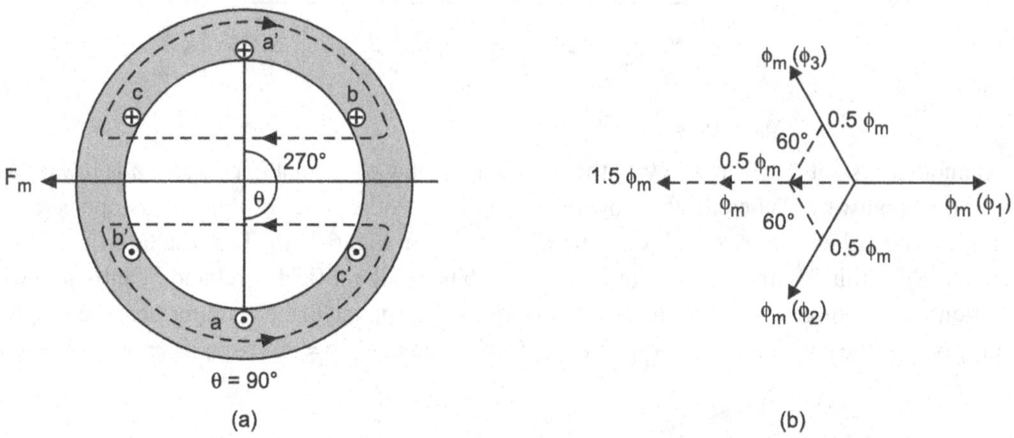

Fig. 6.34 Position of resultant field at instant 't_3'

At this instant, angle θ is 270°, therefore,

$$\phi_1 = \sin 270° = -\ \phi_m$$

$$\phi_2 = \sin (270° - 120°) = 0{\cdot}5\ \phi_m$$

$$\phi_3 = \phi_m \sin (270 - 240°) = 0{\cdot}5\ \phi_m$$

The three fields, ϕ_1, ϕ_2 and ϕ_3 are shown vectorially in Fig. 6.34(*b*). Field ϕ_1 is shown in opposite direction. The resultant field ϕ_R is the vector sum of ϕ_1, ϕ_2 and ϕ_3. Resolving the fields along the axis of field ϕ_1.

$$\phi_R = \phi_1 + \phi_2 \cos 60° + \phi_3 \cos 60°$$

or $$\phi_R = \phi_m + 0{\cdot}5\ \phi_m \times 0{\cdot}5 + 0{\cdot}5\ \phi_m \times 0{\cdot}5$$

or $$\phi_R = 1{\cdot}5\ \phi_m$$

Hence, in one cycle the resultant field rotates through one revolution. It may also be seen that when supply from three phases 1, 2 and 3 is given to coils *a-a'*, *b-b'* and *c-c'* respectively, an anticlockwise rotating field is produced. If supply to the coils *a* – *a'*, *b* – *b'* and *c* – *c'* is given from phase 1, 3 and 2 respectively, the direction of rotation of the resultant field is reversed.

Thus, to reverse the direction of rotation of rotating magnetic field, the connections of any two supply terminals are interchanged.

Conclusion

Hence, it is concluded that when a three-phase supply is given to a three-phase wound stator of an induction motor or synchronous motor, resultant field F_m of magnitude 1·5 ϕ_m is produced which rotates in space at a constant speed called synchronous speed ($N_s = 120f/P$). The direction of rotation of the resultant field depends upon the sequence in which supply in given to the stator winding.

6.20 Ferrari's Principle (Vector Representation of Alternating Field)

An alternating field can be represented by two vectors revolving in opposite directions at constant speed of ω radians per second. It can be well explained with the help of Ferrari's principle.

Ferrari's principle states that a single alternating magnetic field is the resultant of two fields, each of half the magnitude of alternating field, rotating in opposite direction at fixed speed, called synchronous speed.

Consider a field ϕ_m having two components 1 and 2 each of magnitude $\phi_m/2$, which are rotating in oppose direction at a constant angular speed of ω radians/sec.

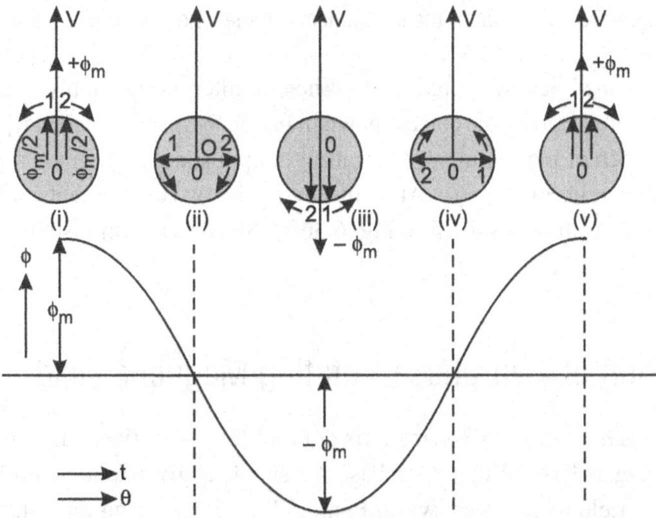

Fig. 6.35 Two-vectors revolving in opposite direction and its wave diagram

Let at start, the two components are vertically upwards i.e., along the vertical axis and the resultant field is $+\phi_m$. When $\omega = 90°$, the resultant field is zero as shown in Fig. 6.35(*ii*), whereas when $\phi =$ 180° the resultant field is $-\phi_m$ and so on.

To determine the value of resultant field at any instant (say after *t* seconds), the two components make an angle $\phi = \omega t$ with the reference (vertical) axis

Then, the resultant field at that instant,

$$\phi = \phi_m \cos \omega t$$

or

$$\phi_R = \phi_m \cos \theta$$

which is an alternating field.

Hence, an alternating field can be represented by two components of field having half the magnitude rotating in opposite direction at a constant speed, called synchronous speed.

According to Ferrari's Principle, an alternating field produced by each phase (coil) can be represented by two components of field each having half the magnitude rotating in opposite direction at a constant speed, called synchronous speed.

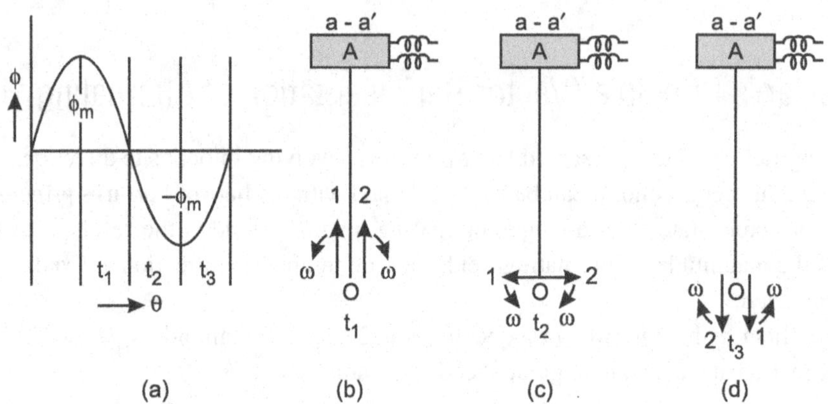

Fig. 6.36 Position of the components of the fields at various instants

Considering a coil *a-a′* having an axis *OA*. When an alternating supply is given to the coil, an alternating field (ϕ) is produced by the coil as shown in Fig. 6.36(*a*). At an instant t_1, the two components of the field 1 and 2 each of magnitude $\phi_m/2$ rotating in opposite direction are represented along the axis of the coil as shown in Fig. 6.36(*b*). At an instant t_2, the two components of the field are rotated through 90° and their position is shown in Fig. 6.36(*c*). Similarly at an instant t_3 the position of the two components is shown in Fig. 6.36(*d*).

6.21 Production of Two-phase Rotating Magnetic Field

Considering two coils *a-a′* and *b-b′* having axis *OA* and *OB* respectively displaced from each other by 90° as shown in Fig. 6.37(*b*). When two-phase AC supply is given to these coils, each one of them produces alternating field (flux) as shown in Fig. 6.37(*a*). Considering an instant t_1, flux produced by coil *a-a′* is + ϕ_m, therefore the component 1 and 2 each of value $\phi_m/2$ are along the axis of coil *a-a′* i.e., *OA*, as shown in Fig. 6.37(*b*).

The flux produced by the coil *b-b′* is zero at this instant and will become + ϕ_m after a rotation of 90°. Accordingly the position of component 3 and 4 each equal to $\phi_m/2$ is shown in Fig. 6.37(*b*).

At this instant, the component 2 and 4 are equal and opposite and rotating in opposite direction, thus neutralising each other. The resultant component is the sum of 1 and 3 equal ϕ_m rotating in anticlockwise direction i.e., the direction of the supply sequence.

Thus, it is concluded that when 2-phase supply is applied (or current flows) to a 2-phase wound stator, resultant field of magnitude ϕ_m is set-up which rotates in space at a synchronous speed in the direction of supply sequence.

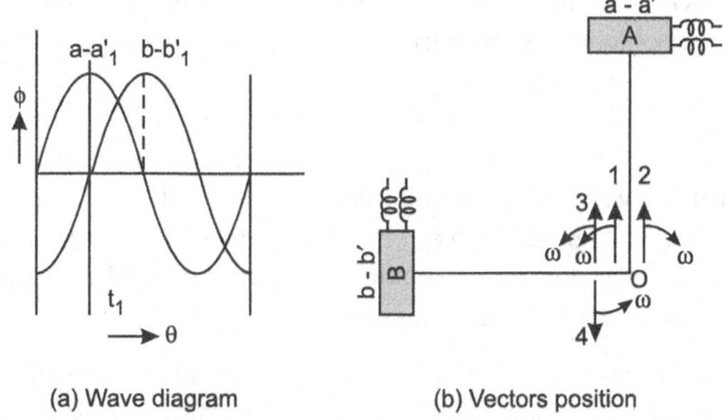

(a) Wave diagram (b) Vectors position

Fig. 6.37 Position of the components of the fields at instant t_1

6.22 Production of Three-phase Rotating Magnetic Field

Considering three coils $a-a'$, $b-b'$ and $c-c'$ having axis *OA, OB* and *OC* respectively as shown in Fig. 6.38(*b*). When three phase supply is given to these three coils, each of them produces alternating field (flux) as shown in Fig. 6.38(*a*). Considering an instant t_1, flux produced by coil $a-a'$ is $+\phi_m$, therefore the components 1 and 2 each of value $\phi_m/2$ are long the axis of coil $a-a'$ i.e., *OA*.

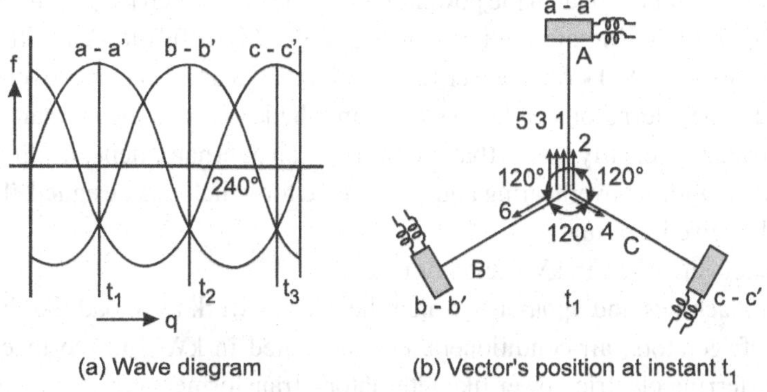

(a) Wave diagram (b) Vector's position at instant t_1

Fig. 6.38 Resultant magnetic field

The flux produced by coil $b-b'$ will become positive maximum $(+\phi_m)$ after an angular rotation of 120 electrical degrees. Therefore, one of the component (3) of the field produced by this coil, rotating in anticlockwise direction, at this instant is along the axis of $a-a'$ i.e., *OA*. Whereas, the other component (4), rotating in clockwise direction is along the axis of coil $c-c'$ i.e., *OC*.

The flux produced by the coil $c-c'$ will become positive maximum $(+\phi_m)$ after an angular rotation of 240 electrical degrees. Therefore, one of the component (5), rotating in anticlockwise direction, at this instant is along the axis of coil $a-a'$ i.e., *OA*. Whereas, the other component (6), rotating in clockwise direction is along the axis of coil $b-b'$ i.e., *OB*.

The resultant of the three components 2, 4 and 6, which are rotating in clockwise direction and displaced by an angle of 120 electrical degrees is zero. The resultant of the remaining three components

1, 3 and 5 is 1·5 ϕ_m which is rotating in anticlockwise direction. It is the direction in which the supply is given to the three phases (coils) of the machine.

Conclusion

Hence, it is concluded that when a three-phase supply is given to the stator of a three-phase wound machine, a resultant field of magnitude 1 · 5 ϕ_m is produced which rotates in space at a constant speed, called synchronous speed. The direction of rotating of the resultant field depends upon the sequence in which supply is given to the three phases (or coils).

6.23 Rating of Alternators

The rating of all the power apparatus whether mechanical (steam or *IC* engines) or electrical (electrical machines) depends upon the power which they can handle safely and efficiently under some specific conditions.

Electrical machines or apparatus are usually rated as per the load they can carry without overheating and damaging the insulation. The rating of electrical machines is governed by the temperature rise caused by the internal losses of the machine. The copper loss (I^2R), in the armature depends upon the magnitude of current passing through the armature conductors and core loss depends upon the operating voltage. Both these losses are independent of the power factor.

The power in kW is proportional to the power factor of the load which may be different for different loads. For instant, let the rating of an alternator be 1000 kVA, at full-load it will be in position to deliver 400, 600, 800 and 1000 kW at power factor 0.4, 0.6, 0.8 and 1.0 respectively. At these loads, current supplied by the alternator will be the same and the losses will also remain the same.

The above explanation clearly shows that the rating of an alternator depends upon, the maximum current it can deliver without overheating and the voltage for which it is designed. The pf of the load is not considered to give its rating.

Hence, alternators are rated in kVA and not in kW.

The electrical machines and apparatus which themselves work as a load like induction motors, ovens, heaters, refrigerators, air-conditioners, etc., are rated in kW. But the machines which are supplying or transferring electric power like alternators, transformers etc., are rated in kVA.

6.24 Armature Resistance

The resistance per phase is armature resistance of an alternator. The resistance of a winding, depending upon resistivity of winding material, length and area of cross-section $\left[R = \rho \frac{l}{a}\right]$ is called *DC* resistance of the winding. The actual or *AC* resistance in nearly 1·25 to 1·75 times the *DC* resistance. This is because of unequal distribution of alternating current (Skin effect) over the cross-section of the conductor. Voltage drop (*IR*) occurs in the winding because of this resistance which is in-phase with the current phasor. However, the value of this resistance is very small as compared to synchronous reactance of the machine, which is why, many times, its voltage drop effect is neglected. Heavy copper losses occur in the machine because of armature resistance.

6.25 Armature Leakage Reactance

When current flows through the armature conductors, local fluxes are set at various places. *The flux which links with the armature winding but not with rotor field winding is called* **leakage flux.** The leakage flux may be divided into the following three components:

(*i*) **Slot leakage flux:** The flux which links or surrounds the armature conductors embedded in the iron or placed in the slots but does not passes through the air gap is called the *slots leakage flux* as shown in Fig. 6.39.

(*ii*) **Air gap leakage flux:** The flux which surrounds the armature conductors and passes through the air gap (i.e., the flux which crosses from tooth to tooth in the air gap) as shown in Fig. 6.39 is called *air gap leakage flux.*

(*iii*) **End-connection leakage flux:** The flux which links with only the end-connections of the armature winding is called *end-connection leakage flux.* It is also called as an *over hang leakage flux.*

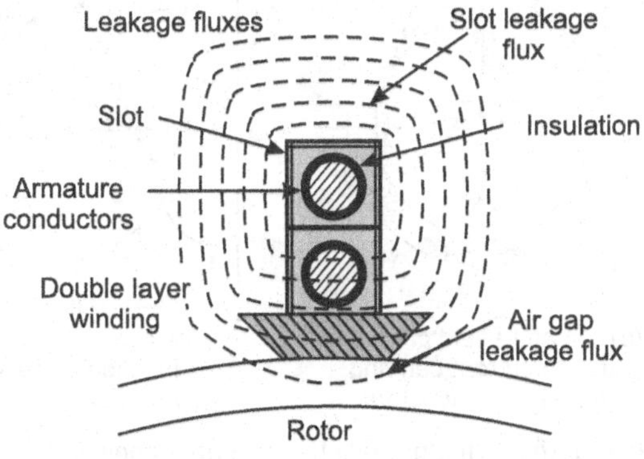

Fig. 6.39 Leakage Fluxes

The total leakage flux which links with the armature winding (conductors) gives rise to inductance. The magnitude of the inductance is given as

$$L = \frac{\phi N}{I} \text{ henry}$$

where ϕ is the leakage flux in weber, N is the number of turns and I is the armature current. This inductance L, when multiplied by ω gives the leakage reactance X_L. Hence,

$$X_L = \omega L = 2\pi f L \text{ ohm}$$

A voltage drop (IX_L) occurs in the winding because of this reactance. This voltage drop is in quadrature to the current vector.

According, $\overline{E} = \overline{V} + \overline{I}\overline{Z} = \overline{V} + \overline{I}(R + jX_L)$

6.26 Armature Reaction

At no-load, the only field (*mmf*) acting in the synchronous machine (alternator) is the main field 'F_m' produced by the exciting winding. When load is connected to the alternator, current flows through the armature conductors and produces a field (*mmf*) called armature field F_a. This field affects the main magnetic field F_m.

Thus, *the effect of the armature field on the main magnetic field is known as* **armature reaction**.

The current flowing through the armature conductors depends upon the power factor of the load. Therefore, the armature reaction will be studied at three extreme conditions of the *p.f.*, i.e., unity, zero lagging and zero leading.

At Unity Power Factor

Consider a two pole alternator with the poles rotating in anticlockwise direction. The three phases are represented by concentric coils *a-a'*, *b-b'* and *c-c'* which are displaced by 120° electrical from each other. Where *a*, *b* and *c* are the start terminals and *a'*, *b'* and *c'* are the finish terminals.

At an instant t_1, depending upon the position of the poles, direction of induced *emfs* in three coils is shown in Fig. 6.40(*a*) and their wave diagrams are shown in Fig. 6.40(*c*) (i.e., $v_{aa'}$, $v_{bb'}$ and $v_{cc'}$).

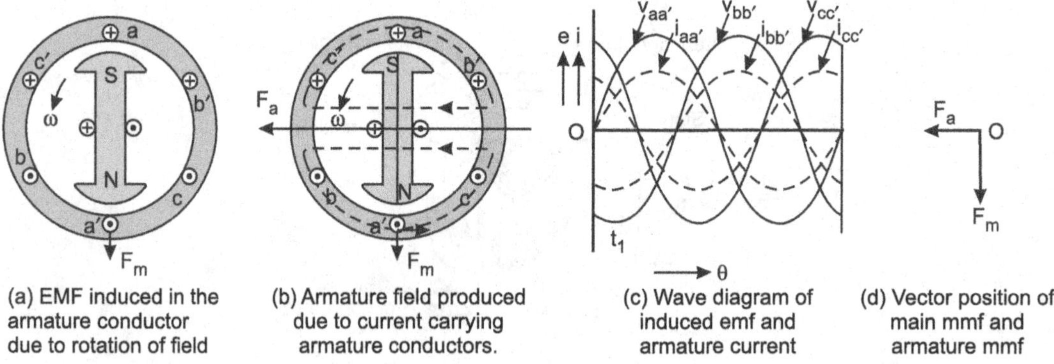

Fig. 6.40 Armature reaction at unity of power factor

The position of the main magnetic field F_m produced by the exciting field coils is shown in Fig. 6.40(*a*) as well as Fig. 6.40 (*b*). When a load of unity power factor is connected to the alternator, current flows through the coils which is in phase with the induced *emf* as shown in Fig. 6.40(*b*) and their wave diagrams are shown in Fig. 6.40 (*c*) (i.e., $i_{aa'}$, $i_{bb'}$ and $i_{cc'}$). A resultant armature field F_a is produced by the current carrying armature conductors of magnitude 1·5 ϕ_m (where ϕ_m is the maximum value of flux produced by the current flowing through each phase) in the direction as shown in Fig. 6.40(*b*). The two fields *mmfs*. are shown vectorially in Fig. 6.40(*d*). The armature field F_a is perpendicular of the main magnetic field F_m and produces cross-magnetising effect. *It is also clear from the position of the two fields that armature field* $\mathbf{F}_a$ *lags behind the main magnetic field* $\mathbf{F}_m$ *by 90° causing the same effect as it is being caused by a* ***pure inductance***.

Thus, for a non-inductive load (at unit p.f.) the effect of armature reaction is **cross-magnetising** i.e., *it distorts the main field.*

At Zero Power Factor Lagging

At an instant, t_2, depending upon the position of the poles, direction of induced *emfs* in three coils is shown in Fig. 6.41(*a*). For simplicity, the wave diagram of induced *emf* in coil $a - a'$ is only shown in Fig. 6.41(*c*). At this instant, induced *emf* in coil-side '*a*' is zero and becoming negative. The position of main magnetic field F_m produced by the exciting field coils is shown in Fig. 6.41(*a*) as well as Fig. 6.41(*b*).

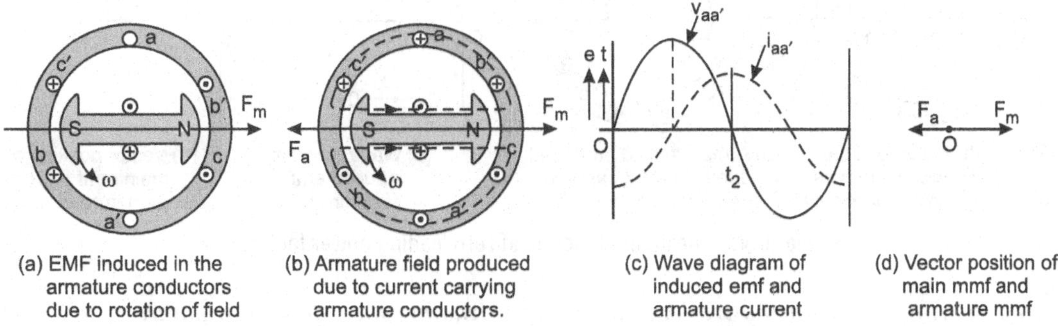

Fig. 6.41 Armature reaction at zero lagging power factor

When a pure inductive load at zero *p.f.* lagging is connected to the alternator, currents will flow through the coils, which lag behind their respective induced *emfs* by 90° electrical. The wave diagram for the phase representing coil *aa'* is shown in Fig. 6.41(*c*).

Accordingly, the direction of flow of current in various coil is shown in Fig. 6.41(*b*). A resultant armature field F_a of magnitude 1·5 ϕ_m is produced by the current carrying armature conductors in the direction as shown in Fig. 6.41(*b*). The two fields (*mmfs.*) are shown vectorially in Fig. 6.41(*d*). The armature field F_a acts in opposite direction to that of main magnetic field ϕ_m and produces demagnetising effect. *It is also clear from the position of the two fields that armature field F_a lags behind the main magnetic field F_m by 180°* i.e., *it lags by 90° because of a pure inductive load and 90° because of armature reaction.*

Thus, with a pure inductive load (at zero p.f. lagging) the effect of armature reaction is **demagnetising** i.e., *it weakness the main field.*

At Zero Power Factor Leading

At an instant t_3, depending upon the position of the poles, direction of induced *emfs* in the three coils is shown in Fig. 6.42(*a*). Only wave diagram of induced *emf* in coil $a - a'$ is shown in Fig. 6.42(*c*). At this instant, induced *emf* in coilside '*a*' is zero and becoming positive. The position of main magnetic field F_m produced by the exciting field coils is shown in Fig. 6.42(*a*) as well as 6.42(*b*).

When a pure capacitive load at zero *p.f.* leading is connected to the alternator, currents will flow through the coils which leads their respective induced *emfs* by 90° electrical. The wave diagram for the phase representing coil *a-a'* is shown in Fig. 6.42(*c*).

Accordingly, the direction of flow of current in various coils is shown in Fig. 6.42(*b*). A resultant armature field F_a of magnitude 1·5 ϕ_m is produced as shown in Fig. 6.42(*b*). The two fields (*mmfs.*) are shown vectorially in Fig. 6.42(*d*). The armature field F_a acts in the same direction as that of the main magnetic field F_m and produces magnetising effect. *It is also clear from the position of the two fields* that the armature field F_a is in phase with the main magnetic field F_m. In fact, the armature field was to lead the main field by 90° due capacitive load but it falls back to its original position due to armature reaction. Hence, the armature field acts in phase with the main field.

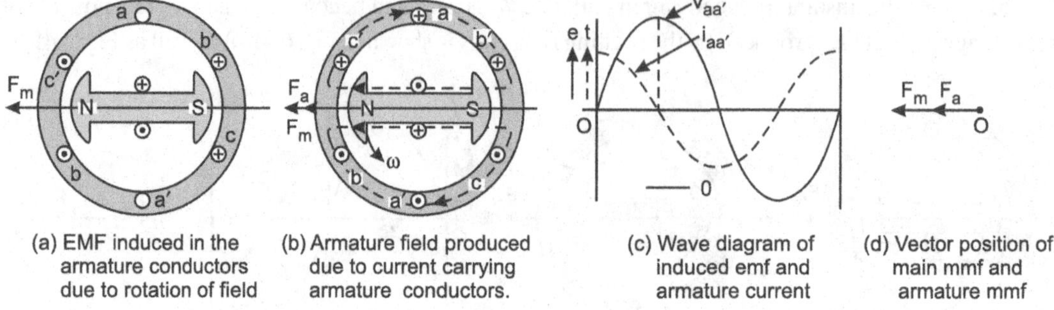

(a) EMF induced in the armature conductors due to rotation of field (b) Armature field produced due to current carrying armature conductors. (c) Wave diagram of induced emf and armature current (d) Vector position of main mmf and armature mmf

Fig. 6.42 Armature reaction at zero leading power factor

Thus, with a pure capacitive load (at zero p.f. leading) the effect of armature reaction is magnetising i.e., *it strengthens the main field.*

Conclusion

The above explanation reveals that

1. The armature reaction flux ($\phi_a = 1.5\ \phi_{arm}$) is constant in magnitude and rotates at synchronous speed.
2. When generator supplies a load at unity *pf*, the armature reaction is *cross-magnetising*.
3. When generator supplies a load at zero *pf* lagging, the armature reaction in *demagnetising*.
4. When generator supplies a load at zero *pf* leading, the armature reaction is *magnetising*.
5. In all the cases, if the armature-reaction flux is assumed to act independently of the main field flux, it induces voltage in each phase which lags the respective phase current by 90°. Hence armature reaction causes an armature reactance drop.

6.27 Effect of Armature Reaction on emf of Alternator

To see the effect of armature reaction on the *emf* induced in the alternator.

Let, E_0 = *emf* induced per phase by the main field (flux) at no-load.

E_a = *emf* induced per phase by the armature field (flux).

E = Net *emf* induced per phase i.e., resultant of E_0 and E_a.

$$\therefore \qquad \bar{E} = \bar{E}_0 + \bar{E}_a$$

When the armature is at no-load, no current flows through the armature. Thus the armature field is zero. The main field F_m (or ϕ_m) will induce an *emf* E_0 which lags behind ϕ_m by 90° as shown in 6.43. Therefore, net induced *emf* $E = E_0$.

When a non inductive load is connected to the alternator, current (I) flows through the armature conductors which is in phase with E_0 and produces an armature field F_a or ϕ_a which lags behind the main field by 90°. An *emf* E_a is induced by this field which lags behind the armature field by 90° as shown in Fig. 6.44. Thus a resultant *emf* E is obtained across the armature. $(\bar{E} = \bar{E}_0 + \bar{E}_a)$

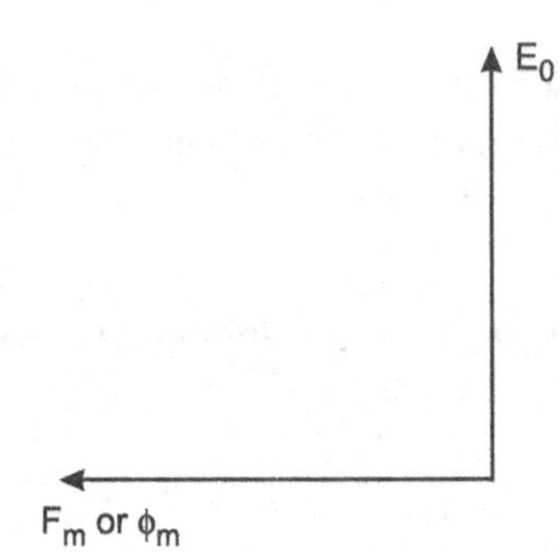

Fig. 6.43 Phasor diagram at no-load

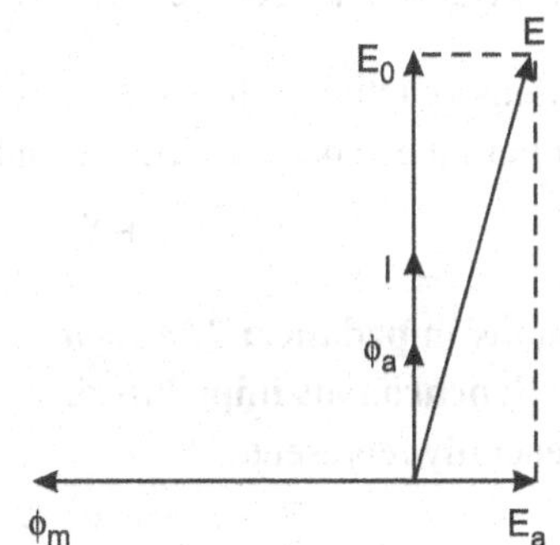

Fig. 6.44 Phasor diagram for unity p.f.

When a pure inductive load is connected to the alternator, a current (I) flows through the armature conductors which lags behind E_0 by 90°. This current produces armature field ϕ_a which further induces an *emf* E_a in the armature lagging behind ϕ_a by 90° as shown in Fig. 6.45. Thus a resultant *emf* $(E = E_0 - E_a)$ is obtained across the armature.

When a pure capacitive load is connected to the alternator, a current (I) flows through the armature conductors which leads the *emf* E_0 by 90°. This current produces armature field ϕ_a which further induces *emf* E_a in the armature lagging behind ϕ_a by 90° but in phase with E_0 as shown in Fig. 6.46. Thus, a resultant *emf* $(E = E_0 + E_a)$ is obtained across the armature.

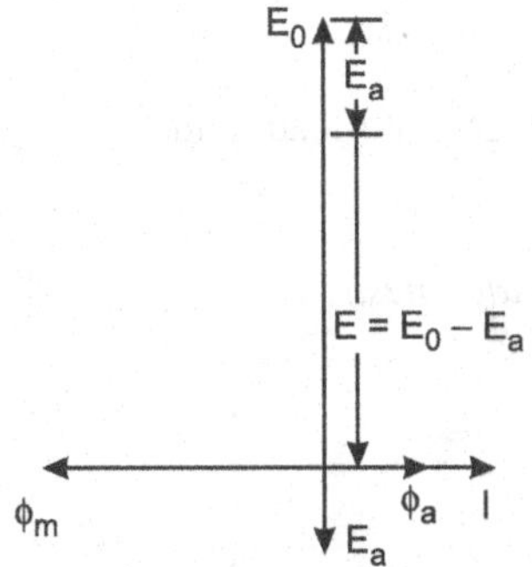

Fig. 6.45 Phasor diagram for zero lagging pf

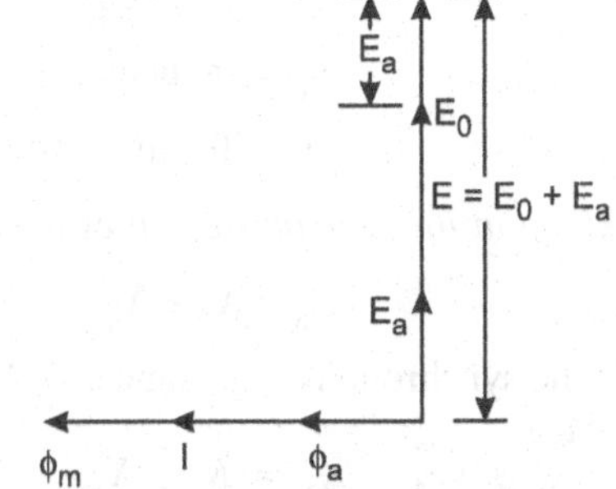

Fig. 6.46 Phasor diagram for zero loading

Conclusion

From the above discussion, it is concluded that E_0 is always in phase quadrature with the load current I and proportional to it. It, therefore, resembles with the *emf* induced in an inductive reactance so that the effect of armature reaction is exactly as if the stator winding has a reactance $X_a = E_a/I$. Whereas X_a is termed as armature reaction reactance.

Thus, an armature reaction produces a reactance effect in the armature winding called **armature reaction reactance.**

6.28 Synchronous Reactance and Synchronous Impedance

Synchronous reactance: It is fictitious reactance which has the effect equivalent to the combined effects of both the leakage reactance and armature reaction reactance. It is represented by X_s.

$$X_s = X_L + X_a$$

Synchronous impedance: *The vector sum of armature resistance and synchronous reactance is known as* **Synchronous impedance.**

It is generally represented by Z_s.

$$\overline{Z}_S = \overline{R} + \overline{X}_S = R + jX_S$$

or
$$Z_S = \sqrt{R^2 + X_S^2}$$

6.29 Equivalent Circuit of an Alternator and Phasor Diagram

The equivalent circuit of an alternator is shown in Fig. 6.47(*a*). Where,

R = Armature resistance,

X_L = Leakage reactance,

X_a = Armature reaction reactance,

I = Load current

E_0 = Induced *emf* in the armature (voltage at no-load) and

V = Terminal voltage

(**Note:** *All quantities are phase values in case of three-phase machines.*)

i.e.,
$$X_S = X_L + X_a$$

and Z_S is the synchronous impedance.

i.e.,
$$\overline{Z}_S = R + jX_s \quad \text{or} \quad Z_S = \sqrt{R^2 + X_S^2}$$

The simplified equivalent circuit is shown in Fig. 6.47(*b*), where X_s is the synchronous reactance.

From circuit diagram:

$$\overline{E}_0 = \overline{V} + \overline{I}\overline{Z}_s$$

Phasor Diagram: From the simplified equivalent circuit shown in Fig. 6.47(*b*), phasor diagram of an alternator can be drawn. The phasor diagram depends upon the type of load. The load may be non-inductive, inductive or capacitive.

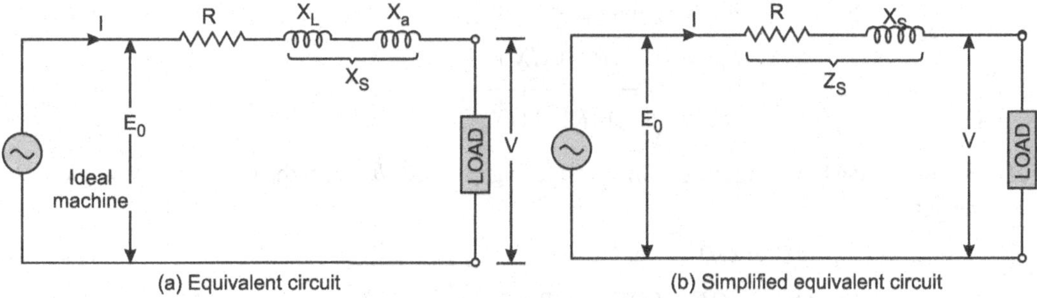

Fig. 6.47 Equivalent circuit for alternator

The phasor diagram for non-inductive, inductive and capacitive load are shown in Fig. 6.48(*a*), (*b*) and (*c*) respectively. While drawing phasor diagram, following steps are followed:

(*i*) Terminal voltage V is taken as the reference vector.

(*ii*) Load current I is drawn in phase with voltage vector V for non-inductive load. For inductive load, it is drawn so that it lags behind the voltage vector by an angle ϕ. Whereas for capacitive load it leads the voltage vector by an angle ϕ. Where ϕ is the power factor angle.

(*iii*) Draw voltage drop vector IR parallel to current vector from point A.

(*iv*) Draw voltage drop vector IX_S from point B perpendicular to vector AB since it is in quadrature to current vector.

(*v*) Join point O and C, where OC is the *emf* E_0 which is induced due to the main field ϕ_m or the terminal voltage at no-load and is the vector sum of V, IR and IX_S.

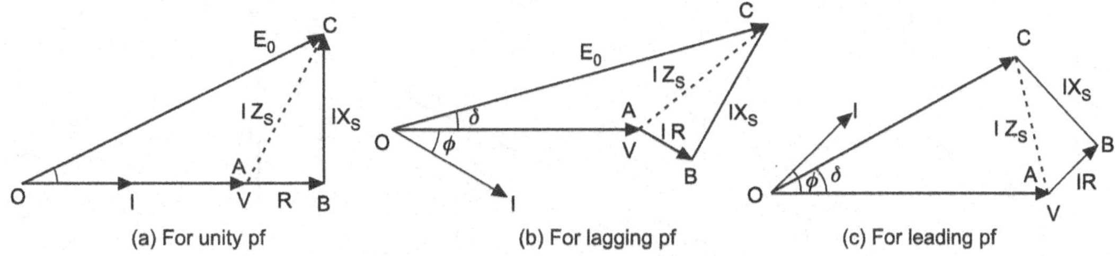

Fig. 6.48 Phasor diagram

6.30 Expression for No-load Terminal Voltage

To deduce an expression for no-load terminal voltage, complete the phasor diagrams for various loads as shown in Fig. 6.49.

The no-load terminal voltage E_0 is the actual induced *emf* in the armature produced by the main magnetic field ϕ_m. It is the vector sum of terminal voltage at load, drop in resistance and drop in synchronous reactance. From phasor diagram of alternator at different load conditions. we can deduce the expression for no-load terminal voltage.

(*i*) **For non-inductive load (unit p.f.):** Considering Fig. 6.49(*a*), from the right angle triangle *OBC*, we get,

$$OC^2 = OB^2 + BC^2$$

or $$OC^2 = (OA + AB)^2 + (BC)^2$$

or $$E_0^2 = (V + IR)^2 + (IX_s)^2$$

or $$E_0 = \sqrt{(V + IR)^2 + (IX_s)^2}$$

(ii) **For inductive load (lagging p.f.):** Considering Fig. 6.49(*b*), from the right angle triangle *ODC*, we get,

$$OC^2 = OD^2 + DC^2$$

or $$OC^2 = (OE + ED)^2 + (DB + BC)^2$$

or $$OC^2 = (OE + AB)^2 + (EA + BC)^2$$

or $$E_0^2 = (V\cos\phi + IR)^2 + (V\sin\phi + IX_s)^2$$

or $$E_0 = \sqrt{(V\cos\phi + IR)^2 + (V\sin\phi + IX_s)^2}$$

Also in right angle triangle *OFC*,

$$OC^2 = OF^2 + FC^2$$
$$= (OA + AH + HF)^2 + (GC - GF)^2$$
$$= (OA + AH + BG)^2 + (GC - BH)^2$$
$$E_0^2 = (V + IR\cos\phi + IX_S\sin\phi)^2 + (IX_S\cos\phi - IR\sin\phi)^2$$
$$E_0 = \sqrt{(V + IR\cos\phi + IX_S\sin\phi)^2 + (IX_S\cos\phi - IR\sin\phi)^2}$$

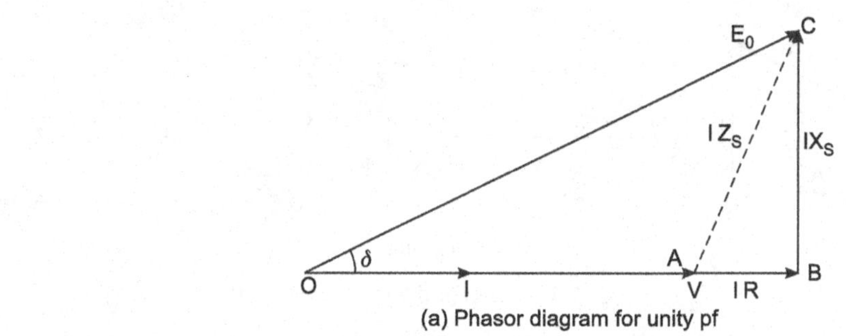

(a) Phasor diagram for unity pf

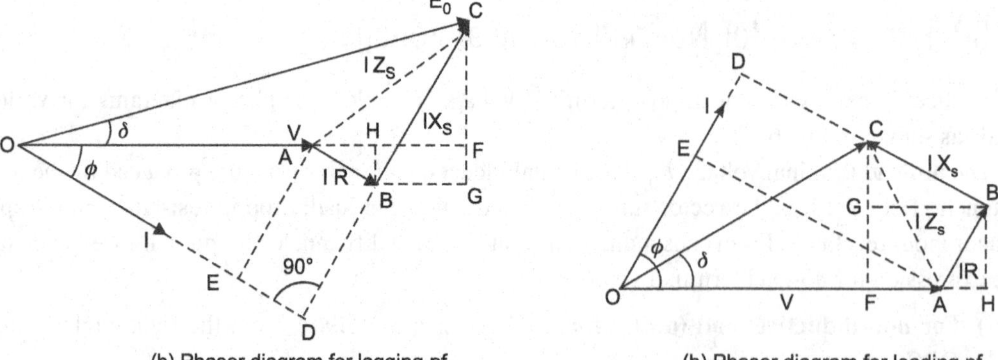

(b) Phasor diagram for lagging pf

(b) Phasor diagram for leading pf

Fig. 6.49 Phasor diagram of different power factors

(*iii*) **For capacitive load (leading p.f.):** Considering Fig. 6.49(*c*), from the right angle triangle *ODC*, we get,

$$OC^2 = OD^2 + DC^2$$

or $$OC^2 = (OE + ED)^2 + (DB - BC)^2$$

or $$OC^2 = (OE + AB)^2 + (EA - BC)^2$$

or $$E_0^2 = (V\cos\phi + IR)^2 + (V\sin\phi - IX_s)^2$$

or $$E_0 = \sqrt{(V\cos\phi + IR)^2 + (V\sin\phi - IX_s)^2}$$

Also in right angle triangle *OFC*,

$$OC^2 = OF^2 + FC^2$$

$$= (OA + AH - HF)^2 + (GC + GF)$$

$$= (OA + AH - BG)^2 + (GC + BH)^2$$

$$E_0^2 = (V + IR\cos\phi - IX_S\sin\phi)^2 + (IX_S\cos\phi + IR\sin\phi)^2$$

$$E_0 = \sqrt{(V + IR\cos\phi - IX_S\sin\phi)^2 + (IX_S\cos\phi + IR\sin\phi)^2}$$

Approximate expression for no-load terminal voltage (considering δ to be very small)

(*i*) *For non-inductive load (unity pf)*

$$OC \cong OB = OA + AB \quad \text{or} \quad E_0 = V + IR$$

(*ii*) *For inductive load (lagging pf)*

$$OC \cong OF = OA + AH + HF \quad \text{or} \quad E_0 = V + IR\cos\phi + IX_s\sin\phi$$

(*iii*) *For capacitive load (leading pf)*

$$OC = OF = OA + AH - HF \quad \text{or} \quad E_0 = V + IR\cos\phi - IX_s\sin\phi$$

6.31 Voltage Regulation

We have seen that because of the voltage drop in the armature resistance and synchronous reactance, the terminal voltage of synchronous generator depends upon the load and its *p.f.*

The rise in terminal voltage when the given load is thrown off, the excitation and speed remaining constant, is called the **voltage regulation** *of a synchronous generator (alternator).*

If E_0 = no-load terminal voltage and V = terminal voltage at a given load.

$$\text{Then voltage regulation} = \frac{E_0 - V}{V} \text{ per unit}$$

$$\text{Percentage regulation, \% Reg} = \frac{E_0 - V}{V} \times 100$$

The voltage regulation is positive both at *unit* and *lagging p.f.* because this causes rise in terminal voltage when the load is thrown off (removed). However, in case of leading *p.f.* the terminal voltage may fall when the load is thrown off. Therefore, at *leading p.f.* voltage regulation may be **negative.**

The voltage regulation may be zero, when load and its pf may be such that

$$^{*}IR \cos \phi = IX_s \sin \phi$$

The voltage regulation of a synchronous generator is usually much high than that of power transformer because of large amount of voltage drop in synchronous reactance.

6.32 Determination of Voltage Regulation

To determine the voltage regulation of smooth cylindrical rotor type alternators, the following methods may be used:

1. Direct load test 2. Indirect Method

1. **Direct load test:** Direct load test is performed only on small alternators (say 5 kVA). In this case, the alternator is run at synchronous speed with the help of a prime-moves and its terminal voltage is adjusted to its rated value V and the load is varied until the ammeter and wattmeter indicate the rated values at given power factor.
   Then the load is removed keeping the speed and field excitation constant. At open circuit i.e., at no-load, the terminal voltage E_0 is recorded. Then voltage regulation is determined as under:
   $$\text{Percentage voltage regulation} = \frac{E_0 - V}{V} \times 100$$
2. **Indirect method:** For large size alternator, indirect methods are used to determine voltage regulation. These methods are:
   (*i*) Synchronous impedance method or *EMF* method
   (*ii*) Ampere-turn method or *MMF* method
   (*iii*) Zero power factor method or Potier method.

6.33 Synchronous Impedance Method or emf Method

This method is based on the concept of replacing the effect of armature reaction by a fictitious reactance.

For a synchronous generator

$$\bar{V} = \bar{E}_0 - \bar{I}\,\bar{Z}_3$$

where $\bar{Z}_3 = R + jX_S$

---

* The approximate expression for E_0 at leading *pf* is

$$E_0 = V + IR\cos\phi - IX_s \sin\phi$$

$$E_0 - V = IR\cos\phi - IX_s \sin\phi$$

Voltage regulation will be zero when $E_0 - V = 0$

$$\therefore \quad IR\cos\phi - IX_s \sin\phi = 0$$

or

$$IR\cos\phi = IX_s \sin\phi$$

To determine synchronous impedance, open-circuit and short circuit tests are performed and to determine synchronous reactance, armature resistance is measured by ammeter-voltmeter method.

By using these parameters, the regulation of the alternator can be determined at any load.

6.33.1 Determination of Synchronous Impedance

To determine the value of synchronous impedance of an alternator experimentally, the following two tests are performed on the machine:

(*i*) Open circuit test (*ii*) Short circuit test.

(*i*) **Open circuit test:** To perform open circuit test, the terminals of the alternator are kept open and is rotated by the prime-mover at synchronous speed. A *DC* supply is given to the field winding through a rheostat. A voltmeter is connected across the terminals of the alternator to measure open circuit voltage E and an ammeter is connected in the field circuit to measure field current I_f as shown in Fig. 6.50. The field current I_f (excitation) is gradually varied (increased in steps) and the voltage across the terminals of the alternator E is recorded for every change in the field current. A graph is plotted taking I_f along abscissa and E along the ordinate called open circuit characteristics (*O.C.C.*). The *O.C.C.* curve so obtained is shown in Fig. 6.52. The curve rises steeply and then flattened due to saturation of the magnetic circuit.

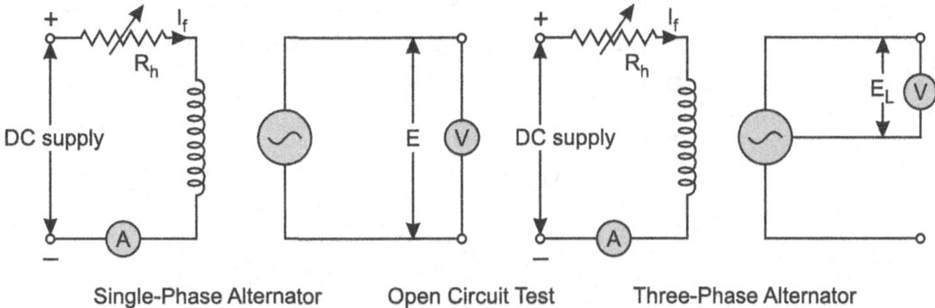

Fig. 6.50 Circuit diagram for open circuit test on single and three phase alternators

Note: In case of three-phase star-connected alternator, to plot the curve phase value of the terminal voltage $E = E_L/\sqrt{3}$ is to be considered.

(*ii*) **Short circuit test:** To perform short circuit test, the terminals of the alternator are short circuited by a thick strip or an ammeter as shown in Fig. 6.51. And its rotor is rotated by the prime-mover at synchronous speed. The field current I_f is gradually increased and the short circuit current I_{sc} is recorded for every change in the field current with the help of ammeter connected across the alternator terminals. A graph is plotted taking I_f along abscissa and I_{sc} along with ordinate called short circuit characteristics (*S.C.C.*). The *S.C.C.* curve so obtained is shown in Fig. 6.52 which is almost a straight line.

It should be noted that both the *O.C.C.* and *S.C.C.* curves are drawn on a common field current I_f as shown in Fig. 6.52.

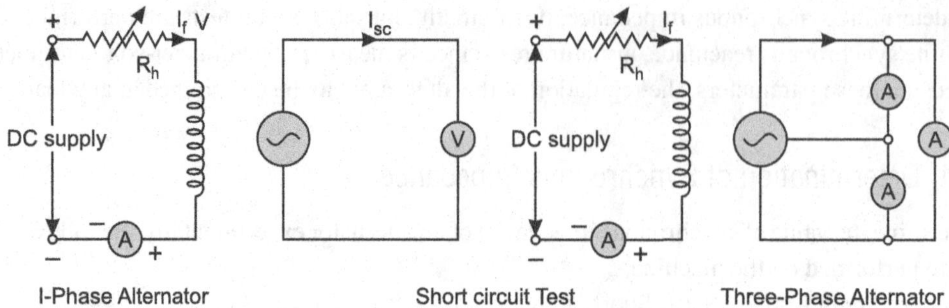

Fig. 6.51 Circuit diagram for short circuit test on single and three phase alternators

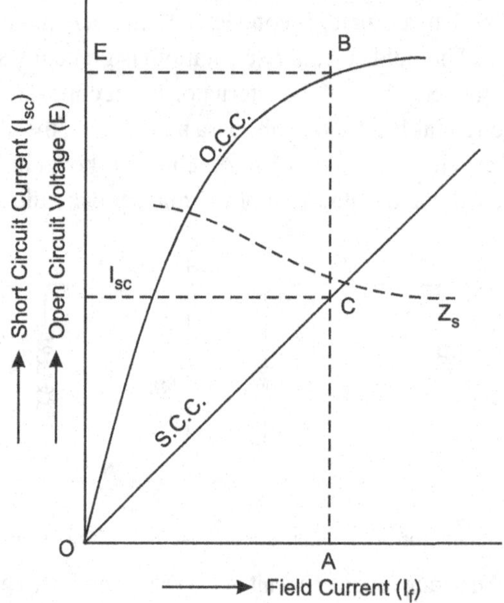

Fig. 6.52 Open circuit and short circuit characteristics

Determination of Synchronous Impedance

To determine synchronous impedance of the alternator, let *OA* be the extension (field current I_f). For this field current *OA*, open circuit voltage is *AB* (i.e., *E*) and for the same field current the short circuit current is *AC* (i.e., I_{sc}). When the alternator is short circuited terminal voltage is zero. Therefore, at short circuit, whole of the induced voltage *E* is being utilised for circulating the short circuit current I_{sc} through the synchronous impedance Z_s.

$\therefore$ $$E = I_{sc} Z_s$$

or $$Z_s = \frac{E}{I_{SC}} = \frac{\text{Open circuit voltage}}{\text{Short circuit current}}$$

at the same field current.

The value of synchronous impedance is not constant. Its value is slightly high at low magnetic saturation. An approximate graph of Z_s against field current. I_f is drawn by the dotted line as shown in Fig. 6.52.

Note: The value of synchronous impedance is usually determined at the field current which provides the rated *emf* of the machine

6.33.2 Determination of Synchronous Reactance

To determine the value of synchronous reactance, first of all armature resistance is calculated by voltmeter–ammeter method, circuit shown in Fig. 6.53. Since the value of armature resistance is very small, a low *DC* supply voltage is connected across the terminals of a one-phase alternator. The value of armature resistance $\left(R_{dc} = \frac{V_{dc}}{I_{dc}}\right)$ is the *DC* resistance. The actual resistance of the armature of an alternator is 1.25 to 1.75 times to that of *DC* resistance.

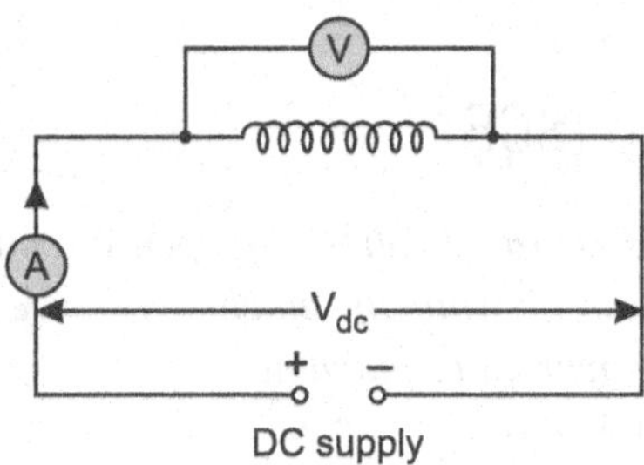

Fig. 6.53 Voltmeter–ammeter test to determine armature resistance

Therefore, $R = 1.25\, R_{dc}$

Now synchronous reactance,

$$X_s = \sqrt{(Z_s)^2 - (R)^2}$$

Imp. Note: In case of three-phase, star connected alternators, DC source, ammeter and voltmeter are connected across the terminals, then the measured resistance comes out to be the resistance of two phases.

DC Resistance per phase,

$$R_{dc} = \frac{\text{measured resistance}}{2}$$

Also, all the vector diagrams are drawn considering the phase quantities whether the alternator is single-phase or three-phase. To determine the regulation, all the quantities must be phase value.

6.34 Modern Alternators

When the terminals of the alternator are short circuited due to any fault (short circuit fault) short circuit current I_{sc} flows through the armature conductors. This current depends upon the induced *emf* E_0 and the synchronous impedance Z_s of the alternator i.e.,

$$I_{sc} = \frac{E_0}{Z_s}.$$

This short circuit current may damage the armature winding of the alternator if Z_S is very small. Therefore, to limit the short circuit current, the modern alternators are designed for higher synchronous impedance. The resistance cannot be increased because it would increase the losses in the machine. *Thus modern alternator are designed for higher synchronous reactance (or leakage reactance) but smaller resistance.*

The value of synchronous reactance is some times more than 20 times the armature resistance. Therefore, for all practical purposes the voltage drop due to armature resistance is neglected as compared to voltage drop due to synchronous reactance.

6.35 Short-Circuit Ratio (SCR)

The ratio of field current to produce rated voltage on open-circuit to the field current required to circulate rated current on short-circuit while the machine is driven at synchronous speed is called **short-circuit ratio (SCR)** *of a synchronous machine.*

From *OCC* and *SCC* shown in Fig. 6.54.

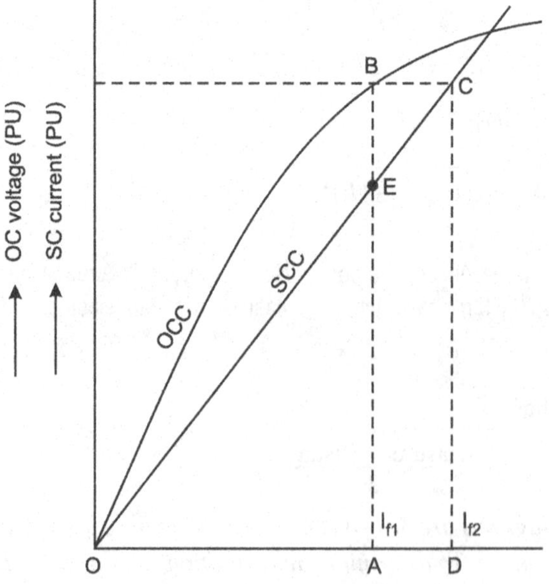

Fig, 6.54 *OCC* and *SCC* for an alternator

$$\text{Short-circuit ratio, } SCR = \frac{I_{f1}}{I_{f2}} = \frac{OA}{OD} = \frac{AE}{DC} = \frac{AE}{AB} = \frac{1}{AB / AE}$$

$$\text{Where } \frac{AB}{AE} = \frac{\text{Per unit voltage on open-circuit}}{\text{Corresponding per unit current on short-circuit}} = X_s$$

$$\therefore \qquad SCR = \frac{1}{X_s}$$

Thus, SCR is just reciprocal of per unit synchronous reactance X_S of the machine. The value of synchronous reactance depends upon saturated conditions of the machine whereas, SCR is specific and defined at rated voltage.

Significance of SCR

Smaller is the value of *SCR*, larger is the value of synchronous reactance which limits the short circuit current to smaller value. But it causes difficulty during parallel operation of the machines owing to smaller value of synchronising power.

Larger value of *SCR* increases the stability of the machine and improves its voltage regulation.

Usually, the *SCR* of a high speed non-salient pole alternators lies between 0.5 and 0.75 whereas it lies between 1.0 and 1.5 for low speed salient pole type alternators.

Therefore, the salient pole type alternators are more stable than non-salient pole type alternators.

Example 6.16

A 1-phase 60 kVA, 220 V, 50 Hz, alternator has an effective armature leakage reactance of 0·07 ohm and negligible armature resistance. Calculate the voltage induced in the armature when the alternator is delivering rated current at a load power factor of 0·7 lagging.

Solution:

Here, Rated power = 60 kVA = 60×10^3 VA

Terminal voltage, V = 220 V

Leakage reactance, $X_L = 0{\cdot}07\ \Omega$

Load power factor, $\cos\phi = 0{\cdot}7$ lag; $\sin\phi = \sin\cos^{-1} 0{\cdot}7 = 0{\cdot}7141$

$$\text{Full load current, } I = \frac{60 \times 10^3}{220} = 272 \cdot 72 \text{ A}$$

$$\text{Voltage induced in the armature, } E_0 = \sqrt{(V\cos\phi + IR)^2 + (V\sin\phi + IX_L)^2}$$

$$= \sqrt{(V\cos\phi)^2 + (V\sin\phi + IX_L)^2}$$

(since value of R is not given)

$$= \sqrt{(220 \times 0 \cdot 7)^2 + (220 \times 0 \cdot 7141 + 272 \cdot 72 \times 0 \cdot 07)^2}$$

$$= \sqrt{(154)^2 + (176 \cdot 19)^2} = \mathbf{234\ V}\ (Ans.)$$

Example 6.17

A single-phase 100 kVA, 600V, 50 Hz alternator has effective armature resistance and leakage reactance of 0·072 and 0·18 ohm respectively. At rated terminal voltage and kVA load, determine internal induced emf at (i) unit p.f. (ii) 0·75 p.f. lagging; (iii) 0·75 p.f. leading.

Solution:

Here, Rated power = 100 kVA = 100×10^3 VA; Terminal voltage, $V = 600$ V

Armature resistance, $R = 0{\cdot}072$ ohm; Leakage reactance, $X_L = 0{\cdot}18$ ohm

$$\text{Rated current, } I = \frac{100 \times 10^3}{600} = 166 \cdot 67 \text{ A}$$

(*i*) When the p.f., $\cos\phi = 1$; $\sin\phi = \sin\cos^{-1} 1 = 0$

∴ Induced *emf* $E_0 = \sqrt{(V\cos\phi + IR)^2 + (IX_L)^2}$

$$= \sqrt{(600 \times 1 + 166 \cdot 67 \times 0 \cdot 72)^2 + (166 \cdot 67 \times 0 \cdot 18)^2}$$

= **612·73 V** (*Ans.*)

(*ii*) When the *p.f.*, $\cos\phi = 0{\cdot}75$ lagging; $\sin\phi = \sin\cos^{-1} 0{\cdot}75 = 0{\cdot}6614$

∴ Induced *emf,* $E_0 = \sqrt{(V\cos\phi + IR)^2 + (V\sin\phi - IX_L)^2}$

$$= \sqrt{(600 \times 0.75 + 166 \cdot 67 \times 0 \cdot 72)^2 + (600 \times 0 \cdot 6614 + 166.67 \times 0 \cdot 18)^2}$$

$$= \sqrt{(462)^2 + (366 \cdot 8)^2}$$

= **629 V** (*Ans.*)

(*iii*) When the *p.f.*, $\cos\phi = 0{\cdot}75$ leading; $\sin\phi = \sin\cos^{-1} 0.75 = 0{\cdot}6614$

∴ Induced *emf* $E_0 = \sqrt{(V\cos\phi + IR)^2 + (V\sin\phi - IX_L)^2}$

$$= \sqrt{(600 \times 0.75 + 166 \cdot 67 \times 0 \cdot 072)^2 + (600 \times 0 \cdot 6614 - 166.67 \times 0 \cdot 18)^2}$$

$= \sqrt{(462)^2 + (366 \cdot 8)^2}$ = **590 V** (*Ans.*)

Example 6.18

A single-phase, 500 V, 50 Hz alternator produces a short-circuit current of 170 A and an open circuit emf of 425 V when a field current of 15A passes through its field winding. If its armature has an effective resistance of 0.2 ohm, determine its full-load regulation at unity pf and at 0.8 pf lagging.

Solution:

Here, Rated power = 50 kVA = 50×10^3 VA

Terminal voltage, $V = 500$ V; Armature resistance, $R = 0{\cdot}2$ ohm

Short circuit current, $I_{sc} = 170$ A; Open circuit *emf,* $E = 425$ V

$$\text{Synchronous impedance, } Z_s = \frac{E}{I_{SC}} = \frac{425}{170} = 2 \cdot 5 \text{ ohm}$$

$$\text{Synchronous reactance, } X_s = \sqrt{(Z_s)^2 - (R)^2} = \sqrt{(2 \cdot 5)^2 - (0 \cdot 2)^2} = 2{\cdot}492 \text{ ohm}$$

$$\text{Full load current, } I = \frac{50 \times 10^3}{500} = 100 \text{ A}$$

When *p.f.* $\cos\phi = 1$; $\sin\phi = 0$

$$E_0 = \sqrt{(V\cos\phi + IR)^2 + (IX_S)^2}$$

$$= \sqrt{(500 \times 1 + 100 \times 0 \cdot 2)^2 + (100 \times 2 \cdot 492)^2}$$

$$= 576.63 \text{ V}$$

$$\% \text{ Reg.} = \frac{E_0 - V}{V} \times 100 = \frac{576 \cdot 63 - 500}{500} \times 100$$

$= \mathbf{15{\cdot}326\%}$ (*Ans.*)

When *p.f.* $\cos\phi = 0{\cdot}8$ lagging; $\sin\phi = \sin\cos^{-1} 0{\cdot}8 = 0{\cdot}6$

$$E_0 = \sqrt{(V\cos\phi + IR)^2 + (V\sin\phi + IX_S)^2}$$

$$= \sqrt{(500 \times 0 \cdot 8 + 100 \times 0 \cdot 2)^2 + (500 \times 0 \cdot 6 + 100 \times 2 \cdot 492)^2}$$

$$= \sqrt{(420)^2 + (549 \cdot 2)^2} = 671 \cdot 62 \text{ V}$$

$$\% \text{ Reg.} = \frac{E_0 - V}{V} \times 100 = \frac{671 \cdot 62 - 500}{500} \times 100$$

$= \mathbf{34{\cdot}35\%}$ (*Ans.*)

Example 6.19

A three-phase star-connected alternator has an armature resistance of 0·1 ohm per phase. When excited to 173·3 V line voltage and on short circuit the alternator gave 200 A. What should be the emf (in line terms) the alternator must be excited to, in order to maintain a terminal potential difference of 400 V with 100 A armature current at 0·8 power factor lagging?

Solution:

Here, open circuit *emf* (line value), $E_L = 173{\cdot}3$ V

Armature resistance per phase, $R = 0{\cdot}1$ ohm

Short circuit current, $I_{sc} = 200$ A

Terminal voltage (line value), $V_L = 400$ V

Armature current, $I = 100$ A

Open circuit *emf* (phase value), $E = \dfrac{173 \cdot 3}{\sqrt{3}} = 100$ V

Synchronous impedance/phase, $Z_s = \dfrac{E}{I_{sc}} = \dfrac{100}{200} = 0 \cdot 5$ ohm

Synchronous reactance/phase, $X_s = \sqrt{Z_s^2 - R^2} = \sqrt{(0 \cdot 5)^2 - (0 \cdot 1)^2} = 0 \cdot 4899$ ohm

Terminal voltage/phase. $V = \frac{V_L}{\sqrt{3}} = \frac{400}{\sqrt{3}} = 231$ V

Load *p.f.*, $\cos\phi = 0{\cdot}8$; $\sin\phi = \sin\cos^{-1} 0{\cdot}8 = 0{\cdot}6$

No-load terminal voltage/phase.

$$E_0 = \sqrt{(V\cos\phi + IR)^2 + (V\sin\phi + IX_s)^2}$$

$$= \sqrt{(231 \times 0.8 + 100 \times 0 \cdot 1)^2 + (231 \times 0 \cdot 6 + 100 \times 0 \cdot 4899)^2}$$

$$= 70.4 \text{ V}$$

No-load terminal voltage (line value) = $\sqrt{3}E_0 = \sqrt{3} \times 270 \cdot 3 =$ **468·4 V** (*Ans.*)

Example 6.20

A three-phase star connected 1200 kVA, 3300 V, 50 Hz, alternator has armature resistance of 0·25 ohm per phase. A field current of 40 A produces a short circuit current of 200 A and an open circuit emf of 1100 V between lines. Calculate regulation on full load 0·8 power factor lagging.

Solution:

Here, Rated power = 1200 kVA = 1200×10^3 VA

Terminal line voltage, V_L = 3300 V (*star connected*)

Armature resistance, $R = 0{\cdot}25\ \Omega$

At field current of 40 A;

Short circuit current, I_{sc} = 200 A

Open circuit *emf* (phase value), $E_{(ph)} = \frac{1100}{\sqrt{3}} = 635 \cdot 1$ V

Synchronous impedance, $Z_s = \frac{E_{(ph)}}{I_{sc}} = \frac{635 \cdot 1}{200} = 3 \cdot 175\ \Omega$

Synchronous reactance, $X_s = \sqrt{Z_s^2 - R^2} = \sqrt{(3 \cdot 175)^2 - (0 \cdot 25)^2} = 3 \cdot 175\ \Omega$

Full load, current, $I = \frac{1200 \times 10^3}{\sqrt{3} \times 3300} = 210$ A

Terminal phase voltage, $V = \frac{V_L}{\sqrt{3}} = \frac{3300}{\sqrt{3}} = 1905 \cdot 2$ V

Power factor, $\cos\phi = 0.8$; $\sin\phi = \sin\cos^{-1} 0{\cdot}8 = 0{\cdot}6$

Open circuit terminal voltage (phase value),

$$E_0 = \sqrt{(V\cos\phi + IR)^2 + (V\sin + IX_s)^2}$$

$$= \sqrt{(1905 \cdot 0 \times 0.8 + 210 \times 0 \cdot 25)^2 + (1905 \cdot 2 \times 0 \cdot 6 + 210 \times 3 \cdot 175)^2}$$

$$= 2400 \text{ V}$$

$$\% \text{ Reg.} = \frac{E_0 - V}{V} \times 100 = \frac{2400 - 1905 \cdot 2}{1905 \cdot 2} \times 100 = \mathbf{25{\cdot}98\%} \ (Ans.)$$

Example 6.21

A three-phase, star connected, 20 MVA, 11 kV, 50 Hz alternator produces a short-circuit current equal to full-load current when a field current of 70 A passes through its field winding. The same field current produces an emf of 1820 V (line to line) on open circuit. If the alternator has a resistance between each pair of terminals as measured by DC is 0.16 ohm and the effective resistance is 1.5 times the ohmic resistance, what will be its full-load regulation at (i) 0.707 pf lagging and (ii) 0.8 pf leading.

Solution:

Here, Alternator is three-phase, star connected;

Rating of alternator = 20 MVA = 20×10^6 VA

Terminal voltage (line value), V_L = 11000 V

Open circuit *emf* (line value), E_L = 1820 V

Resistance between two terminals = 0·16 ohm

$$\text{Resistance measured/phase} = \frac{0 \cdot 16}{2} = 0 \cdot 08 \text{ ohm}$$

Effective resistance/phase, $R = 1{\cdot}5 \times 0{\cdot}08 = 0{\cdot}12$ ohm

$$\text{Open circuit } emf \text{ (phase value), } E = \frac{E_L}{\sqrt{3}} = \frac{1820}{\sqrt{3}} = 1050 \cdot 8 \text{ V}$$

$$\text{Full load circuit, } I = \frac{20 \times 16^6}{\sqrt{3} \times 11000} = 1049 \cdot 7 \text{ A}$$

Short circuit current, $I_{sc} = I = 1049{\cdot}7$ A

$$\text{Synchronous impedance/phase, } Z_s = \frac{E}{I_{sc}} = \frac{1050 \cdot 8}{1049 \cdot 7} = 1 \cdot 001 \text{ ohm}$$

$$\text{Synchronous reactance/phase, } X_s = \sqrt{(Z_s)^2 - (R)^2}$$

$$= \sqrt{(0 \cdot 001)^2 - (0 \cdot 12)^2} = 0{\cdot}994 \text{ ohm}$$

$$\text{Terminal voltage (phase value), } V = \frac{V_L}{\sqrt{3}} = \frac{11000}{\sqrt{3}} = 6351 \text{ V}$$

(*i*) When *p.f.*, $\cos\phi = 0{\cdot}707$ lagging; $\sin\phi = \sin\cos^{-1} 0{\cdot}707 = 0{\cdot}707$

No-load terminal voltage (phase value),

$$E_0 = \sqrt{(V\cos\phi + IR)^2 + (V\sin\phi + IX_s)^2}$$

$$= \sqrt{(6351 \times 0.707 + 1049 \cdot 7 \times 0 \cdot 12)^2 + (6351 \times 0 \cdot 707 + 1049 \cdot 7 \times 0 \cdot 994)^2}$$

$$= \sqrt{(4616)^2 + (5533 \cdot 5)^2} = 7206 \text{ V}$$

$$\% \text{ Reg.} = \frac{E_0 - V}{V} \times 100 = \frac{7206 - 6351}{6351} \times 100 = \mathbf{13{\cdot}46\%} \; (Ans.)$$

(*ii*) When *p.f.* $\cos \phi = 0{\cdot}8$ leading; $\sin \phi = 0{\cdot}6$

No-load terminal voltage (phase value),

$$E_0 = \sqrt{(V \cos\phi + IR)^2 + (V \sin\phi + IX_s)^2}$$

$$= \sqrt{(6351 \times 0.8 + 1049 \cdot 7 \times 0 \cdot 12)^2 + (6351 \times 0 \cdot 6 + 1049 \cdot 7 \times 0 \cdot 994)^2}$$

$$= \sqrt{(4616)^2 + (2767)^2} = 5381 \cdot 8 \text{ V}$$

$$\% \text{ Reg.} = \frac{E_0 - V}{V} \times 100 = \frac{5381 \cdot 8 - 6351}{6351} \times 100 = \mathbf{-15{\cdot}26\%} \; (Ans.)$$

Example 6.22

Estimate the synchronous impedance for an 11 kV, three-phase, 50 Hz, 20 MVA alternator which develops rated emf on no-load with a field current of 20 A. A field current of 12 A produces a short circuit current equal to rated current.

Solution:

Rated power = 20 MVA = 20×10^6 VA

Line voltage, V_L = 11 kV = 11000 V (*three-phase connections*)

At field current of 20 A;

No-load *emf* $E_{(L)}$ = rated voltage = 11000 V (line value)

At field current of 12 A;

$$\text{Short circuit current} = \text{rated full load current} = \frac{20 \times 10^6}{\sqrt{3} \times 11000} = 1049.73 \text{ A}$$

At field current of 20 A;

$$\text{Short circuit current, } I_{sc} = \frac{20}{12} \times 1049 \cdot 73$$

(*SCC is a straight line curve*) = 1749·54 A

$$\text{Phase value of no-load } emf, \; E_0 = \frac{E_{0(L)}}{\sqrt{3}} = \frac{11000}{\sqrt{3}} = 6351 \text{ V}$$

$$\text{Synchronous impedance, } Z_s = \frac{E_0}{I_{sc}} = \frac{6351}{1749.54} = \mathbf{3{\cdot}63 \; \Omega} \; (Ans.)$$

Solution by using j-notation (polar method)

Example 6.23

A three-phase, star-connected, 10 kVA, 230 V alternator has an armature resistance of 0.5 Ω per phase and a synchronous reactance of 1.2 Ω per phase. Calculate the percent voltage regulation at full load at power factors of (a) 0.8 lagging, (b) 0.8 leading, (c) Determine the power factor such that the voltage regulation becomes zero on full load.

Solution:

Here, Rating = 10 kVA; $R = 0.5\ \Omega$; $X_S = 1.2\ \Omega$

$$\text{Full-load current, } I_L = \frac{kVA \times 1000}{\sqrt{3} \times V_L} = \frac{10 \times 10^3}{\sqrt{3} \times 230} = 25.1\,\text{A}$$

$$\text{Phase current, } I = I_L = 25.1\ \text{A}$$

$$\text{Rated voltage per phase, } V = \frac{V_L}{\sqrt{3}} = \frac{230}{\sqrt{3}} = 132.8\ \text{V}$$

Considering V as reference phasor, $\overline{V} = V\angle 0^\circ = 132.8\ \angle 0^\circ = 132.8 + j0$

$$\text{Synchronous impedance, } \overline{Z}_s = R + jX_S = 0.5 + j1.2 = 1.3\ \angle 67.38\text{"}\ \Omega$$

(*a*) *When power factor is 0.8 lagging*

$$\overline{I} = I\angle - \cos^{-1} 0.8 = 25.1\angle - 36.87\text{" A}$$

$$E_0 = \overline{V} + \overline{I}\ \overline{Z}_s$$

$$= (132.8 + j0) + (25.1\ \angle - 36.87\text{"})\ (1.3\angle 67.38\text{"})$$

$$= 132.8 + 32.63\angle 30.51\text{"} = 132.8 + 28.1 + j\ 1.6.56$$

$$= 160.9 + j16.56 = 161.75\ \angle 5.87\text{"}$$

$$\text{Voltage regulation} = \frac{E_0 - V}{V} \times 100 = \frac{161.75 - 132.8}{132.8} \times 100$$

$= \mathbf{21.8\%}$ (*Ans.*)

(*b*) *When power factor is 0.8 leading*

$$\overline{I} = I\ \angle + \cos^{-1} 0.8 = 25.1\ \angle 36.87\text{" A}$$

$$E_0 = \overline{V} + \overline{I}\,\overline{Z_s}$$

$$= 132.8 + (25.1\angle 36.87\text{"})\ (1.3\angle 67.38\text{"})$$

$$= 132.8 + 32.63\angle 104.25\text{"}$$

$$= 132.8 - 8 + j31.62 = 124.8 + j31.62$$

$$= 128.74\ \angle 14.2\text{" V}$$

$$\text{Voltage regulation} = \frac{E_0 - V}{V} \times 100 = \frac{128.74 - 132.8}{132.8} \times 100$$

$= \mathbf{-3.06\%}$ (*Ans.*)

(*c*) For zero regulation, let ϕ be the required power-factor angle.

$\therefore$
$$\overline{I} = I\angle\phi = 25.1\angle\phi\ \text{A}$$

$$E_0 = \overline{V} + \overline{I}\ \overline{Z_s}$$

$$= 132.8 + (25.1\ \angle\ \phi)\ (1.3\ \angle 67.38^\circ)$$

$$= 132.8 + 32.63\ \angle\ (\phi + 67.38^\circ)$$

$$= 132.8 + 32.63\cos(\phi + 67.38^\circ) + j\ 32.63\sin(\phi + 67.38^\circ)$$

$$E_0^2 = [132.8 + 32.63 \cos(\phi + 67.38)]^2 + [32.63 \sin(\phi + 67.38)]^2$$

$$\text{Voltage regulation} = \frac{E_0 - V}{V} \text{ pu}$$

For zero voltage regulation $E_0 = V = 132.8$ V

$$\therefore \quad (132.8)^2 = [132.8 + 32.63 \cos(\phi + 67.38°)]^2 + [32.63 \sin(\phi + 67.38)]^2$$

$$\text{or} \quad (132.8)^2 = (132.8)^2 + 2 \times 132.8 \times 32.63 \cos(\phi + 67.38°) + (32.63)^2 \cos^2(\phi + 67.38°) + (32.63)^2 \sin^2(\phi + 67.38°)$$

$$(132.8)^2 = (132.8)^2 + 2 \times 132.8 \times 32.63 \cos(\phi + 67.38°) + (32.63)^2$$

$$\text{or} \quad \cos(\phi + 67.38°) = \frac{-32.63}{2 \times 132.8} = -0.12285 = \cos 97"$$

$$\therefore \quad \phi = 97° - 67.38° = +29.62$$

$$\text{and} \quad \cos\phi = \mathbf{0.8692\ leading}\ (\textit{Ans.})$$

Example 6.24

A three-phase, star-connected, 10 kVA, 400V 50Hz alternator has armature resistance of 0.5 ohm/phase and synchronous reactance 10 ohm/phase. Determine its torque angle and voltage regulation when it supplies rated load at 0.8 pf lagging.

Solution:

Here, Rated power = 10 kVA; $R = 0.5\ \Omega$; $X_5 = 10\ \Omega$; $\cos\phi = 0.8$ lagging

$$\text{Rated load current, } I_L = \frac{10 \times 10^3}{\sqrt{3} \times 400} = 14.4 \text{ A}$$

$$\text{Rated phase current, } I = I_L = 14.4 \text{ A}$$

$$\overline{Z}_S = R + jX_S = 0.5 + j10 = 10.012\ \angle 87°\ \Omega$$

$$\text{Rated phase voltage, } V = \frac{V_L}{\sqrt{3}} = \frac{400}{\sqrt{3}} = 230.9 \text{ V}$$

Taking phase voltage V as reference phasor,

$$\therefore \quad \overline{V} = V\angle 0° = 230.9\angle 0° = (230.9 \pm j0) \text{ V}$$

At 0.8 lagging power factor

$$\text{Current, } \overline{I} = I\angle -\cos^{-1} 0.8 = 14.4\angle -36.87"\text{A}$$

$$\overline{E}_0 = \overline{V} + \overline{I}\,\overline{Z}_s$$

$$= 230.9 + j0 + (14.4\angle -36.87°)\ (10.012)\angle 87°)$$

$$= 230.9 + 144.2\ \angle 50.13° = 230.9 + 92.4 + j110.6$$

$$= 323.3 + j110.6 = 341.7\ \angle 18.9° \text{ V}$$

$$\therefore \quad E_0 = 341.7 \text{ V}$$

Torque angle between V and E_0 is $\delta = 18.9°$ (leading)

$$\text{Voltage regulation} = \frac{E_0 - V}{V} = \frac{341.7 - 230.9}{230.9} = \mathbf{0.4798\ pu}\ (\textit{Ans.})$$

Section Practice Problems

Numerical Problems

1. A single-phase, 2200 V, 50 Hz, 40 kVA alternator produces a short-circuit current of 200 A and an open circuit emf of 1160 V when a field current 40A passes through its field winding. Calculate the synchronous impedance and reactance if its armature has an effective resistance 0.5 ohm. (**Ans.** 5.8 ohm, 5.77 ohm)

2. A three-phase star connected alternator has an armature resistance of 0.1 ohm per phase. When excited to 860 V line voltage and on short-circuit the alternator gave 200 A. What should be the emf (in line terms) the alternator must be excited to, in order to maintain a terminal p.d. of 400 volt with 100 ampere armature current at 0.8 p.f. lagging. (**Ans.** 452 V)

3. A 600 volt, 60 kVA, single-phase alternator has an effective armature resistance of 0.3 ohm. An exciting current of 5 ampere produces an emf of 400 volt on open circuit and an armature current of 200 ampere on short circuit. Calculate:
   (*i*) the synchronous impedance and synchronous reactance.
   (*ii*) the full load regulation with 0.8 p.f. lagging. (**Ans.** 2 Ω, 2.974 Ω, 24.81%)

4. A three-phase, star-connected, 1000 kVA, 3000 V, 50 Hz alternator produces a short-circuit current of 200 A and an open circuit voltage of 1040 V with the same field current 40 A. if its armature has an effective resistance of 0.2 ohm per phase, calculate full-load percentage regulation of the machine at a pf of 0.8 lagging. (**Ans.** 24·34%)

5. A three-phase, star-connected, 1000 kVA, 3300 V, 50 Hz alternator produces a short-circuit current of 200 A and an open circuit voltage of 1040 V with the same field current 40 A. It its armature has an effective resistance of 0.2 ohm per phase, calculate full-load regulation of the alternator at a pf of 0.8 lagging and 0.8 leading. (**Ans.** 19.9%, –11.8%)

Short Answer Type Questions

Q.1. What are the different modes of operation of a synchronous machine?

Ans. *Operation of synchronous machines*: A synchronous machine may operate independently or in parallel with other machines. Accordingly, their operation is called;

(*i*) *First mode of operation*: The mode of operation in which a synchronous machine works independently is called its *first mode of operation*. In this case, machine works only as a generator.

(*ii*) *Second mode of operation*: The mode of operation in which a synchronous machine works in parallel with other machines or connected to the infinite bus bars is called *second mode of operation*. In this case, a machine may work as a generator or motor.

Q.2. What is armature leakage reactance?

Ans. Due to leakage fluxes like slot leakage flux, air gap leakage flux, end-connection leakage flux etc., the armature winding has some inductance (L) in each phase which offers leakage reactance ($X_L = 2\pi f L$) and is called *armature leakage reactance*.

Q.3. What do you mean by armature reaction reactance?

Ans. In fact, in synchronous generators the armature reaction produces the same effect as if a reactance is connected in series with the winding and hence is called *armature reaction reactance*.

Q.4. Write down the expression for no-load terminal voltage for resistive, inductive and capacitive load.

Ans. *Expression for No-load terminal voltage*

For resistive load, $E_0 = \sqrt{(V + IR)^2 + (IX_s)^2}$

For inductive load, $E_0 = \sqrt{(V\cos\phi + IR)^2 + (V\sin\phi + IX_s)^2}$

For capacitive load, $E_0 = \sqrt{(V\cos\phi + IR)^2 + (V\sin\phi - IX_s)^2}$

Q.5. In short tell how to determine synchronous impedance of an alternator.

Ans. *Determination of synchronous impedance*: It is determined by performing open circuit and short circuit test on the alternators.

By performing open circuit test, a curve between field current (I_f) and open circuit voltage or induced *emf* (E-phase value) is drawn.

By performing short circuit test, a curve between field current (I_f) and short circuit current (I_{sc} – phase value) is drawn.

Then, synchronous impedance, $Z_s = \frac{E}{I_{sc}}$ at the same I_f.

Q.6. Name different methods used to determine voltage regulation of an alternator.

Ans. Voltage regulation of an alternator can be determined by the following methods:

(*i*) Synchronous impedance method or emf method.

(*ii*) Ampere-turn method or mmf method

(*iii*) Zero power factor method or Potier method.

6.36 Assumptions Made in Synchronous Impedance Method

The following assumptions are made in synchronous impedance method:

1. *The synchronous impedance is considered to be constant.*
   In fact, it is constant only when *OCC* an *SCC* are straight line. But above the *Knee-point* of *OCC*, when the saturation starts, the value of synchronous impedance starts decreasing. Hence, the synchronous impedance obtained under test conditions is usually larger than actual value. This is the major source of error of determining voltage regulation of an alternator by synchronous impedance method.
2. *The flux under test conditions is considered to be the same as that under load conditions.*
   The same value of field current is not producing the same flux always. When the armature is short circuited, the armature current lags behind the generated voltage by almost 90°, hence armature reaction produces demagnetising effect. This reduces the degree of saturation further. The actual resultant flux is reduced which reduces the generated emf. These conditions are different from those when the machine is actually loaded. Hence, the synchronous impedance obtained under test condition is usually larger than actual value. This causes a source of error while determining voltage regulation of an alternator by this method.
3. *The effect of armature reaction flux is usually replaced by a voltage drop proportional to the armature current.*

This assumption also causes errors because the shift of armature flux varies with the power factor and the load current.

4. *The magnetic reluctance to the armature flux is considered to be constant regardless of the power factor.*
Although, this assumption is substantially true for non-salient pole type alternators because air-gap in these machines is uniform. But in case of salient pole type alternators, the position of armature flux relative to field poles varies with the power factor. This assumption also introduces considerable error.
Hence, it is found that the regulation determined by using synchronous impedance method is higher than the actual value.

6.37 Ampere-turn (or mmf) Method

The synchronous impedance method is based on the concept of replacing the effect of armature reaction by a fictitious reactance. Accordingly, some assumptions were made. But due to those assumptions, the voltage regulation obtained by that method was higher than the actual value.

In ampere-turn or mmf method, the effect of armature leakage reactance is to be replaced by an equivalent additional armature reaction mmf. This additional mmf is combined with the armature reaction mmf.

To determine the regulation of an alternator by mmf method, the following information is required.

(*i*) *The resistance of the stator winding per phase.*
(*ii*) *Open-circuit characteristics at synchronous speed.*
(*iii*) *Short-circuit characteristics.*

All these informations can be obtained by performing the same tests as preformed in emf method i.e., open circuit test, short circuit test and ammeter–voltmeter method for finding armature resistance.

The open circuit and short circuit characteristics are shown in Fig. 6.54, where field current $I_{f1;}$ is determined to give rated voltage *V* on no-load, neglecting armature resistance drop, the field current I_{f2} is determined to cause short-circuit current, equal to full load current, on short circuit.

On short circuit, the field excitation I_{f_2}, balances the impedance drop in addition to armature reaction on full load. But, as we know that *R* is usually very small and X_L is also small for low voltage on short circuit, so impedance drop can be neglected. Hence *pf* on short circuit is almost zero lagging and the field amp-turns are used entirely to overcome the armature reaction. Therefore, I_{f_2}, gives demagnetising amp-turns at full load.

Now, let us consider that the alternator is supplying full load current at a pf of cos ϕ. Draw a line *OA* representing I_{f_1} to give full load rated voltage, *V* [actually it is equal to $V + I R \cos \phi$] as shown in Fig. 6.55. Then draw *AB* at an angle ($90° \pm \phi$) representing I_{f2} to give full load current on short circuit; + ve sign for lagging pf and -ve sign for leading pf. Now find field current I_f measuring *OB*, which will give open-circuit *EMF* E_0, which can be determined from *OCC*.

Then percentage regulation can be determined from the relation,

$$\% \text{ Regulation} = \frac{E_0 - V}{V} \times 100 .$$

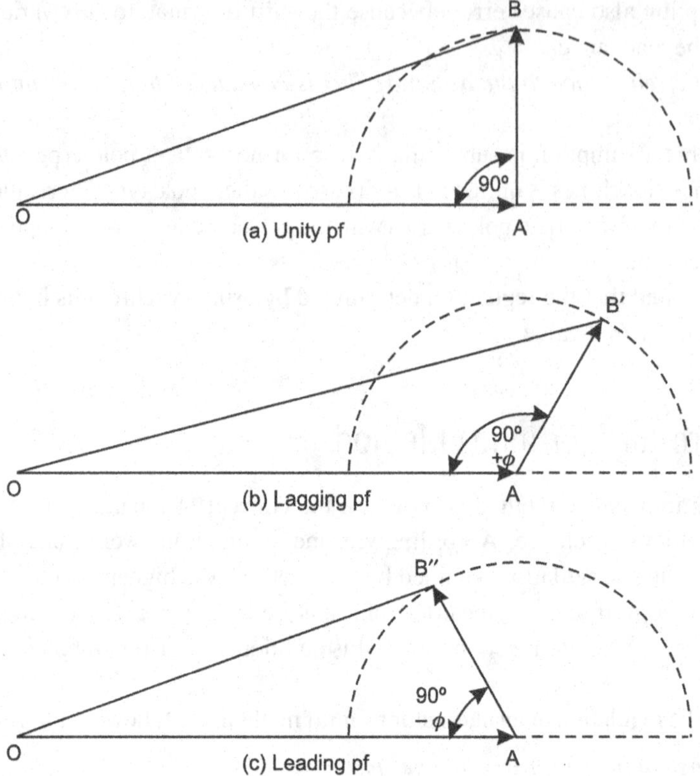

Fig. 6.55 Phasor diagram at different power factors

This method of determining synchronous impedance is known as *optimistic method* since it gives values lower than actual values. It is because the excitation to overcome armature reaction is determined on unsaturated part of the magnetising curve.

Example 6.25

The open-circuit test data of a 500 kVA, 4000 volt, 8 pole, 3-phase, 50 Hz alternator is:

ATs, per pole	*2000*	*3000*	*3560*	*5000*	*6200*	*7000*	*8000*
Terminal voltage	*1990*	*2900*	*3400*	*4000*	*4400*	*4590*	*4800*

The equivalent armature reaction expressed in ampere-turn per pole is 1.1 × ampere conductors per pole per phase. There are 240 conductors per phase in series. If the inductive voltage drop 8% on full load and the resistance drop is negligible. Then determine (i) Short circuit characteristic (ii) field excitation and regulation for full load at 0-8 p. f. lagging.

Solution:

Converting three-phase terminal line voltage into phase values, we have

$$\frac{1990}{\sqrt{3}}, \frac{2900}{\sqrt{3}}, \frac{3400}{\sqrt{3}}, \frac{4000}{\sqrt{3}}, \frac{4400}{\sqrt{3}}, \frac{4800}{\sqrt{3}} = 1150, 1675, 1963, 2310, 2540, 2650, 2770$$

$$\text{Phase voltage, } V = \frac{4000}{\sqrt{3}} = 2310 \text{ V}$$

Open-circuit characteristic is drawn by taking *ATs* per pole along the abscissa and voltage per phase along the ordinate, as shown in Fig. 6.56.

$$\text{Full load current } I = \frac{kVA \times 1000}{\sqrt{3} \times V_L} = \frac{500 \times 1000}{\sqrt{3} \times 4000} = 72 \text{ A.}$$

Armature reaction *ATs* per pole per phase for full load,

$$= 1.1 \times \text{ampere conductors per pole per phase}$$

$$= 1.1 \times I \times \frac{Z}{P} = 1.1 \times 72 \times \frac{240}{8} = 2376$$

$$\text{Inductive drop} = \text{Leakage reactance drop} = \frac{8}{100} \times \frac{4000}{\sqrt{3}} = 185 \text{ volt}$$

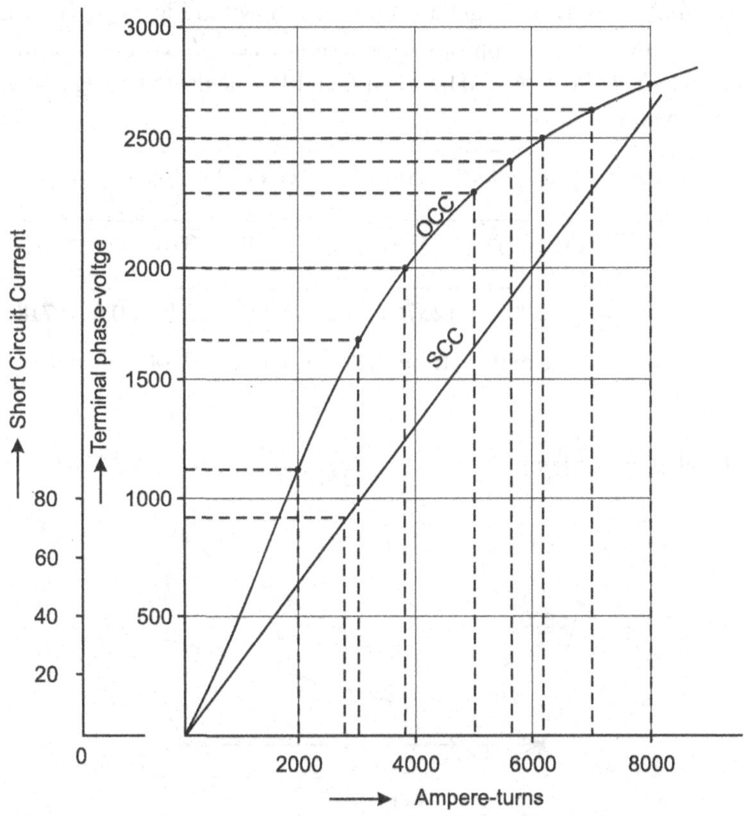

Fig. 6.56 *OCC* and *SCC* as per data

The field *ATs* or simply field current that is obtained from *OCC* is used to overcome the effects of armature reaction and leakage reactance. The *ATs* 2376 are the field *ATs* for balancing the armature reaction. The field *ATs* to balance or overcome the leakage reactance can be read off from the *OCC* graph corresponding to leakage reactance drop of 185 volt and it comes out to be 370 ampere turn.

$\therefore$ Short circuit field *ATs* = 2376 + 370 = 2746 ATs.

So the *SCC* is drawn with two points, one the origin (0,0) and second point is (2746, 72), These two points are joined, hence we get a straight line.

To determine total ampere-turns, proceed as follows:

Draw the phasor diagram as shown in Fig. 6.57. where terminal phase voltage V is taken as reference vector and current lags behind this voltage by an angle 36.87° ($\phi = \cos^{-1} 0.8 = 36.87°$). Here, resistance drop is zero and drop in leakage reactance IX_s is 185 V which leads the current vector by 90°.

$$E = \sqrt{(V\cos\phi + IR)^2 + (V\sin\phi + IX_s)^2}$$

$$= \sqrt{(2310 \times 0.8 + 0)^2 + (2310 \times 0.6 + 185)^2} \quad \text{(since } R = 0\text{)}$$

$$= 2525 \text{ V}$$

From the *OCC* curve the field *ATs* corresponding to 2425 volt are 5500

These field *ATs*, (*oa*) are drawn at right angle to E as shown in Fig. 6.57. The armature reaction *ATs* (2376) only are drawn parallel opposition to current I i.e., *ab* as shown in the Fig. 6.57. The angle between *oa* and *ab*, is $(90 + \phi)$. ($\angle\delta$ between E and V is neglected being small). The resultant vector *ob* is given as below:

$$ob = \sqrt{oa^2 + ab^2 - 2oa \times ab \times \cos(90 + 36.87")}$$

$$= \sqrt{(5500)^2 + (2376)^2 + 2 \times 5500 \times 2376 \times \cos(53.13")}$$

$$= \sqrt{(5500)^2 + (2376)^2 + 2 \times 5500 \times 2376 \times 0.6} = \mathbf{7180\ AT}\ (Ans.)$$

Corresponding to 7180 AT, the emf E_o from the *OCC* curve is 2700 volt and lags the *ob* by 90° as shown.

$$\%\text{age regulation} = \frac{E_0 - V}{V} \times 100 = \frac{2700 - 2310}{2310} \times 100 = \mathbf{16.88\%}\ (Ans.)$$

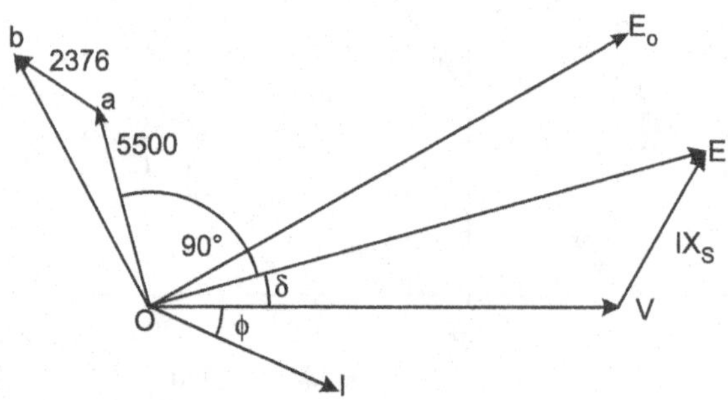

Fig. 6.57 Phasor diagram

<u>**Example 6.26**</u>

The following test results were obtained on a 345 kVA, three-phase, 6600 volt, star connected non salient pole type alternator.

Open circuit characteristics.

Terminal voltage in volt	*1400*	*2100*	*5600*	*6600*	*7240*	*8100*
Exciting current in ampere	*20*	*30*	*46.5*	*58*	*76.5*	*96*

Short circuit characteristics:

Stator current 35 A with an exciting current of 50 A. Leakage reactance on full load is 8% Neglect armature resistance.

Calculate the exciting current (for full load) at p.f. 0-8 lagging and at unity.

Solution:

Reducing all line voltages to phase voltage, for a star connected machine.

$$\frac{1400}{\sqrt{3}}, \frac{2100}{\sqrt{3}}, \frac{5600}{\sqrt{3}}, \frac{6600}{\sqrt{3}}, \frac{7240}{\sqrt{3}}, \frac{8100}{\sqrt{3}} = 808, 1212, 3233, 3810, 4180, 4676$$

Plot the *OCC* as shown in Fig. 6.58

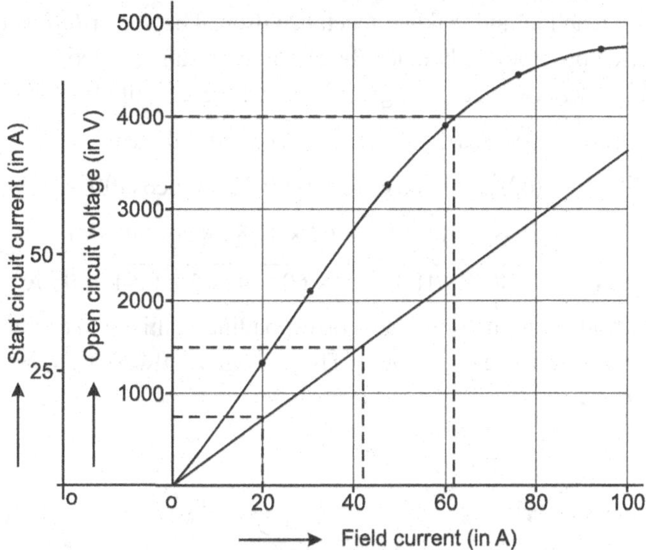

Fig. 6.58 OCC and SCC of the alternator

Plot the *SCC* from the given data. It is a straight line as shown in Fig. 6.58..

$$\text{Leakage reactance drop} = 8\% = \frac{8}{100} \times \frac{6600}{\sqrt{3}} = 304.8 \text{ volt}$$

Terminal voltage (phase value), $V = \frac{6600}{\sqrt{3}} = 3810$ V is taken as reference vector. Since armature resistance is neglected, there is only leakage reactance drop.

$$\therefore \qquad IX_S = 304.8 \text{ volt (leads the current vector by 90°)}$$

Full-load current, $$I = \frac{kVA \times 1000}{\sqrt{3}\,V_L} = \frac{345 \times 1000}{\sqrt{3} \times 6600} = 30 \text{ A}$$

For pf 0.8 lagging

For *pf* 0.8 lagging, current logs behind terminal voltage V by an angle $\phi = \cos^{-1} 0.8 = 36.87°$, as shown in vector diagram

$$E = \sqrt{(V\cos\phi + IR)^2 + (V\sin\phi + IX_S)^2}$$

$$= \sqrt{(3810 \times 0.8 + 0)^2 + (3890 \times 0.6 + 304.8)^2} = 4000 \text{ V}$$

Phase difference between E and V is $\delta = \sin^{-1}\dfrac{IX_S\cos\phi}{E} = \sin^{-1}\dfrac{304.8 \times 0.8}{4000} = 3.5''$

From *OCC* graph, corresponding to 4000 volt, the field current is 62 ampere.

This excitation (I_{f1}) is called normal excitation which is drawn at right angles to E. This is represented by vector *oa*.

From short-circuit characteristic, corresponding to full load current of 30 A, the exciting current is 42.85 A. This excitation (I_{f2}) is in phase opposition with the current I and represents the field excitation of armature reaction and leakage reactance drop. The vector (*ab*) is drawn in parallel to current vector in phase opposition to balance the effect of armature reaction.

Now, the resultant of these two vectors is given by *ob* whose magnitude (I_f) is calculated as follows.

$$ob^2 = oa^2 + ab^2 - 2oa \times ab \times \cos(90 + \phi + \delta)$$

$$= (62)^2 + (42.85)^2 - 2 \times 62 \times 42.85 \times \cos(90 + 36.87° + 3.5°)$$

$$= 3844 + 1800 + 2 \times 62 \times 42.85 \times \cos(49.63°)$$

$$ob = \sqrt{3844 + 1800 + 2 \times 62 \times 42.85 \times 0.64} = 95 \text{ A}$$

If, it is required to find the regulation, then corresponding to this field excitation the open circuit voltage E_o can be read off from the *OCC* curve. This voltage E_o always lags the excitation *ob* by 90° as shown in Fig. 6.59.

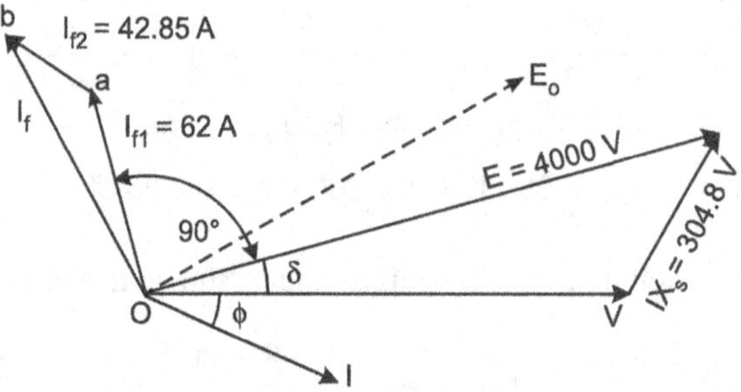

Fig. 6.59 Phasor diagram

For Unity p.f.

For unity *pf*, the current I is in phase with the terminal voltage V and the leakage reactance drop IX_S equal to 304.8 volt is drawn at right angle to the current I or at right angles to voltage vector V, as shown in Fig. 6.60.

Hence from the vector diagram *opt*, we have

$$E = \sqrt{V^2 + (IX_S)^2} = \sqrt{(3810)^2 + (304.8)^2} = 3822 \text{ V}$$

Phase difference between E and V is say δ'

$$\delta' = \sin^{-1}\frac{IX_S}{E} = \sin^{-1}\frac{304.8}{3822} = 4.58°$$

Corresponding to 3822 V, the field excitation from *OCC* curve is 59.8 A

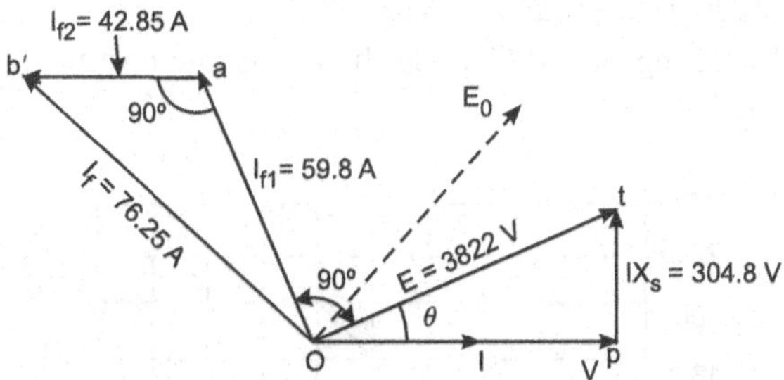

Fig. 6.60 Phasor-diagram

Full load current at unity p.f.

$$I = \frac{345 \times 1000}{\sqrt{3} \times 6600} = 30\text{A}$$

From short circuit characteristic 30 A corresponds to exciting current of 42.85 A

Now *oa'*. is drawn at right angles to *E* and is equal to 59.8 A. Now *a'b'* equal to 42.85 A is drawn parallel to current vector. The vector *ob'* is the vector sum of *oa'* and *a'b'*.

$$ob' = \sqrt{oa^2 + a'b'^2 - 2oa' \times a'b' \times \cos(90 + \theta)}$$

$$= \sqrt{(59.8)^2 + (42.85)^2 + 2 \times 59.8 \times 42.85 \times \cos 85.44°}$$

= **76.25 A** (*Ans.*)

Again if it is required to find the regulation then corresponding to this field current of 76.25 A the open-circuit voltage E_o can be determined which always lags the excitation *ob'* by 90°.

$$\therefore \quad \%\text{ age regulation} = \frac{E_o - V}{V} \times 100$$

Example 6.27

The data for open circuit characteristics of a 3.3 kV, 1500 kVA star-connected, three-phase alternator running at 760 rpm is given below:

Terminal Voltage (V):	*1500*	*2200*	*2700*	*3200*	*3550*	*3800*	*41100*	*4100*	*4150*
Field AT, per pole:	*2000*	*3000*	*4000*	*5000*	*6000*	*7000*	*8000*	*9000*	*10000*

The number of turns per phase is 55. The resistance measured between terminals is 0.5 ohm and leakage reactance per phase is 1.5 ohm. Assume that the armature reaction ampere turns per pole are equal to 1 4 times ampere turns per pole per phase. The ratio of effective resistance to measured resistance is 1.4.

Determine the percentage regulation when full load at normal voltage and at p. f. 0 8 lagging is switched off and the speed of the alternator increases to 770 rpm

Solution:

Changing the line voltages into phase voltages, we have,

$$\frac{1500}{\sqrt{3}}, \frac{2200}{\sqrt{3}}, \frac{2700}{\sqrt{3}}, \frac{3200}{\sqrt{3}}, \frac{3550}{\sqrt{3}}, \frac{3800}{\sqrt{3}}, \frac{4000}{\sqrt{3}}, \frac{4100}{\sqrt{3},}, \frac{4150}{\sqrt{3}}$$

866, 1270, 1560, 1848, 2050, 2190, 2310, 2370, 2400

Plot the *OCC* curve by taking field AT_S along the *x*-axis and terminal phase voltages along the *y*-axis as shown in Fig. 6.61.

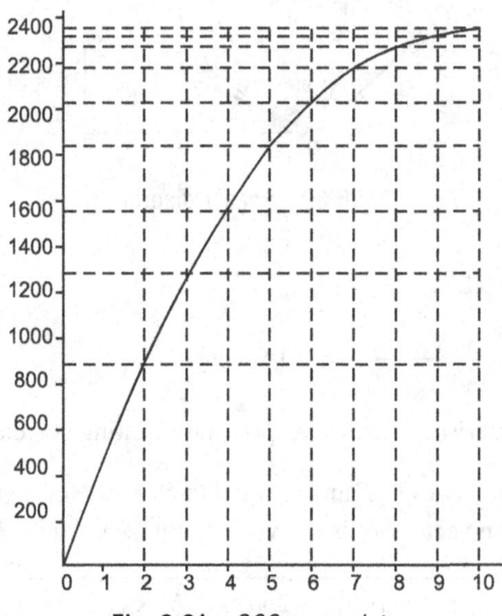

Fig. 6.61 *OCC* as per data

Full load current, $I_{fl} = \dfrac{kVA \times 1000}{\sqrt{3}\, V_L} = \dfrac{1500 \times 1000}{\sqrt{3} \times 3300} = 263 \text{ A}$

Resistance between terminal = 0.5 Ω.

In star connected windings, the resistance per phase = $\dfrac{0.5}{2} = 0.25\ \Omega$.

Effective resistance per phase, $R = 1.4 \times 0.25 = 0.35\ \Omega$

Leakage reactance per phase,

$$X_L = 1.5\ \Omega$$

The phasor diagram is shown in Fig. 6.62. for finding the voltage *E*. The terminal voltage *V* is taken as reference phasor.

Rated phase voltage, $V = \dfrac{3.3 \times 1000}{\sqrt{3}} = 1905 \text{ V}$

$$E = \sqrt{(V \cos \phi + IR)^2 + (V \sin \phi + IX_s)^2}$$

$$= \sqrt{(1905 \times 0.8 + 263 \times (0.35)^2 + (1905 \times 0.6 + 263 \times 1.5)^2}$$

$$= \sqrt{(1616)^2 + (1537.5)^2} = 2230 \text{ V}$$

Corresponding to 2230 volt, the field ampere turns from *OCC* curve is 7150 ATs.

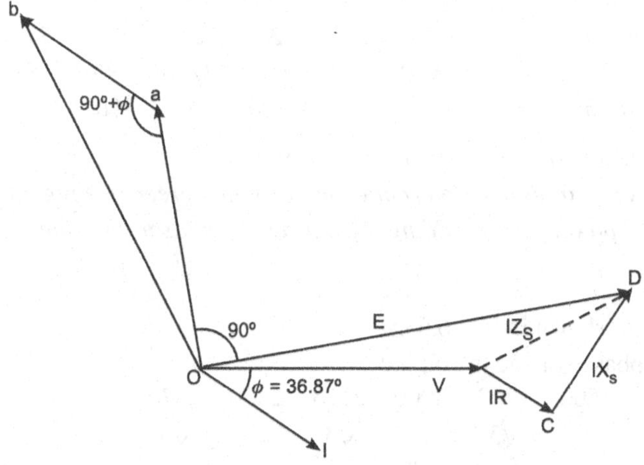

Fig. 6.62 Phasor diagram

Armature reaction *ATs* per pole per phase are = 1.4 × AT_S per pole per phase

$$= 1.4 \times \frac{I \times \text{Turns per phase}}{\text{No. of poles}} = 1.4 \times \frac{263 \times 55}{8} = 2525 \text{ ATs}$$

The field *ATs* as obtained from the *OCC* curve corresponding to normal open circuit voltage are called no-load *ATs*. These ATs (*oa*) are drawn at right angle to the voltage vector *OE*. The armature reaction *ATs* per pole per phase (2525) is drawn parallel opposition to load current *I*. i.e., *ab* is drawn parallel opposition to *I*. The resultant of *oa*, and *ab* is given by *ob*. The angle between *oa* and *ab* is (90 + ϕ)

$$\therefore \quad ob = \sqrt{oa^2 + ab^2 - 2oa \times ab \times \cos(90 + 36.87")}$$

$$= \sqrt{(7150)^2 + (2525)^2 + 2 \times 7150 \times 2525 \times \cos(53.13")}$$

$$= \sqrt{(7150)^2 + (2525)^2 + 2 \times 7150 \times 2525 \times 0.6} = 8900 \text{ ATs}$$

Corresponding to 8900 ATs, the open circuit voltage E_o is 2380 from the *OCC* curve.

Hence when the load is switched off the open circuit voltage is 2380 volt.

Now the speed has increased from 760 to 770 rpm but we need to have open circuit voltage corresponding to 760 rpm, therefore E_0 corresponding to 760 rpm,

$$E_o = \frac{2380 \times 760}{770} = 2350 \text{ V}$$

$$\% \text{ age Regulation} = \frac{E_o - V}{V} \times 100 = \frac{2350 - 1905}{1905} \times 100 = \mathbf{23.4\%} \ (Ans.)$$

Solution by using polar method

Example 6.28

When an open circuit and short circuit tests were performed on a three-phase, star-connected, 1000 kVA, 2000 V, 50 Hz alternator, the following results were obtained:

Field current (in ampere):	*10*	*15*	*20*	*25*	*30*	*35*	*40*	*45*	*50*
O.C. terminal voltage (in volt):	*800*	*1143*	*1500*	*1760*	*2000*	*2173*	*2350*	*2476*	*2600*
S.C. armature current (in ampere):	—	—	*200*	*250*	*300*	—	—	—	—

The armature effective resistance per phase is 0.2 Ω.

Draw the characteristic curves and determine the full-load percentage regulation at (a) 0.8 power factor lagging, (b) 0.8 power factor leading. Also draw the phasor diagrams.

Solution:

The *OCC* and *SCC* are shown in Fig. 6.63.

The open circuit phase voltage in volt are

$$\frac{800}{\sqrt{3}}, \frac{1143}{\sqrt{3}}, \frac{1500}{\sqrt{3}}, \frac{1760}{\sqrt{3}}, \frac{2000}{\sqrt{3}}, \frac{2173}{\sqrt{3}}, \frac{2350}{\sqrt{3}}, \frac{2476}{\sqrt{3}}, \frac{2600}{\sqrt{3}};$$

or $= 462, 660, 866, 1016, 1155, 1255, 1357, 1430, 1501$

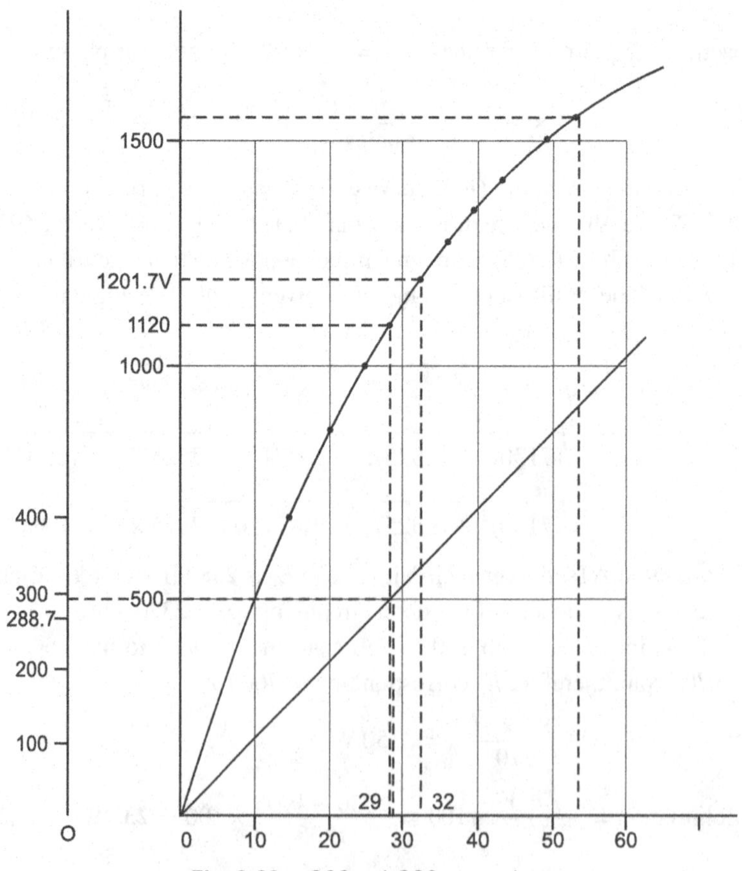

Fig. 6.63 *OCC* and *SCC* as per data

Full-load phase voltage $V = \dfrac{2000}{\sqrt{3}} = 1155$ V

Full load line current, $I_L = \dfrac{kVA \times 1000}{\sqrt{3} \times V_L}$

$$= \frac{1000 \times 1000}{\sqrt{3} \times 2000} = 288.7 \text{ A}$$

Full-load phase current, $I = I_L = 288.7$ A

(a) *At power factor 0.8 lagging* (neglecting leakage reactance since not given)

$$E = \bar{V} + \bar{I}\bar{R} = 1155 + (288.7\angle - \cos^{-1} 0.8) \times 0.2$$

$$= 1155 + (57.74 \times 0.8 - j57.74 \times 0.6)$$

$$= 1155 + 46.2 - j34.64$$

$$= 1201.2 - j34.64 = 1201.7 \angle{-1.65°} \text{ V}$$

Here $\delta = -1.65°$ V

From the *OCC*, the field current required to produce the voltage of 1201.7 V is 32 A. Therefore $oa = I_{f_1} = 32$ A. This current leads the voltage vector *OE* by 90° or leads the terminal voltage vector *OV* by $(90 - \delta = 90 - 1.65° = 88.35°)$ 88.35°.

$$\bar{I}_{f_1} = I_{f_1} \angle 90° - 1.65° = 32\angle 88.35° = (0.92 + i31.98) \text{ A}$$

From the *SCC*, the field current required to produce full-load current of 288.7A is 29 A. Therefore $ob = I_{f_2} = 29$ A. For $\cos\phi = 0.8$, $\phi = 36.87°$

From the phasor diagram shown in Fig. 6.64.

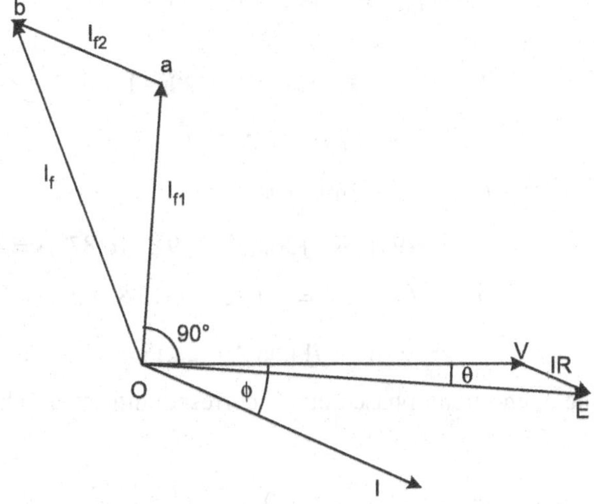

Fig. 6.64 Phasor diagram

$$\bar{I}_{f_2} = I_{f_2} \angle 180° - \phi = 29 \angle 180° - 36.87°$$

$$= 28° \angle 143.13° \text{ A} = -23.2 + j17.4$$

$$\bar{I}_f = \bar{I}_{f_2} + \bar{I}_{f_1} = -23.2 + j17.4 + 0.92 + j31.98$$

$$= -22.28 + j49.38 = 54.18 \angle 114.3° \text{ A}$$

From the *OCC*, the open circuit phase voltage corresponding to the field current of 54.18 A is 1555 V.

$$\therefore \text{ Percentage voltage regulation} = \frac{E_o - V}{V} \times 100 = \frac{1555 - 1155}{1155} \times 100 = \mathbf{34.63\%} \text{ (Ans.)}$$

(*b*) *At power factor 0.8 leading*

$$\bar{E} = \bar{V} + \bar{I}R$$

$$= 1155 + (288.7 \angle +\cos^{-1} 0.8) \times 0.2$$

$$= 1155 + 46.2 + j34.64$$

$$= 1201.2 + j34.64 = 1201.7 \angle +1.65° \text{ V.}$$

From the phasor diagram shown in Fig. 6.65.

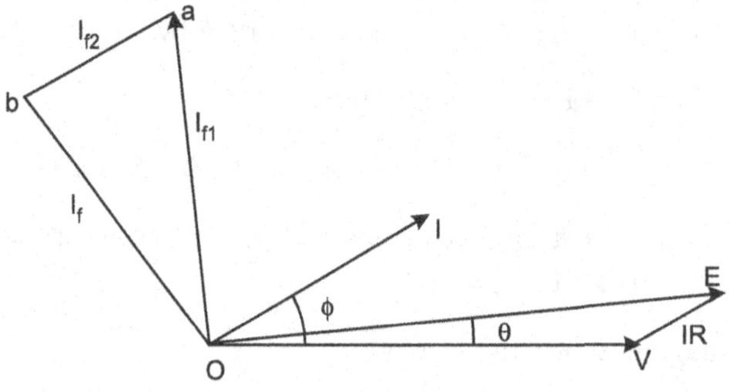

Fig. 6.65 Phasor diagram

$$I_{f_1} = I_{f_1} \angle 90° + \delta = 32 \angle 90° + 1.65 = 32 \angle 91.65° \text{ A}$$

$$= -0.92 + j31.98 \text{ A}$$

$$I_{f_2} = I_{f_2} \angle 180° + \phi$$

$$= 29 \angle 180° + 36.87° = 29 \angle 216.87° \text{ A} = -23.2 - j17.4 \text{ A}$$

$$\bar{I}_f = \bar{I}_{f_2} + \bar{I}_{f_1} = -0.92 + j31.98 - 23.2 - j17.4$$

$$= -24.12 + j14.58 \text{ A} = 28.18 \angle 31.15° \text{ A}$$

From the O.C.C., the open-circuit phase voltage corresponding to a field current of 28.18 A is 1120 V.

$$\text{Percentage voltage regulation} = \frac{E_o - V}{V} \times 100 = \frac{1120 - 1155}{1155} \times 100 = \mathbf{-3.03\%} \text{ (Ans.)}$$

6.38 Zero Power Factor or Potier Method

The regulation obtained by synchronous impedance (or emf) method and ampere-turn (or mmf) method is based on the total synchronous reactance i.e., (the sum of reactance due to armature leakage flux and due to armature reaction effect). This introduces error due to vectorial addition of magnetic fluxes. Whereas, the zero power factor method is based on the separation of reactances due to leakage flux and that due to armature reaction flux, therefore, it gives more accurate results.

For determining voltage regulation by this method, the following data is required:

(*i*) effective resistance of armature winding
(*ii*) open-circuit characteristic
(*iii*) field current to circulate full-load current in the stator and
(*iv*) zero-power factor full-load voltage characteristic—a curve plotted between terminal voltage and field current while the machine is being running on synchronous speed and delivering full-load current at zero power factor.

The first three requirements can be fulfilled by performing open circuit test, short circuit test and resistance measurement test, as discussed earlier. The curve of zero power factor characteristic can be obtained by taking various steps as explained below:

1. The machine is rotated at rated synchronous speed by a prime-mover.
2. A pure inductive load (variable load reactors or an under-excited synchronous motor) is connected across the armature terminals and the excitation or field current of the alternator is raised so as to cause flow of full-load armature current.
   Usually, the alternator is loaded by an under-excited synchronous motor while plotting zero pf full-load curve.
3. The value of the reactance is then increased step by step in such a way that the excitation current is adjusted to a value that causes full-load rated armature current to flow. In the process, the armature terminal-voltages are varied from 125 % to 25 % of the rated voltage in steps, maintaining the speed and rated armature current constant throughout the test.
4. Note down the armature terminal voltages and excitation currents at each step.
5. Draw a curve between terminal voltage and excitation current, as shown in Fig. 6.66. It gives the *zero power factor (lagging) characteristic.*

Now, the zero power factor (lagging) characteristic can be used for obtaining the Potier reactance, it is sufficient to determine the point representing rated armature current and rated voltage. This is indicated by point *B* in Fig. 6.66.

From *OCC* and zero power factor curve, it reveals that there is a definite relationship between the zero power factor (lagging) characteristic and an open-circuit characteristic of an alternator. The zero power factor characteristic curve is of exactly the same shape, as that of *OCC* but it is shifted vertically downward by leakage reactance drop IX_L and horizontally, by the armature reaction mmf.

Zero power factor full-load voltage-excitation characteristic can be drawn by knowing two points *A* and *B*. Point *A* is obtained from a short-circuit test with full-load armature current. Hence *OA* represents excitation (field current) required to overcome demagnetising effect of armature reaction and to balance leakage reactance drop at full load. Point *B* is obtained when full load current flows through the armature but wattmeter reads zero.

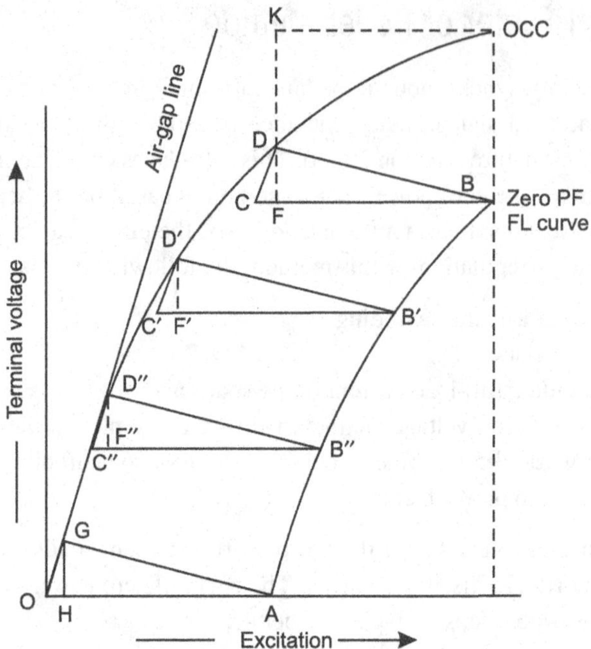

Fig. 6.66 *OCC* of the alternator with zero pf full-load curve

From point *B*, line *BC* is drawn equal and parallel to *AO*. Then a line is drawn through C parallel to initial straight part of *OCC* (parallel to extended *OG*), intersecting the *OCC* at *D*. *BD* is joined and a perpendicular *DF* is dropped on *BC*. The triangle *BFD* is imposed at various points of *OCC* to obtain corresponding points on the zero factor curve.

In triangle *BDF* the length *BF* represents armature reaction excitation and the length *DF* represents leakage reactance drop $(I X_L)$*. This is known as Potier reactance voltage drop and the triangle is known as *Potier triangle.* The Potier reactance is given, as

$$X_P = \frac{DF \text{ (voltage drop per phase)}}{\text{Zero power current per phase}}$$

It is observed that in case of cylindrical rotor machines, Potier reactance is nearly equal to armature leakage reactance, but in case of salient pole machines, the magnetising circuit is more saturated and the armature leakage reactance is smaller than the Potier reactance.

Potier Regulation Diagram

To determine voltage regulation, Potier regulation diagram is drawn as follows:

(*i*) *OV* is drawn horizontally to represent terminal voltage, *V* on full load and *OI* is drawn to represent full load current at a given power factor. (say lagging).

(*ii*) Draw *VE* perpendicular to phasor *OI* and equal to reactance drop (IX_L), neglecting resistance drop.

(*iii*) Join *OE*, where *OE* represents generated emf *E*.

(*iv*) From *OCC* find field excitation I_{f_1} corresponding to generated emf *E*.

(*v*) Draw $oa = I_{f1}$ perpendicular to phasor OE to represent excitation required to induce emf OE on open circuit.

(*vi*) Draw $ab = I_{f2}$ parallel to load current phasor OI to represent excitation equivalent to full-load armature reaction.

(*vii*) Join $ob = I_f$ which gives total excitation required. If the load is thrown off, then terminal voltage will be equal to generated emf corresponding to field excitation $ob = I_f$.
Hence, emf E_0 may be determined from OCC corresponding to field excitation $ob = I_f$. Where, the phasor E_0 will lag behind phasor ob by 90°. Here, EE_0 represents voltage drop due to armature reaction. Now regulation can be obtained from the relation.

$$\% \text{ Regulation} = \frac{E_0 - V}{V} \times 100$$

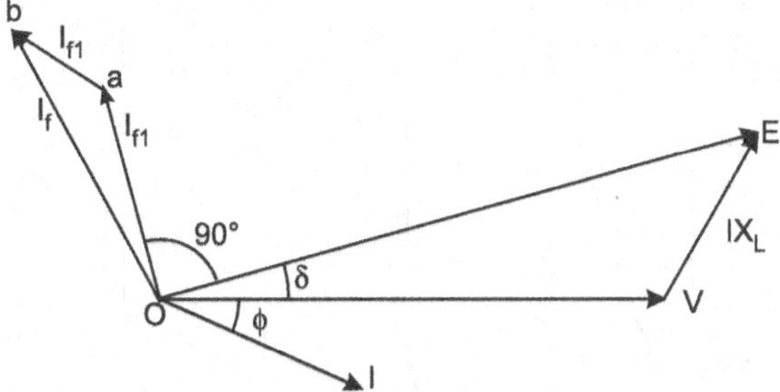

Fig. 6.67 Phasor diagram

Example 6.29

The following test results were obtained when these were performed on a 15 MVA, 11 kV, three-phase, 50 Hz, star-connected alternator:

Field AT per pole in thousand	5	10	15	18	25	30	35	40	45	50
Open-circuit line emf in kV	2.9	5.0	7.0	8.1	10.0	11.1	11.9	12.7	13.3	13.65
Full-load current, zero power factor test, line pd in kV	—	—	—	0	—	—	—	—	10.0	—

Find the armature reaction ampere-turns, the leakage reactance and the regulation for full load at 0.8 pf lagging. Neglect resistance.

Solution:

From the given data, draw OCC between phase voltage and field current. Full-load zero power factor curve is drawn, taking point A (18, 0) and point B (45, $\frac{10.2}{\sqrt{3}}$) being known. From the triangle BDF drawn in Fig. 6.68.

$$\text{Armature reaction ampere turns} = BF = \mathbf{15000\ AT/pole.}\ (\textit{Ans.})$$

$$\text{Full-load reactance drop} = DF = 11.15 - 10 = 1.15 \text{ kV} = 1150 \text{ volt}$$

$$\text{Leakage reactance drop per phase, } IX_L = \frac{1150}{\sqrt{3}} = 664 \text{ V}$$

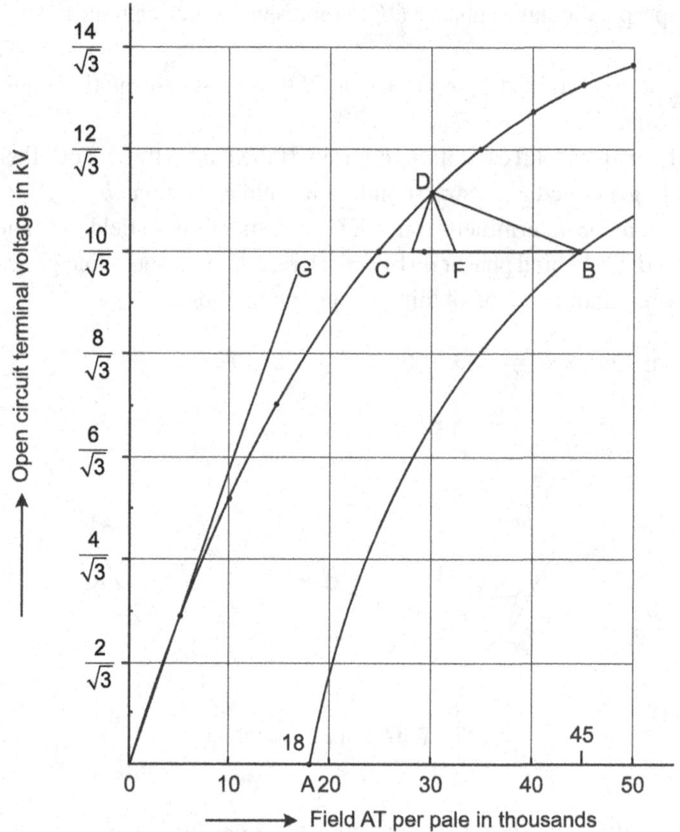

Fig. 6.68 *OCC* and Z pf triangle as per data

$$\text{Full-load current} = \frac{15 \times 10^6}{\sqrt{3} \times 11000} = 787 \text{ A}$$

$$\text{Leakage reactance per phase, } X_L = \frac{IX_L}{I} = \frac{664}{787} = 0.844\ \Omega$$

The phasor diagram is shown in Fig. 6.69, where

$$OV = \text{Terminal phase voltage} = \frac{11000}{\sqrt{3}} = 6351 \text{ V}$$

$$VE = 1150 \text{ volt representing reactance drop}$$

$$OE = E = \sqrt{(V\cos\phi + IR)^2 + (V\sin\phi + IXs)^2}$$

Since $\cos\phi = 0.8$ lag: $\sin\phi = \sin\cos^{-1}0.8 = 0.6$ and $R = 0$

$$E = \sqrt{(6351 \times 0.8)^2 + (6351 \times 0.6 + 664)^2} = 6770 \text{ V}$$

Excitation corresponding to $6770\text{ V}\left(= \frac{11727}{\sqrt{3}}\text{V}\right)$ is 33000 AT

Draw oa equal to 33000 AT, perpendicular to OE.

Draw ab = 15000 AT parallel to current vector OI

Total ampere turns, $ob = \sqrt{(oa^2 + ab^2 - 2oa \times ab \times \cos(90° + 36.87°)}$

$= \sqrt{(33000)^2 + (15000)^2 - 2 \times 33000 \times 15000 \times \cos 126.87°}$

= 43680 AT

Induced emf corresponding to 43680 AT = $\dfrac{13200}{\sqrt{3}}$ V = 7621 V

$$\% R_{eg} = \frac{E_0 - V}{V} \times 100 = \frac{7621 - 6351}{6351} \times 100 = \mathbf{20\%}\ (Ans.)$$

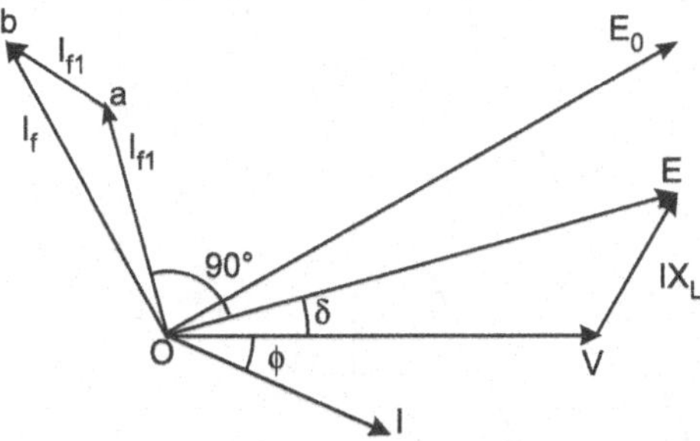

Fig. 6.69 Phasor diagram

Example 6.30

Determine the voltage regulation by zero power factor method of a 500 kVA, 6600V, three-phase, star-connected alternator having a resistance of 0.075 ohm per phase, when delivering a current of 500 A at power factor (i) 0.8 lagging (ii) 0.707 leading and (iii) unity. The alternator has the following open circuit and full-load zero power factor curves:

Field current in A:	*24*	*32*	*50*	*75*	*100*	*125*	*150*
Open circuit terminal voltage in V:	*1400*	—	*4500*	*6400*	*7500*	*8100*	*8400*
Saturated curve, zero pf in V:	*0*	*0*	*1900*	*4200*	*5750*	*6750*	*7100*

Solution:

The *OCC* and *ZPFC* are plotted as shown in Fig. 6.70. At rated terminal voltage of $\dfrac{6600}{\sqrt{3}}$ V, draw a horizontal line at *B*. Take *BC* = *OA* = 32A.

This field current *OA* is the field current required to circulate full-load current on short circuit. Draw a line *CD* parallel to *OG* (the initial slope of *OCC*) to meet *OCC* at *D*. From point *D* draw a perpendicular *DF* on the line *BC*. Here *BCD* is the Potier's triangle.

From Potier's triangle,

Field current required to overcome armature reaction on load = *FB* = 26A and $$FD = \frac{800}{\sqrt{3}} = 462 \text{ V}$$

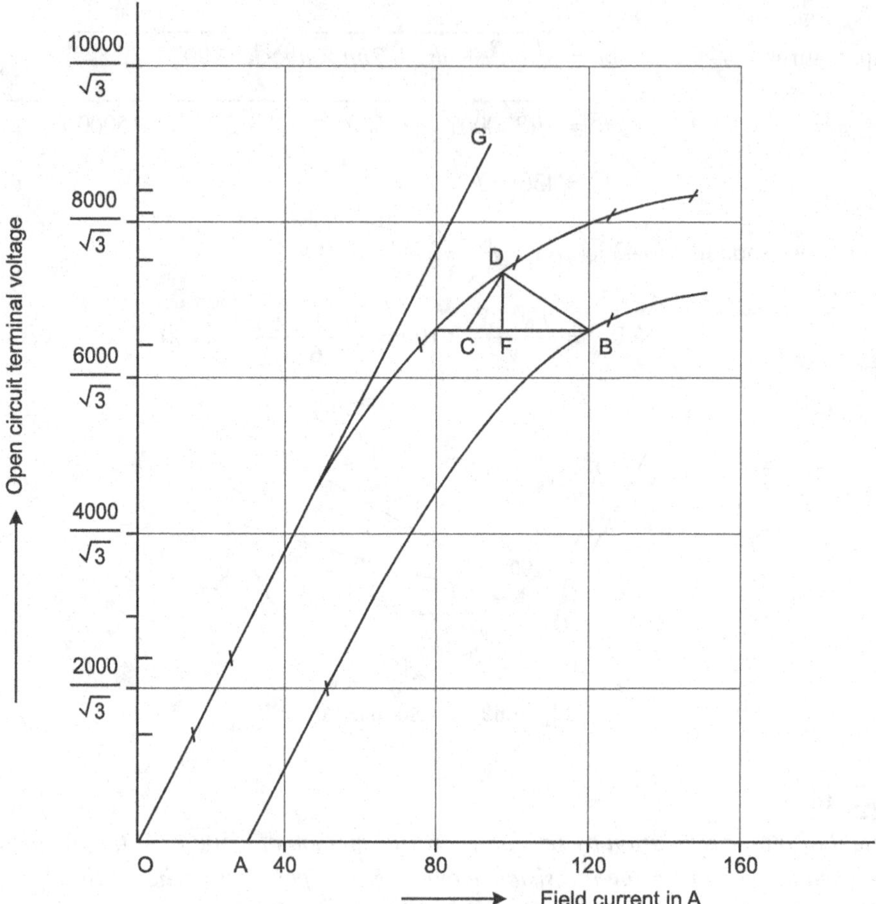

Fig. 6.70 *OCC* and Z pf as per data

Where *FD* represents voltage drop in leakage reactance at full-load current of 500 A (given)

Now, $$IX_L = 462$$

$$\therefore \quad X_L = \frac{462}{500} = 0.924 \ \Omega$$

Draw the phasor diagram, as shown in Fig. 6.71, where,

$$OV = \text{terminal phase voltage, } V = \frac{6600}{\sqrt{3}} = 3810 \text{ V}$$

(*i*) When *pf.* cos $\phi = 0.8$ lagging; $\sin \phi = \sin \cos^{-1} 0.8 = 0.6$

$$OE = E = \sqrt{(V \cos \phi + IR)^2 + (V \sin \phi + 1X_L)^2}$$

$$= \sqrt{(3810 \times 0.8 + 500 \times 0.075)^2 + (3510 \times 0.6 + 500 \times 0.924)^2}$$

$$= \sqrt{(3085)^2 + (2748)^2} = 4131\,\text{V}$$

From *OCC*, the field current corresponding to 4131 V (i.e., $\frac{7156}{\sqrt{3}}$ V)

$$oa = I_{f_1} = 92\,\text{A} \text{ (it leads vector } OE \text{ by } 90°)$$

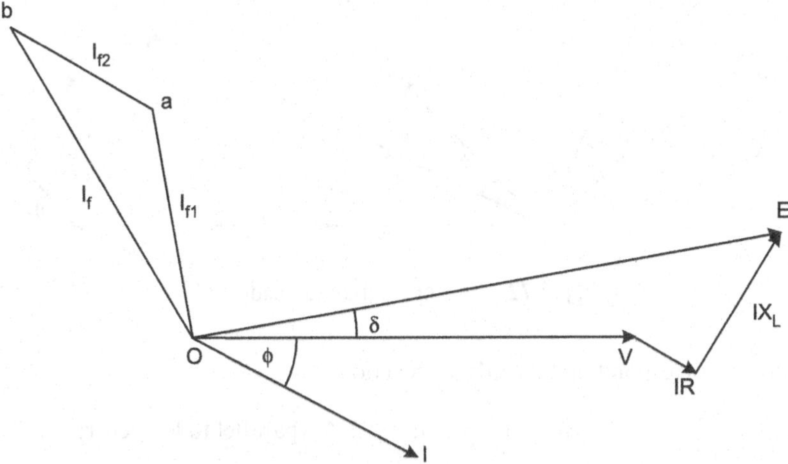

Fig. 6.71 Phasor diagram for lagging pf

Field current corresponding to armature reaction

$$ab = I_{f_2} = BF = 23\text{ A} \qquad \text{(it is parallel to load current } OI)$$

Total field current

$$I_f = Ob = \sqrt{oa^2 + ab^2 - 2oa \times ab \times \cos(90° + 36.87°)}$$

$$= \sqrt{(92)^2 + (23)^2 - 2 \times 92 \times 23 \cos 126.87°}$$

$$= \sqrt{8464 + 529 + 2539} = 107.4\text{ A}$$

Corresponding to this field current of 107.4A, the terminal voltage from *OCC* is 7700 V (line value)

$$E_o = \frac{7200}{\sqrt{3}} = 4446\,\text{V (phase value)}$$

$$\%\,\text{Reg} = \frac{E_0 - V}{V} \times 100 = \frac{4446 - 3810}{3810} \times 100 = \mathbf{16.69\%}\ (Ans.)$$

(*ii*) When *p.f.*, $\cos\phi_1 = 0.707$ leading; $\sin\phi_1 = \sin\cos^{-1} 0.707 = 0.707$; $\phi_1 = 45°$ leading.

The phasor diagram is shown in Fig. 6.72.

$$OE' = E' = \sqrt{(V\cos\phi_1 + IR)^2 + (V\sin\phi_2 - IX_L)^2}$$

$$= \sqrt{(3810 \times 0.707 + 500 \times 0.075)^2 + (3810 \times 0.707 - 500 \times 0.924)^2}$$

$$= \sqrt{(2731)^2 + (2232)^2} = 3527\ \text{V}$$

From *OCC*, the field current corresponding to 3527 V $\left(i.e.\ \frac{6109}{\sqrt{3}}\ \text{V}\right)$

$$oa' = I'_{f_1} = 72\ \text{A}$$

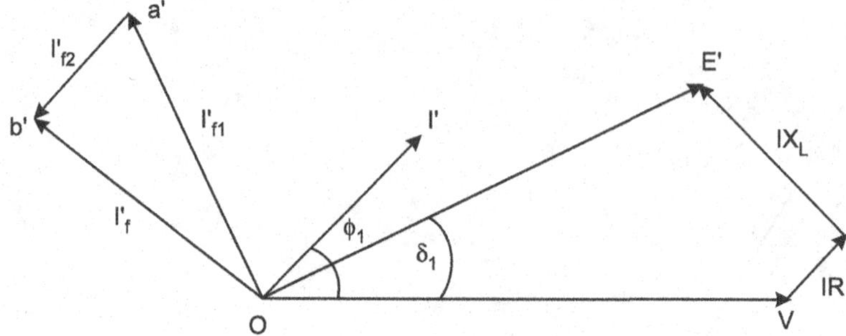

Fig. 6.72 Phasor diagram for leading pf

Field current corresponding to armature reaction

$$a'b' = I'_{f_2} = BE = 23\ \text{A}\ \text{(parallel to load current } OI'\text{)}$$

Total field current,

$$I'_f = ob' = \sqrt{(oa')^2 + (a'b')^2 - 2\,oa' \times a'b' \times \cos\,(90° - 45°)}$$

$$= \sqrt{(92)^2 + (23)^2 - 2 \times 92 \times 23 \times \cos 45°}$$

$$= \sqrt{(8464 + 529 - 1496} = 86.6\ \text{A}$$

Corresponding to this field current of 86.6 A, the terminal voltage from *OCC* is $\frac{5000}{\sqrt{3}} = 2887$ V

$$\%R_{eg} = \frac{B'_0 - V}{V} \times 100 = \frac{2887 - 3810}{3810} \times 100$$

$$= \mathbf{24.23\%}\ (Ans.)$$

(*iii*) When *pf*, $\cos \phi_2 = 1$; $\sin \phi_2 = \sin \cos^{-1} 1 = 0$; $\phi_2 = 0°$

Draw the phasor diagram as shown in Fig. 6.73.

$$OE'' = E'' = \sqrt{(V + IR)^2 + (IX_L)^2}$$

$$= \sqrt{(3810 + 500 \times 0.075)^2 + (500 \times 0.924)^2}$$

$$= \sqrt{(3847)^2 + (462)^2} = 3875\ \text{V}$$

From *OCC*, the field current corresponding to 3875 V $\left(i.e.\ \frac{6712}{\sqrt{3}}\ \text{V}\right)$

$$oa'' = I''_{f_1} = 82\text{A}$$

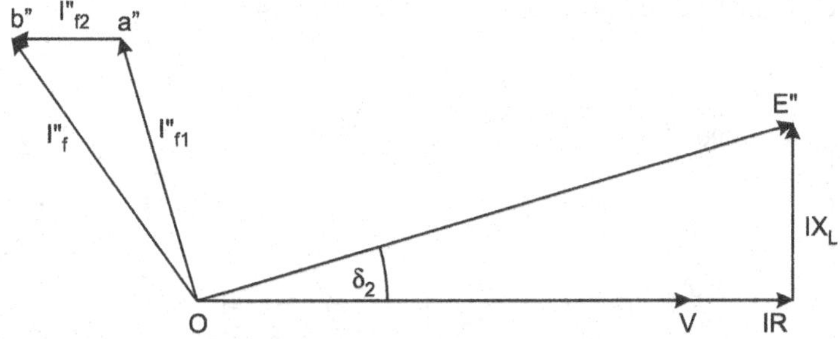

Fig. 6.73 Phasor diagram for unity pf

Field current corresponding to armature reaction

$$a''b'' = I''_{f_2} = BF = 23\text{ A}$$

(parallel to load current OI'' i.e., in phase with OV)

Total field current,

$$I''_f = ob'' = \sqrt{(oa'')^2 + (a''b'')^2 - 2oa'' \times a''b'' \times \cos(90 \pm 0)}$$

$$= \sqrt{(80)^2 + (23)^2 - 0} = 85.2\text{ A}$$

Corresponding to this current of 85.2A, the terminal voltage from OCC is $\frac{6750}{\sqrt{3}} = 3897$ V

$$\%\text{Reg} = \frac{E''_o - V}{V} \times 100 = \frac{3897 - 3810}{3810} \times 100 = \mathbf{2.26\%}\ (Ans.)$$

Example 6.31

An open circuit, short circuit and load zero power factor tests are performed on a 6-pole, 440 V, 50 Hz, three-phase star-connected alternator. The effective ohmic resistance between any two terminals of the armature is 0.4 Ω. and the test results are tabulated below:

Field current (A)	1	2	3	4	5	6	7	8	10	12	14	16	18
O.C. terminal voltage (V)	70	156	220	288	350	396	440	474	530	568	592	610	—
S.C. line current (A)	—	11	—	22	—	34	40	46	57	69	80	—	—
Zero p.f. terminal voltage (V)	—	—	—	—	—	—	0	80	206	314	398	460	504

Determine the regulation at full-load current of 40 A at 0.8 power factor lagging using
(a) synchronous impedance method,
(b) mmf method,
(c) Potier-triangle method

Solution:

$$\text{Armature resistance per phase} = \frac{1}{2} \times 0.4 = 0.2\ \Omega$$

$$\text{Terminal voltage per phase, } V = \frac{440}{\sqrt{3}} = 254\text{ V}$$

The *O.C.C.*, *S.C.C.* and *ZPFC* are plotted as shown in Fig. 6.74.

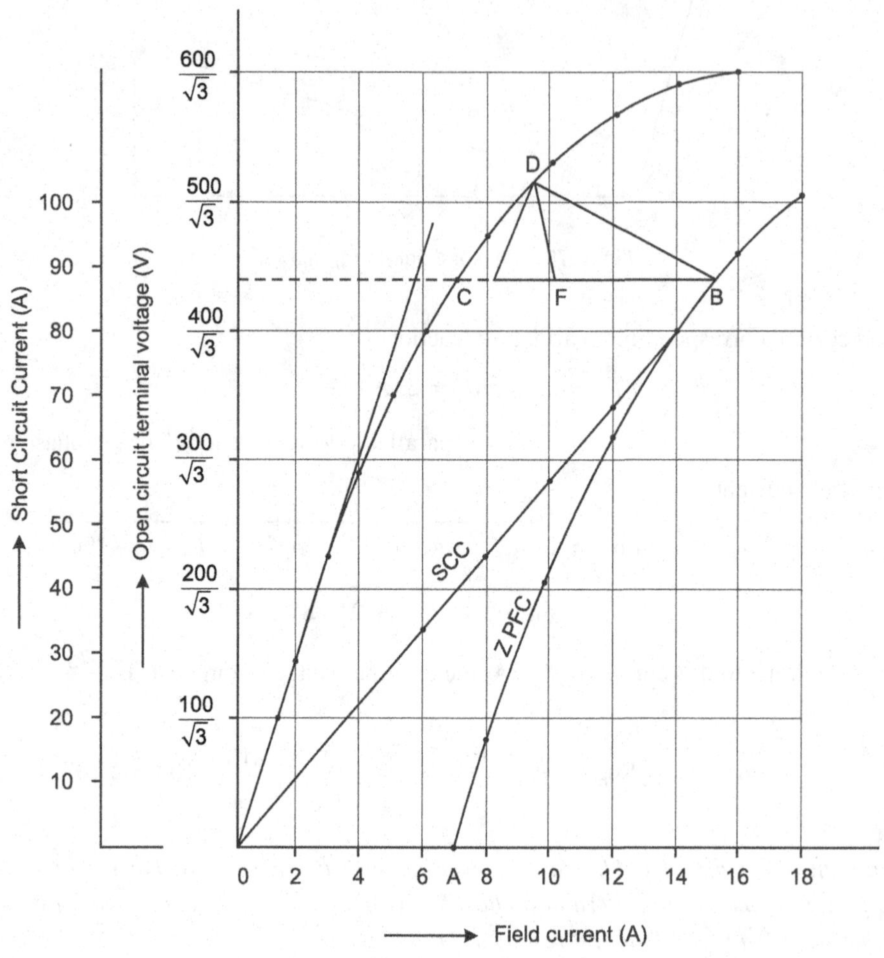

Fig. 6.74 *OCC* and Z pf triangle

(*a*) **Synchronous Impedance Method**

For a field current of 7 A the open-circuit phase voltage is $\frac{440}{\sqrt{3}}$ V and the short circuit current is 40 A. Therefore the synchronous impedance

$$Z_s = \frac{\text{O.C. phase voltage for field current of 7A}}{\text{S.C. current for field current of 7A}}$$

$$= \frac{440/\sqrt{3}}{40} = 6.351\ \Omega$$

$$X_s = \sqrt{Z_s^2 - R^2}$$

$$= \sqrt{(6.35)^2 - (0.2)^2} = 6.348\ \Omega$$

$pf_1 \cos\phi = 0.8$ lagging; $\phi = \cos^{-1} 0.8 = 36.87°$ lag.; $\sin\phi = \sin 36.87° = 0.6$

The phasor diagram in shown in Fig. 6.75.

$$E = \sqrt{(V\cos\phi + IR)^2 + (V\sin\phi + IX_s)^2}$$

$$= \sqrt{(254 \times 0.8 + 40 \times 0.2)^2 + (254 \times 0.6 + 40 \times 6.348)^2}$$

$$= \sqrt{(211.2)^2 + (406.32)^2} = 458 \text{ V}$$

$$\text{Voltage regulation} = \frac{E - V}{V} = \frac{458 - 254}{254}$$

$$= 0.803 \text{ pu} = 80.3\% \ (Ans)$$

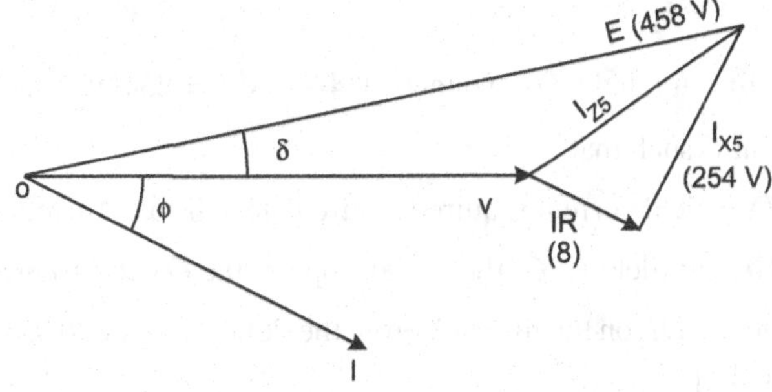

Fig. 6.75 Phasor diagram for lagging pf (syn. impedance method)

(*b*) **MMF Method**

Form the given data, the field current required to give the rated phase voltage of 254 V is 7A. Therefore $oa = I_{f_1} = 7$ A. (perpendicular to vector *OE*) is drawn in phasor diagram shown in Fig. 6.76.

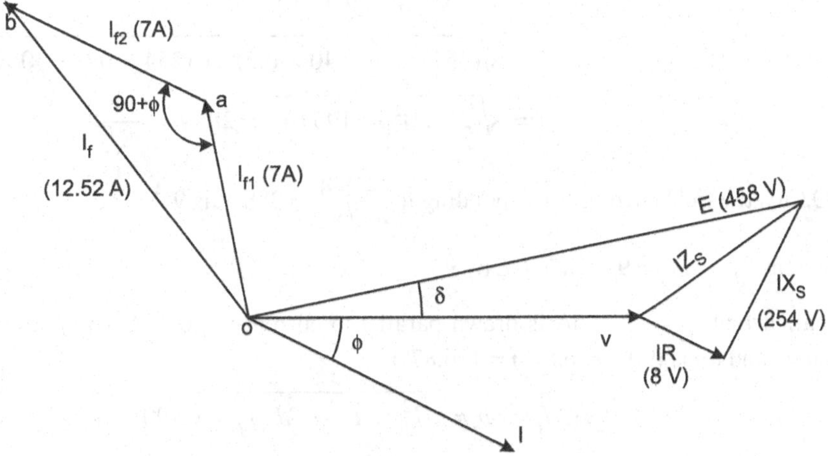

Fig. 6.76 Phasor diagram (mmf method)

Field current required to give full-load current of 40 A on short circuit is $ab = I_{f_2} = 7\text{A}$.

(parallel to vector OI)

Total field current, $$I_f = \sqrt{I_{f_1}^2 + I_{f_2}^2 + 2I_{f_1}I_{f_2}\cos(90 + \phi)}$$

$$= \sqrt{(7)^2 + (7)^2 - 2 \times 7 \times 7 \cos(90 + \phi)} = 12.52 \text{ A}$$

Terminal voltage corresponding to this field current $\dfrac{580}{\sqrt{2}} = 334.86$ V

$$\text{Voltage regulation, Reg} = \frac{Eo - V}{V} = \frac{334.86 - 254}{254} = 0.3183\,\text{pu} = \textbf{(31.83\%) Ans}$$

(*c*) **Zero-Power Factor Method**

Referring to Fig. 6.74.

Draw a horizontal line at rated phase voltage of $440 / \sqrt{3} = 254$ V to meet the *ZPFC* at *B*. On this line take a point *C* such that

$BC = OA = 7\text{A}$ = field current required to circulate full-load current on short circuit.

Through *C* draw *CD* parallel to *OG* (the initial slope of the *O.C.C.*) to meet the *O.C.C.* at *D*.

Draw the perpendicular *DF* on the line *BC*. From the Potier triangle *BCD* we have *DF* = leakage impedance voltage drop

and BF = field current required to overcome armature reaction on load

From curves shown in Fig. 6.74,

$$DF = \frac{75}{\sqrt{3}} = 43.3\,\text{V}, \; BF = 6.0 \text{ A}$$

$$\therefore \quad IX_L = 46, \; X_L = \frac{43.3}{40} = 1.08\ \Omega$$

$$E = \sqrt{(V\cos\phi + IR)^2 + (V\sin\phi + IX_L)^2}$$

$$= \sqrt{(254 \times 0.8 + 40 \times 0.2)^2 + (254 \times 0.6 + 40 \times 1.08)^2}$$

$$= \sqrt{(211.2)^2 + (195.6)^2} = 288\text{V} = \frac{498.8}{\sqrt{3}}\text{ V}$$

From O.C.C. the field current corresponding to $\dfrac{498.8}{\sqrt{3}} = 288$ V is 9 A.

This current $I_{f_1} = oa = 9\text{A}$ leads *OE* by 90°

The field current $I_{f_2} = ab = 6\text{A}$ is drawn parallel to current vector *OI*. The phase difference between *oa* and *ab* is (90° + 36.87°) = 126.87°)

$$\therefore \quad I_f = ob = \sqrt{I_{f_1}^2 + I_{f_2}^2 - 2I_{f_1}I_{f_2}\cos(90° + \phi)}$$

$$= \sqrt{(9)^2 + (6)^2 - 2 \times (9) \times 6 \times \cos 126.87°} = 13.5 \text{ A}$$

From O.C.C., corresponding to a field current of I_f= 13.5 A,

Open circuit terminal voltage is $\frac{590}{\sqrt{3}} = 341$ V

$$\therefore \quad \text{Voltage regulation} = \frac{E_o - V}{V} = \frac{341 - 254}{254} = 0.3455 \text{ pu} = \mathbf{34.55\%} \; (Ans.)$$

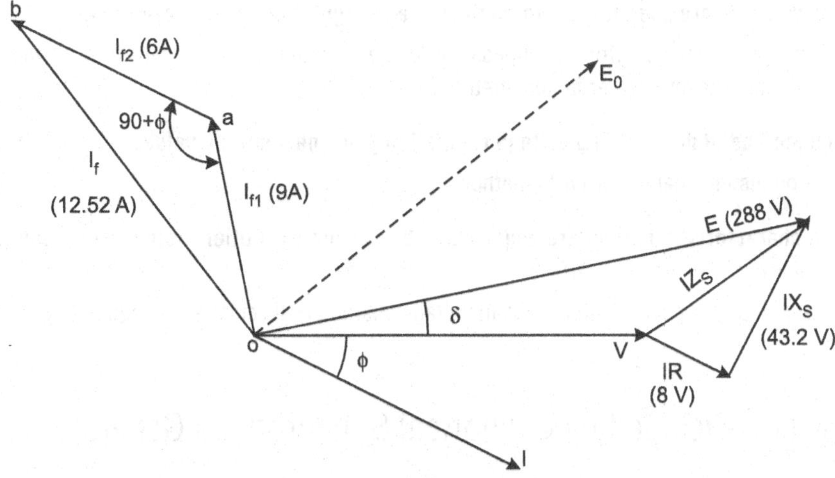

Fig. 6.77 Phasor diagram (Z pf method)

Section Practice Problems

Numerical Problems

1. The open circuit and short circuit test readings of a single-phase 500V,50 kVA alternator having an armature resistance of 0.2 ohm are given below:

Open circuit emf (in V)	125	250	370	480	566	640
Short circuit current (in A)	73	146	220	–	–	–
Field current (in A)	5	10	15	20	25	30

Determine full-load voltage regulation at

(i) Unity p.f. (ii) 0.8 lagging p.f. and

(iii) 0.8 leading p.f. using ampere-turn method (**Ans.** *7.6%; 18.4%; – 9%*)

2. A 4160 V, 3500kVA, 50Hz, three-phase alternator gave the following test results at open circuit:

Field current (in A)	50	100	150	200	250	300	350	400
Open circuit line emf (in V)	1620	3150	4160	4750	5130	5370	5550	5650

Full-load current flows at short circuit when a current of 200A flows through the field winding Neglecting the armature resistance determine full-load voltage regulation of the alternator at 0.8 pf lagging by using

(i) Synchronous impedance method. (ii) Ampere-turn method.

Also comment on the results. (**Ans.** *91.7%; 30.7%*)

3. A 5 MVA, 6600 V, 50Hz, three-phase star connected alternator has the following test data.

Field current (in A)	32	50	75	100	140

OC line voltage (in V)	3100	4900	6600	7500	8300
Line voltage zero pf (in V)	0	1850	4250	5800	7000

Neglecting armature resistance, determine the voltage regulation of the alternator by zero power factor method. (**Ans.** *3.33%*)

Short Answer Type Questions

Q.1. Which methods are used to determine the voltage regulation of an alternation?

Ans. The methods are (*i*) synchronous impedance (or *emf*) methods, (*ii*) Ampere-turn (or *mmf*) methods and (*iii*) Potier (or zero power factor) method.

Q.2. Which method of determining voltage regulation gives pessimistic value?

Ans. Synchronous impedance (or *emf*) method

Q.3. Which characteristic curves are required to be plotted for Potier method of computing voltage regulation?

Ans. *OCC, SCC* and zero power-factor full load voltage characteristic curves are required to be plotted.

6.39 Power Developed by Cylindrical Synchronous Generators

The simplified equivalent circuit of a cylindrical rotor synchronous generator is shown in Fig. 6.78, where

V = Terminal voltage (phase value) E = Excitation voltage (phase value)

I = armature or load current/phase ϕ = *pf* angle (lagging)

δ = load angle between V and E

The phasor diagram for the machine for inductive load is shown in Fig. 6.79.

Synchronous impedance,

$$\overline{Z_s} = R + jX_s = Z_s\angle\theta_s \quad \ldots(i)$$

and

$$\theta_s = \tan^{-1}\frac{X_s}{R}$$

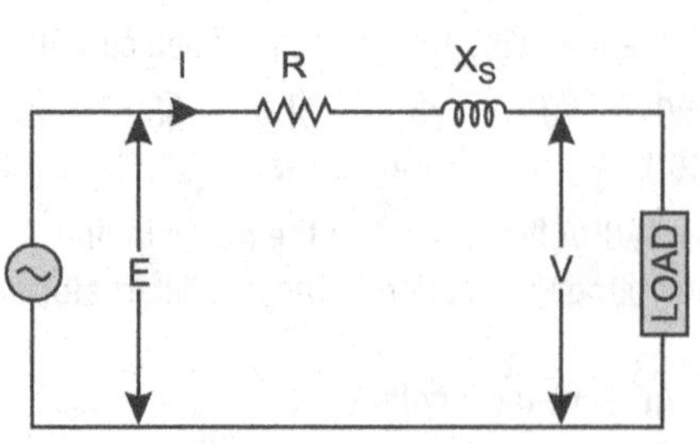

Fig. 6.78 Equivalent circuit

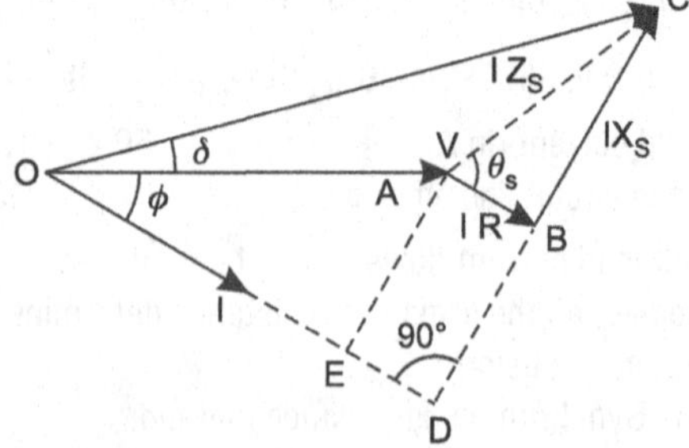

Fig. 6.79 Phasor diagram with lagging pf

Taking terminal voltage V (phase value) as reference vector,

$$\overline{V} = V \pm j0 = V\angle 0 \qquad \text{...(ii)}$$

and

$$\overline{E} = E\angle \delta \qquad \text{...(iii)}$$

$$\overline{E} = \overline{V} + \overline{I}\,\overline{Z_s} \qquad \text{...(iv)}$$

or

$$\overline{I} = \frac{\overline{E} - \overline{V}}{\overline{Z_s}} \qquad \text{...(v)}$$

6.39.1 Power Output of an AC Generator (in Complex Form)

$$S_{og} = P_{og} + jQ_{og} = \overline{V}\,\overline{I}^{\,*} \qquad \text{..... (vi)}$$

$$= \overline{V}\left[\frac{\overline{E} - \overline{V}}{\overline{Z_s}}\right]^{*} = V\angle 0\left[\frac{E\angle_{\delta} - V\angle 0}{Z_s\angle \theta_s}\right]^{*}$$

$$= V\angle 0\left[\frac{E}{Z_s}\angle(\delta - \theta_s) - \frac{V}{Z_s}\angle(-\theta_s)\right]^{*}$$

$$= V\angle 0\left[\frac{E}{Z_s}\angle(\theta_s - \delta) - \frac{V}{Z_s}\angle \theta_s\right]$$

$$= \frac{VE}{Z_s}\angle(\theta_s - \delta) - \frac{V^2}{Z_s}\angle \theta_s$$

$$P_{og} + jQ_{og} = \frac{VE}{Z_s}[\cos(\theta_s - \delta) + j\sin(\theta_s - \delta)] - \frac{V^2}{Z_s}(\cos\theta_s + j\sin\theta_s)$$

$$= \left[\frac{VE}{Z_s}\cos(\theta_s - \delta) - \frac{V^2}{Z_s}\cos\theta_s\right] + j\left[\frac{VE}{Z_s}\sin(\theta_s - \delta) - \frac{V^2}{Z_s}\sin\theta_s\right] \qquad \text{...(vii)}$$

6.39.2 Real Power Output of an AC Generator

Considering real part of the eqn. (*vii*), we get,

$$P_{og} = \frac{VE}{Z_s}\cos(\theta_s - \delta) - \frac{V^2}{Z_s}\cos\theta_s$$

where $\cos\theta_s = \frac{R}{Z_s}$ and $\theta_s = 90 - \alpha_s$ ...(*viii*)

$$P_{og} = \frac{VE}{Z_s}\cos[90 - (\alpha_s + \delta)] - \frac{V^2}{Z_s} \times \frac{R}{Z_s}$$

$$= \frac{VE}{Z_s}\sin(\delta + \alpha_s) - \frac{V^2}{Z_s^2} \times R \qquad \text{...(ix)}$$

$$P_{og} = 3\left[\frac{VE}{Z_s}\sin(\delta + \alpha_s) - \frac{V^2}{Z_s^2}R\right] \qquad \text{(for 3-}\phi\text{ generators)}$$

6.39.3 Reactive Power Output of an AC Generator

From eqn. (*vii*), we get

$$Q_{og} = \frac{VE}{Z_s}\sin(\theta_s - \delta) - \frac{V^2}{Z_s}\sin\theta_s$$

$$= \frac{VE}{Z_s}\sin[90 - (\alpha_s + \delta)] - \frac{V^2}{Z_s} \times \frac{X_s}{Z_s}$$

$$= \frac{VE}{Z_s}\cos(\delta + \alpha_s) - \frac{V^2}{Z_s^2} \times X_s \qquad \dots(x)$$

$$Q_{og} = 3\left[\frac{VE}{Z_s}\cos(\delta + \alpha_s) - \frac{V^2}{Z_s^2} \times X_s\right] \qquad \text{(for 3-}\phi\text{ generators)}$$

6.39.4 Power Input to an AC Generator (in Complex Form)

$$S_{ig} = P_{ig} + jQ_{ig} = \bar{E}\,\bar{I}^{\,*} \qquad \dots(xi)$$

$$= \bar{E}\left[\frac{\bar{E} - \bar{V}}{\bar{Z}_s}\right]^* = E\angle\delta\left[\frac{E\angle\delta - V\angle 0_0}{Z_s\angle\theta_s}\right]^*$$

$$= E\angle\delta\left[\frac{E}{Z_s}\angle(\delta - \theta_s) - \frac{V}{Z_s}\angle -\theta_s\right]^*$$

$$= E\angle\delta\left[\frac{E}{Z_s}\angle(\theta_s - \delta) - \frac{V}{Z_s}\angle\theta_s\right]$$

$$= \frac{E^2}{Z_s}\angle\theta_s - \frac{EV}{Z_s}\angle\theta_s + \delta$$

$$P_{ig} + jQ_{ig} = \frac{E^2}{Z_s}\cos\theta_s + j\frac{E^2}{Z_s}\sin\theta_s - \frac{EV}{Z_s}\cos(\theta_s + \delta) - j\frac{EV}{Z_s}\sin(\theta_s + \delta)$$

$$= \frac{E^2}{Z_s}\cos\theta_s - \frac{EV}{Z_s}\cos(\theta_s + \delta) + j\frac{E^2}{Z_s}\sin\theta_s - j\frac{EV}{Z_s}\sin(\theta_s - \delta) \qquad \dots(xii)$$

6.39.5 Real Power Input to an AC Generator

Considering real part of the eqn. (*xii*), we get

$$P_{ig} = \frac{E^2}{Z_s}\cos\theta_3 - \frac{EV}{Z_s}\cos(\theta_s + \delta)$$

$$= \frac{E^2}{Z_s} \times \frac{R}{Z_s} - \frac{EV}{Z_s}\cos[90" - (\delta - \alpha_s)]$$

$$= \frac{E^2}{Z_s^2}R - \frac{EV}{Z_s}\sin(\delta - \alpha_s) \qquad \dots(xiii)$$

$$P_{ig} = 3\left[\frac{E^2}{Z_s^2}R - \frac{EV}{Z_s}\sin(\delta - \alpha_s)\right] \quad \text{(for 3-}\phi\text{ machine)}$$

6.39.6 Reactive Power Input to an AC Generator

From eqn. (*xii*), we get,

$$Q_{ig} = \frac{E^2}{Z_s}\sin\theta_s - \frac{EV}{Z_s}\sin(\theta_s + \delta)$$

$$= \frac{E^2}{Z_s} \times \frac{X_s}{Z_s} - \frac{EV}{Z_s}\sin[90 - (\delta - \alpha_s)]$$

$$= \frac{E^2}{Z_s^2}X_s - \frac{EV}{Z_s}\cos(\delta - \alpha_s) \quad \text{...(xiv)}$$

$$Q_{ig} = 3\left[\frac{E^2}{Z_s^2}X_s - \frac{EV}{Z_s}\cos(\delta - \alpha_s)\right] \quad \text{(for – 3}\phi\text{ machine)}$$

Mech. power input to the AC generator

$$P_{i(\text{mech.})} = P_{ig} + \text{friction and windage loss} + \text{Core loss}$$

6.39.7 Condition for Maximum Power Output

Output power basically depends upon load (or torque) angle δ. Therefore, condition for maximum power output is obtained, when

$$\frac{dP_{og}}{d\delta} = 0 \quad \text{and} \quad \frac{d^2P_{og}}{P\delta^2} < 0$$

Differentiating the equation of real power output with respect to δ and equating it to zero, we get.

$$\frac{d}{d\delta}\left[\frac{EV}{Z_s}\sin(\delta + \alpha_s) - \frac{V^2}{Z_s^2}R\right] = 0 \quad \text{...(xv)}$$

As E, V, Z_s and R are constant

$$\frac{EV}{Z_s}\cos(\delta + \alpha_s) = 0 \quad \text{or} \quad \cos(\delta + \alpha_s) = 0$$

or $\delta + \alpha_s = 90°$ or $\delta = 90° - \alpha_s$

or $\delta = \theta_s$ ...(*xvi*)

Thus, the output power will be maximum, when load angle, δ = impedance angle θ_s

The maximum output power can be obtained by substituting this value in the given equation, i.e.,

$$P_{o(max.)} = \frac{EV}{Z_s}\sin(90 - \alpha_s + \alpha_s) - \frac{V^2}{Z_s}R$$

$$= \frac{EV}{Z_s} - \frac{V^2}{Z_s}R \quad \text{...(xvii)}$$

6.39.8 Condition for Maximum Power Input

To obtain this condition, put

$$\frac{dP_{ig}}{d\delta} = 0 \quad \text{and} \quad \frac{d^2P_{ig}}{d\delta^2} < 0$$

Differentiating the equation of real power input with respect to δ and equating it to zero, we get,

$$\frac{d}{d_s}\left[\frac{E^2}{Z_s^2}R + \frac{EV}{Z_s}\sin(\delta - \alpha_s)\right] = 0 \qquad \text{...(xviii)}$$

$$\frac{EV}{Z_s}\cos(\delta - \alpha_s) = 0 \qquad \text{or} \qquad \cos(\delta - \alpha_s) = 0$$

or $$\delta - \alpha_s = 90^\circ \qquad \text{or} \qquad \delta = 90^\circ + \alpha_s$$

or $$\delta = 90^\circ + (90 - \theta_s) \qquad \text{or} \qquad \delta = 180^\circ - \theta_s \qquad \text{...(xix)}$$

Thus, the input power will be maximum, when

$$\text{load angle, } \delta = 180^\circ - \text{impedance angle } \theta_s$$

The maximum input power can be obtained by substituting the value of δ in given equation of power input, i.e.,

$$P_{i(\max)} = \frac{E^2}{Z_s^2}R + \frac{EV}{Z_s}\sin[180^\circ - \theta_s - (90^\circ - \theta_s)]$$

$$= \frac{E^2}{Z_s^2}R + \frac{EV}{Z_s}\sin 90^\circ = \frac{E^2}{Z_s}R + \frac{EV}{Z_s} \qquad \text{...(xx)}$$

6.39.9 Power Equations, when Armature Resistance is Neglected

When armature resistance is neglected, $R = 0$; $Z_s = X_s$; $\alpha_s = 0$.

Real power output [Considering equation (*ix*)]

$$P_{og} = \frac{VE}{X_s}\sin\delta \qquad \text{....(xxi)}$$

Reactive power output [Considering equation (*x*)]

$$Q_{og} = \frac{VE}{X_s}\cos\delta - \frac{V^2}{X_s} \qquad \text{...(xxii)}$$

Real power input [Considering equation (*xiii*)]

$$P_{ig} = \frac{EV}{X_s}\sin\delta = p_{og} \qquad \text{...(xxiii)}$$

Reactive power input [Considering equation (*xiv*)]

$$Q_{ig} = \frac{E^2}{X_s} - \frac{EV}{X_s}\cos\delta \qquad \text{...(xxiv)}$$

Also, $$P_{o(\max)} = \frac{EV}{X_s} = P_{i(\max)} \quad \ldots(xv)$$

Example 6.32

A 762 kVA, 2200 V, 50 Hz, three-phase, star connected alternator has an effective resistance of 0.6 ohm per phase. A field current of 30 A produces a full-load current on short circuit and a line to line emf of 1039 V on open circuit. Determine the power angle of the alternator when it delivers full load at 0.8 p.f. lagging. Also determine SCR of the alternator.

Solution:

Here, Rating= 762 kVA; V_L = 2200 V; f = 50 Hz; three-phase

$$R = 0.6\ \Omega,\ I_f = 30\ \text{A};\ I_{sc} = I_{fl};\ E_{oc_{(s)}} = 1039\ \text{V}$$

$$\cos\phi = 0.8;\ \sin\phi = \sin\cos^{-1} 0.8 = 0.6$$

$$I_{sc} = I_{fl} = \frac{762 \times 1000}{\sqrt{3} \times 2200} = 200\ \text{A}\ ;\ V = \frac{V_L}{\sqrt{3}} = \frac{2200}{\sqrt{3}} = 1270\ \text{V}$$

$$E_{oc(\text{phase})} = \frac{1039}{\sqrt{3}} = 600\ \text{V}$$

$$Z_s = \frac{E_{oc}}{I_{sc}} = \frac{600}{200} = 3\Omega\ ;\ X_s = \sqrt{Z_s^2 - R^2} = \sqrt{(3)^2 - (0.6)^2} = 2.94\ \Omega$$

$$E = \sqrt{(V\cos\phi + IR)^2 + (V\sin\phi + IX_s)^2}$$

$$= \sqrt{(1270 \times 0.8 + 200 \times 0.6)^2 + (1270 \times 0.6 + 200 \times 2.94)^2}$$

$$= \sqrt{(1136)^2 + (1350)^2} = 1764\ \text{V}$$

Power developed per phase

$$\frac{EV}{X_s}\sin\delta = \frac{\text{kVA} \times 1000 \times \cos\phi}{3} \text{ or } \frac{1764 \times 1270}{2.94}\sin\delta = \frac{762 \times 1000 \times 0.8}{3}$$

$$\therefore \quad \sin\delta = \frac{762 \times 1000 \times 0.8 \times 2.94}{3 \times 1764 \times 1270} = 0.267$$

$$\delta = \sin^{-1} 0.267 = \mathbf{15.47^\circ}\ (\textit{Ans.})$$

$$SCR \text{ of the alternator} = \frac{1}{X_s} = \frac{1}{2.94} = \mathbf{0.34}\ (\textit{Ans.})$$

6.40 Two-Reactance Concept for Salient Pole Synchronous Machines

In case of a multi-polar cylindrical rotor machine, the airgap is uniform and therefore, its reactance remains the same, irrespective of the rotor position. The effect of armature reaction, fluxes and voltages induced can, therefore, be treated in a simple way with concept of a synchronous reactance and taking it as constant for all positions of field poles with respect to the armature. But in case of a salient pole synchronous machine, the air-gap is non-uniform due to which its reactance varies with the rotor position. Therefore, a salient pole machine possesses two axes of geometric symmetry

(i) field pole axis, called *direct axis* or *d-axis* and
(ii) axis passing through the centres of the inter-polar space, called the *quadrature axis* or *q-axis*, These axes are shown in Fig. 6.80. In case of a cylindrical rotor machines, there is only one axis of symmetry (pole axis or direct axis).

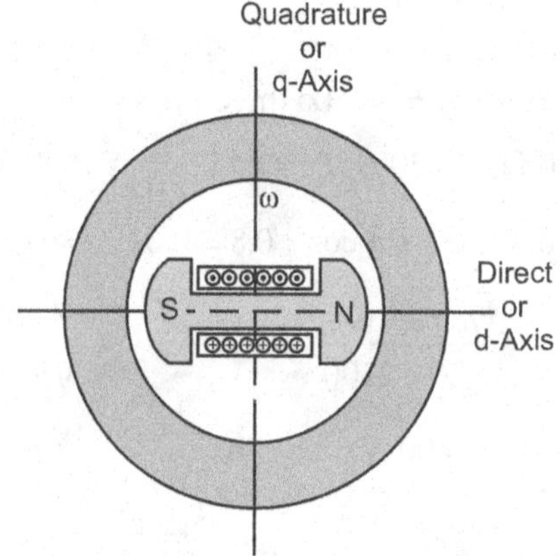

Fig. 6.80 Representation of d-axis and q-axis

Thus, for salient pole machines, the reluctance of the magnetic paths on which the induced emf depends, acts differently along the direct axis and quadrature axis. The reluctance of the direct axis magnetic circuit is due to yoke and teeth of the stator, air-gap, and pole shoe and core of the rotor. In quadrature axis, the reluctance is mainly due to large air-gap in the interpolar space.

Thus it is observed that because of non-uniformity of the reluctance of the magnetic paths, the mmf of the armature is divided into two components namely:

(*i*) a direct acting component and (*ii*) a quadrature (or cross) component.

We have seen that when armature current is in phase with the excitation voltage E the entire mmf of the armature acts at right angles to the axis of the salient poles and, therefore, all the armature mmf is in quadrature. But, if the armature current is in quadrature with the excitation voltage E, the entire mmf of the armature acts directly along the magnetic axis of the salient poles. Hence, all of the armature mmf is either added or subtracted from the mmf of the salient pole field. However when the phase difference between armature current and excitation voltage, is of some angle in between 0 and 90°, the armature mmf will have both a direct acting and a quadrature component. The direct-acting component is proportional to the sine of the phase angle between the armature current and excitation voltage, whereas the quadrature (or cross) component is proportional to the cosine of the phase angle between the armature current and the excitation voltage.

The two reactance concept is similar to the synchronous impedance concept where the effect of armature reaction is taken into account by means of equivalent armature reactance voltage. Since, there is difference in the reluctance of the magnetic paths upon which the two components of the armature mmf act, the value of the equivalent reactance for the direct component of armature mmf is greater than the value of the equivalent reactance for the quadrature component of the armature mmf. Thus the two-reactance concept for salient-pole machines replaces the effect of armature reaction by two fictitious voltages. These reactance voltages are respectively $I_d X_{ad}$ *and* $I_q X_{aq}$, where I_d and I_q are the components of the armature current along direct and quadrature axis respectively. Each of these components of armature current also, produce a leakage-reactance voltage caused by the armature leakage flux. However, the armature leakage reactance exists is assumed to have the same value X_L for both components of the armature current, Therefore, synchronous reactance for each component of the armature mmf is as follows:

Synchronous reactance for direct axis,

$$X_d = X_{ad} + X_L$$

Synchronous reactance for quadrature axis,

$$X_q = X_{aq} + X_L$$

The voltage equation for each phase of the armature based on the two-reactance concept,

$$\overline{V} = \overline{E}_0 - \overline{I}_R - \overline{I}_d X_d - \overline{I}_q X_q$$

Usually, for salient pole synchronous machines X_q = 0.6 to 0.7 times X_d whereas in cylindrical rotor machines $X_q = X_d$

6.40.1 Determination of X_d and X_q by Low Voltage Slip Test

To determine the value of X_d and X_q. a low voltage slip test is performed on the machine, as described below:

Step-I: The circuit is arranged as shown in Fig. 6.81.

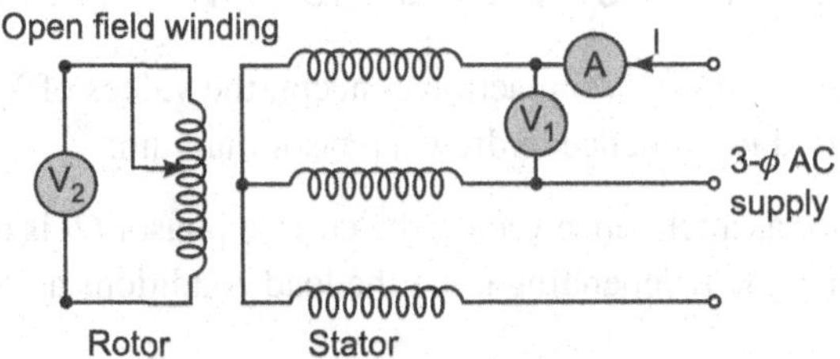

Fig. 6.81 Circuit for low-voltage test

Step-II: A three-phase balanced reduced voltage (say *V* volt) is applied to the stator winding of the unexcited machine operating at a speed little less than its synchronous speed (slip being less than 1%).

Step-III: Using oscillographs, measure and draw the wave shapes of the voltage applied across armature winding, current flowing through it and the voltage induced in the field winding (see Fig. 6.82)

Theory: When low voltage *V* is applied to the stator winding, a current *I* flows through it which produces stator *mmf.* This stator *mmf* moves slowly relative to the poles and induces an *emf* in the circuit at slip frequency.

When the axis of the poles and axis of the armature reaction *mmf* wave coincide, the armature *mmf* acts through the field magnetic circuit. The voltage applied to the armature is then equal to drop caused by the direct component of armature reaction reactance and leakage reactance.

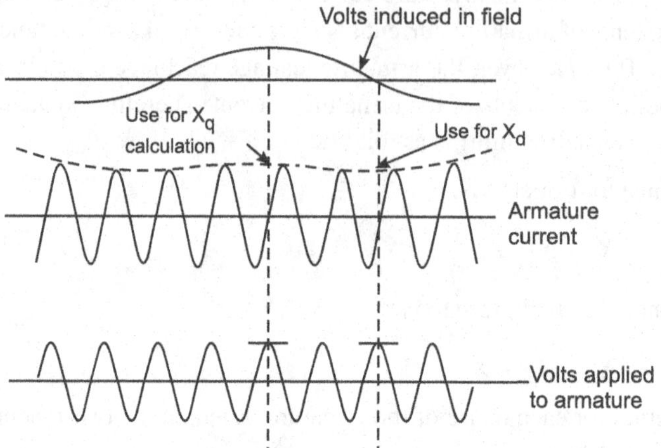

Fig. 6.82 Wave diagrams for applied voltage and armature current

When the armature reaction *mmf* is in quadrature with the field poles, the applied voltage is equal to the leakage reactance drop plus the equivalent voltage drop of the corresponding field component. Accordingly, the value X_d and X_q is determined from the oscillograph record as mentioned below:

$$X_d = \frac{\text{Maximum voltage}}{\text{Minimum current}} \text{ and } X_q = \frac{\text{Minimum voltage}}{\text{Maximum current}}$$

6.41 Construction of Phasor Diagram for Two-Reaction Concept

To construct the phasor diagram for two reaction concept, the values of X_d and X_q must be known. The following steps are used in sequence to draw a phasor diagram:

(*i*) Taking voltage phasor as a reference vector, the current phasor *OI* is drawn lagging behind the voltage vector by an angle ϕ depending upon the load conditions as shown in Fig. 6.83.

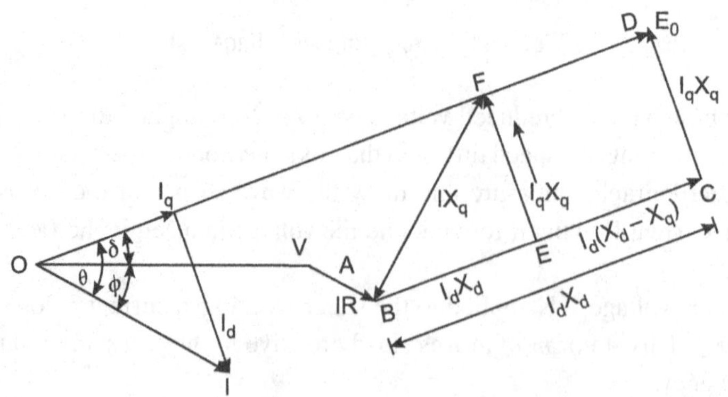

Fig. 6.83 Phasor diagram of a salient pole alternator

(*ii*) From the extreme point of the voltage phasor a line parallel to current phasor *OI* and equal to *IR* is drawn.

(*iii*) From the extreme point of the phasor *IR* a line perpendicular to current phasor *OI* and equal to IX_q is drawn

(*iv*) Draw a line from the origin *O* passing through the extreme point of IX_q phasor. This line gives the direction of excitation voltage E_0.

(*v*) Knowing the direction of E_0 the currents I_d and I_q are drawn and their values are determined.

(*vi*) From extreme point of vector *IR*, draw a vector I_dX_d perpendicular to I_d and I_qX_q from the extreme point of vector I_dX_d, as shown in Fig. 6.83, perpendicular to I_q.

(*vii*) By adding vectorially I_dX_d and $I_q X_q$ to extremity of phasor *IR*, the magnitude of excitation voltage E_0 is determined.

The angle θ between E_0 and *I* is called the *internal power factor angle* whereas, the angle δ between E_0 and *V* is called the *load or power angle.*

The phasor diagram is redrawn in Fig. 6.84 taking E_0 in horizontal direction.

From Fig. 6.84

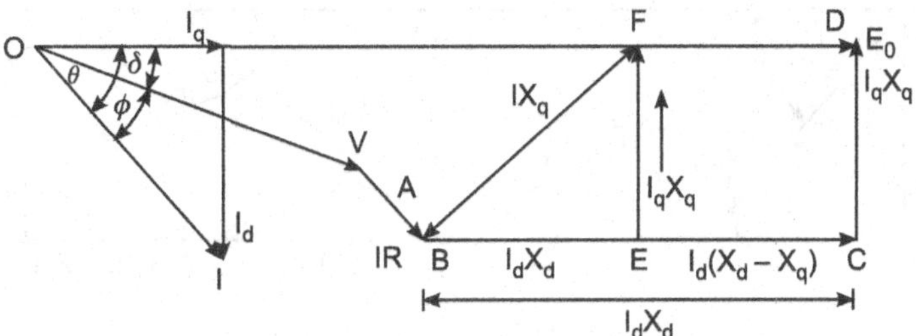

Fig. 6.84 Phasor diagram of an alternator with two reaction concept

$$I_d = I \sin \theta, I_q = I \cos \theta$$

In right $\angle$ triangle *BFE*,

$\angle BFE = \theta$ (since line *BF* is perpendicular to phasor *OI* and line *EF* is perpendicular to phasor E_0. Again redrawing the phasor diagram with extended lines is shown in Fig. 6.85.

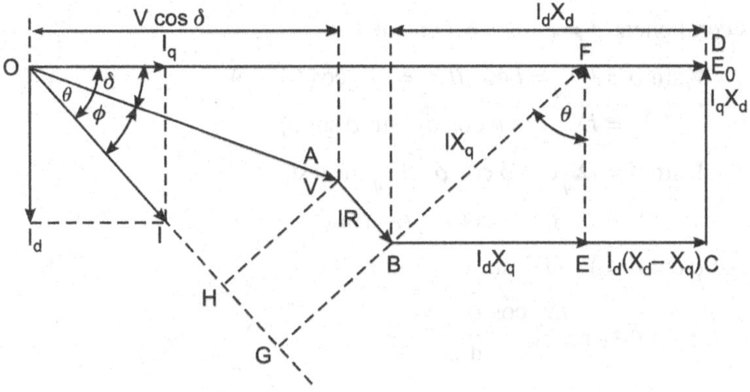

Fig. 6.85 Phasor diagram considering armature

From ΔOGF

$$\tan\theta = \frac{GF}{OG} = \frac{GB + BF}{OH + HG} = \frac{HA + BF}{DF + AE}$$

$$= \frac{V\sin\phi + IX_q}{V\cos\phi + IR} \quad \text{(for* generating action) ...(i)}$$

$$= \frac{V\sin\phi - IX_q}{V\cos\phi - IR_a} \quad \text{or} \quad \theta = \tan^{-1}\frac{V\sin\phi + IX_q}{V\cos\phi + IR_a} \quad \text{...(ii)}$$

Load angle $\delta = \theta - \phi$ (for* generating action)

For lagging power factor, angle ϕ is taken as +ve but for leading power factor it is taken as –ve.

Usually, the value armature resistance is so small as compared to reactance that it is neglected, then the phasor diagram becomes as shown in Fig. 6.86.The load angle δ can be determined directly as below:

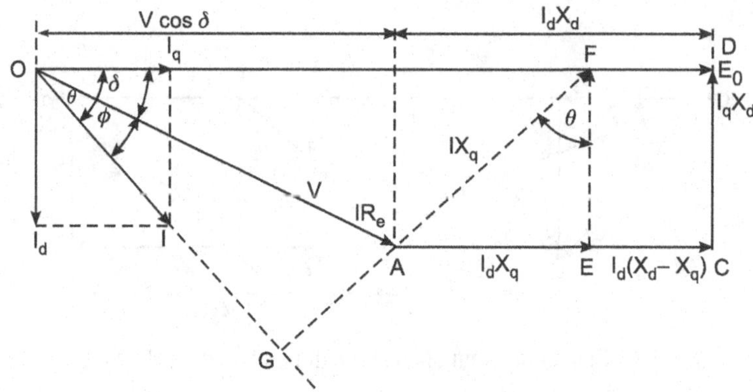

Fig. 6.86 Phasor diagram neglecting armature resistance

$$\delta = \theta - \phi \text{ (generating action)}$$

or $\theta = \phi + \delta$ in general ...(iii)

Direct-axis component, $I_d = I\sin\theta = I\sin(\phi + \delta)$ ...(iv)

Quadrature axis component, $I_q = I\cos\theta = I\cos(\phi + \delta)$ ...(v)

$$V\sin\delta = I_qX_q = I\cos\theta X_q = IX_q\cos(\phi + \delta) \quad \text{...(vi)}$$

$$= IX_q(\cos\phi\cos\delta - \sin\phi\sin\delta)$$

or $V\sin\delta = IX_q\cos\phi\cos\delta - IX_q\sin\phi\sin\delta$

or $V = IX_q\cos\phi\cot\delta - IX_q\sin\phi$

or $IX_q\cos\phi\cot\delta = V + IX_q\sin\phi$

or $$\tan\delta = \frac{IX_q\cos\phi}{V + IX_q\sin\phi} \quad \text{...(vii)}$$

* For motoring action, $\tan\theta = \dfrac{V\sin\phi - IX_q}{V\cos\phi - IR}$

If R is neglected, $E_0 = V\cos\delta + I_d X_d$ ...(viii)

If R is considered, $E_0 = V\cos\delta + IR\cos\theta + I_d X_d = V\cos\delta + I_q R + I_d X_d$ ...(ix)

***Note:** For motoring action, all the equations will be as mentioned below:

$$\delta = \phi - \theta \text{ or } \theta = \phi - \delta$$

$$I_d = I\sin(\phi - \delta)$$

$$I_q = I\cos(\phi - \delta)$$

$$V = IX_q\cos\phi\cot\delta + IX_q\sin\phi$$

$$\tan\delta = \frac{IX_q\cos\phi}{V - IX_q\sin\phi}$$

$$E_0 = V\cos\delta - I_d X_d \text{ (when } R \text{ is neglected)}$$

$$E_0 = V\cos\delta - IR\cos\theta - I_d X_d \text{ (when } R \text{ is considered)}$$

6.42 Power Developed by a Salient Pole Synchronous Generator

(Neglecting mechanical losses)

Power developed per phase,

$$P_d = \text{Power output } (P_{out}) \text{ per phase}$$

$$= VI\cos\phi \text{ considering that } R \text{ or copper loss is negligible}$$

Referring Fig. 6.86

$$I_q X_q = V\sin\delta \quad \text{...(i)}$$

$$I_d X_d = E_0 - V\cos\delta \quad \text{...(ii)}$$

Also $I\cos\phi = I_d\sin\delta + I_q\cos\delta$

Therefore, power developed per phase, $P = VI_d\sin\delta + VI_q\cos\delta$

Substituting values of I_d and I_q, we get,

$$P = V.\frac{E_0 - V\cos\delta}{X_d}\sin\delta + V.\frac{V\sin\delta}{X_q}\cos\delta$$

$$= \frac{E_0 V}{X_d}\sin\delta - \frac{V^2}{X_d}\sin\delta\cos\delta + \frac{V^2}{X_q}\sin\delta\cos\delta$$

$$= \frac{E_0 V}{X_d}\sin\delta + V^2\left[\frac{1}{X_q} - \frac{1}{X_d}\right]\sin\delta\cos\delta$$

$$= \frac{E_0 V}{X_d}\sin\delta + \frac{V^2}{2}\left[\frac{1}{X_q} - \frac{1}{X_d}\right]\sin 2\delta$$

For 3-phase synchronous generators, power developed will be,

$$P = \frac{3E_0V}{X_d}\sin\delta + \frac{3V^2}{2}\left[\frac{1}{X_q} - \frac{1}{X_d}\right]\sin 2\delta$$

From the above expression, it is very clear that power developed consists of two terms, the first term representing power due to excitation and the second term represents the reluctance power i.e., power due to salient rotor construction. In case of a cylindrical rotor machine $X_d = X_q$ and hence the second term becomes zero and the power is given by the first term only.

Example 6.33

A three-phase star-connected salient pole synchronous generator is driven at a speed slightly less then synchronous speed with open circuited field winding. Its stator is supplied from a balanced three-phase supply. Voltmeter connected across the line gave minimum and maximum readings of 2810 and 2830 volt. The line current varies between 365 and 280 ampere. Find the direct and quadrature axis synchronous reactances per phase. Neglect armature resistance.

Solution:

Maximum voltage = 2830 V

Minimum voltage = 2810 V

Maximum current = 365 A

Minimum current = 280 A

Direct-axis synchronous reactance, per phase, X_d

$$= \frac{\text{Maximum voltage / Phase}}{\text{Minimum current}}$$

$$= \frac{2830}{\sqrt{3}\times 280} = \mathbf{5.83\,\Omega}\ (\mathit{Ans})$$

Quadrature synchronous reactance, per phase,

$$X_q = \frac{\text{Minimum voltage / phase}}{\text{Maximum current}}$$

$$= \frac{2810}{\sqrt{3}\times 280} = \mathbf{4.44}\ \Omega\ (\mathit{Ans})$$

Example 6.34

A three-phase, 3300 V, 50 Hz, star-connected alternator has an effective resistance of 0.5 Ω/phase. A field current of 30 A produces full-load current of 180 A on short-circuit and a line to line emf of 1000 V on open circuit. Determine

(i) the power angle of the alternator when it delivers full-load at 0.8 pf (lag)

(ii) the SCR of the alternator.

Solution:

Here, V_L = 3300 V: $I_{fl} = I$ =180 A: I_f = 30 A: E_{OL} =1000 V; R = 0.5 Ω / phase

$$\text{Phase voltage, } V = \frac{V_L}{\sqrt{3}} = \frac{3300}{\sqrt{3}} = 1905\,\text{V}$$

$$\text{Synchronous impedance per phase, } Z_s = \frac{OC \text{ phase voltage}}{SC \text{ current per phase}} \text{ for the same field current}$$

$$= \frac{1000/\sqrt{3}}{180} = 3.21\,\Omega$$

$$\text{Synchronous impedance per phase, } X_s = \sqrt{Z_s^2 - R^2} = \sqrt{(3.21)^2 - (0.5)^2} = 3.165\,\Omega$$

$\cos\phi = 0.8$ lagging; $\sin\phi = \sin\cos^{-1}0.8 = 0.6$

$$\text{Open-circuit voltage per phase, } E = \sqrt{(V\cos\phi + IR)^2 + (V\sin\phi + IX_e)^2}$$

$$= \sqrt{(1905 \times 0.8 + 180 \times 0.5)^2 + (1905 \times 0.6 + 180 \times 3.165)^2}$$

$$= \sqrt{(1614)^2 + (1712.7)^2} = 2353\text{ V}$$

Power output per phase, $P = VI\cos\phi = 1905 \times 180 \times 0.8 = 274320$ W

Power developed per phase due to field excitation

$$P = \frac{EV}{X_S}\sin\delta \text{ neglecting losses}$$

$$\sin\delta = \frac{PX_S}{EV} = \frac{274320 \times 3.165}{2353 \times 1905} = 0.1937$$

(*i*) Power angle of alternator $\delta = \sin^{-1} 0.1937 = \mathbf{11.17°}$ (*Ans.*)

(*ii*) SCR of alternator $= \frac{1}{X_S} = \frac{1}{3.165} = 0.316$ (*Ans.*)

Example 6.35

A three-phase, star-connected salient pole, alternator at 4.5 MVA, 6000 V, has a resistance of 2% and leakage reactance of 10%. A field current of 60 A produces short-circuit armature current equal to full-load current. The armature cross-reaction per armature turn is half of the direct reaction. The open-circuit characteristic is as follows:

Field current in ampere	*30*	*50*	*75*	*100*	*140*
Terminal voltage in volt	*2906*	*4700*	*6600*	*7500*	*8300*

Find the percentage regulation on full load at a power factor of 0.8 (lagging)

Solution:

Converting terminal line voltage to phase voltages:

Field current in A:	*30*	*50*	*75*	*100*	*140*
O.C. phase voltage in V:	$2906/\sqrt{3}$	$4700/\sqrt{3}$	$6600/\sqrt{3}$	$7500/\sqrt{3}$	$8300/\sqrt{3}$
	1678	2714	3810	4330	4792

The OCC is drawn as shown in Fig. 6.87

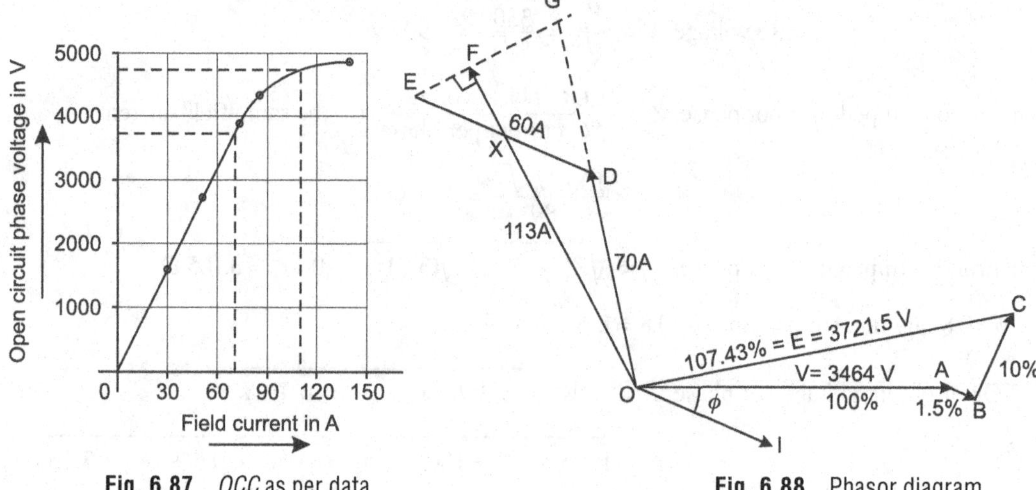

Fig. 6.87 *OCC* as per data

Fig. 6.88 Phasor diagram

The phasor diagram for full load 0.8 power factor (lagging) is drawn as shown in Fig. 6.88,

where, $OA = V = \dfrac{6000}{\sqrt{3}} = 3464$ V (phase value) = 100%

AB = 1.5% drawn parallel to current vector OI

BC = 10% drawn quadrature to OI

OC = is the vector sum of OA, AB and BC

$$E = OC\sqrt{(100 + 1.5\cos\phi + 10\sin\theta)^2 + (10\cos\phi - 1.5\sin\phi)^2}$$

$$= \sqrt{(100 + 1.5 \times 0.8 + 10 \times 0.6)^2 + (10 \times 0.8 - 1.5 \times 0.6)^2}$$

$$= \sqrt{(107.2)^2 + (7.1)^2} = 107.43\% = \frac{3464 \times 107.43}{100} = 3721.5\text{ V}$$

From *OCC* the field excitation corresponding to 3721.5 V, I = 70A.

Now, draw vector OD perpendicular to OC (i.e., E) to represent excitation required to induce E, equivalent to 70A.

Draw DE = 60A, parallel to OI, to represent excitation equivalent to full-load armature reaction on short circuit.

The vector DE is divided at X such that $DX = K.\ DE$ where K is the ratio of the cross- reaction to the direct reaction per ampere-turn. In this case $K = 0.5$ therefore, $DX = EX$.

Now draw a line OX and extend it. Draw a perpendicular from E on this line to meet at F. Then OF is the exciting current required to balance the direct reaction (DG), to balance cross-reaction (GF) and full load exciting field (OD).

By measurement OF = 113 A

The emf generated corresponding to excitation of 113 ampere = 4700V

$$\text{Percentage regulation} = \frac{4700 - 3464}{3464} \times 100 = 35.7\% \ (\textit{Ans.})$$

Example 6.36

A three-phase alternator has a direct axis synchronous reactance of 1.0 pu and a quadrature axis synchronous reactance of 0.65 pu per phase. When the machine is operating at full load at a pf of 0.8 lagging, draw the phase diagram and estimate from there (i) the load angle and (ii) pu no-load emf. Neglect armature resistance.

Solution:

Here, $X_d = 1.0$ pu; $X_q = 0.65$ pu; $\cos\phi = 0.8$ lagging

Terminal voltage, $V = 1.0$ pu

Armature current, $I = 1.0$ pu

Now, $\cos\phi = 0.8$; $\sin\phi = \sin\cos^{-1}0.8 = 0.6$

(*i*) The phasor diagram is (shown in Fig. 6.89)

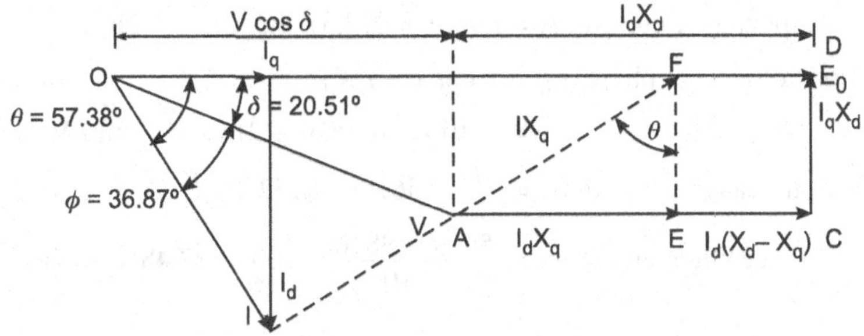

Fig. 6.89 Phasor diagram

From rt. angle triangle *OIF*

$$\tan\theta = \frac{IF}{OI} = \frac{IA + AF}{OI} = \frac{V\sin\phi + IX_q}{V\cos\phi}$$

$$= \frac{1 \times 0.6 + 1 \times 0.65}{1 \times 0.8} = 1.5625$$

and $\theta = \tan^{-1} 1.5625 = 57.38°$

(*ii*) Load angle, $\delta = \theta - \phi = 57.38° - 36.87° = \mathbf{20.51°}$ (*Ans.*)

Direct-axis component of current, $I_d = I\sin\theta = 1 \times \sin 57.38° = 0.8423$ p u

No-load emf, $E_0 = V\cos\delta + I_dX_d$

$= 1.0 \times \cos 20.51° + 0.8423 \times 1.0 = \mathbf{1.7789}$ (*Ans.*)

Example 6.37

The direct and quadrature axis synchronous reactances of a three-phase, 6.6 kV, 4 MVA, 32 salient pole alternator are 9 and 60 ohm respectively. Determine its regulation and excitation emf needed to maintain 6.6 kV at the terminals when supplying a load of 2.5 MVA at 0.8 pf lagging. What maximum power can this alternator supply at the rated terminal voltage if the field becomes open circuited? Neglect armature resistance.

Solution:

Here, $V_L = 6600V$; $\cos\phi = 0.8$ lagging; $X_d = 9\Omega$; $X_q = 6\Omega$

$$\text{Terminal voltage per phase, } V = \frac{6600}{\sqrt{3}} = 3810 \text{ V}$$

$$\text{Armature current, } I = \frac{2.5 \times 10^6}{\sqrt{3} \times 6600 \times 0.8} = 273.37 \text{ A}$$

$$\phi = 0.8; \sin\phi = \sin\cos^{-1}0.8 = 0.6$$

Now,
$$\tan\delta = \frac{IX_q \cos\phi}{V + IX_q \sin\phi} = \frac{273.37 \times 6 \times 0.8}{3810 + 273.37 \times 6 \times 0.6} = 0.2737$$

or
$$\delta = \tan^{-1} 0.2737 = 15.3°$$

$$\text{Angle, } \theta = \phi + \delta = 36.87 + 15.3 = 52.17°$$

$$\text{Direct-axis component of current, } I_d = I \sin\theta = 273.37 \sin 52.17° = 216 \text{ A}$$

$$\text{Excitation voltage per phase, } E_0 = V\cos\delta + I_d X_d$$

$$= 3810 \times \cos 15.3° + 216 \times 9 = 5619 \text{ V}$$

$$\text{Excitation voltage (line value)} = \sqrt{3} \times 5619 = \mathbf{9732} \text{ V } (Ans.)$$

$$\text{Percentage regulation} = \frac{5619 - 3810}{3810} \times 100 = \mathbf{47.48\%} \; (Ans.)$$

When the field is open-circuited, the power developed = $\frac{V_L^2}{2}\left[\frac{1}{X_q} - \frac{1}{X_d}\right] \sin 2\delta$.

The power developed will be maximum for $\sin 2\delta = 1$ and so the maximum power, that the alternator can supply at the rated terminal voltage, with field open-circuited.

$$= \frac{V_L^2}{2}\left[\frac{1}{X_q} - \frac{1}{X_d}\right] = \frac{(6600)^2}{2}\left[\frac{1}{6} - \frac{1}{9}\right] = 1210000 \text{ W or } \mathbf{1.21\ MW} \; (Ans.)$$

Example 6.38

A 3-phase, star connected, 10 kVA, 400 V, salient pole alternator with direct and quadrature axis reactances of 15 ohm and 8 ohm respectively, delivers full-load current at 0.8 power factor lagging. Calculate the excitation voltage, neglecting resistance.

Solution:

Here, $V_L = 400$ V; 10kVA; $X_d = 15\,\Omega$; $X_q = 8\,\Omega$; $\cos\phi = 0.8$ lagging

$$\text{Terminal voltage per phase, } V = \frac{400}{\sqrt{3}} = 231 \text{ V}$$

$$\text{Armature current, } I = \frac{kVA \times 1000}{\sqrt{3} V_L} = \frac{10 \times 1000}{\sqrt{3} \times 400} = 14.43 \text{ A}$$

$\cos\phi = 0.8$; $\sin\phi = \sin\cos^{-1}0.8 = \sin 36.87° = 0.6$

Now, $$\tan\delta = \frac{IX_q \cos\phi}{V + IX_q \sin\phi} = \frac{14.43 \times 8 \times 0.8}{231 + 14.43 \times 8 \times 0.6} = 0.3076$$

$$\text{Load angle, } \delta = \tan^{-1} 0.3076 = 17.1°$$

$$\text{Angle, } \theta = \delta + \phi = 17.1° + 36.87° = 53.97°$$

$$\text{Direct-axis component of current, } I_d = I \sin\theta = 14.43 \times \sin 53.97° = 11.67 \text{ A}$$

$$\text{Excitation voltage, } E_0 = V\cos\delta + I_d X_d = 231 \cos 17.10° + 11.67 \times 15$$

$$= 396 \text{ V}$$

Excitation line voltage, $E_{0L} = \sqrt{3} \times 396 =$ **686 V** (*Ans.*)

Example 6.39

The armature of a three-phase, star-connected, 10 kVA, 400 V, 50 Hz salient pole alternator has a resistance of 1 ohm per phase. Its direct and quadrature axis reactances are 15 ohm and 9 ohm respectively. The machine is delivering rated load at rated voltage and pf 0.8 lagging. If the load angle is 17°, find (i) The direct axis and quadrature axis component of armature current. (ii) excitation voltage of the generator.

Solution:

Here, $V_L = 400$ V; $R = 1\Omega$; $X_d = 15\Omega$; $X_q = 9\Omega$; $\cos\phi = 0.8$ lagging; $\delta = 17°$

$\cos\phi = 0.8$ lagging; $\phi = \cos^{-1}0.8 = 36.87°$; $\sin\phi = \sin 36.87° = 0.6$

$$\text{Terminal voltage per phase, } V = \frac{400}{\sqrt{3}} = 231 \text{ V}$$

$$\text{Armature, } I = \frac{kVA \times 1000}{\sqrt{3}V_L} = \frac{10 \times 1000}{\sqrt{3} \times 400} = 14.43 \text{ A}$$

Phaser diagram is shown in Fig. 6.90.

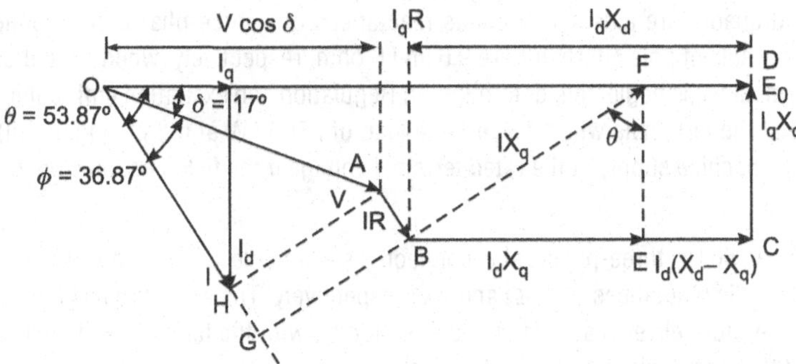

Fig. 6.90 Phasor diagram

Determination of angle θ, $$\tan\theta = \frac{GF}{OG} = \frac{GB + BF}{OH + HG} = \frac{V\sin\phi + IX_q}{V\cos\phi + IR}$$

$$= \frac{231 \times 0.6 + 14.43 \times 9}{231 \times 0.8 + 14.43 \times 1} = 1.3475$$

$$\theta = \tan^{-1} 1.3475 = 53.42°$$

Direct-axis component of armature current, $I_d = I \sin \theta = 14.43 \sin 53.42° =$ **11.59 A** (*Ans.*)

Quadrature-axis component of armature current, $I_q = I \cos \theta = 14.43 \cos 53.42° =$ **8.6 A** (*Ans.*)

$$\text{Excitation voltage, } E_0 = V \cos \delta + I_q R + I_d X_d$$

$$= 231 \cos 16.15° + 8.6 \times 1 + 11.59 \times 15$$

$$= 221.88 + 8.6 + 173.85 = 404.33 \text{ V}$$

$$E_{OL} = \sqrt{3} \times 404.33 = \mathbf{700\ V}\ (Ans.)$$

Section Practice Problems

Numerical Problems

1. A three-phase star-connected salient pole synchronous generator is driven at a speed slightly less than synchronous speed with open circuited field winding. Its stator is supplied from a balanced three-phase supply. Voltmeter connected across the line gave maximum and minimum readings of 2820 and 2800 volt. The line current varies between 275 and 360 ampere. Find the direct and quadrature axis synchronous reactances per phase. Neglect armature resistance. (**Ans.** *5.92 ohm, 4.5 ohm*)

2. A three-phase, star-connected salient pole, alternator at 5000 kVA, 6 kV, has a resistance of 1.5% and leakage reactance of 10%. A field current of 60 A produces short-circuit armature current equal to full-load current. The armature cross-reaction per armature turn is half of the direct reaction. The open-circuit characteristic is as follows:

Field current in ampere	25	50	75	100	140
Terminal voltage in volt	2340	4700	6600	7500	8300

Find the percentage regulation on full load at a power factor of 0.8 (lagging) by mmf method. (**Ans.** *23.94%*)

3. The direct and quadrature axis synchronous reactances of a three-phase, star connected, 3500 kVA, 6600 V, 32-pole salient pole alternator are 9.6 and 6 ohm, respectively, when measured by the slip test. If armature resistance is neglected, determine; (i) Regulation and excitation emf required to maintain the rated voltage at the terminals when delivering a load of 2500 kW at 0.8 pf lagging. (ii) What maximum power can this machine supply at the rated terminal voltage if the field becomes open-circuited? (**Ans.** *1361 kW*)

4. A 10 kVA, 4380 V, 50 Hz, three-phase, star-connected salient pole synchronous generator has direct axis and quadrature axis reactances of 12 Ω and 8 Ω respectively. The armature has a resistance of 1 Ω per phase. The generator delivers rated load at 0.8 pf lagging with the terminal voltage being maintained at rated value. If the load angle is 16.15°, determine.
   (i) the direct axis and quadrature axis components of armature current,
   (ii) excitation voltage of the generator. (**Ans.** *12.14 A, 9.14 A; 633 V*)

Short Answer Type Questions

Q.1. What is the use of slip test performed on an alternator?

Ans. It is performed to determine the value of synchronous reactance for direct axis *Xd* and synchronous reactance for quadrature axis *Xq*.

Q.2. Why two reaction theory is applied only to salient pole synchronous machines?

Ans. In case of salient pole machines, the air-gap is not uniform and its reactance varies with the rotor position. Because of this non-uniformity of the reactance of the magnetic paths, the mmf of the armature is divided into two components called direct-acting component along the field pole axis *i.e* direct-axis and quadrature-component along the axis passing through the centre of the two consecutive salient poles i.e., quadrature axis.

Q.3. What are the factors on which the power angle d depends?

Ans. The power angle d depends upon the following factors:

(*i*) Supply voltage
(*ii*) armature current
(*iii*) load power factor and
(*iv*) quadrature component of synchronous reactance.

Q.4. How the value of *SCR* affect the stability limit?

Ans. We know *Xd* =, smaller the value of *SCR*, larger is the value of *Xd*. Now maximum power output of a machine is inversely proportional to *Xd*.

Hence, larger the value of *Xd* lower will be the stability.

Thus, smaller the value of *SCR* lower will be the stability limit.

6.43 Transients in Alternators

A sudden change in the operating conditions of an alternator causes *transients*. Transients may occur due to

(*i*) Switching
(*ii*) Sudden change of load
(*iii*) Short –circuiting (either line-to line or line to neutral or short circuiting of all the terminals)

The short-circuiting of thermals may develop severe mechanical stresses on the armature winding which may damage the machine or its prime-mover.

Therefore, it is desirable to analyse the synchronous machines under such conditions to predict the possible conditions that may occur due to these abnormal operations.

The complete analysis of transient conditions that may occur due to short – circuiting is quite extensive and beyond the scope of this book. However, we shall limit our discussions regarding this phenomena to the extent that how to determine the short-circuit transient currents and the reactance that limit these currents.

The analysis of transients depends upon two fundamental facts i.e.,

(*i*) The current in an inductive circuit cannot change instantly and
(*ii*) The theory of constant flux linkages which states that flux linkages cannot change with a closed circuit having zero resistance and no source.

Consider a three-phase alternator running at synchronous speed with its DC field excitation, without any load. When a short-circuit occurs at its terminals, the resulting currents in the three phases will develop as shown in Fig. 6.91.

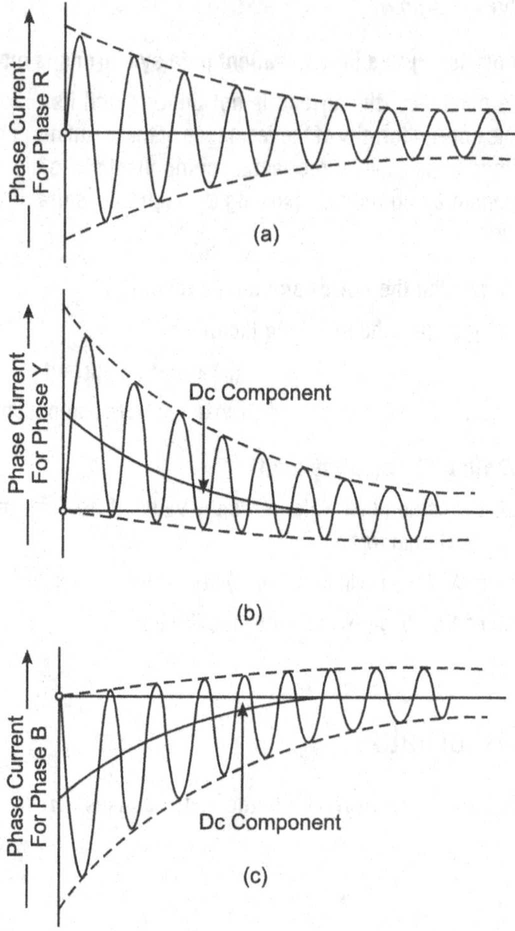

Fig. 6.91 Wave diagram of currents on the three-phases

The armature current in each phase has an AC as well as DC component. The AC component corresponds to the armature current required to oppose the time varying flux produced by the field winding as it rotates and the DC component corresponds to the initial flux linkages exist at the instant of short circuit. The net resultant of these currents produces the armature flux linkages. Each phase of armature keeps its initial flux linkages constant. The similar effects occur in the rotor field winding and these actions occur simultaneously.

Most of the synchronous machines have damper winding which has resistance, self-inductance and mutual inductance with respect to armature and field windings. This winding also affects the short circuit currents.

The transient conditions (currents) do not remain indefinitely. These tend to decay due to resistance of armature and field winding, as shown in Fig. 6.92.

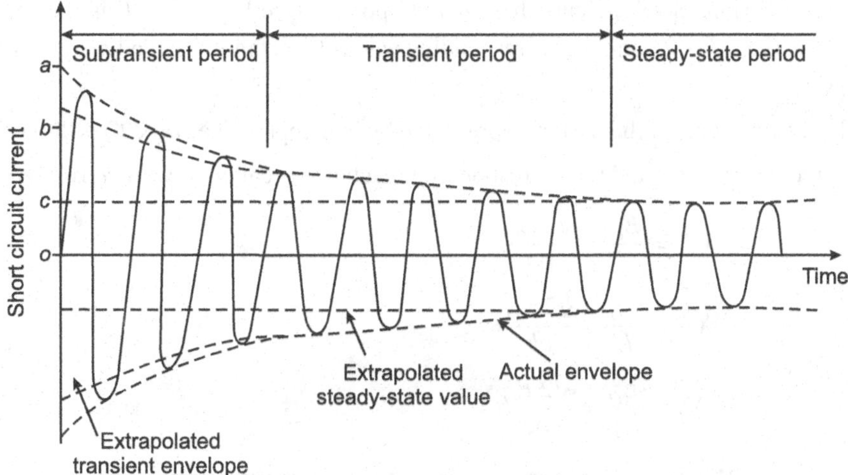

Fig. 6.92 Symmetrical short-circuit stator current in synchronous machine

As the voltage at the three phases are 120° out of phase, the short circuit occur at different voltages at the three phases. Accordingly the AC component of currents, at the instant of short circuit, is different in each phase. As a result the DC component of armature current is also different in each phase. The total initial current in about 1.5 times AC component of current.

As shown in Fig. 6.92, the AC component of current can be divided into three distinct periods

(i) *Subtransient period:* During one cycle or so, the AC current is very large and falls very rapidly. This period of time is called *subtransient period* and the *rms* value of AC current flowing during this period is called *sub-transient current* (I''). This current is developed due to the effect of both damper and field winding. It falls quickly with a time constant T_d''

(ii) *Transient period:* After sub-transient period, the current continues to fall but at slow rate till it attains a steady value. This period of slow decay is called *transient period* and the rms value of AC current during this period is called *transient current (I')*. it is caused in the field winding at the instant of short circuit. The time constant of transient period (T_d') is much longer (nearly 5 times) than the sub-transient time period. It is because the time constant of field winding circuit is much longer than damper winding circuit.

(iii) *Steady state period:* After transient period, the short circuit current reaches a *steady state value* I_{SC} It persists as long the circuit is not opened by some protective device.

It is possible to observe the three periods of current if the *rms* value of AC component is plotted as a function of time on a semi-log scale, as shown in Fig. 6.93. Accordingly, it is possible to determine the time constants T_d'' and T_d' from such a plot.

6.43.1 Sub-transient, Transient and Direct-Reactance

Sub-transient current I'' is the rms value of the initial current which occur at the instant of short circuit (i.e., $oa\ \sqrt{2}$) the corresponding value of reactance is known as *direct axis sub-transient reactance* (X_d'').

The transient current envelope cuts the *y-axis* at point *b* (see Fig. 6.92). I' is the rms value of transient current (i.e., $ob\sqrt{2}$) and the corresponding value of reactance is called *direct axis transient reactance* (X_d').

Similarly, the rms value of the current represented by intercept *oc* (i.e., $oc\sqrt{2}$) is known as steady state short circuit current I_{SC} and the corresponding reactance is called the *direct axis reactance* X_d.

Thus

$$X_d'' = \frac{E_O}{I''} = \frac{\sqrt{2}\,E_O}{oa} \quad \text{... (i)}$$

$$X_d' = \frac{E_O}{I'} = \frac{\sqrt{2}\,E_O}{ob} \quad \text{...(ii)}$$

$$X_d = \frac{E_O}{I_{SC}} = \frac{\sqrt{2}\,E_O}{oc} \quad \text{...(iii)}$$

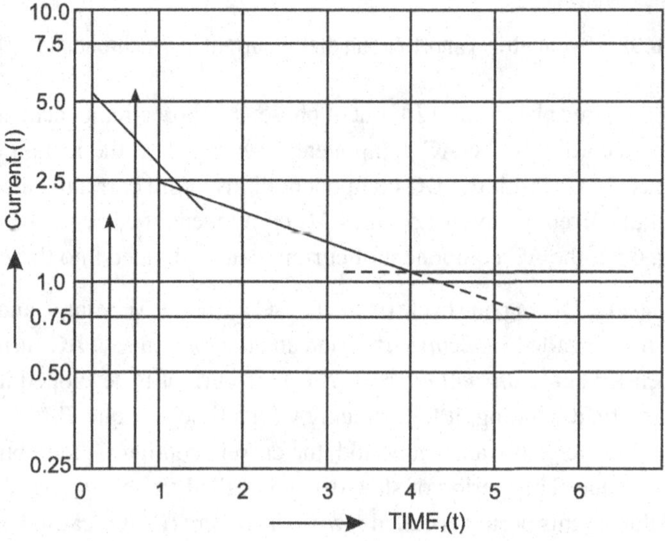

Fig. 6.93 Graph for transient current

Where E_O is the rms value of the open circuit voltage per phase.

Also,

$$I'' = \frac{E_O}{X_d''};\ I' = \frac{E_O}{X_d'} \text{ and } I_{SC} = \frac{E_O}{X_d}$$

The rms value of current at any time *t* [i.e., *I* (*t*)] after short circuit can be determined by the equation

$$I(t) = (I'' - I')e^{-t\,/\,T_d''} + (I' - I_{SC})^{-t\,/\,T_d'} + I_{SC} \quad \text{...(iv)}$$

Example 6.40

A three-phase, 11kV 100 MVA synchronous generator is running at synchronous speed with rated voltage at no-load. Suddenly, a short circuit fault develops at its terminals, the per unit reactances are

$$X_d'' = 0.15;\ X_d' = 0.25;\ X_d = 1.0$$

The time constants are $T_d'' = 0.05$ s and $T_d' = 1.2$ s and the initial DC component in such that the total current is 1.8 times of the initial AC component of current. Find

(i) *AC component of current at the instant of short circuit*

(ii) *total current at the instant of short circuit*

(iii) AC component of current after 2 cycles

(iv) AC component of current after 4 s

Solution:

Here, $V_L = 11$ kV; rating = 100 MVA; $T_d'' = 0.05$ s; $T_d' = 1.2$ s

$$X_d'' = 0.15 \text{ pu}; X_d' = 0.25 \text{ pu}; X_d = 1.0 \text{ pu}$$

$$\text{Base current (full-load)} = \frac{MVA \times 10^6}{\sqrt{3} \times V_L \text{ in } kV \times 10^3} = \frac{100 \times 10^6}{\sqrt{3} \times 11 \times 10^3} = 5248.6 \text{ A}$$

(*i*) Sub-transient current i.e., AC component of current at the instant of short circuit,

$$I'' = \frac{E}{X_d''} = \frac{1.0}{0.15} = 6.667 \text{ pu} = 6.667 \times 5248.6 = \mathbf{34991\ A}\ (\textit{Ans.})$$

(*ii*) Total current (AC plus DC) at the instant of fault

$$= 1.8 \times I'' = 1.8 \times 34991 = \mathbf{62983\ A}\ (\textit{Ans.})$$

$$\text{Transient current, } I' = \frac{E}{X_d'} = \frac{1.0}{0.25} = 4.0 \text{ pu} = 4.0 \times 5248.6 = 20994.4 \text{ A}$$

Steady-state short circuit current,

$$I_{SC} = \frac{E}{X_d} = \frac{1.0}{1.0} = 1.0 \text{ pu}$$

$$= 1.0 \times 5248.6 = 5248.6 \text{ A}$$

(*iii*) The AC component of current after time (t)

$$I(t) = (I'' - I')e^{-t / T_d''} + (I' - I_{SC})e^{-t / T_d'} + I_{SC}$$

$$\text{Here } t = \frac{2}{50} = 0.04 \text{ s}$$

$$I(t) = (34991 - 20994.4)e^{-0.04 / 0.05} + (20994.4 - 5248.6)e^{-0.04 / 1.2} + 5248.6$$

$$= 6288.7 + 15229.3 + 5248.6 = \mathbf{26766.6\ A}\ (\textit{Ans.})$$

(*iv*) When $t = 4$ s

$$I(t) = (34991 - 20994.4)e^{-4 / 0.05} + (20994.4 - 5248.6)e^{-4 / 1.2} + 5248.6$$

$$= 0 + 561.4 + 5248.6 = \mathbf{5810\ A}\ (\textit{Ans.})$$

6.44 Losses in a Synchronous Machine and Efficiency

A synchronous machine is used to convert mechanical energy into electrical energy or vice-versa. While doing so, the whole of input energy does not appear at the output but a part of it is lost in the form of heat in the surroundings. This wasted energy is called *losses in the machine.*

These losses affect the efficiency of the machine. A reduction in these losses leads to higher efficiency. Thus, the major objective in the design of a synchronous machine is to reduce these losses. The various losses occurring in a synchronous machine can be sub-divided as:

1. Copper losses.
2. Iron losses.
3. Mechanical losses
4. Stray losses

1. **Copper losses:** The various windings of the synchronous machine such as armature and field winding are made of copper and have some resistance. When current flows through them, there will be power loss proportional to the square of their respective currents. *These power losses are called* **copper losses.**

   In general, the various copper losses in a synchronous machine are:

   (*i*) Armature copper loss = I^2R

   (*ii*) Field winding copper loss = $I_f^2R_f$

   (*iii*) Brush contact loss – I^2R_b

   The brush contact loss is generally included in field winding copper losses.

2. **Iron losses:** *The losses which occur in the iron parts of the DC machine are called iron losses or core losses or magnetic losses.* These losses consist of the following:

   (*i*) **Hysteresis loss:** Whenever a magnetic material is subjected to reversal of magnetic flux, this loss occurs. It is due to retentivity (a property) of the magnetic material. It is expressed with reasonable accuracy by the following expression:

   $$P_h = K_h\, V f B_m^{1.6}$$

   where, K_h = hysteresis constant in J/m^3 i.e., energy loss per unit volume of magnetic material during one magnetic reversal, its value depends upon nature of material;

   V = volume of magnetic material in m^3.

   f = frequency of magnetic reversal in cycle/second and

   B_m = maximum flux density in the magnetic material in tesla.

   It occurs in the armature (stator core). To minimise this loss, the armature core is made of silicon steel which has low hysteresis constant.

   (*ii*) **Eddy current loss:** When flux linking with the magnetic material changes (or flux is cut by the magnetic material) an emf is induced in it which circulates eddy currents through it. These eddy currents produce eddy current loss in the form of heat. It is expressed with reasonable accuracy by the expression:

   $$P_e = K_e V f^2 t^2 \underline{B}^2{}_m$$

   where, K_e = constant called co-efficient of eddy current, its value depends upon the nature of magnetic material;

   t = thickness of lamination in m;

   V, f and B_m are the same as above.

The major part of this loss occur in the armature core. To minimise this loss, the armature core is laminated into thin sheets (0·3 to 0·5 mm) since this loss is directly proportional to the square of thickness of the laminations.

3. **Mechanical losses:** As the field system of a synchronous machine is a rotating part, some power is required to overcome:
   (*i*) Air friction of rotating field system (windage loss).
   (*ii*) Friction at the bearing and friction between brushes and slip rings (friction loss).

   These losses are known as **mechanical losses.** To reduce these losses proper lubrication is done at the bearings.
4. **Stray losses.** In addition to the iron losses, the core losses are also caused by distortion of the magnetic field under load conditions and losses in insulation of armature and field winding, these losses are called stray lasses. These losses are also included while determining the efficiency of synchronous machines.

Efficiency of a synchronous Generator

The ratio of output power to the input power of a synchronous generator is called its *efficiency.*

$$\text{Efficiency, } \eta = \frac{\text{Output}}{\text{Input}} = \frac{\text{Input} - \text{Losses}}{\text{Input}} = 1 - \frac{\text{Losses}}{\text{Input}}$$

Also,
$$\eta = \frac{\text{Output}}{\text{Output} + \text{Losses}}$$

6.45 Power Flow Diagram

The power flow diagram for a synchronous machine working as a generation is shown in Fig. 6.94

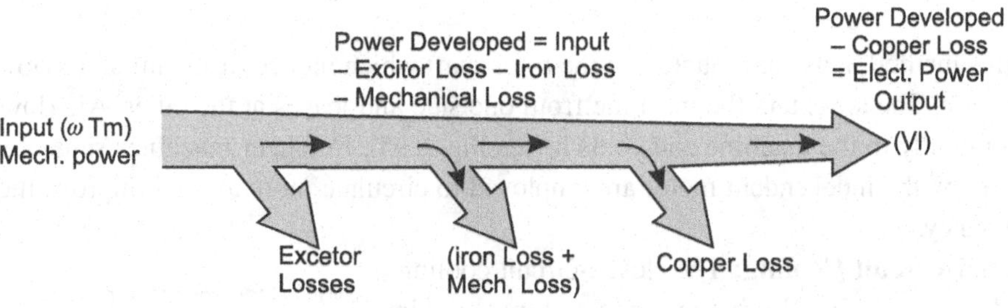

Fig. 6.94 Power flow diagram

6.46 Necessity of Cooling

In the synchronous machine, following are the main losses;

(*i*) Copper losses (*ii*) Iron losses (*iii*) Mechanical losses.

Because of these losses heat is produced which increases the temperature of the machine. This rise in temperature deteriorates the dielectric strength and ability to withstand mechanical stresses of the insulation. Thus it reduces the life of insulation on which life of machine depends. The rise in temperature further increases the losses and reduces the efficiency. Hence the rise in temperature is very harmful for the electrical machines. Therefore, various means are adopted to dissipate this heat into the atmosphere and to reduce the temperature of machine. At a stage when the rate of heat production becomes equal to the rate of heat dissipation, final temperature is achieved. Hence, to keep the temperature of the machine within the limits, efficient cooling method is necessary.

6.47 Methods of Cooling

Cooling methods of synchronous machine are broadly divided into;

(*i*) Open-circuit cooling (*ii*) Closed-circuit cooling

(*i*) **Open-Circuit Cooling:** In an open-circuit cooling system, a machine is cooled by the intake of cold air taken from atmosphere, which passes through the machine and is expelled into the atmosphere.

To prevent clogging of the machine with dust which air always contains, a filter may be mounted at the air intake, but it must be frequently cleaned. However, the filter increases the resistance to air flow, requiring addition fan power.

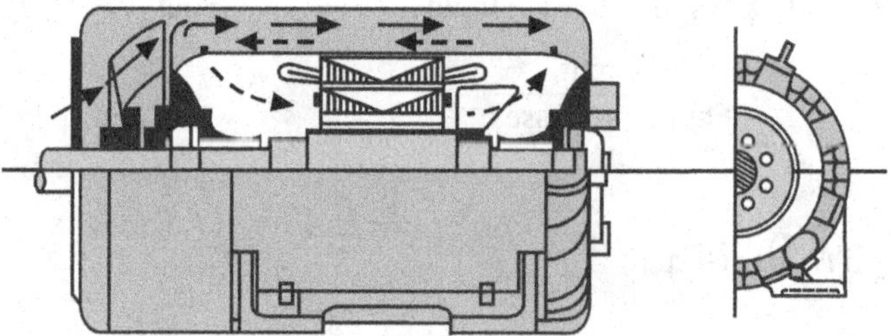

Fig. 6.95 Open circuit cooling

In this method, fans are mounted on the rotor shaft which induces a stream of air flow into the machine. The air enters the machine from one side and leaves at the other. Air flows axially and radially in the machine and cools it (see Fig. 6.95). In bigger machines separate fans are driven by the independent motor are employed to circulate more air and improve the cooling efficiency.

(*ii*) **Closed-Circuit Cooling:** The closed-circuit cooling system is one in which the same volume of air passes through a closed circuit. It passes through the alternator, becomes heated, then it passes through air cooler, where it is cooled down and again circulated through the alternator by a forced fan as shown in Fig. 6.96. The air thus circulates in a closed system. *The closed-circuit cooling system is widely used with large synchronous generators.*

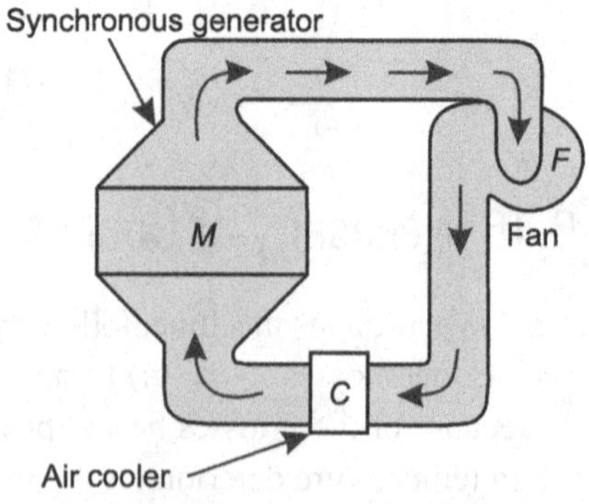

Fig. 6.96 Closed circuit cooling

In closed-circuit cooling system not only air, but also other gases may be used for cooling of large synchronous generators. At present hydrogen at a pressure of 3–4 atmosphere is mainly used.

Hydrogen cooling has a number of valuable advantages over air cooling as mentioned below;

1. The heat conductivity of hydrogen is 7 times greater than that of air and as a result the surface heat-transfer coefficient is 1·4 times greater than air. Hydrogen, therefore, cools a machine more effectively and a machine of higher output can be built with smaller dimensions.
2. Hydrogen is 14 times lighter than air. Hence, the windage losses, which in high speed machines make up the bulk of the total losses, decreases to about one-tenth of the losses when air is used. This results in the increase in efficiency of the machine.
3. With the use of hydrogen, when the corona phenomena occurs ozone gas is not liberated to cause intense oxidation of the insulation. Thus the service life of insulation is improved.

6.48 Preventive Maintenance

Preventive maintenance and routine inspection techniques conserve and prolong the life of electric machinery. Synchronous machines with general purpose ball or roller bearings require periodic lubrication while those equipped with self-lubricating *"lifetime" bearings require no lubrication at all. It may be noted that in lubricating electric machinery (synchronous machines), excessive oiling is just as damaging as insufficient lubrication. Hence, periodic and appropriate lubrication is required.*

In synchronous machines, the brushes and sliprings also require periodic maintenance in addition to lubrication. The sliprings must be checked and cleaned periodically for dust sticking on its surface.

Care must be taken that there should not be any type of oil leakage onto the stator. This may cause insulation breakdown of stator winding.

The body temperature and insulation temperature must be recorded frequently. Its value must fall within the prescribed limits.

Section Practice Problems

Numerical Problems

1. *A three-phase, 11kV 100 MVA synchronous generator is running at synchronous speed with rated voltage at no-load. Suddenly, a short circuit fault develops at its terminals, the per unit reactances are*

$$X_d'' = 0.12;\ X_d' = 0.25;\ X_d = 1.0$$

The time constants are $T_d'' = 0.04$ s *and* $T_d' = 1.1$ s *and the initial DC component in such that the total current is 1.5 times of the initial AC component of current. Find*

(i) AC component of current at the instant of short circuit
(ii) total current at the instant of short circuit
(iii) AC component of current after 2 cycles
(iv) AC *component of current after 5* s (**Ans.** *43738 A: 65608 A: 28800 A: 5416 A*)

Short Answer Type Questions

Q1. What do you mean by transient conditions in alternators?

Ans. A sudden change in the operating conditions of an alternator causes *transients.*

Q.2. When the alternator terminals are suddenly short-circuited due to any fault, what do you mean by subtransient, transient and steady state period

Ans. *Subtransient period:* During one cycle or so, the AC current is very large and falls very rapidly, this period is called *subtransient period* (Td'')

Transient period: After subtransient period, the current continues to fall but at slow rate till it attains steady value, this period between subtransient and steady state period is called *transient period* (Td').

Steady state period: After transient period, the current attains a steady value (ISC) till the fault is removed this period is called *steady state period.*

Q.3. What are the losses in a synchronous machine?

Ans.
1. Copper loss
2. Iron loss
3. Mechanical loss
4. Stray loss

Q.4. What are the methods of cooling of synchronous machines?

Ans. (i) Open circuit cooling (ii) Closed-circuit cooling.

Review Questions

1. Deduce the relation between number of poles frequency and speed of an alternator.
2. A Synchronous machine (generator or motor) is named as synchronous machine, why?
3. Name the various part of a synchronous machine. Give the function and material used for each of them.
4. Give the constructional details of cylindrical rotor alternator.
5. Explain the difference between salient pole and cylindrical pole type of rotor used in alternators. Mention their applications.
6. Explain why the stator core of an alternator is laminated.
7. The Synchronous generators employed at hydro-electric power plant have larger diameter and smaller length, why?
8. The Synchronous generators employed at steam power plants have smaller diameter and larger length, why?
9. List the advantages of making field system rotating and armature stationary in case of a alternator.
10. What is meant by 'full pitch' and 'fractional pitch' windings?
11. What are the advantages of using, 'fractional pitch' windings?
12. What is meant by distribution factor and how does it effects the generated emf of an alternator.
13. What are the advantages of using distributed winding in alternators.
14. Explain distribution factor and pitch factor in *AC* winding.

    or

    Explain the terms '*breadth factor*' and '*Pitch factor*' in case of alternator winding.
15. Why are double layer windings preferred over single layer winding in AC machines?
16. In case of *AC* system, give reasons why special efforts are made to ensure that the generated emf has a sine wave.

17. The magnitude of induced *emf* in a synchronous generator decreases by employing short pitched and distributed winding but still it is preferred, why?
18. The outer frame of a synchronous machine may not be made of magnetic material (cast iron), state why?
19. Derive an expression for induced emf for an alternator.
20. Prove that a three-phase supply when given to a three-phase winding produces a rotating magnetic field of constant magnitude.
21. State what is the effect of armature current in an alternator on the main field:
    (*i*) When it is in phase with the no-load induced emf
    (*ii*) When it lags the no-load emf by 90°.
    (*iii*) When it leads the no-load emf by 90°.
22. Explain the term (*Armature reaction*). Explain armature reaction both at lagging power factor and leading power factor.
23. A pure inductive load is connected to a three-phase synchronous machine. Show by current and flux distribution in the machine and vector diagram, the effect of this load on the terminal voltage.
24. State the causes of voltage drop in an alternator.
25. Explain the term synchronous impedance of an alternator.
26. Draw the open circuit and short circuit characteristics of a synchronous generator. Explain the shape of the characteristics.
27. Draw the equivalent circuit of an alternator.
28. Draw the vector (phasor) diagram of a loaded alternator for unity, lagging and leading power factor.
29. Using phasor diagram, show how will you determine the induced emf in an alternator when the terminal voltage. armature resistance drop and armature reactance drop are known. Consider that the load is delivered at (*i*) unity pf, (*ii*) lagging pf and (*iii*) leading pf.
30. Name the factors responsible for making terminal voltage of an alternator less than induced voltage. Explain them.
31. What do you understand by 'Voltage regulation' in the case of alternator?
32. Explain synchronous reactance and synchronous impedance in case of an alternator. How do they effect the regulation of an alternator?
33. Modern alternators are designed to have large leakage reactance and very small armature resistance why?
34. Alternators do have negative regulation, state and explain why?
35. Draw the load characteristics of an alternator for different load power factor and describe it.
36. What do you mean by short-circuit ratio (*SCR*)? Show that *SCR* is reciprocal of synchronous impedance in pu.
37. Why do the modern alternators are designed with a high value of *SCR?*
38. Describe how *OC* and *SC* tests are performed on an alternator in the laboratory. How will you be in position to determine voltage regulation from these test?
39. Describe the *mmf* method of determining the voltage regulation of an alternator.
40. Make a comparison between synchronous impedance method and ampere-turn method of determining voltage regulation of an alternator which method will you prefer and why?

41. Explain the experimental method of separating stator leakage reactance drop and drop due to armature reaction, when the alternator is loaded.

42. Describe the Potier method of determining the voltage regulation of an alternator.

43. Define and explain two reaction theory applicable to salient pole alternators.

44. Describe the method of determining direct and quadrature axis reactance of a salient pole synchronous generator.

45. What do you understand by the terms direct-axis synchronous reactance and quadrature-axis synchronous reactance of a three-phase salient pole synchronous generator?

46. Describe the experimental method for determining the direct-axis and quadrature axis synchronous reactances of a salient pole alternator.

47. Using two reaction theory, derive an expression for finding voltage regulation of a salient pole alternator. Also draw the phasor diagram.

48. Describe the slip test method for determining the value of direct-axis and quadrature-axis synchronous reactances of a synchronous machine.

49. Draw the phasor diagram of a salient pole synchronous generator delivering a load at lagging pf and explain it.

50. Derive an expression for the power developed by a non-salient pole alternator as a function of power angle, neglecting armature resistance.

51. Derive an expression for power developed by a salient pole alternator as a function of load angle.

52. A salient pole alternator is supplying power at its rated value with lagging power factor, draw and explain its phasor diagram, also show that

$$P = \frac{EV}{X_d} \sin \delta + \frac{V^2}{2}\left[\frac{1}{X_q} - \frac{1}{X_d}\right] \sin 2\delta$$

53. Which losses incur in a synchronous generator? How these losses are determined?

54. Why do we need to cool down the alternators? Describe the different methods by which alternators are cooled down.

55. What are the advantages of hydrogen as a cooling medium as compared to air?

56. What precautions are taken while using hydrogen as a cooling medium for cooling of synchronous generators?

Multiple Choice Questions

1. In an alternator the eddy current and hysteresis losses occur in

(*a*) Iron of field structure only
(*b*) Iron of armature structure only.
(*c*) Both (*a*) and (*b*)
(*d*) None of above.

2. Which of the following prime mover is least efficient?

(*a*) Gas turbine
(*b*) Petrol Engine
(*c*) Steam Engine
(*d*) Diesel Engine.

3. The rotors preferred for alternators coupled to hydraulic turbines are
(*a*) salient pole type
(*b*) cylindrical rotor type
(*c*) solid rotor type
(*d*) any of above.

4. Stator of an alternator consists of
(*a*) an iron core
(*b*) stator winding
(*c*) both *a* and *b* above
(*d*) none above

5. An alternator having 8 poles rotating at 250 rpm will have
(*a*) 60 Hz
(*b*) 50 Hz
(*c*) 25 Hz
(*d*) 50/3 Hz

6. What is the largest size of alternator being manufactured in India.
(*a*) 110 MVA
(*b*) 210 MVA
(*c*) 250 MVA
(*d*) 500 MVA

7. An exciter is nothing but a
(*a*) DC shunt motor
(*b*) *DC* series
(*c*) *DC* shunt generator
(*d*) *DC* series generator

8. The rating of alternators is usually expressed in
(*a*) full load current
(*b*) Horse power
(*c*) kVA
(*d*) kW

9. For a three-phase, 4-pole alternator the winding is placed in 96 slots. The No. of slots/pole/phase will be
(*a*) 2
(*b*) 4
(*c*) 8
(*d*) 16

10. The number of electrical degrees completed in one revolution of a six pole synchronous generator is
(*a*) 120
(*b*) 360
(*c*) 720
(*d*) 1080

11. *RMS* value of voltage generated per phase in an alternator is given by:
(*a*) $E_{ph} = 4{\cdot}44\ K_c K_d N f f$
(*b*) $E_{ph} = 4{\cdot}44\ K_c K_d N f$
(*c*) $E_{ph} = 4{\cdot}44\ K_c K_d N^2 f f$
(*d*) $E_{ph} = 1{\cdot}11\ K_c K_d N f f$

12. If the speed of an alternator is changed from 3000 rpm to 1500 rpm the generated emf/phase will become
(*a*) double
(*b*) unchanged
(*c*) half
(*d*) one fourth

13. In which of the following cases the wave form generated will be close to sine wave form.
(*a*) distributed winding with short pitch coils
(*b*) distributed winding with full pitch coils
(*c*) concentrated winding with full pitch coils
(*d*) same in all above.

14. A alternator has its field winding on the rotor and armature winding on the stator. When running under steady state conditions its air gap field is
(*a*) stationary with respect to rotor.
(*b*) rotating at synchronous speed in the direction of rotor rotation.
(*c*) rotating at synchronous speed with respect to rotor,
(*d*) both a and b.

15. When a balanced three-phase distributed armature winding carries three-phase balanced currents, the strength of the resultant rotating magnetic field is
(*a*) three times the amplitude of each constituent pulsating magnetic field.
(*b*) double to the amplitude of each constituent pulsating magnetic field.

(*c*) equal to the amplitude of each constituent pulsating magnetic field.
(*d*) one and half times the amplitude of each constituent of pulsating magnetic field.

16. Synchronous reactance of an alternator represents
(*a*) armature reaction reactance and leakage reactance.
(*b*) leakage reactance and field winding reactance.
(*c*) field winding reactance and armature reaction reactance
(*d*) a reactance connected in series with a synchronous generator.

17. The armature reaction effect in a synchronous machine depends on
(*a*) Terminal voltage. (*b*) Power factor of the load.
(*c*) Speed of the machine. (*d*) Type of prime-mover

18. For a leading pf load on an alternator implies that its voltage regulation shall be
(*a*) positive. (*b*) negative.
(*c*) zero. (*d*) any one of these.

19. Unbalanced three-phase stator current cause
(*a*) vibrations. (*b*) heating of rotor.
(*c*) double frequency currents in the rotor. (*d*) all of the these.

20. In a synchronous generator
(*a*) the open-circuit voltage (*OC*) lags the terminal voltage by an angle known as power angle.
(*b*) the *OC* voltage leads the terminal voltage by an angle known as power factor angle.
(*c*) the *OC* voltage leads the terminal voltage by an angle known as power angle.
(*d*) the *OC* voltage leads the terminal voltage by an angle known as power factor angle.

21. For Potier diagram, the zero power factor characteristic can be obtained by loading the alternator using
(*a*) lamp load. (*b*) water load.
(*c*) synchronous motor as load. (*d*) DC motor as load.

22. Curve *A* and *B* in the figure denote open circuit and full load zero power factor characteristics of a synchronous generator. Where *Q* is a point on the *Z pf* characteristics at 1.0 pu voltage. The vertical distance *PQ* in the figure gives the voltage drop across

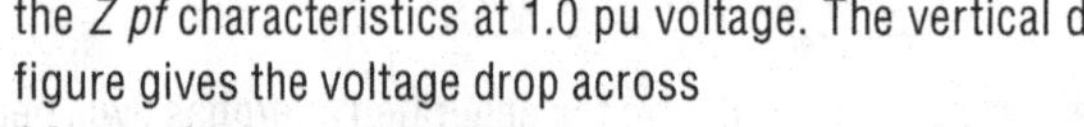

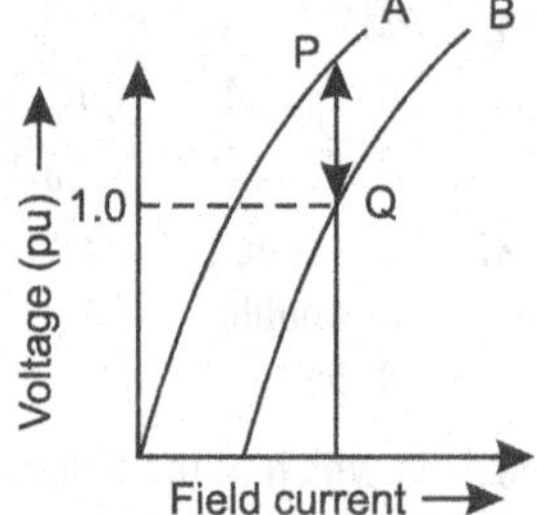

(*a*) synchronous reactance.
(*b*) armature reaction reactance.
(*c*) leakage reactance.
(*d*) Potier reactance.

23. By performing slip test we can determine
(*a*) slip.
(*b*) positive-sequence reactance and negative-sequence reactance.
(*c*) direct-axis reactance and quadrature axis reactance.
(*d*) sub-transient reactance.

24. In a slip test the direct-axis reactance (*Xd*) of a star- connected salient pole alternator is determined by taking
(*a*) *Vmax / Imax* (*b*) *Vmin / Imax*
(*c*) *Vmax / Imin* (*d*) *Vmin / Imin*

25. When a three-phase alternator is suddenly short circuited at its terminals, the initial value of short-circuit current is limited by

(*a*) subtransient reactance Xd''. (*b*) transient reactance Xd'.
(*c*) synchronous reactance XS. (*d*) sum of Xd'', Xd' and XS.

26. If Xd, $X'd$ and Xd'' are steady-state *d*-axis synchronous reactance, transient *d*-axis reactance and subtransient *d*-axis reactance of a synchronous machine respectively, then
(*a*) $Xd < Xd' < Xd''$. (*b*) $Xd'' < Xd' < Xd$.
(*c*) $Xd' > Xd'' > Xd$. (*d*) $Xd > Xd'' > Xd'$.

27. The phase sequence of a three-phase AC generator will be reversed if
(*a*) the field current is reversed keeping the direction of rotation same,
(*b*) the field current remains the same but the direction of rotation is reversed.
(*c*) the field current is reversed and the number of poles is doubled.
(*d*) the number of poles is doubled without reversing the field current.

28. An alternator is delivering load to infinite busbar. Its prime mover suddenly shutdowns. The alternator will
(*a*) continue to work as generator but the direction of rotation will be reversed
(*b*) come to standstill.
(*c*) continue to work as synchronous motor with same direction of rotation.
(*d*) start working as induction motor.

29. A synchronous generator is feeding power to an infinite bus and supplying power at UPF. If its excitation is now decreased it will feed
(*a*) the same power but at leading *pf.*
(*b*) the same power but lagging *pf.*
(*c*) more power at unity *pf.*
(*d*) less power at unity *pf.*

Keys to Multiple Choice Questions

1. b	**2.** c	**3.** a	**4.** c	**5.** d	**6.** d	**7.** c	**8.** c	**9.** c	**10.** d
11. a	**12.** c	**13.** a	**14.** d	**15.** d	**16.** a	**17.** b	**18.** d	**19.** d	**20.** c
21. c	**22.** a	**23.** c	**24.** c	**25.** a	**26.** b	**27.** b	**28.** c	**29.** a	

Parallel Operation of Alternators

CHAPTER

7

Chapter Objectives

After the completion of this unit, students/readers will be able to understand:

- ✓ Why alternators are required to be operated in parallel?
- ✓ What are requirements for parallel operation of alternators?
- ✓ What conditions are required to be fulfilled for proper synchronising?
- ✓ What are the different methods by which alternators are synchronised?
- ✓ How load is shifted from one alternator to the other?
- ✓ How load is shared between the alternators?
- ✓ What is synchronising current, power and torque?
- ✓ What is the effect of change is input power, excitation, reactance and governor's characteristics?
- ✓ What is hunting? What are its ill-effects? And how it can be reduced? Working of synchronous machine depends?

Introduction

To meet with the ever increasing demand of electrical power, it is economical and advisable to run number of generating units in parallel. Although a single larger unit used to meet with the demand is more economical but it reduces the reliability. In fact, there are a number of good reasons to use number of smaller units operating in parallel to meet with the existing demand. In the present scenario, not only the number of units placed at one generating station are operated in parallel rather all the other units placed at the other generating stations which are interconnected are also operating in parallel with each other.

In this chapter, we shall study different aspect of parallel operation of alternators or synchronous generators.

7.1 Necessity of Parallel Operation of Alternators

To meet with the demand of electrical power, a variety of generating stations hawing large capacity have been erected. In the modern generating stations, where huge power is generated, several alternators are operated in parallel.

When number of alternators are connected to same bus-bars, they are called to be ***connected in parallel.***

Such practice is considered necessary for the following reasons:

1. *Physical size*: The output power of modern power stations is so high that it is difficult to build a single unit of that capacity.
2. *Reliability or continuity of service*: Several small units are more reliable than a single large unit because if one unit fails, the continuity of supply can be maintained by operating the other units.
3. *Repair and maintenance*: Repair and maintenance of a unit is more convenient and economical if a large number of smaller units are installed at the power station.
4. *Size and cost of stand-by unit*: Since each unit is of smaller size, the cost of stand by unit is small.
5. *Extension of power plant*: The additional unit can be installed as and when the load demand increases.
6. *Operating efficiency*: Moreover, the load on the power station varies greatly both during day and night as well as during the different seasons. Thus the number of units operating at a particular time can be varied depending upon the load at that time. This keeps the machines loaded upto their rated capacity and hence results in increase in efficiency of operation as the efficiency of an electrical machine is maximum at or near rated capacity.

7.2 Requirements for Parallel Operation of Alternators

The following are the requirements for the parallel operation of alternators:

1. *Output voltage rating*: The output voltage rating of all the alternators must be the same
2. Output frequency: The rated speed of all the machines should be such that they produce the same frequency (f = PN/120).
3. *Output wave shape:* The output wave form of all the alternators must be the same, although their kVA rating may be different.
4. *Speed-load characteristics*: The drooping speed-load characteristics of the prime-movers of the alternators should be the same so that alternators share the load in their proportion as per their output (kVA) rating.
5. *Impedance triangles:* The impedance triangles of the alternators should be identical for successful parallel operation.

7.3 Synchronising Alternators

The procedure of connecting an alternator in parallel with another or with common bus-bars to which a number of alternators are already connected, is called **synchronising of alternators.**

7.4 Conditions for Proper Synchronising

For proper synchronising of alternators, the following conditions must be fulfilled:

1. The terminal voltage of the incoming alternator must be equal to that of the bus-bar voltage.

2. The voltage of the incoming alternator should be in phase opposition to the bus-bar voltage. This implies that there will be no circuiting current between the windings of the alternators already connected to bus-bars and the incoming alternator.
3. The speed of the incoming alternator must be such that its frequency is equal to that of bus-bar frequency.
4. In case of three-phase alternators the phase of the incoming alternator must be identical with the phase of the bus-bars. In other words, the phase sequence of the incoming alternator must be same as that of the bus-bars.

Before studying the practical method of synchronising the alternators, we must be clear about the situation that will arise if the above conditions are not achieved. Considering only two single-phase alternators connected to the common bus-bars operating in parallel as shown in Fig. 7.1. A load is connected to the bus-bars. There are two circuits to which current can be supplied by the alternators. One circuit is the external load and the other is local internal circuit i.e., synchronous impedance of the two alternators.

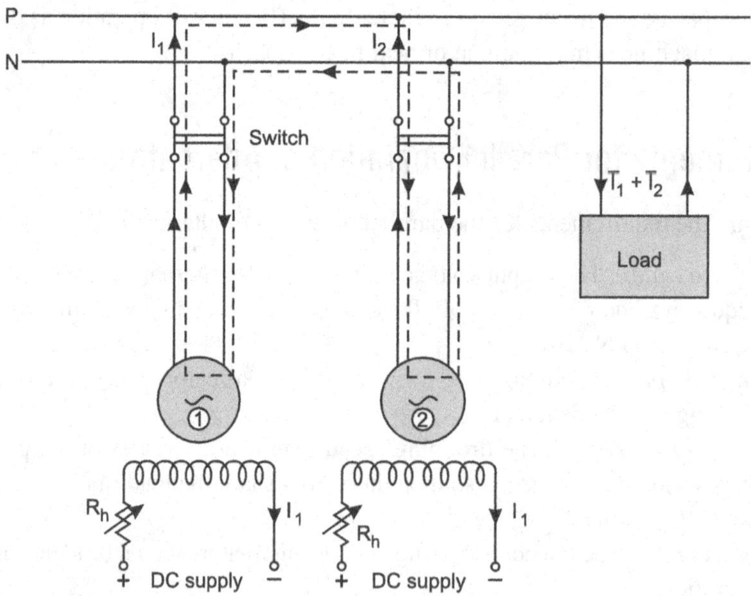

Fig. 7.1 Two 1-phase alternators connected in parallel

For the external circuit (i.e., load), the terminal voltage of the two alternators are equal and in phase with each other as shown in Fig. 7.2 (*a*) and the alternators share the load current as per their respective ratings.

Considering first and second condition, for correct synchronising, the terminal voltage of the incoming alternator must be equal and opposite (180° out of phase) to the voltage of the alternator already connected to the bus-bars with respect to the internal circuit as shown in Fig. 7.2 (*b*). Thus the resultant voltage is zero and no current circulates in the internal circuit. But if these conditions are not fulfilled i.e., the two voltages are not equal to each other in magnitude as shown in Fig. 7.2 (*c*), a resultant voltage E_r appears which will result in a circulating current I_s in the local circuit. This circulating current lags behind the resultant voltage E_r by 90° since the opposition of the local circuit is only synchronous reactance of the two alternators as resistance is neglected. Moreover, the circulating current will load the two alternators, without supplying any power to the external load.

Another condition can be when the voltage of second alternator is equal to the bus-bar voltage in magnitude but it is not out of phase by 180°, as shown in Fig. 7.2 (d). In this case also a resultant voltage E_r acts to circulate current I_s in the internal circuit of the alternators.

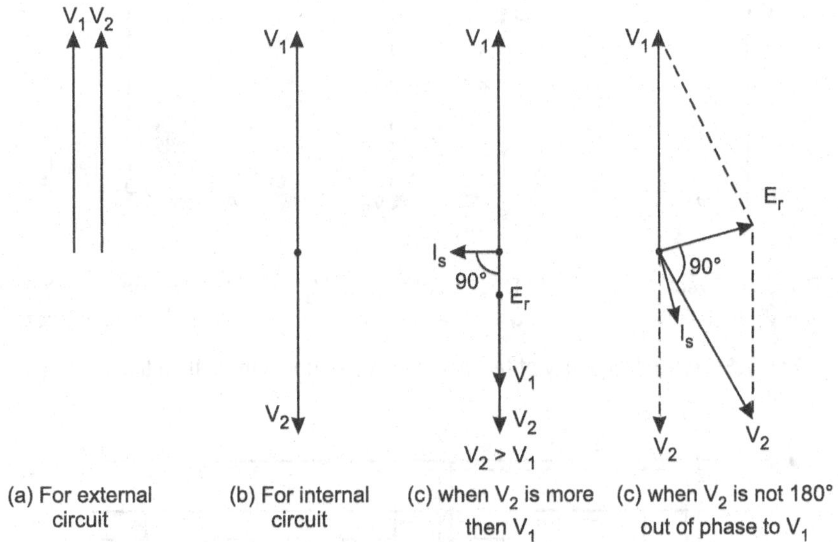

Fig. 7.2 Phasor representation of terminal voltages of two alternators

Considering third condition, for correct synchronising, the frequency of the incoming alternator must be the same as that of the frequency of the alternator already connected to the bus-bars i.e., the vectors representing the two voltages must rotate at the same speed. But if this condition is not fulfilled, one of the vectors rotates at speed higher or lesser than the other vector (say vector V_2 is rotating at a speed higher than vector V_1). The position of the two vectors at different three instants is shown in Fig. 7.3. Thus, the voltages of the two alternators will come in phase and go out of phase. This will happen at the rate equal to the difference of the frequencies of the two alternators. Because of this for local circuit, the resultant voltage E_r varies from zero to double the voltage ($V_1 + V_2$) and circulates variable current in the local circuit.

7.5 Synchronising Single-phase Alternators

Synchronising of single-phase alternators is generally done by using lamps called lamp methods. There are two methods, namely, (*i*) Dark lamp method (*ii*) Bright lamp method.

7.5.1 Dark Lamp Method

An alternator A is already connected to the bus-bars as shown in Fig. 7.4. Alternator B is of the synchronised to the bus-bars. This is done with the help of two lamps L_1 and L_2 connected across the main switch (synchronising switch) of alternator B as shown in Fig. 7.4.

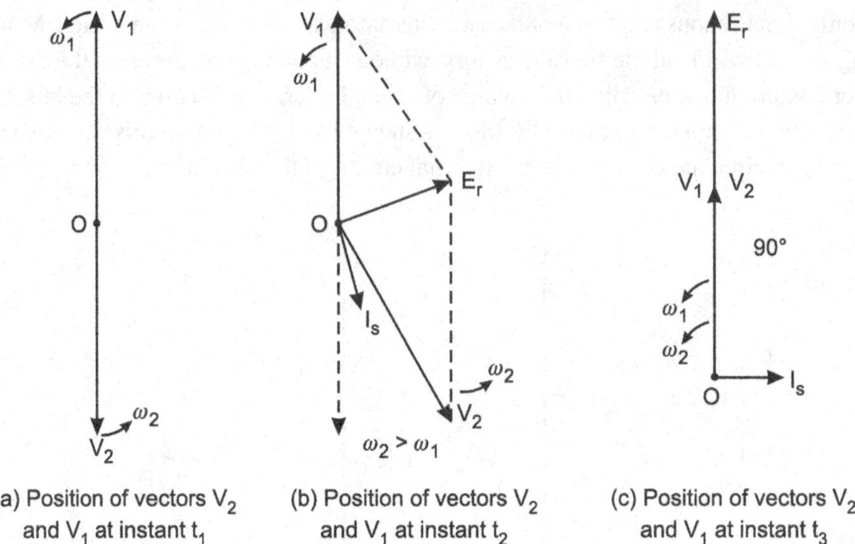

Fig. 7.3 When frequency of the incoming alternator is more than the existing

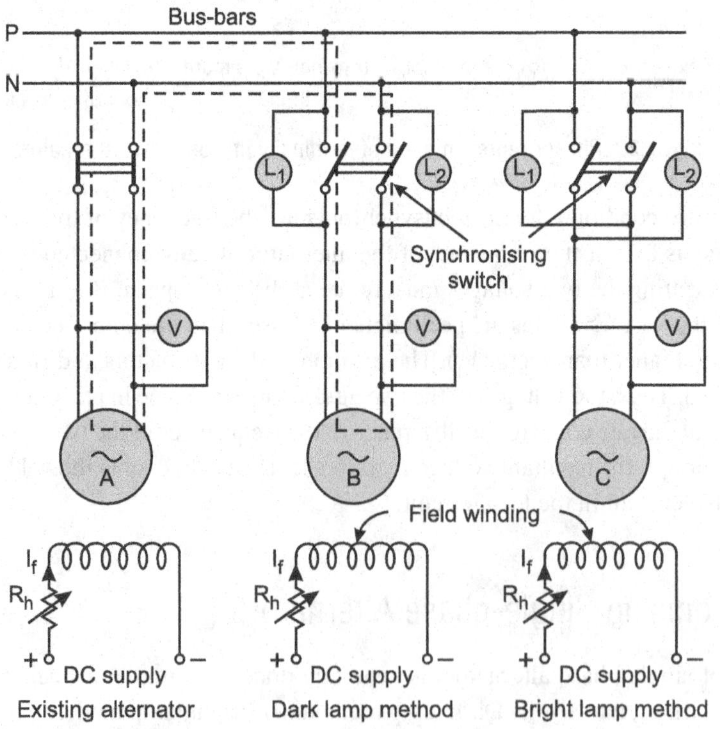

Fig. 7.4 Synchronising of 1-phase alternators

The prime-mover of alternator *B* is started and its speed is brought up near to the rated speed. The field winding of the alternator is then excited by *DC* source. The excitation is then increased to raise the voltage of the alternator equal to that of the bus-bar voltage.

If the frequency of the incoming alternator B is not equal to the frequency of alternator A, already connected to the bus-bars (i.e., $f_b \neq f_a$), a phase difference will exit between the voltages. This phase difference will be continuously changing, and therefore, the current flowing through the local circuit (shown dotted) and the synchronising lamps L_1 and L_2 will go on changing. As a result the lamps will become alternately bright and dark which produces flickering of lamps. The frequency of flickering of lamps will be equal to the difference of frequencies ($f_b - f_a$ or $f_a - f_b$) of the two alternators. The flickering of the lamps will be rapid when there is large difference in frequencies and slow when the frequencies are nearly equal. *To reduce the flickering, the speed of incoming alternator is adjusted..*

If the voltage of incoming alternator is not equal to or in phase opposition to the bus-bar voltage, a resultant voltage will appear across the two lamps L_1 and L_2. Thus, the lamps will not be totally dark. The voltage can also be checked by connecting voltmeters across the two alternators as shown in Fig. 7.4. *The voltage of the incoming alternator is made equal to the bus-bar voltage by adjusting its excitation.*

Hence, to synchronise the alternator to the bus-bars, its speed is adjusted until the lamps flickering reduces to almost zero and, its terminal voltage is made equal to the bus-bar voltage by adjusting the excitation. Then synchronising switch is closed in the middle of the dark period.

Limitations

The lamps can be dark even though a considerable voltage may exist across their terminals; therefore, it is difficult to judge the correct instant of zero voltage in the *dark lamp method*. Keeping in view of this difficulty, some engineers prefer to synchronise the alternators by *bright lamp method* because of the fact that lamps are much more sensitive to change of voltages at their maximum brightness than when they are quite dark.

7.5.2 Bright Lamp Method

In this method, the two lamps L_1 and L_2 are cross connected as shown in Fig. 7.4. When the voltage of the incoming alternator C is equal to the bus-bar voltage, the voltage across the two lamps will be double to that of alternator C. The synchronising switch is closed in this case when the two lamps are equally bright i.e., in the middle of the bright period.

7.6 Synchronising Three-phase Alternators

For synchronising of three-phase alternators, two methods are used i.e., (*i*) By using lamps, (*ii*) By using synchroscope.

Usually, low voltage alternators are synchronised by using three lamps. Depending upon the connections by which three lamps are connected across the terminals of the incoming alternator and bus-bars, the incoming alternator is switched-*ON* to the bus-bars at a particular instant. Accordingly, the lamp methods are named as (*i*) Three dark lamps method and (*ii*) Two bright and one dark lamp method.

7.6.1 Three Dark Lamps Method

An alternator A is already connected to the bus-bars as shown in Fig. 7.5. Alternator B is to be synchronised by three-dark lamps method. In this case lamp L_1 and L_2 and L_3 are connected across $R–R'$, $Y–Y'$ and $B–B'$ terminals of the synchronising switch, respectively, where RYB are the bus-bar terminals and $R'Y'B'$ are the terminals of incoming alternator B.

The voltages of the bus-bars and the incoming alternator are shown vectorially in Fig. 7.6. If the voltages are equal, the frequencies are identical and the phase sequence is correct, the voltage across all the three lamps will be zero. Under this condition the lamps are completely dark. This is the ideal condition for closing the synchronising switch.

When the frequency of the incoming alternator is different from that of bus-bar frequency and the remaining conditions are satisfied, then the three lamps will flicker in unison. At any instant, when the frequency of the incoming alternator is less than the bus-bar frequency, the voltage across the three lamps L_1, L_2, L_3 is shown in Fig. 7.7. The frequency of flickering depends upon the difference between the frequency of alternator B and the bus-bar frequency (i.e., $f_b - f_a$). However, if the phase sequence is not correct the three lamps will flicker one after the other instead of flickering simultaneously. Then the phase sequence should be corrected by interchanging any two leads to the incoming alternator at the synchronising switch. *The flickering of lamps can be minimised by adjusting the speed of the prime mover of the incoming alternator.*

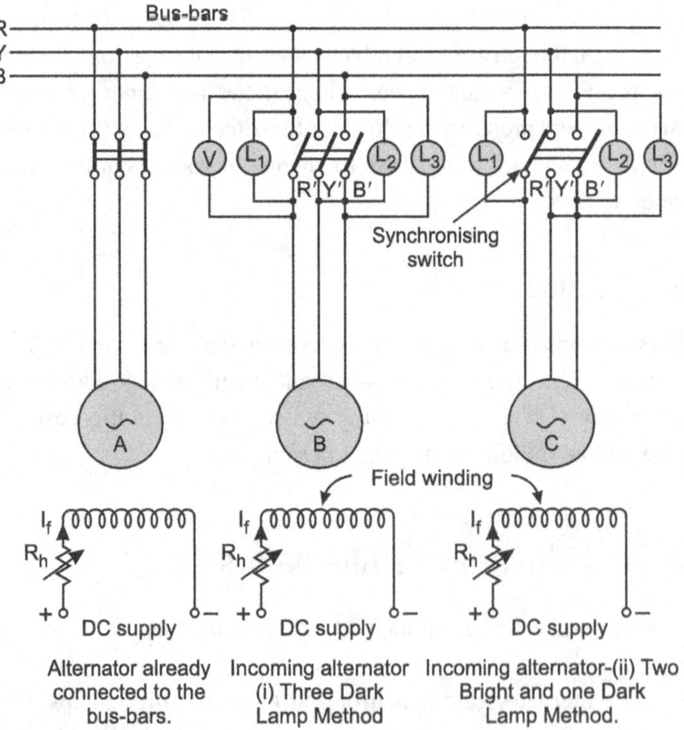

Fig. 7.5 Synchronising of three-phase alternators by using lamps

If the voltage of the incoming alternator is not equal to that of bus-bar voltage and the other conditions are fulfilled, all the lamps will flow with equal brightness and will continue to attain the same brightness. *The ideal condition can be achieved by adjusting the excitation of the incoming alternator B.*

The disadvantage of this method is that a lamp can be dark even though a considerable voltage may exist across its terminals and if the alternators are connected at this instant, this may cause considerable disturbance to high speed turbine-driven alternator. Therefore, in this case voltmeter is also connected across the synchronising switch as shown in Fig. 7.5. The other disadvantage of this method is that the flickering of lamps does not indicate whether the incoming alternator is slow or fast. These difficulties are eliminated in two bright and one dark lamp method.

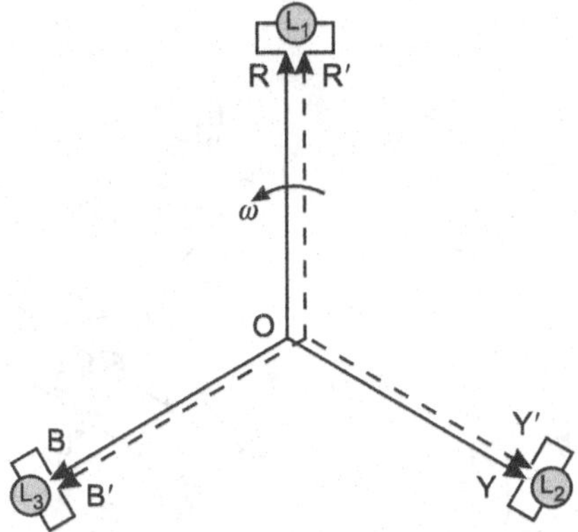

Fig. 7.6 Phasor diagram when $V_2 = V_1$; $f_2 = f_1$ and phase sequence is incorrect

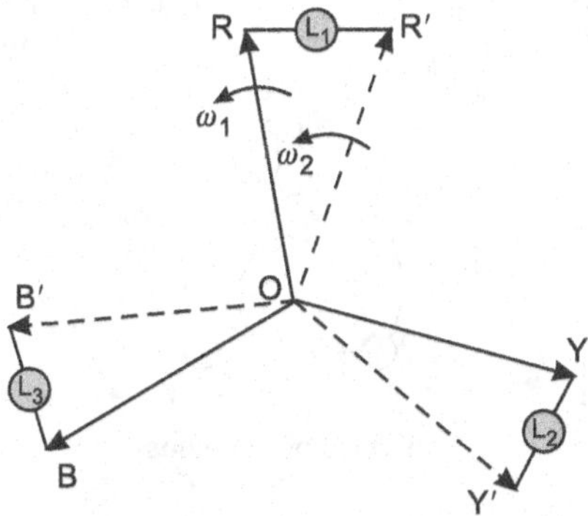

Fig. 7.7 Phasor diagram when $V_2 = V_1$ and phase sequence is correct but $f_2 < f_1$

7.6.2 Two Bright and One Dark Lamp Method

Consider alternator *C* to be synchronised by two bright and one dark lamp method. In this case lamp L_1 is connected across *R* – *R¢*, lamp L_2 is connected across *Y* – *B¢* and lamp L_3 across *B* – *Y¢* as shown in Fig. 7.5.

The bus-bar voltages *OR*, *OY* and *OB* and the voltage of incoming alternator *OR¢*, *OY¢* and *OB¢* are shown vectorially in Fig. 7.8. If the voltages are equal, the frequencies are identical and the phase sequence is correct, the voltage across L_1 will be zero and across L_2 and L_3 will be line value (V_L) as shown in Fig. 7.8. Under this condition lamp L_1 will be completely dark and the lamps L_2 and L_3 will be equally bright. This is the ideal condition for closing the synchronising switch.

When the frequencies of the incoming alternator are different from that of bus-bar frequency and the remaining conditions are fulfilled, then the three lamps will flicker alternately (i.e., one after the other in sequence). There can be two different conditions i.e., either the incoming alternator is running *too fast* or *too slow.*

If the incoming alternator *C* is running *too fast*, then the vector diagram *R¢Y¢B¢* will rotate faster than *RYB*. Because of this voltage across lamp L_1 is increasing, across L_2 is decreasing and across L_3 is increasing as shown in Fig. 7.9. representing two instants (*a*) and (*b*). The lamps will then become bright or flicker one after the other in the order L_3, L_1, L_2, L_3, L_1, L_2, *etc.* To distinguish the sequence of flickering of lamps, the lamps are generally placed on the vertices of an equilateral triangle as shown in Fig. 7.10.

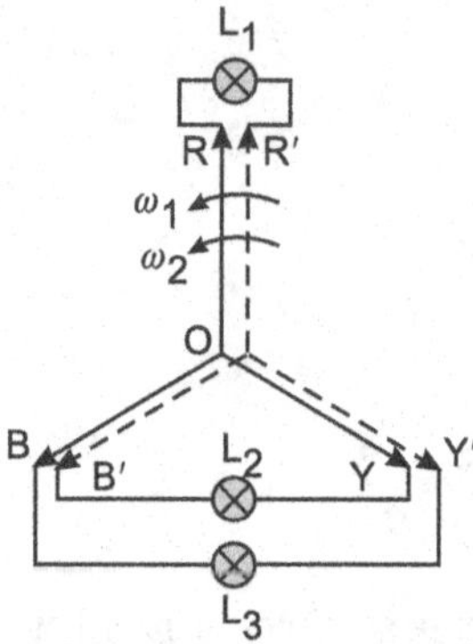

Fig. 7.8 Phasor diagram when $V_2 = V_1$; $f_2 = f_1$ and phase sequence is correct

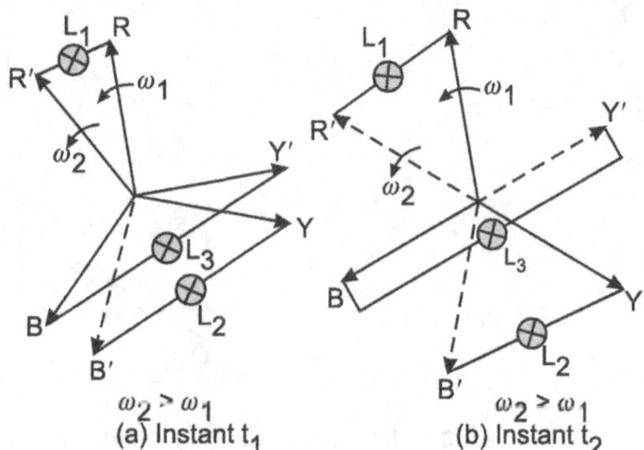

Fig. 7.9 Incoming alternator is running two fast ($f_2 > f_1$)

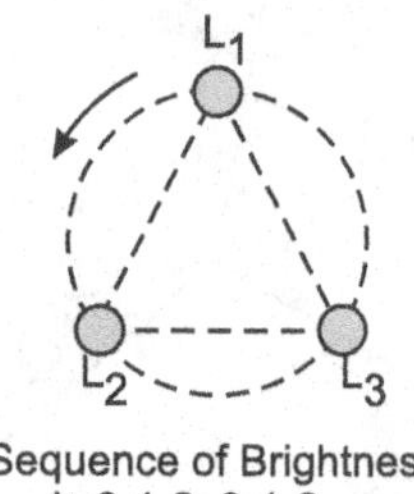

Fig. 7.10 Lamps placed on the vertices of an equilateral triangle

On the other hand, if the incoming alternator is *too slow*, then the vector diagram *R'Y'B'* will rotate slower than *RYB*. Because of this, voltage across lamp L_1 is increasing, across L_2 is increasing and across L_3 is decreasing as shown in Fig. 7.11. But in this case, lamps will become bright or flicker one after the other in the order L_2, L_1, L_3, L_2 L_1, L_3, etc., as shown in Fig. 7.12.

Thus in case of two bright and one dark lamp method, the sequence of flickering of the lamps will indicate whether the incoming alternator is too fast or too slow, accordingly the speed is adjusted to minimise the flickering of the lamps.

However, if the phase sequence is not correct all the three lamps will flicker in unison. Then the phase sequence should be corrected by interchanging any two leads of the incoming alternator at the synchronising switch.

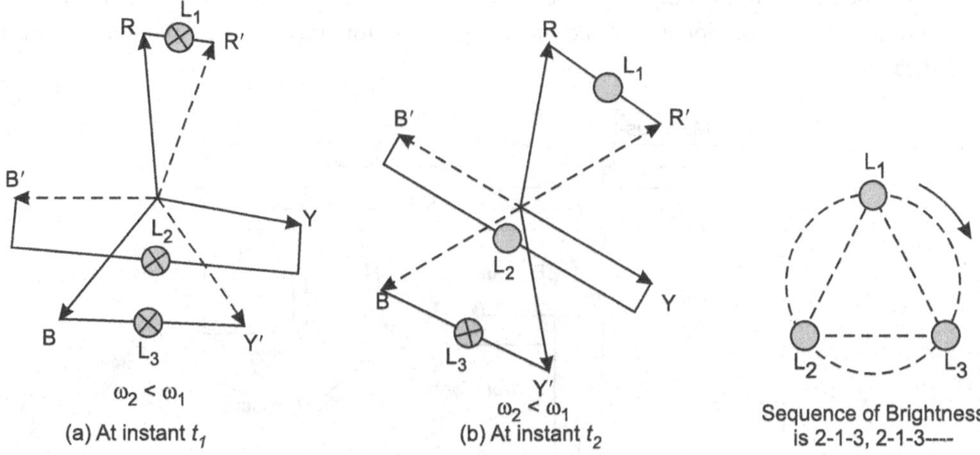

Fig. 7.11 Incoming alternator is running two slow

Fig. 7.12 Lamps placed on the vertices of an equilateral triangle

If the voltage of the incoming alternator is not equal to that of bus voltage and the other conditions are satisfied, all the lamps will flow with different brightness and will continue to attain the same brightness. *The ideal condition can be achieved by adjusting the excitation of the incoming alternator.*

Thus when the flickering frequency is minimised to zero, lamp L_1 is totally dark and L_2 and L_3 are equally bright the synchronising switch should be closed.

The lamp methods are only suitable for small low voltage alternators. For large capacity, high voltage alternators, a sychroscope is almost invariably used for synchronising.

7.7 Synchronising Three-phase Alternators using Synchroscope

High voltage alternators are generally synchronised with the help of *synchroscope.*

A *synchroscope* is an instrument which has been devised, with a rotating pointer and a fixed index. The pointer indicates whether the incoming machine is slow or fast. In fact, it shows the precise instant of synchronisation when the paralleling switch should be closed. Synchroscopes are manufactured in a variety of designs such as *polarised-vane type, moving iron type* and *crossed-coil type.*

In simplest form, a synchroscope utilises a split phase arrangement to give a rotating field which is excited by the incoming alternator.

When the two machines differ in frequencies, the reaction of the two fields produces rotation. A pointer connected to the rotor moves over the dial face and indicates the phase positions of the alternators. In this method of parallel operation two additional bus-bars, known as synchronising bus-bars are necessary by means of which synchroscope is connected through potential transformers (*P.T.*) to the two alternator to be synchronised and to the main bus-bar as shown in Fig. 7.13. The secondary windings of the potential transformers are connected to the synchronising bus-bars.

The incoming alternator is arrowed to run at its rated speed and its voltage is adjusted equal to the main bus-bars voltage. The two **plugs** are inserted in the **secondaries** of the potential transformers so that the synchroscope is acted upon by the voltages of both the incoming alternator and the main bus-bars. The greater the difference in frequency between bus-bars and the incoming alternators.

The greater will be the movement of the pointer. If the incoming alternator runs slower the pointer rotates in anti-clockwise direction and if the incoming alternator runs faster its pointer rotates in clockwise direction.

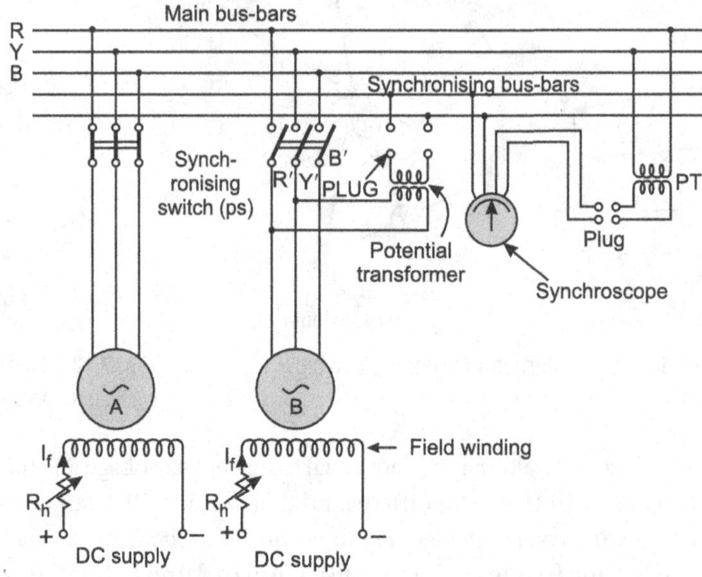

Fig. 7.13 Synchronising of three-phase alternators by using synchroscope

The speed of the incoming alternator is now adjusted until the pointer assumes a vertical position and thus indicates that the two alternators can be put into synchronism. At this instant 3 pole paralleling switch *P.S.* is closed.

7.8 Shifting of Load

After fulfilling all the necessary conditions of synchronising, the synchronising switch is closed and the alternator is said to be connected in parallel with the other alternators already connected to the bus-bars, by closing the synchronising switch. The incoming alternator is connected in parallel with the existing alternator but at this instant, the incoming alternator is not delivering any load (current) to the bus-bars. The load is shifted from the existing alternator to the incoming alternator by increasing the mechanical power input to the prime-mover of the incoming alternator and simultaneously reducing the mechanical power input to the prime-mover of existing alternator. In case of steam machines this may be readily done by admitting more steam (by opening the steam inlet valve) to the steam-turbine of incoming alternator and simultaneously closing the steam valve of existing alternator. Thus, any load can be shifted to incoming alternator from existing alternator. Generally it is shifted as per their respective ratings.

If the existing alternator is to be disconnected from the bus-bars the process continues till whole of the load is shifted to the incoming alternator as indicated by the ammeter and wattmeters in the circuits. Then the line circuit breaker (main switch) and field breaker (field switch) of the existing alternator are opened.

The most important point, to be emphasised, is that load cannot be shifted from one machine to the other by adjusting the excitation. Once the alternator is connected to the bus-bars, the change in excitation only changes the power factor of the alternator it does not affect the load sharing.

7.9 Load Sharing between Two Alternators

Consider two alternators *I* and *II* with identical speed/load characteristics are connected in parallel with a common terminal voltage *V* and delivering power to a load, as shown in Fig. 7.14.

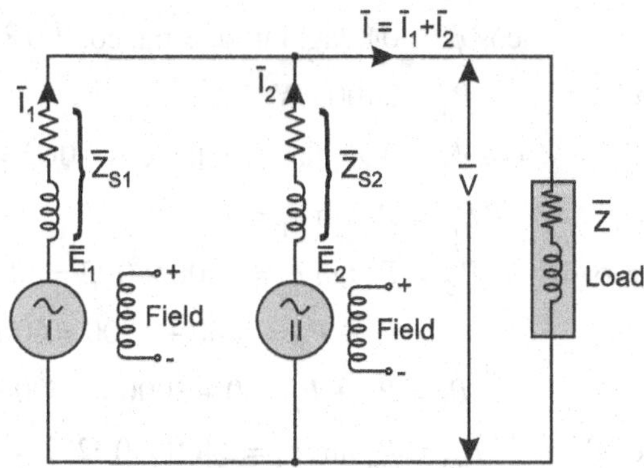

Fig. 7.14 Alternator I and II operating in parallel delivering load

Let $\overline{E}_1$ and $\overline{E}_2$ be the induced emf/phase of the two machines *I* and *II*, respectively, $\overline{Z}_{S1}$ and $\overline{Z}_{S2}$ be the impedance/phase of two machines *I* and *II*, respectively, whereas I_1 and I_2 be the current supplied by the two machines, respectively.

$$\overline{Z} = \text{load impedance/phase}$$

Now,
$$\overline{V} = \overline{E}_1 - \overline{I}_1\overline{Z}_{S1} = \overline{E}_2 - \overline{I}_2\overline{Z}_{S2}$$

$$\overline{I}_1 = \frac{\overline{E}_1 - \overline{V}}{\overline{Z}_{S1}} \text{ and } \overline{I}_2 = \frac{\overline{E}_2 - \overline{V}}{\overline{Z}_{S2}}$$

$$\overline{I} = \overline{I}_1 + \overline{I}_2 = \frac{\overline{E}_1 - \overline{V}}{\overline{Z}_{S1}} + \frac{\overline{E}_2 - \overline{V}}{\overline{Z}_{S2}}$$

Also,
$$V = \overline{I}_1\overline{Z} = (\overline{I}_1 + \overline{I}_2)\overline{Z}$$

Also,
$$\frac{\overline{V}}{\overline{Z}} = \frac{\overline{E}_1 - \overline{V}}{\overline{Z}_{S1}} + \frac{\overline{E}_2 - \overline{V}}{\overline{Z}_{S2}}$$

$$\overline{V}\left[\frac{1}{\overline{Z}} + \frac{1}{\overline{Z}_{S1}} + \frac{1}{\overline{Z}_{S2}}\right] = \frac{\overline{E}_1}{\overline{Z}_{S1}} + \frac{\overline{E}_2}{\overline{Z}_{S2}} \text{ or } \overline{V} = \frac{\dfrac{\overline{E}_1}{\overline{Z}_{S1}} + \dfrac{\overline{E}_2}{\overline{Z}_{S2}}}{\dfrac{1}{\overline{Z}} + \dfrac{1}{\overline{Z}_{S1}} + \dfrac{1}{\overline{Z}_{S2}}}$$

Circulating current on no-load, $\overline{I}_C = \dfrac{\overline{E}_1 - \overline{E}_2}{Z_{S1} + Z_{S2}}$

Example 7.1

A lighting load of 2000 kW and a motor load of 4000 kW at 0.8 pf lagging are supplied by two alternators operating in parallel. If one of them is delivering a load of 2400 kW at 0.95 pf lagging, what will the output and pf of the other alternator.

Solution:

Lighting load, $P_1 = 2000$ kW

Power factor, $\cos\phi_1 = 1; \tan\phi_1 = \tan\cos^{-1} 1 = 0$

Motor load, $P_2 = 4000$ kW

Power factor, $\cos\phi_2 = 0{\cdot}8$ lag; $\tan\phi_2 = \tan\cos^{-1} 0{\cdot}8 = 0{\cdot}75$

Power supplied by machine 'A', $P_A = 2400$ kW

Power factor of machine 'A' $\cos\phi_A = 0{\cdot}95; \tan\phi_A = \tan\cos^{-1} 0{\cdot}95 = 0{\cdot}3287$

Reactive of lightning load, $P_{r1} = P_1 \tan\phi_1 = 0$

Reactive of power of motor load, $P_{r2} = P_2 \tan\phi_2 = 4000 \times 0{\cdot}75 = 3000$ kVAR

Total load, $P = P_1 + P_2 = 2000 + 4000 = 6000$ kW

Total reactive power, $P_r = P_{r1} + P_{r2} = 0 + 3000 = 3000$ kVAR

Reactive power of machine 'A', $P_{rA} = P_A \tan\phi_A = 2400 \times 0{\cdot}3287 = 788{\cdot}84$ kVAR

Power supplied by machine 'B' $P_B = P - P_A = 6000 - 2400 = 3600$ kW

Reactive power of machine 'B', $P_{rB} = P_r - P_{rA} = 3000 - 788{\cdot}84 = 2211{\cdot}16$ kVAR

$$\therefore \quad \tan\phi_B = \frac{P_{rB}}{P_B} = \frac{2211 \cdot 15}{3600} = 0 \cdot 6142$$

Power factor of machine 'B', $\cos\phi_B = \cos\tan^{-1} 0{\cdot}6142 = 0{\cdot}852$ lag

Hence output of second machine is **3600 kW** *at p.f.* **0·852 lagging** (*Ans.*)

***Example* 7.2**

Two synchronous generators are running in parallel and supply a lighting load of 600 kW and a motor load of 707 kW at 0.707 pf lagging. If one of the machine is supplying 900 kW at 0.9 pf lagging, what load at what pf is supplied by the other machine?

Solution:

Lighting load, $P_1 = 600$ kW

Motor load, $P_2 = 707$ kW

Motor load *p.f.*, $\cos\phi_2 = 0{\cdot}707$ lagging

Power supplied by machine 'A', $P_A = 900$ kW

Power factor of machine 'A', $\cos\phi_A = 0{\cdot}9$ lagging

Power factor of lighting load, $\cos\phi_1 = 1$

$\therefore$ Reactive power of lighting load, $P_{r1} = 0$

$$\cos \phi_2 = 0{\cdot}707; \tan \phi_2 = \tan \cos^{-1} 0{\cdot}707 = 1$$

Reactive power of motor load, $P_{r2} = P_2 \tan \phi_2 = 707$ kVAR

Total load, $P = P_1 + P_2 = 600 + 707 = 1307$ kW

Total reactive power, $P_r = P_{r1} + P_{r2}$

$= 0 + 707 = 707$ kVAR

$$\cos \phi_A = 0{\cdot}9; \tan \phi_A = \tan \cos^{-1} 0{\cdot}9 = 0{\cdot}4843$$

Reactive power supplied by alternator '*A*',

$$P_{rA} = P_A \tan \phi_A = 900 \times 0{\cdot}4843 = 435{\cdot}89 \text{ kVAR}$$

Power supplied by alternator '*B*' $P_B = P - P_A = 1307 - 900 =$ **407 kW** (*Ans.*)

Reactive power supplied by alternator '*B*',

$$P_{rB} = P_r - P_{rA} = 707 - 435{\cdot}89 = 271{\cdot}11 \text{ kVAR}$$

$$\therefore \quad \tan \phi_2 = \frac{P_{rB}}{P_B} = \frac{271 \cdot 11}{407} = 0 \cdot 6661$$

Power factor of alternator '*B*', $\cos \phi_B = \cos \tan^{-1} 0{\cdot}6661 =$ **0·8322 lagging** (*Ans.*)

Example 7.3

Two similar 400V, three-phase, 50 Hz, alternators share equal power jointly delivered to a balance three-phase, 50 kW, 0·8 p.f. lagging load, If p.f. of one machine is 0·95 lagging, find the current supplied by the other machine.

Solution:

Total load, $P = 50$ kW

Load *p.f.*, $\cos \phi = 0{\cdot}8$ lagging

Terminal voltage, $V = 400$ V

$$\cos \phi = 0{\cdot}8; \tan \phi = \tan \cos^{-1} 0{\cdot}8 = 0{\cdot}75$$

Reactive power of the load, $P_r = P \tan \phi = 30 \times 0{\cdot}75 = 37{\cdot}5$ kVAR

Load on each machine, $P_A = P_B = \frac{50}{2} = 25$ kW

$$\cos \phi_A = 0{\cdot}95; \tan \phi = \tan \cos^{-1} 0{\cdot}95 = 0{\cdot}3287$$

Reactive power supplied by machine *A*,

$$P_{rA} = P_A \tan \phi_A = 25 \times 0{\cdot}3287 = 8{\cdot}217 \text{ kVAR}$$

Reactive power supplied by machine '*B*'.

$$P_{rB} = P_r - P_{rA} = 37{\cdot}5 - 8{\cdot}217 = 29{\cdot}283 \text{ kVAR}$$

$$\therefore \quad \tan \phi_B = \frac{P_{rB}}{P} = \frac{29 \cdot 283}{25} = 1 \cdot 1713$$

Power factor of machine '*B*', $\cos \phi_B = \cos \tan^{-1} 1{\cdot}1713 = 0{\cdot}6493$ lagging current supplied by machine '*B*', $I_B = \dfrac{P_B}{\sqrt{3}\, V \cos\phi_B} = \dfrac{25 \times 1000}{\sqrt{3} \times 400 \times 0 \cdot 6493}$ = **55·574 A** (*Ans.*)

Example 7.4

Two alternators A and B are operating in parallel, supplying single phase power at 2300 V to a load of 1200 kW whose power factor is unity. Alternator A supplies 200 A at 0·9 p.f. lagging. Determine for alternator B (i) power; (ii) current; (iii) power factor.

Solution:

Total power supplied, P = 1200 kW

Terminal voltage, V = 2300 V

Current supplied by alternator *A*, I_A = 200 A

Power factor of alternator *A*, $\cos\phi_A$ = 0·9 lagging

Power supplied by alternator *A*, $P_A = I_A V \cos\phi = 200 \times 2300 \times 0{\cdot}9$ W = 414 kW

$\cos\phi_A = 0{\cdot}9;\ \tan\phi_A = \tan\cos^{-1} 0{\cdot}9 = 0{\cdot}4843$

Reactive power supplied by alternator *A*,

$$P_{rA} = P_A \tan\phi_A = 414 \times 0{\cdot}4843 = 200{\cdot}5 \text{ kVAR}$$

Total reactive power, $P_r = 0$ (∵ *p.f.* is unity)

Reactive power supplied by alternator *B*,

$$P_{rB} = P_r - P_{rA} = 0 - 200{\cdot}5 = -\,200{\cdot}5 \text{ kVAR}$$

Active power supplied by alternator *B*, $P_B = P - P_A$ = 1200 – 414 = **786 kW** (*Ans.*)

$$\tan\phi_B = \frac{P_{rB}}{P_B} = \frac{-200 \cdot 5}{786} = -0 \cdot 225$$

Power factor of alternator *B*, $\cos\phi_B = \cos\tan^{-1}(-\,0{\cdot}255)$ = **0·969 leading** (*Ans.*)

Current supplied by alternator *B*, $I_B = \dfrac{P_B}{V \cos\phi_B} = \dfrac{786 \times 10^3}{2300 \times 0 \cdot 969}$ = **352·67 A** (*Ans.*)

Example 7.5

Two three-phase, 6600 V, star-connected alternators working in parallel are supplying the following loads:

(i) 1500 kW at p.f. 0·75 lagging;
(ii) 1000 kW at p.f. 0·8 lagging;
(iii) 800 kW at p.f. 0·9 lagging;
(iv) 500 kW at unity power factor.

If one of machine is delivering a load of 1800 kW at 0·85 p.f. lagging, what will be the output, armature current and p.f. of the other machine.

Solution:

Active power of load (*i*), P_1 = 1500 kW

p.f. $\cos \phi_1 = 0{\cdot}75$ lagging; $\tan \phi_1 = \tan \cos^{-1} 0{\cdot}75 = 0{\cdot}8819$

Reactive power of load (*i*), $P_{r1} = P_1 \tan \phi = 1500 \times 0{\cdot}8819 = 1322{\cdot}85$ kVAR

Active power of load (*ii*), $P_2 = 1000$ kW

p.f. $\cos \phi_2 = 0{\cdot}8$ lagging; $\tan \phi_2 = \tan \cos^{-1} 0{\cdot}8 = 0{\cdot}75$

Reactive power of load (*ii*), $P_{r2} = P_2 \tan \phi_2 = 1000 \times 0{\cdot}75 = 750$ kVAR.

Active power of load (*iii*), $P_3 = 800$ kW

p.f. $\cos \phi_3 = 0{\cdot}9$ lagging; $\tan \phi_3 = \tan \cos^{-1} 0{\cdot}9 = 00{\cdot}4843$

Reactive power of load (*iii*), $P_{r3} = P_3 \tan \phi_3 = 800 \times 0{\cdot}4843 = 387{\cdot}44$ kVAR

Active power of load (*iv*), $P_4 = 500$ kW

p.f. $\cos \phi_4 = 1$; $\tan \phi_4 = \tan \cos^{-1} 1 = 0$

Reactive power of load (*v*). $P_{r4} = P_4 \tan \phi_4 = 0$

Total active power, $P = P_1 + P_2 + P_3 + P_4$

$= 1500 + 1000 + 800 + 500 = 3800$ kW

Total active power, $P_r = P_{r1} + P_{r2} + P_{r3} + P_{r4}$

$= 1322{\cdot}85 + 750 + 387{\cdot}44 + 0 = 2460{\cdot}29$ kVAR

Active power of machine *A*, $P_A = 1800$ kW

p.f. $\cos \phi_A = 0{\cdot}85$ lagging; $\tan \phi_A = \tan \cos^{-1} 0{\cdot}85 = 0{\cdot}6197$

Reactive power of machine *A*, $P_{rA} = P_A \tan \phi_A = 1800 \times 0{\cdot}6197 = 1115{\cdot}46$ kVAR

Active power of machine *B*, $P_B = P - P_A = 3800 - 1800 = 2000$ kW (*Ans.*)

Reactive power of machine *B*, $P_{rB} = P_r - P_{rA}$

$= 2460{\cdot}29 - 1115{\cdot}46 = 1344{\cdot}83$ kVAR

$$\tan \phi_B = \frac{P_{rB}}{P_B} = \frac{1344 \cdot 83}{2000} = 0 \cdot 6724$$

Power factor of machine, *B*, $\cos \phi_B = \cos \tan^{-1} 0{\cdot}6724 =$ **0·8298 lagging** (*Ans.*)

$$\text{Armature current, } I_B = \frac{P_B}{\sqrt{3} V_B \cos\phi_B} = \frac{2000 \times 10^3}{\sqrt{3} \times 6600 \times 0 \cdot 8298}$$

= **210·84 A** (*Ans.*)

Example 7.6

Two alternators working in parallel supply the following loads:

(i) *Lighting load of 600 kW;*

(ii) *Inductive load of 800 kW at 0·9 p.f. lagging;*

(iii) *Capacitive load of 800 kW at 0·8 p.f. leading.*

One alternator is supplying 1000 kW at 0·85 p.f. lagging. Calculate the kW output and p.f. of the other alternator.

Solution:

S. No.	**Load**	**Power (P) in kW**	**p.f. cos ϕ**	**tan ϕ**	**Reactive Power $P_r = P \tan \phi$**
(*i*)	*Lighting load*	600	1	0	0
(*ii*)	*Inductive load*	800	0·9 lagging	0·4843	387·44 kVAR
(*iii*)	*Capacitive load*	800	0·8 leading	0·75	– 600 kVAR

$$\text{Total active power, } P = P_1 + P_2 + P_3 = 600 + 800 + 800 = 2200 \text{ kW}$$

$$\text{Total reactive power, } P_r = P_{r1} + P_{r2} + P_{r3} = 0 + 387{\cdot}44 - 600 = -212{\cdot}56 \text{ kVAR}$$

Active power supplied by alternator A, $P_A = 1000$ kW

$$p.f. \cos \phi_A = 0{\cdot}85 \text{ lagging; } \tan \phi_A = \tan \cos^{-1} 0{\cdot}85 = 0{\cdot}61974$$

$$\text{Reactive power alternator } A, P_{rA} = P_A \tan \phi_A = 1000 \times 0{\cdot}61974$$

$$\text{Reactive power of alternator } A, P_{rA} = P_A \tan \phi_A = 1000 \times 0{\cdot}61974 = 619{\cdot}74 \text{ kVAR}$$

Active power supplied by alternator B,

$$P_B = P - P_A = 2200 - 1000 = \mathbf{200\ kW}\ (Ans.)$$

Reactive power supplied by alternator B,

$$P_{rB} = P_r - P_{rA} = -212{\cdot}56 - 619{\cdot}74 = -832{\cdot}3 \text{ kVAR}$$

$$\tan \phi_B = \frac{P_{rB}}{P_B} = -\frac{832 \cdot 3}{1200} = -0 \cdot 6932$$

Power factor of alternator B, $\cos \phi_B = \cos \tan^{-1} (-0{\cdot}6936) =$ **0·8217 leading** (*Ans.*)

Example 7.7

Two single-phase alternators are connected in parallel and supplying current to a load at a terminal voltage of 11000 ∠0°V. Alternator-1 has an induced emf of 13000 ∠20° V and a reactance of 3ohm whereas alternator-II has an emf of 13500 ∠15° V and a reactance of 4 ohm. What will be the current supplied by each alternator.

Solution:

Here, $\overline{V} = 11000 \angle 0°$ V; $\overline{E}_1 = 13000 \angle 20°$ V

$$\overline{E}_2 = 13500\angle 15° \text{ V}; X_{S1} = 3\Omega \text{ and } X_{S2} = 4\ \Omega$$

Current supplied by alternator-I

$$\overline{I}_1 = \frac{\overline{E_1} - \overline{V}}{\overline{X}_{S1}} = \frac{13000\angle 20° - 11000\angle 0°}{3\angle 90°}$$

$$= \frac{12216 + j\,4446 - 11000}{3\angle 90°} = \frac{1216 + j\,4446}{3\angle 90°}$$

$$= \frac{4609 \angle 77.7°}{3\angle 90°} = \mathbf{1536 \angle{-12.3°}\ A}\ (Ans.)$$

Current supplied by alternator-II

$$\bar{I}_2 = \frac{\bar{E}_2 - \bar{V}}{\bar{X}_{S2}} = \frac{13500\angle 15^\circ - 11000\angle 0^\circ}{4\angle 90^\circ}$$

$$= \frac{13500(\cos 15^\circ + j\sin 15^\circ) - 11000(\cos 0^\circ - j\sin 0^\circ)}{4\angle 90^\circ}$$

$$= \frac{13040 + j3494 - 11000}{4\angle 90^\circ}$$

$$= \frac{510 + j3494}{4\angle 90^\circ} \frac{3531\angle 81.69^\circ}{4\angle 90^\circ}$$

$$= \mathbf{882.75 \angle{-8.08^\circ}\ A}\ (Ans.)$$

Example 7.8

Two single phase alternators having induced emf E_1 and E_2 and impedance Z_1 and Z_2 are connected in parallel and supplying a load of impedance Z.

(i) *Represent the terminal voltage in terms of alternators emfs E_1 and E_2 and admittances Y, Y_1 and Y_2.*

(ii) *If* $\bar{E}_1 = 220$ V, $\bar{E}_2 = 230$ V, $\bar{Z} = 3 + j4$ and $\bar{Z}_1 = \bar{Z}_2 = (0.2 + j0.8)$ ohm, *determine the terminal voltage, circulating current at no-load and power delivered by each alternator in kW.*

Solution:

We know, terminal voltage in given by the relation

$$\bar{V} = \frac{\dfrac{\bar{E}_1}{\bar{Z}_{S1}} + \dfrac{\bar{E}_2}{\bar{Z}_{S2}}}{\dfrac{1}{\bar{Z}} + \dfrac{1}{\bar{Z}_{S1}} + \dfrac{1}{\bar{Z}_{S2}}}$$

We also know that admittance, $Y = \frac{1}{\bar{Z}}$

$$\therefore \quad \frac{1}{\bar{Z}_{S1}} = \bar{Y}_{S1};\ \frac{1}{\bar{Z}_{S2}} = \bar{Y}_{S2} \text{ and } \frac{1}{\bar{Z}} = \bar{Y}$$

$$\therefore \text{Terminal voltage, } \bar{V} = \frac{\bar{E}_1\bar{Y}_{S1} + \bar{E}_2\bar{Y}_{S2}}{\bar{Y} + \bar{Y}_{S1} + \bar{Y}_{S2}} \ (Ans.)$$

Give that $\bar{Z} = (3 + j\,4)$ ohm; $\bar{Z}_{S1} = \bar{Z}_{S2} = (0.2 + j\,0.8)$ ohm; $\bar{E}_1 = 220$ V; $\bar{E}_2 = 230$ V

$$\bar{Y} = \frac{1}{\bar{Z}} = \frac{1}{3 + j4} \times \frac{3 - j4}{3 - j4} = \frac{3 - j4}{3^2 + 4^2} = (0.12 - j0.168) \text{ mho}$$

$$\bar{Y}_{S1} = \bar{Y}_{S2} = \frac{1}{\bar{Z}_{S1}} = \frac{1}{0.2 + j0.8}$$

$$= \frac{1}{0.2 + j0.8} \times \frac{0.2 - j0.8}{0.2 - j0.8} = (0.2425 - j0.97) \text{ mho}$$

$$\bar{V} = \frac{\bar{E}_1\bar{Y}_{S1} + \bar{E}_2\bar{Y}_{S2}}{\bar{Y} + \bar{Y}_{S1} + \bar{Y}_{S2}} = \frac{(220 + 230)(0.2425 - j0.97)}{(0.12 - j0.168) + 2(0.2425 - j0.97)}$$

$$= \frac{109.125 - j436.5}{(0.605 - j2.108} = \frac{450\angle -76^\circ}{2.193\angle -74^\circ} = 205.2\angle -2^\circ \text{ volt } (Ans)$$

$$= (205.075 - j7.16) \text{ V}$$

Circulating current at no-load

$$\bar{I}_C = \frac{\bar{E}_2 - \bar{E}_1}{\bar{Z}_{S1} + \bar{Z}_{S2}} = \frac{230 - 220}{2(0.2 + j0.8)} = \frac{10}{0.4 + j1.6} = \frac{10}{1.65\angle 76^\circ}$$

$$= \mathbf{6.06\angle{-76}^\circ A} \; (Ans)$$

Current supplied to the load by alternator I

$$\bar{I}_1 = \frac{\bar{E}_1 - \bar{V}}{\bar{Z}_{S1}} = (\bar{E}_1 - \bar{V})\bar{Y}_{S1} = (220 - 205.075 + j7.16)(0.2425 - j0.97)$$

$$= (14.925 + j\,7.16)\,(0.2425 - j\,0.97) = (16.55\angle{+25.63}^\circ)\,(1\angle{-76}^\circ)$$

$$= 16.55\angle -50.37^\circ \text{A}$$

Current supplied to the load by alternator II

$$\bar{I}_2 = \frac{\bar{E}_2 - \bar{V}}{\bar{Z}_{S2}} = (\bar{E}_2 - \bar{V})\bar{Y}_{S2} = (230 - 205.075 + j7.16)(0.2425 - j0.97)$$

$$= (24.925 + j\,7.16)\,(0.2425 - j\,0.97) = (25.93\angle{+16}^\circ)\,(1\angle{-76}^\circ)$$

$$= 25.93\angle{-50}^\circ \text{ A}$$

Power delivered by alternator I

$$P_1 = \bar{V}\bar{I}_1 = (205.2\angle -2^\circ)\,(16.55\angle{-50.37}^\circ) = 3396\angle -52.37$$

$$= (2073.5 - j\,2.69) \text{ VA} = 2073.5\text{W} = \mathbf{2.0735\ kW} \; (Ans)$$

Power delivered by alternator II

$$P_2 = \bar{V}\bar{I}_2 = (205.2\angle -2^\circ)\,(25.93\angle -50^\circ) = 5321\angle -52^\circ$$

$$= (3276 - j\,4.193) \text{ VA} = 3276 \text{ W} = \mathbf{3.276\ W} \; (Ans)$$

Example 7.9

A three-phase,10 MVA, 11 kV, 50 Hz alternator hawing 10% reactance is connected to a substation by a line having a reactance of 3.8 Ω. Another three-phase 20 MVA, 11 kV, 50 Hz alternator having 15% reactance is connected to the same substation through a line having a reactance of 4.1 Ω. The substation is supplying a load of impedance (50 + j40)Ω.

If the generated emf of alternator I and II is 10 kV and 12 kV, respectively, and the emf of alternator I leads the emf of alternator II by 10° electrical, what will be the current delivered by each alternator.

Solution:

The circuit shown in Fig. 7.14.

$$\text{Phase voltage, } E = \frac{11000}{\sqrt{3}} = 6351 \text{ V}$$

Rated full-load current of alternator-I,

$$I_1 = \frac{10 \times 10^6}{\sqrt{3} \times 11 \times 10^3} = 525 \text{ A}$$

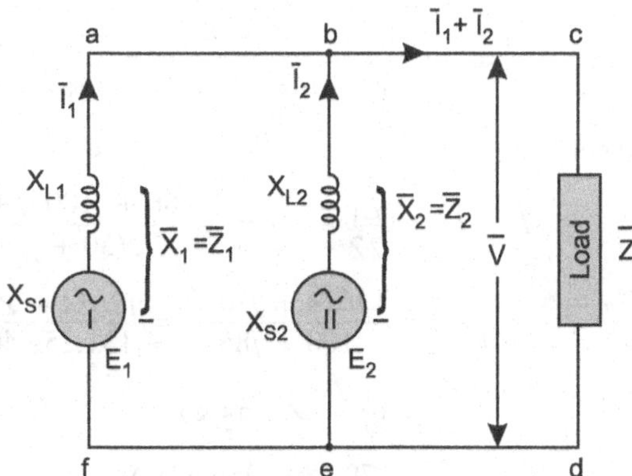

Fig. 7.14 Circuit diagram

Rated full-load current of alternator-II

$$I_2 = \frac{20 \times 10^6}{\sqrt{3} \times 11 \times 10^3} = 1050 \text{ A}$$

Now, $I_1 X_{S1} = 10\%$ of 6351 or $X_{S1} = \frac{10}{100} \times \frac{6351}{525} = 1.2 \ \Omega$

$I_2 X_{S2} = 15\%$ of 6351 or $X_{S2} = \frac{15}{100} \times \frac{6351}{1050} = 0.9 \ \Omega$

Effective impedance/reactance of alternator-*I* and line,

$$Z_1 = X_1 = X_{L1} + X_{S1} = 3.8 + 1.2 = 5 \ \Omega; \ \bar{Z}_1 = j5 \text{ ohm}$$

Effective impedance reactance of alternator-*II* and line,

$$Z_2 = X_2 = X_{L2} + X_{S2} = 4.1 + 0.9 = 5 \ \Omega; \ \bar{Z}_2 = j5 \ ohm$$

Load impedance, $Z = (50 + j\,40)$ ohm $= 64.03 \angle 38.65°$

Considering emf of alternator-*II* i.e., E_2 as reference vector,

$$\bar{E}_2 = \frac{12000}{\sqrt{3}} \angle 0° = 6928 \angle 0° = (6928 + j0) \text{ V}$$

$$\bar{E}_1 = \frac{10000}{\sqrt{3}} \angle 10° = 5774 \angle 10° = (5686 + j1003) \text{ V}$$

Applying Kirchhoff's second law to the closed circuit *a b c d e f a* and *b c d e b*, respectively, we get,

$$\bar{I}_1 \bar{Z}_1 + (\bar{I}_1 + \bar{I}_2)Z = \bar{E} \quad \text{...(i)}$$

$$\bar{I}_2 \bar{Z}_2 + (\bar{I}_1 + \bar{I}_2)Z = \bar{E}_2 \quad \text{...(ii)}$$

Subtracting eqn. (*ii*) from (*i*), we get,

$$\bar{I}_1 - \bar{I}_2 = \frac{\bar{E}_1 - \bar{E}_2}{\bar{Z}_1} \text{(since } \bar{Z}_1 = \bar{Z}_2 \text{ or } X_1 = X_2 = 5\,\Omega)$$

$$= \frac{(5686+j1003) - (6928+j0)}{j5} = \frac{-1242 + j1003}{j5}$$

$$= (200.6 + j248.4) \text{ A} \qquad \ldots\text{(iii)}$$

Adding eqn. (*i*) and (*ii*), we get,

$$\bar{I}_1 + \bar{I}_2 = \frac{\bar{E}_1 + \bar{E}_2}{2\bar{Z} + \bar{Z}_1} = \frac{(5686 + j1003) + (6928 + j0)}{2(50 + j40) + j\,5}$$

$$= \frac{12614 + j1003}{100 + j85} = \frac{12654 + \angle 4.54^\circ}{131.25\angle 40.36^\circ}$$

$$= 96.41\angle - 35.82^\circ$$

$$= (78.18 - j56.42) \text{ A} \qquad \ldots\text{(iv)}$$

Adding eqn.(*iii*) and (*iv*), we get

$$\bar{I}_1 = (139.39 + j98) \text{ A} = 170.4\angle 35^\circ = \mathbf{170.4\ A}\ (\textit{Ans.})$$

$$\bar{I}_2 = (65.7 + j152.4) \text{ A} = 166\angle 66.67^\circ = \mathbf{166\ A}\ (\textit{Ans.})$$

Section Practice Problems

Numerical Problems

1. A lightning load of 2500 kW and a motor load of 5000 kW at a p.f. of 0·707 are supplied by two alternators running in parallel. If one machine is supplying a load of 4000 kW at a p.f. of 0·8 lagging, what is the kW output and p.f. of the other machine? (**Ans.** *3500 kW, 0·8682 lagging*)

2. The following loads are supplied by two alternators running in parallel:
   (i) Lighting load of 400 kW; (ii) 800 kW at p.f. of 0·8 lagging;
   (iii) 800 kW at p.f. of 0·8 leading; (iv) 200 kW at p.f. of 0·9 lagging;

   If one alternator is supplying a load of 1200 kW at 0·9 p.f. lagging, calculate the kW output and p.f. of the other alternator. (**Ans.** *1000 kW, 0·9 leading*)

3. Two single-phase alternators are connected in parallel and supplying current to a load at a terminal voltage of 11000 $\angle 0°$V. Alternator-1 has an induced emf of 13000 $\angle 20°$ V and a reactance of 3 ohm whereas alternator-II has an emf of 13500 $\angle 15°$ V and a reactance of 4 ohm. What will be the current supplied by each alternator. (**Ans.** *2700 $\angle$–21.8°; 2500$\angle$– 0°* A)

Short Answer Type Questions

Q.1. Why are synchronous generators put in parallel?

Ans. Synchronous generators are put in parallel for high efficiency of operation, better reliability, convenience in repair and maintenance, extension of power plant capacity, etc.

Q.2. What conditions are required to be fulfilled for parallel operation of alternators?

Ans. Terminal voltage, frequency and the phase (or phase sequence) of the incoming alternator must be the same as that of the existing alternator or the bus-bars.

Q.3. For synchronising of three-phase alternators; which method is preferred, dark lamp method or one-dark and two-bright lamp method?

Ans. One-dark and two-bright lamp method is preferred.

Q.4. How load in shifted to the incoming alternator?

Ans. After synchronising the incoming alternator to the bus-bars of the existing alternator, the input power to the incoming alternator is increased whereas to the existing alternator is decreased gradually.

7.10 Two Alternators Operating in Parallel

When the two alternators are operating in parallel as shown in Fig. 7.15 under stable equilibrium, they share the total load in proportion to their ratings or equally if both are identical. Any tendency on the part of one to pull out of synchronism because of change in load or excitation is immediately counteracted by the production of a synchronising current (hence synchronising torque) which pulls it back to synchronism.

Let us study the effect of change in input power change in excitation or change in speed of the alternators.

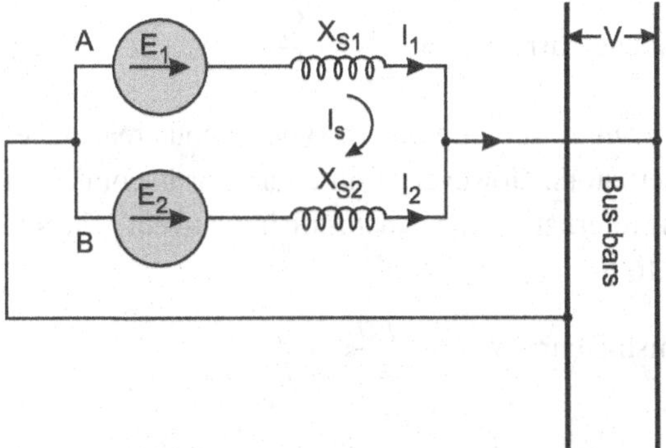

Fig. 7.15 Two alternators operating in parallel

7.11 Synchronising Current, Power and Torque

When the two identical alternators are in exact synchronism, the *emfs* of two alternators E_1 and E_2 are equal and in exact phase opposition, as shown in Fig. 7.16 (*a*), as far as the local circuit of the alternators in concerned. No circulating current flows through the local circuit.

If the induced *emfs.* of the two alternators are equal but due to some reason the *emf* of the second alternator E_2 fall back by an angle θ (electrical radians) as shown in Fig. 7.16 (*b*), then a resultant *emf* E_r acts in the local circuit and causes flow of current known as *synchronising current.*

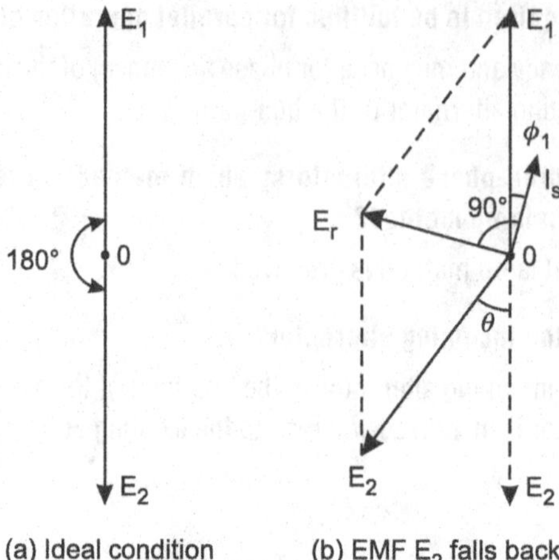

(a) Ideal condition (b) EMF E_2 falls back

Fig. 7.16 Phasor diagram of two alternators operating in parallel

$$\text{Resultant } emf,\ E_r = 2E\cos\left[\frac{180-\theta}{2}\right] \qquad (\because E_1 = E_2 = E)$$

$$= 2E\cos\left(90-\frac{\theta}{2}\right) = 2E\sin\frac{\theta}{2}$$

$$= 2E\frac{\theta}{2} = E\cdot\theta \qquad \text{(since } \theta \text{ is very small)}$$

$$\text{Synchronising current, } I_s = \frac{E_r}{Z_s} = \frac{E_r}{X_s}$$

Since resistance is very small as compared to synchronous reactance. Here Z_S is the combined impedance of the two alternators. However, if an alternator is connected to infinite bus-bars, it is the impedance of only one alternator. The synchronising current I_s lags behind the resultant *emf* E_r by 90° as shown in Fig. 7.16 (*b*).

$$\therefore \qquad \text{Synchronising current, } I_s = \frac{E\theta}{X_s}$$

Synchronising Power

In the above case, because of synchronising current, first alternator will supply extra power known as *synchronising power.*

$$\text{Synchronising power, } P_s = E_1 I_s \cos\phi_1 \qquad \text{(per phase)}$$

$$= E_1 I_s \qquad (\because \phi_1 \text{ is very small})$$

$$= E I_s \qquad \text{(since } E_1 = E)$$

$$= E\frac{E\theta}{X_s} = \frac{E^2\theta}{X_s}$$

$$\text{Synchronising power for three-phase} = 3P_s = 3\frac{E^2\theta}{X_s} \text{ watt}$$

Synchronising Torque

Let T_s be the synchronising torque in *Nm* and N_s be the synchronous speed in *rpm*, then total synchronising power;

$$3P_s = T_s \times \frac{2\pi N_s}{60}$$

$\therefore$ Synchronising torque, $T_s = \dfrac{3P_s \times 60}{2\pi N_s}$ Nm

7.12 Effect of Change in Input Power of One of the Alternators

When two identical alternators are operating in parallel, they supply equal current to the load at the same power factor. Both the alternators have same reactance and induced *emfs*. Under such conditions, all the quantities are shown vectorially in Fig. 7.17 for the local circuit of the two alternators.

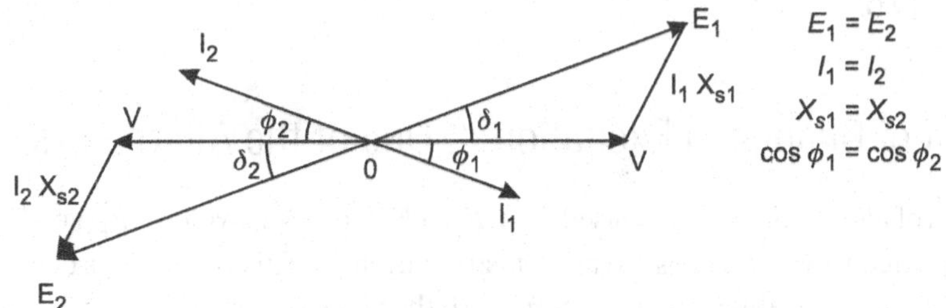

Fig. 7.17 Phasor diagram with ideal conditions,

If input power (steam supply or water supply or fuel supply) to alternator 'A' is increased, it will increase the torque angle δ_1 to δ_1' (since input power is proportional to δ neglecting losses). Thus the *emf* E_1 of the alternator 'A' attains a new position as shown in Fig. 7.18 and a resultant *emf* E_r appears across the local circuit which circulates synchronising current I_s. This synchronising current lags behind the resultant *emf* E_r by 90°. The current-delivered by alternator 'A' increases to I_1' and the current delivered by alternator 'B'' decreases to I_2'. The power factor of the two alternators also changes to cos ϕ_1' and cos ϕ_2', respectively.

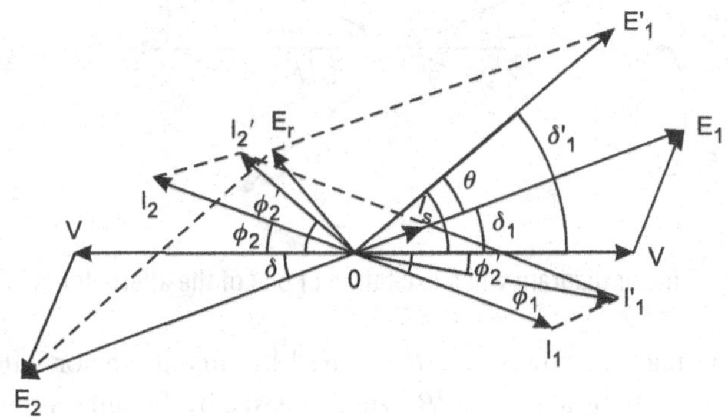

Fig. 7.18 Phasor diagram when input to one of the alternator is increased

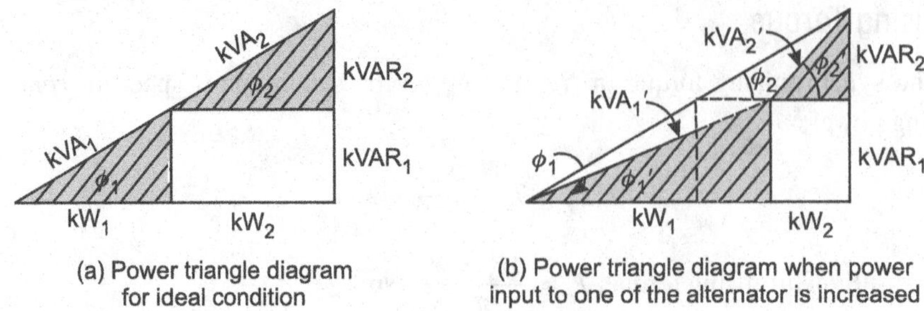

(a) Power triangle diagram for ideal condition

(b) Power triangle diagram when power input to one of the alternator is increased

Fig. 7.19 Power triangle representation

Thus, the increase in power input to prime-mover of alternator 'A', increases the kW output of alternator 'A' and decreases that of alternator 'B' by the same amount. If this process is continued, a point may reach where output power of alternator 'A' becomes more than the total load supplied, then the alternator 'B' instead of supplying load will start drawing power from alternator 'A'. Hence, the alternator A will start working as a motor. The load sharing diagram of the two alternators is shown in Fig. 7.19 (b).

7.13 Effect of Change in Excitation of One of the Alternators

If the excitation of alternator 'A' is increased from E_1 to E_1', a resultant voltage E_r (i.e., $E_1' - E_2$) appears across the local circuit and circulates a synchronising current I_s. This current lags behind the resultant voltage E_r by 90° and acts almost in quadrature with the voltage vector. Therefore, it does not affect the active component of current delivered by the two alternators. The resultant currents delivered by the two alternators are I_1' and I_2' as shown in Fig. 7.20. The power factor of the two alternators also changes in cos ϕ_1' and cos ϕ_2', respectively.

Thus, the increase in excitation of alternator 'A' does not affect the kW loading of the two alternators but it affects the power factor and reactive power of the two alternators. When the excitation of alternator

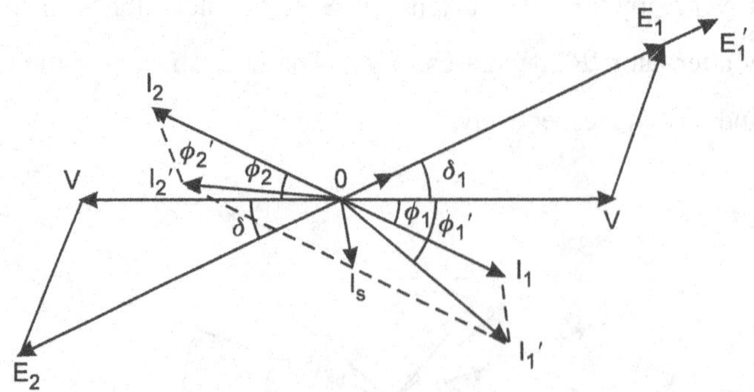

Fig. 7.20 Phasor diagram when excitation of one of the alternator is increased.

'A' is increased, the reactive power ($kVAR_1$) shared by this alternator will increase while the reactive power ($kVAR_2$) shared by alternator 'B' will decreased by the same amount. The load sharing diagram of the two alternators is shown in Fig. 7.21.

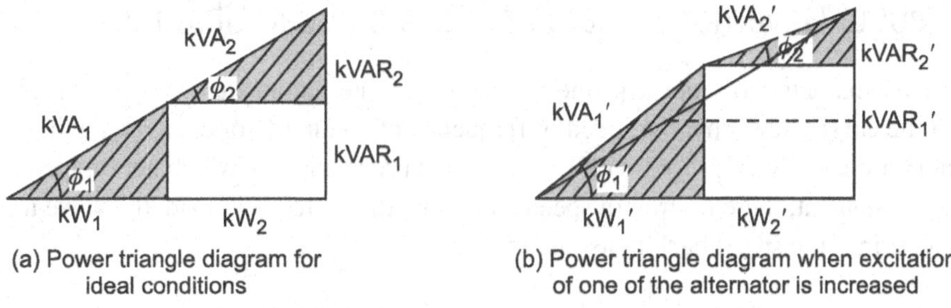

(a) Power triangle diagram for ideal conditions

(b) Power triangle diagram when excitation of one of the alternator is increased

Fig. 7.21 Power triangle representation

7.14 Effect of Reactance

When two alternators are running in parallel, the *emf* of one alternator acts opposite to the *emf* of the other alternator with respect to their local circuit. As such one machine runs as a synchronous motor with respect to the other. Due to any reason, if the input supply to one of the machine is cut-off, it is quite evident that it would draw a wattful current from the other machine to run as a motor.

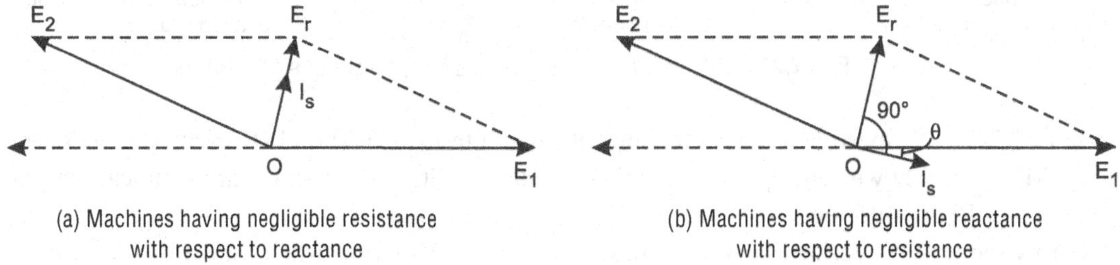

(a) Machines having negligible resistance with respect to reactance

(b) Machines having negligible reactance with respect to resistance

Fig. 7.22 Pharos diagram for the internal circuit of two synchronous machine operating in parallel

Considering that the two machines are having only resistance and negligible reactance. Their *emfs* E_1 and E_2 are practically in phase opposition, as shown in Fig. 7.22(a) and their resultant emf E_r is in quadrature with E_1 and E_2. The synchronising current I_S [$I_S = E_r/(R_1 + R_2)$] will be in phase with E_r but in quadrature with E_1 and E_2. Hence, no real power is transferred from one machine to the other since the synchronising current is the wattless current (needs help for synchronisation).

Now, considering that the two machines are having large reactance as compared to resistance. The synchronising current I_S will lag behind the resultant voltage E_r by 90° as shown in Fig. 7.22 (b), this synchronising current I_S will develop a real power $E_1 I_S$ cos q which will be delivered by one machine to the other, Hence, one machine works as a generator and the other works as a motor. This causes the machine to restore synchronism.

Conclusion

From the above discussions, it is concluded that machines must have high reactance for satisfactory parallel operation of synchronous generators.

7.15 Effect of Governors' Characteristics on Load Sharing

The speed-load characteristics of the prime-movers of the alternators are drooping curves as shown in Fig. 7.23 the curve shows that the speed or frequency of an alternator decreases with the load.

Governors are usually employed with the generator prime-movers which act on the centrifugal principle. The moment, there in drop in speed caused by the increase in load, the valve mechanism operates and brings the speed back to its rated value.

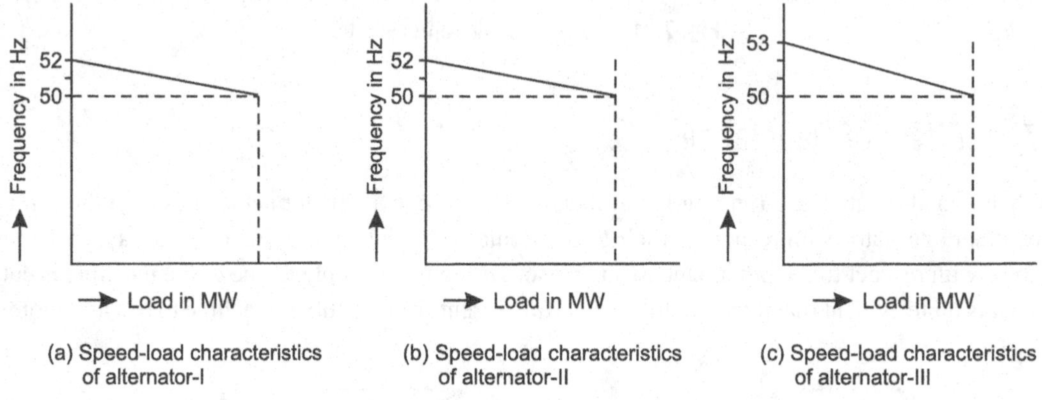

Fig. 7.23 Speed-load characteristics of synchronous generators

In Fig. 7.23 The speed-load characteristics of three alternator I, II and III each of same rating (say 5 MVA), are shown. The speed-load characteristics of alternator I and II are identical but that of alternator III are different.

Let us see how speed-load characteristics affect the load sharing when they are connected is parallel. Consider that alternator I and II are operating in parallel and both of them are supplying their rated full load of 4MW (say the pf of the load is 0.8 lagging). If the load on the system decreases (say from 8MW to 6MW), as shown in Fig. 7.23 (a) the frequency of both the alternators increases momentarily but equal load (3MW each) is shared by them.

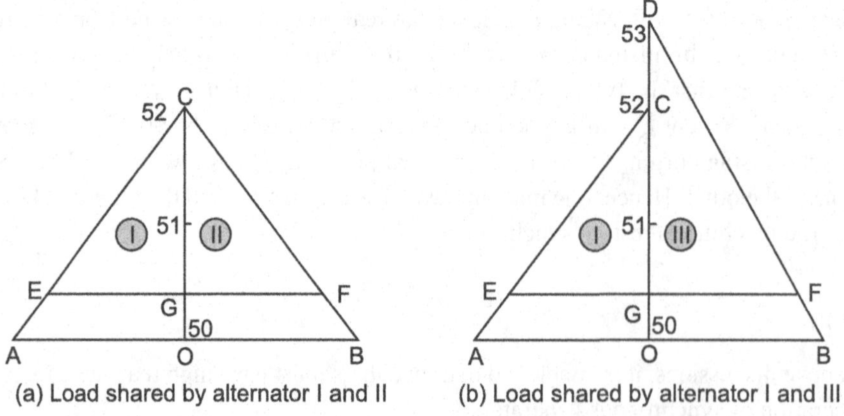

Fig. 7.24 Load shared by two alternators running in parallel.

Now, consider that alternator I and III are operating in parallel and both of them are supplying their rated full load of 4MW (say the pf of the load is 0.8 lagging). If the load on the system decreases (say from 8 MW to 6MW), as shown in Fig. 7.24(b), the frequency of both the alternators increases momentarily but they are not sharing the load equally, alternator I is sharing less load than the alternator III as shown in Fig. 7.24 (b).

Hence, it is concluded that the speed-load characteristics play their role. Therefore, the governess placed on the alternators must be very sensitive so that they bring the alternators to operate at their rated frequency and share the load as per their rating.

7.16 Hunting

When a synchronous machine is loaded, the rotor poles slightly fall back in position with respect to the stator field (poles) by an angle δ known as power angle or torque angle or retarding angle. As the load is gradually increased, this angle δ also increases gradually so as to produce more torque for coping with the increased load. If the load is suddenly thrown off, angle δ decreases suddenly and the rotor poles are pulled into almost exact opposition to the stator poles, but due to inertia of rotor and rotor poles travel too far. They are then pulled back again, and so on, thus oscillations are set up around the equilibrium position, corresponding to new load.

The oscillation of the rotor about its equilibrium position is known as **hunting**.

Hunting effect produces heavy mechanical tresses in the machine parts particularly on the bearings. Therefore effort are made to prevent hunting

The hunting (oscillations) can be prevented by providing damper winding on the rotor pole faces in case of salient pole alternators. This damper winding consists of number of copper or aluminium bars embedded into the slots provided on the outer periphery of the pole shoes and then short circuited by end rings. When hunting takes place, there is relative motion of the rotor with respect to the stator field, which sets up eddy currents in this winding which flow in such a way that it suppresses the oscillations. Hunting can also be reduced by placing heavy flywheels on the shaft and putting dash-pots on the engine governors.

However, non-salient pole of alternators used in steam-fed power plants do not have the tendency to hunt.

The hunting also occurs when the machine operates as a motor. In this case also because of sudden change in load oscillations are set up in the rotor called hunting; which can be prevented by providing damper winding on the rotor.

Example 7.10

Determine the synchronising power and synchronising torque per mechanical degree of rotor displacement at no-load of a 2.5 MVA, three-phase, six-pole synchronous generator connected to 6000V, 50Hz bus-bars. The synchronous reactance of the machine is 5 ohm per phase.

Solution:

Here, rating of alternator = 2.5MVA; V_L = 6000V; P = 6; X_S = 5 Ω; f = 50Hz

$$\text{Phase voltage, } E = V = \frac{V_L}{\sqrt{3}} = \frac{6000}{\sqrt{3}} = 3464 \text{ V}$$

Rotor displacement in electrical degree = Mech. angle $\times \frac{P}{2} = 1 \times \frac{6}{2} = 3°$

Electrical angle in radians, $\theta \times \frac{3 \times \pi}{180} = 0.0524$

$$\text{Synchronous speed, } N_s = \frac{120f}{P} = \frac{120 \times 50}{6} = 1000 \text{ rpm}$$

Synchronising power for three-phase alternator,

$$P_s = \frac{3E^2\theta}{X_s} = \frac{3 \times (3464)^2 \times 0.0524}{5} = 377258 \text{ W}$$

$$= \mathbf{377.258\ kW}\ (Ans.)$$

$$\text{Synchronising torque, } T_s = \frac{P_s \times 60}{2\pi N_s} = \frac{377258 \times 60}{2\pi \times 1000}$$

$$= \mathbf{3603\ Nm/mech.degree}\ (Ans.)$$

Example 7.11

A three-phase, 6 MVA, 50Hz alternator has a synchronous reactance of 0.4 pu. it is running at 1500 rpm and excited to give 11kV. Calculate the synchronising torque for per degree mechanical displacement.

Solution:

Here, Rating of alternator = 6 *MVA*; f = 50 *Hz*; E_L = 11 *kV* N_S= *1500 rpm*; X_S= *0.4 pu*

$$\text{Phase voltage, } E = \frac{E_L}{\sqrt{3}} = \frac{11000}{\sqrt{3}} = 6351 \text{ V}$$

$$\text{Full-load current, } I = \frac{MVA \times 10^6}{\sqrt{3}\ E_L \times 10^3} = \frac{6 \times 10^6}{\sqrt{3} \times 11 \times 10^3} = 315 \text{ A}$$

Now, $IX_S = 0.4$ of 6351

$\therefore$ $X_S = \frac{0.4 \times 6351}{315} = 8.065\ \Omega$

$$\text{No. of poles of the machine, } P = \frac{120f}{N_S} = \frac{120 \times 50}{1500} = 4$$

Rotor displacement in electrical degree = displacement in mechanical degree $\times \frac{P}{2} = 1 \times \frac{4}{2} = 2°$

$$\text{Rotor displacement in radian, } \theta = 1 \times \frac{2 \times \pi}{180} = 0.0349 \text{ electrical}$$

synchronising power for three-phase machine

$$P_S = \frac{3E^2\theta}{X_S} = \frac{3 \times (6351)^2 \times 0.0349}{8.065} = 523731 \text{ W}$$

$$\text{Synchronising torque, } T_S = \frac{P_S \times 60}{2\pi N} = \frac{523731 \times 60}{2\pi \times 1500} = \mathbf{3334\ Nm}\ (Ans.)$$

Example 7.12

What will be synchronising power developed in a three-phase synchronous generator for one mechanical degree of displacement from its equilibrium position. The synchronous generator is connected to 11 kV infinite bus-bars and having the following data: Rated capacity =5 MVA; Frequency, f = 50 Hz; Number of poles, P = 8; synchronous reactance = 25%

Also determine the corresponding value of the synchronising torque.

Solution:

$$\text{Phase voltage, } E = \frac{E_L}{\sqrt{3}} = \frac{11000}{\sqrt{3}} = 6351 \text{ V}$$

$$\text{Full-load current, } I = \frac{MVA \times 10^6}{\sqrt{3} \times E_L \times 10^3} = \frac{5 \times 10^6}{\sqrt{3} \times 11 \times 10^3} = 263 \text{ A}$$

Now,

$$IX_S = 25\% \text{ of } 6351 \quad \text{or} \quad X_S = \frac{25 \times 6351}{100 \times 263} = 6.05\ \Omega$$

$$\text{Rotor displacement in electrical degree} = \text{Mech. degree} \times \frac{P}{2} = 1 \times \frac{8}{2} = 4$$

$$\text{Rotor displacement in radian, } \theta = \frac{4 \times \pi}{180} = 0.0698$$

$$\text{Synchronous power, } P_S = \frac{3E^2\theta}{X_S} = \frac{3 \times (6351)^2 \times 0.0698}{6.05}$$

$$= \mathbf{1396328\ W}\ (Ans.)$$

$$\text{Synchronous speed, } N_S = \frac{120f}{P} = \frac{120 \times 50}{8} = 750 \text{ rpm}$$

$$\text{Synchronising torque, } T_S = \frac{P_S \times 60}{2\pi N_S} = \frac{1396328 \times 60}{2\pi \times 750} = \mathbf{17778\ Nm}\ (Ans.)$$

Example 7.13

A three-phase, star-connected, 3300 V alternator is connected to the bus-bars. Another 3000 kVA, six-pole, three-phase, star-connected alternator running at 1000 rpm with 30% reactance is to be synchronised to the same bus-bars. Calculate the synchronising current and power per one mechanical degree of displacement and the corresponding synchronising torque.

Solution:

Alternator is three-phase, star connected

$$\text{Rating of alternator,} = 3000 \text{ kVA} = 3000 \times 10^3 \text{ VA}$$

$$\text{No. of poles, } P = 6$$

$$\text{Synchronous speed, } N_s = 1000 \text{ rpm}$$

$$\text{Terminal voltage (line value), } V_L = 3300 \text{ V}$$

$$\text{Terminal voltage (phase value), } V_{-} = \frac{3300}{\sqrt{3}} = 1905 \cdot 26 \text{ V}$$

$$\text{Full load current, } I = \frac{3300 \times 10^3}{\sqrt{3} \times 3300} = 524 \cdot 86 \text{ A}$$

$$\text{Synchronous impedance, } Z_s = \frac{1905 \cdot 26}{524 \cdot 86} = 3 \cdot 63 \text{ ohm}$$

$$\text{Synchronous reactance, } X_s = 30\% \text{ of } Z_s = \frac{30}{100} \times 3 \cdot 63 = 1 \cdot 089 \text{ ohm}$$

$$\text{Angular displacement in elect. degrees, } \theta = \text{mech. degree} \times \text{pair of poles} = 1^\circ \times \frac{6}{2} = 3^\circ$$

$$\text{Angular displacement in radians, } \theta = \frac{3 \times \pi}{180} = 0 \cdot 05236$$

$$\text{Synchronising current, } I_s = \frac{E\theta}{X_s} = \frac{1905 \cdot 26 \times 0 \cdot 5236}{1 \cdot 089} = \mathbf{91{\cdot}6\ A}\ (Ans.)$$

$$\text{Synchronising power} = 3\, I_s\, V = 3 \times 91{\cdot}6 \times 1905{\cdot}26 = \mathbf{523{\cdot}6\ kW}\ (Ans.)$$

$$\text{Synchronising torque, } T_s = \frac{\text{Synchronising power}}{2\pi N_s / 60}$$

$$= \frac{523 \cdot 6 \times 10^3}{2\pi \times 1000} \times 60 = \mathbf{5000\ Nm}\ (Ans.)$$

Example 7.14

The governors of the two 50 MVA, three-phase alternators operating in parallel. are set in such a way that the rise in speed from full-load to no-load is 2% in one machine and 3% in the other. The characteristics being straight lines in both cases. If each machine is fully loaded when the total load is 100 MW (unity pf), what will be the load on each machine when load is reduced to 75 MW.

Solution:

For parallel operation of two alternators having same number of poles, the speed of both should be the same. When each machine is fully loaded, the operating points are A and B (see Fig. 7.25). When the total load is reduced to 75 MW, let the speed rises to some value $x\%$ *of* the full load speed. Now the operating points of alternator I and II are shifted to F and E.

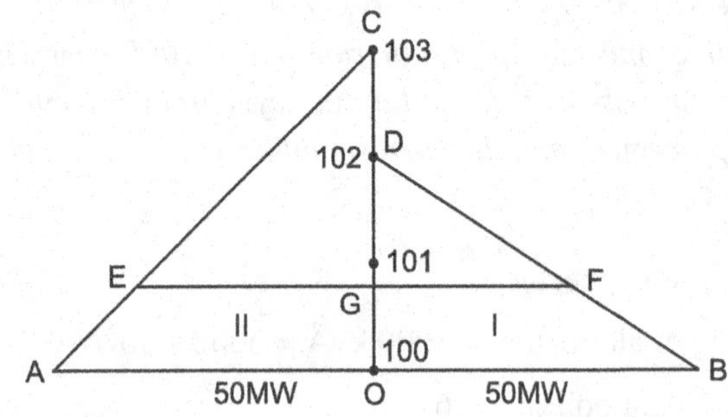

Fig. 7.25 Frequency-load characteristics of two alternators

From similar triangles ACO and ECG

$$\frac{EG}{AO} = \frac{GC}{OC}$$

$$\therefore \qquad EG = AO \times \frac{OC - OG}{OC}$$

$$= 50 \times \frac{3 - x}{3} = 50 - \frac{3 - x}{3} \qquad ...(i)$$

From similar triangles *BDO* and *FDG*

$$\frac{FG}{BO} = \frac{GD}{OD}$$

or

$$FG = BO \times \frac{OD - OG}{OD} = 50 \times \frac{2 - x}{2} = 50 - 25x \qquad ...(ii)$$

Adding (*i*) and (*ii*)

$$(EG + GF) = 50 - \frac{50x}{3} + 50 - 2x$$

or

$$75 = 100 - \frac{125x}{3}$$

or

$$x = 0{\cdot}6$$

From equation (*i*), load shared by alternator II

$$EG = 50 - \frac{50 \times 0 \cdot 6}{3} = 50 - 10 = \textbf{40 MW} \ (Ans.)$$

From equation (*ii*), load shared by alternator I.

$$FG = 50 - 25 \times 0{\cdot}6 = \textbf{35 MW} \ (Ans.)$$

Example 7.15

The two alternators I and II having a full-load capacity of 500 kW each are operating in parallel. Their speed regulation is 100% to 104% and 100% to 105% from full-load to no-load, respectively. How they will share a load of 600 kW. Also find the load at which one machine ceases to supply any load?

Solution:

When each alternator is operating at full-load, the operating points are *A* and *B*, as shown in Fig. 7.26 When the total load is reduced to 600 kW, let the speed be raised to (100 + *x*)% Now the operating points of alternator *I* and *II* are *E* and *F*.

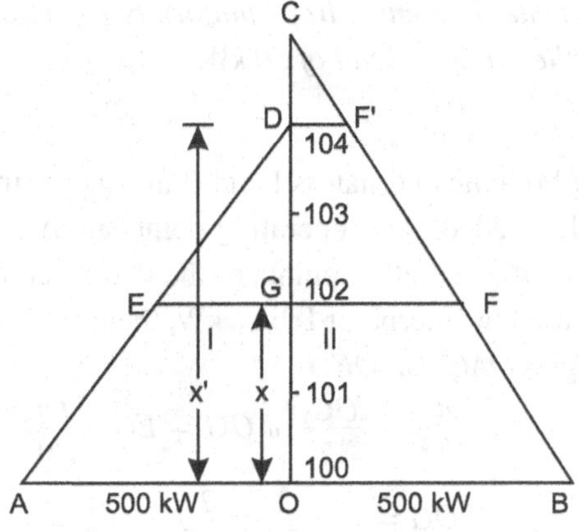

Fig. 7.26 Speed-load curves as per data

From similar triangle *AOD* and *EGD*

$$\frac{EG}{AO} = \frac{GD}{OD} \text{ or } EG = AO \times \frac{OD - OG}{OD}$$

or
$$EG = 500 \times \frac{4 - x}{4} = 500 - 125\,x \qquad \ldots(i)$$

From similar triangle *AOD* and *EGD*

$$\frac{FG}{BO} = \frac{GC}{OC} \text{ or } FG = BO \times \frac{OC - OG}{OC}$$

or
$$FG = 500 \times \frac{5 - x}{5} = 500 - 100\,x \qquad \ldots(ii)$$

Adding eqn.(*i*) and (*ii*), we get,

$$EG + FG = 500 - 125\,x + 500 - 100\,x$$

$$600 = 1000 - 225\,x$$

$$225\,x = 1000 - 600 \text{ or } x = \frac{400}{225} = 1.78$$

From eqn. (*i*), load shared by alternator I

$$EG = 500 \times \frac{4 - 1.78}{4} = \mathbf{277.78\ kW}\ (Ans)$$

Load shared by alternator II, $FG = 600 - 277.78 = \mathbf{322.22\ kW}\ (Ans)$

When load reduces to *DF′*, one of the alternator i.e., alternator *I* will cease to deliver load.

From similar triangles *OBC* and *DF′C*, we get,

$$\frac{DF'}{OB} = \frac{DC}{OC}$$

or
$$DF' = \frac{DC}{OC} \times OB = \frac{1}{5} \times 500 = \mathbf{100\ kW}\ (Ans)$$

Example 7.16

The governors of two identical three-phase alternators operating in parallel are set such that the frequency of first alternator drops uniformly from 50 Hz on no-load to 48 Hz on full-load of 25 kW, whereas, in second alternator the frequency drops uniformly from 50 Hz on no-load to 47 Hz on full-load of 25 kW. How will they share a load of 30 kW.

Solution:

The frequency-load curves of both the alternators I and II are shown in Fig. 7.27.

When each machine is fully loaded, their operating points are *A* and *B*, respectively. When the load of the system is 30 kW, let the operating points of the two machines be *E* and *F*, respectively. At this instant, let the load shared by alternator I be *x* kW, then;

Considering similar triangles *OAC* and *OEG*

$$\frac{OG}{EG} = \frac{OC}{AC} \text{ or } OG = EG \times \frac{OC}{AC}$$

or
$$OG = x \times \frac{2}{25} = \frac{2x}{25} \qquad \ldots(i)$$

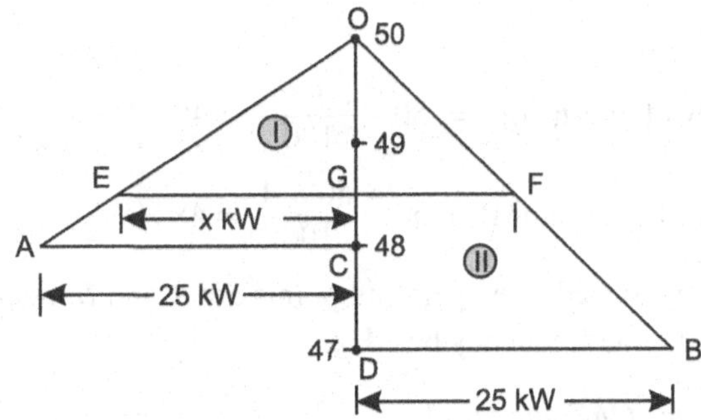

Fig. 7.27 Frequency-load curves of the alternators as per data

Considering similar triangles *ODB* and *OGF*

$$\frac{OG}{GF} = \frac{OD}{DB} \text{ or } OG = GF \times \frac{OD}{DB}$$

or
$$OG = (30 - x) \times \frac{3}{25} = \frac{90 - 3x}{25} \qquad \ldots(ii)$$

Equating eqn. (*i*) and (*ii*), we get

$$\frac{2x}{25} = \frac{90 - 3x}{25} \text{ or } 2x = 90 - 3x$$

or
$$2x + 3x = 90 \text{ or } x = \frac{90}{5} = 18 \text{ kW}$$

Load shared by alternator I = **18 kW** (*Ans.*)

Load shared by alternator II = 30 – 18 = 12 **kW** (*Ans.*)

Example 7.17

The drop characteristics of the governors of two alternators having rating 300 MW and 400 MW operating in parallel are of 3% and 4%, respectively. If the generators operate at no-load at 50 Hz. What will be the frequency at which they would operate with a total load of 600 MW. Also comment on the results.

Solution:

The frequency-load curves of both the alternators I and II are shown in Fig. 7.28

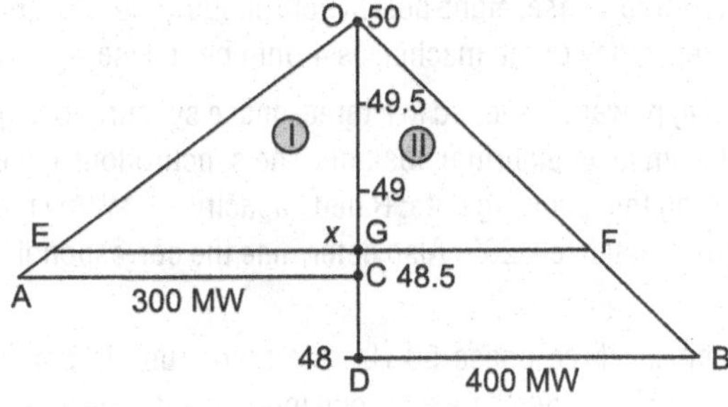

Fig. 7.28 Frequency-load curves of the alternators as per data

Al full-load,

$$\text{Frequency of alternator I} = 50 - \frac{50 \times 3}{100} = 48.5 \text{ Hz}$$

$$\text{Frequency of alternator II} = 50 - \frac{50 \times 4}{100} = 48 \text{ Hz}$$

When each machine is fully loaded, the operating points are *A* and *B*, respectively. When the load on the system is 600 MW let the frequency be *x* Hz.

Considering similar triangles *OAC* and *OEG*

$$\frac{EG}{OG} = \frac{AC}{OC} \text{ or } EG = OG \times \frac{AC}{OC}$$

or

$$EG = (50 - x) \times \frac{300}{(50 - 48.5)} = 10000 - 200\,x \qquad \ldots(i)$$

Considering similar triangles *ODB* and *OGF*

$$\frac{GF}{OG} = \frac{DB}{OD} \text{ or } GF = OG \times \frac{DB}{OD}$$

or

$$GF = (50 - x) \times \frac{400}{(50 - 48)} = 10000 - 200x \qquad \ldots(ii)$$

Adding eqn. (*i*) and (*ii*), we get

$$EG + GF = 10000 - 200\,x + 10000 - 200\,x$$

or $$600 = 20000 - 400\,x \qquad (\text{Since } EG + GF = 600 \text{ MW})$$

or $$400x = 19400$$

or $$x = \mathbf{48.5\ Hz}\ (Ans.)$$

With the given drooping characteristics, the two alternators will not be in position to take-up load more than 600 MW

Section Practice Problems

Numerical Problems

1. Determine the synchronising power and synchronising torque per mechanical degree of rotor displacement at no-load of a 2000 kVA, three-phase, eight-pole synchronous generator connected to 6000V, 50Hz bus-bars. The synchronous reactance of the machine is 4 ohm per phase. (**Ans.** 628.64 kW, 8000 Nm)
2. What will be synchronising power developed in a three-phase synchronous generator for one mechanical degree of displacement from its equilibrium position. The synchronous generator is connected to 3.3 kV infinite bus-bars and having the following data: Rated capacity = 3 MVA; Frequency, *f* = 50 Hz; Number of poles, *P* = 6; synchronous reactance = 25% Also determine the corresponding value of the synchronising torque. (**Ans.** 6.28.4 kW, 6000Nm)
3. A 750 kVA, 3300 V, 8 poles, three-phase 50 Hz, alternator runs in parallel with other alternator. Its synchronous reactance is 24%. Find the synchronising current, power and torque of one mechanical degree displacement. (**Ans.** 38·16A, 218·12 kW, 2777·23 Nm)

4. Two three-phase 40 MW synchronous generator I and II are operating in parallel. Their speed regulation is 100% to 104% and 100% to 103% from full-load to no-load, respectively. How they will share a load of 60 MW. Also find the load at which one machine ceases to supply any load?
(**Ans.** 31.43 MW, 28.57 MW, 10 MW)

5. The settings of governors of two identical three-phase alternators are operating in parallel are such that the frequency of first alternator drops uniformly from 50 Hz on no-load to 48 Hz on full-load of 20 kW, whereas, in second alternator the frequency drops uniformly from 50 Hz on no-load to 47.5 Hz on full-load of 20 kW. How will they share a load of 30 kW. (**Ans.** 13.33 kW, 16.67 kW)

Short Answer Type Questions

Q.1. The synchronous generators with larger air-gap have higher synchronising power, why?

Ans. A large air-gap offers a large reluctance to the armature mmf and reduces the armature reaction. This reduces the value of synchronous reactance. Since synchronising power is inversely proportion to synchronous reactance, the smaller synchronous reactance will develop higher value of synchronising power.

Q.2. Will an increase in excitation of an alternator connected to infinite bus-bars increase its real power generation?

Ans. No, real power generation will not be affected, it will remain the same.

Q.3. What do you understand by synchronising power and what is its significance?

Ans. When the equilibrium of a synchronous generator connected to infinite bus-bars in disturbed due to any reason, a synchronising current flows through it which develops *synchronising power.* The tendency of the synchronising power is to bring the machine back in to synchronism.

Q.4. How can the load sharing between the alternators operating in parallel be changed?

Ans. The load sharing between the alternators operating in parallel can be changed by adjusting the input power to their prime-movers.

Q.5. Do the non-salient pole type of alternators employed at steam power plants have a tendency to hunt?

Ans. No, they don't have the tendency to hunt.

Q.6. What are the causes of hunting in synchronous machines?

Ans. In synchronous machines, hunting occurs due to sudden change in load on the machine.

Review Questions

1. Explain what you understand by Infinite Bus-bars.
2. What are the advantages of connecting alternators in parallel? What conditions are required to be fulfilled before connecting an alternator in parallel with the existing alternators?
3. What is the necessity of parallel operation of alternators? What conditions are required to be satisfied before connecting an alternator to the infinite bus-bars.
4. What are necessary conditions for parallel operation of three-phase alternators?

or

State the necessary conditions which must be satisfied before an incoming alternator is switched *ON* with other alternator already working in parallel.

5. What is meant by synchronising? What are the various methods of synchronisation. Explain any one of them.

6. Describe "One dark and two bright" lamp method of synchronising two three-phase alternators.

7. Explain the term synchronising. Discuss any one methods of synchronising of 1-phase alternators.

8. Explain with diagram any one method of synchronising of two three-phase alternators

9. What are the advantages of connecting the three synchronising lamps in a manner so as to give one dark and two birght instead of all dark at a time while synchronising an alternator?

10. Discuss the use of synchroscope in the parallel operation of three-phase alternators.

11. An alternator has just been synchronised and is floating on the bus-bars. What steps will you take to make it share load? Justify your answer with the help of a phasor diagram.

12. What steps should be taken before an alternator is switched off from infinite bus-bars?

13. Derive an expression for synchronising power and torque when two alternators are running in Parallel.

14. Two identical alternators are running in parallel sharing equal load. What is the effect on their load sharing if:

(*i*) Field excitation of one of the alternators is increased.
(*ii*) Input power of one of the alternators is increased.

15. Explain the effect of change in excitation of a synchronous generator connected to an infinite bus-bars.

16. If two alternators are operating in parallel, what is the effect on the phasor of their induced emf of increasing the torque of one of the prime mover driving one of them. Show that the resultant emf produces a circulating current called synchronising current and the action of this current is such that it makes the parallel operation of alternators a conditions of stable equilibrium.

17. Two alternators with negligible synchronous reactances as compared to their resistances are Connected in parallel. Will they be in position to run successfully? Justify you answer.

18. Explain the factors that affect the load sharing between the two synchronous generators running in Parallel.

19. What do you understand by 'power angle' of a synchronous generator? Will it increase or decrease if the input to the prime-mover of the machine connected to infinite bus-bar is Increased? Justify your answer.

20. Drive the condition for maximum power output of a synchronous generator connected to an infinite bus and operating at constant excitation.

21. "Synchronous reactance is necessary for the flow of synchronising power'. Justify the statement.

22. How does change in the excitation of the alternator connected to infinite bus-bars affect the operation?

23. How does change in the driving torque of the alternator connected to infinite bus-bars affect the operation.

24. "It is desirable to adjust the frequency of the incoming machine slightly higher than that of the bus with which it is to be connected". Justify the statement.

25. What do you understand by hunting in synchronous generators? What are the ill-effects of hunting. How hunting can be minimised?

Multiple Choice Questions

1. The speed of prime mover for an alternator is maintained constant by
(*a*) filters (*b*) coupling
(*c*) brakes (*d*) governors

2. Two alternators are operating in parallel. What will happen when the excitation of *A* is increased?
(*a*) alternator will burn out (*b*) wattless component will change
(*c*) power output will reduce (*d*) the machines will stop

3. Two alternators *A* and *B* have the same % age impedance. They will share any additional load.
(*a*) equally (*b*) in proportion to their pf
(*c*) in proportion to their kVA rating (*d*) in proportion to their speed.

4. Which of the following is not an essential conditions for parallel operation of two alternators?
(*a*) identical terminal voltage for two machines
(*b*) same phase sequence for two machines
(*c*) same kVA rating for two machines
(*d*) same frequency of voltage generated.

5. A common load is supplied by two identical alternators *A* and *B* running is parallel. Both of them are having identical excitations and steam supplies to their prime-movers. Now, if the steam supply to the prime-mover of generator *A* is increased compared to *B* with field excitation kept unchanged, then.
(*a*) Active power of *A* will remain the same but the reactive power will increase.
(*b*) Active power of *A* will decrease while the reactive power will increase.
(*c*) both active and reactive components of power of generator *A* will increase.
(*d*) active power contribution of *A* will increase but reactive power contributions of both will remain unchanged.

6. Synchronising current means
(*a*) The total current supplied to the load by the alternators operating in parallel.
(*b*) The current circulating in the local circuit of two alternators operating in parallel which brings the alternators in synchronism once they are out of it.
(*c*) The current supplied by the synchronous generator.
(*d*) All of the above.

7. An inductive load is shared by two identical synchronous generators *A* and *B* equally, if the excitation of alternator *A* is increased.
(*a*) Alternator *A* will deliver more current and alternator *B* will deliver less current.
(*b*) Alternator *B* will deliver more current and alternator *A* will deliver less current.
(*c*) both will continue to deliver same current.
(*d*) both will deliver more current.

8. The governor droop characteristics of two 300 MW and 400 MW alternators appearing in parallel are 4% and 5%, respectively. If the alternators operate at no-load at 50 Hz, the frequency at which they would operate with a total load of 600 MW is
(a) 48.06 Hz. (b) 47.06 Hz.
(c) 49.66 Hz. (d) 47.66 Hz.

9. Two alternators each having 4% speed regulation are working in parallel. Alternator 1 is rated 12 MW and alternator 2 is rated 8 MW. When the total load is 12 MW, the load shared by alternators 1 and 2 would be, respectively

(*a*) 4 MW and 8 MW.
(*b*) 8MW and 4 MW.
(*c*) 4.8 MW and 7.2 MW.
(*c*) 12 MW and zero.

10. The following data pertaining to two alternators working in parallel and supplying a total load of 75 MW:
Machine 1: 50 MVA with 5% speed regulation
Machine 2: 75 MVA with 5% speed regulation
The load sharing between machine 1 and 2 will be

(*a*) 45 MW and 30 MW
(*b*) 40 MW and 35 MW
(*c*) 35 MW and 40 MW
(*d*) 30 MW and 45 MW

11. An infinite bus-bar has

(*a*) constant voltage.
(*b*) constant frequency.
(*c*) constant current
(*d*) both (a) and (b).

Keys to Multiple Choice Questions

1. d	**2.** b	**3.** c	**4.** c	**5.** d	**6.** b	**7.** a	**8.** a	**9.** c	**10.** d
11. d									

CHAPTER

8

Synchronous Motors

Chapter Objectives

After the completion of this unit, students/readers will be able to understand:

- ✓ What is the basic principle of operation of a synchronous motor?
- ✓ How synchronous motor operates on load?
- ✓ How to determine relation between supply voltage *V* and excitation voltage *E*.
- ✓ What are the different torques developed in a synchronous motor?
- ✓ How to draw a power flow diagram of a synchronous motor?
- ✓ What is the effect of change in excitation when load on the machine is kept constant?
- ✓ What is the effect of change in load when excitation is kept constant?
- ✓ What are V-curves and inverted V-curves.
- ✓ What are the different methods by which a synchronous motor is made self-starting?
- ✓ How a synchronous motor is used to improve the pf of a system?
- ✓ What are the important characteristics and applications of synchronous motors?
- ✓ How synchronous motors are compared with induction motors?

Introduction

The same synchronous machine can be used as a generator or as a motor. When it converts mechanical power or energy into electric power or energy, it is called a synchronous generator. On the other hand, when it converts electric power or energy into mechanical power or energy, it is called a synchronous motor.

For instance, if two alternators *A* and *B* are operating in parallel and the power input to one of the alternator say '*A*' is increased, it starts delivering more power. Since the demand of the load is unaltered, the alternator '*B*' will thus be relieved off load by an equal amount. If this process is continued till output power of alternator '*A*' becomes more than the total load supplied, then the alternator '*B*' instead of supplying power starts receiving power from the bus-bars. Thus, the machine starts working as a motor. However, when two alternators *A* and *B* (or any number of alternators) are operating in parallel and the driving force (power input) of any one of them is removed, it continues to run as a motor by receiving power from the other alternator (or bus bars).

Thus, the machine starts working as a motor and is called **synchronous motor.**

Hence, a synchronous machine can be operated as a generator (called alternator or synchronous generator) or a motor (called synchronous motor).

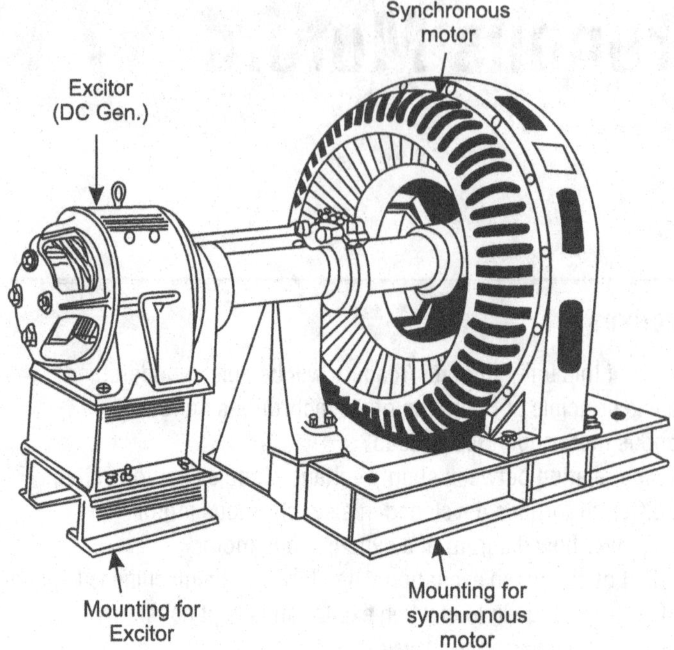

Fig. 8.1 Synchronous motor with exciter

Note: The construction of synchronous motor is same as that of synchronous generator or alternator. In fact, the same machine can be operated as a generator or as a motor.

8.1 Working Principle of a Three-phase Synchronous Motor

When a 3-phase supply is given to the stator of a 3-phase wound synchronous motor, a revolving field is set up (say in anticlockwise) which rotates at a synchronous speed ($N_S = \frac{120f}{P}$). This field is represented by the imaginary stator poles. At an instant as shown in Fig. 8.2(*a*), the opposite poles of stator and rotor are facing each other (for simplicity two-pole machine is considered). As there is a force of attraction between them, an anticlockwise torque is produced in the rotor as the rotor poles are dragged by the stator revolving poles or field.

After half a cycle, polarity of the stator poles is reversed whereas the rotor poles could not change their position due to inertia. Thus, like poles are facing each other and due to force of repulsion a clockwise torque is produced in the rotor as shown in Fig. 8.2 (*b*).

Hence, the torque produced in a 3-phase synchronous motor is not unidirectional and as such this motor is **not self-starting**.

However, if rotor of synchronous motor is rotated by some external means at the start so that it also reverses its polarity as the polarity of stator poles is reversed after half a cycle as shown in Fig.

8.2(*c*). A continuous force of attraction between stator and rotor poles exists. This is called magnetic locking. Once the magnetic locking is obtained, the rotor poles are dragged by the stator revolving field (imaginary poles) and a continuous torque is obtained. As the rotor poles are dragged by the stator revolving field, hence the rotor rotates at the same speed as that of stator revolving field, i.e., synchronous speed.

Thus, a synchronous motor only runs at a constant speed called synchronous speed.

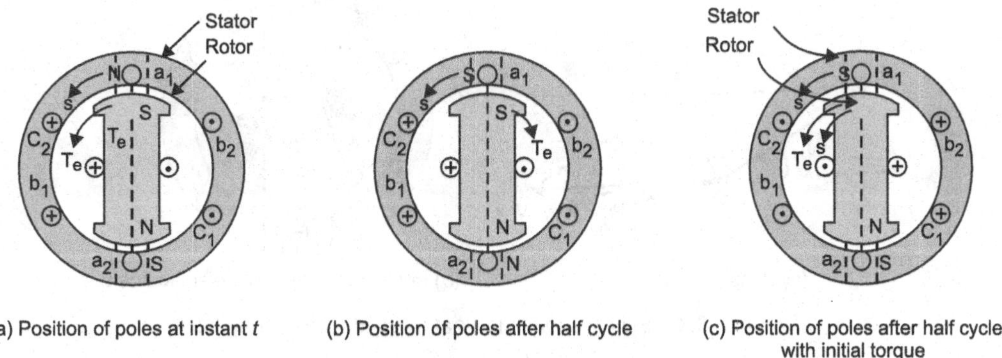

Fig. 8.2 Representing working principle with initial torque

8.2 Effect of Load on Synchronous Motor

When a synchronous motor is connected to the lines and started by some external means, it starts rotating at synchronous speed. If the motor is running at no-load and has no losses (ideal condition), then the induced *emf* E is equal and opposite to applied voltage V as shown in Fig. 8.3(*a*) and the stator and rotor poles are in line with each other as shown in Fig. 8.3(*b*). The resultant *emf* and hence the current drawn by the motor is zero. Thus, the motor is said to be floating on the lines.

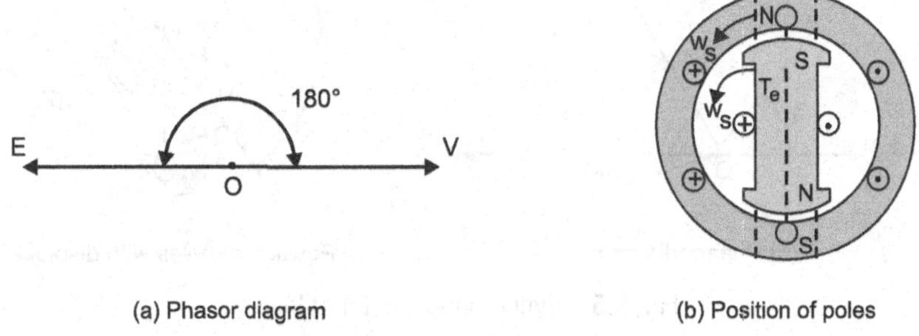

Fig. 8.3 Synchronous motor on no-load (ideal condition)

However, in actual machine some losses are always present with the result induced *emf* E falls back by an angle δ_0 relative to the stator poles as shown in Fig. 8.3(*b*). This causes a resultant voltage E_r across the armature circuit and motor draws no-load current I_0 ($I_0 = E_r/Z_s$) from the mains. This no-load current lags behind the resultant voltage by an angle θ where $\theta = \tan^{-1}\frac{X_S}{R}$; X_S is the synchronous reactance and R is the resistance of armature (stator winding). Since resistance is very small as compared to synchronous reactance, therefore angle θ is nearly 90°. The power drawn by the motor at no-load is $VI_0 \cos \phi_0$ which is sufficient to meet with the losses and make the motor running continuously at synchronous speed.

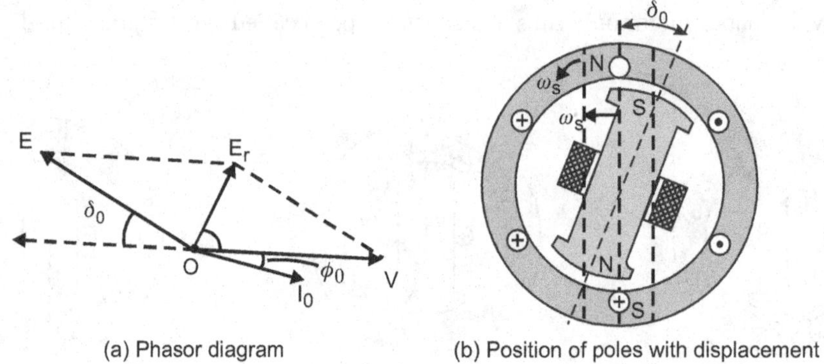

(a) Phasor diagram (b) Position of poles with displacement

Fig. 8.4 Synchronous motor at no-load (considering losses)

However, when load is applied through the shaft on the motor, the rotor poles fall back a little more (angle δ) relative to stator poles as shown in Fig. 8.5(*b*). Hence the torque angle increases to δ with the increase in load. This increases the resultant voltage E_r which in turn increases the current I ($I = E_r/Z_s$) drawn by the motor from the mains.

Thus, a synchronous motor is able to supply power to the increased mechanical load, not by decrease in speed, but by shifting the position of the rotor poles with respect of the stator poles or field.

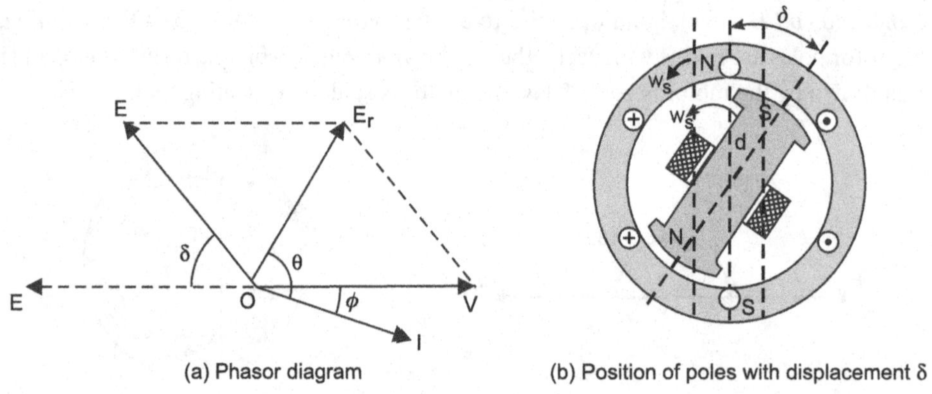

(a) Phasor diagram (b) Position of poles with displacement δ

Fig. 8.5 Synchronous motor at load

When load applied on the shaft of the motor is further increased, the induced, *emf* E falls back further. Hence load angle (torque angle) δ increases with the increase in load. When δ increases the resultant voltage E_r increases and so the armature current I. If too great mechanical load is applied to the synchronous motor, the rotor is pulled out of synchronism, after which it comes to stand still.

This maximum value of torque that a motor can develop without losing its synchronism is called **pull-out torque.**

8.3 Equivalent Circuit of a Synchronous Motor

The equivalent circuit of a synchronous motor is the same, in all respects, as that of a synchronous generator, the only difference is that the direction of power flow is reversed in this case, as shown in Fig. 8.6. In the equivalent circuit

V = Voltage applied to the armature (phase value)

I = armature current (phase value)

R = effective armature resistance in ohm per phase

X_S = synchronous reactance per phase

E = excitation voltage (phase value)

Z_S = synchronous impedance per phase

I_f = exciting or field current

where, $\overline{Z}_S = R + jX_S = \sqrt{R^2 + X_S^2}\ \angle \tan^{-1}\dfrac{X_S}{R}$ ohm

For synchronous motor:

$$\overline{V} = \overline{E} + \overline{I}\overline{Z}_s = \overline{E} + \overline{I}(R + jX_S) = \overline{E} + \overline{I}R + j\overline{I}X_S \quad \text{...(i)}$$

or

$$\overline{E} = \overline{V} - \overline{I}\overline{Z}_s = \overline{V} - \overline{I}(R + jX_S) = \overline{V} - \overline{I}R - j\overline{I}X_S \quad \text{...(ii)}$$

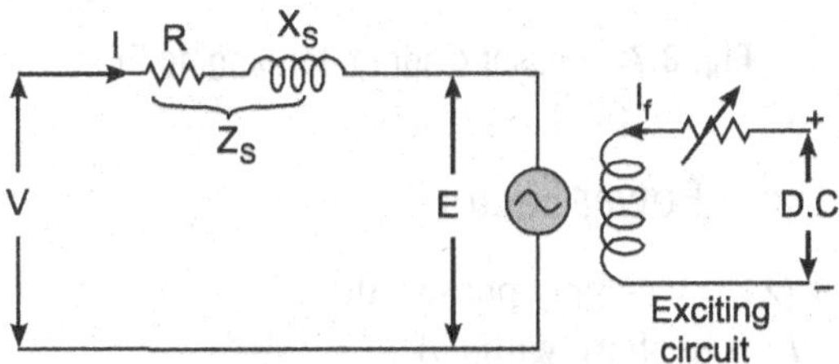

Fig. 8.6 Equivalent circuit

8.4 Phasor Diagram of a Synchronous Motor (Cylindrical Rotor)

A 3-phase cylindrical rotor synchronous motor may operate at different power factors i.e., lagging, unity or leading. Accordingly its phasor diagram is drawn with the help of above equations.

(i) Phasor Diagram for Lagging Power Factor

The phasor diagram may be drawn by considering equation $\overline{E} = \overline{V} - \overline{I}R - j\overline{I}X_s$.

The phasor diagrams are shown in Fig. 8.7(*a*) and (*b*).

To draw the phasor diagram, proceed as follows:

1. Supply voltage V (phase value) is taken as reference vector such that $OA = V$.
2. Armature current I lags behind the voltage vector by an angle ϕ for lagging p.f. such that $OI = I$.

3. Draw phasor $AB = -IR$ in phase opposition to current phasor as shown in Fig. 8.7(*a*) or draw phasor $AD = -IX_S$ perpendicular to current phasor but is opposite direction as shown in Fig. 8.7(*b*).
4. Draw phasor $BC = -IX_S$ perpendicular to current phasor but in opposite direction as shown in Fig. 8.7(*a*) or draw phasor $DC = -IR$ in phase opposition to current phasor as shown in Fig. 8.7(*b*).
   Hence, $\bar{E} = \bar{V} - \bar{I}\,\bar{Z}_S$ or $\bar{V} = \bar{E} + \bar{I}\bar{Z}_S$
5. Join OC, which represents phasor E i.e., induced emf per phase.
6. The phase difference between V and E i.e., angle δ is known as power angle or torque angle.

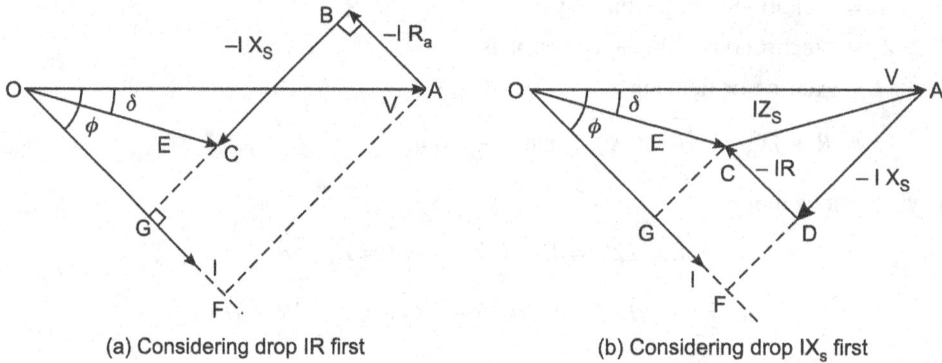

(a) Considering drop IR first (b) Considering drop IX$_s$ first

Fig. 8.7 Phasor diagram for lagging PF

(ii) Phasor Diagram at Unity Power Factor

1. Draw reference phasor $OA = V$ (supply phase value)
2. Draw current phasor $OI = I$ in phase with OA.
3. Draw phasor $AB = -IR$ in phase opposition to current phasor as shown in Fig. 8.7(*a*) or draw phasor $AD = -IX_S$ perpendicular to current phasor but in opposite direction as shown in Fig. 8.7(*b*).
4. Draw phasor $BC = -IX_S$ perpendicular to current phasor but in opposite direction as shown in Fig. 8.8(*a*) or draw phasor $DC = -IR$ in phase opposition to current phasor as shown in Fig. 8.8 (*b*).
5. Join OC, which represents phasor E i.e., induced emf per phase.

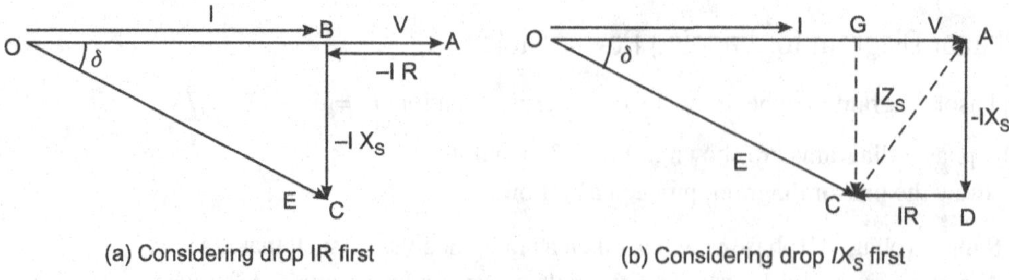

(a) Considering drop IR first (b) Considering drop IX_S first

Fig. 8.8 Phasor diagram for unity *pf*

6. The phase difference between V and E i.e., angle δ is known as power angle or torque angle.
   Hence, $\bar{E} = \bar{V} - \bar{I}\,\bar{Z}_S$ or $\bar{V} = \bar{E} + \bar{I}\,\bar{Z}_S$

(iii) Phasor Diagram for Leading Power Factor

1. Draw a reference phasor $OA = V$ (supply voltage phase value)
2. Draw current phasor $OI = I$ leading the phasor OA (i.e., V) by an angle ϕ since *pf* $\cos\phi$ is leading.
3. Draw phasor $AB = -IR$ i.e., drop in armature resistance per phase, parallel to current phasor but in phase opposition as shown in Fig. 8.9(*a*) or draw phasor $AD = -IX_S$ i.e., drop in synchronous reactance per phase, perpendicular to current phasor but in opposite direction as shown in Fig. 8.9(*b*).
4. Draw phasor $BC = -IX_S$ i.e., drop in synchronous reactance per phase, perpendicular to current phasor but in opposite direction as shown in Fig. 8.9(*a*) or draw $DC = -IR$ i.e., drop in armature resistance per phase parallel to current phasor but in phase opposition as shown in Fig. 8.9(*b*).
5. Join OC, which represents phasor value of E i.e., induced emf per phase.
6. The phase difference between V and E i.e., angle δ is known as power angle or torque angle. Hence, $\bar{E} = \bar{V} - \bar{I}\,\bar{Z}_S$ or $\bar{V} = \bar{E} + \bar{I}\,\bar{Z}_S$

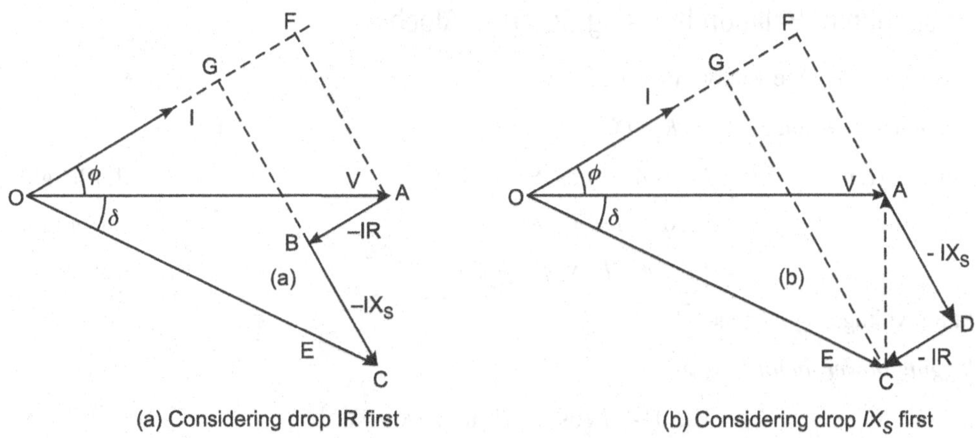

(a) Considering drop IR first (b) Considering drop IX_S first

Fig. 8.9 Phasor diagram for leading *pf*

8.5 Relation between Supply Voltage V and Excitation Voltage E

The relation between excitation voltage E (phase value) and supply voltage V (phase value) can be determined for different power factors either by using phasor diagram or complex algebra.

Determination of a Relation by using Phasor Diagram

For lagging power factor, consider phasor diagram shown in Fig. 8.7(*b*).

From right angle triangle OCG

$$OC^2 = OG^2 + CG^2 = (OF - FG)^2 + (DF)^2 = (OF - DC)^2 + (AF - AD)^2$$

$$E^2 = (V\cos\phi - IR)^2 + (V\sin\phi - IX_S)^2$$

or
$$E = \sqrt{(V\cos\phi - IR)^2 + (V\sin\phi - IX_S)^2}$$

For unity power factor, consider phasor diagram shown in Fig. 8.8(*b*).

From right angle triangle *OCG*

or
$$OC^2 = OG^2 + CG^2 = (OA - AG)^2 + CG^2 = (OA - DC)^2 + DA^2$$

$$E^2 = (V - IR)^2 + (IX_S)^2$$

or
$$E = \sqrt{(V - IR)^2 + (IX_S)^2}$$

For leading power factor, consider phasor diagram shown in Fig. 8.9(*b*)

From right angle triangle *OCG.*

$$OC^2 = OG^2 + CG^2 = (OF - FG)^2 + DF^2 = (OF - DC)^2 + (FA + AD)^2$$

$$E^2 = (V\cos\phi - IR)^2 + (V\sin\phi + IX_S)^2$$

or
$$E = \sqrt{(V\cos\phi - IR)^2 + (V\sin\phi + IX_S)^2}$$

Determination of Relation by using Complex Algebra

Since V is the reference vector, $\overline{V} = V\angle 0° = V \pm j0$

Synchronous impedance, $\overline{Z}_S = R + jX_S$

Armature current, $\overline{I} = I\angle -\phi = I\cos\phi - jI\sin\phi$ (for lagging *pf*)

$= I\angle 0° = I \pm j0$ (for unity *pf*)

$= I\angle +\phi = I\cos\phi + jI\sin\phi$ (for leading *pf*)

Excitation voltage is given as, $\overline{E} = \overline{V} - \overline{I}\,\overline{Z}_S$

For lagging power factor ($\cos\phi$)

$$\overline{E} = V\angle 0° - (I\cos\phi - jI\sin\phi)(R + jX_S)$$

$$= V \pm j0 - [IR\cos\phi + j\,IX_S\cos\phi - j\,IR\sin\phi + IX_S\sin\phi]$$

$$= (V - IR\cos\phi - IX_S\sin\phi) - j\,(IX_S\cos\phi - IR\sin\phi)$$

Now,
$$\overline{E} = E\angle\delta°$$

$\therefore$
$$E = \sqrt{(V - IR\cos\phi - IX_S\sin\phi)^2 + (IX_S\cos\phi - IR\sin\phi)^2}$$

$$\angle\delta = -\tan^{-1}\frac{IX_S\cos\phi - IR\sin\phi}{V - IR\cos\phi - IX_S\sin\phi} \quad \text{(lagging)}$$

Similarly, for leading power factor ($\cos\phi$)

$$E = \sqrt{(V - IR\cos\phi + IX_S\sin\phi)^2 + (IX_S\cos\phi + IR\sin\phi)^2}$$

$$\angle\delta = -\tan^{-1}\frac{IX_S\cos\phi + IR\sin\phi}{V - IR\cos\phi + IX_S\sin\phi} \quad \text{(lagging)}$$

For unity power factor ($\cos\phi = 1$; $\sin\phi = 0$)

$$E = \sqrt{(V - IR)^2 + (IX_S)^2}$$

$$\angle\delta = -\tan^{-1}\frac{IX_S}{V - IR} \qquad \text{(lagging)}$$

8.6 Different Torques in a Synchronous Motor

The following torques of a synchronous motor are usually considered while selecting it for a particular application:

1. **Locked-rotor torque.** The minimum torque at any angular rotor position that is developed with the rotor locked (that is, stationary) and rated voltage at rated frequency is applied to the terminals is called *locked rotor torque.* This torque is provided by the stator windings.
2. **Running Torque.** The torque developed by the motor under running conditions is called *running torque.* It is determined by the power rating and speed of the driven machine.
3. **Pull-in torque.** A synchronous motor is usually started as induction motor. When it runs 2 to 5 percent below the synchronous speed, DC excitation is applied and the rotor pulls into step with the synchronously rotating stator field. The *pull-in torque* is the maximum constant torque at rated voltage and frequency under which a motor will pull a connected load into synchronism when DC excitation is applied.
4. **Pull-out torque.** The maximum value of torque which a synchronous motor can develop at rated voltage and frequency without losing synchronism is called *pull-out torque.*

8.7 Power Developed in a Synchronous Motor (Cylindrical Rotor)

Considering circuit diagram of a cylindrical rotor synchronous motor shown in Fig. 8.6 and its phasor diagram for lagging power factor shown in 8.7(*b*). Here E lags behind V by an angle δ, then

$$\overline{V} = V\angle 0° \text{ and } \overline{E} = E\angle -\delta$$

Also

$$\overline{V} = \overline{E} + \overline{I}\,\overline{Z}_S \qquad \ldots(8.7.1)$$

$$\overline{I} = \frac{\overline{V} - \overline{E}}{\overline{Z}_S} = \frac{V\angle 0°}{Z_S\angle\theta} - \frac{E\angle -\delta}{Z_S\angle\theta}$$

$$= \frac{V}{Z_S}\angle -\theta - \frac{E}{Z_S}\angle -(\delta + \theta)$$

or

$$\overline{I}^* = \frac{V}{Z_S}\angle\theta - \frac{E}{Z_S}\angle(\delta + \theta) \qquad \ldots(8.7.2)$$

Power Input to Motor Per Phase

Apparent power = Active power + j Reactive power

$$S_{im} = P_{im} + jQ_{im} = \bar{V}\bar{I}^* \quad \text{...(8.7.3)}$$

$$= (V\angle 0°)\left[\frac{V}{Z_S}\angle\theta - \frac{E}{Z_S}\angle(\delta+\theta)\right]$$

$$= \frac{V^2}{Z_S}\angle\theta - \frac{VE}{Z_S}\angle(\delta+\theta)$$

$$= \left[\frac{V^2}{Z_S}\cos\theta + j\frac{V^2}{Z_S}\sin\theta\right] - \left[\frac{VE}{Z_S}\cos(\delta+\theta) + j\frac{VE}{Z_S}\sin(\delta+\theta)\right]$$

$$= \left[\frac{V^2}{Z_S}\cos\theta - \frac{VE}{Z_S}\cos(\delta+\theta)\right] + j\left[\frac{V^2}{Z_S}\sin\theta - \frac{VE}{Z_S}\sin(\delta+\theta)\right] \quad \text{...(8.7.4)}$$

Real Power Input to the Motor (Per Phase)

Taking real part of eqn. 8.7.4, we get,

$$P_{im} = \frac{V^2}{Z_S}\cos\theta - \frac{VE}{Z_S}\cos(\delta+\theta) \quad \text{...(8.7.5)}$$

$$= \frac{V^2}{Z_S^2}R - \frac{VE}{Z_S}\cos(\delta+\theta)$$

where, $\theta_S = 90° - \alpha$

$$\therefore \quad \cos(\delta+\theta) = \cos[90 + (\delta-\alpha)] = -\sin(\delta-\alpha)$$

$$\therefore \quad P_{im} = \frac{V^2}{Z_S^2}R + \frac{VE}{Z_S} - \sin(\delta-\alpha) \quad \text{...(8.7.6)}$$

Reactive Power Input to the Motor (Per Phase)

Taking imaginary part of the eqn. (8.7.4), we get,

$$Q_{im} = \frac{V^2}{Z_S}\sin\theta - \frac{VE}{Z_S}\sin(\delta+\theta) \quad \text{...(8.7.7)}$$

$$Q_{im} = \frac{V^2}{Z_S^2}X_S - \frac{VE}{Z_S}\sin(\delta+\theta)$$

where $\theta = 90 - \alpha$; $\sin(\delta+\theta) = \sin[90 + (\delta-\alpha)] = \cos(\delta-\alpha)$

$$\therefore \quad Q_{im} = \frac{V^2}{Z_S^2}X_S - \frac{VE}{Z_S}\cos(\delta-\alpha) \quad \text{...(8.7.8)}$$

Power Output of Motor Per Phase

$$S_{om} = P_{om} + jQ_{om} = \bar{E}\bar{I}^* \quad \text{...(8.7.9)}$$

$$= (E\angle -\delta)\left[\frac{V}{Z_S}\angle\theta - \frac{E}{Z_S}\angle(\delta+\theta)\right]$$

$$= \frac{VE}{Z_S}\angle(\theta-\delta) - \frac{E^2}{Z_S}\angle\theta$$

$$= \left[\frac{VE}{Z_S}\cos(\theta-\delta) + j\frac{VE}{Z_S}\sin(\theta-\delta)\right] - \left[\frac{E^2}{Z_S}\cos\theta + j\frac{E^2}{Z_S}\sin\theta\right]$$

$$= \left[\frac{VE}{Z_S}\cos(\theta-\delta) - \frac{E^2}{Z_S}\cos\theta\right] + j\left[\frac{VE}{Z_S}\sin(\theta-\delta) - \frac{E^2}{Z_S}\sin\theta\right] \quad ...(8.7.10)$$

Real Power Output of the Motor Per Phase

Taking real part of the eqn. (8.7.10), we get,

$$P_{om} = \frac{VE}{Z_S}\cos(\theta-\delta) - \frac{E^2}{Z_S}\cos\theta \quad ...(8.7.11)$$

$$= \frac{VE}{Z_S}\cos(\theta-\delta) - \frac{E^2}{Z_S^2}R \quad ...(8.7.12)$$

where $\theta = 90° - \alpha$; $\cos(\theta - \delta) = \cos[90° - (\delta + \alpha)] = \sin(\delta + \alpha)$

$$\therefore \quad P_{om} = \frac{VE}{Z_S}\sin(\delta+\alpha) - \frac{E^2}{Z_S^2}R \quad ...(8.7.13)$$

In large machines, usually the value of armature resistances R is so small than the synchronous reactance X_S that it is neglected, then $\theta = 90°$. Considering eqn. 8.7.11,

$$P_{om} = \frac{VE}{Z_s}\cos(q-\delta) - \frac{E^2}{Z_s}\cos\theta = \frac{VE}{X_s}\cos(90-\delta) - \frac{E^2}{X_s}\cos 90°$$

$$= \frac{VE}{X_S}\sin\delta \quad ...(8.7.14)$$

This is also known as synchronising power of the motor. For 3-phase machine,

$$\text{Synchronising power, } P_{syn} = 3\frac{VE}{X_S}\sin\delta \quad ...(8.7.15)$$

Reactive Power Output of the Motor Per Phase

Taking imaginary part of the eqn. (8.7.10), we get,

$$Q_{om} = \frac{VE}{Z_S}\sin(\theta-\delta) - \frac{E^2}{Z_S}\sin\theta \quad ...(8.7.16)$$

$$= \frac{VE}{Z_S}\sin(\theta-\delta) - \frac{E^2}{Z_S^2}X_S$$

where $\theta = 90° - \alpha$; $\sin(\theta - \delta) = \sin[90° - (\alpha + S)] = \cos(\alpha + \delta)$

$$\therefore \qquad Q_{om} = \frac{VE}{Z_S}\cos(\alpha + \delta) - \frac{E^2}{Z_S^2}X_S \qquad \text{...(8.7.17)}$$

For a synchronous motor, actual power developed in the motor is P_{om} and the power available at the shaft will be

Power at the shaft = Mechanical power developed – rotational losses

where Rotational losses include friction, windage and core loss.

Maximum Power Output of the Motor

For maximum power output, differentiate P_{om} with respect to the variable i.e., torque angle δ and put that equal to zero, i.e.,

$$\frac{d}{d\delta}P_{om} = 0 \text{ and } \frac{d^2}{d\delta^2}P_m < 0$$

$$\frac{d}{d\delta}P_{om} = \frac{d}{d\delta}\left[\frac{VE}{Z_S}\sin(\delta + \alpha) - \frac{E^2}{Z_S^2}R\right] = 0 \qquad \text{...(8.7.18)}$$

$$\therefore \qquad \frac{VE}{Z_S}\cos(\delta + \alpha) = 0$$

or $\qquad \cos(\delta + \alpha) = 0$ or $\delta + \alpha = 90°$ or $\delta = 90° - \alpha$

or $\qquad \delta = \theta \qquad$ ...(8.7.19)

Maximum power output

$$P_{om\,(\max)} = \frac{VE}{Z_S}\sin 90° - \frac{E^2}{Z_S^2}R \text{ or } P_{om(max)} = \frac{VE}{Z_S} - \frac{E^2}{Z_S^2}R \qquad \text{...(8.7.20)}$$

Neglecting armature resistance,

$$P_{om(\max)} = \frac{VE}{X_S} \qquad \text{...(8.7.21)}$$

8.8 Phasor Diagrams of a Salient-pole Synchronous Motor

The voltage equation for a salient-pole synchronous motor is

$$\bar{V} = \bar{E} + \bar{I}R + jI_dX_d + jI_qX_q$$

where

V = terminal voltage (phase value);	E = induced emf (phase value);
I = armature current (phase value);	I_d = direct axis current;
I_q = quadrature current;	R = armature resistance;
X_d = direct axis reactance;	X_q = quadrature reactance;

(a) For Lagging Power Factor cos ϕ

The phasor diagram at lagging pf cos ϕ is shown in Fig. 8.10.

$$OD = OA + AC + CD$$
$$V\cos\delta = E + I_qR + I_dX_d$$
$$BH = BC + CH$$
$$I_qX_q = I_dR + V\sin\delta$$
$$\psi = \phi - \delta$$
$$I_d = I\sin\psi = I\sin(\phi - \delta)$$
$$I_q = I\cos\psi = I\cos(\phi - \delta)$$

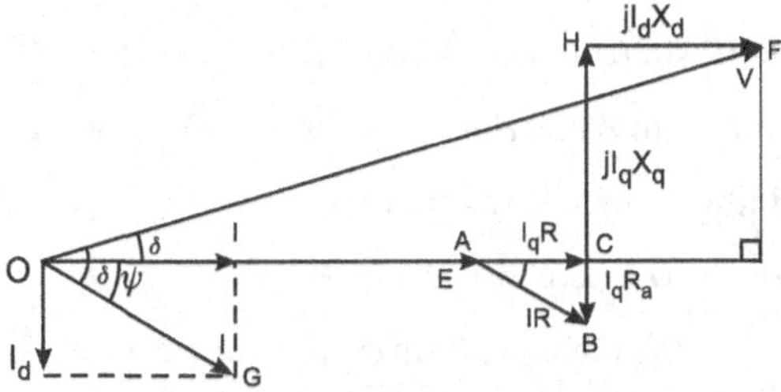

Fig. 8.10 Phasor diagram of a salient-pole synchronous motor at lagging power factor

After combining the above equations, we get

$$IX_q\cos(\phi - \delta) = IR\sin(\phi - \delta) + V\sin\delta$$
$$IX_q(\cos\phi\cos\delta + \sin\phi\sin\delta) = IR(\sin\phi\cos\delta - \cos\phi\sin\delta) + V\sin\delta$$
$$(V - IX_q\sin\phi - IR\cos\phi)\sin\delta = (IX_q\cos\phi - IR\sin\phi)\cos\delta$$
$$\therefore \qquad \tan\delta = \frac{IX_q\cos\phi - IR\sin\phi}{V - IX_q\sin\phi - IR\cos\phi}$$

(b) For Leading Power Factor cos ϕ

The phasor diagram at leading pf cos ϕ is shown in Fig. 8.11.

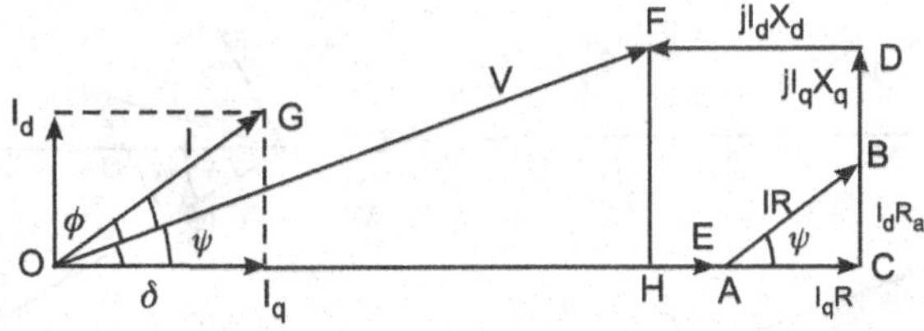

Fig. 8.11 Phasor diagram of a salient-pole synchronous motor at leading power factor

As per phasor diagram shown in Fig. 8.11.

$$OA = E,\ AC = IR,\ BD = I_qX_q,\ DF = I_dX_d$$

$$OF = V,\ AC = AB\cos\psi = IR\cos\psi = I_qR$$

$$BC = AB\sin\psi = IR\sin\psi = I_dR$$

Also, $$OA = OH + HA = OH + HC - AC = OH + FD - AC$$

$$E = V\cos\delta + I_dX_d - I_qR$$

$$HF = CD = CB + BD$$

$$V\sin\delta = I_dR + I_qX_q$$

$$\psi = \phi + \delta$$

$$I_d = I\sin\psi = I\sin(\phi + \delta)$$

$$I_q = I\cos\psi = I\cos(\phi + \delta)$$

Substituting the values of I_d and $I_{q,}$ we get,

$$V\sin\delta = IR\sin(\phi + \delta) + IX_q\cos(\phi + \phi)$$

$$= IR(\sin\phi\cos\delta + \cos\phi\sin\delta) + IX_q(\cos\phi\cos\delta - \sin\phi\sin\delta)$$

Separating the terms containing $\sin\delta$ and $\cos\delta$, we get

$$(V + IX_q\sin\phi - IR\cos\phi)\sin\delta = (IX_q\cos\phi + IR\sin\phi)\cos\delta$$

$$\therefore \qquad \tan\delta = \frac{IX_q\cos\phi + IR\sin\phi}{V + IX_q\sin\phi - IR\cos\phi}$$

(c) Unity Power Factor

$$\cos\phi = 1,\ \phi = 0°,\ \sin\phi = 0$$

$$\therefore \qquad \tan\delta = \frac{IX_q}{V - IR}$$

Phasor Diagrams of Salient-pole Synchronous Motor Neglecting Armature Resistance

In the modern synchronous machines or motors, the value of resistance in so small in comparison to reactance that it is neglected. In such cases, the phasor diagram of synchronous motor are modified as shown in Fig. 8.12.

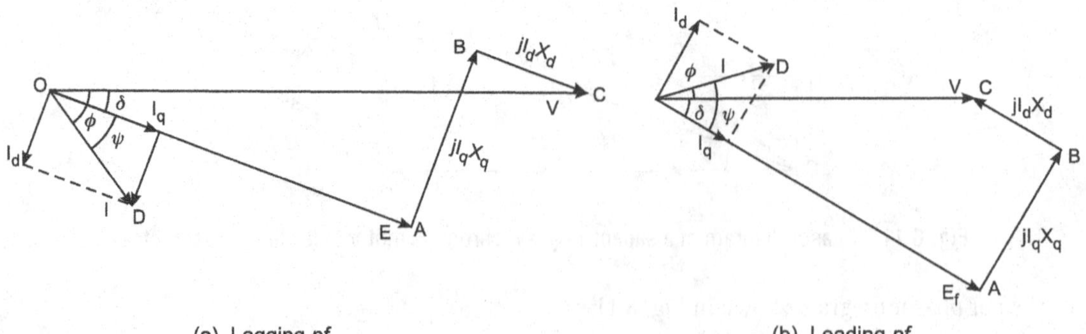

(a) Lagging *pf* (b) Leading *pf*

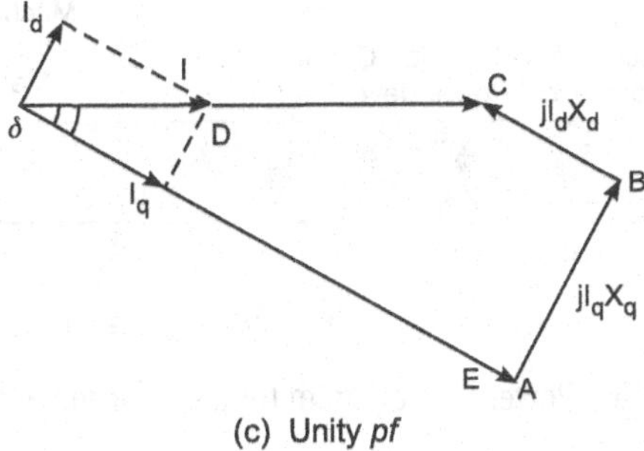

(c) Unity *pf*

Fig. 8.12 Phasor diagram of a salient-pole synchronous motor, neglecting *R*

8.9 Power Developed in a Salient-pole Synchronous Motor

The expressions for the power developed by a salient-pole synchronous generator derived in chapter 3 also apply to a salient-pole synchronous motor.

Real power developed per phase in watt,

$$P_{1\phi} = \frac{VE}{X_d}\sin\delta + \frac{V^2}{2}\left(\frac{1}{X_q} - \frac{1}{X_d}\right)\sin 2\delta$$

Real power developed by three phases in watt $P_{3\phi} = 3P_{1\phi}$

Reactive power per phase in VAR

$$Q_{1\phi} = \frac{VE}{X_d}\cos\delta - \frac{V^2}{2X_dX_d}[(X_d + X_q) - (X_d - X_q)\cos 2\delta]$$

Reactive power for three phases in VAR, $Q_{3\phi} = 3Q_{1\phi}$

All the power equations are true for both salient-pole synchronous generators and synchronous motors. The torque angle δ is positive for the generator and negative for the motor.

The expression $\frac{VE}{X_d}\sin\delta$ represents excitation power,

whereas the expression $\frac{V^2}{2}\left[\frac{1}{X_q} + \frac{1}{X_d}\right]\sin 2\delta$ represents reluctance power. Thus,

$$\text{Excitation power per phase} = \frac{VE}{X_d}\sin\delta$$

$$\text{Reluctance power per phase} = \frac{V^2}{2}\left(\frac{1}{X_q} - \frac{1}{X_d}\right)\sin 2\delta$$

8.10 Power Flow in a Synchronous Motor

The power flow diagram for a synchronous machine working as a motor is shown in Fig. 8.13.

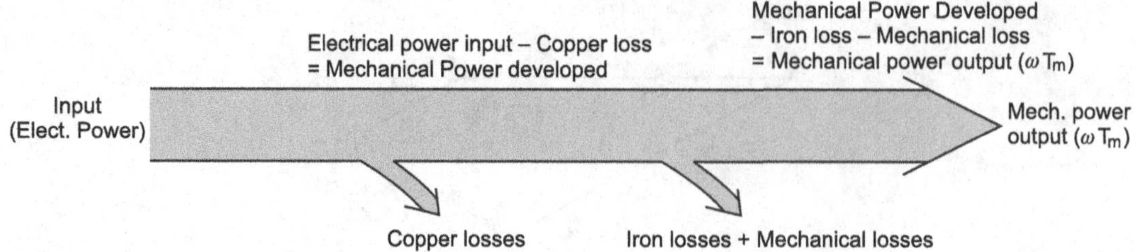

Fig. 8.13 Power flow diagram for a synchronous motor

Example 8.1

A 2.3 kV, 3-ϕ star-connected synchronous motor has Z_S = (0.2 + j2.2) ohm per phase. The motor is operating at 0.5 power factor leading with a line current of 200 A. Determine the generated emf per phase.

Solution by using phasor algebra

Supply voltage (phase value), $V = \dfrac{2.3 \times 1000}{\sqrt{3}} = 1328$ V

Taking V as reference vector, $\overline{V} = V\angle 0° = V \pm j0$

$p.f.$, $\cos\phi = 0.5$; $\sin\cos^{-1} 0.5 = 0.866$; $\phi = \cos^{-1} 0.5 = 60°$ leading.

Current, $\overline{I} = I\angle + \phi' = 200\angle 60°$

Impedance, $\overline{Z} = R + jX_S = (0.2 + j2.2)$ ohm

$$= \sqrt{(0.2)^2 + (2.2)^2}\ \angle\tan^{-1}\frac{2.2}{0.2} = 2.21\angle 84.8°$$

Now,

$$\overline{E} = \overline{V} - \overline{I}\,\overline{Z}_S$$
$$= (V \pm j0) - [(200\angle 60°)(2.21\angle 84.8°)]$$
$$= 1328 - [442\angle 144.8°]$$
$$= 1328 - [442\cos 144.8° + j\sin 144.8°]$$
$$= 1328 - [442 \times (-0.8171) + j0.5764)]$$
$$= 1328 - [-361.16 + j254.77]$$
$$= 1689.16 - j254.77 = 1708\angle -8.58°$$

Generated emf/phase, E = **1708 V** (*Ans.*)

Torque angle, δ = **8.58° lagging** (*Ans.*)

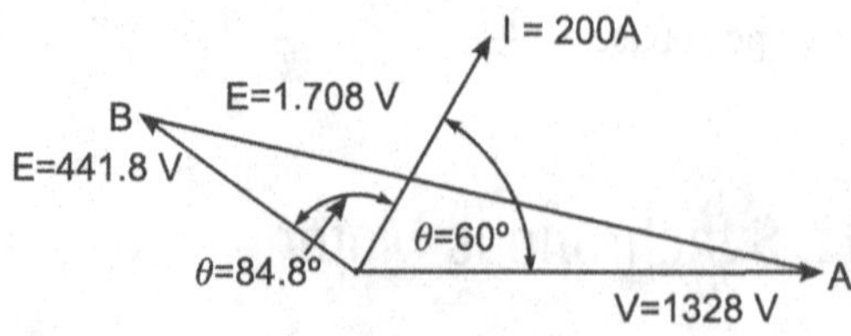

Fig. 8.14 Phasor diagram

Solution by vector method

Supply voltage (phase value), $V = \dfrac{2.3 \times 1000}{\sqrt{3}} = 1328$ V

Input current, $I = 200$ A

Power factor, $\cos\phi = 0.5$; $\phi = \cos^{-1} 0.5 = 60°$; $\sin\phi = 0.866$

Angle between E_r and I, $\theta = \tan^{-1}\dfrac{X_S}{R} = \tan^{-1}\dfrac{2.2}{0.2} = 84.8°$

Synchronous impedance/phase, $Z_S = \sqrt{R^2 + X_S^2} = \sqrt{(0.2)^2 + (2.2)^2} = 2.21\ \Omega$

Voltage drop in impedance, i.e., $E_r = IZ_S = 200 \times 2.21 = 442$ V

From phasor diagram shown in Fig. 8.14.

$$\text{Induced emf, } E = \sqrt{V^2 + E_r^2 - 2VE_r\cos(\theta + \phi)}$$

$$= \sqrt{(1328)^2 + (442)^2 - 2 \times 1328 \times 442 \times \cos(84.8° + 60°)}$$

$$= \mathbf{1708\ V}\ (Ans.)$$

New, from phasor diagram,

$$\frac{E}{\sin(\theta + \phi)} = \frac{E_r}{\sin\delta} \text{ or } \sin\delta = \frac{E_r}{E} \times \sin(\theta + \phi)$$

or

$$\sin\delta = \frac{442}{1708} \times \sin(84.8° + 60°)$$

$$= \frac{442}{1708} \times 0.5764 = 0.1492$$

Torque angle, $\delta = \sin^{-1} 0.1492 =$ **8.58° lagging** (*Ans.*)

Note: All the numerical problems can be solved by either of the two methods.

Example 8.2

A 3-phase, 400 V synchronous motor takes 60 A at a power factor of 0.8 leading. Determine the induced emf and the power supplied. The motor impedance per phase is (0.25 + j3.2) Ω.

Solution:

Here, Phase voltage, $V = \dfrac{400}{\sqrt{3}} = 231$ V ; $I = 60$ A; $\cos\phi = 0.8$ leading; $Z = (0.25 + j3.2)\ \Omega$.

For leading power factor $E^2 = (V\cos\phi - IR)^2 + (V\sin\phi + IX_S)^2$

$$= (231 \times 0.8 - 60 \times 0.25)^2 + (231 \times 0.6 + 60 \times 3.2)^2$$

$$= (170)^2 + (330)^2 \text{ or } E = \sqrt{(170)^2 + (330)^2}$$

$$E = 371.2 \text{ V}$$

$$\text{Line emf} = \sqrt{3} \times 371.2 = \mathbf{643\ V}\ (Ans.)$$

$$\text{Power supplied, } P_i = \sqrt{3}\, V_L I \cos\phi$$

$$= \sqrt{3} \times 400 \times 60 \times 0.8 = \mathbf{33255\ W}\ (Ans.)$$

Example 8.3

A 3-phase synchronous motor of 10 kW at 1100 V has synchronous reactance of 8 Ω per phase. Find the minimum current and the corresponding induced emf for full-load condition. The efficiency of the machine is 0.8. Neglect armature resistance.

Solution:

Here, Power output, $P_0 = 10$ kW; $V_L = 1100$ V; $X_S = 8\ \Omega$; $\eta = 0.8$

The current in the motor is minimum when the power factor is unity, that is, $\cos\phi = 1$.

$$\text{Motor input} = \frac{\text{motor output}}{\text{efficiency}}$$

$$P_i = \frac{P_0}{\eta} = \frac{10}{0.8} = 12.5\,\text{kW}$$

$$I_L = \frac{P_i}{\sqrt{3}\, V_L \cos\phi} = \frac{12.5 \times 10^3}{\sqrt{3} \times 1100 \times 1} = \mathbf{6.56\ A}\ (Ans.)$$

Phase voltage,

$$V = \frac{V_L}{\sqrt{3}} = \frac{1100}{\sqrt{3}} = 635\ \text{V}$$

For unity power factor

$$E^2 = V^2 + (IX_S)^2 = (635)^2 + (6.56 \times 8)^2$$

$\therefore$

$$E = \mathbf{637.25}\ \text{V per phase}\ (Ans.)$$

Example 8.4

A 6600 V, 3-phase, star-connected synchronous motor draws a full-load current of 80 A at 0.8 p.f. leading. The armature resistance is 2.2 Ω and synchronous reactance 22 Ω per phase. If the stray losses of the machine are 3200 W, determine: (a) the emf induced; (b) the output power; (c) the efficiency.

Solution:

$$V = \frac{6600}{\sqrt{3}} = 3810.6\ \text{V (phase value)}$$

$$I = 80\ \text{A},\ \cos\phi = 0.8,\ \sin\phi = 0.6;\ R = 2.2\ \Omega;\ X = 22\ \Omega$$

For leading power factor

$$\overline{E} = (V - IR\cos\phi + IX_S \sin\phi) - j\,(IX_S\cos\phi + IR\sin\phi)$$

$$= (3810.6 - 80 \times 2.2 \times 0.8 + 80 \times 22 \times 0.6) - j(80 \times 22 \times 0.8 + 80 \times 2.2 \times 0.6)$$

$$= 4725.8 - j1513.6 = 4962\ \angle{-17.76^\circ}\ \text{V}$$

Induced line emf = $\sqrt{3} \times 4962$ = **8594 V** (*Ans.*)

Power input = $\sqrt{3} \times V_L I \cos\phi = \sqrt{3} \times 6600 \times 80 \times 0.8 = 731618$ W

Total copper loss = $3I^2 R = 3 \times 80^2 \times 2.2 = 42240$ W

Stray loss = 3200 W

Power output = power input – copper losses – stray loss

= 731618 –42240 – 3200 = **686178 W** (*Ans.*)

Efficiency, $\eta = \dfrac{\text{output}}{\text{input}} = \dfrac{686178}{731618}$. = **0.9379 p.u.** (*Ans.*)

Example 8.5

A 1000 kVA, 11 kV, 3-phase star-connected synchronous motor has an armature resistance and reactance per phase of 3.5 Ω and 40 Ω, respectively. Determine the induced emf and angular retardation of the rotor when fully loaded at (a) unity power factor, (b) 0.8 power factor lagging,] (c) 0.8 power factor leading.

Solution:

$$V = \frac{11 \times 1000}{\sqrt{3}} = 6351; \; I_L = \frac{kVA \times 1000}{\sqrt{3}\, V_L} = \frac{1000 \times 1000}{\sqrt{3} \times 11000} = 52.5 \text{ A}$$

(a) At unity power factor

$$\overline{V} = V\angle 0° = V \pm j0$$

$$\cos\phi = 1.0, \; \phi = 0°, \; \overline{I} = 52.49 \angle 0° \text{ A}$$

$$\overline{E} = \overline{V} - \overline{I}\,(R + jX_S)$$

$$= 6351 - (52.49 \angle 0°)(3.5 + j40) = 6351 - (183.7 + j2099.6)$$

$$E\angle\delta = 6167.3 - j2099.6 = 6515 \angle{-18.8°} \text{ V}$$

$E = 6515$ V per phase; $E_L = \sqrt{3} \times 6515$ = **11.284 kV** (*Ans.*)

δ = **–18.8°** (*Ans.*)

(b) At 0.8 power factor lagging

$$\cos\phi = 0.8, \; \sin\phi = 0.6; \; \overline{I} = \overline{I} \angle -\phi$$

$$\overline{E} = \overline{V} - \overline{I}\,\overline{Z}_S = V - (I\angle{-\phi})(R + jX_S) = V - (I\cos\phi - j\,I\sin\phi)(R + jX_S)$$

$$= (V - I\,R\cos\phi - I\,X_S\sin\phi) - j(IX_S\cos\phi - IR\sin\phi)$$

$$= (6351 - 52.49 \times 3.5 \times 0.8 - 52.49 \times 40 \times 0.6)$$

$$- j(52.49 \times 40 \times 0.8 - 52.49 \times 3.5 \times 0.6)$$

$$E\angle\delta = 4944 - j1569.5 = 5187 \angle{-17.6°} \text{ V}$$

∴ E = **5187 V per phase**, δ = – **17.6°** (*Ans.*)

Induced line voltage, $E_L = \sqrt{3} \times 5187$ = **8984 V** (*Ans.*)

(c) At 0.8 power factor leading

$$\bar{I} = I\angle+\phi$$

$$\bar{E} = \bar{V} - \bar{I}\bar{Z}_S = V - (I\angle+\phi)\,(R + jX_S)$$

$$= (V - IR\cos\phi + IX_S\sin\phi) - j\,(IX_S\cos\phi + IR\sin\phi)$$

$$= (6351 - 52.49\times3.5\times0.8 + 52.49\times40\times0.6)$$

$$-j(52.49\times40\times0.8 + 52.49\times3.5\times0.6)$$

$$E\angle\delta = 7463.8 - j1790 = 7675\ \angle{-13.48°}\ \text{V}$$

E = **7675 V per phase** (*Ans.*)

δ = **–13.48°** (*Ans.*)

Induced line voltage,

$E_L = \sqrt{3}\times7675$ = **13.3 kV** (*Ans.*)

Example 8.6

The excitation corresponding to no-load voltage of a 3-phase synchronous motor running at 1500 rpm is kept constant. Determine the power input, power factor and torque developed for an armature current of 200 A if the synchronous reactance is 5 Ω per phase and armature resistance is neglected.

Solution:

Supply voltage per phase, $V = \dfrac{3000}{\sqrt{3}} = 1732\ \text{V}$

Induced emf per phase, $E = \dfrac{3000}{\sqrt{3}} = 1732\ \text{V}$

$$\bar{Z}_S = R + jX_S = 0 + j5 = 5\ \angle90°\ \Omega$$

$$\bar{E} = \bar{V} - \bar{I}\bar{Z}_S$$

If V is taken as reference phasor, then for lagging power factor,

$$\bar{I} = \bar{I}\angle-\phi$$

$$\bar{E} = V - (I\angle-\phi)\,(5 < 90°) = V - 5\,I\angle(90° - \phi)$$

$$= V - 5\times200\ \angle90 - \phi$$

$$= V - 1000\,[\cos(90° - \phi) + j\sin(90° - \phi)]$$

$$= V - 1000\,(\sin\phi + j\cos\phi)$$

$$= (V - 1000\sin\phi) - j\,1000\cos\phi$$

$$E^2 = (V - 1000\sin\phi)^2 + (1000\cos\phi)^2$$

$$= V^2 - 2V\times1000\sin\phi + (1000\sin\phi)^2 + (1000\cos\phi)^2$$

$$1732^2 = 1732^2 - 2\times1732\times1000\sin\phi + (1000)^2$$

$$2 \times 1732 \times 1000 \sin \phi = (1000)^2$$

$$\sin \phi = \frac{1000}{2 \times 1732} = 0.2887 \; ; \; \phi = \sin^{-1} 0.2887 = 16.78°$$

$$\cos \phi = \mathbf{0.9574 \text{ (lagging)}} \; (Ans.)$$

Input power $$p_i = \sqrt{3}\, V_L I \cos \phi = \sqrt{3} \times 3000 \times 200 \times 0.9574$$

$$= 994959 \text{ W} = \mathbf{994.96 \text{ kW}} \; (Ans.)$$

Also, $$P_i = \frac{2\pi\, N_S T}{60}$$

$\therefore$ $$\text{Torque } T = \frac{P_i \times 60}{2\pi\, N_S} = \frac{994959}{2\pi \times 1500} = \mathbf{7712.5 \text{ Nm}} \; (Ans.)$$

Example 8.7

A 3-ϕ, star-connected 6600 V synchronous motor has synchronous reactance per phase of 20 Ω. For a certain load the input is 900 kW at normal voltage and the induced line emf is 8500 V. Determine the line current and power factor.

Solution:

$$V_L = 6600 \text{ V}, V = \frac{6600}{\sqrt{3}} = 3810.5 \text{ V} \; ; X_S = 20 \; \Omega, R = 0, Z_S = jX_S = j20 \text{ ohm}$$

$$\text{Input, } P_i = \sqrt{3}\, V_L I \cos \phi \quad \text{or} \quad I \cos \phi = \frac{900 \times 10^3}{\sqrt{3} \times 6600} = 78.73 \text{ A}$$

$$\text{Induced emf/phase } E = \frac{8500}{\sqrt{3}} = 4907.5 \text{ V}$$

Since $E > V$, the power factor is leading

$$\overline{E} = \overline{V} - \overline{I}\,\overline{Z}$$

$$= (V + j0) - (I \angle \phi)(20 \angle 90°) = V - 20\, I \angle 90° + \phi$$

$$= V - 20\, I\, [-\sin \phi + j \cos \phi]$$

$$\overline{E} = (V + 20\, I \sin \phi) - j\,(20\, I \cos \phi)$$

$$E^2 = (V + 20\, I \sin \phi)^2 + (20\, I \cos \phi)^2$$

$$(4907.5)^2 = (3810.5 + 20\, I \sin \phi)^2 + (20 \times 78.73)^2$$

$$(3810.5 + 20\, I \sin \phi) = \sqrt{(4907.5)^2 - (20 \times 75.73)^2} = 4648$$

$$I \sin \phi = \frac{4648 - 3810.5}{20} = 41.876$$

$\therefore$ $$I = I \cos \phi + jI \sin \phi = 78.73 + j41.876 = 89.17 \angle 28° \text{ A}$$

$$\cos \phi = \cos 28° = \mathbf{0.8829 \text{ leading}} \; (Ans.)$$

and $$I = \mathbf{89.17 \text{ A}} \; (Ans.)$$

Example 8.8

The resistance and synchronous reactance per phase of a 75 kW, 400 V, 4-pole, 3-phase, star-connected synchronous motor is 0.04 Ω and 0.4 Ω, respectively. Determine for full load 0.8 pf leading the open-circuit emf per phase and gross mechanical power developed. Assume an efficiency of 92.5%.

Solution:

Here, Motor output = 75 kW; $V_L = 400$ V; $P = 4$; $R = 0.04\ \Omega$; $X_s = 0.4\ \Omega$

$$\text{Motor input} = \frac{\text{Motor output}}{\eta} = \frac{75}{0.925} = 81.081 \text{ kW or } 81081 \text{ watt}$$

$$\text{Armature current, } I = \frac{\text{Motor input}}{\sqrt{3}\, V_L \cos\phi} = \frac{81081}{\sqrt{3} \times 400 \times 0.8} = 146.3 \text{ A}$$

$$\text{Supply voltage/phase, } V = \frac{400}{\sqrt{3}} = 231 \text{ V}$$

$$\text{Resultant voltage, } E_r = IZ_s = 146.3\sqrt{(0.04)^2 + (0.4)^2} = 58.81 \text{ V}$$

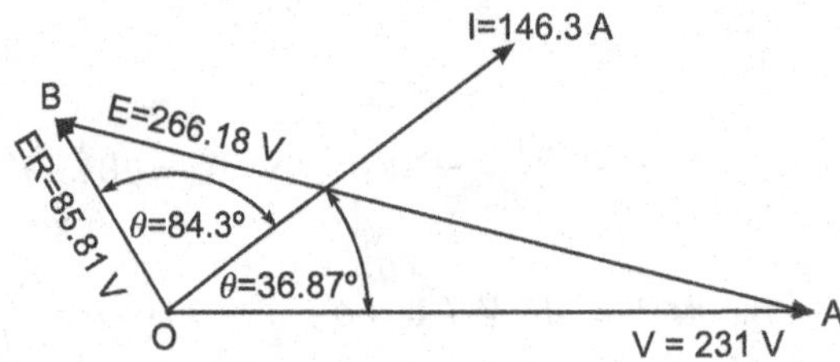

Fig. 8.15 Phasor diagram

$$\text{Internal phase angle, } \theta = \tan^{-1}\frac{X_s}{R} = \tan^{-1}\frac{0.4}{0.04} = 84.3°$$

From phasor diagram shown in Fig. 8.15, open circuit emf/phase,

$$E = \sqrt{V^2 + E_r^2 - 2VE_r \cos(\theta + \phi)}$$

$$= \sqrt{(231)^2 + (58.81)^2 - 2 \times 231 \times 58.81 \times \cos(84.3° + 36.87°)}$$

$$= \sqrt{70855} = \mathbf{266.18\ V}\ (Ans.)$$

Gross mechanical power developed for all the 3 phases,

$$P_m = P_{in} - 3I^2R = 81081 - 3 \times (146.3)^2 \times 0.04$$

$$= 78512 \text{ watt or } \mathbf{78.512\ kW}\ (Ans.)$$

Example 8.9

The effective resistance and synchronous reactance of a 2000 V, 3-phase, star-connected synchronous motor is 0.2 Ω and 2.2 Ω, respectively. The input is 800 kW at normal voltage and the generated line emf is 2500 V. Calculate the line current and power factor.

Solution:

$$\text{Supply voltage/phase, } V = \frac{2000}{\sqrt{3}} = 1155 \text{ V}$$

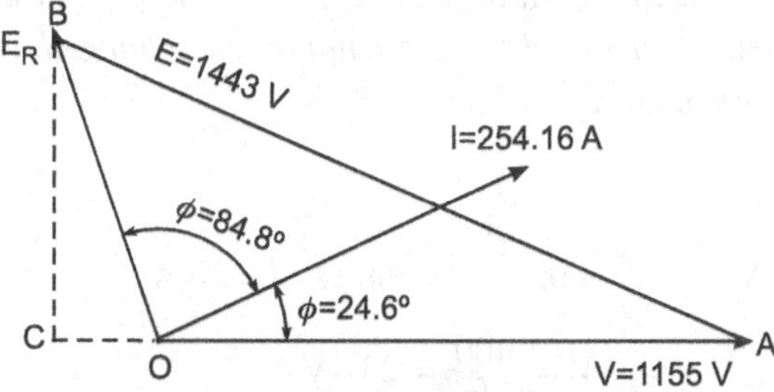

Fig. 8.16 Phasor diagram

$$\text{Induced emf/phase, } E = \frac{2500}{\sqrt{3}} = 1443 \text{ V}$$

$$\text{Internal phase angle of the motor, } \theta = \tan^{-1}\frac{X_s}{R} = \tan^{-1}\frac{2.2}{0.2} = 84.8°$$

Since induced emf is greater than supply voltage, the motor must be operating with a leading power factor.

$$\bar{I} = I\angle\phi \text{ or } \bar{I} = (I\cos\phi + jI\sin\phi)$$

$$\text{Since power input} = \sqrt{3}\,V_L I_L \cos\phi$$

or
$$I\cos\phi = \frac{\text{Power input}}{\sqrt{3}\,V_L} = \frac{800 \times 1000}{\sqrt{3} \times 2000} = 231 \text{ A}$$

$$\text{Impedance voltage/per phase, } \bar{E}_r = \bar{I}\bar{Z}_s = (I\cos\phi + jI\sin\phi)(0.2 + j2.2)$$

$$= (231 + jI\sin\phi)(0.2 + j2.2.)$$

$$= (46.2 - 2.2\,I\sin\phi) + j(508.2 + 0.2\,I\sin\phi)$$

Now from phasor diagram shown in Fig. 8.16

$$OA = 1155 \text{ V}; AB = 1443 \text{ V}; \theta = 84.8°$$

$$OC = 46.2 - 2.2\,I\sin\phi;\ BC = 508.2 + 0.2\,I\sin\phi$$

From the right angled ΔABC we have

$$AB^2 = BC^2 + AC^2 = BC^2 + (AO + OC)^2$$

or
$$(1443)^2 = (508.2 + 0.2\,I\sin\phi)^2 + [1155 - (46.2 - 2.2\,I\sin\phi)]^2$$

or
$$I(\sin\phi) = 106 \text{ A}$$

$$\text{Line current, } I_L = I = \sqrt{(I\cos\phi)^2 + (I\sin\phi)^2} = \sqrt{(231)^2 + (106)^2}$$

$$= 254.16 \text{ A } (Ans.)$$

$$\text{Power factor, } \cos\phi = \frac{I\cos\phi}{I} = \frac{231}{254.16} = \mathbf{0.909\ (leading)}\ (Ans.)$$

Example 8.10

The resistance and synchronous impedance of a 400 V, 6-pole, 3-phase, 50 Hz, star-connected synchronous motor is 0.5 Ω and 4Ω per phase, respectively. It takes a current of 20 A at unity power factor when operating with a certain field current. If the load torque is increased until the line current is increased to 60 A, the field current remaining unchanged, calculate the gross torque developed and the new power factor.

Solution:

Here $R = 0.5\ \Omega$; $Z_s = 4\ \Omega$; $X_s = \sqrt{4^2 - 0.5^2} = 3.968\ \Omega$; $I = 20$ A

$$V = \frac{400}{\sqrt{3}} = 231\ \text{V};$$

At unity p.f., i.e., $\cos\phi = 1$, $\sin\phi = 0$

$$\bar{E} = V - IR - jIX_s$$

$$= 231 - 20 \times 0.5 - j20 \times 3.968 = 221 - j79.36$$

$$= 224.8\ \angle{-19.75^\circ}\ \text{V}$$

When load on the motor is increased, but the field current is kept constant, E remains the same. For lagging power factor $\cos\phi$,

$$E^2 = (V\cos\phi + IR)^2 + (V\sin\phi + IX_s)^2$$

$$= V^2 + (IZ_s)^2 - 2VIZ_s\cos(\theta - \phi)$$

$$(224.8)^2 = 231^2 + (60 \times 4)^2 - 2 \times 231 \times 60 \times 4 \times \cos(\theta - \phi)$$

$$\cos(\theta - \phi) = \frac{231^2 + (60 \times 4)^2 - (224.8)^2}{2 \times 231 \times 60 \times 4} = 0.545$$

$$\theta - \phi = \cos^{-1} 0.545 = 56.97^\circ$$

$$\tan\theta = \frac{X_s}{R} = \frac{3.968}{0.5} = 7.936;\ \theta = \tan^{-1} 7.936 = 82.82^\circ$$

$$\therefore \quad \phi = \theta - 56.97^\circ = 82.82^\circ - 56.97^\circ = 25.85^\circ$$

New power factor $\cos\phi = \cos 25.85^\circ =$ **0.9 (lag)** (*Ans.*)

$$\text{Motor input, } P_i = \sqrt{3}\,V_L I\cos\phi = \sqrt{3} \times 400 \times 60 \times 0.9 = 37412\ \text{W}$$

$$\text{Total armature copper loss} = 3I^2R = 3 \times (60)^2 \times 0.5 = 5400\ \text{W}$$

Electrical power converted into mechanical power,

$$P_m = P_i - 3I^2R$$

$$= 37412 - 5400 = 32012\ \text{W}$$

$$\text{Synchronous speed, } N_s = \frac{120f}{P} = \frac{120 \times 50}{6} = 1000\ \text{r.p.m.}$$

$$\frac{2\pi N_s T}{60} = P_i$$

or $$T = \frac{P_i \times 60}{2\pi N_s} = \frac{32012 \times 60}{2\pi \times 1000}$$

$= \mathbf{305.7\ Nm}$ (*Ans.*)

Example 8.11

The excitation to a 2000 V, 3-phase, 4-pole, Y-connected synchronous motor running at 1500 rpm is kept constant to produce an open-circuit voltage of 2000 V. The resistance is negligible as compared to synchronous reactance of 3 Ω/phase. Determine the power input, power factor and torque developed for an armature current of 200 A.

Solution:

Here, $V_L = 2000$ V; $E_L = 2000$ V; $I = 200$ A

Supply voltage per phase, $V = \frac{2000}{\sqrt{3}} = 1154.7$ V

Induced emf per phase, $E = \frac{2,000}{\sqrt{3}} = 1154.7$ V

Impedance drop per phase, $E_r = IZ_s = 200 \times 3 = 600$ V

Internal angle, $\theta = 90°$ ∵ resistance is negligible

Assuming armature current lagging behind the supply voltage by an angle ϕ, as shown in phasor diagram (Fig. 8.17).

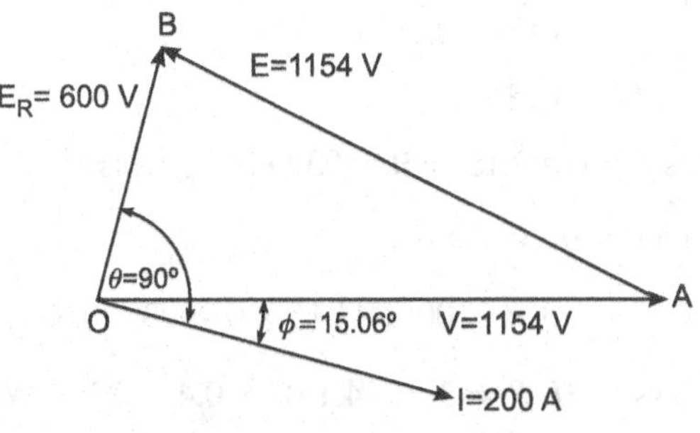

Fig. 8.17 Phasor diagram

In ΔAOB we have $$E^2 = V^2 + E_r^2 - 2VE_r \cos(90° - \phi)$$

or $$1154.7^2 = 1154.7^2 + 600^2 - 2 \times 1154.7 \times 600 \sin \phi$$

or $$\sin \phi = \frac{600^2}{2 \times 1154.7 \times 600} = 0.26$$

$$\phi = \sin^{-1} 0.26 = 15.06°$$

Power factor = cos 15.06° = **0.966 lagging** (*Ans.*)

Power input $= \sqrt{3} V_L I_L \cos \phi = \sqrt{3} \times 2000 \times 200 \times 0.966$

$= 669029$ W

or $$\text{Torque developed} = \frac{\text{Power input} - \text{copper losses in armature}}{2\pi N_s / 60}$$

Since armature resistance is neglected, copper losses in the armature are considered to be zero

$$\therefore \qquad T = \frac{669029}{2\pi \times 1500 / 60} = \mathbf{4259\ Nm}\ (Ans.)$$

Example 8.12

A 2200 V, 3-phase, star-connected, 8-pole synchronous motor has impedance per phase equal to (0.4 + j6) Ω. When the motor runs at no-load, the field excitation is adjusted so that E is made equal to V. When the motor is loaded, the rotor is retarded by 3° mechanical. Calculate the armature current, power factor and power of the motor. What is the maximum power the motor can supply without falling out of step?

Solution:

$$E = V = \frac{2200}{\sqrt{3}} = 1270.2 \text{ V} = \overline{Z}_s = (0.4 + j6) = 6.0133 \angle 86.18°$$

$$\delta = 3° \text{ mechanical} = 3 \times \frac{P}{2} = \frac{3 \times 8}{2} = 12° \text{ (elec.)}$$

$$\overline{I} = \frac{\overline{V} - \overline{E}}{\overline{Z}_s} = \frac{V\angle 0° - E\angle -\delta}{Z_s \angle \theta} = \frac{1270.2 \angle 0° - 1270.2 \angle -\delta}{6.0133 \angle 86.18°}$$

$$= \frac{1270.2\,[1 - (\cos 12° - j \sin 12°)]}{6.0133 \angle 86.18°} = \frac{1270.2 \times 0.2090 \angle 84°}{6.0133 \angle 86.18°}$$

$$= 44.15 \angle -2.18° \text{A}$$

Armature current, I = **44.15 A**

Power factor, $\cos \phi = \cos 2.18°$ = **0.9993 lagging** (*Ans.*)

$$\text{Total power input} = \sqrt{3}\, V_L I \cos \phi$$

$$= \sqrt{3} \times 2200 \times 44.15 \times 0.9993 = 168116 \text{ W} = \mathbf{168.116\ kW}\ (Ans.)$$

$$\text{Total copper loss} = 3I^2R = 3 \times (44.15)^2 \times 0.4 = 2339 \text{ W}$$

Power developed by the motor = motor input – copper losses

= 168116 – 2339 = 165777 W = 165.777 kW

Maximum power $$P_{max} = \frac{EV}{Z_S} - \frac{E^2}{Z_S^2} R = \frac{1270.2 \times 1270.2}{6.0133} - \frac{(1270.2)^2}{(6.0133)^2} \times 0.4$$

= 250459 W = **250.459** kW (*Ans.*)

Example 8.13

The excitation supplied to a 3-phase, star-connected, 30 kW, 660 V, 50 Hz, 20-pole synchronous motor operating at rated voltage is such that it generates the same emf per phase as that of the supply voltage per phase. When loaded the motor is retarded by 3 mechanical degrees from its synchronous position. The synchronous reactance and armature resistance are 10 Ω and 1 Ω per phase. Calculate (i) armature current per phase (ii) the power per phase and the total power drawn by the motor from the bus and (iii) the developed power.

Solution:

Here, $V_L = 660$ V; $X_S = 10\ \Omega$; $R = 1\ \Omega$; $P = 20$; $f = 50$ Hz.

$$\text{Supply voltage per phase, } V = \frac{660}{\sqrt{3}} = 381 \text{ V}$$

Induced emf per phase, $E = 381$ V (given)

$$\text{Load or torque angle, } \delta = \frac{P}{2} \times \text{angle of retardation in mechanical degrees}$$

$$= \frac{20}{2} \times 3 = 30° \text{ electrical}$$

$$\text{Impedance per phase, } Z_s = \sqrt{1^2 + 10^2} = 10.05\ \Omega$$

Now from ΔAOB of phasor diagram shown in Fig. 8.18

$$E_r = \sqrt{V^2 + E^2 - 2VE \cos \delta}$$

$$= \sqrt{(381)^2 + (381)^2 - 2 \times 381 \times 381 \cos 30°} = 197.2 \text{ V}$$

(*i*) Armature current per phase, $I = \dfrac{E_r}{Z_s} = \dfrac{197.2}{10.05} = 19.62$ A (*Ans.*)

From ΔAOB of phasor diagram shown in Fig. 8.18.

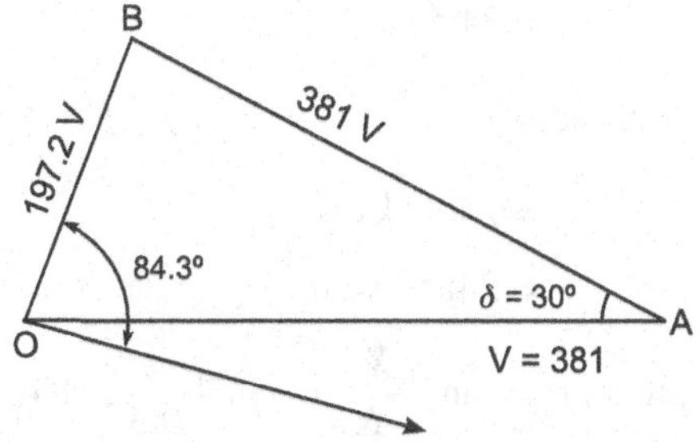

Fig. 8.18 Phasor diagram

$$\frac{AB}{\sin \angle AOB} = \frac{OB}{\sin \angle OAB}$$

$$\frac{381}{\sin (\theta + \phi)} = \frac{197.2}{\sin 30°}$$

or $$\sin (\theta + \phi) = \frac{381}{197.2} \sin 30° = 0.966$$

or $$\theta + \phi = \sin^{-1} 0.966 = 75°$$

and $$\theta = \tan^{-1} \frac{X_s}{R_e} = \tan^{-1} \frac{10}{1} = 84.3°$$

$$\phi = 75° - \theta = 75 - 84.3° = -9.3° \text{ i.e., } 19.3° \text{ lagging}$$

(*iii*) Power per phase $= VI\cos\phi = 381 \times 19.62 \times \cos 9.3°$

$= 7377$ W or **7.377 W** (*Ans.*)

Total power drawn $= 3 \times 7.377 = 22.313$ kW (*Ans.*)

Power developed = Total power drawn $- 3\,I^2 R$

$$= 22.313 - \frac{3 \times (19.62)^2 \times 1}{1000}$$

$= 22.313 - 1.155 =$ **21.158 kW** (*Ans.*)

Example 8.14

A 500 V, 6-pole, 3-phase, 50 Hz, star-connected synchronous motor has a resistance and synchronous *reactance of 0.3 Ω and 3 Ω per phase, respectively. The open-circuit voltage is 600 V. If the friction and core losses total 1 kW, calculate the line current and power factor when the motor output is 100 hp.*

Solution:

$$\text{Supply voltage per phase, } V = \frac{500}{\sqrt{3}} = 288.7 \text{ V}$$

$$\text{Induced emf per phase, } E = \frac{600}{\sqrt{3}} = 346.41 \text{ V}$$

Total power developed, $3\,P_{mech}$ = output + mech. loss

$= 100 \times 0.7355 + 1$

$= 74.55$ kW

$$\text{Power developed per phase} = \frac{74.55}{3}$$

$= 24.85$ kW

or $= 24850$ watt

$$\text{Internal angle, } \theta = \tan^{-1}\frac{X_s}{R} = \tan^{-1}\frac{3}{0.3} = 84.3°$$

$$\text{Synchronous impedance/phase, } Z_s = \sqrt{0.3^2 + 3^2} = 3.015\ \Omega$$

Power developed per phase is given as

$$P_{\text{mech}} = \frac{EV}{Z_s}\cos(\theta - \delta) - \frac{E^2}{Z_s}\cos\theta$$

or $$24850 = \frac{346.41 \times 288.7}{3.015}\cos(84.3° - \delta) - \frac{(346.41)^2}{3.015}\cos 84.3°$$

or $$24850 = 33170.3\cos(84.3° - \delta) = 3953$$

or $$\cos(84.3° - \delta) = 0.8684$$

or $$84.3° - \delta = \cos^{-1} 0.8684 = 29.73°$$

or $$\delta = 84.3° - 29.73° = 54.57° \text{ (lag)}$$

From phasor diagram shown in Fig. 8.19.

$$E_r = \sqrt{V^2 + E^2 - 2VE \cos \delta}$$

$$= \sqrt{288.7^2 + 346.41^2 - 2 \times 288.7 \times 346.41 \cos 54.57°}$$

$$= 295.6 \text{ V}$$

$$\text{Line current} = \text{Phase current, } I = \frac{E_R}{Z_s} = \frac{295.6}{3.015} = \mathbf{98.1\ A} \text{ (}Ans.\text{)}$$

Again considering phasor diagram

$$\sin(\theta + \phi) = \frac{E}{E_r} \sin \delta = \frac{346.41}{295.6} \sin 54.57° = 0.9542$$

or $$\phi + \theta = \sin^{-1} 0.9542 = 72.7°$$

$$\text{or } \theta = 72.7° - 84.3° = -11.6° \text{ lagging}$$

$\therefore$ $$\text{Power factor} = \cos 11.6° = \mathbf{0.98\ lagging} \text{ (}Ans.\text{)}$$

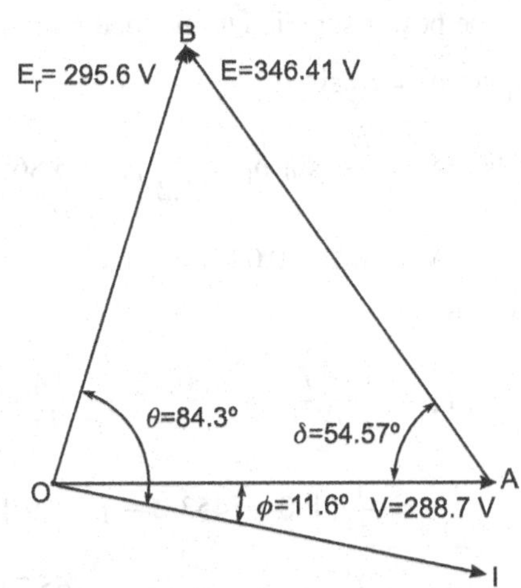

Fig. 8.19 Phasor diagram

Example 8.15

The 400 V, 50 kVA, 0.8 power factor leading delta connected synchronous motor has synchronous reactance of 3 ohm, resistance neglected. It is supplying a 12 kW load with initial power factor of 0.86 lagging. The windage and friction losses are 2.0 kW and the core losses are 1.5 kW. Determine the line current, armature current and excitation voltage. If the flux of the motor is increased by 20 percent determine the excitation voltage, armature current and the new power factor.

Solution:

$$P_i = P_0 + all\ losses;\ P_i = P_o + P_{mech} + P_{core} + P_{elec} = 12 + 2 + 1.5 + 0 = 15.5 \text{ kW}$$

Now, $$P_i = 3\,VI_1 \cos\phi$$

$\therefore$ $$I_1 = \frac{P_i}{3V\cos\phi} = \frac{15.5\times10^3}{3\times400\times0.86} = 15\text{ A}$$

Since the power factor of the motor is 0.86 lagging, the phasor armature current is given by

$$\bar{I}_1 = I_1\angle-\cos^{-1}0.86 = 15\angle-30.68°\text{A}$$

$$\text{Line current } I_L = \sqrt{3}\,I_1 = \sqrt{3}\times15 = \mathbf{25.98\ A}\ (\textit{Ans.})$$

The excitation voltage E is given by

$$\bar{E}_1 = \bar{V} - \bar{I}_1\bar{Z}_s = V - jI_1X_s$$
$$= 400\angle0° - j3\,(15\angle-30.68°)$$
$$= 400\angle0° - 45\angle90° - 30.68° = 400 - 45\angle59.32°$$
$$= 400 - (22.96 + j38.7) = 377.04 = j38.7 = 379\angle-5.86°\text{ V}$$

If the flux is increased by 20%, then E will also increase by 20%.

$\therefore$ $$E_2 = 1.2\,E_1 = 1.2\times379 = 454.8\text{ V}$$

When flux is increased by 20%, the power supplied to the load remains constant,

$\therefore$ $$E_1\sin\delta_1 = E_2\sin\delta_2$$

$$\sin\delta_2 = \frac{E_1}{E_2}\sin\delta_1 = \frac{1}{1.2}\sin(-5.86°) = -0.085$$

$$\delta_2 = \sin^{-1}(-0.085) = -4.88°$$

The new armature current is given by

$$\bar{I}_2 = \frac{\bar{V} - E_2}{jX_s} = \frac{400\angle0° - 454.8\angle-4.88°}{j3}$$

$$= \frac{1}{j3}[400 - (453.2 - j38.69)]$$

$$= \frac{1}{j3}(-53.2 + j38.69) = \frac{65.78\angle148°}{3\angle90°} = 21.93\angle58°\text{ A}$$

New power factor of the motor, $\cos\phi_2 = \cos 58° =$ **0.53 leading** (*Ans.*)

Example 8.16

A 3-phase, 11 kV, 5000 kVA, 50 Hz, 1000 rpm star-connected synchronous motor operates at full load at a power factor of 0.8 leading. The synchronous reactance is 60% and the resistance may be neglected. Calculate the synchronising power and torque per mechanical degree of angular displacement. What is the value of maximum torque and the ratio of maximum to full-load torque?

Solution:

Here, Rating = 5000 kVA; f = 50 Hz; N_s = 1000 rpm; $\cos\phi$ = 0.8 leading;

$$X_s = 60\%;\ \delta_{\text{mech}} = 1^\circ;\ V_L = 11 \text{ kV}$$

$$I = \frac{kVA \times 10^3}{\sqrt{3} V_L} = \frac{5000 \times 10^3}{\sqrt{3} \times 11 \times 10^3} = 262.4 \text{ A}$$

$$\text{Phase voltage, } V = \frac{V_L}{\sqrt{3}} = \frac{11 \times 10^3}{\sqrt{3}} = 6351 \text{ V}$$

$$\text{Now, } \% X_s = \frac{IX_s}{V} \times 100 \text{ or } X_s = \frac{\% X_s}{100} \times \frac{V}{I} = \frac{60}{100} \times \frac{6351}{262.4} = 14.52\ \Omega$$

$$\bar{E} = \bar{V} - \bar{I}\bar{Z}_s = V - (1 \angle \phi)(X_s \angle 90^\circ)$$

$$= (V - IX_S \sin \phi) - jIX_s \cos \phi$$

$$= (6351 + 262.4 \times 14.52 \times 0.6) - j\, 262.4 \times 14.52 \times 0.8$$

$$= (6351 + 2286) - j\, 3048 = 8637 - j3048$$

$$= 9159 \angle -19.44^\circ \text{V}$$

$\therefore$ $$E = 9159 \text{ V},\ \delta = -19.44^\circ$$

Now, $$f = \frac{PN_s}{120}$$

or $$P = \frac{120 \times f}{N_s} = \frac{120 \times 50}{1000} = 6;$$

or $$\text{pair of poles, } p = \frac{P}{2} = \frac{6}{2} = 3$$

$$\text{Synchronising power, } P_{syn} = \left(\frac{3VE}{X_s} \cos \delta\right) p \frac{\pi}{180}$$

$$= \frac{3 \times 6351 \times 9159}{14.52} (\cos 19.44^\circ) \times \frac{3\pi}{180} = \mathbf{593404\ W}\ (\textit{Ans.})$$

$$\text{Synchronising torque, } T_{syn} = \frac{P_{syn}}{2\pi N_s / 60} = \frac{593404}{2\pi \times \frac{1000}{60}}$$

$$= \mathbf{5666.6\ Nm\ per\ mech\ degree}\ (\textit{Ans.})$$

$$\text{Maximum power, } P_{max} = \frac{3VE}{X_s} = \frac{3 \times 6352 \times 9159}{14.52} = 12018349 \text{ W}$$

$$\text{Maximum torque} = \frac{P_{\max}}{2\pi N_s / 60} = \frac{12018349}{2\pi \times \frac{1000}{60}} = \mathbf{114767\ Nm}\ (\textit{Ans.})$$

We know that full-load torque = maximum torque × sin δ

$\therefore$ Ration of max. and full-load torque, i.e.,

$$\frac{\text{maximum torque}}{\text{full load torque}} = \frac{1}{\sin \delta} = \frac{1}{\sin 19.44^\circ} = \mathbf{3}\ (\textit{Ans.})$$

Example 8.17

A 3-phase, 100 hp, 440 V, star-connected synchronous motor has a synchronous impedance per phase of (0.1 + j 1) Ω. The excitation and torque losses are 4 kW and may be assumed constant. Calculate the current, power factor and efficiency when operating at full load with an excitation equivalent to 500 line volt. (Assume 1 hp = 746 W)

Solution:

Here $Z_S = 0.1 + j\,1 = 1.005\angle 84.28°\ \Omega$, $\theta_S = 84.28°$;

$$\text{Phase value and } V = \frac{V_L}{\sqrt{3}} = \frac{440}{\sqrt{3}} = 254 \text{ V}$$

and
$$E = \frac{E_L}{\sqrt{3}} = \frac{500}{\sqrt{3}} = 288.7 \text{ V}$$

$$\text{Output} = 100 \times 746 = 74600 \text{ W}$$

$$\text{Power developed} = \text{output + excitation and torque losses}$$
$$= 74600 + 4000 = 78600 \text{ W}$$

$$\text{We know, } P_o = \frac{3VE}{Z_S}\cos(\theta - \delta) - \frac{3E^2}{Z_s^2}R$$

or
$$78600 = \frac{3(254)\,(288.7)}{1.005}\cos(84.28° - \delta) - \frac{3(288.7)^2 \times 0.1}{1.005}$$

$$78600 + 24875 = 218895\cos(84.28° - \delta)$$

$$\cos(84.28° - \delta) = \frac{103475}{218895} = 0.4727 = \cos 61.79°$$

$$84.28° - \delta = 61.79°,\ \delta = 84.28° - 61.79° = 22.49°$$

$$\bar{E} = E\angle -\delta = 288.7\angle -22.49° = 266.74 - j110.4$$

$$\bar{V} = V\angle 0° = V \pm j0 = 254 \pm j0$$

$$\bar{I} = \frac{\bar{V} - \bar{E}}{\bar{Z}_s} = \frac{254 + j0 - 266.74 + j110.4}{1.005\angle 84.28°}$$

$$= \frac{-12.74 + j110.4}{1.005\angle 84.28°} = \frac{111.13 + \angle -83.41°}{1.005\angle 84.28°}$$

$$\bar{I} = \mathbf{111.13\angle{+12.3°}\ A}\ (\textit{Ans.})$$

Power factor $\cos\phi = \cos 12.3° =$ **0.977 leading** (*Ans.*)

$$\text{Efficiency, } \eta = \frac{\text{output}}{\text{input}} = \frac{100 \times 746}{\sqrt{3} \times 440 \times 11.13 \times 0.977} = \mathbf{0.9016\ pu}\ (\textit{Ans.})$$

Example 8.18

A 10 MVA, 3-phase, star-connected, 11-kV, 12-pole, 50-Hz salient-pole synchronous motor has reactance of $X_d = 4\ \Omega$, $X_q = 2\ \Omega$. At full-load, unity power factor and rated voltage determine.

(i) *the excitation voltage,*

(ii) *active power,*

(iii) *synchronising power per electrical degree and the corresponding torque,*
(iv) *synchronising power mechanical degree and the corresponding torque.*

Solution:

Here, $V_L = 11$ kV; Rating = 10 MVA = 10×10^6 VA; $P = 12; f = 50$ Hz

$$X_d = 4\ \Omega;\ X_q = 2\ \Omega;\ N_s = \frac{120 \times 50}{12} = 500 \text{ rpm}$$

$$\text{Phase voltage, } V = \frac{V_L}{\sqrt{3}} = \frac{11000}{\sqrt{3}} = 6351 \text{ V}$$

$$I = \frac{MVA \times 10^6}{\sqrt{3}\,V_L} = \frac{10 \times 10^6}{\sqrt{3} \times 11 \times 10^3} = 524.86 \text{ A}$$

From the phasor diagram of a salient pole motor at unity p.f.

$$V \sin \delta = I_q X_q \text{ and } I_q = I \cos \delta;\ I_d = I \sin \delta$$

∴ $$V \sin \delta = (I \cos \delta)\, X_q$$

or $$\tan \delta = \frac{IX_q}{V} = \frac{524.26 \times 2}{6351} = 0.1653$$

or $$\delta = \tan^{-1} 0.1653 = 9.38°$$

∴ $$I_q = I \cos \delta = 524.86 \times \cos 9.38° = 517.8 \text{ A}$$

$$I_d = I \sin \delta = 524.86 \times \sin 9.38° = 85.54 \text{ A}$$

(*i*) Excitation voltage/phase,

$$E = V \cos \delta + I_d X_d = 6351 \times 0.9866 + 85.54 \times 4 = 6266 + 342 = 6608 \text{ V}$$

Line value, $E_L = \sqrt{3}\,E = \sqrt{3} \times 6608 = 11445 \text{ V}$ = **11.445 kV (Ans.)**

(*ii*) Active power per phase of the motor

$$P = \frac{3VE}{X_d} \sin \delta + \frac{3V^2}{2} \frac{(X_d - X_q)}{X_d\, X_q} \sin 2\delta$$

$$= \frac{3 \times 6351 \times 6608}{4} \sin 9.38 + \frac{3 \times (6351)^2}{2} \frac{(4-2)}{4 \times 2} \sin 2 \times 9.38$$

= (5130 + 4930) kW = **10060 kW** (*Ans.*)

(*iii*) Synchronising power per electrical degree

$$P_{syn} = \left[\frac{3EV}{X_d} \cos \delta + 3V^2 \left[\frac{X_d - X_q}{X_d X_q} \right] \cos 2\delta \right] \frac{\pi}{180}$$

$$= \left[\frac{3 \times 6608 \times 6351}{4} \cos 9.38 + 3 \times (6351)^2 \left[\frac{4-2}{4 \times 2} \right] \cos 2 \times 9.38 \right] \frac{\pi}{180}$$

= (31054 + 28644) $\frac{\pi}{180}$ kW = **1042 kW** (*Ans.*)

Synchronising torque per electrical degree

$$T_{syn} = \frac{P_{syn}}{2\pi N_s / 60} = \frac{1042 \times 10^3 \times 60}{2\pi \times 500} = \mathbf{19900\ Nm}\ (Ans.)$$

(*iv*) Synchronising power per mechanical degree

$$\text{Electrical angle} = \frac{P}{2} \times \text{Mechanical angle}$$

$$\therefore \quad 1^\circ \text{ Mech.} = \left(\frac{P}{2}\right) \times 1 = \frac{12}{2} \times 1 = 6^\circ \text{ elect.}$$

$\therefore \quad P_{syn}$/mech. degree = 6 × 1042 = 6252 kW

Corresponding synchronising torque

T_{syn}/mech. degree = 6 × 19900 = **119400 Nm** (*Ans.*)

Example 8.19

An 11 kV, 3-phase, star-connected synchronous motor is running in parallel with an infinite bus. Its direct-and quadrature-axis synchronous reactances are 10 Ω and 5 Ω, respectively. If the field current is reduced to zero, find the maximum load that can be put on the synchronous motor. Also calculate the armature current and the maximum power. Neglect armature resistance.

Solution:

Here, $$V_L = 11 \text{ kV}, V = \frac{V_L}{\sqrt{3}} = \frac{11 \times 10^3}{\sqrt{3}} = 6351 \text{ V}$$

$$P = \frac{EV}{X_d} \sin \delta + \frac{V^2}{2}\left(\frac{1}{X_q} - \frac{1}{X_d}\right) \sin 2\delta$$

When the field current becomes zero, $E = 0$

$$\therefore \quad P = \frac{V^2}{2}\left(\frac{1}{X_q} - \frac{1}{X_d}\right) \sin 2\delta$$

For maximum reluctance power,

$$\sin 2\delta = 1, 2\delta = 90 \text{ or } \delta = 45^\circ$$

$$\therefore \quad P_{max} = \frac{V^2}{2}\left(\frac{1}{X_q} - \frac{1}{X_d}\right)$$

$$= \frac{(6351)^2}{2}\left(\frac{1}{5} - \frac{1}{10}\right) = 2017 \times 10^3 \text{ W per phase}$$

Total maximum power for all the three phases

$$= 3 \times 2017 \times 10^3 \text{ W} = 6050 \text{ kW}$$

For maximum power, $\delta = 45^\circ$ and

$$I_d = \frac{V \cos \delta}{X_d} = 6351 \times \frac{\cos 45^\circ}{10} = 449 \text{ A}$$

$$I_q = \frac{V \sin \delta}{X_q} = 6351 \times \frac{\sin 45°}{10} = 898 \text{ A}$$

Armature current at maximum power

$$I = \sqrt{I_d^2 + I_q^2} = \sqrt{(449)^2 + (898)^2} = 1004 \text{ A}$$

Example 8.20

A 125 MVA, 3-phase, star-connected 11 kV, 4-pole, 50 Hz synchronous motor has a reactance of 0.15 pu and negligible armature resistance. Calculate the synchronising power per mechanical degree when it supplies full load at 11 kV and 0.8 power factor leading.

Solution:

Here, Rating = 125 MVA; $V_L = 11$ kV; $X_{spu} = 0.15$; $P = 4$; $f = 50$ Hz

$$I = \frac{125 \times 10^6}{\sqrt{3} \times 11 \times 10^3} = 6561 \text{ A}$$

and
$$V = \frac{V_L}{\sqrt{3}} = \frac{11 \times 10^3}{\sqrt{3}} = 6351 \text{ V}$$

$$X_{s\,pu} = \frac{X_s \text{ in ohm}}{V / I}$$

or
$$X_s = X_{s\,pu} \frac{V}{I} = 0.15 \times \frac{6351}{6561} = 0.14518 \ \Omega$$

$$\begin{aligned}
\overline{E} &= \overline{V} - \overline{I}\,\overline{Z_S} \\
&= V - (I \angle \phi)(X_s \angle 90°) = V - IX_s \angle 90° + \phi \\
&= V - IX_s [\cos (90 + \phi) + j \sin (90 + \phi)] \\
&= (V + IX_s \sin \phi) - jIX_s \cos \phi \\
&= 6350 + 6561 \times 0.14518 \times 0.6 - j6561 \times 0.14518 \times 0.8 \\
&= 6921.5 - j762 = 6963.4 \angle 6.2828°
\end{aligned}$$

$$E = 6963.4 \text{ V}, \ \delta = 6.2826°$$

Synchronising power per mechanical degree

$$P_{syn} = \left(\frac{d}{d\delta} P\right) p \frac{\pi}{180} \text{ and P} = \frac{3VE}{X_s} \sin \delta$$

$$P_{syn} = \left(\frac{3VE}{X_s} \cos \delta\right) p \frac{\pi}{180}$$

$$= \left(\frac{3 \times 6350 \times 6963.4}{0.14518} \cos 6.2826°\right) \times \frac{6\pi}{180}$$

$= 95109087$ W – **95.11 MW** (*Ans.*)

Section Practice Problems

Numerical Problems

1. A 50 kW, 400 V, 3-phase synchronous motor is operating at full load with an efficiency of 92%. If the field current is adjusted to make its power factor 0.8 leading, estimate the armature current. (**Ans.** *98A*)

2. The effective resistance and synchronous reactance of a 2000 V, 3-phase, star-connected synchronous motor are 0.2 Ω and 2.2 Ω per phase, respectively. The input is 800 kW at normal voltage and the induced line emf is 2500 V. Calculate the line current and power factor. (**Ans.** *254.4A; 0.91 leading*)

3. A 660 V, 3-phase, star-connected synchronous motor draws 50 kW at power factor 0.8 lagging. Find the new current and power factor when the back emf increases by 50%. The machine has synchronous reactance of 3Ω and effective resistance is negligible. (**Ans.** *49.2A, 0.89 leading*)

4. A 3-phase, 400 V, star-connected synchronous motor draws 24 A at full-load unity power factor. If the machine is operating at 95% efficiency. What will be the induced emf and total mechanical power developed at full load and 0.9 p.f. leading? The synchronous impedance per phase is (0.2 + j2) Ω.
(**Ans.** *441 V;16.2 kW*)

5. A 6-pole, 2200 V, 50 Hz, 3-phase, star-connected synchronous motor has armature resistance of 0.4 Ω per phase and synchronous reactance of 4 Ω per phase. While running on no-load, the excitation has been adjusted so as to make the emf numerically equal to and antiphase with the terminal voltage. With a certain load torque applied, if the rotor gets retarded by 3 mechanical degrees, calculate the armature current and power factor of the motor. (**Ans.** *49.57 A; 0.999 leading*)

6. A 3-phase, 415 V, 6-pole, 50 Hz star-connected synchronous motor has emf of 520 V (L-L). The stator winding has a synchronous reactance of 2 Ω per phase, and the motor develops a torque of 220 Nm. The motor is operating at 415 V, 50 Hz bus-bar (*a*) Calculate the current drawn from the supply and its power factor (b) draw the phasor diagram showing all the relevant quantities. (**Ans.** *42.12 A; 0.76 leading*)

7. The effective resistance and synchronous reactance of a 3-phase, 11 kV star-connected synchronous motor are 1 ohm and 30 ohm per phase, respectively. If it draws 50 A from the main, calculate the induced emf for a power factor of (a) 0.8 lagging (b) 0.8 leading and (c) the power supplied to the motor.
(**Ans.** *5536 V; 7315 V; 762 V*)

8. A 3-phase, star-connected synchronous motor is connected to 693 V lines and draws 48 kW at 0.8 pf lagging. If its induced emf is increased by 30% without changing the power input, what will be the new current and pf. Z_s, equals (0 + *j*2) ohm/phase. (**Ans.** *46.366 A; 0.863 leading*)

9. A 3-phase, star-connected synchronous motor takes 20 kW at 400 V from the mains. The synchronous reactance is 4 Ω and the effective resistance is negligible. If the exciting current is so adjusted that the back emf is 550 V, calculate the line current and the power factor of the motor.
(**Ans.** *33.1 A, 0.872 leading*)

10. A 3-phase, 11-kV, 50-Hz, 10-pole, 200-kW star-connected salient-pole synchronous motor has X_d = 1.2 pu and X_q = 0.8 pu. It operates at rated power at 0.98 power factor leading. Determine
(*a*) the internal emf and the load angle.
(*b*) the maximum power developed. (**Ans.** *18553 V, 34°; 298 kW*)

Short Answer Type Questions

Q.1. Why the synchronous motor is not self-starting?

Or

Why can't a synchronous motor start by itself?

Ans. When 3-phase supply is given to the stator of a 3-phase wound synchronous motor, a revolving field is set-up in the stator. Simultaneously, DC supply is given to the rotor which excites the poles. The stator revolving field tries to drag the rotor poles (or field) along with it but fails to do so due to rotor inertia.

Thus, a synchronous motor is not self-starting or in other words, a synchronous motor can't start by itself.

Q.2. What is the speed regulation of a synchronous motor?

Ans. Speed regulation of a synchronous motor is zero, because

$$\text{Speed regulation} = \frac{\text{No load speed-full load speed}}{\text{No-load speed}} = \frac{N_s - N_s}{N_s} = \text{zero}$$

Q.3. What is pull-out torque?

Ans. ***Pull-out*** **torque:** The maximum value to torque which a synchronous motor can develop at rated voltage and frequency without losing synchronism is called pull-out torque.

Q.4. Why a synchronous motor runs only at synchronous speed?

Ans. Synchronous motor can run only at synchronous speed because rotor poles are magnetically locked with the stator revolving field and the these rotor poles are dragged by the stator revolving field.

8.11 Effect of Change in Excitation

Consider a synchronous motor loaded with a constant mechanical load and normal (100%) excitation i.e., having induced *emf* equal to applied voltage V in magnitude. At the given load it takes a current of I ampere lagging behind the applied voltage V by an angle ϕ. As load on the machine is constant,

$$P = VI\cos\phi = \frac{VE}{X_S}\sin\delta \text{ is constant}$$

Here, V and X_S are constant, therefore, for constant power output $E\sin\delta$ and $I\cos\phi$ should remain constant, i.e.,

$$I\cos\phi = \text{Constant}$$

$$E\sin\delta = \text{Constant}$$

Since $I\cos\phi$ is constant, while drawing phasor diagram draw a horizontal dotted line passing through the extreme point of vector OI. On the other hand, as $E\sin\delta$ is constant, draw a vertical line passing through the extreme point of vector OE as shown in Fig. 8.20(*a*) and (*b*).

When the excitation is decreased, the induced *emf* decreases to E' in magnitude so that $E\sin\delta$ remains constant i.e., E decreases to E' angle delta increases to δ as shown in Fig. 8.20(a). This increases the resultant voltage Er to E_r', which is also shifted in clockwise direction. With the increase in resultant voltage, current increases to I' ($I' = E_r'/Z_S$). Since the phase angle θ between the resultant voltage and current is constant, therefore, current is also shifted in clockwise direction. This increases the phase angle between voltage and current to ϕ' which in turn decreases the lagging power factor cos ϕ to cos ϕ' more lagging but the active component of current I' cos ϕ' remains the same i.e., $i \cos \phi' = I \cos \phi$,

Hence *with the decrease in excitation, synchronous motor draws more current from the supply mains at* **lower (lagging) power factor.** *Thus, the Motor operates as an* **inductor.**

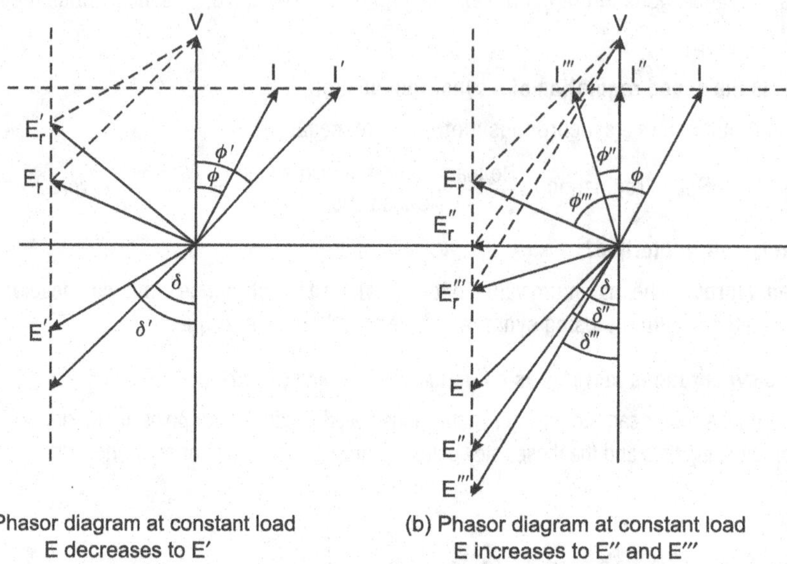

(a) Phasor diagram at constant load E decreases to E′

(b) Phasor diagram at constant load E increases to E″ and E‴

Fig. 8.20 Phasor diagram at constant load

When the excitation is increased, the induced *emf* increases in magnitude to E'' keeping E'' sin δ' to be the same as that of E sin δ for constant load. This decreases the resultant voltage which is also shifted in anti-clockwise direction. Since the phase angle θ between the resultant voltage and current is constant, current is also shifted in anti-clockwise direction and decreases in magnitude till it comes in phase with the voltage vector as shown in Fig. 8.20(*b*). *At this instant, the current drawn by the synchronous motor is minimum and the power factor is maximum* i.e., *one.*

Now, if the excitation is further increased, the induced *emf* increases to E''' in magnitude. This increases the resultant voltage to E_r''' which is also shifted further in anti-clockwise direction. Therefore, current is also shifted in anti-clockwise direction and its magnitude increases to I'''. This increases the phase angle between voltage and current to ϕ''' in opposite direction as shown in Fig. 8.19(*b*), which makes the power factor leading.

Thus, when synchronous motor is over excited, it draws more current at a **leading power factor** *from the supply mains. A synchronous motor operating under this condition is also called* **synchronous condenser.**

It may be noted that when excitation voltage E is changed, the torque angle δ changes such that

$$E \sin \delta = E' \sin \delta = E'' \sin \delta' = E''' \sin \delta''$$

Similarly, the active component of armature current also remains the same, i.e.,

$$I \cos \phi = I' \cos \phi' = I'' \cos \phi'' = I''' \cos \phi''$$

It is also observed that when the value of E increases, the magnitude of armature current first decreases and then increases. The armature current is minimum at unity *pf* and more at lagging or leading power factors.

Minimum Excitation

From Fig. 8.20, it is very clear that as excitation decreases, the torque angle increases. The minimum permissible excitation, E_{min} corresponds to the stability limit, i.e., $\delta = 90°$.

Therefore, $$E_{min} = \frac{PX_S}{V} \text{ (where } P \text{ is the power input)}$$

8.12 V-Curves and Inverted V-Curves

While changing the excitation of a 3-phase synchronous motor, keeping the load constant, the *curve plotted between field current I_f and armature or load current I is called* **V-curve,**

It is named as V-curve because its shape resembles with the shape of English alphabet '*V*'.

When the excitation of a three-phase synchronous motor taking constant power is varied, it changes the operating power factor of the motor. If

$$P = 3\,VI \cos \phi$$

where, P = Power input.
V = Terminal voltage (phase value).
I = Armature current (phase value) and
$\cos \phi$ = Power factor.

then for constant power input P and terminal voltage V, only increase in power factor causes decrease in armature current I and vice versa. Armature current will be minimum at unity power factor and increases when the power factor decreases on either side (lagging or leading).

Hence variation in excitation (field current) causes the variation in armature current I. *If we plot a curve taking field current I_f on X-axis and the armature current I on Y-axis, the curve so obtained is called V-curve because of its shape.* The V-curve at different power inputs are shown in Fig. 8.21.

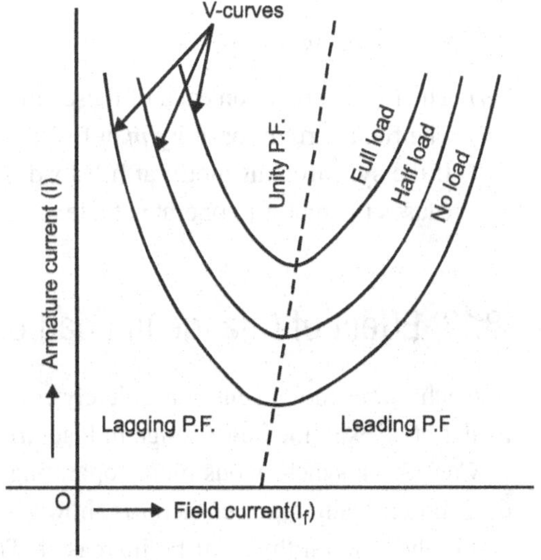

Fig. 8.21 V-curves

Significance of V-Curves

V-curves are useful in adjusting the field current. By increasing the field current I_f beyond the level of minimum armature current I, we can obtain leading *pf*. On the other hand by decreasing the field current I_f below the level of minimum armature current I, we can obtain lagging *pf*. Therefore by controlling the field current of a synchronous motor, the reactive power supplied to or consumed from the power system can be controlled.

Thus, by changing the field current (or excitation), a synchronous motor can be used as a condenser or inductor.

Inverted V-Curves

If we plot a family of curves between power factor and field current (I_f), the curves so obtained are called **invested V-curves,** *as shown in Fig. 8.22, because of their shape.*

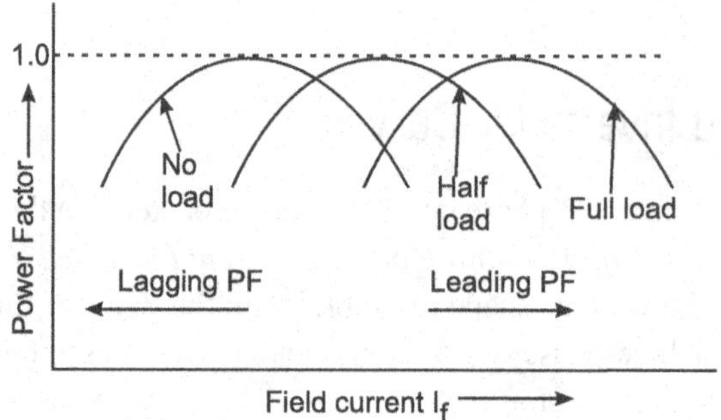

Fig. 8.22 Inverted V-curves

These curves reveal that:

(*i*) The highest point on each of these curves indicate unity p.f..
(*ii*) The field current for unity *pf* at full-load is more than the field current for unity *pf* at no-load.
(*iii*) If the synchronous motor at full-load is operating at unity p.f., then the removal of shaft load causes the motor to operate at a leading power factor.

8.13 Effect of Change in Load on a Synchronous Motor

A synchronous motor runs at absolutely constant speed called synchronous speed, regardless of the load. Let us examine how change in load affects its performance.

Consider a synchronous motor operating initially at a leading power factor cos ϕ_1. The phasor diagram for leading power factor is shown in Fig. 8.23.

Let the load on the shaft be increased. The rotor slows down momentarily since it takes some time to settle down as per the increased power drawn from the lines. In other words, although still rotating at synchronous speed, the rotor slips back in space as a result of increased loading as shown in Fig. 8.23. In this process the torque angle δ becomes larger and therefore the induced torque $\left(T_{ind} = \dfrac{VE \sin \delta}{\omega X_S}\right)$ increases. The increased torque increases the rotor speed and the motor again picks up the synchronous speed but with a larger torque angle δ. Thus, with the increase in load, the torque angle δ increases. But the magnitude of excitation voltage E remains constant which depends upon field current I_f and speed and both of them are constant.

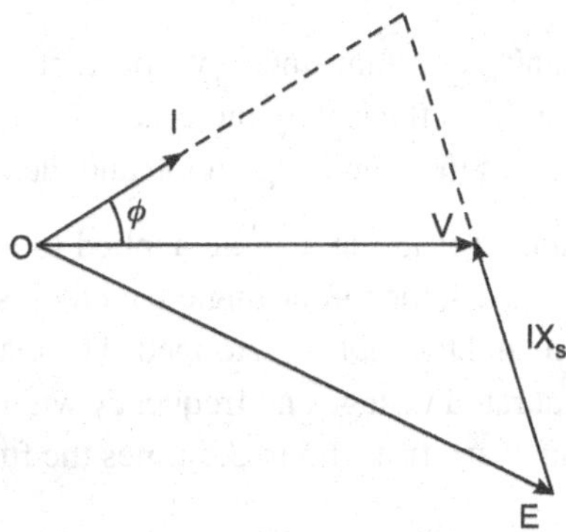

Fig. 8.23 Phasor diagram for leading power factor

$$\text{Now,} \qquad P = \frac{VE \sin \delta}{X_s} = VI \cos \phi,$$

$$\therefore \qquad E \sin \delta = \frac{X_s}{V} P = KP \text{ where } K = \frac{X_s}{V} = a \text{ constant.}$$

These expressions show that the increase in P increases $E \sin \delta$ and $I \cos \Phi$. The locus of E is shown in Fig. 8.24. It is seen from Fig. 8.24 that with the increase of the load, the quantity jIX_s goes on increasing so that the relation $V = E + jIX_S$ is satisfied and therefore the armature current I also increases. It is also seen from Fig. 8.24 that the power factor angle ϕ also changes. It becomes less and less leading and then becomes more and more lagging.

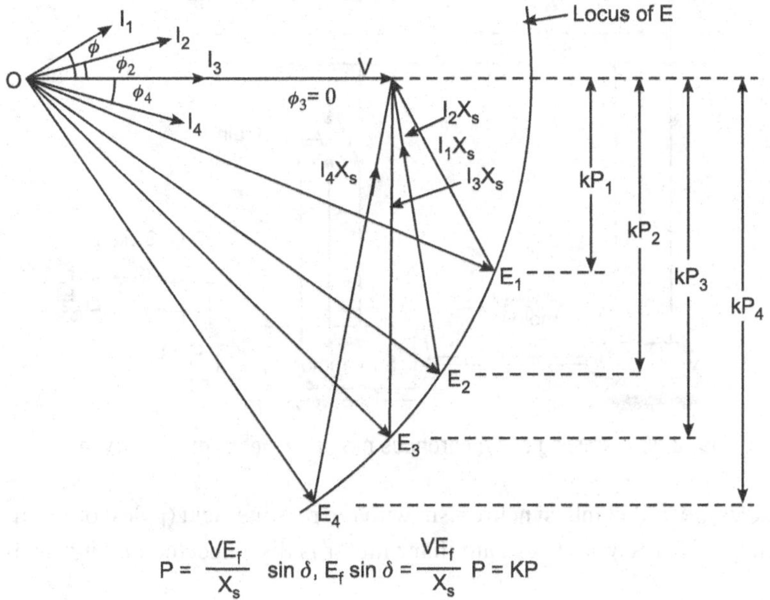

Fig. 8.24 Effect of increase in load on the operation of a synchronous motor

Thus, when the load on a synchronous motor is increased,

(*i*) the motor speed does not change, it continues to run at synchronous speed.
(*ii*) the torque angle δ increases.
(*iii*) the excitation voltage E remains constant since I_f is constant
(*iv*) the armature current I drawn from the supply increases.
(*v*) the phase angle ϕ first decreases in the leading region and then increases in the lagging region.

There is a limit to the mechanical load that can be applied to a synchronous motor depending upon its rating. As the load is increased, the torque angle δ increases till a stage is reached when the rotor is pulled out of synchronism and the motor is stopped. The maximum value of torque which a synchronous motor can develop at rated voltage and frequency without losing synchronism is called the **pull-out torque.** Its value may vary from 1.5 to 3.5 times the full-load torque.

8.14 Methods of Starting of Synchronous Motors

Since a synchronous motor is inherently not self-starting, the following methods are generally adopted to start the synchronous motor:

1. By means of Auxiliary Motor: A small induction motor called the pony motor (auxiliary motor) is mounted on the same shaft or coupled to synchronous motor as shown in Fig. 8.25. The auxiliary motor should have the same number of poles as that of synchronous motor or preferably one pole pair less so that it can rotate the motor nearly at synchronous speed. First of all supply is given to the pony motor. When it rotates the rotor of the synchronous motor near to the synchronous speed the main switch and DC switch of the main synchronous motor are closed.

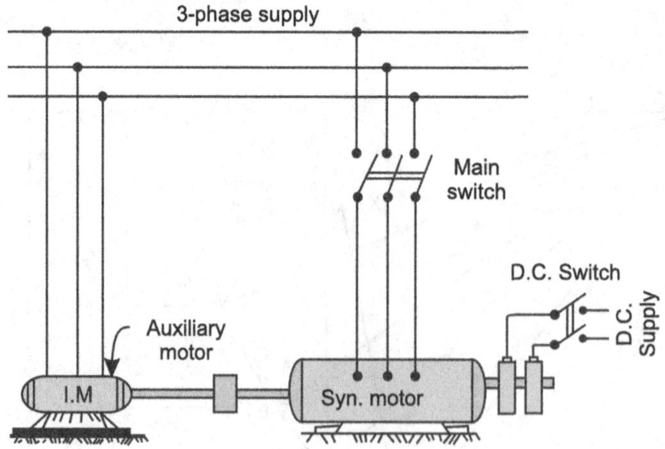

Fig. 8.25 Starting of synchronous motor by means of auxiliary motor

The rotor poles are pulled into synchronism with the rotating field (poles) of the armature (stator) of the main motor. Then supply to the auxiliary motor is disconnected and it acts as a load on the main motor.

2. By providing damper winding: This is a most common method of starting a synchronous motor. In this method, the motor is first started as a squirrel cage induction motor by providing a special winding on the rotor poles, known as *damper* or *squirrel cage winding.* This damper winding consists of number of copper bars embedded into the slots or holes provided on the outer periphery of the pole shoes, where salient poles are employed, and then short circuiting these bars by brazing them to end rings as shown in Fig. 8.26. In a non-salient pole machine, the damper winding conductors are placed in the rotor slots above the main field winding and short circuited by the end rings.

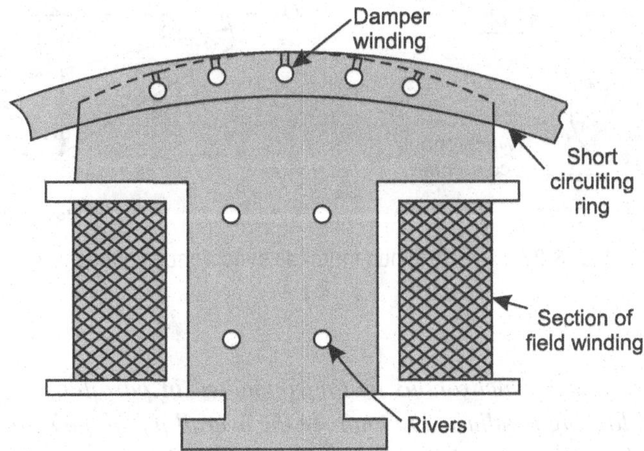

Fig. 8.26 Starting of synchronous motor by providing damper winding

When the synchronous motor (armature) is connected to 3-phase supply mains, a revolving field is set up which causes the rotor to rotate as a squirrel cage induction motor. As soon as motor attains about 65% synchronous speed, the rotor winding is connected to DC mains (exciter) and the rotor field is magnetically locked with the stator rotating field and the motor starts running runs as a synchronous motor.

8.15 Synchronous Condenser

The power factor of a synchronous motor can be controlled over a wide range by adjusting its excitation. At no-load, when the motor is over-excited it may draw the current from mains which leads the voltage by large angle nearly 90°. Hence, the motor acts like a static capacitor and is known as a *synchronous condenser.*

Thus, an over excited synchronous motor operating at no-load is called a **synchronous condenser or synchronous capacitor.**

When an over excited motor is operated on the same electrical system to which some industrial load (induction motors, induction furnaces, arc furnaces, etc.) is operating at lagging power factor, the leading reactive power supplied by the synchronous motor compensates for the lagging reactive power of industrial load and improves the overall power factor of the system. In large industrial plants, which have a low lagging power factor load, it is often found economical to install an over excited synchronous motor (synchronous condenser), even though the motor is not required to drive a load.

Consider an industrial load P_L operating at a power factor cos ϕ_1. When an over excited motor drawing power P_m is connected in parallel with the existing load as shown in Fig. 8.27(*a*), some of the lagging reactive power of the industrial load in compensated by the leading reactive power of the motor (i.e., P_{rm}) which improves the over-all power factor to cos ϕ_2 as shown in Fig. 8.27(*b*).

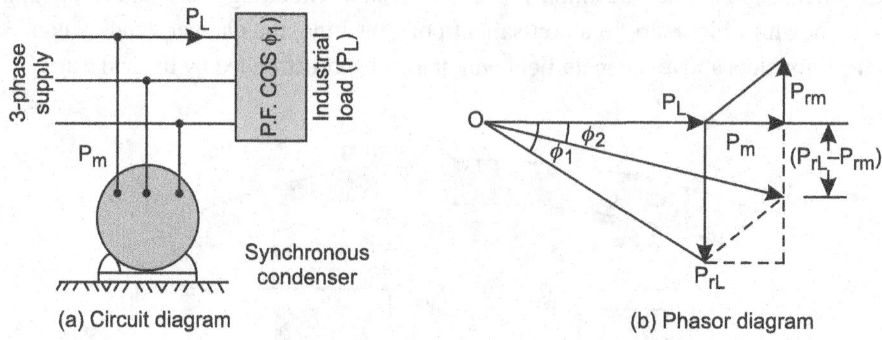

Fig. 8.27 Synchronous motor as synchronous condenser

Example 8.21

The excitation of a 3-phase synchronous motor connected in parallel with a load of 500 kW operating at 0·8 p.f. lagging is adjusted to improve the overall p.f. of the system to 0.9 lagging. If the mechanical load on the motor including losses is 125 kW, calculate the kVA input to the synchronous motor and its p.f.

Solution:

Industrial load, $P_L = 500$ kW

Load *p.f.*, $\cos \phi_L$ = lagging; $\tan \phi_L = \tan \cos^{-1} 0{\cdot}8 = 0{\cdot}75$

Reactive power of the industrial load,

$$P_{rL} = P_L \tan \phi_L = 500 \times 0{\cdot}75 = 375 \text{ kVAR}$$

Motor load, $P_m = 125$ kW

Total active power, $P = P_L + P_m = 500 + 125 = 625$ kW

Power factor of total load, $\cos \phi = 0{\cdot}9$ lag

$\tan \phi = \tan \cos^{-1} 0{\cdot}9 = 0{\cdot}4843$

Total reactive power, $P_r = P \tan \phi = 625 \times 0{\cdot}4843 = 302{\cdot}7$ kVAR

Reactive power supplied by synchronous motor,

$$P_{rm} = P_r - P_{rL} = 302{\cdot}7 - 375 = -72{\cdot}3 \text{ kVAR}$$

Input of the motor in kVA, $P_{am} = \sqrt{P_m^2 + P_{rm}^2}$

$$P_{am} = \sqrt{(125)^2 + (72 \cdot 3)^2}$$

$= \mathbf{144{\cdot}4\ kVA}$ (*Ans.*)

$$\text{Power factor of the motor, } \cos\phi_m = \frac{P_m}{P_{am}} = \frac{125}{144 \cdot 4} = \mathbf{0{\cdot}8656\ leading}\ (\textit{Ans.})$$

Example 8.22

The electric loads connected to the supply are:

(i) A lighting load of 500 kW;

(ii) A load of 400 kW at p.f. 0·707 lagging;

(iii) A load of 800 kW at p.f. 0·8 leading;

(iv) A load of 500 kW at p.f. 0·6 lagging;

(v) A synchronous motor driving a 540 kW DC generator and having an overall efficiency of 90%.

Determine the p.f. of the synchronous motor to improve the overall p.f. to unity.

Solution:

The load chart is given below:

S. No.	Load in kW	p.f. $\cos\phi$	$\tan\phi$	Reactive power $P_r = P\tan\phi$
1.	500	1	zero	zero
2.	400	0·707 lag.	1	400 kVAR
3.	800	0·8 lead.	0·75 – ve	– 600 kVAR
4.	500	0·6 lag.	1·333	666·6 kVAR

$$\text{Industrial load, } P_L = 500 + 400 + 800 + 500 = 2200 \text{ kW}$$

Reactive power of the industrial load,

$$P_{rL} = 0 + 400 - 600 + 666{\cdot}6 = 466{\cdot}6 \text{ kVAR}$$

$$\text{Input power to the motor, } P_m = \frac{\text{output}}{\eta} = \frac{540}{0 \cdot 9} = 600 \text{ kW}$$

$$\text{Total active power, } P = P_L + P_m = 2200 + 600 = 2800 \text{ kW}$$

$$p.f.\ \cos\phi = 1;\ \tan\phi = \tan\cos^{-1} 1 = 0$$

$$\text{Reactive power of total load, } P_r = 0$$

$$\text{Reactive power supplied by motor, } P_{rm} = P_r - P_{rL} = 0 - 466{\cdot}6 = -466{\cdot}6 \text{ kVAR}$$

$$\tan\phi_m = \frac{P_{rm}}{P_m} = -\frac{466 \cdot 6}{600} = -0 \cdot 7777$$

$$\text{Power factor of the motor, } \cos\phi_m = \cos\tan^{-1} = 0{\cdot}777 = \mathbf{0{\cdot}7893\ leading}\ (\textit{Ans.})$$

Example 8.23

A 3-phase synchronous motor takes a load of 50 kW is connected in parallel with a factory load of 250 kW operating at a lagging power factor of 0.8. If the p.f. of overall load is required to be improved to 0.9 lagging, what is the value of the leading kVAR supplied by the motor and at what power factor is it working?

Solution:

$$\text{Factory load, } P_L = 250 \text{ kW}$$

$$\text{Power factor, } \cos\phi_L = 0.8 \text{ lagging; } \tan\phi_L = \tan\cos^{-1} 0.8 = 0.75$$

$$\text{Load in kVAR, } P_{rL} = 250 \tan\phi_L = 250 \times 0.75 = 187.5 \text{ kVAR}$$

$$\text{Synchronous motor load, } P_m = 50 \text{ kW}$$

$$\text{Total load, } P = P_L + P_m = 250 + 50 = 300 \text{ kW}$$

$$\text{Overall power factor, } \cos\phi = 0.9 \text{ (lagging)}$$

$$\text{Phase angle, } \phi = \cos^{-1} 0.9 = 25.84^\circ$$

$$\text{Total kVAR, } P_r = P\tan\phi = P\tan 25.84^\circ$$

$$= 300 \times 0.4843 = 145.3 \text{ (lagging)}$$

Leading kVAR supplied by the motor,

$$P_{rm} = P_{rL} - P_r = 187.5 - 145.3 = \mathbf{42.2\ kVAR}\ (Ans.)$$

$$\text{k}VA \text{ supplied by the motor, } P_{am} = \sqrt{P_m + P_{rm}} = \sqrt{50^2 + (42.2)^2} = 65.43 \text{ kVA}$$

$$\text{Power factor of the motor, } \cos\phi_m = \frac{P_m}{P_{am}} = \frac{50}{65.43} = 0.764 \text{ (leading) } (Ans.)$$

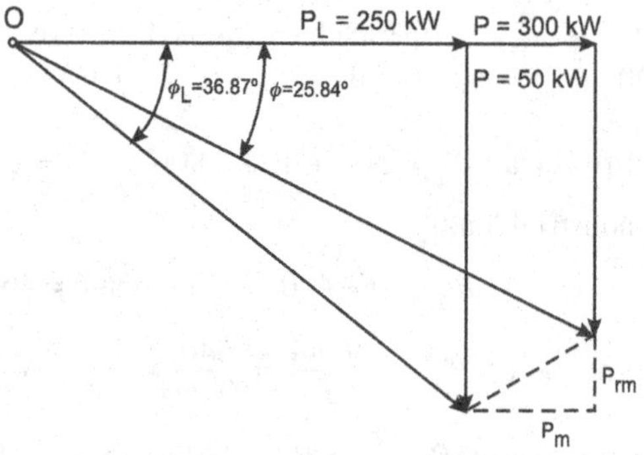

Fig. 8.28 Phasor diagram

8.16 Characteristics of Synchronous Motor

A synchronous motor has the following important characteristics;

1. It is inherently *not a self-starting* motor.
2. For a given frequency, it operates only at one speed called *synchronous speed* given by the expression $N_S = 120 f / P$.
3. It can be operated under *a wide range of power factors* both lagging and leading.
4. In addition to the motor being used for mechanical load, it is also used as a *power factor improvement equipment* and is known as *synchronous condenser.*
5. At no-load it draws a very small current from the mains to meet the internal losses of the motor. With the increase in load, the torque angle δ increases due to which motor draws more current from the mains. After the input current reaches maximum (torque angle δ in nearly 90°) no further increase in load is possible. If the motor is further loaded it goes out of synchronism and stops.

8.17 Hunting

When a synchronous motor is loaded, the rotor poles slightly fall back in position with respect to the stator field (poles) by an angle δ known as power angle or torque angle or retarding angle. As the load is gradually increased, this angle δ also increases gradually so as to produce more torque for coping with the increased load and the motor remains in equilibrium. If the load is suddenly thrown off, angle δ decreases suddenly and the rotor poles are pulled into almost exact opposition to the stator poles, but due to inertia of rotor and rotor poles travel too far. They are then pulled back again, and so on, thus oscillations are set up around the equilibrium position, corresponding to new load. If these oscillations are too large they may throw the motor out of synchronism and stops.

The oscillation of the rotor about its equilibrium position is known as **hunting**.

The hunting (oscillations) can be prevented by providing damper winding or squirrel cage winding on the rotor pole faces. This damper winding consists of number of copper bars embedded into the slots provided on the outer periphery of the pole shoes and then short circuited by end rings. When hunting takes place, there is relative motion of the rotor with respect to the stator field, which sets up eddy currents in this winding which flow in such a way that it suppresses the oscillations.

The hunting may also occur when the machine is operating as an alternator. In this case also because of sudden change in electrical output or mechanical input oscillations are set up in the rotor called hunting; which can be prevented by providing damper winding on the rotor.

From the above discussions, the following factors are evolved:

Causes of Hunting

1. Sudden change in the load.
2. Sudden change in the field current.
3. Cyclic variation of the load torque.
4. Faults occurring in the system.

Effects of Hunting

1. The machine (generator or motor) may go out of synchronism. This is the most important phenomenon to be avoided.
2. Large mechanical stresses may develop in the rotor shaft.
3. It increases the possibility of resonance. It happens when the frequency of torque component becomes equal to that of the transient oscillations of the synchronous machine.
4. It increases the machine losses which may overheat the machine.
5. It may cause variations in supply voltage (generator action) producing lamp flickering.

Reduction of Hunting

1. *By using damper winding*: The eddy currents developed in the damper winding, damped down the oscillations.
2. *By using flywheels:* The prime-mover is provided with a large and heavy flywheel which increases the inertia of the prime-mover and helps in maintaining the motor speed constant.

8.18 Applications of Synchronous Motors

The following are the important applications of synchronous motors:

1. These are used to improve the power factor of large industries.
2. These are used at the substations to improve the power factor.
3. These are used to control the voltage at the end of transmission lines by varying their excitation. When operated in this manner, it is called a *synchronous phase modifier or reactor.*
4. These are used to textile mills, cement factories, mining industries and rubber mills for power applications.

These motors are mostly used to drive equipment which are operated at constant speed continuously such as centrifugal pumps, centrifugal fans, air compressors, motor-generator sets, blowers, etc.

8.19 Comparison between Three-phase Synchronous and Induction Motors

Se. No.	Particulars	Synchronous motor	Induction motor
1.	Magnetic excitation	A synchronous motor is a doubly *excited* machine. Its armature winding is energised from an AC source, whereas, its field winding is excited from a DC source	An induction motor is a *singly-excited* machine. Its stator winding is energised from an AC source and rotor is excited due to induction.
2.	Speed	It always runs at *synchronous speed* and its speed is independent of load.	Its speed falls with the increase in load and is always *less than the synchronous speed.*
3.	Starting	It is *not self-starting.* It has to be run upto synchronous speed by some means before it can be synchronised to AC supply.	An induction motor has got *self-starting* torque.
4.	Power factor	A synchronous motor can be operated under *wide range of power factors, both lagging and leading* by changing its excitation.	An induction motor operates at only *lagging power factor,* which becomes very poor at light loads.
5.	Major application	It is usually used for *power factor correction* in addition to supplying torque to drive mechanical loads.	An induction motor is used for *driving mechanical loads* only.
6.	Efficiency	It is *more efficient* than induction motor of the same output and voltage rating.	Its *efficiency is lesser* than that of a synchronous motor of the same output and voltage rating.
7.	Cost	A synchronous motor is *costlier* than an induction motor of the same output and voltage rating.	The cost of an induction motor is *less* than a synchronous motor of the same output and voltage rating.

8.20 Merits and Demerits of Synchronous Motor

Merits

(*i*) The ease with which the power factor can be controlled. An over-excited synchronous motor having a leading power factor can be operated in parallel with induction motors and other power apparatus operating at lagging power factors, thereby improving the power factor of the supply system.

(*ii*) The speed is constant and independent of load. This characteristic is mainly of use when the motor is required to drive another alternator to generate a supply at a different frequency as in frequency changers.

(*iii*) Electro-magnetic power varies linearly with the voltage.

(*iv*) These motors can be constructed with wider air gaps than induction motors, which make them better mechanically.

(*v*) These motors usually operate at higher efficiencies, especially in the low speed unity pf ranges.

Demerits

(*i*) The cost per kW output is generally higher than that of an induction motor.

(*ii*) It requires DC excitation which must be supplied from external source.

(*iii*) The synchronous motor is inherently not self-starting motor and needs some arrangement for its starting and synchronising.

(*iv*) It cannot be used for variable speed jobs as there is no possibility of speed adjustment.

(*v*) It cannot be started under load. Its starting torque is zero.

(*vi*) It has a tendency to hunt.

(*vii*) It may fall out of synchronism and stop when over-loaded.

(*viii*) Collector rings and brushes are required.

(*ix*) For some purposes synchronous motors are not desirable as for driving shafts in small workshops having no other power available for starting and in cases where frequent starting or strong starting torque is required.

Section Practice Problems

Numerical Problems

1. A 500 V, 50 Hz, 3-phase load draws 20 A at 0.8 power factor lagging. A synchronous motor is used to raise the power factor to unity. Calculate the kVA input to the motor, and its power factor when driving a mechanical load of 7.5 kW. The motor has an efficiency of 85%. (**Ans.** *13.63 kVA; 0.6472 (leading)*)

2. An alternator supplying a load of 500 kW at 0.7 power factor lagging. If its power factor is required to be raised to unity by means of an over-excited synchronous motor. At a constant armature current how much input power is required for the synchronous motor? Find the power factor of the synchronous motor. (**Ans.** *510 kVAR; Zero leading*)

3. A substation is operating at its full load of 1000 kVA supplies a load of power factor 0..71 lagging. Calculate the permissible additional load at this power factor and the rating of synchronous condenser to raise the power factor to 0.87 lagging. (**Ans.** *225.35 kVA; 370 kVAR*)

Short Answer Type Questions

Q.1. What will happen, if the excitation of a synchronous motor connected to an infinite bus-bar is varied, load remaining constant?

Or

What happens if the excitation of a synchronous motor is varied at constant load.

Ans. Effect of change in excitation keeping load constant on a synchronous motor

The change in excitation changes the power factor of the machine.

(*i*) If excitation is decreased, the power factor becomes lagging and is further reduced. The machine behaves as an inductor and the machine is known as *synchronous inductor.*

(*ii*) If excitation is increased, the lagging power factor of the machine is improved and becomes unity. If it is further increased, the machine starts operating at leading power factor. The machine behaves as a capacitor and it is known as *synchronous condenser.*

Q.2. How a synchronous motor can be used as a synchronous condenser?

Ans. An over-excited synchronous motor draws power at leading power factor, under this condition it can be used as a synchronous condenser.

Q.3. Why dampers are used in a synchronous motor?

Ans. In synchronous motors, dampers are used to suppress the oscillations of the rotor around its find position.

Q.4. How is a synchronous motor started.

Or

How a synchronous motor is made self-starting?

Ans. A synchronous motor is not self-starting. It can be started by using either of the following methods:

1. *By using an auxiliary motor:* This motor is coupled to the shaft of main motor and rotates it almost at synchronous speed. Then DC excitation is given to rotor poles of the main motor which are magnetically locked with the stator revolving field and motor starts rotating at synchronous speed.

2. *By using damper winding*: Due to damper winding the synchronous motor stars rotating as a squirrel cage induction motor. Once, it picks-up speed near to synchronous speed, DC excitation is given to the rotor poles. Thus, the rotor poles are magnetically locked with the stator revolving field and the motor starts rotating at synchronous speed.

Q.5. Define the term over-excitation and under-excitation with reference to synchronous machines.

Ans. *Over-excitation:* The excitation (field current) at which a synchronous motor operates at leading power factor is called over-excitation.

Under-excitation: The excitation (field current) at which a synchronous motor operates at lagging power factor is called under-excitation.

Q.6. What could be reasons if a synchronous motor fails to start?

Ans. A synchronous motor may fail to start because of the following reasons:

1. Faulty connections
2. Supply voltage too low
3. One phase open circuited or short circuited
4. Wrong connections
5. High friction at the bearings
6. Field winding is not excited properly
7. Over-loaded at the start
8. Some fault in the armature or field winding
9. Poor contact of brushes at the sliprings.

Q.7. Mention the characteristic features of synchronous motor.

Ans. The main characteristic features of a synchronous motor are:

1. Synchronous motor operates only at synchronous speed.
2. It is not a self-starting motor.
3. It can be operated in a wide range of power factor both lagging and leading.
4. At no-load it draws very small current to meet with the mechanical losses.
5. It can be used to regulate the voltage of transmission lines.

Q.8. What happens when the field current of a synchronous motor is increased beyond the normal value at constant input?

Ans. When field current of a synchronous motor is increased beyond normal it starts working at leading power factor and is called as a synchronous condenser.

Q.9. Why synchronous motor is designated as P.F. improving device?

Ans. An over excited synchronous motor operates at leading power factor, under this condition it can improve the power factor of the system. Hence, designated as power factor improving device.

Q.10. Show the inverted V-curve for synchronous machine.

Ans. Inverted V-Curves

If we plot a family of curves between power factor and field current (I_f), the curves so obtained are called **invested V-curves,** *as shown in Fig. Q.19, because of their shape.*

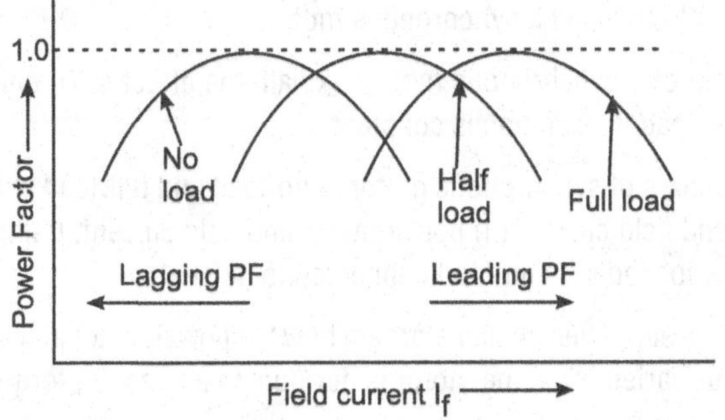

Fig. Q.19 Inverted V-curves

Q.11. Enumerate methods employed for cooling of 3-phase synchronous motors.

Ans. Small machines are cooled with the help of blast of air by providing a fan on the shaft. For large size machines closed circuit hydrogen cooling is provided.

Review Questions

1. Explain the construction and principle of operation of a synchronous motor.
2. Explain why the speed of a 3-phase synchronous motor remains constant at variable loads when it is fed from a constant frequency supply.
3. Give the principle of operation of a synchronous machine. How speed is related to frequency.
4. Explain with neat sketches the principle of operation of a 3-phase synchronous motor. Also explain why it never runs at a speed other than synchronous speed.
5. Describe with the help of neat sketches, the principle of operation of a 3-phase synchronous motor.
6. Explain why a synchronous motor runs at synchronous speed or not at all. How can the speed of such motor be varied?
7. What is the effect on the speed of a synchronous motor if load on it is increased?
8. Explain the action of synchronous motor when loaded.
9. Explain clearly the effect of excitation on the performance of a synchronous motor.
10. Describe briefly the effect of varying excitation upon the armature current and power factor of a 3-phase synchronous motor when input power to the motor is kept constant.
11. With the help of proper phasor diagrams, explain the operation of a 3-phase synchronous machine with normal excitation under following conditions.
    (*i*) The machine is floating on the supply bus.
    (*ii*) The machine is working as a synchronous motor at no-load and also with load.
12. Explain effects of varying excitation on armature current and power factor in a synchronous motor. Also draw 'V' curves.
13. Show with the help of phasor diagram, how change in excitation of a synchronous motor causes it to work (*i*) as an inductor (*ii*) as a condenser.
14. Draw and explain the 'V'-curves of a synchronous motor.
15. Draw a phasor diagram of a synchronous motor. Explain the effect of (i) change in excitation if load is constant (ii) change in load if excitation is constant.
16. Draw the families of curves of synchronous motor at no-load and full-load showing the relation between (i) armature current and field current (ii) power factor and field current. Using these curves explain how the motor may be over loaded with no-load connected to its shaft
17. Explain with the help of a simplified phasor diagram how the power factor at which a 3-phase synchronous motor operates may be varied when the motor is developing a constant torque.
18. Explain V and inverted V curves.
19. Draw and explain equivalent circuit of a synchronous motor.

20. Draw and explain the phasor diagrams of a 3-phase synchronous motor when (i) it is overexcited (ii) it is under excited

21. Draw the phasor diagrams of synchronous motor for lagging, leading and unity power factor conditions. Name all the phasors.

22. Derive the commonly used expression for the power developed by a synchronous motor.

23. What is meant by constant power circle for synchronous motor? How is it derived?

24. Show that the locus of stator current for a constant output of 3-ϕ synchronous motor connected to a constant voltage, constant frequency bus-bars is a circle.

25. How does a synchronous machine act as a capacitor? Explain it with phasor diagrams.

26. What do you understand by a synchronous condenser? Explain with the help of phasor diagrams its operation and application.

27. What is a synchronous condenser? Show the region of operation of the condenser on V-curves. Where are synchronous condensers used?

28. Briefly describe the phenomenon of 'hunting' in a synchronous machine. How is it remedied?

29. With the help of a neat sketch discuss any one method of starting a synchronous motor.

30. Name different methods of starting a synchronous motor, explain any one in detail.

31. Explain the functions of a damper winding in a synchronous motor.

32. Explain two important functions served by damper winding in a synchronous motor. State applications of synchronous motor.

33. A synchronous motor will operate at a constant speed on every load, why? Explain in detail. Make a list of applications of a synchronous motor.

Multiple Choice Questions

1. Synchronous motors are not self-starting because
(*a*) starters cannot be used on these machines.
(*b*) starting winding is not providing on these machines
(*c*) the direction of rotation is not reversed.
(*d*) the direction in instantaneous torque reverse after half cycle.

2. A pony motor is basically a:
(*a*) DC series motor
(*b*) DC shunt motor
(*c*) double winding AC/DC motor
(*d*) small induction motor.

3. A Synchronous Motor can be started by
(*a*) providing damper winding
(*b*) pony motor
(*c*) DC compound motor
(*d*) any of above

4. As compared to an Induction Motor of same size the air gap in a synchronous motor is
(*a*) less than half
(*b*) half to three fourth
(*c*) same
(*d*) three to five times.

5. A syn. motor is switched on to supply with its field winding shorted on themselves. It will
(*a*) not start
(*b*) start and continue to run as an induction motor
(*c*) start as Induction motor and then run as Syn. motor
(*d*) none of these.

6. When V is the applied voltage, the breakdown torque of a syn. motor varies as
(*a*) V (*b*) $\sqrt{V}$
(*c*) V^2 (*d*) $1/V$

7. If one phase of a 3-phase synchronous motor is short circuited, the motor will:
(*a*) not start (*b*) run at half of synchronous speed
(*c*) run with excessive vibrations (*d*) Develops no torque.

8. A synchronous motor can develop synchronous torque
(*a*) only at Syn. speed. (*b*) while over excited
(*c*) when under loaded (*d*) Below or above Synchronous speed.

9. A three phase Syn. motor will have:
(*a*) no slip rings (*b*) two slip rings
(*c*) three slip rings (*d*) four slip rings.

10. Cage winding in a Syn. motor carries
(*a*) high starting and running current (*b*) no starting current
(*c*) no running current (*d*) no starting as well as running current

11. Syn. Watt is
(*a*) a unit to express the rating of syn. motors.
(*b*) kW as applicable to a Syn. motor
(*c*) the torque which under Syn. speed would develop a power of one watt
(*d*) none of these

12. Slip rings in a synchronous motor carry:
(*a*) DC (*b*) AC
(*c*) both *a* and *b* (*d*) no current.

13. The efficiency of a properly designed Syn. motor will usually fall in the range:
(*a*) 99% to 99·5% (*b*) 85% to 95%
(*c*) 75% to 80% (*d*) 60% to 70%

14. A Syn. motor working on leading p.f. at no-load is known as
(*a*) condenser (*b*) Syn. condenser
(*c*) inverter (*d*) convertor.

15. The maximum value of torque that a syn. motor can develop without losing its synchronism is called:
(*a*) slip torque (*b*) pullout torque
(*c*) breaking torque (*d*) syn. torque.

16. The armature current in a syn. motor will be least when p.f. is
(*a*) zero (*b*) unity
(*c*) leading (*d*) lagging

17. When the field of a syn. motor is under excited, the p.f. will be
(*a*) zero (*b*) unity
(*c*) lagging (*d*) leading.

18. Operating speed of a Syn. motor can be changed to new fixed value by
(*a*) changing the load (*b*) changing the supply voltage
(*c*) changing frequency (*d*) using brakes.

19. Back emf set up in the stator of Syn motor depends on
(*a*) rotor speed (*b*) load
(*c*) rotor excitation (*d*) coupling angle.

20. The % age slip in case of syn. motor is
(*a*) 1% (*b*) 100 %
(*c*) 0·5% (*d*) zero.

21. The speed regulation of a syn. motor is always:
(*a*) 1% (*b*) 0·5%
(*c*) positive (*d*) zero.

22. A synchronous motor can be made self-starting by providing
(*a*) damper winding on rotor pole (*b*) damper winding on stator
(*c*) damper winding on stator as well as rotor (*c*) none of above.

23. Synchronous speed for a syn. motor is given by
(*a*) 200 *f/p* (*b*) 120 *f/p*
(*c*) 120 *p/f* (*d*) 120 *f.p.*

24. Oscillations in a Syn. motor can be damped out by:
(*a*) maintaining constant excitation
(*b*) running the motor on leading power factor
(*c*) providing damping bars on the rotor pole faces
(*d*) oscillations cannot be damped.

25. A Syn. motor has its field winding shorted and stator is supplied variable voltage. This is likely to result in:
(*a*) burning or rotor (*b*) motor running as Induction motor
(*c*) crawling of rotor (*d*) magnetic locking of rotor.

26. Torque angle for a Syn. Motor is
(*a*) the angle through which the motor lags behind the Syn. speed.
(*b*) the angle of lag from no-load to full load conditions.
(*c*) the angle between the rotating stator flux and the field produced by rotor poles.
(*d*) none of these.

27. A Syn. motor can be used a Syn. Capacitor when it is
(*a*) under loaded (*b*) over loaded
(*c*) under excited (*d*) over excited

28. Power developed by a syn. motor will be maximum when the load angle is
(*a*) zero (*b*) 45°
(*c*) 90° (*d*) 120°

29. A 3-phase induction motor draws 1000 kVA at a pf of 0.8 lag. A synchronous condenser is connected in parallel to draw an additional 750 kVA at a power factor of 0.6 lead. The pf of the total load supplied by the mains is

(*a*) unity. (*b*) zero.
(*c*) 0.6 lag. (*d*) 0.707 lead.

30. Squirrel cage winding is provided on a synchronous motor to make it

(*a*) noise free (*b*) self-starting
(*c*) cheap. (*d*) quick start

31. A pony motor is basically a

(*a*) small induction motor. (*b*) DC series motor,
(*c*) DC shunt motor. (*c*) double winding AC/DC motor.

32. While starting a synchronous motor its field winding should be

(*a*) connected to a DC source. (*b*) short-circuited.
(*c*) kept open. (*d*) none of the above.

Keys to Multiple Choice Questions

1. d	**2.** d	**3.** d	**4.** d	**5.** b	**6.** a	**7.** a	**8.** a	**9.** b	**10.** c
11. c	**12.** a	**13.** b	**14.** b	**15.** b	**16.** b	**17.** c	**18.** c	**19.** c	**20.** d
21. d	**22.** a	**23.** b	**24.** c	**25.** b	**26.** c	**27.** d	**28.** c	**29.** a	**30.** b
31. a	**32.** b								

Three-Phase Induction Motors

CHAPTER

9

Chapter Objectives

After the completion of this unit, students/readers will be able to understand:

- ✓ What are the various parts of 3-phase induction motor?
- ✓ What is basically a squirrel cage and phase wound induction motor?
- ✓ How a rotating field is developed in the stator core of a 3-phase induction motor?
- ✓ What is the basic principle of operation of this motor and why it is named as induction motor?
- ✓ Why it is also called as asynchronous motor?
- ✓ How the direction of rotation of a 3-phase induction motor is reversed?
- ✓ What are the parameters of the rotor of an induction motor?
- ✓ How to draw an equivalent circuit of rotor circuit?
- ✓ What are stator parameters and how to draw exact equivalent circuit of a 3-phase induction motor?
- ✓ How to draw its phasor diagram?
- ✓ What are the similarities and dissimilarities of a 3-phase induction motor and a transformer?
- ✓ How power flows in an induction motor?
- ✓ What are the factors on which torque produced depends in an induction motor?
- ✓ Under what condition, torque developed in an induction motor is maximum.
- ✓ How the performance of an induction motor is affected by the addition of resistance in its rotor circuit?
- ✓ How to perform various tests on an induction motor to determine its parameters and performance?
- ✓ How to determine performance of a 3-phase induction motor using circle diagram?
- ✓ What are various harmonics which affect the performance of an induction motor?
- ✓ What are cogging and crawling phenomena and how these can be eliminated (or minimised) in case of induction motors?
- ✓ What are deep-bar and double-cage induction motors?

Introduction

Induction machines are also called asynchronous machines i.e., the machines which never run at a synchronous speed. Whenever we say induction machine we mean to say induction motor. Induction motors may be single-phase or three-phase. The single phase induction motors are usually built in small sizes (upto 3 H.P). Three phase induction motors are the most commonly used AC motors in the industry because they have simple and rugged construction, low cost, high efficiency, reasonably good power factor, self-starting and low maintenance cost. Almost more than 90% of the mechanical power used in industry is provided by three phase induction motors.

In this chapter, we shall deal with all the important aspects of a three phase induction motor.

9.1 Constructional Features of a Three-phase Induction Motor

A 3-phase induction motor consists of two main parts, namely stator and rotor.

1. **Stator:** It is the stationary part of the motor. It has three main parts, namely. (*i*) Outer frame, (*ii*) Stator core and (*iii*) Stator winding.

   (*i*) **Outer frame**: It is the outer body of the motor. Its function is to support the stator core and to protect the inner parts of the machine. For small machines the fame is casted but for large machines it is fabricated.

   To place the motor on the foundation, feet are provided in the outer frame as shown in Fig. 9.1.

   (*ii*) **Stator core***:* When AC supply is given to the induction motor, an alternating flux is set -up in the stator core. This alternating field produces hysteresis and eddy current loss. To minimise these losses, the core is made of high grade silicon steel stampings. The stampings are assembled under hydraulic pressure and are keyed to the frame. Each stamping is insulated from the other with a thin varnish layer. The thickness to the stamping usually varies from 0.3 to 0.5 mm. Slots are punched on the inner periphery of the stampings, as shown in Fig. 9.2, to accommodate stator winding.

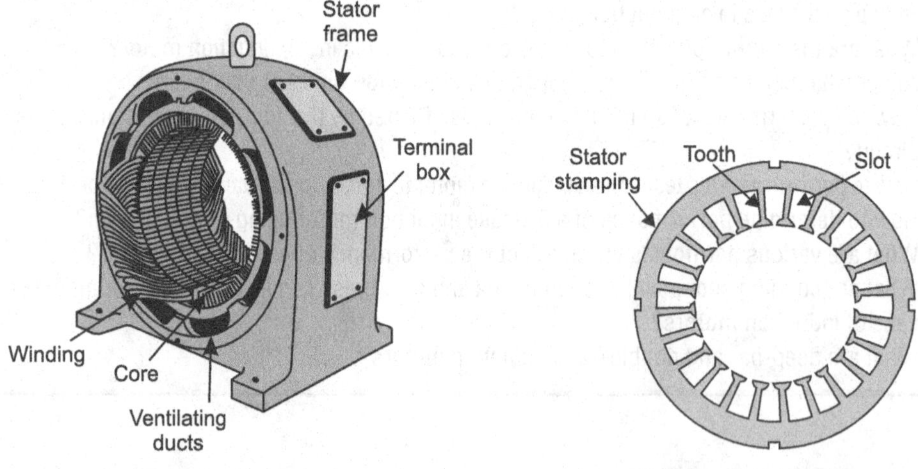

Fig. 9.1 Stator

Fig. 9.2 Stator stamping

   (*iii*) **Stator winding:** The stator core carries a three phase winding which is usually supplied from a three phase supply system. The six terminals of the winding (two of each phase) are connected in the terminal box of the machine. The stator of the motor is wound for definite number of poles, the exact number being determined by the requirement of speed. It will be seen that greater the number of poles, the lower is the speed and vice-versa, since $N_S \propto \frac{1}{P}\left(\because N_S = \frac{120f}{P}\right)$. The three- phase winding may be connected in star or delta externally through a starter.

2. **Rotor:** The rotating part of the motor is called *rotor.* Two types of rotors are used for 3-phase induction motors. (*i*) Squirrel cage rotor (*ii*) Phase wound rotor.

   (*i*) **Squirrel cage rotor:** The motors in which these rotors are employed are called *Squirrel cage induction motors*. Because of simple and rugged construction, the most of the induction motors employed in the industry are of this type. A squirrel cage rotor consists of a laminated cylindrical core having semi-closed circular slots at the outer periphery. Copper or aluminium bar conductors are placed in these slots and short circuited at each end by copper or aluminium rings, called short circuiting rings, as shown in Fig. 9.3. Thus, in these rotors, the rotor winding is permanently short-circuited and no external resistance can be added in the rotor circuit. Figure 9.3 clearly shows that the slots are not parallel to the shaft but these are skewed. The skewing provides the following advantages:

   (*a*) Humming is reduced, that ensures quiet running.
   (*b*) At different positions of the rotor, smooth and sufficient torque is obtained.
   (*c*) It reduces the magnetic locking of the stator and rotor,
   (*d*) It increases the rotor resistance due to the increased length of the rotor bar conductors.

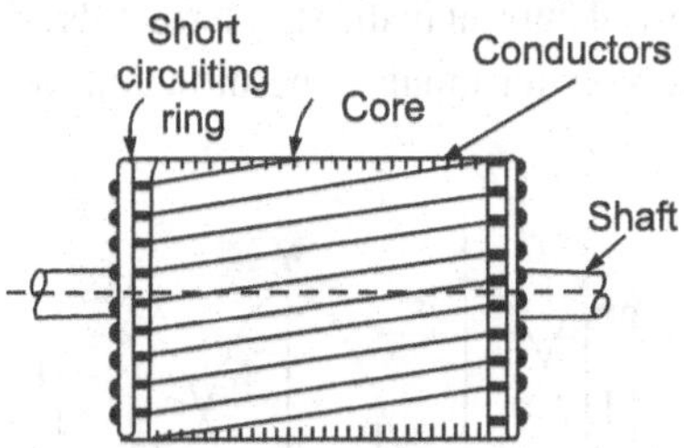

Fig. 9.3 Squirrel cage rotor

   (*ii*) **Phase wound rotor***:* It is also known as slip ring rotor and the motors in which these rotors are employed are known as *phase wound* or *slipring induction motors.* This rotor is also cylindrical in shape which consists of large number of stampings. A number of semi-closed slots are punched at its outer periphery. A 3-phase insulated winding is placed in these slots. The rotor is wound for the same number of poles as that of stator. The rotor winding is connected in star and its remaining three terminals are connected to the slip rings. The rotor core is keyed to the shaft. Similarly, slip-rings are also keyed to the shaft but these are insulated from the shaft. (see Fig. 9.4).

   In this case, depending upon the requirement any external resistance can be added in the rotor circuit. In this case also the rotor is skewed.

   A mild steel shaft is passed through the centre of the rotor and is fixed to it with key. The purpose of shaft is to transfer mechanical power.

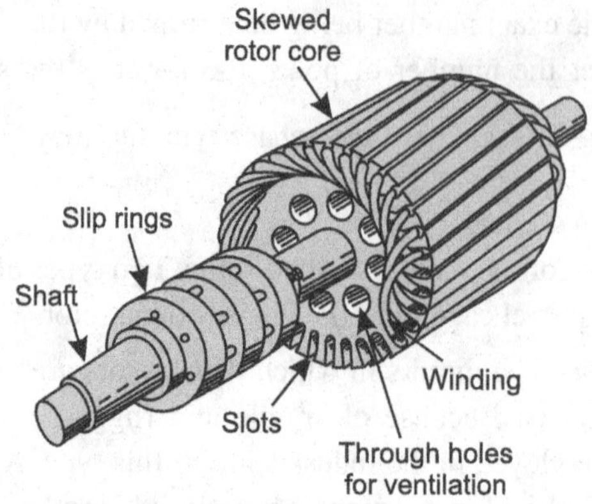

Fig. 9.4 Phase wound rotor

9.2 Production of Revolving Field

The detail description of production of revolving field has been discussed in chapter 6 please refer the same. However, here we will go through the topic quickly.

Consider a stator on which three different windings represented by three concentric coils a_1a_2, b_1b_2 and c_1c_2 respectively are placed 120° electrically apart.

Let a 3-phase supply, as shown in Fig. 9.5, is applied to the stator. Three phase currents will flow through the three coils and produce their own magnetic fields. The positive half cycle of the alternating current is considered as inward flow of current in the start terminals and negative half cycle is considered as outward flow of current in the start terminals. The direction of flow of current is opposite in the finish terminals of the same coil.

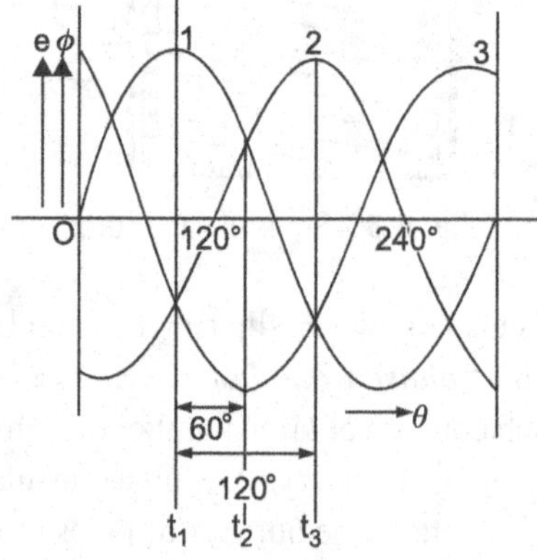

Fig. 9.5 Wave diagram of 3- phase supply

Let at any instant t_1, current in coilside a_1 be inward and in b_1 and c_1 outward. Whereas, the current in the other sides of the same coils is opposite i.e., in coil side a_2 is outward and b_2 and c_2 is inward. The resultant field and its direction (Fm) is marked in Fig. 9.6.

At instant t_2 when θ is 60°, current in coil sides a_1 and $b_1$is inward and in c_1 is outward. Whereas, the current in the opposite sides is opposite. The resultant field and its direction is shown in Fig. 9.7, which is rotated through an angle $\theta = 60°$ from its previous position.

At instant t_3 when θ is 120°, current in coilside b_1 is inward and in c_1 and a_1 is outward. The resultant field and its direction is shown in Fig. 9.8. Which is rotated through an angle θ = 120° electrical from its first position.

Thus, in one cycle, the resultant field completes one revolution. Hence, we conclude that when 3-phase supply is given to a 3-phase wound stator, a resultant field is produced which revolves at a constant speed, called synchronous speed ($N_s = 120°f/P$).

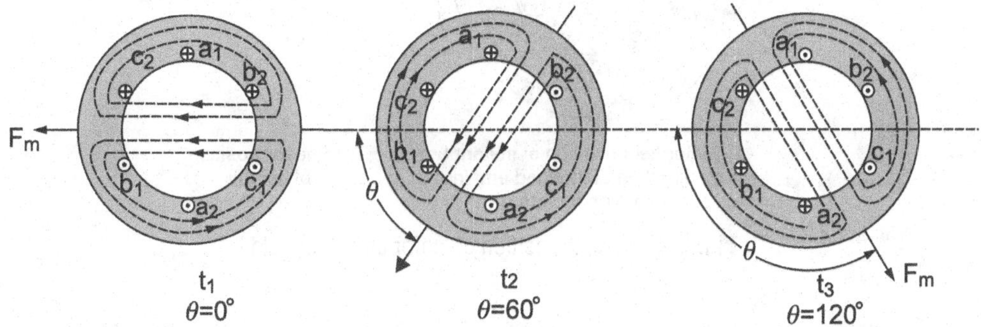

Fig. 9.6 Stator field position at instant t_1 **Fig. 9.7** Stator field position at instant t_2 **Fig. 9.8** Stator field position at instant t_3

In this case, we have seen that when supply from phase 1,2 and 3 is given to coil a_1a_2, b_1b_2 and c_1c_2, respectively, an anticlockwise rotating field is produced. If the supply to coil a_1a_2, b_1b_2 and c_1c_2 is given from phase 1, 3 and 2 respectively, the direction of rotating field is reversed. Thus, to reverse the direction of rotation of rotating field the connections of any two supply terminals are inter changed.

9.3 Principle of Operation

When 3-phase supply is given to the stator winding of a 3-phase wound induction motor, a revolving field is set up in the stator core. The resultant magnetic field set-up by the stator core, at any instant, is shown in Fig. 9.9. The direction of the resultant field is marked by an arrow head *Fm*. As per the supply sequence, let this field is rotating in an anti-clockwise direction at synchronous speed ωs radian per second.

The revolving field is cut by the stationary rotor conductors and an emf is induced in the rotor conductors. Since the rotor conductors are short circuited, current flows through them in the direction as marked in Fig. 9.9(a). A resultant field *Fr* is set-up by the rotor current carrying conductors. This field tries to come in line with the stator revolving field *Fm*, due to which an electromagnetic torque *Te* develops and rotor starts rotating in same direction as that of stator revolving field.

Alternately: Reproducing section X of Fig. 9.9(*a*) as shown in Fig. 9.10, when the revolving stator field see Fig. 9.10 (*a*) cuts the stationary rotor conductors, an emf is induced in the conductors by induction. As rotor conductors are short circuited, current flows through them, as marked in Fig. 9.10 (*b*), which sets up field around them. A resultant field is set up, as shown in Fig. 9.10 (*c*), which exerts force on the rotor conductors. Thus, the rotor starts rotating in the same direction in which stator field is revolving.

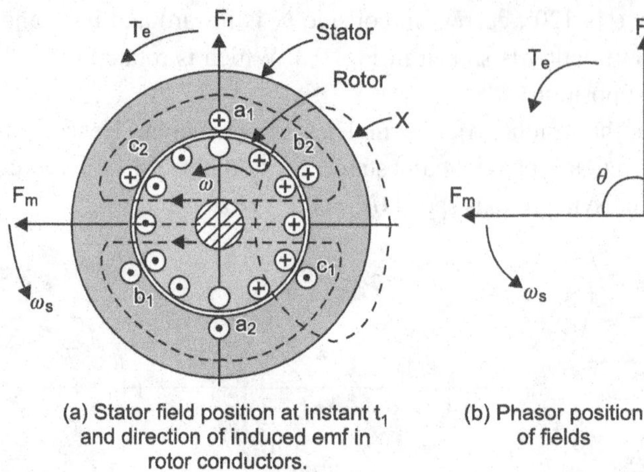

(a) Stator field position at instant t_1 and direction of induced emf in rotor conductors.

(b) Phasor position of fields

Fig. 9.9 Phaser position of stator and rotor field

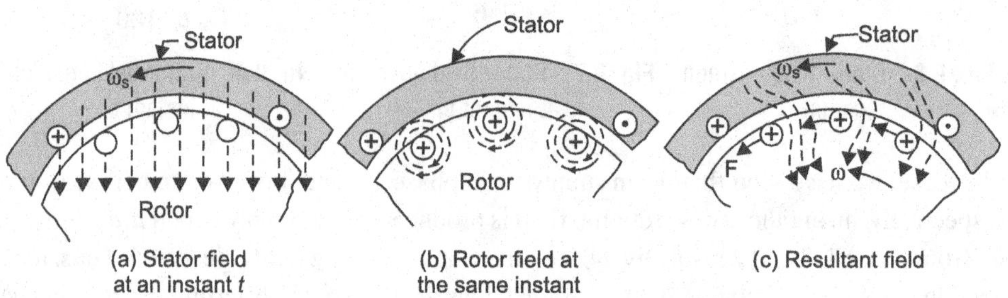

(a) Stator field at an instant *t*

(b) Rotor field at the same instant

(c) Resultant field

Fig. 9.10 Working principle of induction motor

The rotor picks up speed and tries to attain the synchronous speed but fails to do so. It is because if the rotor attains the synchronous speed then the relative speed between revolving stator field and rotor will be zero, no emf will be induced in rotor conductors. No emf means no current, no rotor field F_r and hence no torque is produced. Thus, *an induction motor can never run at synchronous speed.* It always runs at a speed less than synchronous speed.

Since, *the principle of operation of this motor depends upon electromagnetic induction, hence the name* ***induction motor***.

9.4 Reversal of Direction of Rotation of Three-phase Induction Motors

In section 9.2 it has been seen that a revolving field is set up in the stator of a 3-phase induction motor when 3-phase supply is given to its winding and the direction of rotation depends upon the supply sequence.

In section 9.3, it has been see that rotor of a three phase induction motor rotates in the same direction as that of the revolving field.

The direction of rotation of the revolving field or that of the rotor can be reversed if the sequence of supply is reversed. The supply sequence can be reversed by interchanging the connections of any two supply leads at the stator terminals.

Hence, *the direction of rotation of a 3-phase induction motor can be reversed by interchanging the connections of any two supply leads at the stator terminals.*

9.5 Slip

In an induction motor, the speed of rotor is always less than synchronous speed.

The difference between the speed of revolving field (N_s) and the rotor speed (N) is called **slip**.

The slip is usually expressed as a percentage of synchronous speed (N_s) and is represented by symbol S.

Mathematically,

% slip, $$\% S = \frac{N_s - N}{N_s} \times 100$$

or Fractional slip, $$S = \frac{N_s - N}{N_s}$$

Rotor speed, $$N = N_s (l - S)$$

The difference between synchronous speed and rotor speed is called *slip speed* i.e.,

$$\text{Slip speed} = N_s - N$$

The value of slip at full load varies from about 6% small motors to about 2% for large motors.

Importance of Slip

Slip plays an important role in the operation of an induction motor. We have already seen that the difference between the rotor speed and synchronous speed of flux determine the rate at which the flux is cut by rotor conductors and hence the magnitude of induced emf i.e., $e_2 \propto N_S - N$

Rotor current, $i_2 \propto e_2$ and torque, $T \propto i_2$

$$\therefore \quad T = K (N_S - N)$$

or $$T = KN_S \left(\frac{N_s - N}{N_s} \right)$$

or $$T = K_1 S$$

Hence, $T \propto S$

Thus, greater the slip greater will be the induced emf or rotor current and hence larger will be the torque developed.

At no-load, induction motor requires small torque to meet with the losses only such as mechanical, iron and other losses. Therefore rotor speed at no-load is very high and the slip is very small. When the motor is loaded, greater torque is required to drive the load, therefore, the slip increases and rotor speed decreases slightly.

Thus, it is seen that ship in an induction motor adjusts itself to such a value so as to meet the required driving torque under normal operation.

9.6 Frequency of Rotor Currents

The frequency of rotor currents depends upon the relative speed between rotor and stator field. When the rotor is stationary, the relative speed between stator revolving field and stationary rotor conductors is ($Ns - 0 = Ns$) the frequency of rotor currents is the same as that of the supply frequency. But once the rotor starts rotating, the frequency of rotor currents decreases depends upon relative speed or slip speed ($N_s - N$).

Let at any speed N, the frequency of rotor currents be f_r. Then

$$f_r = \frac{(N_s - N)P}{120} = \frac{(N_s - N)}{N_s} \cdot \frac{N_s P}{120} = S \times f$$

Example 9.1

A 3-phase, 4 pole induction motor is connected to 400 V, 50 Hz supply. Determine: (i) Synchronous speed. (ii) Actual speed of the motor when running at 4% slip (iii) Frequency of emf induced in rotor.

Solution:

Synchronous speed, $N_S = \frac{120f}{P} = \frac{120 \times 50}{4}$ **= 1500 rpm** (*Ans.*)

Actual speed of motor, $N = N_S(1 - S)$ where, $S = 0{\cdot}04$

$N = 1500\,(1 - 0{\cdot}04)$ **= 1440 rpm** (*Ans*)

Frequency of rotor emf, $f_r = Sf = 0.04 \times 50 =$ **2 Hz** (*Ans*)

Example 9.2

A 3-phase, 10 HP squirrel cage induction motor is wound for 6 poles. When the motor is connected to 230 V, 50 Hz supply, at full-load, it operates at 4% slip. Determine (i) full loud speed. (ii) full load torque in Newton-metre. (iii) frequency of rotor current under this condition and (iv) speed of rotation of the stator mmf

Solution:

Synchronous speed, $N_S = \frac{120f}{P} = \frac{120 \times 50}{6} = 1000$ rpm

Full load speed, $N = N_S(1 - S) = 1000\,(1 - 0.04)$ **= 960 rpm** (*Ans*)

Output $= \omega T = 10 \times 735.5$ W (where $\omega = 2\pi N/60$)

Or $T = \frac{10 \times 735{\cdot}5 \times 60}{2\pi \times 960}$ **= 73·16 Nm** (*Ans*)

Rotor current frequency, $f_r = S f = 0.04 \times 50 =$ **2 Hz** (*Ans*)

Speed of rotation of stator mmf $= N_S =$ **1000 rpm** (*Ans*)

Example 9.3

Power to an induction motor is supplied by a 12 pole, 3-phase, 500 rpm alternator. The full load speed of the motor is 1440 rpm Find the percentage slip and number of poles in the motor.

Solution:

Speed of the alternator, $N_{sa} = 500$ rpm

No. of poles of the alternator, $P_a = 12$

$$\text{Generated or supply frequency, } f = \frac{P_a N_{sa}}{120} = \frac{12 \times 500}{120} = 50 \text{ Hz}$$

Motor Speed, $N = 1440$ rpm

$$\text{No. of poles of motor, } P = \frac{120f}{N} = \frac{120 \times 50}{1440} = 4{\cdot}16 = \mathbf{4} \ (Ans)$$

$$\text{Synchronous speed, } N_S = \frac{120f}{P} = \frac{120 \times 50}{4} = 1500 \text{ rpm}$$

$$\text{Percentage slip, } S = \frac{N_s - N}{N_s} \times 100 = \frac{1500 - 1440}{1500} \times 100 = \mathbf{4\%} \ (Ans)$$

<u>**Example 9.4**</u>

A 500 HP, 3-Phase, 440 V, 50 Hz induction motor has a speed of 950 rpm at full load. The machine has 6 poles. Calculate the slip. How many complete alternations will the rotor emf make per minute.

Solution:

Here, $f = 50$ Hz, $P = 6$, $N = 950$ rpm

Synchronous speed, $$N_S = \frac{120f}{P} = \frac{120 \times 50}{6} = 1000 \text{ rpm}$$

Slip, $$S = \frac{N_s - N}{N_s}$$

$$= \frac{1000 - 950}{1000} = \mathbf{0{\cdot}05 \ or \ 5\%} \ (Ans)$$

Frequency of rotor emf $$f_r = Sf = 0.05 \times 50 = 2.5 \text{ Hz or } 2.5 \text{ c/s}$$

Alternations of rotor emf per min = 2.5×60 = **150 c/ min** (*Ans*)

9.7 Speed of Rotor Field or mmf

When three-phase currents are supplied to the stator winding of a poly phase induction motor, a resultant field is set up which rotates at a constant speed called synchronous speed ($N_S = 120f/P$).

This rotating field induces poly phase emfs. in the rotor winding and if rotor winding is closed, poly phase currents circulate in it. These currents set up a revolving field in the rotor which rotates at a speed

$$N_r = 120 f_r / P \text{ with respect to rotor.}$$

Now $$N_r = 120 \times Sf/P = SN_S$$

When rotor itself is rotating at a speed N rpm in the space.

$\therefore$ Speed of rotor field in space $= N + N_r = (1 - S) N_S + SN_S = N_S - SN_S + SN_S = N_S$

Thus, rotor magnetic field also rotates, in space, at the same speed and in the same direction as that of stator field. Hence, the two fields are magnetically locked with each other and are stationary with respect to each other.

9.8 Rotor emf

The revolving magnetic field set up in the stator by poly phase currents is common to both stator and rotor winding. This field induces emfs. in both the windings. The stator induced emf per phase is given by the relation;

$$E_1 = 4.44\, kw1\, T_1 f f_m \qquad ...(i)$$

Where kw_1 = winding factor i.e., product or coil span factor kc and distribution factor kd.

T_1 = No. of turns/phase of stator winding;

f = stator or supply frequency and

f_m = maximum value of flux.

The rotor induced emf/phase, $E_2 = 4.44\, kw_2\, T_2 f_r f_m$ ...(ii)

Where f_r is the rotor current frequency, and under stationary condition i.e., at the start $f_r = f$. Therefore, rotor induced emf/phase at stand still or start, $E_{2s} = 4.44\, kw_2\, \mathrm{T}_2 f f_m$

Dividing eq. (*ii*) by (*i*), we get,

$$\frac{E_{2s}}{E_1} = \frac{4{\cdot}44\, kw_2 T_2 f \phi_m}{4{\cdot}44\, kw_1 T_1 f \phi_m} = \frac{T_2}{T_1} = K \quad \text{(i.e., transformation ratio)}$$

From eq. (*ii*), induced emf in the rotor under running condition,

$$E_2 = 4{\cdot}44\, kw_2\, T_2\, (Sf)\, \phi_m = SE_{2S}$$

The induced emf in the rotor circuit is maximum at the start and varies according to the value of slip under running condition. Since, the value of normal slip under loaded condition is nearly 5% therefore, the rotor induced emf is nearly 5% of the maximum value.

9.9 Rotor Resistance

Since the rotor winding is made of some conducting material (copper or aluminium), it has a definite resistance ($R = \rho\, l/a$). Its value remains constant and is denoted by R_2.

9.10 Rotor Reactance

Whole of the flux produced by the rotor currents does not link with the stator winding. The part of rotor flux which links the rotor conductors but not with the stator winding is called leakages flux and hence develops leakage inductance (L_2). The leakage flux and hence the inductance is very small if the rotor conductors are placed at the outermost periphery of the rotor as shown in Fig. 9.11. Depending upon the rotor current frequency, rotor reactance will be developed.

$$\text{Rotor reactance, } X_2 = 2\pi fr\, L_2 = 2\pi\, Sf\, L_2 = S\,(2\pi f L_2)$$

When the rotor is standstill i.e., at the start, when slip, $S = 1$

$$\text{The value of rotor reactance} = X2s = 2\pi f L_2$$

Thus, under normal running, rotor reactance, $X_2 = SX2s$

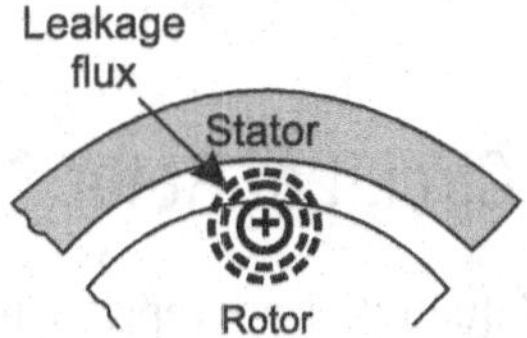

Fig. 9.11 Rotor leakage flux

9.11 Rotor Impedance

The total opposition offered to the flow of rotor current by the rotor circuit is called the rotor impedance.

$$\text{Rotor impedance,} = R_2 + j X_2 = R_2 + j SX_2s$$

$$\text{Magnitude of rotor impedance, } Z_2 = \sqrt{(R_2)^2 + (SX_{2s})^2}$$

9.12 Rotor Current and Power Factor

The rotor circuit diagram of an induction motor is shown in Fig. 9.12.

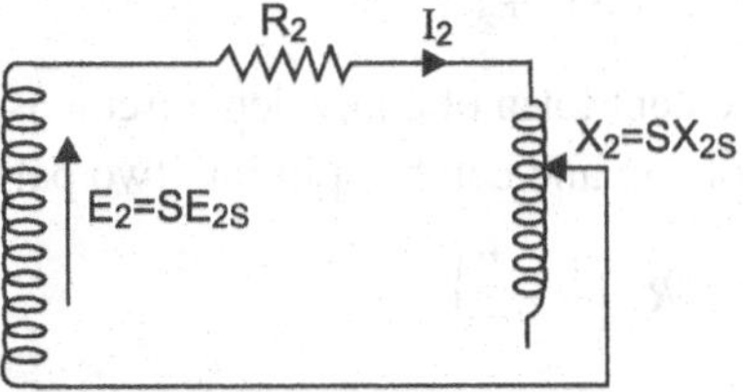

Fig. 9.12 Rotor equivalent circuit

Under running condition;

$$\text{Rotor induced emf} = E_2 = SE_2\text{s}$$

$$\text{Rotor impedance, } Z_2 = \sqrt{R_2^2 + X_2^2} = \sqrt{(R_2)^2 + (SX_{2s})^2}$$

$$\text{Rotor current, } I_2 = \frac{E_2}{Z_2}$$

$$= \frac{E_2}{\sqrt{(R_2)^2 + (X_2)^2}}$$

$$= \frac{SE_{2s}}{\sqrt{(R_2)^2 + (SX_{2s})^2}}$$

Rotor power factor, $\cos\phi_2 = \frac{R_2}{Z_2} = \frac{R_2}{(R_2)^2 + (SX_{2s})^2}$

9.13 Simplified Equivalent Circuit of Rotor

The various parameters and electrical quantities are represented on the circuit diagram, as shown in Fig. 9.13. The rotor current is given by the expression:

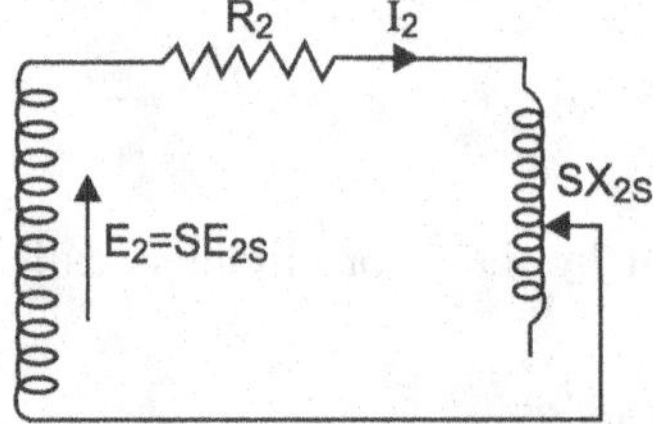

Fig. 9.13 Rotor equivalent circuit

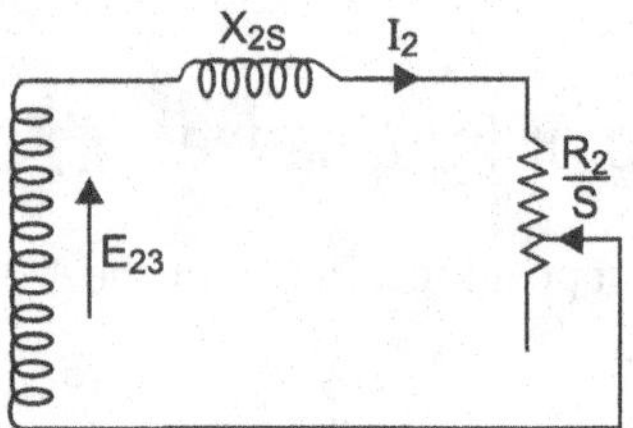

Fig. 9.14 Rotor equivalent circuit

$$I_2 = \frac{SE_{2s}}{\sqrt{(R_2)^2 + (SX_{2s})^2}}$$

The other expression for the rotor current is

$$I_2 = \frac{E_{2s}}{\sqrt{(R_2 / S)^2 + (X_{2s})^2}}$$ (dividing the numerator and denominator by S)

This expression gives a convenient form of equivalent circuit as shown in Fig. 9.14.
The resistance is a function of slip and can be split into two parts;

$$\frac{R_2}{S} = R_2 + R_2\left(\frac{1-S}{S}\right)$$

Where $R_2\left(\frac{1-S}{S}\right)$ represents electrical load on the rotor, Say R_L

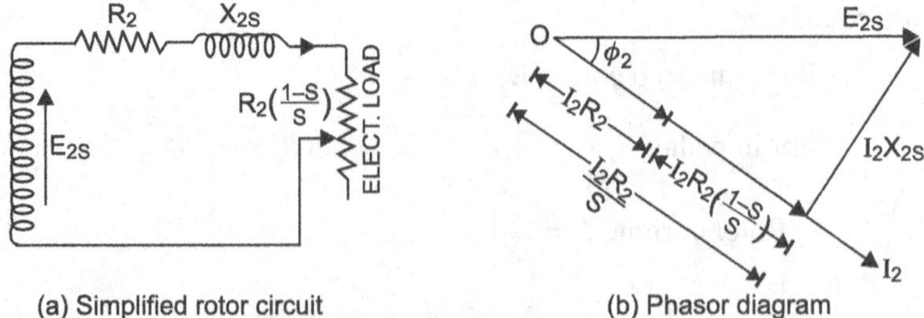

(a) Simplified rotor circuit (b) Phasor diagram

Fig. 9.15 Simplified rotor equivalent circuit with phasor diagram

Thus, the final simplified equivalent rotor circuit is shown in Fig. 9.15 (*a*). Where R_2 is rotor resistance and X_{2s} is standstill leakage reactance. The resistance $R_2\left(\frac{1-S}{S}\right)$ is fictitious resistance representing load (R_L).

The power consumed by this fictitious resistance i.e., $I_2^2 R_2\left(\frac{1-S}{S}\right)$ is the electrical power which is converted into mechanical power to pick the load. After subtracting the mechanical losses, we get the output power available at the shaft.

Thus, electrical power converted into mechanical power, $= I_2^2 R_2\left(\frac{1-S}{S}\right)$ watt

From the simplified equivalent circuit the phasor diagram of rotor circuit is drawn as shown in Fig. 9.15 (*b*).

Rotor current I_2 lags behind the rotor standstill induced emf E_{2s} by an angle f.

The voltage drop across R_2 i.e., I_2R_2 and across $R_2\left(\frac{1-S}{S}\right)$ i.e., $I_2R_2\left(\frac{1-S}{S}\right)$ are in phase with current I_2, whereas the voltage drop in X_{2s} i.e., I_2X_{2s} leads the current I_2 by 90°.

The vector sum of all the three drops is equal to E_{2s} i.e.,

$$E_{2s} = I_2\sqrt{(R_2 / S)^2 + (X_{2s})^2}$$

Power factor of rotor circuit,

$$\cos\phi_2 = \frac{R_2 / S}{\sqrt{(R_2 / S)^2 + (X_{2s})^2}}$$

Example 9.5

An 8 HP, 3-phase, 4-pole squirrel cage induction motor is connected to 400 V, 50 Hz supply. The motor is operating at full-load with 5% slip. Calculate the following: (i) The speed of the revolving field relative to the stator structure;(ii) The frequency of the rotor currents; (iii) The speed of the rotor mmf relative to the rotor structure; (iv) The speed of the rotor mmf relative to the stator structure; (v) The speed of the rotor mmf relative to the stator field distribution; (vi) Are the conditions right for the development of the net unidirectional torque?

Solution:

Here, $P = 4$; $f = 50$ Hz; $S = 0{\cdot}05$

(*i*) The speed of the revolving field relative to the stator structure

i.e., $N_s = \frac{120f}{P} = \frac{120 \times 50}{4}$ **= 1500 rpm** (*Ans*)

(*ii*) $f_r = Sf = 0{\cdot}05 \times 50$ **= 2·5 Hz** (*Ans*)

(*iii*) The speed of rotor mmf relative to the rotor structure,

$N_r = \frac{120f_r}{P} = \frac{120 \times 2{\cdot}5}{4}$ **= 75 rpm** (*Ans*)

(*iv*) Rotor speed, $N = N_s(1 - S) = 1500(1 - 0{\cdot}05) =$ **1425 rpm** (*Ans*)

The speed of the rotor mmf relative to the stator structure

$$= N + N_r = 1425 + 75 = \mathbf{1500\ rpm}\ (Ans)$$

(v) The speed of the rotor *mmf* relative to the stator field distribution

$$= N_s - (N + N_r) = 1500 - 1500 = \mathbf{zero}\ (Ans)$$

(vi) **Yes,** the given conditions fully satisfy for the development of net unidirectional torque.

Example 9.6

The resistance and stand-still reactance per phase of a 3-phase induction motor is 0.1 ohm and 0.4 ohm respectively. If 100 V per phase is induced in the rotor circuit at start then calculate rotor current and rotor p.f. (i) when rotor is stationary and (ii) when running with a slip of 5%.

Solution:

Here, $R_2 = 0{\cdot}1\ \Omega$; $X_{2s} = 0{\cdot}4\ \Omega$; $E_{2s} = 100$ V

(i) *When the rotor is stationary*

$$\text{Rotor current,}\quad I_{2s} = \frac{E_{2s}}{\sqrt{(R_2)^2 + (X_{2s})^2}} = \frac{100}{\sqrt{(0{\cdot}1)^2 + (0{\cdot}4)^2}} = \mathbf{242{\cdot}5\ A}\ (Ans)$$

Rotor power factors,

$$\cos f_{2s} = \frac{R_2}{\sqrt{(R_2)^2 + (X_{2s})^2}} = \frac{0{\cdot}1}{\sqrt{(0{\cdot}1)^2 + (0{\cdot}4)^2}} = \mathbf{0.2425\ lag}\ (Ans)$$

(ii) *When rotor is running with a slip of 5%* i.e., $S = 0{\cdot}05$

$$\textit{Rotor current,}\quad I_2 = \frac{SE_{2s}}{\sqrt{(R_2)^2 + (SX_{2s})^2}} = \frac{0{\cdot}04 \times 100}{(0{\cdot}1)^2 + (0{\cdot}05 \times 0{\cdot}4)^2} = \mathbf{49\ A}\ (Ans)$$

Rotor power factor,

$$\cos f_2 = \frac{R_2}{\sqrt{(R_2)^2 + (SX_{2s})^2}} = \frac{0{\cdot}1}{\sqrt{(0{\cdot}1)^2 + (0{\cdot}05 \times 0{\cdot}4)^2}} = \mathbf{0{\cdot}98\ lag}\ (Ans)$$

Example 9.7

The resistance and stand-still reactance per phase of a 3-phase, 4-pole, 50 Hz induction motor is 0.2 ohm and 2 ohm respectively. The rotor is connected in star and emf induced between the slip-rings at start is 80 V. If at full-load motor is running at a speed of 1440 rpm, calculate (i) the slip, (ii) rotor induced emf per phase, (iii) the rotor current and power factor under running condition and (iv) rotor current and p.f. at standstill when the slip rings are short circuited.

Solution:

No. of poles, $P = 4$

Supply frequency, $f = 50$ Hz

Rotor resistance/ phase, $R_2 = 0.2$ ohm

Rotor standstill reactance/ phase, $X_{2s} = 2$ ohm

Rotor speed, $N = 1440$ rpm

Rotor induced emf at start (line value),

$$E_{2s(L)} = 80 \text{ V}$$

$$\text{Synchronous speed, } N_s = \frac{120\,f}{P} = \frac{120 \times 50}{4} = 1500 \text{ rpm}$$

(*i*) Slip,

$$S = \frac{N_2 - N}{N_s} = \frac{1500 - 1440}{1500} = \mathbf{0{\cdot}04} \; (Ans)$$

(*ii*) Rotor induced emf/phase at standstill,

$$E_{2s} = \frac{E_{2s(L)}}{\sqrt{3}} = \frac{80}{\sqrt{3}} = \mathbf{46{\cdot}2 \text{ V}} \; (Ans)$$

(*iii*) *Under running condition:*

$$\text{Rotor current, } I_2 = \frac{S\,E_{2s}}{\sqrt{(R_2)^2 + (SX_{2s})^2}}$$

$$= \frac{0{\cdot}04 \times 46{\cdot}2}{\sqrt{(0{\cdot}2)^2 + (0{\cdot}04 \times 2)^2}} = \mathbf{8{\cdot}58 \text{ A}} \; (Ans)$$

$$\text{Rotor } p.f., \cos\phi_2 = \frac{R^2}{\sqrt{(R_2)^2 + (SX_{2s})^2}}$$

$$= \frac{0{\cdot}2}{\sqrt{(0{\cdot}2)^2 + (0{\cdot}04 \times 2)^2}} = \mathbf{0{\cdot}9285 \text{ lagging}} \; (Ans)$$

(*iv*) At standstill condition i.e., at start;

$$\text{Rotor current, } I_{2s} = \frac{E_{2s}}{\sqrt{(R_2)^2 + (X_{2s})^2}} \qquad (since\ S = 1)$$

$$= \frac{46{\cdot}2}{\sqrt{(0{\cdot}2)^2 + (2)^2}} = \mathbf{22{\cdot}98 \text{ A}} \; (Ans)$$

$$\text{Rotor p.f., } \cos\phi_{2s} = \frac{R^2}{\sqrt{(R_2)^2 + (X_{2s})^2}}$$

$$= \frac{0{\cdot}2}{\sqrt{(0{\cdot}2)^2 + (2)^2}} = \mathbf{0{\cdot}0995 \text{ lag}} \; (Ans)$$

Example 9.8

The resistance and stand-still reactance per phase of a 3-phase star-connected rotor of a phase wound induction motor is 1 ohm and 4 ohm respectively. If emf induced across the slip-rings is 80 V at start then calculate the current per phase and power factor when (i) sliprings are short-circuited (ii) sliprings are connected to a star-connected rheostat of 3 ohm per phase.

Solution:

Rotor induced emf (line value), $E_{2s(L)} = 60$ V

Rotor resistance/phase, $R_2 = 1$ ohm

Rotor reactance/phase at standstill, $X_{2s} = 4$ ohm

Rotor induced emf at standstill (phase value),

$$E_{2s} = \frac{E_{2s(L)}}{\sqrt{3}} = \frac{60}{\sqrt{3}} - 34 \cdot 64 \text{ V}$$

Rotor impedance/phase at stand-still $Z_{2s} = \sqrt{(R_2)^2 + (X_{2s})^2}$

$$= \sqrt{(1)^2 + (4)^2} = 4 \cdot 123 \text{ ohm}$$

(*i*) *When sliprings are short circuited:*

$$\text{Rotor current, } I_{2s} = \frac{E_{2s}}{Z_{2s}} = \frac{34 \cdot 64}{4 \cdot 123} = \mathbf{8{\cdot}4\ A}\ (Ans.)$$

$$\text{Rotor p.f. } \cos f_{2s} = \frac{R_2}{Z_{2s}} = \frac{1}{4 \cdot 123} = \mathbf{0{\cdot}2425\ lagging}\ (Ans.)$$

(*ii*) When sliprings are connected to a star-connected rheostat of 3 ohm per phase as shown in Fig. 9.16.

Rotor impedance/phase, $Z'_{2s} = \sqrt{(R_2 + R)^2 + (X_{2s})^2}$

$$= \sqrt{(1+3)^2 + (4)^2} = 5 \cdot 657 \text{ ohm}$$

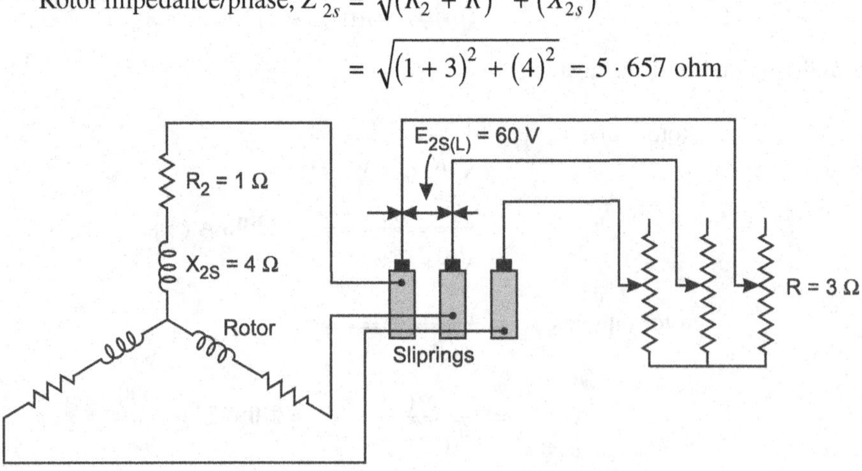

Fig. 9.16 Rotor winding with external resistors

$$\text{Rotor current, } I'_{2s} = \frac{E_{2s}}{Z'_{2s}} = \frac{34 \cdot 64}{5 \cdot 657} = \mathbf{6{\cdot}123\ A}\ (Ans.)$$

$$\text{Rotor p.f., } \cos\phi'_{2s} = \frac{R_2 + R}{Z'_{2s}} = \frac{1+3}{5 \cdot 657} = \mathbf{0{\cdot}707\ lagging}\ (Ans.)$$

Example 9.9

A 3-phase induction motor with star connected rotor has an induced emf per phase of 60 V with the slip rings open circuited and normal voltage applied to stator. The resistance and standstill reactance of each rotor phase are 0·6 ohm and 0·4 ohm respectively. Calculate the rotor current per phase:

(i) at stand still when the rotor circuit is connected through rheostat having a resistance of 5 ohm and reactance 2 ohm per phase.

(ii) when running with slip rings short circuited with slip of 4%.

Solution:

Here, $E_{2s} = 60$ V; $R_2 = 0{\cdot}6\ \Omega$; $X_{2s} = 0{\cdot}4\ \Omega$; $R = 5\Omega$; $X = 2\Omega$

(i) *At standstill, when a rheostat is connected with rotor circuit* (see Fig. 9.17):

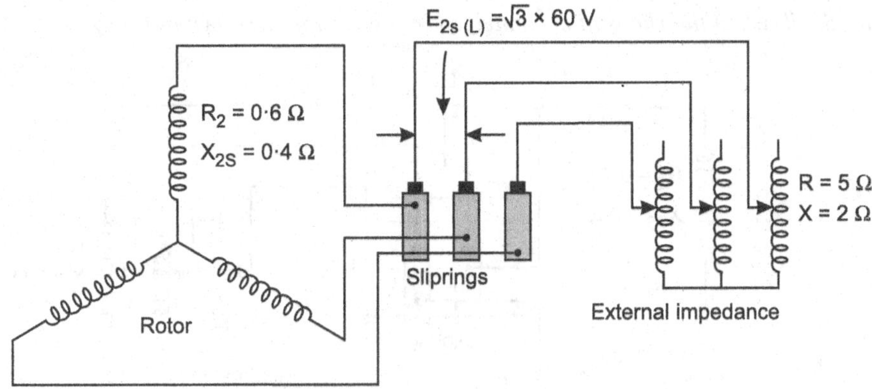

Fig. 9.17 Rotor with external impedance

$$\text{Rotor current, } I_{2s} = \frac{E_{2s}}{\sqrt{(R_2 + R)^2 + (X_{2s} + X_2)^2}}$$

$$= \frac{60}{\sqrt{(0 \cdot 6 + 5)^5 + (0 \cdot 4 + 2)^2}} = \mathbf{9{\cdot}85\ A}\ (Ans.)$$

(ii) *When the slip rings are short circuited and the rotor is running at a slip of 4%* i.e., $S = 0{\cdot}04$

$$\text{Rotor current, } I_2 = \frac{SE_{2s}}{\sqrt{(R_2)^2 + (SX_{2s})^2}} = \frac{0 \cdot 04 \times 60}{\sqrt{(0 \cdot 6)^2 + (0 \cdot 04 \times 0 \cdot 4)^2}} = \mathbf{6{\cdot}66\ A}\ (Ans.)$$

Example 9.10

The standstill impedance of a 3-phase, star-connected rotor of a phase-wound induction motor is (0.4+j4) ohm. When normal supply is connected to the stator, an emf of 80 V appears across the two slip-rings of the motor on open circuit. If a rheostat having impedance per phase of (4+j2) ohm is connected in the rotor, determine.

(a) the rotor current at standstill with the rheostat is in the circuit;

(b) when running short-circuit with slip of 3%.

Solution:

Rotor induced emf at standstill (line value),

$$E_{2s(L)} = 80 \text{ V}$$

Rotor resistance/phase, $R_2 = 0{\cdot}4$ ohm

Rotor reactance/phase at standstill, $X_{2s} = 4$ ohm

Rheostat resistance/phase, $R = 4$ ohm

Rheostat reactance/phase, $X = 2$ ohm

Rotor induced emf at standstill (phase value),

$$E_{2s} = \frac{E_{2s(L)}}{\sqrt{3}} = \frac{80}{\sqrt{3}} = 46 \cdot 2 \text{ V}$$

(*a*) *At standstill, when the rheostat is connected in the circuit as shown in Fig. 9.18.*

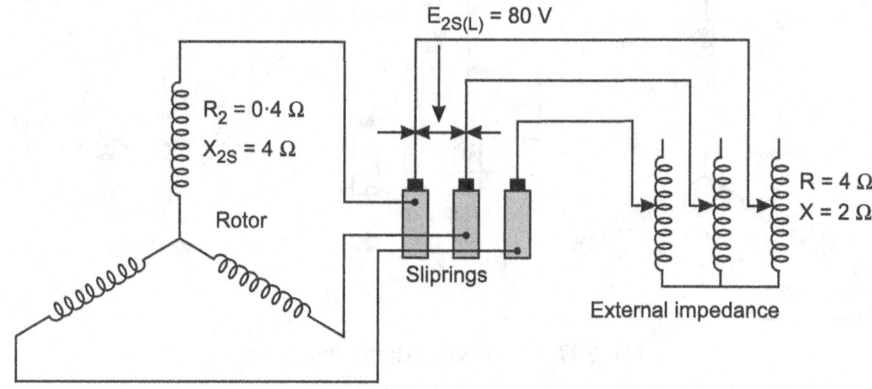

Fig. 9.18 Rotor winding with external impedance

Total rotor circuit resistance/phase, $R'_2 = R_2 + R = 0{\cdot}4 + 4 = 4{\cdot}4$ ohm

Total rotor circuit reactance/phase, $X'_{2s} = X_{2s} + X = 4 + 2 = 6$ ohm

Rotor circuit impedance/phase, $Z'_{2s} = \sqrt{(R_2')^2 + (X_{2s}')^2} = \sqrt{(4 \cdot 4)^2 + (6)^2} = 7.44$ ohm

Rotor current, $I'_{2s} = \dfrac{E_{2s}}{Z_{2s}'} = \dfrac{46 \cdot 2}{7 \cdot 44} = \mathbf{6{\cdot}2\ A}$ (*Ans.*)

(*b*) *When the sliprings are short-circuited, and the rotor is running at a slip, S = 0·03*

$$\text{Rotor impedance, } Z_2 = \sqrt{(R_2)^2 + (SX_{2s})^2}$$

$$= \sqrt{(0 \cdot 4)^2 + (0.03 \times 4)^2} = 0 \cdot 4176 \text{ ohm}$$

Rotor induced, emf, $E_2 = S\,E_{2s} = 0{\cdot}03 \times 46{\cdot}2 = 1{\cdot}386$ V

Rotor current, $I_2 = \dfrac{E_2}{Z_2} = \dfrac{1 \cdot 386}{0 \cdot 4176} = \mathbf{3{\cdot}32\ A}$ (*Ans.*)

Section Practice Problems

Numerical Problems

1. An 8-pole induction motor is supplied at 50 Hz. At full-load its rotor frequency is 1.5 Hz, what will be its speed and slip. (**Ans.** *727 × 5 rpm, 0 × 03*)

2. A -3-phase induction motor is required to be operated at about 700 rpm. What will be the number of poles of the machine if supply frequency is (i) 60 Hz (ii) 25 Hz. Also determine its actual speed if slip is 5%. (**Ans.** *10; 684 rpm, 4; 712 × 5 rpm*)

3. A 3-phase, 6-pole induction motor is supplied from a 3-phase, 400 V, 50 Hz supply. If it is operating at full-load with 0.04 slip, determine:
   (*i*) The speed of the revolving field relative to the stator structure.
   (*ii*) The frequency of the rotor currents.
   (*iii*) The speed of the rotor mmf relative to the rotor structure.
   (*iv*) The speed of the rotor mmf relative to the stator structure.
   (*v*) The speed of the rotor mmf relative to the stator field distribution.
   (*vi*) Are the conditions right for the development of the net unidirectional torque?
   (**Ans.** *1000 rpm; 2 Hz; 40 rpm; 1000 rpm; zero; yes*)

4. The rotor of a 3-phase, phase-wound induction motor has resistance and stand-still reactance of 0.5 ohm and 2 ohm per phase respectively. When normal rated supply is fed to the stator, 80 V is induced across two slip-rings at start on open circuit. Determine the current per phase and p.f. when (*i*) sliprings are short circuited (*ii*) sliprings are connected to a star connected rheostat of 4 ohm per phase. (**Ans.** *95 × 22 A, 0 × 2425 lagging; 9 × 38 A, 0 × 9138 lagging*)

5. A 3-phase induction motor, with star-connected rotor, has an induced emf per phase of 60 V with the sliprings open circuited and normal voltage applied to stator. The resistance and standstill reactance of each rotor phase are 0·6 ohm and 0·4 ohm respectively. Calculate the rotor current per phase:
   (*a*) at standstill when the rotor circuit is connected to a star-connected rheostat having a resistance of 5 ohm and reactance 2 ohm per phase;
   (*b*) when running with slip rings short-circuited at 4% slip. (**Ans.** *5·686 A; 2·31 A*)

Short Answer Type Questions

Q.1. How does the name of an induction motors is derived? **(PTU)**

Ans. The name of an induction motor is derived from the fact that it work on the basic principle of mutual induction. The current in the rotor conductions is induced by the motion of rotor conductor relative to the magnetic field produced by the stator currents.

Q.2. Classify 3-phase induction motors on the basis of their construction. Which one is generally preferred and why?

Ans. Three-phase induction motors may be classified as (*i*) squirrel cage induction motors and (*ii*) phase wound or slip-ring induction motor.

Squirrel cage induction motor is generally preferred due to its low construction cost, low maintenance, high pf, high efficiency, robust construction etc.

Q.3. Usually semi-closed slots are preferred is small motors, why?

Ans. In case of semi-closed slots, the reluctance becomes more uniform and improves the power factor of the motor.

Q.4. Why the rotor slots of an induction motor are skewed?

Ans. The rotor slots of an induction motor are skewed to reduce humming noise and ensuring quiet running, reduce magnetic locking and for smooth and uniform torque.

Q.5. Why the rotor conductors of the squirrel cage rotor are short-circuited in the case of slip-ring induction motors, the rotor circuit is closed through resistors?

Ans. In induction motors, torque develops by the interaction of stator and rotor fields. The rotor field is developed only if current flows through the rotor conductors which is only possible if rotor circuit is closed or short-circuited.

Q.6. What is slip in an induction motor?

Ans. The difference between the synchronous speed of stator revolving field and the rotor speed expressed as a fraction of synchronous speed is known as a slip in an induction motor.

Q.7. What is the value of slip at start in an induction motor?

Ans. 1

Q.8. If the full load speed of a 3-phase, 50 Hz induction motor is 1460 rpm, what will be its synchronous speed?

Ans. 1500 rpm.

9.14 Stator Parameters

Like rotor the stator winding of the motor also has resistance R_1. The flux produced by stator winding linking with its own turns only (leakage flux) produces leakage reactance X_1.

Of the total voltage V applied to the stator, a part of it is consumed by stator resistance (I_1R_1) and leakage reactance (I_1X_1) and the remaining is utilised in establishing mutual flux which links with stator and rotor winding both. When it links with the stator winding it produces self-induced emf E_1.

$$\therefore \qquad \bar{E}_1 = \bar{V}_1 - \bar{I}_1\bar{R}_1 - \bar{I}_1\bar{X}_1$$

9.15 Induction Motor on No-load

(Rotor circuit open)

In slip ring induction motor rotor circuit can be opened. Under this condition, when stator is connected to 3-phase supply, it draws a very small current called no-load current I_0. This current has two components i.e., working component I_w and magnetising component I_{mag}. Working component is in phase with the supply voltage and it supplies the stator iron losses. Whereas, magnetising component lags behind the supply voltage V by 90° and produces the mutual flux which links with stator and rotor winding and induces E_1 and E_{2s}.

The equivalent circuit and phasor diagram of the motor under this condition is shown in Fig. 9.19.

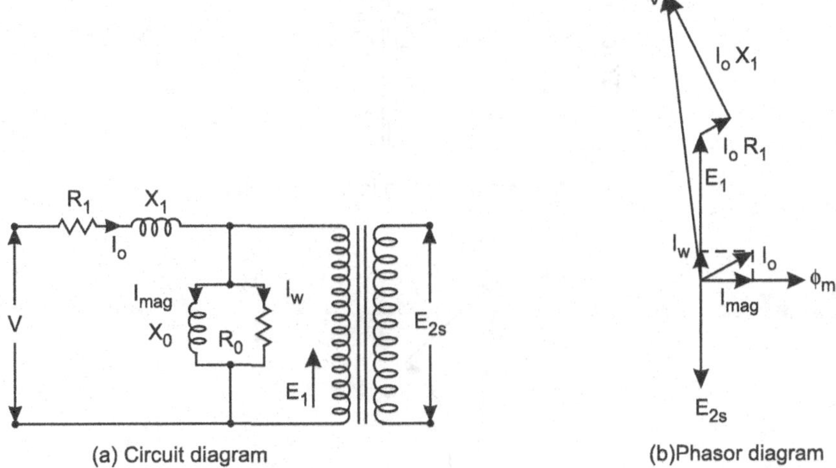

Fig. 9.19 Induction motor at no-load

9.16 Induction Motor on Load

When load is applied on the induction motor its speed decreases slightly and slip increases. Thus rotor current I_2 increases. Simultaneously, to meet with this load, motor draws extra current from the supply mains *similar to that of a transformer.* In fact, power is transferred through magnetic field or flux.

The complete circuit diagram and phasor diagram of a loaded induction motor is shown in Fig. 9.20(a) and (b), respectively.

Here, X_0 – exciting reactance $= \dfrac{V}{I_{mag}}$

R_0 – exciting resistance $= \dfrac{V}{I_w}$

all other abbreviation have their usual meaning.

It may be noted that in the equivalent circuit (per phase) of a 3-phase induction motor, the mechanical load has been replaced by an equivalent electrical resistance R_L given as:

$$R_L = R_2\left(\frac{1-S}{S}\right)$$

The circuit shown in Fig. 20(a) is similar to the equivalent circuit of a transformer with secondary load equal to R_L. The rotor emf in the equivalent circuit only depends on the transformation ratio K (= E_2/ E_1)

Hence, an induction motor can be represented as an equivalent transformer connected to a variable load $R_L\left(= R_2\dfrac{1-S}{S}\right)$.

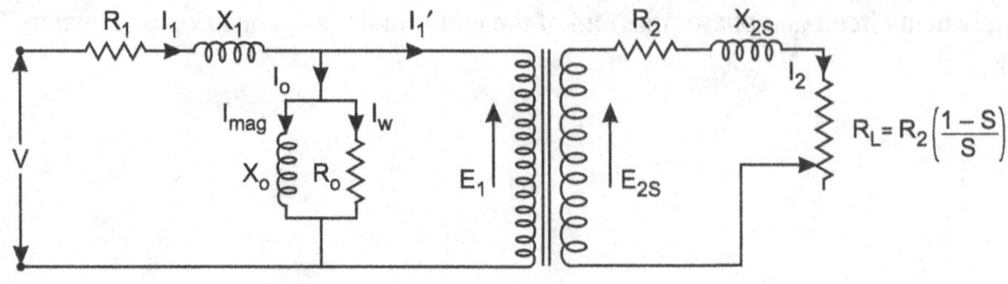

(a) Circuit diagram of an induction motor

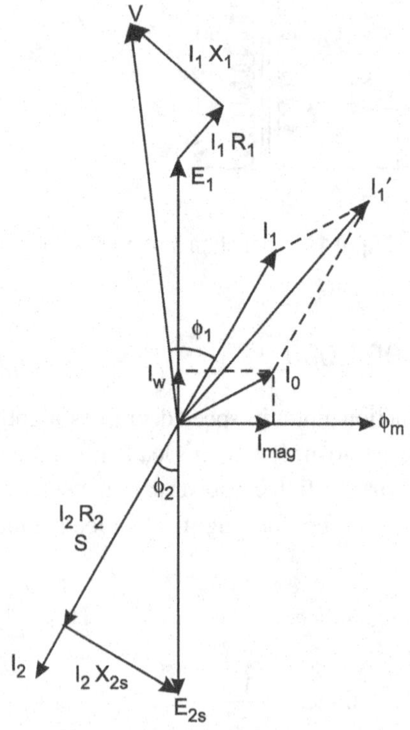

(b) Phasor diagram of an induction motor

Fig. 9.20 Circuit diagram and phasor diagram of an induction motor on-load

9.17 Induction Motor vs Transformer

An induction motor can be compared with a transformer considering the following facts:

Similarities

1. *Induction devices:* Both induction motor and transformer are the induction devices. the stator and rotor of an induction motor can be compared to the primary and secondary of a transformer.
2. *Performance at no-load:* When a 3-phase supply in fed to the stator of a 3-phase induction

motor, a revolving field is set-up in the stator core and an emf is induced in the rotor as well as in the stator winding (*please refer to Art. No 9.8*) similarly as it is induced in the secondary and primary winding of a transformer but in this case field is pulsating not revolving. If we refer to the equations, the only difference is that in case of induction motors, the winding factor k_w ($k_w = k_c \times k_d$) is considered since the winding is distributed over the stator core as per the design requirement. Thus, an induction motor at standstill exhibits the characteristics of a transformer on no-load. In induction motors also the leakage fluxes exist as in a transformer.

3. *Performance on-load:* Under load conditions also the behaviour of both the devices is similar. The flow of current in the rotor conductors produces rotor mmf. The rotor mmf and stator mmf both are rotating at synchronous speed and hence stationary with respect to each other, thus can be combined. This produces the same effect as it is produced in a transformer where mmf produced by primary and secondary are combined. Because of this combined mmf, when load on the transformer (secondary) increases, the primary draws extra current from the supply and the load current is transferred through mmf depending upon transformation ratio. Similarly, when mechanical load on an induction motor increases, rotor current increases and simultaneously motor draws extra current from the supply and the transfer of load occurs through mmf,
4. *Performance under short circuit:* An induction motor with blocked rotor is similar to a transformer under short circuit. The stator and rotor windings have resistance and leakage reactances just as transformer primary and secondary windings have and hence, the equivalent circuit and phasor diagram of an induction motor is similar to a transformer.

Dissimilarities

Although, as per performance, both are similar devices as explained above but at the same time there are some dissimilarities too as mentioned below:

1. *Winding placement and designing:* As per the need, induction motor winding is well distributed over its core, where as in a transformer, the winding is concentrated.
2. *Operating conditions:* An induction motor is a rotating device; basically it converts electrical energy into mechanical energy. Whereas a transformer is a static device; basically it transfers electrical energy from one circuit to the other and while doing so the voltage level is raised or lowered.
3. *Leakage reactance and magnetising current:* since there is air gap between rotor and stator, the leakage reactances in an induction motor are higher than in a transformer. Moreover, induction motor draws more magnetising current which reduces its power factor.
4. *Losses and efficiency:* induction motor has more losses (i.e., mechanical losses because of rotating parts) and lower efficiency than a transformer.

9.18 Reasons of Low Power Factor of Induction Motors

The basic principle of operation of an induction motor is mutual induction. When 3-phase supply is given to a 3-phase wound stator of an induction motor, a revolving field is set up in the stator. This field (flux) is also set up in the air between stator and rotor which links with rotor conductors and an emf is induced in them by mutual induction.

To set up the mutual flux, induction motor draws magnetising current (I_{mag}) from the mains which lags behind the voltage by 90° as shown in Figs. 9.19(*a*) and 9.20(*b*). The magnitude of this current is quite large because of high reluctance of the air gap between stator and rotor.

The power factor of the induction motor is minimum at no-load as shown in Fig. 9.19 (*a*) since the magnetising current has its dominating effect. However, the power factor increases with increase in load on the induction motor and is maximum at full load as shown in Fig. 9.20 (*b*).

Therefore, it is advised to operate the induction at full load.

Thus, because of air gap, induction motor draws large magnetising current and operates at low lagging power factor.

9.19 Main Losses in an Induction Motor

The major losses in an induction motor are:

1. **Stator losses:** The losses which occur in the stator of an induction motor are called stator losses.
   (*i*) Stator copper losses – $I_1^2 R_1$ (per phase)
   (*ii*) Stator iron losses- These are the hysteresis and eddy current losses.
2. **Rotor losses:** which occur in the rotor of an induction motor are called rotor losses.
   (*i*) Rotor copper losses – $I_2^2 R_2$ (per phase)
   (*ii*) Rotor iron losses- Since under normal running condition rotor frequency is very small, therefore. These losses are so small that they are neglected.
3. **Mechanical losses:** The sum of windage and friction losses is called mechanical losses.

9.20 Power Flow Diagram

Electrical power input is given to the stator. There are stator copper and iron losses and the remaining power i.e.,stator output is transferred to the rotor through magnetic flux called rotor input. In the rotor there are rotor copper losses and the remaining power is converted into mechanical power called mechanical power developed in the rotor. Then there are mechanical losses and the remaining power is available at the shaft called mechanical power output.

The power flow diagram is shown in Fig. 9.21.

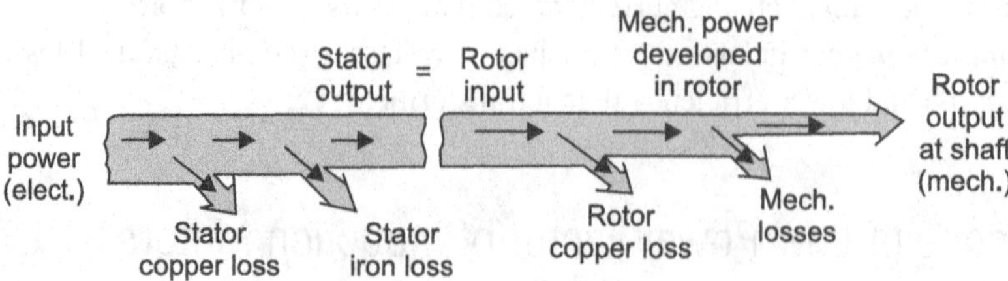

Fig. 9.21 Power flow diagram

9.21 Relation between Rotor Copper Loss, Slip and Rotor Input

We have seen that the electrical power developed in the rotor is converted into mechanical power which is given by the relation:

$$\text{Mech. power developed in the rotor} = I_2^2 R_2 \left(\frac{I - S}{S}\right) \quad \ldots(i)$$

$$\text{The rotor copper losses} = I_2^2 R_2 \quad \ldots(ii)$$

From power flow diagram:

$$\text{Rotor input} = \text{Mech. power developed} + \text{rotor copper losses}$$

$$= I_2^2 R_2 \left(\frac{I - S}{S}\right) + I_2^2 R_2 = \left(\frac{I_2^2 R_2}{S}\right) \quad \ldots(iii)$$

From eq. (*i*) and (*ii*), we get,

$$\frac{\text{Rotor copper loss}}{\text{Mech. power developed}} = \frac{I_2^2 R_2}{I_2^2 R_2 \left(\frac{1 - S}{S}\right)}$$

$\therefore$ $$\textbf{Rotor copper loss} = \left(\frac{\mathbf{S}}{\mathbf{1 - S}}\right) \textbf{Mech. power developed}$$

From. eq. (*ii*) and (*iii*), we get,

$$\frac{\text{Rotor copper loss}}{\text{Rotor input}} = \frac{I_2^2 R_2}{I_2^2 R_2 / S}$$

$\therefore$ **Rotor copper loss = S × Rotor input**

Note: All the values are the phase values.

9.22 Rotor Efficiency

The ratio of rotor output (i.e., mech, power developed in rotor neglecting mechanical losses) to the rotor input is called the rotor efficiency.

$$\text{Rotor efficiency} = \frac{\text{Mech. Power developed}}{\text{Rotor input}} = \frac{I_2^2 R_2 \left(\frac{1 - S}{S}\right)}{I_2^2 R_2 / S} = (1 - S)$$

Example 9.11

The power input to a 3-phase induction motor is 80 kW. The stator losses total 1×5 kW. Find the total mechanical power developed if the motor is running with a slip of 4%.

Solution:

$$\text{Stator output or rotor input} = \text{Stator input} - \text{stator losses} = 80 - 1{\cdot}5 = 78{\cdot}5 \text{ kW}$$

$$\text{Rotor copper losses} = S \times \text{Rotor input} = 0{\cdot}04 \times 78{\cdot}5 = 3{\cdot}14 \text{ kW}$$

$$\text{Mechanical power developed} = \text{Rotor input} - \text{Rotor copper losses}$$

$$= 78{\cdot}5 - 3{\cdot}14 = \mathbf{75{\cdot}36\ kW}\ (Ans)$$

Example 9.12

A 10 H.P., 4 pole, 25 Hz, 3-phase, wound rotor induction motor is taking 9100 watt from the line. Core loss is 290 watt, stator copper loss is 568 watt, rotor copper loss in 445 watt, friction and windage losses are 100 watt. Determine; (a) power transferred across air gap; (b) mechanical power in watt developed by rotor; (c) mechanical power output in watt; (d) efficiency; (e) slip.

Solution:

$$\text{Power input to motor or stator} = 9100 \text{ watt}$$

$$\text{Power transferred across air gap} = \text{Stator input} - \text{Stator core loss} - \text{Stator copper loss}$$

$$= 9100 - 290 - 568 = \mathbf{8242\ W}\ (Ans)$$

$$\text{Mechanical power developed in rotor} = \text{rotor input} - \text{Rotor copper loss} = 8242 - 445 = \mathbf{7797}$$

$$\text{Rotor output} = \text{Mechanical power developed} - \text{Mechanical loss}$$

$$= 7797 - 100 = \mathbf{7697\ W}\ (Ans.)$$

$$\text{Motor efficiency} = \frac{\text{Output}}{\text{input}} \times 100 = \frac{7697}{9100} \times 100 = \mathbf{84{\cdot}58\ \%}\ (Ans)$$

$$\text{Slip, } S = \frac{\text{Rotor copper loss}}{\text{Rotor input}} = \frac{445}{8242} = \mathbf{0{\cdot}05399}\ (Ans)$$

Example 9.13

A 50 H.P., 3-f, 6-pole induction motor delivers full load output at 960 rpm at 0·8 p.f. when supplied with 400V, 50Hz supply. Losses due to windage and friction come out to be 2H.P. and stator losses are 2 kW. Find out. (a) total rotor Cu loss; (b) efficiency and (c) line current.

Solution:

Here, Output = 50 HP, 3-phase, $P = 6$; $N = 960$ rpm;

$$\cos\phi = 0{\cdot}8;\ V_L = 400\text{ V};\ f = 50\text{ Hz};$$

$$\text{Mech. loss} = 2\text{ HP; stator losses} = 2\text{ kW}$$

$$\text{Power developed in rotor} = \text{rotor output} + \text{mech. loss}$$

$$= 50 + 2 = 52\text{ HP} = 52 \times 735{\cdot}5 = 38246\text{ W}$$

$$\text{Synchronous speed, } N_S = \frac{120f}{P} = \frac{120 \times 50}{6} = 1000\text{ rpm}$$

$$\text{Fractional slip, } S = \frac{N_S - N}{N_S} = \frac{1000 - 960}{1000} = 0{\cdot}04$$

$$\text{Rotor copper loss} = \frac{S}{1-S} \times \text{power developed in rotor}$$

$$= \frac{0{\cdot}04}{1 - 0{\cdot}04} \times 38246 = \mathbf{1594\ W}\ (Ans.)$$

$$\text{Input to motor} = \text{power developed in rotor + Cu loss + stator loss.}$$

$$= 38246 + 1594 + 2000 = 41840 \text{ W}$$

$$\text{Motor efficiency, } \eta = \frac{\text{output}}{\text{input}} \times 100 = \frac{50 \times 735{\cdot}5}{418040} \times 100 = \mathbf{87{\cdot}9\%}\ (Ans.)$$

$$\text{Line current, } I_L = \frac{\text{input}}{\sqrt{3}\, V_L \cos\phi} = \frac{41840}{\sqrt{3} \times 400 \times 0{\cdot}8} = \mathbf{75{\cdot}49\ A}\ (Ans.)$$

Example 9.14

A 4 pole, 3-phase, 50 Hz, 400 V induction motor has a delta connected stator and a star connected rotor. Each phase of rotor winding carries one-fourth of the number of turns on each phase of stator winding. The full load speed is 1455 rpm The rotor resistance is 0×3 ohm, and rotor standstill reactance is 1 ohm per phase. The stator and rotor windings are similar. Stator losses equal 100 W. Friction and windage losses are equal to 50 W. Calculate.

(i) blocked rotor voltage per phase *(ii) rotor current per phase at full load*
(iii) total rotor power input at full load *(iv) rotor power loss at full load*
(v) efficiency.

Solution:

Here, $P = 4; f = 50$ Hz; $V_2 = 400$ V; $\frac{N_2}{N_1} = \frac{1}{4}$;

$$N = 1455 \text{ rpm}; R_2 = 0{\cdot}3\ \Omega; X_{2s} = 1\ \Omega;$$

$$\text{Stator loss} = 100 \text{ W; Mech loss} = 50 \text{ W}$$

$$\text{Stator induced emf per phase, } E_1 = V_L = 400 \text{ V } (\textit{delta connected})$$

Now,

$$\frac{E_{2s}}{E_1} = \frac{N_2}{N_1} = \frac{1}{4}$$

$\therefore$ Blocked rotor voltage per phase, $E_{2s} = \frac{400}{4} = \mathbf{100\ V}\ (Ans)$

$$\text{Synchronous speed, } N_s = \frac{120f}{P} = \frac{120 \times 50}{4} = 1500 \text{ rpm}$$

$$\text{Slip, } S = \frac{N_s - N}{N_s} = \frac{1500 - 1455}{1500} = 0 \times 03$$

$$\text{Rotor current, } I_2 = \frac{SE_{2s}}{\sqrt{(R_2)^2 + (SX_{2s})^2}} = \frac{0 \cdot 03 \times 100}{\sqrt{(0 \cdot 3)^2 + (0 \cdot 03 \times 1)^2}}$$

$$= \mathbf{9 \cdot 95\ A}\ (Ans)$$

$$\text{Rotor copper loss} = 3I_2^2 R2 = 3 \times (9{\cdot}95)^2 \times 0{\cdot}3 = \mathbf{89\ W}\ (Ans)$$

$$\text{Power input to rotor} = \frac{\text{Rotor copper loss}}{\text{Slip}} = \frac{89}{0 \cdot 03} = \mathbf{2967\ W}\ (Ans)$$

$$\text{Input to the motor} = \text{rotor input + stator losses} = 2967 + 100 = 3067 \text{ W}$$

$$\text{Output at the shaft} = \text{rotor input} - \text{rotor copper loss} - \text{mech. loss}$$
$$= 2967 - 89 - 50 = 2828 \text{ W}$$
$$\text{Efficiency} = \frac{\text{output}}{\text{Input}} \times 100 = \frac{2828}{3067} \times 100 = \mathbf{92 \cdot 2\ \%}\ (Ans)$$

Example 9.15

A 3-phase, 6 pole, 400 V, 50 Hz induction motor develops 20 HP including mechanical losses when running at 965 rpm, the power factor being 0 × 87 lagging. Calculate (i) the slip (ii) rotor copper losses (iii) the total input if the stator losses are 1500 watt (iv) line current and (v) the number of cycles made per minute by the rotor emf

Solution:

Here V_L =400 V; P = 6; f = 50 Hz; N = 965 rpm

$$\cos\phi = 0 \times 4 \text{ lag; Stator copper loss} = 1500 \text{ W}$$
$$\text{Synchronous speed, } N_s = \frac{120f}{P} = \frac{120 \times 50}{6} = 1000 \text{ rpm}$$
$$\text{Slip, } S = \frac{N_s - N}{N_s} \times 100 = \frac{1000 - 965}{1000} \times 100 = \mathbf{3 \cdot 5\ \%}\ (Ans)$$
$$\text{Mechanical power developed} = 20 \text{ HP} = 20 \times 735{\cdot}5 = 14710 \text{ W}$$
$$\text{Rotor copper losses} = \left(\frac{S}{1-S}\right) \text{mech. power developed}$$
$$= \frac{0 \cdot 035}{1 - 0 \cdot 035} \times 14710 = \mathbf{533 \times 5\ W}\ (Ans)$$
$$\text{Input to stator} = \text{mech. power developed} + \text{rotor copper loss} + \text{stator copper loss}$$
$$= 14710 + 533\ 5 + 1500 = \mathbf{16743 \cdot 5\ W}\ (Ans)$$
$$\text{Line current, } I_L = \frac{\text{Input}}{\sqrt{3}\, V_L \cos\phi} = \frac{16743 \cdot 5}{\sqrt{3} \times 400 \times 0 \cdot 87} = \mathbf{27 \cdot 78\ A}\ (Ans)$$
$$\text{Rotor frequency, } f_r = S \times f = 0 \cdot 035 \times 50 = 1 \cdot 75 \text{ Hz or c/s}$$

No. of cycles made per minute by rotor emf = 1 · 75 × 60 = **105 cycle/min.** (*Ans*)

Example 9.16

A 60 H.P., 6-pole, 3-phase induction motor delivers load output at 960 rpm at 0·8 power factor when supplied with 400 V. 50 Hz supply. Losses due to windage and friction come to 3 H.P. and the stator losses are 2kW. Find out: (a) Total rotor copper loss; (b) efficiency, and (c) line current.

Solution:

$$\text{Rotor output} = 60 \text{ H.P} = 60 \times 735{\cdot}5 = 44130 \text{ W}$$
$$\text{Mechanical losses} = 3 \text{ H.P.} = 3 \times 735{\cdot}5 = 2206 \text{ W}$$
$$\text{Mechanical power developed} = \text{Rotor output} + \text{Mechanical losses}$$
$$= 44130 + 2206 = 44336 \text{ W}$$

$$\text{Synchronous speed, } N_s = \frac{120f}{P} = \frac{120 \times 50}{6} = 1000 \text{ r.p.m.}$$

$$\text{Slip, } S = \frac{N_s - N}{N_s} = \frac{1000 - 960}{1000} = 0 \cdot 04$$

$$\text{Rotor copper loss} = \frac{S}{1-S} \text{ mech. power developed}$$

$$= \frac{0 \cdot 04}{1 - 0 \cdot 04} \times 46336 = 1931 \text{ W}$$

Stator losses = 2kW = 2000W

Input to stator or motor = Mech. power developed + Rotor *Cu* loss + Stator losses

= 46336 + 1931 = 2000 = 50267W

$$\text{Efficiency, } \eta = \frac{\text{Output}}{\text{Input}} = \frac{44130}{50267} \times 100 = \mathbf{87{\cdot}79\%} \textit{ (Ans.)}$$

$$\text{Line current, } I_2 = \frac{\text{Input power}}{\sqrt{3} \times \text{Line voltage} \times \text{ p.f.}} = \frac{50267}{\sqrt{3} \times 400 \times 0 \cdot 8} = \mathbf{90{\cdot}67 \text{ A}} \textit{ (Ans.)}$$

Example 9.17

A 4-pole, 3-phase, 50 Hz induction motor supplies a useful torque of 159 Newton-metre. Calculate at 4% slip: (i) The rotor input; (ii) Motor input; (iii) Motor efficiency, if the friction and windage losses total 500 watt and stator losses are 1000 watt.

Solution:

No. of poles, P = 4; Frequency, f = Hz

Torque at shaft, T_m = 159 Nm; Slip, S = 4% = 0·04

Mechanical losses = 500 W; Stator losses = 1000 W

$$\text{Synchronous speed, } N_s = \frac{120f}{P} = \frac{120 \times 50}{4} = 1500 \text{ r.p.m.}$$

Rotor speed, $N = N_s (1 - S) = 1500 (1 - 0{\cdot}04) = 1440$ rpm

$$\text{Angular speed, } \omega = \frac{2\pi N}{60}$$

$$\text{Rotor output} = \omega T_m = \frac{2\pi N}{60} T_m = \frac{2\pi \times 1440}{60} \times 159 = 23977 \text{ W}$$

Mech. power developed in rotor = Rotor output + Mech. losses = 23977 + 500 = 24477 W

$$\text{Rotor Cu loss} = \frac{S}{1-S} \text{ mech. power developed} = \frac{0 \cdot 04}{1 - 0 \cdot 04} \times 24477 = 1020 \text{ W}$$

∴ Rotor input = Mech. power developed + Rotor Cu loss

= 24477 + 1020 = **25497 W** *(Ans.)*

Motor input = Rotor input + Stator losses = 25497 + 1000 = **26497 W** *(Ans.)*

$$\text{Motor efficiency, } \eta = \frac{\text{Output}}{\text{Input}} \times 100 = \frac{23977}{26497} \times 100 = \mathbf{90{\cdot}49\%} \textit{ (Ans.)}$$

Example 9.18

A 4-pole, 50 Hz, 3-phase induction motor has an efficiency of 85% for useful output power at the shaft of 17 kW. For this load the total stator losses are 900 W and the windage and friction losses are 1100 W. Calculate the slip, torque developed by the rotor and torque available at the rotor shaft.

Solution:

No. of poles, $P = 4$; Supply frequency, $f = 50$ Hz

Motor efficiency, $\eta = 85\% = 0{\cdot}85$; Output power = 17 kW = 17000 W.

Stator losses = 900 W; Mechanical losses = 1100 W

$$\text{Input power} = \frac{\text{Output power}}{\eta} = \frac{17000}{0\cdot 85} = 20000 \text{ W}$$

Stator output or rotor input = Input power – Stator losses = 20000 – 900 = 19100 W

Mech. power developed = Output power + Mech. losses = 17000 + 1100 = 18100 W

Rotor Cu losses = Rotor input – Mech. power developed = 19100 – 18100 = 1000 W

$$\text{Slip, } S = \frac{\text{Rotor Cu loss}}{\text{Rotor input}} = \frac{1000}{19100} = 0\cdot 05235$$

$$\text{Synchronous speed, } N_s = \frac{120f}{P} = \frac{120 \times 50}{4} = 1500 \text{ r.p.m.}$$

$$\text{Rotor speed, } N = N_s(1 - S) = 1500\,(1 - 0{\cdot}05235) = 1421{\cdot}47 \text{ rpm}$$

$$\text{Angular speed, } \omega = \frac{2\pi N}{60} = \frac{2\pi \times 1421\cdot 47}{60} = 148{\cdot}85 \text{ rad/sec.}$$

Mech. power developed in the rotor = ωT

$$\therefore \quad \text{Torque developed, } T = \frac{\text{Mech. power developed}}{\omega} = \frac{18100}{148\cdot 85} = \mathbf{121{\cdot}6\ Nm}\ (Ans.)$$

$$\text{Torque at the shaft, } T_m = \frac{\text{Rotor output}}{\omega} = \frac{17100}{148\cdot 85} = \mathbf{114{\cdot}2\ Nm}\ (Ans.)$$

Example 9.19

A 3-phase induction motor has an efficiency of 90% and runs at a speed of 480 rpm The motor is supplied from 400 V mains and it takes a current of 75 A at 0·77 p.f. Calculate the B.H.P. (metric) of the motor and pull on the belt when driving the line shaft through pulley of 0·75 m diameter.

Solution:

Supply voltage, $V_L = 400$ V; Rotor speed, $N = 480$ rpm

Motor efficiency, $\eta = 90\% = 0{\cdot}9$; Current drawn from mains, $I_L = 75$ A

Motor p.f., $\cos\phi = 0{\cdot}77$ lag. Diameter of pulley, $d = 0{\cdot}75$ m

$$\text{Radius of pulley, } r = \frac{0\cdot 75}{2} = 0\cdot 375 \text{ m}$$

$$\text{Input power} = \sqrt{3}\, V_L I_L \cos\phi = \sqrt{3} \times 400 \times 75 \times 0{\cdot}77 = 40010 \text{ W}$$

$$\text{Output power} = \text{Input power} \times \eta = 40010 \times 0{\cdot}9 = 36009 \text{ W}$$

$$\text{B.H.P. of the motor} = \frac{\text{Output power}}{735 \cdot 5} = \frac{36009}{735 \cdot 5} = \mathbf{48 \cdot 958}\ (Ans.)$$

$$\text{Angular speed, } \omega = \frac{2\pi N}{60} = \frac{2\pi \times 480}{60} = 16\pi$$

$$\text{Torque at the shaft, } T_m = \frac{\text{Ouptut power}}{\omega} = \frac{36009}{16\pi} = 716{\cdot}376 \text{ Nm}$$

Now, torque, T_m = Pull on the belt × radius of pulley

$$\therefore \quad \text{Pull on the belt} = \frac{T_m}{r} = \frac{716 \cdot 376}{0 \cdot 375} = 1910 \cdot 34 \text{ N} = \frac{1910 \cdot 34}{9 \cdot 81} = \mathbf{194 \cdot 73\ kg}\ (Ans.)$$

Example 9.20

A 440 V, 50 Hz, 6-pole, 3-phase induction motor draws an input power of 81 kW from the mains. The rotor emf makes 120 complete cycle per minute. Its stator losses are 1 kW and rotor current per phase is 65 ampere. Calculate:

(i) Rotor copper losses per phase; *(ii) Rotor resistance per phase;*

(iii) Torque developed.

Solution:

Supply voltage, V_L = 440 V; Supply frequency, f = 50 Hz

No. of poles, P = 6; Input power = 81 kW = 81000 W

Stator losses = 1 kW = 1000 W

$$\text{Rotor frequency, } f_r = 120 \text{ c/m} = \frac{120}{60} = 2 \text{ c/s}$$

Rotor current per phase, I_2 = 65 A

Stator output or rotor input = Stator input – Stator losses = 81000 – 1000 = 80000 W

$$\text{Slip, } S = \frac{f_r}{f} = \frac{2}{50} = 0 \cdot 04$$

Total rotor Cu loss = S × Rotor input = 0·04 × 80000 = 3200 W.

$$\text{Rotor Cu losses per phase} = \frac{3200}{3} = \mathbf{1067\ W}\ (Ans.)$$

$$\text{Rotor Cu losses per phase} = I_2^2 R_2$$

$$\therefore \quad \text{Rotor resistance } R_2 = \frac{1067}{(65)^2} = \mathbf{0 \cdot 2525\ ohm}\ (Ans.)$$

Mech. power developed in the rotor

= Rotor input – Rotor Cu losses = 80000 – 3200 = 76800 W

$$\text{Synchronous speed, } N_s = \frac{120f}{P} = \frac{120 \times 50}{6} = 1000 \text{ r.p.m.}$$

Rotor speed, $N = N_s(1 - S)$ = 1000 (1 – 0·04) = 960 rpm

$$\text{Angular speed, } \omega = \frac{3\pi N}{60} = \frac{2\pi \times 960}{60} = 30\ \pi \text{ rad/sec.}$$

Mech. power developed = ωT

$$\therefore \quad \text{Torque developed, } T = \frac{\text{Mech. power developed}}{\omega} = \frac{76800}{32\pi} = \mathbf{763{\cdot}94\ Nm}\ (Ans.)$$

Section Practice Problems

Numerical Problems

1. A 3-phase induction motor draws 40 kW from the main, the corresponding stator losses are 1·5 kW. Calculate:

(*a*) The total mechanical power developed and the rotor I^2R losses when the slip is 0 · 04.

(*b*) The output power of the motor if the friction and windage losses are 0 · 8 kW.

(*c*) The efficiency of the motor. Neglect the rotor iron losses.

(**Ans.** *36·96 kW; 1·54 kW; 36·16 kW; 90·4%*)

2. A 400 V, 6-pole, 50 Hz, 3-phase induction motor develops 20 H.P. inclusive of mechanical losses when running at 995 rpm, the power factor being 0 · 87. Calculate:

(*i*) Slip (*ii*) The rotor copper losses (*iii*) The line current. The stator copper loss is 1500 W.

(**Ans.** *0 · 005; 73 · 92 W; 27 A*)

3. A 6-pole, 3-phase induction motor runs at a speed of 960 rpm and the shaft torque is 135· 7 Nm. Calculate the rotor copper loss if the friction and windage losses amount of 150 watt. The frequency of supply is 50 Hz. (**Ans.** *574 · 67 W*)

Short Answer Type Questions

Q.1. What are the major parameters of the stator of an induction motor?

Ans. Major parameters of the stator of an induction motor are stator winding resistance R_1, stator winding reactance (leakage reactance) X_1, exciting resistance R_0 and exciting reactance X_0.

Q.2. Why an induction motor is considered similar to a transformer?

Ans. Since both are induction devices and power is transferred from one circuit to the other through magnetic field.

Q.3. Which are the major losses in an induction motor?

Ans. The major losses in an induction motor are copper losses, iron losses and mechanical losses.

9.23 Torque Developed by an Induction Motor

We have already seen that the electrical power of 3-phase induction motor converted into mechanical power is given by the relation;

$$P_0 = 3I_2^2R_2\left(\frac{1-S}{S}\right) \quad \ldots(i)$$

also,

$$P_0 = \omega T \quad \ldots(ii)$$

Where, ω = angular speed of the rotor in rad/sec. and

T = torque developed by an induction motor in Nm.

Equating eq. (*i*) and (*ii*), we get

$$\omega\, T = 3I_2{}^2R_2\left(\frac{1-S}{S}\right) = 3\frac{I_2^2R_2}{S}\cdot\frac{1-S}{\omega}$$

or $$T = \frac{3}{\omega_2}\cdot\frac{I_2^2\,R_2}{S}$$ [since $\omega = \omega_s\,(1-S)$]

where ω_s = angular synchronous speed in rad/sec.

As $$I_2 = \frac{SE_{2s}}{\sqrt{(R_2)^2+(SX_{2s})^2}}$$

∴ $$T = \frac{3}{\omega_2}\left(\frac{SE_{2s}}{\sqrt{(R_2)^2+(SX_{2s})^2}}\right)^2\cdot\frac{R_2}{S}$$

or $$T = \frac{3}{\omega_s}\,\frac{SE_{2s}^2R_2}{[(R_2)^2+(SX_{2s})^2]} = \frac{3}{\omega_s}\,\frac{E_{2s}^2R_2\,/\,S}{[(R_2\,/\,S)^2+(X_{2s})^2]}$$

This is the expression for *full load torque.*

9.24 Condition for Maximum Torque and Equation for Maximum Torque

The full load torque developed in an induction motor is given by the relation:

$$T = \frac{3}{\omega_2}\,\frac{SE_{2s}^2R_2}{[(R_2)^2+(SX_{2s})^2]}$$

or $$T \propto \frac{SR_2}{R_2^2+S^2X_{2x}^2}$$ (since $\frac{3}{\omega_s}E_{2s}^2$ is constant)

The torque developed will be maximum at a particular value of slip. As, slip (*S*) is a variable quantity, therefore, to obtain the condition for maximum torque, the above expression for torque is differentiated with respect to *S* and equated to zero.

$$\frac{dT}{dS} = \frac{(R_2^2+S^2X_{2s}^2)\,R_2 - SR_2(0+2SX_{2s}^2)}{(R_2^2+S^2X_{2s}^2)^2} = 0$$

or $$(R_2{}^2+S^2\,X_{2s}^2\,)\,R_2 = 2R_2S^2\,X_{2s}{}^2$$

or $$R_2{}^2 = (SX_{2s})^2 \text{ or } R_2 = SX_{2s}$$

or $S = R_2/X_{2s}$ is the slip at which torque is maximum.

To obtain the expression for maximum torque substitute the value of $R_2 = SX_{2s}$ in the expression for full load torque, we get,

$$\text{Maximum torque, } T_m = \frac{3}{\omega_s} \frac{SE_{2s}^2\,(SX_{2s})}{[(SX_{2s})^2 + (SX_{2s})^2]} = \frac{3E_{2s}^2}{2\omega X_{2s}}$$

Thus, the maximum torque is independent of rotor resistance but it is inversely proportional to rotor reactance at standstill (i.e., X_{2s}). Therefore, to achieve higher value of maximum torque, the leakage reactance of the rotor should be kept minimum. This is achieved (*i*) *by placing the rotor conductors very near to the outer periphery of the rotor and* (*ii*) *by reducing the air gap between stator and rotor to smallest possible value.*

9.25 Starting Torque

At start rotor is stationary and the value of slip is one i.e., $S = 1$.

Thus, to obtain the expression for starting torque, substitute the value of slip, $S = 1$ in the expression of full load torque;

$$\therefore \quad \text{Starting torque, } T_s = \frac{3}{\omega_s} \frac{E_{2s}^2 R_2}{[(R_2)^2 + (X_{2s})^2]}$$

Sometimes maximum torque is required at start. In that case, in the condition for maximum torque substitute the value of $S = 1$.

$$R_2 = SX_{2s} = X_{2s} \qquad \text{(since } S = 1 \text{ at start)}$$

Thus, to obtain maximum torque at start, the value of rotor resistance must be equal to rotor leakage reactance at standstill. Therefore, at start some external resistance is added in the rotor circuit. This is only possible in case of slip ring induction motors. This is the reason, why slip ring induction motors are applied where heavy loads are required to be picked up at start such as in lifts, cranes, elevators etc. Once the motor picks up the load the external resistance is gradually reduced to zero.

In case of squirrel cage induction motors, the rotor resistance is fixed and is kept quite low in comparison to rotor reactance, otherwise the rotor copper losses would be high and the efficiency of the motor would fall to low value. However to obtain higher starting torque in case of squirrel cage induction motors another cage is embedded in the rotor and the motor is called a *double cage induction motor.*

9.26 Ratio of Starting to Maximum Torque

Starting torque is given by the expression:

$$T_s = \frac{3}{\omega_s} \frac{E_{2s}^2 R_2}{\left[R_2^2 + X_{2s}^2\right]} \qquad \ldots(i)$$

Maximum torque is given by the expression.

$$T_m = \frac{3E_{2s}^2}{2\omega X_{2s}} \qquad \ldots(ii)$$

The ratio of starting to maximum torque is obtained by dividing equation (*i*) by (*ii*);

$$\frac{T_s}{T_m} = \frac{2R_2 X_{2s}}{R_2^2 + X_{2s}^2} = \frac{2R_2 / X_{2s}}{(R_2 / X_{2s}) + 1}$$

Putting $\dfrac{R_2}{X_{2s}} = a;\ \dfrac{T_s}{T_m} = \dfrac{2a}{a^2 + 1}$

9.27 Ratio of Full Load Torque to Maximum Torque

Full load torque is given by the expression:

$$T = \frac{3}{\omega_s} \cdot \frac{SE_{2s}^2 R_2}{\left[(R_2)^2 + (SX_{2s})^2\right]} \qquad \ldots(i)$$

Maximum torque is given by the expression:

$$T_m = \frac{3E_{2s}^2}{2\omega_s X_{2s}}$$

To obtain the ratio of full load torque to maximum torque divide equation (*i*) by the (*ii*);

$$\frac{T}{T_m} = \frac{2R_2 X_{2s} S}{\left[(R_2)^2 \times (SX_{2s})^2\right]} = \frac{2S R_2 / X_{2s}}{(R_2 / X_{2s})^2 + (S)^2}$$

Putting $\dfrac{R_2}{X_{2s}} = a;\ \dfrac{T}{T_m} = \dfrac{2Sa}{a^2 + S^2}$

9.28 Effect of Change in Supply Voltage on Torque

The torque developed by the induction motor, when it is running with slip S, is given by the expression;

$$T = \frac{3}{\omega_s} \frac{SE_{2s}^2 R2}{\left[R_2^2 + (SX_{2s}^2)\right]} = \frac{KSE_{2s}^2 R2}{R_2^2 + (SX_{2s})2} \qquad \left(\text{constant } K = \frac{3}{\omega_s}\right)$$

or

$$T = \frac{K' SR_2 V^2}{R_2^2 + (SX_{2s})2} \qquad \left[\begin{array}{c}\text{since } E_{2s} \propto \phi \propto V(\text{applied voltage})\\ \text{and } K' \text{ is another constant)}\end{array}\right]$$

At full load, the slip S is very low; therefore, the value of $(S^2 X_{2s}^{\ 2})$ is so small that it can be neglected in comparison to R_2.

$$\therefore \quad T = \frac{K' R_2 SV^2}{R_2^2} = \frac{K'}{R_2} SV^2 = K'' SV^2 \qquad \left(\text{where constant } K'' = \frac{K'}{R_2}\right)$$

or $T \propto SV^2$

Thus, when the supply voltage V is changed, it changes the torque T developed by the motor under running condition. With the decrease in supply voltage, torque decreases abruptly and in order to maintain the same torque to pick up the load slip increases or speed decreases. Hence, the motor draws extra current from the supply mains which may overheat the motor. If the motor is operated continuously under this condition, it may burn.

9.29 Torque-slip Curve

The full load torque developed by an induction motor is given by the expression;

$$T = \frac{3}{\omega_s} \frac{SE_{2s}^2 R_2}{[R_2^2 + (SX_{2s})^2]}$$

To draw the torque-slip or torque-speed curve the following points are considered:

(*i*) At synchronous speed (N_s); slip, $S = 0$ and torque $T = 0$.

(*ii*) When rotor speed is very near to synchronous speed i.e., when the slip is very low the value of the term $(SX_{2s})^2$ is very small in comparison to R_2^2 [i.e., $(SX_{2s})^2 << R_2^2$)] and is neglected. Therefore, torque is given by the expression;

$$T = \frac{3}{\omega_s} \frac{SE_{2s}^2 R_2}{R_2^2} = KS, \text{ or } T \propto S$$

Thus, at low values of slip, torque is approximately proportional to slip S and the torque-slip curve is a straight line, as shown in Fig. 9.22.

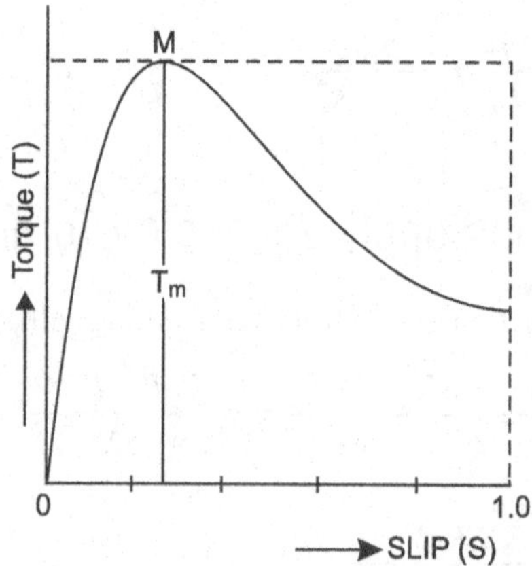

Fig. 9.22 Torque-slip curve

(*iii*) As the slip increases torque increases and attains its maximum value when $S = R_2/X_{2s}$. *This maximum value of torque is also known as* **break down** *or* **pull out torque.**

(*iv*) With further increase in slip due to increase in load beyond the point of maximum torque i.e., when slip is high, the value of term $(SX_{2s})^2$ is very large in comparison to R_2^2 [i.e., $(SX_{2s})^2 >> R_2^2$]. Therefore, R_2^2 is neglected as compared to $(SX_{2s})^2$ and the torque is given by the expression.

$$T = \frac{3}{\omega_s} \frac{SE_{2s}^2 R_2}{X^2 X_{2s}^2} = K' \frac{1}{S} \text{ or } T \propto \frac{1}{S}$$

Thus, at higher value of slip (i.e., the slip beyond that corresponding to maximum torque), torque is approximately inversely proportional to slip S and the torque-slip curve is a rectangular hyperbola, as shown in Fig. 9.22.

Thus, with the increase of slip beyond the point of maximum torque, due to increase in load, torque decrease. The result is that the motor could not pick-up the load and slows down and eventually stops. This results in *blocked rotor or short circuited motor.*

9.30 Torque-speed Curve and Operating Region

The torque-speed curve of an induction motor is shown in Fig. 9.23. It is the same curve which is already drawn, the only difference is that speed is taken on the abscissa instead of slip.

From the curve, it is clear that induction motor develops the same torque at point X and Y. However at point X the motor is unstable because with the increase in load speed decreases and the torque developed by the motor also decreases. Therefore, the motor could not pick up the load and the result is that the motor slows down and eventually stops. The miniature circuit breakers will be tripped open if the circuit has been so protected.

At point Y, the motor is stable because in this region with the increase in load speed decreases but the torque developed by the motor increases. Thus the motor will be in position to pick up the extra load effectively.

Thus, on the torque-speed curve region BC is the unstable region and **region AB is the stable or operating region** *of the induction motor as shown in Fig. 9.23.*

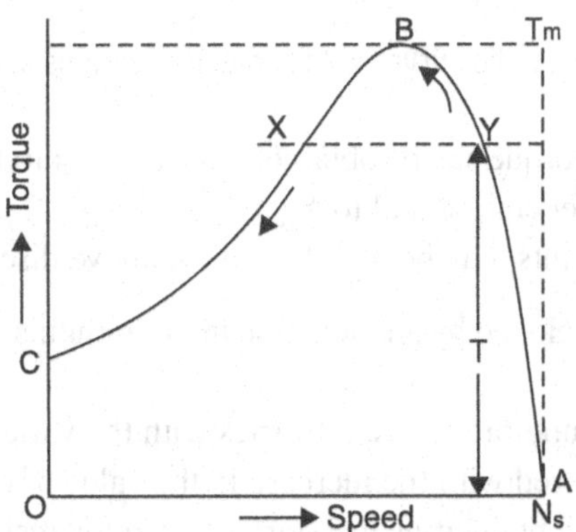

Fig. 9.23 Torque-speed curve

9.31 Effect of Rotor Resistance on Torque-slip Curve

To see the effect of rotor resistance on torque-slip curve, consider a slip ring induction motor in which additional resistance in the rotor circuit can be introduced through slip rings. The rotor reactance X_{2s} remains constant.

It has already been seen that maximum value of the torque developed by an induction motor is independent of rotor resistance R_2, but is inversely proportional to rotor standstill reactance X_{2s}. Therefore, the effect of change in rotor resistance will change the value of slip at which this maximum torque occurs; Greater the rotor resistance, greater the value of slip at which the maximum torque occurs since

$$S = \frac{R_2}{X_{2s}} \text{ at which torque is maximum}$$

The torque-slip curves are shown in Fig. 9.24. for various values of rotor resistance R_2 keeping rotor reactance X_{2s} constant. When R_2 is $0 \cdot 1$ times of X_{2s}, the maximum torque will occur at slip $S = \frac{R_2}{X_{2s}} = \frac{0.1\,X_{2s}}{X_{2s}} = 0 \cdot 1$. Now, if the rotor resistance R_2 is increased, by adding some resistance externally, to the value so that it becomes $0 \cdot 2$ times of X_{2s} then maximum torque would occur at a slip $S = \frac{R_2}{X_{2s}} = \frac{0.2X_{2s}}{X_{2s}} = 0 \cdot 2$ and so on.

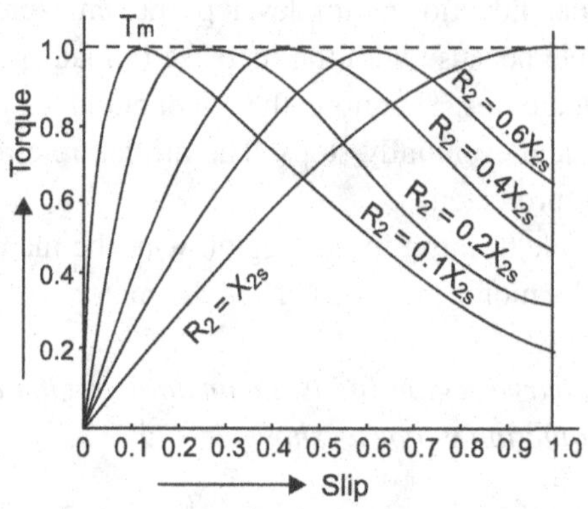

Fig. 9.24 Torque-slip curves with addition of various external resistors

The maximum value of the torque can be obtained even at the start by adding that much resistance in the rotor circuit so that R_2 becomes equal to X_{2s}.

The following important points may be noted from the above discussions:

(*i*) The maximum torque developed by an induction motor remains constant since it is independent of the rotor resistance.

(*ii*) The slip at which maximum torque occurs varies with the variation of the rotor resistance.

(*iii*) The starting torque increased with the increase in the value of rotor resistance.

(*iv*) The maximum torque is obtained at the start when rotor resistance is made equal to rotor reactance at standstill i.e., $R_2 = X_{2s}$

Example 9.21

A 3-phase induction motor has a 4-pole star-connected stator winding. The motor runs at a line voltage of 200 V, 50 Hz supply. The rotor resistance and standstill reactance per phase are 0.1 and 0.9 ohm respectively. The ratio of rotor to stator turns is 0.67. Calculate the total torque at 4% slip.

Solution:

Here, $P = 4$; $f = 50$ Hz; $R_2 = 0{\cdot}1\ \Omega$; $X_{2s} = 0{\cdot}9\ \Omega$; $S = 4\%$

Supply voltage (line value), $V_L = 200$ V

Stator induced voltage (phase value), $E_1 = \frac{V_L}{\sqrt{3}} = \frac{200}{\sqrt{3}} = 115{\cdot}47$ V

Ratio of rotor to stator turns, $\frac{T_2}{T_1} = 0{\cdot}67 = \frac{E_{2s}}{E_1}$

$\therefore$ Rotor induced emf at standstill, $E_{2s} = 0{\cdot}67\, E_1 = 0{\cdot}67 \times 115{\cdot}47 = 77{\cdot}365$ V

Synchronous speed, $N_s = \frac{120f}{P} = \frac{120 \times 50}{4} = 1500$ rpm

Synchronous angular speed, $\omega_s = \frac{2\pi N_s}{60} = \frac{2\pi \times 1500}{60} = 50\,\pi$ rad/sec.

$$\text{Torque developed, } T = \frac{3}{\omega_s} \frac{SE_{2s}^2\, R_2}{[R_2^2 + (SX_{2s})^2]}$$

$$= \frac{3}{50\pi} \times \frac{0 \cdot 04 \times (77 \cdot 365)^2 \times 0 \cdot 1}{[(0 \cdot 1)^2 + (0 \cdot 04 \times 0 \cdot 9)^2]} = \mathbf{40 \cdot 48\ Nm}\ (Ans)$$

Example 9.22

The rotor resistance and standstill rotor reactance of a 3-phase, 4-pole, 50 Hz, phase-wound induction motor is 0.21 ohm and 0.7 ohm per phase respectively. Calculate the speed at which maximum torque is developed.

Solution:

Here, $P = 4$; $f = 50$ Hz; $R_2 = 0 \cdot 21\ \Omega$; $X_{2s} = 0 \cdot 7\ \Omega$

Condition for maximum torque is $R_2 = SX_{2s}$

$\therefore$ Torque will be maximum at a slip, $S = \frac{R_2}{X_{2s}} = \frac{0 \cdot 21}{0 \cdot 7} = 0{\cdot}03$

Synchronous speed, $N_s = \frac{120f}{P} = \frac{120 \times 50}{4} = 1500$ rpm

Speed at which the torque will be maximum,

$N = N_s\,(1 - S) = 1500\,(1 - 0{\cdot}03) =$ **1455 rpm** (*Ans*)

Example 9.23

A 400 V, 3-phase motor has a rotor resistance of 0·02 ohm and standstill reactance of 0·1 ohm. It has stator to rotor turns ratio of 5. What must be the value of total resistance of a starter to be used in series with rotor circuit for maximum torque to be exerted at starting. Also find the rotor current under this condition.

Solution:

Here, $V_L = 400$ V; $R_2 = 0{\cdot}02\ \Omega$; $X_{2s} = 0{\cdot}1\ \Omega$

Ratio of stator to rotor turns, $\frac{T_1}{T_2} = 5$

Let us consider that stator be connected in star;

$\therefore$ Stator supply voltage (phase value), $V = \dfrac{V_L}{\sqrt{3}} = 230{\cdot}94$ V,

$$E_1 = V = 230{\cdot}94 \text{ V}$$

$$\text{Transformation ratio, } K = \frac{T_2}{T_1} = \frac{E_{2s}}{E_1} = \frac{1}{5}$$

$$\therefore \quad \text{Rotor induced emf at standstill, } E_{2s} = \frac{E_1}{5} = \frac{230 \cdot 94}{5} = 46{\cdot}19 \text{ V}$$

Condition for maximum torque; $R_2 = SX_{2s}$

At start slip, $S = 1$ and let the additional rotor resistance per phase be R ohm,

$$\therefore \quad R_2 + R = X_{2s}$$

or

$$R = X_{2s} - R_2 = 0{\cdot}1 - 0{\cdot}02 = \mathbf{0{\cdot}08 \text{ ohm}} \ (Ans)$$

Under this condition; rotor current

$$I_2 = \frac{E_{2s}}{\sqrt{(R_2 + R)^2 + (X_{2s})^2}}$$

$$= \frac{46 \cdot 19}{\sqrt{(0 \cdot 1)^2 + (0 \cdot 1)^2}} = \mathbf{326{\cdot}6 \text{ A}} \ (Ans.)$$

Example 9.24

The impedance of the rotor circuit at standstill of a 1000 HP, 3-phase, 16-pole induction motor is (0.02 + j0.15) ohm. It develops full-load torque at 360 rpm what will be.

(i) The ratio of maximum to full load torque;
(ii) The speed at maximum torque;
(iii) The rotor resistance to be added to get maximum starting torque. **(UPTU)**

Solution:

Here, $f = 50$ Hz; $P = 16$; $R_2 = 0{\cdot}02\ \Omega$; $X_{2s} = 0{\cdot}15\ \Omega$; $N = 360$ rpm

$$\text{Synchronous speed, } N_s = \frac{120f}{P} = \frac{120 \times 50}{16} = 375 \text{ r.p.m.}$$

$$\text{Full load slip, } S = \frac{N_s - N}{N_s} = \frac{375 - 360}{375} = 0 \cdot 04$$

Full load torque is given by the expression:

$$T = \frac{3}{\omega_s} \times \frac{SE_{2s}^2 R_2}{[R_2^2 + (SX_{2s})^2]} \quad \ldots(i)$$

Maximum torque is given by the expression:

$$T_m = \frac{3E_{2s}^2}{2\omega_s X_{2s}} \quad \ldots(ii)$$

From (*i*) and (*ii*) equation; Ratio of maximum to full load torque,

$$\frac{T_m}{T} = \frac{\left[R_2^2 + (SX_{2s})^2\right]}{2SR_2X_{2s}}$$

$$= \frac{(0\cdot02)^2 + (0\cdot04 \times 0\cdot15)^2}{2 \times 0\cdot04 \times 0\cdot02 \times 0\cdot15} = \mathbf{1{\cdot}817}\ (Ans.)$$

The condition for maximum torque is $R_2 = SX_{2s}$

∴ The slip at which maximum torque occurs,

$$S = \frac{R_2}{X_{2s}} = \frac{0\cdot02}{0\cdot15} = 0{\cdot}1333$$

For maximum torque rotor speed, $N = N_s(1 - S) = 375\,(1 - 0{\cdot}1333) =$ **325 rpm** (*Ans.*)

To obtain maximum torque at the start, let R be the addition resistance connected in series with each rotor phase, then

$$R_2 + R = X_{2s}$$

∴ $$R = X_{2s} - R_2 = 0{\cdot}15 - 0{\cdot}02 = \mathbf{0{\cdot}13\ ohm}\ (Ans.)$$

Example 9.25

A 440 V, 50 Hz, 4-pole, 3-phase, star-connected induction motor running at 1440 rpm on full load has a slip ring rotor of resistance 0·01 ohm and standstill reactance of 0·167 ohm, per phase. Calculate (i) the ratio of standing torque to maximum torque (ii) the ratio of maximum torque to full load torque.

Solution:

Here, $P = 4$; $f = 50$ Hz; $R_2 = 0{\cdot}01\ \Omega$; $X_{2s} = 0{\cdot}167\ \Omega$; $N = 1440$ rpm

$$\text{Synchronous speed, } N_s = \frac{120f}{P} = \frac{120 \times 50}{4} = 1500 \text{ r.p.m.}$$

$$\text{Full load slip, } S = \frac{N_s - N}{N_s} = \frac{1500 - 1440}{1500} = 0\cdot04$$

(*i*) Ratio of starting torque to maximum torque:

$$\frac{T_s}{T_m} = \frac{2a}{a^2 + 1}$$

where $a = \dfrac{R_2}{X_{2s}} = \dfrac{0\cdot01}{0\cdot167} = 0\cdot06$

∴ $$\frac{T_s}{T_m} = \frac{2 \times 0\cdot06}{(0\cdot06)^2 + 1} = \mathbf{0{\cdot}12}\ (Ans.)$$

(*ii*) Ratio of maximum torque to full load torque:

$$\frac{T_m}{T} = \frac{a^2 + S^2}{2aS} = \frac{(0\cdot06)^2 + (0\cdot04)^2}{2 \times 0\cdot06 \times 0\cdot04} = \mathbf{1{\cdot}083}\ (Ans.)$$

Example 9.26

Two squirrel cage induction motors A and C are identical in all respect except that rotor of A is made of aluminium having a resistivity of 3 × 10⁻⁸ ohm-m and that of motor C is made of copper having a resistivity of 2 × 10⁻⁸ ohm-m. The starting torque of motor A is 120 Nm. Predict the starting torque of motor C.

Solution:

Resistivity of aluminium, $\rho_a = 3 \times 10^{-8}$ ohm-m

Resistivity of copper, $\rho_c = 2 \times 10^{-8}$ ohm-m

Starting torque of motor *A*, $T_{s(a)} = 120$ Nm

Starting torque of an induction motor is given by the expression:

$$T_s = \frac{3}{\omega_s} \frac{E_{2s}^2 R_2}{\left[R_2^2 + X_{2s}^2\right]} = K \frac{R_2}{R_2^2 + X_{2s}} = K' \frac{1}{R_2}$$

or

$$T_s \propto \frac{1}{R_2} \propto \frac{1}{\rho \frac{l}{a}} \propto \frac{a}{pl} \propto \frac{1}{\rho}$$

Thus, the starting torque is just inversely proportional to resistivity.

$$\therefore \qquad T_{s(a)} \propto \frac{1}{\rho_c}$$

and

$$T_{s(c)} \propto \frac{1}{\rho_c}$$

Therefore, $\dfrac{T_{s(c)}}{T_{s(a)}} = \dfrac{\rho a}{\rho c}$

Thus, starting torque developed by motor *C*,

$$T_{s(c)} = \frac{\rho a}{\rho c} \times T_{s(a)} = \frac{3 \times 10^{-8}}{2 \times 10^{-8}} \times 120 = \mathbf{180\ Nm}\ (Ans.)$$

Example 9.27

A 4-pole, 3-phase, 50 Hz induction motor has resistance and stand still reactance of 0·03 Ω and 0·12 Ω per phase respectively. Find the amount of rotor resistance per phase to be inserted to obtain 75% of maximum torque at start.

Solution:

Rotor resistance per phase, $R_2 = 0{\cdot}03\ \Omega$

Rotor reactance/phase at stand still, $X_{2s} = 0{\cdot}12\ \Omega$

To obtain 75% of maximum torque at start let a resistance of *R* ohm be connected in each phase of rotor circuit, then

Rotor resistance, $R'_2 = R_2 + R$

The ratio of starting torque to maximum torque is given by the relation;

$$\frac{T_s}{T_m} = \frac{2a}{a^2+1} = \frac{75}{100}$$

where $a = \dfrac{R_2'}{X_{2s}}$

or $$3a^2 + 3 - 8a = 0$$

or $$a = \frac{8 \pm \sqrt{64-36}}{6} = \frac{8 \pm 5 \cdot 29}{6}$$

or $$a = 2{\cdot}215 \text{ or } 0{\cdot}4517$$

or $$\frac{R_2'}{X_{2s}} = 2{\cdot}215 \text{ or } 0{\cdot}4517$$

or $$R'_2 = 0{\cdot}2658 \text{ or } 0{\cdot}0542 \text{ ohm} \qquad (\text{since } X_{2s} = 0{\cdot}12)$$

Higher value of R'_2 i.e., 0·2658 is not considered because of higher losses.

∴ $$R'_2 = 0{\cdot}0542 \text{ or } R_2 + R = 0{\cdot}0542$$

or $$R = 0{\cdot}0542 - 0{\cdot}03 = \mathbf{0{\cdot}0242\ ohm}\ (Ans.)$$

Example 9.28

400 V, 3-phase, 50 Hz, 6 pole induction motor is supplying a load of 20 kW, when the frequency of rotor currents is 2Hz. Estimate (i) Slip and speed at this load (ii) Rotor copper loss (iii) Speed of motor when supplying 30 kW load assuming torque-slip curve to be a straight line.

Solution:

Here, f = 50 Hz; P = 6; f_r = 2Hz; Output = 20 kW

$$\text{Synchronous speed, } N_s = \frac{120f}{P} = \frac{120 \times 50}{6} = 1000 \text{ r.p.m.}$$

$$\text{Slip, } S = \frac{f_r}{f} = \frac{2}{50} = 0 \cdot 04 \text{ or, } \mathbf{4\%}\ (Ans.)$$

$$\text{Rotor speed, } N = N_s - SN_s \qquad \left(\text{since } S = \frac{N_s - N}{N_s}\right)$$

$$= 1000 - 0{\cdot}04 \times 1000 = \mathbf{960\ rpm}\ (Ans.)$$

$$\text{Rotor copper loss} = \frac{S}{1-S} \times \text{mech. power developed}$$

$$= \frac{0 \cdot 04}{1 - 0 \cdot 04} \times 20 = 0 \cdot 833 \text{ kW} = \mathbf{833\ W}\ (Ans.)$$

It is given that torque-slip curve to be straight line.

∴ $$T \propto S \text{ as well as } T \propto \text{output}$$

Hence, $S \propto$ output

$\therefore$ $$S_1 \propto \text{output}_1 \text{ and } S_2 \propto \text{output}_2$$

or $$\frac{S_2}{S_1} = \frac{30 \text{ kW}}{20 \text{ kW}}$$

or $$S_2 = \frac{30}{50} \times 0 \cdot 04$$

or $$S_2 = 0{\cdot}06$$

$\therefore$ $$\text{Speed, } N = N_s (1 - S) = 1000 (1 - 0{\cdot}06) = \mathbf{940 \text{ rpm}} \ (Ans.)$$

Section Practice Problems

Numerical Problems

1. A 3-phase induction motor has a rotor resistance of 0.02 ohm and standstill reactance of 0.1 ohm per phase. What must be the value of total resistance of a starter for the rotor circuit for maximum torque to be exerted at starting. (**Ans.** *0.08 ohm*)

2. The rotor resistance and stand-still reactance per phase of a 3-phase, 8-pole, 50 Hz, phase-wound induction motor is 0.01 ohm and 0.1 ohm respectively. At full-load the machine is operating at 4% slip. Find the ratio of maximum torque to full load torque. Also find the speed at which this torque occurs. (**Ans.** *1·45, 675 rpm*)

3. A 3-phase induction motor has a 4-pole, star-connected stator winding. The motor runs at a ling voltage of 400 V, 50 Hz supply. The motor resistance and stand-still reactance per phase are 0·1 ohm and 1 ohm respectively. The ratio of stator to rotor turns is 4. Calculate: (*i*) Starting torque; (*ii*) Full load torque, when the slip is 4%; (*iii*) Slip at which maximum torque occurs and rotor speed; (*iv*) Maximum torque; (*v*) Ratio of starting to maximum torque and full load torque to maximum torque; (*vi*) Value of additional rotor resistance per phase to obtain maximum torque at start. (**Ans.** *6·3 Nm; 21·95 Nm; 0·1, 1350 rpm; 31·83 Nm; 0·198, 0·69; 0·9 ohm*)

Short Answer Type Questions

Q.1. What parameters determine the torque of an induction motor?

Ans. Torque depends upon rotor induced emf, E_{2S} rotor resistance, R_2, rotor reactance, X_{2S}, and slip (or rotor speed N)

Q.2. Can we get maximum torque at start in an induction motor, how?

Ans. Yes, we can.by adding external resistance in the rotor circuit (in case of phase-wound induction motor) so that $X_{2S} = (R_2 + R)$

Q.3. Which type of motor develops higher starting torque?

Ans. Phase-wound (or slip ring) induction motor.

Q.4. Does rotor resistance affect the maximum torque developed in the induction motor?

Ans. No, maximum torque developed in an induction motor is independent of rotor resistance.

Q.5. How does the supply voltage affect the torque developed by an induction motor?

Ans. $T \alpha V^2$ i.e., torque developed varies as the square of the supply voltage.

Q.6. What is the effect on the slip when the load on an induction motor increases?

Ans. When load increases, slip also increases.

Q.7. Why the rotor conductors are placed at the outermost periphery of the rotor?

Ans. This arrangement minimise the rotor reactance X_{2s} and increases the maximum torque produced by the motor, because $T \alpha \frac{1}{X_{2s}}$

9.32 Constant and Variable Losses in an Induction Motor

Various losses which occur in an induction motor during energy conversions are given below:

(*i*) Constant losses (*ii*) Variable losses.

(*i*) **Constant losses:** The losses which are independent of the load and remain constant irrespective of the load variation are called *constant losses*. These losses may be:

(*a*) *Core losses*: These include hysteresis and eddy current losses in stator as well as in rotor core. Eddy current losses in rotor core is negligible since rotor current frequency is very small of the order of 0·5 to 2 Hz. These losses are constant or fixed losses since these depend upon voltage and frequency which is practically constant.

(*b*) *Friction and windage losses*: These losses are also constant as these losses depend upon the speed of the induction motor. The speed of induction motor is approximately constant (for normal running, slip is very small). These losses occur in the machine because of power loss due to friction at the bearings and to overcome wind resistance. Additional sliding friction loss occur in the slip ring induction motor.

(*ii*) **Variable losses:** The losses which depend on the load and change with the variation in load are called *variable losses*. These losses are:

(*a*) I^2R loss in stator winding.

(*b*) $I^2 R$ loss in rotor winding.

There losses occur due to the resistance of stator winding as well as resistance of rotor winding. This loss is also called copper loss. It is proportional to the square of the current following in the stator as well as in rotor winding.

(*c*) *Brush contact loss*: This loss occurs only in slip ring induction motors. This is occurring because of contact resistance between brushes and slip rings. Its magnitude is very small since contact resistance is made minimum.

(*iii*) *Stray losses*: These losses are occurring in iron as well winding of the machine. These cannot be determined exactly but are accounted for when the efficiency of the machine is calculated, by suitable factor.

$$\text{Efficiency, } \eta = \frac{\text{output}}{\text{input}} = \frac{\text{output}}{\text{output + losses}}$$

9.33 Main Tests Performed on an Induction Motor

To determine losses, power factor, efficiency and other characteristics of an induction motor various tests *viz* stator resistance test, no-load test, blocked rotor test, heat run etc., are performed on an induction motor

9.33.1 Stator Resistance Test

This test is performed to determine the resistance of stator winding. In this test, resistance between two terminals of the stator winding is measured by voltmeter-ammeter method using DC supply as shown in Fig. 9.25 this gives the resistance of two phases in series. To obtain the resistance of each phase, the measured value is divided by 2. Let the resistance measured between two terminal be R', where

$$R' = \frac{\text{voltmeter reading (V)}}{\text{ammeter reading (I)}}$$

$$\text{Resistance of each phase, } R = \frac{R'}{2}$$

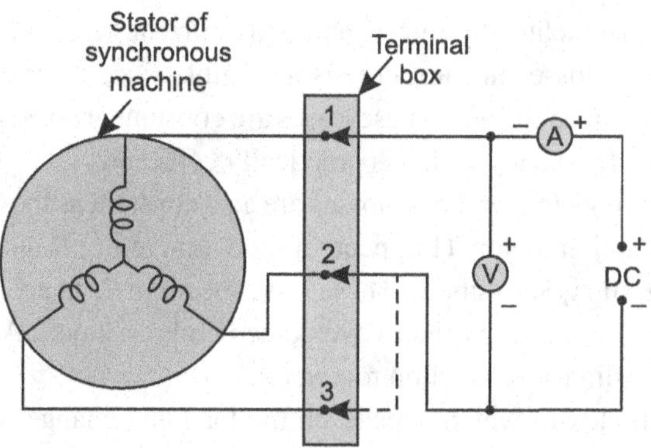

Fig. 9.25 Circuit for stator resistance test

To obtain more accurate results, the resistance may be measured between terminals 1-2,2-3 and 3-1 as R'_1, R'_2, R'_3, respectively. Then their average value is considered i.e., $R' = \frac{R'_1 + R'_2 + R'_3}{3}$. Since the effective AC resistance is higher than the DC resistance, the average value R' is multiplied by a is multiplied by 1.2 to obtain the effective value.

Hence, Winding resistance/phase, $R = 1.2\,R'$

9.33.2 Voltage-ratio Test

This test can only be performed on a phase-wound induction motors. While performing this test, brushes are lifted from the slip-rings and stator is excited by 3-phase AC supply through a 3-phase variac at normal frequency. The ratio of rotor to stator voltage can be measured by means of voltmeters connected on the both sides, as shown in Fig. 9.26. to obtain more accurate results, various reading across different phases near to rated voltages are taken and the average value is considered.

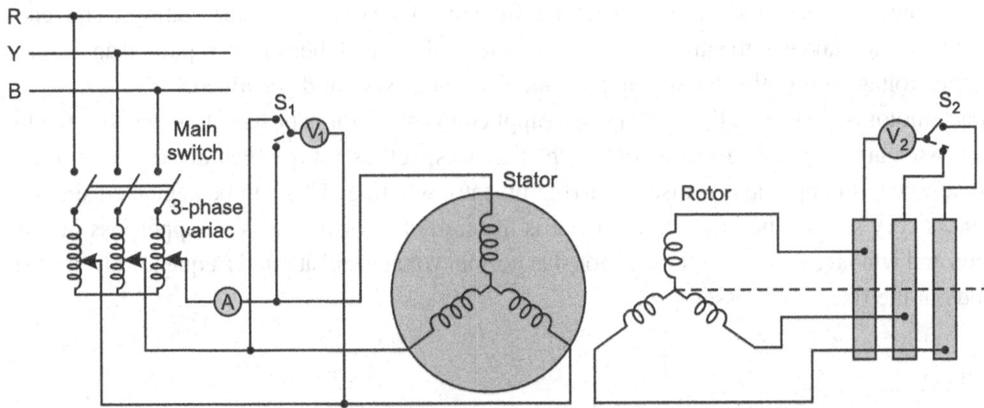

Fig. 9.26 Circuit for voltage ratio test

In this case, it may be noted that the rotor current frequency is the same as that of the supply frequency since the rotor is stationary.

9.33.3 No-load Test

This test, as the name indicate, is performed on induction motor when it is running without load. This test tells us the magnitude of constant losses occurring in the motor.

The machine is started in the usual way and runs without load at normal voltage. On the stator side, suitable instruments are connected between supply mains and motor terminals to measure power, line current and line voltage. as shown in Fig. 9.27. For power and power factor measurement wattmeter readings W_1 and W_2 are taken keeping the switch S_1 at stud-1 and then at stud-2. Since motor is running without load, p.f. of the motor is low (less than 0·5), so once the wattmeter will give down-scale reading. To obtain up-scale reading, the connections of the potential coil of the wattmeter are reversed by changing the position of switch-2 from stud-1 to stud-2, but the reading is considered as negative.

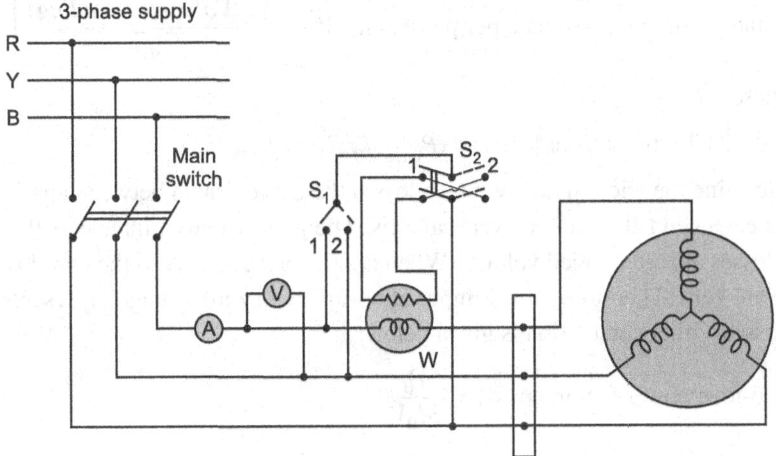

Fig. 9.27 Circuit diagram for no-load test

Total power drawn by the motor is then the difference of the two wattmeter readings. The readings of wattmeters are taken with various values of applied voltage and then curve is plotted against power and input voltage. From this curve windage and friction losses are determined.

Since motor is at no-load the input power supplied to the motor is utilised to meet with the losses only. Losses are occurring in iron core of the stator as well as rotor which are called core losses. Small amount of copper loss is also occurring in stator winding. This can be neglected since stator current is very small. Therefore, total power is the sum of rotor iron loss + copper loss in stator + friction and windage loss. The readings noted at normal voltage and at rated frequency are considered to find out the fixed core losses.

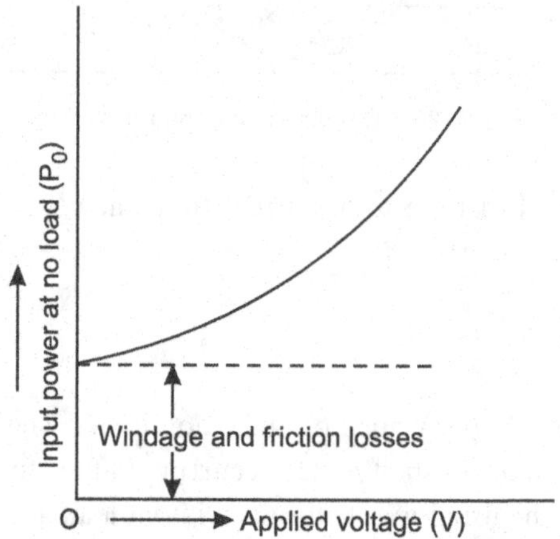

Fig. 9.28 Curve between applied voltage and power input to the motor at no-load

The procedure to separate the various losses is given below:

Total power measured by two readings of the wattmeter = $P_0 = (W_1 - W_2)$ watt,

$$\text{Copper losses in stator winding} = 3I_0^2R_1$$

where I_0 is the no-load current measured by ammeter at normal voltage and rated frequency of supply and R_1 is the stator winding resistance per phase. and $V = \left[\dfrac{V_L(\textit{Voltmeter reading})}{\sqrt{3}}\right]$ is the applied voltage per phase.

$$\therefore \quad \text{Total constant losses} = (P_0 - 3I_0^2R_1) \text{ watt}$$

Now to determine the friction and windage losses the curve drawn between applied voltage and input power is extended till it cuts the vertical axis. The point where it intersects the vertical axis, represent the losses at zero applied voltage. When applied voltage is zero the core losses and stator copper losses are zero. Therefore, power input at no-load and zero voltage represents the windage and friction losses. Other calculation is given below:

$$\text{No-load power factor, } \cos\phi_0 = \frac{P_0}{3I_0V}$$

$$\text{No-load exciting resistance, } R_0 = \frac{V}{I_0\cos\phi_0} = \frac{V}{I_w}$$

No-load exciting reactance, $X_0 = \dfrac{V}{I_0 \cos \phi_0} = \dfrac{V}{I_{mag}}$

The equivalent circuit of an induction motor at no-load is shown in Fig. 9.29.

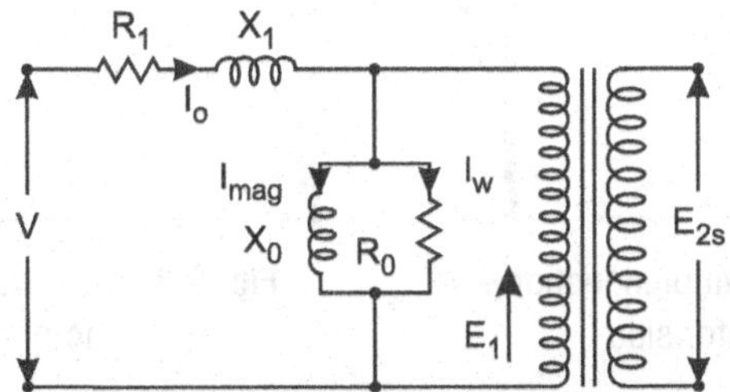

Fig. 9.29 Equivalent circuit of an induction motor at no-load

9.33.4 Blocked Rotor Test

This test is performed by locking the rotor (by holding the rotor not to rotate). This is carried out to know the copper loss at short circuit, power factor at short circuit and total equivalent resistance and reactance of the motor referred to stator side. This test is just equivalent to short circuit test of a transformer.

Keeping switch $S_1$and S_2 at stud-1 and starting with zero voltage across the stator, the applied voltage is gradually increased in steps till motor draws the full-load current. The readings of voltmeter, ammeter and wattmeters are noted. the wattmeter reading is taken as W_1 it is again noted by changing the position of switch S_2 from stud-1 to stud-2 and recorded as W_2. While performing this test the following points are taken care of.

1. The mechanism used to hold (or block) the rotor should be of proper strength.
2. The direction of rotation of rotor should be established prior to start the test. And direction of force which is to keep the rotor blocked (unmoved) should be in opposite direction.
3. As the windings get heated while performing the test, therefore this test should be carried out quickly.
4. The short circuit current should not be allowed to rise more than one and half times the full load current.

Calculations from Test Results

Let P_{sc} be the total power measured when I_{sc}/phase be the circulating current and V_{sc} is the voltage applied per phase. where V_{sc} = voltmeter reading / $\sqrt{3}$.

$$\therefore \qquad P_{sc} = W_1 + W_2$$

Then equivalent impedance/phase of the motor referred to stator side $Z_{eql} = \dfrac{V_{sc}}{I_{sc}} \Omega$

Power factor, $\cos \phi_{sc} = \dfrac{P_{sc}}{3V_{sc}I_{sc}}$

Let R_{eql} be the equivalent resistance of the motor referred to stator side.

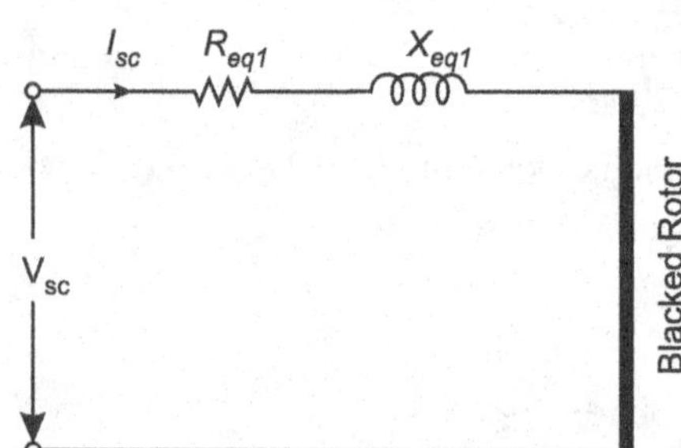

Fig. 9.30 Equivalent circuit of induction motor referred to stator side

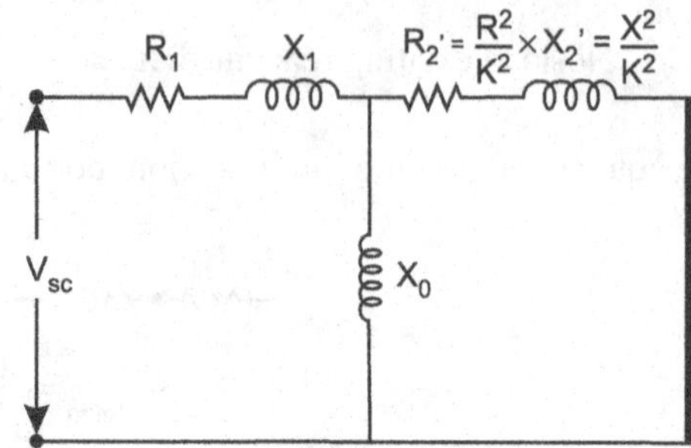

Fig. 9.31 Equivalent circuit of induction motor on stator side

$$R_{eql} = \frac{P_{sc}}{3I_{sc}^2}\,\Omega$$

∴ Equivalent reactance, of the motor referred to the stator side,

$$X_{eql} = \sqrt{\left(Z_{eql}\right)^2 - \left(R_{eql}\right)^2}$$

The equivalent circuit of the induction motor referred to stator side is shown in Fig. 9.30

Usually, the stator reactance per phase X_1 is assumed to be equal to rotor reactance per phase as referred to stator side, i.e.,

$$X_2' = X_1 = \frac{X_{eql}}{2} \text{ and } X_2' = \frac{X_2}{K^2}$$

In case of phase wound motors, the stator and rotor winding resistances per phase can be measured separately and the resistance determined by the test is divided in the same ratio.

In case of squirrel cage motor, the rotor resistance per phase, as referred to stator, can be determined by subtracting R_1 (calculated from stator resistance test) from R_{eql}, i.e.,

$$R_2' = \left(R_{eql} - R_1\right) = \frac{R_2}{K^2}$$

The equivalent circuit is shown in Fig. 9.31.

The total power input to motor when rotor is locked is absorbed as total copper losses in the motor as well as small iron losses (usually neglected). The iron losses are small since small voltage only 10 to 15% of normal voltage is applied to circulate full load current in the motor.

∴ P_{sc} = total Cu loss in motor.

If I_{sc} = full load current of motor then P_{sc} is total copper loss on full load.

9.33.5 Heat Run Test

When a motor is operated at load, it is heated-up due to various losses which occur in different parts of the machine. Maximum temperature is obtained by the machine at a certain load. The life of the winding insulation depends upon the temperature.

Therefore, this test is performed on the motor to check the rise in temperature which must remain with-in the prescribed limits.

The prescribed limits depend upon the type of insulation material used and the duty cycle of the motor. The temperature is measured both while the motor is operating and after it is shut-down. The motor should be stopped as soon as the motor attains the thermal equilibrium.

Example 9.29

No-load and blocked rotor test were performed on a 400 V, 3-phase delta-connected inductions motor and the following results were obtained:

No-load test: 400 V; 2.5A; 600 W

Blocked-rotor test: 200 V; 12.5 A; 1500 W

Determine energy (or working)component and magnetising component of no-load current, no-load power factor, exciting resistance and reactance per phase referred to stator (primary) side assuming that friction and winding lasses are 180 W.

Also determine equivalent resistance and reactance of the motor referred to stator side as well as current and power factor on short circuit with normal rated voltage of 400 V assuming that the stator winding resistance per phase is 5 ohm.

Solution:

Here, Motor is delta connected.

From no-load test: $V_L = 400$ V; $I_{0(L)} = 2.5$ A; $P_0 = 600$ W and $R_1 = 5$ ohm

$$\text{Phase voltage, } V = V_L = 400 \text{ V}$$

$$\text{No-load phase current, } I_o = \frac{I_{o(L)}}{\sqrt{3}} = \frac{2.5}{\sqrt{3}} = 1.443 \text{ A}$$

$$\text{Stator copper loss, } P_{cl} = 3I_o^2 R_1 = 3 \times (1.443)^2 \times 5 = 31.25 \text{ W}$$

$$\text{Windage and friction loss, } P_{wf} = 180 \text{ W (given)}$$

$$\therefore \quad \text{Stator iron lass, } P_i = P_o - P_{Cl} - P_{wf}$$

$$= 600 - 31.25 - 180 = 388.75 \text{ W}$$

$$\text{No-load power factor, } \cos\phi_o = \frac{Pi}{3VI_o} = \frac{388.75}{3 \times 400 \times 1.443} = \mathbf{0.2245} \text{ lag. } (Ans)$$

Energy component of no-load current,

$$I_w = I_o \cos\phi_o = 1.443 \times 0.2245 = \mathbf{0.324\ A}\ (Ans.)$$

Magnetising component of no-load current,

$$I_{mag} = \sqrt{I_o^2 - I_w^2} = \sqrt{(1.443)^2 - (0.324)^2} = \mathbf{1.406\ A}\ (Ans.)$$

$$\text{Exciting resistance, } R_o = \frac{V}{I_w} = \frac{400}{0.324} = \mathbf{1235\ ohm}\ (Ans.)$$

$$\text{Exciting reactance, } X_o = \frac{V}{I_{mag}} = \frac{400}{1.406} = \mathbf{284.5\ ohm}\ (Ans.)$$

From blocked-rotor test: $V_{sc(L)} = 200$ V; $I_{sc(L)} = 12.5$ A; $P_{sc} = 1500$ W

Phase voltage on short circuit, $V_{sc} = V_{sc(L)} = 200$ V

Phase current on short-circuit, $I_{sc} = \frac{I_{sc(L)}}{\sqrt{3}} = \frac{12.5}{\sqrt{3}} = 7.217$ A

Equivalent impedance referred to stator side,

$$Z_{eq1} = \frac{V_{sc}}{I_{sc}} = \frac{200}{7.217} = 27.7 \text{ ohm}$$

Equivalent resistance referred to stator side,

$$R_{eq1} = \frac{P_{sc}}{3I_{sc}^2} = \frac{1500}{3 \times (7.217)^2} = \mathbf{9.6\ \Omega}\ (Ans.)$$

Equivalent reactance referred to stator side,

$$X_{eq1} = \sqrt{Z_{eq1}^2 - R_{eq1}^2} = \sqrt{(27.7)^2 - (9.6)^2} = \mathbf{25.98\ \Omega}\ (Ans.)$$

Short circuit current at rated voltage of 400 V, $\frac{V}{V_{sc}} \times I_{sc(L)} = \frac{400}{200} \times 12.5 = \mathbf{25\ A}$ (*Ans.*)

Power factor on short circuit, $\cos \phi_{sc} = \frac{P_{sc}}{3\,V_{sc}\,I_{sc}} = \frac{1500}{3 \times 200 \times 7.217} = \mathbf{0.3464\ lag}$ (*Ans.*)

Example 9.30

A 400 V, 3-phase, star connected induction motor draws 40 A at a line voltage of 80 V with rotor locked. The power drawn by the motor under this condition is 480 W. the DC resistance measured between a pair of stator terminals is 0.2 Ω. if the core loss is 80 W and the ratio of AC to DC resistance is 1.4, determine the equivalent leakage reactance per phase of the motor as well as stator and rotor resistance per phase.

Solution:

Here, $V_{sc(L)} = 80$ V; $I_{sc} = 40$ A; $P_{sc} = 480$ W

Phase voltage at short circuit, $V_{sc} = \frac{V_{sc(L)}}{\sqrt{3}} = \frac{80}{\sqrt{3}} = 46.2$ V

Equivalent impedance referred to stator side/phase,

$$Z_{eq1} = \frac{V_{sc}}{I_{sc}} = \frac{46.2}{40} = 1.115\ \Omega$$

At short-circuit copper loss $P_c = P_{sc} - P_i = 480 - 80 = 400$ W

Equivalent resistance referred to stator side/phase,

$$Z_{eq1} = \frac{P_c}{I_{sc}^2} = \frac{400}{(40)^2} = 0.25\ \Omega$$

Equivalent reactance referred to stator side/phase,

$$X_{eq1} = \sqrt{Z_{eq1}^2 - R_{eq1}^2} = \sqrt{(1.115)^2 - (0.25)^2}$$

$$= \mathbf{1.087\ \Omega}\ (Ans.)$$

DC resistance of the stator winding/phase,

$$R_{dc} = \frac{0.2}{2} = 0.1\Omega$$

AC resistance of the stator winding/phase,

$$R_{ac} = R_1 = 0.1 \times 1.4 = \mathbf{0.14\ \Omega}\ (Ans.)$$

Rotor resistance referred to stator side/phase,

$$R'_2 = 0.25 - 0.14 = \mathbf{0.11\ \Omega}\ (Ans.)$$

9.34 Equivalent Circuit of an Induction Motor

It is seen that when stator winding of a 3-phase induction motor is connected to 3-phase supply, the rotor circuit being closed (or short circuited), torque is developed and rotor rotates. Thus, energy is transferred from stator winding to rotor winding through magnetic flux. Therefore, an induction motor is similar to a transformer with a rotating secondary. The induction motor may be viewed as shown in Fig. 9.32. where per phase values have been considered.

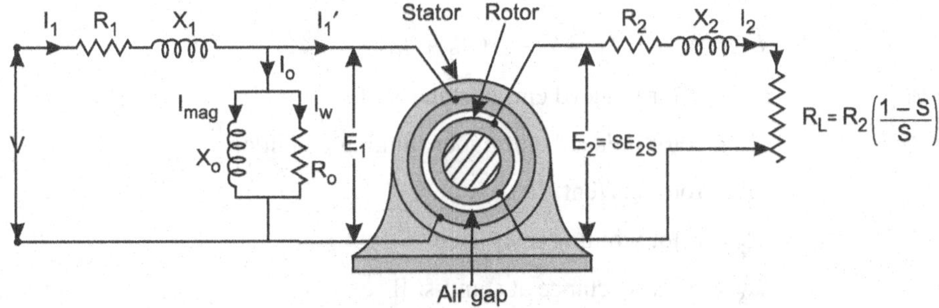

Fig. 9.32 Induction motor circuit

This circuit can be further represented as shown in Fig. 9.33. Here, stator and rotor winding is shown just as it is the primary and secondary of a transformer.

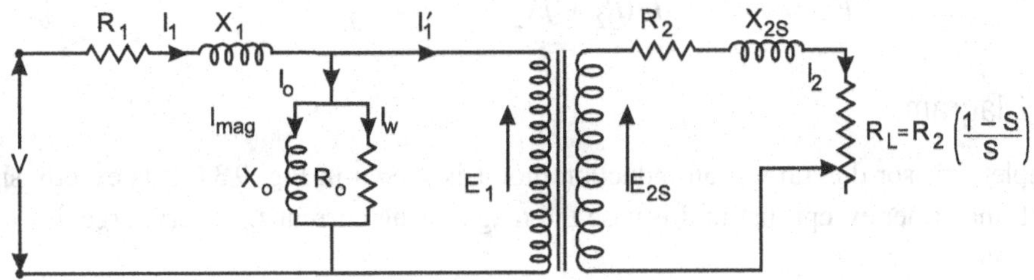

Fig. 9.33 Equivalent circuit of induction motor

The various quantities (all phase values) in the circuit are:

Stator side:

V = Supply voltage

R_1 = Stator winding resistance

X_1 = Stator reactance

I_1 = Stator current

I_0 = Stator no-load current

I_{mag} = magnetising component of no-load current

I_w = working component of no-load current

X_0 = exciting reactance

R_0 = exciting resistance

I_2' = rotor current referred to stator (primary) side

Here,

$$I_{mag} = \frac{V}{X_0} \text{ or } X_0 = \frac{V}{I_{mag}}$$

$$I_w = \frac{V}{R_0} \text{ or } R_0 = \frac{V}{I_w}$$

$$\bar{I}_o = \bar{I}_{mag} + \bar{I}_w$$

$$\overline{I'}_2 = KI_2 (K = \text{transformation ratio if not it is taken as 1})$$

$$\bar{I}_1 = \overline{I'}_2 + \bar{I}_0$$

$$\bar{E}_1 = \bar{V} - \bar{I}_1\bar{Z}_1 = \bar{V} - \bar{I}_1(R_1 + jX_1)$$

Rotor side:

E_{2S} = rotor induced emf at stand-still

E_2 = rotor induced emf under running condition

I_2 = rotor current

R_2 = rotor winding resistance

X_{2S} = rotor reactance at stand-still

X_2 = rotor reactance under running condition

X_L = fictitious resistance representing load

Here,

$$E_{2S} = KE_1;\ R_L = R_2\left(\frac{1-S}{S}\right);\ E_2 = SE_{2S};\ X_2 = SX_{2S}$$

$$\bar{E}_2 = \bar{I}_2\bar{Z}_2 = \bar{I}_2(R_2 + jX_2)$$

Phasor Diagram

The complete phasor diagram of an induction motor is shown in Fig. 9.34. It is exactly similar to that of a transformer except that in this case the magnetising current I_0 is very large due to air gap between stator and rotor.

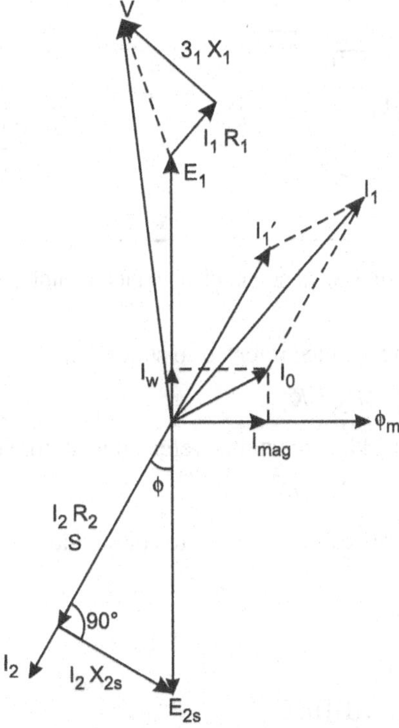

Fig. 9.34 Complete phasor diagram of an induction motor

9.35 Simplified Equivalent Circuit of an Induction Motor

To simplify the circuit, all the quantities which are represented on the rotor side can be referred to the stator side in the similar method as these are transferred (or referred) from secondary to primary for a transformer. Then, the simplified equivalent circuit of the induction motor will become as shown in Fig. 9.35. Here, R_L' represents the equivalent electrical resistance represent gross mechanical load on the motor (or mechanical power developed in the rotor)

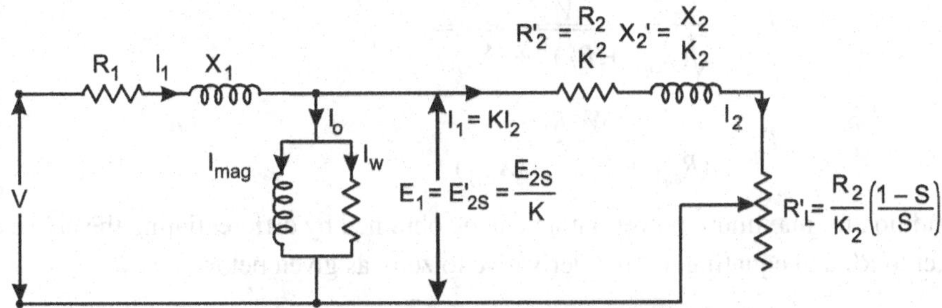

Fig. 9.35 Simplified equivalent circuit (all quantities referred to stator side)

The equivalent circuit of an induction motor can further be simplified as shown in Fig. 9.36.

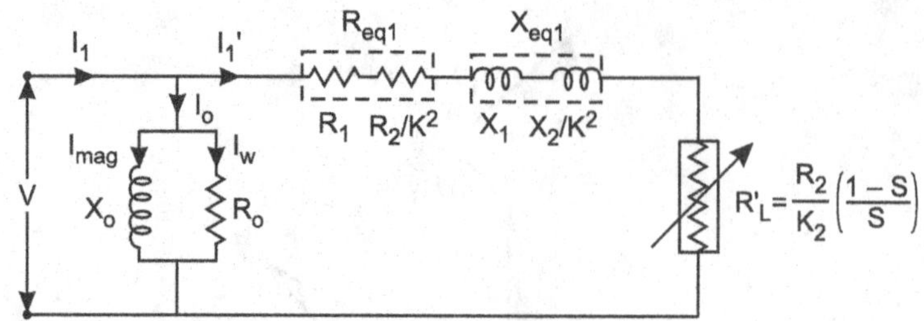

Fig. 9.36 Simplified equivalent circuit of an induction motor referred to stator side

Where, R_{eql} = overall motor resistance referred to stator side.

$= R_1 + R_2' = R_1 + R_2 / K^2$

X_{eql} = overall motor leakage reactance referred to stator side.

$= X_1 + X_2' = X_1 + X_2 / K^2$

R_L' = Fictitious load resistance referred to stator side.

$$= \frac{R_L}{K^2}\left(\frac{1-S}{S}\right)$$

9.36 Maximum Power Output

A simplified equivalent circuit of a 3-phase induction motor (representing one phase only) is shown in Fig. 9.36.

Where, $$R_{eql} = R_1 + \frac{R_2}{K^2} \text{ and } X_{eql} = X_1 + \frac{X_2}{K^2}$$

Load resistance, $$R'_L = \frac{R_2}{K^2}\left(\frac{1-S}{S}\right)$$

Power developed in the motor or gross power output of the motor,

$$P_g = 3I_1'^2 R_L'$$

Where, $$I'_1 = \frac{V}{\sqrt{(R_{eql} + R_L')^2 + (X_{eql})^2}}$$

$\therefore$ $$P_g = \frac{3V^2 R_L'}{(R_{eql} + R_L')^2 + (X_{eql})^2}$$

The condition for maximum power output can be obtained by differentiating the above equation with respect to R'_L and equating its first derivative to zero, as given below:

$$\frac{d}{dR_L'} P_g = \frac{\left[\left(R_{eql} + R_L'\right)^2 + \left(X_{eql}\right)^2\right] 3V^2 - 3V^2 R_L'\left[2\left(R_{eql} + R_1'\right)\right]}{\left[(R_{eql} + R_L')^2 + (X_{eql})^2\right]^2} = 0$$

or $3V^2 R'_L\left[2(R_{eq1}+R'_L)\right] = \left[(R_{eq1}+R_L')^2+(X_{eq1})^2\right]3\,V^2$

or $$2R_{eq1}R'_L + R'_L = R_{eq1}^{\;2} + R_L^{\;2} + 2\,R_{eq1}\,R'_L + X_{eq1}^{\;2}$$

or $$R_L'^{2} = R_{eq1}^{\;2} + X_{eq1}^{\;2}$$

or $$R'_L = \sqrt{R_{eq1}^{\;2} + X_{eq1}^{\;2}} = Z_{eq1}$$

The above equation shows that an induction motor will develop maximum power when load equivalent resistance R'_L will be equal to standstill leakage impedance (Z_{eq1}) of the motor.

Accordingly, the gross maximum power output of the motor,

$$P_{g(\max)} = \frac{3\,V^2\,Z_{eq1}}{(R_{eq1}+Z_{eq1})^2+(X_{eq1})^2} \qquad \text{(Putting } R'_L = Z_{eq1}\text{)}$$

$$= \frac{3\,V^2\,Z_{eq1}}{R_{eq1}^{\;2}+Z_{eq1}^{\;2}+2\,R_{eq1}\,Z_{eq1}+X_{eq1}^{\;2}} = \frac{3\,V^2\,Z_{eq1}}{(R_{eq1}^{\;2}+X_{eq1}^{\;2})+Z_{eq1}^{\;2}+2\,R_{eq1}\,Z_{eq1}}$$

$$= \frac{3\,V^2\,Z_{eq1}}{Z_{eq1}^{\;2}+Z_{eq1}^{\;2}+2\,R_{eq1}\,Z_{eq1}} = \frac{3\,V^2\,Z_{eq1}}{2\,Z_{eq1}\,(R_{eq1}+Z_{eq1})} = \frac{3\,V^2}{2\,(R_{eq1}+Z_{eq1})}$$

The slip corresponding to maximum power developed (i.e., gross power output) can be obtained by putting the condition, i.e.,

$$Z_{eq1} = R'_L = \frac{R_2}{K^2}\left(\frac{1-S}{S}\right) = \frac{R_2}{K^2}\left(\frac{1}{S}-1\right)$$

or $$Z_{eq1} + \frac{R_2}{K^2} = \frac{R_2}{SK^2}$$

or $$S = \frac{R_2 / K^2}{Z_{eq1} + (R_2 / K^2)}$$

Example 9.31

A 3-phase induction motor with a unity turn ratio has the following data/phase referred to stator side:

Stator impedance $\bar{Z}_1 = (1.0 + j\,3.0)$ ohm;

Rotor standstill impedance $\bar{Z}_{2S}' = (1.0 + j\,2.0)$ ohm;

No-load or exciting impedance, $Z_O = (10 + j\,50)$ ohm

Supply voltage, $V = 240$ V

Estimate the stator current, equivalent rotor current, mechanical power developed, power factor, slip and efficiency when the machine is operating at 4% slip.

Solution:

The equivalent circuit for the motor as per data is shown in Fig. 9.37.

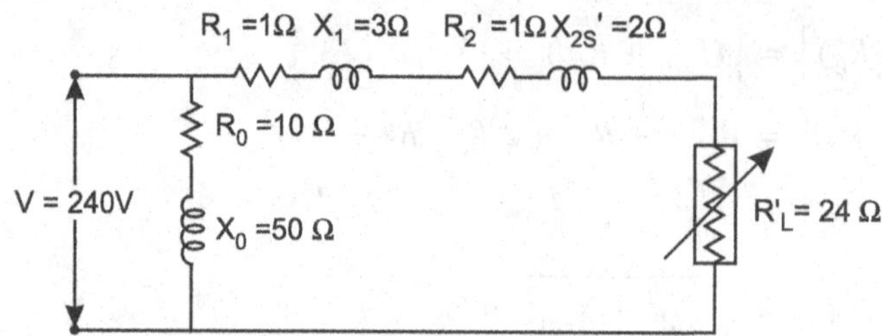

Fig. 9.37 Equivalent circuit as per data

$$\text{Equivalent load resistance, } R_L' = \frac{R_2}{K^2}\left(\frac{1-S}{S}\right) = \frac{1}{(1)^2}\left(\frac{1-0.04}{0.04}\right) = 24\ \Omega$$

$$\text{Effective impedance per phase, } \bar{Z}_{eff} = \left(R_1 + \frac{R_2}{K^2} + R_L'\right) + j\left(X_1 + \frac{X_{2S}}{K^2}\right)$$

$$= \left(1 + \frac{1}{(1)^2} + 24\right) + j\left(3 + \frac{2}{(I)^2}\right) = (26 + j5)\ \Omega$$

$$\text{Stator load current, } \bar{I}_1' = \frac{\bar{V}}{\bar{Z}_{eq}} = \frac{240}{(26 + j5)}$$

$$= \frac{240}{26.47\angle 10.9^\circ} = (9.06\angle -10.9^\circ)\ \text{A}$$

$$= 9.06(\cos 10.9^\circ - j\sin 10.9^\circ) = (8.896 - j1.713)\ \text{A}$$

$$\text{Equivalent rotor current, } I_2 = \frac{I_1'}{K} = \frac{9.06}{1} = \mathbf{9.06\ A}\ (Ans.)$$

$$\text{No-load current, } \bar{I}_0 = \frac{\bar{V}}{\bar{Z}_0} = \frac{240}{(10 + j\,50)} = \frac{240}{51\angle 78.7^\circ}$$

$$= 4.7\angle -78.7^\circ = (0.921 - j4.61)A$$

$$\text{Stator equivalent current, } \bar{I}_1 = \bar{I}_1' + \bar{I}_0 = (8.896 - j1.713) + (0.921 - j4.61)$$

$$= (9.817 - j6.323) = (11.7\angle -32.78^\circ)\ \text{A}$$

$$\therefore \quad I_1 = \mathbf{11.7\ A}\ (Ans.)$$

$$\text{Power factor, } \cos\phi_1 = \cos 32.78^\circ = \mathbf{0.84\ lagging}\ (Ans.)$$

$$\text{Mechanical power developed} = 3\,(I_1')^2 R_L' = 3\,(9.06)^2 \times 24 = \mathbf{5910\ W}\ (Ans.)$$

$$\text{Motor efficiency, } \eta = \frac{\text{Output}}{\text{input}} = \frac{5910}{3 \times 240 \times 11.7 \times 0.84}$$

$$= 0.8352 = \mathbf{83.52\%}\ (Ans.)$$

Example 9.32

A 3-phase 4-pole, 50 Hz, 400 V, 8 kW, star connected squirrel cage induction has the following data:

$R_1 = 0.4\ \Omega$/phase; $R_2 = 0.25\ \Omega$/phase; $X_1 = X_{2S} = 0.5\ \Omega$/phase; $X_m = 15.5\ \Omega$/phase.

The motor develops full-load internal torque at a slip of 4%.

Assume that the shunt branch is connected across the supply, determine (i) slip at maximum torque (ii) maximum torque developed at rated voltage and frequency (iii) torque developed at the start at rated voltage and frequency.

Solution:

Here, $R_1 = 0.4\ \Omega$; $X_1 = 0.5\ \Omega$; $R_2 = 0.25\ \Omega$; $X_{2S} = 0.5\Omega$

$X_m = 15.5\ \Omega$; $S = 0.04$; $V_L = 400$ V; $f = 50\ Hz$.

$$\text{Phase voltage, } V = \frac{V_L}{\sqrt{3}} = \frac{400}{\sqrt{3}} = 231 \text{ V}$$

$$\text{Slip to develop maximum torque, } S_m = \frac{R_2}{X_{2S}} = \frac{0.25}{0.5} = \mathbf{0.5} \ (Ans.)$$

For maximum torque development

Equivalent impedance of the motor at S_m

$$\overline{Z}_{eq1(m)} = \left(R_1 + \frac{R_2}{S_m}\right) + j\left(X_1 + X_{2S}\right) = \left(0.4 + \frac{0.25}{0.5}\right) + j\left(0.5 + 0.5\right)$$

$$= 0.9 + j\,1 = 1.345 \angle 48° \text{ ohm}$$

$$\text{Rotor current, } \overline{I}_2 = \frac{\overline{E}_{2S}}{\overline{Z}} = \frac{231}{1.345\angle 48°}\left(\text{here, } E_2 = \text{V since } K = 1\right)$$

$$= 171.7 \angle{-48°}$$

$$I_2 = \mathbf{171.7\ A}\ (Ans.)$$

$$\text{Rotor copper loss} = 3\,\overline{I}_2^{\,2}\,R_2 = 3 \times (171.7)^2 \times 0.25 = 22122 \text{ W}$$

$$\text{Power input to rotor, } P_2 = P_2 = \frac{\text{Rotor copper loss}}{S_m} = \frac{22122}{0.5} = 44244 \text{ W}$$

Synchronous speed of revolving field,

$$N_S = \frac{120f}{P} = \frac{120 \times 50}{4} = 1500 \text{ rpm}$$

$$\text{Maximum torque developed, } T_m = \frac{P_2}{\omega_S} = \frac{P_2 \times 60}{2\,\pi\,N_S} = \frac{44244 \times 60}{2\,\pi \times 1500} = \mathbf{281.67\ Nm}\ (Ans.)$$

For starting torque:

At start, slip, $S_S = 1.0$

$$\text{Equivalent motor impedance, } \overline{Z}_{eq1(S)} = \left(R_1 + \frac{R_2}{S_S}\right) + j\left(X_1 + X_{2S}\right)$$

$$= \left(0.4 + \frac{0.25}{1}\right) + j\left(0.5 + 0.5\right) = \left(0.46 + j1\right) \text{ohm}$$

$$= (1.1 \angle 65.3°) \text{ ohm.}$$

$$\text{Rotor current at start, } \bar{I}_{2S} = \frac{\bar{E}_{2S}}{\bar{Z}_S} = \frac{231}{1.1\angle 65.3^\circ} = 210\angle -65.3^\circ$$

$$I_{2S} = \mathbf{210\ A}\ (Ans.)$$

$$\text{Rotor copper loss} = 3\, I_{2S}^2\, R_2 = 3 \times (210)^2 \times 0.25 = 33075 \text{ W}$$

$$\text{Power input to rotor, } P_{2S} = \frac{33075}{1} = 33075 \text{ W}$$

$$\text{Starting torque developed, } T_S = \frac{P_{2S}}{\omega_S} = \frac{33.75 \times 60}{2\pi \times 1500} = \mathbf{210.6\ Nm}\ (Ans.)$$

Example 9.33

A 3- phase 4-pole, 50 Hz 400 V, star-connected squirrel cage induction has the following data:

$$\text{Stator impedance, } \bar{Z}_1 = (0.07 + j\,0.25) \text{ ohm}$$

$$\text{Equivalent rotor impedance, } \bar{Z}_2' = (0.07 + j0.3) \text{ ohm}$$

Determine the maximum power developed (gross output), torque and the slip at which it occurs. Neglect exciting reactance.

Solution:

Here, $V_L = 400\ V$; $P = 4$; $f = 50\ Hz$; $\bar{Z}_1 = (0.07 + j\,0.25)$ ohm

$\bar{Z}_2' = (0.07 + j\,0.3)$ ohm

$$\text{Phase voltage, } V = \frac{V_L}{\sqrt{3}} = \frac{400}{\sqrt{3}} = 231 \text{ V}$$

motor equivalent impedance referred to stator side

$$\bar{Z}_{eq1} = (0.07 + j0.25) + (0.07 + j0.3) = (0.14 + j0.55) \text{ ohm}$$

$$= (0.5675\angle 75.7^\circ) \text{ ohm}$$

$$Z_{eq1} = 0.5675 \text{ ohm}$$

Slip corresponding to maximum power developed

$$S_m = \frac{R_2 / K^2}{Z_{eq1} + R_2 / K^2} \qquad \text{(where } K\text{=1 not given)}$$

$$= \frac{0.07}{0.5675 + 0.07} = 0.11 = \mathbf{11\%}\ (Ans.)$$

$$\text{Maximum power developed, } P_m = \frac{3\,V^2}{2(R_{eq1} + Z_{eq1})} = \frac{3 \times (231)^2}{2(0.14 + 0.5675)}$$

$$= 113132 \text{ W} \cong \mathbf{113\ kW}\ (Ans.)$$

$$\text{Synchronous speed, } N_S = \frac{120\,f}{P} = \frac{120 \times 50}{4} = 1500 \text{ r.p.m.}$$

Maximum torque developed, $T_m = \frac{P_m \times 60}{2\pi N_S} = \frac{113132 \times 60}{2\pi \times 1500} = \mathbf{2263\ Nm}$ (*Ans.*)

9.37 Circle Diagram

The performance of an induction motor can be studied, at a glance, with the help of its circle diagram. The construction of a circle diagram is based on the simple equation of a circle.

To understand it more clearly, consider an *RL* series circuit, as shown in Fig. 9.38. The circuit is fed at constant voltage *V*. It contains a constant reactance *X* but a variable resistance *R*.

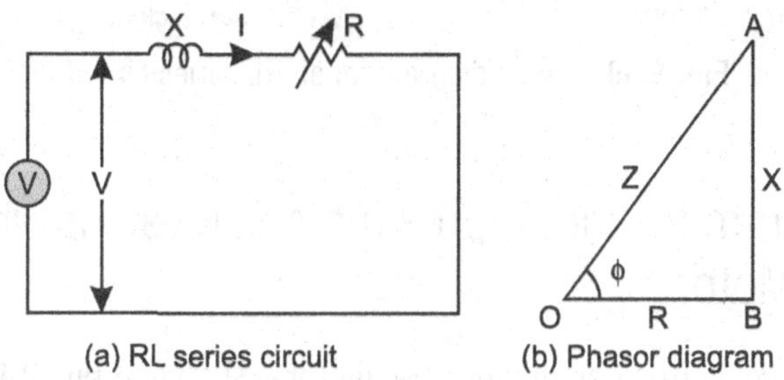

(a) RL series circuit (b) Phasor diagram

Fig. 9.38 RL series circuit

From the circuit,

$$I = I = \frac{V}{\sqrt{R^2 + X^2}}$$

$$= \frac{V}{X} \times \frac{X}{\sqrt{R^2 + X^2}} = \frac{V}{X} \sin\phi$$

as $\sin\phi = \frac{X}{Z} = \frac{X}{\sqrt{R^2 + X^2}}$ (see phasor diagram)

∴ $\mathrm{I} = \frac{V}{X}\sin\phi$ ...(i)

where $\frac{V}{X}$ is constant

The equation (*i*) represents the equation of a circle in polar from having diameter (*V/X*). Such a circle is shown in Fig. 9.39 (a). While drawing this circle the magnitude of the current and power factor angle are taken as polar co-ordinates of the point *A*.

The circuit represents the locus of point *A* which is the extreme point of vector *OA*= *I* since, *R* is variable, its magnitude varies which also changes the phase angle ϕ

$\phi = \sin^{-1}\frac{X}{\sqrt{R^2 + X^2}}$) is such a way that point *A* lies on the circle having diameter equal to *V/X* (constant value). Usually, for lagging power factors, the circle diagrams are drawn by taking diameter on horizontal position (X-axis) and voltage vector on vertical (Y-axis) position, as shown in Fig. 9.39 (b). All relations are true for both the positions.

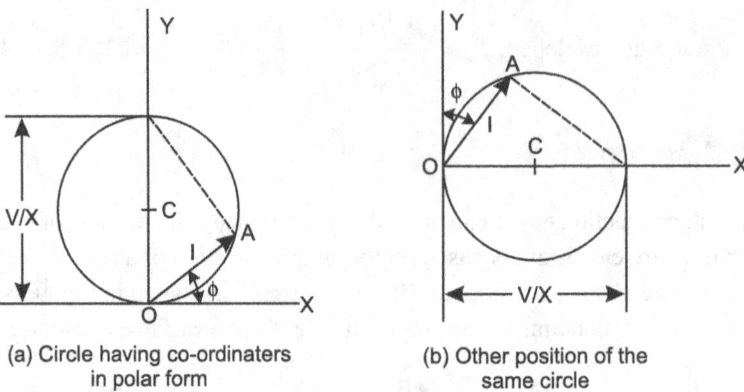

Fig. 9.39 Circle diagram for an RL series circuit

9.38 Circle Diagram for the Approximate Equivalent Circuit of an Induction Motor

The approximate equivalent circuit of an induction motor is shown in Fig. 9.40. It clearly shows that the circuit beyond (on the right hand side of) point *ab* is similar to an *RL* series circuit having constant input voltage *V* and reactance X_{eq1} but the value of resistance varies depending upon the slip (or load).

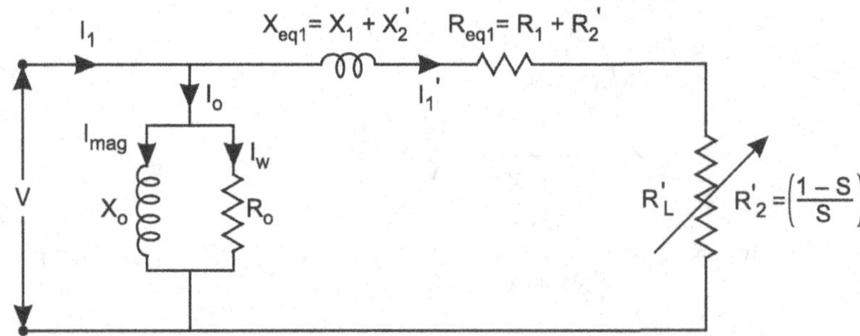

Fig. 9.40 Simplified equivalent circuit of an induction motor

Hence, the extreme point *A* of the current vector I_1' i.e., rotor current referred to the stator side, will lie on a circle with a diameter of V/X_{eq1} (i.e., *OB*) as shown in Fig. 9.41, this is the rotor current referred to the stator side. When I_1' is lagging and $\phi_2 = 90°$ then position of vector for I_1' will be along *OB* i.e., at right angles to vertical axis *OY*.

I_0 is the no-load current which is the vector sum of I_w and I_{mag}. If it is assumed that exciting resistance R_o and exciting reactance X_0 as constant, then I_0 and cos ϕ_o will be constant.

The end of current vector I_1 i.e., vector sum of current I_0 and I_1' $\left(\bar{I}_1 = \bar{I}_0 + \bar{I}_1'\right)$ will also lie on another circle which is displaced from the dotted circle by an amount I_0. Its diameter will be V/X_{eq1} and parallel to X-axis.

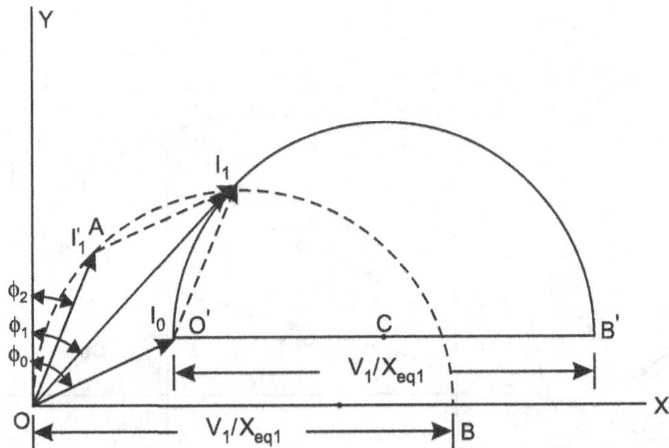

Fig. 9.41 Circle diagram of an induction motor for its simplified equivalent circuit

9.39 Construction of a Circle Diagram for an Induction Motor

The circle diagram of an induction motor may be drawn by using the data obtained from (*i*) stator resistance test, (*ii*) no-load test and (*iii*) blocked rotor test. From stator resistance test, the stator winding resistance per phase R_1 is determined.

From no-load test, the phase voltage $V\left(V=\frac{V_L}{\sqrt{3}}\right)$, no-load current I_0 and no-load power input P_0 is measured and from this data no-load power factor angle is determined.

From blocked rotor test, the corresponding value of short circuit current at normal stator applied voltage $I_{SN}\left(I_{SN}=\frac{V}{V_{SC}}\times I_{SC}\right)$ and short circuit phase angle ϕ_{SC} are determined.

From this data, a circle diagram for an induction motor is drawn by taking the following steps:

1. Draw the voltage phasor *OV* along y-axis.
2. Choose a convenient scale for current and draw a line $OO' = I_0$ at an angle of ϕ_0 with phasor *OV*.
3. Draw a line *OX* perpendicular to voltage phasor *OV*. Also draw a line *O'X* from *O'* parallel to line *OX*.
4. From *O* draw a line *OA* equal to blocked rotor current I_{SN} corresponding to normal rated voltage, with the same scale used for I_0, lagging behind voltage vector *OV* by an angle ϕ_{SC}. Then phasor *O'A* will represent the rotor current referred to stator side i.e., I_2'. The line *O'A* is known as output line.
5. It is obvious that both point *O'* and *A* will lie on the circle. To determine the centre of circle *C*, divide the line *O'A* equally. Extend the dividing line on both sides and the point where it meets the line *O'X'* that will be the centre *C* of the circle (see Fig. 9.42).
6. Taking *C* as centre and *CO'* as radius, draw a semicircle *O'AB*.
7. Draw vertical lines *O'P* and *AQ* from *O'* and *A* respectively parallel to *Y*-axis or voltage vector *OV*.

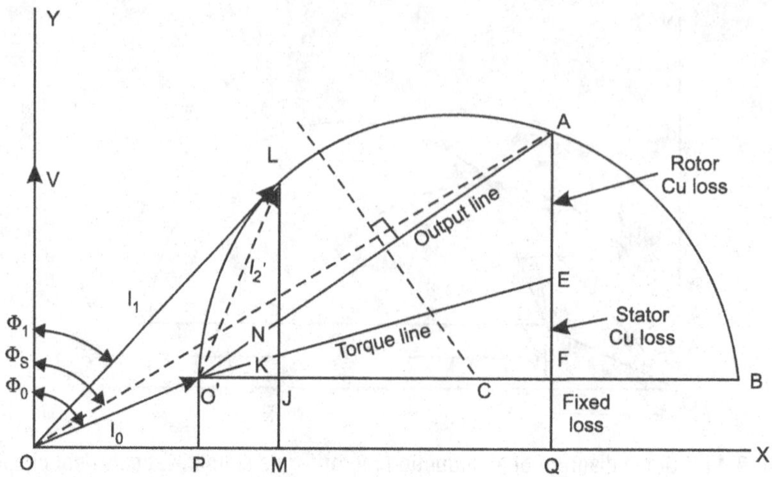

Fig. 9.42 Circle diagram for an induction motor

It may be noted that all the vertical distances represent the active or power components of current since they are in phase with the voltage vector *OV*. Accordingly, *O′P* represents working component of no-load current (i.e., $O'P = OO' \cos \phi_O = I_0 \cos \phi_O = I_w$); otherwise it represents no-load input which includes core losses, friction and windage losses and very small amount of stator copper loss. Similarly, *AQ* i.e., vertical component of *OA* is proportional to motor input on short circuit which includes rotor copper loss, stator copper and fixed losses i.e.,stator iron loss and mechanical loss.

Where $FQ = O'P$ = Stator iron loss + windage and friction loss

AE = rotor copper loss

EF = stator copper loss

The rotor and stator copper losses are represented by the torque line *O′E* and '*E*' is located as follows:

(*i*) *In case of squirrel cage machines:* The stator resistance per phase R_1 is determined by stator resistance test. Then

$$\text{Rotor copper loss} = \text{power input at short circuit -stator copper loss}$$

$$= P_{SC} - 3\, I_{SC}^2\, R_1$$

$$\therefore \qquad \frac{AE}{EF} = \frac{P_{SC} - 3I_{SC}^2 R_1}{3I_{SC}^2 R_1}$$

(*ii*) *In case of phase- wound machines:* In this case, the stator and rotor resistances per phase i.e., R_1 and R_2 can be measured separately for any value of stator and rotor currents, say I_1 and I_2.

$$\frac{AE}{EF} = \frac{I_2^2 R_2}{I_1^2 R_1} = \frac{R_2}{R_1} \times \left(\frac{I_2}{I_1}\right)^2 = \frac{R_2}{R_1} \times \frac{1}{K^2} \qquad \left(\text{Since } \frac{I_1}{I_2} = K\right)$$

$$\frac{AE}{EF} = \frac{R_2 / K^2}{R_1} = \frac{R_2'}{R_1} = \frac{\text{Rotor resistance referred to stator}}{\text{Rotor resistance}}$$

Value of *K* can be determined by voltage ratio test or by using two ammeters on stator and rotor circuit during blocked rotor test.

9.40 Results Obtainable from Circle Diagram

Let us assume that the motor is drawing a current of I_1 ampere at any load. Draw an arc with radius $OL = I_1$ with O as its centre. From L, draw a line LM parallel to y-axis, as shown in Fig. 9.42 which intersects various lines at points N, K and J. Then JM represents fixed losses, JK as stator copper loss, KN as rotor copper loss, KL as rotor input, Nl as rotor output and LM as total motor input.

Various analysis can be done by using the following relations:

1. Motor input = $3V(LM)$ watt
2. Fixed losses = $3V(JM)$ watt
3. Stator Cu loss = $3V(JK)$ watt
4. Rotor Cu loss = $3V(NK)$ watt
5. Total losses = $3V(NM)$ watt
6. Mech. power developed = $3V(NL)$ watt
7. Rotor input = $3V(KL)$ watt
8. Torque developed = $3V\dfrac{(\text{KL})}{\omega}$ Nm
9. Motor efficiency = $\dfrac{\text{LN}}{\text{LM}} = \dfrac{\text{output}}{\text{input}}$
10. Slip = $\dfrac{\text{NK}}{\text{LK}} = \dfrac{\text{Rotor copper loss}}{\text{Rotor input}}$
11. Power factor = $\dfrac{\text{LM}}{\text{OL}}$
12. $\dfrac{\text{Rotor Output}}{\text{Rotor input}} = 1 - \text{S} = \dfrac{\text{N}}{\text{N}_\text{S}} = \dfrac{\text{NK}}{\text{LK}}$

$$\therefore \qquad \text{Rotor speed} = N_S \times \left(\frac{\text{NK}}{\text{LK}}\right) \text{ rpm.}$$

9.41 Maximum Quantities

The maximum quantities can be determined from the circle diagram shown in Fig. 9.43, as mentioned below:

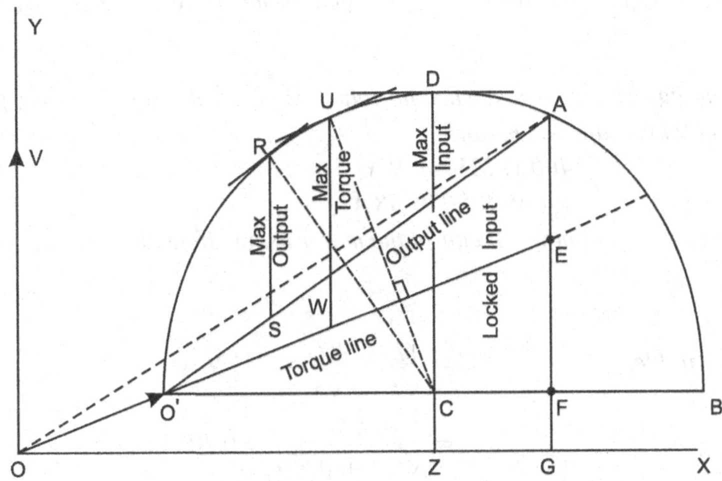

Fig. 9.43 Maximum quantities on a circle diagram

1. *Maximum output*: From centre point C drop a perpendicular on output line $O'A$ and extend it to meet the circle at R. From R draw a vertical line intersecting the output line $O'A$ at S. Then maximum output is represent by RS, such that

$$\text{Maximum output} = 3\,V\,(RS) \text{ watt.}$$

2. *Maximum torque or rotor input*: From centre point *C* drop a perpendicular on torque line *O'E* and extend it to meet the circle at *U*. From *U* draw a vertical line intersecting the torque line *O'E* at *W*. Then maximum torque is represent by *UW*, such that

$$\text{Maximum torque} = 3\,V\left(\frac{UW}{\omega}\right) \text{ Nm}$$

3. *Maximum input power:* Draw a vertical line parallel to y-axis passing through the centre of circle *C*. It intersects the circle at *D* and the x-axis at *Z*. The point *D* where the tangent to the circle is horizontal represents the maximum input power which is proportional to *DZ*, such that

$$\text{Maximum power} = 3V\,(DZ) \text{ watt.}$$

But the motor will be unstable here since the point *D* occur beyond the point of maximum torque *U*. However, maximum input indicates the ability of the motor to carry short-time over loads. Usually. the maximum input power is 2 to 3 times the rated power.

9.42 Significance of Some Lines in the Circle Diagram

Although, many lines have been drawn in the circle diagram but some of them (given below) are important and have special significance.

1. *Input line OX:* The vertical distance between any point on the circle and this line represents the input power and hence called *input line.*
2. *Output line O'A:* The vertical distance between any point on the circle and this line represents the output power and hence called *output line.*
3. *Torque line or Air-gap power line O'E:* Since *KL* represents mechanical power developed in the rotor i.e., air-gap power, the line *O'E* is called *torque line* or *air-gap power line.*

Example 9.34

The following readings were obtained when no-load and blocked rotor tests were performed on a 3-phase, 400 V, 14.9 kW induction motor:

No-load test: *400 V, 1250 W, 9 A*

Blocked rotor test: *150 V, 4000 W, 38 A*

Find full-load current and power factor of the motor using circle diagram.

Solution:

Using the data available,

$$V = \frac{V_L}{\sqrt{3}} = \frac{400}{\sqrt{3}} = 231 \text{ V}$$

$$\cos\phi_0 = \frac{1250}{\sqrt{3}\times 400\times 9} = 0.2004 \text{ lagging}$$

$$\phi_0 = \cos^{-1} 0.2004 = 78.5° \text{ lagging}$$

$$\cos\phi_{SC} = \frac{4000}{\sqrt{3}\times 150\times 38} = 0.405 \text{ lagging}$$

$$\phi_{SC} = \cos^{-1}0.405 = 66.1° \text{ lagging}$$

Short circuit current with normal voltage,

$$I_{SN} = I_{SN} = \frac{V_{SN}}{V_{SC}} \times I_{SC} = \frac{400/\sqrt{3}}{150/\sqrt{3}} \times 38 = 101.3 \text{ A}$$

Power drawn with normal voltage would be

$$= \left(\frac{400/\sqrt{3}}{150/\sqrt{3}}\right)^2 \times 4000 = 28440 \text{ W}$$

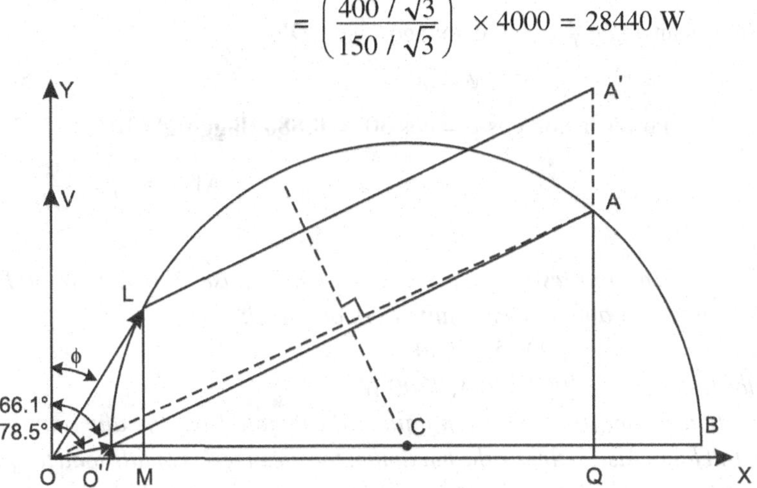

Fig. 9.44 Circle diagram as per data

Construction of Circle Diagram

Draw a vector *OV* along y-axis, as shown in Fig. 9.44.

Let the scale for current be 5 A = 1cm.

Draw vector *OO′* i.e., $I_o = 9$ A $\left(OO' = \frac{9}{5} = 1.8 \text{ cm}\right)$ lagging behind the voltage vector *OV* by an angle ϕ_0 ($\phi_0 = 78.5°$).

Draw vector *OA* i.e., $I_{SN} = 101.3$A ($OA = \frac{101.39}{5} = 20.26$ cm) lagging behind the voltage vector *OV* by an angle ϕ_{SC} $\left(\phi_{SC} = 66.1°\right)$.

Note: To adjust the figures on the page, the size of the figures are reduced.

Draw horizontal x-axis (*OX*) perpendicular to *OY* and a line *O′B* parallel to x-axis.

Draw a line *O′A* and its bisector which meets the line *O′B* at *C*. Draw the semicircle *O′AB* from centre *C* with radius *CO′*.

Draw a line *AQ* parallel to y-axis which represents 28440 W, as calculated above from blocked rotor test. Measure this line which comes out to be 8.1 cm.

Hence the scale for power is 1 cm = $\frac{28440}{8.1} = 3511$ W

Now full-load motor output = 14.9 kW = 14900 W

$$14900 \text{ W} = \frac{14900}{3511} = 4.245 \text{ cm}$$

For locating full-load power point on the circle, extend *QA* to *QA* such that *AA′* = 4.245 cm. From *A′* draw a line parallel to output line *O′A* which intersects the semicircle at *L*. From *L* drop a perpendicular on x-axis which meets the line *OX* at *M*.

Line current = OL = 6 cm = 6 × 5 = **30 A** (*Ans.*)

Phasor *OL* makes an angle ϕ with the voltage vector *OV*.

$$\phi = 30° \qquad \text{(by measurement)}$$

Power factor, cos ϕ = cos 30° = **0.886 (lagging)** (*Ans.*)

$$\left(\text{Also } \cos\phi = \frac{OM}{OL} = \frac{5.3}{6} = 0.863\right)$$

Example 9.35

The no-load and blocked rotor tests were performed on a 3-phase, 200 V, 4 kW, 50 Hz, 4 pole, star connected induction motor and follows results were obtained:

No-load test: *200 V, 5 A, 350 W*

Blocked-rotor test: *100 V, 26 A, 1700 W*

Draw the circle diagram from the given data and estimate from the diagram line current and power factor at full load. Also estimate the maximum torque in terms of full-load torque considering that rotor copper loss at standstill is half the total copper loss.

Solution:

Using data available from no-load test;

$$V = V = \frac{V_L}{\sqrt{3}} = \frac{200}{\sqrt{3}} = 115.5 \text{ V}; I_o = 5\text{A}$$

$$\cos\phi_0 = \frac{350}{\sqrt{3} \times 200 \times 5} = 0.202 \text{ lagging;}$$

$$\phi_0 = \cos^{-1} 0.202 = 78.34° \text{ lagging}$$

Using data available from blocked-rotor test,

$$\cos\phi_{SC} = \frac{1700}{\sqrt{3} \times 100 \times 26} = 0.378 \text{ lagging}; \phi_{SC} = \cos^{-1} 0.378 = 67.82° \text{ lagging}$$

Short circuit current with normal voltage

$$I_{SN} = \frac{V_{SN}}{V_{SC}} \times I_{SC} = \frac{200/\sqrt{3}}{100/\sqrt{3}} \times 26 = 52 \text{ A}$$

Power drawn with normal voltage at short circuit would be

$$= \left(\frac{200/\sqrt{3}}{100/\sqrt{3}}\right)^2 \times 1700 = 6800 \text{ W}$$

Construction of Circle Diagram

Draw line OX and OY as horizontal (x-axis) and vertical (y-axis) axis respectively. Draw a vector OV along y-axis, as shown in Fig. 9.45.

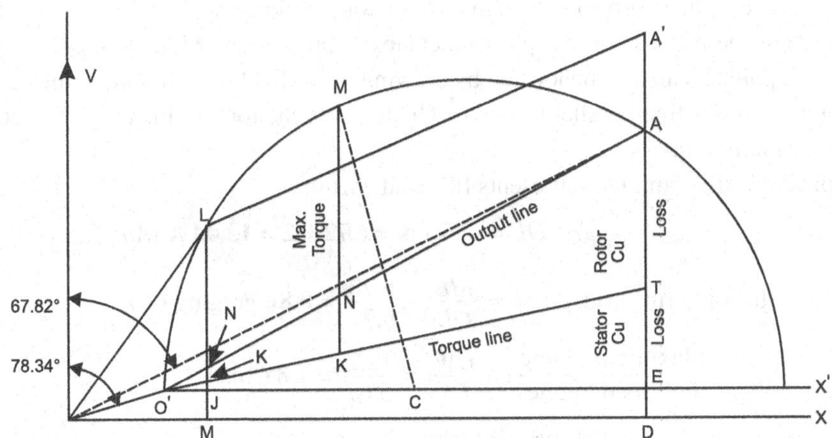

Fig. 9.45 Circle diagram as per data

Let scale for the current be 2 A = 1 cm

Draw vector OO' i.e., $I_0 = 5$ A ($OO' = \frac{5}{2} = 2.5$ cm) lagging behind the voltage phasor OV by an angle ϕ_0 ($\phi_0 = 67.82°$)

Draw phasor OA i.e., $I_{SN} = 52$ A ($OA = \frac{52}{2} = 26$ cm) lagging behind the voltage phasor OV by an angle ϕ_{SC} ($\phi_{SC} = 67.82°$)

Note: To adjust the figure on the page, the size of the figure has been reduced.

From O' draw a line $O'X'$ parallel to x-axis.

Draw a line $O'A$ and its bisector which meets the line $O'X$ at C. Then draw a semicircle $O'AB$ taking centre C and radius CO'. This line $O'A$ is called the output line.

Draw a line AQ parallel to y-axis which represents 6800 W, as calculated above from blocked rotor test, represents total losses. Measure this line which comes out to be 9.8 cm.

Hence scale for power is 1 cm $= \frac{6800}{9.8} = 694$ W

Out of total loss AQ, the portion QF represents no-load loss i.e., 350 W, whereas, the remaining portion AF represents total copper loss.

Find the centre point of line AF at E since at blocked-rotor, the stator copper loss is equal to rotor copper loss i.e., $AE = EF$

Join O' and E, the line $O'E$ represents torque line.

Now motor output = 4 kW = 4000 W $= \frac{4000}{694} = 5.76$ cm

Thus, locate the point L as follows:

Extend line FA to A' so that $AA' = 5.76$ cm and from A' draw a line parallel to output line $O'A$ which cuts the circle at L. From L drop a perpendicular on OX which meets OX at M. Join point L with origin O.

To locate point U which corresponds to maximum torque, proceed as follows:

From centre of the circle C drop a perpendicular on torque line $O'E$ and extend it to meet the circle at U. This point U can also be located by drawing a parallel line with torque line $O'E$ from A'.

From point U draw a line parallel to y-axis which meets the torque line at W, The section UW represents maximum torque.

Now in the circle diagram, OL represents full load current,

$$OL = 6.72 \text{ cm} = 6.72 \times 2 = \mathbf{13.44\ A}\ (Ans.)$$

$$\text{Power factor at full load, } \cos\phi = \frac{ML}{OL} = \frac{5.78}{6.72} = \mathbf{0.86\ lagging}\ (Ans.)$$

$$\frac{\text{Maximum torque}}{\text{Full-load torque}} = \frac{UW}{LK} = \frac{10.2}{5.67} = \mathbf{1.8}\ (Ans.)$$

i.e., Maximum torque = 180 of full-load torque.

Example 9.36

When no-load and blocked rotor tests were performed on a 3-phase, 400 V, 50 Hz, star connected induction motor, the following results were obtained:

No-load test: *400 V, 8.5 A, 1100 W*

Blocked-rotor test: *180 V, 45 A, 5700 W*

Draw the circle diagram and estimate the line current and power factor of the motor when operating at 4% slip. The stator resistance per phase is measured as 0.5 ohm.

Solution:

From no-load test;

$$\text{Phase voltage, } V = \frac{V_L}{\sqrt{3}} = \frac{400}{\sqrt{3}} = 231 \text{ V}; \ I_o = 8.5 \text{ A}$$

$$\cos\phi_o = \frac{1100}{\sqrt{3} \times 400 \times 8.5} = 0.187 \text{ lagging}; \ \phi_o = \cos^{-1} 0.187 = 79.2° \text{ lagging}$$

From blocked-rotor test:

$$\cos\phi_{SC} = \frac{5700}{\sqrt{3} \times 180 \times 45} = 0.4063 \text{ lagging}; \ \phi_{SC} = \cos^{-1} 0.4063 = 66° \text{ lagging}$$

Short circuit current at rated voltage,

$$I_{SN} = \frac{V_{SN}}{V_{SC}} \times I_{SC} = \left(\frac{400/\sqrt{3}}{180/\sqrt{3}}\right) \times 45 = 100 \text{ A}$$

Short-circuit power input at normal voltage would be

$$P_{SN} = \left(\frac{400/\sqrt{3}}{180/\sqrt{3}}\right)^2 \times 5700 = 28150 \text{ W}$$

Construction of circle diagram

Draw horizontal and vertical axis OX and OY respectively, as shown in Fig. 9.46.

Draw a phasor OV along y-axis.

Let scale for the current be 10A = 1cm

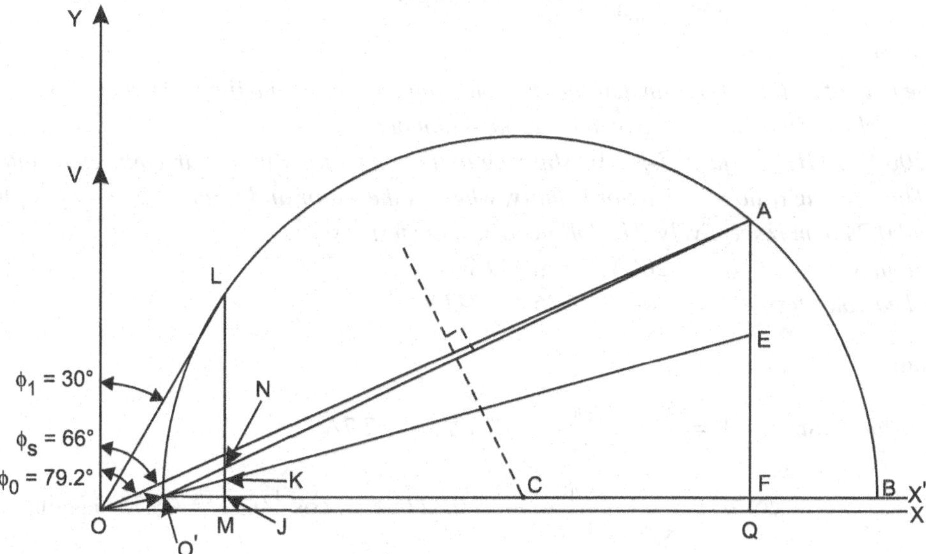

Fig. 9.46 Circle diagram

Draw OO' i.e., $I_o = 8.5\text{A}$ ($OO' = \frac{8.5}{10} = 0.85\text{cm}$) lagging behind the phasor OV by an angle of ϕ_0(ϕ_0 = 79.2°).

Draw phasor OA i.e., $I_{SN} = 100$ A ($OA = \frac{100}{10}$ =10cm) lagging behind the phasor OV by an angle ϕ_{SN} ($\phi_{SN} = 66°$).

From O' draw a line $O'X'$ parallel to x- axis.

Draw a line $O'A$ and its bisector which meets the line $O'X$ at C. taking C as centre draw a semicircle $O'AB$ having radius CO', this line $O'A$ is called output line.

Draw a line AQ parallel to y-axis which represents total losses at short circuit with normal voltage

i.e., $P_{SN} = 28150$ W. Measure this line which comes out to be 4.05 cm.

$$\text{Hence, scale for power is 1 cm} = \frac{28150}{4.05} = 6950 \text{ W}$$

$$\text{Stator copper loss on short circuit} = 3I_{SN}^2 R_1 = 3 \times (100)^2 \times 0.5 = 15000 \text{ W}$$

$$\therefore \quad EF = \frac{15000}{6950} = 2.16 \text{ cm}$$

Hence, point E is located on the line AQ.

Draw a line $O'E$ by joining point E with O'.

Now, $$\text{slip, } S = \frac{\text{Rotor copper loss}}{\text{Rotor input}} = 0.04 = \frac{NK}{LK}$$

Using this ratio $\left(\text{i.e. } \frac{LK}{NK} = \frac{1}{0.04} = 25\right)$ locate point L on the circle by hit and trial method.

Thus, line current, $I_L = OL = 2.42 \text{ cm} = 10 \times 2.42 = \mathbf{24.2\ A}$ (*Ans.*)

Power factor, $\cos \phi = \frac{OM}{OL} = \frac{2.1}{2.42} = \mathbf{0.866\ lagging}$ (*Ans.*)

Example 9.37

With the help of a circle diagram determine (i) starting torque, (ii) maximum torque, (iii) maximum output (iv) slip for maximum torque and (v) maximum output

For a 200 V, 50 Hz, 3-phase, 7.46 kW, slip ring induction motor with a star-connected stator and rotor. The winding ratio of the motor is unity, whereas the stator and rotor resistance per phase is 0.38 and 0.24 ohm respectively. The following are the test results:

No-load test: *200 V, 7.7 A, 874 W*

Blocked-rotor test: *100 V, 39.36 A, 3743 W*

Solution:

From no-load test: $V = \frac{V_L}{\sqrt{3}} = \frac{200}{\sqrt{3}} = 115.5 \text{ V}; I_o = 7.7\text{A}$

$$\cos \phi_o = \frac{874}{\sqrt{3} \times 200 \times 0.77} = 0.327\,; \phi_o = \cos^{-1} 0.327 = 78.75° \text{ lagging}$$

From blocked-rotor test:

$$\cos \phi_{SC} = \frac{3743}{\sqrt{3} \times 100 \times 39.36} = 0.549; \phi_{SC} = \cos^{-1} 0.549 = 63° \text{ lagging}$$

Short circuit current with normal voltage

$$I_{SN} = \frac{V_{SN}}{V_{SC}} \times I_{SC} = \left(\frac{200/\sqrt{3}}{100/\sqrt{3}}\right) \times 39.36 = 78.72 \text{ A}$$

Power drawn with normal voltage at short circuit would be

$$P_{SN} = \left(\frac{200/\sqrt{3}}{100/\sqrt{3}}\right)^2 \times 3743 = 14972 \text{ W}$$

Let scale for the current be 5 A = 1 cm.

Construct the circle diagram as usual, as shown in Fig. 9.47.

Here, no-load current, $I_o = 7.7$ A;

$$OO' = \frac{7.7}{5} = 1.54 \text{ cm lagging behind } OV \text{ by } 78.75°$$

$$\text{Phasor } OA = I_{SN} = \frac{78.72}{5} = 15.74 \text{ cm lagging behind phasor } OV \text{ by } 63°.$$

The vertical line AQ measures the power input on short circuit with normal voltage i.e.,14972 W and its measurement is 8.7 cm.

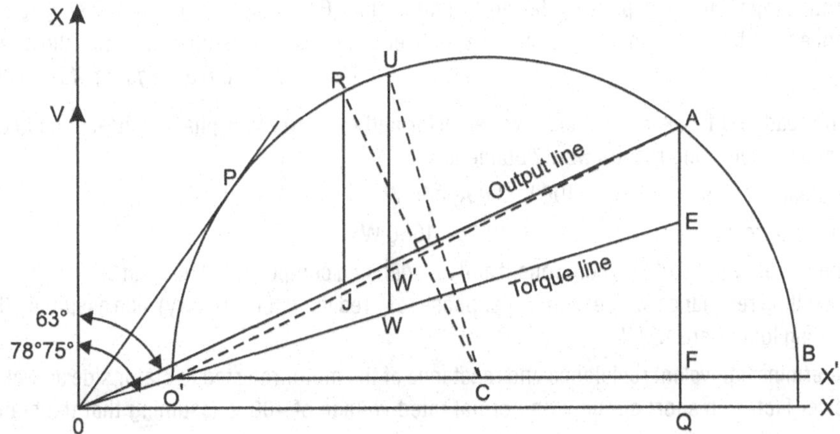

Fig. 9.47 Circle diagram

$\therefore$ Power scale is 1 cm $= \dfrac{14972}{8.7} = 1721$ W

Total copper loss, $AF = 14972 - 874 = 14098 = 8.2$ cm

Ratio of rotor copper loss and total copper loss i.e.,

$$\frac{AE}{AF} = \frac{\text{Rotor resistance}}{\text{Rotor + stator resistance}} = \frac{0.24}{0.24 + 0.38}$$

Or $$AE = \frac{0.24}{0.62} \times 8.2 = 3.18 \text{ cm}$$

(*i*) Starting torque = $AE = 3.18{\times}1721$ = **5473 synch. watt** (*Ans.*)

(*ii*) From centre *C* draw a perpendicular to torque line *O'E* and extent it to meet the circle at *U*. from *U* draw a line parallel to y-axis which meets *O'E* at *W*, then *UW* represents maximum torque.

$$UW = 7.15 \text{ cm} = 7.15 \times 1721 = \mathbf{12305 \text{ synch.watt}} \ (Ans.)$$

(*iii*) For maximum power output, drop a perpendicular on output line *O'A* from centre *C* and extend it to meet the circle at *R*. Draw a line from *R* parallel to y-axis which meets the line *O'A* at *J*. Then *RJ* represents maximum power output.

Maximum power output, $P_{max} = RJ = 5.9$ cm $= 5.9{\times}1721$ = **10154 W** (*Ans.*)

(*iv*) The slip for the maximum torque $= \dfrac{WW'}{WU} = \dfrac{1.4}{7.15} = 0.195$

(*v*) For maximum power factor, draw a tangent on circle from *O* which touches the circle at *P*. Then maximum power factor = cos $\angle POV$ = cos 28.5° = **0.879 lagging** (*Ans.*)

Section Practice Problems

Numerical Problems

1. Determine stator current, stator power factor, torque developed and efficiency of a 400 V, 50 Hz, 4-pole, 3-phase, star connected induction motor having the following data. Stator impedance per phase (0.07+*j*0.3) ohm, rotor impedance per phase referred to stator side (0.08+*j*0.3) ohm, magnetising reactance and resistance per phase 10 ohm and 50 ohm respectively. The machine is operating at a slip of 4%.

(**Ans.** *119.7 A, 0.89 lagging, 437.35 Nm, 89.4%*)

2. When no-load and blocked rotor tests were performed on a 400 V, 3-phase delta-connected induction motor and the following results were obtained:

   No-load test: 400 V; 3.0 A; 645 W

   Blocked-rotor test: 200 V; 12.0 A; 1660 W

   Determine energy (or working) component and magnetising component of no-load current, no-load power factor, exciting resistance and reactance per phase referred to stator (primary) side assuming that friction and winding losses are 183 W.

   Also determine equivalent resistance and reactance of the motor referred to stator side as well as current and power factor on short circuit with normal rated voltage of 400 V assuming that the stator winding resistance per phase is 5 ohm.

   (**Ans.** *0.3464 A, 2.94 A, 0.2 lagging, 1155 ohm, 236 ohm, 11.53 ohm, 26.5 ohm, 24 A, 0.4 lagging)*

3. Draw the circle diagram and determine the line current and power factor at rated output and maximum torque developed for a 415 V, 29.84 kW, 50 Hz delta connected motor. The following test data is available:

   No-load test: 415 V; 21 A; 1250 W

   Blocked-rotor test: 100 V; 45 A; 2730 W

   Assume that stator copper losses are equal to rotor copper losses at standstill.

   (**Ans.** *104A, 0.819 lagging, 51980 synch.watt*)

Short Answer Type Questions

Q.1. What are constant and variable losses in an induction motor?

Ans. The losses which remain constant from no-load to full-load are called constant losses. The losses which vary with the load are called variable losses.

Q.2. Which tests are performed to determine parameters of an induction motor?

Ans. Stator resistance test, no-load test and blocked-rotor test.

Q.3. Draw simplified equivalent circuit of a 3-phase induction motor referred to stator side.

Ans.

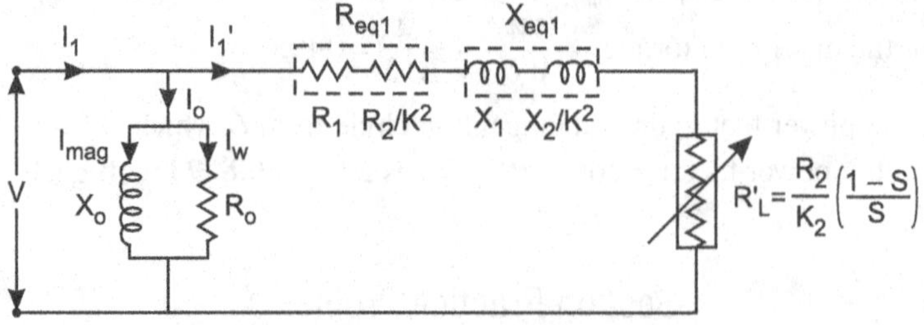

Fig. Q.3

Q.4. List out the tests necessary to perform on a 3-phase induction motor to draw its circle diagram.

Ans. No-load and blocked-rotor test.

9.43 Effect of Space Harmonies

While discussing the performance of an induction motor, so far, it is assumed that the flux in the air-gap is distributed sinusoidally which produces a revolving field which rotates at synchronous speed in space. But in actual practice, it is not so. Either *line harmonics* or *space harmonics are there to affect the torque.*

These harmonics may have components that rotate at speeds other than that correspond to the fundamental frequency.

The time harmonic voltages are usually small in comparison to fundamental voltages, moreover the motor reactance to the time harmonics is high and consequently the torques developed by these harmonies are very small. Thus, the performance of an induction motor is not much affected by time harmonics.

However, the space harmonic fields are developed by the windings, slotting, magnetic saturation and irregularity in air-gap length. These harmonic fields induce emfs which circulate harmonic currents in the rotor windings. These harmonic currents interact with the harmonic fields and develop harmonic torques which produce vibrations and noise.

With certain relationship between the number of pole, stator and rotor slots in cage motors peculiar behaviour may be observed when the motor is started.

9.43.1 Cogging in Three-phase Induction Motors

In an induction motor, the magnetic circuit is completed through stator and rotor. It is the tendency of a magnetic circuit to align itself in a position of minimum reluctance. Thus, if the number of stator slots is equal to the number of rotor slots, there exists a position of minimum reluctance when the teeth of rotor and stator are aligned opposite to each other. The radial alignment forces become very strong when the machine is at rest and rotor and stator teeth are aligned. These forces may exceed the tangential forces of acceleration thereby preventing the motor from starting.

Thus, *the phenomenon by which the radial alignment forces exceed the tangential accelerating forces and the machine (induction motor) refuses to start is called* **cogging.**

Due to cogging, the rotor of induction motor is locked which may overheat the motor winding.

To avoid cogging the number of rotor slots are never made equal to the number of stator slots. Moreover, the rotor is always skewed for smooth running of the rotor.

9.43.2 Crawling in Three-phase Induction Motors

In a three-phase induction motor, the fundamental frequency produces the synchronous torque due to which rotor starts rotating at its rated speed. However, in the induction motors, harmonics are developed out of which 5th and 7th harmonics are more important. These harmonics generate rotor emfs, currents and torques of the same general torque/speed shape as that of the fundamental but with synchronous speeds 1/5th (backward) and 1/7th (forward) of the fundamental synchronous speed.

The 7th harmonic torque reaches its maximum value just before 1/7th of synchronous speed, but beyond this speed the 7th harmonic torque becomes negative since the slip in the harmonic field is negative. The resultant torque-speed curve combined with 7th harmonic and fundamental frequency is shown in Fig. 9.48. Torque due to harmonic flux causes dip in the torque-speed characteristics.

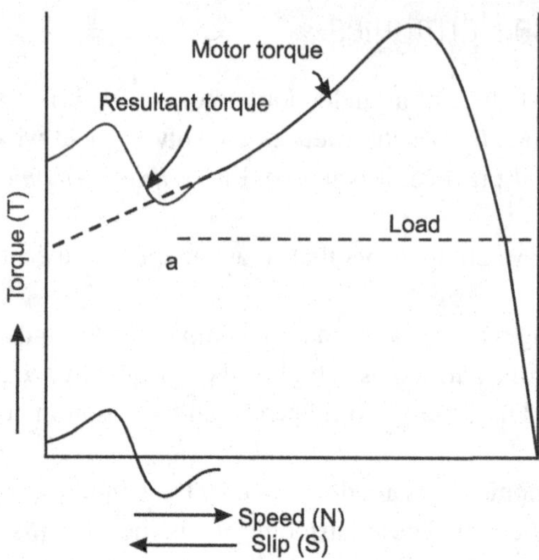

Fig. 9.48 Effect of harmonics on starting torque of an induction motor

When dip in the torque occur due to 7th harmonic and is insufficient to pick the load, the motor can not accelerate to its full speed but starts *crawling* at the speed corresponding to point '*a*'.

Thus, *when a 3-phase induction motor continues to rotor at a speed little lower than the 1/7th synchronous speed, it is said to be* **crawling.**

Such operation of a motor is unstable, a momentary reduction in load may permit the motor to accelerate to the rated speed. In fact, the crawling effect can be eliminated by proper choice of coil pitch and distribution of coils while designing the winding. This reduces the harmonic flux in the air gap to very low value.

9.44 Performance Curves of Induction Motors

The behaviour of an induction motor at different loads is different. The graphical representation of various quantities of an induction motor with respect to its output represents its performance curve.

1. **Speed-output characteristics:** The speed–output (or slip-output) characteristics of a 3-phase induction motor are shown in Fig. 9.49. It shows that the speed of the motor decreases slightly from no-load to full-load but slip increases. Hence, it is almost a constant speed motor, similar to a DC shunt motor.
2. **Power factor characteristics:** The power factor–output characteristics of a 3-phase induction motor are shown in Fig. 9.49. At no-load, motor draws magnetising current I_{mag} which is very large component due to high reluctance of the air-gap and no-load current I_o lags behind the voltage by a large angle and thus power factor of the motor is very low at no-load. As the load increases, motor draws more current from the supply to meet with the load and the percentage of magnetising current component decreases with respect to the total load current drawn by the motor and thus power factor improves and goes on improving with the load, as shown in Fig. 9.49.

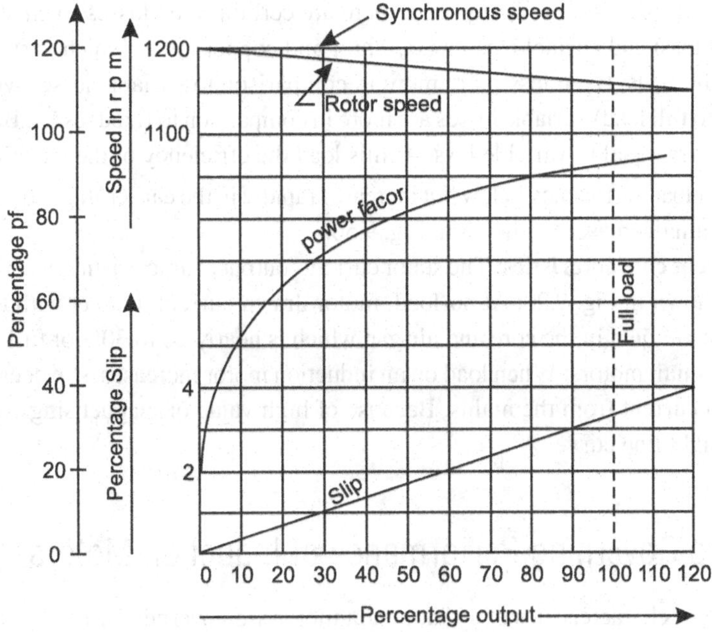

Fig. 9.49 Performance Curves of a 3-phase, 50 Hz squirrel cage motor

3. **Efficiency characteristics:** The efficiency–output characteristics of a 3-phase induction motor are shown in Fig. 9.50.

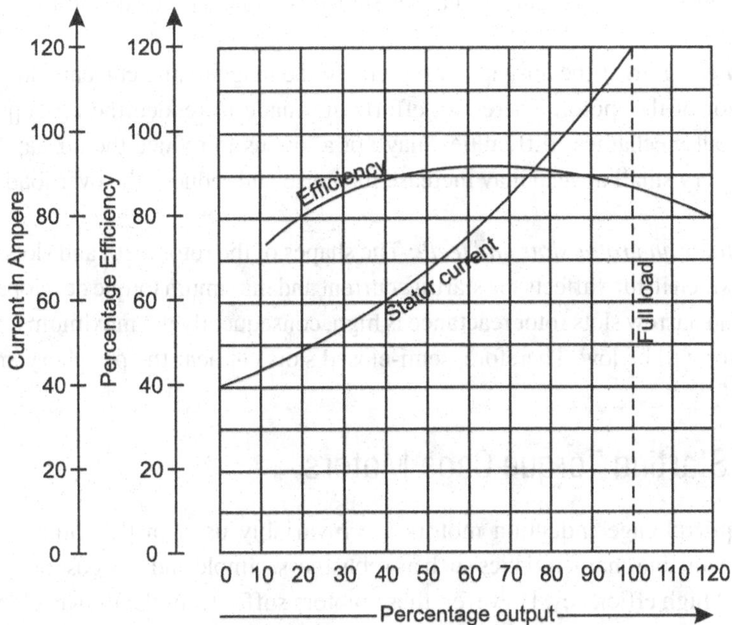

Fig. 9.50 Performance Curves of a 3-phase, 50 Hz squirrel cage motor

Under operating conditions, at any load, there are certain fixed losses (such as iron losses and mechanical loss) and variable losses i.e., stator and copper losses which vary as square of the load. At light loads, fixed losses are more in comparison to variable losses whereas, at heavy loads (near to full load) variable losses are more in comparison to fixed losses. But at certain load fixed losses are equal to variable loss. At this load the efficiency of the machine is maximum. Thus, at light loads efficiency in low but increases rapidly in the earlier stages becomes maximum and then again decreases as shown in Fig. 9.50.

4. **Stator current characteristics:** The stator current–output characteristics of a 3-phase induction motor are shown in Fig. 9.20. At no-load, motor drawn current to meet with the losses and to set-up magnetic field in the core and air-gap which is nearly 20 to 30% of full-load which may be more for small motors. When load on an induction motor increases, slip decreases and motor drawn extra current from the mains. Because of high value of magnetising current this curve is not a straight line curve.

9.45 Factors Governing Performance of Induction Motors

The performance or characteristics of an induction motor are governed by its

(*i*) Rotor resistance, (*ii*) air-gap length (*iii*) shape of stator and rotor slots and teeth.

(*i*) *Rotor resistance:* Smaller value of rotor resistance improves the efficiency of the motor but the starting torque becomes very poor. To improve the starting torque rotor resistance cannot be increased in squirrel cage rotors. Hence these motors have poor starting torque. But in case of slip ring induction motors, an external resistance is added at the start to improve the starting torque.

(*ii*) *Air-gap length:* Larger the air-gap, larger will be the magnetising current and poor will be the power factor of the motor. Therefore, efforts are made to reduce the air-gap between stator and rotor, but mechanical difficulties may not allow us to reduce the air-gap to large extent. Moreover, very small air-gap may increase the noise and reduces the over-load capacity of the motor.

(*iii*) *Shape of stator and rotor slots and teeth:* The shapes of the rotor teeth and slots affect the rotor reactance which in turn affects the starting current and maximum torque developed by the motor. For deep and narrow slots rotor reactance is high, consequently the maximum torque developed by the motor will be low. Therefore, semi-closed slots cut near the periphery are preferred.

9.46 High Starting Torque Cage Motors

Three-phase squirrel-cage induction motors are invariably used in the industrial applications because of their valuable characteristics such as robustness, simple and low cost of construction, low maintenance and high efficiency. However, these motors suffer from the drawback of poor starting torque. Starting torque of an induction motor can be increased by increasing the rotor resistance at the start which is possible in case of slip-ring induction motors. The rotor resistance of the squirrel cage induction motor is fixed since their cages are short circuited. If their rotor resistance is increased by designing their rotor circuit, then their efficiency would become very poor under running conditions. However, two types of rotors i.e., deep bar rotors and double cage rotors with short circuited cages are suggested to improve the starting torque without affecting the efficiency of the motor.

9.46.1 Deep Bar Cage Rotor Motors

A deep bar cage rotor is a rotor which carries deep bar conductors i.e.,its bar conductors are 20 to 25 mm high with a depth to width ratio of 10:1 or 12:1, as shown in Fig. 9.51 (a).

In deep bar cage rotor motors, skin effect is dominating factor to increase the rotor resistance at the start to obtain higher starting torque. Let us see how this skin effect occurs in deep bar cage rotor motors.

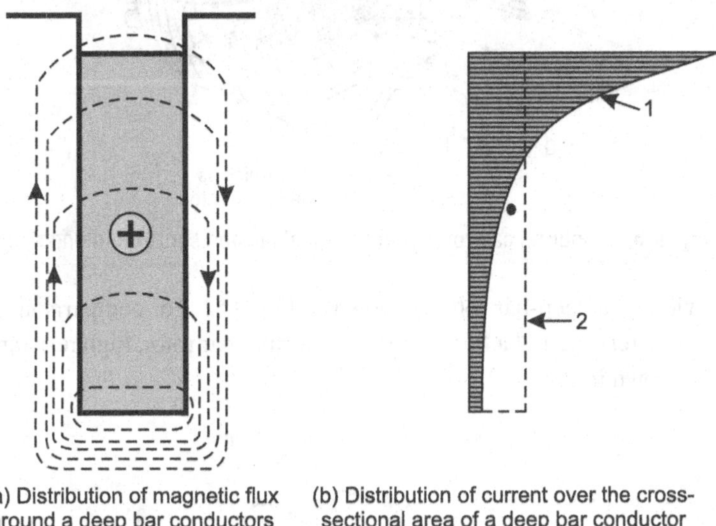

(a) Distribution of magnetic flux around a deep bar conductors (b) Distribution of current over the cross-sectional area of a deep bar conductor

Fig. 9.51 Magnetic flux and current distribution in a deep bar cage rotor conductors

Consider a slot and conductor of a deep bar rotor. At the start, the rotor current frequency in equal to supply frequency. The flux will be set-up around the rotor conductions as shown in Fig. 9.51 (a). the concentration of magnetic field is more in the lower part than the upper.

Therefore, more emf is induced in the lower part of the bar conductor than the upper part. This result is more current to flow in the upper part than the lower. The current density in the conductor is depicted as shown in Fig. 9.51 (b). This effect is called skin effect. The non-uniform current distribution causes greater ohmic loss. Thus, the apposition offered by conductor increases at the start. The AC resistance is substantially high ($R_{ac} >> R_{dc}$). As a result, the starting torque increases and the starting current decreases.

Once motor picks up speed, the frequency of rotor currents decreases to the values corresponding to steady state operating conditions. The skin effect gradually diminishes and current is distributed uniformly over the area of cross-section of the conductor as shown by the dotted line -2, in Fig. 9.51 (b). Thus, the resistance of the rotor gradually decreases and becomes normal under operating conditions.

Thus, in these motors, high starting torque with lesser starting current is obtained without losing the efficiency of the motor. However, in this case the rotor reactance will be high as compared to the normal cage motor because of more leakage flux. Hence, the maximum torque obtained by these motors will be smaller. It reduces the over-load capacity. of the motor. Moreover, the power factor of these motors is also reduced.

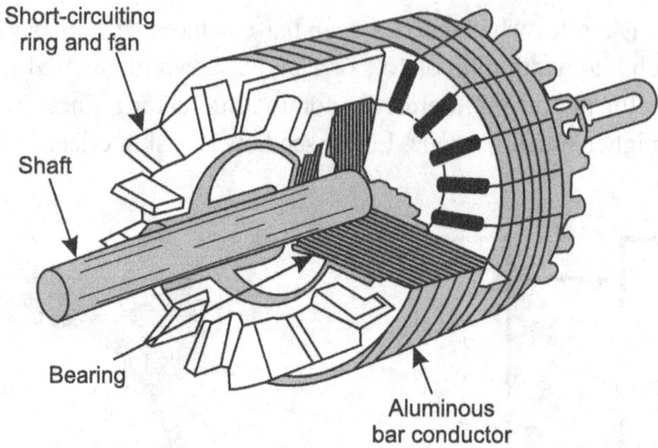

Fig. 9.52 Squirrel cag rotor with sectional fan and short-circuiting rings

The cut-away view of a deep-bar rotor is shown in Fig. 9.52. For comparison, the torque-speed characteristics of squirrel cage induction motor with normal bar rotor, high resistance-bar rotor and deep-bar rotor are shown in Fig. 9.53.

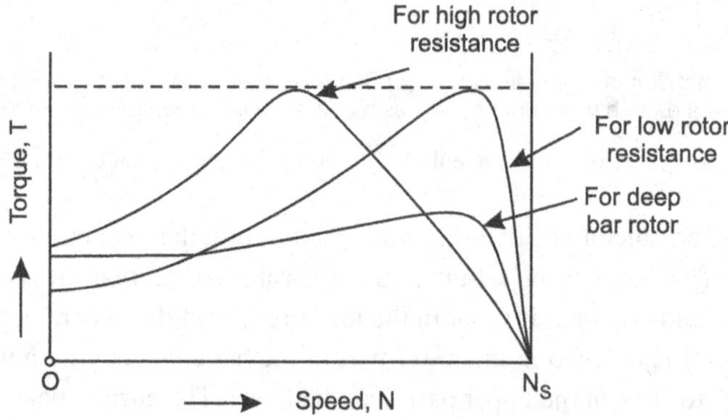

Fig. 9.53 Torque-speed characteristics of low- resistance and beep-bar rotor motor

9.46.2 Double Cage Induction Motor

The rotor design of a double squirrel cage induction motor is somewhat expensive than deep-bar rotor design but it provides better starting and running characteristics. The rotor of this motor is so designed that it provides a high starting torque (200 to250 percent) with a low starting current (restricting to 400 to 600 percent of full-load value). The major advantage of this design is that a high resistance rotor circuit comes into action during starting and a low resistance rotor circuit during running which improves the efficiency of motor during normal running conditions.

Construction

A double cage induction motor consists of a rotor which has two independent cages one above the other in the same slot. A double cage rotor is shown in Fig. 9.54. The upper slot conductors form the outer cage and the lower slot conductors form the inner cage. The outer cage consists of bars of high resistance (resistance of outer cage in nearly 5 to 6 times the resistance of inner cage), generally bars of brass alloy of smaller area of cross-section, are employed. The inner cage consists of bars of low resistance, generally bars of copper metal of larger area of cross section, are employed. Since the inner cage is embedded deep in the iron, it has relatively more leakage flux as shown in Fig. 9.55 and has high reactance.

Thus, the outer cage has high resistance and low reactance whereas, the inner cage has low resistance and high reactance. A slit is provided between upper and lower slot so that it offers high reluctance for the stator field. Thus the stator field instead of linking with only outer cage, links both the cages simultaneously.

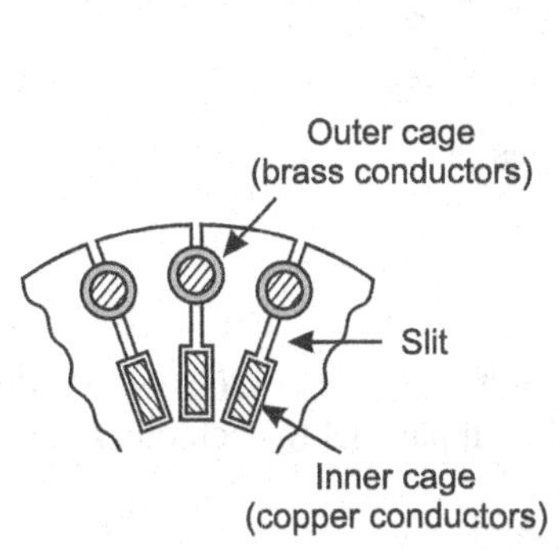

Fig. 9.54 Double-cage rotor conductors

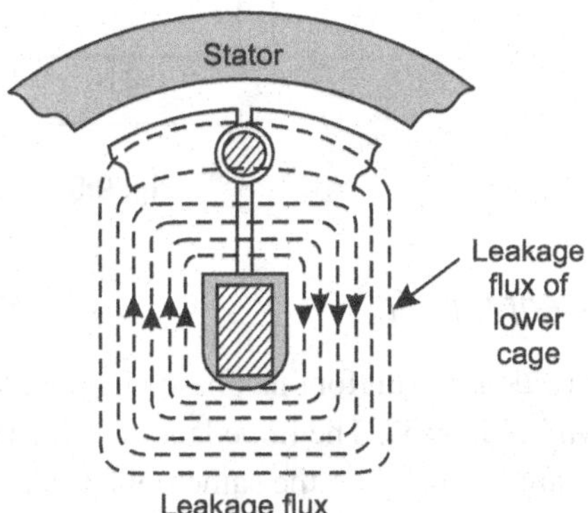

Fig. 9.55 Rotor flux distribution

Operation

To study the operation of a double-cage induction motor, the following points must be kept in mind:

1. Since the maximum torque produced by each cage is independent of their resistance but is inversely proportional to their stand still reactance. Therefore, the value of the maximum torque developed by the inner cage is more than the outer cage as shown in Fig. 9.56.
2. At start, the rotor current frequency is equal to the supply frequency, therefore, the reactances ($2\pi f L$) of the two cages are high as compared to their resistances. The divisions of the rotor current between the cages is determined by their relative impedance (reactance). Therefore, at start, comparatively a small current flows through inner cage bars while a large current flows through the outer cage in spite of its high resistance.

   Thus, at start the torque is mainly produced by the outer cage as shown in Fig. 9.56.

3. As the motor picks up speed, the rotor current frequency decreases and is very low at rated speed. Under this condition the reactance of the two cages are very low as compared to their resistances. Therefore, the division of the rotor current between the cages is determined by their relative resistance and most of the current flows through the inner cage while a small current flows through the outer cage.

   Thus, under normal running condition, the torque is mainly produced by the inner cage as shown in Fig. 9.56.

The torque-speed curves of outer and inner cages as well as double cage induction motor are shown in Fig. 9.56 separately.

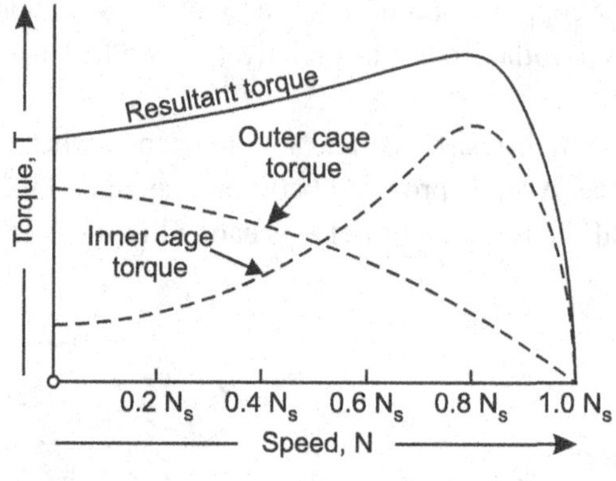

Fig. 9.56 Torque-speed curves

Equivalent Circuit

The equivalent circuit for one phase of a double squirrel cage induction motor referred to stator side is shown in Fig. 9.57. The impedances of the two cages are shown parallel to each other since both of them are linking with the same field produced by the stator.

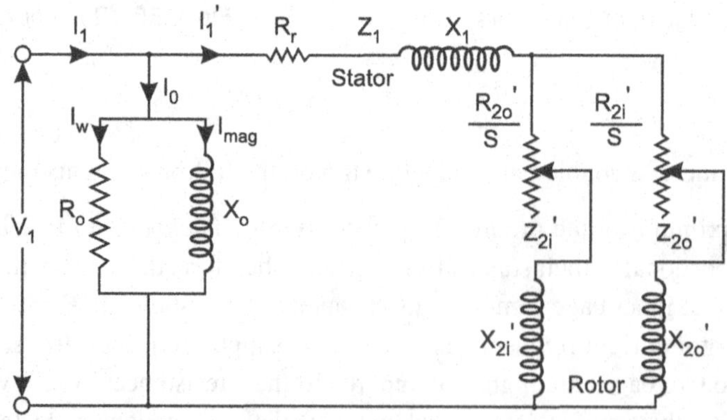

Fig. 9.57 Equivalent circuit of double squirrel cage induction motor referred to stator side with magnetic circuit

Here, R_1 = Stator resistance; X_1= Stator reactance;

R_0 = Exciting resistance; X_0= Exciting reactance;

$\frac{R'_{20}}{S}$ = Rotor circuit resistance of outer cage referred to stator under running condition.

$$\frac{R'_{20}}{S} = \frac{R_{20}}{K^2 \times S}$$

X'_{20} = Rotor circuit reactance of outer cage referred to stator under running condition.

$$X'_{20} = \frac{X_{20}}{K^2}$$

$\frac{R'_{2i}}{S}$ = Rotor circuit resistance of inner cage referred to stator under running conditions.

$$\frac{R'_{2i}}{S} = \frac{R_{2i}}{K^2 \times S}$$

X'_{2i} = Rotor circuit reactance of inner cage referred to stator under running conditions.

$$X'_{2i} = \frac{X_{2i}}{K^2}$$

A simplified equivalent circuit neglecting the magnetising circuit is shown in Fig. 9.58.

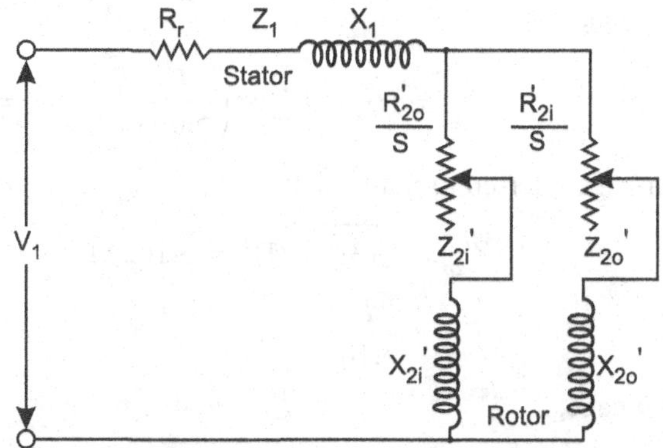

Fig. 6.58 Simplified equivalent circuit of double squirrel cage induction motor

Example 9.38

The standstill impedances per phase of the inner cage and outer cage of a 3-phase, 400 V, 50 Hz, double squirrel cage induction motor referred to stator side are given as;

$$\bar{Z}_{2i} = (0.6 + j5)\ \Omega; \bar{Z}_{2o} = (3 + j1)\ \Omega$$

Determine the ratio of the currents and torques of the two cages (i) at stand-still (ii) at a slip of 4%. Neglect magnetising current.

Solution:

The simplified equivalent circuit of the motor as per data is shown in Fig. 9.59.

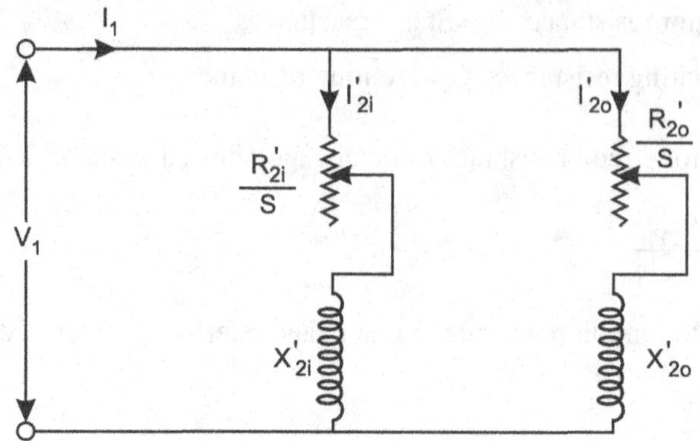

Fig. 9.59 Equivalent circuit

(*i*) *At standstill;* slip, $S = 1$

Impedance of inner-cage referred to stator side;

$$Z'_{2i} = \sqrt{(0.6)^2 + (5)^2} = 5.036\ \Omega$$

Let the voltage impressed across the cage be E_1, then

$$I'_{2i} = \frac{E_1}{5.036} \text{ ampere}$$

Power input to inner-cage,

$$P'_{2i} = (I'_{2i})^2 R'_{2i} = \left(\frac{E_1}{5.036}\right)^2 \times 0.6 = 0.02366\ E^2$$

Impedance of outer-cage referred to stator side;

$$Z'_{2o} = \sqrt{(3)^2 + (1)^2} = 3.162\ \Omega$$

$$I'_{2o} = \frac{E_1}{3.162}$$

Power input to outer-cage,

$$P'_{2o} = (I'_{2o})^2 R'_{2o} = \left(\frac{E_1}{3.162}\right)^2 \times 3 = 0.3\ E^2$$

Ratio of outer to inner cage current,

$$\frac{I'_{2o}}{I'_{2i}} = \frac{E_1}{3.162} \times \frac{5.036}{E_1} = \mathbf{1.593} \text{ (Ans)}$$

Ratio of outer to inner-cage torques or power

$$\frac{T_{2o}}{T_{2i}} = \frac{P'_{2o}}{P'_{2i}} = \frac{0.3E_1^{\ 2}}{0.02366E_1^{\ 2}} = \mathbf{12.68} \text{ (Ans)}$$

(*ii*) *At a slip of 4%;* Slip, $S = 0.04$

Impedance of inner-cage referred to stator side;

$$Z'_{2i} = \sqrt{\left(\frac{R'_{2i}}{S}\right)^2 + (X_{2i})^2} = \sqrt{\left(\frac{0.6}{0.04}\right)^2 + (5)^2} = 15.81\ \Omega$$

Current, $$I'_{2i} = \frac{E_1}{Z_{2i}} = \frac{E_1}{15.81} \text{ ampere}$$

Power input, $$P'_{2i} = (I'_{2i})^2 \times \frac{R'_{2i}}{S} = \left(\frac{E_1}{15.8}\right)^2 \times \frac{0.6}{0.04} = 0.06\ E_1^{\ 2}$$

Impedance of outer-cage referred to stator side;

$$Z'_{2i} = \sqrt{\left(\frac{R'_{2o}}{S}\right)^2 + (X_{2o})^2} = \sqrt{\left(\frac{3}{0.04}\right)^2 + (1)^2} = 75\ \Omega$$

Current, $$I'_{2o} = \frac{E_1}{Z_{2o}} = \frac{E_1}{75} \text{ ampere}$$

Power input to outer-cage,

$$P'_{2o} = (I'_{2o})^2 \times \frac{R'_{2o}}{S} = \left(\frac{E_1}{75}\right)^2 \times \frac{3}{0.04} = 0.01333\ E_1^{\ 2}$$

Ratio of outer to inner-cage current,

$$\frac{I'_{2o}}{I'_{2i}} = \frac{E_1}{75} \times \frac{15.81}{E_1} = \mathbf{0.2108}\ (Ans)$$

Ratio of outer to inner-cage torques or power

$$\frac{T'_{2o}}{T'_{2i}} = \frac{P'_{2o}}{P'_{2i}} = \frac{0.01333\ E_1^{\ 2}}{0.06\ E_1^{\ 2}} = \mathbf{0.222}\ (Ans)$$

Conclusion: During starting major part of the power is developed by the outer-cage but during running condition major power is developed by the inner-cage.

Example 9.39

The standstill impedances per phase of the inner and outer cage of a double cage induction motor is given below:

$$Z_{2i} = (0.5 + j2)\ \Omega;\ Z_{2o} = (2 + j0.5)\ \Omega$$

Determine the slip at which the two cages develop equal torques.

Solution:

Let S be the slip at which the two cages develop equal torques.

During running condition,

$$Z_{2i} = \sqrt{\left(\frac{0.5}{S}\right)^2 + (2)^2} \text{ and } Z_{2o} = \sqrt{\left(\frac{2}{S}\right)^2 + (0.5)^2}$$

Now,
$$\frac{I_{2i}}{I_{2o}} = \frac{Z_{2o}}{Z_{2i}}$$

or
$$\left(\frac{I_{2i}}{I_{2o}}\right)^2 = \left(\frac{Z_{2o}}{Z_{2i}}\right)^2 = \frac{(2/S)^2 + (0.5)^2}{(0.5/S)^2 + (2)^2}$$

Power developed in the two cages

$$P_{2i} = I_{2i}^2 \times \frac{R_{2i}}{S} = I_{2i}^2 \times \frac{0.5}{S};\ P_{2o} = I_{2o}^2 \times \frac{R_{2o}}{S} = I_{2o}^2 \times \frac{2}{S}$$

$T_{2i} \propto P_{2i}$ and $T_{2o} \propto P_2$

and
$$T_{2i} = T_{2o}$$

or
$$\frac{T_{2i}}{T_{2o}} = 1$$

or
$$\frac{[(2/S)^2 + (0.5)^2](0.5/S)}{[(0.5/S)^2 + (2)^2](2/S)} = 1 \quad \text{or} \quad \frac{\left(\frac{4}{S^2} + 0.25\right) \times 0.5}{\left(\frac{0.25}{S^2} + 4\right) \times 2} = 1$$

or
$$\frac{4}{S^2} + 0.25 = \left(\frac{0.25}{S^2} + 4\right)4 \quad \text{or} \quad \frac{4}{S^2} + 0.25 = \frac{1}{S^2} + 16$$

or
$$\frac{3}{S^2} = 15.75 \quad \text{or} \quad S = \sqrt{\frac{3}{15.75}} = 0.4364 = \mathbf{43.64\%}\ (Ans)$$

Example 9.40

The impedance per phase of a double cage induc*tion motor are:*

Inner cage-(0.06 +j0.5) ohm; Outer cage – (0.6 + j0.12) ohm,

Estimate the toque in synchronous watt per phase at standstill and 4% slip considering that the rotor equivalent induced emf per phase is 110 V at standstill.

Solution:

The equivalent circuit of a double-cage rotor is shown in Fig. 9.60.

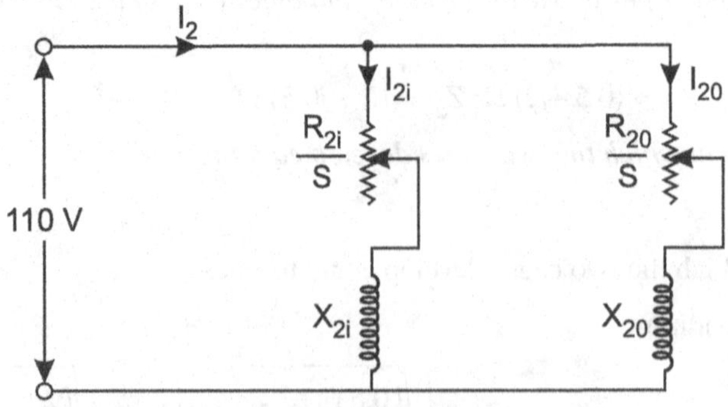

Fig. 9.60 Equivalent circuit as per data

Considering voltage as a reference phasor, $E_{2S} = 110 \pm j0 = 110 \angle 0°$

$$\overline{Z}_{2i} = (0.06 + j0.5)\ \Omega;\ \overline{Z}_{2o} = (0.6 + j0.12)\ \Omega$$

At standstill

$$\overline{Z}_{2i} = (0.06 + j0.5) = \sqrt{(0.06)^2 + (0.5)^2} \angle \tan^{-1}\frac{0.5}{0.06} = 0.5036\angle 92.4°$$

$$\overline{Z}_{2o} = (0.6 + j0.12) = \sqrt{(0.6)^2 + (0.12)^2} \angle \tan^{-1}\frac{0.12}{0.6} = 0.612\angle 12.57°$$

$$\overline{Z}_{2i} + \overline{Z}_{2o} = (0.06 + 0.6 + j0.5 + j0.12) = (0.66 + j0.62) = 0.906\angle 48°$$

Total impedance of the two cages

$$\overline{Z}_2 = \frac{\overline{Z}_{2i}\ \overline{Z}_{2o}}{(\overline{Z}_{2i} + \overline{Z}_{2o})} = \frac{(0.5036 \angle 92.4°)(0.612 \angle 12.57°)}{(0.906 \angle 48°)}$$

$$= 0.34 \angle 56.97° = (0.34 \cos\ 56.97° + j\ 0.34 \sin\ 56.93°)$$

$$= (0.2127 + j\ 0.2652)\text{ohm}$$

Rotor current at standstill

$$\overline{I}_{2S} = \frac{\overline{E}_{2S}}{\overline{Z}_2} = \frac{110 \angle 0°}{0.34 \angle 56.97°} = 323.5 \angle -56.97°$$

Synchronising power or torque developed per phase in the rotor at start

$$T_{syn} = I^2{}_{2S}\ R_{2S} = (323.5)^2 \times 0.2127 = \mathbf{22260\ syn.\ watt}\ (Ans)$$

At a slip of 4%

$$\overline{Z}_{2i} = \left(\frac{0.06}{0.04} + j0.5\right) = (1.5 + j0.5) = 1.58 \angle 20.48°$$

$$\overline{Z}_{2o} = \left(\frac{0.6}{0.04} + j0.12\right) = (15 + j0.12) = 15 \angle 0.5°$$

$$\overline{Z}_{2i} + \overline{Z}_{2o} = (1.5 + j0.5 + 15 + j0.12) = (16.5 + j0.62) = 16.51 \angle 2.39°$$

Total impedance of the two cages

$$\overline{Z}_2 = \frac{\overline{Z}_{2i}\ \overline{Z}_{2o}}{(\overline{Z}_{2i} + \overline{Z}_{2o})} = \frac{(1.58 \angle 20.48°)(15 \angle 0.5°)}{16.51 \angle 2.39°} = 1.435 \angle 18.59°$$

$$= 1.435 (\cos 18.59° + j \sin 18.59°) = (1.374 = j\ 0.413)\ \text{ohm}$$

Rotor current during running

$$\overline{I}_2 = \frac{S\overline{E}_{2S}}{\overline{Z}_2} = \frac{0.04 \times 110 \angle 0°}{1.435 \angle 18.59} = 3.07 \angle -18.59°$$

Synchronous power or torque developed per phase in the rotor,

$$T_{syn} = I_2^2\ R_2 = (3.07)^2 \times 1.374 = \mathbf{12.95\ syn.\ watt}\ (Ans)$$

9.47 Motor Enclosures

Various types of enclosure of the motor have been designed to protect the windings, bearings, and other mechanical parts from *moisture, chemicals, mechanical damage* and *abrasion from grit.*

NEMA (National Electrical Manufacturers Association) has suggested 7 Most Common Motor Enclosure. These types are defined by NEMA Standards.

Under the categories of open machines, totally enclosed machines and with sealed windings, more than 20 types of enclosures have been defined in NEMA standards MGl-1.25 to 1.27. Out of which **the 7 most common types of enclosures are:**

1. Open Drip Proof (ODP) Enclosure

An open Drip Proof (ODP) enclosure for a premium efficient super-E motor is shown in Fig. 9.61. This enclosure prevents the liquid drops to enter the motor when they fall with 15 degree angle from the vertical but it allows air to circulate through the internal winding for cooling. These enclosures are typically used for **indoor applications in relatively clean, dry locations**.

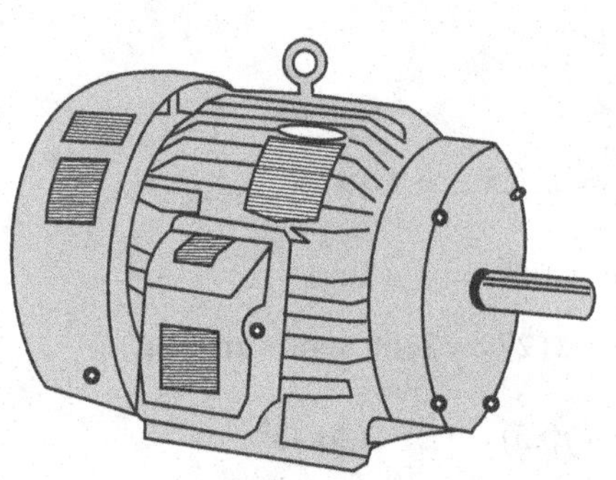

Fig. 9.61 Open Drip Proof (ODP) Enclosure

Fig. 9.62 Totally Enclosed Fan Cooled (TEFC) Enclosure

2. Totally Enclosed Fan Cooled (TEFC) Enclosure

The NEMA Premium Efficiency – Three Phase TEFC Enclosure Motor is shown in Fig. 9.62.

This enclosure prevents the free exchange of air between the inside and outside of the frame, but the frame is not made completely air-tight. To increase the cooling efficiency, a fan is attached to the motor shaft which circulates the air during operation. Ribs or fins are attached on the frame to increase the surface area which further improves cooling. These enclosures are more versatile and are used with motors for *pumps, fans, compressors, general industrial belt drive and direct connected equipment.*

3. Totally Enclosed Non-Ventilated (TENV) Enclosure

A1/2 HP, 3-Phase, 1140 RPM Motor with Totally Enclosed Non-Ventilated (TENV) Enclosure is shown in Fig. 9.63. It is almost similar to a TEFC enclosure but in these enclosures, fan is not employed and heat is dissipated only by convection. It is a tightly enclosed enclosure hence prevents free exchange of air. Although, no vent is provided in the enclosure but it is not air-tight

These are suitable for uses which are exposed to dirt or dampness, **but not very moist or hazardous (explosive) locations**.

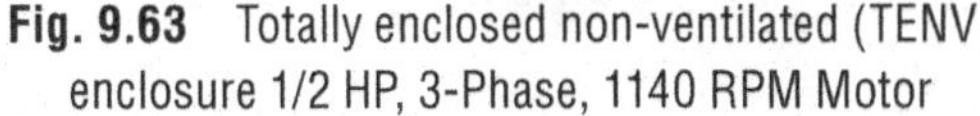

Fig. 9.63 Totally enclosed non-ventilated (TENV) enclosure 1/2 HP, 3-Phase, 1140 RPM Motor

Fig. 9.64 US Totally enclosed air over (TEAO) enclosure. 1/3 Hp, 1800 rpm, 24 VDC

4. Totally Enclosed Air Over (TEAO) Enclosure

A1/3 HP, 24 V DC, 1800 RPM Motor with Totally Enclosed Air Over (TEAO) Enclosure is shown in Fig. 9.64.

Dust-tight fan and blower duty motors designed for shaft mounted fans or belt driven fans. The motor must be mounted within the airflow of the fan.

5. Totally Enclosed Wash Down (TEWD) Enclosure

A Motor with Totally Enclosed Wash Down (TEAO) Enclosure is shown in Fig. 9.65.

The Washdown Duty Motors are used for food processing, packaging, pharmaceuticals, or in the applications where motors are regularly exposed to high pressure washdown. While designing these enclosures one must take care of high wash-down pressure, humidity and wet environment. Various enclosures have been designed for such (wet) conditions.

Fig. 9.65 Totally Enclosed Wash down (TEWD) motor

Fig. 9.66 Explosion-proof enclosures (EXPL) motor

6. Explosion-proof Enclosures (EXPL)

A Motor with Explosion-proof enclosures (EXPL) is shown in Fig. 9.66.

The Explosion-proof enclosure (EXPL) motors are used in chemical plants, the oil industry, in gas works, wood and plastic processing industry or in agriculture. Where there is hazardous environment, the enclosure of explosion proof motors is designed in such a way that it can withstand heavy pressures built inside the motor. However, if pressure is built-up more than the designed value that must be released through the safety valve and switch off the supply simultaneously.

These motors are designed for **specific hazardous purposes**, such as atmospheres containing gases or hazardous dusts. For safe operation, the maximum motor operating temperature must be kept below the ignition temperature of surrounding gases or vapours.

Explosion proof motors are designed, manufactured and tested for particular requirements.

7. Hazardous Location (HAZ) Enclosures

3-Phase, 5 HP Motor with Hazardous Location (HAZ) Enclosure is shown in Fig. 9.67. It meets with the requirements of National Electrical Code for Hazardous Locations.

Fig. 9.67 Explosion-proof enclosures (EXPL) motor

These motors are used to operate Power Fans, Blowers, Pumps or Air Compressors in the hazardous locations.

While selecting the enclosure and motor, the environment of the location is determined and accordingly the specific material is used for the enclosure so that the enclosure's strength may not deteriorate.

The format used to define this information is a class, group, division and temperature code structure as defined by the **National Electric Code (NFPA-70)**.

9.48 Standard Types of Squirrel Cage Motor

With the slight variation in design as per the need, the squirrel cage induction motors have been standardised. According to their electric characteristics and design, these are designated as design *A, B, C, D, E and F.*

The commercial simple squirrel-cage induction motor with shallow slots are designated as class *A*. These motors are considered to have normal slip and hence used as reference.

9.48.1 Class A Motors

Normal starting torque, normal starting current, normal slip.

Characteristics

- Have relatively normal (low) resistance and normal (low) reactance
- Rotor slots are semi-closed cut near the periphery to have small reactance.
- At rated voltage, the locked rotor current is generally more than 6 times the rated full-load current.
- For the machines having smaller sizes and number of poles, at rated voltage, the starting torque is twice the full-load torque
- For the machines having larger sizes and number of poles, at rated voltage, the starting torque is 1.1 times the full-load torque.
- Full-load slip is less than 5%.

Applications

These motors can pick-up low inertia loads very quickly and can be started and stopped frequently therefore, these are generally used for fans, pumps and conveyors etc.

9.48.2 Class B Motors

Normal starting torque, normal starting current, normal slip.

Characteristics

- Have deep and narrow bars (see Fig. 9.62) therefore offer high reactance at start.
- Can be started at full-load. Then it draws relatively low starting current but develops normal starting torque.
- Generally, locked-rotor current draws 5 to 6 times the full-load current at rated voltage..
- Full-load slip is less than 6%.

Applications

These motors can pick-up almost full-load while developing normal starting torque with relatively low starting current and best suited for compressors, heavy fans having high, machine tool applications centrifugal pumps etc.

9.48.3 Class C Motors

High starting torque, low starting current, normal slip.

Characteristics

- Have two separate cages and are called double squirrel-cage motors (see Fig. 9.63).
- Locked-rotor current with rated voltage is generally 5 to 6 times the full-load current as in case of class-B motors.
- Starting torque with rated voltage is usually 2.75 times the full-load torque.
- Full-load slip is normal less than 6%.

Applications

These motors develop high torque at start nearly 2.75 times the full-load torque with low starting current at rated voltage. Therefore, these are best suited to pick heavy loads at start and are generally used for crushers, compression pumps, large refrigerators, conveyors, textile industry, boring mills and wood working equipment.

9.48.4 Class D Motors

High starting torque, low starting current, high slip

Characteristics

- Have relatively high resistance to obtain high starting torque with low starting current.
- The bar conductors of the cage have relatively smaller area of cross-section to obtain high resistance. These conductors are located very near the periphery to have low leakage reactance.
- Locked-rotor current with rated voltage is generally 5 to 6 times the full-load current similar to class-*C*.
- Full-load slip varies from 5 to 20%, depending upon the application.

Applications

These motor have the ability to develop extremely high torque at start, therefore, these are best suited for shearing machines, pinching processes, stamping machines, bulldozers, laundry equipment and metal drawing equipment.

9.48.5 Class E Motors

Low starting torque, low starting current, low slip

Characteristics

- Have relatively very low resistance accordingly have low starting torque and low reactance.
- Locked-rotor current with normal rated voltage is generally 8 to 10 times the rated full-load current, for larger sizes it may be more than this even.
- Their power factor and efficiency is relatively high.

Applications

These motor have low resistance and operate at higher power factor and efficiency, therefore, best suited for continuous running such as exhaust fans.

9.48.6 Class F Motors

Low starting torque, low starting current, normal slip

Characteristics

- Have relatively low resistance and to reduce the starting current, rotor reactance is made very high by embedding the rotor bar conductors deep in the slots for increasing the leakage flux.
- Locked-rotor current with rated voltage is generally 5–6 times the rated full-load current.
- The starting torque with full voltage is only 1.25 times the full-load torque.
- Full-load slip less than 6% depending upon the load.

9.49 Advantages and Disadvantages of Induction Motors

9.49.1 Squirrel Cage Induction Motors

Advantages

1. Cost of construction is low comparatively.
2. These motors are rugged in construction and mechanically robust.
3. Copper losses in cage rotor are low hence, their efficiency in high.
4. There is no overhang of rotor winding hence, their length is smaller.
5. Slip-rungs and brushes are not used, therefore, their maintenance cost is low and there is no risk of sparking.
6. These can be cooled down comfortably by using large fans on the sides since more space is available.
7. These motors have almost constant speed, high over-load capacity and starting arrangement.

Disadvantages

1. These motors have low starting torque (1.5 to 2 times the full-load torque) with large starting current (5 to 9 times the full-load current). Their power factor is also low.
2. Speed cannot be controlled comfortably.
3. These motor are very sensitive to supply voltage fluctuations.
4. During starting, there is more energy loss in comparison to slip-ring induction motors. Therefore these motors are not used where frequent starting is required.

9.49.2 Slip-ring Induction Motors

Advantages

1. In these motors, external resistance can be added in the rotor circuit at start. Hence heavy starting torque at low starting current can be obtained. The maximum torque can be obtained at the start by making rotor resistance equal to standstill rotor reactance (i.e., $R_2 = X_{2S}$).
2. Speed of the motor can be controlled or adjusted as per need by simply inserting external resistance in the rotor circuit.

Disadvantages

1. Cost of construction is high comparatively.
2. Rotor copper losses are more in these motors and hence the efficiency is low in comparison to squirrel cage induction motors.
3. The size of these motors is larger due to overhang or rotor winding and slip-rings.
4. Repair and maintenance charges of these motors are more because of slip-rings and brushes.
5. Because of overhang of rotor winding and slip-rings larges fans cannot be accommodated which affects its cooling.

9.50 Applications of Three-phase Induction Motors

The applications of squirrel cage induction motors and slip-ring (phase wound) induction motors are given below:

1. **Squirrel cage induction motors:** These motors are mechanically robust and are operated almost at constant speed. These motors operate at high power factor and have high over load capacity. However, these motors have low starting torque. (i.e., these motors cannot pick-up heavy loads) and draw heavy current at start. On the bases of these characteristics, these motors are best suited for:

   (*i*) Printing machinery
   (*ii*) Flour mills
   (*iii*) Saw mills
   (*iv*) Shaft drives of small industries
   (*v*) Pumps
   (*vi*) Prime-movers with small generators etc.

2. **Slip-ring (or phase-wound) induction motors:** These motors have all the important characteristics (advantage) of squirrel cage induction motors and at the same time have the ability to pick-up heavy loads at start drawing smaller current from the mains. Accordingly these motors are best suited for;

   (*i*) Rolling mills
   (*ii*) Lifts and hoists
   (*iii*) Big flour mills
   (*iv*) Large pumps
   (*iv*) Line shafts of heavy industries
   (*vi*) Prime-moves with medium and large generators.

9.51 Comparison of Squirrel Cage and Phase Wound Induction Motors

For comparison of two types of induction motors, the same output is considered. The comparison given below is made on the basis of construction, cost, losses, maintenance, starting, performance etc.

Se. No.	*Particulars*	Squirrel Cage induction motor	*Phase Wound Induction motor*
1.	Construction	It is *simple in construction* and mechanically robust.	Its rotor is phase wound which *needs care.*
2.	Cost	It is *cheaper* in construction cost.	It is more *costly.*
3.	Maintenance	It requires *less maintenance.*	Because of slip rings, brushes etc. it requires *more maintenance.*
4.	Utilisation of slot space	*Space of rotor slots* is better utilised since rotor conductors are in shape of bars.	Whole of the *slots space* is not utilised since winding is placed in the slots.
5.	Coper losses	It has got *small copper losses* because of low rotor resistance.	It has got *more copper losses.*
6.	Efficiency	Its *efficiency* is *high.*	It has *lower efficiency.*
7.	Cooling	The heat can be dissipated more efficiently because the end rings are bare thus there is more space for providing a *cooling fan.*	In this case, overhang of the winding occupies space and less space is left for the provision of a good cooling fan. Thus *cooling* is not quite efficient.
8.	Starting torque	It develops very *poor starting torque* due to low resistance which cannot be increased by ordinary means.	In this case an external resistance can be added in the rotor circuit at the start to *improve its starting torque.*
9.	Starter	Starters are applied on stator side and *starting methods* are quite simple.	Starter is applied on rotor side through *sliprings and brush gears.*

9.52 Comparison between Induction Motor and Synchronous Motor

The comparison between induction motor and synchronous motor is given below:

Se. No.	*Particulars*	*Induction motor*	*Synchronous motor*
1.	Starting	It is a *self-starting* motor.	It is inherently *not a self-starting* motor.
2.	DC Excitation	Its basic principle is mutual induction and *no excitation* is required.	It *requires* DC for *field excitation.*
3.	Speed	Its speed is always *less than synchronous* speed and speed decreases with the increase in load.	It only runs at a *synchronous speed* and its speed is not affected by load. With the increase in load the torque angle δ increases.

Se. No.	Particulars	Induction motor	Synchronous motor
4.	Power factor	It runs at *lagging p.f.* which may become very low at light loads.	It can be operated under wide range of power factors *both lagging and leading* by changing its excitation.
5.	Speed control	Its *speed* can be *controlled.*	It runs only at *synchronous speed.* The only way to change its speed is to vary the supply frequency.
6.	Applications	Its *application* is limited to the supply of mechanical load.	It is also used to supply mechanical load and in addition it *improves the p.f.* of the system.
7.	Cost	It is very *cheap to manufacture* and mechanically robust.	It is *costlier* and complicated to manufacture.
8.	Maintenance	It requires *less maintenance.*	It requires *more maintenance.*

9.53 Installation of Induction Motors

An electrical engineer must have the general idea regarding the installation and preventive maintenance of induction motors. The general information regarding installation and preventive maintenance of induction motors is available in the sources like Indian Standards (*IS*: 900-1965), manufacturers manuals, reference books, hand books etc.

Installation of induction motors involves the following points:

1. Inspection of the motor on its arrival and its storage.
2. Section of location for the installation of motor and control gear.
3. Preparation of foundation and arrangement of levelling.
4. Checking for proper alignment.
5. Fitting of loading system (pulleys, couplings etc.)
6. Earthing etc.

1. Inspection of the Motor on its Arrival and its Storage

Unload the wooden crates containing motors with the help of cranes, chain and pulleys, ropes etc. Care should be taken that under no circumstances motor be damaged while unloading. After unloading, the motor is carried to the storage site. While handling, hook on the frame should be used for its lifting.

Inspection and storage: If there is visible sign of damage on the crate, open delivery from carrier be taken and loss/damage certificate be got signed. Otherwise open the crate by keeping the motor in upright position with the help of screw drivers, main pullers, pliers, hammers etc. Care should be taken that the motor may not be damaged while opening the crate. On opening the crate, the material should be checked by comparing it with the packing test. For any type of loss or damage during transit the supplier and the insurance company should be informed immediately.

The motor should be stored in a clean and dry place. After storing, the rotor should be rotated 90° and relocked periodically (say every fortnight). If motor is stored at locations susceptible to vibrations, rubber pads are provided underneath to avoid stationary indentation.

2. Selection of Location for the Installation of Motor and Control Gear

The following points may be considered while selecting a location for the installation of moto:

(*i*) The motor should be installed at a place, where sufficient space around the motor is available. It will facilitate the movement of operating personal.

(*ii*) It should be installed at a place where it is easily accessible for carrying out repair and maintenance. Sometimes, the lifting arrangement is also provided with the structure at the building.

(*iii*) The motor and control gear should be installed at a place so that they may not be exposed to water, corrosive liquid, oil, steam, metallic dust, dirt etc. However, if it cannot be avoided, the motor control gears must be properly enclosed.

3. Preparation of Foundation and Arrangement of Leveling

Depending upon the size of the motor, a strong and rigid cement concrete foundation is provided. The foundation block should be at least 20 cm longer and 20 cm wider than the motor feet. The anchor or rag bolts which are to hold the motor are embedded in concrete. Their location is determined with great accuracy.

Where the motor rests directly on foundation, a great care is needed in leveling the foundation otherwise the motor will be strained when the mounting bolts are tightened. After leveling, the motor is also carefully aligned with the driven equipment to avoid extra load or strain on the motor.

4. Checking for Proper Alignment

The motor should be properly aligned with the driven equipment to avoid undue loading. The alignment is generally checked by free rotation of the motor with or without directly coupled load. Proper alignment for direct coupled drives is achieved by ensuring three operations i.e., axial positioning of shaft, paralleling of shaft and centering at shaft.

5. Fitting of Loading System

The loading system (pulleys, coupling etc.) is fitted on the motor shaft and is properly aligned so that the motor may not be unduly overloaded.

6. Earthing

Each motor frame is connected at two places with two distinct earthing systems. The earth wire should have proper size as per the size of the motor.

9.54 Preventive Maintenance of Three-phase Induction Motors

It is always better to prevent faults to occur on the motor than to repair it afterwards. This can only be achieved by giving periodic attention to the motor. The periodic attention prevents the faults to occur and provides trouble free service for long time.

Some of the important aspects of maintenance of an induction motor are given below:

(*i*) The motor frame or outer body should be repainted periodically.
(*ii*) Motor should be cleaned periodically with the help of blowers and vacuum cleaners.
(*iii*) Motor should be kept dry. The stator and rotor winding should be kept free from oil, grease, dampness, dirt, smoke etc.
(*iv*) Terminals should be kept clean and tight to maintain good connections.
(*v*) Air-gap between stator and rotor should be kept clean i.e., free from any accumulation of dirt.
(*vi*) Bearings should be greased in case they make noise. Every three years, bearings should be washed with petrol and new grease of good quality be filled.
(*vii*) Dust, oil, moisture etc., should not be allowed to accumulate on the surface of slip rings and brush-gear.
(*viii*) Spring tension and brush pressure on the slip rings should be checked periodically.
(*ix*) Insulation of winding should be tested periodically.
(*x*) To verify the effectiveness of earthing, the earth resistance should be measured periodically.
(*xi*) The control and switching devices should be cleaned and checked periodically.

Recommended Maintenance Schedule

Refer IS: 900-1965

1. **Daily Maintenance**
   (*i*) Check lead connections and earth connections visually.
   (*ii*) Check motor winding temperature just by touching the outer body. It should be such so that hand can bear it.
   (*iii*) Check the control equipment.
2. **Weekly Maintenance**
   (*i*) Blow out dust with the help of blowers if the motor is situated in dust locations.
   (*ii*) Check the contacts of the starter where motor is started and stopped frequently.
   (*iii*) Check lubricating oil in case of oil-ring lubricated bearings for contamination of dust, dirt etc.
3. **Monthly Maintenance**
   (*i*) Inspect and clean the control and switch gear system.
   (*ii*) Renew oil in high-speed bearings in damp and dusty locations.
   (*iii*) Check and clean the brushes and brush-holder of slip-ring motors.
4. **Half-Yearly Maintenance**
   (*i*) Clean windings of motors subjected to corrosive and damp locations. Dip it in varnish and bake it if necessary.
   (*ii*) Check and clean the slip-rings and brush gear in case of slip-ring induction motors.
   (*iii*) Check grease in ball and roller bearings and fill grease if necessary but avoid overfilling.
   (*iv*) In case of oil-ring lubricating motors, drain-off the oil, clean the bearing with petrol and refill with clean and fresh oil.

5. Yearly Maintenance

(*i*) Clean the motor winding by blowing clean and dry air.
(*ii*) Clean and varnish dirty and oily windings. After varnishing do baking.
(*iii*) Check all high-speed bearings and replace if necessary.
(*iv*) Check oil and renew if necessary.
(*v*) Replace the switch and fuse contacts, if damaged.
(*vi*) Check insulation resistance of the winding.
(*vii*) Check resistance of earth connections.
(*viii*) Check air-gap between stator and rotor.
(*ix*) Check roller and ball bearings, if giving noise replace them.

Section Practice Problems

Numerical Problems

1. The standstill impedances per phase of the inner-cage and outer cage of a double squirrel cage induction motor are given as

$$Z_{2i} = (0.4 + j2) \text{ ohm and } Z_{2o} = (2 + j0.4)\text{ohm}$$

Determine the ratio of torques produced by the two cages

(*i*) At standstill (*ii*) at a slip of 5%. (***Ans.*** T_0; T_i;: 5:1; T_0:T_i::0.21:1)

2. The standstill impedances per phase of the inner and outer cage of a double cage induction motor is given as

$$Z_{2i} = (0.5 + j3.5) \text{ ohm and } Z_{2o} = (2 + j1.2)\text{ohm}$$

Determine the slip at which the two cages develop equal torque. (**Ans.** 25.1%)

3. The impedances per phase of a double cage induction motor are inner cage: (0.05 + *j*0.4) ohm; Outer-cage: (0.5 + *j*0.1) ohm Estimate the torque in synchronous watt per phase at standstill and 5% slip considering that the rotor equivalent induced emf per phase is 100V at standstill.

(***Ans.*** 26000 syn.watt; 22.47 syn.watt)

Short Answer Type Questions

Q.1. What do you mean by cogging in 3-phase induction motors?

Ans. The phenomenon by which the radial alignment forces exceed the tangential accelerating forces and the machine refuses to start is called *cogging*.

Q.2. What do you mean by crawling in 3-phase induction motors?

Ans. When a 3-phase induction motor continues to rotate at a speed little lower than 1/7th of its synchronous speed, then it is said to be crawling.

Q.3. How load affects the speed of an induction motor.

Ans. With the increase in load, the speed of an induction motor decreases slightly. However, in general, it is considered as a constant speed motor.

Q.4. What is the effect of load on the efficiency of an induction motor?

Ans. The speed or slip of an induction motor changes with the change in load. The efficiency of an induction motor is maximum at a slip when $S = R_2/X_{2S}$. the load at which $S = R_2/X_{2S}$, the efficiency is maximum. The value of slip decreases when the load on the motor is more or less than the given condition (i.e., $R_2 = S \times X_{2S}$).

Q.5. Which are the high starting torque cage motors?

Ans. (*i*) Deep bar cage rotor motors and (*ii*) double cage induction motors.

Q.6. Name seven-common enclosures of induction motor suggested by NEMA

Ans. These are

1. Open drip proof (*ODP*),
2. Totally enclosed fan cooled (*TEFC*),
3. Totally enclosed non-ventilated (*TENV*),
4. Totally Enclosed Air Over (*TEAO*),
5. Totally Enclosed Wash Down (*TEWD*),
6. Explosion-proof (*EXPL*) and
7. Hazardous Location (*HAZ*) enclosure.

Q.7. State three major advantages of squirrel cage induction motor over phase-wound induction motor.

Ans. 1. Cheaper is cost 2. Rugged is construction and 3. Low maintenance.

Q.8. What is the major difference between a synchronous motor and an induction motor?

Ans. Synchronous motor runs only at synchronous speed irrespective of the load. Induction motor can never run at the synchronous speed, it runs at a speed slightly less than the synchronous speed.

Review Questions

1. Which is the most commonly used 3-phase motor in the industry? Justify its name.
2. Differentiate between squirrel cage and phase wound rotor construction with the help of sketches.
3. Explain the construction of a 3-phase squirrel cage induction motor.
4. Explain, with the help of sketches, the construction of a 3-phase phase-wound induction motor.
5. Why the rotor of an induction motor is skewed?
6. Discuss how a rotating field is produced in a 3-phase induction motor. How does the rotating field help in the production of torque.
7. Explain the working of a 3-phase induction motor.
8. Explain with the help of suitable diagram, how rotating magnetic field is developed in a 3-phase induction motor when 3-phase AC supply is given to it.
9. Explain with neat sketches, the principle of operation of a 3-phase induction motor. Explain clearly how torque develops.
10. Can induction motor (3-phase) run at synchronous speed? Explain your answer.
11. How much torque is developed in an induction motor at synchronous speed? What do you understand by slip? What is the normal slip of an induction motor? How can you reverse the direction of rotation of a 3-phase induction motor.

12. Explain why the rotor of a squirrel cage induction motor rotates in the same direction as that of the stator revolving field. Why this motor does not operate at synchronous speed?
13. The stator winding of a 3-phase, phase-wound induction motor may be connected either in delta or star but the rotor winding is always connected in star, why? How does an induction motor adjust its current with the change in mechanical load?
14. What do you mean by slip, slip frequency, cage rotor and wound rotor?
15. Derive a relation between rotor current frequency and supply frequency.
16. Derive the relationship between the frequency of rotor currents and supply frequency in case of a 3-phase induction motor.
17. Derive an expression for the torque developed in a 3-phase induction motor and the condition for maximum torque.
18. Derive an expression for an induction motor for (*i*) full-load torque (*ii*) starting torque (*iii*) maximum torque.
19. Develop a relation between (*i*) Starting torque and full-load torque (*ii*) full-load torque and maximum torque and (*iii*) starting torque and maximum torque.
20. If some resistance is added in the rotor circuit of an induction motor, how does it affect the starting torque and maximum torque.
21. What is the effect on the starting and maximum torque of an induction motor if a resistance is inserted in its rotor circuit? What should be the maximum limit of additional resistance?
22. Derive the condition for maximum torque in a 3-phase induction motor under normal operation.
23. Draw a torque-speed characteristic of a 3-phase induction motor, mark the operating region on it and show the effect of change in rotor resistance.
24. Derive an expression for full-load torque in a 3-phase induction motor. Also obtain relation between full-load torque and maximum torque i.e., T_{fe}/T_{max}.
25. For a 3-phase induction motor, derive a relation between maximum torque and full-load torque.
26. For a 3-phase induction motor, derive a relation between starting torque and maximum torque.
27. List the different types of losses in an induction motor.
28. What are the various losses in an induction motor? On what factors do they depend?
29. Derive an expression for the rotor copper loss in terms of slip and input to the rotor.
30. Show that the ratio of rotor speed to synchronous speed is equal to the ratio of rotor output to rotor input. Also obtain an expression for rotor copper loss in terms of rotor input.
31. Explain the term air-gap power or power developed in the rotor and shaft power. How these terms are related to each other?
32. Draw the equivalent circuit and phasor diagram of a 3-phase induction motor.
33. Develop the equivalent circuit of a 3-phase induction motor. Explain how the mechanical power developed in the motor is taken care in this circuit.
34. Derive an exact equivalent circuit of a 3-phase induction motor. How it is converted to simplified equivalent circuit. From simplified equivalent circuit find rotor output, power developed and slip.

35. Describe how no-load test and blocked rotor test is performed on a 3-phase induction motor.

36. Describe the test by which one can determine the parameters of a 3-phase induction motor.

37. Draw the equivalent circuit of a 3-phase induction motor operating under following conditions:
(*i*) Blocked rotor condition. (*ii*) light load condition. (*ii*) Rated load condition.

38. Draw the equivalent circuit of a 3-phase induction motor. What is significance of each element, haw their values can be determined in electrical laboratories.

39. What is a circle diagram of a 3-phase induction motor? Draw and explain the circle diagram. What experimental test are to be performed on an induction motor to draw the circle diagram.

40. How will you determine the motor characteristics from the circle diagram of a 3-phase induction motor?

41. What are the space harmonic fields? How they affect the performance of a 3-phase induction motor? What design construction can minimise these effects?

42. Explain the phenomenon of cogging and crawling is squirrel cage induction motors.

43. What do you mean by deep bar cage construction? How starting torque is improved by using this construction?

44. Describe the construction and working of a 3-phase double squirrel cage induction motor.

45. Describe the main features of a double cage induction motor.

46. State the difference between squirrel cage rotor and wound rotor type of induction motors.

47. Compare relative advantages of cage rotor type induction motor and wound rotor type induction motor.

48. Give a list of advantages and disadvantages of 3-phase induction motors.

49. Compare cage and wound type induction motors.

50. Compare a cage induction motor with a slip-ring induction motor with reference to construction, performance and applications point of view.

51. Make a comparison between 3-phase induction motor and 3-phase synchronous motor.

Multiple Choice Questions

1. Stator core of an induction motor is built of
(*a*) laminated cast iron
(*b*) mild steel
(*c*) silicon steel stampings
(*d*) soft wood.

2. The stator winding of an induction motor can be designed for
(*a*) any number of pole
(*b*) any even number of poles
(*c*) any odd number of poles
(*d*) only for four poles.

3. The true statement associated with a 3-phase induction motor is.
(a) the cage rotor is made of copper.
(b) 3-phase AC supply is usually connected to squirrel cage rotor.
(c) the rotating field is produced by the stator winding.
(d) the rotor stampings should be properly insulated from each other.

4. Usually, in induction motors, die-cast aluminium rotors are used because aluminium is
(a) easy to cast owing to its low melting point and is easily available.
(b) lighter in weight.
(c) of low resistivity.
(d) having less cost.

5. The rotors of squirrel cage induction motors are provided with blades to
(a) provide cooling to motor
(b) balance the rotor dynamically.
(c) eliminate noise.
(d) eliminate harmonics.

6. The rotor conductors of 3-phase squirrel cage induction motors are
(a) short-circuited through end rings.
(b) short-circuited through slip rings.
(c) kept open.
(d) short circuited through resistors.

7. A wound rotor induction motor can be distinguished from squirrel cage induction motor by
(a) frame's structure.
(b) presence of slip-rings
(c) shaft diameter.
(d) all of the above.

8. In a large induction motor usually the value of full load slip is
(*a*) 0 · 4%
(*b*) 20%
(*c*) 3 to 5%
(*d*) 6 to 15%

9. Uneven air gap in an induction motor may cause
(a) heating of motor.
(b) unbalancing of motor shaft.
(c) over loading.
(d) all of these.

10. The stator and rotor of a 3-phase 6-pole squirrel cage induction motor has 54 and 42 slots respectively. The number of phases in the rotor is
(a) 1.
(b) 3.
(c) 6.
(d) 9.

11. At start, the slip of the induction motor is
(*a*) zero
(*b*) 0 · 5
(*c*) one
(*d*) infinite.

12. In squirrel-cage induction motors the rotor slots are slightly skewed to.
(a) increase the mechanical strength of rotor bars.
(b) reduce the cost of construction
(c) reduce the magnetic hum and locking tendency of rotor.
(d) provide balancing.

13. In a 3-phase induction motor, skewing of rotor slots is done to
(a) increase mechanical strength of rotor.
(b) reduce vibration and noise.
(c) improve motor efficiency.
(d) decrease rotor resistance and save copper.

14. The rotor of an induction motor never pick-up the speed up to synchronous speed, if it would run at synchronous speed then the relative speed between rotor conductors and the rotating field will be.
(a) zero and hence, torque will be maximum.
(b) zero and hence, torque will be zero.
(c) synchronous speed and hence, torque will be maximum.
(d) synchronous speed and hence, torque will be zero.

15. The direction of rotation of a 3-phase induction motor is clockwise when it is supplied with phase sequence R-Y-B. If its direction of rotation is to be reversed. The phase sequence of the power supply should be
(a) Y-R-B (b) R-B-Y
(c) B-Y-R (d) all of these

16. A squirrel cage induction motor is, in general analogous to
(a) auto-transformer.
(b) two winding transformer with secondary open-circuited.
(c) two winding transformer with secondary short-circuited.
(d) none of the above.

17. The stator and rotor of an induction motor behave like
(a) an ordinary two winding transformer.
(b) an auto-transformer.
(c) a variable voltage constant frequency transformer
(d) constant voltage variable frequency transformer.

18. The full load slip of a 60 Hz, 6-pole squirrel cage induction motor is 4%. Its full load speed is
(a) 960 rpm. (b) 1152 rpm.
(c) 1140 rpm. (d) 950 rpm.

19. The synchronous speed of an induction motor is 1000 rpm. What will be slip when it is running at a speed of 960 rpm?
(a) +5%. (b) –5%
(c) +4% (d) –3%. [A.M.I.E]

20. A 6-pole, 3-phase induction motor is running at 5 percent slip at full load. If the speed of the motor is 1140 rpm, the supply frequency is
(a) 30 Hz. (b) 25 Hz.
(c) 50 Hz. (d) 60 Hz.

21. A 4-pole, 3-phase alternator running at 1500 rpm supplies to a 6-pole, 3-phase induction motor which has a rotor current of frequency 2 Hz. The speed of the motor is
(a) 1440 rpm. (b) 960 rpm.
(c) 840 rpm. (d) 720 rpm.

22. If E_{2S} is the standstill rotor phase emf, I_2 is the standstill rotor phase current and cos ϕ_2 is the rotor power factor then torque developed by a 3-phase induction motor varies as
(a) $E_2 I_2$. (b) $E_{2S} I_2/\sin \phi_2$
(c) $E_2 I_2 \cos \phi_2$. (d) $E_{2S} I_2 \sin \phi_2$

23. If the rotor resistance is increased in a slip-ring induction motor, then
(a) both starting torque and pf will increase. (b) both starting torque and pf will decrease.
(c) starting torque decreases but pf increases. (d) starting torque increases but pf decreases.

24. In a 3-phase induction motor, maximum torque is developed when the rotor circuit resistance per phase is equal to
(a) rotor leakage reactance per phase at standstill.
(b) stator leakage reactance per phase.
(c) slip times the rotor leakage reactance per phase at standstill.
(d) starting current will increase but starting torque decrease.

25. In a 3-phase slipring induction motor, if some resistance is added in the rotor circuit.
(*a*) its starting torque will decrease and maximum torque will increase.
(*b*) its both starting torque and maximum torque will increase
(*c*) its starting torque will increase but the maximum torque will remain the same.
(*d*) its starting torque will remain the same but maximum torque will increase.

26. In a 3-phase induction motor if the leakage reactance is increased by using deep slots
(a) starting torque and starting current will decrease but power factor will increase.
(b) starting torque and starting current both will increase but power factor will decrease.
(c) pull-out torque will decrease.
(d) starting current will increase but starting torque decrease.

27. Breakdown torque of a 3-phase induction motor is
(a) inversely proportional to the rotor resistance.
(b) directly proportional to rotor resistance.
(c) inversely proportional to the rotor leakage reactance
(d) directly proportional to the rotor reactance.

28. The power input in blocked-rotor test performed on a 3-phase induction motor is approximately equal to
(*a*) iron loss in the core.
(*b*) hysteresis loss in the core
(*c*) eddy current loss in the core
(*d*) copper loss in the windings

29. In a 3-phase induction motor iron loss mainly occurs in
(a) rotor core and rotor teeth.
(b) stator and rotor core.
(c) stator and rotor winding.
(d) stator core and rotor teeth.

30. The power input to a 400 V, 50 Hz, 4-pole, 3-phase induction motor running at 1440 rpm is 40 kW. The stator losses are 1 kW, the rotor copper loss will be
(a) 1600 W.
(b) 1560 W.
(c) 1500 W.
(d) 1440 W.

31. The torque developed by an induction motor is
(*a*) directly proportional to the square of the rotor resistance
(*b*) directly proportional to the square of the rotor reactance
(*c*) directly proportional to the square of the supply voltage
(*d*) directly proportional to the square of the slip.

32. A 3-phase, 400 V, 50 Hz, 4 pole induction motor cannot run at 1500 rpm because
(*a*) at this speed motor will draw such a heavy current which may damage the motor.
(*b*) at this speed motor bearings may be damaged.
(*c*) at this speed, emf will not be induced in the rotor circuit and hence no torque will be developed.
(*d*) all of these.

33. A 400 V, 12 kW, 4-pole, 50 Hz induction motor has full-load slip of 5%. Its torque at full-load will be
(a) 8.04 Nm.
(b) 80.4 N.
(c) 7.64 Nm.
(d) 76.4 Nm.

34. The equivalent circuit per phase of a three phase induction motor is similar to that of a three phase transformer but the transformer does not develop any torque. It is because of
(a) insufficient power input.
(b) insufficient supply frequency.
(c) insufficient voltage.
(d) non-fulfilment of winding placement conditions.

35. Squirrel cage induction motors will have low pf at
(a) heavy loads only.
(b) light loads only.
(c) both light and heavy loads.
(d) rate load only.

36. Torque developed by a 3-phase, 400 V, induction motor is 200 Nm. If the supply voltage is reduced to 200 V, the developed torque will be
(*a*) 100 Nm
(*b*) 50 Nm
(*c*) 75 Nm
(*d*) 200 Nm

37. Under locked rotor condition, the rotor circuit frequency of a 3-phase induction motor connected to a 50 Hz supply will be
(a) 3 Hz.
(b) 30 Hz.
(c) 50 Hz.
(d) 60 Hz.

38. While performing no-load test on a 3-phase induction motor, a curve is plotted between input power and applied voltage. When this curve is extended backward to intersect the y-axis. This intersection point yields
(a) core loss.
(b) stator iron loss.
(c) stator copper loss.
(d) friction and windage loss.

39. The phenomenon in squirrel cage induction motors due to which they show a tendency to run at a very low speed in known as
(*a*) crawling
(*b*) cogging
(*c*) skewing
(*d*) humming

40. The crawling in an induction motor is caused by
(*a*) improper design of the machine
(*b*) low supply voltage
(*b*) both *a* and *b*
(*d*) harmonics developed in the motor.

41. In an induction motor crawling may occur due to
(*a*) harmonic synchronous torques.
(*b*) slip torques.
(*c*) vibration torques.
(*d*) all of these.

42. The crawling in an induction motor is caused due to
(*a*) improper design of stator laminations.
(*b*) improper winding design.
(*c*) harmonics developed in motor.
(*d*) low supply voltage.

43. When the number of stator slots of an induction motor is equal to an integral multiple of rotor slots.
(*a*) a high starting torque will be available.
(*b*) the maximum torque will be high.
(*c*) the machine will fail to start.
(*d*) the motor picks-up speed quickly.

44. The crawling in a 3-phase induction motor may be due to
(*a*) 7th space harmonics of air-gap field.
(*b*) 7th time harmonics of voltage wave.
(*c*) 5th space harmonics.
(*d*) 5th times harmonics.

45. If the air-gap in an induction motor is increased, its
(*a*) speed will reduce.
(*b*) power factor will be lowered.
(*c*) efficiency will improve.
(*d*) breakdown torque will reduce.

46. In induction motors cogging occurs when
(*a*) number of stator teeth are more than number of rotor teeth.
(*b*) number of stator teeth are less than number of rotor teeth.
(*c*) number of stator teeth are equal to number of rotor teeth.
(*d*) number of stator teeth are double to that of number of rotor teeth.

47. In induction motors semi-closed slots are used essentially to
(*a*) improve maximum torque developed.
(*b*) improve pull-out torque.
(*c*) improve efficiency.
(*d*) improve power factor and reduce magnetising current.

48. In double cage rotors
(*a*) the inner cage has high resistance and outer cage has low resistance.
(*b*) both the cages have low resistances.
(*c*) both the cages have high resistances.
(*d*) the inner cage has low resistance and outer cage has high resistance.

49. The most widely used motor is
(a) double cage induction motor.
(*b*) slip-ring induction motor.
(*c*) squirrel cage induction motor.
(*d*) synchronous motor.

Keys to Multiple Choice Questions

1. c	**2.** b	**3.** c	**4.** a	**5.** a	**6.** a	**7.** b	**8.** c	**9.** a	**10.** b
11. c	**12.** c	**13.** b	**14.** b	**15.** d	**16.** c	**17.** a	**18.** b	**19.** c	**20.** d
21. b	**22.** c	**23.** a	**24.** c	**25.** c	**26.** b	**27.** c	**28.** d	**29.** d	**30.** b
31. c	**32.** c	**33.** b	**34.** d	**35.** b	**36.** b	**37.** c	**38.** d	**39.** b	**40.** d
41. a	**42.** c	**43.** c	**44.** a	**45.** b	**46.** c	**47.** d	**48.** d	**49.** c	

Starting Methods and Speed Control of Three-Phase Induction Motors

CHAPTER 10

Chapter Objectives

After the completion of this unit, students/readers will be able to understand:

- ✓ Why a starter is employed to start an induction motor?
- ✓ Where we can apply a direct-on-line starter?
- ✓ What are the functions of over-current and no-volt relays?
- ✓ What is a stator resistance starter and what are its limitations?
- ✓ What is a star-delta starter and why it is not used with heavy (more than 20 kW) induction motor?
- ✓ How an auto-transformer starter works and why it is more suitable for heavy motors?
- ✓ Why rotor resistance starting method is preferred over stator starting methods.
- ✓ How step resistors are designed for a rotor resistance starter?
- ✓ What are the factors on which speed of an induction motor depends?
- ✓ How can we change the speed of squirrel cage and slip-ring induction motors?
- ✓ What are the merits and limitations of various speed control methods.
- ✓ How speed of heavy slip-ring induction motors is controlled?

Introduction

In the previous chapter, it has been found that a 3-phase induction motor is a self-starting motor. When a 3-phase supply is given to the 3-phase wound stator of an induction motor, a rotating field is developed. By induction, an emf is induced in the rotor conductors, torque develops and rotor starts rotating. But, if the motor is directly switched on to the supply, it draws heavy current from the mains due to inertia of rotor. Thus, a device is required to limit the inrush flow of current at the start, called starter.

Once the motor picks-up speed and takes-up the load, there are many applications where regulated speed is required. To fulfil this requirement various means and methods have been devised.

In this chapter, we shall discuss various methods and devices which are used for starting and speed control of 3-phase induction motors.

10.1 Necessity of a Starter

The current drawn by a motor from the mains, depends upon the rotor current. The rotor current under running condition is given by the expression:

$$I_2 = \frac{SE_{2s}}{\sqrt{R_2^2 + (SX_{2s})^2}}$$

At start slip $S = 1$, therefore, rotor current.

$$I_{2s} = \frac{E_{2s}}{\sqrt{R_2^2 + X_{2s}^2}}$$

This current is very large as compared to its full load current. Thus, when a squirrel cage induction motor is directly connected to the supply mains, it draws very large current (nearly 5 to 7 times of the full load current) from the mains. This heavy current may not be dangerous for the motor because it occurs for a short duration of time, but it causes the following affects:

(*i*) It produces large voltage drop in the distribution lines and thus affects the voltage regulation of the supply system.

(*ii*) It adversely affects the other motors and loads connected to the same lines.

Hence it is not advisable to start large capacity induction motors by direct switching. Rather, such motors should be started by means of some starting device known as *starter.*

The function of a starter is to limit the initial rush of current to a predetermined value.

A starter also has some protective devices to protect the induction motors against over loading.

10.2 Starting Methods of Squirrel Cage Induction Motors

The various starters which are employed to restrict the initial rush of current in squirrel cage induction motors are given below:

1. Direct On Line (D.O.L.) Starter; 2. Primary resistance (or inductance) starter; 3. Star/Delta Starter; 4. Auto-transformer Starter.

10.2.1 Direct on Line (D.O.L.) Starter

It is a starter by which the motor is switched *ON* direct to the supply mains by switching conductor. With normal industrial motors this operation results in a heavy rush of current of the order of five to seven times of the normal full load current. This high current rapidly decreases as the motor picks up speed but it is at a very low power factor and thus tends to disturb the voltage of the supply in the distribution lines. For this reason, the supply authorities limit the size of motor upto 5 H.P. which can be started by this starter. An automatic *D.O.L.* Starter is shown in Fig. 10.1.

A direct on line starter essentially consists of a contactor having four normally open (*N.O.*) contacts and a contactor coil also known as no-volt coil or no volt release. There are two push buttons *ON* and *OFF* which are used to start and stop the motor. To protect motor against overload, thermal or magnetic over-load coils are connected in each phase.

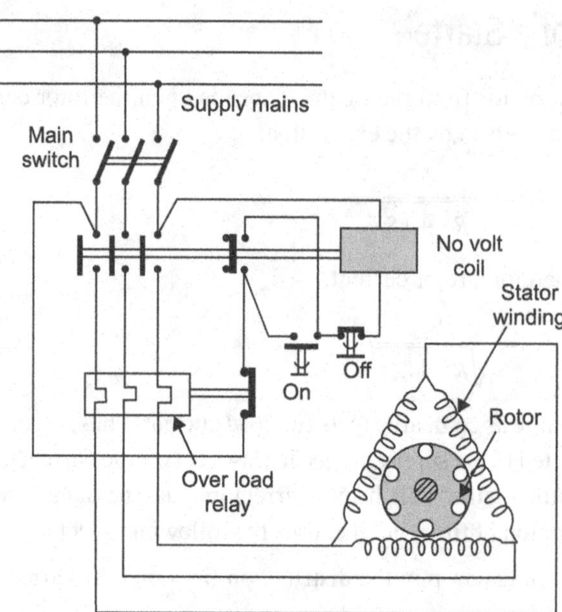

Fig. 10.1 Direct-on-line starter

To start the motor, the *ON* push button (green) is pressed which energies the no-volt coil by connecting it across two phases. The no-volt coil pulls its plunger in such a direction that all the normally open (*NO*) contacts are closed and motor is connected across supply through three contacts. The fourth contact serves as a hold on contact which keeps the no-volt coil circuit closed even after the *ON* push button is released. To stop the motor, *OFF* push button (red) is pressed momentarily which de-energises the no volt coil opening the main contacts.

When the motor is over loaded, the thermal overload relay contact, connected in the control circuit opens thus disconnecting the No-volt relay from the supply. Overload protection is achieved by thermal element overload relay.

Torque developed by the motor when started by direct on line starter:

Power developed in the rotor or rotor input = ωT

$$\text{Rotor copper loss} = S \times \text{rotor input}$$

$$= S\omega T \quad \ldots(i)$$

$$\text{Also rotor copper loss} = 3\, I_2^{\,2}\, R_2 \quad \ldots(ii)$$

Where S is the slip, ω is the angular velocity, T is the torque developed, I_2 is the rotor current per phase and R_2 is the rotor resistance per phase.

Equating the eqn. (*i*) and (*ii*), we get,

$$S\omega T = 3\, I_2^{\,2}\, R_2$$

$$T = \frac{3\, I_2^{\,2}\, R_2}{S\omega}$$

i.e., $T \propto \frac{I_2^{\,2}}{S}$ since rotor resistance and supply frequency is constant.

As stator current is proportional to rotor current,

i.e.,
$$T \propto \frac{I_1^2}{\omega}$$

or
$$T = K\frac{I_1^2}{\omega} \qquad \text{(where } K \text{ is a constant)}$$

At start, slip, $S = 1$

$\therefore$ Starting torque, $T_{st} = K(I_{st})^2$ (where I_{St} is the starting current)...(*iii*)

Full load torque, $T_{fl} = \dfrac{KI_{Sl}^2}{S_{fl}}$ ...(*iv*)

Where I_{fl} is the full-load current and S_{fl} is the full-load slip.

From eqn. (*iii*) and (*iv*), we get

$$\frac{T_{st}}{T_{fl}} = \frac{KI_{st}^2}{KI_{fl}^2 / S_{fl}} = \left(\frac{I_{st}}{I_{fl}}\right)^2 S_{fl}$$

or
$$\text{Starting torque, } T_{st} = T_{fl} \times \left(\frac{I_{st}}{I_{fl}}\right)^2 \times S_{fl}$$

When motor is connected to the mains by direct on line starter, the starting current of the motor will be equal to the short-circuit current I_{sc}

$\therefore$
$$\text{Starting torque, } T_{st} = T_{fl} \times \left(\frac{I_{st}}{I_{fl}}\right)^2 \times S_{fl} \qquad \text{...(v)}$$

10.2.2 Stator Resistance (or Reactance) Starter

In this method, a variable resistor (or inductor) is connected in series with each phase of the stator winding of a 3-phase squirrel cage induction motor, as shown in Fig. 10.2.

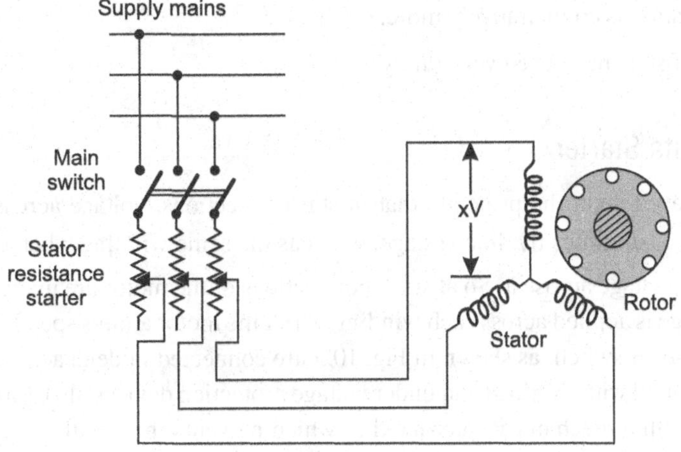

Fig. 10.2 Stator resistance starter

Very low voltage is supplied to each phase of the winding at the start by inserting more resistance (or inductance) in series with the winding. As the motor picks-up the speed, the resistance (or inductance) is taken out of circuit.

Let the voltage applied across each phase be reduced to a fraction of x

Voltage applied across each phase at start = xV

$$\text{Starting current, } I_{St} = x\, I_{SC}$$

Hence, the current is reduced to fraction x.

$$\text{Starting torque, } T_{st} = T_{fl} \times \left(\frac{I_{st}}{I_{fl}}\right)^2 \times S_f$$

$$= T_{fl} \times \left(\frac{x\, I_{sc}}{I_{fl}}\right)^2 \times S_f$$

$$= x^2\, T_{fl} \left(\frac{I_{sc}}{I_{fl}}\right)^2 \times S_f$$

Starting torque = x^2 × torque developed by direct switching.

Merits

(*i*) Smooth acceleration since the resistance (or inductance) is reduced gradually.
(*ii*) Simple and less expansive
(*iii*) Higher pf during starting

Demerits

(*i*) Current is reduced by a fraction of x but the torque is reduced to fraction x^2, hence starting torque is poor.
(*ii*) More energy loss in the resistors, therefore, inductors are preferred. But still loss is more.
(*iii*) Starting duration is comparatively more.

This method of starting is used very rarely.

10.2.3 Star-Delta Starter

This method is based upon the principle that in star connections, voltage across each winding is phase voltage i.e., $1/\sqrt{3}$ times the line voltage, whereas the same winding when connected in delta will have full line voltage across it. So at start, connections of the motor are made in star fashion so that reduced voltage is applied across each winding. After the motor attains speed the same windings through a change-over switch, as shown in Fig. 10.3 are connected in delta across the same supply. The starter is provided with overload and under voltage protection devices also. Moreover, the starter is also provided with a mechanical inter-locking which prevents the handle to put in run position first. Simplified connections of a star delta starter are shown in Fig. 10.4.

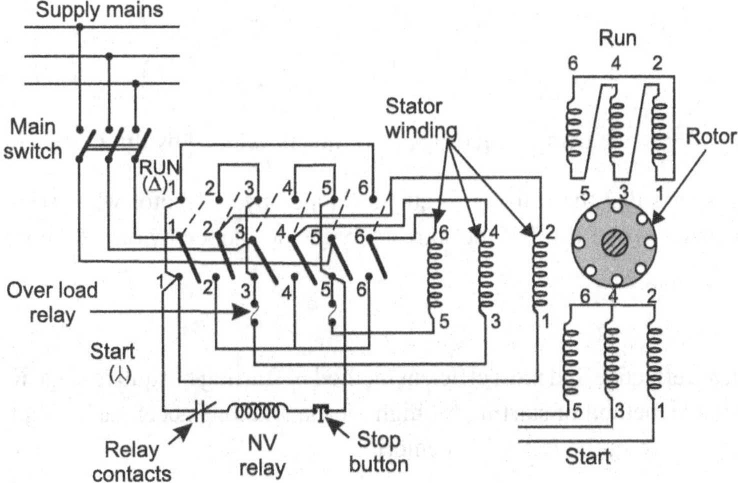

Fig. 10.3 Star-delta starter

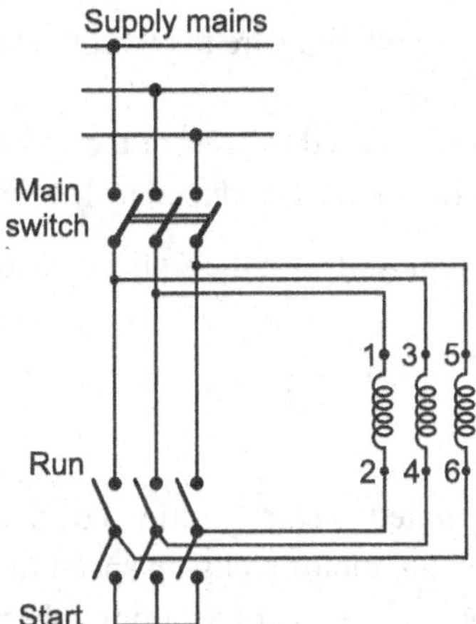

Fig. 10.4 Simplified star-delta starter

Since at start stator windings are connected in star connection, so voltage across each phase winding is reduced to $1/\sqrt{3}$ of line voltage, therefore, starting current/phase becomes equal to

$$I_{sc}/\sqrt{3} = \text{Starting line current}$$

Starting line current by direct switching with stator winding connected in delta = $\sqrt{3}\,I_{sc}$

$$\therefore \quad \frac{\text{Line current with star delta starter}}{\text{Line current with direct switching}} = \frac{I_{sc}/\sqrt{3}}{\sqrt{3}I_{sc}} = \frac{1}{3}$$

Thus, it concludes that when a 3-phase motor is started by a star/delta starter, the current drawn by it is limited to 1/3rd of the value that it would draw without starter.

Torque developed by motor when started by star-delta starter

$$\text{Starting torque, } T_{st} = T_{fl}\left(\frac{I_{st}}{I_{fl}}\right)^2 \times S_{fl}$$

$$= T_{fl}\left(\frac{I_{st}/\sqrt{3}}{I_{fl}}\right)^2 \times S_{fl} = \frac{1}{3}T_{fl}\left(\frac{I_{sc}}{I_{fl}}\right)^2 \times S_{fl}$$

$$\text{Starting torque} = \frac{1}{3} \times \text{torque developed by direct switching}$$

The equation shows that the starting torque developed by the motor when started by star-delta starter is also reduced to one-third of the starting torque developed by direct on line switching.

Merits

It is a simple, cheap, effective and most efficient method of starting of squirrel cage induction motors. It is the most suitable method of starting for high inertia and long acceleration loads.

However, it suffers from the following demerits.

Demerits

1. All the six terminals of the stator winding are to be brought out to starter through six leads and the motor has to be operated in delta.
2. The starting current can only be limited to $1/3^{rd}$ of the short circuit current.
3. Starting torque developed by the motor is reduced to $1/3^{rd}$ which is very low to pick the load.

Accordingly, such starters are employed with the squirrel cage induction motors having capacity 4 kW to 20 kW.

Precautions

To start the motor, the handle is operated to start position i.e., to connect the motor winding in star. Keep the handle in this position till the motor picks-up speed to 70% of its rated value. Only after that shift the handle to run position (i.e., delta connection), otherwise there will be heavy sparking at the starter terminals.

10.2.4 Auto-transformer Starter

In the previous method, the current can only be reduced to 1/3 times the short circuit current. Whereas, in this method, the voltage applied across the motor and hence current can be reduced to a very low value at the time of start. At the time of start, the motor is connected to supply through auto-transformer by a 6 pole double throw switch. When the motor is accelerated to about full speed, the operating handle is moved to run position. By this, motor is directly connected to the line as shown in Fig. 10.5.

Overload protection and under voltage protection is provided as explained in the first method.

Although this type of starter is expensive but is most suitable for both the star-connected and delta-connected induction motors. It is most suitable for starting of large motor.

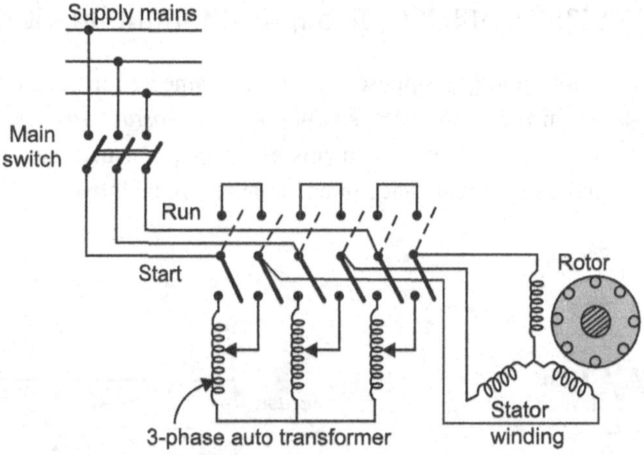

Fig. 10.5 Auto-transformer starter

Large size motors draw huge amount of current from the mains if they are connected to mains without starter. However, if they are connected to the mains through star/delta starter, the current is limited to 1/3rd value which is still, so large that it would disturb the other loads connected to the same lines. Hence, to limit the initial rush of current to low values *auto-transformer starters* are employed. With the help of auto-transformer starters, we can limit the starting current to any predetermined value as explained below:

Let the motor be started by an auto transformer having transformation ratio K.

If I_{sc} is the starting current when normal voltage is applied.

$$\text{Applied voltage to stator at start} = KV$$

$$\text{Then motor input current } I_{st} = KI_{sc}$$

$$\text{Supply current} = \text{Primary current of Auto transformer}$$

$$= K \times \text{Secondary current of Auto-transformer}$$

$$= KKI_{sc} = K^2 I_{sc}$$

If 20% (i.e., one-fifth) voltage is applied to the motor through auto-transformer starter, the current drawn from the mains is reduced to $\left(\frac{1}{5}\right)^2$ i.e., 1/25th times.

Torque developed by motor when started by an auto-transformer starter

$$\text{Starting torque, } T_{st} = T_{fl}\left(\frac{I_{st}}{I_{fl}}\right)^2 \times S_{fl}$$

$$= T_{fl}\left(\frac{KI_{sc}}{I_{fl}}\right)^2 \times S_{fl} = K^2 T_{fl}\left(\frac{I_{sc}}{I_{fl}}\right)^2 \times S_{fl}$$

$$\text{Starting torque} = K^2 \times \text{torque developed by direct switching}$$

10.3 Rotor Resistance Starter for Slip Ring Induction Motors

To start a slip ring induction motor, a 3-phase rheostat is connected in series with the rotor circuit through brushes as shown in Fig. 10.6. This is called *rotor rheostat starter*. This is made of three separate variable resistors joined together by means of a 3-phase armed handle which forms a star point. By moving the handle equal resistance in each phase can be introduced.

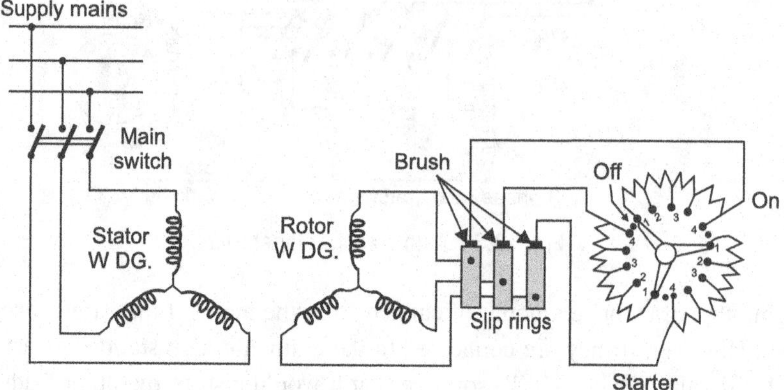

Fig. 10.6 3-phase slip-ring induction motor starter

At start, whole of the rheostat resistance is inserted in the rotor circuit and the rotor current is reduced to

$$I_{2s} = \frac{E_{2s}}{\sqrt{(R_2 + R_S)^2 + (X_{2s})^2}}$$

Correspondingly it reduces the current drawn by the motor from the mains at start.

When the motor picks up speed the external resistance is reduced gradually and ultimately whole of the resistance is taken out of circuit and slip rings are short-circuited.

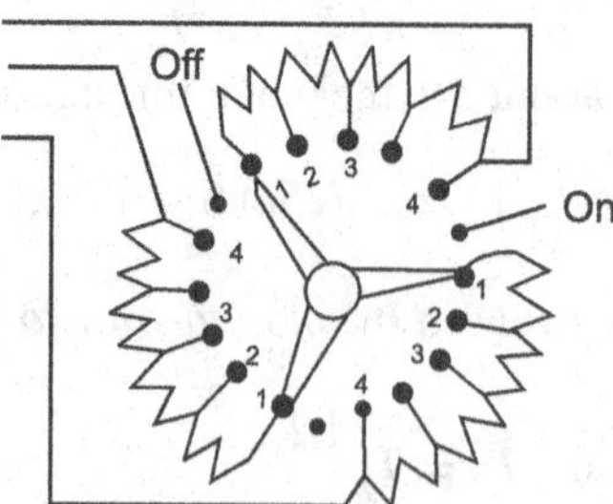

Fig. 10.7 3-phase slip-ring induction motor starter

By inserting external resistance in the rotor circuit, not only the starting current is reduced but at the same time starting torque is increased due to improvement in power factor:

At starts:

$$\text{Power factor without starter, } \cos \phi_s = \frac{R_2}{\sqrt{(R_2)^2 + (X_{2s})^2}}$$

Power factor with starter, $\cos' \phi_s = \dfrac{(R_2 + R_S)}{\sqrt{(R_2 + R_S)^2 + (X_{2s})^2}}$

Hence, $\cos' \phi_s >> \cos \phi_s$

Calculation of Starting Resistance at Various Steps

Assumptions: consider that (*i*) the motor starts against a constant load or torque and (*ii*) the rotor current fluctuates between fixed maximum and minimum values of current i.e., I_{2max} and I_{2min}

For calculations, consider only one phase of a 3-phase slip-ring induction motor having R_2 as rotor resistance and X_{2S} as rotor reactance at stand-still, as shown in Fig. 10.8.

Let, 1, 2, ...*n* be the number of studs of the starter

$r_1, r_2 \ldots r_{n-1}$ be the resistance of different steps in each phase.

$R_{S1}, R_{S2\ldots} R_{Sn}$ be the total resistance per phase in the rotor circuit at stud 1, 2...and *n*, respectively.

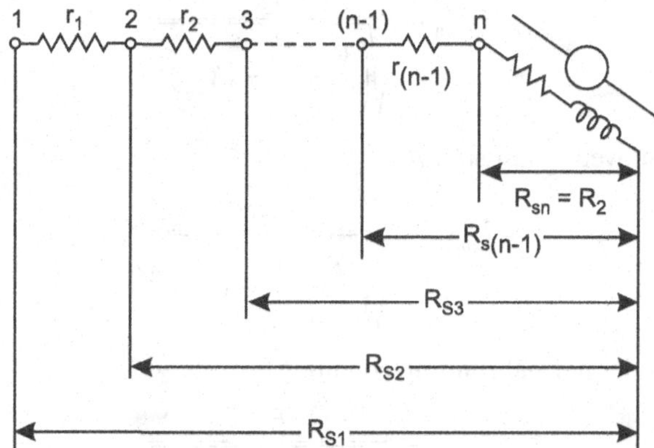

Fig. 10.8 Resistance of various steps of a slip-ring induction motor starter (For one-phase)

When the handle comes in contact with stud -1, the current rises to its maximum value, i.e.,

$$I_{2\max} = \frac{S_1 E_{2S}}{\sqrt{(R_{S1})^2 + (S_1 X_{2S})^2}} = \frac{E_{2S}}{\sqrt{\left(\dfrac{R_{S1}}{S_1}\right)^2 + (X_{2S})^2}} \quad \ldots(i)$$

When motor is stationary and handle is moved to stad-1, at that instant slip $S_1 = 1$. The rotor starts rotating and it picks-up the speed so that the slip reaches to S_2. Before moving the handle to stud-2, the slip of the motor is S_2 and the current obtains its minimum value, i.e.,

$$I_{2\min} = \frac{E_{2S}}{\sqrt{\left(\dfrac{R_{S1}}{S_2}\right)^2 + (X_{2S})^2}} \quad \ldots(ii)$$

When the handle in moved to stud-2, the current again rises to its maximum value, the speed momentarily remains the same, i.e.,

$$I_{2max} = \frac{E_{2S}}{\sqrt{\left(\frac{R_{S2}}{S_2}\right)^2 + (X_{2S})^2}} \quad ...(iii)$$

At this (2nd) stud, the speed of the motor rises, slip becomes S_3 and current reduces to $I_{2\,min}$, i.e.,

$$I_{2min} = \frac{E_{2S}}{\sqrt{\left(\frac{R_{S2}}{S_3}\right)^2 + (X_{2S})^2}}. \quad ...(iv)$$

Similarly, when the handle is moved to stud-3,

$$I_{2max} = \frac{E_{2S}}{\sqrt{\left(\frac{R_{S3}}{S_3}\right)^2 + (X_{2S})^2}} \quad ...(v)$$

and

$$I_{2min} = \frac{E_{2S}}{\sqrt{\left(\frac{R_{S3}}{S_4}\right)^2 + (X_{2S})^2}} \quad ...(vi)$$

When the handle is moved to last stud-*n*,

$$I_{2max} = \frac{E_{2S}}{\sqrt{\left(\frac{R_{Sn}}{S_{max}}\right)^2 + (X_{2S})^2}}$$

where S_{max} is the slip under normal running condition

$$= \frac{E_{2S}}{\sqrt{\left(\frac{R_2}{S_{max}}\right)^2 + (X_{2S})^2}} \quad ...(vii)$$

From eqn. (*i*), (*iii*), (*v*) and (*vii*), we get,

$$= \frac{E_{2S}}{\sqrt{\left(\frac{R_{S1}}{S_1}\right)^2 + (X_{2S})^2}} = \frac{E_{2S}}{\sqrt{\left(\frac{R_{S2}}{S_2}\right)^2 + (X_{2S})^2}} = \frac{E_{2S}}{\sqrt{\left(\frac{R_{S3}}{S_3}\right)^2 + (X_{2S})^2}} = \frac{E_{2S}}{\sqrt{\left(\frac{R_2}{S_{max}}\right)^2 + (X_{2S})^2}}$$

or

$$\frac{R_{S1}}{S_1} = \frac{R_{S2}}{S_2} = \frac{R_{S3}}{S_3} ... = \frac{R_{S(n-1)}}{S_{(n-1)}} = \frac{R_2}{S_{max}} \quad ...(viii)$$

Similarly from eqn. (*ii*), (*iv*), (*vi*)..., we get,

$$= \frac{E_{2S}}{\sqrt{\left(\frac{R_{S1}}{S_2}\right)^2 + (X_{2S})^2}} = \frac{E_{2S}}{\sqrt{\left(\frac{R_{S2}}{S_3}\right)^2 + (X_{2S})^2}} = \frac{E_{2S}}{\sqrt{\left(\frac{R_{S3}}{S_4}\right)^2 + (X_{2S})^2}} = \frac{E_{2S}}{\sqrt{\left(\frac{R_{S\,(n-1)}}{S_{max}}\right)^2 + (X_{2S})^2}}$$

or

$$\frac{R_{S1}}{S_2} = \frac{R_{S2}}{S_3} = \frac{R_{S3}}{S_4} ... = \frac{R_{S(n-1)}}{S_{max}} \quad ...(ix)$$

Combining eqn. (*viii*) and (*ix*), we get,

$$\frac{S_2}{S_1} = \frac{S_3}{S_2} = \frac{S_4}{S_3} \cdots = \frac{S_{max}}{S_{n-1}} = \frac{R_{S2}}{R_{S1}} = \frac{R_{S3}}{R_{S2}} = \frac{R_{S4}}{R_{S3}} = \frac{R_2}{R_{S(n-1)}}\ K(say) \qquad \ldots(x)$$

From eqn. (*viii*), total rotor circuit resistance per phase at stud-1,

$$R_{S1} = R_2 \times \frac{S_1}{S_{max}} = \frac{R_2}{S_{max}} \qquad \text{(Since } S_1 = 1) \ldots(xi)$$

as well as, $R_{S2} = K\,R_{S1}$; $R_{S3} = K\,R_{S2} = K^2\,R_{S1}$

Also, $$R_2 = KR_{S(n-1)} = K^{(n-1)}R_{S1} = K^{(n-1)} . \frac{R_2}{S_{max}}$$

or $$K = (S_{max})^{1/(n-1)}$$

where (n-1) represents the number of sections of the starter.

Hence, the resistance of various sections of the starter are determined as,

$$r_1 = R_{S1} - R_{S2} = R_{S1}\,(1 - K)$$

$$r_2 = R_{S2} - R_{S1} = R_{S1}\,(K - K^2) = K\,r_1$$

$$r_3 = R_{S3} - R_{S2} = R_{S1}\,(K^2 - K^3) = K^2\,r_1$$

Example 10.1

Calculate the ratio of starting torque to full-load torque of a 3-phase induction motor having short-circuit current 4 times the full-load current with a full-load slip of 3%.

Solution:

Hence, $I_{sc} = 4\,I_{fl}$; $S_f = 3\% = 0.03$

Ratio of starting to full-load torque, $$\frac{T_{st}}{T_{fl}} = \left(\frac{I_{sc}}{I_{fl}}\right)^2 \times S_f = (4)^2 \times 0.03 = \mathbf{0.48}\ (Ans.)$$

Example 10.2

Determine the ratio of starting current to full-load current and starting torque to full-load torque for a small 3-phase squirrel cage induction motor having short-circuit current 5 times the full-load current. The motor is started with a stator resistance starter which reduces the impressed voltage to 70% of rated value. The full-load slip of the motor is 4%.

Solution:

Here, $I_{sc} = 5\,I_{fl}$; $x = 70\% = 0.7$; $S_f = 4\% = 0.04$

Starting current, $$I_{st} = xI_{sc} = 0.7 \times 5 \times I_{fl} = 3.5I_{fl}$$

or $$\frac{I_{st}}{I_{fl}} = \mathbf{3.5}\ (Ans.)$$

$$\text{Starting torque, } T_{st} = T_{st} = \left(\frac{I_{st}}{I_{fl}}\right)^2 \times S_f \times T_{fl}$$

or
$$\frac{T_{st}}{T_{fl}} = (3.5)^2 \times 0.04 = 0.49 = \mathbf{49\%}\ (Ans.)$$

Example 10.3

A 10 H.P. 3-phase induction motor with full load efficiency and p.f. of 0·83 and 0·8 respectively has a short circuit current of 3·5 times full load current. Estimate the line current at the instant of starting the motor from a 500 V supply by means of star delta starter.

Solution:

Output of the motor = 10 H.P. = 7355 W

Power factor, $\cos\phi = 0{\cdot}8$

Efficiency, $\eta = 0{\cdot}83$

Supply voltage (line value), $V_L = 500$ V

$$\text{Input to the motor} = \frac{\text{output}}{\eta} = \frac{7355}{0.83} = 8861.45\text{W}$$

$$\text{Full load current, } I_{fl} = \frac{\text{Input}}{\sqrt{3}\, V_L \cos\phi} = \frac{8861 \cdot 45}{\sqrt{3} \times 500 \times 0 \cdot 8} = 12 \cdot 79 \text{ A}$$

$$\text{Ratio of, } \frac{I_{sc}}{I_{fl}} = 3{\cdot}5$$

∴ Short circuit current, $I_{sc} = 3{\cdot}5 \times 12{\cdot}79 = 44{\cdot}76$ A

$$\text{Staring current, } I_s = \frac{I_{sc}}{3} = \frac{44.76}{3} = \mathbf{14{\cdot}92\ A}\ (Ans.)$$

Example 10.4

Determine the starting torque of a 3-phase induction motor in terms of full load torque when started by means of:

(i) Star delta starter; and

(ii) An auto-transformer starter with 50% tapings.

The motor draws a starting current of 5 times the full load current when started direct on line. The full load slip is 4 percent.

Solution:

Ratio of short circuit current to full load current,

$$\frac{I_{sc}}{I_{fl}} = 5$$

Full load slip, $S = 0{\cdot}04$

(*i*) *For star-delta starter.*

Ratio of starting torque to full load torque,

$$\frac{T_{st}}{T_{fl}} = K^2\left(\frac{I_{sc}}{I_{fl}}\right)^2 \times S_{fl} = \frac{1}{3}(5)^2 \times 0 \cdot 04 = 0 \cdot 3333$$

∴ **Starting torque = 33·33% of full load torque** (*Ans.*)

(*ii*) *For Auto-transformer starter:*

Transformation ratio, or tapings,

$$K = 50\% = \frac{1}{2}$$

Ratio of starting torque to full load torque

$$\frac{T_{st}}{T_{fl}} = K^2\left(\frac{I_{sc}}{I_{fl}}\right)^2 \times S_{fl} = \left(\frac{1}{2}\right)^1 \times (5)^2 \times 0.04 = 0.25$$

∴ **Starting torque = 25% of full load torque** (*Ans.*)

Example 10.5

A 20 H.P., 3-phase, 6-pole, 50 Hz, 400 V induction motor runs at 960 rpm on full load. If it takes 120 A on direct starting, find the ratio of starting torque to full load torque with a star-delta starter. Full load efficiency and p.f. are 90% and 0·85.

Solution:

Output power = 20 *H.P.* = 20 × 735·5 = 14710 W

No. of poles, $P = 6$

Supply voltage (line value), $V_L = 400$ V

Supply frequency, $f = 50$ Hz

Short circuit current, $I_{sc} = 120$ A

Efficiency, $\eta = 90\% = 0{\cdot}9$

Power factor, $\cos\phi = 0{\cdot}85$

Rotor speed, $N = 960$ rpm

$$\text{Full load current, } I_{fl} = \frac{14170}{\sqrt{3} \times 400 \times 0 \cdot 9 \times 0 \cdot 85} = 27{\cdot}75 \text{ A}$$

$$\text{Synchronous speed, } N_s = \frac{120f}{p} = \frac{120 \times 50}{6} = 1000 \text{ r.p.m}$$

$$\text{Full-load slip, } S_{fl} = \frac{N_2 - N}{N_2} = \frac{1000 - 960}{1000} = 0.04$$

Ratio of starting torque to full load torque,

$$\frac{T_{st}}{T_{fl}} = \frac{1}{3}\left(\frac{I_{sc}}{I_{fl}}\right)^2 \times S$$

$$= \frac{1}{3}\left(\frac{120}{27.75}\right)^2 \times 0.04 = \mathbf{0{\cdot}2493}\ (Ans.)$$

Example 10.6

Find the suitable tapping on an auto-transformer starter for an induction motor required to start the motor with 36% of full load torque. The short circuit current of the motor is 5 times the full load current and the full load slip is 4%. Determine also the starting current in the supply mains as a percentage of full load current.

Solution:

Ratio of starting torque to full load torque,

$$\frac{T_{st}}{T_{fl}} = 0.36$$

Ratio of short-circuit current to full load current,

$$\frac{I_{sc}}{I_{fl}} = 5$$

Full load slip, $S = 0{\cdot}04$

Now
$$\frac{T_{st}}{T_{fl}} = K^2\left(\frac{I_{sc}}{I_{fl}}\right)^2 \times S$$

$$0{\cdot}36 = K^2 \times (5)^2 \times 0{\cdot}04$$

∴ Transformation ratio or tapping of auto-transformer,

$$K = \sqrt{\frac{0.036}{5 \times 5 \times 0.04}} = \mathbf{0{\cdot}6} \text{ or } \mathbf{60\%}\ (Ans.)$$

In an auto-transformer starter, ratio of starting current to short-circuit current,

$$\frac{I_s}{I_{sc}} = K^2 \text{ where } I_{sc} = 5\, I_f$$

∴
$$\frac{I_s}{I_f} = 5K^2 = 5 \times (0{\cdot}6)^2 = \mathbf{1{\cdot}8} \text{ or } \mathbf{180\%}\ (Ans.)$$

Example 10.7

Find the ratio of starting to full load current for 10 kW input, 415 V, 3-phase induction motor with star-delta starter at full load efficiency 0·9 and the full load p.f. 0·8. The short circuit current is 40 A at 210 V and the magnetising current is negligible.

Solution:

Input power, $P = 10$ kW = 1000 W

Supply voltage (line value), $V_L = 415$ V

Power factor, $\cos\phi = 0{\cdot}8$

Short circuit current at 210 V = 40 A

$$\text{Short-circuit current at 415 V, } I_{sc} = \frac{415 \times 40}{210} = 79 \text{ A}$$

$$\text{Full load current, } I_{fl} = \frac{P}{\sqrt{3}\, V_L \cos\phi} = \frac{10000}{\sqrt{3} \times 415 \times 0 \cdot 8} = 17 \cdot 39 \text{ A}$$

In case of star-delta starter;

$$\text{Starting current, } I_{st} = \frac{I_{sc}}{3} = \frac{79}{3} = 26 \cdot 33 \text{ A}$$

$$\text{Ratio of starting to full load current, } \frac{I_{st}}{I_{fl}} = \frac{26 \cdot 33}{17 \cdot 39} = \mathbf{1{\cdot}514} \text{ (Ans.)}$$

Example 10.8

The full-load slip of a 3-phase, 400 V, 50 Hz slip-ring induction motor is 2% and its starting current is 1.5 times the full-load current. Design a five-section 6-stud rotor starter if the resistance of rotor circuit per phase is.03 ohm.

Solution:

Here, full-load slip, $S_{fl} = 2\% = 0.02$

$$\text{Rotor current at full-load, } I_{2fl} = I_{2fl} = \frac{S_{fl}\, E_{2S}}{\sqrt{R_2^{\,2} + (S_{fl} X_{2S})^2}}$$

When slip is very small, the value of $(S_{fl}\, X_{2S})^2$ is very small and can be neglected, therefore,

$$I_{2fl} = \frac{S_{fl} E_{2S}}{R_2}$$

Thus, slip $\propto$ Rotor current

At the starting current, slip is maximum, $S_{max} = 1.5 \times S_{fl} = 1.5 \times 0.02 = 0.03$

Total resistance in the rotor circuit per phase on stud-1,

$$R_{S1} = \frac{R_2}{S_{\max}} = \frac{0.03}{0.03} = 1.0 \text{ ohm}$$

and $$K = (S_{\max})^{\frac{1}{n-1}} = (0.03)^{\frac{1}{5}} = 0.496$$

Resistance of various steps,

$r_1 = (1-K)\, R_{S1} = (1-0.496) \times 1 = \mathbf{0.504\ ohm}$ (*Ans.*)

$r_2 = K \times r_1 = 0.496 \times 0.504 = \mathbf{0.245\ ohm}$ (*Ans.*)

$r_3 = K^2 \times r_1 = (0.496)^2 \times 0.504 = \mathbf{0.124\ ohm}$ (*Ans.*)

$r_4 = K^3 \times r_1 = (0.496)^3 \times 0.504 = \mathbf{0.0615\ ohm}$ (*Ans.*)

$r_5 = K^4 \times r_1 = (0.496)^4 \times 0.504 = \mathbf{0.0305\ ohm}$ (*Ans.*)

Section Practice Problems

Numerical Problems

1. Calculate the starting torque to full-load torque of a 3-phase induction motor having short-circuit current 4 times the full-load current with a full-load slip of 2.5%. (**Ans.** *0.4*)

2. Calculate the reduction in starting current and starting torque when the motor is started by stator resistance starter which reduces the impressed voltage to 80%. (**Ans.** *20%; 36%*)

3. A 3-phase squirrel cage induction motor has a full-load slip of 4%. When it is started direct on line, it draws starting current which is 5 times the full-load current. What should be the transformation ratio of the auto-transformer to enable the motor to start with a current which is twice the full-load current. Also determine the starting torque under the conditions.

   (**Ans.** *63.25%; 40% of* T_{fl})

4. A 3-phase slip-ring induction motor has a full-load slip of 2% and a starting current of 1.5 times the full-load current. If the resistance of rotor per phase is 0.02 ohm, design a five step 6-stud rotor starter.

   (**Ans.** *0.336, 0.1667, 0.0826, 0.041, 0.0203 ohm*)

Short Answer Type Questions

Q.1. What is the need of a starter to start an induction motor?

Ans. Starter is required to limit the starting current to predetermined value and to provide necessary protection to the motor.

Q.2. Which starter has the minimum cost?

Ans. Direct on line starter.

Q.3. How much starting current is reduced if the motor is started with star-delta starter?

Ans. it is reduced to 1/3rd.

Q.4. What protections are provided in 3-phase induction motor starters?

Ans. Over-load protection and no-volt protection.

Q.5. Which method is adopted to start a slip-ring induction motor?

Ans. Rotor resistance method is used to start slip-ring induction motors?

10.4 Speed Control of Induction Motors

The speed of an induction motor is given by the relation

$$N = N_s (1 - S) \text{ or } N = \frac{120f}{P}(1 - S)$$

Hence, the speed of an induction motor depends upon three factors i.e., frequency, slip and number of poles for which the motor is wound. It also depends upon the supply voltage. Thus, the speed of an induction motor can be controlled by changing or controlling any one of these quantities.

10.5 Speed Control by Changing the Slip

The speed of an induction motor can be changed by changing its slip, and the slip can be changed (1) by changing the rotor circuit resistance (2) by changing the supply voltage and (3) by injecting voltage in the rotor circuit.

10.5.1 Speed Control by Changing the Rotor Circuit Resistance

In the wound type motor the slip may be changed by introducing resistance in the rotor circuit and hence speed is changed.

Torque developed in an induction motor is given by the expression:

$$T = \frac{3}{\omega_s} \frac{E_{2s}^2 R_2 / S}{[(R_2 / S)^2 + (X_{2s})^2}$$

The torque will remain constant if $\frac{R_2}{S}$ is constant. For a given torque, the slip at which a motor works is proportional to the rotor resistance.

The torque-speed curve (dotted) of a slip ring induction rotor is shown in Fig. 10.9. When an external resistance is added in the rotor circuit, speed decreases for the same torque T, so that ratio $\frac{R_2}{S}$ remains constant.

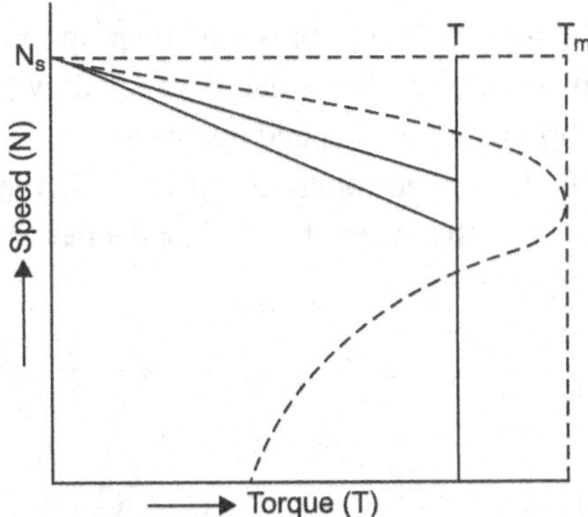

Fig. 10.9 Torque-speed curve when speed is controlled by changing the rotor circuit resistance

The disadvantages of this method of speed control are

(*i*) *Poor efficiency:* By the introduction of external resistance in the rotor circuit, there is extra power loss ($I_2{}^2R$) in the rotor circuit which reduces the overall efficiency of the motor.

(*ii*) *Poor speed regulation:* When speed of the induction motor is controlled by adding some external resistance in the rotor circuit, the change in speed is larger when load on the machine changes from one value to the other. Hence the machine operates at a *poor regulation.*

For illustration, refer to Fig. 10.10. When the rotor resistance is R_1 and the load on the machine changes from half-load to full-load, the speed of the motor decreases from N_1 to N_2. However, when some resistance is added in the rotor circuit so that its value becomes R_2 (i.e., $R_2 > R_1$), then the speed changes from N_3 to N_4 when load on the motor changes from half-load to full-load. It is very clear that $N_3 - N_4$ is larger than $N_1 - N_2$. Hence the machine operates at a *poor regulation.*

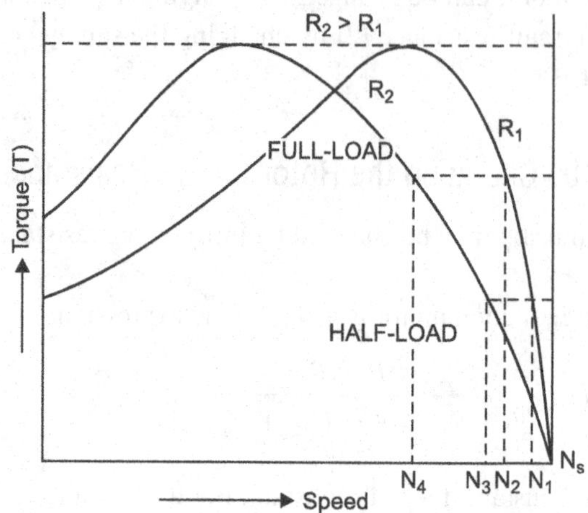

Fig. 10.10 Torque-speed curve representing the effect on maximum torque when speed is changed by changing the rotor circuit resistance

10.5.2 Speed Control by Controlling the Supply Voltage

Slip or speed of a motor can also be changed by controlling the voltage fed to the motor. We have already seen that the torque developed by the motor is directly proportional to the square of the supply voltage. If the supply voltage is decreased, the torque developed by the motor decreases rapidly ($T \propto V^2$) and to pick-up the load slip increases or speed decreases.

For illustration, look at Fig. 10.11. At rated voltage and given load, the speed of the motor is N_1. If the supply voltage is reduced (say to 90%), the speed of the motor decreases to N_2 to pick-up the given load.

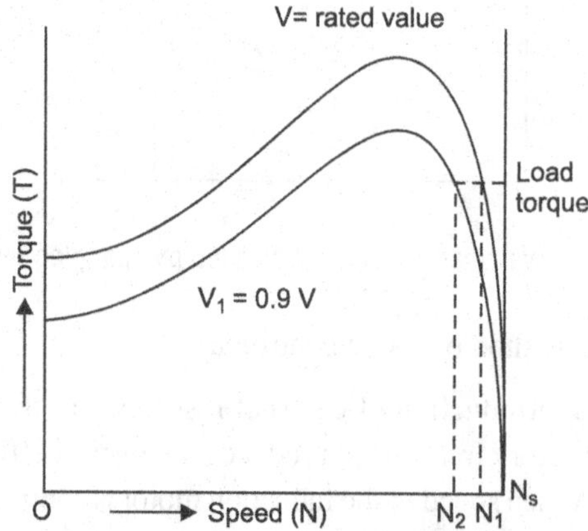

Fig. 10.11 Torque-speed curve when speed is changed by changing the supply voltage

This method is never used for the speed control of three-phase large induction motors because the voltage control devices are very costly and bulky. However, this method is usually employed with single-phase induction motors e.g., ceiling fans, etc.

10.5.3 Speed Control by Injecting Voltage in the Rotor Circuit

The speed of an induction motor can also be controlled by injecting a voltage at slip frequency directly into the rotor circuit. This method of speed control is discussed in detail in Art.no.10.9. First of all, this method was introduced by K.H. Schrage of Sweden and the motor in which this method is employed is called *Schrage motor.* (shall be dealt in detail in the next chapters) If the injected emf has a component directly opposite to the rotor induced emf, the motor speed decreases. On the other hand, if the injected emf has a component in phase with the rotor induced emf the motor speed increases and may rises beyond the synchronous speed.

Now-a-days, Schrage motors are not preferred because of their heavy cost and bulky construction but these are still employed in large printing presses like newspaper printing.

10.6 Speed Control by Changing the Supply Frequency

The frequency of the power supply is constant, therefore, to control the speed of an induction motor by this method, the induction motor is connected to the alternator operating independently. To control the speed, the frequency of the alternator is changed. This is a costly affair.

Recent improvements in the capabilities of controlled rectifiers (*SCR*) and continued decrease in the cost of their manufacturing, it has made it possible to control the speed of induction motor by controlling the supply frequency fed to the motor. By this method 5 to 10% of rated speed of induction motors can be controlled. However, if the speed is to be controlled beyond this value, the motor design has to be changed accordingly.

Further, If the speed of the motor is changed by changing the frequency, to keep the flux level constant in the stator core, supply voltage is also changed so that v/f is kept constant.

10.7 Speed Control by Changing the Poles

By means of suitable switch, the stator winding connections can be changed in such a manner that the number of stator poles is changed. This changes the actual speed of motor since actual speed of the motor is approximately inversely proportional to the number of poles.

By suitable connections one winding can give two different speeds.

Suppose there are four coils per phase. If these are connected in such a way that they carry current in same direction then it will form eight poles altogether as shown in Fig. 10.12(*a*). Now if the connections are such that the alternate coils carry current in opposite directions, we get four poles altogether as shown in Fig. 10.12(*b*).

If more than two speeds are required. Two separate winding are housed in same slots and if each is arranged to give two speeds then two windings can give four different speeds.

In squirrel cage motors, the rotor poles are adjusted automatically. However, in wound type motors, care has to be taken to change the rotor poles accordingly. Moreover, due to flux distribution, the pf of the motor is also affected by changing the number of poles to obtain different speeds. If the speed is decreased by increasing the number of poles the power factor of the motor decreases.

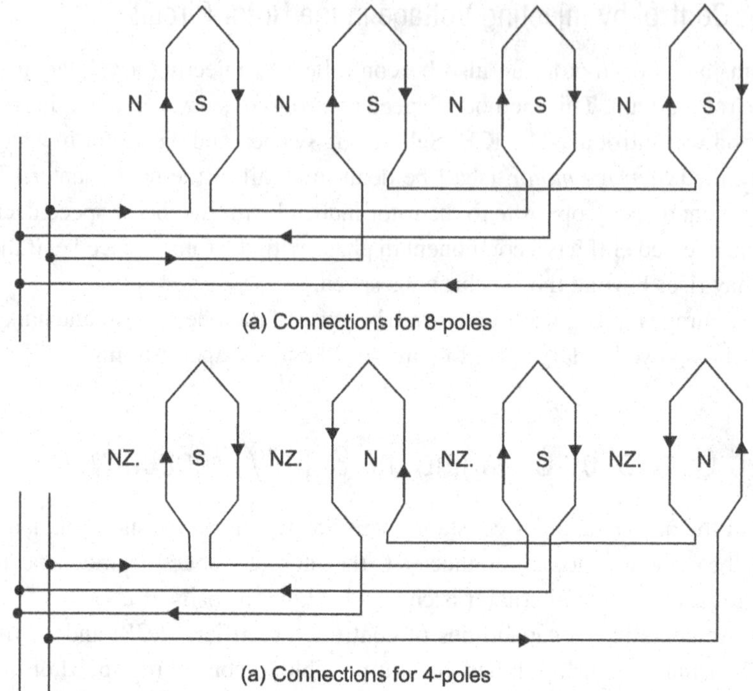

Fig. 10.12 Speed control by changing the number of poles.

10.8 Speed Control by Cascade Method

A method of speed control of induction motor involving two or more motors is known as *cascade method of speed control.* The two machines are mechanically coupled with each other, one of them must be a slip ring induction motor. Mostly both slip ring motors are used having transformation ratio equal to unity. In this case supply is connected to the stator of one of the induction motor and the induced emf of the rotor from slip rings is fed to the stator or rotor of the second induction motor, as shown in Fig. 10.13.

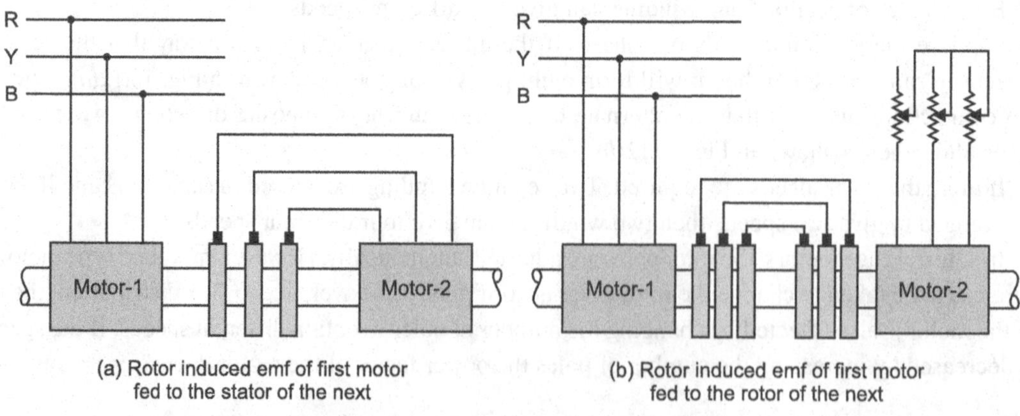

Fig. 10.13 Motors connected in cascade

If P_1 and P_2 are the number of poles of the two machines and f is the supply frequency, then the set can give the following different speeds:

(*i*) When machine I works alone:

$$\text{The synchronous speed} = \frac{120f}{P_1}$$

(*ii*) When machine II works alone;

$$\text{The synchronous speed} = \frac{120f}{P_2}$$

(*iii*) When machine I and II are connected in cumulative cascade i.e., the torque of the two motors are in same direction;

$$\text{The synchronous speed of the set} = \frac{120f}{P_1 + P_2}$$

(*iv*) When machine I and II are connected in differential cascade i.e., the torque of the motors are in opposite direction;

$$\text{The synchronous speed of the set} = \frac{120f}{P_1 - P_2}$$

To prove the above relations, proceed as follows:

Let, f be the supply frequency and P_1 and P_2 be the number of poles of machine I and II respectively, then

$$\text{Synchronous speed of motor -1, } N_{S1} = \frac{120f}{P_1}$$

$$\text{Speed of motor-1 with slip } S_1, N_1 = (1 - S_1) = \frac{120\,f}{P_1}(1 - S_1) \qquad \ldots(i)$$

Frequency of the emf induced in the rotor of motor-1 fed to the motor-2, $f_2 = S_1 \times f$

$$\text{Synchronous speed of motor-2, } N_{S2} = \frac{120\,f_2}{P_2} = \frac{120\,S_1 f}{P_2}$$

$$\text{Speed of motor-2 with slip } S_2, N_2 = \frac{120\,S_1 f}{P_2}(1 - S_2) \qquad \ldots(ii)$$

Since the two motors are mechanically coupled with each other N_1 must be equal to N_2.

Therefore, equating eqn. (*i*) and (*ii*), we get

$$\frac{120\,f}{P_1}(1 - S_1) = \frac{120\,S_1\,f}{P_2}(1 - S_2)$$

$$\frac{1 - S_1}{P_1} = \frac{S_1\,(1 - S_2)}{P_2}$$

$$\frac{1 - S_1}{P_1} = \frac{S_1 - S_1\,S_2}{P_2} \qquad \ldots(iii)$$

Neglecting $S_1 S_2$ being very small, we get

$$\frac{1-S_1}{P_1} = \frac{S_1}{P_2}$$

or
$$\frac{1-S_1}{S_1} = \frac{P_1}{P_2}$$

or
$$\frac{1}{S_1} - 1 = \frac{P_1}{P_2}$$

or
$$\frac{1}{S_1} = \frac{P_1 + P_2}{P_2}$$

or
$$S_1 = \frac{P_2}{P_1 + P_2}$$

$$\text{Thus, cascaded speed, } N = \frac{120f}{P_1}(1 - S_1) = \frac{120f}{P_1}\left(1 - \frac{P_2}{P_1 + P_2}\right)$$

$$N = \frac{120f}{P_1 + P_2}. \qquad \text{...(iv)}$$

Now, if the electrical connections are such that the torques of two motors are acting in opposite direction, then from eqn. (*iii*),

$$\frac{1-S_1}{P_1} = \frac{-S_1 + S_1 S_2}{P_2} \qquad \text{...(v)}$$

or
$$\frac{1-S_1}{P_1} = \frac{-S_1}{P_2} \quad \text{or} \quad \frac{1-S_1}{P_1} = \frac{-P_1}{P_2}$$

or
$$\frac{1}{S_1} - 1 = \frac{-P_1}{P_2} \quad \text{or} \quad \frac{1}{S_1} = \frac{P_2 - P_1}{P_2}$$

or
$$S_1 = \frac{P_2}{P_2 - P_1}$$

$$\text{Thus, cascaded speed, } N = \frac{120\,f}{P_1}(1 - S_1) = \frac{120\,f}{P_1}\left(1 - \frac{P_2}{P_2 - P_1}\right)$$

or
$$N = \frac{-120\,f}{P_2 - P_1} \quad \text{or} \quad N = \frac{120\,f}{P_1 - P_2} \qquad \text{...(vi)}$$

10.9. Speed Control by Injecting an emf in the Rotor Circuit

In this method, the speed of a 3-phase slip-ring induction motor is controlled by injecting a voltage (or emf) in the rotor circuit. While injecting emf in the rotor circuit, we are to take care for its frequency which must be the same as that of the rotor frequency of the induction motor. Although there is no restriction as to the phase of the injected voltage.

If the injected voltage is in phase with the rotor induced emf, it amounts to *decrease* in the rotor resistance, on the other hand, if the injected voltage is in phase opposition to the rotor induced emf, it amounts to *increase* in rotor resistance.

Hence, by changing the phase of the injected emf, which in effect changes the rotor resistance, the speed of the motor can be changed.

The following systems are used to control the speed of slip-ring induction motors:

10.9.1 Kramer System of Speed Control

This system of speed control was suggested by Kramer as illustrated below:

It consists of a rotary converter *C* which converts the low slip frequency AC power into DC. This DC power is used to drive a DC shunt motor *D* mechanically coupled to the main motor *M*, as shown in Fig. 10.14.

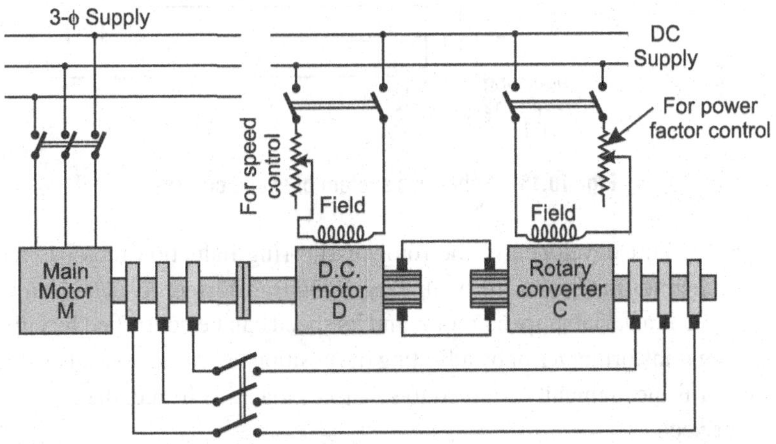

Fig. 10.14 Kramer system of speed control

When 3-phase AC supply is given to the main slip-ring induction motor *M*, an emf at slip frequency is induced in the rotor. This emf induced in the rotor at slip frequency is given to the rotary converter *C* through slip. rings. The DC output of rotary converter *C* is given to DC motor *D*. Both DC motor *D* and rotary converter *C* are excited from DC bus-bars (or from exciters). The field of DC motor *D* can be regulated which regulates its back emf (E_b) and hence the DC potential at the commutator of *C* which further controls the slip-ring voltage and therefore, the speed of motor *M*.

This method of speed control is used for the speed control of large motors of capacity 4000 kW or above. The major advantage of this method is that any speed, within the working range, can be obtained instead of only in two or three steps, as with other methods of speed control.

Another major advantage of this method is that the power factor of the main motor can be improved by over-exciting the rotary converter.

10.9.2 Scherbius System of Speed Control

The other method employed to control the speed of large size slip-ring induction motors is scherbius system of speed control. In this system, the slip frequency energy is not converted into DC and then fed to a DC motor, rather it is fed directly to a specially designed 3-phase (or 6-phase) commutator motor C called a scherbius machine, as shown in Fig. 10.15.

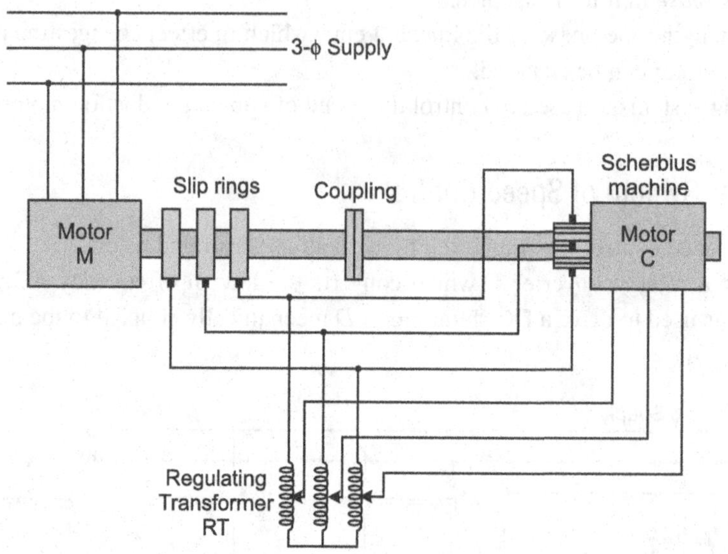

Fig. 10.15 Scherbius system of speed control

The slip frequency emf developed in the rotor of slip-ring induction motor is fed to the poly-phase winding of scherbius machine *C* through a regulated transformer *RT*. The commentator motor (scherbius machine) *C* is a variable speed motor and its speed can be controlled by either varying the tapings on regulating transformer *RT* or by adjusting the position of brusher placed on its commentator. Since main motor *M* is mechanically coupled to scherbius machine hence, the speed of main motor is controlled as per need.

This system of speed control of heavy slip-ring induction motors is preferred over Kramer system since it is more economical because DC motor is not required in this case. Moreover, it provides smooth operation.

Example 10.9

A 3-phase, 4-pole, 50Hz slip-ring induction motor is rotating at a speed of 1440 rpm at full-load. Its rotor resistance is 0.25 ohm per phase. What external resistance should be added in the rotor circuit to reduce the speed to 1320 rpm, the torque is to be kept the same.

Solution:

Here, $R_2 = 0.25$ ohm ; $f = 50$ Hz ; $P = 4$; $N_1 = 1440$ rpm ; $N_2 = 1320$ rpm

$$\text{Torque developed, } T = \frac{S\,E_{2S}\,R_2}{[R_2^2 + (S\,X_{2S})^2]} = \frac{S\,E_{2S}\,R_2}{R_2^2}$$

(Rejecting $S\,X_{2S,}$ not given)

$$T = \frac{K\,S}{R_2}$$

(Where K is a constant)

In first case, $$T_1 = \frac{K\,S_1}{R_2} \quad \ldots(i)$$

In second case, $$T_2 = \frac{K\,S_2}{(R_2 + r)} \quad \ldots(ii)$$

$$N_S = \frac{120\,f}{P} = \frac{120 \times 50}{4} = 1500 \text{ rpm}$$

$$S_1 = \frac{1500 - 1440}{1500} = 0.04\ ;\ S_2 = \frac{1500 - 1320}{1500}\ 0.12$$

Since $T_1 = T_2$ equating equation (*i*) and (*ii*), we get,

$$\frac{S_1}{R_2} = \frac{S_2}{R_2 + r} \quad \text{or} \quad R_2 + r = \frac{S_2}{S_1} \times R_z$$

$\therefore$ $$r = \frac{0.12}{0.04} \times 0.25 - 0.25 = \mathbf{0.5\ ohm}\ (Ans.)$$

Example 10.10

The slip of a 3-phase, 6-pole, 50 Hz induction motor is 4% at full-load. Assuming rotor resistance per phase as 0.3 ohm, find the value of external resistance to be connected in series with each phase of the rotor to reduce the speed by 10%. Assume torque to remain the same.

Solution:

$$P = 6;\ f = 50 \text{ Hz};\ S_1 = 4\% = 0.04;\ R_2 = 0.3\ \Omega$$

$$N_S = \frac{120f}{P} = \frac{120 \times 50}{6} = 1000 \text{ rpm}\ ;\ N_1 = N_S\ (1 - S) = 1000\ (1 - 0.04) = 960 \text{ rpm}$$

$$N_S = N_1 - \frac{10}{100} \times N_1 = 960 - \frac{10}{100} \times 960 = 864 \text{ rpm}$$

$$S_2 = \frac{N_S - N_2}{N_S} = \frac{1500 - 864}{1500} = 0.424$$

Since, $$T_1 = T_2;\ \frac{S_1}{R_2} = \frac{S_2}{R_2 + r} \quad \text{or} \quad R_2 + r = \frac{S_2}{S_1} \times R_2$$

or $$r = \frac{S_2}{S_1} \times R_2 - R_2 = \frac{0.424}{0.04} \times 0.3 - 0.3 = \mathbf{2.88\ ohm}\ (Ans.)$$

Example 10.11

The rotor of a 4-pole, 50 Hz, slipring induction motor has resistance of 0.25 ohm per phase and runs at 1440 rpm at full load. Calculate the external resistance per phase which must be added to lower the speed to 1200 rpm, torque remaining constant.

Solution:

No. of poles, $P = 4$

Supply frequency, $f = 50$ Hz

Rotor resistance, $R_2 = 0{\cdot}25$ ohm

$$\text{Full load speed, } N = 1440 \text{ rpm}$$

$$\text{Controlled speed, } N' = 1200 \text{ rpm}$$

$$\text{Synchronous speed, } N_s = \frac{120\,f}{P} = \frac{120 \times 50}{4} = 1500 \text{ rpm}$$

$$\text{Full load slip, } S = \frac{N_s - N}{N_s} = \frac{1500 - 1400}{1500} = 0{\cdot}04$$

Let the external resistance per phase added in the rotor circuit to control the speed be r.

Then total rotor circuit resistance $= R_2 + r$

Slip at 1200 rpm $$S_2 = \frac{N_s - N_2}{N_s} = \frac{1500 - 1200}{1500} = 0{\cdot}2$$

For constant torque, ratio of $\frac{R_2}{S_1} = \text{constant} = \frac{R_2 + r}{S_2}$

$$\therefore \quad R_2 + r = \frac{R_2}{S} \times S_2 = \frac{0 \cdot 25}{0 \cdot 04} \times 0 \cdot 2 = 1{\cdot}25 \text{ ohm}$$

$$\therefore \quad r = 1.25 - R_2 = 1{\cdot}25 - 0{\cdot}25 = \mathbf{1\ ohm}\ (Ans.)$$

Section Practice Problems

Numerical Problems

1. A 3-phase, 400 V, 6-pole, 50 Hz slip-ring induction motor has rotor resistance 0.2 ohm per phase and runs at 960 rpm at full-load. What value of external resistance is required to be added in the rotor to lower the speed to 800 rpm, torque remaining the same. (**Ans.** 0.8 ohm)

Short Answer Type Questions

Q.1. What are the factors on which speed of an induction motor depends?

Ans. Speed of an induction motor depends upon supply voltage, supply frequency, slip and number of poles of the stator.

Q.2. If the speed of an induction motor is decreased by increasing the number of poles, how does it affect the pf of the motor?

Ans. The pf of the motor will reduce.

Q.3. If two slip-ring induction motors having poles P_1 and P_2 are cascaded, how many step-speeds can be obtained?

Ans. $N_{S1} = \frac{120\,f}{P_1}$; $N_{S2} = \frac{120\,f}{P_2}$; $N_{S3} = \frac{120\,f}{P_1 + P_2}$; $N_{S4} = \frac{120\,f}{P_1 - P_2}$

Review Questions

1. What are the various methods of speed control of 3-phase induction motors? Explain the working of a direct-on-line starter with neat sketch.

2. How over-load relay and no-volt relay protect the induction motor?
3. Explain with the help of a neat sketch the working of a star-delta starter.
4. Why a starter is required to start a 3-phase induction motor? Explain how an auto-transformer starter limits the starting current?
5. Explain with the help of a neat sketch the working of an auto-transformer starter.
6. Explain how a stator resistance starter limits the starting current when used to start a 3-phase squirrel cage induction motor. What are its disadvantages?
7. Explain the working of a rotor rheostat starter used for a 3-phase slip-ring induction motor with the help of neat sketches.
8. Compare the star-delta and auto-transformer method of starting of 3-phase squirrel cage induction motors on the basis of starting torque and starting current, adoptability to motor ratting and starting conditions.
9. What are the factors which determine the speed of a 3-phase induction motor? Name various methods of speed control of 3-phase induction motor.
10. Draw and explain the speed torque characteristics of a 3-phase induction motor when its speed is controlled by (*i*) changing the rotor circuit resistance and (*ii*) controlling the supply voltage.
11. Discuss any two methods of speed control of 3-phase squirrel cage induction motors.
12. Discuss any two methods of speed control of 3-phase slip-ring induction motors.
13. Discuss the torque-speed characteristics of induction motor under *V/f* speed control.
14. Discuss the method of speed control of 3-phase squirrel cage induction motor by changing the number of poles.
15. What is the basic principle of speed control of 3-phase induction motor by pole changing method? Explain how two different torque-speed characteristics are obtained for pole changing method.
16. Explain the method of speed control of 3-phase slip-ring induction motor by changing the rotor circuit resistance. What are advantages and limitations of this method?
17. Explain cascade arrangement for controlling the speed of three-phase slip-ring induction motors.
18. Explain Kramer system of speed control of 3-phase slip-ring induction motors.
19. Explain Scherbius system of speed control of 3-phase slip-ring induction motors.
20. Explain the principle of speed control of 3-phase induction motor by injecting voltage. Explain Kramer system of speed control of 3-phase induction motors.

Multiple Choice Questions

1. An induction motor will draw largest starting current when employed with,
   (*a*) reduced voltage starting.
   (*b*) star-delta starting
   (*c*) auto-transformer starting.
   (*d*) direct-on-line starting.
2. If an induction motor is started by a star-delta starter what tapping of an auto transformer starter will provide the same results.
   (*a*) 50%.
   (*b*) 33.33%.
   (*c*) 57.7%.
   (*d*) 73.2%.

3. The star-delta starting of a 3-phase squirrel cage induction motor in comparison to DOL starting shall have
   (*a*) more starting current
   (*b*) reduced starting current
   (*c*) more starting torque
   (*d*) more acceleration.
4. In V/f speed control technique
   (*a*) air gap flux is kept constant.
   (*b*) power developed is kept constant.
   (*c*) torque developed is kept constant.
   (*d*) rotor speed is kept constant.
5. Rotor resistance speed control method is used in
   (*a*) synchronous motor
   (*b*) DC motor
   (*c*) slip-ring induction motor
   (*d*) squirrel cage induction motor.

Keys to Multiple Choice Questions

1. d	**2.** c	**3.** b	**4.** a	**5.** c

CHAPTER

11

Single-Phase Motors

Chapter Objectives

After the completion of this unit, students/readers will be able to understand:

- ✓ What are the various types of 1-phase motors?
- ✓ What is two-field revolving theory?
- ✓ How a rotating field is produced when two-phase AC supply is given to a two-phase wound motor?
- ✓ How a 1-phase induction motor is made self-starting (split-phase motor).
- ✓ How a capacitor affects the starting torque of a single-phase induction motor when placed in the starting winding?
- ✓ What is a shaded pole motor?
- ✓ What is a reluctance motor?
- ✓ What is a 1-phase synchronous motor?
- ✓ What is a hysteresis motor?
- ✓ What is a universal motor and what are its charactistics?
- ✓ How is a 1-phase induction motor selected for an application?

Introduction

Although, 3-phase induction motors are invariably employed in the industry for bulk power conversion from electrical to mechanical. But for small power conversions 1-phase induction motors are mostly used.

These motors, usually have output less than one horse-power or one kilowatt, hence are called *fractional horse-power* or *fractional kilowatt motors.* AC single-phase, fractional kilowatt motors perform variety of services in the homes, offices, business concerns, factories etc. Almost in all the domestic appliances such as refrigerators, fans, washing machines, hair driers, mixer grinders etc., only 1-phase induction motors are employed. In this chapter, we shall focus our attention on the general principles, operation and performance of single-phase induction motors.

11.1 Classification of Single-phase Motors

Usually, single-phase motors are built in fractional horse-power range and are called as fractional horse-power (*FHP*) or fractional kilo-watt (*FKW*) motors. These motors may be classified into the following four, basic categories:

1. Single-phase induction motors
   (*i*) split-phase type (*ii*) capacitor type (*iii*) Shaded-pole type
2. AC series motors or universal motors
3. Repulsion motors
   (*i*) Repulsion-start induction-run motors (*ii*) Repulsion-induction motors
4. Synchronous motors
   (*i*) Reluctance motors (*ii*) Hysteresis motors.

11.2 Single-phase Induction Motors

A single-phase induction motor is very similar to a 3-phase squirrel cage induction motor in construction. The pictorial disassembled view of a single phase induction motor is shown in Fig. 11.1. Similar to 3-phase induction motor it consists of two main parts namely stator and rotor.

Fig. 11.1 Dismantled view of a 1-phase induction motor

1. **Stator:** It is the stationary part of the motor. It has three main parts, namely. (*i*) Outer frame, (*ii*) Stator core and (*iii*) Stator winding.
   (*i*) ***Outer frame:*** It is the outer body of the motor. Its function is to support the stator core and to protect the inner parts of the machine. Usually, it is made of cost iron.
   To place the motor on the foundation, feet are provided in the outer frame as shown in Fig. 11.2(a).

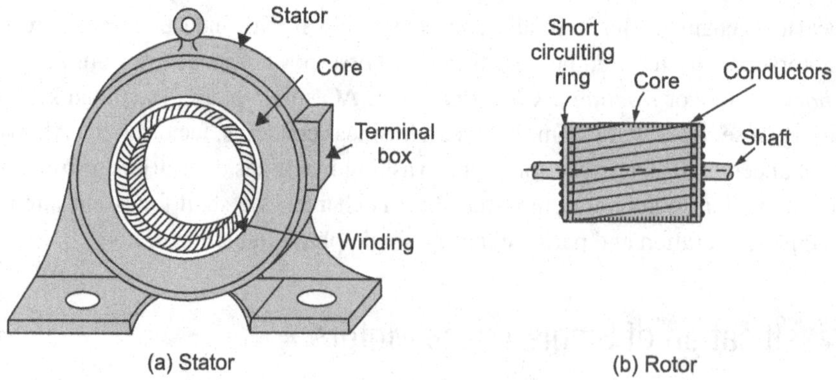

Fig. 11.2 Single-phase induction motor

(*ii*) ***Stator core:*** The stator core is to carry the alternating magnetic field which produces hysteresis and eddy current losses. To minimise these losses high grade silicon steel stampings are used to build core. The stampings are assembled under hydraulic pressure and are keyed to the outer frame. The stampings are insulated from each other by a thin varnish layer. The thickness of the stamping usually varies from 0.3 to 0.5 mm. Slots are punched on the inner periphery of the stampings to accommodate stator winding.

(*iii*) ***Stator winding*:** The stator core carries a single phase winding which is usually supplied from a single phase AC supply system. The terminals of the winding are connected in the terminal box of the machine. The stator of the motor is wound for definite number of poles, as per the need of speed.

2. **Rotor:** It is the rotating part of the motor. A squirrel cage rotor is used in single phase induction motors.

   It consists of a laminated cylindrical core of some high quality magnetic material. Semi-closed circular slots are punched at the outer periphery. Aluminium bar conductors are placed in these slots and short circuited at each end by aluminium rings, called short circuiting rings, as shown in Fig. 11.2(b). Thus, the rotor winding is permanently short circuited.

   The rotor slots are usually not parallel to the shaft but are skewed. Skewing of rotor has the following advantages:

   (*a*) It reduces humming thus ensuring quiet running of a motor,
   (*b*) It results in a smoother torque curves for different positions of the rotor,
   (*c*) It reduces the magnetic locking of the stator and rotor,
   (*d*) It increases the rotor resistance due to the increased length of the rotor bar conductors.

   The other miscellaneous parts of a 1-phase induction motor are shaft, bearings, end-rings, fan, nut-bolts, etc. as shown in disassembled view of the motor (see Fig. 11.1)

11.3 Nature of Field Produced in Single Phase Induction Motors

The field produced in a single-phase induction motor can be explained by *double revolving field* theory which is given below:

This theory is based on the "Ferraris Principle" that pulsating field produced in single phase motor can be resolved into two components of half the magnitude and rotating in opposite direction at the same synchronous speed.

Thus the alternating flux which passes across the air gap of single phase induction motor at stand still consists of combination of two fields of same strength which are revolving with same speed, one in clockwise direction and the other in anticlockwise direction. The strength of each one of these fields will be equal to one half of the maximum field strength of the actual alternating field as shown in Fig. 11.3 (*a*).

Let ϕ_m be the pulsating field which has two components each of magnitude $\phi_m/2$. Both are rotating at the same angular speed ω_s rad/sec but in opposite direction as shown in Fig. 11.3 (*a*). The resultant of the two fields is $\phi_m \cos\theta$. This shows that resultant field varies according to cosine of the angle θ. The wave shape of the resultant field is shown in Fig. 11.3 (*b*).

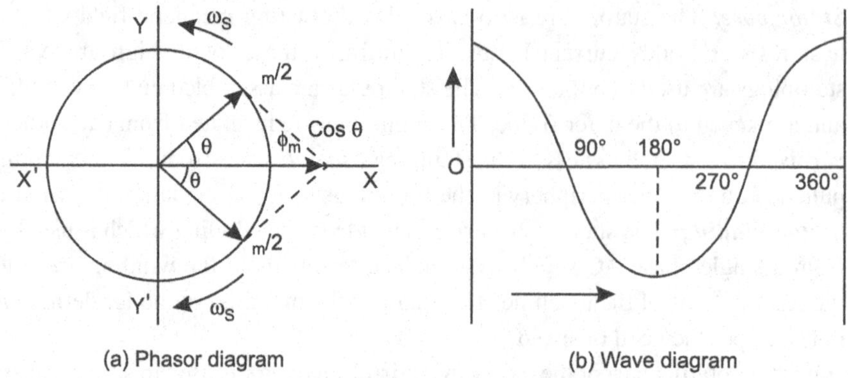

(a) Phasor diagram (b) Wave diagram

Fig. 11.3 Two field revolving theory

Mathematically

Consider the phasor diagram shown in Fig. 11.3 (a), where two magnetic fluxes each of magnitude $\phi_m/2$ are revolving in opposite direction. At any instant t, the two fluxes have been rotated through angle $\theta\,(\theta = \omega t)$. To determine the resultant value at this instant, resolve the flux vectors along *x-axis* and *y-axis*;

$$\text{Total value of flux along } x\text{-axis} = \frac{\phi_m}{2}\cos\omega t + \frac{\phi_m}{2}\cos\omega t = \phi_m \cos\omega t$$

$$\text{Total value of flux along } y\text{-axis} = \frac{\phi_m}{2}\sin\omega t - \frac{\phi_m}{2}\sin\omega t = 0$$

$$\text{Resultant flux, } \phi = \sqrt{(\phi_m \cos\omega t)^2 + (0)^2} = \phi_m \cos\omega t$$

Thus an alternating field can be represented by the two fields each of half the magnitude rotating at same angular (synchronous) speed of ω_s radians/sec but in opposite direction.

11.4 Torque Produced by Single-phase Induction Motor

The two revolving fields will produce torques in opposite directions. Let the two revolving fields be field No. 1 and field No. 2 revolving in clockwise and anticlockwise direction. The clockwise field produces torque in clockwise direction, whereas, the anticlockwise field produces torque in anticlockwise direction. The clockwise torque is plotted as positive and anticlockwise as negative as shown in Fig. 11.4. At stand still, slip for both fields is one. At synchronous speed, for clockwise direction, the field-1 will give condition of zero slip but it will give slip = 2 for field No. 2. Similarly, at synchronous speed in a counter clockwise direction, will give condition of zero slip for field -2 but slip = 2 for field No. 1. The resultant torque developed in the rotor is shown by the curve passing through zero position as shown in Fig. 11.4.

Now, if we examine the resultant torque it is observed that the starting torque (torque at slip = 1) is zero. And except at starting there is always some magnitude of resultant torque, (see at position 1-1′, the torque developed by field-1 is dominating, therefore, motor will pick-up the speed in clockwise direction, similarly, at position 2-2′, the torque developed by field-2 is dominating, therefore, motor will pick-up the speed in anti-clockwise direction). This shows that if this type of motor is once started (rotated) in either direction it will develop torque in that direction and rotor will pick-up the required speed.

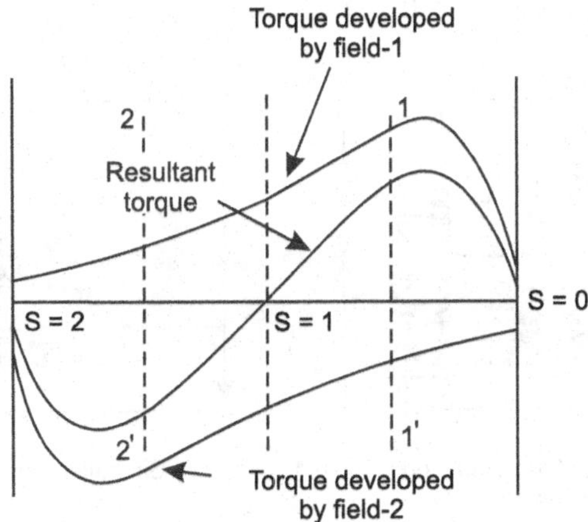

Fig. 11.4 Graphical representation of torque developed

The above analysis shows that single phase induction motor with single winding develops no starting torque but if the rotor is rotated in any direction by some auxiliary means in will develop torque in the same direction and will start rotating in that direction. So the problem is to find out the auxiliary means to give the starting torque to the motor.

11.5 Equivalent Circuit of Single-phase Induction Motor

The equivalent circuit of a single-phase induction motor may be drawn on the basis of two revolving field theory. Accordingly, each of the field is producing emf in the rotor by induction. Therefore, a single-phase induction motor may be imagined to have common stator but two rotors revolving in opposite directions. Where, each rotor has resistance and reactance half the actual rotor values:

Let R_1 be the resistance of stator winding

X_1 be the leakage reactance of stator winding

X_m be the total magnetising reactance

R_m be the total magnetising resistance

R_2' be the resistance of rotor referred to stator

X_2' be the reactance of rotor referred to stator

While developing the equivalent circuit, it is considered that the stator is having only one winding. The equivalent circuit can be developed under stand-still (at start) and running (operating) conditions.

At Standstill Conditions

At standstill, the motor is considered simply as a transformer with its secondary short-circuited. The only difference is that in this case, two fields are considered revolving in opposite direction. Therefore, for each field rotor resistance and reactance is considered to be half the value, i.e., $R_2'/2$

and $X_2'/2$. Moreover, each rotor is associated with half the total magnetising reactance and resistance i.e., $X_m/2$ and $R_m/2$ respectively.

The equivalent circuit of a single-phase induction motor at standstill is shown in Fig. 11.5(a), Its simplified circuit is shown in Fig. 11.5(b) where core loss component R_m has been neglected.

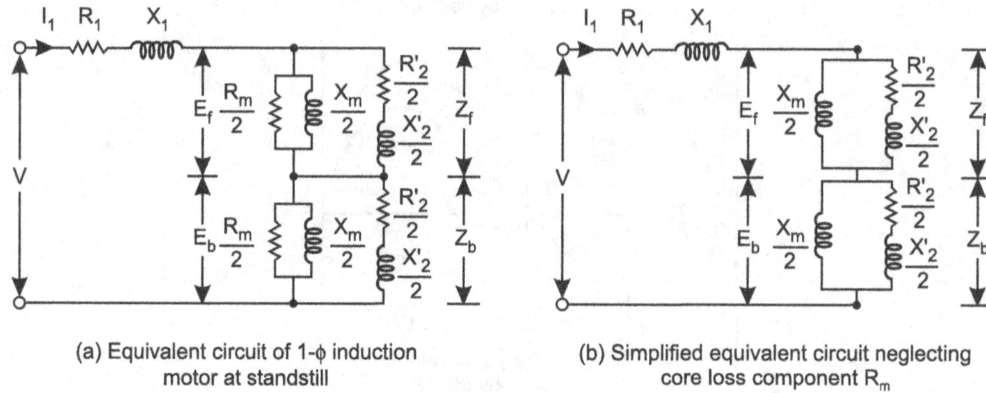

Fig. 11.5 Equivalent circuit of 1-phase induction motor at standstill.

Here, $E_f = 4.44 f N \phi_f$

and $E_b = 4.44 f N \phi_b$

At standstill, $\phi_f = \phi_b$,

therefore, $E_f = E_b$

Now, as per circuit, $\overline{V} = \overline{E}_f + \overline{E}_b = \overline{I}_1\overline{Z}_f + \overline{I}_1\,\overline{Z}_b$

Where Z_f = impedance of forward parallel branch

Z_b = impedance of backward parallel branch

At Running Condition

At running condition, if the rotor is rotating in the direction of the forward revolving field with the slip *S*, the rotor current produced by the forward field will have a frequency *Sf*. At the same time the rotor current produced by the backward field will have a frequency *(2-S) f*. Accordingly, the equivalent circuit of single-phase induction motor at running condition is drawn and shown in Fig. 11.6(a). Whereas, it is simplified in Fig. 11.6(b) where core loss component R_m has been neglected.

Total circuit impedance,

$$\overline{Z}_T = \overline{Z}_1 + \overline{Z}_f + \overline{Z}_b$$

Where, $\overline{Z}_1 = R_1 + jX_1$

$$\overline{Z}_f = \frac{j\dfrac{X_m}{2}\left(\dfrac{R_2'}{2S} + j\dfrac{X_2'}{2}\right)}{\dfrac{R_2'}{2S} + j\left(\dfrac{X_m}{2} + \dfrac{X_2'}{2}\right)}$$

$$\overline{Z}_b = \frac{j\frac{X_m}{2}\left(\frac{R_2'}{2(2\text{-}S)} + j\frac{X_2'}{2}\right)}{\frac{R_2'}{2(2\text{-}S)} + j\left(\frac{X_m}{2} + \frac{X_2'}{2}\right)}$$

and $\overline{I}_1 = \frac{\overline{V}_1}{\overline{Z}_T}$

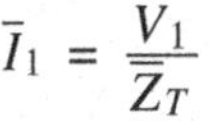

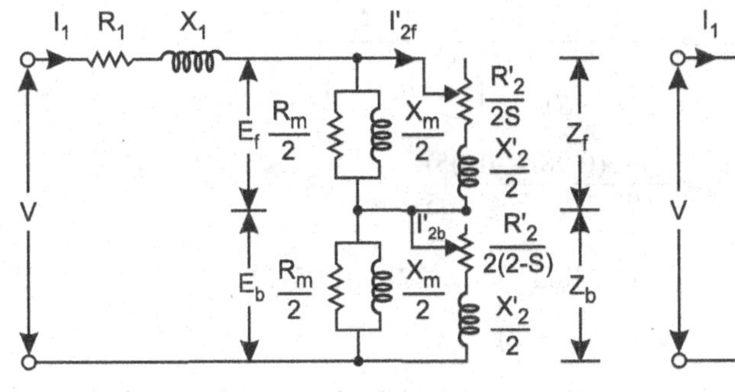

(a) Equivalent circuit of 1-φ induction motor at running condition

(b) Simplified equivalent circuit neglecting core loss component

Fig. 11.6 Equivalent circuit of a 1-ϕ induction motor at running condition

Example 11.1

The following are the parameters of a 230V, 50 Hz, 4-pole, single phase induction motor $R_1 = 2.2\Omega$; $X_1 = 3.0\Omega$; $R'_2 = 3.8\Omega$; $X'_2 = 2.1\Omega$; $X_m = 86\Omega$.

Calculate the input current and power when the motor is operating at full-load speed of 1410 rpm.

If the mechanical losses (iron and friction loss) are 60 W, determine gross power developed, useful power at the shaft, shaft torque and efficiency.

Solution:

Here, $V = 230$ V; $f = 50$ Hz; $P = 4$; $N = 1410$ rpm

$$R_1 = 2.2\ \Omega;\ X_1 = 3.0\ \Omega;\ R_2' = 3.8\ \Omega;\ R_2' = 2.1\ \Omega;\ X_m = 86\ \Omega$$

For equivalent circuit, refer to Fig. 11.6 (b).

$$\text{Synchronous speed, } N_S = \frac{120f}{P} = \frac{120 \times 50}{4} = 1500 \text{ rpm}$$

$$\text{Slip, } S = \frac{N_S - N}{N_S} = \frac{1500 - 1410}{1500} = 0.06$$

Now,

$$\frac{R_2'}{2S} = \frac{3.8}{2 \times 0.06} = 31.67\ \Omega;\ \frac{X_2'}{2} = \frac{2.1}{2} = 1.05\ \Omega$$

$$\frac{R_2'}{2(2-S)} = \frac{3.8}{2(2-0.06)} = 0.98\ \Omega;\ \frac{X_m}{2} = \frac{86}{2} = 43\ \Omega$$

$$\bar{Z}_f = \frac{j\dfrac{X_m}{2}\left(\dfrac{R_2'}{2S} + j\dfrac{X_2'}{2}\right)}{\dfrac{R_2'}{2S} + j\left(\dfrac{X_m}{2} + \dfrac{X_2'}{2}\right)} = \frac{j43(31.67 + j1.05)}{31.67 + j(43 + 1.05)}$$

$$= (25.22 \angle 37.61°) \text{ ohm} = (19.98 + j\,15.39)\ \Omega$$

$$\bar{Z}_b = \frac{j\dfrac{X_m}{2}\left(\dfrac{R_2'}{2\,(2-S)} + j\dfrac{X_2'}{2}\right)}{\dfrac{R_2'}{2\,(2-S)} + j\left(\dfrac{X_m}{2} + \dfrac{X_2'}{2}\right)}$$

$$= \frac{j43(0.98 + j1.05)}{0.98 + j(43 + 1.05)}$$

$$= 1.40\angle 48.3° = (0.933 + j1.05)\ \Omega$$

Total impedance, $\bar{Z}_T = \bar{Z}_1 + \bar{Z}_f + \bar{Z}_b$

$$= (2.2 + j3) + (19.98 + j15.39) + (0.933 + j1.05)$$

$$= (23.11 + j19.44)\ \Omega = 30.2 \angle 40.1°$$

Input current, $\bar{I}_I = \dfrac{\bar{V}}{\bar{Z}_T} = \dfrac{230 \angle 0°}{30.2 \angle 40.1°} = 7.616 \angle -40.1°$ A

$I_I = \mathbf{7.616\ A}$ (*Ans.*)

Input power, $P = I_I^2\, R_T = (7.616)^2 \times 23.11 = \mathbf{1340\ W}$ (*Ans.*)

Alternately, $P = V I_I \cos\phi = 230 \times 7.616 \times \cos 40.1° = \mathbf{1340\ W}$ (*Ans.*)

The voltage across *AB*, i.e.,

$$\bar{V}_f = \bar{I}_1 \bar{Z}_f = 7.616\angle -40.1° \times 25.22\angle 37.61°$$

$$= 200.5 \angle -2.5° \text{ V}$$

Current in the forward parallel branch, i.e.,

$$\bar{I}_f = \frac{\bar{V}_f}{\bar{Z}_2'} = \frac{200.5 \angle -2.5°}{31.67 + j\,1.05} = \frac{200.5 \angle -2.5°}{31.69 \angle 1.9°}$$

$$= 6.327 \angle -4.4° \text{ A}$$

The voltage across *BC*, i.e.,

$$\bar{V}_b = \bar{I}_1\, \bar{Z}_b = 7.616 \angle -40.1° \times 1.40 \angle 48.3°$$

$$= 10.66 \angle 8.2°$$

Current in the backward parallel branch, i.e.,

$$\bar{I}_b = \frac{\bar{V}_b}{\bar{Z}_2'} = \frac{10.66 \angle 8.2°}{0.98 + j\,1.05} = \frac{10.66 \angle 8.2°}{1.436 \angle 47°} = 7.42 \angle -38.8° \text{ A}$$

Effective resistance of the forward branch $= \dfrac{R_2'}{2S} = \dfrac{3.8}{2 \times 0.06} = 31.67\ \Omega$

The actual resistance existing in the branch = $\frac{3.8}{2} = 1.9\ \Omega$

It depicts that (31.67 – 1.9 = 29.77 Ω) is the electrical equivalent resistance of mechanical load.

∴ Power developed in the forward branch

$$P_{mf} = I_f^2 \times 29.77 = (6.327)^2 \times 29.77 = 1192\ \text{W}$$

Effective resistance of the backward branch = $\frac{3.8}{2\,(2-S)} = \frac{3.8}{2\,(2-0.06)} = 0.98\ \Omega$

Actual resistance existing in the branch = $\frac{3.8}{2} = 1.9\ \Omega$

Electrical equivalent resistance representing mechanical load

$$= (0.98 - 1.9) = -0.92\ \Omega$$

The negative sign indicates that the power developed produces a backward torque.

Power developed in the backward branch,

$$P_{mb} = I_b^2 \times (-0.92) = (7.42)^2 \times (-0.92) = -51\ \text{W}$$

Hence, total mechanical power developed,

$$P_m = P_{mf} - P_{mb} = 1192 - 51 = 1141\ \text{W}$$

Power at the shaft, $P_{Sh} = P_m$ – mechanical loss = 1141 – 60 = **1081 W** (*Ans.*)

Shaft torque = $\frac{P_{sh}}{2\pi N / 60} = \frac{60 \times 1081}{2\pi \times 1410}$ = **7.32 Nm** (*Ans.*)

Efficiency = $\frac{P_{out}}{P_{in}} = \frac{1081}{1340}$ = 0.8067 = **80.67%** (*Ans.*)

Example 11.2

The gross power absorbed by the forward and backward field of a 230V, 4-pole, 50 Hz, single-phase induction motor is 180 W and 30 W respectively at a motor speed of 1425 rpm. Find the shaft torque if the no-load friction losses are 50W.

Solution:

Here, $V = 230$ V; $P = 4$; $f = 50$ Hz; $N = 1425$ rpm

$P_f = 180$ W; $P_b = 30$ W; Mechanical loss = 50 W

Synchronous speed, $N_S = \frac{120f}{P} = \frac{120 \times 50}{4} = 1500$ rpm

Slip, $S = \frac{N_S - N}{N_S} = \frac{1500 - 1425}{1500} = 0.05$

Power input to the rotor, $P_{i(r)} = P_f - P_b = 180 - 30 = 150$ W

Mechanical power developed, $P_m = P_{i(r)}(1-S) = 150\,(1-0.05) = 142.5$ W

Power at the shaft, $P_{sh} = P_m$ – Mechanical loss = 142.5 – 50 = 92.5 W

$$\text{Shaft torque, } T_{sh} = T_{sh} = \frac{P_{sh}}{\omega} = \frac{P_{sh}}{2\pi N / 60} = \frac{92.5 \times 60}{2\pi \times 1425} = \mathbf{0.62\ Nm}\ (Ans.)$$

11.6 Rotating Magnetic Field from Two-phase Supply

In case of 3-phase induction motors, we have seen that a revolving field of magnitude 1.5 ϕ_m is produced in the stator core when 3-phase AC supply is given to its 3-phase stator winding. Similarly, a 2-phase balanced AC supply also produces a rotating magnetic field of constant magnitude ϕ_m.

Let us see how?

Consider the stator of a 2-pole machine having two phase winding represented by the concentric coils *a-a'* and *b-b'* respectively as shown in Fig. 11.7.

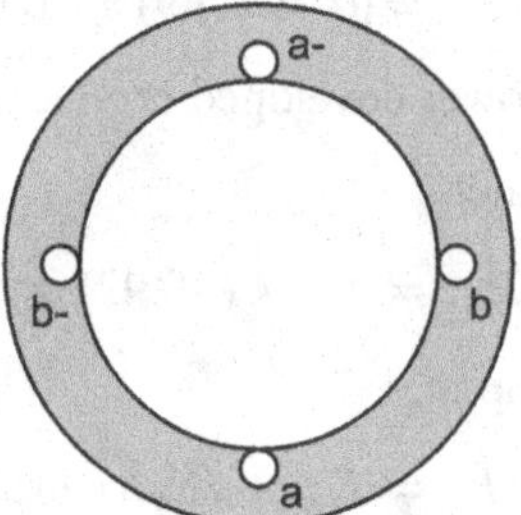

Fig. 11.7 Stotar carrying 2-phase winding

Let a 2-phase supply having wave diagram shown in Fig. 11.8(a) is applied to the stator winding. Phase-1 is connected to coil *a-a'* and phase-2 is connected to coil *b-b'*. Alternating currents having the same wave shape as that of supply voltage start flowing through the coils and produce their own magnetic field ϕ_m (each). The phasor diagram of the fields at an instant is shown in Fig. 11.8(b). The positive half cycle of the alternating current is considered as inward flow of current [cross in a circle ⊕] and negative half cycle as outward flow of current (dot in a circle ⊙). This abbreviation is marked in the start terminals of the two coils i.e., *a* and *b*, respectively. Whereas, the direction of flow of current is opposite in the other two ends of the coils.

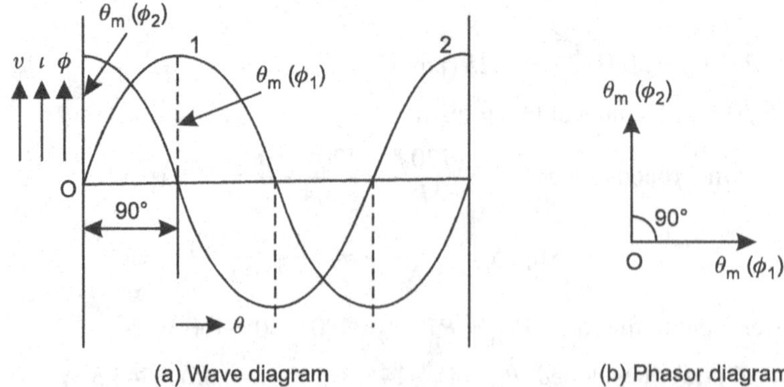

(a) Wave diagram (b) Phasor diagram

Fig. 11.8 Wave diagram and phasor diagram of fields produced by 2-phase supply

The two fields are given by the equations:

$$\phi_1 = \phi_m \sin \theta \, (or\ \phi_m \sin \omega t)$$

$$\phi_2 = \phi_m \sin (\theta + 90°)$$

Considering instants:

$t_1 - \theta = 0°$ $\quad$ $t_4 - \theta = 180°$

$t_2 - \theta = 45°$ $\quad$ $t_5 - \theta = 225°$

$t_3 - \theta = 90°$ $\quad$ $t_6 - \theta = 270°$

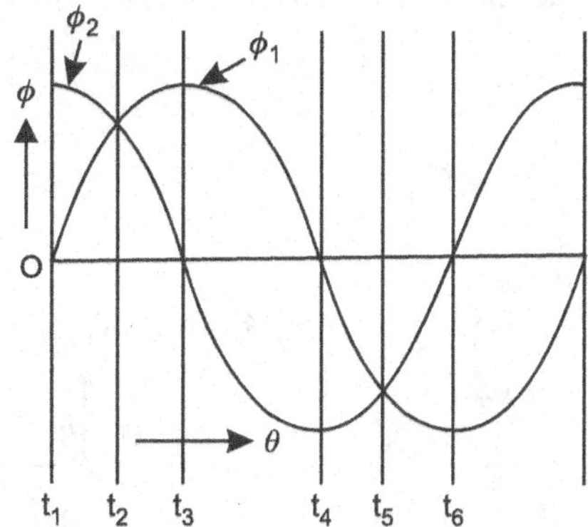

Fig. 11.9 Various instants marked on the wave diagram

Let us see the position of the resultant magnetic field at various instants marked in Fig. 11.9

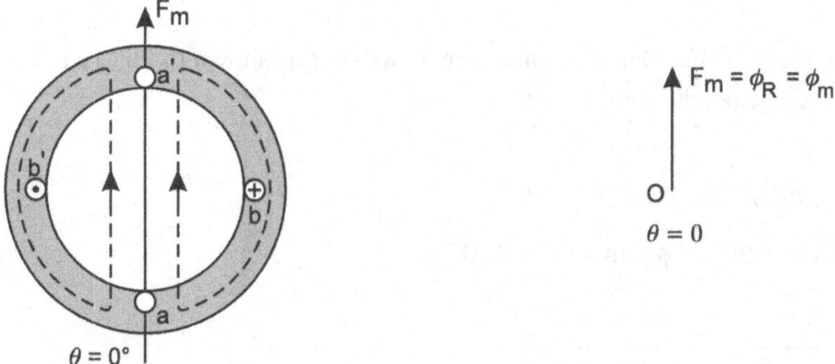

(a) Position of the resultant field (b) Phasor position of the resultant field

Fig 11.10 Flow of current in coil-sides and position of resultant field at instant t_1

At an instant t_1, when $\theta = 0$, the flow of current in the start terminal of coil *a-a'* i.e., *a* is zero, whereas the start terminal of *b-b*, i.e., *b* it is positive. Accordingly, current in terminal *a* is zero and in *a'* also zero but in terminal *b* it is ⊕, whereas in *b'* it is ⊙ as shown in Fig. 11.10(a). The magnetic field produced in the stator core at instant t_1 is shown in Fig. 11.10(a). F_m represents the direction of axis of the resultant field.

When $\theta = 0$

$$\phi_1 = \phi_m \sin\theta = \phi_m \sin 0° = 0$$
$$\phi_2 = \phi_m \sin(\theta + 90°) = \phi_m \sin 90° = \phi_m$$
$$\bar{\phi}_R = \bar{\phi}_1 + \bar{\phi}_2 = 0 + \bar{\phi}_m = \bar{\phi}_m \text{ or } \phi_R = \phi_m$$

The phasor representation of the resultant field F_m or $\phi_R = \phi_m$ is shown in Fig. 11.10(b). The magnitude of the resultant field is ϕ_m.

At an instant t_2, when $\theta = 45°$, the flow of current is start terminal of coil *a-a'* i.e., *a* is positive as well as in start terminal of coil *b-b;* i.e., *b* is also positive as shown in wave diagram Fig. 11.9(a). Accordingly, the direction of flow of current in start terminals of the two coils i.e., *a* and *b* is marked as ⊕ but the flow of current in the finish terminals of the two coils i.e., *a'* and *b'* is ⊙ as shown in Fig. 11.11(a).

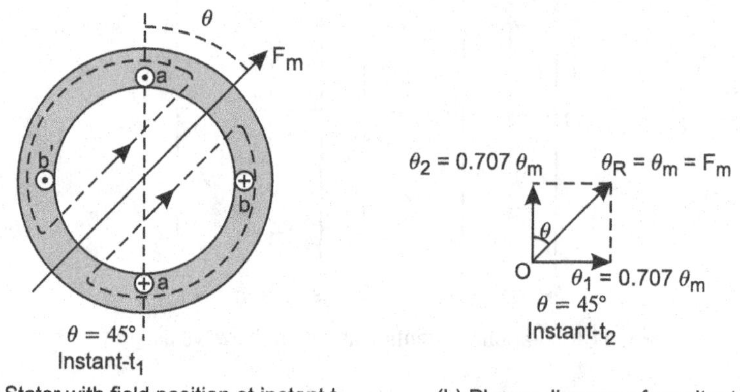

(a) Stator with field position at instant t_2 (b) Phasor diagram of resultant field

Fig. 11.11 Axis of resultant field in the stator core at instant t_2

The magnetic field produced in the stator core at instant t_2 is shown in Fig. 11.11(a). F_m represents the direction of axis of resultant main field.

When $\theta = 45°$

$$\phi_1 = \phi_m \sin\theta = \phi_m \sin 45° = 0.707\,\phi_m$$
$$\phi_2 = \phi_m \sin(\theta + 90°) = \phi_m \sin 135° = 0.707\,\phi_m$$
$$\bar{\phi}_R = \bar{\phi}_1 + \bar{\phi}_2$$
$$\phi_R = \sqrt{(0.707\,\Phi_m)^2 + (0.707\,\Phi_m)^2} = \Phi_m$$
$$\theta = \tan^{-1}\frac{\phi_1}{\phi_2} = \tan^{-1}\frac{0.707\phi_m}{0.707\phi_m} = 45°$$

The phasor representation of the resultant field F_m or $\phi_R = \phi_m$ is shown in Fig. 11.9(b). The magnitude of the resultant field is ϕ_m but it has rotated through an angle $\theta = 45°$ with respect to its first position.

At an instant t_3, when $\theta = 90°$, the flow of current in the start terminal of coil *a-a'*, i.e., *a* is positive, whereas, in the start terminal, of coil *b-b* i.e., '*b*' is zero. Accordingly, current in terminal *a* is ⊕ and in *a'* it is ⊙ whereas, in conductor *b* and *b'* it is zero as shown in Fig. 11.12(a). The magnetic field produced in the stator core at an instant t_3 is shown in Fig. 11.11(a). F_m represents the direction of axis of the resultant field.

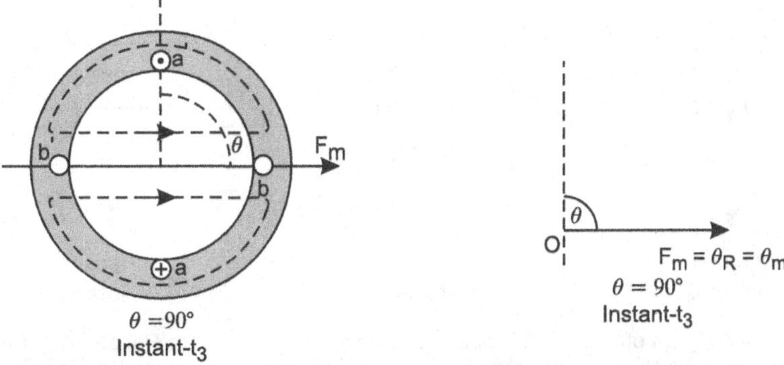

(a) Stator with field position at instant t_2 (b) Phasor diagram of resultant field

Fig. 11.12 Axis of resultant field in the stator core at instant t_3

When $\theta = 90°$

$$\phi_1 = \phi_m \sin \theta = \phi_m \sin 90° = \phi_m$$

$$\phi_2 = \phi_m \sin (\theta + 90°) = \phi_m \sin 180° = 0$$

$$\bar{\phi}_R = \bar{\phi}_1 + \bar{\phi}_2 = \bar{\phi}_m + 0 = \bar{\phi}_m \text{ or } \phi_R = \phi_m$$

The phasor representation of the resultant field F_m or $\phi_R = \phi_m$ is shown in Fig. 11.12(b). The magnitude of the resultant field is ϕ_m but it has been rotated through an angle 90° with respect to its first position.

Similarly, the position of the axis of the resultant field set-up in the stator at various instants (t_4, t_5 and t_6) are shown in Fig. 11.13, whereas, the phasor representation and magnitude of resultant field at various instants (t_4, t_5 and t_6) are shown in Fig. 11.14. Thus, the resultant phasor completes one revolution in one cycle.

From the above discussion, it is revealed that when a 2-phase AC supply is given to a 2-phase wound stator of a machine, a resultant field of magnitude ϕ_m is produced which rotates in the space of the stator at synchronous speed $\left(N_S = \dfrac{120f}{P}\right)$.

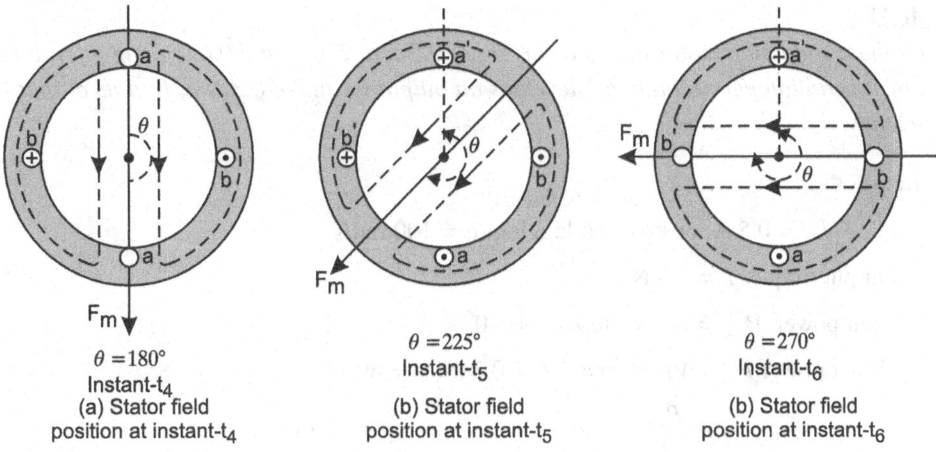

(a) Stator field position at instant-t_4 (b) Stator field position at instant-t_5 (b) Stator field position at instant-t_6

Fig. 11.13 Stator with field position at various instants (t_4, t_5 and t_6)

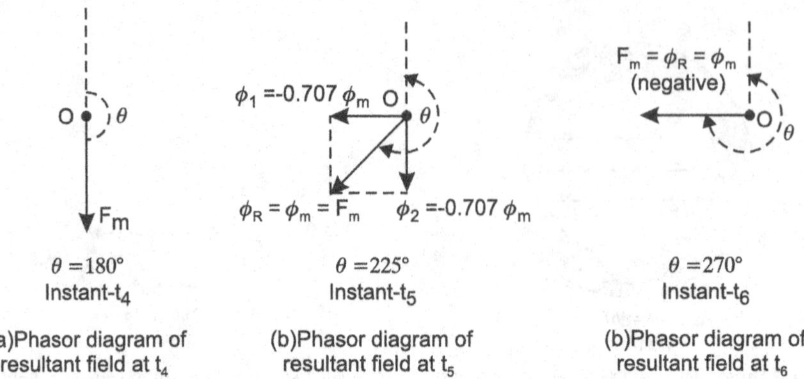

Fig. 11.14 Phasor position of the resultant field at various instants (t_4, t_5 and t_6)

11.7 Methods to make Single-phase Induction Motor Self-starting

As discussed, a single-phase induction motor inherently is not self-starting. To make it self-starting, some method is required to be evolved to produce a revolving magnetic field in the stator core. This may be obtained by converting a single-phase supply into two-phase supply which can be achieved by using an additional winding (this additional winding may be or may not be disconnected once the motor starts and picks-up the speed) or by creating another field (by induction) using a short circuiting band or ring. Accordingly, depending upon the method used to make a 1-phase induction motor self-starting, single-phase induction motors can be classified as:

1. **Sprit-phase motors:** These motors are started by employing two-phase motor action through the use of an auxiliary winding called starting winding.
2. **Capacitor motors:** These motors are started by employing two-phase motor action through the use of an auxiliary winding with capacitor.
3. **Shaded-pole motors:** These motors are started by the interaction of the field produced by a shading band or short circuiting ring placed around a portion of the pole structure.

Example 11.3

A single-phase induction motor draws a current of 0.5 A at 230 V and 0.6 lagging p.f. If it runs at a speed of 100 radian per second and develops an output torque of 0.3 Nm, find its output power and efficiency.

Solution:

Here, $V = 230$ V; $I = 0.5$ A; $\cos\phi = 0.6$ lagging; $\omega = 100$ rad/s

Output torque, $T = 0.3$ Nm

Output power, $P_{out} = \omega T = 100 \times 0.3 = 30$ W

Input power $P_{in} = VI\cos\phi = 230 \times 0.5 \times 0.6 = 69$ W

$$\text{Efficiency, } \eta = \frac{P_{out}}{P_{in}} \times 100 = \frac{30}{69} \times 100 = \mathbf{43.48\%} \text{ (Ans.)}$$

11.8 Split Phase Motors

Construction

The outer frame and stator core of a split-phase motor is similar to the outer frame and stator core of a 3-phase induction motor as shown in Fig. 11.2. It is provided with an auxiliary stator winding called starting winding in addition to main winding. These windings are placed in the stator slots. Both the windings are put in parallel as shown in Fig. 11.15(*a*). The purpose is to get two different currents sufficiently displaced from each other so that a revolving field is produced. The main winding which is highly inductive is connected across the line in the usual manner. The auxiliary or starting winding has a greater resistances and lesser reactance as compared to main winding.

The current in the starting winding I_s lags the supply voltage by lesser angle ϕ_s whereas the current in the main winding I_m being highly inductive lags the supply voltage by greater angle ϕ_m as shown in Fig. 11.15(*b*). The two currents have a phase difference of $\theta°$ electrical. Thus, a revolving field is set-up in the stator and a starting torque is developed in the rotor.

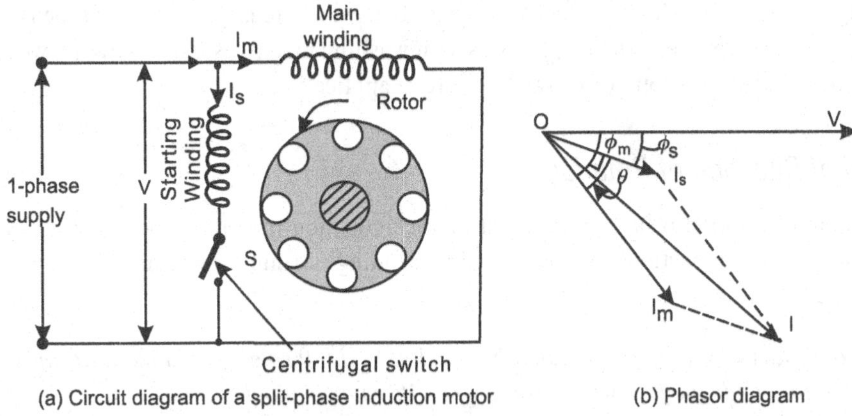

(a) Circuit diagram of a split-phase induction motor (b) Phasor diagram

Fig. 11.15 Split-phase induction motor

Consequently rotor starts rotating and picks up the speed. A centrifugal switch which is normally closed is incorporated in series with the starting winding. When the motor attains a speed about 75% of synchronous speed, the centrifugal switch is opened automatically with the help of centrifugal force and puts the starting winding out of circuit. It is important that the centrifugal switch should open otherwise the auxiliary winding being made of thin wire will be over heated and may damage.

Performance and Characteristics

A typical torque speed characteristics are shown in Fig. 11.16, the starting torque is about twice the full load torque. The current at start is about 6 to 8 times. The speed falls with increase in load by only about 5% to 7% otherwise it is a constant speed motor. Speed is governed by the relation

$$N_S = \frac{120\,f}{P} \text{ rpm}$$

Actual speed is less than synchronous speed N_S. For the same weight its rating is about 60% to that of the poly phase induction motor. It has lower p.f. and lesser efficiency. P.f. is about 0.6 and efficiency is also about 60%.

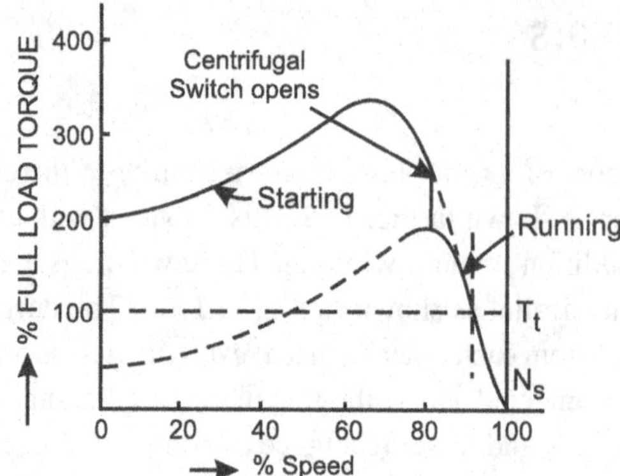

Fig. 11.16 Torque-speed characteristics of a split-phase induction motor

Applications

As starting torque is not so high so this machine is not used where large starting torque is required. It is used for smaller sizes about 0.25 h.p. It is used in washing machines, fans, blowers, wood working tools, grinders and various other low starting torque applications.

Reversal of Direction of Rotation

The direction of rotation of a 1-phase (split phase) induction motor can be reversed by reversing (interchanging) the connections of either starting winding or running winding.

Example 11.4

The main and starting winding of a 230V, 50 Hz, 190 W, 1750 rpm, resistance split-phase (1-phase) induction motor yields the following results when its rotor is blocked:

	Applied voltage	*Current*	*Active Power*
Main winding:	*45 V*	*4 A*	*120 W*
Starting winding:	*45 V*	*1.5 A*	*60 W*

Calculate (i) The phase angle between I_m and I_S

(ii) The locked-rotor current drawn from the lines at 230 V.

Solution:

Here, $V = 230$ V; $f = 50$ Hz; $N = 1750$ rpm

At block rotor, for main winding, $I_m = 4$A; $V_{sc} = 45$V; $P_{sc(m)} = 120$ W

For starting winding, $I_S = 4$A; $V_{sc} = 45$V; $P_{sc(s)} = 60$ W

(*i*) Let us first calculate the phase angle between I_m and the applied voltage V for the main winding.

$$\text{Apparent power, } P_{a(m)} = V_{sc} I_m = 45 \times 4 = 180 \text{ VA}$$

$$\therefore \quad \text{Power factor, } \cos \phi_m = P_{sc(m)} / P_{a(m)} = 120 / 180 = 0.667 \therefore \phi_m = 48.2°$$

Therefore, I_m lags behind V by an angle of 48.2°

Now, we calculate the phase angle between I_S and V for the starting winding.

$$\text{Apparent power, } P_{a(s)} = V_{sc} I_s = 45 \times 1.5 = 67.5 \text{ VA}$$

$$\therefore \quad \text{Power factor, } \cos\phi_S = P_{sc(s)} / P_{a(s)} = 60 / 67.5 = 0.89 \therefore \phi_s = 27.3°$$

Therefore, I_S lags behind V by an angle of 27.3°.

The phase angle between I_m and I_S, $\theta = \phi_m - \phi_s = 8.2° - 27.3° = \mathbf{20.9°}$

(ii)

$$\text{Total active power, } P = P_m + P_s = 120 + 60 = 180 \text{ W}$$

$$\text{Total reactive power, } P_r = P_{rm} + P_{rs} = P_m \tan\phi_m + P_s \tan\phi_s$$

$$= 120 \tan 48.2° + 60 \tan 27.3° = 165.18 \text{ VAR}$$

$$\text{Total apparent power, } P_a = \sqrt{P^2 + P_r^2} = \sqrt{(180)^2 + (165.18)^2} = 244.3 \text{ VA}$$

$$\text{Short circuit line current at 45V} = \frac{P_a}{V_{sc}} = \frac{244.3}{45} = 5.43 \text{ A}$$

$$\text{Short circuit current at 230V} = 5.43 \times \frac{230}{45} = \mathbf{27.75\ A}\ (Ans.)$$

Example 11.5

The main winding of a 230V, 0.5 HP single-phase induction motor (split-phase) draw a current of 6 A which lags behind the voltage vector by 45°, whereas the starting winding draw a current of 4A which lags being the voltage by 15°. Determine

(i) Current drawn by the motor at start and its pf.
(ii) Current draw by the motor during running and its pf.
(iii) Phase angle between the current drawn by the main winding and starting winding.
(iv) Power drawn by the motor during starting and running.

Solution:

Here, $V = 230$ V; $I_m = 6$A; $\phi_m = 45°$; $I_S = 4$A; $\phi_S = 15°$; output = 0.5 HP

Taking supply voltage as a reference vector,

$$\text{Total current drawn by the motor, } \bar{I} = \bar{I}_m + \bar{I}_S = 6\angle -45° + 4\angle -15°$$

$$= 6(\cos 45° - j\sin 45°) + 4(\cos 15° - j\sin 15°)$$

$$= (4.243 - j\,4.243) + 4\,(3.864 - j\,1.035)$$

$$= (8.107 - j\,5.278) = 9.67 \angle -33° \text{ A}$$

(i) Current drawn by motor, I = **9.67A** (*Ans.*)

Power factor of the motor, $\cos\phi = \cos 33°$ = **0.838 lagging** (*Ans.*)

(ii) During running, the starting winding is disconnected

$\therefore$ Current drawn by the motor, I_m = **6A** (*Ans.*)

Power factor during running, $\cos\phi_m = \cos 45°$ = **0.7.7 lagging** (*Ans.*)

(iii) Phase difference between vector I_m and I_S,

$$\theta = \phi_m - \phi_s = 45° - 15° = \mathbf{30°}\ (Ans.)$$

(iv) Power drawn by motor during starting,

$$P_S = VI\cos\phi = 230 \times 9.67 \times \cos 33° = \mathbf{1865.3\ W}\ (Ans.)$$

Power drawn by motor during running.

$$P = VI_m \cos\phi_m = 230 \times 6 \times \cos 45° = \mathbf{975.8\ W}\ (Ans.)$$

Example 11.6

The parameters of the main and starting winding of a 240 V, 50 Hz split-phase induction motor are given as:

Main winding: $R_m = 6\ \Omega;\ X_m = 8\ \Omega$
Starting winding: $R_S = 8\ \Omega;\ X_S = 6\ \Omega$

Determine:

(i) *Current in the main winding*
(ii) *Current in the starting winding*
(iii) *Phase angle between I_m and I_S*
(iv) *Line current and pf of the motor*
(v) *Power drawn by the motor.*

Solution:

Here, $V = 240$ V ; $f = 50$ Hz

$R_m = 6\ \Omega;\ X_m = 8\ \Omega;\ R_S = 8\ \Omega;\ X_S = 6\ \Omega$

Taking supply voltage V as reference vector, $\bar{V} = V \angle 0°$

(i) Impedance of main winding, $\bar{Z}_m = (6 + j8)$ ohm $= 10 \angle 53.13°$

$$\text{Current, } \bar{I}_m = \frac{\bar{V}}{\bar{Z}_m} = \frac{230\angle 0°}{10\angle 53.13°} = 23\angle -53.13\text{A}$$

$$\bar{I}_m = \mathbf{23\ A;\ \phi_m = 53.13°\ lagging}\ (Ans.)$$

(ii) Impedance of starting winding, $\bar{Z}_S = (8 + j\,6) = 10 \angle 36.87°\ \Omega$

$$\text{Current, } \bar{I}_S = \frac{\bar{V}}{\bar{Z}_S} = \frac{230 \angle 0°}{10 \angle 36.87°} = 23 \angle -36.87°\text{ A}$$

$$I_S = \mathbf{23\ A;\ \phi_S = 36.87°\ lagging}\ (Ans.)$$

(iii) Phase angle between I_m and I_S,

$$\theta = \phi_m - \phi_s = 53.13° - 36.87° = \mathbf{16.26}°\ (Ans.)$$

(iv)

$$\text{Line current, } \bar{I} = \bar{I}_m + \bar{I}_s = 23\angle -53.13° + 23\angle 36.87°$$

$$= 13.8 - j18.4 + 18.4 - j13.8 = (32.2 - j32.2)\text{ ohm}$$

$$= 45.54 \angle -45°$$

Power factor of the motor $= \cos\phi = \cos 45° = \mathbf{0.707\ lagging}$ (*Ans.*)

(v) Power drawn by the motor, $P = VI\cos\phi$

$$P = VI\cos\phi = 230 \times 45.54 \times 0.707 = \mathbf{7405\ W}\ (Ans.)$$

11.9 Capacitor Motors

It is also a split phase motor. In this motor, a capacitor is connected in series with the starting winding. This is an improved form of the above said split phase motor. In these motors, the angular displacement between I_S and I_m can be made nearly 90° and high starting torques can be obtained since starting torque is directly proportional to sine of angle θ. The capacitor in the starting winding may be connected permanently or temporarily. Accordingly, capacitor motors may be

1. Capacitor start motors. 2.Capacitor run motors. 3. Capacitor start and capacitor run motors.

1. **Capacitor start motors:** In the capacitor start induction motor capacitor C is of large value such that the motor will give high starting torque since torque $T \times \sin \phi$ and in this case, the phase angle between I_m and I_S is made near to 90°, as shown in Fig. 11.17(b). Capacitor employed is of short time duty rating. Capacitor is of electrolytic type. Electrolytic capacitor C is connected in series with the starting winding along with centrifugal switch S as shown in Fig. 11.17(*a*). When the motor attains the speed of about 75% of synchronous speed starting winding is cut-off. The construction of the motor and winding is similar to usual split phase motor. It is used where high starting torque is required such as refrigerators.

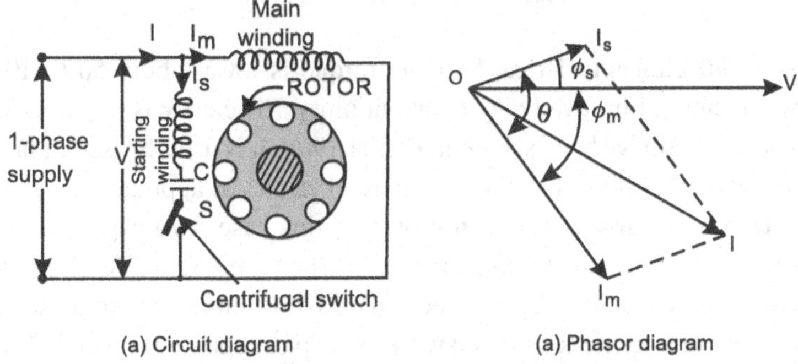

Fig. 11.17 Capacitor start motor

Performance and characteristics: Speed is almost constant with in 5% slip. This type of motor develops high starting torque about 4 to 5 times the full load torque. It draw low starting current. A typical torque speed curve is shown in Fig. 11.18. The direction of rotation can be changed by interchanging the connection of either starting or running winding.

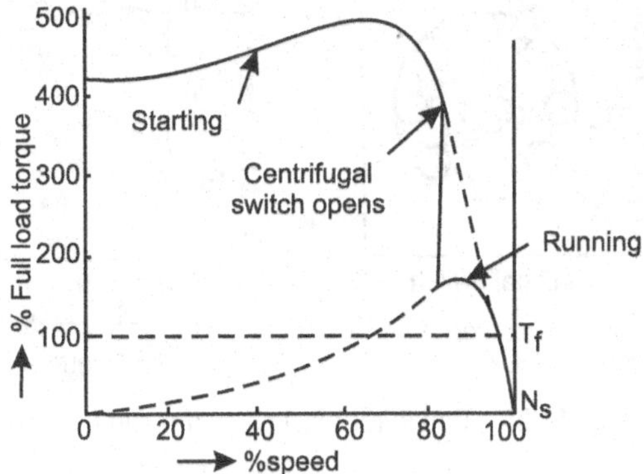

Fig. 11.18 Torque-speed characteristics

2. **Capacitor run motors (fan motors):** In these motors, a paper capacitor is permanently connected in the starting winding, as shown in Fig. 11.19(*a*). In this case, electrolytic capacitor cannot be used since this type of capacitor is designed only for short time rating and hence cannot be permanently connected in the winding. Both main as well as starting winding is of equal rating.

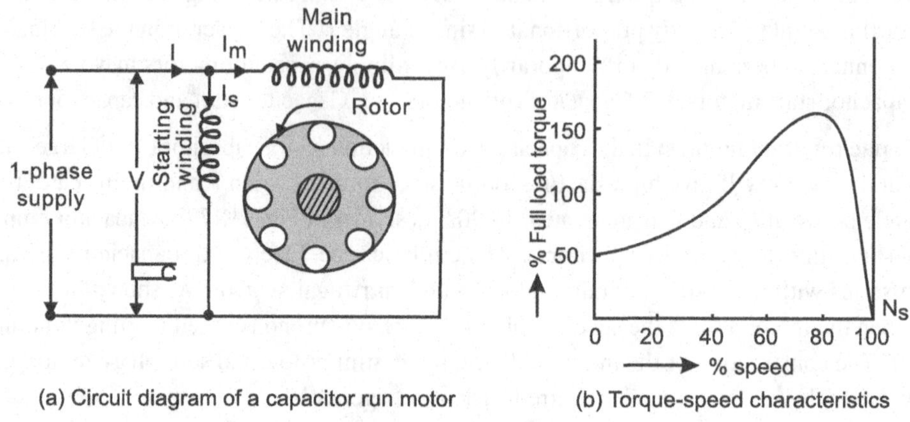

(a) Circuit diagram of a capacitor run motor (b) Torque-speed characteristics

Fig. 11.19: Capacitor run motor

Performance and characteristics. Starting torque is lower about 50 to 100% of full load torque. Power factor is improved may be about unity. Efficiency is improved to about 75%. A typical characteristics have been shown in Fig. 11.19(*b*). It is usually used in fans, room coolers, portable tools and other domestic and commercial electrical appliances.

3. **Capacitor start and capacitor run motors:** In this case, two capacitors are used one for starting purpose and other for running purpose as shown in Fig. 11.20(*a*). For starting purpose *an* electrolytic type capacitor (C_s) is used which is disconnected from the supply when the motor attains 75% of synchronous speed with the help of centrifugal switch *S*. Whereas, a paper capacitor C_R is used for running purpose which remains in the circuit of starting winding during running conditions. This type of motor gives best running and starting operation. Starting capacitor C_S which is of higher value than the value of running capacitor C_R.

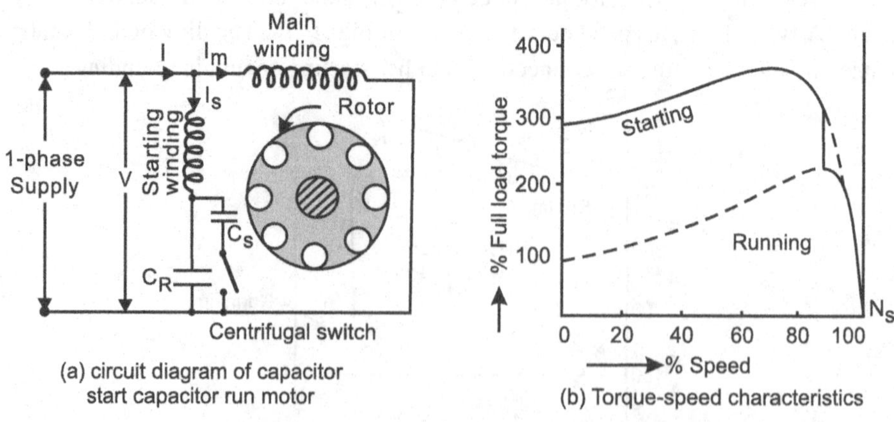

(a) circuit diagram of capacitor start capacitor run motor (b) Torque-speed characteristics

Fig. 11.20 Capacitor start capacitor run motor

Performance and characteristics: Such motors operate as two phase motors giving best performance and noiseless operation. Starting torque is high, starting current is low and give better efficiency and higher p.f. The only disadvantage is high cost. A typical torque speed curve is shown in Fig. 11.20(*b*).

Example 11.7

The main winding of a 110V, 60 Hz, 0.25 hp, single phase capacitor start motor carries a current of 6-ampere which lags behind the applied voltage by 42°. The starting winding carries a current of 4A which leads the voltage vector by 40°. Calculate (i) the total starting current and the power factor. (ii) the phase angle between the main winding current and starting winding current.

Solution:

Here, $I_m = 6$ A; $\phi_m = 42°$ lagging; $I_s = 4$ A; $\phi_s = 40°$ leading

$V = 110$ V; $f = 60$ Hz; 0.25 HP

Considering supply voltage as a reference phasor

$$\bar{V} = V\angle 0° = 110\angle 0° = (110 - \text{j}0)\text{V}$$

$$\text{Main winding current, } \bar{I}_m = 6\angle -42°\text{A} = (4.6 - \text{j}3.86)\text{ A}$$

$$\text{Starting winding current, } \bar{I}_s = 4\angle 40°\text{A} = (2.97 + \text{j}2.68)\text{ A}$$

$$\text{Total starting current, } \bar{I} = \bar{I}_m + \bar{I}_s = (4.6 - \text{j}3.86) + (2.97 + \text{j}2.68)$$

$$= (7.57 - j1.18)\text{ A} = 7.66\angle -8.9°\text{ A}$$

Starting current, I = **7.66 A** (*Ans.*)

Power factor, $\cos\phi = \cos 8.9° = 0.988$ **lagging** (*Ans*)

Phase difference between I_m and I_s, $\theta = \phi_s - (-\phi_m) = 40° - (-42°) =$ **82°** (*Ans.*)

Example 11.8

The winding impedances of the main winding and starting winding of a 230 V, 50 Hz, 250 W capacitor start motor are (5 + j4) ohm and (10 + j4) ohm respectively. What value of capacitor is required to be connected in series with the starting winding to obtain maximum torque at the start.

Solution:

Here, $V = 230$V; $f = 50$Hz; $P_o = 250$ W; $Z_m = (5 + j4)$ ohm ; $Z_s = (10 + j4)$ ohm

$$\phi_m = \tan^{-1}\frac{4}{5} = 38.66° \text{ lagging}$$

$$\phi_s = \tan^{-1}\frac{4}{10} = 28.8° \text{ lagging}$$

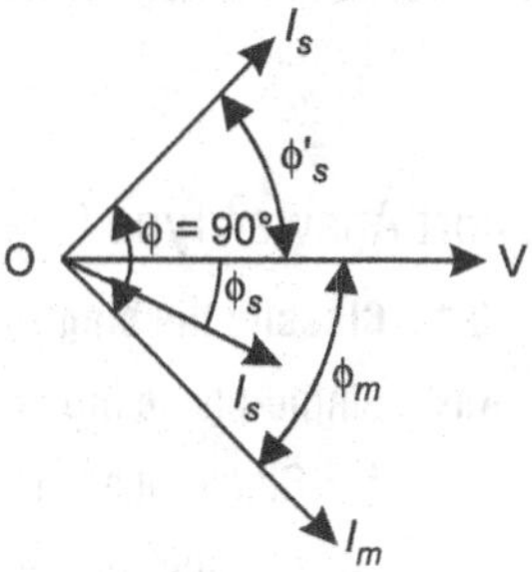

Fig. 11.21 Phasor diagram

The phasor diagram of the motor is shown in Fig. 11.21. Here, the main winding current I_m lags behind the voltage vector by $\phi_m = 38.66°$, whereas, the starting winding current lags behind the voltage vector by $\phi_s = 21.8°$. The starting torque will be maximum when phase angle between I_s and I_m i.e., $\theta = 90°$.

Let C farad be the value of capacitor to be connected in series with the starting winding to obtain maximum torque at start.

$$\phi'_s = 38.66° - 90° = -51.34°$$

From phasor diagram; $\tan(-51.34°) = \dfrac{4 - X_c}{10}$

$$-1.25 = \frac{4 - X_c}{10}$$

$$X_c = 16.5 \text{ ohm}$$

Capacitance of capacitor, $C = \dfrac{1}{2\pi f X_c} = \dfrac{1}{2\pi \times 50 \times 16.5}$ = **193 μF** (*Ans.*)

Section Practice Problems

Numerical Problems

1. The following are the parameters of a 230V, 50 Hz, 2-pole, single phase induction motor: R_1= 2.2Ω; X_1= 3.0Ω; R'_2 = 3.8Ω; X'_2 = 2.1Ω; X_m = 86Ω. Calculate the input current and power when the motor is operating at full-load speed of 2820 rpm. If the mechanical losses (iron and friction loss) are 50 W, determine gross power developed, useful power at the shaft, shaft torque and efficiency. (**Ans.** 7.95A; 1460W; 1138 W; 3.67 Nm; 74.5%)

2. A single-phase induction motor draws a current of 0.6 A at 230 V and 0.6 lagging p.f. If it runs at a speed of 100 radian per second and develops an output torque of 0.25 Nm, find its output power and efficiency. (**Ans.** 25 W; 30.19%)

3. The main winding of A 230V, 0.5 HP Single-phase induction motor (split-phase) draw a current of 6.2 A which lags behind the voltage vector by 40°, whereas the starting winding draw a current of 4.2 A which lags being the voltage by 10°.
   (i) Current drawn by the motor at start and its pf.
   (ii) Current draw by the motor during running and its pf.
   (iii) Phase angle between the current drawn by the main winding and starting winding.
   (iv) Power drawn by the motor during starting and running.
   (**Ans.** 10.6 A; 0.88 log; 6.2 A; 0.766 log; 30°; 2043 W; 1092.3 8W)

4. The main winding of a 110V, 60 HZ, 0.25 hp. Single phase capacitor start motor carries a current of 6-ampere which lags behind the applied voltage by 40°. The starting winding carries a current of 4A which leads the voltage vector by 15°. Calculate the total starting current and the power factor starting winding current. (**Ans.** 9.77A; 0.866 lag.)

Short Answer Type Questions

Q.1. Classify the single-phase motor.

Ans. Single-phase motors may be classified as

1. Single-phase induction motors
   (*i*) split-phase type (*ii*) capacitor type (*iii*) Shaded-pole type
2. AC series motors or universal motors

3. Repulsion motors
   (*i*) Repulsion-start induction-run motors (*ii*) Repulsion-induction motors
4. Synchronous motors
   (*i*) Reluctance motors (*ii*) Hysteresis motors.

Q.2. What is two-field revolving theory?

Ans. Two-field revolving theory states that a sinusoidally varying magnetic field can be represented by two vectors of field having half the magnitude revolving in opposite direction at synchronous speed.

Q.3. What is magnitude and speed of the resultant revolving field when 2-phase AC supply is given to a 2-phase wound stator winding?

Ans. Magnitude of resultant field $\phi_R = \phi_m$ and speed of the resultant revolving field, $= \omega_s$ or N_s

Q.4. How a split-phase method helps to make a 1-phase induction motor self-starting.

Ans. By splitting the phase, the flux is set-up in such a way that a revolving field is set-up in the core as if it is produced by a 2-phase AC supply

11.10 Shaded Pole Motor

Construction

Shaded pole motor is constructed with salient poles in stator. Each pole has its own exciting winding as shown in Fig. 11.22(*a*). A 1/3rd portion of each pole core is surrounded by a copper strip forming a closed loop called the shading band as shown in Figs 11.22(*a*) and (*b*). Rotor is usually squirrel cage type.

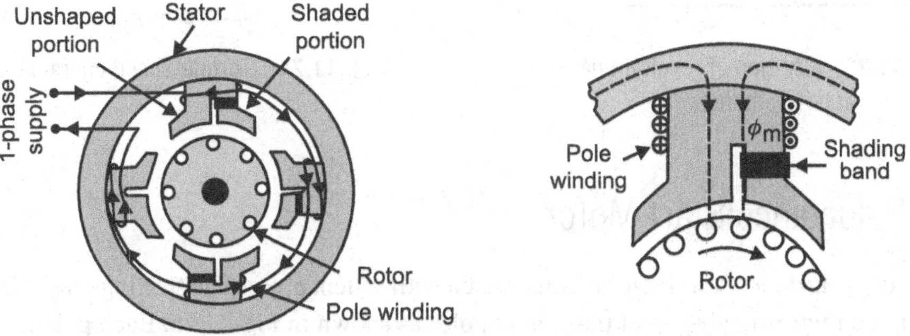

(a) Position of main winding and shading band (b) Field produced by main winding

Fig. 11.22 Shaded pole induction motor

When a single phase supply is given to the stator (exciting) winding, it produces alternating flux. When the flux is increasing in the pole, a portion of the flux attempts to pass through the shaded portion of the pole. This flux induces an emf and hence current in the shading band or copper ring. As per Lenz's law the direction of this current is such that it opposes the cause *which produces it* i.e., *increase* of flux in shaded portion. Hence in the beginning, the greater portion of flux passes through unshaded side of each pole and resultant lies on unshaded side of the pole. When the flux reaches its maximum value, its rate of change is zero, thereby the emf and hence current in the shading coil becomes zero. Flux is uniformly distributed over the pole face and the resultant field lies at the centre of the pole. After this the main flux tends to decrease, the emf and hence the current induced in the shading coil now tends to increase the flux on the shaded portion of the pole and resultant lies on the shaded portion of the pole as shown in Fig. 11.23.

Hence, a revolving field is set up which rotates from unshaded portion of the pole to the shaded portion of the pole as marked by the arrow head in Fig. 11.23. Thus, by electromagnetic induction, a starting torque develops in the rotor and the rotor starts rotating. After that its rotor picks up the speed.

Performance and Characteristics

A typical speed torque characteristics is shown in Fig. 11.24. The starting torque of this motor is very small about 50% of full load torque. Efficiency is low because of continuous power loss in shading coil. These motors are used for small fans, electric clocks, gramophones etc.

Its direction of rotation depends upon the position of the shading coil, i.e., which portion of the pole is wrapped with shading coil. The direction of rotation is from unshaded portion of the pole to the shaded portion. Its direction of rotation cannot be reversed unless the position of the poles is reversed.

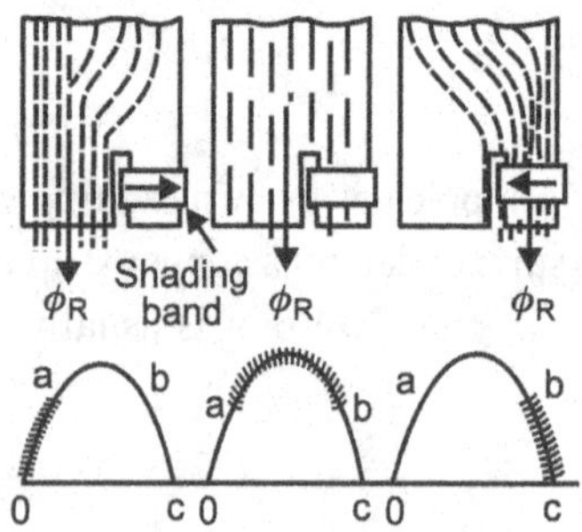

Fig. 11.23 Position of resultant field at different instants

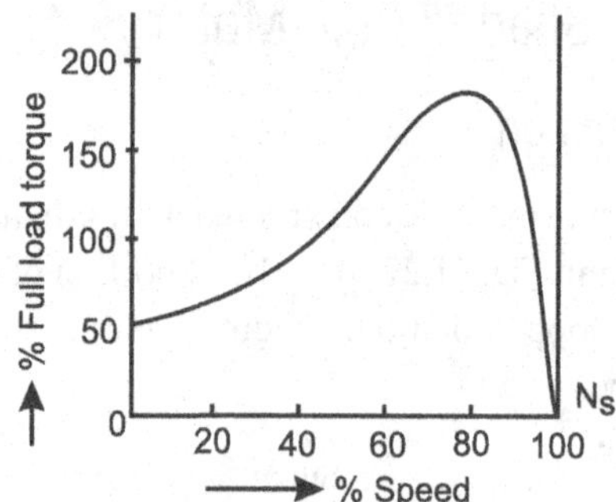

Fig. 11.24 Torque-speed characteristics

11.11 Reluctance Start Motor

The stator of a reluctance start motor is constructed with salient poles. The starting torque is achieved by creating non-uniform air-gap of the salient poles as shown in Fig. 11.25. Each pole is excited by its own winding carrying the same current. The rotor is usually squirrel cage type.

The non-uniform air-gap between stator poles and rotor offers different reluctance to the magnetic lines of force. The flux (ϕ_A) set up in the portion having greater air gap will be more in phase with the current than the flux (ϕ_B) set-up in the portion having smaller air-gap. This can be illustrated more clearly by considering two coils one having air-core and the other having iron core as shown in Fig. 11.27. In air-cored coil, there is no core loss, I lags behind the voltage vector by 90°. This current is the magnetising current and sets up field ϕ_A in phase with I. Whereas, in iron-cored coil, there are core losses, I lags behind the voltage vector by an angle α (less than 90°). This current has two components, I_e (energy component) in phase with voltage to meet with losses and I_m (magnetising component) in quadrature to voltage vector to set-up field ϕ_i in the core. Hence, field ϕ_i lags behind the current vector I by an angle θ.

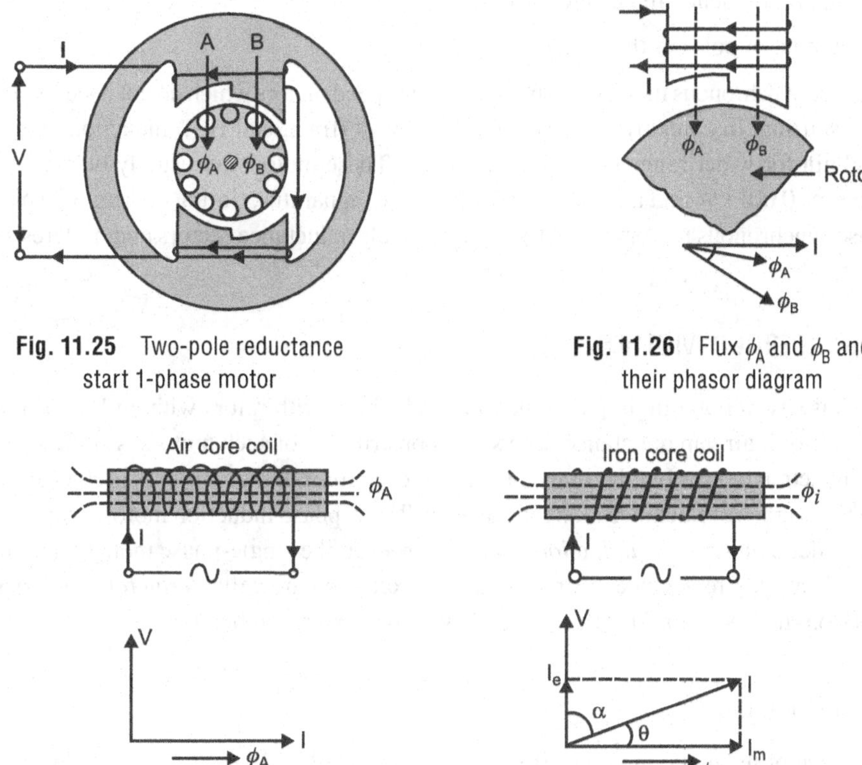

Fig. 11.25 Two-pole reductance start 1-phase motor

Fig. 11.26 Flux ϕ_A and ϕ_B and their phasor diagram

(a)Field produced in air core

(b) Field produced in iron core

Fig. 11.27 Effect of reluctance on the field.

Hence, the fluxes set-up by the two portions of the poles will lag behind the current by different angles and are displaced in time from one another. Thus, the resultant magnetic axis will shift across the poles from the longer air-gap region to the shorter air-gap region since ϕ_A is in more phase with current I *than* ϕ_B. Consequently, the rotor starts rotating in the same direction. Once the rotor starts rotating, it will continue to rotate like other types of single-phase inductor motors.

It is evident that the direction of rotation of these motors is fixed by the construction (i.e., stator poles) and cannot be reversed.

Reluctance-start motors have very small starting torque, low efficiency and poor power factor, therefore, their applications are limited. For most of the small power applications, shaded pole motors are preferred.

11.12 Single-phase Synchronous Motors

Synchronous motors are constant speed motors which rotate only at synchronous speed. The speed depends upon supply frequency and the number of poles for which the machine is designed. It is given by the relation

$$N_S = \frac{120\,f}{P} \text{ rpm}$$

where f = supply frequency in Hz and

P = number of poles of the motor

Single-phase synchronous motors are small, constant speed motors which do not need DC excitation and are self-starting. In comparison to fractional kilowatt (fractional H.P.) induction motors, these motors are built for wider range of output and speed. These motors are usually built in miniature ratings as low as 0.001 kW and are used in clocks, control apparatus, timing devices, CD players etc. Single-phase synchronous motors are of two types namely, reluctance motors and hysteresis motors.

11.13 Reluctance Motors

The three-phase as well as single-phase motors can be built with rotors without DC excitation but having non-uniform air-gap reluctance. The stator construction of such motors is similar to 3-phase or 1-phase induction motors but the shape of squirrel cage rotor of these motors is changes such that it has variable magnetic reluctance along the air-gap. The 3-phase induction motors with such rotors are usually called *synchronous induction motors*. However, the single-phase induction motors built with a variable air-gap reluctance rotor without DC excitation are called *reluctance motors*. These motors work on the basis of reluctance principle which is explained below:

Reluctance Principle

Consider a piece of magnetic material (iron), free to rotate, placed in a magnetic field as shown in Fig. 11.28(a). A torque will act (by the alignment of two fields) on the piece of magnetic material (see Fig. 11.28(a)] which will bring it to the position shown in Fig. 11.28(b), so that it offers minimum reluctance to the magnetic flux.

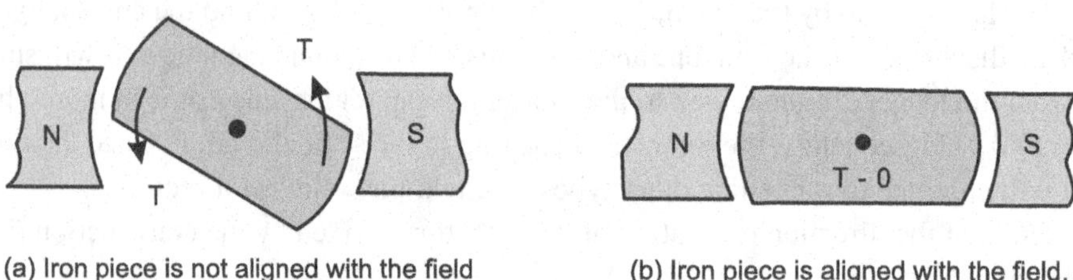

(a) Iron piece is not aligned with the field (b) Iron piece is aligned with the field.

Fig. 11.28 Reluctance principle

Construction

The stator of a reluctance motor can be of any type of single-phase induction motor. To obtain starting torque, the *stator may have an auxiliary winding with the main winding. The rotor is a modified squired cage rotor. The variation of the air-gap reluctance is created either by removing some of the rotor teeth or by shaping the rotor stampings as shown in Fig. 11.29 (a) and (b), respectively.

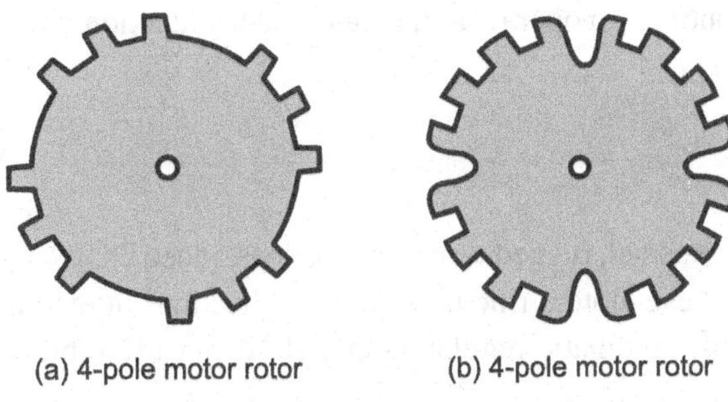

Fig. 11.29 Rotor stampings of reluctance motor

Working

When supply is given to the stator of a single-phase (split-phase) reluctance motor, a revolving field is set-up in the stator. The rotor starts rotating as an induction motor rotor and picks-up the speed near to synchronous speed. Then the rotor is pulled into synchronous speed with the stator field since reluctance torque is developed at the salient poles which have smaller air-gap reluctance. Thus, the rotor starts rotating at synchronous speed once the reluctance torque is developed.

Torque-Speed Characteristics

The torque-speed characteristics of a typical reluctance motor is shown in Fig. 11.30. The motor starts at any where from 300 to 400 percent of its full-load torque (depending upon the rotor position of the unsymmetrical rotor with respect to the field windings) as a split-phase motor.

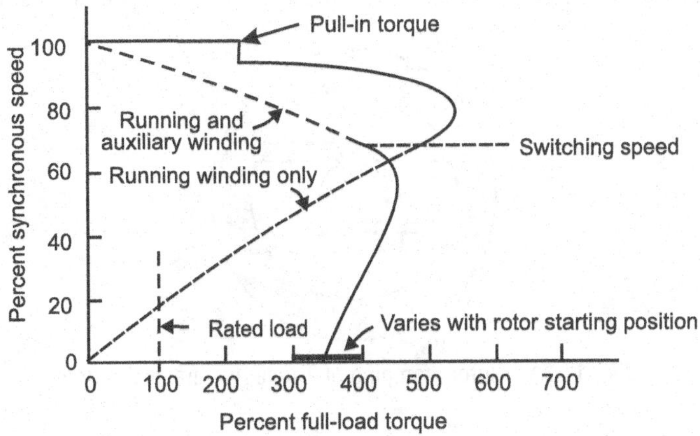

Fig. 11.30 Torque-speed curve of reluctance motor

At about 70% of synchronous speed, the starting winding is opened by a centrifugal switch but the motor continues to develop a single-phase torque produced by its running winding only. As it approaches synchronous speed, the reluctance torque is sufficient to pull the rotor into synchronism with stator field. The motor operates as a constant speed, single-phase, non-excited synchronous motor upto a little over 200 percent of its full-load torque. If it is loaded beyond the value of synchronous pull-out torque, it will continue to operate as a single-phase induction motor upto over 500 percent of its rated torque.

Applications

These motors have constant speed, rugged construction, do not need DC excitation and need minimum maintenance. Therefore, these motors find their wide applications in recording instruments, timing devices, control apparatus, automatic regulators etc. However, these motors have poor efficiency and power factor.

11.14 Hysteresis Motors

A hysteresis motor is a single-phase cylindrical (non-salient pole type) synchronous induction motor. The difference between this motor and reluctance motor is in (i) the shape of the rotor and (ii) the nature of the torque produced.

Construction

The stator construction of a hysteresis motor is either split-phase type or shaded-pole type which produces a revolving field in the stator when single phase AC supply is given to it.

The rotor of hysteresis motor is specially designed and is made of *Hysteresis-type laminations* of the shape shown in Fig. 11.31. These are usually made of hardened high-retentivity steel rather than commercial low-retentivity dynamo steel. During operations, the cross arms of the rotor are permanently magnetised due to high retentivity of the steel used for its construction.

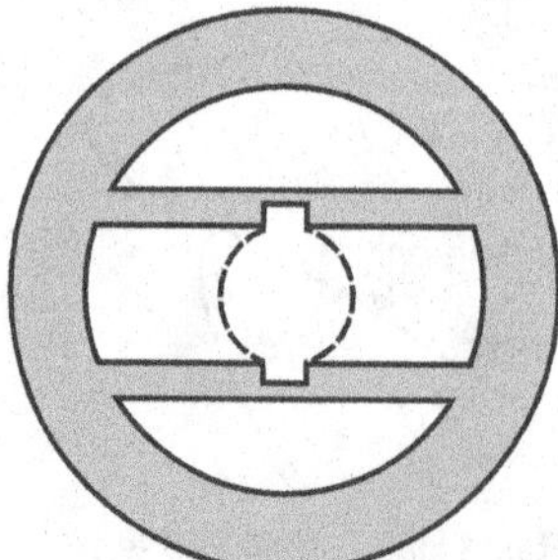

Fig. 11.31 Rotor stamping of 2-pole Hysteresis motor

Principle and Working

When single-phase supply is given to the stator (split-phase type or shaded pole type), a revolving magnetic field is set-up by it. Eddy currents are induced in the rotor. These eddy currents set up the rotor magnetic field which causes rotor to rotate. A high starting torque is produced as a result of the high rotor resistance (proportional to the hysteresis loss). As the motor approaches synchronous speed, the frequency of current reversal in the cross bars decreases and the rotor becomes *permanently magnetised in one direction* through cross-arms as a result of high retentivity of the steel used for the construction of rotor. With the two permanently set field poles, the rotor (shown in Fig. 11.31) will develop a speed of 3000 rpm at 50 Hz. Thus, *the motor runs as a hysteresis motor on hysteresis torque because the rotor is permanently magnetised.*

Applications

The amount of torque produced as a result of rotor magnetisation is not as great as reluctance torque. But hysteresis torque is *extremely steady* in both amplitude and phase in spite of fluctuations in supply voltage. As a result of this, these motors are extremely popular in driving high-quality cassette players, compact disc players, record players, tape recorders, clocks, etc.

11.15 AC Series Motor or Commutator Motor

When a 1-phase AC supply is given to a DC series motor, a unidirectional torque is developed in it. In fact, during positive half cycle, same current flows in the series field winding and armature winding which develops a torque in one direction (say clockwise). During the negative half cycle, current flowing through series field winding is reversed and at the same time current flowing through the armature also reverses, therefore, torque is developed in the same direction (i.e., clockwise direction). Thus, a continuous rotation is obtained.

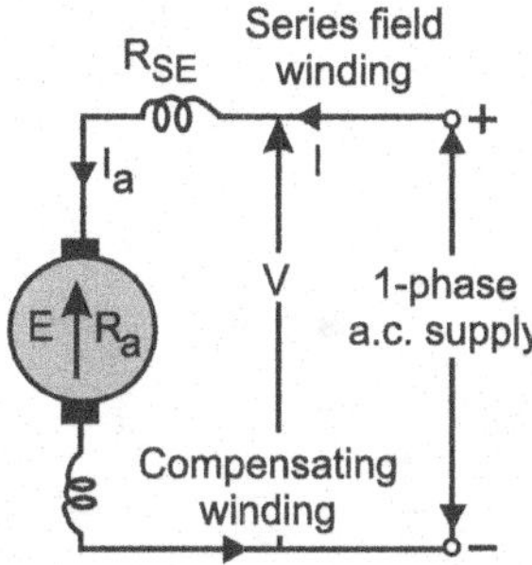

Fig. 11.32 AC series motor

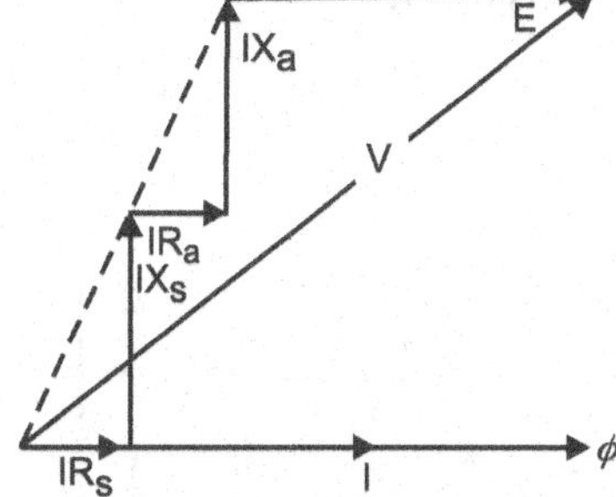

Fig. 11.33 Phesor diagram for AC series motor

Mathematically we know:

Torque in DC series motors, $T \propto \phi_{se} I_a$ where ϕ_{se} is the series field winding flux and I_a is the armature current.

Now when AC supply is given to series motor $T \propto \phi_{se} I_a$ for positive half cycle. For negative half cycle $T \propto (-\phi_{se})(-I_a) \propto \phi_{se} I_a$. Thus same torque is produced during positive and negative half cycle. However, some modifications are necessary in a DC series motor for its satisfactory operation on AC which are given below:

1. The iron structure of field and yoke are laminated.
2. The series field winding is so designed that it will produce relatively small mmf as compared to DC This is done by reducing the number of turns. The smaller field mmf would result in reduced air gap flux. Therefore, in order to develop the necessary torque the number of armature conductors have to be increased proportionately.

3. Increase in armature conductors would result in increased inductive reactance of the armature so the net inductive reactance may not be reduced. In order to overcome this difficulty compensating winding is connected in series with armature as shown in Fig. 11.32. This completely neutralises the inductive effect of the armature winding.
4. The reluctance of the magnetic circuit is reduced to have high flux with reduced mmf So magnetic material used should be of high permeability and air gap should be small.

As shown in the vector diagram (Fig. 11.33) large voltage drop occurs in resistance and reactance of the armature and field winding. Voltage left for operation is only *E*.

Where $\overline{E} = \overline{V} - (\overline{IR_S} + \overline{IX_S} + \overline{IR_a} + \overline{IX_a})$

Where IR_S = voltage drop in series winding resistance
IX_S = voltage drop in series winding reactance
IR_a = voltage drop in armature resistance
IX_a = voltage drop in armature reactance.

Performance and Characteristics

The speed-torque characteristic for DC and AC series motors are shown in Fig. 11.34. The torque varies as square of the current and the speed varies inversely proportional to the current approximately. The efficiency will not be good as that of corresponding DC machine because of greater eddy current loss and effects of p.f.

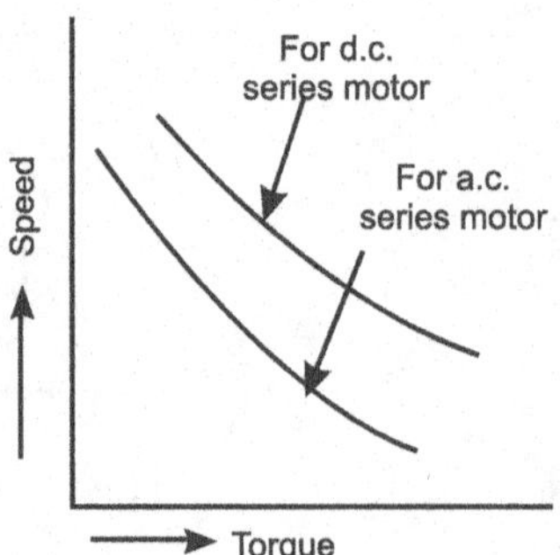

Fig. 11.34 Torque-speed characteristics of DC and AC series motors

These motors have their wide applications where high speed (20000 rpm) is required e.g., mixer grinders, blowers, hair dryers, etc.

11.16 Universal Motor

A motor which can be operated on AC as well as on DC supply at the rated voltage is called **universal motor.**

Basically, universal motor is an AC series motor. It is just an improved from of a DC series motor. The core size of an universal motor is more than the core size of a DC series motor of the same rating.

Construction

The motor has two main parts namely.

1. Stator and 2. Rotor

1. **Stator.** It is the stationary part of the motor. It consists of *magnetic frame* (or yoke), *pole core* and *pole shoe* and *field* or *exciting winding* as shown in Fig. 11.35.

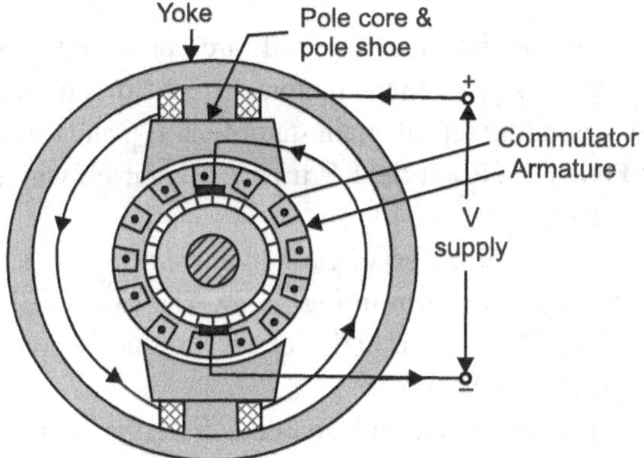

Fig. 11.35 Side-sectional view of an universal motor

The magnetic frame, pole core and pole shoe are made of silicon steel stampings. The stempings are insulated from each other by varnish layer. The hysteresis losses are very small in silicon steel and eddy current losses are reduced due to stampings. The field winding made of enamelled copper is wound around the poles to produce the required flux.

2. **Rotor.** It is the rotating part of the motor. It consists of *shaft, armature, armature winding* and *commutator,* as shown in Fig. 11.36.

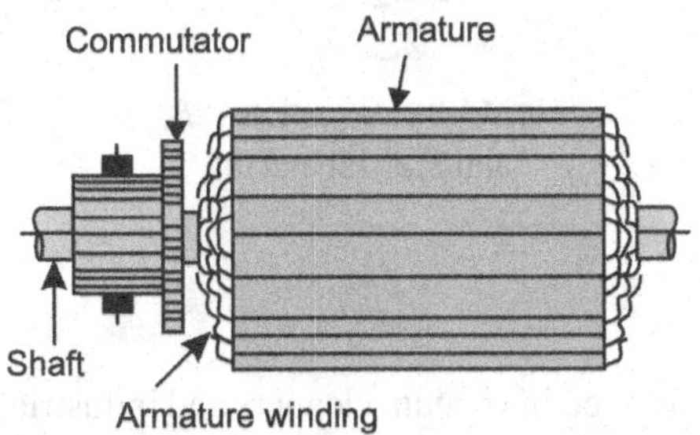

Fig. 11.36 Rotor of a universal motor

Shaft is a part of rotor which transfers mechanical power or energy to the load. It is made-up of mild steel. Armature is made-up of stempings of silicon steel material since it is to carry the magnetic field. It is keyed to the shaft. Slots are cut at its outer periphery to accommodate armature winding. The ends of armature winding are braced to the commutator segments. Commutator is made-up of wedge shaped segments forming a ring. The wedges are insulated from each other by an insulating layer of micanite. The commutator is also keyed to the shaft. Carbon brushes are pressed over the commutator surface to deliver current to the machine.

Principle

When a current carrying conductor is placed in the magnetic field, a force is exerted on it and torque develops. In other words, when the rotor field produced by the rotor current carrying conductors, tries to come in line with the main field, torque develops and rotor rotates.

Working

The armature winding and stator field winding both are connected in series, as shown in Fig. 11.37. When 1-phase, AC supply is given to the motor, current flows through the field winding and armature winding. The field winding sets up main stator field F_m, and the armature winding sets up rotor field F_r as shown in Fig. 11.38. Rotor field F_r tries to align itself with the main field F_m and an anticlockwise torque in produced.

During negative half cycle, the direction of flow of current in the field winding as well as in the armature winding is reversed, as shown in Fig. 11.39. The two windings set up their fields in the direction as shown in Fig. 11.39, again anticlockwise torque is produced in the rotor. Thus, unidirectional torque is produced in the motor.

To obtain continuous torque, commutator reverses the direction of flow of current in the coil or conductors which cross the magnetic neutral axis (MNA).

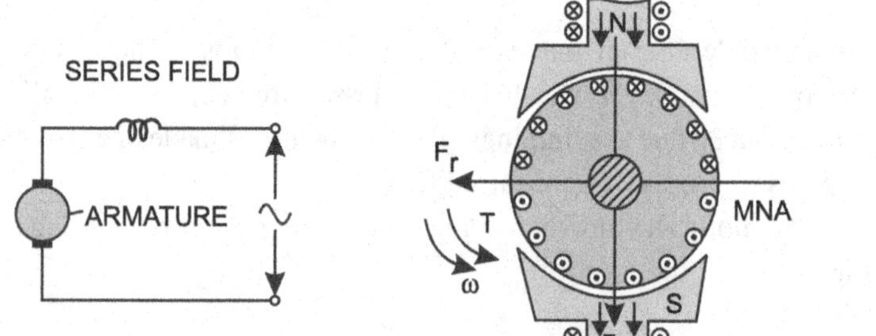

Fig. 11.37 Circuit diagram

Fig. 11.38 Position of F_m and F_r at 1st instant

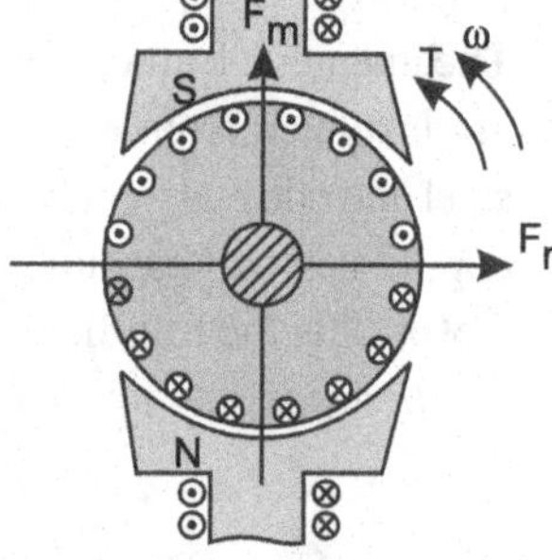

Fig. 11.39 Position of F_m and F_r at 2nd instant

Applications

In large sizes of $\frac{3}{4}$ HP, these are used in vacuum cleaners and industrial sewing machines. In smaller sizes of $\frac{1}{4}$ HP or less, these are used in electric hand drills, mixers, can openers, blenders, electric shavers, hair dryers, etc.

11.17 Comparison of Single-phase Motors

The comparison of single-phase motors is given in the Table 11.1.

Table 11.1 Comparison of Single-Phase Motors on the basis of Characteristics, Ranges, and Applications

Type of motor Torque	*Starting Torque (% Rated)*	*Pull-Out (% Rated)*	*Pull-Out Torque (% Rated)*	*Rated Power Factor (%)*	*Rated Efficiency (%)*	*Rage (hp)*	*Application*
General-purpose split-phase	90–200 medium normal	200–250	185–250	55–65	62–67	1/29 to 3/4	Fans, blowers, office appliances, food-preparation machinery. Low or medium starting torque, low inertia, continuous-operation loads.
Permanent-split, capacitor run (high torque)	up to 200 normal	200	260	80–95	55–65	1/6 to 3/4	Belt-driven or direct-drive fans and blowers, centrifugal pumps, oilburners, washing machines. Moderate-starting torque loads.
General-purpose, capacitor-start	up to 435 very high	265 high	upto 400	80–95	55–65	1/8 to 3/4	Dual voltage. Compressors, stokers, conveyors, pumps. Belt-driven loads with high static friction.
Capacitor-start, capacitor-run	380 high	260	up to 260	80–95	55–65	1/8 to 3/4	Compressors, stokers, conveyors,pumps. High torque loads. High power factor.
Shaded-pole	50 very low	50	150	40–50	30–40	1/300 to 1/3	Fans, toys, hair dryers, unit heaters, blowers. Low-starting torque loads.
Series (universal)	400–500	very high	400–500	85–95	40–60	1/150 to 1	High speed (3000–11000) rpm). Vacuum cleaners, snow blowers, centrifugal fans, mixers.

11.18 Trouble Shooting in Motors

Table 11.2 Reference Guide to Probable Causes of MotorTroubles

Motor type symptom or trouble	*AC single phase*				*AC polyphase (two or three phase)*	*Brush-type (universal, series, shunt, or compound)*
	Split-phase	*Capacitor start*	*Capacitor start and run*	*Shaded pole*		
Will not start	1, 2, 3, 5	1, 2, 3, 4, 5	1, 2, 4, 7, 17	1, 2, 7, 16, 17	1, 2, 9	1, 2, 12, 13

Will not start even at no-load, but will run in either direction when started manually	3, 5	3, 4, 5	4, 9	–	9	
	6, 8	6, 8	4, 8	8	8	8
Starts, but heats up rapidly	8	8	4, 8	8	8	8
Starts, but runs too hot	3, 5, 8	3, 4, 5, 8	4, 89	8, 9		
Will not start, but will run in either direction when started manually-overheats	—	—	—	—	—	10, 11, 12, 13, 14
Sluggish – sparks severely at the brushes	—	—	—	—	—	15
Abnormally high speed – sparks severely at the brushes	8, 16, 17	8, 16, 17	8, 16, 17	8, 16, 17	8, 16, 17	13, 16, 17
Reduction in power – motor gets too hot	8, 18	8, 18	8, 18	8, 18	8, 18	18, 19
Motor blows fuse, or will not stop when switch is turned to off position	—	—	—	—	—	10, 11, 12, 13, 19
Jerky operation – severe vibration						

1. Open in connection to line.
2. Open circuit in motor winding
3. Contact of centrifugal switch not closed.
4. Defective capacitor.
5. Starting winding open.
6. Centrifugal starting switch not opening.
7. Motor overloaded
8. Winding short-circuited or grounded.
9. One or more windings open.
10. High mica between commutator bars.
11. Dirty commutator or commutator is out of round.
12. Worn brushes and/or annealed brush springs.
13. Open circuit or shot circuit in the armature winding.
14. Oil-soaked burshes.
15. Open circuit in shunt winding
16. Sticky or tight bearings.
17. Interference between stationary and rotating members.
18. Grounded near switch end of winding.
19. Shorted or grounded armature winding.

Section Practice Problems

Short Answer Type Questions

Q.1. Can you reverse the direction of rotation of a single-phase (Split-phase) induction motor How?

Ans. Yes, by reversing the leads of either starting winding on main winding, but not both.

Q.2. Which motor, capacitor-start or resistance-start motor has a larger starting torque?

Ans. Capacitor start motor.

Q.3. How much starting torque is produced by the shaded pole motor in comparison to split-phase induction motors?

Ans. The starting torque of a shaded pole motors is very poor.

Q.4. State some of the important applications of shaded pole motors

Ans. Since their starting torque is very poor, these are used with small fans, desert-water cooler pumps, clocks, toy-motors, heat radiators etc.

Q.5. Why the yoke of an AC series motor is laminated?

Ans. In AC series motors, an alternating flux is set-up in the yoke due to this heavy eddy current loss would occur in case of solid yoke. To reduce eddy current loss in the yoke it is laminated.

Q.6. Why series motor are not operated at no-load?

Ans. At no-load series motors obtain dangerously high speed.

Review Questions

1. Explain that basically a single-phase induction motor is not a self-starting motor.
2. How will you make a single-phase induction motor self-starting?
3. Show that a single-phase sinusoidal field can be replaced by two fields rotating around the air gap in opposite directions. Draw a torque-slip curve due to each of these two fields. Explain that single-phase induction motor has no starting torque, How will you explain the fact of the motor accelerating in the direction in which it is started?

   State the manner in which the starting torque of a single-phase motor may be obtained by splitting the phase.
4. Explain the construction and working of a single-phase capacitor start induction motor.
5. What is the use of a capacitor employed with a ceiling fan?
6. If a ceiling fan is switched on to a single-phase AC supply and it does not rotate, what may be the reasons?
7. Explain the construction (with sketch) and working of a capacitor-start capacitor-run single-phase induction motor. What are its advantages and practical applications?
8. How can you change the direction of rotation of a single-phase induction motor?
9. Compare the various types of split-phase single-phase induction motor.
10. Explain the construction and working of a single-phase shaded pole induction motor.
11. Describe the working of a single-phase series motor. Name two electrical gadgets where these motors are used.
12. (*a*) Suggest suitable single-phase induction motors for following applications and give reasons for your choice:

    (*i*) Ceiling fan; (*ii*) Portable drilling machine;
    (*iii*) Domestic Refrigerators; (*iv*) Sewing machine.

    (*b*) Name five applications where you have seen single-phase induction motors being used.
13. Explain the construction and working of a reluctance start induction motor.
14. Explain the principle, construction and characteristics of a single-phase reluctance motor.
15. Explain the construction and working of a hysteresis motor.

Multiple Choice Questions

1. Stator core of a single-phase, split winding induction motor is made of
(*a*) laminated cast iron
(*b*) mild steel
(*c*) silicon steel stampings
(*d*) soft wood.

2. Two field revolving theory is base upon.
(*a*) Ferrari's principle
(*b*) Lenz's Law.
(*c*) Blondal's theorem
(*d*) Newton's law.

3. In a single phase induction motor at start, the two revolving fields produce
(*a*) unequal torques in the rotor conductors
(*b*) no torque in the rotor conductors
(*c*) equal and opposite torque in the rotor conductors
(*d*) equal torques in same direction in the rotor conductors

4. During running condition if the starting winding of a split phase induction motor is disconnected
(*a*) the motor will stop
(*b*) the motor winding will burn
(*c*) the main winding will be damaged
(*d*) the motor will continue to rotate

5. The function of the centrifugal switch is to disconnect the
(*a*) main winding during starting
(*b*) auxiliary winding during starting.
(*c*) main winding during run.
(*d*) auxciliary winding during running.

6. The direction of rotation of a split-phase induction motor can be reversed by
(*a*) reversing the connections of the supply terminals
(*b*) reversing the connections of the main winding only
(*c*) reversing the connections of the starting winding only
(*d*) either (*b*) or (*c*).

7. The direction of rotation of the shaded pole induction motor.
(*a*) cannot be reversed
(*b*) can be reversed with the help of regulator
(*c*) either a or b
(*d*) none of these

8. The DC series motor.
(*a*) can be operated on AC
(*b*) is a high speed motor
(*c*) cannot be started at no-load

9. Which motor is best suited for domestic refrigerator?
(*a*) 3-phase induction motor
(*b*) Universal motor
(*c*) Capacitor start motor
(*d*) Shaded pole motor.

10. The material used for the construction of rotor of a hysteresis motor is
(*a*) cast iron
(*b*) copper
(*c*) hardened high-retentivity steel
(*d*) low-retentivity steel

Keys to Multiple Choice Questions

1. c	**2.** a	**3.** c	**4.** d	**5.** d	**6.** d	**7.** a	**8.** d	**9.** c	**10.** c

CHAPTER

Special Purpose Machines 12

Chapter Objectives

After the completion of this unit, students/readers will be able to understand:

- ✓ What is servomechanism?
- ✓ What are the servomotors and how these are used in control system?
- ✓ What is brush-less synchronous generator?
- ✓ What is brush-less DC motor?
- ✓ What is construction, working, characteristics and applications of
  - (i) Schrage motor
  - (ii) Stepper motor
  - (iii) Linear induction motor
  - (iv) Permanent magnet DC motor
- ✓ What is an induction generator?
- ✓ How a submersible pumps and motor work?
- ✓ What are energy efficient motors?

Introduction

So for, we have discussed about synchronous machines, induction machines, fractional kW machines etc. These machines are the basic machines which are employed at various places as per requirement. However, in industries, at commercial places and in houses there are variety of jobs for which these general purpose basic machines are not suitable. But these machines with suitable amendments in their design can be employed to perform these jobs. Such specially designed machines are called *special purpose machines.*

The machines which are designed specially to perform a particular job are called **special purpose machines.**

12.1 Feedback Control System

A control system which tends to maintain a prescribed relationship between a controlled variable and a reference input by comparing relationship between a controlled variable and a reference and difference as a means of control is call a *feedback control system.*

Electrical feedback control systems rely upon mainly electrical energy for operation. The essential of this type of control are (i) an error detecting device, (ii) an amplifier and (iii) an error-correcting device as shown in Fig. 12.1. Each one of them serves a functional purpose in matching the controlled variable to the reference input.

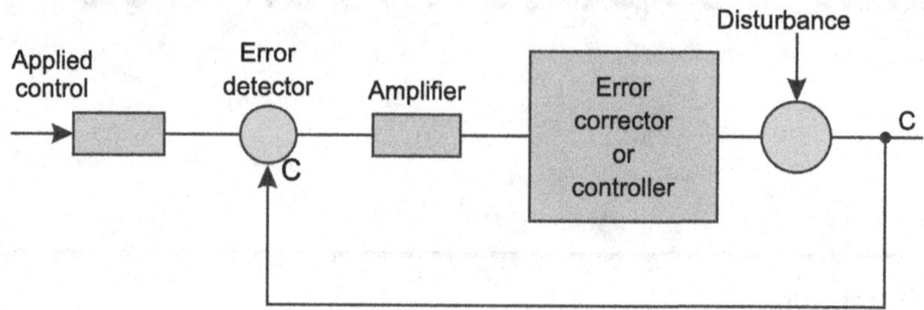

Fig. 12.1 Feedback control system (Black diagram)

The error-detecting device determines when the controlled variable is different from the reference input. It then sends out on error signal to the amplifier, which is turn supplies power the error correcting device changes the controlled variable so that it matches the reference input. The closed loop at feedback control system is shown in Fig. 12.1.

Thus, a system in which output is compared with the input and error is used to control the operating system is called **feedback control system.**

12.2 Servomechanism

The mechanism in which the control variable is adjusted by the error served by comparing output and input is called **servomechanism.**

Any quantity e.g., voltage, speed, temperature, position, direction or torque be controlled by providing appropriate feedback. Any quantity can be used as a reference input. Moreover, the reference input need not have the same units as the controlled variable provided that the proper error detecting device is employed.

When the controlled variable is mechanically positioned or adjusted of its time derivatives, the feedback control system is commonly called a *servomechanism.*

12.3 Servomotors

The motors which respond to the error signal abruptly and accelerate the load quickly are called **servomotors.** Servomotors are usually employed with control system.

Servomotors may be either DC (shunt or series) motors or AC (2-phase) induction motors. DC motors are preferred because of their high torque to inertia ratio and high starting torque. On the other hand AC motors are known for their reliability and freedom from commutation problems such as noise and wearing of brushes etc. Servomotors vary in size from 0.05 HP to 1000 HP.

The fundamental characteristics to be sought in any servomotor (DC or AC) are

(i) The motor output torque should be proportional to its applied control voltage (developed by the amplifier in response to an error signal).
(ii) The direction of the toque is determined by the (instantaneous) polarity of the control voltage.

12.4 DC Servomotors

The DC motors which are employed in the control system are called **DC servomotors.**

There are four types of DC servomotors which are used in the control system. These are (1) the field-controlled shunt motor, (2) the armature controlled shunt motor, (3) the series split-field motor, and (4) the permanent-magnet (fixed field excitation) shunt motor.

12.4.1 Field-controlled DC Servomotors

In this motor armature current is always kept constant and the field is excited by the DC error amplifier, as shown in Fig. 12.2. The torque produced by the motor is zero when no field excitation is supplied by the DC error amplifier. Torque produced by the motor varies directly as the field flux and also as the field current up to saturation ($T = k\,\phi\,I_a$). The direction of the motor (i.e., rotation) is reversed if polarity of the field is reversed.

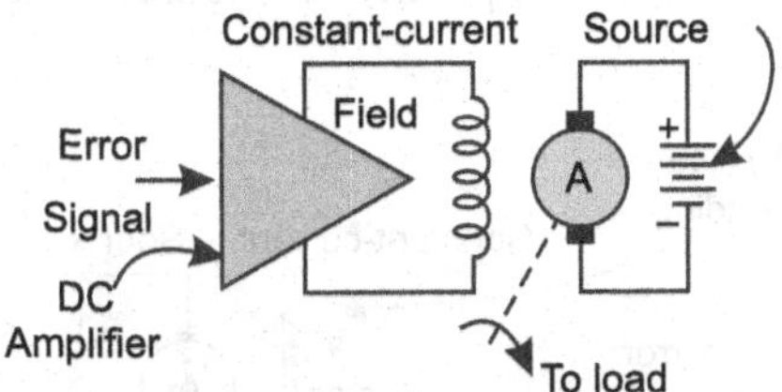

Fig. 12.2 Field control DC servomotor

The control of field current by this method is used only in small servomotors because

(i) it is undesirable to supply a large and fixed armature current as would be required for large DC servomotors and
(ii) its dynamic response is slower than the armature-controlled motor because of the longer time constant of the highly inductive field circuit.

12.4.2 Armature-controlled DC Servomotors

This servomotor employs a fixed DC field excitation furnished by a constant-current source, as shown in Fig. 12.3. As sated, this type of control possesses certain dynamic advantages not possessed by the field-control method. A sudden large or small change in armature voltage produced by an error signal will cause an almost immediate response in torque because the armature circuit is essentially resistive as compared to the highly inductive field circuit.

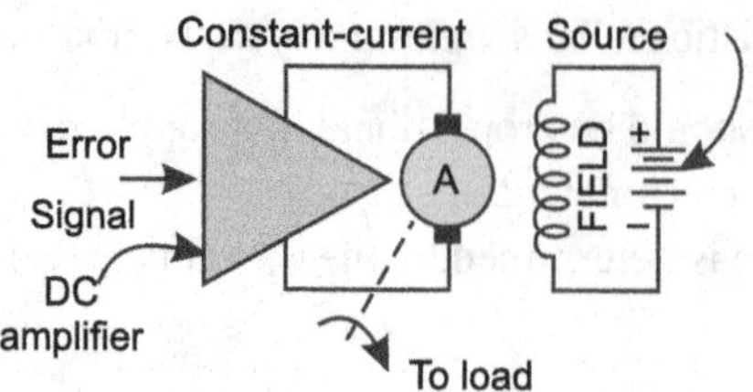

Fig. 12.3 Armature control DC servomotor

The field of this motor is normally operated well beyond the knee of the saturation curve to keep the torque less sensitive to slight changes in voltage from the constant-current source. In addition, a high field flux increases the torque sensitivity to the motor ($T = k\phi I_a$) for the same small change in armature current. DC motors up to 1000 hp are driven by armature voltage control in this manner. If the error signal polarity is reversed, the motor reverses its direction.

12.4.3 Series Split-field DC Servomotors

Small fractional-horsepower, series split-field DC motors may be operated as separately excited field-controlled motors, as shown in Fig. 12.4. One winding is called the main winding and the other is called the auxiliary winding, although they are generally equal in *MMF* and are wound about the field poles in such a direction as to produce reversal of rotation with respect to each other. As shown in Fig. 12.3 the motors may be separately excited, and the armature may be supplied by a constant-current source.

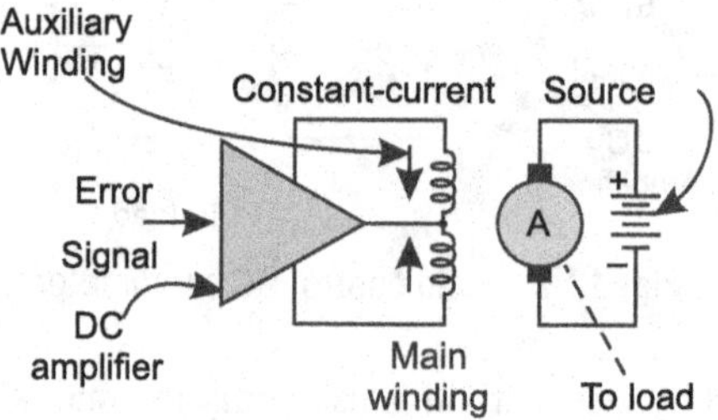

Fig. 12.4 Series split-field DC servomotor

The advantages of the split-field method of field control are the (1) the dynamic response of the armature is improved since the fields are always excited (there is no delay due to inductive time constant) and (2) a finer degree of control is obtained because the direction of rotation is more responsive to extremely small differences in current between the main and auxiliary windings.

12.4.4 Permanent-magnet Armature-controlled DC Servomotor

This type of extremely popular servomotor uses permanent magnets (either Alnico or ceramic) for constant field excitation, as shown in Fig. 12.5. It is usually manufactured in 6V and 28 V ratings in fractional-horsepower sizes and in 150 V rating in integral-horsepower sizes up to 2 hp. The field structure for this type of motor usually consists of Alnico VI alloy, cast in the form of a circular ring, completely surrounding the armature and providing a strong, constant flux. The permanent-magnet motors are well compensated by means of commutating windings to avoid demagnetisation of the field magnets whenever the DC armature voltage is suddenly reversed. Eddy currents and hysteresis effects in these motors are generally negligible, and the pole pieces are usually laminated to reduce arcing at brushes whenever a rapid change of signal voltage occurs. These devices are also controlled by armature voltage control in the same manner as the armature-controlled shunt motor.

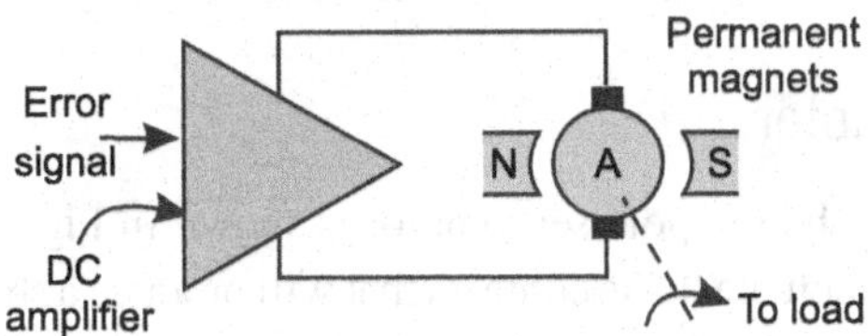

Fig. 12.5 Permanent magnet DC servomotor

The *PM* DC servomotor is used in two modes of operation: (1) position control and (2) velocity control. *PM* DC servomotors used for position control are sometimes called "torque motors" because they develop extremely high torques at standstill or low speeds. Conversely, at high speeds, since, for a given horsepower, torque varies inversely as speed the developed torque is extremely small. The developed torque is also a function of the voltage applied to the armature.

High speed velocity control servomotors are usually totally enclosed and have large frame sizes to permit adequate heat dissipation. Depending on the velocity to be controlled and the torque requirements of the load driven, a motor may also require a built-in fan to improve the heat dissipation and cool the armature.

12.5 AC Servomotors

The AC motors which are employed in the control system are called **AC servomotors.**

Along with smaller DC stepping motors, most of the smaller AC servomotors are of the AC two-phase or the shaded-pole induction motor-type.

Two-phase AC Servomotors

Figure 12.6 shows the schematic diagram of a two-phase servomotor. This motor is a true two-phase motor, having two stator windings displaced 90° in space on the stator. The reference windings in constantly and usually excited through a capacitor by the fixed AC supply. With no error signal, the squirrel-cage rotor is at standstill. A small error signal of some particular instantaneous polarity with respect to the reference winding is amplified by the AC amplifier and fed to the control winding. Torque i.e., rotation is produced in such a direction so as to reduce the error signal, and the motor ceases to rotate when a null (zero error signal) is produced at the control winding.

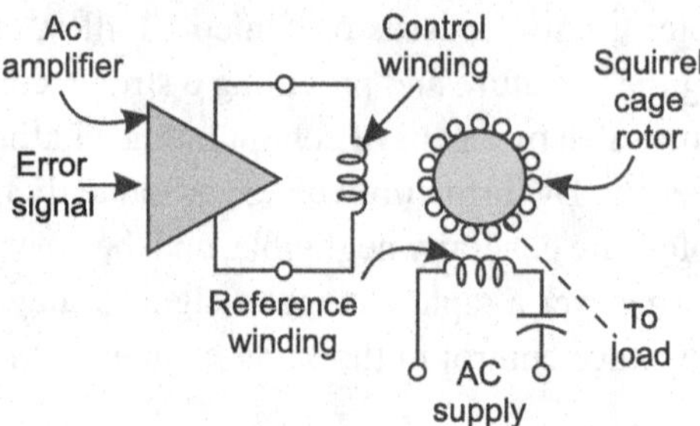

Fig. 12.6 Two-phase AC servomotor

Shaded-pole AC Servomotor

The schematic diagram of a shaded-pole servomotor is shown in Fig. 12.7. In this motor, a phase-sensitive relay is employed to actuate those contacts that will produce a short circuit of the shaded-pole winding to develop rotation in the desired direction. As with all shaded-pole windings, a single-phase AC field winding is connected to the AC supply. In the presence of an error signal sufficient to actuate the relay, one pair of shaded-pole windings is shorted; there upon, the servomotor rotates until the null is produced (at which point the relay drops out) and the motor stops. An error signal of opposite polarity will actuate the phase-sensitive relay to short-circuit a different pair of windings, causing rotation of the servomotor in the reverse direction.

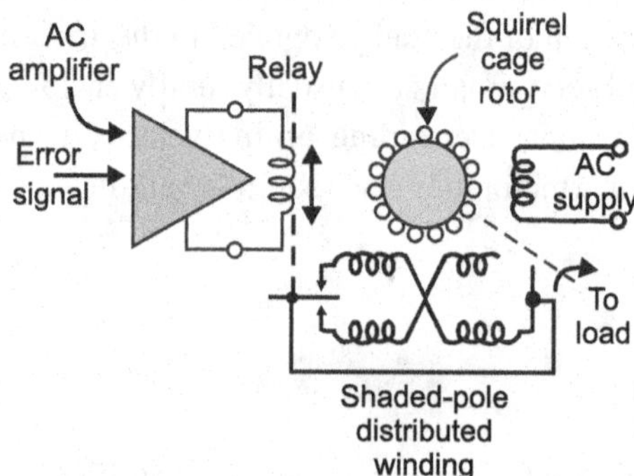

Fig. 12.7 Shaded pole AC servomotor

It is fairly evident that the two-phase motor design of Fig. 12.6 is better of the two types since it is capable of responding to small error signals. A shade-pole servomotor will respond only when the amplified error signal is of sufficient magnitude to cause the relay to actuate. The response of the two-phase servomotor to very small control signals is further improved by reducing the weight and inertia of the motor. **Shaded-pole AC servomotor**

12.6 Schrage motor

The demand of the industry to have a motor of moderate size having desirable features of speed control and power factor improvement without the auxiliary devices was met with when **K H Schrage** of *Sweden* developed such a motor and was named as *Schrage motor.*

Schrage motor *is a motor that contains both power factor correction arrangement and speed control in one and the same motor.*

These motors are generally built in sizes from 3 hp to 50 hp, however these can also be built in large sizes upto 150 hp. But beyond this these are not economical. These motors are designed for 3-phase, 50 Hz, 400 V supply. A simple diagram of a two-pole Schrage motor is shown in Fig. 12.8. Since the position of the brushes placed on commutator can be adjusted to obtain the desired speed, therefore, this motor is also known as *brush shift AC motor.* This motor provides characteristics similar to a DC shunt motor, hence it is also called as *AC shunt motor.*

Construction

The winding or electrical connection diagram of a 2-pole Schrage motor is shown in Fig. 12.8. This motor is an inverted wound rotor induction motor in which the rotor winding connected in star or delta is the **Primary** and is supplied from a 3 phase supply through slip rings as shown in Fig. 12.8

The motor has another winding known as compensating or adjusting or regulating winding, entirely similar to an ordinary DC armature winding, and is tapped to segments of a commutator in the usual manner. The conductors of the regulating or adjusting winding are placed in the same slots in which the conductors of the rotor winding or primary winding are housed. The conductors of the regulating winding are placed above the conductors of the primary winding as shown in Fig. 12.9. Brushes held by brush rockers are placed on the surface of the commutator. The stator winding forms the secondary and is divided into 3 isolated phases as shown in Fig. 12.9. The two ends of each phase of the stator winding are connected to two brushes *A* and *B* placed on the commutator as shown in Figs. 12.8 and 12.10.

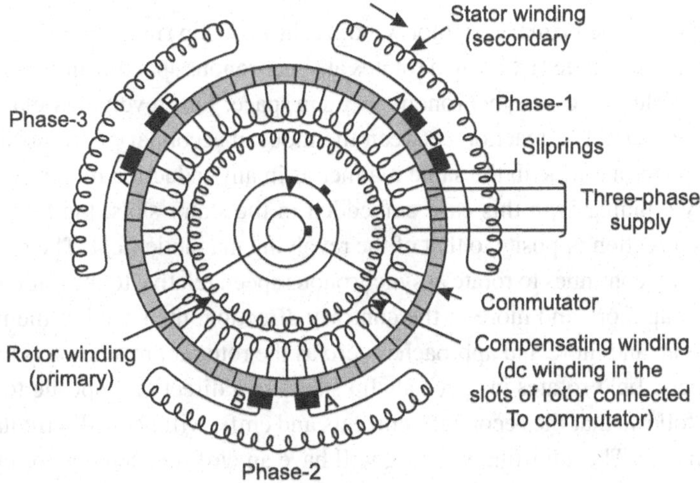

Fig. 12.8 Conventional diagram of Schrage motor

Two **sets of brushes** (i.e., a set of three A-brushes and a set of three *B*-brushes) are supported on two brush rockers. Each rocker carrying three brushes are located at opposite ends of the commutator and are so arranged that they can be rotated relative to each other in either direction by means of a hand wheel. Thus by turning the hand wheel one brush rocker or set of brushes moves clockwise and the other moves anticlockwise, thereby allowing the brushes of the two rockers to be displaced with respect to each other or to be placed in line with one another. The three brushes of one rocker or set of brushes are connected to the respective starts of the three phases of the stator or secondary winding, while the finishes of these winding are connected to the corresponding brushes on the second rocker or set of brushes.

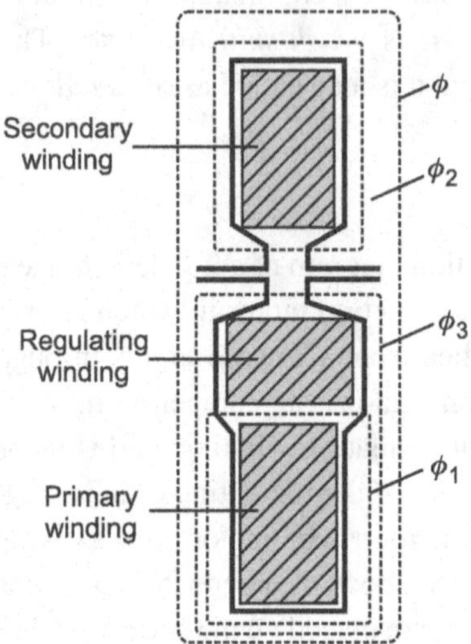

Fig. 12.9 Slots for stator and rotor

Working

When supply is given to the primary or rotor winding through slip rings, the primary (rotor) current will set up a resultant magnetic field which rotates at synchronous speed with respect to the rotor. At start this magnetic field rotates at synchronous speed in space. This revolving field induces *emfs* and currents in the secondary. The reaction between the induced secondary currents and the revolving magnetic field develops a torque in the same manner as in any induction motor. Because of the fact that the secondary winding is in this case embedded in the stator slots, the torque will cause the rotor to rotate in a direction opposite to that of the revolving magnetic field. The revolving magnetic field itself necessarily continues to rotate at synchronous speed relative to the rotor and consequently as the rotor speeds up more and more in the contrary direction, the speed of the magnetic field in space decreases more and more and approaches zero as the rotor approaches synchronous speed. In general, the magnetic field rotates in space at slip speed in a direction opposite to that of the rotor core. From this it follows that the secondary currents and emfs. will be of slip frequency (*Sf*) just as in any induction motor. The adjusting winding will have an emf induced due to transformer action from the primary winding provided in the same slots. Hence the frequency of the voltages in both the windings is the same as that of the supply. With constant supply voltage, the r.m.s. values of induced voltages in the adjusting winding are nearly constant at all loads.

Thus it is concluded that when the primary is fed from the supply, emfs. are induced in the regulating winding at **supply frequency** *and as the rotor rotates, emfs. induced in the secondary winding are at* **slip frequency.**

Speed Control by Changing the Brush Position

From the Fig. 12.10 it is clear that for any position of brushes connected to each phase of the secondary, there exists a voltage, the value of which depends upon the number of turns of the regulatingwinding which is contained between these brushes. This means that at any instant the voltage of the Regulating winding between the two brushes is in series with secondary emf The voltage of the regulating winding is zero when the two brushes touch the same bar and its value is maximum when the two brushes are 180° electrical apart. Thus the magnitude of the voltage of the regulating winding is varied by changing the separation between the brushes of the same phase. Thus, voltage of the regulating winding in injected into each phase of the secondary winding and if this injected emf is in direct opposition to secondary induced emf the speed will be below normal. On the other hand if it is in phase with the secondary induced emf will be above normal. There are three possible positions of brushes as illustrated in Figs. 12.10 (*a*), (*b*) and (*c*).

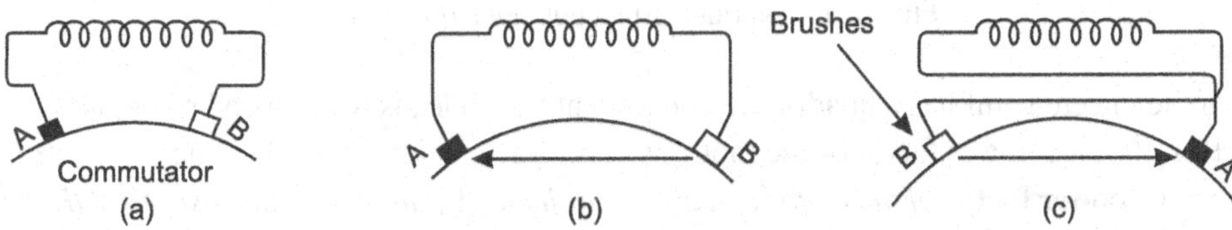

Fig. 12.10 Various brush positions for (speed control)

(*i*) Figure 12.10 (*a*) shows when the brushes *A* and *B* carried by two rockers are in line with one another, so that at every instant there will be direct electrical contact through a commutator segment. Hence each secondary phase is short circuited and the motor will run like a plain induction motor.

In fact, at this instant, there is no **regulating voltage** injected into the secondary circuit and the only emf is due to secondary winding and thus the motor runs at normal speed.

(*ii*) Figure 12.10 (*b*) shows when the brushes are opened out. In this case, the regulating voltage equivalent to the voltage induced in the turns contained between the brushes *A* and *B* is injected into each phase of the secondary circuit so that it opposes the secondary induced emf This reduces its magnetic effect with the result the speed of the motor decreases.

(*iii*) Figure 12.10 (*c*) shows when the brushes are moved apart in the opposite direction i.e., when the position of the brushes *A* and *B* is reversed. In this case, the regulating voltage will be in phase with secondary induced emf The total emf acting in the secondary circuit increases which in turn increases the current in the secondary winding thus the motor speed increases.

Hence, it is concluded that if the brushes are separated, regulating voltage will appear between the brushes *A* and *B*, which is injected into each phase of the secondary winding. This injected voltage may aid or oppose the secondary induced voltage, depending upon the direction of shifting of brushes. Thus, for any position of the brushes there will be definite speed of rotation either above or below synchronous speed. In practice a hand wheel is fitted which operates the brushes in either direction through differential gearing.

Power Factor Control by Changing the Brush Position

Although the magnitude of the emf between the brushes *A* and *B* is dependent on their relative angular position, the phase of the emf depends upon the position of the centre line with respect to the centre of the rotor winding. Hence if the brushes as a whole are moved round the commutator as shown in Fig. 12.11 the phase of the emf, injected into the secondary circuit will be varied relative to the rotor induced emf and therefore power factor of the motor will be varied.

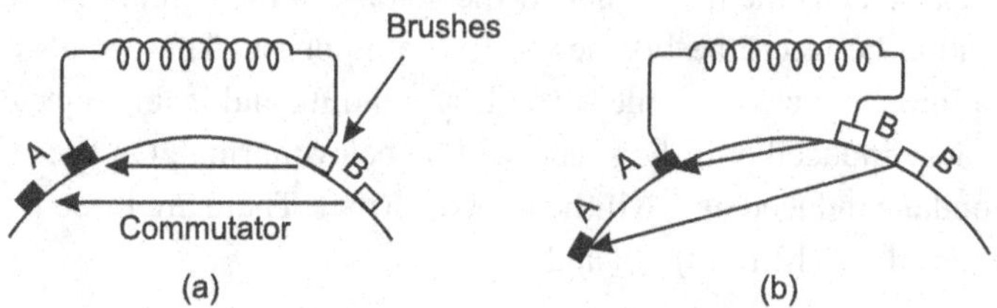

Fig. 12.11 Various brush positions (pf control)

When the injected emf has a quadrature component which leads the induced rotor *emf* [as shown in Fig. 12.11 (*b*)] the power factor of the motor will improve and becomes leading.

Hence, the **power factor** *of the motor can also be adjusted by adjusting the position of the brushes* i.e., *by changing the position of brush A and B simultaneously in the same direction.*

Conclusion

From the above discussions we conclude that the speed of motor depends on the angular distance between the individual brushes *A* and *B* while the p.f. depends upon the angular position of the brushes as a whole. Owing to the changes in power factor over the speed range where motor is to be used for one direction only, the brushes are set to give best p.f. over the whole range. A motor should note be reversed without making an adjustment in the setting of the brushes. The power factor is very high at speeds above synchronous and relatively low at sub-synchronous speeds. The low power factor at reduced speed can be improved by unsymmetrical brush displacement. This can be achieved by allowing one brush rocker to move faster than the other, so that the centre line of the brushes is rotated round the commutator. Motor which are designed with unsymmetrical brush displacement can be run in one direction only, but by a simple reconnection or readjustment of the brushes the direction of rotation can be changed. Reversible machines such as crane and hoisting motor, are designed with symmetrical brush displacement. To reverse the direction of rotation of such a machine it is preferred to inter change two primary supply leads.

Advantage

(*i*) It has a continuous speed regulation with the required range so that the driven machinery may be run at the best possible speeds.

(*ii*) It has high average efficiency and power factor.

Applications

There are many industrial applications of the Schrage motor, some of them are given below.

(*i*) It is extensively used in *printing industry* in which variable speed is essential. The motor has proved equally suitable for driving small presses as well as large rotary presses used in the newspaper production.

(*ii*) In the *paper industry* this motor has been employed for paper-machine drives.

(*iii*) In *rubber industry* this motor is extensively employed for driving rubber cylinders.

(*iv*) In Steam boiler installations, Schrage motor are commonly employed for boiler fans and strokes drives.

(*v*) The other important applications are in cement kilns, cranes, lifting pumps, large machine tools, sugar refining machinery, textile industry, belt conveyors etc.

Section Practice Problems

Short Answer Type Questions

Q.1. What do you mean by feedback control system?

Ans. A system in which output is compared with the input and error is used to control the operating system is called *feedback* control system

Q.2. What do you mean by servomotors?

Ans. The motors which respond to the error signal abruptly and accelerate the load quickly are called **servomotors.**

Q.3. Torque to inertia ratio of servomotors should be high, why?

Ans. Higher the torque to inertia ratio smaller will be the response time.

Q.4. What is a Schrage motor?

Ans. Schrage motor: It is a motor that contains both power factor correction arrangement and speed control arrangement in one and the same motor.

12.7 Brushless Synchronous Generator

The elimination of the exciter on the shaft of a synchronous generator eliminates the problems associated with commutation (like sparking at the brushes, excessive heating, wear and tear, maintenance etc.) of a DC generator. But it is still necessary to supply DC to the field via brushes and slip rings. In order to eliminate slip rings and brushes. The brushless *synchronous generators* were developed.

A block diagram of one type of brushless synchronous generator is shown in Fig. 12.12. Its explanation is given below:

1. **Block-1:** It is a stationary part, its main function is to provide the necessary field current (or excitation) to the field winding of the auxiliary alternator. The function of each part of this block is given below:
   (i) *Single-phase variac*: When 1-phase AC supply is given to variac, it provides variable voltage at its output which is fed to a full wave rectifier.
   (ii) *Full-wave rectifier*: It converts AC into DC. As the output of rectifier contains ripples, a filter circuit is provided which removes the ripples. Pure DC is fed to a potentio-divider.
   (iii) *Potentio-divider*: By changing the position of jockey, we can change the magnitude of field current fed to the field winding of auxiliary alternator.
   (iv) *Field winding of auxiliary alternator*: This winding provides the necessary magnetic field so that a required voltage is induced in the armature winding of the auxiliary alternator.

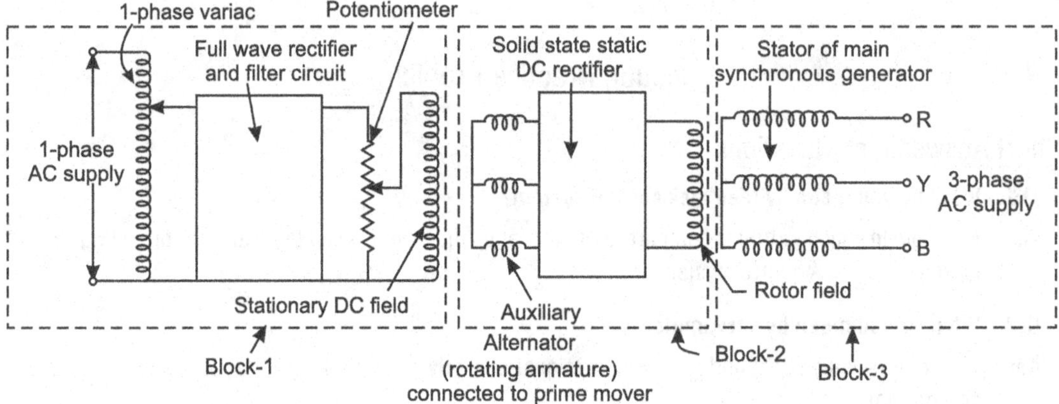

Fig. 12.12 Brushless synchronous generator (block diagram)

2. **Block-2:** It is a rotating part and is coupled with the prime-mover. Its main function is to provide necessary field (or flux) to the main synchronous generator. The function of each section of this block is given below:
   (i) *Armature of auxiliary alternator*: It is a part in which necessary *emf* is induced. This part, ultimately provides the required field current for the field winding of main synchronous generator. The output of auxiliary alternator is given to the solid state static DC rectifier.
   (ii) *Solid state static DC rectifier*: It is an electronic circuit containing transistors, *SCRs*, resistors and capacitors. This circuit converts AC into DC which is further fed to the field winding of the main synchronous generator.
   (iii) *Field winding of main synchronous generator*: It is a winding which produces the magnetic flux. The magnitude of flux produced by it depends upon the current flowing through it. The value of this field current can be varied by varying the voltage supplied to rectifier in block-1 or by varying the value of potentiometer.
3. **Block-3:** It is the stationary armature of the main synchronous generator. The three-phase AC supply is available at its output terminals.

The above means of controlling the DC field excitation of the synchronous generator does not need exciter, slip rings or brushes of any kind.

The reduced cost and increased reliability of solid state rectifier packages has encouraged a variety of brushless generators in the low and medium power ratings in both poly-phase and single-phase types. This trend is expected to continue because of the advantages such as no commutation problems, no sparking and no maintenance on brushes.

12.7.1 Brushless DC Generator

When the output of a brushless synchronous generator is converted into DC with the help of a solid state rectifier, the complete set-up is called a brushless DC generator.

12.8 Brushless Synchronous Motor

The elimination of the exciter on the shaft of the synchronous motor eliminated the problems associated with commutation of a DC generator and sparking at the brushes connected to the commutator. In order to completely eliminate slip rings and brush maintenance, the *brushless synchronous motor* was developed.

A block diagram of one type of brushless synchronous motor is shown in Fig. 12.13.

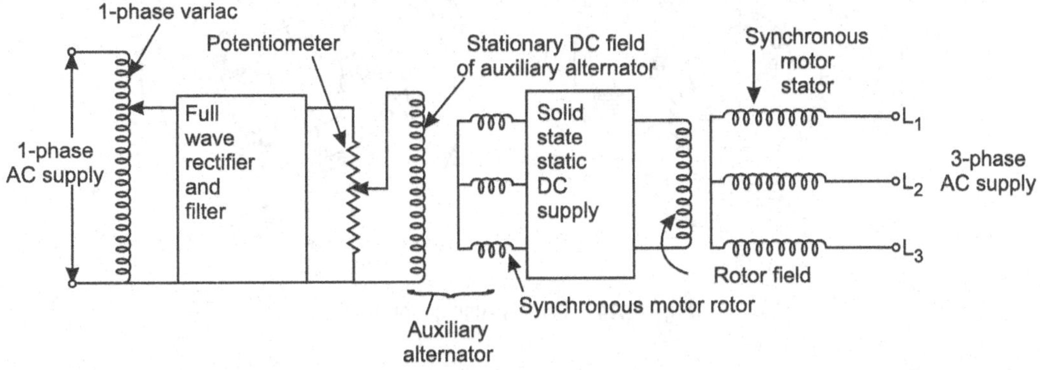

Fig. 12.13 Brushless synchronous motor (block diagram)

This arrangement requires an AC generator (auxiliary alternator) having a stationary DC field and a rotating poly-phase armature in which AC voltages are generated. The DC excitation of the synchronous motor is controlled by the single-phase variac exciting the stationary DC field of the poly-phase alternator, which is on the same rotor shaft as the rotor field of the synchronous motor, as shown in Fig. 12.13.

The rotor of the synchronous motor, as shown in Fig. 12.13, carries the alternator armature, the static DC control and rectification system, consisting of transistors and *SCRs* and the rotor field of the synchronous motor.

The arrangement shown in Fig. 12.13, provides a means of controlling the DC field excitation of a synchronous motor without the need for exciter or slip rings or brushes of any kind.

The reduced cost and increased reliability of solid-state rectifier packages has encouraged a variety of brushless synchronous motors in the low to medium horse-power ranges in both poly-phase and single-phase types. This trend is expected to continue because of the advantages of this type of motor, i.e., no commutation problems, no sparking, and no maintenance on brushes, as found in conventional synchronous motors.

12.9 Three-brush (or Third-brush) Generator

In some types of small generators, particularly those which operate at variable speed, such as automobile generators, advantage is taken of armature reaction to regulate the current. If an ordinary shunt or compound generator used in an automobile to charge the battery, the generator would be lightly loaded at low speeds and too heavily loaded at the higher speeds.

A common method of regulating automobile generators is the third-brush method, in which the shunt field is connected between one of the main brushes and a small third brush located between the two main brushes. In Fig. 12.14, *A* is the positive main brush, and *B* is the negative main brush, which is usually grounded to the frame of the car or engine.

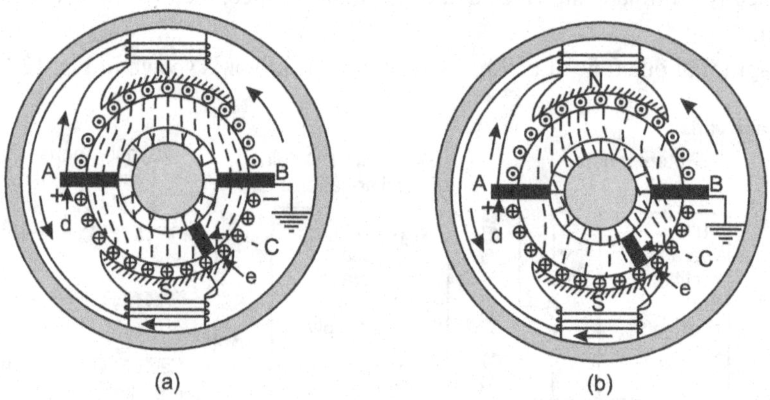

Fig. 12.14 Three-brush generator

The auxiliary, or third, brush *C* is placed at an angle of about 60° from brush *B* in the direction opposite to the direction of rotation. The shunt field is connected between the brush *C* and the positive brush. In Fig. 12.14 (*a*) the flux through the armature under light-load condition is shown diagrammatically. There is negligible armature reaction. The voltage across the field will be substantially the emf induced in the conductors connected in series between *d* and *e*, which cut the flux included between brushes *A* and *C*.

If the speed of the generator increases, other factors remaining constant, the emf will increase and the generator will deliver a greater current. This current, however, distorts the magnetic field in the direction of rotation, as shown in Fig. 12.14 (*b*). The effect is to transfer the flux from between conductors *d* and *e* to the region between brushes *C* and *B*. Since the total emf between conductors *d* and *e*, and hence the voltage across the field, is proportional to the flux between conductors *d* and *e*, the effect of the increased speed will be compensated in part at least by the lesser flux and the current will increase only slightly if at all. A study of Fig. 12.14 (*b*) shows that the current may be increased by moving the third brush *C* in the direction of rotation and the current may be decreased by moving if against the direction of rotation.

When such generators are used to charge batteries, as in automobiles, there should be a cutout relay which connects the battery to the generators only after the generator has built up to a voltage slightly higher than that of the battery. If the generator shows down, this relay opens when the current reverses and begins to flow from battery to generator.

12.10 Brushless DC Motors

A conventional DC motors suffers from the following two major problems:

(*i*) They require more maintenance and need to replace brushes periodically

(*ii*) Their operating voltage and speed is limited because of commutation difficulties

To over-come these difficulties, we have to eliminate commentator and brushes.

Thus, *a motor that retains the characteristics of a DC motor but eliminates the commutator and brushes is called a* **brushless DC motor.**

The brushless DC motors are equipped with electronic circuits and devices which perform the same function as that of a mechanical commutator.

Operating Principle

The schematic diagram of a brushless DC motor is shown in Fig. 12.15. It consists of a multi-phase winding wound on a non-salient stator and a permanent magnet (*PM*) rotor. The required voltage is applied to the individual phase winding through a sequential switching operation such that the necessary commutation is achieved to impart rotation to the motor. The necessary switching operation is achieved by using electronic circuit and devices such as transistors, thyristors, etc.

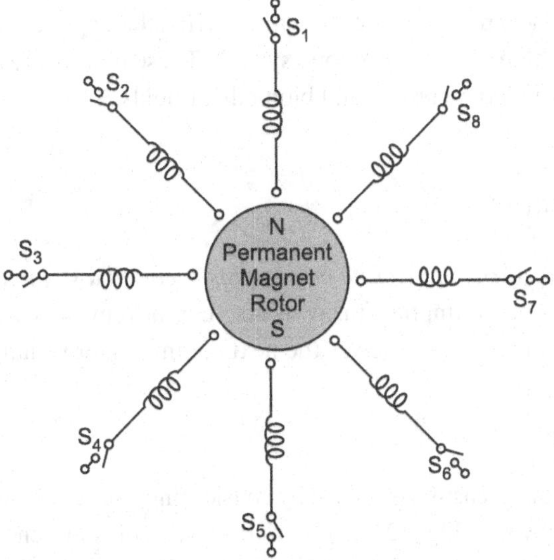

Fig. 12.15 Conventional diagram of a brush-less DC motor

Let us see how it operates

When switch S_1 is closed, winding-1 is energised, the *PM* rotor is aligned with its magnetic field. When switch S_1 is opened and S_2 is closed, winding-2 is energised, the *PM* rotor is aligned with its magnetic field and turn through a particular angle. When a number of such phase windings are energised sequentially, the rotor rotates.

In such motors, the windings can be designed for required voltage and very high speeds, as high as 40000 rpm.

Advantages

(*i*) Require little or no maintenance
(*ii*) Have longer operating life
(*iii*) Less losses, more operating efficiency
(*iv*) No sparking, hence can be used in the vicinity of combustible fluids and gases.
(*v*) These are very reliable and efficient (efficiency is more than 75%).
(*vi*) They are capable to run at very high speeds (more than 40000 rpm)

Disadvantages

(i) More expressive then conventional DC motors
(ii) Additional electronic circuit and devises are required that increases the overall size of the machine

Application

Due to high reliability and low maintenance these motors find their applications in aerospace industry. satellites, gyroscope and high efficiency robotic system. These motors are also suitable for artificial heart pumps, disc drives, video recorders and biomedical fields.

12.11 Stepper Motors

A motor in which the rotor turns in discrete movements is called a ***stepper motor****.*

A stepper motor, as its name implies, turns in discrete movements called steps. After the rotor makes a step, it stops turning until it receives the next command (or signal).

Principle of Operation

Stepper motor operation can be easily visualised by considering a series of electromagnets or solenoids arranged in a circle as shown in Fig. 12.16. When these solenoids are energised in sequence, their fields interact with the rotor, causing it to turn either clock wise or counter clockwise, depending upon the input commands (or signals). The stepping angle (α) is determined by the design of the motor, but it should not be greater than 180° in any case.

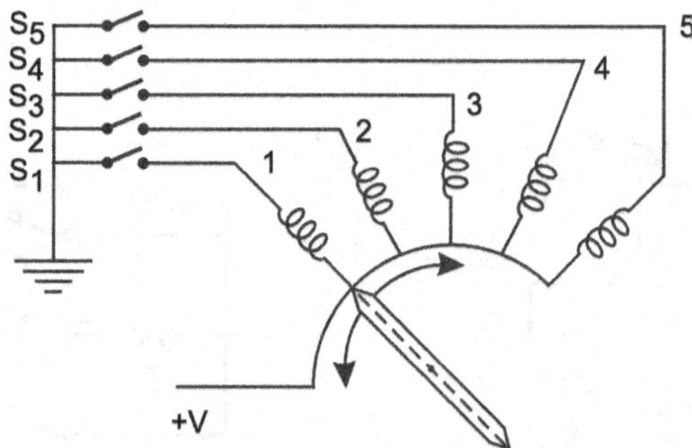

Fig. 12.16 Rotor movement

Types of Stepper Motors

There are two basic types of magnetic stepper motors

(i) Permanent-magnet (*PM*) stepper motor

(ii) Variable-reluctance (*VR*) stepper motor.

12.11.1 Permanent-magnet (PM) Stepper Motor

A stepper motor in which rotor is made of permanent magnet having even number of poles is called **a permanent-magnet (PM) stepper motor.**

Construction

The stator of a permanent-magnet stepper motor carries the similar winding as that of a conventional two-phase, three-phase or higher poly-phase induction or synchronous motor. The end terminals of all the windings are brought out to the terminal box for DC excitations. The rotor of this motor has even number of poles made of high retentivity steel alloy (Alnico) producing a multipole permanent magnet. Both rotor and stator may employ salient or non-salient pole construction. Usually, the stepper motors having small stepping angles are of non-salient pole type construction.

Working

Figure 12.17 (a) shows a typical *PM* stepper motor. With stator winding $A_1 - A_2$ energised only, for the polarity shown in Fig. 12.17(b) the *PM* rotor is "locked" in the position shown. If stator coil $A_1 - A_2$ is de-energized and $B_1 - B_2$ is energised, the polarity of south (*S*) is produced at B_1 and north (*N*) at B_2, causing the *PM* rotor to rotate 90° in response to the *excitation torque* produced by the winding $B_1 - B_2$. The excitation torque is maximum when the angle between the *PM* rotor and stator winding is 90°.

Reversing the current in windings *A* and *B*, consecutively, in turn, results in continuous clock wise rotation of the *PM* rotor. Instead of reversing the supply voltages feeding each phase, a simplified switching arrangement using two solid-state (transistors or thyristors) switches accomplishes the current reversals in each phase of the two-phase windings, as shown in Fig. 12.17 (b).

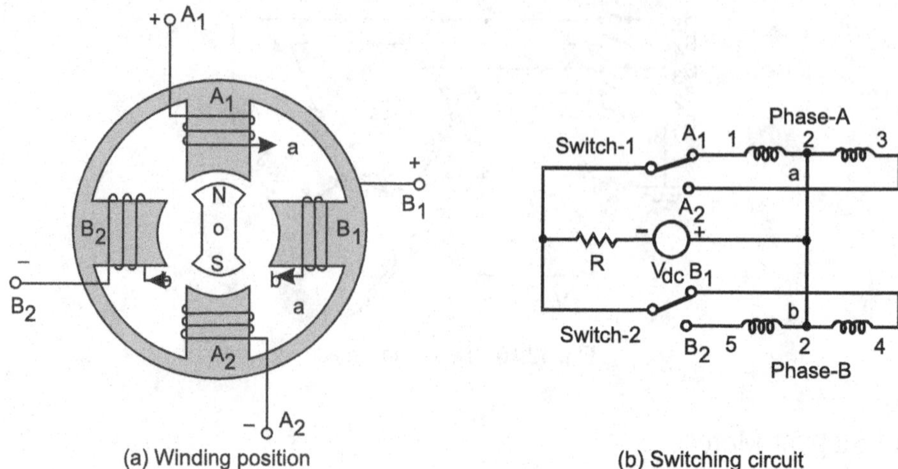

(a) Winding position (b) Switching circuit

Fig. 12.17 Permanent magnet stepper motor with switching circuit

The adjoining table shows the switching sequence for the four-step stepper motor. Note that for a two-pole, two-phase stepper, the pole-phase product is 4. If 360° is divided by this product, we obtain the *stepping length, or stepping angle,* for any *PM* or *VR* stepper motor.

i.e.,
$$\alpha = \frac{360^\circ}{nP} \textit{ degrees}$$

where n = the number of phases or stacks

P = the number of rotor poles (or teeth)

Switching step	*Switch 1*	*Switch 2*	*Angle of rotation*
1	1	4	90°
2	3	4	180°
3	3	5	270°
4	1	5	360° (0°)
1	1	4	90°

Since the number of teeth (or poles) on a rotor of given diameter is limited, it might appear that the solution to smaller stepping lengths is to increase the number of phases. But, as shown in Fig. 12.17 (b), increasing the number of phases (or stacks) results in a corresponding increase in the number of solid-state driven circuits. Since increasing the number of phases produces no particular performance advantages, steppers are rarely found having more than three phases or stacks.

Finally, if reversal of the stepping motor is desired, the switching steps shown in the table (reading from top the bottom) may be reversed by performing the sequences steps 4-3-2-1 (reading from bottom to top).

A new type of *PM* stepper motor has overcome the rotor size and weight problems that limit the maximum speed which motor can achieve. The rotor of this new type of stepper motor is a disc rather than the more typical cylinder. The rotor is a thin disc made from rare-earth magnetic material

(Fig. 12.18). Because the disc is thin, it can be magnetised up to a hundred individual tiny magnets, evenly spaced around the edge of the disc. Conventional *PM* steppers are generally limited to a minimum step angle of 30°, for a maximum of 12 steps per revolution. The new thin-disc motors are generally half the size of the ordinary stepper motors and weigh 60% less.

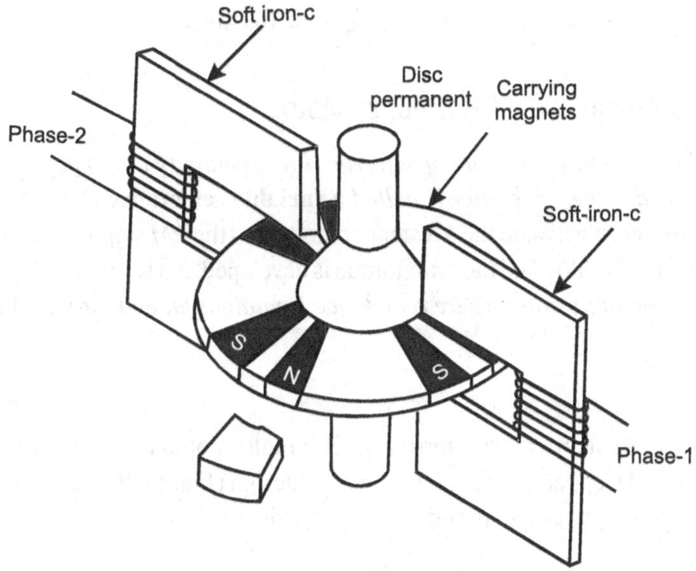

Fig. 12.18 Permanent magnet stepper motor

The disc of the stepper motor is supported on a nonmagnetic hub, the disc and hub together form the rotor. The disc magnets are polarised with alternating north and south poles, as shown in Fig. 12.18. *A* simple *C*-shaped electromagnet forms the field poles. When one of the phases is energised the rotor will align itself with the electromagnetic field generated. Then, when the first phase is turned off and the second is turned on, the rotor will turn by one-half of a half (or one-quarter) of a rotor pole to align itself with the field from the second phase. So that the rotor keeps turning in the same direction, the second phase is turned off and the first phase is turned on again.

Example 12.1

Determine the stepping angle for a 3-phase, 24 pole PM stepper motor.

Solution:

Stepping angle, $\alpha = \dfrac{360^\circ}{nP}$

Where $n = 3 \text{ and } P = 24$

$\therefore$ $\alpha = \dfrac{360^\circ}{3 \times 24} = \mathbf{5^\circ/step}$ (*Ans.*)

Example 12.2

To obtain a stepping angle of 7.5° for a 3-phase PM stepper motor, how many poles should the rotor have?

Solution:

Stepper angle, $\alpha = \dfrac{360^\circ}{nP}$

where, $n = 3$ and $\alpha = 7{\cdot}5^\circ/step$

$\therefore$ $7{\cdot}5^\circ = \dfrac{360^\circ}{3 \times P}$ or $P = \dfrac{360^\circ}{3 \times 7{\cdot}5^\circ} = \mathbf{16}$ (*Ans.*)

12.11.2 Variable-reluctance (VR) Stepper Motor

A stepper motor in which rotor is made of soft iron (low retentivity alloy) having teeth at the outer periphery to obtain variable reluctance is called a **variable reluctance (VR) stepper motor.**

The essential difference between the *VR* stepper motor and the *PM* stepper is that the *VR* rotor uses soft-iron or low-retentivity alloy and the rotor torque is developed as a result of *reluctance torque. That is, the rotor moves to that position where reluctance is minimised,* and air-gap flux is maximised.

Construction

Variable-reluctance (*VR*) stepper use a ferromagnetic multi-toothed rotor with an electromagnetic stator similar to the *PM* stepper. A typical three-phase design (Fig. 12.19). has 12 stator poles spaced 30° apart; the rotor has eight poles spaced at 45° intervals.

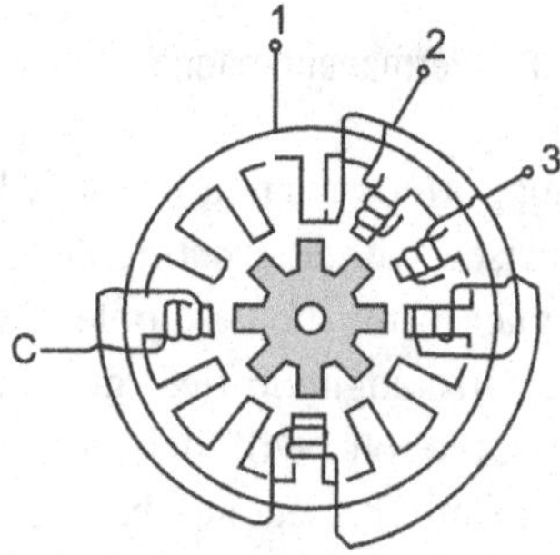

Fig. 12.19 Variable reluctance stepper motor

Working

The stator poles are energised sequentially by the three-phase winding. When current is supplied to phase-1, the rotor teeth closest to the four energised (magnetised) stator poles are pulled into alignment. The four remaining rotor teeth align midway between the non-energised reluctance between rotor and stator field.

Energising phase-2 produces an identical response. The second set of four stator poles magnetically attracts the four nearest rotor teeth, causing the rotor to advance along the path of minimum reluctance into a position of alignment. This action is repeated as the stator's electromagnetic field is sequentially shifted around the rotor. Energising the poles in a definite-sequence produces either clockwise or counterclockwise stepping motion.

The exact increment of motion (step angle) is the difference in angular pitch between stator and rotor teeth, in this case 30° and 45°, respectively, for a net difference of 15°. The *VR* stepper's step angles are small, making possible finer resolution that can be obtained with the *PM* type. Maximum stepping rates generally are higher than that in the *PM* stepper. Also, because of the non-retentive rotor, *VR* steppers do not have detent torque when unenergised.

A typical *VR* stepper uses a stator with 12 fields. Poles are set about 30° apart and grouped for three-phase operation where each phase has four coils set 90° apart. The rotor has eight teeth spaced 45° apart. *VR* steppers have a maximum stepping speed of about 18,000 steps/s, much higher than the PM stepper can produce. At high speeds, however, the *VR* stepper tends to overshoot and must be damped.

Applications

Stepper motors are often used as output devices for microprocessor-based control system as in paper drives on printers and X – Y graphical plotters. For example, (*i*) in a graphical plotter pen driven by X-axis and Y-axis steppers is controlled by a microprocessor whose phase-controlled signals operate the drive circuits producing phase currents for each stepper motor. (*ii*) These motors are also used in closed-loop servo-systems, replacing conventional DC servomotors in DC-operated servo-mechanisms, to position machine tools and valves. (*iii*) Further, since the signals fed to a stepper consist of a digital pulse train (of 0s and 1s), at a given repetition rate, the stepper may also serve as a digital-to-analog converter.

Example 12.3

Calculate the stepping angle for a 3-stack, 16-tooth rotor VR stepper.

Solution:

Stepping angle, $\alpha = \dfrac{360^\circ}{nP}$

where $n = 3$ and $P = 16$

$\therefore \quad \alpha = \dfrac{360^\circ}{3 \times 16} = 7{\cdot}5^\circ\ /\ step$ (*Ans.*)

Section Practice Problems

Numerical Problems

1. Determine the stepping angle for a 2-phase, 18 pole PM stepper motor. (**Ans.** *10°/step*)
2. To obtain the stepping angle of 10° for a 3-phase PM stepper motor, how many poles should the rotor have? (**Ans.** *12*)
3. Calculate the stepping angle for a 3-stack, 20 pole rotor VR stepper. (**Ans.** *6°/step*)

Short Answer Type Questions

Q.1. Why brushless synchronous generators (or DC generators) are developed?

Ans. To eliminate the problems associated with commutation (like sparking at the brushes, excessive heating, wear and tear, maintenance etc.) and brushes, brushless synchronous generators (or DC generators) have been developed.

Q.2. What is a third-brush generator?

Ans. It is a DC generator which contains three brushes instead of two. The third brush provides a regulated (constant) voltage irrespective to the speed.

Q.3. What do you mean by brushless DC motor?

Ans. A motor that retains the characteristics of a DC motor but eliminates the commutator and brushes is called a **brushless DC motor.**

Q.4. What do you mean by a stepper motor?

Ans. A motor in which rotor turns in discretc movement is called a *stepper motor.*

Q.5. What are the factors on which the angular movement of the rotor of permanent magnet stepper motor depends?

Ans. The angular movement of a stepper motor depends upon (i) number of phases or stacks (n) and (ii) the number of rotor poles (P)

$$\theta = \frac{360^\circ}{nP}$$

Q.6. Why a thire-disc stepper motor is preferred over a conventional *PM* stepper motor?

Ans. Because in these motors the size and weight of the rotor is very less comparatively.

12.12 Switched Reluctance Motor (SRM)

Construction

The schematic diagram of a switched reluctance motor (SRM) is shown in Fig. 12.20. Both stator and rotor of the motor have salient poles. Exciting coils are placed on the stator poles wherein the diametrically opposite coils are connected in series to from a pair of poles. In Fig. 12.20. Four coils are grouped to form two phases (or pair of poles) *A* and *B*. The laminated stator poles are attached to a laminated yoke. The rotor is also laminated and made of some magnetic material.

Operating Principle and Working

The operating principle of a switched reluctance motor is the same as that of a variable reluctance motor. When stator coils are energised sequentially with a single pulse, a reluctance torque is developed due to the attraction between the rotor and stator poles. To obtain continuous rotation, the position of rotor and sensor's timings are well synchronised (designed). High speeds can be developed by using high frequency pulses.

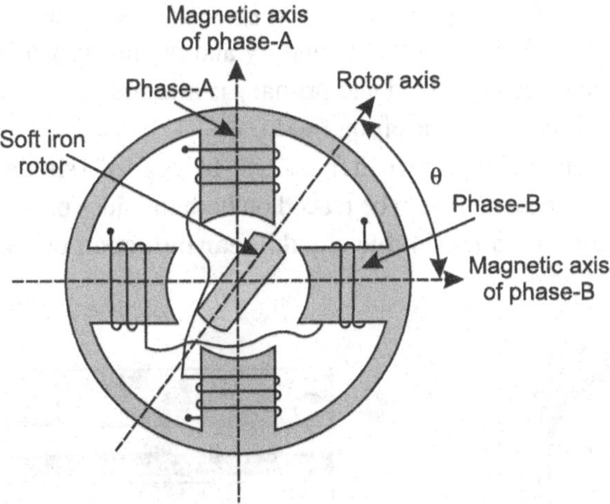

Fig. 12.20 Switched reluctance motor (SRM)

Advantages

(*i*) Robust construction since no winding on the rotor or slip-rings are used
(*ii*) Efficiency is high (about 75%)
(*iii*) Can be operated at high speeds (10000 to 30000 rpm)
(*iv*) Also perform well at higher temperatures.

12.13 Linear Induction Motor (LIM)

A conventional induction motor gives a circular motion whereas, a linear induction motor gives a linear or translational motion.

Thus, a **linear induction motor (LIM)** *is a motor which gives a linear or translational motion instead of rotational motion as is obtained from a conventional induction motor.*

Construction

To understand its construction, consider a 3-phase conventional induction motor as shown in Fig. 12.21(a). Cut it along the axis A A' and spread out flat as shown in Fig. 12.21(b). The stator winding is placed in the stator core called primary, whereas, the rotor short circuited ring or rotor conductor's spread over the rotor core is called secondary. In the linear induction motors the secondary consists of a flat aluminium conductor with ferromagnetic core.

Working

As in case of conventional 3-phase induction motor, when 3-phase supply is given to the stator, a resulting rotating flux is produced. Similarly, in case of linear 3-phase induction motor when 3-phase supply is given to its primary, a resultant travelling flux is produced which travels from one end to the other and repeats the same. EMF or current is induced in the aluminium conductor due to relative motion between the travelling primary field and stationary aluminium conductor or strip. BY the interaction of field produced by the secondary and the primary a linear force (or thrust) F is produced. If the secondary is kept fixed and primary is allowed to move, the force or thrust will move the primary in the direction of travelling field.

The linear induction motor (LIM) shown in Fig. 12.21 (b) is called a **single-sided linear induction motor (*SLIM*).** However, the primary can be placed on both the sides of the secondary, as shown in Fig. 12.22(a). then the motor is called **double-sided linear induction motor (*DLIM*)**

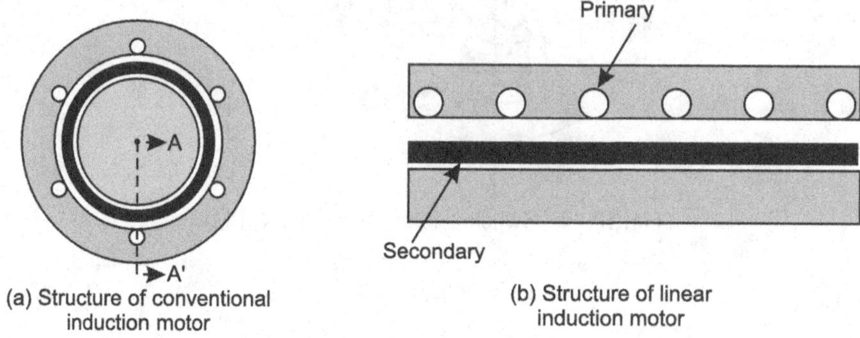

(a) Structure of conventional induction motor

(b) Structure of linear induction motor

Fig. 12.21 Constructional features of Linear induction motor

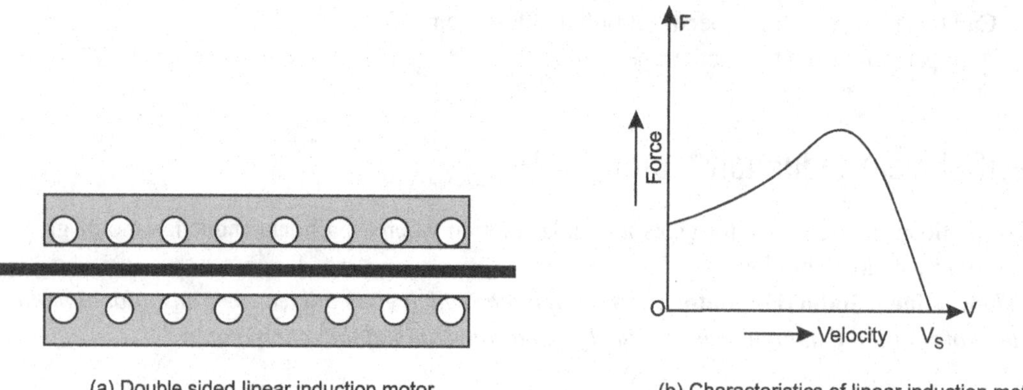

(a) Double sided linear induction motor

(b) Characteristics of linear induction motor

Fig. 12.22 Double sided linear induction motor and its characteristics

Performance

The linear synchronous speed of the travelling field,

$$v_s = 2f \times pole\ pitch \text{ metre/second}$$

The linear speed of the secondary

$$v = v_s\,(1 - S)$$

Where S is the slip of the linear induction motor

$$S = \frac{v_s - v}{v_s} \text{ pu}$$

The thrust or linear force acting on the secondary or primary

$$F = \frac{\text{air gap power}}{\text{linear synchronous velocity i.e., } v_s}$$

The thrust and velocity curve of the linear induction motor is shown in Fig. 12.22(b). It is similar to the torque-speed curve of a conventional rotary induction motor

Applications

The main application of this motor is in transportation where primary is mounted on the vehicle and secondary is laid along the track.

There are employed in cranes for material handily pumping of liquid metals, actuators for door movements, actuators for high voltage circuit breakers, etc.

Comparison of Conventional (rotary) and Linear Induction Motors

Se. No	*Particulars*	*Conventional (or rotary) induction motor*	*Linear induction motor*
1.	Type of movement	Angular rotation is obtained	Linear movement is obtained.
2.	Torque or Thrust	Torque develops in the rotor	Thrust (force) is exerted on the secondary (or primary)
3.	Parts	Main parts are called stator and rotor	Main parts are called primary and secondary
4.	Magnetising current	Air-gap is very small, magnetising current is small and pf is reasonably high.	Air-gap in more and magnetising current is more which reduces the pf
5.	Transients	Flux distribution is regular and transients are less	Due to shorter secondary, flux distribution is not regular, transients are more

12.14 Permanent Magnet DC Motors

A permanent magnet (*PM*) DC motor is similar to an ordinary DC shunt motor, except that its poles are made of permanent magnets and no excitation is required for the poles. Although these motors can be designed upto 75 hp but their actual field is confined to smaller sizes (fractional hp motors).

Construction

In these motors, the permanent magnets are supported by a cylindrical steel stator which provides an easy path for the magnetic flux. The cross-sectional view of a 2-pole permanent magnet DC motor is shown in Fig. 12.23. The permanent magnets occupy comparatively less space which reduces the overall size of the machine. The armature with conductors in its slots, commentator and brushes, etc, are the same as in a conventional DC machine.

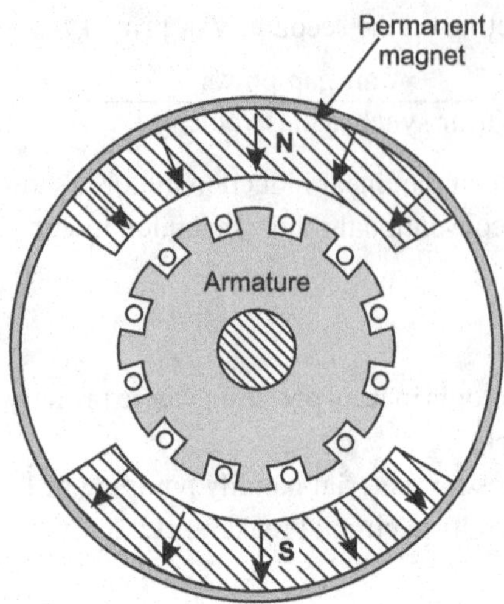

Fig. 12.23 Permanent magnet DC motor

Working

Usually, these motors are operated on 12V (or less) DC supply which may be obtained from batteries or rectifiers. When supply is given to the armature, torque develops and motor picks up the speed as in ordinary motors. These motors are put directly on line without starter since their stalling torque is very high (5 to 6 times full load torque).

The schematic diagram of the motor is shown in Fig. 12.24. It may be noted that, the supply is given to the armature and there is no field connections.

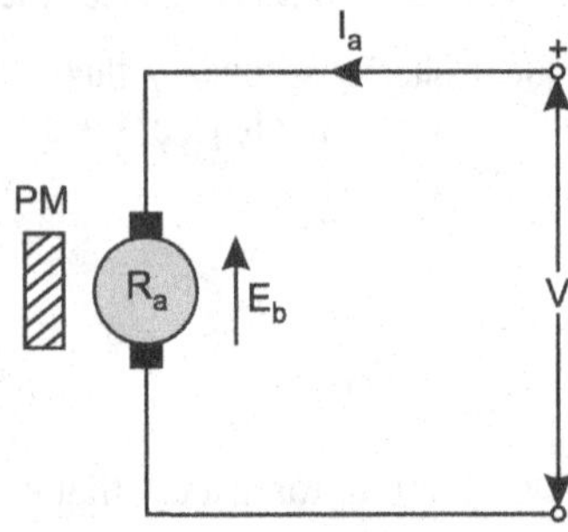

Fig. 12.24 Schematic diagram

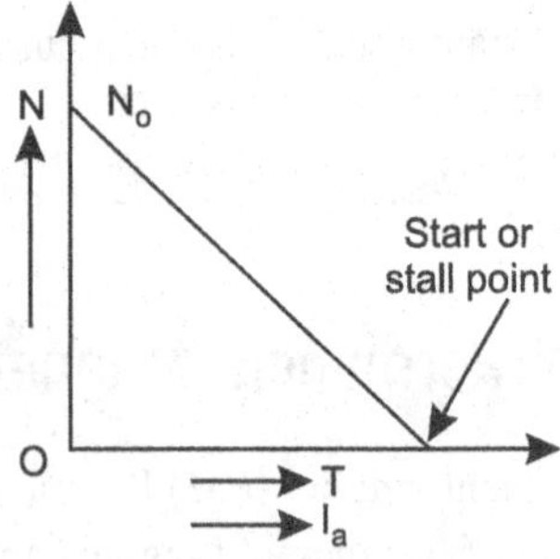

Fig. 12.25 Speed-torque characteristics.

$$E_b = V - I_a R_a$$

Motor speed, $N \alpha \frac{E_b}{\phi}$ Or $N \alpha \frac{V - I_a R_a}{\phi}$

$$N \alpha \; V - I_a R_a \qquad \text{(since } \phi \text{ is constant)}$$

$$T \alpha I_a \qquad \text{(as } T \alpha I_a \phi \text{ and } \phi \text{ is constant)}$$

The above relation shows that torque-armature current as well as torque-speed characteristics fallow the straight line, as shown in Fig. 12.25

The characteristics of a permanent magnet (*PM*) DC motor can be changed by changing either

(*i*) the supply voltage, *V* or

(*ii*) resistance of the armature circuit

By changing the supply voltage, the no-load speed N_O of the motor is changed without affecting the slope of the characteristic, as shown in Fig. 12.26. Accordingly, a set of parallel speed-torque characteristics are obtained for a set of supply voltage (V_1, V_2, V_3, ...)

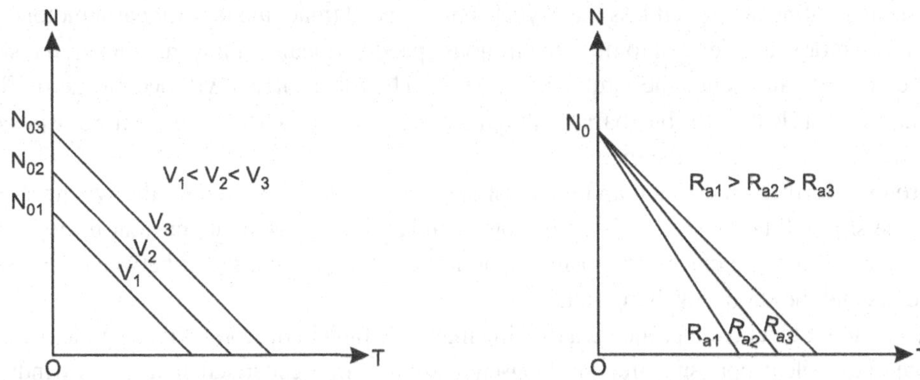

Fig. 12.26 Effect of change in supply voltage **Fig. 12.27** Effect of change in armature circuit resistance

By change the effective resistance of the armature circuit (inserting some resistance in series with the armature), the slope of the speed-torque characteristic changes without affecting the no-load speed N_O of the motor, as shown in Fig. 12.27..

Salient Features of PM DC Motor

(*i*) Generally, the size of this motor, for the same rating, is smaller than the excited field wound motors.

(*ii*) Since these motors do not have exciting field winding, losses are reduced and improved efficiency is obtained.

(*iii*) Usually, these motors are used for low voltage (less than 12V) and smaller power rating (0.1 to 5W).

(*iv*) Smaller life for small motors which is basically determined by the commutator and brushes used. Usually, it is limited to 1000 to 2000 hour.

While designing such motors, a compromise is made between cost, life span and efficiency. Since these motors are operated by battery, these are designed for an efficiency of 80%.

Applications

These motors are used for toys, windshield wipers, cordless tools, shavers, portable vacuum cleaners, automobile heaters, fans, wheel-chairs etc. For all these applications, millions of small *PM* DC motors with the ratings of 0.1 to 5W, 12 V or less than 12 V are produced every year.

12.15 Induction Generator

A three-phase induction motor converts electrical energy into mechanical energy. When the same machine starts converting mechanical energy in to electrical energy, it is called as an induction generator.

Let us see how this condition is achieved (**Principle of operation**).

- When 3-phase AC supply in given to the stator winding of a 3-phase wound induction machine, a revolving field is produced in the stator core which rotates at synchronous speed. Current is induced in the rotor conductors and torque develops by the interaction of rotor and stator fields. Rotor starts rotating at a speed less then synchronous speed (machine is acting as a motor).
- Now, if the rotor is accelerated to the synchronous speed by means of a prime mover, the slip will become zero and hence the torque will be zero. The rotor current will become zero. The machine will just float on the bus-bars. At this stage, the machine is neither receiving any power nor delivering any power to the mains.
- If the rotor is further accelerated and the rotor speed is increased more than the synchronous speed, the slip will become negative. The rotor conductors will generate current in opposite direction since they are cutting stator magnetic field due to relative speed between the rotor conductors and the stator revolving field.
- This generated rotor current produces a rotating magnetic field in the rotor which applies thrust on to the stator field in opposite direction. Consequently, an emf is generated in the stator winding which is larger than the applied voltage and acts in opposite direction. This pushes the current to flow out of the machine, *thus the machine starts working as an induction generator.*

The induction machine working as a generator is not a self-excited machine. In order to develop rotating magnetic field, it requires magnetising current and draws reactive power from the mains. At the same instant, it supplies active power to the mains as shown in Fig. 12.28, proportional to the slip above the synchronous speed. It shows that an induction generator cannot work in isolation because it continuously requires reactive power from the supply mains.

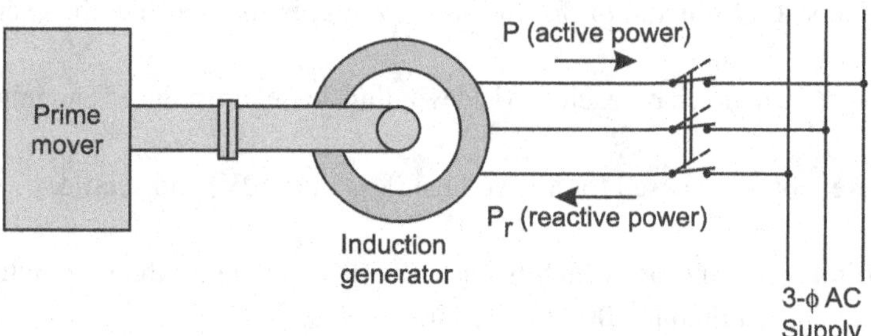

Fig. 12.28 Induction machine working as a generator

However, the problem can be solved by connecting a *capacitor bank* to supply reactive power equivalent to the reactive power which was earlier supplied by the mains.

Isolated Induction Generator

Isolated induction generator is also called a self-excited induction generator. It is called a self-excited because in isolation when the rotor of induction generator is rotated at an enough speed, a small voltage is generated across the stator terminals due to residual magnetism. This small voltage produces current to pass through the capacitor bank connected across the stator terminals. The capacitor bank provides the further reactive power for magnetisation.

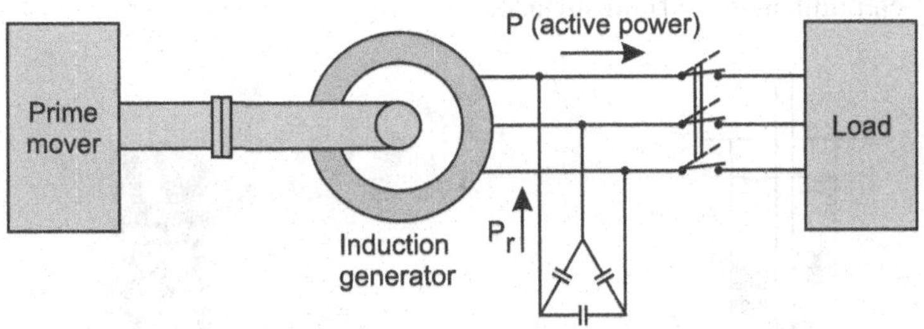

Fig. 12.29 Induction Generator in Isolated

In induction generators, the reactive power required to establish air-gap flux is provided by a bank of capacitors in case of stand-alone system. But is case of grid connected system, the reactive power is met by the grid. In case of grid-connected system, the voltage and frequency of the machine is taken care by the grid. But in case of stand-alone system, the frequency and voltage of the machine depend upon machine parameters, capacitors used for excitation and the type and value of load applied.

Advantages

1. Induction generators do not require commutator and brush arrangement as it is needed for synchronous generators and hence these are more rugged.
2. These are more simple in construction (mechanically and electrically) than other generators

Disadvantages

They take quite a large amount of reactive power.

Limitations

The requirement of large amount of reactive power puts limit to its common applications. These are only used for some special applications.

Applications

These generators are best suited where rotor speed is varying occasionally such as in wind turbines and in micro-hydro power plants

12.16 Submersible Pumps and Motors

A **submersible pump** or **electric submersible pump** *(ESP) is a device which has a **hermetically sealed motor coupled to a pump body as shown in Fig. 12.30(*a*). The whole assembly is submerged in the fluid or liquid to be pumped. Submersible pumps push fluid to the surface instead of pulling it as it is being done in jet pumps. Submersibles are more efficient than jet pumps. The main advantage of this type of pump is that it prevents pump cavitation, a problem associated with a high elevation difference between pump and the fluid surface.

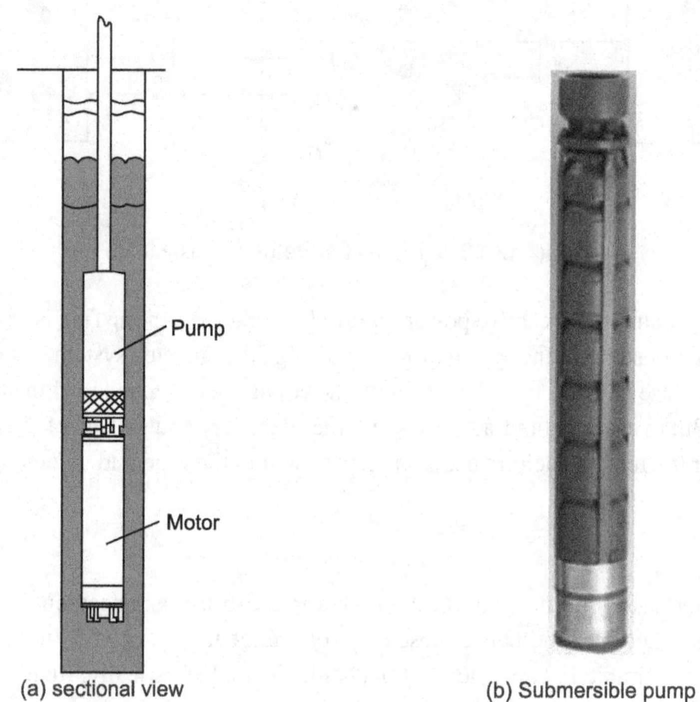

(a) sectional view (b) Submersible pump

Fig. 12.30 Submersible pump and motor assembly

Working Principle of Pump

The submersible pumps used in ESP installations are multistage centrifugal pumps operating in a vertical position. Its external view is shown in Fig. 12.30(*b*). The basic principle of a submersible pump is that the surrounding liquid (water or oil) after being subjected to great centrifugal forces caused by the high rotational speed of the impeller, lose their kinetic energy in the diffuser where a conversion of kinetic energy to pressure energy takes place. This is the main operational mechanism of radial and mixed flow pumps.

* ESP means Electric Submersible Pump

** Hermetical real means airtight seal

The pump shaft is connected to the gas separator or the protector by a mechanical coupling at the bottom of the pump. The surrounding fluid enters the pump through an intake screen and is lifted by the pump stages. Other parts include the radial bearings (bushings) distributed along the length of the shaft providing radial support to the pump shaft turning at high rotational speeds. An optional thrust bearing takes up part of the axial forces arising in the pump but most of those forces are absorbed by the protector's thrust bearing.

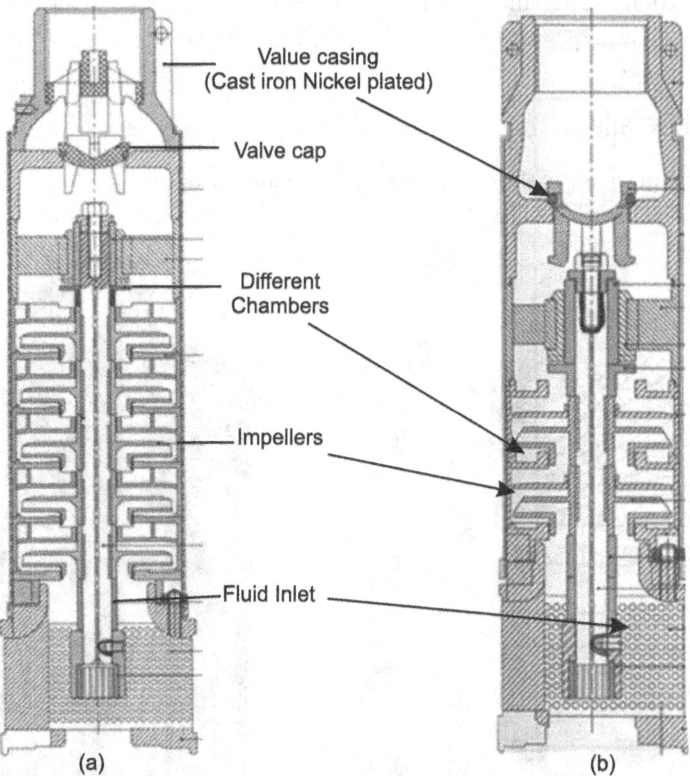

Fig. 12.31 Sectional view of pump

The pump itself is a multi-stage unit, as shown in Fig. 12.31, with the number of stages being determined by the work requirements. Each stage consists of a driven impeller and a diffuser which directs flow to the next stage of the pump. Pumps come in diameters from 90 mm (3.5 inch) to 254 mm (10 inch) and their length varies between 1 metre (3 ft) and 8.7 metre (29 ft). The motor used to drive the pump is typically a three phase, squirrel cage induction motor, with a nameplate power rating in the range 7.5 kW to 560 kW (at 50 or 60 Hz).

The ESP system consists of a number of components that turn a staged series of centrifugal pumps to increase the pressure of the well fluid and push it to the surface. The energy to turn the pump comes from a high-voltage (3 to 5 kV) alternating-current source to drive a special motor that can work at high temperatures of up to 300°F (149°C) and high pressures of up to 5000 psi (34 MPa), from deep wells of up to 12000 feet (3.7 km) deep with high energy requirements of up to about 1000 horsepower (750 kW). ESPs have dramatically lower efficiencies with significant fractions of gas, greater than about 10% volume at the pump intake. Given their high rotational speed of up to 4000 rpm (67 Hz) and tight clearances, they are not very tolerant of solids such as sand.

Submersible Motor

A submersible pump coupled to a motor, as shown in Fig. 12.30(*a*), is used to lift fluid (or water) is called *submersible motor.* It may be a single-phase capacitor start capacitor run motor or a 3-phase squirrel cage induction motor which is hermetically sealed and coupled to the pump. The size of the motor varies from 75 mm (3 inch) to 250 mm (10 inch) in diameter.

Constructional Details

The sectional view of motor is shown in Figs. 12.32(*a*) and 12.32(*b*)

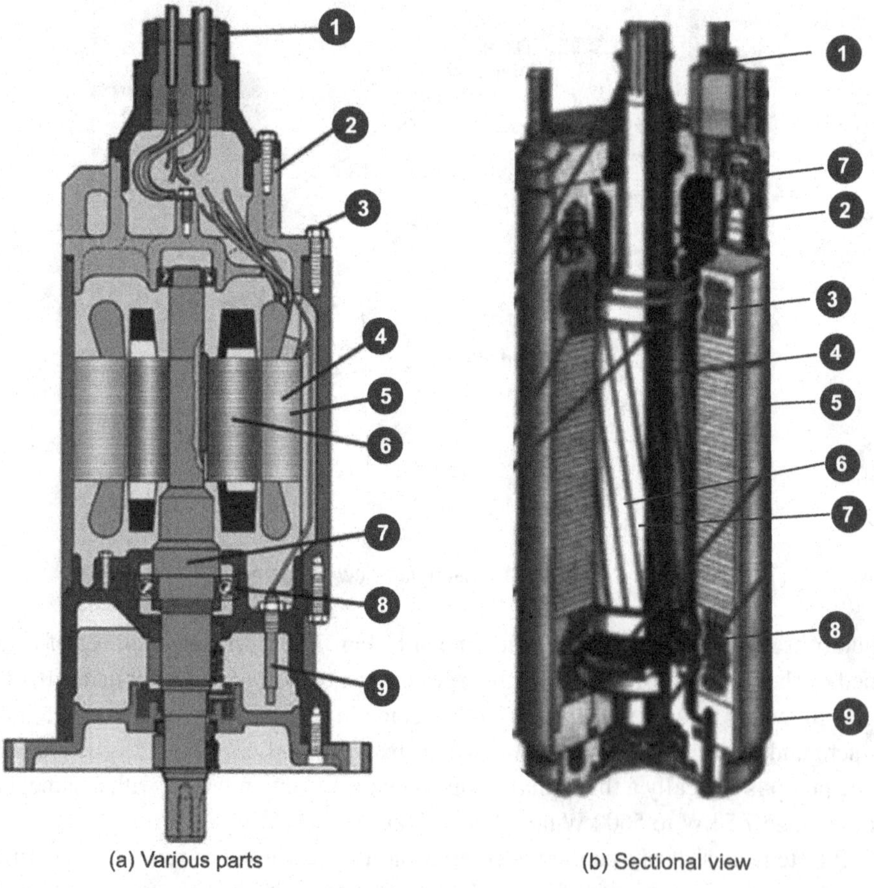

Fig. 12.32 Submersible motor

1. **Cable Entry System.** Epoxy sealed, cable connections are being made.
2. **Frame.** Unlimited run dry capability, optional continuous-in-air duty allows extended exposure of the motor under full load without overheating. UL explosion proof and rust proof available.
3. **Hardware.** Nuts and bolts of different sizes.

4. **Stator core.** It is made of high grade silicon steel stampings.
5. **Stator Insulation.** Specially treated Class "F" and "H" with special coating no hydroscopic insulation.
6. **Rotor.** Die cast aluminium, dynamically balanced. As per NEMA (National Electrical Manufacturer Association) standards.
7. **Shaft.** High strength, 416 stainless steel.
8. **Bearings.** Specially adapted for vertical assembly. Both upper and lower bearing are prefixed and sealed for life with special high temperature grease.
9. **Monitoring System.** Two wire, two probe monitoring system constantly monitors oil conditions.

Technical Specifications of Oil Lubricated Submersible Motor

4" Oil Lubricated Motors

Make: XYZ (rewireable)

Upper/Lower bracket and motor base are made of cast iron with nickel coating.

Motor shaft is of SS 420 grade or above

Stator shell is of SS 304

Winding wire is dual coated with Class-B insulation

Flange and shaft size as per NEMA (National Electrical Manufacturer Association) standard

Degree of protection : IP 58 (IP 58 is an International Protection Rating Code)

Max oil temperature : 50°C

Start/H : Max30

Allowable voltage variation : +6% / -10%

Max depth immersion : 250 m

Mounting : vertical / horizontal

Single phase motors are capacitor start capacitor run

Motor cable length : 3m/3 core

Coolant : Dielectric non-toxic.

Versions

Single Phase 0.25 kW to 2.2 kW, 220-230 volt/50 Hz

Three Phase 0.55 kW to 5.5 kW, 380-400 volt/50 Hz

Other voltage and frequencies are also available on demand

4" Water Lubricated Motors

Make: XYZ (rewireable)

Upper/Lower bracket and motor base are cast iron with nickel coating

Motor shaft is of SS 420 grade or above

Stator shell is of SS 304

Winding wire poly propylene with Class-B insulation

Flange and shaft size as per NEMA standard

Water lubricated radial and thrust bearing

Degree of protection	:	IP 68
Max water temperature	:	50°C
Start/H	:	Max 30
Allowable voltage variation	:	+6%/-10%
Max depth immersion	:	250 m
Mounting	:	vertical / horizontal

Single phase motors are capacitor start capacitor run

Motor cable length	:	3 m/3 core

Versions

Single Phase 0.25 kW to 2.2 kW, 220-230 volt/50Hz

Three Phase 0.55 kW to 5.5 kW, 380-400 volt/50 Hz

Other voltage and frequencies are also available on demand.

Applications

Submersible pumps are found in many applications. Single stage pumps are used for drainage, sewage pumping, general industrial pumping and slurry pumping. They are also popular with aquarium filters. Multiple stage submersible pumps are typically lowered down a borehole and used for water abstraction, water wells and in oil wells.

12.17 Energy Efficient Motors

There is a large gap between the demand and supply of electrical energy. This gap is widening at the rate of 3%. To reduce this gap either we are to increase the generation of electrical energy or we are to conserve it. It is very difficult to increase the generation of electrical energy as per the *increasing demand*. Therefore, it is suggested to employ energy efficient electric motors.

For conversion of electrical energy into mechanical energy, the basic need of industry is electric motors. Electric motors consume around 70% of the total electricity used in the industrial sector. As motors are the largest users of electrical energy, even small efficiency improvements can produce very large savings across the country. Energy conservation measure taken by individual consumers in this direction can improve the national economy and benefit the environment on global scale.

An **energy efficient motor** *produces the same shaft output power but draws less input power than a standard motor. Hence energy efficient motor consumes less electricity than comparable standards motor.*

Fig. 12.33 Pictorial view of induction motor

Energy-efficient motors, also called premium or high-efficiency motors, are 2 to 8% more efficient than standard motors. Motors qualify as "energy-efficient" if they meet or exceed the efficiency levels listed in the National Electric Manufacturers Association's (NEMA's)

Energy-efficient motors possess their higher performance characteristics because of their improved design and more accurate manufacturing tolerances, as mentioned below:

1. *More copper,* using wire of larger diameter in the stator saves energy by reducing the resistance of the stator winding consequently reducing copper losses.
2. *Thinner steel laminations*, decrease eddy current losses
3. *Longer stator,* lowers magnetic density and increases cooling capacity. As a result, both magnetic and load losses are reduced.
4. *Efficient cooling fan design*, improves air flow and reduces power required to drive the fan.
5. *Premium grade steel core*, reduces hysteresis losses.
6. *Modified stator slot design*, helps to decrease magnetic losses and makes room for larger diameter wire.
7. *Larger rotor conductor bars and end rings*, reduce rotor resistance losses

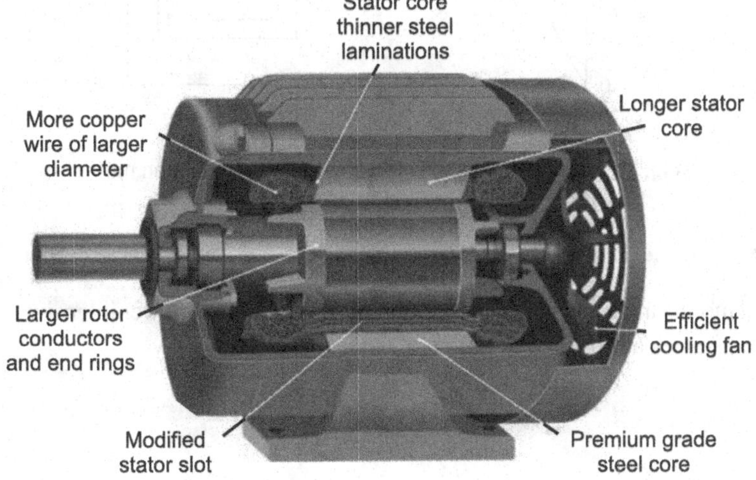

Fig. 12.34 Sectional view of an induction motor

Features:

Standards	:	IS: 12615, IS: 325, IS: 1231 and IS: 2223
Mounting	:	Foot (B3), flange (B5), face (B14) and combinations
Voltage	:	415 ± 10% or as required
Frequency	:	50 Hz ± 5% or as required
Ambient	:	45°C.
Altitude	:	upto 1000 m above m.s.l.
Enclosure	:	Totally enclosed fan cooled (TEFC)
Protection	:	IP55
Insulation class	:	Class F insulation with temp. rise limited to class B.

Advantages of energy efficient motors

- Save energy and money
- Near uniform efficiency from 50% to 100% of full load ensuring energy savings even at part load conditions also
- Short payback period
- Substantial savings after payback period

Range

Energy efficient motors manufactured by major industries are designed as per the values specified in IS: 12615. They have near uniform efficiencies between 50 and 100% of load as shown in Fig. 12.17. The motors are available from 2.2 kW to 200.0 kW in 2 and 4 pole variety.

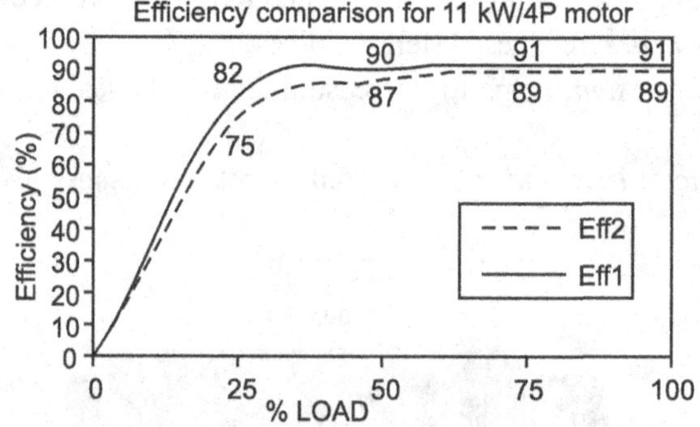

Fig. 12.35: Characteristics of energy efficient induction motor

Applications

Energy efficient motors are especially suited for industries which are power intensive and equipments which run on constant load for long duration.

Section Practice Problems

Short Answer Type Questions

Q.1. What is linear induction motor (LIM)?

Ans. A motor which gives a linear or translational motion instead of rotational motion is called a linear induction motor.

Q.2. Which type of motor is usually applied in toys?

Ans. Usually, permanent magnet DC motor is employed in toys.

Q.3. What is an induction generator?

Ans. When a 3-phase induction motor is driven by the prime-mover at a speed higher than synchronous speed, it is called as an induction generator.

Q.4. What is the major difference in the construction of submersible pump motor and a conventional induction motor?

Ans. A submersible pump motor is larger in length but smaller in diameter.

Review Questions

1. What is a servomotor? Classify DC servomotors.
2. Explain the working of an armature-controlled DC servomotors.
3. How a two-phase servomotor works.
4. Explain the constructional details of a Schrage motor.
5. Explain, how speed control and p.f. control is possible in Schrage motors.
6. Give at least five applications of a Schrage motor.
7. With the help of a block diagram explain the working of a brushless synchronous generator.
8. Explain, how a brushless synchronous motor works.
9. With the help of neat diagram explain how a third brush generator is most suitable for charging of batteries in automobiles.
10. What is a brushless DC motor, how it operates? What are its advantages and disadvantage as compared to a conventional DC motor?
11. What is a stepper motor? Explain the constructions and working of a PM stepper motor.
12. With a neat sketch explain the constructions and working of a VR stepper motor. How it differs from a PM stepper motor?
13. Explain the construction and working of a disk type PM stepper motor. Why it is preferred over a conventional stepper motor?
14. Mention some of the applications of a stepper motor.
15. Explain the construction and working of a switched reluctance motor.
16. What is a linear induction motor? Explain its working and applications.

17. Make a comparison between a linear induction motor and a conventional (rotary) induction motor.

18. Explain the construction and working of a permanent magnet DC motor.

19. Why a permanent magnet DC motor is best suited for toys?

20. What is an induction generator? What are its limitations?

21. Can an induction generator work as a stand-alone unit? Justify your answer.

22. Describe the working principle of submersible pump.

23. Describe the construction and working of submersible motor-pump.

24. Describe the main design characteristics and applications of a submersible motor.

25. What is the necessity of employing energy efficient motors? Its initial cost is more still it is preferred, justify.

26. To improve the efficiency of an induction motor, what adjustments are required to be made in its designing? Explain.

27. By improving the design of an induction motor to make it "energy efficient", its cost will increase, how will you justify that it will be more economical for a consumer in the long run?

Multiple Choice Questions

1. The motors which are used in the control system are called
(*a*) quick motors (*b*) servomotors
(*c*) toy motors (*d*) all of these

2. When the brushes of a Schrage motor are moved apart, the speed of the motor
(*a*) decreases (*b*) increases
(*c*) both *a* and *b* (*d*) none of these

3. When the brushes of a Schrage motor are move apart symmetrically, the p.f. of the motor
(*a*) improves (*b*) become poor
(*c*) is not affected (*d*) nothing can be said

4. A brushless generator is preferred over generator having brushes because it eliminates
(*a*) heavy sparking at brushes (*b*) excessive heating
(*c*) wear and tear (*d*) all of these

5. In PM stepper motor, the rotor contains
(*a*) electromagnets
(*b*) permanent magnet having even number of poles
(*c*) permanent magnet having odd number of poles
(*d*) none of these

6. The alloy used for the construction of rotor of a PM stepper motor is
(*a*) Alnico (*b*) Brass
(*c*) Bronze (*d*) none of these

7. A 3-phase induction motor is called a 3-phase induction generator
   (*a*) when its speed is equal to synchronous speed
   (*b*) when it is driven at a speed higher than synchronous speed
   (*c*) when its slip is negative
   (*d*) both b and c.

8. The submersible pumps used in ESP installations are placed in ...
   (*a*) horizontal position
   (*b*) vertical position
   (*c*) inclined at 60° position
   (*d*) any desired position

9. To improve efficiency of induction motors, the following improvements in design are made....
   (*a*) larger diameter of winding wire is used
   (*b*) stempings are made thin
   (*c*) length of stator core is increased
   (*d*) all of these

Keys to Multiple Choice Questions

1. b	**2.** a	**3.** c	**4.** d	**5.** b	**6.** a	**7.** d	**8.** b	**9.** d

Open Book Questions

Test Questions			Refer Art. No.
		Chapter 1	
I			
	1.	What do you mean by magnetic field? Give its significance.	**1.1**
	2.	What do you mean by ohm's law for magnetic circuits? How it is different to the ohm's law for electric circuits?	**1.2**
	3.	What are the similarities of magnetic and electric circuits?	**1.4**
	4.	Analyze a series magnetic circuit.	**1.6**
	5.	What do you mean by leakage flux, useful flux and leakage co-efficient?	**1.8**
	6.	Draw and explain magnetisation curve of a magnetic material.	**1.9**
	7.	What do you mean by hysteresis loss? Why do we prefer to use silicon steel material for the construction of core of a transformer?	**1.11 and 1.12**
II			
	1.	Define and explain electro-magnetic induction.	**1.13**
	2.	State and explain Fleming's right hand rule and Lenz's law.	**1.15**
	3.	What do you mean by statically induced emf? How mutually induced emf is different to self induced emf?	**1.18**
	4.	State and explain how mutual inductance develops between the two coils.	**1.20**
	5.	What will be the total induction when two inductances are connect in series (i) when their field are set-up in the same direction (ii) When their field are set-up in opposite direction.	**1.22**
	6.	What are the factors on which energy stored in the magnetic field depends?	**1.23**
III			
	1.	How will you define a generator and motor? Can the same machine work as a generator and motor.	**1.26**
	2.	By using a permanent magnet placed in the magnetic field, explain how torque is developer by the alignment of two magnetic fields?	**1.27.2**
	3.	Draw and explain the wave diagram of an induced emf in a coil, when coil is rotated in a uniform constant magnetic field.	**1.30**
	4.	Differentiate between generator and motor action.	**1.31.2**

Test Questions			Refer Art. No.
Chapter 2			
I			
	1.	What is the basic principle underlying the operation of practically all transformers? How are current, voltage and number of turns related to one another in the primary and secondary of the transformer.	**2.2 and 2.5**
	2.	How will you classify the concentric windings used in large transformers?	**2.3.3**
	3.	What are transformer bushings? How will you differentiate between oil filled and condenser bushings?	**2.3.5**
	4.	Explain the construction and working principle of a transformer and derive its emf equation.	**2.4, 2.2 and 2.7**
	5.	Derive an expression for the emf induced in a transformer winding. Show that the emf induced per turn in primary is equal to the emf per turn in secondary winding.	**2.7**
	6.	A transformer is analogous to mechanical gear drive, justify.	**2.5**
II			
	1.	Explain the principle of working of a transformer. Draw its phasor diagram on no-load.	**2.2 and 2.8**
	2.	What do you know about no-load current of a transformer and upon what factors does it depend?	**2.8**
	3.	What is the effect of saturation on exciting current of single-phase transformer?	**2.9**
	4.	With the help of wave diagram explain the effect of magnetisation on no-load current drawn by transformer.	**2.9**
	5.	If the magnetising current of a single-phase transformer is assumed a sine wave then show that the flux wave must be flat-topped.	**2.9**
	6.	Describe the inrush current phenomenon in a transformer.	**2.10**
III			
	1.	Explain how flux in the transformer core remains fairly constant from no-load to full-load (ampere-turns balance).	**1.11**
	2.	Draw and explain the phasor diagram of a transformer on load, neglecting winding resistance and reactance, at unity and lagging power factor.	**2.11**
	3.	Explain the working of a transformer at no-load and on load conditions.	**2.8 and 2.11**
	4.	What should be the value of primary resistance when referred to secondary side.	**2.13**
	5.	What is the effect of leakage flux in a transformer?	**2.14**
	6.	Derive the value of total impedance of a transformer referred to primary side.	**2.13, 2.15 and 2.17**
IV			
	1.	Draw the phasor diagram of a single-phase transformer on-load and describe it.	**2.16**

Test Questions			Refer Art. No.
	2.	Draw and briefly explain about transformer equivalent circuit.	**2.16**
	3.	Draw a complete phasor diagram for a transformer when the load p.f. is lagging.	**2.16**
	4.	Draw approximate equivalent circuit of a transformer referred to primary side and indicate how it differs from the exact equivalent circuit.	**2.18**
	5.	Define voltage regulation of a transformer and deduce expression for voltage regulation.	**2.19 and 2.20**
	6.	What is voltage regulation of transformer and derive expression for the real transformer supplying load at lagging power factor?	**2.19 and 2.20**
	7.	Draw Kapp's regulation diagram and explain how the regulation of the transformer is determined from the diagram.	**2.22**
V			
	1.	What are the various losses in a transformer? Where do they occur and how do they vary with load?	**2.23**
	2.	Why does all the transformers produce buzzing noise? Can it be eliminated? How can it be minimised?	**2.23**
	3.	State and prove the condition for maximum efficiency of a transformer.	**2.26**
	4.	How load power factor affects the efficiency of transformer? Explain.	**2.28**
	5.	What is all-day efficiency of a transformer? How does it differ from ordinary efficiency?	**2.29**
VI			
	1.	How can you determine the polarity of leads of a single-phase transformer?	**2.31**
	2.	Draw and explain the circuit diagram of open circuit test in single-phase transformer. What is the use of this test?	**2.33**
	3.	Explain how can you determine the parameters of equivalent circuit of a transformer no-load and short circuit test.	**2.33 and 2.35**
	4.	What are load losses occurring in a single-phase transformer? Explain the test used to determine variable losses.	**2.33 and 2.35**
	5.	Describe a back-to-back test for separation of losses in two identical transformers.	**2.36**
VII			
	1.	Discuss parallel operation of single-phase transformers.	**2.38 and 2.39**
	2.	What are the conditions necessary for parallel operation of single-phase transformers? How to achieve them?	**2.40**
	3.	Explain the necessary and sufficient conditions for parallel operation of transformers.	**2.40**
	4.	Obtain the load division between transformers of unequal voltage ratio in parallel operation of the transformers.	**2.41**

Test Questions			Refer Art. No.
VIII			
	1.	Write short notes on an Auto Transformer	**2.42**
	2.	What is an autotransformer? How does it differ from conventional two winding transformer? State its one application.	**2.42 and 2.52**
	3.	Make a comparison between an auto-transformer and a potential divider	**2.43**
	4.	Write down the working principle, construction and operation of auto transformer. Draw the phasor diagram also.	**2.42 and 2.47**
	5.	What are the advantages of using an autotransformer over two winding transformer?	**2.45**
	6.	What are the disadvantages of using a two winding transformer over an autotransformer?	**2.45**
	7.	What are the disadvantages of auto-transformer in comparison to two-winding transformer?	**2.46**
	8.	Compare the weights of copper (G) in an auto-transformer and a two winding transformer of the same VA rating. Show that: $\frac{G_{auto}}{G_{TW}} = 1 - \frac{V_2}{V_1}$ where $\frac{V_2}{V_1}$ = Voltage ratio of transformer.	**2.44**
	9.	Discuss the advantage, disadvantages and applications of auto-transformer as compared to two winding transformer.	**2.45 and 2.52**
	10.	Draw and explain phasor diagram of an auto-transformer.	**2.47**
	11.	Obtain the required voltage expressions for the equivalent circuit of an auto transformer and draw the equivalent circuit of auto transformer.	**2.48 and 2.49**
	12.	Compare the operation of autotransformer with two winding transformer.	**2.41**
	13.	Compare the characteristics of auto transformers and two winding transformers.	**2.41**
	14.	Can a 2-winding transform be converted into an auto-transformer, how?	**2.40**
	15.	Why is auto-transformer not used as a distribution transformer?	**2.46**
	16.	Show that in case of an auto-transformer $\frac{\text{Inductively transferred power}}{\text{Total power}} = \frac{\text{High voltage} - \text{Low voltage}}{\text{High voltage}}$	**2.42**
		Chapter 3	
I			
	1.	Explain core type and shell type construction of 3-phase transformers.	**3.2**
	2.	Draw and explain the connections of 3-phase transformer with symbolic notation Dd0.	**3.7.2**
	3.	What are different winding connections normally used in 3-phase transformers? What is the effect of the connections on exciting current?	**3.6**
	4.	Write a short note on 3-phase transformer.	**3.2 and 3.3**

Test Questions			Refer Art. No.
II			
	1.	What are the necessary conditions to be fulfilled before connecting two transformers in parallel? If these conditions are not fulfilled what happens?	**3.10**
	2.	How stabilisation is provided by tertiary winding in star-star transformer?	**3.12.1**
	3.	What is the use of tap changes provided on the transformers?	**3.13**
	4.	Explain various types of on load tap changing methods.	**3.14**
	5.	Explain about on-load Tap changing transformer	**3.14.2**
III			
	1.	Draw and explain open delta connection. Compare its capacity with delta-delta connection.	**3.16**
	2.	Explain open-delta or V-V connections with the help of phasor diagrams.	**3.16**
	3.	What are Scott connections? Discuss with appropriate connections.	**3.18**
	4.	In a 3-phase transformer, discuss the 3-phase to 2-phase conversion and vice-versa.	**3.19**
	5.	What are the methods of cooling of 3-phase transformers? Explain two methods in detail.	**3.21**
		Chapter 4	
	1.	Name major parts of a DC machine. Draw the sketch and show the path of magnetic flux in a 4-pole DC machine.	**4.2**
	2.	Discuss the difference between single layer and double layer winding.	**4.7**
	3.	Compare lap and wave winding. Where each type of winding is used and why?	**4.1 and 4.3**
	4.	Explain why a commutator and brush arrangement is necessary for the operation of a DC machine.	**4.1 and 4.2**
	5.	With the help of sketches, describe the main parts and working principle of a DC generator.	**4.2**
	6.	Explain in brief the construction of a DC machine.	**4.3**
	7.	Draw neat diagram of a commutator segment and explain its functions.	**4.5**
	8.	Explain what do you understand by (*i*) lap and wave windings (*ii*) duplex windings.	**4.8**
	9.	Illustrate with suitable diagrams the patterns of (*i*) lap winding and (*ii*) wave winding of the armature of a DC machine. Explain the relative merits and the applications of the two types of windings.	**4.8 and 4.13**
	10.	Explain why equaliser connections are used in lap windings and dummy coils are sometimes used in wave windings.	**4.8 and 4.13**
	11.	Highlight the factors on which the choice of type of armature winding of a DC machine will depend.	**4.10 (6)**

Test Questions			Refer Art. No.
	12.	Write short notes on the following:	
	(i)	Principle of operation of DC generator.	**4.1**
	(ii)	Construction and function of commutator.	**4.2**
	(iii)	Lap winding.	**4.8**
	(iv)	Wave winding-merits and demerits over lap winding.	**4.13**
II			
	1.	Derive the EMF equation of a DC machine?	**4.16**
	2.	Derive an equation for emf in a DC machine.	**4.16**
	3.	Derive the emf equation from fundamentals and explain why lap winding is preferred for low voltage high current machines.	**4.16**
	4.	Derive torque equation of a DC machine and mention the factors on which it depends.	**4.17**
	5.	What do you understand by the term 'armature reaction'? How does it affect the main field flux? Define GNP and MNP.	**4.18**
	6.	For a DC generator working without interpoles, explain the necessity of the brushes having a positive/forward lead.	**4.18**
	7.	What are the effects of armature reaction of a non-interpole DC generator? What are the remedial measures taken to counter the effects of armature reaction?	**4.18**
	8.	Discuss briefly armature reaction and commutation in DC machines.	**4.18 and 4.20**
	9.	"As brushes in a DC machines are shifted, magnetising/demagnetising ampere-turns appears". Explain with the help of a diagram.	**4.19**
	10.	What are the types of commutation possible in a DC machine? Explain them with graph.	**4.20**
	11.	Explain the process of commutation in a DC machine and discuss the methods to improve it.	**4.20**
	12.	What is meant by commutation process in DC machines? How is sparkless commutation achieved in DC machines?	**4.20**
	13.	How inductance of a coil affects commutation in a DC machine? How its effect can be neutralised?	**4.21**
	14.	What do you mean by good commutation and bad commutation?	**4.23**
	15.	What are commutating poles? Why are they used? Why are these poles (*i*) wound with series turns, and (*ii*) tapered?	**4.24**
	16.	Why are compensating winding and interpoles provided in a DC machine? Describe their role in improving the performance of the machine as a generator.	**4.24 and 4.25**
	17.	What is commutation? Give causes of sparking on the commutator and state how it can be avoided.	**4.20, 4.24 and 4.25**
	18.	Discuss the methods in details to achieve sparkless commutation.	**4.24 and 4.25**

Test Questions			Refer Art. No.
	19.	Describe the effects of armature reaction on the operation of DC machines. Describe also the remedies employed for decreasing the effects of armature reaction.	4.18, 4.24 and 4.25
III			
	1.	Describe with suitable diagram various types of DC machines.	4.27, 4.28 and 4.29
	2.	Write a short note on DC generator characteristics.	4.31
	3.	Describe the open-circuit characteristics of DC generators.	4.32
	4.	Explain the process of voltage build up in a DC shunt generator. What is the field circuit critical resistance?	4.33
	5.	What could be the causes for the failure of voltage build-up of DC self excite generator? How can the problem be remedied?	4.33
	6.	A DC shunt generator fails to build up voltage when it is operated at the rated speed. What may be the possible reasons for this? Explain.	4.33
	7.	Why is the shunt generator characteristic on load drooping and turns back as it is overloaded? Explain.	4.35
	8.	A DC shunt generator can operate under short-circuit condition but series and cumulative compound generator cannot operate under this condition. Explain the reason.	4.35, 4.36 and 4.37
	9.	Differentiate over, level and differential compounding with the help of their external characteristics.	4.37
	10.	What is degree of compounding? How does it affect the external characteristic of a DC generator? How may the series field ampere-turns be estimated for a desired degree of compounding?	4.37
	11.	Give the applications of different types of DC generators with reasons and justification.	4.39
Chapter 5			
I			
	1.	Explain the principle of operation of DC motors. What is back emf in DC motors? What is its significance?	5.2, 5.3
	2.	What is significance of back emf in DC machines and derive the expression for it.	5.3
	3.	What is back emf? Is the back emf greater or lesser than the applied voltage? Why? By what amount do the two voltages differ. Write voltage equation of a motor.	5.3
	4.	Derive an expression for the speed of a DC motor in terms of back emf and flux per pole.	5.3
	5.	How may the direction of rotation of a DC shunt motor be reversed? What is the effect of reversing the line terminals?	5.4

Test Questions			Refer Art. No.
	6.	Define an expression for the torque developed by a DC motor in terms of ϕ, Z, P, A, L_a where the symbols have usual meaning.	5.4
	7.	Mention different types of DC motors.	5.7
	8.	Differentiate between different types of DC motors.	5.7
	9.	Explain the speed-current, torque-current and speed-torque characteristics of a DC shunt motor.	5.9
	10.	What are the important characteristics of DC motors? Sketch all these characteristics for DC series motors.	5.10
	11.	Derive an expression for torque of a DC motor. Hence draw torque vs current (or load) characteristics of DC shunt and DC series motors.	5.10
	12.	Compare the speed-current characteristics of various types of DC motors.	5.9,5.10,5.11
	13.	Explain the speed-armature current characteristics of series and shunt motors.	5.9, 5.10
	14.	Draw speed-torque characteristics of (*i*) DC shunt motor (*ii*) DC series motor (*iii*) DC cumulatively compounded motor. Mention one industrial application of each one of these motors.	5.11 and 5.12
	15.	List applications of DC shunt, DC series and DC compound motors.	5.12
	16.	Explain the function of no-volt and over load release coil in a two-point series motor starter?	5.19
	17.	Name various starters used for DC motor and explain working of any one of them with a suitable diagram.	5.14, 5.15
	18.	Explain why a starter is required for starting a DC motor. Describe a 3-point starter having no-volt and over-load protections for starting a DC shunt motor. What modification is made in a 4 point starter?	5.14, 5.16, 5.13
	19.	What may happen to a DC shunt motor connected to a three-point starter if the field excitation is kept minimum at the time of starting?	5.16
	20.	Way a DC motor draws heavy current at start? How does a 4-point starter of DC motor works?	5.14, 5.17
II			
	1.	What are the different methods of speed control of a DC motor? give the advantages and disadvantages of each in brief. Why is it dangerous to uncouple the mechanical load of a DC series motor?	5.10, 5.20
	2.	Make a list of different speed control methods for DC motor. Discuss merits and demerits of each method.	5.20
	3.	What are various speed control methods of a DC motor? Explain each of them in brief with neat diagram.	5.20
	4.	How may the speed of a DC shunt motor be controlled by armature voltage and field flux? Explain the principles and state the limitations of the two methods.	5.21
	5.	For a DC motor the field flux-speed control method is called a constant power drive method. Explain.	5.21

Test Questions			Refer Art. No.
	6.	Explain with neat sketch how speed control of a DC shunt motor is obtained by Ward Leonard control system. How the direction of rotation of the motor is usually reversed in this method of speed control?	**5.22**
	7.	Define speed regulation and percentage speed regulation of a DC motor.	**5.23**
	8.	Explain the series-parallel control of two identical series motors. How many speeds are possible? Are the motors run faster in series combination or parallel combination?	**5.24.3**
	9.	With relevant circuit diagrams, describe different methods of speed control of DC series motor.	**5.24**
	10.	What is meant by braking of DC motors? Describe, briefly, the various methods of braking of DC shunt motors.	**5.25**
	11.	What is regenerative braking? Why it cannot be accomplished with a series wound DC motor without modifications in the circuit diagram? Give merits and demerits?	**5.26.3**
III			
	1.	What are the various losses occurring in rotating machines? Mention the method to reduce them.	**5.27**
	2.	Describe all types of losses in a shunt and compound wound motor. State which comprise constant loss.	**5.27, 5.28**
	3.	Draw a power flow diagram of a DC motor and define commercial, mechanical and electrical efficiency of a motor.	**5.30, 5.31**
	4.	What are various losses in a DC machine? How the efficiency is estimated? Write the condition for maximum efficiency.	**5.27, 5.32**
	5.	What are the various tests performed on a DC Machine? Explain any two.	**5.33, 5.34**
	6.	Explain why Swinburne's test cannot be used to determine the efficiency of DC series machines.	**5.35**
	7.	Explain the Hopkinson Test of DC machine in details with the help of connection diagram.	**5.36**
	8.	What is Hopkinson test? Draw a diagram and explain the procedure of Hopkinson test.	**5.36**
	9.	Classify various losses in a DC machine and indicate the factors on which these depend. Explain the regenerative method of determining the efficiency of a DC machine. List the merits and demerits of the method.	**5.27, 5.36**
	10.	With a neat circuit diagram discuss how the efficiency of DC series motors can be determined, by conducting field's test.	**5.37**
		Chapter 6	
I			
	1.	Define and explain the working principle of a synchronous machine working as (*i*) a generator, (*ii*) a motor.	**6.2**

Test Questions			Refer Art. No.
	2.	How an alternating emf is generated by an alternates? How will you relate speed and frequency in a synchronous machine?	**6.4 and 6.5**
	3.	Explain the working of a brushless excitation system.	**6.8.3**
II			
	1.	Explain what do you mean by concentric (or spiral), lap and wave winding used in the armature of an alternator.	**6.10**
	2.	What are the different types of armature winding used in the alternators?	**6.10**
	3.	Define and explain the following terms: (i) Electrical and Mechanical angle (ii) Single-turn and multi-turn coil (iii) Coil-span (iv) Pole-pitch and coil-pitch (v) slot pitch.	**6.11**
III			
	1.	Define and explain coil span factor and derive a relation for this factor.	**6.12**
	2.	Define and explain distribution factor for a distributed armature winding. Derive the necessary relation.	**6.13**
	3.	Derive emf equation of an alternator.	**6.15**
IV			
	1.	How leakage reactance develops in the armature of an alternator and what is its significance?	**6.25**
	2.	What is synchronous impedance? Draw an equivalent circuit of an alternator and simplify it.	**6.28 and 6.29**
	3.	Draw a phasor diagram of an alternator and derive a relation between terminal voltage and induced emf at no-load for a capacitive load.	**6.30**
	4.	What are the different methods of determining voltage regulation of an alternator? Explain synchronous impedance method or emf method.	
V			
	1.	What assumptions are made while determining voltage regulation of an alternator by synchronous impedance method?	**6.36**
	2.	How will you plot open circuit characteristics of an alternator?.	**6.33.1**
	3.	Describe mmf method for determining voltage regulation of an alternator.	**6.37**
VI			
	1.	Derive a relation for the power output of a non-salient pole type of alternator.	**6.39.1**
	2.	Derive a relation for the reactive power output of an AC generator.	**6.39.3**
	3.	Derive a relation for the reactive power input to an AC generator.	**6.39.6**
	4.	Explain two reaction theory for salient pole synchronous generators.	**6.40**
	5.	Describe the slip test method for determining X_d and X_q of synchronous machines.	**6.40.1**

Test Questions			Refer Art. No.
	6.	Derive an expression for power developed by a salient pole alternator delivering power at lagging p.f. neglecting the effect of armature resistance.	**6.41.1**
VII			
	1.	What do you mean by transient in alternators? Draw and explain the wave-shapes of the transient currents.	**6.43**
	2.	Explain what are the various losses which way occur in synchronous machines?	**6.44**
	3.	Why synchronous machines are cooled? Explain the methods by which these machines are cooled down?	**6.46 and 6.47**
		Chapter 7	
	I		
	1.	What is the necessity of connecting the alternators is parallel? What conditions are required to be fulfilled for successful parallel operation of alternators?	**7.1 and 7 .3**
	2.	If the terminal voltage of incoming alternator is more than the terminal voltage of existing alternator connected to the bus-bars to which incoming alternator is connected, what would you expect? How the terminal voltage of the incoming alternator is adjusted.	**7.4**
	3.	Explain one-dark and two-bright lamp method of synchronising of two 3-phase alternators.	**7.6**
	4.	Discuss the use of synchroscope in synchronising the alternators.	**7.7**
	5.	What steps are taken while disconnecting an alternator from the bus-bars?	**7.8**
II			
	1.	What do you mean by synchronising current, power and torque? What are the factors on which these depend?	**7.11**
	2.	What will be the effect of change in excitation of one of the alternator which is running in parallel with another alternator?	**7.13**
	3.	How do governors characteristics effect the load sharing of alternators operating in parallel?	**7.15**
		Chapter 8	
I			
	1.	Discuss only a synchronous motor is not self-starting.	**8.1**
	2.	Draw the equivalent circuit of a synchronous motor and derive the commonly used expression for the power developed by a synchronous motor.	**8.3 and 8.7**
II			
	1.	Explain hunting phenomenon in synchronous machines. Why it is objectionable? What are its causes and how can it be reduced?	**8.17**
	2.	What is synchronous condenser? What are the practical benefits of using one?	**8.15**
	3.	Write short notes on hunting.	**8.17**

Test Questions			Refer Art. No.
	4.	Write short note on V-curves and inverted V-curves.	**8.12**
		Chapter 9	
I			
	1.	Explain the construction of 3-phase squirrel cage induction motor.	**9.1**
	2.	The rotor of an inductor motor is skewed, why?	**9.1**
	3.	An induction motor is also called asynchronous motor, why?	**9.1**
	4.	Can an induction motor run at synchronous speed, state why?	**9.3**
	5.	Explain the principle of working of a three-phase induction motor and give the expression of percentage slip.	**9.3**
	6.	Derive the relationship between the frequency of rotor currents and supply frequency in case of a 3-phase induction motor.	**9.5**
	7.	Draw the rotor equivalent circuit of an induction motor and simplify it.	**9.13**
II			
	1.	Draw the stator equivalent circuit and phasor diagram of an induction motor.	**9.15**
	2.	Why the power factor of an induction motor is low at light loads than full-load.	**9.18**
	3.	List the different types of losses in an induction motor.	**9.19**
	4.	Derive an expression for the rotor copper loss in terms of slip and input to the rotor.	**9.21**
III			
	1.	Obtain an expression for torque under running condition, for a 3-phase induction motor and then deduce the condition for maximum torque.	**9.23 and 9.24**
	2.	Derive the simplified equation for torque for an induction motor.	**9.23**
	3.	Draw and explain the slip-torque characteristics of a 3-phase slip ring induction motor. Mark on it starting and maximum torque.	**9.29**
	4.	Explain briefly the effect of increasing the rotor resistance of an induction motor on (*i*) Starting torque (*ii*) Running torque.	**9.31**
IV			
	1.	Explain how stator resistance test and voltage ratio test is performed on a 3-phase induction motor.	**9.33.1 and 9.33.2**
	2.	How blocked- rotor test is performed on a 3-phase induction motor? What information do we get from this test?	**9.33.4**
	3.	Draw equivalent circuit and phasor diagram of a 3-phase induction motor.	**9.34**
	4.	Derive a relation for maximum power output.	**9.36**
	5.	Describe the circle diagram for the approximate equivalent circuit of an induction motor.	**9.38**
	6.	What results can be obtained from the circle diagram of an induction motor.	**9.40**

Test Questions			Refer Art. No.
V			
	1.	Explain the phenomenon of cogging in a 3-phase induction motor. How can it be avoided?	**9.43.1**
	2.	Explain the phenomenon of cogging and crawling in squirrel cage induction motors.	**9.43.1 and 9.43.2**
	3.	What is the effect of change in load on speed, power factor and efficiency of an induction motor	**9.44**
	4.	Explain with diagram the performance characteristics of an induction motor.	**9.44**
	5.	What do understand by high starting torque cage motors?	**9.46**
	6.	Describe double squirrel cage induction motor.	**9.46.2**
	7.	How will you distinguish the motor enclosures as per the guide-lines of NEMA?	**9.47**
	8.	What are the advantages and disadvantages of squirrel cage induction motors over phase-wound induction motors?	**9.49**
	9.	Mention some of the important applications of squirrel-cage and phase-wound induction motors.	**9.50**
	10.	Compare relative advantages of cage rotor type induction motor and wound rotor type induction motor.	**9.49**
	11.	Compare cage and wound type induction motors.	**9.51**
	12.	Compare a squirrel cage motor with a slip ring induction motor with reference to construction, performance and application.	**9.52**
Chapter 10			
I			
	1.	What is the necessity of a starter? Explain the working of direct-on-line starter?	**10.1**
	2.	When a 3-phase squirrel cage induction motor is started by using a direct-on-line starter, show that $\frac{T_{st}}{T_{fl}} = \left(\frac{I_{st}}{I_{fl}}\right)^2 \times S_{fl}$	**10.2.1**
	3.	When a 3-phase squirrel cage induction motor is started by a stator resistance starter, show that (*i*) $I_{st} = x I_{sc}$ and (*ii*) $T_{st} = x^2 T_{fl}(I_{sc} / I_{sl})^2 \times S_{fl}$. Also mention the demerits of a stator resistance starter	**10.2.2**
	4.	Explain how a star-delta starter reduces the starting current to one third of its value if the motor is started directly on line.	**10.2.3**
	5.	Explain the working of an auto-transformer starter with the help of a neat sketch. Also explain why this starter is preferred over a star-delta starter to start induction motors of larger capacity.	**10.2.4**
	6.	Explain the working of a rotor resistance starter used to start 3-phase slip-ring induction motors. Why it is preferred over other starters used to start squirrel cage induction motors.	**10.3**

Test Questions			Refer Art. No.
II			
	1.	What are the various methods of speed control of 3-phase induction motors? Explain any one method in detail.	10.4 and10.5.1
	2.	Describe any two methods of speed control of 3-phase squirrel cage induction motors. Also draw the torque speed curves.	10.5
	3.	Explain cascade method of speed control of 3-phase induction motors.	10.8
		Chapter 11	
I			
	1.	Prove that a single-phase AC supply produces two equal and oppositc revolving fields when supplied to a 1-phase wound induction motor.	11.3
	2.	Show that starting torque produced in a 1-phase induction motor is zero.	11.4
	3.	Develop an equivalent circuit of a single-phase induction motor with the help of two-field revolving theory.	11.5
	4.	Show that a revolving field is developed if 2-phase supply is given to the stator of a 2-phase wound induction motor.	11.6
	5.	Draw the circuit diagram of a chapacitor start induction motor and explain its working.	11.9
II			
	1.	Describe the constructional features and operating principle of a shaded pole1-phase induction motor.	11.10
	2.	Describe the performance characteristics and applications of a shaded-pole 1-phase induction motor.	11.10
	3.	How will you differentiate between a shaded-pole and reluctance start 1-phase induction motors?	11.10 and 11.11
	4.	Describe the constructional features and operating principle of a 1-phase reluctance motor.	11.13
	5.	What is hysteresis motor? Explain its construction and principle of operation.	11.14
	6.	Why series motors are not operated at no-load?	11.15
		Chapter 12	
I			
	1.	Explain feedback control system with the help of a block diagram.	12.1
	2.	Explain the working of armature-controlled and field controlled DC servomotors. Why armature-controlled servomotors are preferred over field-controlled servomotors?	12.4
	3.	Explain how the speed of a Schrage motor is controlled.	12.6

Test Questions			Refer Art. No.
II			
	1.	Explain the working of a brushless synchronous generator with the help of block diagram	**12.7**
	2.	Explain the construction and working of a brushless DC motor.	**12.10**
	3.	Explain the basic working principle of a stepper motor.	**12.11**
	4.	Explain the construction and working of a variable reluctance (*VR*) stepper motor	**12.11.2**
III			
	1.	Explain the construction and working principle of a linear induction motor.	**12.12**
	2.	Explain the construction and working of a permanent magnet DC motor.	**12.13**
	3.	Explain the construction and working of a submersible pump motor.	**12.15**

Index

Printed in the United States
by Baker & Taylor Publisher Services

Printed in the United States
by Baker & Taylor Publisher Services